Manual of numerical methods in concrete

Modelling and applications validated by experimental and site-monitoring data

M. Y. H. Bangash

Thomas Telford

Published by Thomas Telford Publishing, Thomas Telford Ltd,
1 Heron Quay, London E14 4JD.

URL: http://www.thomastelford.com

Distributors for Thomas Telford books are
USA: ASCE Press, 1801 Alexander Bell Drive, Reston,
VA 20191-4400, USA
Japan: Maruzen Co. Ltd, Book Department, 3–10 Nihonbashi 2-chome,
Chuo-ku, Tokyo 103
Australia: DA Books and Journals, 648 Whitehorse Road,
Mitcham 3132, Victoria

First published as **Concrete and concrete structures; numerical modelling** and applications by MYH Bangash (Elsevier Science), in 1989.

This edition first published 2001

Also available from Thomas Telford Books
Prototype bridge structures, by MYH Bangash, 1999, ISBN 0 7277 2778 8
Prototype building structures, by MYH Bangash, 1999, ISBN 0 72772777 X

A catalogue record for this book is available from the British Library

ISBN: 0 7277 2946 2

Typeset by Academic + Technical, Bristol
Printed and bound in Great Britain by MPG Books, Bodmin

Preface

The problem of modelling the constitutive equations representing the isotropic and anisotropic behaviour of concrete in the analysis of concrete structures remains one of the most difficult tasks in the field of structural engineering. At present analyses are being carried out on concrete structures for elastic, plastic and cracking conditions under static and dynamic loads. The behaviour of concrete under such complex loads depends on the state of stress, relative Poisson's ratios and Young's moduli, and the nature of cracking. All these vary in three dimensions. The state of stress or cracking may also depend on the instant load combinations such as compression–compression–compression, tension–compression–compression, tension–tension–tension, etc. All these cases at present cannot easily be handled by codes of practice or simplified empirical formulae.

The basic aim of this book is to present a unified approach for the available mathematical models of concrete. These are then linked to finite element analysis and to computer programs in which special provision are made for concrete plasticity, cracking and crushing with and without concrete aggregate interlocking. Creep, temperature and shrinkage formulations are included and also geared to various concrete constitutive models. Their influence is taken into consideration in the operational and overloading behaviour of concrete structures.

The book is divided into three parts. In the first part Chapter I gives a literature review of the basic properties of concrete as seen by various researchers. Useful data are included on the stress–strain criteria of concrete under monotonic, cyclic and thermal loads. The effects of creep, shrinkage and cryogenic temperatures on the material properties of concrete are clearly established. Equations and graphs are provided to support the data for concrete under uniaxial, biaxial and triaxial conditions with and without the changing influence of Young's modulus and Poisson's ratio. They are arranged in such a way that they can be used in the proposed finite element analysis. A comprehensive literature review and data are presented on dense concrete and shotcrete using heavy steel aggregate, steel shots and steel fibres. Tables are provided for material properties. Experimental tests on beams and slabs with dense concrete have been reported. A validation work using numerical modelling and finite element analysis has been included in detail. Results such as compression strength with age, time dependent stress–strain relation, comparative studies of material properties and force versus elastic/fracture energy have been plotted and fully validated. An appendix supporting this work is also included in the text.

Chapter II gives various mathematical models for concrete. They include the one-parameter to five-parameter models; the hypoelastic model; the shear and bulk moduli models; the endochronic theory; fracture modelling of concrete is discussed in detail together with cracking criteria for single and multiple cracks; blunt crack band propagation; and concrete penetration, perforation, scabbing and spalling. A reference is given on explosion effect on concrete structures. The blast effects on concrete is given in Appendix XVII. Tables and graphs are provided along with diagrams to support these numerical models. Chapter III gives several constitutive creep models including time and strain hardening concepts related to creep models. They are followed by a comprehensive treatment of the three-dimensional thermal creep model. Shrinkage and thermal models are discussed under separate headings within this chapter. The chapter is supported by relevant data. All concrete models in these chapters are presented in such a way that they can also be easily incorporated into the finite element analysis.

The bond and bond–slip phenomenon is one of the most difficult concepts in the field of concrete. Many areas have not been properly understood because of several variables and problems encountered in the measurement of bond. A comprehensive treatment of the bond and bond–slip phenomenon is given in Chapter IV with an emphasis on models which are related to the finite element analysis. Both conventional steel and prestressed steel are examined for bond and bond–slip conditions. For the first time a three-dimensional bond linkage element has been developed by Dr M. Ahmed under the supervision of the author. The element is known as the Ahmlink element. Various constitutive equations and the final stiffness matrix are given for this element to be included in the global finite element analysis. Analytical bond–slip curves obtained have been validated by experimental results and by those available in published data. Chapter V gives a summary of the empirical formulae for the fatigue of concrete. A number of computer subroutines have been developed on the basis of these formulae. They are linked to the global finite element analysis and to the computer program ISOPAR. A comprehensive fatigue analysis is extended and results are fully corroborated in Appendix XV.

Part II contains two major chapters on finite element analysis and the simulation techniques of various concrete numerical models. Chapter VI gives a step-by-step formulation of the finite element analysis of concrete structures in which provision is made for elastic, elasto-plastic cracking and crushing of concrete structures. The influence of creep, shrinkage, bond and fatigue is included. The analysis caters for both static and dynamic loads with special emphasis on the effects of impact, blast and earthquakes. Solid isoparametric elements, panel/shell elements and line elements are included to represent concrete, steel liners/membranes and reinforcement/ prestressing tendons respectively. Other popular elements are introduced as an alternative choice for finite element analysis. This chapter includes details of solution procedures for the finite element equations applied to a number of concrete structures. Chapter VI is supported by a number of appendices which include additional analyses, element types, material compliance matrices and some computer subroutines.

Chapter VII gives material modelling of major constitutive equations given in earlier chapters. A procedure is given for the simulation of these models in the finite element analysis. The numerical models described in Chapter II along with additional numerical models are expanded and a link is established between them and the relevant equations of Chapter VI. This linking procedure forms the basis on which various subroutines are written for the computer program ISOPAR. The linking procedure is also appropriate for other similar finite element packages and paves the way for the effective use of this technique in other research areas for concrete.

Chapter VIII is devoted entirely to seismic criteria and numerical modelling of soil–concrete structure interaction with and without isolators. Structural design requirements for concrete structures have been thoroughly discussed. Structural framing systems are introduced. Seismic performance categories are dealt with in detail with various concrete building configurations. Sections exist on combination of load effects which generally cause deflection and drifts. For these various limits are included. Equivalent lateral force procedures are systematically introduced with full evaluations of horizontal shear distributions. Vertical distribution of seismic forces are then brought into play. They are followed by drift determination and P–Δ effects. These then pave the way for the modal analysis procedure. A vast section exists on soil–concrete structure interaction effects. Methods of analysis such as *response spectrum* and *time-history* are introduced for any concrete structures interacting with soils. Characteristics of interaction are clarified. The site response problems including vertical seismic response and P–Δ effect are fully dealt with. Frequency domain analysis is fully exposed to the reader with and without sub-structure method. Numerical modelling for mode superposition method is now introduced. A full blown analysis is given on *Modal Analysis of Concrete Cyclic Structures* with and without modal dumping. Energies are introduced which can be included in the dynamic non-linear finite element analyses.

A vast section exists on modelling of isolators. The advantages of isolators are given. Isolation systems are discussed. Prior to selection of any one of them, a clear cut approach is given for the isolator uses in constructed facilities. After giving the mechanical characteristics, numerical modelling with isolators is introduced in detail. Equations of motion with isolators are then introduced. Response spectra with discrete energy expressions have been determined. The entire numerical work is now ready to form the basis for the finite element analysis of concrete structures in the seismic environments, taking into consideration the interaction effects with ground.

Part III is devoted to the application of this work. Case studies have been selected from various areas of engineering. They include structures for building works, hydro-electric and nuclear power stations, offshore, defence, shelters and storage facilities etc. Care is taken to include those existing structures which are initially supported by experimental test results and/or site monitoring. In this way the work reported in the text can assume a degree of validation.

Chapter IX covers various case studies which include beams, slabs, concrete-filled tubes, tunnels, nuclear reactor pressure and containment vessels, underground nuclear shelters, gravity platforms and liquefied natural gas tanks. The analytical results compared well with those obtained from the experiments, site monitoring and some published results. Various concrete structures–foundation interaction problems are solved.

A total number of seventeen appendices are included which act as supporting analyses to the main analyses given in the text. They include relevant element library, transformation matrices, material and cracking matrices with computer subroutines, Gaussian quadrature and hourglassing, acceleration and convergence criteria, experimental and numerical modelling of cubes and reinforcing bars, stress–strain curves for conventional and prestressing steel, computer subroutines for cracking of concrete leading to three dimensional cases, data for concrete offshore structures with computer subroutines of individual cases, geometry and forces for domes, large strain analysis, collapse analysis for bridge decks, interface or gap elements, data on the Kobe earthquake and computer subroutines for concrete–soil structure interaction, fatigue analysis data, ductility of beams and blast effects on concrete structures.

In order to acquaint the reader with such specialised topics, each case study is supported by a brief introduction to the background of the relevant areas. In this way the reader becomes familiar with the problem prior to its solution. In addition, the text gives a comprehensive bibliography for those who wish to study these topics in greater depth.

When writing the book the author relied on a wide range of published references and data provided by organisations, but he also sought to incorporate the results of his own research papers which have appeared in various journals, periodicals and conference proceedings but have not been published in book form. The book has the flexibility to include additional topics and case studies. The book can be used by engineers, technologists, mathematicians and specialists in computer-aided techniques who are involved in numerical modelling of materials such as concrete and the finite element modelling of concrete structures. This book can also be used by practising engineers and designers in the field of concrete for nuclear power, concrete in the ocean, concrete under extreme temperatures and loads and concrete in conventional structures such as buildings, bridges, dams and heavy foundations. The book is also relevant to postgraduate courses on concrete structures subject to static, impact, dynamic, blast and earthquake loads.

M. Y. H. BANGASH

Acknowledgements

The author wishes to express his appreciation and gratitude to many civil/structural, nuclear, offshore, petroleum and aerospace engineers, engineering consortia and research establishments, whose names are mentioned in the text, for helpful discussions, criticisms and guidance. The author is also indebted to all the authors of the research papers and reports mentioned in the Bibliography for generous help in the acquisition of data, graphs and computer outputs.

Special thanks to Professor Z. P. Bazant, Professor F. H. Wittmann, Dr G. L. England, Dr J. Stevenson, Dr J. Swanson, Dr D. W. Murray, Professor O. C. Zienkewitz, Professor J. H. Argyrus and Dr N. S. Ottoson.

Thanks to Professor E. Wilson, Dr K. J. William, Mr Janson, Mr A. Nielsen and Mr D. Olsen for generously providing their research materials.

It will not be out of place to mention the help of many students notably, Dr M. Ahmad, Mr A. Aziz, Mr C. Lai, Mr N. Y. Lau, Mr V. Salashoori, Mr J. Tang, Mr A. Ghorbani and Mr Thirugonosothy, in the development of some computer subroutines.

The author is indebted to the following:

Mr Mike Chrimes and his staff, the Institution of Civil Engineers Library, London
Mrs Anita Witten and her staff, the Institution of Structural Engineers Library, London
The American Society of Civil Engineers, New York
The American Concrete Institute, Detroit
The Concrete Society, London
The Royal Institute of British Architects, London
The British Cement Association, UK
Patent Office Library, Chancery Lane, UK
The Home Office, Croydon, Surrey, UK
The Ulster Constabulary, Belfast
The Institution of Chemical Engineers Library, Derby, UK
The MUTO Institute, Tokyo
The Japan Society of Civil Engineers, Tokyo
The Chinese Society of Civil Engineers, Beijing
The Pakistan High Commission, London
The Institution of Mechanical Engineers, London

Der Bauingenieur, Germany
Times Index, Research Publications, Reading, UK
The American Association of State Highway and Transportation Officials (AASHTO), California
The Ontario Highways Department, Canada
The European Union Library, Brussels, Belgium
The Asean Countries, Central Library, Korea
Beratende Ingenieure VBI, Hamburg, Germany
Dr Ing Fritz Notzold, Consulting Engineers, Langen, Germany
Transport and Road Research Laboratory, Crowthorne, Berkshire, UK
Flint & Niel, Consulting Engineers, London
Leonhardt, Andra & Partners, Stuttgart, Germany
Swindell Fresseler Corp., Pittsburgh
The Indian Concrete Journal and the Indian Concrete Society, New Dehli
The Punjab Engineering Congress, Lahore, Pakistan
DIANA Analysis, Delft, The Netherlands
Abacus, Manchester
FIP Industriale, SpA, Italy
BBR Systems Ltd, Switzerland
Weidlinger Associates Ltd, New York
Acrow Corporation of America, New Jersey
The D. S. Brown & Co, Baltimore, Ohio, USA
The International Association of Bridge and Structural Engineering c/o the Institution of Structural Engineers, London
J. Muller International, San Diego, California, USA
Parsons Brinckerhoff, New York
Yapi Merkezi Inc., Istanbul
Nippon Steel, Tokyo
Takenaka Corporation, Tokyo
Nihon Sekki Ltd, Tokyo
Hyder Consultants, Surrey, UK
Alan Grant & Partners, Cobham, Surrey, UK
Mott Macdonald, Croydon, Surrey, UK
Gifford & Partners, UK
US Agency for International Development, Washington DC
Civil and Structural Computer Services Ltd, Newtonabbey, Co. Antrim, UK
Taylor Woodrow Construction, Southall, UK

This acknowledgement would be incomplete without praising the enormous help given by the City of London Police and the UK Highways Agency, and the computer bureaux STRUCOM Ltd, London, and the FEA, Kingston, Surrey, UK.

The author acknowledges that owing to a unified approach, many symbols given in original sources had to be replaced by new ones.

Major notation

$[B]$	Strain–displacement matrix
$[B_i]$	Strain–displacement matrix at node i
$[C]$	Inverse of material matrix
d, ϕ	Diameter of the line element
d, D	Missile diameter or diameter for structures
$\det J$	Determinant of Jacobian
ds	Differential surface area
d vol	Differential volume
$[D]$	Material compliance matrix
$[D_e]$	Elastic material matrix for steel
$[D_{ep}]$	Elasto-plastic material matrix for steel
$[D_p]$	Plastic material matrix for steel
$\{D_E\}$	Elastic material matrix for membrane element
$[D_T]$	Tangent constitutive matrix in global co-ordinate system
$[D^*]$	Constitutive matrix in crack co-ordinate system
E_b	Bond linkage constitutive matrix
E_c	Modulus of elasticity for concrete
E_1, E_2, E_3	Moduli of elasticity in three principal directions
F	Vector of external element loads
$\{F\}$	Global load vector
G	Shear modulus
G'	Shear modulus for cracked concrete
$[J]$	Jacobian matrix
k_H, k_V, k_E (or k_h, k_v, k_e)	Bond–slip moduli in horizontal, vertical and lateral directions
$[K]$	Global structural stiffness matrix
$[K_b]$	Bond linkage stiffness matrix
$[K_T]$	Total stiffness matrix
$[K_0]$	Effective stiffness matrix
$[K^e]$	Element stiffness matrix
l, m, n	Direction cosines relating local to global axes

L	Length of the line elements
n_1, n_2	Unit vectors
N	Element shape function matrix
N_i	Shape function of node i
$\{P_G\}$	Vector of body forces per unit volume
P_s	Vector of surface pressure
q	Starting iteration vector
$\{R_i\}$	Residual force vector
$\{R_t\}$	External load vector
$\{R_t\}$	Projected load vector
s	Spalling thickness/scabbing thickness
t	Time during analysis
$[T]''$	Transformation matrix (relating local to global displacements at nodes)
$\{T''_\varepsilon\}$	Strain transformation matrix
$\{T''_\sigma\}$	Stress transformation matrix
$\{U\}$	Global nodal displacement vector
U_i, V_i, W_i	Displacements at node i
$\{U^e\}$	Element nodal displacement vector
V	Impact velocity
w, W	Missile weight or any other weights
W_e	External work
w_i	Nodal displacement in the z-direction at node i
X	Constant
X, Y, Z	Global co-ordinate system
X', Y', Z'	Local Cartesian co-ordinate system
X^*, Y^*, Z^*	Crack co-ordinate system
α_{ij}	Total translation of the yield surface
β'	Shear interlocking factor
δ, u	Displacement
$\dot{\delta}$, $\dot{u}$	Velocity
$\ddot{\delta}$, $\ddot{u}$	Acceleration
δ_{ij}	Kronecker delta
$\{\delta\}_0$	First displacement vector
δ_1, δ_2, δ_3	Displacement increments
δ^*	Virtual displacement
$\{\Delta S\}$	Incremental slips
$\Delta\varepsilon$	Incremental uniaxial strain
$\Delta\varepsilon_i$, $\Delta\sigma_i$	Incremental strain/stress vectors
$\{\Delta\varepsilon_0\}$	Plastic strain increment
$\{\Delta\varepsilon_t\}$	Incremental thermal strains
$\{\Delta\varepsilon^*\}\{\Delta\sigma^*\}$	Incremental strain/stress vectors in crack directions
$\{\Delta\sigma_b\}$	Incremental bond stress vector

ε	Equivalent strain
$\{\varepsilon_{cu}\}$	Concrete crushing strain
ε_i	Vector of Cartesian strain components at point i
ε_0	Initial strain vector
ε^c	Creep strain vector
$\{\varepsilon^*\}$	Strain vector in crack directions
ν_c	Poisson's ratio for concrete
ξ, η, ζ	Local curvilinear co-ordinates
σ	Equivalent stress
$\{\sigma_b\}$	Bond stress vector
σ_c	Concrete cylinder compressive strength
σ_i	Principal stresses, $i = 1, 2, 3$
σ_i	Vector of Cartesian stress components at point i
σ_{ij}	Stress tensor
σ''_{ij}	Deviatoric component of the stress tensor
σ_m	Mean stress
σ_0	Initial stress
σ_0	Initial stress vector
$\{\sigma^*\}$	Stress vector in crack directions
σ_t	Concrete limiting tensile strength
τ	Increase in time t

Conversion factors

Imperial units	*SI units*	*Additional*
1 in	= 254 mm	
1 in^3	= 0.003785 (in^3)	
1 ft	= 30.48 cm	
1 ft^2	= 0.09290 (in^2)	1 ft^3 = 0.02832 (m^3)
10 ga	= 3.57 mm	1 cu yard = 0.765 (m^3)
18 ga	= 1.27 mm	
1 lb	= 0.454 kg	
1 ton	= 9.964 kN	
1 sq ft	= 929 $(cm)^2$	
1 cubic ft	= 16.4 $(cm)^3$	
1 psi	= 6.89 kPa	= 6895 Pascal (Pa)
20 T/ft^2	= 1915.2 kN/m^2	
1 lb/sq ft	= 992.16 kPa	= 47.88 Pascal (Pa)
1 lb/ft^3	= 16.02 kg/m^3	
1 ft/lb	= 1.356 N.m	
1 ft/sec	= 0.3048 M/s	
1 slug	= 14.59 kg	
1 in lb	= 0.1129848 Nm	
1 kip/in	= 175.1268 kN/m	
1 bar	= 100 kN/m^2	
1 kip	= 1000 lb	= 4.448 kN
1 short ton (2000 lb)	= 0.9072 Megagram (Mg)	

MKS units	*SI units*
1 Pascal (Pa)	= 1 N/m^2
1 kgm	= 9.807 Nm
1 kgf	= 9.807 N

Temperature in °C (Celsius)

1°F (Fahrenheit) = $t_f = \frac{5}{9}$K		$t_c = (t_f - 32)/118$
1°C		$t_f = 1.8t_c + 32$
1 BTU = 1055 J		

1 m^2/g	= 1000 m^2/kg
1 mm^2/mg	= 1 m^2/kg
1 radian	= 57.296 deg
1 ft/sec	= 0.3048 m/sec

Contents

PART I

This part contains a literature review of the material properties and strengths of concrete. Detailed work is presented on the numerical modelling of concrete strength and the failure of concrete. Numerical modelling of creep, shrinkage temperature, bond and fatigue is also covered in Parts II and III for the finite element analysis of concrete structures.

1. Material properties and strength of concrete — a literature survey of numerical modelling

1.1 Introduction

This chapter includes a literature survey of the material properties and strength of concrete. Various concrete strength models are discussed. Material data collected for concrete subjected to monotonic loads, cyclic loads and temperature has been used elsewhere in the analysis of concrete structures. The effects of creep and shrinkage are discussed within the text. A comparative study of the behaviour of concrete is given for such numerical models.

1.2 Stress–strain curves

1.2.1 Uniaxial compression

Experimental tests show that concrete behaves in a highly nonlinear manner in uniaxial compression. Figure 1.1 shows a typical stress–strain relationship subjected to uniaxial compression. This stress–strain curve is linearly elastic up to 30% of the maximum compressive strength. Above this point the curve increases gradually up to about 70–90% of the maximum compressive strength. Eventually it reaches the peak value which is the maximum compressive strength σ_{cu}. Immediately after the peak value, this stress–strain curve descends. This part of the curve is termed *softening*. After the curve descends, crushing failure occurs at an ultimate strain ε_{cu}. A numerical expression has been developed by Hognestad [1] which treats (Fig. 1.1) the ascending part as a parabola and the descending part as a straight line. This numerical expression is given as:

$$\text{for } 0 < \varepsilon < \varepsilon'_0 \qquad \frac{\sigma}{\sigma_{cu}} = 2\frac{\varepsilon}{\varepsilon'_0}\left(1 - \frac{\varepsilon}{2\varepsilon'_0}\right) \tag{1.1}$$

$$\text{for } \varepsilon'_0 < \varepsilon < \varepsilon_{cu} \qquad \frac{\sigma}{\sigma_{cu}} = 1 - 0.15\left(\frac{\varepsilon - \varepsilon'_0}{\varepsilon_{cu} - \varepsilon'_0}\right) \tag{1.2}$$

where ε'_0 is the strain at the peak stress, ε_{cu} is the ultimate strain and σ is the stress.

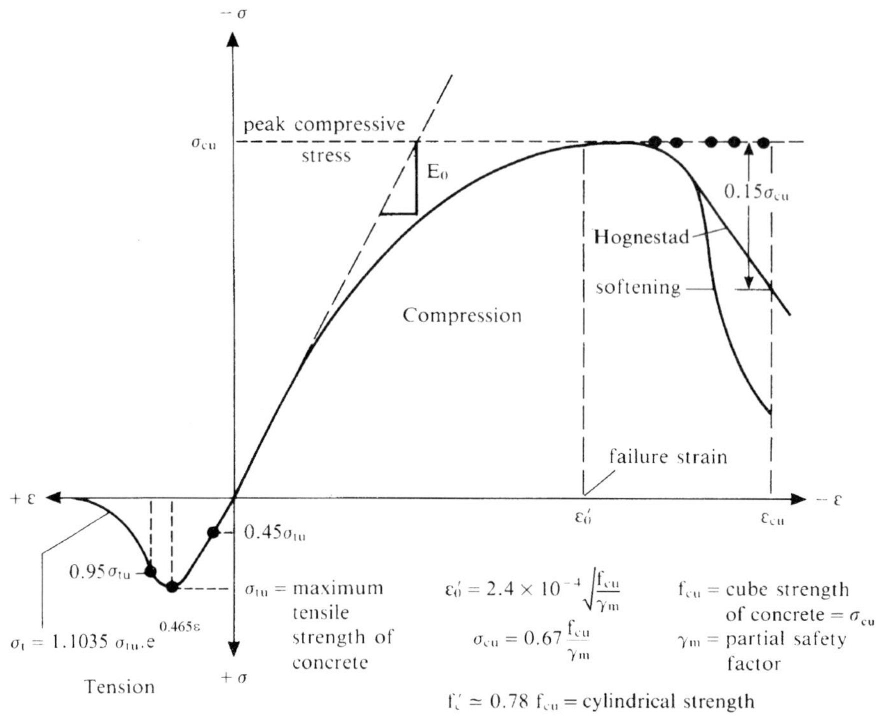

Fig. I.1. Typical uniaxial compression or tension stress–strain curve for concrete.

The initial tangent modulus E_0 is then computed as

$$E_0 = \left.\frac{d\sigma}{d\varepsilon}\right|_{\varepsilon'=0} = 2\frac{\sigma_{cu}}{\varepsilon'_0} \tag{I.3}$$

and equals twice the secant modulus taken at the peak stress σ_{cu}.

Hognestad suggested that the modulus of elasticity E_0 in psi is

$$E_0 = 1.8 \times 10^6 + 460(0.85 \times \text{the compressive strength of standard cylinders}) \tag{I.4}$$

Desayi and Krishnan [2] give a general expression for both the ascending and descending parts of the curve as

$$\sigma = \frac{E_0 \varepsilon}{1 + \left(\dfrac{\varepsilon}{\varepsilon'_0}\right)^2} \tag{I.5}$$

The European Concrete Committee (CEB) for short-term loading gives a parabola and a straight line up to ultimate strain ε_{cu} as shown in Fig. I.1. In numerical terms this is given as

$$\frac{f'_c}{\sigma_{cu}} = \frac{k\eta - \eta^2}{1 + (k-2)\eta} \tag{I.6}$$

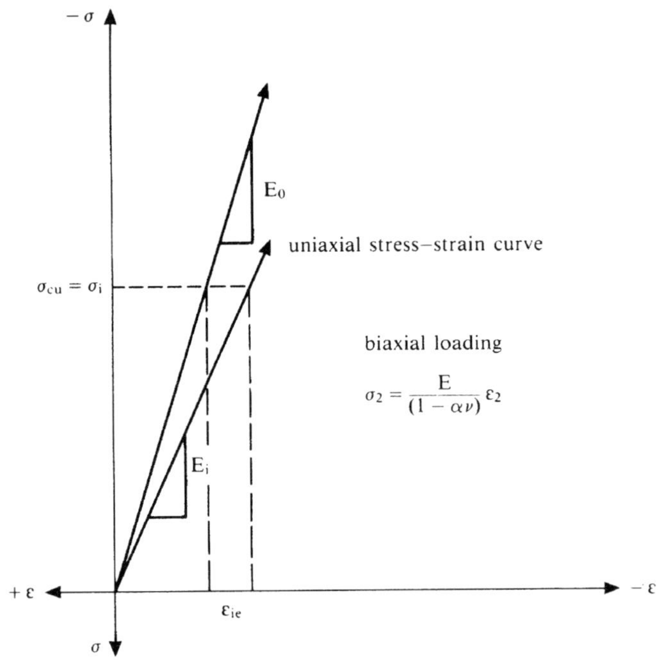

Fig. I.2. Stress–strain curve and equivalent strain.

where

$$\eta = \frac{\varepsilon_{cu}}{0.002} \qquad k = \frac{[0.0022(1.1E)]}{\sigma_{cu}}$$

The value of ε_{cu} is given as between 0.003 and 0.0035.

The concept of equivalent uniaxial strain for any stress has been developed [3] so as to assess, at any stage, the degradation of stiffness and strength of plain concrete and also to allow the actual biaxial stress–strain curves to be duplicated from such uniaxial curves. This concept provides a method of separating the Poisson effect from the cumulative strain. Figure I.2 shows the equivalent uniaxial strain ε_{ie} for concrete as a linear material. The value of ε_{ie} for nonlinear material is written as

$$\varepsilon_{ie} = \sum \mathrm{d}\varepsilon_{ie} = \int \frac{\mathrm{d}\sigma_i}{E_{ti}} = \sum \frac{\Delta\sigma_i}{E_{ti}} \tag{I.7}$$

where $\mathrm{d}\varepsilon_{ie}$ and $\mathrm{d}\sigma_i$ are, respectively, differential changes in stress and equivalent uniaxial strain in the ith direction. $\sum$ is for all load increments. E_{ti} is tangent modulus in the ith direction.

Saenz [4] selected curves for compressive loading and has defined them numerically (Fig. I.3) as

$$\sigma_{cu} \simeq \sigma_i = \frac{\varepsilon_{ie} E_0}{1 + \left(\dfrac{E_0}{E_{sec}} - 2\right)\dfrac{\varepsilon_{ie}}{\varepsilon_0'} + \left(\dfrac{\varepsilon_{ie}}{\varepsilon_0'}\right)^2} \tag{I.8}$$

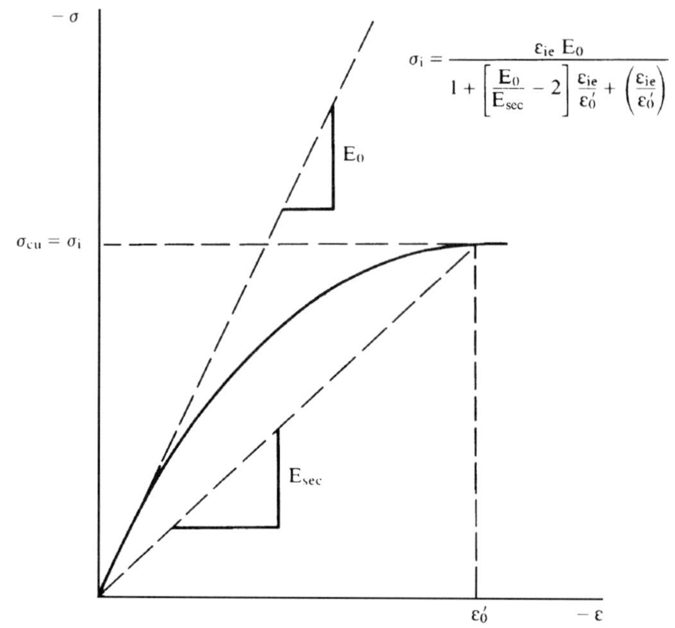

Fig. I.3. Equivalent uniaxial stress–strain curve.

where E is the second modulus. Elwi and Murray [5] have expressed eqn (I.8) as

$$\varepsilon_{ie}E_0 \Big/ 1 + \left(\frac{E_0}{E_{sec}} + R - 2\right)\frac{\varepsilon_{ie}}{\varepsilon'_0} - (2R-1)\left(\frac{\varepsilon_{ie}}{\varepsilon'_0}\right)^2 + R\left(\frac{\varepsilon_{ie}}{\varepsilon'_0}\right)^3 \qquad \text{(I.8a)}$$

where

$$R = \frac{E_0}{E_{sec}}\left(\frac{R_\sigma - 1}{R_\varepsilon - 1}\right) - \frac{1}{R_\varepsilon} \qquad R_\sigma = \frac{1}{0.85} \qquad R_\varepsilon = \frac{\varepsilon'_0}{\varepsilon_{cu}}$$

Typical stress–strain curves for concrete in compression have been produced by Bangash [6] for different concrete grades, and are shown in Fig. I.4 along with the BS 8110 actual stress curves. The idealised curves are also plotted in the same figure. Figure I.5 shows a comparative study of concrete stress–strain diagrams in compression under various age conditions.

Figure I.6 shows a typical stress–strain curve for concrete undergoing several cycles of load reversals in uniaxial compression. In such a case many values of strain occur for a single value of stress. An idealised model curve is proposed by Darwin and Pecknold [7a,b] and is plotted on such cyclic curves as given in Fig. I.6. This curve is based on several experimental results from Karson and Jirsa [8]. Equation (I.8) is used to describe the ascending part of the idealised curve, and the descending part of the curve is a straight line ending at $0.2f'_c$, where f'_c is the cylindrical compressive strength of concrete. Figure I.7 shows some of Karson and Jirsa's experimental results. An empirical relation is established between ε_{en} (the strain on the envelope curve prior to unloading) and the residual strain remaining after the

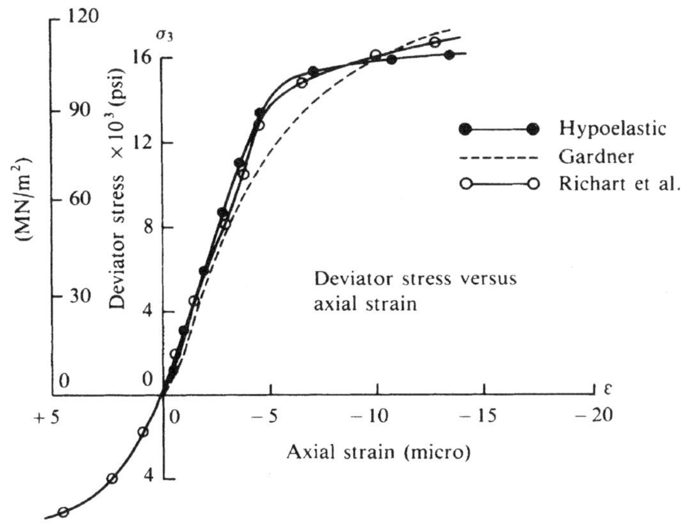

Fig. I.4(a). Deviator stress versus axial strain.

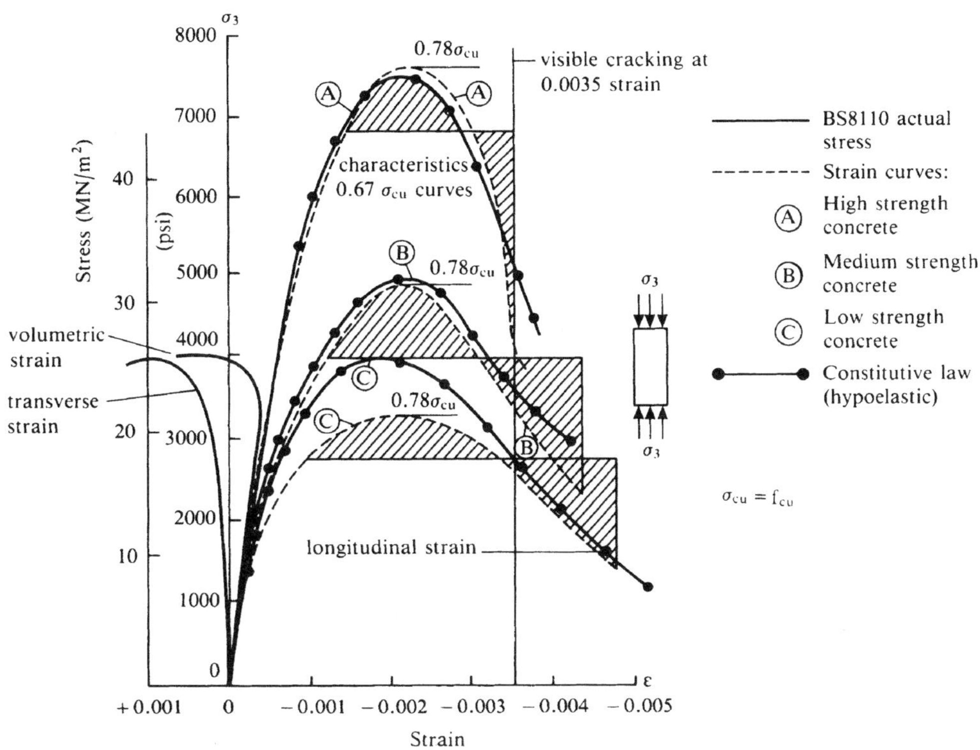

Fig. I.4(b). Typical stress–strain curves for concrete compression: comparison with BS 8110.

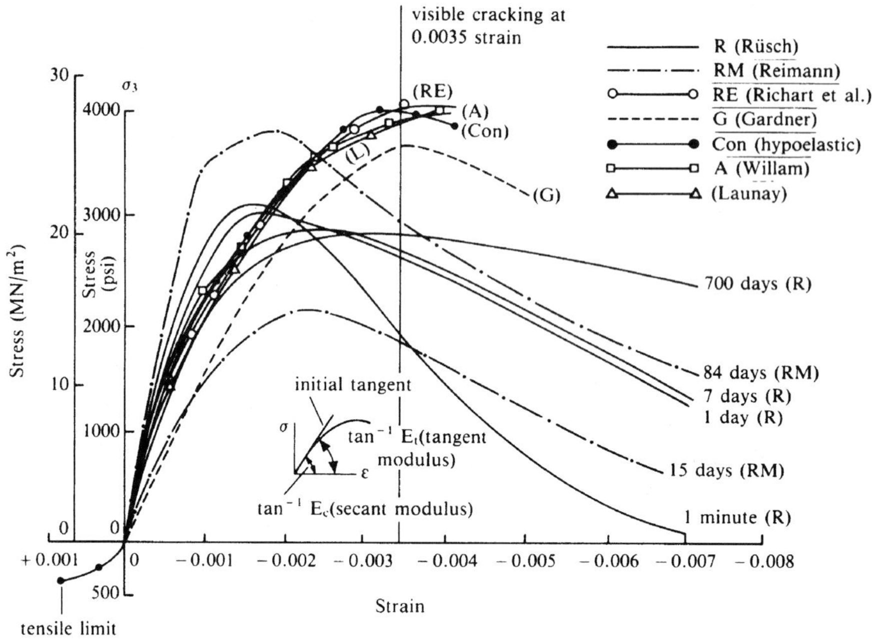

Fig. I.5. Stress–strain curve for concrete.

entire load is released, ε_p (the plastic strain). This is given as

$$\frac{\varepsilon_p}{\varepsilon_{cu}} = 0.145\left(\frac{\varepsilon_{en}}{\varepsilon_{cu}}\right)^2 + 0.13\left(\frac{\varepsilon_{en}}{\varepsilon_{cu}}\right) \tag{I.9}$$

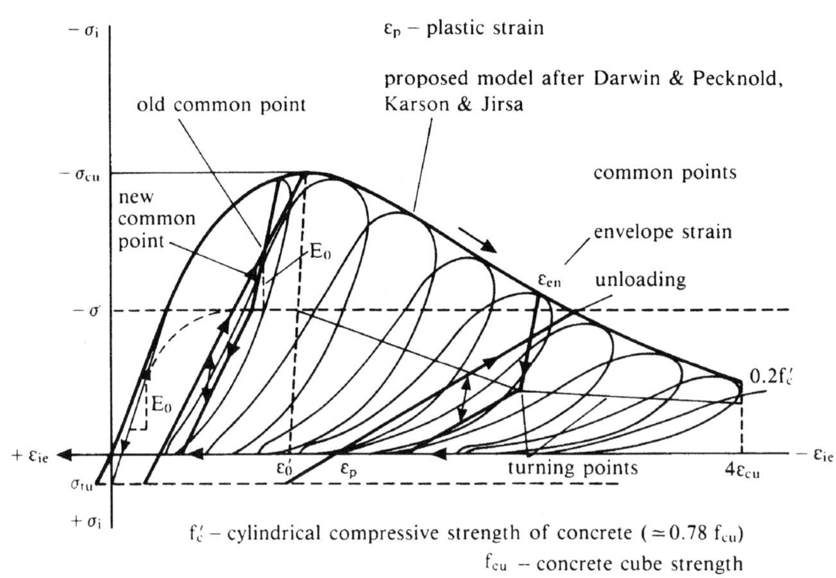

Fig. I.6. Cyclic loading of concrete.

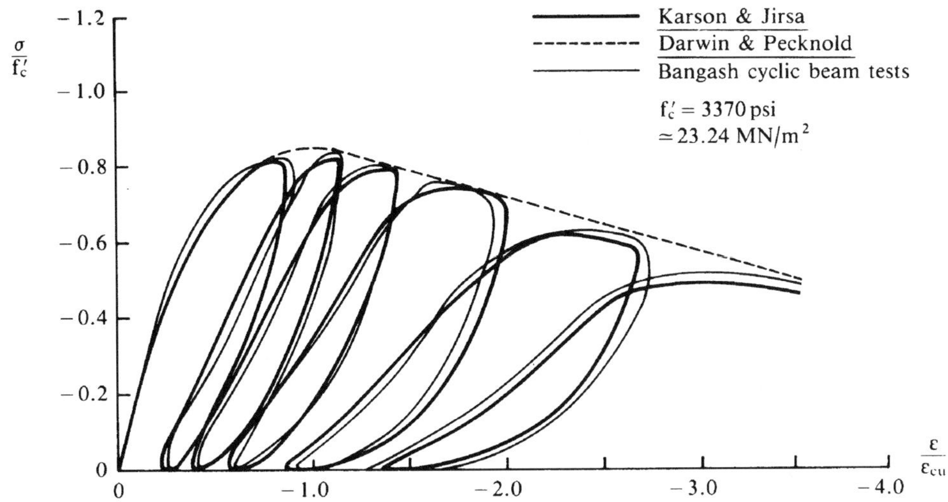

Fig. I.7. Concrete under cyclic load — a comparative study.

Karson and Jirsa found a band of points on the stress–strain plane which controls the degradation of the concrete under continued load cycles. Below this band is the 'stability limit' and above it is the 'common point' limit. The band is reduced to a single curve which is then called the *locus of common points*. As shown in Fig. I.6, if the load is cycled above the stability limit and peak stress maintained between cycles, additional permanent strain would accumulate. The common point limit shows the maximum stress at which a reloading curve may intersect the original unloading curve. Each new envelope point defines a common point and a new plastic strain. The location of the common point with respect to the envelope curve can be adjusted in order to control the number of cycles leading to failure. Continued cycles above this common point can eventually intersect the envelope curve. If the locus of common points is lowered, fewer cycles are needed to intersect the envelope for a given value of the peak stress. A typical locus of common points from Darwin and Pecknold [7a,b] is given in Fig. I.8. The energy dissipated for each cycle is controlled by these turning points. The lower the turning point, the greater the energy dissipated per cycle. Figure I.8 gives a simple satisfactory energy for all but low values of ε_{en}. For low values of ε_{en} the unloading and reloading take place on a single line when the slope is E_0 and when

$$|\varepsilon_{en}| \leq \tfrac{1}{4}|\varepsilon_{cu}| \qquad \text{(I.10)}$$

I.2.2 Uniaxial tension

Tensile behaviour is evaluated by either

(a) the split cylinder test or
(b) the modulus of rupture or bending test

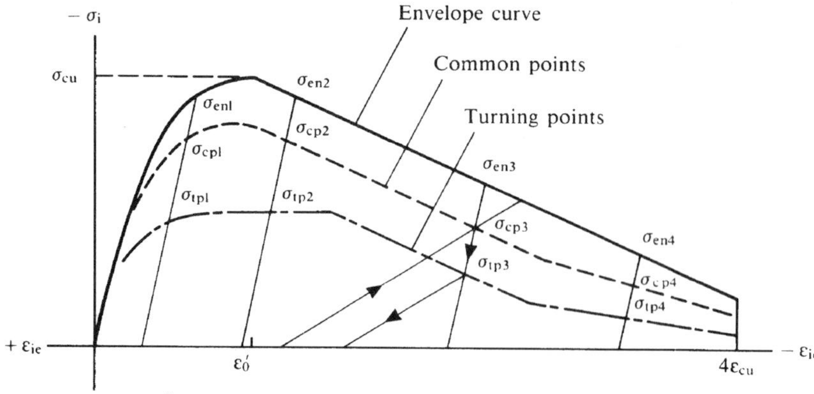

σ_i 1: $\sigma_{cp1} = \frac{5}{6}\sigma_{en1}$

$\sigma_{tp1} = \frac{1}{2}\sigma_{en1}$

2: $\sigma_{cp2} = \sigma_{en2} - \text{Min*}\left(\frac{1}{6}\sigma_{en2}, \frac{1}{6}f_c'\right)$

$\sigma_{tp2} = \text{Min*}\left(\frac{1}{2}\sigma_{en2}, \frac{1}{2}f_c'\right)$

* Largest in absolute magnitude

3: $\sigma_{cp3} = \sigma_{en3} - \frac{1}{6}f_c'$

$\sigma_{tp3} = \sigma_{en3} - 2(\sigma_{en3} - \sigma_{cp3})$

$= \sigma_{en3} - \frac{1}{3}f_c'$

4: $\sigma_{cp4} = \frac{2}{3}\sigma_{en4}$

$\sigma_{tp4} = \frac{1}{3}\sigma_{en4}$

Fig. I.8. Envelope curve, common points and turning points [7].

In both these methods, when the tensile strength σ_{cu} decreases, the compressive strength increases. For any stress less than $0.6\sigma_{cu}$, microcracks are ignored. Welch [9] and Evans [10] give a value of $0.75\sigma_{cu}$ for the onset of unstable cracks. The tensile strength can be expressed in terms of the cylindrical compressive strength of concrete. Hughes and Chapman [11] have carried out several tests. They found that the ratio between uniaxial tensile strength and compressive strength ranges from 0.05 to 0.1. The modulus of elasticity under uniaxial tension is higher, and Poisson's ratio lower, than those obtained from the uniaxial compression curves. For the tensile splitting case, σ_{tu} is given by

$$\sigma_{tu} = 0.55\sqrt{f_c'} \quad \text{(MPa)} \tag{I.11}$$

For the modulus of rupture case, the tensile strength σ_t is given by

$$\sigma_{tu} = 0.95\sqrt{f_c'} \quad \text{(MPa)} \tag{I.12}$$

where f_c' is the cylindrical strength of concrete in compression. Figures I.1 and I.5 show the tensile 'cut-off'. Data on the complete stress–strain curves for concrete subjected to direct tension have been reported by Evans [10].

Johnson [13] carried out tests and found that the ratio of uniaxial tensile strength to compressive strength ranges from 0.05 for high-strength concrete to about 0.1 for a concrete of medium strength, thus endorsing the figures given by Hughes and Chapman. He also found that the type of aggregate influences the tensile strength of concrete. Siliceous aggregates decrease the tensile strength, and calcareous aggregates increase the tensile strength of concrete.

A general form of the tensile stress–strain curve is given in Fig. I.1.

I.3 Behaviour of concrete under biaxial stress

A number of investigators [1–31] have attempted to evaluate concrete strength under biaxial compression, biaxial tension–compression, or under any combination of loads. Kupfer *et al.* [29] and Liu *et al.* [30] have conducted biaxial loading tests. Three detailed experiments have been carried out and the results are summarised in Figs I.9 and I.10. For biaxial compression, Kupfer and Gerstle [31] have closely approximated the strength curve by eqn (I.13). Where $\sigma_1 \geq \sigma_2$,

$$\left(\frac{\sigma_1}{f_c'}+\frac{\sigma_2}{f_c'}\right)^2-\left(\frac{\sigma_2}{f_c'}\right)-3.65\left(\frac{\sigma_1}{f_c'}\right)=0 \tag{I.13}$$

Equation (I.13) can be rewritten for $\alpha = \sigma_{1c}/\sigma_{2c}$ to give σ_{2c}. The maximum compressive strength of concrete as a function of the uniaxial compressive strength f_c' and α (Fig. I.11) is

$$\sigma_{2c}=\frac{1+3.65\alpha}{(1+\alpha)^2}f_c' \tag{I.14}$$

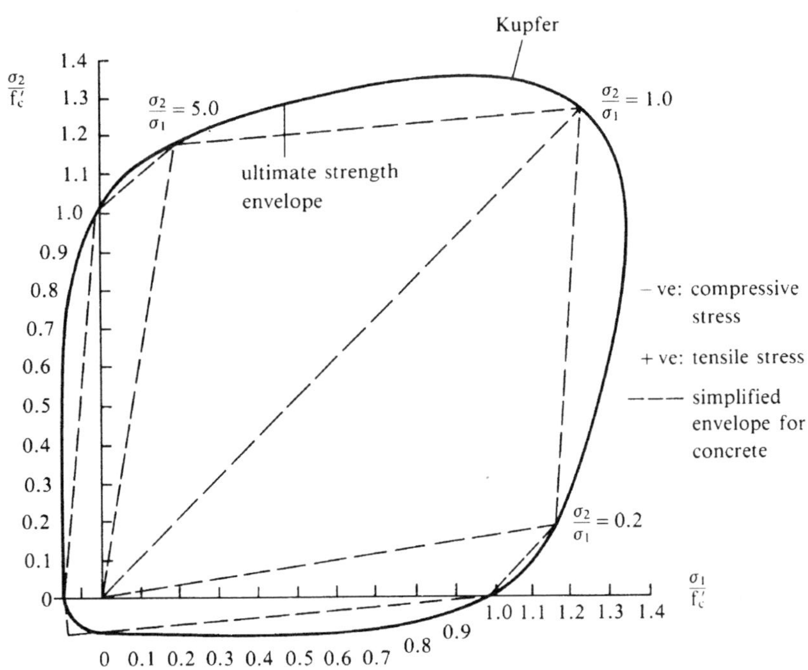

Fig. I.9. Concrete under biaxial stress state in terms of ultimate uniaxial cylinder crushing strength.

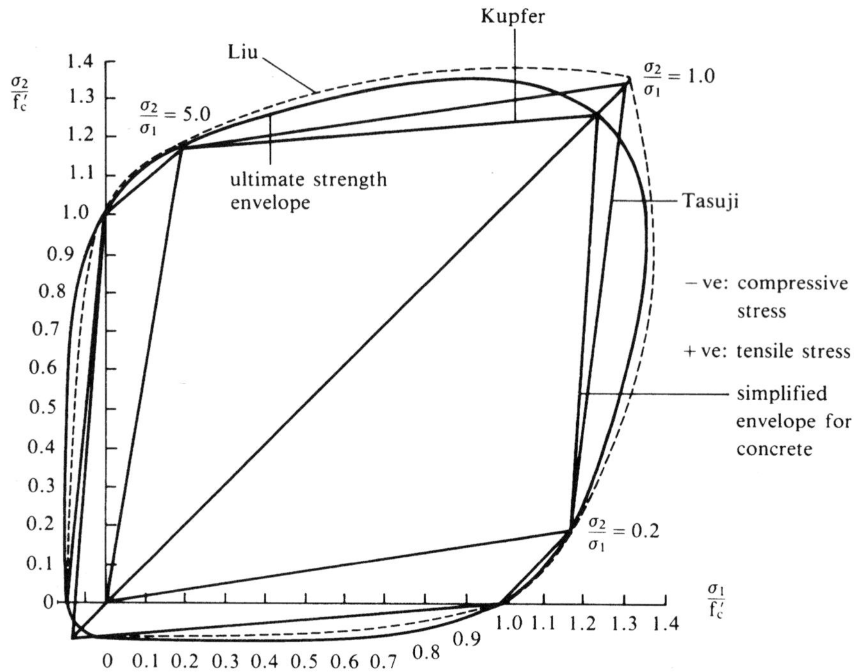

Fig. I.10. Concrete under biaxial stress state in terms of ultimate uniaxial cylinder crushing strength.

The peak stress σ_{cu} is given by

$$\sigma_{1c} = \sigma_{cu} = \alpha\sigma_{2c} \tag{I.15}$$

For the tension–compression case, Kupfer and Gerstle suggest a straight line reduction in tensile strength, taking into consideration the increased compressive stress

$$\sigma_{pt} = \left(1 - 0.8\frac{\sigma_2}{f'_c}\right)\sigma_{tu} \tag{I.16}$$

For tension–compression, Darwin and Pecknold [7a] modified eqn (I.14) and presented it as

$$\sigma_2 = \left[\frac{1 + 3.28\alpha}{(1+\alpha)^2}\right]f'_c \tag{I.17}$$

For tension–tension, Kupfer and Gerstle suggest a constant tensile strength equal to the uniaxial tensile strength of concrete.

The biaxial compression stress–strain curve ($\alpha = 1$) can determine the effect on increased ductility. In this case the real strain at maximum compressive stress, i.e. the peak stress σ_{cu}, is converted to $\varepsilon'_0/(1-\nu)$, thus eliminating the Poisson effect from the strain value, as stated in Section I.2.

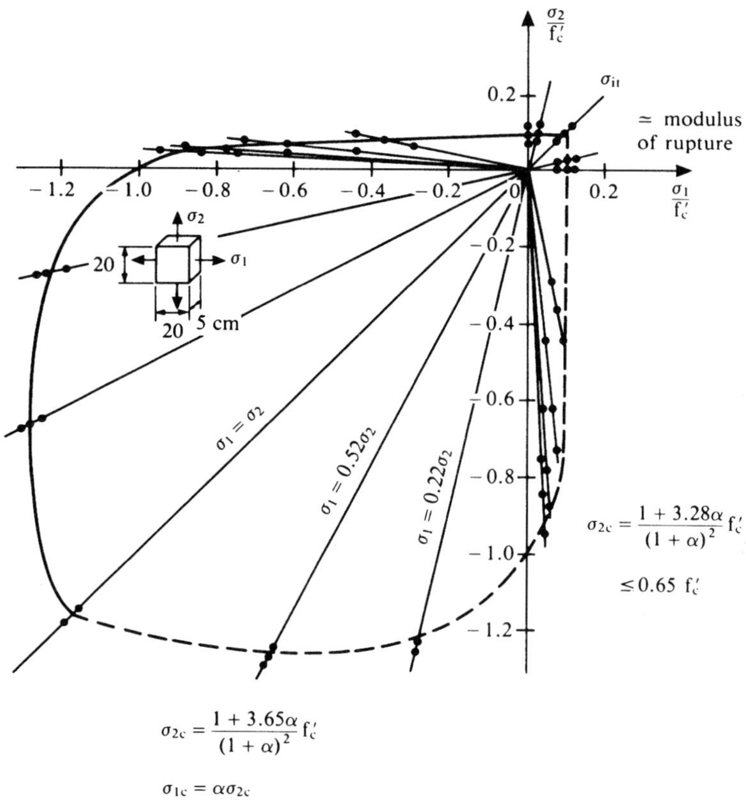

Fig. I.11. Biaxial stress state strength [7a,b].

If the value of ε'_0 varies with σ_{cu}, then ε'_0 is given by

$$\varepsilon'_0 = \varepsilon_{cu}\left[1.6\left(\frac{\sigma_{cu}}{f'_c}\right)^3 + 2.25\left(\frac{\sigma_{cu}}{f'_c}\right) + 0.35\left(\frac{\sigma_{cu}}{f'_c}\right)\right] \tag{I.18}$$

The tensile ductility is greater under biaxial compression than under uniaxial compression. It is estimated that a 25% increase in maximum strength occurs if $\alpha = 2$, and a decrease of approximately 15.5% occurs when $\alpha = 1$. It is interesting to note that when a failure point is approached, with increasing compressive stress, an increase in volume occurs; this inelastic volume is known as *dilatancy*. This is the onset of progressive growth of microcracks in concrete.

Many attempts have been made to produce graphically the biaxial ultimate strength envelope. Figure I.9 shows the biaxial stress state in terms of ultimate cylindrical strength. Using data provided by Kupfer and Gerstle [31], Fig. I.10 shows a comparative study of the biaxial ultimate strength envelopes of Kupfer and Gerstle [31], Tasuji and co-workers [73, 74], and Liu and co-workers [69, 70].

Failure occurs when non-dimensionalised stress lies outside the failure envelope. The maximum compressive stress σ_{cu} in directions 1 and 2 is then determined by

$$\sigma_{cu1} = f_c' \left[1 + \frac{\left(\frac{\sigma_2}{\sigma_1}\right)}{1.2 - \left(\frac{\sigma_2}{\sigma_1}\right)} \right] \quad \text{when } \frac{\sigma_2}{\sigma_1} < 0.2 \tag{I.19}$$

$$\sigma_{cu1} = 1.2 f_c' \quad \text{when } 1.0 \geq \frac{\sigma_2}{\sigma_1} \geq 0.2$$

$$\sigma_{cu2} = f_c' \left[1 + \frac{\left(\frac{\sigma_1}{\sigma_2}\right)}{1.2 - \left(\frac{\sigma_1}{\sigma_2}\right)} \right] \quad \text{when } \frac{\sigma_2}{\sigma_1} > 5.0 \tag{I.20}$$

$$\sigma_{cu2} = 1.2 f_c' \quad \text{when } 1.0 \leq \frac{\sigma_2}{\sigma_1} \leq 5.0$$

Throughout the analysis, tensile stresses are considered to be positive, and compressive stresses to be negative. For tensile failure it is assumed that the load may be carried in the direction of the crack and may not be transferred across the crack. For a case of compressive failure it is assumed that no load may be carried in any direction. In order to comply with these limitations, the following boundaries for corresponding regions are established:

(a) Tension–tension region: σ_{tu}/f_c' as non-dimensionalised values for directions 1 and 2.

(b) Compression–tension region: the boundary is defined as

$$\frac{\sigma_1}{f_c'} = \frac{\sigma_{tu}}{f_c'}\left(1 - \frac{\sigma_1}{f_c'}\right) \qquad \frac{\sigma_2}{f_c'} = \frac{\sigma_{tu}}{f_c'}\left(1 - \frac{\sigma_2}{f_c'}\right) \tag{I.21}$$

(c) Compression–compression region:

Direction 1

$$\frac{\sigma_1}{f_c'} = 1 + \left[\frac{\left(\frac{\sigma_2}{\sigma_1}\right)}{1.2 - \left(\frac{\sigma_2}{\sigma_1}\right)} \right] \quad \text{when } \frac{\sigma_2}{\sigma_1} < 0.2$$

$$\frac{\sigma_2}{f_c'} = 1 + \left[\frac{1}{1.2\left(\frac{\sigma_1}{\sigma_2}\right) - 1} \right]\left(\frac{\sigma_2}{\sigma_1}\right) \quad \text{when } \frac{\sigma_2}{\sigma_1} < 0.2 \tag{I.22}$$

$$\frac{\sigma_1}{\sigma_2} = 1.2 \quad \text{when } 1.0 \geq \frac{\sigma_2}{\sigma_1} \geq 0.2$$

$$\frac{\sigma_2}{f_c'} = 1.2\left(\frac{\sigma_2}{\sigma_1}\right) \quad \text{when } 1.0 \geq \frac{\sigma_2}{\sigma_1} \geq 0.2$$

Direction 2

$$\frac{\sigma_2}{f_c'} = 1 + \left(\frac{\sigma_1}{\sigma_2}\right) \Big/ 1.2\left(\frac{\sigma_1}{\sigma_2}\right) \qquad \text{when } \frac{\sigma_2}{\sigma_1} > 5.0$$

$$\frac{\sigma_1}{\sigma_2} = \left[1 + \frac{1}{1.2\left(\frac{\sigma_2}{\sigma_1}\right) - 1}\right]\left(\frac{\sigma_1}{\sigma_2}\right) \qquad \text{when } \frac{\sigma_2}{\sigma_1} > 5.0 \qquad \text{(I.23)}$$

$$\frac{\sigma_2}{\sigma} = 1.2 \qquad \text{when } 5.0 > \frac{\sigma_2}{\sigma_1} > 1.0$$

$$\frac{\sigma_1}{\sigma} = 1.2\left(\frac{\sigma_1}{\sigma_2}\right) \qquad \text{when } 5.0 > \frac{\sigma_2}{\sigma_1} > 1.0$$

I.3.1 Biaxial state adopting the octahedral stress approach

In the biaxial state of stress, the value of the shear modulus is assumed to be a function of the confined pressure and the second invariants of the stress and strain. The yield criterion in general is represented by $\tau_0 = f(\sigma_0)$, in which the hydrostatic pressure σ_0, as explained later, is given by $\sigma_0 = \frac{1}{3}J_1$. Again using Kupfer *et al.*'s experimental results [29], a good approximation is obtained from a yield function with a linear expression of the form

$$\tau_0 + A\sigma_0 - B = 0 \qquad \text{(I.24)}$$

in which A and B are constants, and τ_0 and σ_0 are the corresponding octahedral stresses. As a result, eqn (I.24) will have two forms — one representing the biaxial compression region and the other representing the biaxial tension and compression–tension regions. It is interesting to note that eqn (I.24) can now be extended to the triaxial state as well, in which case a surface of revolution with an axis $\sigma_1 = \sigma_2 = \sigma_3$ can easily be formed. The linear form of eqn (I.24) is a cone. Equation (I.24) has two forms, and hence two different cones can be plotted.

I.3.1.1 Determination of constants A and B

(a) Biaxial compression stress state. Under this condition, if $\sigma_1 = \sigma_2$, the concrete, as non-homogeneous material, will yield when it reaches its biaxial yield strength in compression $R_c f_c'$. The numerical value of R_c will simply be the ratio of biaxial/uniaxial strength of the concrete.

Using

$$\tau_0 = \frac{\sqrt{2}}{3}\sigma_{cu}$$

together with $R_c f_c'$, eqn (I.24) becomes

$$+\tfrac{2}{3}R_c f_c'(A - 1) - B = 0 \qquad \text{(I.25)}$$

For a uniaxial value at yield with $\sigma_1 = \sigma$ and $\sigma_2 = \sigma_3 = 0$ eqn (I.24) leads to

$$+\tfrac{1}{3}f_c'(A - \sqrt{2}) - B = 0 \qquad \text{(I.26)}$$

Solving eqns (I.24) and (I.26), the values for the constants A and B are

$$A = \sqrt{2}L_1 \qquad B = \frac{\sqrt{2}}{3} L_1 f_c' \tag{I.27}$$

where

$$L_1 = \left(\frac{R_c - 1}{2R_c - 1} \right)$$

Equation (I.24) now represents a simplified function of the type given in eqn (I.28):

$$\tau_0 + \sqrt{2}L_1\sigma_0 + \frac{\sqrt{2}}{3} L_1 f_c' = 0 \tag{I.28}$$

The intersection of the yield surface, based on eqn (I.28) for the σ_1–σ_2 plane, is now represented by

$$\sigma_1^2 L_1^2 - \sigma_1^2 + 2\sigma_1\sigma_2 L_1^2 + \sigma_1\sigma_2 + \sigma_2^2 L_1^2 - \sigma_2^2 - 2L_1^2\sigma_1 f_c' - 2L_1^2\sigma_2 f_c' + L_1^2 f_c'^2 = 0 \tag{I.29}$$

Equation (1.28) is a general equation for the intersection of yield surfaces. The value of R_c can be obtained by experiment or from the test results of the several investigators mentioned earlier. This general equation can become the basis of comparison and can test the sensitivity of the investigator's results. For example, Kupfer's tests give the value of $R_c = 1.16$, therefore eqn (I.28) leads to an expression which indicates that eqn (I.28) represents an ellipse:

$$\left(\frac{2L_1^2 + 1}{2} \right)^2 - (L_1^2 - 1)^2 < 0 \tag{I.30}$$

The construction of such an ellipse is a sure test of the authenticity of these equations.

At the centre of the ellipse it is obvious that $\sigma_1 = 0 = \sigma_2$. The direction of axes of such an ellipse can be determined in the manner given below.

If the inclination of σ_1, say, is α', then

$$\tan 2\alpha' = \frac{(2L_1^2 + 1)}{(L_1^2 - 1) - (L_1^2 - 1)} = \infty \tag{I.31}$$

from which $\alpha' = \pi/4$ or $3/4\pi$.

(b) Biaxial tension and combined compression–tension stress states. In this case two of the three stresses are less than zero and the third stress is zero. For example, $\sigma_1 < 0$, $\sigma_2 < 0$ and $\sigma_3 = 0$ will give a biaxial tension state of stress. Under the uniaxial tension condition the concrete will reach its cracking condition with a limiting

value of $R_t f'_c$, where $R_t = \sigma_{tu}/f'_c$. When f'_c is less than zero and σ_{tu} is greater than zero, the value of R_t is normally around 10%, which is in full agreement with the findings of several investigators [29–39]. As with eqn (I.25) and eqn (I.26), the yield function becomes

$$-\tfrac{1}{3}R_t f'_c(A + \sqrt{2}) - B = 0 \tag{I.32}$$

Equation (I.29) is available for uniaxial compression. The second set of values for the constants A and B is determined simultaneously:

$$A = \sqrt{2}L_3 \qquad B = \frac{2\sqrt{2}}{3}L_4 \tag{I.33}$$

The yield function for a single constitutive equation defined by eqn (I.24) can now be expressed as

$$\tau_0 + \sqrt{2}L_3\sigma_0 + \frac{2\sqrt{2}}{3}L_4 = 0 \tag{I.34}$$

where

$$L_3 = \frac{1 - R_t}{1 + R_t} \qquad L_4 = \left(\frac{R_t}{1 + R_t}\right)f'_c$$

The yield surface intersection of eqn (I.35) for the $\sigma_1 - \sigma_2$ plane becomes

$$\sigma_1 L_3^2 - \sigma_1 + 2L_3^2\sigma_1\sigma_2 + \sigma_1\sigma_2 + \sigma_2 L_3^2 - \sigma_2 + 2\sigma_1 L_3 L_4 \tag{I.35}$$

The polynomial of eqn (I.35) reduces to a simplified equation, eqn (I.36), by using the accepted value of $R_t = 0.1$:

$$\left[\left(\frac{2L_3^2 + 1}{2}\right)^2 - (L_3 - 1)^2\right] > 0 \tag{I.36}$$

Equation (I.36) is very similar to eqn (I.30), but represents a hyperbola for which σ_1 and σ_2 at the centre are

$$\sigma_1 = \frac{4L_3L_4}{(1 - 4L_3^2)} = \sigma_2 \tag{I.37}$$

The direction of this hyperbola with respect to the σ_1 axis, say at an angle of α', is still $\pi/4$ or $\frac{3}{4}\pi$ since

$$\tan 2\alpha' = \frac{2L_1^2 + 1}{(L_3^2 - 1) - (L_3^2 - 1)} = \infty \tag{I.38}$$

I.4 The incremental constitutive laws for concrete under a multiaxial state of stress

Numerous attempts have been made [32–125] to define the mechanics of failure of concrete under a complex state of stress. Some investigators [57–125] have also made attempts to derive a single failure criterion for concrete. Due to crude assumptions and limited boundary conditions, most of these laws or criteria lack consistency in their results and, in some cases, they are diametrically opposed to one another. Although most of them [32–125] do recognise concrete as a redundant structure, able to carry loads even when cracks have extended through the mortar mix, advances in the use of prestressing techniques for prestressed concrete reactor vessels (PCRV) have necessitated a thorough investigation of the existing failure criteria. In PCRV, due to variations in loading conditions, the principal stresses at any time and any point may vary from biaxial and triaxial compression to compression–tension–tension in any combination. These instant changes in the principal stress profile are not directly covered by these laws or criteria. However, with the advent of computers, these instant changes can be detected and simulated by iteration procedures. In the following pages a systematic review is given of the work of some well-known investigators.

I.4.1 General review of several failure criteria and strength theories (existing failure theories)

Failure criteria proposed for various materials have been reviewed by Nadai [18], Timoshenko [101] and Marin [102].

Many of the proposed criteria are given in empirical form, e.g. the maximum principal stress of Lamé [103], Clapeyron [104] and Rankine [105]; the maximum shear stresses of Tresca [108], Coulomb [109], Navier [110] and Mohr [111]; and the octahedral shear and normal stresses of Balmer [16], Nadai [18], Wastlung [22], Bresler and Pister [23, 42], Kupfer and Gerstle [31], Vile [40] and Bellamy [43], etc. Another independent theory expressed in terms of shear strain energy has been suggested by Von Mises [35]. Most of the elasto-plastic and crack analyses do make use of these criteria. Nevertheless, they are better suited to metallic structures because they do not predict characteristics such as concrete in tension and compression, and material instability leading to a cracking phenomenon. In order to understand such characteristics, some investigators [48–125] modify their failure criteria in order genuinely to predict the behaviour of concrete under a complex state of stress.

The Tresca criterion is a one-parameter theory based on the assumption that failure occurs when the maximum shear stress reaches a limiting value. The yield surface for a cohesive material is represented by a regular hexagon with its axis on the space diagonal. For concrete with cohesionless ingredients the Tresca criterion gives a yield surface which is represented by a regular hexagonal pyramid with its axis on the space diagonals and its apex at the origin. Both these cases are shown in Fig. I.12a. The maximum shear stress τ_{max} is generally $(\sigma_3 - \sigma_1)/2$. For simple tension when $\sigma_2 = \sigma_1 = 0$ the maximum shear stress at yield will be $\frac{1}{2}\sigma_{cu}$, where σ_{cu} is the ultimate or yield stress. The Tresca criterion then asserts that yielding will occur when any one of the following conditions is met:

$$\sigma_3 - \sigma_2 = \pm\sigma_{cu} \qquad \sigma_2 - \sigma_1 = \pm\sigma_{cu} \qquad \sigma_1 - \sigma_3 = \pm\sigma_{cu} \tag{I.39}$$

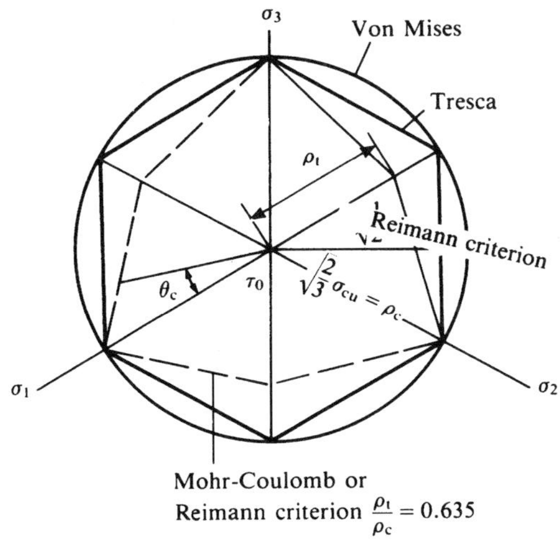

Fig. I.12a. View along space diagonal. Principal stress space for different failure theories. (Note: $\sigma_{cu} = f_{cu}$.)

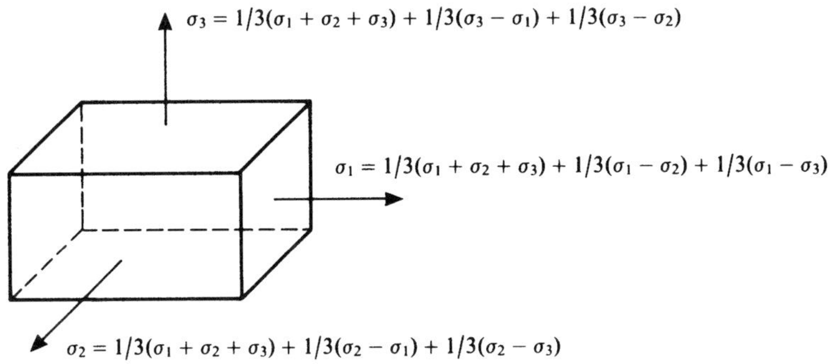

Fig. I.12b. Values of stress resultants.

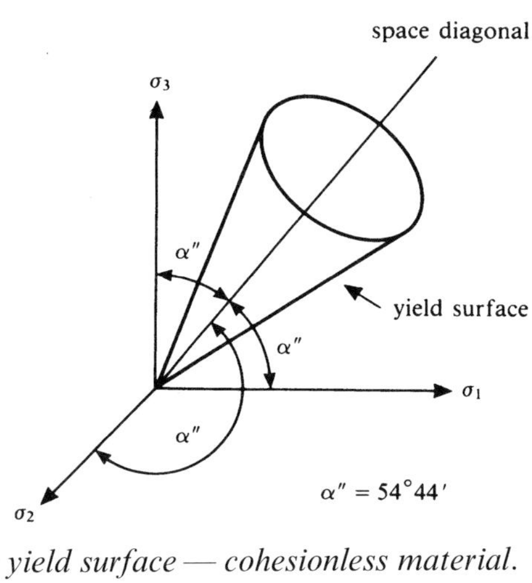

Fig. I.12c. Von Mises yield surface — cohesionless material.

For the biaxial state of stress when $\sigma_3 = 0$, the following conditions hold good:

$$\begin{aligned}
\sigma_1, \sigma_2 &= \sigma_{cu} && \text{for } \sigma_1 > 0, \sigma_2 < 0 \\
\sigma_1, \sigma_2 &= -\sigma_{cu} && \text{for } \sigma_1 < 0, \sigma_2 > 0 \\
\sigma_2 &= \sigma_{cu} && \text{for } \sigma_2 > \sigma_1 > 0 \\
\sigma_1 &= \sigma_{cu} && \text{for } \sigma_1 > \sigma_2 > 0 \\
\sigma_1 &= -\sigma_{cu} && \text{for } \sigma_1 < \sigma_2 < 0 \\
\sigma_2 &= -\sigma_{cu} && \text{for } \sigma_2 < \sigma_1 < 0
\end{aligned} \tag{I.39a}$$

For $\sigma_1 = -\sigma_2 = \bar{y}$, then $\bar{y} = \frac{1}{2}\sigma_{cu}$

As this criterion is in fair agreement with the experimental results [48–125] it is used to a considerable extent by the designers. It suffers, however, from one major drawback, i.e. it will always be necessary to know in advance the maximum and minimum principal stresses. It is extremely difficult to forecast these stresses in advance in structures.

Von Mises theory, on the other hand, is a distortion energy theory, and failure is assumed to occur when the shear strain energy density exceeds a critical value or when the octahedral shear stress exceeds a limiting value. For three-dimensional systems the three principal stresses are expressed in Fig. I.12b, and the first term in each expression is the hydrostatic mean stress:

$$\sigma_0 = \tfrac{1}{3}(\sigma_1 + \sigma_2 + \sigma_3) = \tfrac{1}{3}J_1 \tag{I.40}$$

which acts on all faces of the cube and which produces a volumetric strain only. The second and third terms are proportional to the maximum shear stress values and it is these stresses that cause distortion or energy of distortion. The volumetric strain ε_ν and volumetric strain energy U_ν can easily be computed from Figs I.12b and I.12c, and are given in eqn (I.41):

$$\varepsilon_\nu = \left(\frac{\sigma_1 + \sigma_2 + \sigma_3}{E}\right)(1 - 2\nu) \qquad U_\nu = \frac{\varepsilon_\nu}{6} \tag{I.41}$$

For a three-dimensional system the strains in three directions, ε_1, ε_2, ε_3, which are dependent on the Poisson value of ν and Young's modulus E, are difficult to determine. Using constant values of ν and E as assumed by Von Mises, the shear strain energy can be computed by subtracting the volumetric strain energy from the total strain energy:

$$\begin{aligned}
&\tfrac{1}{2}/E[\sigma_1^2 + \sigma_2^2 + \sigma_3^2 - 2\nu(\sigma_1\sigma_2 + \sigma_2\sigma_3 + \sigma_1\sigma_3)] \\
&\quad - \tfrac{1}{6}/E[(\sigma_1 + \sigma_2 + \sigma_3)^2(1 - 2\nu)]
\end{aligned} \tag{I.42}$$

Substituting $(1 + \nu) = E/2G$, eqn (I.42) becomes

$$\tfrac{1}{12}/G[(\sigma_1 - \sigma_2)^2 + (\sigma_2 - \sigma_3)^2 + (\sigma_3 - \sigma_1)^2] \tag{I.43}$$

or

$$(\sigma_1 - \sigma_2)^2 + (\sigma_2 - \sigma_3)^2 + (\sigma_3 - \sigma_1)^2 = \text{a constant}$$

For a case of pure shear when $\sigma_3 = -\sigma_1$ and a $\sigma_2 = 0$, Von Mises general yield condition becomes (Figs I.12a and I.12c)

$$\tfrac{1}{2}[(\sigma_1 - \sigma_2)^2 + (\sigma_2 - \sigma_3)^2 + (\sigma_3 - \sigma_1)^2] = \sigma_{\text{cu}}^2 = \bar{y}^2 \tag{I.44}$$

Using tensor notation, the second invariant I_2 of the stress deviator at the yield point becomes

$$I_2 = \tfrac{1}{3}\sigma_{\text{cu}}^2 = \tfrac{1}{6}(\sigma_1 - \sigma_2)^2 + (\sigma_2 - \sigma_3)^2 + (\sigma_3 - \sigma_1)^2 = \bar{y}^2 \tag{I.45}$$

A yield stress in pure shear $\bar{y}$ becomes

$$\bar{y} = \frac{1}{\surd 3}\sigma_{\text{cu}} \tag{I.46}$$

Von Mises criterion predicts a pure shear yield stress which is about 15% higher than predicted by the Tresca criterion. The criterion is widely used since no knowledge is needed of the relative magnitude of the principal stresses. Equations (I.40) and (I.46) are unable to cover areas such as instant tension–compression and unstable cracking phenomena. However, for elasto-plastic problems involving nonlinearity arising from the nonlinear material and geometric behaviour, the stress–strain relation can be expressed in incremental form by combining the yield criteria of the Von Mises and Prandtl–Reuss [35, 36] equations. This idea is used in the following chapters for the solution of various problems.

Hencky and Nadai [37] later showed that it was convenient to assume that yielding can take place when the octahedral shear stress τ_0 reaches the octahedral shear stress at yield in simple tension. Referring to eqn (I.44) this then becomes

$$\tau_0 = \frac{\surd 2}{3}\sigma_{\text{cu}}^2 = \tfrac{1}{3}\surd[(\sigma_1 - \sigma_2)^2 + (\sigma_2 - \sigma_3)^2 + (\sigma_3 - \sigma_1)^2] \tag{I.47}$$

For a simple tension at yield, Fig. I.12a shows the value of τ_0, which is given by

$$\tau_0 = \frac{\surd 2}{3}\sigma_{\text{cu}}^2 = \frac{\surd 2}{3}J_2 \tag{I.48}$$

Alternatively, this can be looked upon as stating that yielding will occur when the second invariant J_2 of the stress deviator tensor reaches a critical value. This is the stage where the importance of the octahedral shear stress becomes imminent. If a suitable relationship between a normal stress and this octahedral shear stress can be found, then it could be possible to define the mechanics of failure criteria. In general terms this relationship can be written as

$$F(\sigma_0, \tau_0) = 0 \tag{I.49}$$

$$\tau_0 = f(\sigma_0) \tag{I.50}$$

Whatever the type of loading on structures, using eqns (I.49) and (I.50) should give rise to normal or shear stresses, or both, in the concrete of structures. As concrete is a non-homogeneous and anisotropic material, the simulation of some boundary conditions under a complex state of stress in eqns (I.49) and (I.50) is, by itself, a complex phenomenon as these boundary conditions depend on several dependent variables. The important variables amongst them are the loading history and time, temperature and creep dependent values of E and ν at any axis of the failure surface.

Mohr [111] simplifies eqns (I.49) and (I.50) by stating a failure criterion in which the limiting shear stress τ_0 in any plane is dependent upon the normal stress σ_0 of eqn (I.50) in the same plane. This means that the intermediate principal stress has no influence on the failure. This may present problems if it is applied to the behaviour of concrete in complex structures. A special version of the Mohr criterion is Coulomb's criterion, which is a two-parameter theory in which the coefficient of friction μ and cohesion value c are used. All this involves is the achievement of a limiting Mohr envelope for all possible Mohr circles that can be drawn for different states of stress. The value of τ_0 will then become

$$|\tau_0| = c + \mu\sigma \tag{I.51}$$

It is interesting to note that for $\mu = 0$ Coulomb's criterion corresponds to Tresca's criterion when $\sigma_3 \geq \sigma_2 \geq \sigma_1$. This theory, as is now evident, can be the basis of a satisfactory criterion for triaxial compression only when the two lesser principal stresses are equal. However, this fails when one of them is zero or becomes negative, which may well be the case in complex structures under extreme loads.

This point has been supported by Bresler and Pister [23], Bellamy [43], Balmer [16], and Karni and McHenry [83]. Vile [40] later took this further to indicate that the limiting tensile strain at discontinuity depends on the state of stress, and increases with the degree of hydrostatic compression shown by the value of mean normal stress. This means that the tensile failures due to compressive forces (mainly shear bond failures) occur at higher strains than those due to direct tensile forces (mainly tensile bond failures). This observation is supported by Sturman *et al.* [38]. It is interesting to note that, although failure due to slip lines in the Mohr–Coulomb case is widely accepted because of, amongst other things, the creation of cones in the uniaxial compressive stress, it was found by Wastlung [22] and Kupfer [113] that the cone creation can easily be explained by the friction on the loaded surface, and hence that the compressive failure in the biaxial state of stress is generally followed by splitting perpendicular to the direction of the greater principal strain.

Rüsch [55] indicated the significance of creep and creep strain, and stated that the behaviour and ultimate strength of concrete are affected by the rate of loading. At slow rates the creep strain becomes significant at a small proportion of ultimate load. He therefore suggested that the static modulus of elasticity should be defined in terms of secant value at an appropriate stress level. Using such a relation he showed that the ultimate compressive strength of concrete under long-term loading is around 80% of the short-term ultimate strength. This is relevant to the short- and long-term concrete behaviour in some structures. It was not clear whether or not his results took into consideration short- and long-term thermal effects — a vital case in certain complex structures. In any event, the behaviour of concrete under creep and thermal effects, which is very much the environment in reactor vessels, is something that was not considered in earlier criteria.

Newman and Newman [21] have dealt more fully with the relationship expressed in eqns (I.43), (I.48) and (I.50) by averaging data obtained by Richart *et al.* [14], Balmer [16], Bresler and Pister [23], Karni and McHenry [83], Bellamy [43], and Weigher and Beeker [57], and have plotted the values at failure of mean normal stress σ_0 against the value of shear strain. They have concluded that a small tensile stress combined with compression can, to a greater extent, reduce the strength of concrete to a value well below the biaxial compression line. The value of octahedral normal stress was kept as σ_0, as given in eqn (I.47). They have discovered that a short cylindrical specimen will fail under a triaxial state of stress if the height/width ratio is less than about $1\frac{1}{2}$ and will fail under a uniaxial compressive state of stress ($\sigma_1 = \sigma_2 = 0$) if the height/width ratio is greater than $2\frac{1}{2}$. They consider a true uniaxial tensile stress to be equal to one of the values of σ. Throughout their work they have considered concrete to be a homogeneous and isotropic material.

Hannant and Frederick [58] have suggested a failure criterion for concrete which is valid only for multiaxial compressive stresses. Simple equations are presented which take into consideration the form of the failure surface as a three-lobed curve which is not rotationally axisymmetric.

Two cases have been reviewed by Hannant and Frederick. For isotropic material conditions these cases are

$$\begin{aligned} &\sigma_3 > \sigma_1 = \sigma_2 \\ &\qquad \sigma_3 = \sigma_{cu} + 4\sigma_2 \qquad \text{(Case 1)} \\ &\qquad \sigma_3 = 1.75\sigma_{cu} + 3\sigma_2 \quad \text{(Case 2)} \end{aligned} \tag{I.52}$$

Based on short-term loading conditions, they arrived at the conclusion that for a safe design when the maximum principal stress is increased by a factor of two while other stresses remain constant, failure will not occur when

$$\begin{aligned} \sigma_1 = \sigma_2 \quad &\text{and} \quad \sigma_3 = \tfrac{1}{2}\sigma_{cu} + 2\sigma_2 \\ \sigma_3 = \sigma_1 \quad &\text{and} \quad \sigma_2 = \tfrac{1}{2}\sigma_{cu} + 2\sigma_2 \\ \sigma_3 = \sigma_2 \quad &\text{and} \quad \sigma_1 = \tfrac{1}{2}\sigma_{cu} + 2\sigma_2 \end{aligned} \tag{I.53}$$

For a maximum stress of $4\sigma_{cu}$, the factor of safety is found to be 1.75. The introduction of a safety factor in the stress space is plausible. This work is partially recommended in the new British code of practice [89]. Figure I.13 shows a failure surface covering both the cases given above.

Although this criterion is quite an improvement upon the previous criteria, nevertheless concrete has been considered as a material behaving isotropically. It might be

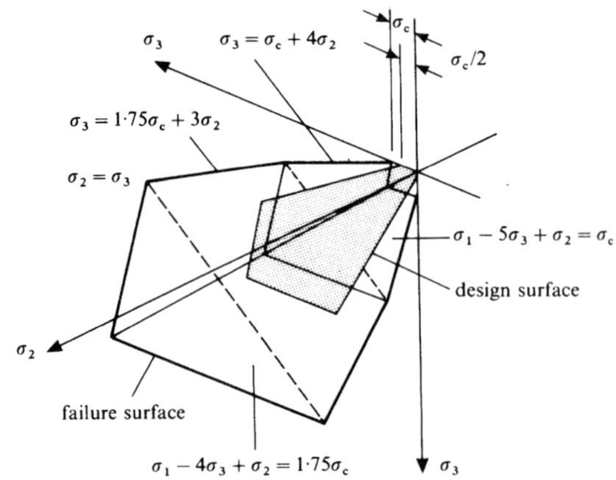

Fig. I.13. Concrete failure surface under triaxial compression (after Hannant and Frederick [58]). (Note: $\sigma_c = f'_c \sigma_c = \sigma_{cu}$.)

interesting to look at the numerical work done by Link [59], who has presented a numerical analysis of the failure criterion of concrete incorporating a material behaviour law for concrete, in the elastic range. The analysis is based only on the biaxial test results of Kupfer and co-workers [29,31] and Rüsch [55].

An isotropic stress–strain relation has been derived with two Poisson's ratios ν_1 and ν_2 and two Young's moduli E_1 and E_2 in the entire biaxial stress field covering the domains of tension–tension, tension–compression and compression–compression. Although such an analysis is best suited to ultimate load analyses using finite elements, the relationship of initial and new values of ν and E are based on lightweight concrete. The use of a true relationship between ν and E would certainly produce an accurate failure criterion under a biaxial state of stress.

A criterion for concrete failure suggested by Baker [60] is based on the fact that concrete has a heterogeneous system, and the principal causes of cracking and failure are the differential stiffness between the aggregate matrix and interface, and the bond interaction and the weakness in tension of the mortar mix. He has modelled the failure surface by using a tetrahedron (Fig. I.14) in which the Poisson thrust ring, plastic flow and crack effects are represented by the relative stiffnesses of rods, which, in turn, depend on the changing load behaviour. He has developed expressions assuming that cracks form first in mortar pockets and then eventually extend around the stone interfaces. It is an impressive contribution to improving representation of the situation in which even the Poisson's ratio and the Young's modulus E change rapidly as the failure approaches. Using the convention for co-ordinate axes adopted by Baker, the two equations obtained are

$$\frac{\sigma_3 + \sigma_1}{f'_c} = 1 + 4.5\left(\frac{\sigma_2}{f'_c}\right) \quad \text{for } \sigma_3 > \sigma_1 = \sigma_2 \tag{I.54}$$

$$\frac{\sigma_3 + \sigma_1}{f'_c} = 2.2 + 10\left(\frac{\sigma_2}{f'_c}\right) \quad \text{for } \sigma_3 = \sigma_1 > \sigma_2 \tag{I.55}$$

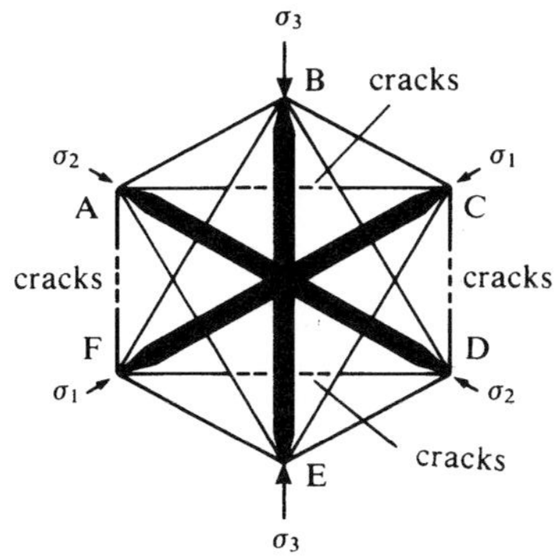

Fig. I.14. Isometric of tetrahedral model (after Baker).

For $\sigma_2 = 0$, the factor of safety

$$\frac{\sigma_3}{f'_c} = \frac{\sigma_1}{f'_c} = 1.1 \tag{I.56}$$

The test case in eqn (I.56) proves what Vile [40] has established. It is interesting to note that, using eqns (I.54) and (I.55) throughout the triaxial compression quadrant of stress space, the following relationships are achieved, respectively:

$$\sigma_3 = f'_c + 3.50\sigma_2 \qquad \sigma_3 = 1.1f'_c + 5\sigma_2 \tag{I.57}$$

When they are plotted in the stress space, as shown in Fig. I.15, the failure envelope agrees both in shape and magnitude, as suggested by Hannant and Frederick [58] in eqns (I.56) and (I.57). These relationships are extremely sensitive to the assumed values of ν and E. The safety factors thus depend both on values of ν and values

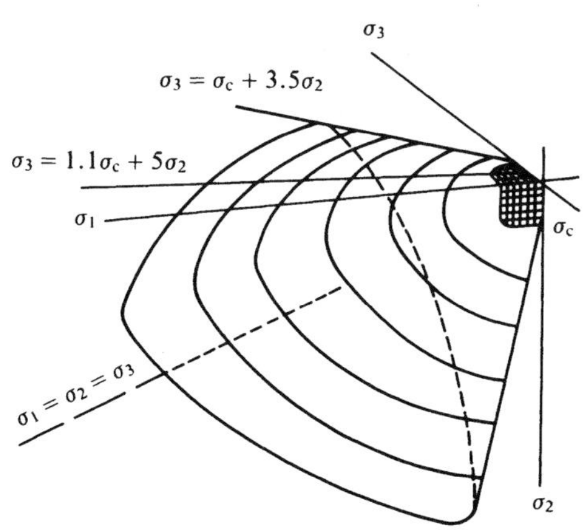

Fig. I.15. Failure surface for concrete in the compression quadrant of stress space based on eqns (II.8a and 8b) (after Hannant and Frederick [58]). (Note: $\sigma_c = f'_c$.)

Table I.1. Pressure vessel parameters

O

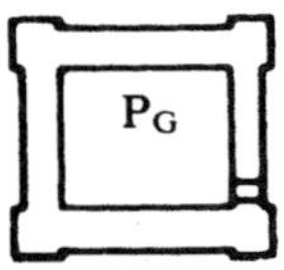

P = gas pressure
= 2.45 MN/m^2
Oldbury

P = 2.64 MN/m^2
Wylfa

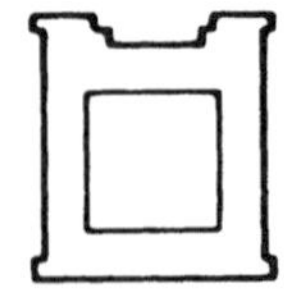

P = 4.03 MN/m^2
Hinkley point

H

P_G = 4.03 MN/m^2
Hartlepool

D

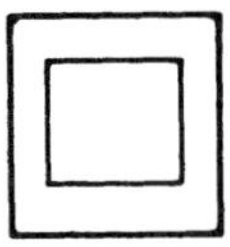

P_G = 3.30 MN/m^2
Dungeness B

FT

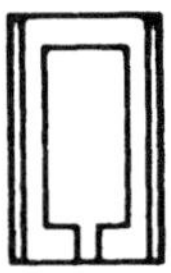

P_G = 4.86 MN/m^2
Fort St Vrain

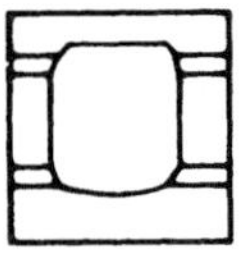

P = 3.04 MN/m^2
Chinon

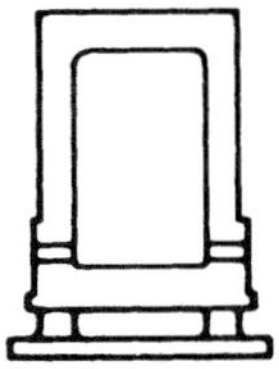

P = 2.76 MN/m^2
St Laurent

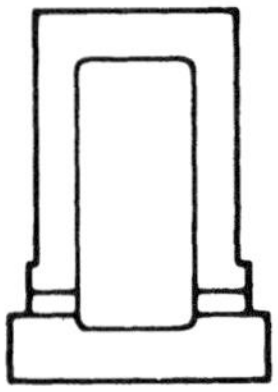

P = 4.48 MN/m^2
Bugey

Parameter	O	D	H	FT
$2R_1$ = Internal diameter	18.3m	19.95m	13.1048m	10.67m
$2H_1$ = Internal height	23.5m	17.70m	29.261m	21.95m
t = Wall thickness	4.6m	3.82m	6.40m	2.5m
D_1 or D_2 = Cap thickness	6.7m	6.2T 5.94B	5.5m	5.0m

Concrete

σ_{cu} =	minimum cube strength of concrete at 28 days	-41.34 kN/mm^2
σ_y =	yield strength	$-0.66\sigma_{cu}$
σ_t =	tensile strength	$+0.1\sigma_{cu}$
E_e =	elastic modulus	$+41.4$ kN/mm^2
E_p =	plastic modulus	$+0.476E$
ε_{cu} =	ultimate strains	0.0035

ν = Poisson's ratio	0.18
α_T = coefficient of linear thermal expansion	8.0 μM/m °C
K = thermal conductivity	1.75 M/m °C
d' = coefficient for aggregates	0.65, 0.68, 0.87, 0.87
Conventional steel	
$\bar{\gamma} = \sigma_y$ = yield strength	4516 N/mm^2 (50∅) 4400 (25∅)
E_e = elastic modulus	200 kN/mm^2
E_p = plastic modulus	$0.1E = 20$ kN/mm^2
Liner	
t_s = thickness	12 mm + 10%, 12 mm + 10%, 19 mm + 5%, 19 mm + 5% Up to 40 mm max, up to 40 mm, up to 25 mm, up to 25 mm
$\bar{Y} = \sigma_y$ = yield strength	3.4×10^5 kN/m^2
α_T = coefficient of linear thermal expansion	10 μM/m °C
k = thermal conductivity	41.6 W/m °C

Loadings

Case 1 Prestress and ambient temperature
Case 2 Prestress + 1.15 × design pressure + ambient temperature
Case 3[a] Prestress + design pressure + temperature
Case 4[a] Overload (prestress + increasing pressure + temperature)

[a]Short- and long-term conditions apply [126]

of E, and could vary from 1.1 to 2.5. Further research is needed in Baker's equations for $(\sigma_1 + \sigma_2 + \sigma_3)$ in excess of 16. Chawalla [61], very similarly to other investigators, shows the three-dimensional envelope which defines the onset of failure as a surface of revolution having an axis of symmetry, the meridian of which can be represented by σ_0–τ_0. The surface is similar to a nose-cone of a rocket with a peak partially flattened to a lesser or greater extent, depending upon material properties. The only difference is that $\sigma_1 = \sigma_2 = \sigma_3 = \sigma_{cu}$ defines the stress in the material at rupture caused by equal tensile forces, and the failure is anticipated if the point is outside the σ_0–τ_0 envelope.

Chinn and Zimmerman have reported their own work on [62] this subject, and it appears they are the only researchers who have investigated the failure of concrete when the cases of both axial pressure and confining pressure predominate. The outcome of their investigation is that the maximum principal stress, say σ_3, is equal to $\sigma_{cu} + 4\sigma_1$ (or σ_2). Akroyd [48] has found that this is only true up to about $\sigma_{cu} + 3\sigma_1$ (or σ_2). These findings again are very much in line with those found by Hannant and Frederick [58], Baker [60] and Garas [68].

Reimann [63] gives a failure criterion, and using Fig. I.12a the compressive meridian ρ_c is expressed by

$$\frac{\frac{1}{\sqrt{3}} J_1}{f_c'} = A \frac{\rho_c^2}{f_c'} + B \frac{\rho_c}{f_c'} + C \tag{I.58}$$

Equation (I.58) is a parabolic equation where A, B and C are constants, and $J_1 = \sigma_1 + \sigma_2 + \sigma_3$ is the first invariant of the stress tensor. The other meridians are assumed to be affined with the compressive meridians $\rho = \phi(\theta_c)\rho_c$ for $-60° < \theta_c < 60°$. This corresponds to a failure curve in the deviatoric plane consisting of straight lines which are tangents to a circle of which the radius corresponds to the tensile meridian. As shown in Fig. I.12a, besides having a constant value, the surface is not smooth and it is only valid for compressive stresses.

Argyris [64] suggested the following three-parameter criterion involving all the stress invariants:

$$A\frac{J_1}{f_c'} + (B - C\cos 3\theta_c)\left(\frac{\sqrt{I_2}}{f_c'}\right) - 1 = 0 \quad \text{(straight meridians)} \tag{I.59}$$

where A, B and C are constants, and f_c' is the uniaxial compressive cylindrical strength, i.e. straight meridians. The failure curves are similar in all deviatoric planes and concavity exists when $\rho_t/\rho_c < 0.777$.

Willam and Wanke [130] suggested a five-parameter criterion in which the tensile and compressive meridians are expressed by

$$\begin{aligned} \sqrt{I_{2,t}} &= A_{01} + A_{11}J_1 + A_{22}\left(\frac{J_1^2}{f_c'}\right) \quad \text{tensile meridian } (\theta_c = 0°) \\ \sqrt{I_{2,c}} &= B_{01} + B_{11}J_1 + B_{22}\left(\frac{J_1^2}{f_c'}\right) \quad \text{compressive meridian } (\theta_c = 60°) \end{aligned} \tag{I.60}$$

where A_{01}, A_{11}, A_{22}, B_{01}, B_{11} and B_{22} are constants. The hydrostatic axis intersects these two meridians at the same point. They succeeded in finding an expression for the failure curve in the deviatoric plane, which is convex for all ratios $\frac{1}{2} < \rho_t/\rho_c < 1$. This was obtained by prescribing this curve as part of an elliptic curve.

Referring to Fig. I.12a, Launay and Gachon [66] suggested the following relation to express the failure curve in the deviatoric plane:

$$\rho^2\left[\frac{\cos^2\frac{3}{2}\theta_c}{\rho_c\Big/\left(\frac{J_1^2}{3}\right)} + \frac{\sin^2\frac{3}{2}\theta_c}{\rho_t\Big/\left(\frac{J_1^2}{3}\right)}\right] = 1 \tag{I.61}$$

where

$$\rho = \frac{\sqrt{2\rho_c\rho_t}}{\sqrt{(\rho_t^2 + \rho_c^2) + (\rho_t^2 - \rho_c^2)\cos 3\theta_c}} \tag{I.61a}$$

where

$$\cos 3\theta_c = \frac{3\sqrt{3}}{2}[(I_3)/(I_2)^{3/2}]$$

and

$$I_3 = \tfrac{1}{3}(\sigma_1^3 + \sigma_2^3 + \sigma_3^3)$$

However, this gives convexity only for $\rho_t/\rho_c > 0.745$, and for nearly all practical applications $\rho_t/\rho_c > 0.745$. The tensile and compressive meridians have not been prescribed in this equation.

The details of these several parameter models have been discussed later on in relation to numerical modelling. Bangash has reported [126] triaxial stress results using statistical results based on the mutual relationship of stresses at various incremental stages of deformations of several existing vessels listed in Table I.1. A computer program was developed to compare these stresses individually or in combination and evaluate them in terms of a uniaxial compressive stress. Figures I.15 and I.16 show a relationship between these stresses for a case of triaxial compression–compression–tension. They also indicate a relationship for various increments of

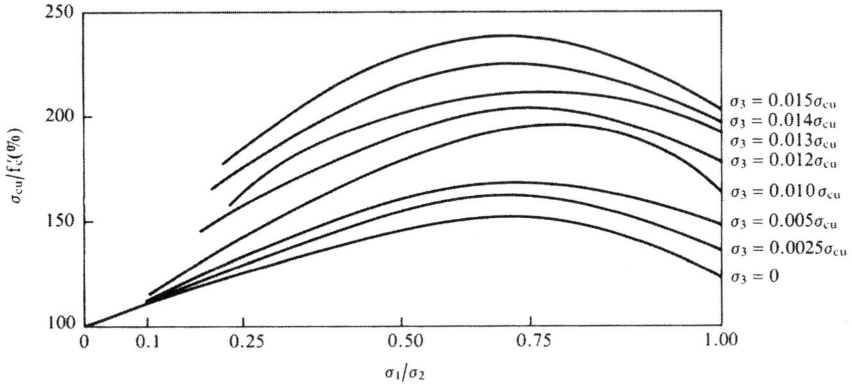

Type	2H (m)	2R (m)	T (m)	t (m)	Design gas pressure P_{GD} (MN/m^2)
Oldbury	18.28	23.46	6.7T &B	4.57	2.8
Dungeness	17.66	19.95	6.2T 5.94B	3.82	3.45
Hinkley 'B' / Hunterston 'B'	19.40	18.90	8.38T 8.07B	5.02	4.41
EDF 3	20.70	18.74	5.33	4.19	3.09
EDF 4	36.28	19.00	5.64T 5.94B	4.75	3.15
Bugey I / Bugey II	38.25	17.06	7.46	5.48	4.69
Enel/Is	18.28	16.00	6.60	5.41	3.40
Feasibility Study-HTGCR	21.93	15.85	6.70	5.18	5.36

Fig. I.16. Triaxial compression/compression/compression from the vessel operational analysis.

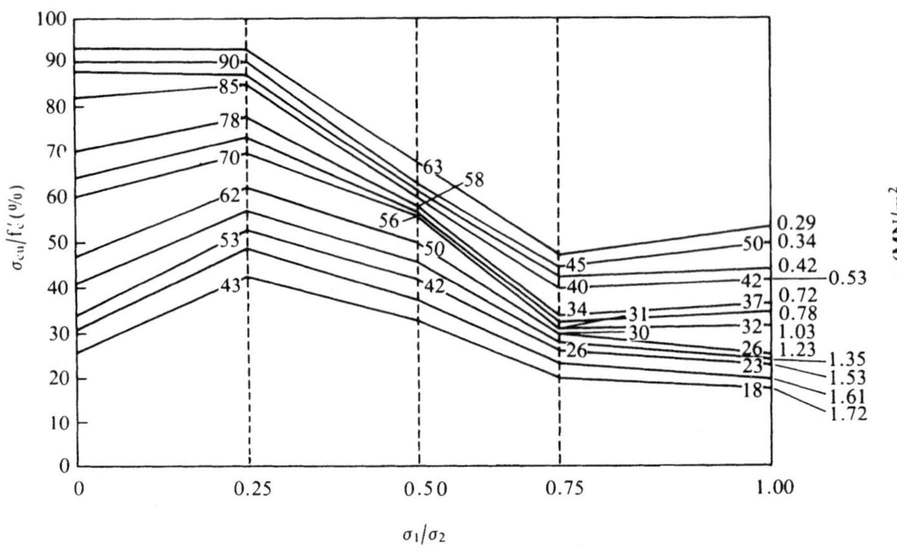

σ_3 = tensile stress component
σ_1, σ_2 = compressive stresses $\sigma_1 < \sigma_2$
σ_c = uniaxial strength of concrete = f_c'
σ_{cu} = computed ultimate load in the direction of high compressive stress = f_{cu}

Fig. I.17. Triaxial compression/compression/tension with specified tensile stress.

the stress components. Results can be similarly obtained for triaxial compression. If any two stresses are equal and the third is decreased by 1/10 to 3/10 of the value of the others, the values of (maximum stress)/(uniaxial compressive stress 100%) = 200 to 260%.

The triaxial compression results have been reported by Bangash [126] using the hypoelastic concept discussed later in this book. These results have been compared with those obtained by Richart *et al.* [14] in Fig. I.18. Similarly, hypoelastic analysis is compared with the experimental results of Kupfer [113] and summarised in Fig. I.19.

Hobbs [131] has carried out a series of tests for the failure criteria of concrete under triaxial conditions for various ages of concrete. He has obtained an ultimate strength equation of

$$\frac{\sigma_1}{f_c'} = 1 + \frac{4.8\sigma_3}{f_c'} \tag{I.62}$$

where σ_1 and σ_3 are the principal stresses at failure along the 1, 3 axes.

Hobbs has carried out a special triaxial test, contrary to the conventional triaxial tests in which intermediate principal stress is equal to the minor principal stress, in which he has managed to equate the intermediate principal stress to the major principal stress. This is known as the 'extension test'. This test has the flexibility

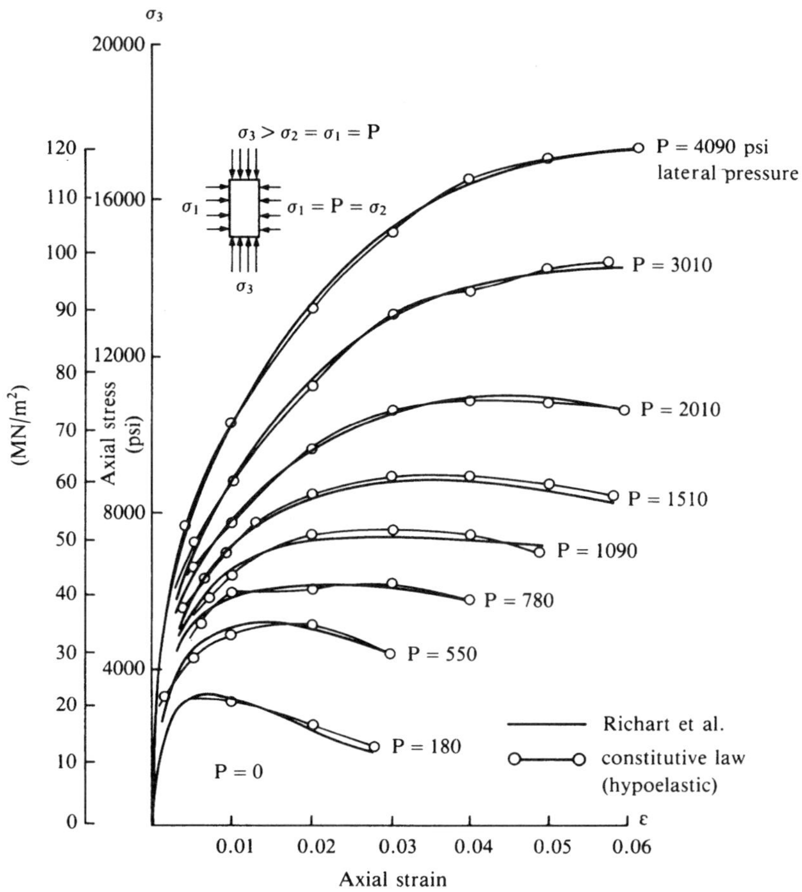

Fig. I.18. Stress–strain curves for concrete in triaxial compression.

so that the biaxial compression plus tension can be included. The results of Hobbs [131] and Kotsovos [132] are summarised in Fig. I.20. Hobbs comes to the following major conclusions:

$$\sigma_u = 0.45 f_{cu} + 3\sigma_3 \tag{I.63}$$

for the ultimate collapse condition where $\sigma_1 > \sigma_2 > \sigma_3 \geq 0$

$$\sigma_s = 0.3 f_{cu} + 2\sigma_3$$

for the serviceability condition where $\sigma_1 > \sigma_2 > \sigma_3 \geq 0$ (I.64)

The other important theories, namely hypoelastic [91, 96, 5], bulk modulus [44, 47], endochronic [172, 388], etc., are discussed later in this book. They are of special interest to researchers involved in the analysis and design of complex concrete structures, such as nuclear reactor pressure and containment vessels, offshore platforms, underground shelters, silos, bunkers, LPG/LNG tanks, etc.

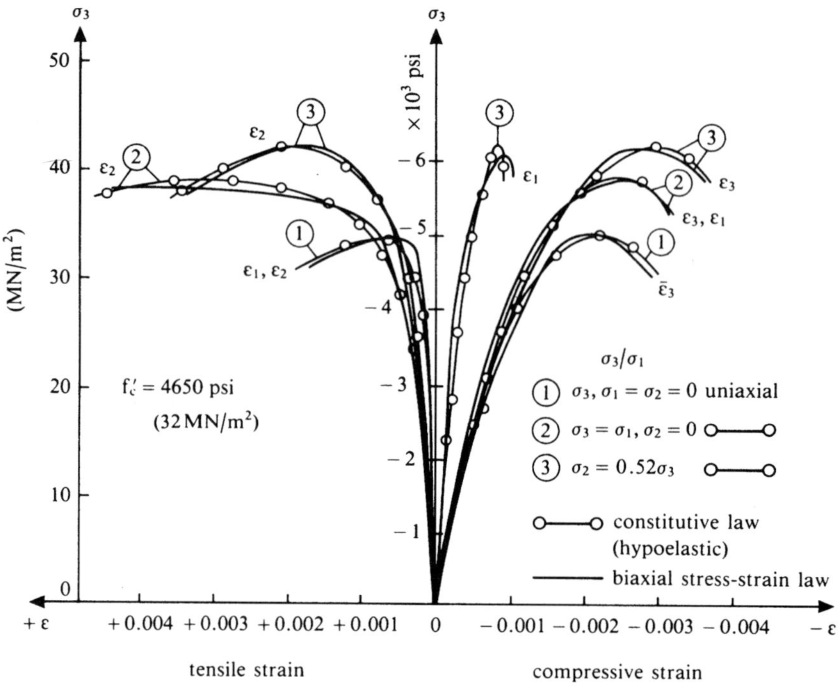

Fig. I.19. Biaxial stress–strain curves.

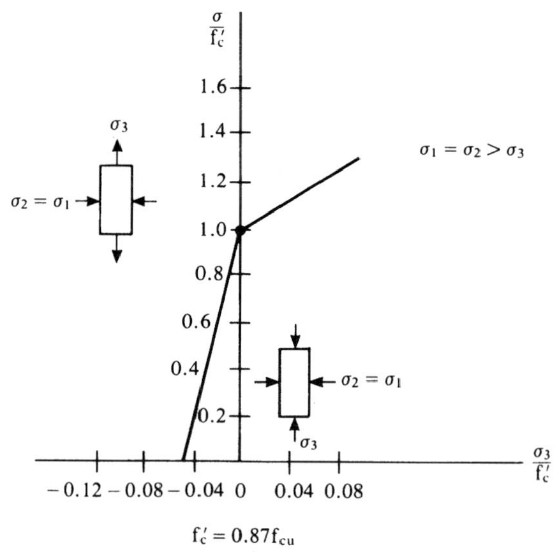

Fig. I.20. Concrete under triaxial stress — extension test results (after Hobbs [131] and Kotsovos [132]).

I.5 Data on Poisson's ratio and Young's modulus

I.5.1 Poisson's ratio

Some investigators [68–87] have carried out experiments on the triaxial behaviour of concrete, with special emphasis on the study of Poisson's value ν and Young's modulus E in the directions of the principal stresses. Launay and Gachon [66] have produced diagrams which relate stresses to Young's moduli; for example, the secant modulus E_3 and Poisson's ratios ν_{13} and ν_{23} in the 1, 3 and 2, 3 directions can be computed as

$$E_{sec} = E_3 = \frac{\sigma_3}{\varepsilon_3} \qquad \nu_{13} = -\frac{\varepsilon_1 \sigma_3}{\varepsilon_3 \sigma_3} \qquad \nu_{23} = -\frac{\varepsilon_2 \sigma_3}{\varepsilon_3 \sigma_3} \tag{I.65}$$

The values in eqn (I.65) were derived for $\sigma_3 > \sigma_2 > \sigma_1$ on 140-mm cubes subject to temperatures of 20, 40 and 60°C at pressures up to 500 bar. The value of ν_{23} (average value = 0.2) remained unchanged for any value $\sigma_1 \leq \sigma_2 \leq 400$ bar and $\sigma_3 \leq 500$ bar. The value of ν_{13} varied erratically and, according to Launay and Gachon [66], this occurs in the following manner:

$$\begin{array}{ll} \text{for } \sigma_1 = \sigma_2, & \nu_{13} = \nu_{23} = 0.2 \text{ for any value of } \sigma_3 \\ \text{for } \sigma_1 < \sigma_2, & \nu_{13} > \nu_{23}; \text{ at } \sigma_{1_{max}} = 100 \text{ bar } \nu_{13} = \nu_{23} \\ \text{for } \sigma_{1_{max}} = 0, & \nu_{13} = 0.2 \text{ to } 0.4 \text{ for any value of } \sigma_2; \\ & \nu_{13} = 0.4 \text{ for } \sigma_3 \text{ up to } 500 \text{ bar} \\ \text{for } \sigma_2 = \sigma_3 > \sigma_1, & \nu_{13} > \nu_{23} \end{array} \tag{I.65a}$$

The suggested stabilising values of ν_{23} and ν_{13} are 0.2 and 0.3, respectively. The above values of E and ν are not far from those of other researchers.

York *et al.* [67] have studied creep under elevated temperatures suitable for vessel concrete and found that Poisson's ratio for biaxially loaded specimens is 55% higher than for uniaxially loaded specimens. Garas [68] found on wire-wound cylindrical specimens that concrete fails under equal biaxial compression at a stress equivalent to 1.2 to $1.5f'_c$. With no axial restraints, it fails at a stress equal to between 0.5 and $0.62f'_c$, which in turn is equivalent to a limiting tensile strain of about 150 micro-strains. In these experiments Poisson's ratio increases from an initial value of 0.15 to a maximum value of 0.42. Hussain and Saugy [76], whilst assuming the concrete surface at failure to be a cone in tension and a truncated cone in compression, arrived at a figure of 1.25 as the ratio of the biaxial compression strength to uniaxial compression, and $\nu = 0.25$.

Linse *et al.* [77] have investigated the strength of concrete under multiaxial all-state stresses of σ_1, σ_2 and σ_3, and under elevated temperatures. They have found that Poisson's ratio of creep at a temperature of 80°C is about 35–50% higher than at normal temperatures. The values suggested range between 0.21 and 0.235.

Gardner tested specimens using the theory developed by Row [78], which relates the stress in the concrete to the instantaneous value of ν. The theory relates the ratio of the minimum work done by the axial stress to the instantaneous geometry of a

Table I.2. Modulus of elasticity and Poisson's ratio at cryogenic temperatures

Mix no.	Aggregates	Cement content (kg/m^3)	Condition	E_0 (kN/mm^2)						ν
				24°C	2°C	−18°C	−59°C	−101°C	−157°C	
1	sand/gravel	415	moist	45.5	45.5	48.3	55.2	62.1	62.1	0.22
2	sand/gravel	326	moist	37.9	40.0	45.5	52.5	58.0	58.7	0.22
2	sand/gravel	326	50% relative humidity	34.5	35.2	35.2	35.2	36.6	37.3	0.23
2	sand/gravel	326	dry	29.7	29.7	29.7	29.7	29.7	29.7	0.25
3	sand/gravel	237	moist	35.9	35.9	42.8	46.2	56.6	57.3	0.18
4	expanded shale	355	moist	20.7	20.7	22.8	26.2	29.0	29.7	0.21

regular packing of uniform spheres. Comparison of this theory with triaxial test results on sand showed good agreement.

Under monotonic loading in tension–tension and compression, Poisson's ratio of 0.2 is quite effective. This value of 0.2 is adequate for uniaxial compression and compression–tension at very low values of stress. Around $0.8f'_c$, the value of 0.2 (as pointed out) is too low, and according to Darwin and Pecknold [7a,b] should be computed as

$$\nu = 0.2 + 0.6\left(\frac{\sigma_2}{f'_c}\right)^4 + 0.4\left(\frac{\sigma_1}{f'_c}\right)^4 \tag{I.66}$$

for uniaxial compression and compression–tension cases.

Elwi and Murray [5, 135], using a least square fit of a cubic polynomial, evaluated an expression for Poisson's ratio. They used the uniaxial compression data of Kupfer [113]. This expression is

$$\nu = \nu_0[1 + 1.3763(\varepsilon_R) - 5.36(\varepsilon_R)^2 + 8.586(\varepsilon_R)^3] \tag{I.67}$$

where

ν_0 = initial value of Poisson's ratio

$\varepsilon_R = \varepsilon_i/\varepsilon'_0$

ε_i = uniaxial strain in the ith direction

Poisson's ratio under cryogenic temperatures for various conditions [136] are given in Table I.2.

The German Specification DIN 4227 gives the following Poisson ratios:

E_0 (kN/mm^2):	24	30	35	40
ν:	0.15–0.18	0.17–0.20	0.20–0.25	0.25–0.30

(I.68)

I.5.2 Young's modulus

Young's modulus does not change substantially with time. After 28 days, in general, about 86% of the final value is reached and, after three months, it attains 99.7% of its final value. The values of E_0 defined in some codes are given below:

British Standard BS 8110:

$$E_0 = 5.5\sqrt{f_{cu}/\gamma_m} \quad kN/mm^2 \tag{I.69}$$

Indian Standard IS 456–1979:

$$E_0 = 5688\sqrt{f_{cu}} \quad N/mm^2 \tag{I.70}$$

The European Concrete Committee (CEB):

$$E_0 = 6000\sqrt{f_{cu}} \quad N/mm^2 \tag{I.71}$$

The American Concrete Institute (ACI):

$$E_0 = 4800\sqrt{f_{cu}} \quad N/mm^2 \tag{I.72}$$

For lightweight concrete these values are multiplied by $(D_c/2300)^2$, where D_c is the density of lightweight concrete in kg/m^3.

The German Code DIN 4227 gives E_0 values already given in eqn (I.68). The E_0 values for concrete under cryogenic conditions are given in Table I.2.

The US Bureau of Reclamation [102] quoted by Marin expresses E_0 values in the form of

$$E_0 = E(\bar{\tau}) = [4.4 \times 10^6(1 - e^{-0.07\bar{\tau}})\,lbf/in^2] \times 6.895\,kN/m^2 \tag{I.73}$$

Hobbs reported a theoretical expression for E_0:

$$E(\bar{\tau}) = E_{cm}(\bar{\tau})\left\{1 + \frac{2V_a[E_a - E_{cm}(\tau)]}{E_a + E_{cm} - V_a[E_a - E_{cm}(\tau)]}\right\} \tag{I.74}$$

where $\bar{\tau}$ = number of days, E_{cm} = Young's modulus of cement paste, E_a = Young's modulus of aggregate, and V_a = aggregate volume concentration.

Green and Swanson [145] give an empirical equation:

$$E(\bar{\tau}) = E\left\{0.4 + 0.6\left[\frac{f(\bar{\tau})}{f_{cu}}\right]\right\} \tag{I.75}$$

where E = Young's modulus at 28 days, $f(\bar{\tau})$ = cube strength of concrete at time $\bar{\tau}$, and f_{cu} = cube strength of concrete at 28 days.

I.5.3 Temperature effects on Young's modulus

The influence of temperature distribution on the Young's modulus is much more substantial.

Directly predicted thermal stresses are dependent on the values assumed for Young's modulus and, therefore, the stresses are altered whether or not the change in the elastic properties is the same throughout. Therefore Young's moduli used in constitutive equations need modification in order to suit a particular temperature distribution for given vessel parameters. Three variations of modulus with temperature are plotted in Fig. I.21 for data from Ross [92, 141], Browne [84] and Kasami [80]. In the analyses that follow these variations are considered for specific temperature distributions.

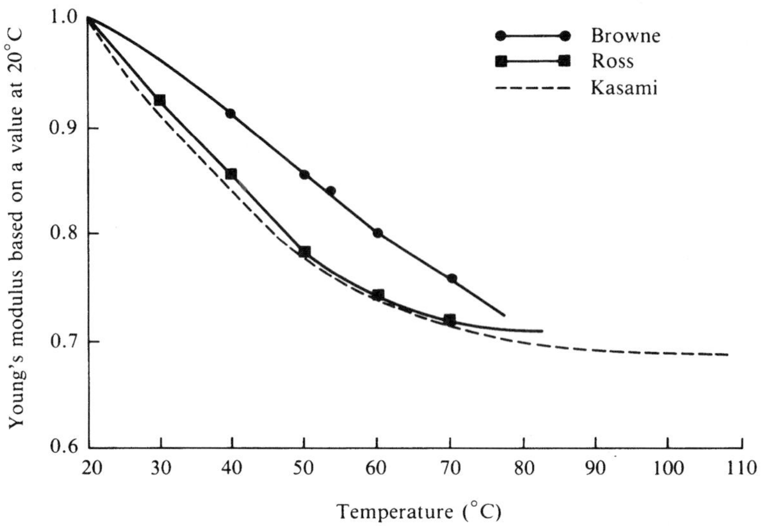

Fig. I.21. Variation of Young's modulus with temperature for concrete.

Equations (I.71) and (I.72) also give expressions for Young's modulus. From the work of Elwi and Murray [5, 135] eqn (I.8a) is differentiated. The tangent moduli with respect to equivalent uniaxial strains are given by

$$E = \frac{d\sigma_i}{d\varepsilon_{0_i}} = \frac{E_0[1 + (2R - 1)\varepsilon_R^2 - 2R\varepsilon_R^3]}{\left\{1 + \left[\left(\frac{E_0}{E_{sec}}\right) + R - 2\right]\varepsilon_R - (2R - 1)\varepsilon_R^2 + R\varepsilon_R^2\right\}^2} \quad (i = 1, 2, 3) \tag{I.76}$$

where $\varepsilon_R = \varepsilon_i/\varepsilon_0'$.

Bangash and England [94] suggested two simplified expressions for temperature-dependent E:

$$E_T = E\left(1 - \frac{T - 20^\circ C}{137}\right) \quad \text{for } 20^\circ C \leq T \leq 50^\circ C$$
$$E_T = 0.78E\left(1 - \frac{T - 50^\circ C}{341}\right) \quad \text{for } 50^\circ C \leq T \leq 85^\circ C \tag{I.77}$$

where E is taken to be at 20°C and T is any temperature not exceeding 85°C.

A similar equation has been produced by Dibsi [147] describing E_{sec} due to temperature effects as

$$E_{sec}(T) = E\left[1 - \left(\frac{T - 125}{1000}\right)^{0.48}\right] \tag{I.78}$$

where E is at 20°C.

1.6 Dense concrete and shotcrete

1.6.1 Introduction

Considerable literature is available [1040–1208] on dense and shot concrete design mix. The most important factor in producing quality shotcrete construction is the skill and ability of the nozzleman, who ensures that no improper encasement of reinforcement, no voids behind steel, no excessive cracking caused by shrinkage, no sand pockets and no cases of overspray and rebound can occur. Dense concrete with high strength, required for biological shields, can be obtained by adding aggregates such as ilmenite, magnetites, and others. Steel fibres are commonly used in the manufacture of shotcrete, producing high densities. Successful high density shielding concrete has also been made using barite as an aggregate.

In many instances, these particular ores have been combined with various forms of manufactured heavy aggregates as well as with other heavy ores. In all instances, the cost of the heavy aggregate and the expense involved in its transportation have been a significant consideration.

Iron-serpentine concrete is a special type of radiation-shielding concrete made with iron aggregate, crushed serpentine rock and (hydrated) Portland cement. The iron provides high density, while the latter two materials are sources of chemically bound water, required for efficient neutron attenuation. By increasing concrete density and water content, shield thickness can be reduced. Serpentine, a hydrous magnesium-silicate mineral, is unique in that it holds most of its bound water at temperatures as high as 45°C.

Data are presented on the physical and nuclear shielding properties of two types of iron–serpentine concrete. Specimens were tested under standard conditions and after being heated at elevated temperatures up to about 320°C. These high-density concretes were designed for use in the construction of nuclear reactor shields which will operate at elevated temperatures. Densities of 35 and 42 kN/m^3 were obtained by using steel aggregates in combination with serpentine. The latter aggregate retains most of its chemically bound water at temperatures as high as 450°C, and thus is a reliable source of hydrogen for attenuating neutrons when shield temperatures are high. Only certain types of serpentine rock are suitable for making shielding concrete. Therefore, care must be exercised in order to obtain rock which is high in bound water content, structurally strong, and relatively free of fibre.

Use of limonite or magnetite ores crushed to fine aggregate gradation and graded steel scrap as coarse aggregate results in a concrete of approximately double the unit weight of conventional concrete. The shielding properties of limonite or magnetite matrix concretes are obtained primarily from this increased density. In certain cases, high density is the only consideration; in others, both density and structural strength are necessary. For structural purposes these concretes are reinforced in the same ways as ordinary concrete. In all cases, the heavy concretes require close laboratory control.

Material properties, experimental and analytical test results are given in this section for these types of concrete.

I.6.2 Material properties

For dense concrete and shotcrete, the choice of material properties and aggregate mix are important as they generally define the strength and density. Table I.3 gives the material mixes for magnetite and ilmenite dense concrete. Table I.4 shows the concrete tests carried out on serpentine aggregate used in beams and cubes. The design mix is given in Table I.5.

I.6.3 Heavy steel aggregate concrete

The ingredients used in making this heavy concrete are

(a) water,
(b) cement,
(c) admixture,
(d) steel punchings, and
(e) steel shot.

Table I.3. Properties of magnetite and ilmenite aggregate concretes for four different mixtures

	Mixture no.	Maximum size of coarse aggregate (mm)	Water/ cement ratio by weight	Theoretical cement content bags per m^3	Sand/ aggregate, percent (fine aggregate as volume percent of total aggregate)	Actual unit weight (MN/m^3)	Slump (mm) range of tests*
Magnetite	1	38	0.35	5.9	39	6.65	12–18
coarse	2	38	0.30	7.0	38	6.56	12–18
and fine	3	12	0.35	6.55	41	6.55	12–18
aggregates	4	12	0.30	7.75	40	6.42	18
Ilmenite	5	38	0.35	5.9	39	6.60	6–12
coarse	6	38	0.30	7.0	38	6.42	6–12
and fine	7	12	0.35	6.55	41	6.48	6–18
aggregates	8	12	0.30	7.75	40	6.38	6–12

*Each result is the average of three determinations.
6″ (152 mm) × 12″ (304.8 mm) cylinders.
Compressive strength:

$52.4\,N/mm^2$ to $64.8\,N/mm^2$ } 7 days

$62\,N/mm^2$ to $75.8\,N/mm^2$ } 28 days

$69\,N/mm^2$ to $84\,N/mm^2$ } 90 days

Table I.4. Properties of serpentine aggregate for iron–serpentine concrete

The mineral serpentine[a]	
Composition	A magnesium silicate, $3MgO.2SiO_2.2H_2O$ or $H_4Mg_3Si_2O_9$
	Silica 44.1%; magnesia 43.0%; water 12.9%
	Ferrous iron and nickel may be present in small amounts.
Structure	Usually massive, but microscopically fibrous and felted.
	The fibrous variety is called chrysotile and is the asbestos of commerce.
Physical properties	Hardness 3.5–5.0; specific gravity 2.5–2.65.
	Lustre is greasy, wax-like in massive varieties.
	Colour is olive to blackish green, yellowish green or white. Often variegated, showing mottling in lighter and darker shades of green.

[a]Petrographic analyses by Mr Arthur W. Schmidt, Chief of the Petrography Branch, NPD Materials Laboratory.

W/C gal. per bag	Cement factor bags per m^3	Theoretical density (kN/m^3)	Actual density (kN/m^3)	Compressive strength (N/mm^2)			Flexural strength (N/mm^2)
				7 day	14 day	21 day	
$4\frac{1}{2}$	6.3	381	376	2860			
$4\frac{1}{2}$	7	372	376	4310			
$4\frac{1}{2}$	8.3	353	353	4330			
$4\frac{1}{2}$	9	342	352	3730			
$3\frac{1}{2}$	7	384	371	5390	5460	3220	615
4	7	377	377	5270	4490	6080	462
$4\frac{1}{4}$	7	370	375	4440	4030	5430	425
$3\frac{1}{2}$	8	372	372	4880	6400	6380	384
4	8	364	370	4690	5760	5290	354
$4\frac{1}{2}$	8	357	361	3960	4600	4260	462
$3\frac{1}{2}$	9	361	356	4930	4850	5320	567
4	9	352	353	3980	4190	4190	405
$4\frac{1}{2}$	9	342	343	2760	3130	3220	329

E value ranges from 20 GN/m^2 to 60 GN/m^2.
Cylinder and beams 152 mm × 304.8 mm, 150 × 150 × 750 mm.

The steel punchings and shot are referred to as the inert ingredients because they do not contribute to the chemical reaction producing the final hardened concrete mass.

The steel punchings used as coarse aggregate consist of a random mix of waste material from steel fabricating shops. For the most part, the pieces are flat and circular, having a maximum diameter of approximately 25 mm and maximum thickness of 6 mm. Intermixed with the flat and circular pieces are many oblong and other variously shaped flat pieces. The specific gravities and density of these ingredients are given in Table I.6.

With shots and punchings as aggregates, theoretical maximum density is calculated establishing the upper limit as 10.71 kN/m^2 for heavy concrete as shown in Table I.7.

I.6.4 Steel fibre reinforced concrete (SFRS)

Generally, most of the design information developed for plain shotcrete may be applied to fibrous shotcrete, with some modifications. The purpose of the

Table I.5. Mix data for shielded concrete

Plastic mortar data	
air content (%)	2.7
unit weight (kN/m^3)	7.4
Consistency	
flow (%)	32
flow (s)	
Batch weight (kg)	
sand	15.9
steel shot, SAE 330	28.2
steel shot, SAE 1320	56.9
cement	12.7
water	6.8
fluidifer	0.13
Water/cement ratio	0:54
Cement factor, s_k/c_y	8.02
Sand grading, percent passing	
No. 4	98.7
8	52.5
16	43.3
30	13.1
50	5.4
100	1.4
200	–
Fineness modulus	3.86
Specific gravity, SSD	2.59
unit weight (kN/m^3)	7.785
Weight change (%)	
28 days at 20°C	0.25
after heating at 85°C	−4.03
200°C	−4.12
350°C	−4.31
Length change (%)	
28 days at 20°C	0.007
350°C	−0.154
Dyn. Mod. Elast. (psi $\times 10^{-6}$)	
28 days at 20°C	40.0 (100%)
after heating at 85°C	16.5 (41%)
200°C	15.26 (38%)
350°C	13.12 (33%)
Coeff. at therm. expansion (in./in.°C)	
28 days at 20°C	13.19
after heating at 85°C	14.63
200°C	14.32
350°C	13.18

Table I.5. (continued)

Flexural strength (kN/m^2)	
28 days at 20°C and after after heating at 350°C	5.79×10^3
Modified cube compression after heating at 350°C	47.6×10^3
Thermal conductivity, BTU/ft/hr/°F	
moist 20°C	1.063
dry 85°C	0.168
200°C	0.196
350°C	0.284
Compressive strength (kN/m^2)	
28 days at 20°C	62.64
after heating at 85°C	84.37 (100%)
200°C	79.27 (113%)
350°C	51.35

Table I.6. Specific gravities and density of ingredients

Material	Specific gravity	Dry packed density (kN/m^3)
Water	1.00	
Cement	3.10	
Steel punchings	7.56	7.0
Steel shot	7.46	8.38
Wrought iron	7.70	13.42
Common steel	7.83	13.64
Armco ingot	7.86	13.17
Ferro-nickel steel	8.10	14.27

Table I.7. Material proportions for theoretical maximum concrete density

Material	Specific gravity	Specific weight	Weight of concrete (kN/m^3)	Volume (m^3)
Shot	7.46	13	4.91	0.01070
Punchings	7.56	13.17	4.91	0.01060
Cement	3.10	5.40	0.68	0.00368
Water	1.00	1.74	0.21	0.00342
Concrete			10.71	0.02832

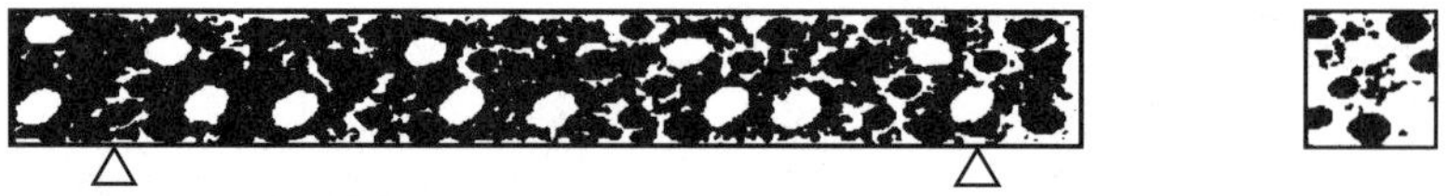

Fig. I.22. RC beams with steel shots as aggregates.

reinforcement is to absorb shrinkage stresses and thermal stresses, improve the shear and tensile strength and distribute concentrated loads over a wide area. Even the densest shotcrete will remain impermeable to water only if it does not crack. Cracks are caused not only by excessive stressing or shifting of the substrate surface, but also because of shrinkage or thermal stresses. Shotcrete should therefore be provided with adequate reinforcement wherever tensile stresses may be expected.

Tensile stresses are also the result of the external loading or rock pressure. Anchors, as used in the NATM, press against the shotcrete layer in a point loading pattern. The immediate vicinity of the anchor head is subjected to a shear load, while a bending moment can occur in the zone between the anchors.

Steel fibres, therefore, provide a homogeneous reinforcement to the shotcrete layer, absorbing all tensile stresses wherever produced and whatever the origin may be.

I.6.5 Experimental tests on beams and slabs with dense concrete

Tests carried out in Section IX.2, Case Study No. 1 and in Section IX.3, Case Study No. 2 are repeated. This time, the aggregates were steel shots (serpentine type) and steel fibres. The beams are reinforced and similar failure tests have been carried out. Various stress–strain curves have been obtained. These are compared later with those obtained from the finite element analysis. Figure I.22 details the beam showing aggregates and Fig. I.23 shows the concrete cube with only three layers for the finite element analysis.

I.6.6 Numerical modelling and finite element analysis

Figures I.24 and I.25 are the optimum finite element mesh schemes including dense concrete aggregates. First, the cubes are tested as shown in Appendix VI.

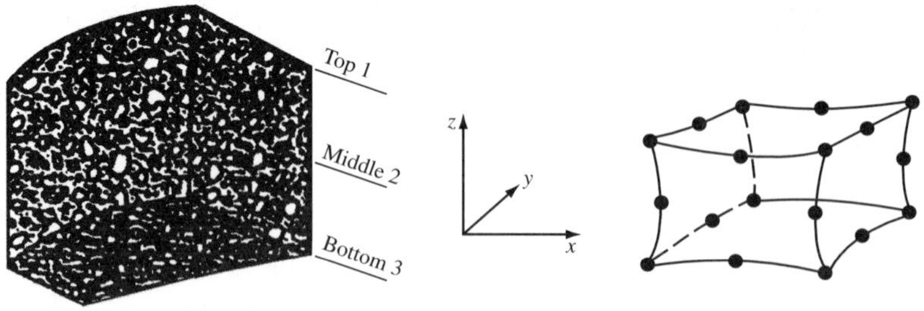

Fig. I.23. 20-noded isoparametric finite element for a concrete cube with steel shot and steel fibre aggregates.

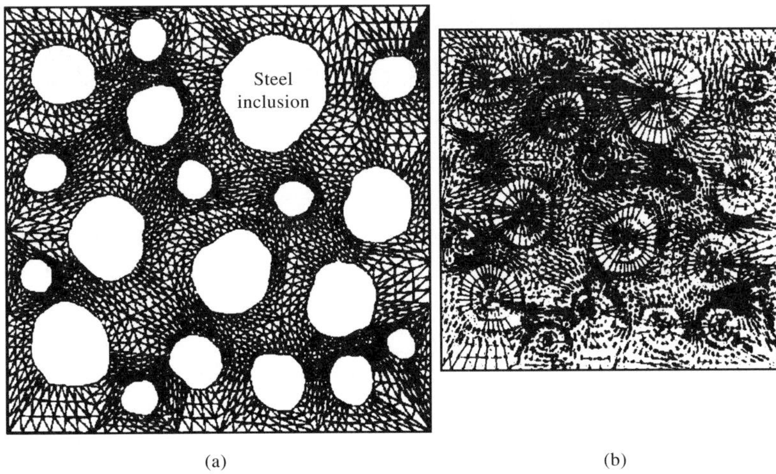

Fig. I.24. FE mesh schemes for first and third layers: (a) first and third layer FE mesh scheme — aggregate with individual boundary conditions; (b) damage scenario — aggregate part of mesh scheme.

The analysis is repeated using the steel aggregates. The same procedure is adopted and the cube compression tests are given in Fig. I.26. Here the data given in Table I.3 are used. Where reinforcement is used, the criteria given in Appendix VII have been adopted. Chapter VI and Appendix IA give technical details about the finite element.

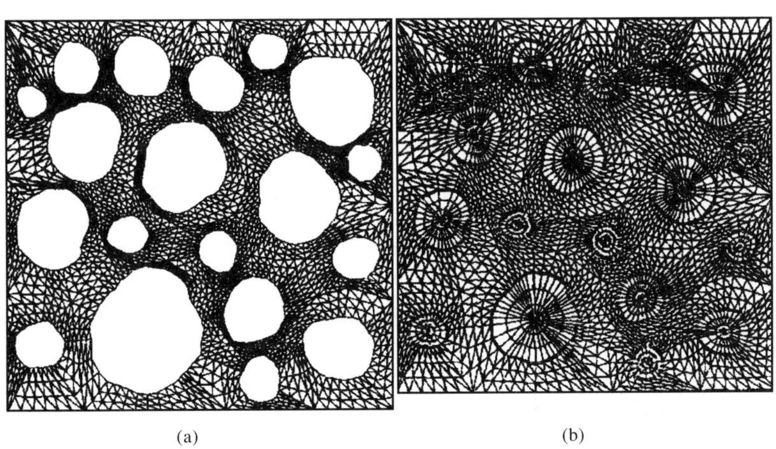

Fig. I.25. Mid-layer finite element mesh scheme with and without aggregate mesh idealisation: (a) second layer at mid-point, FE mesh scheme without aggregate as part of the mesh scheme; (b) second layer at mid-point, FE mesh scheme with aggregate as part of the mesh.

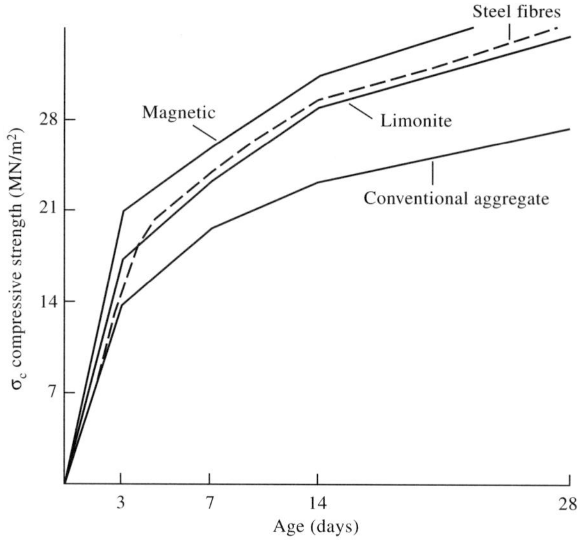

Fig. I.26. A comparative study of compression strength with age.

I.6.6.1 Data for dense cubes and dense reinforced concrete beams

(*a*) 150 × 150 × 150 mm dense concrete cubes with steel fibres as aggregates — FE

E_s = elastic modulus of 28 days = 30 GPa — 1800 elements without shots and fibres

σ_c = compressive strength of 28 days = 32 MPa

ν = Poisson's ratio = 0.25; w/c = 0.45 — 2800 nodes

d = thickness = 150 mm — depth of each element = 50 mm

θ = friction angle (average) = 300

c = cohesion — 3 layers = 150 mm total depth

σ_{ss} = steel shots or fibres = 0.25 MPa

ψ = dilation angle = 200

f = dilation factor = 2

Loads as in the previous case studies: Section IX.2, No. 1, Section IX.3, No. 2 and Appendix VI.

(*b*) Beam with steel shot Fig. I.22
Concrete data as above
Increment of loads 5 kN
L = 1985 mm
f_{cu} = 30 N/mm^2
Loads (kN) 10, 20, 30, 40, 50, 60, 70–100
All data given in Section 1.6.6.1(a)
Total no. of elements = 20 Half beam symmetrical for analysis
Total no. of nodes = 340
Depth of each element = 95 mm

I.6.6.2 Additional criteria for the finite element analysis with steel shot

The following major areas are highlighted here as they form the basis for the analysis criteria.

(a) *Constitutive law for concrete.* Ottosan criteria with strain hardening and softening material. Blunt crack propagation for the formulation of cracking direction.
(b) *Steel shot and fibres.* Steel shot stress–strain criteria multi-linear type fibre are assumed to be distributed in different directions.
(c) *Bond.* Bond-slip phenomenon indicated in the text is assumed for both ordinary and steel aggregate gap element given in Appendix XIII.

During the iteration procedure in the analysis, compression–compression regions and tension–tension or tension–compression regions are examined, taking into consideration the first invariant of the stress tensor and the second invariant of the deviatoric stress tensor.

A non-linear iterative method using the initial stiffness approach is adopted.

I.6.6.3 Cube tests evaluation

The equivalent days chosen for the analysis are 2, 5, 7, 10, 14, 21 and 28 days. They are converted to minutes/seconds for the finite element analysis. Maximum grain size is 12 mm.

Finite element layered schemes for the dense cubes and RC beams. The finite element mesh schemes are shown with steel shot. Three layers are chosen in both the cubes and in the assumed beam. Top and bottom layers are shown by Fig. I.24. The results in general are similar to those given in Appendix VI with a difference indicated by Figs I.26 to I.29. In general, the strength of cubes is enhanced by 30–40% above those with ordinary concrete.

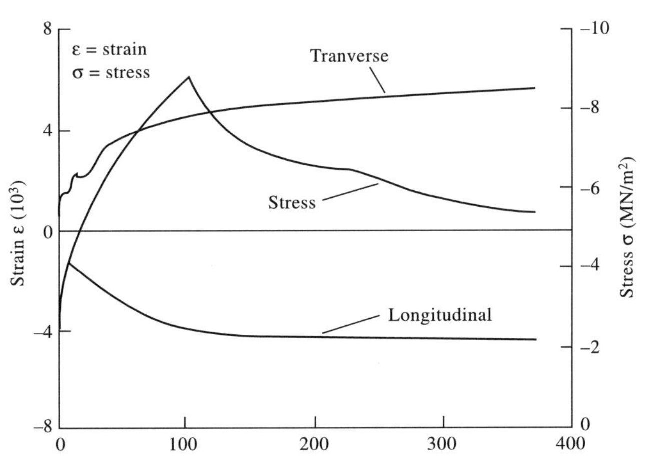

Fig. I.27. Time-dependent stress–strain relations.

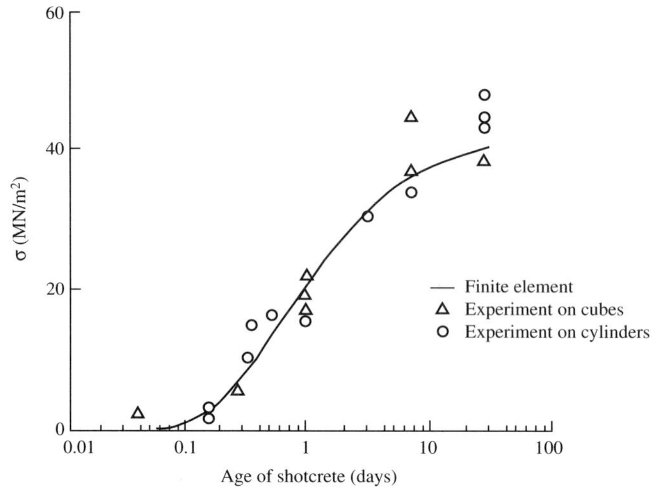

Fig. I.28. A comparative study of stress at ages of concrete using steel shots.

The force–displacement can now be compared with various aggregates in Fig. I.30. Here, the comparative study shows that the behaviour of a comparable beam using steel shot has been improved by at least 30% with regard to failure load. The load at failure for the beam based on the criteria explained earlier, is 77 kN as opposed to 55 kN at crush stages. The strain profiles given in Fig. I.27 have been deflected by under 30%, i.e. there is an improvement to all those values. The cracking and crushing effects of the same beam at 77 kN are identical to those shown later on in Figs IX.8 and IX.10 for the same finite element mesh

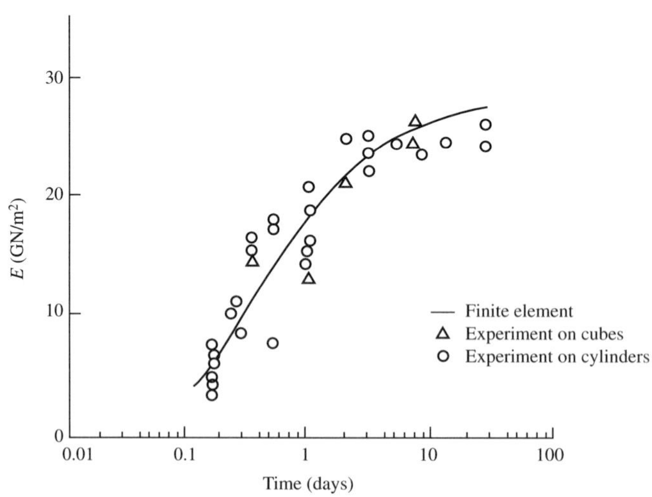

Fig. I.29. Young's modulus versus time in days.

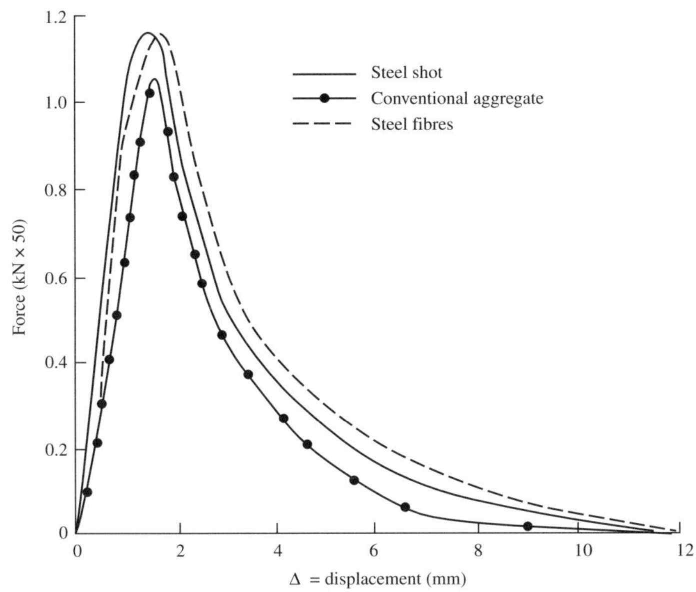

Fig. I.30. Force–displacement relations using conventional aggregates and shots or fibres.

in Fig. IX.9. Fracture energy has been evaluated for the dense concrete for the beam given in Fig. IX.9, assuming the same quantity of the reinforcing steel. The analytical results are compared with those using the RILEM, Bazant and Eurocode methods and they are plotted in Fig. I.31.

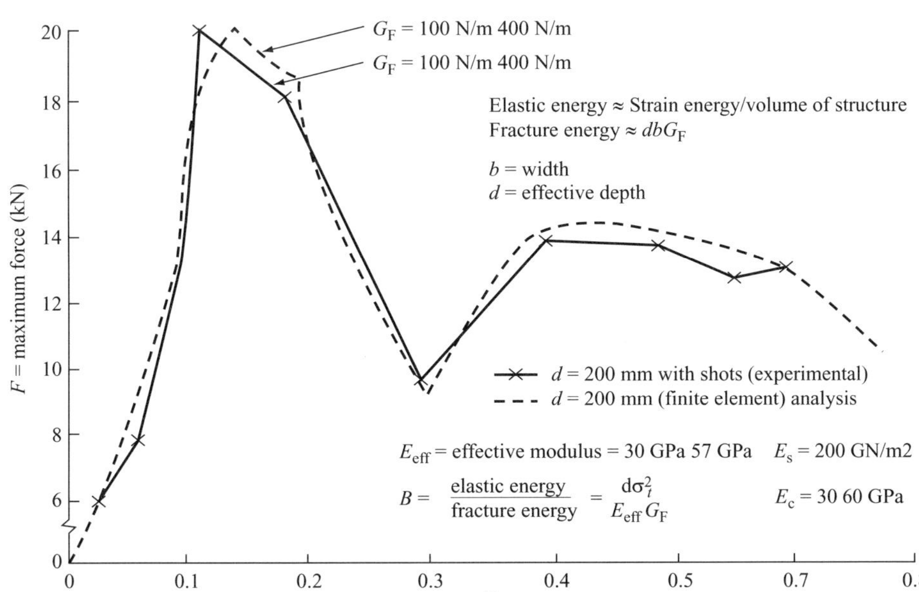

Fig. I.31. Force versus ratio of elastic/fracture energy.

II. Numerical modelling of concrete strength and failure

A literature review is given for the constitutive and numerical modelling of concrete. Concrete is characterised by a materially nonlinear deformation behaviour. The material nonlinearity is assumed to occur due to cracking of concrete in tension and plasticity of concrete in compression. All numerical models discussed in this chapter include the possibility of such effects. They are written specifically in a format suitable for finite element analysis. They have the flexibility to include creep, shrinkage and temperature changes, and also fatigue — discussed later in this book.

II.1 Failure models for concrete

II.1.1 One-parameter model

The Tresca and Von Mises shear stress criteria are one-parameter models and have already been discussed in Chapter I, Section I.4. For numerical analysis the Tresca model is written as

$$f(J_2, \theta) = \sqrt{J_2} \sin\left(\theta + \frac{\pi}{3}\right) - \bar{y} = 0 \tag{II.1}$$

In terms of invariants J_2 and J_3, eqn (II.1) can also be written as

$$f(J_2, J_3) = J_2^3 - 6.75J_3^2 - 9\bar{y}^2 J_2^2 + 24\bar{y}^4 J_2 - 16\bar{y}^6 = 0 \tag{II.2}$$

If the material is isotropic and equal yield stress exists in tension and in compression, eqn (II.2) converges to

$$f(J_2, J_3) = J_2^3 - 2.25J_3^2 - \bar{y}^6 = 0 \tag{II.3}$$

Here $\bar{y}$ can also be approximated as $(1/\sqrt{3})f_{cu}$.

An alternative to the yield stress in pure shear, the octahedral shear stress concept, can be substituted, which is given in the form

$$f(J_2) = J_2 - \bar{y}^2 = 0 \tag{II.4}$$

where J_2 is related to τ_0 as given in eqn (I.48).

Using finite element analysis, the Von Mises surface can be combined with the tension cut-off surface in order to take into consideration the limited tensile capacity of concrete. Similarly the same Von Mises surface can be combined with the ultimate compressive strain criterion ε_{cu} for the compressive ductility and crushing of concrete.

II.1.2 Two-parameter model

Equation (II.5) assumes that yielding will occur when the shear stress τ_0 reaches the value that depends on the cohesion stress C and the normal stress σ. Figure II.1 shows a yield surface in a three-dimensional stress space. This is graphically represented by a hexagonal pyramid, the axis of which lies along the space diagonal, known as the hydrostatic stress axis. The deviatoric plane (π-plane) passes through the origin and perpendicular to this hydrostatic axis. Graphical representation shows that it is an irregular right hexagon and the yield stress in tension differs from that in compression. In numerical form the yield criterion of eqn (II.5) can be expressed in a number of ways. Chen [148] expresses it as

$$f(I_1, J_2, \theta) = \tfrac{1}{3}I_1 \sin\phi + \sqrt{J_2}\sin\left(\theta + \frac{\pi}{3}\right) + \frac{\sqrt{J_2}}{\sqrt{3}}\cos\left(\theta + \frac{\pi}{3}\right)\sin\phi - C\cos\phi = 0 \qquad \text{(II.5)}$$

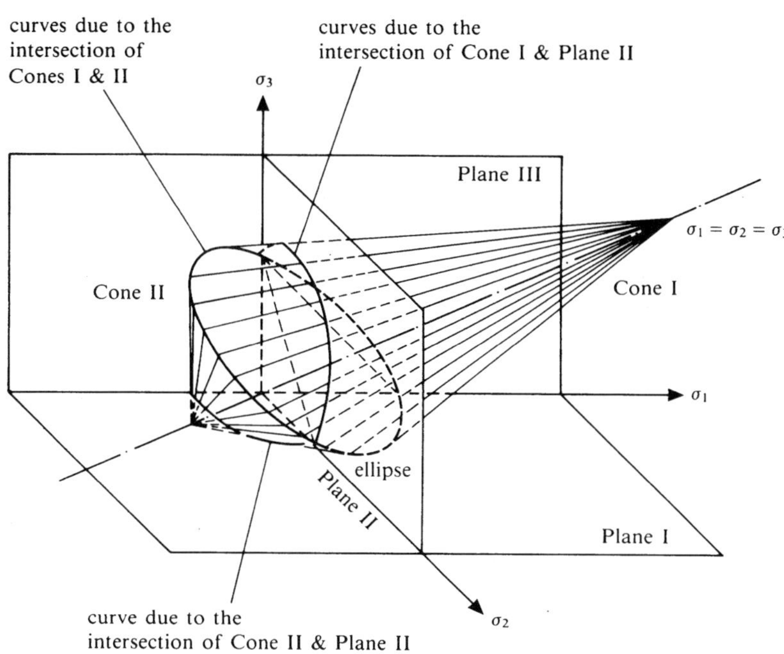

Fig. II.1. Graphical representation of octahedral shearing stress yield criteria of concrete.

The final expressions developed by Kojic and Cheathem [149] are written as

$$f(I_1, J_2, \theta) = \tfrac{1}{3}I_1 \sin\phi + \sqrt{J_2}\cos\theta - \frac{\sqrt{J_2}}{\sqrt{3}}\sin\theta + \sin\phi - C\cos\phi = 0 \qquad \text{(II.6)}$$

in which I_1 is the first invariant of stresses and J_2 is the second invariant of stress deviations. The angle θ is the lode angle which is an alternative form of the third stress invariant.

Bangash [126, 750] has analysed a containment vessel using a simplified version of the two-parameter model as a function of three co-ordinates

$$f(z, \gamma, \delta) = 0 \qquad \text{(II.7)}$$

where

$$z = \frac{1}{\sqrt{3}}(\sigma_1 + \sigma_2 + \sigma_3)$$

$$\gamma = \frac{1}{\sqrt{3}}\sqrt{(\sigma_1 - \sigma_2)^2 + (\sigma_2 - \sigma_3)^2 + (\sigma_1 - \sigma_3)^2}$$

$$\delta = \cos^{-1}[\sigma_1 + \sigma_2 - 2\sigma_3/\sqrt{2}\sqrt{3}\gamma]$$

Suh [152] modified the Mohr–Coulomb yield criterion to take into account both dilatancy and compaction using a closed surface of the failure cone. Mathematically, this is represented as

$$f(I_1, J_2, \theta) = \left[\left(-\frac{I_1}{3} + K\right)^2 + \gamma J_2\right]^{1/2} - \beta\left[\cos\frac{\pi\theta}{2\phi}\right]^n = 0 \qquad \text{(II.8)}$$

where

$$\tan\theta = \sqrt{\gamma J_2}\Big/\left(\frac{I_1}{3} + K\right) \qquad \text{(II.8a)}$$

γ is generally taken as 2/3, and K, n, β are positive constants, depending on concrete properties.

The yield surface is the revolution of a lemniscate. Equation (II.8) does include the effects of strain hardening and porosity. Since the Mohr–Coulomb surface is angular in the plane and has a singular vertex point in the principal stress space, a greater computational advantage can be achieved by eliminating angularity and vertex singularity. Zienkiewicz and Pande [153] have introduced hyperbolic and parabolic approximations of Mohr–Coulomb with an added strain dependent elliptical surface. A general surface without corners has been developed and is given in the form

$$f(I_1, J_2, \theta) = \alpha'\left(\frac{I_1}{3}\right)^2 + \beta\frac{I_1}{3} + \gamma' + \left\{\frac{J_2}{g(\theta)}\right\}^2 = 0 \qquad \text{(II.9)}$$

where α', β', γ' are material constants. The function $g(\theta)$ is such that $g(\pi/6) = 1$ so that it determines the shape of the π-section. Various forms of $g(\theta)$ have been suggested in order to avoid singularities.

In the Drucker–Prager criterion [154] it was discovered that in the case of a Mohr–Coulomb surface where information is not available in advance about the use of any one of the six sides or the non-fixation of the principal directions at each point, numerical computation presents problems. It gives a smooth approximation of the Mohr–Coulomb surface and presents it in a mathematical form:

$$f(I_1, J_2) = \alpha I_1 + \sqrt{J_2} - K = 0 \tag{II.10}$$

where α and K are positive and constant at each point of the material. Drucker and Prager have shown that eqn (II.9) gives a right circular cone which possesses a vertex and in which the corners have been smoothed. The axis of the cone, as shown in Fig. II.1, lies along the hydrostatic stress. If $\alpha = 0$, eqn (II.10) reduces to the Von Mises yield criterion stress. Where $\alpha \neq 0$ the plastic deformation is accompanied by an increase in volume. The values of α and K are given as

$$\alpha = 2\sin\phi/\sqrt{3}(3 \pm \sin\phi)$$
$$K = 6C\cos\phi/\sqrt{3}(3 \pm \sin\phi) \tag{II.10a}$$

ϕ is used when the inner cone passes through the tensile meridian ρ_t when $\theta = 0$; ϕ is used when two surfaces of the model agree along the compressive meridian ρ_c when $\theta = 60°$.

II.1.3 Three-parameter model

Three-parameter models developed by Reimann [63] and Argyris [64] are briefly described in Chapter I. Willam [65] has given a three-parameter failure surface for concrete in a low compression regime and in tension cross-sections. The failure model (Fig. II.2) which is convex and smooth has been achieved by the curve fitting of an ellipse, which is a portion of the failure curve. Owing to symmetry, smoothness and convexity, the elliptic form can easily degenerate to a circle which, in turn, shows that both the Von Mises and the Drucker–Prager criteria are special cases of this model.

The standard form of an ellipse is

$$\frac{x^2}{a^2} + \frac{y^2}{b^2} = 1 \tag{II.11}$$

where a and b are half axes.

Sampling this equation at a point $P_2(m, n)$ yields

$$\frac{m^2}{a^2} + \frac{n^2}{b^2} = 1$$

where these values have been defined as

$$\begin{aligned} m &= \frac{\sqrt{3}}{2}\rho_c \qquad & n &= b - (\rho_t - \tfrac{1}{2}\rho_c) \\ a &= \sqrt{\frac{\rho_c(\rho_t - 2\rho_c)^2}{(5\rho_c - 4\rho_t)}} \qquad & b &= (2\rho_t^2 - 5\rho_t\rho_c + 2\rho_c^2)/(4\rho_t - 5\rho_c) \end{aligned} \tag{II.12}$$

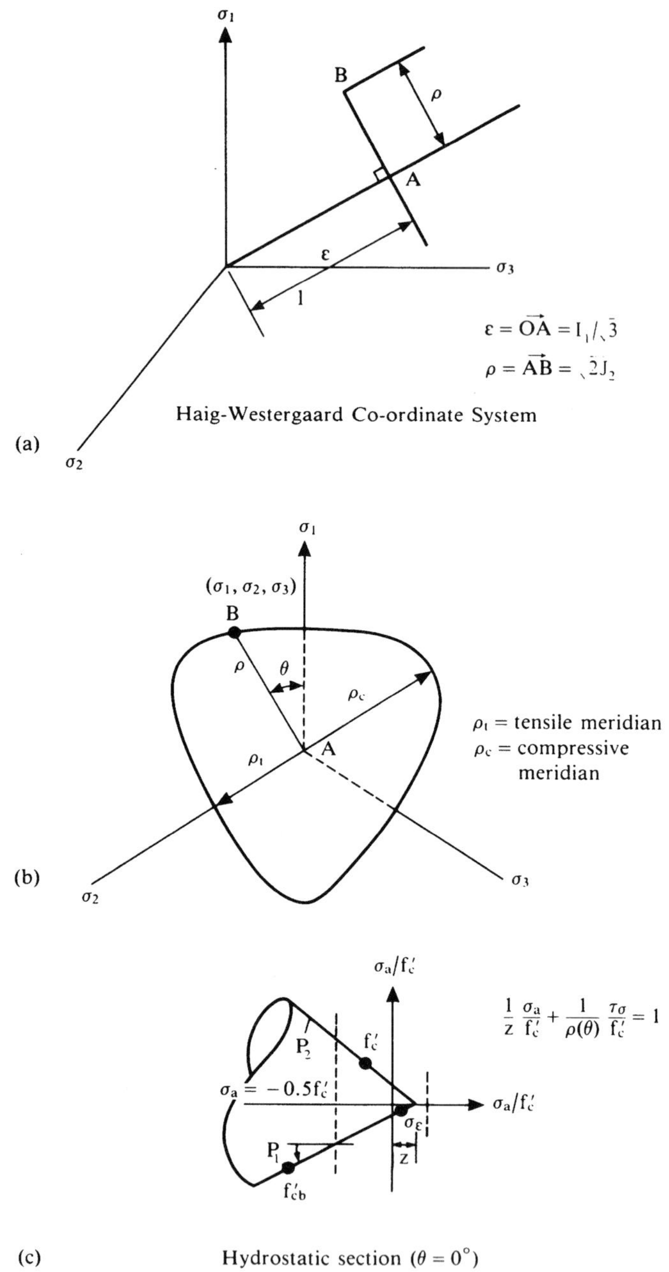

Fig. II.2. Three-parameter model strength ratios, $\alpha_u = 1.3$, $\alpha_z = 0.1$ [63, 64].

Using $x = \rho \sin\theta$, $y = \rho\cos\theta - (\rho_t - b)$, the final surface form in polar coordinates ρ, θ for $0 \le \theta \le 60°$ is given as

$$\rho(\theta) = \frac{2\rho_c(\rho_c^2 - \rho_t^2)\cos\theta + \rho_c(2\rho_t - \rho_c)[\alpha' + 5\rho_t^2 - 4\rho_t\rho_c]^{1/2}}{\alpha' + (\rho_c - 2\rho_t)^2} \qquad \text{(II.12a)}$$

where

$$\alpha' = 4(\rho_c^2 - \rho_t^2)\cos^2\theta$$

The angle of similarity θ is computed as

$$\theta = \cos^{-1}\left[\frac{2\sigma_1 - \sigma_2 - \sigma_3}{\sqrt{2}[(\sigma_1 - \sigma_2)^2 + (\sigma_2 - \sigma_3)^2 + (\sigma_3 - \sigma_1)^2]^{1/2}}\right] \quad \text{(II.12b)}$$

If $\frac{1}{2} \leq \rho_t/\rho_c \leq 1$ satisfies the position vector ρ, it shows that the failure curve has both convexity and smoothness. This three-parameter model predicts failure if the average stresses σ_a, τ_a, and the angle of similarity θ satisfy the boundary conditions of the following equation:

$$f_c(\sigma_a, \tau_a, \theta) = \frac{1}{z}\frac{\sigma_a}{f_c'} + \frac{1}{\rho(\theta)}\frac{\tau_a}{f_c'} - 1 = 0 \quad \text{(II.13)}$$

where

$$\sigma_a = \tfrac{1}{3}I_1 \qquad \tau_a = \sqrt{\tfrac{3}{5}} \qquad \tau_0 = \tfrac{2}{5}J_2 = \tfrac{1}{5}\rho^2 \quad \text{(II.13a)}$$

$$z = \frac{\alpha_u \alpha_z}{\alpha_u - \alpha_z} \qquad \alpha_z = \sigma_t/f_c' \qquad \alpha_u = \frac{\sigma_{cb}}{f_c'} \quad \text{(II.13b)}$$

σ_{cb} = stress at biaxial conditions

The values for the constituents are given as:

$$\begin{array}{cccc}
\dfrac{\sigma_a}{f_c'} & \dfrac{\tau_a}{f_c'} & \theta & \rho(\theta) \\[2ex]
\dfrac{1}{3}\alpha_z & \sqrt{\dfrac{2}{15}}\alpha_z & 0^\circ & \rho_1 = \dfrac{\alpha_u\alpha_z}{2\alpha_u - \alpha_z}\sqrt{\dfrac{6}{5}} \\[2ex]
-\dfrac{1}{3} & \sqrt{\dfrac{2}{15}} & 60^\circ & \rho_2 = \dfrac{\alpha_u\alpha_z}{3\alpha_u\alpha_z + \alpha_u - \alpha_z}\sqrt{\dfrac{6}{5}} \\[2ex]
-\dfrac{2}{3}\alpha_u & \sqrt{\dfrac{2}{15}}\alpha_u & 0^\circ & \rho_1
\end{array} \quad \text{(II.13c)}$$

II.1.4 Four-parameter model

A failure criterion proposed by Ottoson [24–26], known as the four-parameter model, is given here. This failure surface contains all three stress invariants and has the following characteristics:

(a) The surface is smooth and convex with curved meridians, determined by the constants a and b.
(b) It is open in the negative direction of the hydrostatic axis.
(c) The trade in the deviatoric plane changes from an almost triangular to a circular shape with increasing hydrostatic pressure, defined by λ on the deviatoric plane.
(d) The surface is in good agreement with experimental results over a wide range of stress states, including those where tensile stresses occur.

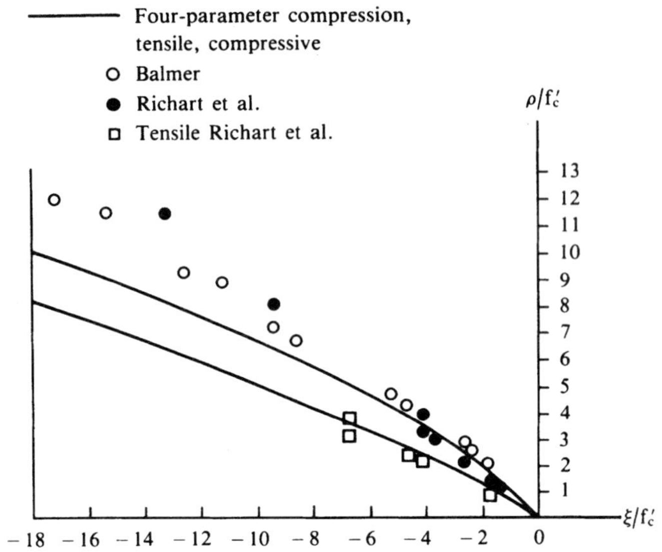

Fig. II.3. Comparison for the four-parameter model.

Figures II.3 and II.4 show the surface in the principal stress co-ordinate system in which the compressive meridian ρ_c ($\theta = 60°$, $\sigma_1 = \sigma_2 > \sigma_3$) and the tensile meridian ρ_t ($\theta = 0°$, $\sigma_1 = \sigma_2 < \sigma_3$) are defined. The meridians are curved, smooth and convex, and ρ increases with increasing hydrostatic pressure.

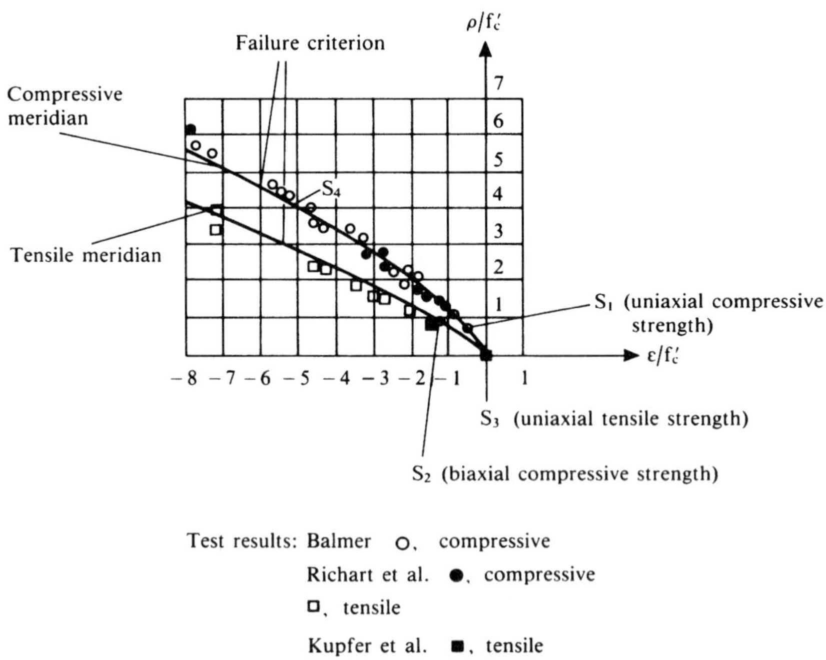

Fig. II.4. Determination of material parameters (S_1, S_2, S_3, S_4 = failure stresses).

An analytical failure surface containing all the above characteristics is defined in the following form by Ottoson

$$f(I_1, J_2, J) = a\frac{J_2}{(f'_c)^2} + \lambda\frac{\sqrt{J_2}}{f'_c} + b\frac{I_1}{f'_c} - 1 = 0 \tag{II.14}$$

where

$$I_1 = \sigma_x + \sigma_y + \sigma_z = \text{the first invariant of the stress tensor} \tag{II.14a}$$

$$J_2 = \text{the second invariant of the stress deviator tensor}$$

$$= \tfrac{1}{2}(S_x^2 + S_y^2 + S_z^2) + \tau_{xy}^2 + \tau_{yz}^2 + \tau_{zx}^2 \tag{II.14b}$$

$$J = \cos 3\theta = 1.5\sqrt{3}\frac{J_3}{\sqrt{J_2}} \tag{II.14c}$$

$$J_3 = \text{the third invariant of the stress deviator tensor}$$

$$= S_x S_y S_z + 2\tau_{xy}\tau_{yz}\tau_{zx} - S_x\tau_{yz}^2 - S_y\tau_{xz}^2 - S_z\tau_{xy}^2 \tag{II.14d}$$

$$S_x = \sigma_x - I_1/3$$
$$S_y = \sigma_y - I_1/3 \tag{II.14e}$$
$$S_z = \sigma_z - I_1/3$$

$\lambda = \lambda(\cos 3\theta) > 0$ $\quad$ a and b are constant

$\lambda = K_1 \cos(\frac{1}{3}\cos^{-1}(K - 2\cos 3\theta))$ for $\cos 3\theta > 0$

$\lambda = K_1 \cos(\pi/3 - \frac{1}{3}\cos^{-1}(-K_2 \cos 3\theta))$ for $\cos 3\theta \leq 0$

K_1, K_2, a and b are material parameters ($0 \leq K_2 \leq 1$)

f'_c = uniaxial compressive cylinder strength for concrete

σ_t = uniaxial tensile strength for concrete

For $a > 0$, $b > 0$, the meridians become curved (non-affine), smooth and convex and the surface opens in a negative direction of the hydrostatic axis. When $a = 0$, λ = constant, this criterion comes closer to the Drucker–Prager criterion. When $a = 0$, $b = 0$ and λ = constant, it represents the Von Mises criterion. The relationship of $\sqrt{J_2}/f'_c$ is given by

$$\frac{\sqrt{J_2}}{f'_c} = \frac{1}{2a}\left[-\lambda + \sqrt{\lambda^2 - 4a\left(b\frac{I_1}{f'_c} - 1\right)}\right] \tag{II.15}$$

Corresponding to the hydrostatic tension, when the vertex of the failure surface is approached, $\sqrt{J_2} \rightarrow 0$, which leads to

$$\frac{\sqrt{J_2}}{f'_c} \rightarrow \frac{1}{\lambda}\left(1 - b\frac{I_1}{f'_c}\right) \tag{II.16}$$

Table II.1. Four material parameters ($k = \sigma_t/\sigma_c$)

k	a	b	k_1	k_2
0.08	1.8076	4.0962	14.4863	0.9914
0.10	1.2759	3.1962	11.7365	0.9801
0.12	0.9218	2.5969	9.9110	0.9647

i.e.

$$\frac{\rho_t}{\rho_c} \rightarrow \frac{\lambda_c}{\lambda_t} \text{ for } \sqrt{J_2} \rightarrow 0$$

in which $\lambda_c = \lambda(-1)$ and $\lambda_t = \lambda(1)$ corresponding to the compressive and tensile meridians respectively.

The value of λ_c/λ_t is inside the range 0.54–0.58. For $\rho_t/\rho_c \rightarrow 1$ for $I_1 \rightarrow -\infty$, the deviatoric plane becomes nearly circular for very high compressive stresses.

Equation (II.14) defines the failure at a point if $f \geq 0$. For $f < 0$, the point is situated inside the failure surface.

II.1.4.1 Parameters a, b, K_1 and K_2 (Tables II.1 and II.2)

The four material parameters are determined using the biaxial test results of Kupfer [113] and the triaxial results of Balmer [16] and Richart *et al.* [14]. The following three failure states were represented:

1. Uniaxial compressive strength, f'_c ($\theta = 60°$). Uniaxial tensile strength, σ_t ($\theta = 0°$) $= Kf'_c$.
2. Biaxial compressive strength, $\sigma_1 = \sigma_2 = -1.16\sigma_c$; $\sigma_3 = 0$ ($\theta = 0°$) (test results of Kupfer [113]).
3. The triaxial state $(\xi/f'_c, \rho/f'_c) = (-5, 4)$ on the compressive meridian ($\theta = 60°$).

Hereafter the method of least squares is adopted to obtain the best fit of the compressive meridian for $\xi/f'_c = -5.0$ to test the results of Balmer [16] and Richart *et al.* [14]. Figure II.3 shows the process where the compressive meridian passes through a point (where ξ/σ_c, $\rho/\sigma_c = (-5, 4)$). With this procedure, the values of material parameters are determined as given in Tables II.1 and II.2. From these tables it is clear that the material parameters show considerable dependence on $k = \sigma_t/f'_c$, but the failure stresses in the compressive regime are only slightly affected.

II.1.4.2 σ_{cu} value

It remains now to assess σ_{cu} ($i = 1, 2, 3$), the peak stress, for the calculation of tangent moduli (eqn (I.75)) for various principal stress ratios. Under uniaxial conditions σ_{cu} is equal to the compressive cylinder strength. However, under multiaxial stress conditions the compressive strength of concrete increases. To

Table II.2. Values of the function ($k = \sigma_t/\sigma_c$)

k	λ_t	λ_c	λ_c/λ_t
0.08	14.4925	7.7834	0.7378
0.1	11.7109	6.5315	0.5577
0.12	9.8720	5.6979	0.5772

obtain σ_{cu} in three directions for the principal stress ratio, a surface in the stress space is used. First of all, current principal stresses are established: let these be σ_{p1}, σ_{p2}, σ_{p3}, where $\sigma_{p1} > \sigma_{p2} > \sigma_{p3}$. It is then assumed that σ_{p1} and σ_{p2} are held constant while the third principal stress is changed such that it reaches to the failure surface. This establishes that the ultimate stress is σ^c_{p3}. Similarly, σ^c_{p1} and σ^c_{p2} are calculated by increasing their values while the other two stresses remain constant. This means that the principal stresses are substituted in the failure surface (eqn (II.14)) and then one of the stresses (the more compressive) is increased while the other two remain constant until eqn (II.14) is satisfied.

II.1.5 Five-parameter model

The three-parameter model developed by Willam [65] has been extended by adding two additional degrees of freedom for describing curved meridians so that the failure surface model can be applied to low as well as high compression regions. The more general failure condition is suggested in contrast to eqn (II.13) which determines the average stress components using a linear relationship. This failure condition is given by

$$f(\sigma) = f(\sigma_a, \tau_a, \theta) = \frac{1}{\rho(\sigma_a, \theta)} \frac{\tau_a}{f'_c} - 1 \qquad \text{(II.17)}$$

The average shear stress is restricted for $f(\sigma) = 0$ to

$$\frac{\tau_a}{f'_c} = \rho(\sigma_\theta, \theta)$$

such that τ_a is now a single function of σ_a and θ. The proposed model now removes the affinity of deviatoric planes and sections built into previous theories.

The failure model is constructed by first obtaining the two meridians at $\theta = 0°$ and $\theta = 60°$ by two second-order parabolas which are connected by an ellipsoidal surface (Fig. II.5). These parabolas, whose tensile and compressive meridians exist, are given by

$$\rho'_1(\sigma_a) = \frac{\rho_t}{\sqrt{5}f'_c} = a_0 + a_1 \frac{\sigma_a}{f'_c} + a_2 \left(\frac{\sigma_a}{f'_c}\right)^2 \quad \text{at } \theta = 0° \qquad \text{(II.18)}$$

$$\rho'_2(\sigma_a) = \frac{\rho_c}{\sqrt{5}f'_c} = b_0 + b_1 \frac{\sigma_a}{f'_c} + b_2 \left(\frac{\sigma_a}{f'_c}\right)^2 \quad \text{at } \theta = 60° \qquad \text{(II.18a)}$$

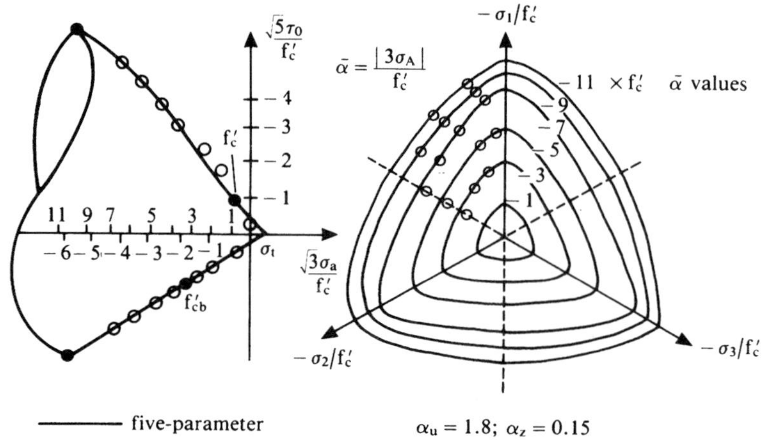

Fig. II.5. Five-parameter model [65].

The surface is defined on the basis of the average normal stress σ_a $(0^\circ \leq \theta \leq 60^\circ)$ as

$$\rho(\sigma_a, \theta) = \frac{2\rho_c(\rho_c^2 - \rho_t^2)\cos\theta + \rho_c(2\rho_t - \rho_c)[\sqrt{\alpha' + 5\rho_t^2 - 4\rho_t\rho_c}]}{\alpha' + (\rho_c - 2\rho_t)^2} \tag{II.19}$$

which is the right-hand-side of eqn (II.12a).

The six degrees of freedom a_0, a_1, a_2, b_0, b_1 and b_2 require evaluation. When the following adjustments are made, the Von Mises, Drucker–Prager, Willam three-parameter and Ottoson criteria are established

$$
\begin{aligned}
&a_0 = b_0 \quad \text{and} \quad a_1 = b_1 = b_2 = a_2 = 0 && \text{leads to Von Mises}\\
&a_0 = b_0 \quad a_1 = b_1 \quad a_2 = b_2 = 0 && \text{leads to Drucker–Prager}\\
&\frac{a_0}{b_0} = \frac{a_1}{b_1} \quad a_2 = 0 = b_2 && \text{leads to Willam three-parameter}\\
&\frac{a_0}{b_0} = \frac{a_1}{b_1} = \frac{a_2}{b_2} && \text{leads to Ottoson}
\end{aligned} \tag{II.20}
$$

When the two parabolas pass through a common apex $\bar{\sigma}_a$ at the equisectrix or hydrostatic axis, this imposes a constraint

$$\rho_t(\xi_0) + \rho_c(\xi_0') = 0 \tag{II.21}$$

where

$$\xi_0' = \frac{\bar{\sigma}_a}{f_c'}$$

The stress state of five sets together with the constraint conditions of the common apex for $\theta = 0^\circ$ and $\theta = 60^\circ$ are given in Table II.3.

Table II.3. Parameters and their values

Test	σ_a/f'_c	τ_a/f'_c	$\theta°$	$\rho(\sigma_a, \theta)$
$\sigma_1 = \sigma_t$	$\frac{1}{3}\alpha_z$	$\sqrt{\frac{2}{15}}\alpha_z$	0	$\rho_t(\sigma_a)$
$\sigma_2 = \sigma_3 = \sigma_{cb}$	$-\frac{2}{3}\alpha_u$	$\sqrt{\frac{2}{15}}\alpha_u$	0	$\rho_t(\sigma_a)$
	$-\zeta'$ [a]	ρ_1 [b]	0	$\rho_t(\sigma_a)$
$\sigma_3 = f'_c$	$-\frac{1}{3}$	$\sqrt{\frac{2}{15}}$	60	$\rho_c(\sigma_a)$
	$-\zeta'$	ρ_2 [c]	60	$\rho_c(\sigma_a)$
	ξ'_0	0	60	$\rho_c(\sigma_a)$

[a] $\zeta' = -\sigma_a/f'_c$.
[b] At $\theta = 0°$; $\rho_1 = \tau_a/f'_c$.
[c] $\rho_2 = \tau_a/f'_c$; at $\theta = 60°$.

Substituting the three strength values in Table II.3 into eqn (II.18a) at the tensile meridian $\theta = 0°$, the three parameters a_0, a_1 and a_2 are given as

$$a_0 = \tfrac{2}{3}\alpha_u a_1 - \tfrac{4}{9}\alpha_u^2\alpha_z + \sqrt{\tfrac{2}{15}}\alpha_u$$

$$a_1 = \tfrac{1}{3}(2\alpha_u - \alpha_z)a_2 + \sqrt{\tfrac{6}{5}}\,\frac{\alpha_z - \alpha_u}{2\alpha_u + \alpha_z} \tag{II.22}$$

$$a_2 = \frac{\sqrt{\frac{6}{5}}\zeta'(\alpha_z - \alpha_u) - \sqrt{\frac{6}{5}}\alpha_z\alpha_u + \rho_1(2\alpha_u + \alpha_z)}{(2\alpha_u + \alpha_z)(\zeta'^2 - \frac{2}{3}\alpha_u\zeta' + \frac{1}{3}\alpha_z\zeta' - \frac{2}{9}\alpha_z\alpha_u)}$$

The apex of the failure surface follows from the condition $\rho_t(\bar{\sigma}_a) = 0$. Hence the following equation is obtained

$$a_2\xi_0'^2 + a_1\xi_0' + a_2 = 0 \quad \text{and} \quad \xi_0'^2 = -\frac{a_1 - \sqrt{a_1 - 4a_0a_2}}{2a_2} \tag{II.23}$$

Substituting the second set of three strength values of Table II.3 into eqn (II.18a) at the compressive meridian $\theta = 60°$, the remaining three parameters b_0, b_1 and b_2 are established:

$$b_0 = -\xi_0' b_1 - \xi_0'^2 b_2$$

$$b_1 = (\zeta' + \tfrac{1}{3})b_2 + \frac{\sqrt{\frac{6}{5}} - 3\rho_2}{3\zeta' - 1} \tag{II.23a}$$

$$b_2 = \frac{\rho_2(\xi_0' + \frac{1}{3}) - \sqrt{\frac{2}{15}}(\xi_0' - \zeta')}{(\zeta' + \xi_0')(\zeta - \frac{1}{3})(\xi_0' + \frac{1}{3})}$$

The five-parameter model is illustrated in Fig. II.5 and is corroborated by the experimental data from Launay and Gachon [66]. The five-parameter model features a smooth surface and produces the principal features of the triaxial failure surface of concrete. It is periodical in the deviatoric plane (with a period of 120°) and has symmetry at 60°. The cross-section is non-circular and the meridians are second-order parabolas with a failure surface in the deviatoric plane described by a part of an elliptic curve. In order to produce convexity in both the deviatoric plane and along the meridians, the following boundary conditions or constraints are applied:

$$a_0 > 0 \quad \text{and} \quad a_1 \leq 0 \qquad a_2 \leq 0$$
$$b_0 > 0 \quad \text{and} \quad b_1 \leq 0 \qquad b_2 \leq 0 \qquad \text{(II.23b)}$$
$$\rho_t(\sigma_a)/\rho_c(\sigma_a) > \tfrac{1}{2}$$

The failure surface intersects the hydrostatic axis if, as in eqns (II.23b), $a_2 \leq 0$, $b_2 \leq 0$.

II.1.6 Hypoelastic model

The hypoelasticity element may be a specific material which can undergo softening or hardening in strain, but, at the same time, has neither a preferred state nor a preferred stress and its response to deformation is smooth and gradual. The hypoelastic material differs from most materials employed in plasticity theory in that it is dynamically sound. If it is possible to judge the behaviour of the hypoelastic material under large strains, large rotations and abrupt stress changes, then the Truesdell [90], Rivlin and Erickson [91] and Noll [166] constitutive equations will be subject to certain additional restrictions.

Such a suggestion obviously frees concrete from the need to choose any of the specific failure criteria discussed earlier. It is admitted in the development of such a constitutive law that the hypoelastic element is appropriate only to the isotropic case unless a preferred initial state is specified. Here, for the sake of argument, the anisotropy concept is appropriate only to materials which have a preferred state. The equations to be developed in this case must be supported by experimental results in order to given an initial preferred state. Since, in hypoelasticity, equations of motion are involved and initial stresses are satisfied by these equations, the situation herein is not the same as that of ordinary elasticity in which special classes of initial stresses obtained from elastic strains are admissible. Using the equations of motion the concept of hypoelasticity offers better treatment of initially stressed configurations of models or prototypes under large deformations. Truesdell [90] has shown that his equations do give large deformations for various specimens. Since these deformations are not covered in the classical theory of finite strains, the opportunity can be taken to extend the existing work to include terms for elasto-plasticity.

In addition, the model contains stability criterion and provides the most general representation for time independent incremental behaviour and accommodates stress-induced anisotropy. More discussions are given by Coons and Evans [159] and Bangash [126].

II.1.6.1 Concrete stress–strain criteria using hypoelasticity

A time dependent law by which an increment of stress is related to the stress is given by Truesdell [90] as

$$\text{rate of stress} = f(\text{stress, rate deformation}) \qquad \text{(II.24)}$$

where f is the tensor function. The purpose is to obtain a concept of the elastic behaviour of the material expressed in terms of rates. In other words, this means that the initial state of the material is arbitrarily stressed, which is different from what was believed earlier, and that the unstrained state of material can also be considered as unstressed. If eqn (II.24) is given a three-dimensional status, the difference between the results based on the constitutive law concept of expressing material behaviour in rates, and the concept of material being unstressed whilst unstrained, will be greater. This is due to the fact that the three-dimensional equations bring about the inter-relationship of several phenomena and parameters, noted amongst these are: extension, shear, hydrostatic pressure, variable elastic properties, etc.

In eqn (II.24) Green and Swanson [145] retained the features of plasticity when they applied them to the deformation of metals. This idea is extremely attractive for predicting the short-term response of the frictional material. Equations developed on this basis, when stability criteria are included, can reasonably predict failure criteria for plain concrete. This is the theme which will be followed throughout in the development of the theoretical model.

Equation (II.24) can easily be written in a combined tensor form in three dimensions as

$$f(\sigma, \varepsilon, \dot{\sigma}, \dot{\varepsilon}) = 0 \tag{II.25}$$

in which $f(\sigma, \varepsilon)$ and $f(\dot{\sigma}, \dot{\varepsilon})$ are related to the elastic behaviour, and $f(\sigma, \dot{\varepsilon})$ defines the onset of plasticity zones. It is now clearly established that eqn (II.25) is a general representation of the constitutive relationship in which classical stress (σ) and strain (ε) tensors and their scalar parameters represented by dots are included. It is initially assumed that the increment of stress in eqn (II.25) is not linearly dependent on the increment of strain. Within eqn (II.25) the incremental stress $\mathrm{d}\sigma_{ij}$ is given as in a 'trilinear' relationship as

$$\mathrm{d}\sigma_{ij} = S_{ijkl}(\sigma_{ij})\,\mathrm{d}\sigma_{kl} \tag{II.26}$$

where S_{ijkl} is the material property tensor, σ_{ij} is the stress, and $\mathrm{d}\sigma_{kl}$ is the rate of deformation as stated in eqn (II.24).

Truesdell [90] suggests for isotropic material that the values of S_{ijkl} and $\mathrm{d}\sigma_{ij}$ are

$$\begin{aligned} S_{ijkl} = {} & (A_{01} + A_{11}\sigma_{rr})\delta_{ij}\delta_{kl} + \tfrac{1}{2}(A_{02} + 2A_{11}\sigma_{rr})(\delta_{ik}\delta_{jl} + \delta_{jk}\delta_{il}) \\ & + A_{13}(\sigma_{ij}\delta_{kl} + \sigma_{kl}\delta_{ij}) + \tfrac{1}{2}A_{14}(\sigma_{jk}\sigma_{li} + \sigma_{jl}\delta_{ki} + \sigma_{ik}\delta_{lj} + \sigma_{il}\delta_{kj}) \end{aligned} \tag{II.26a}$$

$$\begin{aligned} \mathrm{d}\sigma_{ij} = {} & (A_{01}\dot{\varepsilon}_{kk}\delta_{ij} + A_{02}\dot{\varepsilon}_{ij} + A_{11}\sigma_{pp}\dot{\varepsilon}_{kk}\delta_{ij}) - 2A_{11}\sigma_{kk}\dot{\varepsilon}_{ij} \\ & + A_{13}(\sigma_{ij}\dot{\varepsilon}_{kk} + \sigma_{kl}\dot{\varepsilon}_{kl}\delta_{ij}) + A_{14}(\sigma_{jk}\dot{\varepsilon}_{ik} + \sigma_{ik}\dot{\varepsilon}_{jk}) \end{aligned} \tag{II.26b}$$

where δ represents Kronecker deltas and A represents various hydrostatic constants to be determined by experiments on the lines suggested by Gardner [78]. Hence eqn (II.26a) exhibits a stress-induced anisotropicity if its last three terms are retained. This phenomenon occurs in most nonlinear isotropic behaviour. As stated earlier, Richart *et al.*'s data [14] suggest a small amount of asymmetry about the hydrostatic stress axis, and Bresler and Pister [42] have indicated that the failure of concrete is affected by the third invariant of stress J_3; it therefore becomes necessary to retain the constant A_{14} in eqn (II.26a) and eqn (II.26b) in order to represent the asymmetry. Coons and Evans [159], in their general failure equations, have seven material properties: $A_{01}, A_{11} \ldots A_{13} \ldots$ etc. These have been reduced to five in eqns (II.26a) and (II.26b) by making $A_{13} = A_{15}$ and $A_{12} = A_{11}$ in order to guarantee the entire behaviour of concrete as hypoelastic. It is interesting to note the two constants A_{01} and A_{02}. The hypoelastic constant A_{01} is incorporated in the equations (II.26a and II.26b) so that, for its zero value, it should give a zero value for Poisson's ratio at very small loads. Similarly, the constant A_{02} is essential in order to give the slope at the origin of the stress–strain curve for zero lateral pressure. The values of the remaining constants can easily be obtained by using failure data from a number of investigations [77–176] mentioned earlier.

For any combination of material properties, a stress state is sought for which an incremental deformation may be obtained without changing the stress state itself. Coons and Evans [159] suggest that, if constants other than A_{01} and A_{02} are zero, the hypoelastic case will be brought back to an elastic case represented by Hooke's law with an additional freedom that the initial stress can be prescribed for the zero strain rate. For the hypoelastic case suggested by eqns (II.26a) and (II.26b) the five sets of conditions suggested for the kind of stress state for which the incremental deformation is required, are written in modified forms as

$$\begin{aligned} A_{14} &\neq 0 \\ -2A_{11}(3A_{11} - A_{11}) \text{ or } -2A_{11}^2 &= 0 \\ -2A_{11}A_{13} &= 0 \\ A_{13}(3A_{11} + A_{13}) &= 0 \\ -6A_{01}A_{11} + A_{02}(-2A_{11} - A_{13}) &= 0 \end{aligned} \tag{II.27}$$

Equation (II.27) must therefore be satisfied if the stress–strain criterion assumed is initially independent of the loading path. The stress condition for failure after fully satisfying eqn (II.27) will be

$$S_{ijkl}\,\mathrm{d}\varepsilon_{kl} = 0 \tag{II.28}$$

Equation (II.28) is satisfied for the incremental deformation and condition of failure when the determinant of the material constants is zero, i.e.

$$|S_{ijkl}| = 0 \tag{II.29}$$

In order to evaluate the determinant in eqn (II.29), eighty-one components of the material tensor for a homogeneous anisotropic concrete element must have 36 coefficients. From considerations based on the law of conservation of energy and on the study of potential energy stored in the concrete element by the strains, Love [160] and Lekhnitskii [161] have shown that the relation $K_{ij} = K_{ji}$ of $|S_{ijkl}|$,

Table II.4. Direction cosines C

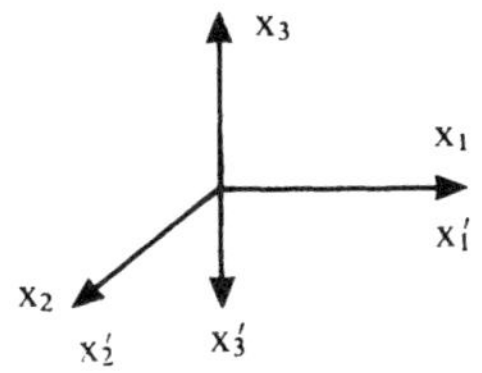

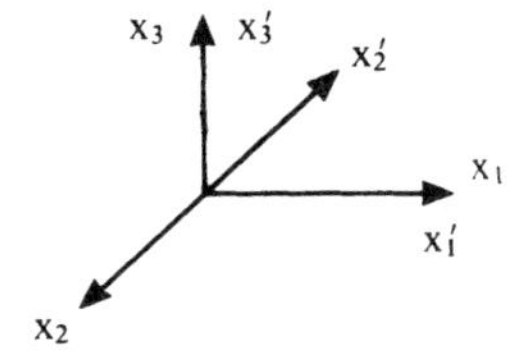

Materials with one plane of symmetry				Orthotropic materials			
	x_1	x_2	x_3		x_1	x_2	x_3
x_1'	0	1	0	x_1'	0	1	0
x_2'	1	0	0	x_2'	−1	0	0
x_3'	0	0	−1	x_3'	0	0	1

for $i, j = 1, 2, \ldots, 6$, holds between these constants. Since the original matrix $[S_{ijkl}]$ is symmetric around the main diagonal, all 36 coefficients of this matrix therefore involve a maximum of 21 constants. Nye [162] thus concludes that the relationship given in eqn (II.29) also exists amongst the material stiffnesses, and hence it is easy to write

$$S_{ijkl} = S_{ijlk} = S_{jikl} = S_{klij} \tag{II.29a}$$

Consider a concrete element that is symmetric in properties with respect to one plane, say x_1–x_2. This symmetry can be expressed by transforming the invariants such that $x_1 = x_1'$, $x_2 = x_2'$ and $x_3 = x_3'$. The direction cosines given in Table II.4 are used for this transformation. The stresses and strains of the transformed co-ordinate systems are related to those of the original co-ordinate system by:

$$\sigma_{kl}' = c_{ki}c_{lj}\sigma_{ij} \qquad \varepsilon_{kl}' = c_{ki}c_{lj}\varepsilon_{ij} \tag{II.29b}$$

Therefore, for $i = 1, 2, 3, 6$, $\sigma_i' = \sigma_i$ and $\varepsilon_i' = \varepsilon_i$. However, $\varepsilon_{23}' = c_{22}c_{33}$, $\varepsilon_{23} = -\varepsilon_{23}$; therefore $\varepsilon_4' = -\varepsilon_4$. Similarly, $\sigma_4' = -\sigma_4$. Likewise $\varepsilon_{31}' = c_{33}c_{11}$, $\varepsilon_{31} = -\varepsilon_{31}$. Therefore $\varepsilon_5' = -\varepsilon_5$ and also $\sigma_5' = -\sigma_5$.

Now eqn (II.26) can be written both in terms of the original co-ordinate system and the transformed co-ordinate systems as:

$$\begin{aligned} \sigma_1 &= D_{11}'\varepsilon_1 + D_{12}'\varepsilon_2 + \cdots + D_{16}'\varepsilon_6 \\ \sigma_1' &= D_{11}'\varepsilon_1' + D_{12}'\varepsilon_2' + \cdots + D_{16}'\varepsilon_6' \end{aligned} \tag{II.29c}$$

Since $\varepsilon_4' = -\varepsilon_4$ and $\varepsilon_5' = -\varepsilon_5$, then $\sigma_1' = \sigma_1$ requires that $D_{14}' = D_{15}' = 0$. By similarly considering the other equations of eqn (II.26) such as $\sigma_2' = \sigma_2$; $\sigma_3' = \sigma_3$; $\sigma_4' = \sigma_4$; $\sigma_5' = \sigma_5$; and $\sigma_6' = \sigma_6$, it is found that $D_{24}' = D_{25}' = D_{34}' = D_{35}' = D_{64}' = D_{65}' = 0$.

Knowing that Lekhnitskii [161] has established the relation $D_{ij} = D_{ji}$, and with the material having one plane of symmetry, the elasticity tensor $[S_{ijkl}] = [D]$ has the following array which involves 13 independent components:

$$[D] = \begin{bmatrix} D'_{11} & D'_{12} & D'_{13} & 0 & 0 & D'_{16} \\ D'_{21} & D'_{22} & D'_{23} & 0 & 0 & D'_{26} \\ D'_{31} & D'_{32} & D'_{33} & 0 & 0 & D'_{36} \\ 0 & 0 & 0 & D'_{44} & D'_{45} & 0 \\ 0 & 0 & 0 & D'_{54} & D'_{55} & 0 \\ D'_{61} & D'_{62} & D'_{63} & 0 & 0 & D'_{66} \end{bmatrix} \qquad \text{(II.29d)}$$

Concrete as a material is generally orthotropic, i.e. it contains three mutually orthogonal planes of symmetry. This also means that this material must be symmetric with respect to the x_1–x_3 and the x_2–x_3 planes. Proceeding as before, the diagram shown in Table II.4 for this case can be made for symmetry with respect to the x_1–x_3 plane. Using the transformation, it becomes clear that $D'_{16} = D'_{26} = D'_{36} = D'_{45} = 0$ from the x_1–x_3 plane of symmetry. For concrete as an orthotropic (orthogonally anisotropic) material the S_{ij} tensor can be modified by substituting these values into eqn (II.29d), and thus the property of this material is described by nine independent components.

Sometimes concrete as a material is assumed to be *transversely isotropic*, i.e. like the orthotropic case, it also has three mutually orthogonal planes but in addition is *isotropic* with respect to one of the three planes. For example, if the material isotropicity is in the x_1–x_2 plane, then this means that $D'_{11} = D'_{22}$, $D'_{13} = D'_{23}$, $D'_{44} = D'_{55}$. Similarly, the tensor coefficients for the x_2–x_3 planes are: $D'_{13} = D'_{12}$, $D'_{31} = D'_{21}$ and $D'_{23} = D'_{32}$.

II.1.6.2 Concrete as an orthotropic material

After modification of the tensor array given in eqn (II.29d)

$$[D] = \begin{bmatrix} D'_{11} & D'_{12} & D'_{13} & 0 & 0 & 0 \\ D'_{21} & D'_{22} & D'_{23} & 0 & 0 & 0 \\ D'_{31} & D'_{32} & D'_{33} & 0 & 0 & 0 \\ 0 & 0 & 0 & D'_{44} & 0 & 0 \\ 0 & 0 & 0 & 0 & D'_{55} & 0 \\ 0 & 0 & 0 & 0 & 0 & D'_{66} \end{bmatrix} \qquad \text{(II.29e)}$$

If the co-ordinates are aligned with the direction of the principal stresses and using specific equations for the invariants of the stress tensor and stress deviator tensor, the *determinant* in eqn (II.29) will simply be the determinant of eqn (II.29e)

$$\begin{aligned} I_1 &= \sigma_1 + \sigma_2 + \sigma_3 \\ J_2 &= \tfrac{1}{2}(\sigma_1^2 + \sigma_2^3 + \sigma_3^2) \\ J_3 &= \tfrac{1}{3}(\sigma_1^3 + \sigma_2^3 + \sigma_3^3) \end{aligned} \qquad \text{(II.30)}$$

Using eqns (II.26), (II.26a), (II.26b), (II.27) and (II.30), the values of the tensor coefficients of eqn (II.29) are written as:

$$D'_{11} = A_{01} + A_{02} - A_{11}I_1 + 2(A_{13} + A_{14})\sigma_1$$

D'_{22}, D'_{33} have all the terms of D_{11} except σ_1 is changed to σ_2 for D'_{22} and σ_3 for D'_{33}

$$\begin{aligned}
D'_{44} &= (A_{02} - 2A_{11}I_1) + A_{14}(J_1 - \sigma_1)\\
D'_{55} &= (A_{02} - 2A_{11}I_1) + A_{14}(J_1 - \sigma_2)\\
D'_{66} &= (A_{02} - 2A_{11}I_1) + A_{14}(J_1 - \sigma_3)\\
D'_{12} &= (A_{01} + A_{11}I_1) + A_{13}(\sigma_1 + \sigma_2) = \bar{D} + A_{13}(\sigma_1 + \sigma_2)\\
D'_{13} &= \bar{D} + A_{13}(\sigma_1 + \sigma_3) \qquad \text{(II.30a)}\\
D'_{23} &= \bar{D} + A_{13}(\sigma_2 + \sigma_3)\\
D'_{21} &= \bar{D} + A_{13}(\sigma_2 + \sigma_1) \quad \text{where } \bar{D} = A_{01} + A_{11}I_1\\
D'_{31} &= \bar{D} + A_{13}(\sigma_3 + \sigma_1)\\
D'_{32} &= \bar{D} + A_{13}(\sigma_3 + \sigma_2)
\end{aligned}$$

The determinant in eqn (II.30) after expansion leads to:

$$\begin{aligned}
&[\![[(A_{02} - 2A_{11}I_1)]\{-3A_{13}^2(2I_2 - J_1^3/3) + [3A_{01} + A_{02} + (A_{11} + 2A_{13})I_1]\}\\
&\quad + 2A_{14}\{(A_{02} - 2A_{11}I_1)^2 I_1 + 2(A_{02} - 2A_{11}I_1)I_1(A_{01} + A_{11}I_1)\\
&\quad + (2J_1^2/3 - (J_2 - I_{1/3}^3))[A_{14}(A_{01} + A_{11}I_1) + (A_{02} - 2A_{11}I_1)(2A_{13} + A_{14})]\\
&\quad + ((J_{3/3} - I_1(J_2 - I_1^3/3))/2)[2A_{14}(6A_{13} + 2A_{14}) + 6A_{13}^2]\\
&\quad - A_{13}^2[J_1^3/3 + I_1(J_2 - I_1^3/3) - J_3]\}]\!]\\
&\quad \times [(A_{02} - 2A_{11}I_1) + A_{14}(I_1 - \sigma)] \times [(A_{02} - 2A_{11}I_1) + A_{14}(I_1 - \sigma_2)]\\
&\quad \times [(A_{02} - 2A_{11}I_1) + A_{14}(I_1 - \sigma_3)] = 0 \qquad \text{(II.31)}
\end{aligned}$$

Equation (II.31) in the expanded form geometrically represents a surface in the principal stress space comprising two shapes — the first is pyramidal resulting from the last three products of eqn (II.31); whilst the second, the curved shape, results from the determinant whose elements are D_{12}, D_{13}, D_{23}, D_{21}, D_{31} and D_{32} in eqn (II.30a). In terms of invariants of the stress deviator tensor, I_1, J_2 and J_3 values are substituted into eqns (II.30), (II.30a) and (II.31).

Looking at eqn (II.30) and (II.31) it is clear that there is a class of deformation associated with stresses, and when plastic yield occurs two important properties emerge.

(a) The incremental plastic strain vector $d\varepsilon_p$ is normal everywhere to the yield surface except the pyramidal part.
(b) On the pyramidal part, the determinant in eqn (II.29), the directions of the principal stresses in this case do not coincide with those of the principal strains.
(c) The principal directions of the stress–strain rate do coincide for the part of the failure surface where $D_{ij} = 1, 2, 3, i \neq j$.

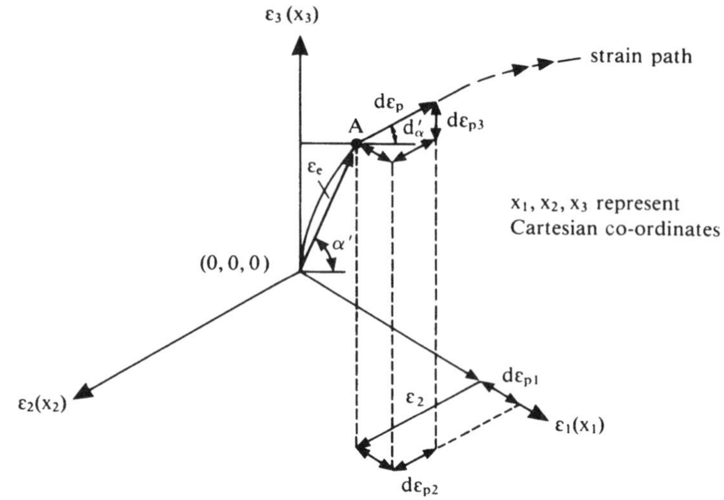

Fig. II.6. Strain space.

If a strain history is regarded as a strain path drawn in this space (Fig. II.6) the values of ε_1, ε_2, and ε_3 may all be varied.

At any point A the equivalent strain ε_e is given by the radius as shown in Fig. (II.6), and $d\varepsilon_p$ is its incremental value along the strain path. For example, for two intervals it can be computed as follows:

$$\frac{d\varepsilon_{p_2}}{d\varepsilon_{p_3}} = \frac{-D'_{11}D'_{23} + D'_{21}D'_{13}}{D'_{12}D'_{23} - D'_{22}D'_{13}}$$

and (II.32)

$$\frac{d\varepsilon_{p_1}}{d\varepsilon_{p_2}} = \frac{-D'_{12}D'_{21} + D'_{11}D'_{22}}{D'_{12}D'_{23} - D'_{22}D'_{13}}$$

The inclination $d\alpha'$ of $d\varepsilon_p$ at any instant is computed from

$$d\alpha' = \tan^{-1}\frac{1}{\sqrt{3}}\frac{(d\varepsilon_{p1} - d\varepsilon_{p2}) + (d\varepsilon_{p1} - d\varepsilon_{p3})}{d\varepsilon_{p2} - d\varepsilon_{p3}} \qquad \text{(II.33)}$$

A_{01} (at infinitesimal load for zero Poisson's ratio) $= 0$

A_{02} (determined from the slope at the origin of the stress–strain curve with zero lateral pressure) $= 8.99 \times 10^5$

A_{14} (for symmetry of hydrostatic stresses) $= 0$

The coefficients A_{11} and A_{13} are found from the stress–strain increment vector and are plotted on the basis of data from Richart *et al.* [14] and Gardner [78]. Again for this specific case the following values are evaluated:

$A_{11} = 43.5 \qquad A_{13} = -1200$

These values can be substituted into equations for orthotropic, transversely isotropic, isotropic and one-dimensional tension or compression cases. Similar data can easily be obtained from the stress–strain curves of other experimental tests.

Similarly, by using eqns (II.30), (II.31) and (II.32) and following the above principles, $d\sigma_3/d\varepsilon_3$ for an *orthotropic* case can be computed. The procedure for evaluating $d\sigma_1/d\varepsilon_1$ and $d\sigma_2/d\varepsilon_2$ for orthotropic and transversely isotropic cases is exactly the same. Only the appropriate tensor coefficients from the respective matrices $[D]$ have to be chosen to replace those of $d\sigma_3/d\varepsilon_3$. For any iteration procedure the tangent moduli can be determined by holding lateral stresses σ_1 and σ_2 constant for prescribed loading conditions. In the case of $d\sigma_3/d\varepsilon_3$ it will give the tangent modulus E_3, which can be prescribed as a function of the current values of σ_1, σ_2 and σ_3. This function must reflect not only the magnitudes of σ_1, σ_2 and σ_3, but also their signs.

The procedure for evaluating $d\sigma_1/d\varepsilon_1$ and $d\sigma_2/d\varepsilon_2$, and the corresponding tangent moduli E_1 and E_2, is exactly the same for both orthotropic and transversely isotropic cases, except that the appropriate tensor coefficients from the respective matrices $[D]$ will replace those of $d\sigma_3/d\varepsilon_3$.

II.1.6.3 Isotropic case

Having established values of constants consistent with both initial and ultimate loading, it is then possible to compare observed and predicted behaviour of the concrete over the complete load range of the concrete structure vessel. Results from the standard triaxial tests of various authors [35–133] can be used for this purpose. For any prescribed loading path of the vessel, the above hypoelasticity equations can be solved by integration. Where integration is avoided, initial slopes of the stress–strain curves are used to provide the relationship between various material constants in eqns (II.29)–(II.32). For example, for an isotropic case, the strain increments are given as

$$\frac{d\varepsilon_2}{d\varepsilon_3} = \frac{[A_{01} + (A_{11} + A_{13})\sigma_3 + (2A_{11} + A_{13})]}{2[A_{01} + \frac{1}{2}A_{02} + (2A_{13} + A_{14})]} \tag{II.34}$$

For $\sigma_3 > \sigma_1$ or σ_2, the stress–strain criterion will simply be $d\sigma_3/d\varepsilon_3$, and

$$\begin{aligned}\frac{d\sigma_3}{d\varepsilon_3} = \Bigg\{ & [(A_{01} + A_{02}) + (-A_{11} + 2A_{13} + 2A_{14})\sigma_3 - 2A_{11}\sigma_1] \\ & \times [A_{01} + \tfrac{1}{2}A_{02} + (2A_{13} + A_{14})\sigma_1] \\ & - \frac{[A_{01} + (A_{11} + A_{13})\sigma_3 + (2A_{11} + A_{13})\sigma_1]^2}{A_{01} + \frac{1}{2}A_{02} + (2A_{13} + A_{14})} \Bigg\}\end{aligned} \tag{II.35}$$

$$\begin{aligned}\frac{d(\sigma_3 - \sigma)}{d\varepsilon_3} = \Bigg\{ & A_{01} + A_{02} + [-A_{11} + 2(A_{13} + A_{14})\sigma_3 - 2A_{11}\sigma] \\ & - \frac{[A_{01} + (A_{11} + A_{13})\sigma_3 + (2A_{11} + A_{13})\sigma]^2}{A_{01} + \frac{1}{2}A_{02} + (2A_{13} + A_{14})\sigma} \Bigg\}\end{aligned} \tag{II.36}$$

$$\frac{d\varepsilon_v}{d\varepsilon_3} = \frac{\frac{1}{2}A_{02} + (-A_{11} + A_{13})\sigma_3 + (-2A_{11} + A_{14} + A_{13})\sigma}{A_{01} + \frac{1}{2}A_{02} + A_{13}\sigma} \tag{II.37}$$

where ε_v is *the volumetric strain under the triaxial state of stress.*

Initial slopes are obtained by setting $\sigma_3 = \sigma$ in eqns (II.36) and (II.37). These initial slopes are related to the constants C_1 and C_2 evaluated in Table II.5 for the isotropic compression case. Hence

$$\left[\frac{\mathrm{d}(\sigma_3 - \sigma)}{\mathrm{d}\varepsilon_3} \Big/ \frac{\mathrm{d}\varepsilon_\mathrm{v}}{\mathrm{d}\varepsilon_3}\right]_{\sigma_3=\sigma} = (C_1 + C_2\sigma) \tag{II.38}$$

A simple case for the material constants is to use only the initial slope of eqn (II.36). For isotropic conditions eqn (II.36) may be written as

$$A_{01} + (3A_{11} + 2A_{13})\sigma = \frac{[\mathrm{d}(\sigma_3 - \sigma)/\mathrm{d}\varepsilon_3]_{\sigma_3=\sigma} - (A + B\sigma)}{\left[\dfrac{\mathrm{d}(\sigma_3 - \sigma)}{\mathrm{d}\varepsilon_3}\right]_{\sigma_3=\sigma} \sigma - 3(A + B\sigma)}(C_1 + C_2\sigma) \tag{II.39}$$

Equation (II.39) gives a relationship between the lateral pressure σ, material constants and the initial slope of the stress–strain curves. At the peak stress

Table II.5. Isotropic compression and plane strain cases

(a) Isotropic condition

$$\sigma_1 = \sigma_2 = \sigma_3 = \sigma \qquad \varepsilon_1 = \varepsilon_2 = \varepsilon_3 = \varepsilon$$

Final differential equations:

$$\mathrm{d}\sigma = (C_1 + C_2\sigma)\,\mathrm{d}\varepsilon$$

$$C_1 = 3A_{01} + A_{02}$$

$$C_2 = 3A_{11} + 6A_{13} + 2A_{14}$$

Upon integration, the stress–strain relation becomes

$$\sigma = (C_1/C_2 + \sigma_\mathrm{c})\,e^{C_2\varepsilon} - C_1/C_2$$

(b) Plain strain condition
Constant pressure (confined) σ.
Differential equations pertaining to this condition:

$$\frac{\mathrm{d}\varepsilon_\mathrm{v}}{\mathrm{d}\varepsilon_3} = \frac{A_{02} + (A_{12} + A_{13})\sigma_3 - 2A_{11}\sigma_2 + (-2A_{11} + A_{13} + 2A_{14})\sigma}{A_{01} + A_{02} - A_{11}(\sigma_3 + \sigma_2) + (-A_{11} + 2A_{13} + 2A_{14})}$$

$$\begin{aligned}\frac{\mathrm{d}\sigma_3}{\mathrm{d}\varepsilon_3} = {} & A_{01} + A_{02} + (-A_{11} + 2A_{13} + 2A_{14})\sigma_3 + (-A_{11})(\sigma_2 + \sigma) \\ & - [A_{01} + (A_{11} + A_{13})\sigma_3 + A_{11}\sigma_2 + (A_{11} + A_{13})\sigma] \\ & \times \frac{[A_{01} + (A_{13} + A_{11})\sigma_3 + A_{11}\sigma_2 + (A_{11} + A_{13})\sigma]}{A_{01} + A_{02} - A_{11}(\sigma_3 + \sigma_2) + (-A_{11} + 2A_{13} + 2A_{14})}\end{aligned}$$

$$\begin{aligned}\frac{\mathrm{d}\sigma_2}{\mathrm{d}\varepsilon_3} = {} & A_{01} + (A_{11} + A_{13})\sigma_3 + A_{11}\sigma_2 + (A_{11} + A_{13})\sigma \\ & - [A_{01} + A_{11}\sigma_3 + (A_{11} + A_{13})\sigma_2 + (A_{11} + A_{13})\sigma] \\ & \times \frac{[A_{01} + (A_{11} + A_{13})\sigma_3 + A_{11}\sigma_2 + (A_{11} + A_{13})\sigma]}{A_{01} + A_{02} - A_{11}(\sigma_3 + \sigma_2) + (-A_{11} + 2A_{13} + 2A_{14})}\end{aligned}$$

condition under the triaxial state of stress, the following relations between the material parameters are obtained by equating eqn (II.36) equal to zero:

$$[3C_1 + C_2(\sigma_3 + 2\sigma)][3A_{02} + (-6A_{11} + 2A_{14})(\sigma_3 + 2\sigma) + 2A_{14}(\sigma_3 - \sigma) - 2(\sigma_3 - \sigma)^2(3A_{13} + 2A_{14})^2] = 0 \quad \text{(II.40)}$$

Equation (II.40) is a simplified version under isotropic conditions of the general eqn (II.31).

II.1.6.4 Deduction of a one-dimensional tension or compression

The application of the boundary conditions and the loading path corresponding to one-dimensional tension or the compression test gives the two related differential equations

$$\frac{d\sigma_3}{d\varepsilon_3} = A_{01} + A_{02} + (-A_{11} + 2A_{13} + 2A_{14})\sigma_3 + (-A_{11})\sigma \quad \text{(II.41)}$$

$$\frac{d\sigma}{d\varepsilon_3} = A_{01} + (A_{11} + A_{13})\sigma_3 + (2A_{11} + A_{13})\sigma \quad \text{(II.42)}$$

assuming $K_0\sigma$ is related to σ_3 by

$$\sigma = K_0\sigma_3 = (1 - \sin\phi)\sigma_3 \quad \text{(II.43)}$$

where ϕ is the angle of frictional resistance for concrete.

Using eqns (II.41)–(II.43), two relations are obtained between the material parameters

$$A_{01} = K_0(A_{01} + A_{02}) \quad \text{(II.44)}$$

$$A_{11}(1 - K_0)(1 + 2K_0) - K_0(1 + 2K_0)(-2A_{11}) - 2K_0A_{14} + (1 + K_0)A_{13} = 0 \quad \text{(II.45)}$$

Substituting eqn (II.43) into eqn (II.41) gives the general equation

$$\frac{d\sigma_3}{d\varepsilon_3} = (\bar{\alpha} + \bar{\beta}\sigma_3) \quad \text{(II.46)}$$

where

$$\bar{\alpha} = A_{01} + A_{02}$$
$$\bar{\beta} = -A_{11}(1 + 2K_0) + 2(A_{13} + A_{14}) \quad \text{(II.47)}$$

Integrating eqn (II.46) leads to a stress–strain relation similar to the one obtained in Table II.5 for the isotropic case:

$$\sigma_3 = \left(\frac{\bar{\alpha}}{\bar{\beta}} + \sigma_{3c}\right) e^{\bar{\beta}\varepsilon_3} - \bar{\alpha}/\bar{\beta} \quad \text{(II.48)}$$

The constants $\bar{\alpha}$ and $\bar{\beta}$, as usual, are computed from the initial slope and one point of the experimental stress–strain curve from a one-dimensional test.

Table II.5 gives differential equations pertaining to plain strain conditions. A similar integration procedure is required in order to arrive at suitable relations for the stress–strain conditions. For example, the test conducted by Gardner [78] gives good agreement with the experimental data of Richart *et al.* [14]. Table I of Gardner [78] with an average of four tests, and Tables 8 and 9 of Richart *et al.* [14] with an average of four tests each, can now be correlated for the mix 1:1:2. Data from Richart *et al.* [14] for this mix gives a good agreement with those of Gardner [78].

II.1.7 Shear and bulk moduli models

Nonlinear incremental elastic models are proposed by Phillips *et al.* [167] in which the bulk modulus K is assumed constant, and the tangential shear modulus G is assumed to be a function of the octahedral shear stress only. A similar approach has been adopted by Cedolin *et al.* [169] who consider the bulk and shear moduli to depend on all the stress invariants. The proposed model is applicable to triaxial compressive states only.

Another triaxial model of nonlinear type was proposed by Saugy *et al.* [170] in which the bulk modulus was considered as constant while the shear modulus was assumed to vary as a logarithmic function of the second stress invariant. This model has been used for three-dimensional work.

In this work the elastic bulk and shear moduli are treated as scalar functions of the invariants of the stress or strain tensor. The deviatoric and hydrostatic (volumetric) components are written as

$$S_{ij} = 2G\varepsilon_{ij} = 2G\,\Delta\varepsilon_{ij} \tag{II.49a}$$

$$\sigma_{\mathrm{m}} = \sigma_0 = K\varepsilon_{kk} \tag{II.49b}$$

in which

$$\sigma_0 = \frac{\sigma_{kk}}{3} \quad \text{or } \Delta\sigma_{\mathrm{m}} = 3K\varepsilon_{kk}$$

The stress–strain relation is then

$$\begin{aligned}
\Delta\sigma_{ij} &= \Delta\delta_{ij} + \delta_{ij}\,\Delta\sigma_{\mathrm{m}} \\
\Delta\sigma_{\mathrm{m}} &= (\Delta\sigma_{11} + \Delta\sigma_{22} + \Delta\sigma_{33})/3 \\
\Delta\varepsilon_{ij} &= \Delta\varepsilon_{ij} - \delta_{ij}\,\Delta\varepsilon_{\mathrm{m}} \\
\Delta\varepsilon_{\mathrm{m}} &= \frac{(\Delta\varepsilon_{11} + \Delta\varepsilon_{22} + \Delta\varepsilon_{33})}{3}
\end{aligned} \tag{II.50}$$

δ_{ij} (the Kronecker delta) $= 1$ for $i = j$ and $= 0$ for $i \neq j$.

Equation (II.50) can also be written as

$$\begin{aligned}
\Delta\sigma_{ij} &= 2G\,\Delta\varepsilon_{\mathrm{d}ij} + K\,\Delta\varepsilon_{kk}\delta_{ij} \\
&= 2\varepsilon\,\Delta\varepsilon_{\mathrm{d}ij} + (K - \tfrac{2}{3}G)\,\Delta\varepsilon_{kk}\delta_{ij}
\end{aligned} \tag{II.51}$$

In the case of the biaxial model and using the approach adopted by Kupfer and Gerstle [31], K and G can be expressed as a function of octahedral shear strain $\gamma_0 = \tau_0/G$. Using E and ν as

$$E = \frac{9KG}{(3K+G)} \qquad \nu = \frac{(3K-2G)}{2(3K+G)} \tag{II.52}$$

Equation (II.51) is written in matrix form as

$$\{\sigma\} = [D]\{\varepsilon\} \tag{II.53}$$

where $[D]$ for the *biaxial case* is given below

$$[D] = 4GD_2 \begin{bmatrix} D_1 & D_3/2 & 0 \\ D_3/2 & D_1 & 0 \\ 0 & 0 & D_2/4 \end{bmatrix} \tag{II.54}$$

where

$$D_1 = G + 3K \qquad D_2 = 4G + 3K \qquad D_3 = 3K - 2G$$

In three dimensions Bangash [126] adopts the stress–strain relation for the triaxial case as:

$$\begin{pmatrix} \Delta\sigma_x \\ \Delta\sigma_y \\ \Delta\sigma_z \\ \Delta\tau_{xy} \\ \Delta\tau_{yz} \\ \Delta\tau_{zx} \end{pmatrix} \begin{bmatrix} (K+\frac{4}{3}G), (K-\frac{2}{3}G), (K-\frac{2}{3}G) & 0 & 0 & 0 \\ (K+\frac{4}{3}G), (K-\frac{2}{3}G) & 0 & 0 & 0 \\ (K+\frac{4}{3}G) & 0 & 0 & 0 \\ & G_{12} & 0 & 0 \\ \text{sym} & & G_{23} & 0 \\ & & & G_{13} \end{bmatrix} \begin{pmatrix} \Delta\varepsilon_x \\ \Delta\varepsilon_y \\ \Delta\varepsilon_z \\ \Delta\gamma_{xy} \\ \Delta\gamma_{yz} \\ \Delta\gamma_{zx} \end{pmatrix} \tag{II.55}$$

or in short

$$\{\Delta\sigma\} = [D]\{\Delta\varepsilon\}$$

where $[D]$ is the required material matrix.

During loading of the structure, the shear modulus of eqn (II.55) is assumed to vary as a logarithmic function of the second stress invariant and is expressed as follows:

$$G_{12} = G_{23} = G_{13} = G = G_e - \alpha \log_e \frac{J_2}{J_2^e} \quad \text{for } J_2 > J_2^e \tag{II.56}$$

$$G = G_e \qquad \text{for } J_2 \leq J_2^e \tag{II.57}$$

and $K = K_e$ is constant throughout, and where

$$G_e = \frac{E}{2(1+\nu)} \quad \text{is the linear elastic shear modulus} \tag{II.58}$$

$$K_e = \frac{E}{3(1-2\nu)} \quad \text{is the linear elastic bulk modulus} \tag{II.59}$$

$$J_2 = \tfrac{1}{2}(S_x^2 + S_y^2 + S_z^2) + \tau_{xy}^2 + \tau_{yz}^2 + \tau_{zx}^2 \quad \text{is the second stress invariant} \tag{II.60}$$

$$S_x = \sigma_x - \sigma_m \qquad S_y = \sigma_y - \sigma_m \qquad S_z = \sigma_z - \sigma_m \tag{II.61}$$

J_2^e is the limit of linearity $= \frac{1}{3} f_c'$.

When E, ν and G change, the material matrix (II.55) is modified on the lines suggested by Bangash [126] and in Appendix I. The values of G_{12}, G_{23} and G_{13} are given as

$$\begin{aligned} G_{12} &= \frac{1}{2}\left[\frac{E_1}{2(1+\nu_{12})} + \frac{E_2}{2(1+\nu_{21})}\right] \\ G_{23} &= \frac{1}{2}\left[\frac{E_2}{2(1+\nu_{23})} + \frac{E_3}{2(1+\nu_{32})}\right] \\ G_{13} &= \frac{1}{2}\left[\frac{E_3}{2(1+\nu_{31})} + \frac{E_1}{2(1+\nu_{13})}\right] \end{aligned} \tag{II.62}$$

The values of ν are given in Chapter I.

Due to symmetry

$$E_1\nu_{21} = E_2\nu_{12} \qquad E_2\nu_{32} = E_3\nu_{23} \qquad E_3\nu_{13} = E_1\nu_{31} \tag{II.62a}$$

In terms of the uniaxial stress–strain relation in compression

$$\varepsilon_{cu} = \frac{\sigma_{cu}}{3}\left[\frac{1}{3K_e} - \frac{1}{G}\right] \leq 0.0035 \qquad \sigma_{cu} = f_{cu}$$

Using eqn (II.56) Figs II.7a and b show variation of G with E and with J_2.

II.1.8 Concrete model based on endochronic theory

This type of model was initially developed for steel by Valanis [171] and has subsequently been modified by Bazant and Bhat [172] and Bazant and Shieh [173]. In this theory the inelastic strains are shown by means of increments of intrinsic time, a non-decreasing scalar variable whose increments depend on strain increments. The difference between the formulations for metal and steel is only due to the hydrostatic pressure sensitivity of inelastic strain from inelastic dilatancy caused by large deviatoric strains. The theory applicable to concrete takes into consideration improved description of strain softening with and without cyclic loading, unloading and reloading. Bazant and Bhat [172] give this theory in an incremental form. Bazant and Shieh [173], using further test data, give a significant refinement of this theory.

The basic incremental constitutive equations of the endochronic theory are:

$$\mathrm{d}\varepsilon_{dij} = \mathrm{d}\varepsilon_{dij}^{E} + \mathrm{d}\varepsilon_{dij}^{P} \tag{II.63}$$

$$= \frac{\mathrm{d}S_{ij}}{2G} + \frac{S_{ij}}{2G}\,\mathrm{d}z \tag{II.63a}$$

$$d\varepsilon_m = d\varepsilon_m^E + d\varepsilon_m^P \tag{II.63b}$$

$$= \frac{d\sigma_m}{3K} + d\lambda \tag{II.63c}$$

$$d\varepsilon_{ij} = d\varepsilon_{dij} + \delta_{ij}\, d\varepsilon_m$$

where

$d\varepsilon_{ij}$ = strain increment

$d\varepsilon_{dij}$ = deviatoric strain increment

$d\varepsilon_m$ = volumetric strain increment

$d\sigma_m$ = volumetric stress increment

dS_{ij} = deviatoric stress increment

$d\varepsilon_{dij}^E$, $d\varepsilon_{dij}^P$ = elastic and inelastic deviatoric strain increments

$d\varepsilon_m^E$, $d\varepsilon_m^P$ = elastic and inelastic volumetric strain increments (II.63d)

δ_{ij} = Kronecker delta ($\delta_{ij} = 1$ for $i = j$, otherwise $\delta_{ij} = 0$)

K = bulk modulus

G = shear modulus

dz = intrinsic time increment

$d\lambda$ = inelastic dilatancy increment

$dS_{ij} = d\sigma_{ij} - \delta_{ij}\, d\sigma_m$

The increment of intrinsic time is defined by Bazant and Bhat [172] as

$$dz = \frac{d\eta}{z_1 F_2 \left\{1 + \dfrac{\beta_1\eta + \beta_2\eta^2}{1 + F_1 a_7}\right\}} \tag{II.64}$$

Bazant and Shieh [47] give the value of dz as

$$dz = \left[\left(\frac{d\zeta}{z_1}\right)^2 + \left(\frac{dt}{\tau_1}\right)^2\right]^{1/2}$$

$$d\zeta = d\eta/f(\eta, \varepsilon\sigma) = \frac{d\eta}{1 + \dfrac{\beta_1\eta + \beta_2\eta^2}{1 + F_2/a_7}} F_3 \tag{II.65}$$

For eqn (II.64) the parameters defined are:

$$d\eta = \left\{\frac{a_0}{1 - (a_6 I_3(\sigma))^{1/3}} + F_1\right\} d\xi \tag{II.65a}$$

$$d\xi = \sqrt{\tfrac{1}{2}\, d\varepsilon_{dij}\, d\varepsilon_{dij}} \tag{II.65b}$$

$$F_1 = \frac{a_2[1 + a_5 I_2(\sigma)]\sqrt{J_2(\varepsilon)}}{\{1 - a_1 I_1(\sigma) - (a_3 I_3(\sigma))^{1/3}\}(1 + a_4 I_2(\sigma)\sqrt{J_2(\varepsilon)})} \tag{II.65c}$$

$$F_2 = 1 + \frac{a_8}{\left(1 + \dfrac{a_9}{\eta^2}\right) J_2(\boldsymbol{\varepsilon})} \tag{II.65d}$$

$I_1(\boldsymbol{\sigma})$ = first stress invariant

$I_2(\boldsymbol{\sigma})$ = second stress invariant

$I_3(\boldsymbol{\sigma})$ = third stress invariant

$J_2(\boldsymbol{\varepsilon})$ = second invariant of the deviatoric strain tensor

In the case of Bazant and Shieh [47] additional factors are included

$$F_3 = 1 + \frac{a_{10}}{(1 + a_9/\eta^2) J_2(\varepsilon)} \tag{II.66}$$

$$F_1 = a_0(1 - g_1)/[1 - a_5(I_3(\sigma)^{1/3}(1 + g_2)] \tag{II.67}$$

$$F_2 = [a_2\sqrt{J_2(\varepsilon)}(1 + [a_6 I_2(\sigma)]^{1/4} + F_5]/[1 - a_1 I_1(\sigma) + \bar{L}] \tag{II.68}$$

$$\bar{L} = [[a_8 I_2(\sigma)^{1/4} F_4 - a_3 I_3(\sigma)[J_2(\sigma)]^{1/8}(1 + g_2)]$$

$$g_1 = g_{11} g_{12}$$

$$g_2 = g_{21} g_{22} g_{23}$$

$$g_{12} = 1 - \left[1 + \left(\frac{\sigma_{min}}{a_{17}(\sigma_{max} - a_{23})}\right)^4\right]^{-1} \tag{II.69}$$

$$g_{11} = a_{14}[J_2(\varepsilon)]^{1/4} \frac{\sigma_{med} - \sigma_{min}}{\sigma_{max} - a_{23}} \times \left[a_{15}\left(\frac{\sigma_{med} - \sigma_{min}}{\sigma_{max} - \sigma_{23}}\right)^{4/3} - a_{16}\right]$$

$$g_{21} = \left[\frac{a_{18}\left(\dfrac{\sigma_{med} - \sigma_{min}}{\sigma_{med} - a_{23}}\right) - 1}{a_{19}\left(1 - a_{20}\dfrac{|\sigma_{min}|}{\sigma_{max} - a_{23}}\right)(\sigma_{min} - a_{23})}\right]^{5/4}$$

$$g_{22} = \left[1 + a_{21}\left(\frac{\sigma_{min}}{\sigma_{max} - a_{23}}\right)^4\right]^{-1}$$

$$g_{23} = \left[\frac{[J_2(\varepsilon)]^{1/4}}{a_{22} + [J_2(\varepsilon)]^{1/2}}\right]^3$$

$$F_4 = \left[\frac{[J_2(\varepsilon)]^{1/4}}{a_4 + [J_2(\varepsilon)]^{1/2}}\right]^3$$

$$F_5 = a_{11}\sigma_{\min}(1 + a_{12}\sigma_{\min})\left[\frac{[J_2(\varepsilon)]^{1/4}}{|a_{13}\sigma_{\min}|^{1/4} + [J_2(\varepsilon)]^{1/2}}\right]^3 \tag{II.70}$$

The values of G and K are

$$G = \frac{E_0}{2(1+\nu)}\left(\frac{1}{1+c_5\lambda}\right) \qquad K = \frac{E_0}{3(1-2\nu)}\left(\frac{1}{1+c_5\lambda}\right)$$

The values of a, β and c are given in Table II.6.

The shear and bulk moduli are assumed to be dependent on λ (inelastic dilatancy) and are given in the case of Bazant and Bhat as

$$G = \frac{E_0}{2(1+\nu)}\left(1 - 0.25\frac{\lambda}{\lambda_0}\right) \tag{II.71}$$

$$K = \frac{E_0}{3(1-2\nu)}\left(1 - 0.25\frac{\lambda}{\lambda_0}\right) \tag{II.72}$$

E_0 = initial modulus of elasticity

$$l(\lambda) = 1 - \frac{\lambda}{\lambda_0} \tag{II.72a}$$

The increment of the inelastic dilatancy variable is defined as:

$$\frac{1-\lambda/\lambda_0}{1-c_1I_1(\sigma)}\left\{\left(\frac{J_2(\varepsilon)}{c_2^2 + J_2(\varepsilon)}\right)^3 + \left(\frac{\lambda}{\lambda_0}\right)^2\right\}\mathrm{d}\xi \tag{II.73}$$

and $\mathrm{d}\lambda$ is accumulated along the load path as

$$\lambda = \sum_{\text{all load increments}} \mathrm{d}\lambda \tag{II.73a}$$

Bazant and Shieh [47] give λ' (shear compaction), and the corresponding intrinsic time for compaction is z'.

Similar to eqns (II.65a), (II.73) and (II.73a), the new expressions are:

$$\mathrm{d}z' = \left[\left(\frac{\mathrm{d}\zeta'}{z_2}\right)^2 + \left(\frac{\mathrm{d}t}{\tau_1}\right)^2\right]^{1/2} \tag{II.74}$$

$$\mathrm{d}\xi' = \frac{\mathrm{d}\eta'}{h(\eta')} \qquad \mathrm{d}\eta' = H(\sigma)\,\mathrm{d}\xi'$$

$$\mathrm{d}\xi' = \sqrt{[I_1(\mathrm{d}\varepsilon)]^2} = [\mathrm{d}\varepsilon_{11} + \mathrm{d}\varepsilon_{22} + \mathrm{d}\varepsilon_{33}] \tag{II.75}$$

$$h(\eta') = 1 + \frac{\eta'}{\beta_3} + \left(\frac{\eta'}{\beta_4}\right)^2 \qquad H(\sigma) = b_1\left[\frac{I(\sigma)}{b_2 - I_1(\sigma)}\right]^2 \tag{II.76}$$

$l(\lambda)$ is as in eqn (II.72a):

$$l(\lambda, \varepsilon, \sigma) = \frac{c_3}{1 - c_1I_1\sigma}\left[\left(\frac{\lambda}{\lambda_0}\right)^2 + \left(\frac{c_4J_2(\varepsilon)}{c_2^2 + J_2(\varepsilon)}\right)^2\right] \tag{II.76a}$$

$$\mathrm{d}\lambda' = l'(\lambda)L'(\lambda', \varepsilon, \sigma)$$

$$l'(\lambda') = c_6\left(1 - \frac{|\lambda'|}{\lambda'_0}\right) \qquad L'(\lambda', \varepsilon, \sigma) = \frac{\sigma_{\min} g_3^{1/3}}{1 + |g_3/c_8|^3} \tag{II.77}$$

$$g_3 = |c_7\sigma_{\min}| \times 0.93 - [J_2(\varepsilon)]^{1/2} \tag{II.78}$$

J_2 is the second invariant of the deviator of the tensor; J_3 is the third invariant of the tensor; $\sigma_{\max}$, $\sigma_{\min}$ and σ_{med} are the maximum minimum and medium principal stresses; and $H(\sigma)$ is the softening function.

The additional parameters defined above are also given in Table 11.6.

Table II.6. Coefficients for endochronic theory

	1[a]	2[b]		1[a]	2[b]
a_0	0.7	0.7	a_{20}	—	14
a_1	$0.6/f'_c$	$0.6/f'_c$	a_{21}	—	1 000
a_2	1400	$1400\left(\frac{f'_c}{4650\text{ psi}}\right)^{1/2}$	a_{22}	—	0.04
			a_{23}	—	$0.2f'_c$
a_3	$500/f'^3_c$	$\frac{90}{f'_c}\left(\frac{3600\text{ psi}}{f'_c}\right)$	b_1	—	$9.1\left(\frac{f'_c}{7200\text{ psi}}\right)$
a_4	$475/f'^2_c$	0.045	b_2	—	$1.0f'_c$
			c_1	$100/f'_c$	$2.0/f'_c$
a_5	$0.8/f'^2_c$	$\frac{0.6}{f'_c}\left(\frac{3600\text{ psi}}{f'_c}\right)$	c_2	0.000 5	0.003
			c_3	—	0.5
a_6	$0.055/f'^3_c$	$0.15/f'^2_c$	c_4	—	2.0
a_7	20.0	0.05	c_5	—	150
a_8	0.000 125	$\frac{15}{f'_c}\left(\frac{f'_c}{3600\text{ psi}}\right)^{1.5}$	β_1	30	30
			β_2	3500	3 500
			β_3	—	0.08
a_9	0.001 5	0.001 5	β_4	—	0.23
a_{10}	—	0.000 125	z_1	0.001 5	0.001 5
a_{11}	—	$0.2/f'_c$	z_2	—	0.012 5
a_{12}	—	$0.8/f'_c$	λ_0	0.001	0.003
a_{13}	—	$2.2 \times 10^{-5}/f'_c$	λ'_0	—	0.003
a_{14}	—	25	c_6	—	$\frac{0.002}{\text{psi}}$
a_{15}	—	1.095			
a_{16}	—	1.216	c_7	—	$\frac{1.05 \times 10^6}{\text{psi}}$
a_{17}	—	0.055			
a_{18}	—	0.94			
a_{19}	—	$\frac{1.0}{f'_c}\left(\frac{3600\text{ psi}}{f'_c}\right)$	c_8	—	0.001
			E_0	$(0.565 + 0.000\,1f'_c) \times 5\,700\sqrt{f'_c}$	

[a] 1: Bazant and Bhat—f'_c shall be in kN/m^2, except E_0 where f'_c shall be in N/mm^2.
[b] 2: Bazant and Shieh—f'_c shall be in kN/m^2, except where the 6.89 factor is applied and f'_c shall be in lbs/in^2.

II.2 Fracture modelling of concrete

II.2.1 Introduction

Concrete in tension is modelled as a linear elastic strain softening material. The principal stresses and their directions are computed initially for an uncracked concrete. If the maximum principal stress for some reason exceeds a limiting value, a crack is assumed to form in a plane orthogonal to this stress. After this, the behaviour of that zone of concrete becomes orthotropic. The local material axes in the given zone coincide with the principal stress direction. If this direction is fixed, the procedure is known as 'fixed crack approach'. If the principal material axes can be given a rotation to coincide with the principal stress and strain directions, the model is known as the 'rotating crack model'. It is also possible in computational models to keep the directions of the first set of cracks fixed. A search is then performed to determine the maximum stress value in the plane parallel to the existing crack. If this maximum stress exceeds the limiting value, then new sets of cracks can be formed which may be perpendicular to the previous one. The limiting values to define the onset of cracking are given.

A new relationship between the fracture energy of concrete, the tensile strength, and the ultimate crack width has been developed by considering the post-cracking strength or softening of concrete. Linear, power, and exponential mathematical models are used to describe the descending branch of the stress–strain diagram. A nonlinear finite element model is developed, using smeared cracks for a layered model approach. This approach is useful, especially for the convenient representation of steel/concrete composite sections. Any fracture model can then easily be incorporated to predict deflection in elastic and inelastic range of reinforced concrete slabs and continuous concrete beams. This section also covers the models related to perforation, scabbing and spalling of concrete.

Fracture mechanics is the study of the behaviour of the material and the conditions in the vicinity of a crack and at the crack tip. The application of fracture mechanics to cementitious materials has opened up a new field for modelling of phenomena that have often been treated empirically in the past.

II.2.1.1 Review of previous studies

Cementitious conglomerates like concrete are made up of different materials in complex arrangements of inclusions of different shapes and sizes. These inclusions are pores, aggregate particles, and pre-existing microcracks. The behaviour of the concrete under load is very much influenced by the amount of inclusion and by the size, distribution and roughness of the aggregates. Stress concentrations prevail at these irregularities in the matrix of the material and will lead to microcracking, which in turn produces progressive concrete failure as the cracks grow.

The use of different parameters in modelling the fracture process of concrete can be examined together with the loss of stiffness due to cracking.

A number of approaches were used by researchers [1082–1109] over the years to describe and quantify crack propagation. However, significant advances were only made following the development of nonlinear finite element methods. Early simulations of concrete fracture were based on the use of the tensile strength only as the governing parameter, whereby the stress normal to the crack was assumed to drop to zero immediately after crack formation [1083]. However, subsequent testing found that single parameter models could not be verified. Bazant and Oh [1083] concluded that strength parameters are 'inobjective', since the results of the analysis can be strongly affected by the size of the finite element mesh.

Other studies normally extended the principles of linear and elastoplastic fracture mechanics to concrete by using the stress intensity factor approach by Walsh [1108], Hawkins *et al.* [1088] and Testa and Stubbs [1107].

Other linear elastic fracture mechanics approaches for cementitious materials have generally proved to be inadequate since the pursuit of unique critical stress intensity factors, K integrals and R-curves, were unsuccessful. The J-integral is the path-independent integral that is an average measure of the elastic–plastic stress–strain field ahead of a crack, and the R-curve is defined as the procedure to characterise the resistance to fracture of the material during incremental slow-stable crack extension. This is primarily due to the fact that concrete cracks throughout the entire fracture zone.

Linear elastic fracture mechanics is meaningful only when the fracture zone is surrounded by an elastic region that can be characterised by stress intensity factors (in case of linear) of J-integral (then nonlinearity is more suitable). This is partly prompted by the nonductile response to concrete since the microcracking is not accompanied by any substantial contraction, as is found when metals yield. Therefore, no essential difference exists between plane–stress and plane–strain conditions for concrete. Consequently, the width of the specimen, i.e. the length of the crack front, is not of major importance for concrete.

II.2.1.2 Concrete tensile strength and cracking

Failure of concrete in tension is induced by the initiation of cracks and their propagation through all or part of the element or member in question. The cracking is governed by the tensile properties of the material. Traditionally, strength parameters have been used to define the initiation of facture — specifically, the tensile and compressive strength of concrete. These are used to position a tension cut-off condition in the principal stress pace. When the combination of principal stresses violates this condition, the crack is initiated. Many failure criteria have been based on these parameters.

For concrete strength and cracking, the following initial assessment is carried out using triaxial tension and tension–compression in combinations:

(a) $\sigma_{\mathrm{p}i} = \sigma_{\mathrm{tu}i} \qquad i = 1, 2, 3$ for triaxial tension (II.79)

(b) tension–tension–compression and tension–compression–compression, linear decreasing tensile strength expressions

$$\begin{aligned} \sigma_{\mathrm{p}i} &= \sigma_{\mathrm{tu}i}\left(1 + \frac{\sigma_{i+1}}{f_{\mathrm{c}}'}\right) && \text{for } \sigma_{i+1} \leq 0 \\ \sigma_{\mathrm{p}i} &= \sigma_{\mathrm{tu}i}\left(1 + \frac{\sigma_{i+1}}{f_{\mathrm{c}}'}\right)\left(1 + \frac{\sigma_{i+2}}{f_{\mathrm{c}}'}\right) && \text{for } \sigma_{i+1}, \sigma_{i+2} \leq 0 \end{aligned} \qquad \text{(II.80)}$$

where $\sigma_{\mathrm{tu}i}$ is the tensile strength of concrete.

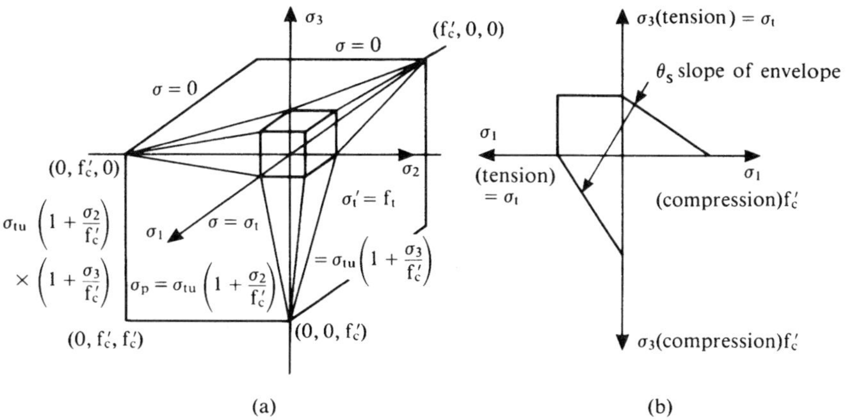

Fig. II.7. Concrete tensile strength and cracking; (a) three-dimensional cracking criteria; (b) criterion in σ_3–σ_1 plane.

Figure II.7 shows a typical three-dimensional tensile failure envelope of a concrete model [751].

II.2.1.3 The strain softening rule

Concrete can be considered as an elastic strain softening material in tension. The problem with this concept is the extent to which the tension stiffening can be incorporated. Alternatively, concrete can be assumed to be elastic–brittle in tension. When a crack occurs, the stress normal to it can be immediately released and drop to zero.

The general strain softening model for concrete is given by Petersson [752] as a bilinear curve. Figure II.8 shows the strain softening model. The stress σ is defined as

$$\sigma = \sigma_{tu}\{\exp[-(\varepsilon - \varepsilon_{cu})/\alpha]\} \tag{II.81}$$

Petersson found that the typical values of fracture energy of concrete are in the range of 60–100 N/m where ε is the nominal tensile strain in the cracked zone, ε_{cu} is the cracking and α, the softening parameter, is given by

$$\alpha = (G_f - 1/2\sigma_{tu}\varepsilon_{cu}l_c)/\sigma_{tu}l_c > 0 \tag{II.82}$$

where

$$l_c = \frac{\text{volume containing a crack}}{\text{area of crack}} = \frac{V}{A_{cs}} \approx (dV)^{1/3}$$

and dV is the volume of concrete represented by the sampling point

$$G_f = \int_0^\infty \sigma(\omega)\,d\omega = \text{fracture energy needed to separate two cracks} \tag{II.83}$$

$$\approx 50\text{–}200\ \text{N/m}$$

$$\omega = \text{crack width} = l_c\varepsilon_{cu}$$

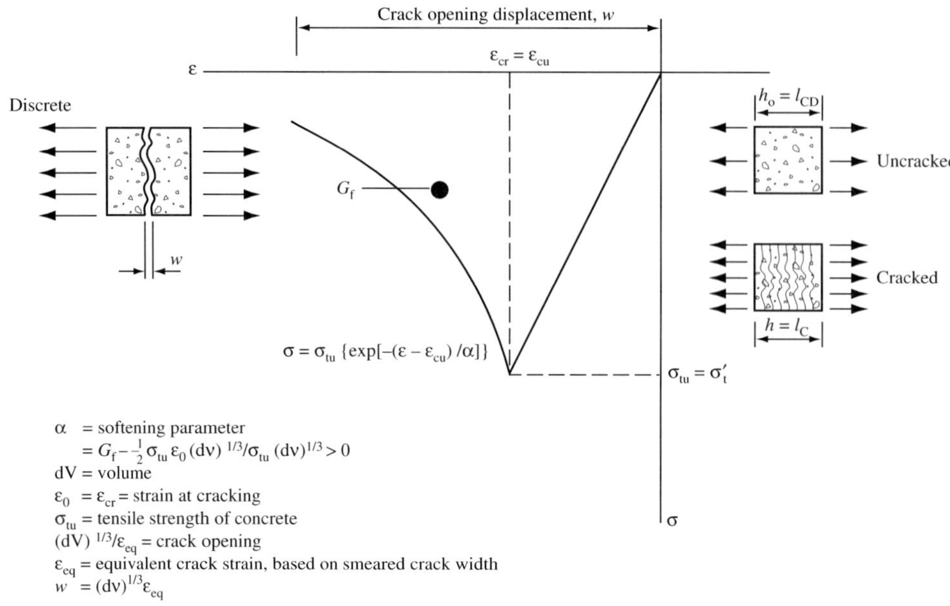

Fig. II.8. Tensile strain softening model for concrete.

The area under the softening curve represents the amount of energy consumed by a crack to propagate through the concrete element depth. By knowing the cross-sectional area of the element, the fracture energy can be calculated. Rots *et al.* [1106] used a better representation by taking a bilinear stress–strain softening diagram to determine the stiffness of cracked concrete. Neither of these approaches are accurate, since tensile tests have shown that the concrete stress–strain softening diagram is highly nonlinear.

The equations given in the following are those proposed by different researchers for nonlinear stress–strain softening curves.

Reinhardt [1103]

$$\sigma/\sigma_{tu} = 1 - (\varepsilon/\varepsilon_0)^k \tag{II.84}$$

where $k = 0.31$; σ = tensile stress normal to the crack; ε = the strain in the direction of the normal stress; and ε_0 = the ultimate strain of the strain-softening branch.

Reinhardt and Cornelissen [1104]

$$\frac{\sigma}{\sigma_{tu}} = \left[1 + c_1\left(\frac{\varepsilon}{\varepsilon_0}\right)^4\right]\exp\left(-c_2\frac{\varepsilon}{\varepsilon_0}\right) \tag{II.85}$$

where c_1 = a material constant = 9.0; and c_2 = a material constant = 5.0.

Gopalaratnam and Shah [1087]

$$\sigma/\sigma_{tu} = \exp(-kw^{\lambda}) \tag{II.86}$$

where λ = a constant = 1.01; k = a constant = 0.063; and w = the crack width in micrometers.

Cornelissen *et al.* [1084]: these researchers conducted several deformation controlled uniaxial tension tests on normal-weight and lightweight concrete specimens to determine the actual stress–strain softening characteristics. For the normal weight concrete, river gravel with a maximum grain size of 8 mm was used, and sintered expanded clay was used for the lightweight concrete. It was concluded that there is a unique relationship between crack opening and ultimate stress for a given concrete; this relationship is not affected by stress or strain history. The following exponential mathematical model was proposed.

$$\sigma/\sigma_{tu} = f(w) - (w/w_0)f(w = w_0) \tag{II.87}$$

where $f(w)$ = a displacement function given by

$$f(w)\left[1 + \left(\frac{c_1 w}{w_0}\right)^3\right] \exp\left(-\frac{c_2 w}{w_0}\right) \tag{II.88}$$

where w = the crack-opening displacement; w_0 = the crack opening at which stress can no longer be transferred; and c_1 and c_2 = material constants (for normal weight concrete; $c_1 = 1.0$ and $c_2 = 5.64$; for lightweight concrete; $c_1 = 3.0$ and $c_2 = 6.93$). This model fits the data points of the tensile-strain softening diagram for aggregate size and type, water to cement ratio, age of concrete at loading and loading rate. The experimental values of w_0 are 160 and 140 μm for normal and lightweight concrete respectively. Equation (II.82) gives the value of $l_c = h$ shown in Fig. II.8. The fracture energy of concrete is equal to the area under the stress crack opening softening diagram. G_f can therefore be expressed as

$$G_f = l_c g_f = hj\sigma\, d\varepsilon \tag{II.89}$$

where g_f = the area under the stress–strain softening diagram. This is the same as eqn (II.83). The strain is in the ratio of $(h - h_0)/h$ with the g_f for smear cracks. Since the fracture energy and the tensile strength are material properties of the concrete, it is important to predict the cracks under various stress levels.

Bazant and Oh [1083] analysed 22 experimental data sets, conducted by several researchers on different sizes of concrete specimens, to predict the fracture energy. The finite element method was used in the analysis, and the fracture of concrete was modelled as a blunt smeared crack band. In this model, the microcracks are assumed to be distributed or smeared over a certain width within the finite element mesh, called the crack band width h. Only mode 1 fracture was considered in the analysis by Barsom and Rolf [1082] in their research for a detailed description of the various modes of fracture. The material fracture properties were characterised by three parameters: the fracture energy, the uniaxial tensile strength and the crack band width over which the cracks were smeared out. The strain softening modulus was a function of these parameters.

An approximate formula was developed to predict the fracture energy of concrete, which was based on the tensile strength and the maximum aggregate size:

$$G_f \approx (2.72 + 0.0214 f_t) f_t^2 \tfrac{3}{4} d_a / E_{c0} \tag{II.90}$$

where G_f = the estimated fracture energy of concrete, f_t = the tensile strength of concrete, d_a = the maximum aggregate size, and E_{c0} = the modulus of elasticity of concrete. Equation (II.90) was based on a linear tensile stress–strain softening diagram. The optimum value of the crack band width was about three times the maximum aggregate size.

Hillerborg [1089, 1090], Hordijk *et al.* [1091], Rots and Blaauwendraad [1105] and Petersson [1100–1102] conducted several experiments on concrete specimens to determine the fracture energy. They concluded that the fracture energy of concrete is a material property and its value is influenced by the aggregate relationship between these material properties and the ultimate crack opening w_0 at which stress can no longer be transferred. Morcos and Bjorhovde [1098, 1099]. By using the Reinhardt [1103, 1104] mathematical model express the term G_f as

$$G_f = h\sigma_{tu} \int_0^{\varepsilon_0} \left[1 - \left(\frac{\varepsilon}{\varepsilon_0} \right)^{0.31} \right] d\varepsilon \tag{II.91}$$

By performing the integration and substituting for the limits it is found that

$$G_f = h\sigma_{tu} \left[\varepsilon_0 - \frac{1}{1.31} \left(\frac{1}{\varepsilon_0} \right)^{0.31} \varepsilon_0^{1.31} \right] = 0.23664\varepsilon_0 h\sigma_{tu} \tag{II.92}$$

The relationship between the ultimate crack opening, the fracture energy and the tensile strength is written as

$$w_0 = 4.226(G_f/\sigma_{tu}) \tag{II.93}$$

By using the Reinhardt [1103, 1104] and Cornelissen *et al.* [1084] exponential mathematical model, the same development leads to the expression

$$w_0 = 4.05(G_f/\sigma_{tu}) \tag{II.94}$$

The Cornelissen *et al.* [1084] exponential mathematical model gives, for normal-weight concrete,

$$w_0 = 5.618(G_f/\sigma_{tu}) \tag{II.95}$$

and for lightweight concrete

$$w_0 = 4.505(G_f/\sigma_{tu}) \tag{II.96}$$

Equations (II.93–II.96) show that the relationship between the fracture energy, the tensile strength, and the ultimate crack width is linear. Further, the relationship between G_f, f_z and w_0 is independent of the mathematical expression used to describe the descending curve of the strain softening diagram. Although different nonlinear models are used, the relationship between the material properties of concrete and the ultimate crack opening is linear.

II.2.2 A step-by-step formulation for cracking

Bangash [126] has presented cracking criteria, assuming that under incremental loads, progressive and well-disposed cracks occur. The following assumptions are made concerning the initiation, closure and reopening of cracks.

(a) Where cracks occur the direct tensile stress cannot be supported in the direction normal to the cracks.
(b) A concrete element parallel to the crack is assumed to carry stress given by the constitutive relationship consistent with biaxial or triaxial conditions prevailing in the plane parallel to the crack. On further loading new cracks occur at an angle to the previous ones.
(c) The opposite faces of a crack opening in a normal direction will interlock when they are subjected to a parallel differential movement. As a result, restraint of movement and volume changes will cause forces which will only be transmitted across the crack. With large crack widths, these surfaces would completely separate and the aggregate interlock phenomenon would cease. If the strain across the crack is positive, the crack is considered open and rigidity for that direction is kept zero.
(d) Where interlocking exists, the shear stress τ_0 along the crack will not be zero and instead it is linearly related to the strain caused by the differential movement.
(e) During the unloading procedure of the structure the interlocking faces of a crack and the surface deterioration will not interfere with the closing of a crack. A crack closes if the normal with negative quantity across the crack is compressive.

The best way to handle these cracks is to adjust material laws in zones where they occur with their natural directions. The material matrix $[D]$ can easily be modified to initiate cracks in specific directions using direction cosines.

II.2.2.1 Step-by-step analysis of single and multiple cracks

The cracks are initiated, closed and re-opened using the following steps.

(i) Assess initial strains $\{\varepsilon_0\}$, initial stress $\{\sigma_{\text{in}}\}$ and adjusted material property matrix $[D]$.
(ii) Compute incremental stresses

$$\begin{aligned} \delta\{\sigma_1\} &= [D]\delta\{\varepsilon_{\text{i}}\} && \text{for no crack} \\ \delta\{\sigma_{\text{i}}\} &= [D]\delta\{\varepsilon_{\text{i}}\} && \text{for an already initiated crack} \end{aligned} \qquad \text{(II.97)}$$

(iii) Calculate total stresses and strains

$$\{\sigma_{\text{i}}\}_{\text{tot}} = \{\sigma_{\text{in}}\}^* + \delta\{\sigma\}_{\text{i}}^* \qquad \{\varepsilon\}_{\text{tot}} = \{\varepsilon_{\text{in}}\}^* + \delta\{\varepsilon\}_{\text{i}}^* \qquad \text{(II.98)}$$

(* is excluded if cracks do exist).

(iv) For the initially uncracked condition, compute principal stresses $\{\sigma\}$ corresponding to total stresses in point step (iii) and check for the initiation of cracking

$$\{\sigma\}_{\text{p}i} \geq \sigma_{\text{tu}} \qquad i = 1, 2, 3, \text{triaxial tension zone} \qquad \text{(II.99)}$$

If the above inequality is satisfied, calculate the angle of cracks and their directions. Principal stresses are calculated and converted into a global system using a standard transformation matrix $[T]$

$$\{\sigma_g\} = [T]\{\sigma\}_{pi} \tag{II.100}$$

(v) The stresses are converted into nodal loads

$$\{F\} = \int [B]^{T''}\{\sigma_g\}\text{dvol} \tag{II.101}$$

(vi) Residual forces are examined and accumulated for elements and are checked at the end of each iteration. The final iteration terminates when the values of these forces are less than a predetermined constant of 0.001 or any other acceptable value.

(vii) At this stage, a thorough check is required of the crushing of concrete and the yielding of steel. When the principal compressive strain $\varepsilon_i \geq \varepsilon_{cu}$, ε_i is such that where ε_{cu} is the limiting compressive strain (assumed 0.0035), concrete is assumed to crush in the compressive zones at the integration points. The material matrix $[D]^*$ is set to zero. The stresses are then zero. For subsequent load increments these points are ignored. Stresses at these points are converted into equivalent nodal forces as shown in eqn (II.100). The steel elements, whether lying within or on the solid element, yield on the basis of a uniaxial stress–strain relationship. The crushing failure condition of concrete can also be on the basis of

$$3J_2' = \varepsilon_{cu}^2 \tag{II.102}$$

where J_2' is the second deviatoric strain invariant.

(viii) This criterion for the crack closure is based on the inequality of strains

$$\varepsilon_{cu} \leq \varepsilon_c \tag{II.103}$$

where

ε_{cu} = strain normal to the crack

$\varepsilon_{cu} = (\nu^*/E_c^*)\sigma_{cu}$ or strain from the stress–strain curve

ν^* and E_c^* = relevant concrete Poisson's ratio and Young's modulus.

If $\varepsilon_{cu} < \varepsilon_c$, compressive strains are induced which may not be compatible with the concrete strains. When cracks are closed at an intermediate load within the increment then at the end of the load increment stresses may occur at the contact faces of the crack.

(ix) The opening of the crack is similar to new crack initiation. The only difference is the tensile resistance offered by concrete. The closed crack will re-open if

$$\sigma_{cu} \geq 0 \tag{II.104}$$

The shear stresses τ_{cu} on the crack interface and the normal stresses σ_{cu} must be redistributed such that

$$\hat{t}_{fv} = \sigma_{cu} \qquad \hat{t}_{f\tau} = \tau_{cv} \tag{II.105}$$

When these stresses are equated, the rest of the procedure should then be the same as given earlier.

(x) The crack nodes are not identical from one iteration to another at any specific Gaussian point. The most obvious reasons are the improvement in $[D]^*$ and the effect of interlocking with the shear stress along the crack. These changes from one crack node to another can be included. The following linear relationship is adopted for shear τ^*

$$\tau^* = \beta' G \gamma^* \tag{II.106}$$

where G is the shear modulus of the uncracked material; β' is the interlocking factor $= 1 - (\varepsilon_t/0.005)^{K_1}$; ε_t is the tensile strain normal to the crack plane; and K_1 ranges from 0.3 to 1.

II.2.3 Blunt crack band propagation

The smeared crack concept is far simpler than the alternative approach of isolated sharp inter-element cracks in which one has to cope with topological changes in element connectivity. The smeared crack requires an adjustment of element stiffness. Cedolin *et al.* [169] have developed the blunt crack approach and a refined version is given by Pfiffer *et al.* [175]. The smeared crack band of a blunt front is that in which one can easily choose cracks of any direction without paying a penalty if the crack direction is unknown. On the contrary, a change in the direction of a sharp inter-element crack requires changes in the co-ordinates of the finite element mesh. The smeared crack concept suffers from the drawback that the magnitude of stress depends on the sharpness of the front of the crack band, particularly the width of the front element of the blunt crack band. The smaller the width, the larger is the stress in the element just ahead of the front under smaller applied load. If the element width is zero, the stresses ahead of the crack front become infinite for any finite load value. This indicates that cracks would propagate at nearly zero load, which cannot be true. To overcome this problem of fracture mechanics, a criterion in conjunction with the highly efficient smeared crack concept is adopted.

The equivalent strength and energy variation are utilised for crack propagation once it is initiated within the element. The equivalent strength criterion is used for crack propagation by specifying an equivalent stress within the surrounding elements of an existing crack at which cracking should be propagated. The expression for the equivalent strength σ_{eq} is given as

$$\sigma_{eq} = C[EG_f/W(1 - 2\nu\sigma_2^0/\sigma_1^0)]^{1/2} \tag{II.107}$$

where

C = a constant dependent on the choice of elements

E = elastic modulus

ν = Poisson's ratio

$W = A/\Delta a = A/r\cos\alpha$

A = area of the element at the front (Fig. II.9)

r = centroidal distance from the last cracked elements in the front element

α = angle measured from the established crack to the line between the centroids

G_f = the energy release rate $= \partial U(P_i, a)/\partial a = \Delta U(P_i, a)/\Delta a$

U = total strain energy in a cracked body

A = cracked area

P_i = loads

ΔU = energy released by the structure into the element which cracks.

The band length is specified as $a + \Delta a/2$.

In the initial state prior to cracking, the strain energy U_0 is based on the principal stresses σ_2^0, where σ_1^0 is the largest tensile stress. After cracking σ_1^1 becomes zero and $\sigma_2^1 \neq 0$, which is used for the current value of U_1. The change of strain energy $\Delta U = U_0 - U_1$ is equated to crack length $\Delta a \times G_f$ defined above. If $\sigma_1^1 = 0$, cracks relieve the σ_1 stress. The energy flow into the crack front depends on γ and α as indicated in Fig. II.9.

II.2.3.1 Formulation of cracking directions

In this method one specifies the cracking direction within an arbitrary grid, given by

$$\theta_A = 1/3(\theta_{n-1} + \theta_n + \theta_{n+1}) \tag{II.108}$$

where θ_A is the average crack direction, θ_{n-1} and θ_n are the cracking angles of the next to last cracked element, and θ_{n-1} is the impending cracking angle of the element adjacent to the crack front. Since in the arbitrary grid the cracking direction is specified by the accumulated error of the crack direction, the accumulated cracking error θ_E is given by

$$\theta_E = \sum_i^n (\theta_{Gi} - \theta_{Ai}) \tag{II.109}$$

where n is the number of cracked elements and θ_G is the actual average crack propagation angle within each element.

Gambarova and Karakoc [177] introduce the relation between the interface shear and the crack displacements in Bazant's formulations. A better description is given of the effects of aggregate size, and a better formulation is given of the tangent shear modulus G^{CR} whenever cracking is initiated.

$$G^{CR} = \frac{\sigma_{nt}^c}{\varepsilon_{nn}^{CR}} k \frac{1}{r(a_3 + a_4|r|^3)} \tag{II.110}$$

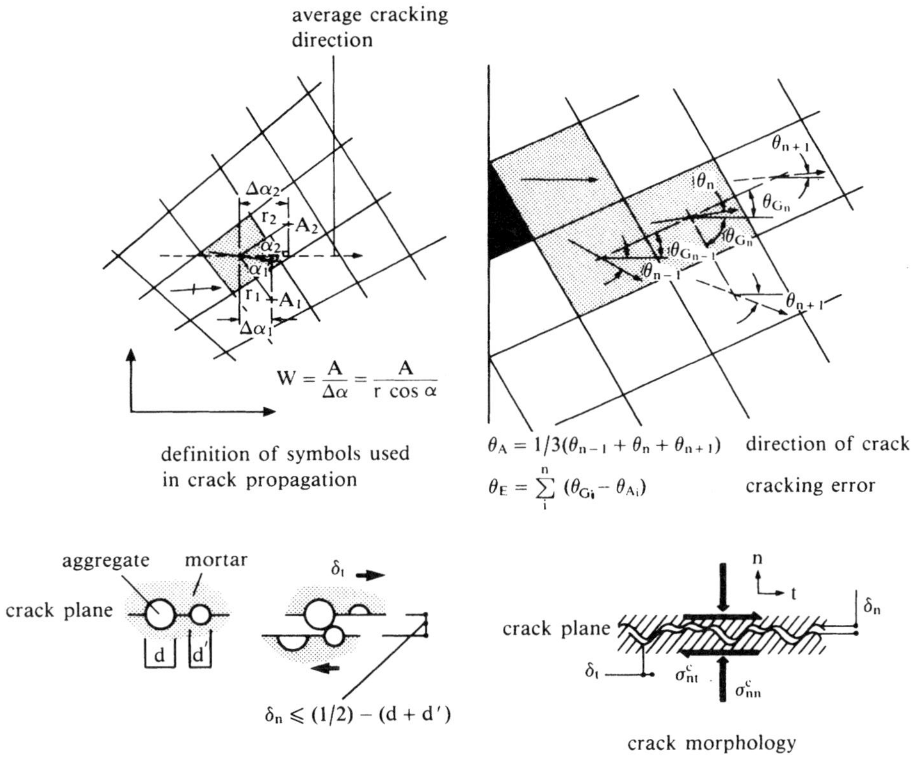

Fig. II.9. Blunt crack propagation.

or

$$G^{CR} = \frac{\tau_0}{\varepsilon_{nn}^{CR}} k\{1 - [2(P/D_a)\varepsilon_{nn}^{CR}]^{1/2}\} \tag{II.111}$$

where

$$K = \frac{a_3 + 4a_4|\gamma|^3 - 3a_3a_4\gamma^4}{(1 + a_4\gamma^4)^2} \tag{II.111a}$$

and

a_3, a_4 = coefficients as a function of the standard cylindrical strength f_c'

τ_0 = crack shear strength (ranging from 0.25 to $0.7f_c'$)

D_a = maximum aggregate size (up to 4 mm)

P = large percentage of crack asperities

$\gamma = \delta_t/\delta_n$

δ_t, δ_n = crack displacements along the normal and tangential directions (Figs II.9 and II.10).

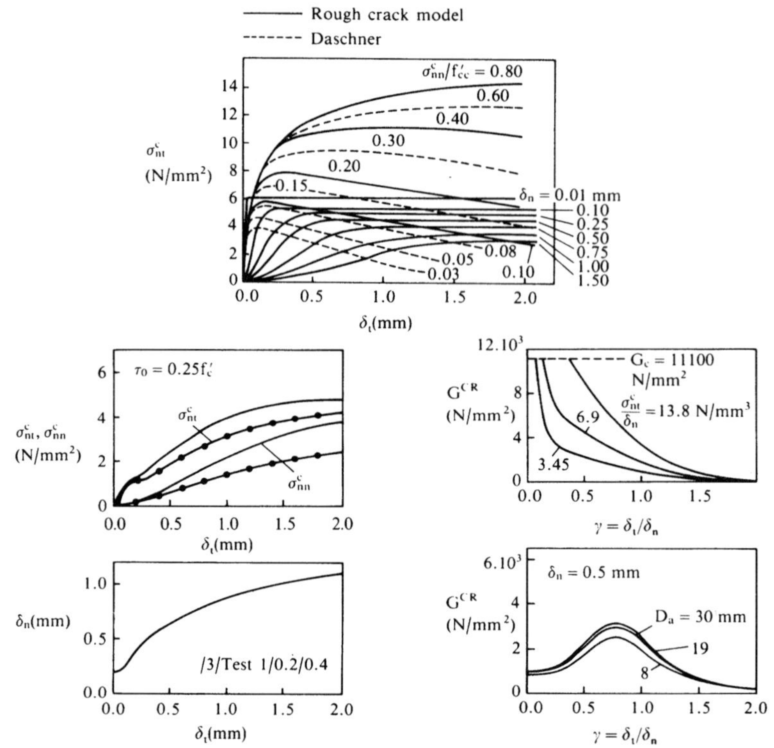

Fig. II.10. Crack displacement versus tangent shear modulus.

Curves have been plotted showing a decrease in the value of G^{CR} when a crack opening linearly increases at increasing shear. The paper [177] gives a constitutive law in which a confinement stress within the rough crack model is given by

$$\sigma^c_{nn} = -a_1 a_2 \frac{\delta_t \sigma^c_{nt}}{(\delta_n^2 + \delta_t^2)^q} \tag{II.112}$$

where

σ^c_{nn} = interface normal stress

σ^c_{nt} = interface shear stress

a_1, a_2 = constants $(a_1 a_2 = 0.62)$

q = a function of crack opening; taken to be 0.25.

For different types of cracking dilatancy δ_n/σ^c_{nt} (Fig. II.9) the tangent shear modulus G^{CR} is plotted against the ratio r of the crack displacement. The value of σ^c_{nt} is given by

$$\sigma^c_{nt} = \tau_0 \left(1 - \sqrt{\frac{2p}{D_a}\varepsilon^{CR}_{nn}}\right) r \frac{a_3 + a_4|r|^3}{1 + a_4 r^4} \qquad r = \gamma^{CR}_{nt}/\varepsilon^{CR}_{mn} \tag{II.113}$$

where p is the crack spacing and CR is cracked concrete

$$\varepsilon_{\mathrm{nn}}^{\mathrm{CR}} = \frac{\delta n}{p} = \text{strain against } \sigma_{\mathrm{nn}}^{\mathrm{c}} \tag{II.114}$$

The $\sigma_{\mathrm{nn}}^{\mathrm{c}}$ values have been computed using points common to the curve of crack opening and constant confinement stress. The authors claim that, using these expressions, the relationship between stresses transferred across the crack and the strains corresponding to the interface displacements can be found. The solution of this work is based on finite element methods, by updating the stiffness matrix at each load increment.

II.2.4 Penetration, perforation, scabbing and spalling

It is important first to define these terms prior to their influence on concrete structures. This section pertains only to local effects largely independent of the dynamic characteristics of the concrete structures. Local effects consist of missile penetration into and perforation through concrete structures, and spalling and scabbing of such concrete structures. They are defined below.

(a) Penetration — the measure of the depth of the crater formed at the level of impact.
(b) Perforation — full penetration through the structure.
(c) Scabbing — the peeling off of the material from the opposite face to that at which the impact occurred.
(d) Spalling — the ejection of target material from the face at which the impact occurs.

Analytical and rational mathematical prediction of these effects is extremely difficult. This is due to the complex nature of the transient loads. Various papers have been published in recent years which have described new ideas and methods for analysing and designing against impact on concrete structures connected with nuclear plant facilities. Owing to complexities in evaluating structural damage due to impact loading, design criteria so far developed have been mainly dependent on experimental tests and empirical formulae.

There are various empirical formulae that used to be employed in connection with missile impact problems on nuclear plant structures, but most of them have been discarded because over the years new, more relevant test data have been made available and these empirical formulae did not give good or realistic results in comparison.

However, there are some older and newer formulae which are still very much applicable. Of these, some empirical formulae are given in the next few paragraphs.

II.2.4.1 The National Defence Research Committee (NDRC) Formula

This formula is really the only one of the older formulae still applicable to nuclear plant facilities for design against impacts. It was put forward in 1946 by the National Defence Research Committee (USA), and proposed a theory of penetration for a non-deforming missile penetrating a massive concrete target. This theory enabled one to calculate the penetration, scabbing and perforation thickness for given missile data, and also gave the thickness that would be required to prevent scabbing and perforation.

The modified NDRC formula for penetration is given below

$$x = \left[4KNW_{\mathrm{d}}\left(\frac{V_0}{1000d}\right)^{1.8}\right]^{1/2} \qquad \text{for } \frac{x}{d} \leq 2.0 \tag{II.115}$$

$$x = \left[KNW\left(\frac{V_0}{1000d}\right)^{1.8} + d\right] \qquad \text{for } \frac{x}{d} \geq 2.0 \tag{II.116}$$

where all the symbols are defined below.

The NDRC formula for perforation thickness is given by

$$\frac{e}{d} = 3.19\left(\frac{x}{d}\right) - 0.718\left(\frac{x}{d}\right)^2 \qquad \text{for } \frac{x}{d} \leq 1.35 \tag{II.117}$$

This formula is applicable for $x : d$ ratios less than 1.35 only, and for other $x : d$ ratios the results are very conservative. For other $x : d$ ratios the US Army Corps of Engineers put forward the formula

$$\frac{e}{d} = 1.32 + 1.24\left(\frac{x}{d}\right) \qquad \text{for } 1.35 \leq \frac{x}{d} \leq 13.5 \tag{II.118}$$

where

W = missile weight (lb)

x = penetration depth (in.)

e = perforation thickness

e_{sc} = scabbing thickness

d = missile diameter (in.)

V_0 = impact velocity (ft/s)

f'_{c} = concrete cylinder compressive strength (psi)

$K = 180/\sqrt{f'_{\mathrm{c}}}$

N = missile shape factor

= 0.72 for flat-nosed bodies

= 0.84 for blunt-nosed bodies

= 1.00 for spherical-ended bodies

N = 1.14 for very sharp-nosed bodies.

The formula for scabbing and perforation thickness x_{sc} and x_{p} respectively for a solid cylindrical steel missile and infinite thickness of the target are given below

$$x_{\mathrm{sc}}/d = 2.12 + 1.36x_{\mathrm{p}}/d \qquad \text{for } 3 \leq t_{\mathrm{sc}}/d \leq 18 \tag{II.119}$$

$$x_{\mathrm{p}}/d = 1.32 + 1.24x_{\mathrm{p}}/d \qquad \text{for } 3 \leq t_{\mathrm{p}}/d \leq 18 \tag{II.120}$$

$$x_{\mathrm{sc}}/d = 7.91(x_{\mathrm{p}}/d) - 5.06(x_{\mathrm{p}}/d)^2 \qquad \text{for } t_{\mathrm{sc}}/d < 3 \tag{II.121}$$

$$x_p/d = 3.19(x_p/d) - 0.718(x_p/d)^2 \qquad \text{for } t_p/d < 3 \tag{II.122}$$

Modified NDRC formula. A modification of the NDRC formula to take into account the finite thickness of a target is proposed. For large diameter missiles impacting on targets of finite thickness the perforation and scabbing thicknesses are given below

$$\begin{aligned} x_p/d &= 3.19(x/d) - 0.718(x/d)^2 \qquad &\text{for } x/d \le 1.35 \\ x_{sc}/d &= 7.91(x/d) - 5.06(x/d)^2 \qquad &\text{for } x/d \le 0.65 \end{aligned} \tag{II.123}$$

Equations will have a range of $0.65 \le x/d \le 11.75$ and $1.35 \le x/d \le 13.5$, respectively. In any case, $t/d \le 3$ gives comfortable results.

It should be stressed that the above formula for perforation within the stated $x : d$ ratios are not strictly NDRC formulae but are based on the penetration thickness x, as obtained from the original NDRC formula.

The values of x and x_{sc} in the formulae are the calculated values of the target thickness to prevent perforation and scabbing, respectively. It has been suggested that a factor of safety be used with these values so that the design thickness of the concrete target is 20% or 30% greater than the value calculated.

For example, if a 20% safety factor is used, the design thickness to prevent perforation, t_4 would be given by

$$t_4 = 1.2x \tag{II.124}$$

Similarly, the thickness to prevent scabbing is given by

$$t_4 = 1.2x_{sc} \tag{II.125}$$

where x_{sc} = scabbing thickness.

It is up to the designer what he or she considers most important for design: the perforation thickness or the scabbing thickness. Usually the larger of two values obtained is used.

In most cases, if the structure is designed against perforation then it is automatically made adequately safe against punching, and further analysis for punching shear is not necessary. This is strictly true for non-deformable missiles such as steel rods but also holds good for most deformable missiles.

II.2.4.2 The Bechtel formula

This formula is sometimes used to determine the thickness to prevent scabbing. It was developed by the Bechtel Corporation and is based on recent test data applicable to missile impacts on nuclear plant structures. However, the formula is restricted to essentially non-deformable missiles such as solid steel slugs and rods. It is only moderately applicable to hollow pipe missiles. There are two formulae for scabbing thickness which are given below.

The Bechtel formula for scabbing thickness for solid steel missiles

$$x_{sc} = \frac{15.5}{\sqrt{f'_c}}\left(\frac{W^{0.4}V_0^{0.5}}{d^{0.2}}\right) \tag{II.126}$$

The Bechtel formula for steel pipe missiles (scabbing)

$$x_{sc} = \frac{5.42}{\sqrt{f'_c}}\left(\frac{W^{0.4}V_0^{0.65}}{d^{0.2}}\right) \tag{II.127}$$

The variables given above are in imperial units. The Bechtel formulae are also written for scabbing and perforation thickness in metric units using a new reference velocity v^* (60.96 m/s):

$$t_{sc} = 1.75\left(\frac{v^*}{v}\right)^{0.13}\frac{(mv^2)^{0.4}}{(d)^{0.3}(f'_c)^{0.4}} \tag{II.128}$$

$$t_p = 0.90\left(\frac{v^*}{v}\right)^{0.25}\frac{(mv^2)^{0.5}}{(df'_c)^{0.5}} \tag{II.129}$$

where $v = V_0$.

The variables given above are in imperial units. The symbols again are defined earlier. d is the nominal missile diameter. The formula was strictly developed with 200 mm (8 in.) diameter pipes but also holds true for other diameters.

Again, x_{sc} obtained from the formula is the target thickness to prevent scabbing. A factor of safety may also be used to convert to design-target thickness.

II.2.4.3 The IRS formulae for penetration and complete protection

The IRS formula for penetration is expressed as

$$x = 1183f_c'^{-0.5} + 1038f_c'^{-0.18}\exp(-0.82f_c'^{0.18}) \tag{II.130}$$

The IRS formula for total protection of a target against penetration, perforation and scabbing is:

$$\text{SVOLL} = 1250f_c'^{-0.5} + 1673f_c'^{-0.18}\exp(-0.82f_c'^{0.18}) \tag{II.131}$$

where SVOLL is the minimum wall thickness to provide complete protection. (Note: the units of the penetration and concrete strength are not imperial. The penetration is in units of centimetres and f'_c is in the units kg force/cm^2 — 1 kg force/cm^2 $\approx$ 100 N/m^2).

In the IRS formula the value of SVOLL is the equivalent of design thickness as obtained from the other formulae. The penetration depth is the same as those in the other above-mentioned formulae.

II.2.4.4 The ACE formulae to prevent penetration or perforation

The formulae for penetration and perforation are as follows.

(a) The formula for penetration

$$x = \frac{282}{\sqrt{f'_c}}\left(\frac{W}{d^2}\right)d^{0.215}\left(\frac{V_0}{1000}\right)^{1.5} + 0.5d \tag{II.132}$$

(b) The formulae for perforation and spalling thicknesses are

$$\begin{aligned} e &= 1.23d + 1.07x \quad &\text{(i)} \\ e_{sp} &= 2.12d + 1.36x \quad &\text{(ii)} \end{aligned} \tag{II.133}$$

The symbols used are those previously defined. The units of the variables are imperial. The ACE formula for perforation gives the thickness of target to prevent perforation. Together with a factor of safety of, say, 1.2, the value of e can be used to determine the target design thickness.

$$t_a = 1.2e \qquad \text{(II.134)}$$

This is because e is the minimum wall thickness of the target to prevent perforation.

II.2.4.5 The Stone and Webster formula

$$\text{Scabbing thickness } (s) = (W\nu_0/C)^{1/3} \qquad \text{(II.135)}$$

C is a coefficient (Table II.7)

II.2.4.6 The CKW–BRL formula for penetration and perforation

$$x = \frac{6Wd^{0.2}}{d^2}\left(\frac{V_0}{1000}\right)^{4/3} \quad \text{(imperial units)} \qquad \text{(II.136)}$$

The perforation thickness is given by

$$e = 1.3x \qquad \text{(II.137)}$$

e is the minimum thickness to prevent perforation.

Table II.7 shows comparisons of scabbing thicknesses.

Table II.7. Comparison of predicted scabbing thickness and known low velocity test results

Missile	Missile velocity (m/s)	Test results scabbing thickness (mm)	Calculated scabbing thickness	
			NDRC (mm)	Bechtel (mm)
200 mm slug 0.95 kN	60	305	360	315
75 mm pipe schedule-40 0.35 kN	61	300	285	343
300 mm pipe 3.3 kN	60	450–600	607	627
75 mm slug 0.06 kN	30	127	129	129
75 mm diameter pipe 0.005 kN	59	133	141	151

Stone and Webster: C coefficients lb/in./s

s/d	1	1.5	2.0	2.5	3.0
Solid	C	←	900	—	950 →
Hollow	C				
$2t/d =$	0.125	2.250	2.450	2.500	2.550
=	0.08	3.00	3.250	3.350	3.400
=	0.06	3.600	3.750	3.900	4.050

II.2.4.7 Modified Ballistic Research Laboratory formula

The modified Ballistic Research Laboratory formula gives the perforation thickness of infinitely thick targets impacted by a non-deformable missile with high velocity as

$$t_p = (427Wd^{0.2})/(d^2\sqrt{f'_c})V_0/1000)^{1.33} \tag{II.138}$$

where

t_p = perforation thickness (in.)

W = missile weight (lb)

d = missile diameter (in.)

f'_c = concrete compressive strength (psi)

V_0 = missile velocity (ft/s)

$x_{sp} = 2x$ is the spalling thickness.

II.2.4.8 Chalapathi, Kennedy and Wall (CKW)–BRL formula

The penetration depth is calculated using the CKW–BRL method as

$$t_p = (6Wd^{0.2}/d^2)(V_0/1000)^{1.33} \tag{II.139}$$

where t_p = thickness to prevent perforation $= 1.3x_p$. All units are imperial, as defined earlier.

II.2.4.9 Dietrich, Furste (DF)–BRL formula

The formula [Dietrich F., personal communication] gives the thickness to prevent perforation as

$$t_p = (3 \times 10^{-4}/\sqrt{f_{ci}})(W/d^{1.8})(\nu/1000)^{4/3} \tag{II.140}$$

where d and t_p are in m, W is in kN, f_{ci} is in kN/cm^2 and v is in km/h.

II.2.4.10 Modified Stone and Webster

The scabbing thickness is given, as for the infinitely thick concrete targets, by

$$t_{sc} = (Wv^2/c'_0) \tag{II.141}$$

where the values of W and v are in lb and ft/s respectively and c'_0 is a coefficient depending on the ratio of t_{sc}/d. The range for which this formula is considered is 3000 psi $\leq f'_c \leq$ 4500 psi and $1.5 \leq t_{sc}/d \leq 3$.

II.2.4.11 Modified Kar formula

The equation for predicting the penetration depth, X, in concrete structures is given as

$$G(X/d) = [\bar{\alpha}/(f'_c)^{0.5}]N_2(E^m/E_s)^{1.25}[W/D(d^{1.8})(V/1000)^{1.8}] \tag{II.142}$$

where

$G(X/d) = (X/2d)^2$ for $X/d \leq 2.0$

$G(X/d) = X/d - 1$ for $X/d > 2.0$

$N = 0.72$ for flat-nosed solid bodies, and

$N = 0.72 + 0.25(n - 0.25)^{0.5} \leq 1.17$ for missiles with special nose shapes; n is the ratio of the radius of the nose to the diameter of the missile, and

$N = 0.72 + [(D/d) - 1](0.0306) \leq 1.17$, for hollow pipe or irregular sections

E^{m} is the modulus of elasticity of the material of the missile (psi)

E_s is the modulus of elasticity of mild steel (psi)

f_c' is the ultimate compressive strength of the concrete test cylinder (psi)

d is the projectile diameter (in.)

D is the outside diameter of a circular section (in.)

W is the weight of the missile (lbs)

V is the impact velocity of the missile (ft/s)

$\bar{\alpha}$ is a constant depending on the type of units used ($\bar{\alpha} = 180.0$ when using imperial units)

The thickness x_a to prevent perforation, and the thickness x_{sc} to prevent backface scabbing are determined by

$$(x_a - a_1)/d = 3.19(X/d) - 0.718(X/d)^2 \quad \text{for } X/d < 1.35 \tag{II.143}$$

$$\rho(x_{sc} - a_1)/d = 7.91(X/d) - 5.06(X/d)^2 \quad \text{for } X/d \leq 0.65 \tag{II.144}$$

in which $\rho = (E_s/E^{m})(\rho = 1$ for steel missiles), and a_1 is half the aggregate size in concrete (in.) $x_a = e$ in this case.

For X/d ratios larger than those shown in the previous equations, the perforation depth and the scabbing thickness can be calculated using the following equations:

$$(x_a - a_1)/d = 1.32 + 1.24(X/d) \quad \text{for } 3 \leq x_a/d \leq 18 \tag{II.145}$$

$$\beta'(x_{sc} - a_1)/d = 2.12 + 1.36(X/d) \quad \text{for } 3 \leq x_{sc}/d \leq 18 \tag{II.146}$$

II.2.4.12 Petry formula

Petry's formula is used for predicting the penetration depth x_p for infinitely thick concrete targets. This formula is derived from tests concerning high velocity impact on infinitely thick concrete targets. Where thickness governs the failure mode or the impact response is influenced by the size and shape of the missile and the presence of the reinforcement, this formula gives accurate assessments.

$$x_p = K_m A V' R \tag{II.147}$$

where

x_p = depth of penetration in a concrete slab of thickness h

K_m = material property constant (L^3/F)
$= 4.76 \times 10^{-3}\,\text{ft}^3/\text{lb} = 2.97 \times 10^{-4}\,\text{m}^3/\text{kg}$

A = sectional mass weight of the missile per unit cross-sectional area of contact (F/L^2)

V' = velocity factor $= \log_{10}[1 + (V_0'^2/V^{*2})]$

V_0' = initial velocity of missile at impact

V^{*2} = reference velocity equal to $215\,000\,\text{ft}^2/\text{s}^2 (19\,973\,\text{m}^2/\text{s}^2)$

R = thickness ratio

$$x_p/D_p = 1 + \exp[-4(\alpha' - 2)]$$

where $\alpha' = h/D_p = h/K_m A V'$ and D_p is the penetration depth of an infinitely thick slab. The penetration depth is restricted to less than h, to satisfy the inequality below. In order to prevent penetration and spalling

$$h \geq C_1 A \times 10^{-5}\,\text{ft or an equivalent value in SI units}$$

where C_1 is taken from Fig. II.11. The time required for penetration is derived from the modified Petry formula

$$F = mx = mv(\mathrm{d}v/\mathrm{d}x) = -1.15(V^{*2}/K_m A)\exp(2.3x_p/K_m A) \qquad \text{(II.148)}$$

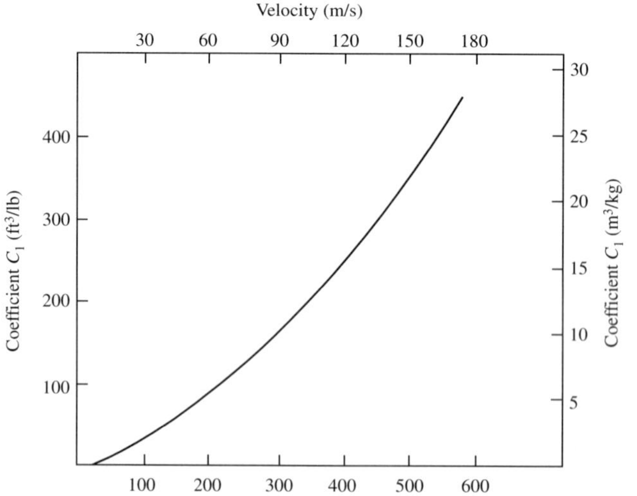

Fig. II.11. Minimum thickness needed to prevent penetration and spalling.

where

F = the resisting force, equal to R_m

x = penetration at any time

V^* = missile velocity at any time

h = penetration depth.

Due to the nonlinear nature of the equation of motion, a numerical integration is necessary to determine the velocity as a function of distance. Then

$$Y = -1.15(V^{*2}/K_m A)\exp[2.3(x - \delta)/K_m A] \quad \text{(II.149)}$$

hence

$$m_x = m_t\delta + k\delta \quad \text{(II.150)}$$

where

x = missile displacement

δ = target displacement

k = target stiffness

m, m_t = mass of the missile and target respectively.

II.2.4.13 Chang formulae

Chang has proposed two semi-analytical formulae for predicting perforation and scabbing thickness of concrete targets impacted by hard steel missiles of non-deformable type:

$$t_p = (u/v)^{0.25}(mv_0^2/df_c')^{0.5} \quad \text{(II.151)}$$

$$t_{sc} = 1.84(u/v_0)^{0.18}(f_c')^{0.4}(d)^{0.2}(mv_0^2)^{0.4} \quad \text{(II.152)}$$

These formulae are probably validated over the following range based on random variable by Bayesian statistics:

$16.7\,\text{m/s} < v_0 < 311.8\,\text{m/s}$

$0.11\,\text{kg} < W < 343\,\text{kg}$

$2.0\,\text{cm} < d < 30.4\,\text{cm}$

$232\,\text{kg/cm}^2 < f_c' < 464.2\,\text{kg/cm}^2$

$5.0\,\text{cm} < h < 60.9\,\text{cm}$

where u is a reference velocity (200 ft/s or 60.96 m/s) and f_c', d, m and v_0 are defined in other sections. The scabbing velocity V_{sc} ft/s, is written as

$$V_{sc} = [(1/2 \times 469)(d^{0.2}f_c'h/W^{0.4})]^{3/2} \quad \text{(II.153)}$$

II.2.5.14 Modified Barr, Carter, Howe and Neilson formulae

Experiments were undertaken on circular reinforced concrete slabs at three different scales (1×, 0.38× and 0.128× prototype). Transient loads, deflections and impactor velocities were measured during impact. The bending reinforcement in the targets varied between 0 and 0.5% each way on each face and the missile velocity varied from 50 m/s to 300 m/s. Based on these tests, a modification of the CEA–EDF formula was suggested. Thus, the velocity to cause perforation V can be calculated by

$$V = 1.3\sigma^{1/2} D_c^{1/6} (\mathrm{d}e^2/m_5)^{2/3} r_b^{0.27} \tag{II.154}$$

Because of the limited test data, it was proposed that the formula be restricted to the following range:

$160\,\mathrm{kg/m^3} <$ reinforcing steel

$20\,\mathrm{m/s} <$ impact velocity $< 230\,\mathrm{m/s}$

$0.5 \leq$ slab thickness/projectile diameter.

II.2.4.15 Modified CEA–EDF formula

Tests were carried out for the French Atomic Energy (CEA) and Electricité de France by Berriaud on a series of slabs subject to impactors with varying velocities (from 20 m/s to 200 m/s), thickness, concrete strength and reinforcement quantities. The empirical formula for a thickness to resist perforation is given by

$$x = 0.82(f_c')^{-3/8}(\rho_c)^{-1/8}(W/d)^{0.5} V_0^{3/4} \tag{II.155}$$

where ρ_c is the density of concrete and the following ranges apply

$30\,\mathrm{MPa} < f_c' < 45\,\mathrm{MPa}$

$0.3 < x_p/d < 4$

$75\,\mathrm{kg/m^3} < p_1 < 300\,\mathrm{kg/m^3}$

where p is the reinforcement quantity. The perforation velocity v_p for the target thickness is given by

$$v_p 1.7 f_c'(\rho_c)^{1/3}(\mathrm{d}h^2/W)^{4/3} \text{ (metric)} \tag{II.156}$$

The CEA–EDF residual velocity formula v_r based on several tests with a correction factor K_0 is given below with all values in imperial units:

$$v_r = \left[\frac{1}{1 + \dfrac{W_1}{W}}(v^2 - v_p^2)\right] \tag{II.157}$$

$$K_0 v_p = \left[v_p^2 - v_r^2\left(1 + \frac{W_1}{W}\right)\right]^{1/2} \tag{II.158}$$

where v_r, v_p, K_0 and W_1 are the residual velocity of the missile, perforation velocity, correction factor and the weight of the target material removed by impact, respectively. The mean and minimum values of K_0 are 1.45 and 1.225 respectively.

II.2.4.16 Haldar, Miller and Hatami method

The impact formula of the NDRC type associated with a non-dimensional impact factor I are presented in imperial units as

$$I = WNv^2 gd^3 f_c' \tag{II.159}$$

where $g = 32.2\,\text{ft/s}^2$ and, when substituted, the above equation becomes

$$I = 12NWv^2/32.2d^3 f_c'$$

where N is the missile nose-shaped factor and all other notations are as defined earlier.

For various impact factors, the NDRC test results were examined using linear regression analysis for x_p/d and t_{sc}/d ratios.

$$\begin{aligned}
t_{sc}/d &= -0.0308 + 0.2251I \qquad 0.3 \le I \le 4.0 \qquad \text{(a)}\\
t_{sc}/d &= 0.6740 + 0.0567I \qquad 4.0 \le I \le 21.0 \qquad \text{(b)}\\
t_{sc}/d &= 1.1875 + 0.0299I \qquad 21.0 \le I \le 455 \qquad \text{(c)}\\
t_{sc}/d &= 3.3437 + 0.0342I \qquad 21.0 \le I \le 385 \qquad \text{(d)}
\end{aligned} \tag{II.160}$$

II.2.4.17 Takeda, Tachikawa and Fujimoto formula

Takeda *et al.* proposed a formula for predicting the penetration depth into reinforced concrete slabs subject to hard missiles:

$$x_0 = [\alpha/(\beta+1)](v_0)^{\beta+1} \tag{II.161}$$

where

$$\alpha = 2^n m^{1-n}/c'\psi$$

$$\beta = 1-2n$$

x_p = maximum depth of penetration (cm)

m = mass of projectile (kg^{-s^2}/cm)

v_0 = impact velocity (cm/s)

ψ = circumference of projectile (cm)

c', n = constants.

Since the formula is based on the kinetic energy as input, it is valid for an energy range from 20×100 kg/cm to 200×10^5 kg/cm.

II.2.4.18 Hughes formulae

These formulae have been developed using the dimensional analysis and test analysis results of NDRC and ACE described earlier. Front and back faces are reinforced (front face 0–0.15%; back face 0.3–1.7% each way). The penetration depth calculated for a concrete barrier, assuming no scabbing or perforation occurs, is given by

$$x_p/d = 0.19NI/s' \tag{II.162}$$

where

N = nose-shaped coefficient = 1, 1.12, 1.26 and 1.39 for flat, blunt, spherical and very sharp noses, respectively

I = impact parameter = $(mv^2/0.63\sqrt{f'_c})d^3$

s' = strain-rate factor = $1 + 12.3\log(1 + 0.03I)$.

The thickness of the concrete target necessary to prevent scabbing and perforation is written as

$$x_{sc}/d = 1.74(x_p/d) + 2.3 \tag{II.163}$$

$$x_p/d = 1.58(x_p/d) + 1.4 \tag{II.164}$$

II.2.4.19 Modified Kar formulae

It is claimed that most of the formulae described earlier do not include dimensions, shapes of the missiles, material properties of the missiles and targets and the size of the coarse aggregate in concrete. Kar gives the penetration depth (in inches) in the concrete targets as

$$G(x_p/d) = \alpha K_p N(W/D)(v/10\,000d)^{1.8} \tag{II.165}$$

where the range for $G(x_p/d)$ is given as

$$\begin{cases} (x_p/2d)^2 & \text{for } x_p/d \leq 2.0 \\ (x_p/d) - 1 & \text{for } x_p/d \geq 2.0 \end{cases} \tag{II.166}$$

and where D = diameter of the actual missile in the case of a circular section or is equal to the projectile diameter d in the case of a rectangular section.

II.2.4.20 Perry and Brown formulae

For solid missiles on concrete targets, the penetration depth is given in Fig. II.12 by

$$x_p/d = 9.605\sqrt{m}[v\sqrt{(E_0 d^3)}] + 0.06 \tag{II.167}$$

where x_p, d and v are as defined earlier and m is the missile mass. If $L < 50$, the scabbing thickness does not occur. If $L > 70$ the scabbing thickness certainly occurs.

$$L = [\sqrt{m}(v/\sqrt{d})h(1 + h/d)]\sqrt{E_c\sigma_s} \tag{II.168}$$

where

h = target thickness

E_c = Young's modulus

σ_s = maximum value of nominal shear stress before damage
$= \sigma_{tm}\sqrt{[1 + (\sigma_{cm}/\sigma_{tm})]}$

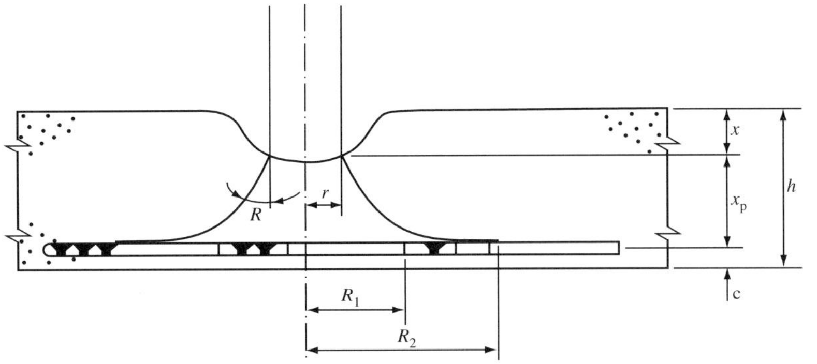

Fig. II.12. Assumed barrier failure mechanism and derivation of hinge radius.

σ_{tm} = mean tensile stress in concrete

σ_{cm} = compressive stress

$$\sigma_{cr} = 0.9\sigma'_{cr}$$

$$= 0.9[0.52(1-\nu^2)/(1-\nu_p^2)]^{1/2}(E_t/E)^{1/2}E_s(t_p/r)]$$

E = Young's modulus

E_t = tangent modulus

E_s = secant modulus

t_p = pipe thickness (in.)

r = pipe radius (in.)

ν, ν_p = Poisson's ratios for elastic and plastic, respectively.

Geometry

$$\tan\theta = R_1 \times r/x = 2$$

$$2x = R_1 \times r$$

$$R_1 = 2(htx_p) + r$$

$$R_2 = C_L t_3$$

$$R_u = M_u D_h$$

where

M_u = ultimate curvature $= eu/C_h \approx x_p/C_h$

D_h = hinge length (or $R_2 R_1$)

$$(R_2 \text{ or } R_1)_{max} = r_u/M_u = 0.07/0.003 = 23.3$$

$$(R_2)_{max} = R_1 + 23$$

The thickness to prevent spalling is given by

$$t_{sp} = x_p + x + c \tag{II.169}$$

$$x(x+r)^2 > \left[\frac{6W^2V_m^2 - W}{2g\pi r_u M_u}\right]\left(\frac{1728}{\rho_c \pi}\right) - R_2^2 C \tag{II.170}$$

where

$\rho_c = 0.15\,\text{kips/ft}^3$

r_u = ultimate rotational capacity at hinge (rad)

C = cover (in.)

M_u = ultimate moment capacity at hinge (in kips/in.) (1 kip = 1000 lbf)

$R_2 = R_1 + 23$

$\nu = \nu_2 \quad \text{if } \nu_2 > \nu_2'$

$\nu = \nu_2' \quad \text{if } \nu_2 < \nu_2' < \nu_1$

if $\nu_2' > \nu_1$ not applicable

the spalling equations are not applicable if $t_{sp} = h < 12$ in.

For pipe missiles on concrete targets, the penetration depth is given by

$$x_p/d = 8[\sqrt{mv}/\sqrt{(Ed^3)}] + 0.24 \tag{II.171}$$

where d is not outside the diameter of the pipe missile. The scabbing thickness can be achieved if $L > 60$ and is unlikely to occur if $L < 50$. The value of L in this case is expressed by

$$L = \sqrt{m}\,\mathrm{d}v/[\mathrm{d}h(1 + h/d)](\sqrt{E_c}/\sigma_s) \tag{II.172}$$

where the various elements of the equations have been defined previously.

II.2.4.21 Formulae for deformable missiles impacting on concrete targets — McMahon, Meyers, Sen model

The model evaluates local damage including penetration and back face spalling of reinforced concrete targets subject to the impact of deformable, tornado-generated missiles such as pipes, etc. The total penetration is given by

$$x_p = x_1 + x_2 + x_3 \tag{II.173}$$

where

x_1 = penetration during time $t_1 = Ft_1^2/6m) + v_0 t_1$

F = interface body $= \sigma_{cr} \times A$

t_1 = rise time $= 3.2 \times 10^{-6} F$

A = pipe area

v_0 = missile velocity at the initial time t_0

m = mass of the missile

$$x_2 - x_1 = x_c \text{ (plastic missile deformation during } t_2 - t_1)$$

$$= \frac{m}{2F}(v_2 - v_1)^2 + \frac{mv_1}{F}$$

$$= \frac{m}{2F}(v_2^2 - v_1^2) \tag{II.174}$$

v_1, v_2 = missile velocity at times t_1 and t_2

$$v_1 = (Ft_1/2m) + v_2$$

$$v_2 = (F/m)(t_2 - t_1)tv_1 \tag{II.175}$$

$$x_3 = x_2 \text{ (penetration during } t_3 - t_2)$$

$$= -mv_2^2/2F \tag{II.176}$$

$$t_2 = \underset{\text{plastic wave}}{l/\sqrt{(E_T/\rho)}} + \underset{\text{elastic wave}}{l/\sqrt{(E/\rho)}} = \text{wrinkling duration} \tag{II.177}$$

$$t_3 = (-mv_2/F) + t_2 = \text{final time} \tag{II.178}$$

where

ρ = material density

l = missile length

II.2.4.22 Rotz damage model

Rotz predicted scabbing thickness using Bayesian estimators as

$$t_{sp} = \bar{K}_p(W^{0.14}v^{0.65}/\sqrt{f'_c}d^{0.2}) \tag{II.179}$$

where

$\bar{K}_p$ = empirical constant = 5.42

v = impact velocity in imperial units (ft/sec)

d = missile diameter (in)

f'_c = concrete compression strength (lb/in^2)

W = missile weight (lbf)

II.2.5 Explosion on concrete structures

Bangash, M. Y. H. [1620] has given a comprehensive treatment of blast effects on structures. A summary of numerical modelling of concrete beams under blast effects is given in the Appendix XVII.

III. Numerical models for creep, shrinkage and temperature

III.1 Introduction

This chapter deals with various constitutive models for creep, shrinkage and temperature. As far as possible up-to-date information is included for all three areas so that it can be used in the finite element analyses presented later in the text.

Concrete exhibits time-dependent strains due to creep shrinkage and temperature, which profoundly affect the properties of concrete as a structural material. The literature covering concrete creep is comprehensive and it is not therefore intended to review the entire literature in great depth here. However, a brief discussion of the results obtained by various researchers is included.

III.2 Creep

III.2.1 A brief survey of creep

Creep in concrete represents the dimensional change in the material under the influence of sustained mechanical loading. Quite small loads will cause the concrete to deform. The phenomenon of creep occurs at elevated and at ambient temperatures. The rate of creep is increased at elevated temperatures. Various experimental tests [204–228] have been conducted to identify the effect of temperature on concrete creep. England and Phok [86] and Ross [207] have presented results for sealed and unsealed cylinders up to a temperature of 140°C and for a test duration of 60 days. The results for sealed cylinders show that the creep at 80 and 140°C was about 3.5 and 4.2 times the value at 20°C. Nasser and Neville [204] reported their observations from experimental tests at temperatures ranging from 21.1 to 96.1°C with stress/strength ratios from 0.35 to 0.7. The concrete was cured at 41.34 N/mm^2 (6000 psi) and tested after 24 h of casting. They found that the pronounced maximum for the creep rate is at a temperature of about 71°C. This creep rate was based on creep measurements made during a period from 21 to 91 days after loading. Hannant [205] conducted creep tests on sealed 104.775 mm ($4\frac{1}{8}$ in) by 305 mm (12 in) cylinders of (approximately 62 N/mm^2 (9000 psi)) limestone aggregate after curing them for five months in water and for an additional month in a sealed and saturated condition. Results showed a nearly linear increase of specific creep with a range of 27 to 77°C

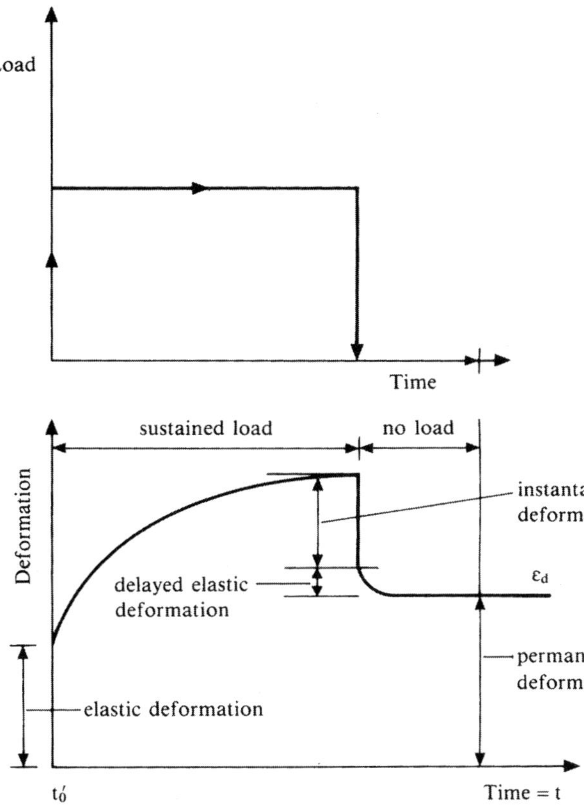

Fig. III.1. Typical deformation versus time curves for concrete under constant load.

for loading periods of two years. The creep at 77°C was approximately 4–4.8 times that at 27°C. Up to stresses of 12.78 N/mm^2 (2000 psi) creep remained proportional to stress. The Poisson ratios of creep determined for sealed specimens were similar in magnitude to their elastic values.

In order to understand the influence of creep, a typical deformation–time curve is shown in Fig. III.1. A concrete specimen loaded under uniaxial compression shows an immediate elastic deformation. If this load is sustained, additional deformation due to creep occurs. The rate of the deformation decreases with time (Fig. III.1). If this load is removed, there is an immediate recovery of deformation and, following this, a recovery of creep deformation (delayed recovery) rate occurs, which rapidly decreases with time, as shown in Fig. III.2. At the end of this, a residual deformation is left which is greater than the initial elastic deformation. Figure III.2 shows a time-dependent strain curve for ambient and elevated temperatures between T_1 and T_2. The creep recovery strain occurs immediately after the instantaneous elastic strain, which is extensive at first but reduces after a short period of time. The creep recovery is essentially independent of temperature [204].

At a low level of stress, the concept of specific creep is introduced. Specific creep or creep strain per unit of stress is a useful indicator of creep effects. It is also sometimes

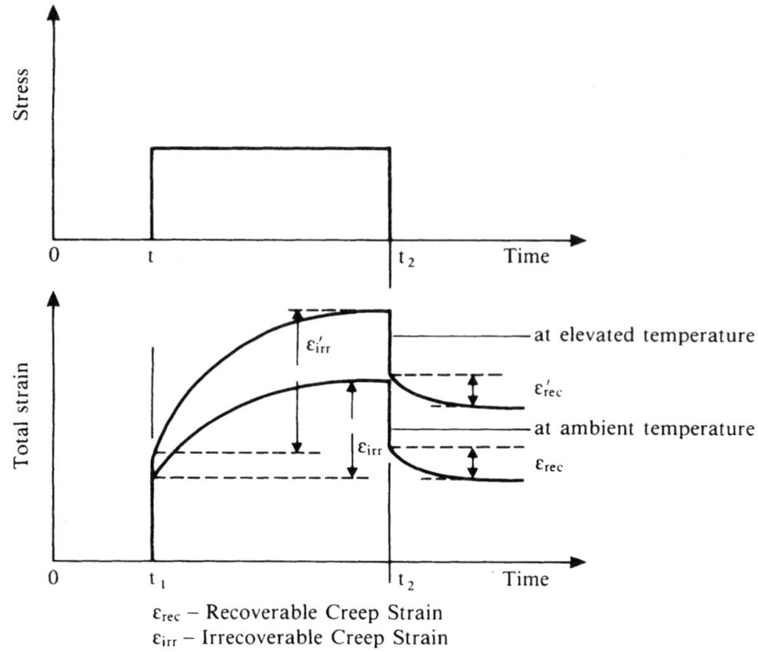

Fig. III.2. Time-dependent strains in concrete.

useful to normalise creep strain data with respect to stress and temperature. This quantity is known as specific thermal creep.

The creep rate, creep–time curves and the Young's modulus of concrete are all temperature dependent. The creep rate is also dependent on the type of aggregate and the mix proportions, and the age of the concrete. The average values and increases in creep rate are subject to +30% due to the different qualities of aggregate. The creep of concrete at normal temperatures has been investigated by several authors [80–96, 207–234, 418]. Some tests indicate that even very old concrete is still liable to creep of considerable magnitude. The creep and time curves of old concrete approach very nearly those of young concrete [80–96, 207–234]. This is a very important phenomenon and is very relevant to the analysis and design of the reactor vessels since these vessels are loaded when concrete has attained several years of age. For prototype vessels the maximum temperature level permitted on concrete [82, 94] is about 80°C, and hence creep is generally high, as suggested by Ross [207]. Bangash and England have carried out an optimum analysis, based on a statistical approach, to plot long-term specific creep values. The long-term specific creep values have been determined by extrapolating short-term (Fig. III.3) tests by Hannant and Frederick [58] and Browne [84] based on a linear relationship between creep rate and log time existing immediately after the initial load application. Water/cement and aggregate/cement ratios are (0.47:0) and (4.5:4.4), respectively, giving (7800:7300) psi (53.78 MN/m^2 – 50.33 MN/m^2). The difference lies in the maximum aggregate size, that used by Hannant and

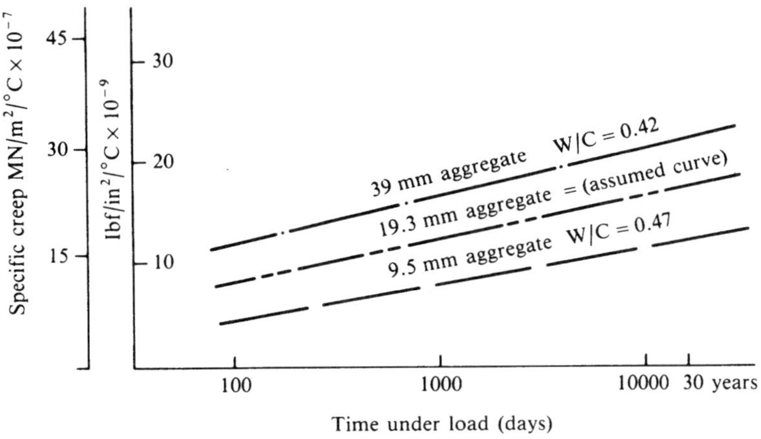

$\log_s = \log a + n \log t'$

$n = C\sigma e^{-E/RO}$

a = factor decreasing with age

t' = time under loads

O = absolute temperature

E = activation energy

R = Boltzmann constant

C = constant

Fig. III.3. Specific creep curves (after Refs [84, 88, 205, 909]).

Frederick being 9.5 mm ($\frac{3}{8}$ in) and that of Browne 39 mm ($1\frac{1}{2}$ in). The specimens are sealed to represent conditions in the vessel wall or cap thickness in the long term. The two curves when plotted would give an upper and lower bound to the creep loss to be expected in any vessel when a maximum aggregate size of 18 mm ($\frac{3}{4}$ in) is used. Assuming loss to be proportional to aggregate size, linear interpolation between the two curves has been carried out in order to obtain a curve for 19.3 mm ($\frac{3}{4}$ in) aggregate which is given in Fig. III.3.

For the long term, under normal operating conditions, creep loss has been determined for nuclear vessels using the specific creep value from this curve at thirty years in conjunction with the normal operating stresses. For a hot shut-down period, Hannant [205] has shown that temperature cycling under constant long-term load does not normally increase the resulting total creep strain. Since the maximum hot shut-down period for vessels is six weeks, it is assumed to be a short-term shut-down, which will not cause a significant temperature reduction within the wall and slab thicknesses of the vessel from that of normal operation.

The shut-down periods last for six weeks and occur every second year after the first four years throughout the vessel life. During these six-week periods there is an increase in stress in the vessel and a corresponding increase in creep strain. These increases in creep strain have been superimposed on the 30-year normal operating creep strain assuming that there is no creep recovery on return to normal operating

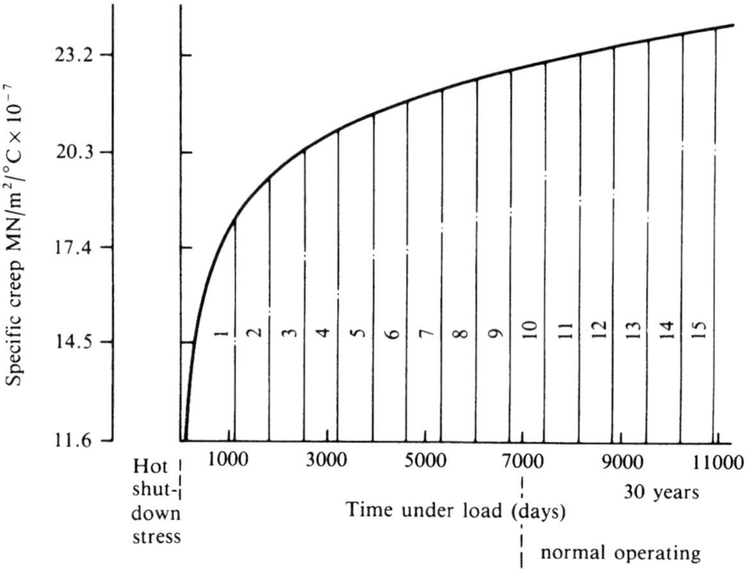

Fig. III.4. Specific creep curve for hot shut-down case of the reactor vessel.

conditions. This, therefore, gives a conservative estimate of creep loss. Figure III.4 gives the relationship between specific creep and time for a hot shut-down case for up to the entire predicted life of the vessel. Similar curves are used for structures in different operating temperature conditions.

Various methods of creep analysis have been used over the last 50 years; they may be classified into two main categories:

(a) direct methods
(b) iterative or step-by-step methods.

The direct methods allow the calculation of creep effects in a single time step. The effective modulus [222] and steady-state [222] methods are examples of this category. Other direct methods, in refined form, have been presented by England and co-workers [86, 93].

In the iterative methods, the period of time (over which creep is sought) is divided into a number ofsteps, and separate calculations are carried out for each step. An assumption is made that stress is constant during each time step while strain is being calculated. The accuracy and stability of the solution depends on the length of the time step chosen and successive calculations depend on those in previous time steps. The types of iterative creep solution methods are:

1. method of superposition
2. rate of creep
3. strain hardening
4. rate of flow.

III.2.2 Creep constitutive models

III.2.2.1 Indian Standard Code IS 1343 [212]

The ultimate creep strain ε_r^u is computed as

$$\varepsilon_r^u = 17.5\left(\frac{1.25 - H}{0.25}\right)\left(\frac{400}{f_{cu}}\right)\left(\frac{f_{cu}}{f_{ci}}\right) \times 10^{-6} \tag{III.1}$$

where

H = the average ambient humidity of the environment (0.5–0.2 dry; 0.75–0.9 humid; 1.0 average)

f_{cu} = cube strength of concrete at 28 days

f_{ci} = cube strength of concrete at transfer

III.2.2.2 The Hilsdorf and Muller method [213]

This method gives the general creep function $\varepsilon_c(t, t')$:

$$\varepsilon_c(t, t') = \frac{1}{E_c(t')} + \frac{\phi(t, t')}{E_c(t')} \tag{III.2}$$

where

$E_c(t')$ = modulus of elasticity at time t'

$\phi(t, t')$ = creep coefficient = $\varepsilon_c/\varepsilon_e$ for concrete at time t

III.2.2.3 The ACI method

The American Concrete Institute ACI 209 method is based mainly upon the method proposed by Branson [214]. The hyperbolic function $\phi(t, t')$ is given by

$$\phi(t, t') = \frac{(t - t')^{0.6}(\phi_\infty)}{10 + (t - t')^{0.6}} \tag{III.3}$$

where ϕ_∞, the creep coefficient, is given by

$\phi_\infty = 2.35\beta_{t'}\beta_H\beta_d\beta_S\beta_F\beta_{AC}$

$\beta_{t'}$ = a coefficient for the age of the concrete t' when load is applied

$$\beta_{t'} = \{1.25t'\}^{-0.118} \quad \text{for } t' \geq 7 \text{ days} \tag{III.3a}$$

β_H = a coefficient to estimate the relative humidity H

$$\beta_H = 1.27 - 0.0067H \quad \text{for } H \geq 0.4 \tag{III.3b}$$

β_d = a coefficient for the average thickness of the concrete member

(i) for $50 \leq d \leq 150$ mm

d (mm):	50	75	100	125	150
β_d:	1.3	1.17	1.11	1.04	1.0

(III.3c)

(ii) for $150 \leq d < 380$ (all in mm)

$$\beta_d = 1.14 - 2.093 \times 10^{-5} \rightarrow t - t' \leq 365 \text{ days}$$

or (III.3d)

$$\beta_d = 1.10 - 1.139 \times 10^{-5} d^2 \rightarrow t - t' > 365 \text{ days}$$

(iii) $d \geq 375$ mm

$$\beta_d = 0.67(1 + 1.13e^{-0.212V/A})(1.13e^{-0.54V/A}) \qquad \text{(III.3e)}$$

(V = volume of the member/structure and A = area of the member/structure)

β_S = a coefficient for the consistency of fresh concrete

$$\beta_S = 0.82 + 1.7688 \times 10^{-4} S \qquad \text{(III.3f)}$$

(S = slump in concrete (mm))

β_F = a coefficient for fine aggregate particles

$$\beta_F = 0.88 + 24 \times 10^{-4} F \qquad \text{(III.3g)}$$

(F = particles/total aggregate content × 100)

β_{AC} = a coefficient for the air content of the concrete

$$\beta_{AC} = 0.46 + 0.09AC \geq 1.0 \qquad \text{(III.3h)}$$

(AC is the percentage air content (in volume of concrete))

III.2.2.4 The CEB–FIP method [215]

This method suggests the following creep function for the initial load application at 28 days:

$$\phi(t, t') = \frac{1}{E_c(t')} + \frac{\alpha_{28}(t, t')}{E_c(t_{28})} \qquad \text{(III.4)}$$

where

$$\alpha_{28}(t, t') = \beta'_H \beta'_{t'} \beta'_F \beta'_d \beta_{td} \qquad \text{(III.4a)}$$

(i) β'_H = a creep coefficient as a function of relative humidity (III.4b)

H:	0.3	0.4	0.5	0.6	0.7	0.8	0.9	1.0
β_H:	3.25	3.15	2.90	2.70	2.35	1.90	1.5	1.0

(ii) $\beta'_{t'}$ = an age coefficient at loading of normal cement (III.4c)

t' (days):	1	3	7	14	28	90	180	360
$\beta'_{t'}$:	1.8	1.6	1.4	1.2	1.0	0.75	0.65	0.3

(iii) β'_F = a coefficient for concrete composition (Table III.1) (III.4d)

Table III.1. β'_F values for different water/cement ratios and unit weights

Water/cement ratio	Unit weight (kg/m^3)			
	500	400	300	200
0.4	1.0	0.8	0.45	–
0.5	1.4	1.2	0.9	–
0.6	–	1.6	1.19	0.75
0.7	–	–	1.5	1.0
0.8	–	–	–	1.3

Table III.2. β_{td} values for different ages and values of d

d (mm)	Age (days)				
	1	10	100	1000	10 000
50	0.15	0.38	0.8	0.99	–
100	–	0.21	0.65	0.95	–
200	–	0.1	0.43	0.90	–
400	–	–	0.24	0.78	–
800	–	–	0.1	0.5	0.85

(iv) β'_d = a coefficient based on cross-sectional dimension (III.4e)

Effective thickness, *d* (cm):	100	200	300	400	500
β'_d:	1.0	0.85	0.75	0.71	0.7

(v) β_{td} = a coefficient based on theoretical thickness versus age of concrete at loading (Table III.2) (III.4f)

III.2.2.5 Hansen's formula as expressed by Nasser and Neville [81, 216]

$$\frac{C}{\sigma}=\frac{B(0.31h_K+W/C)V_1}{(NK_1+0.31)h_K}(1-e^{-mt})+\alpha\frac{W}{C}V_{cp}\log_e\frac{t+t'}{t'} \qquad \text{(III.5)}$$

where

C = creep value

σ = applied stress

W/C = water/cement ratio by weight (correction made for bleeding)

V_{cp} = volume concentration of cement paste

$$=\frac{1}{27}\left[\frac{\text{weight of cement}}{\rho_{cm}\times\rho_w}+\frac{\text{weight of water}}{\rho_w}\right]$$

(ρ_w = specific gravity of water, ρ_{cm} = specific gravity of cement)

h_K = degree of hydration for loading

$$K_1 = \frac{\text{weight of non-evaporable water}}{\text{weight of cement at full hydration}}$$

t' = aggregate loading

$N = 0.75(1 + 4t')$

t = time under load

Hansen reported the values of α, B and m to be

$$\alpha = 5.7 \times 10^{-6} \qquad B = 1.76 \times 10^{-6} \qquad m = 0.33$$

The first right-hand term in eqn (III.5) is the term for delayed elastic deformation and the second right-hand term is the term for viscous (long-term) deformation. The value of $C \times 10^{-6}$ gives creep in inches per inch, and this is then converted to SI units.

III.2.2.6 The Dibsi strain hardening model [147]

The creep rate at any instant of time is defined as

$$\mathrm{d}\varepsilon_c/\mathrm{d}t = f(\bar{\sigma}, T, \varepsilon_c) \tag{III.6}$$

where $\mathrm{d}\varepsilon_c/\mathrm{d}t$ is the creep rate

$$\bar{\sigma} = \text{effective stress} \qquad \varepsilon_c = \text{thermal creep} = (1/10)\sigma_t^{0.31}F(T) \tag{III.6a}$$

$$F(T) = \left\{1 \pm \left[\frac{125 - T}{1000}(t)\right]^{0.6}\right\} \tag{III.6b}$$

The creep rate at any instant of time is written as (time hardening)

$$\mathrm{d}\varepsilon_c/\mathrm{d}t = 0.31\sigma_t^{-0.61}F(T) \tag{III.7}$$

In terms of strain hardening, eqn (III.7) is written as

$$\mathrm{d}\varepsilon_c/\mathrm{d}t = 0.31[\sigma F(T)]^{3.225}\varepsilon_c^{-2.225} \tag{III.8}$$

Equation (III.8) can be used successfully in the finite element creep analysis of structures.

III.2.2.7 The Zienkiewicz and Watson creep formulation [87]

England and Phok [86], for variable stress conditions, suggested a strain rate equation for total strain including elastic and creep. Zienkiewicz and Watson [87] give a modified version of the equation of England and Phok [86] in which the value of E is considered as a function of time and temperature, and is written as

$$\varepsilon_s = \phi(T, t')\log[1 + (t - t')] + (1/E)(T, t') \tag{III.9}$$

where

$$\varepsilon_s = \text{specific strains}$$
$$\phi(T, t') = \text{constant depending upon time and temperature}$$
$$t' = \text{age of loading}$$
$$t = \text{time at which strains are to be considered}$$
$$T = \text{temperature}$$

The last term in eqn (III.9) is the elastic component. The specific value of the creep strain increment can be written in terms of elastic and creep strains:

$$\{\varepsilon_{0,c}\} = [D_0]^{-1}\left[\sum_{t=0}^{t=t_n} \frac{1}{E(T,t')} \frac{\Delta\{\sigma\}}{\Delta t'} \Delta t'\right] \quad \text{(elastic strains)}$$
$$+ \sum_{t=0}^{t=t_n} \phi(T,t') \log[1 + (t - t')] \frac{\Delta\{\sigma\}}{\Delta t'} \Delta t' \quad \text{(creep strains)}$$

$$[D_0]\{\varepsilon_c\}_1 = \Delta\{\sigma_0\} \log(1 + t_1 - t'_0) \tag{III.10a}$$

$$\begin{aligned}[D_0]\{\varepsilon_c\}_2 &= \Delta\{\sigma_0\}\phi(T_1, t'_0) \log(1 + t_2 - t'_0) \\ &\quad + \Delta\{\sigma_1\}\phi(T_2, t'_1) \log(1 + t_2 - t'_1)\end{aligned} \tag{III.10b}$$

$$\begin{aligned}[D_0]\{\varepsilon_c\}_{n-1} &= \Delta\{\sigma_0\}\phi(T_{n-1}, t'_0) \log(1 + t_{n-1} - t'_0) \\ &\quad + \Delta\{\sigma_1\}\phi(T_{n-1}, t'_1) \log(1 + t_{n-1} - t'_1) \\ &\quad + \Delta\{\sigma_{n-2}\}\phi(T_{n-2}, t'_{n-2}) \log(1 + t_{n-1} - t'_{n-2})\end{aligned} \tag{III.10c}$$

$$\begin{aligned}[D_0]\{\varepsilon_c\}_n &= \Delta\{\sigma_0\}\phi(T_n, t'_0) \log(1 + t_n - t'_0) \\ &\quad + \Delta\{\sigma_1\}\phi(T_n, t'_1) \log(1 + t_n - t'_1) \\ &\quad + \Delta\{\sigma_{n-1}\}\phi(T_{n-1}, t'_{n-1}) \log(1 + t_n - t'_{n-1})\end{aligned} \tag{III.10d}$$

Thus any specific value of creep strain increment can be computed from eqns (III.10c) and (III.10d) as

$$\Delta\{\varepsilon_c\}_n = \{\varepsilon_c\}_n - \{\varepsilon_c\}_{n-1} \tag{III.11}$$

The time intervals between $t_1, \ldots, t_n$ may be chosen quite arbitrarily.

III.2.2.8 The Bangash incremental creep formulation

The creep strain increment $\Delta\varepsilon_c$ is written in the form

$$\phi_1(\Delta\varepsilon_c)\phi'(t, t') \tag{III.12}$$

where ϕ' is a function which determines whether time hardening, strain hardening or strain dependent relations, respectively, are being used; and ϕ_1 is a standard function defining stress.

Two forms of eqn (III.12) are given below:
Time hardening:

$$\begin{aligned}
\frac{\Delta\varepsilon_{cx}}{\Delta t} &= \bar{L}(\sigma_x - \sigma_y) + (\sigma_x - \sigma_z) \quad & \frac{\Delta\varepsilon_{xy}}{\Delta t} &= \bar{L} \times 2\tau_{xy} \\
\frac{\Delta\varepsilon_{cy}}{\Delta t} &= \bar{L}(\sigma_y - \sigma_x) + (\sigma_y - \sigma_x) \quad & \frac{\Delta\varepsilon_{xz}}{\Delta t} &= \bar{L} \times 2\tau_{xz} \\
\frac{\Delta\varepsilon_{cz}}{\Delta t} &= \bar{L}(\sigma_z - \sigma_y) + (\sigma_z - \sigma_x) \quad & \frac{\Delta\varepsilon_{zy}}{\Delta t} &= \bar{L} \times 2\tau_{zy}
\end{aligned} \tag{III.13}$$

where $\bar{L} = \frac{1}{2}\phi_1[(\sigma_{eq})/\sigma_{eq}]\phi'(t)\phi(t')$, and where equivalent stress σ_{eq} is defined by

$$\begin{aligned}
\sigma_{eq} &= \sigma_{cu} + \sqrt{3\tau^2} \\
\sigma_{cu} &= f_{cu} = \sqrt{(\sigma_x - \sigma_z)^2 + (\sigma_x - \sigma_y)^2 + (\sigma_y - \sigma_z)^2}
\end{aligned} \tag{III.13a}$$

If σ_{eq} and t' are independent of time then

$$\sigma_x = \sigma_z = \sigma_{xy} = \sigma_{zx}$$

Equation (III.13) reduces to eqn (III.12) since $\varepsilon_c t = 0$, and hence the field of equivalent creep strain can be written as

$$\begin{aligned}
\varepsilon_c = \sum_0^{t_n} \Bigg\{ \frac{9}{2} \Bigg[\left(\frac{\Delta\varepsilon_{cx}}{\Delta t} - \frac{\Delta\varepsilon_{cy}}{\Delta t} \right)^2 + \left(\frac{\Delta\varepsilon_{cx}}{\Delta t} - \frac{\Delta\varepsilon_{cz}}{\Delta t} \right)^2 \\
+ \left(\frac{\Delta\varepsilon_{cz}}{\Delta t} - \frac{\Delta\varepsilon_{cy}}{\Delta t} \right)^2 + \frac{1}{2} \left(\frac{\Delta\varepsilon_{cx}}{\Delta t} + \frac{\Delta\varepsilon}{\Delta t} + \frac{\Delta\varepsilon_{cz}}{\Delta t} \right) \Bigg] \Bigg\}
\end{aligned} \tag{III.14}$$

Strain hardening:
The term $\phi'(t)$ in eqn (III.13) is replaced by $\phi'(t'')$ such that

$$\frac{\Delta\varepsilon_{cx}}{\Delta t} = \frac{1}{2} \frac{\phi'(\sigma_{eq})}{\sigma_{eq}} \phi'(t'')\phi(t')[(\sigma_x - \sigma_y) + (\sigma_x - \sigma_z)], \text{ etc.} \tag{III.15}$$

The equivalent time t'' is calculated from the creep strain

$$\phi(t'') = \varepsilon_c / \phi_1(\sigma_{eq})\phi(\sigma) \tag{III.16}$$

For a large range of stresses a suitable creep strain function $\varepsilon_{c_{eq}}$ can be chosen from any creep model described above.

III.2.2.9 The three-dimensional thermal creep model

Bangash and England [1004] have presented a three-dimensional formulation for creep which allows independent variation of irreversible and delayed recoverable strains with respect to temperature and time. Simultaneous adoption of a pseudo-time concept simplifies the mathematical form of the modelling and allows the four-element Burger representation to be employed in pseudo-time. When the loads are sustained for a long time, the solution of the equation can be governed by the Maxwell elements of the model at elevated temperatures with small creep recovery.

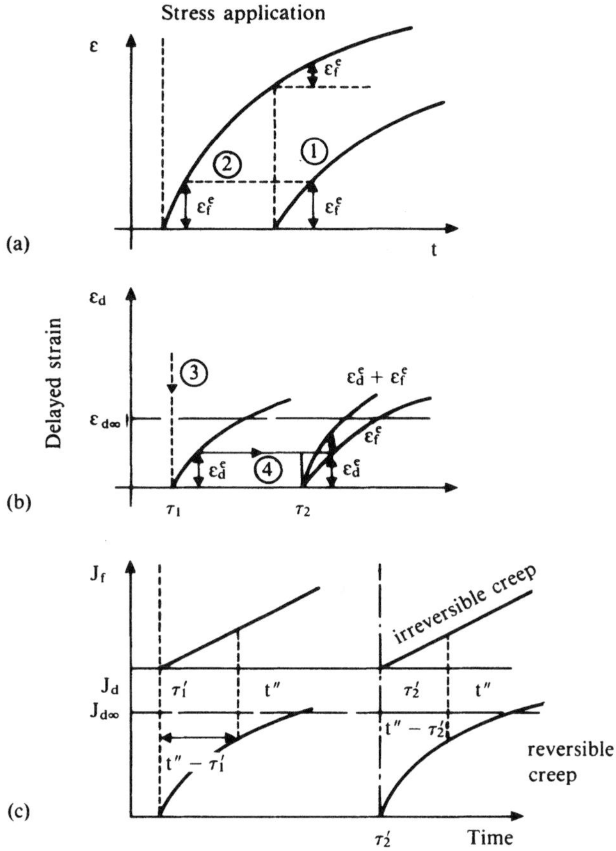

Fig. III.5. Creep components. (a) Influence of age at loading on flow component of strain; (b) creep strain at different age of loading; (c) component of creep strain in pseudo-time.

The constitutive relation for concrete under uniaxial stress is established by using the rate of flow method. The total time dependent strain in concrete may be written (Fig. III.5) as

$$\varepsilon(t) = \varepsilon_{\mathrm{e}}(t) + \varepsilon_{\mathrm{f}}^{\mathrm{e}}(t) + \varepsilon_{\mathrm{d}}^{\mathrm{e}}(t)$$

$$= \frac{\sigma(t')}{E(t')} + \int_0^{t'} \frac{\partial \sigma(\tau)}{\partial \tau} J_{\mathrm{f}}(t'-\tau)\,\mathrm{d}\tau + \int_0^{t'} \frac{\partial \sigma(\tau)}{\partial \tau} J_{\mathrm{d}}(t'-\tau)\,\mathrm{d}\tau \qquad \text{(III.17)}$$

where

$\varepsilon'(t)$ = total strain

$\varepsilon_{\mathrm{f}}(t')$ = irreversible or flow component of creep strain

$$= \sum_0^{t'} J_{\mathrm{f}}(t'-\tau)\Delta\sigma^*$$

$\varepsilon_e(t')$ = elastic strain

$\varepsilon_d(t)$ = reversible or delayed elastic component of creep strain

$$= \sum_0^{t'} J_d(t' - \tau)\Delta\sigma^*$$

$J_f(t')$ = specific flow (flow strain per unit of stress)

$J_d(t')$ = delayed elastic strain per unit of stress

t' = age of concrete

τ = time under load

$\Delta\sigma^*$ represents incremental stress

Creep compliance for concrete is written as

$$J(t', \tau) = (1/E) + J_f(t' - \tau) + J_d(t' - \tau) \tag{III.18}$$

A parameter is introduced at this stage known as pseudo-time, t'', which itself is a specific flow component; and this may be used in place of actual time, t', together with the representation of a non-ageing visco-elastic material. The pseudo-time concept transforms the age dependent creep relationship in real time to the simpler non-ageing Maxwell law in pseudo-time. The time transformation eases the analytical or numerical computation which leads to a solution without changing the basic creep equation. Now, with this transformation, J_f may be written as

$$J_f(t'' - \tau') = t'' - \tau' \tag{III.19}$$

and also the delayed elastic strain:

$$J_d(t'' - \tau') = J_{d\alpha_i}[1 - e^{-\beta(t'' - \tau')}] \tag{III.20}$$

When a Maxwell unit is connected in series with a Kelvin unit (Fig. III.6), a Burger model is obtained. In this model the dashpot and the Kelvin unit correspond to flow and delayed elastic component. Equation (III.18) is written as

$$J(t'' - \tau') = (1/E) + (t'' - \tau') + J_{d\alpha_i}[1 - e^{-\beta(t'' - \tau')}] \tag{III.21}$$

The components of creep strains in pseudo-time having special provisions for reversible and irreversible creep effects can now be written as

$$J_d(t'') = \sum_{i=1}^{n} J_{d\alpha_i}(1 - e^{-J_f/Q_i}) = Q[1 - e^{-\beta(t'' - \tau')}]\ \mathrm{mm^2/N} \tag{III.22}$$

The incremental flow component is written as

$$\Delta J_f(t'') = \Delta t''\rho(T) = [(t'' + \Delta t'') - t'']\rho(T) \tag{III.23}$$

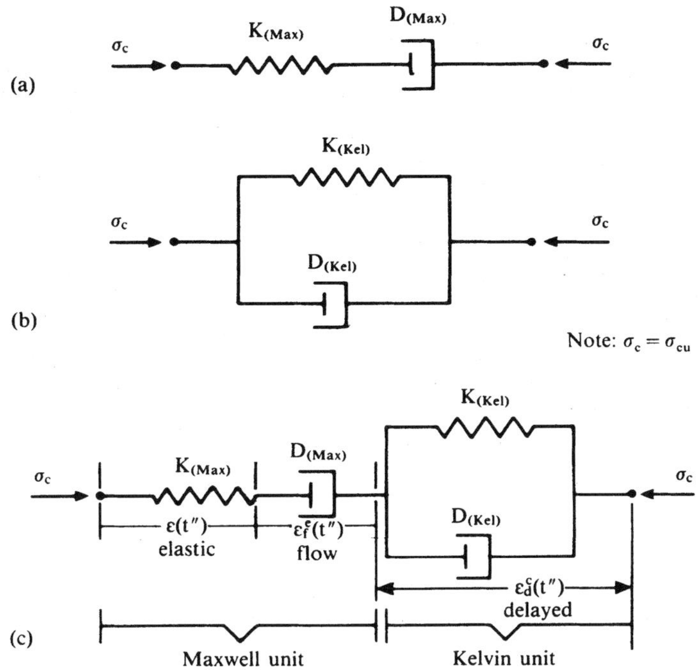

Fig. III.6. Visco-elastic models for concrete creep. (a) Model A — the Maxwell model of creep strain (flow component); (b) Model B — the Kelvin model of creep strain (delayed elastic component); (c) Model C — the Burger model (models A + B).

where

$\Delta t''$ = incremental pseudo-time = 1/50

$\rho(T)$ = temperature function = $1/50(T + 30^\circ)\rho(T_\phi)$

T_ϕ = temperature at ambient level (normally 20°C)

The incremental delayed elastic strain component is given by

$$\Delta J_d(t'') = [J_d(t'' + \Delta t'') - J_d(t'')] = (1 - e^{-\beta\Delta t''})[Q - J_d(t'')] \tag{III.24}$$

For limestone concrete

$$J_f = t'' = [-15.37346 + 7.88977 \log_e(t')] \times 10^{-6} \text{ per N/mm}^2 \tag{III.25}$$

$$Q = 14.21457 \times 10^{-6} \text{ per N/mm}^2, \quad \beta = 0.2813578 \times 10^{6} \text{ N/mm}^2 \tag{III.26}$$

Young's modulus is

$$E(T) = E_{20}\left[1 - \frac{T - 20^\circ}{137}\right] \quad \text{for } 20^\circ \le T \le 50^\circ$$
$$E(T) = E_{20}(0.78)\left[1 - \frac{T - 50^\circ}{341}\right] \quad \text{for } 50^\circ \le T \le 85^\circ \tag{III.27}$$

This concept is used for many problems solved in this text. Further explanations are given later on in the finite element formulations (Chapters VI and VII).

III.3 Shrinkage

Shrinkage is considered to be a time dependent deformation in concrete which occurs due to the movement of the gel water from within the gel pores. The evaporation process of the drying out of the concrete is sufficient to cause such shrinkage strains. Most of the time, shrinkage due to drying out and creep due to stress occur in the same concrete. As moisture seepage takes place in the case of both shrinkage and creep, it is difficult to observe the two cases separately within the same concrete specimen. The type of cement definitely has an influence on the shrinkage of concrete. Although the function of shrinkage is to reduce the volume of the unloaded concrete at constant temperature, it induces stresses acting over extended time periods. These stresses are generally reduced by creep. Hence this is the reason why the shrinkage phenomenon is taken into account with the problem of creep.

III.3.1 Shrinkage models

III.3.1.1 The Hilsdorf model [218]

Hilsdorf found that as the water/cement ratio increases and the relative humidity drops, shrinkage increases. His relationship can be expressed as

$$\varepsilon_{\text{sh}_\infty} = \beta_1 p \sqrt{\frac{100 - H}{100}} \tag{III.28}$$

where

$\varepsilon_{\text{sh}_\infty}$ = final shrinkage of concrete
β_1 = proportionality coefficient
p = total porosity
H = relative humidity

III.3.1.2 The ACI 209 method [219]

The final shrinkage strain is estimated to be

$$\varepsilon_{\text{sh}_\infty} = 780 \times 10^{-6}(\beta^{\text{s}}_{cp}\beta^{\text{s}}_{H}\beta^{\text{s}}_{d}\beta^{\text{s}}_{S}\beta^{\text{s}}_{F}\beta^{\text{s}}_{ce}\beta^{\text{s}}_{AC}) \tag{III.29}$$

(i) β^{s}_{cp} = a coefficient for curing periods different from seven days (III.29a)

Days:	1	3	7	14	28	90
β^{s}_{cp}:	1.2	1.1	1.0	0.93	0.86	0.75

(ii) β^{s}_{H} = a coefficient to estimate the contribution from the relative humidity (III.29b)

$$\beta^{\text{s}}_{H} = 1.40 - 0.01H \qquad 40 \le H \le 80\%$$
$$\beta^{\text{s}}_{H} = 3.0 - 0.03H \qquad 80 \le H \le 100\%$$

(iii) β_d^s = a coefficient for the average thickness of the concrete member (III.29c)

when $50 \leq d \leq 150$ (all mm):

d (mm):	50	75	100	125	150
β_d^s:	1.35	1.25	1.17	1.08	1.0

when $150 \leq d \leq 375$ (all mm):

$$\beta_d = 1.23 - 5.70 \times 10^{-8} d^2 \quad \text{for } t' - t_1 \leq 365 \text{ days}$$
$$\beta_d = 1.17 - 3.19 \times 10^{-5} d^2 \quad \text{for } t' - t_1 \leq 365 \text{ days}$$

when $d > 375\,\text{mm}$:

$$\beta_d^s = 1.2e^{-0.047V/A}(1.2e^{-0.12V/A})$$

where V/A is the volume surface area ratio.

(iv) β_S^s = a coefficient for concrete consistency (III.29d)

$$\beta_S^s = 0.89 + 8.56 \times 10^{-5} S$$

where S is the slump of concrete.

(v) β_F^s = a coefficient for the content of fine aggregates (III.29e)

$$\beta_F^s = 0.30 + 14 \times 10^{-4} F \qquad F \leq 50\%$$
$$\beta_F^s = 0.90 + 2 \times 10^{-3} F \qquad F > 50\%$$

where F is the fine aggregate content (<4.8 mm).

(vi) β_{ce}^s = a coefficient to take into account the cement content (III.29f)

$$\beta_{ce}^s = 0.75 + 2.196 \times 10^{-7} {}^\circ\text{C}$$

(vii) β_{AC}^s = a coefficient for estimating air content (III.29g)

$$\beta_{AC}^s = 0.95 + 8 \times 10^{-3} \qquad A \geq 1.0$$

where A is the air content by volume in %.

The development of shrinkage over time is also generally expressed as

$$\varepsilon_{st} = \frac{(t - t')}{35 + (t - t')} \varepsilon_{sh_\infty} \tag{III.29h}$$

III.3.1.3 The CEB–FIP 70 method [220]

This code proposes prediction of shrinkage by

$$\varepsilon_{sh_\infty} = \beta_H^{s'} \beta_{t'}^{s'} \beta_d^{s'} \beta_p^{s'} \beta_{td}^{s'} \tag{III.30}$$

(i) $\beta_H^{s'}$ = a coefficient for relative humidity (III.30a)

H:	40	50	60	70	80	90	100
$\beta_H^s \times 10^{-6}$:	410	380	320	270	199	100	0

Table III.3. $\beta^{s}_{t'}$ values for different water/cement ratios and unit weights

Water/cement ratio	Unit weight (kg/m^3)			
	500	400	300	200
0.3	–	–	–	–
0.4	1.0	0.80	–	–
0.5	1.43	1.20	0.90	–
0.6	–	1.60	1.19	0.70
0.7	–	–	1.45	1.0
0.8	–	–	1.80	1.25

Table III.4. $\beta^{s'}_{td}$ values for different ages and values of d

d (mm)	Age (days)				
	1	10	100	1000	10 000
50	0.17	0.4	0.80	0.90	–
100	–	0.26	0.66	0.88	–
200	–	0.10	0.43	0.85	–
400	–	–	0.27	0.78	–
800	–	–	0.08	0.50	0.87

(ii) $\beta^{s'}_{t'}$ represents the age of concrete at loading (Table III.3) (III.30b)

(iii) $\beta^{s'}_{d}$ = a coefficient based on cross-sectional dimensions (III.30c)

d (cm):	100	200	300	400	500
$\beta^{s'}_{d}$:	1.10	0.80	0.65	0.55	0.45

(iv) $\beta^{s'}_{p}$ = a coefficient based on longitudinal reinforcement ($p = 100A_{st}/A_c$) (III.30d)

p:	0	1	2	3	4	5
$\beta^{s'}_{p}$:	1.0	0.85	0.70	0.62	0.3	–

(v) $\beta^{s'}_{td}$ = a coefficient based on theoretical thickness versus the age of concrete at loading (Table III.4) (III.30e)

III.4 Temperature

Detailed discussions have been given earlier of the influence of temperature on the material properties of concrete. Bangash [221] has given a comprehensive analysis of movements caused by creep shrinkage and temperature. Temperature analysis using finite elements is given later in this book. Prior to such an analysis, it is essential to know the influence of various conditions on α_T, the thermal coefficient of expansion of concrete. The thermal coefficient of expansion varies with moisture content. It decreases with age, but it increases with temperature and richness of mix. Hatt [229] carried out tests on oven-dried gravel concrete and evaluated coefficients of $7.2 \times 10^{-6}/°C$ at 15.6°C and $11.7 \times 10^{-6}/°C$ at 66°C. Very little difference in these values was found for saturated and partially dried samples chosen for this concrete. Davis [225] carried out high- and low-temperature tests on 100 samples within the range of −9°C to +54°C and discovered a decrease in the coefficient of about 14%. Mitchell [226] and Walker *et al.* [230] reported some improvement of the thermal coefficient by using air-entraining admixtures. Berwanger [231] suggested that within a temperature range of −40°C to +27°C the coefficient reduces at an estimated rate of $0.02 \times 10^{-6}/°C/day$ for specimens at between 28 to 200 days. After 200 days no specific changes occur in the magnitude of the thermal coefficient. Meyers [227] reported that the maximum coefficient of limestone concrete for partially dry specimens is 56% higher than for saturated specimens. Many others [231–234] have reported a 15% increase in the coefficient of the specimen in the partially dry state compared with in the saturated state. Emanuel and Huisey [232] have reported, on the basis of extensive tests, an empirical formula which takes into consideration moisture content, age and mix type:

$$\alpha_T = f_T(f_M f_A B_P \alpha_s + B_{FA}\alpha_{FA} + B_{CA}\alpha_{CA}) \qquad \text{(III.31)}$$

where

f_T = a correction factor for temperature alterations (1.0 for a controlled environment; 0.86 for exposed conditions)

f_M = a correction factor for moisture

moisture content (%):	0	20	40	60	80	100
f_M:	1.2	1.2	1.5	1.8	1.6	1.0

f_A = a correction factor for age

moisture content (%):	0	20	40	60	80	90	100
f_A (0–6 months):	1.0	1.0	1.0	1.0	1.0	1.0	1.0
f_A (12–18 months):	1.0	1.0	0.99	0.90	0.97	1.0	–
f_A (old):	1.0	1.0	0.96	0.80	0.78	1.0	–

B_P = proportion by volume of paste

B_{FA} = proportion by volume of fine aggregate

B_{CA} = proportion by volume of coarse aggregate

α_s = thermal coefficient of expansion for saturated cement paste

$= 10.8 \times 10^{-6}/°C$ for evaluating B_P

α_{FA} = thermal coefficient of expansion for fine aggregate

α_{CA} = thermal coefficient of expansion for coarse aggregate

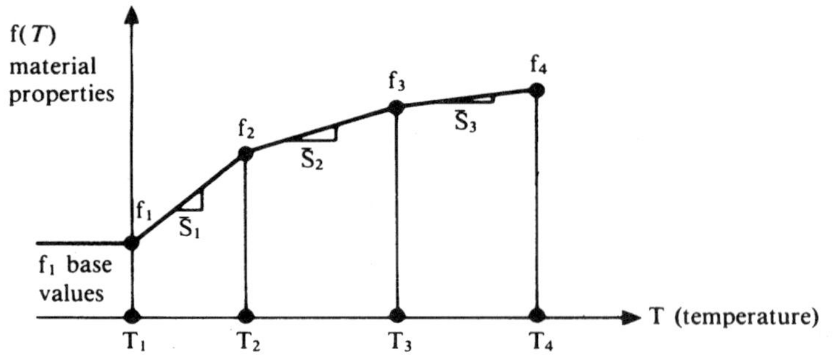

Slope	*Break point*	*Function*	*Temperature*
$\bar{S}_1 = (f_2 - f_1)/(T_2 - T_1)$	T_1	f_1	T_1
$\bar{S}_2 = (f_3 - f_2)/(T_3 - T_2)$	T_2	f_2	T_2
$\bar{S}_3 = (f_4 - f_3)/(T_4 - T_3)$	T_3	f_3	T_3
		f_4	T_4

Fig. III.7. Piecewise linear representation of temperature-dependent material properties.

The strains are given by

$$(\mathrm{d}\varepsilon_{\mathrm{T}})_{ij} = (\alpha_{\mathrm{T}})_{ij}\,\mathrm{d}T \tag{III.32}$$

For isotropic cases eqn (III.32) is written as

$$(\mathrm{d}\varepsilon_{\mathrm{T}})_{ii} = \alpha_{\mathrm{T}}\,\mathrm{d}T = \left[\bar{\alpha}_{\mathrm{T}} + \frac{\mathrm{d}\alpha_{\mathrm{T}}}{\mathrm{d}T}(T - T_0)\right]\mathrm{d}T \tag{III.33}$$

where T_0 is the initial temperature.

Terms inside the brackets of eqn (III.33) can be modelled using piecewise linear slopes and breaking points representing the temperature dependent material properties of concrete. This is shown in Fig. III.7. For details concerning heat conduction and thermal analysis of concrete, additional material can be found in Chapters VI and VII.

IV. Numerical modelling of bond and bond–slip

IV.1 Introduction

A great deal of work has been carried out on the bond between concrete and steel. It remains one of the most difficult topics in the study of concrete, and is still not properly understood. This is due to a number of theoretical and experimental problems encountered in the correct assessment of bond. Nevertheless it is understood that the bond between concrete and steel is due to the combination of adhesion, bearing action and friction. For a small slip there is an adhesion surface between the concrete and steel, and the resistance to slip is given by the shear strength of the fine particles of concrete filling the indentation of the surface of reinforcement. The value of the adhesion is zero when slip takes place. Bearing action is caused by the deformation of the reinforcing bars or cables in contact with concrete. Slip takes place when the concrete surrounding the reinforcement bars or cables is crushed. At the failure of the structure, bond generally depends on the friction at the failure surface. This friction is destroyed when large slip occurs. The bond–slip relationship is vital for bonded structures subjected to incremental and ultimate loads.

The variables which might affect bond between concrete and steel are as follows.

(a) Concrete mix, temperature and humidity.
(b) The age and tensile/compressive strength at test.
(c) Bar or cable type, and the embedment length and depth.
(d) The loading speed and repeated loading.

The most commonly adopted methods of testing bond are the pull-out test and the tensile test. In the pull-out test, the steel in tension is in contact with concrete in compression. In this way meaningful results can be obtained for different types of bar or different concrete mixes and for the transfer zone of prestressing cables or tendons.

This chapter gives several numerical models for bond. A detailed bond–slip model is given which suits the finite element analysis presented in this text. The references given for bond [238–274] will help the reader understand details not covered in this book.

IV.2 Simple models for bond between steel and concrete

IV.2.1 BS 8110 model for bond

Two types of bond, namely anchorage and local bond, are recognised. Both plain and deformed bars are considered. In cases of a simple pull-out, the anchorage bond stress is given by

$$\sigma_b = \sigma_s \phi / 4L \tag{IV.1}$$

where σ_b = bond stress; σ_s = stress in the reinforcing bar; ϕ = the diameter of the bar; and L = the length of the embedded bar, or the development length.

The bond resistance at ultimate strength of the bar is computed as

$$L = 0.87\sigma_{yt}\pi\phi^2/4/\sigma_b\pi\phi L \tag{IV.2}$$

where σ_{yt} is the yield strength of bar.

In the case of flexural and local bond, the code bases its conclusions on the ultimate shear. For a beam of uniform depth, the bond stress is given by

$$\sigma_b = V/d\sum O \tag{IV.3}$$

where V = the ultimate shear; d = the depth of the beam; and $\sum O$ = the sum of the effective parameters of the bars in tension.

BS 8110 suggests that ultimate local bond must not exceed the following values when no slip is considered:

Plain bars	Grade 25 concrete	2.0	(IV.3a)
	Grade 30 concrete	2.2	
Deformed bars	Grade 25 concrete	2.5	
	Grade 30 concrete	2.8	

IV.2.2 Brice, Saliger and Wastlund models

Figure IV.1 shows distributions of steel and concrete and bond stress distributions. It is assumed that two consecutive cracks occur and that the section in between has a constant bending moment. At each crack, the tensile stress in concrete $\sigma_t = 0$, and the maximum stress in the bars has a magnitude σ_{yt}. Away from the cracks, the bond between concrete and steel is introduced, whereby the bars transfer an increasing proportion of the total tensile force to the surrounding concrete. The maximum stress σ_{cu} in concrete reaches $x = L/2$ when the cracking takes place, if and when σ_t has a specific value σ_{tu}.

At a section x, the bond between the steel and concrete is given by

$$\sum O \int_0^x \sigma f(x/L)\,dx \tag{IV.4}$$

where

$$\sigma_{bx} = \sigma_b f(x/L) \tag{IV.4a}$$

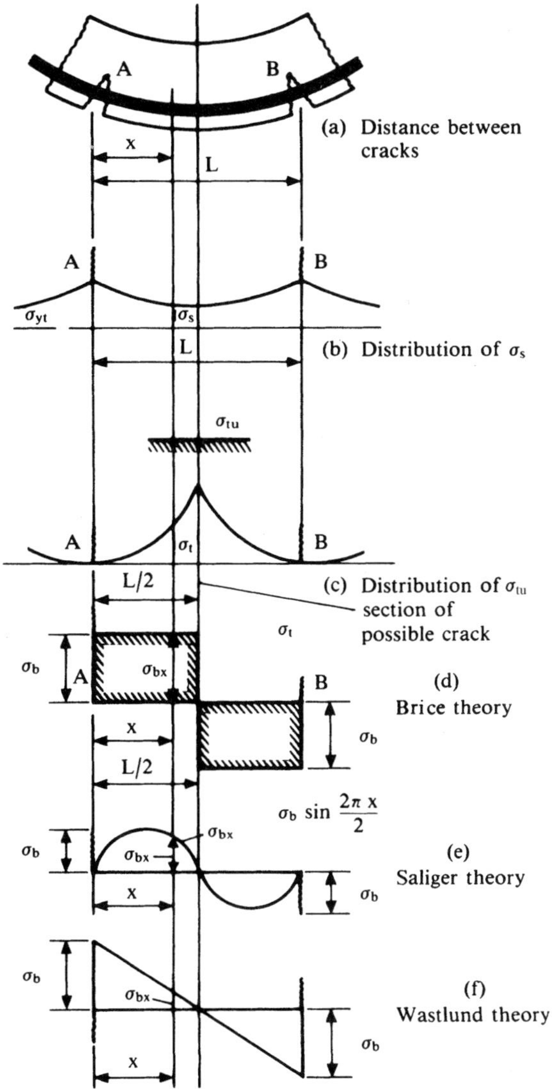

Fig. IV.I. Distribution of various stresses between cracks — based on different theories.

$$\sigma_t, \sigma_s = \sigma_{yt} - (p/A)\sigma_b \int_0^x f(x/L)\,dx \tag{IV.5}$$

$$M_x = \text{moment at } x = \sigma_x A_s z + \sigma_t z_c \tag{IV.6}$$

From eqn (IV.5)

$$\sigma_t = (p\sigma_b z/z_c) \int_0^x f(x/L)\,dx \tag{IV.7}$$

where z_c is the section modulus; and z is the lever arm.

At $L/2$, if $\sigma_t = \sigma_{tu}$ then the developing length L is given by

$$L = B_l \frac{1}{p} \frac{z_c}{z} \frac{\sigma_{tu}}{\sigma_b} \tag{IV.8}$$

where

$$\frac{1}{B_l} = \int_0^{1/2} f(x/L)d(x/L)$$

(a) When $B_l = 2$ and $f(x/L) = 1$, the well-known Brice Theory [239] emerges

$$\sigma_{bx} = \sigma_b = \frac{2K_B}{1 + 3\dfrac{\Sigma\phi}{\Sigma d}} \tag{IV.9}$$

where $K_B = 1.6$; $\phi =$ the diameter of the bars; and $d =$ the thickness of the concrete crossing the line of possible cracking, encountering the largest number of bars.

(b) When $B = \pi$ for $f(x/L) = 1$, Saliger's expression [240] is written as

$$\sigma_{bx} = \sigma_b \sin\frac{2\pi x}{L} \tag{IV.10}$$

For flexure

$$f_c'/4.4\,\text{kg/cm}^2 \tag{IV.11}$$

and for tension

$$\sigma_{tu} = f_c'/12$$

For deformed bars $\sigma_{ca} = \sigma_{tu}$ ranges between $f_c'/3$ and $f_c'/4$.

The corresponding crack width w is given by

$$w = \frac{\pi}{4} \frac{\sigma_{yt}}{E_c} \frac{\sigma_{tu}}{\sigma_b} \frac{\phi}{A_s} bd \tag{IV.12}$$

where $b =$ width and $d =$ depth of the tension zone.

(c) When $\beta_l = 4$ for $f(x/L) = 1$, it gives the equation of Wastlund [cited in Ref. 241]. The distribution of bond stress is given by

$$\sigma_{bx} = \left(1 - \frac{2x}{L}\right) K\sigma_{tu}\sqrt{\frac{w}{\phi}} \tag{IV.13}$$

$$\sigma_b = K\sigma_{tu}\sqrt{\frac{w}{\phi}} \tag{IV.13a}$$

where $\sigma_{tu} = f_{cu}/10$; K is for high tensile steels; and $\sqrt{\phi/w} \simeq 14$.

IV.3 Somayaji and Shah [242] model

This model considers [242, 243] the relationship between σ_{bx} and the second derivative of the local slip S_x. An exponential bond stress distribution function is assumed as well. The value of S_x

$$S_x = Ae^x + Be^{-x} + C\frac{x^2}{2} + Dx + E \tag{IV.14}$$

where A–E are constants

$$A = (1 - e^{-L_t})\varepsilon_s^a/\bar{S}_D$$
$$B = (e^{L_t} - 1)\varepsilon_s^a/\bar{S}_D$$
$$C = (e^{-L_t} - e^{L_t})\varepsilon_s^a/\bar{S}_D$$
$$D = (e^{-L_t} + e^{L_t} - L_t e^{-L_t} + L_t e^{L_t} + 2)\varepsilon_s^a/\bar{S}_D$$
$$E = \varepsilon_s^a[-e^{L/2}(1 - e^{-L_t}) - e^{-L/2}(e^{L_t} - 1)L^2/8(e^{-L_t} - e^{L_t}) - L/2 \times (e^{-L_t} - e^{L_t} - L_t e^{-L_t} + L_t e^{L_t} + 2)]/\bar{S}_D$$
$$\bar{S}_D = (2e^{L_t} + 2e^{-L_t} + L_t e^{-L_t} - L_t e^{L_t} - 4)$$
$$\varepsilon_s^a = \text{applied strain}$$
$$L_t = \text{transfer length} \tag{IV.15}$$

Figure IV.2 shows typical distributions. The value of

$$P_{\text{tran}} = (P/1 + \bar{\alpha}_e p) \tag{IV.16}$$

where P is the applied load, $\bar{\alpha}_e$ the modular ratio and p the reinforcement ratio.

The transmission length given by Hoyer's method [245] is given by

$$L_t = \frac{\phi}{2\mu}(1 + \nu_c)\left(\frac{\bar{\alpha}_e}{\nu_s} - \frac{f_{pi}}{E_c}\right)\left(\frac{f_{pe}}{2f_{pi} - f_{pe}}\right) \tag{IV.17}$$

where μ is the coefficient of friction between steel and concrete; ν_c, ν_s = Poisson's ratio for concrete and steel, respectively; f_{pi} = initial prestress; and f_{pe} = effective prestress in steel.

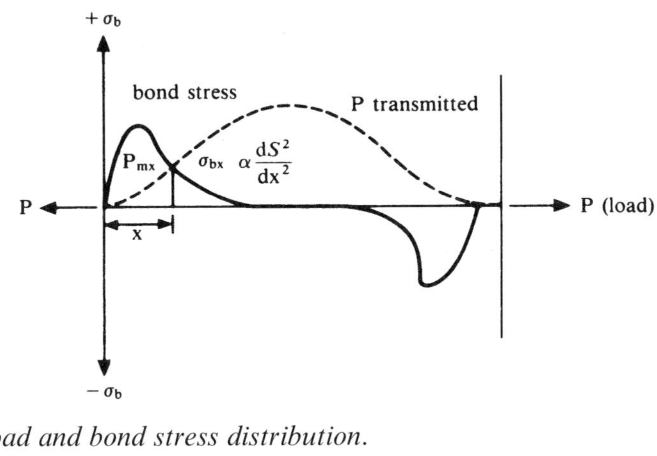

Fig. IV.2. Load and bond stress distribution.

IV.4 Marshall and Krishnamurthy model

Marshall and Krishnamurthy [244] have predicted the transmission length for the prestressing cable as

$$L_t = \sqrt{f_{cu} \times 10^3/\bar{\beta}} \tag{IV.18}$$

where $\bar{\beta}$ is constant depending upon the strand and wire

wire/strand (mm)	$\bar{\beta}$
2 mm wire	0.144
5 mm wire	0.0235
7 mm wire	0.0174
10 mm/7 wire strand	0.144
12.5 mm/19 wire strand	0.058
19 mm/7 wire strand	0.0235

C. G. Marshall [246] and W. T. Marshall [247] then propose a bond stress formula

$$\sigma_{bx} = \sigma_{b_{max}} e^{-4\psi_x/\phi} \tag{IV.19}$$

where σ_{bx} is the bond stress at a distance x from the free end; ψ = constant $\approx 0.007\,25$; $\sigma_{b_{max}} \not> 7.42\ \mathrm{N/mm^2}$ for $f_{cu} = 80\ \mathrm{N/mm^2}$; and ϕ is the diameter of wire.

IV.5 Bond and bond–slip models

Bangash reported [248] the use of finite element analysis for bond and bond–slip conditions. Bonded tendons in concrete vessels were analysed under incremental loads. Figures IV.3–IV.4b show comparisons between the behaviour of bonded and unbonded prestressing tendons in concrete vessels. Similar work was done by

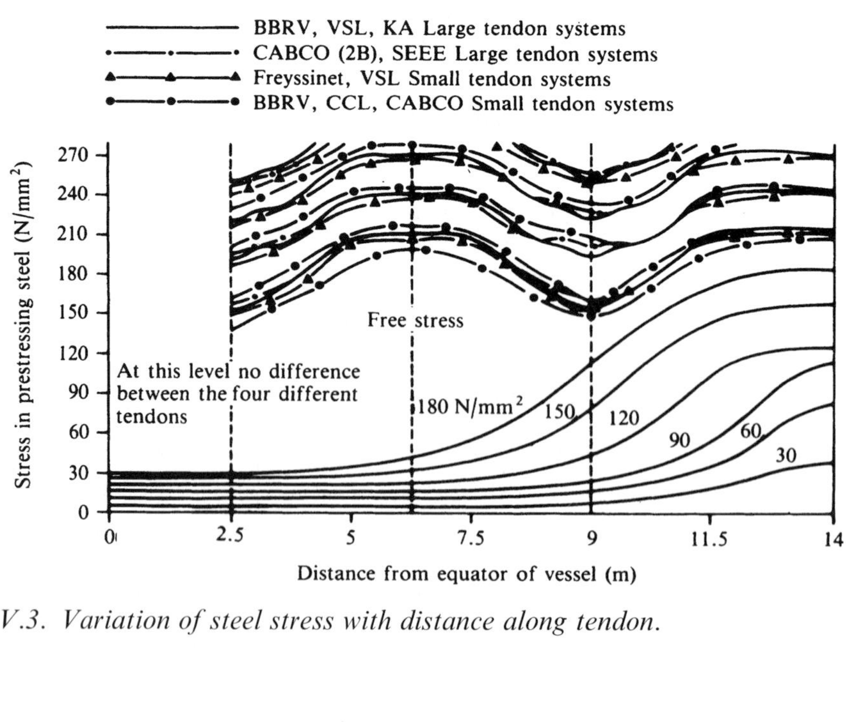

Fig. IV.3. Variation of steel stress with distance along tendon.

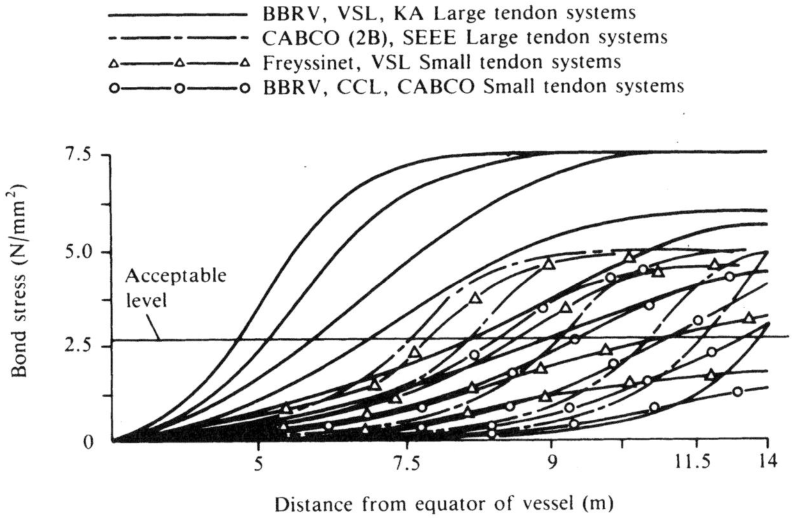

Fig. IV.4a. Variation of stress before cracking.

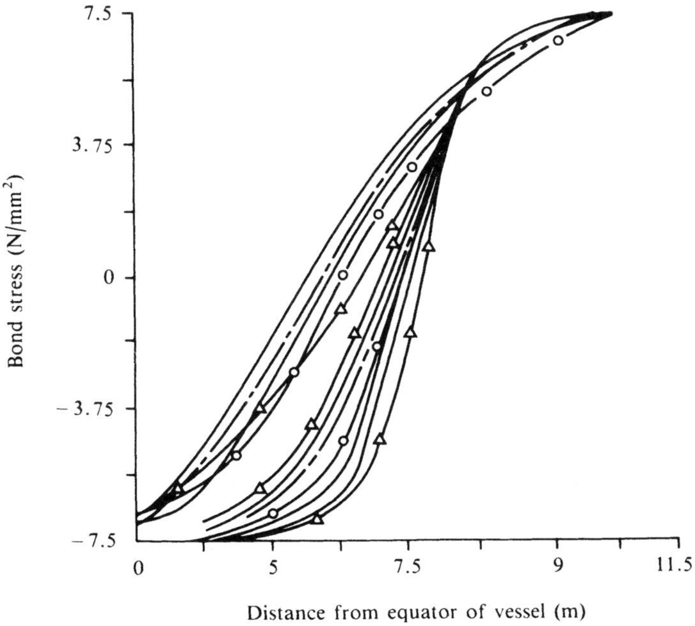

Fig. IV.4b. Variation of stress after primary cracking.

Nilson [251] who used in finite element analysis the relation between local bond-stress and local slip (Fig. IV.5). He used a deformed bar in contact with concrete and obtained the data indirectly for bond stress to local slip using results obtained from several tests on axially loaded tensile specimens. A bond–slip curve was also obtained by Evans and Robinson [253] which is shown in Fig. IV.6. Stocker and Sozen [265] reported the results of an extensive series of pull-out tests on strands

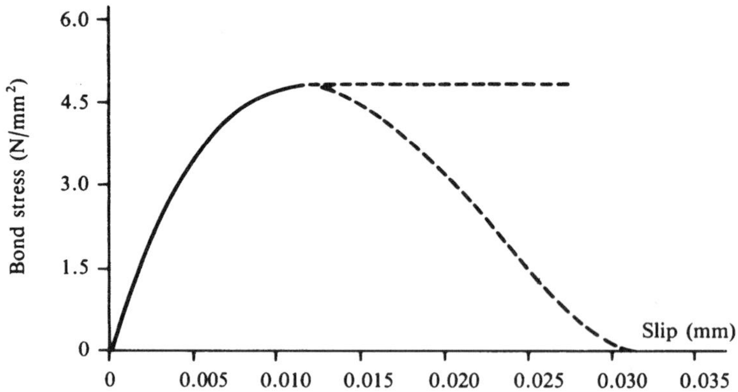

Fig. IV.5. Bond–slip curve for deformed bar [249].

with an embedment length of 25 mm. These results are confirmed also in Figs. IV.7 and IV.8 [249, 250], where each curve represents the average results derived from many curves. From these results, the relationship seems to indicate that bond stress decreases with increasing strand diameter and increases with increase in concrete strength. The unit bond force increases by approximately 10% for every 7.0 N/mm^2 increase in concrete strength. It is concluded that after a slip of 0.0025 mm, further values of slip increase rapidly and the bond strength continues to increase at a small rate. Nilson [251], using the finite element technique, analysed crack formation in a concentrically loaded tension member using the plane stress idealisation and the bond–slip curve for the deformed bars. The bond stress and steel stress distribution are shown in Fig. IV.5. Prior to cracking, both bond stress and steel stress decrease rapidly with distance from the member faces. As the load increases, bond deterioration occurs near the concrete faces and the maximum bond stress moves inward. Primary and secondary cracking produce destruction

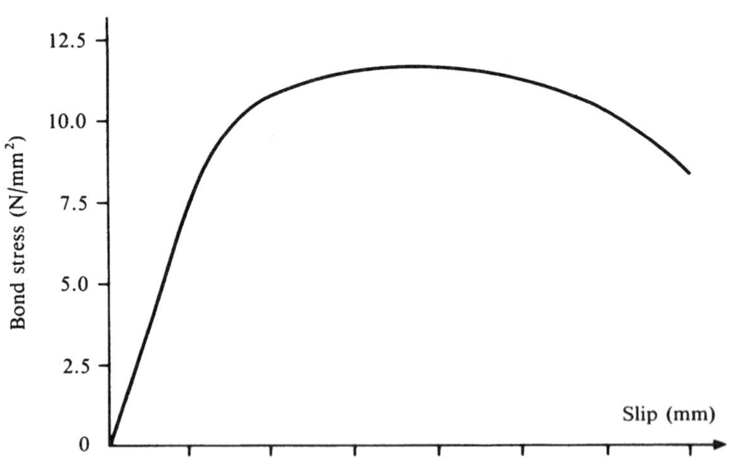

Fig. IV.6. Bond–slip curve for plain wire.

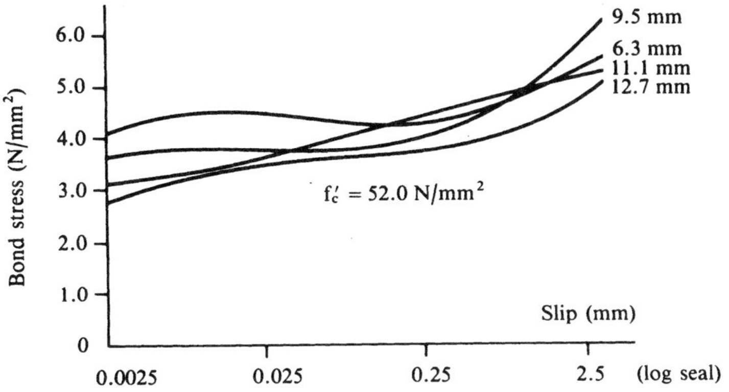

Fig. IV.7. Bond–slip curves for strands of different diameters [249, 250, 262].

of bond. It was concluded that the distribution of longitudinal concrete stresses and the magnitude of the transverse stresses found were different from those given by elastic analysis. The difference is due to the gradual transfer of the tensile forces to concrete by bond. The finite element analysis carried out by Lutz and Gergely [266] concluded that with the assumed perfect bond after cracking, high interfacial tensile stresses existed between steel and concrete at the level of the crack. The separation in the case of plain bars would mean a total loss of bond. The crack width at the bar would be the same as the width of the concrete member. In the case of deformed bars, due to prevention of the complete unloading of the concrete, a great deal of variation of crack width from the bar to the concrete surface existed.

The effects of repeated loading on bond have been reported in detail by Hanson and Kaar [254], Hanson and Hulsbos [263], and Edward and Picard [249]. Hanson and Hulsbos showed that the static bond strength decreased when preceded by repeated loads. When the ratio of the repeated load and static load is higher, the static bond strength (according to Hanson) became lower for the same number of load cycles. Hanson and Kaar [254] showed that under repeated loading, two

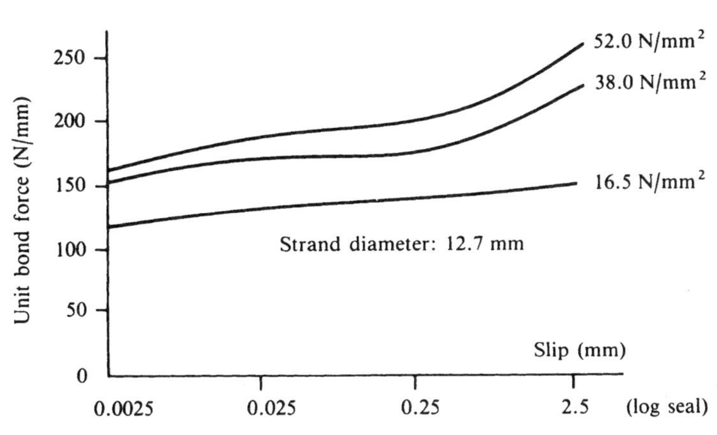

Fig. IV.8. Variation of bond strength with concrete strength [249, 250, 262].

beams with wires failed in bond after 7200 cycles with an average bond stress of 85% of the average value determined from the static test. After 654 000 cycles, the average bond stress was 71% of the value from the static test.

Edwards and Picard [249] reported bond pull-out and tension tests on 12.7 mm wire strand. They used a 38 mm embedded length with three independent concrete covers (12.5 mm, 25 mm and 38 mm). Bond–slip curves showing elasto-plastic type of behaviour have been produced. These results indicate that with the increase in the concrete cover the average maximum bond strength decreases. Empirical equations for crack width and spacings are also given.

Morris [270, 271] carried out tests on unbonded tendons in containment vessels. A 54-strand post-tensioning tendon embedded in a concrete beam was stressed to 76% of guaranteed ultimate tensile stress (g.u.t.s). After 31 days of grouting, the tendon was cut and the bond transfer length was found to be 3.1–3.7 mm. Naus [274] studied both grouted and non-grouted tendons in prestressed concrete pressure vessels. Flexural tests were performed on beams (3 m length × 0.15 m width × 0.31 m depth) prestressed with a 12.5 mm diameter seven-wire prestressed strand. The prestressing load was between 0.5 and 0.7 g.u.t.s with a loading of 0.074–74 kN/s. The results showed that grouted tendons had increased cracking and ultimate load.

Ngo and Scorcelis [272] used the finite element method on several singly reinforced concrete beams with different idealised cracking patterns. A bond linkage element (Fig. IV.9) is used for the bond between steel and concrete. The stiffness matrix for the linkage element is derived later on in this chapter and is similar to the one reported by Bangash [248].

Ahmed [262] developed the Ahmlink element which is a three-dimensional linkage element. He has tested his finite element analysis on a scale model of an octagonal slab, post-tensioned with 5 mm and 7 mm high-tensile wires placed in the two orthogonal directions. In addition, pull-out tests have been carried out on bonded

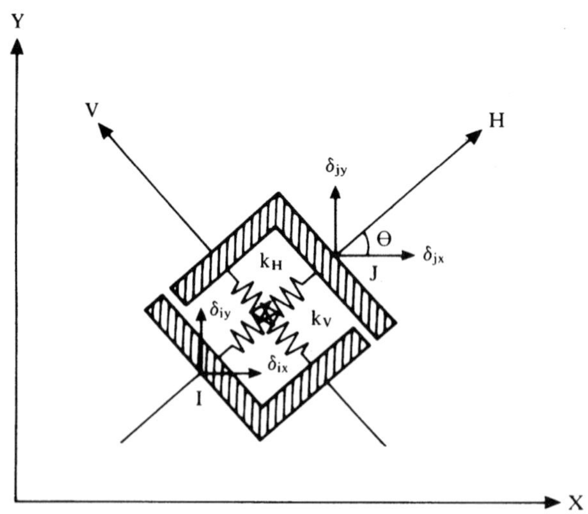

Fig. IV.9. Linkage element [272].

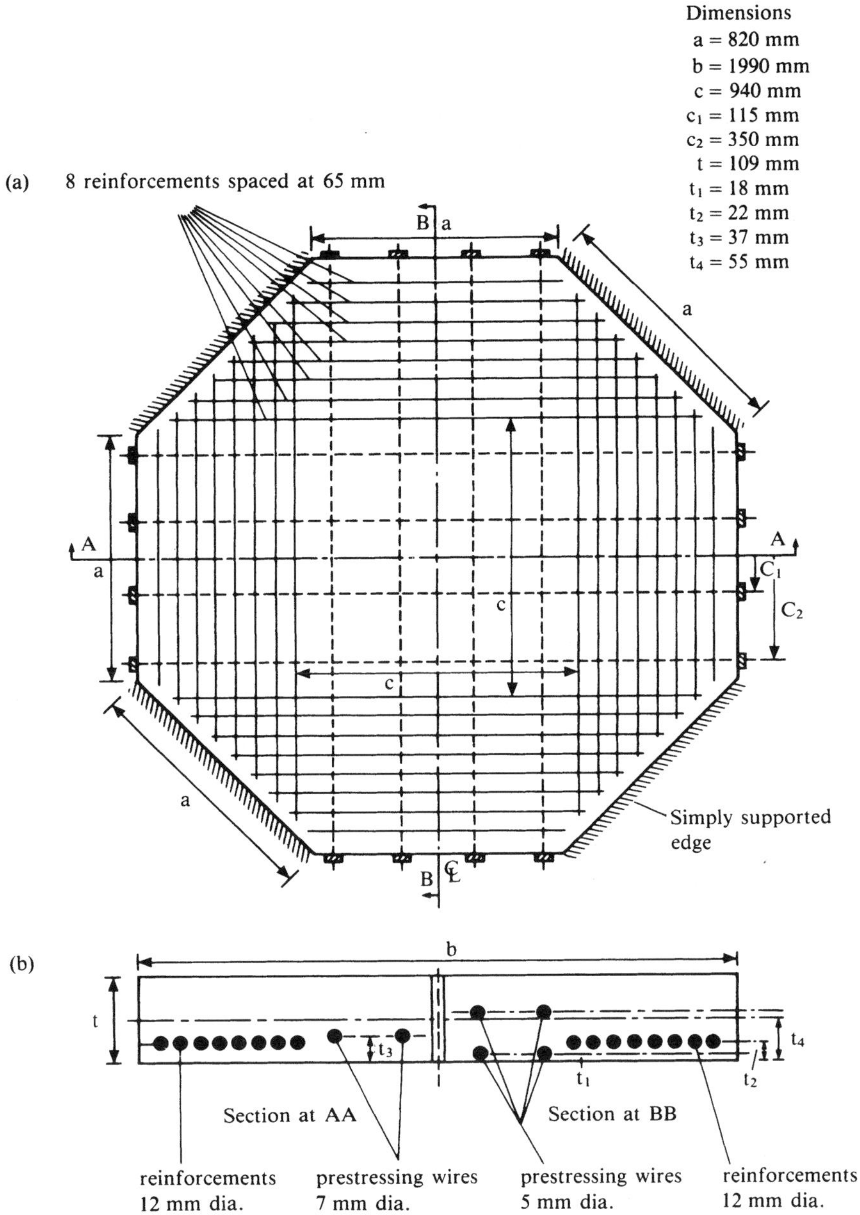

Fig. IV.10. (a) Prestressed concrete slab; and (b) a detailed section (see (a) for dimensions) [262].

prestressed concrete specimens in order to obtain local bond–slip data. This element is used along with other finite elements for the analysis of prestressed concrete vessels for a high-temperature reactor.

Figure IV.10 shows typical details of the pull-out specimens and the concrete slab. Figures IV.11–IV.14 show bond stress–slip curves and prestressing wires.

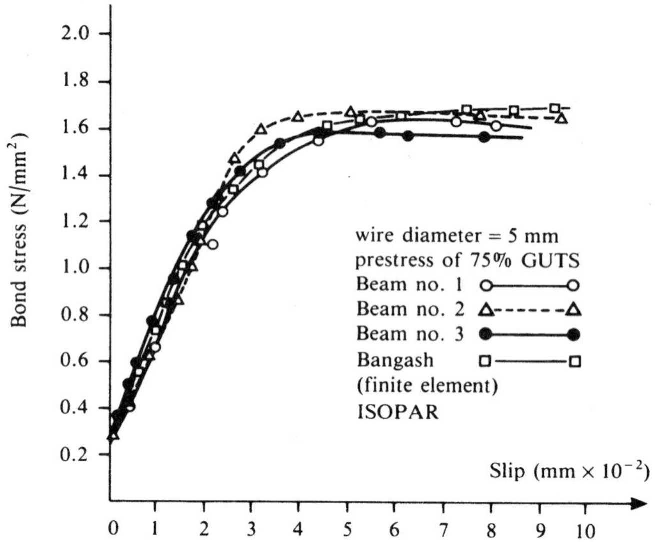

Fig. IV.11. Experimental bond stress–slip curve of 5 mm diameter prestressing wire.

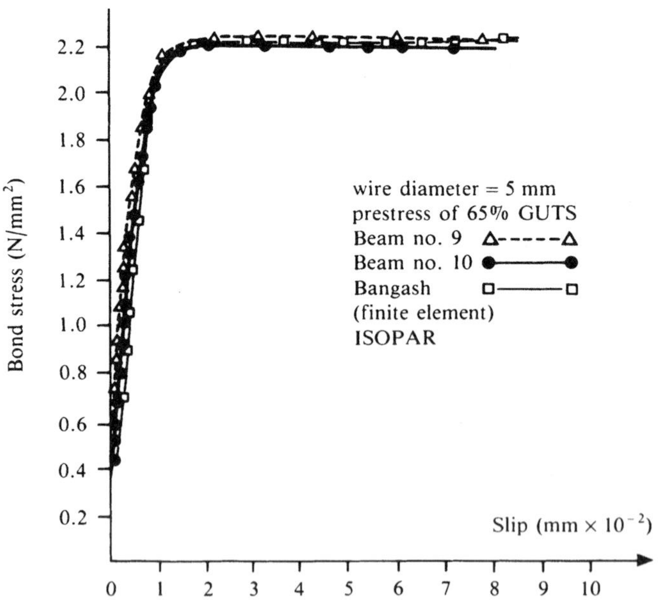

Fig. IV.12. Experimental bond stress–slip curve of 5 mm diameter prestressing wire [262].

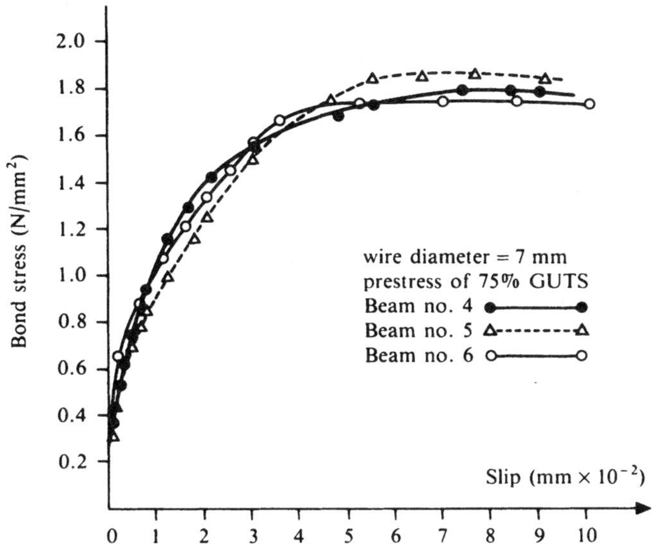

Fig. IV.13. Experimental bond stress–slip curve of 7 mm diameter prestressing wire.

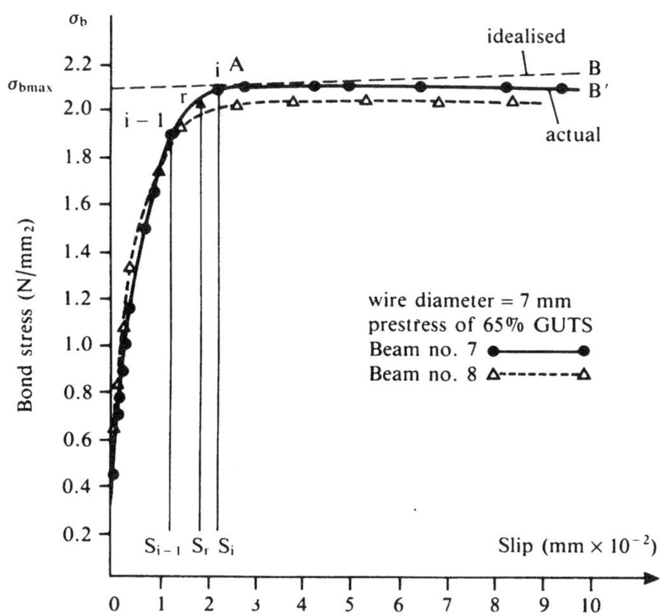

Fig. IV.14. Experimental bond stress–slip curve of 7 mm diameter prestressing wire [262].

IV.5.1 Bond stress–slip constitutive models

IV.5.1.1 Ngo–Scordelis model

Figure IV.9 shows a two-dimensional linkage element to be incorporated into the finite element analysis. The springs are denoted by k_H and k_V respectively. The value of k_H is given by

$$k_H = \mathrm{d}\sigma_\mathrm{b}/\mathrm{d}S \tag{IV.20}$$

where k_H is the slope of the local bond slip curve; and $\mathrm{d}\sigma_\mathrm{b}$ = incremental local bond stress, $\mathrm{d}S$ = incremental local slip (both in the H direction).

For a bar surface area, the right-hand side of eqn (IV.20) is multiplied by $\pi\phi L/2$ which corresponds to one spring. The factor 2 for the bar surface representation indicates two-dimensional idealisation. In the axisymmetric case, when only one spring exists, this factor of 2 disappears. If the H and V axes have an angle θ with the x and y axes, the relationship between the spring deformation and the linkage force is given in a matrix form by

$$\begin{Bmatrix} F_H \\ F_V \end{Bmatrix} = \begin{bmatrix} k_H & 0 \\ 0 & k_V \end{bmatrix} \begin{Bmatrix} \delta_H \\ \delta_V \end{Bmatrix} \tag{IV.21}$$

or the stress–strain relationship is written as

$$\begin{Bmatrix} \sigma_{\mathrm{b}H} \\ \sigma_{\mathrm{b}V} \end{Bmatrix} = \begin{bmatrix} k_H & 0 \\ 0 & k_V \end{bmatrix} \begin{Bmatrix} \varepsilon_{\mathrm{b}H} \\ \varepsilon_{\mathrm{b}V} \end{Bmatrix} \tag{IV.22}$$

or

$$\{\sigma_\mathrm{b}\} = [k_L]\{\varepsilon_\mathrm{b}\} \tag{IV.22a}$$

The strains and displacements are related by the displacement transformation matrix T'' (the H–V and x–y axes system)

$$\begin{Bmatrix} \varepsilon_{\mathrm{b}H} \\ \varepsilon_{\mathrm{b}V} \end{Bmatrix} = \begin{bmatrix} -C & -S & C & S \\ S & -C & -S & C \end{bmatrix} \begin{Bmatrix} \delta_{ix} \\ \delta_{iy} \\ \delta_{jx} \\ \delta_{jy} \end{Bmatrix} \tag{IV.23}$$

$$C = \cos\theta \qquad S = \sin\theta \tag{IV.23a}$$

or

$$\{\varepsilon\} = [T_\mathrm{b}]\{\delta\} \tag{IV.23b}$$

Bangash [221] included the nodal forces in the x–y system as

$$\begin{Bmatrix} F_{ix} \\ F_{iy} \\ F_{jx} \\ F_{jy} \end{Bmatrix} = [T_\mathrm{b}]^{T''}\{F_\mathrm{b}\} \tag{IV.24}$$

where

$$F_b = \begin{Bmatrix} F_H \\ F_V \end{Bmatrix}$$

The stiffness matrix $[K_L]_{xy}$ for the linkage element is thus given by:

$$[K_L]_{xy} = [T_b]^{T''}[k_L][T_b] \tag{IV.25}$$

Table IV.1 shows the matrix $(K_L)_{xy}$ for the linkage element in two dimensions.

Table IV.1. Two-dimensional linkage element matrix $(K_L)_{xy}$

$k_H C^2 + k_V S^2$	$k_H SC - k_V SC$	$-k_H C^2 - k_H S^2$	$-k_H SC + k_V SC$
$k_H SC - k_V SC$	$k_H S^2 + k_V C^2$	$-k_H SC + k_V SC$	$-k_H S^2 - k_V C^2$
$-k_H C^2 - k_V S^2$	$-k_H SC + k_V SC$	$k_H C^2 + k_V S^2$	$k_H SC - k_V SC$
$-k_H SC + k_V SC$	$-k_H S^2 - k_V C^2$	$k_H SC - k_V SC$	$k_H S^2 + k_V C^2$

where $S = \sin\beta$, $C = \cos\beta$. (IV.26)

When $\theta = 0$, i.e. the H and V axes coincide with x and y axes, the matrix $(K_L)_{xy}$ is given by

$$[K_L]_{xy} = \begin{bmatrix} k_H & 0 & -k_H & 0 \\ 0 & k_V & 0 & -k_V \\ -k_H & 0 & k_H & 0 \\ 0 & -k_V & 0 & k_V \end{bmatrix} \tag{IV.26a}$$

IV.5.1.2 Ahmed's constitutive model for bond and bond–slip

Ahmed used eqn (IV.20) in incremental form

$$k_H = \Delta\sigma_b/\Delta S \tag{IV.26b}$$

In order to model various types of interface characteristics, an experimental bond–slip curve idealised as shown in Fig. IV.5 is used. The nonlinear curve is idealised by a series of bond stress and slip points joined linearly. The slope (k_H) at point i of the bond–slip curve is given as

$$k_{Hi} = \frac{\sigma_{bi+1} - \sigma_{bi}}{S_{i+1} - S_i} = \Delta\sigma_{bi}/\Delta S_i \tag{IV.26c}$$

and the other two spring coefficients in the vertical and lateral directions of the steel are taken as

$$\begin{aligned} k_{Vi} &= \alpha_m \times k_{Vi} \\ k_{Li} &= \alpha_m \times k_{Hi} \end{aligned} \tag{IV.26d}$$

where α_m is a multiplying factor. Let I and J (Fig. IV.15) be line element nodes. x, y, z and x', y', z' are global and local co-ordinates. The direction cosines of the

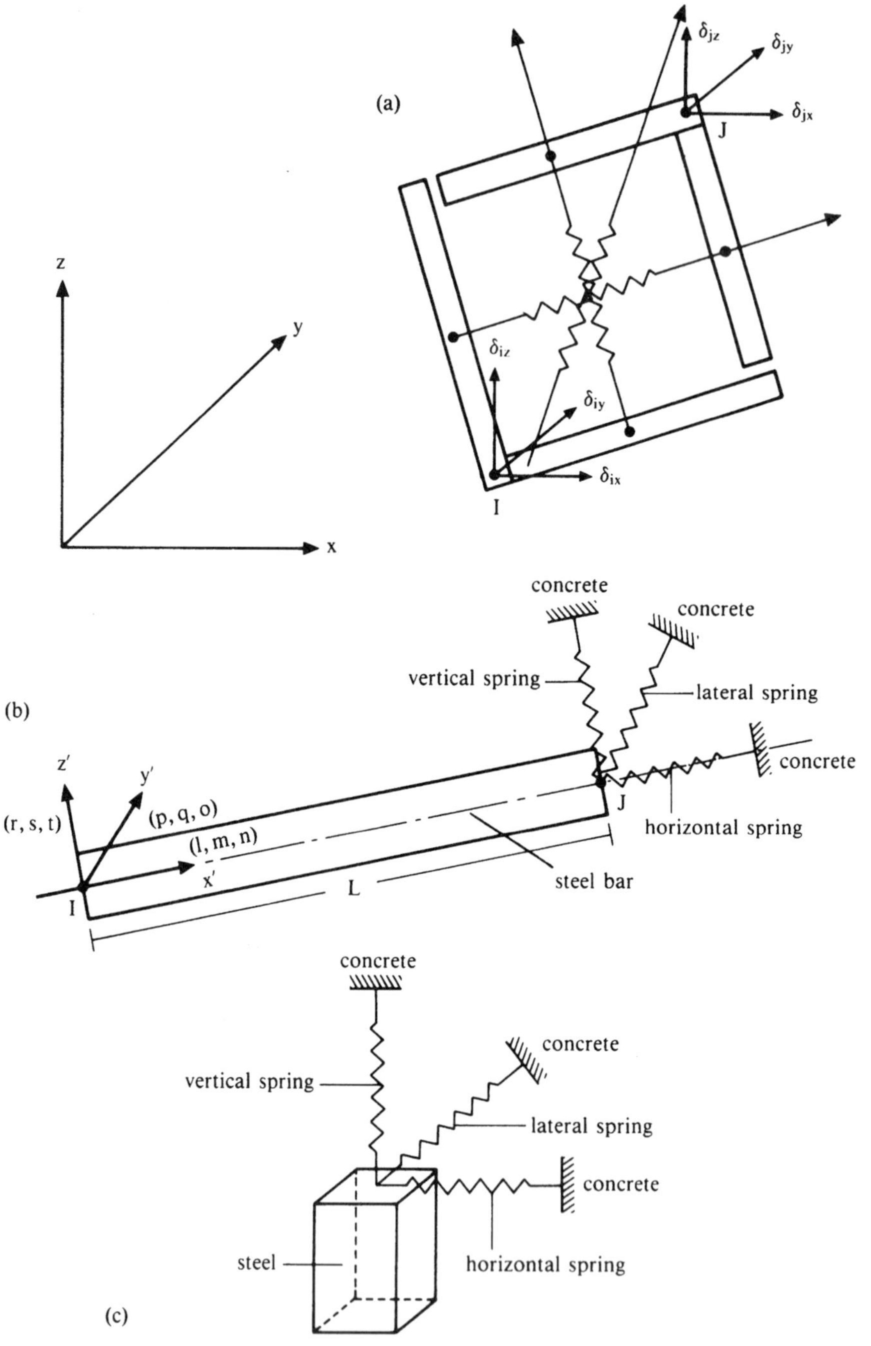

Fig. IV.15. Details of bond linkage elements [262]. (a) Element; (b) direction cosines; (c) bond representation. xzy = global co-ordinate system; x′, y′, z′ = local co-ordinate system; l, m, n, p, q, r, s, t = direction cosines; U, V, W = global displacement variables.

local axes (x', y', z') with respect to the global axes (x, y, z) are (l, m, n), (p, q, θ), (r, s, t). Let (I, J) be the line element nodes.

$$\begin{aligned} l &= (y_J - x_I)/L & m &= (y_J - y_I)/L \\ n &= (z_J - z_I)/L & p &= -m/\sqrt{1-n^2} \\ q &= l/\sqrt{1-n^2} & r &= -ln/\sqrt{1-n^2} \\ s &= -\,mn/\sqrt{1-n^2} & t &= \sqrt{1-n^2} \\ L &= \sqrt{(x_J - x_I)^2 + (y_J - y_I)^2 + (z_J - z_I)^2} \end{aligned} \tag{IV.27}$$

Let ΔS_H, ΔS_V and ΔS_L be the incremental slip in the horizontal, vertical and lateral directions of the steel element. The incremental relationship between the slip and the nodal displacements can be written as

$$\begin{Bmatrix} \Delta S_H \\ \Delta S_V \\ \Delta S_L \end{Bmatrix} = \begin{bmatrix} -l & -m & -n & l & m & n \\ -p & -q & -\theta & p & q & \theta \\ -r & -s & -t & r & s & t \end{bmatrix} \begin{Bmatrix} \delta_{ix} \\ \delta_{iy} \\ \delta_{iz} \\ \delta_{jx} \\ \delta_{jy} \\ \delta_{jz} \end{Bmatrix} \tag{IV.28}$$

or

$$\{\Delta S\}_{H,V,L} = [T_{\mathrm{b}}]^{r,s,t}_{l,m,n} \times \{\Delta U^{\mathrm{e}}\} \tag{IV.28a}$$

where (T_{b}) is the transformation matrix and $\{\Delta U^{\mathrm{e}}\}$ are the global element displacements.

The local incremental bond stress and bond–slip may be written

$$\begin{Bmatrix} \Delta\sigma_{Hi} \\ \Delta\sigma_{Vi} \\ \Delta\sigma_{Li} \end{Bmatrix} = \begin{bmatrix} k_{Hi} & 0 & 0 \\ 0 & k_{Vi} & 0 \\ 0 & 0 & k_{Li} \end{bmatrix} \begin{Bmatrix} \Delta S_{Hi} \\ \Delta S_{Vi} \\ \Delta S_{Li} \end{Bmatrix} \tag{IV.29}$$

or

$$\Delta\sigma^{\mathrm{e}}_{\mathrm{b}i} = K_{\mathrm{b}i}\,\Delta S^{\mathrm{e}}_{Hi} \tag{IV.29a}$$

The element nodal force vector given can be rewritten as

$$\{\Delta P^{\mathrm{e}}_i\} = \pi\,\mathrm{d}L[T_{\mathrm{b}}]^{T''}\{\Delta\sigma^{\mathrm{e}}_{\mathrm{b}i}\} \tag{IV.30}$$

and the element stiffness matrix assumes the form

$$\underset{6\times 6}{[K_L]} = \pi\,\mathrm{d}L[T_{\mathrm{b}}]^{T''}k_{\mathrm{b}i}[T_{\mathrm{b}}]^{T''} \tag{IV.31}$$

where $[T_{\mathrm{b}}]$ is the transformation matrix given in eqn (IV.28).

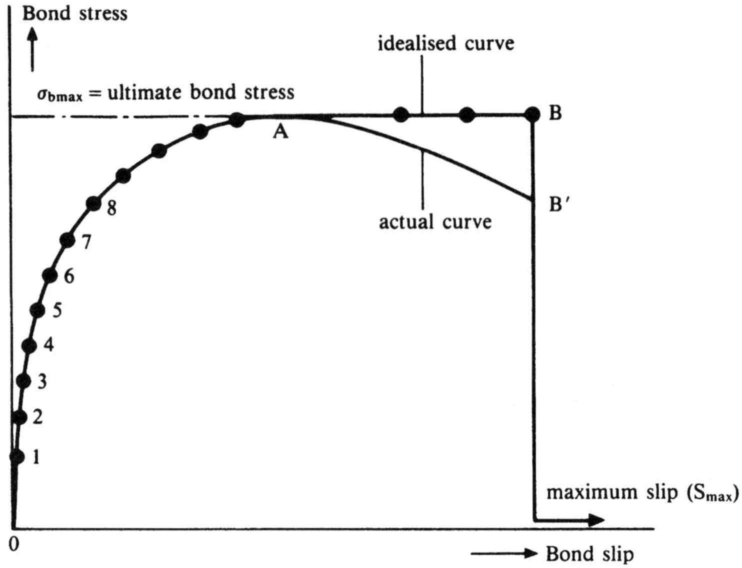

Fig. IV.16a. A typical idealised bond–slip curve.

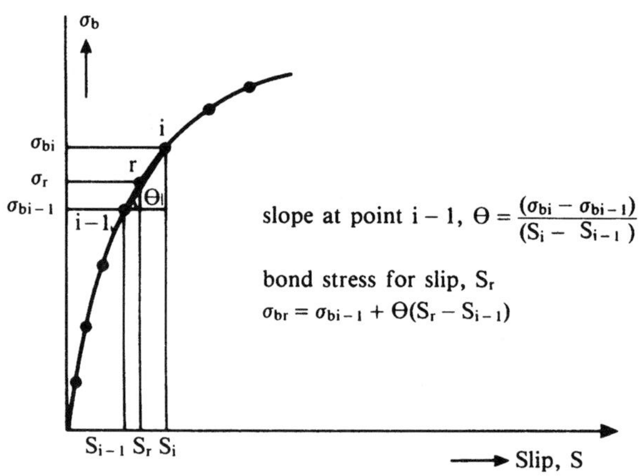

Fig. IV.16b. Linear interpolation of nonlinear bond–slip curve [262].

The constitutive relations (IV.29) are valid between points O and A (Figs IV.15 and IV.16), therefore the bond stress increment becomes zero (region AB), i.e. the slip occurs at constant bond stress. The bond is assumed to fail at point B (Fig. IV.16) when maximum allowable slip has been reached. At this point, the total bond stress is released. This creates a non-equilibrium state which is corrected by performing equilibrium iterations using the initial stress method.

Note: *H*, *V*, *L* are replaced by *h*, *v*, *l* in order to differentiate between these matrices and the previous ones. The directions are the same as defined earlier.

Now a scheme is suggested to calculate correctly bond stress from a specified bond–slip curve (Fig. IV.14). Let S_r be the total slip reached at any point in the calculation. This slip lies between the points $i-1$ and i on the specified curve. Bond stresses are calculated which are compatible with the total slip (S_r) by linear interpolation as

$$\sigma_{br} = \sigma_{bi-1} + \theta(S_r - S_{i-1}) \tag{IV.32}$$

where θ = the slope of the specified curve at point $i-1$

$$\theta = \left(\frac{\sigma_{bi} - \sigma_{bi-1}}{S_i - S_{i-1}}\right) \tag{IV.33}$$

Let the total bond stress calculated using the constitutive equation (IV.29) be σ_{bi}. The difference $(\sigma_{bi} - \sigma_{bv})$ is treated as initial stress and is corrected by performing equilibrium iterations.

Table IV.2 shows the bond linkage matrix K. The stresses in terms of nodal displacements can be written as

$$\{\Delta\sigma_b\} = \{K_b\}[T_b]\{\Delta U^e\} \tag{IV.34}$$

Table IV.2. Bond-linkage stiffness matrix [262]

The explicit form of the bond-linkage stiffness is:

$$\underset{6\times 6}{K_b} = \begin{bmatrix} K_{b11} & K_{b12} \\ K_{b21} & K_{b22} \end{bmatrix}$$

where

$$K_{b12} = K_{b21} = -K_{b11}$$

$$K_{b22} = K_{b11}$$

$$\underset{3\times 3}{K_{b11}} = \begin{bmatrix} K_{b11} & K_{b12} & K_{b13} \\ & K_{b22} & K_{b22} \\ \text{symmetry} & & K_{b33} \end{bmatrix}$$

$$K_{b11} = \pi dL(l^2 E_h + p^2 E_v + r^2 E_l)$$

$$K_{b22} = \pi dL(m^2 E_h + q^2 E_v + s^2 E_l)$$

$$K_{b33} = \pi dL(n^2 E_h + t^2 E_v)$$

$$K_{b12} = \pi dL(lmE_h + pqE_v + rsE_l)$$

$$K_{b13} = \pi dL(lnE_h + rtE_v)$$

$$K_{b23} = \pi dL(mnE_h + stE_l)$$

$$\left.\begin{matrix} l, m, n \\ p, q, r \\ s, t \end{matrix}\right\} = \text{direction cosines}$$

πdL = perimeter of the steel

Table IV.3. Stiffness matrix load vector for bond-linkage element [749]

$$
\underset{6\times 6}{[K_{\mathrm{b}}]} = \pi dL
\begin{bmatrix}
l^2k_h + p^2k_v + r^2k_l & lmk_h + pqk_v + rsk_l & lnk_h + rtk_v & -l^2k_h - p^2k_v - r^2k_l & -lmk_h - pqk_v - rsk_l & -lnk_h - rtk_v \\
 & m^2k_h + q^2k_v + s^2k_l & mnk_h + stE_l & -lmk_h - pqk_v - rsk_l & -m^2k_h - q^2k_v - s^2k_l & -mnk_h - stk_l \\
 & & h^2kh + t^2k_v & -lnk_h - rtk_v & -mnk_h - stk_l & -h^2k_h - t^2k_v \\
 & & & l^2k_h + p^2k_v + r^2k_l & lmk_h + pqk_v + rsk_l & lnk_h + rtk_v \\
 & \text{symmetrical} & & & m^2k_h + q^2k_v + s^2k_l & mnk_h + stk_l \\
 & & & & & n^2k_h + t^2k_v
\end{bmatrix}
$$

$$
\underset{6\times 1}{\Delta\mathbf{P}^{\mathrm{e}}} = \pi dL \begin{Bmatrix} -l\Delta\sigma_h - p\Delta\sigma_v - r\Delta\sigma_l \\ -m\Delta\sigma_h - q\Delta\sigma_v - s\Delta\sigma_l \\ -n\Delta\sigma_h - t\Delta\sigma_l \\ l\Delta\sigma_h - p\Delta\sigma_v + r\Delta\sigma_l \\ m\Delta\sigma_h + q\Delta\sigma_v + s\Delta\sigma_l \\ n\Delta\sigma_h + t\Delta\sigma_l \end{Bmatrix}
$$

πdL = perimeter of the steel

$$
\left.\begin{matrix} l, m, n, \\ p, q, r, \\ s, t \end{matrix}\right\} = \text{direction cosines}
$$

A step-by-step procedure for computing bond stresses using three-dimensional finite elements and the above numerical model is given later in this book.

Bangash [749] carried out similar tests to Ahmed but with more tendons and a steel plate fastened by lugs to a concrete slab. Ahmed's linkage matrix was modified to include the bonds due to the lugs as well. The final matrix is given in Table IV.3.

V. Numerical models for fatigue of concrete

V.1 Introduction

Substantial research has been carried out in recent years [276–285] on the fatigue of concrete structures and their components. The subject of fatigue is more important in concrete structures such as offshore gravity platforms, nuclear reactor vessels, heavy machine foundations, etc., where higher strength materials are used and at the same time the structures are expected to perform successfully under higher stress levels. Fatigue analysis is based on a period of time t_f, which is the planned life of the concrete structures. The value of t_f is taken to be 30 years plus.

Fatigue loading is generally a sequence of repetitive loads. This type of loading is distinctly divided between high-cycle low-amplitude and low-cycle high-amplitude. The distinction is based on an arbitrary number of cycles, normally within the range of 100–1800 cycles. Lower than 1800 cycles of high-amplitude fatigue loading can occur due to earthquakes or due to various sea states. For machine foundations the cycles could go up to 10^7.

A number of investigators [276–285] have reported that concrete specimens under cyclic compressive loadings receive progressive changes within their concrete matrix due to the growth of microcracking. Shah [275] reported that for all types of concrete in different environments and under controlled conditions, for a stress level equivalent to 50% of the compressive strength of concrete, the volume of concrete increases rather than decreases and microcracking occurs at the aggregate–paste interface. Reinforcement plays a great part in the deformation of concrete sections under repeated loading.

Holman [276] gives a relationship between strain ratio $\varepsilon_{max}/\varepsilon_0'$ and the cycle ratio N (number of cycles)/N_{TF} (total life in cycles). This relationship indicates that there is a gradual increase in strain between 10% and 80% of the N_{TF}; however, between 0 and 10% and over 80% the strain increment is fairly rapid. Wilson [285] concluded that the slow-cycle fatigue caused the bond cracks to expand, resulting in a decrease in the concrete durability. His results were based on first freezing concrete and then subjecting it to slow-cycle fatigue loading.

Some of the models for fatigue analysis which can easily be included in the global finite element analysis are given in the following paragraphs.

V.2 Fatigue models

V.2.1 Kakuta et al. model [277]

Kakuta *et al.* [277] reported the following expression for the fatigue of concrete

$$\log N = 17\left[1 - \frac{(\sigma_{\text{max}} - \sigma_{\text{min}})/f_c'}{1 - (\sigma_{\text{min}}/f_c')}\right] \tag{V.1}$$

where N = the number of cycles; σ_{max} = the maximum stress of the cycle; σ_{min} = the minimum stress of the cycle; and f_c' = the cylindrical compressive strength of concrete.

V.2.2 Aas-Jacobson model [278]

This model [278] is based on a linear relationship between log N and the cycles of stresses using a factor β_f based on statistical data

$$\log N = \frac{1}{\beta_f}\left[\frac{1 - (\sigma_{\text{max}}/f_c')}{1 - (\sigma_{\text{max}}/\sigma_{\text{min}})}\right] \tag{V.2}$$

A value of $\beta_f = 0.064$ was given by Aas-Jacobson.

V.2.3 Miner model [279]

The main aspect of the Miner model [279], known as Miner's Rule, is the estimation of the cumulative effect of the stress history by adopting a simplified linear damage relationship. The cumulative usage factor $\bar{\eta}$ is given by

$$\bar{\eta} = \sum_{i=1}^{S} N_i/N_{fi} \tag{V.3}$$

where S = the number of stress blocks considered; N_i = the actual number of stress cycles for the stress block i; and N_{fi} = the permissible number of stress cycles if the stress block i is considered only, or the number of cycles that will cause failure at stress level i.

Stresses occur with variable amplitude in random order and they are substituted by a number of 'stress blocks'. Each stress block represents a number of constant-amplitude stress cycles.

Equation (V.3) is written for a fatigue damage as

$$\bar{\eta}_1 = \sum_{i=1}^{S} \frac{N_i}{K} S_i^{\bar{m}} \tag{V.4}$$

where K is an empirical constant in the S–N equation; S = a fatigue stress cycle; N = the number of cycles to failure; $\bar{m}$ = an empirical constant in the S–N equation; and $S_i^{\bar{m}}$ = the expected value of $S^{\bar{m}}$ in the ith state. Failure occurs when $\bar{\eta} \geq 1$ (see Appendix IX for offshore platform fatigue results).

V.2.4 Waagaard model [280]

Waagaard [280] uses a value of $\bar{\eta}$, the cumulative usage factor, of 0.2 rather than 1 for offshore structures. He represents the fatigue of concrete in compression in offshore structures as

$$\log N = 10\left[\left\{1.0 - \frac{\sigma_{\max}}{\alpha_g(f_c'/\gamma)}\right\} \Big/ \left\{1.0 - \frac{\sigma_{\min}}{\alpha_g(f_c'/\gamma)}\right\}\right] \tag{V.5}$$

where α_g is the factor to account for flexural gradient, $1 \leq \alpha_g < 10$; γ is the partial safety factor for materials $= 1.25$; and $\sigma_{\min}$ is around $0.1f_c'$.

V.2.5 JSCE model [281]

The fatigue of concrete in compression is given by the Japan Society of Civil Engineers [281] as

$$\log N = 15\left[1 - \frac{(\sigma_{\max} - \sigma_{\min})}{(0.9K'f_c' - \sigma_{\min})}\right] \tag{V.6}$$

where $K' =$ a coefficient taken as equal to 0.85.

The characteristic fatigue strength for concrete F_{cr} is given by

$$F_{cr} = (0.9K'f_c' - \sigma_{cp})\left(1 - \frac{\log N}{15}\right) \tag{V.7}$$

where $\sigma_{cp} =$ permanent compressive stress.

For deformed bars of diameter $\not> 32$ mm, the fatigue strength is given by the value of F_{sr}.

$$F_{sr} = \bar{C} \times 10^{-0.2(\log N - 6)} \quad \text{for } N < 6$$

or

$$= \bar{C} \times 10^{-0.1(\log N - 6)} \quad \text{for } N > 6 \tag{V.8}$$

where $\bar{C} = 160\,\text{Mps} - (\sigma_{tp}/3)$, $\sigma_{tp} =$ permanent tensile stress in steel.

V.2.6 CEB–FIP model [282]

According to CEB–FIP [282] recommendations, the stresses corresponding to fatigue strength must be determined by elastic analysis in which provision should be made for dynamic effects, creep and shrinkage, etc. The following conditions are checked:

(a)

$$\Delta\sigma = \sigma_{\max} - \sigma_{\min} \leq \Delta\sigma_{rep}/\gamma_{fat} \tag{V.9}$$

Both $\sigma_{\max}$ and $\sigma_{\min}$ have maximum values corresponding to 2×10^6 cycles. $\sigma_{rep} =$ strength under repeated loads; $\gamma_{fat} = 1.25$.

The fatigue strength of concrete is the 50% of the fractile value obtained from the series of tests.

(b) *Fatigue strength of reinforcement*: For 2×10^6 cycles, the characteristic strength of conventional steel and prestressing steel are 10% fractile and 50% fractile respectively.

For conventional steel

$$\sigma_{max} = 0.7\sigma_{yt}$$

For prestressing tendons

$$\sigma_{max} = 0.85\sigma_{yt}$$

The ratio

$$\Delta\sigma_{rep}/\gamma_{fat} = \Delta\sigma_{sf}/\gamma_{fat} \tag{V.10}$$

where $\gamma_{fat} = 1.15$ for conventional steel and $\gamma_{fat} = 1.25$ for prestressed anchorages. The values of $\Delta\sigma_{sf}$ are given below:

Plain bars: 250 MPa
Deformed bars: 150 MPa
Prestressing tendons: 150–200 MPa

For welding zones a reduction of 0.4 of the above values is taken into consideration.

V.2.7 AASHTO model [283]

The American Association of State Highway and Transportation Officials (AASHTO) [283] has recommended a limitation of stresses for fatigue for concrete, the maximum compressive stress must not exceed $0.5f'_c$. The stress range for the reinforcement is given by

$$\sigma_{sr} = 145 - 0.33\sigma_{min} + 54.9(r/h) \tag{V.11}$$

where r = the base radius of the rolled bar; h = the height radius of the rolled bar; σ_{min} = the minimum stress level; and σ_{sr} = the stress range.

Where prestressing tendons are used the tendons shall offer 60–66% of the minimum tensile strength without failure under 5×10^5 cycles. The period of each cycle is counted from lower stress level to the upper stress level and back to the original lower stress level.

V.2.8 NCHR model [284]

The National Cooperative Highway Research (NCHR) of the US Transportation Research Board studied bars from several manufacturers and as a result an empirical equation was developed [284] in order to predict the long-term limiting stress range below which fatigue damage will not occur. For hot-rolled bars, eqn (V.11) is validated. Above the fatigue limit of eqn (V.11), the safe fatigue life for all stress ranges is represented by eqn (V.12).

$$\begin{aligned}\log N = {} & 6.1044 - 4.07 \times 10^{-5}\sigma_{rs} - 1.38 \times 10^{-5}\sigma_{min} \\ & + 0.71 \times 10^{-5}\sigma_{yt} - 0.0566A_s + 0.3233d\end{aligned} \tag{V.12}$$

where σ_{rs} and σ_{min} are in psi; A_s = the area of the bar in (inches)^2; d = the diameter of the bar in inches; and σ_{yt} = ultimate steel strength.

Using SI units, eqn (V.12) is given by

$$\log N = 6.1044 - 590.8 \times 10^{-5}\sigma_{rs} - 199 \times 10^{-5}\sigma_{min} + 103 \times 10^{-5}\sigma_{yt} - 8.771 \times 10^{-5}A_s + 0.01265d(r/h) \quad \text{(V.13)}$$

where σ_{rs} and σ_{min} are in Mpa, A_s in mm^2, and d in mm.

V.2.9 Wilson model [285]

Based on the ASTM freezing and thawing tests Wilson [285] concluded that the durability factor for concrete under cyclic loads can be calculated as

$$E_{DY} = n_i^2/n_0^2 \times 100 \quad \text{(V.14)}$$

where E_{DY} = the relative dynamic modulus of elasticity in percent after N cycles; n_0 = the fundamental transverse frequency at zero cycles; and n_i = the fundamental transverse frequency after N_i cycles.

The durability factor of the test specimens can be obtained by

$$DF = E_{DY}N_i/N_{fi} \quad \text{(V.15)}$$

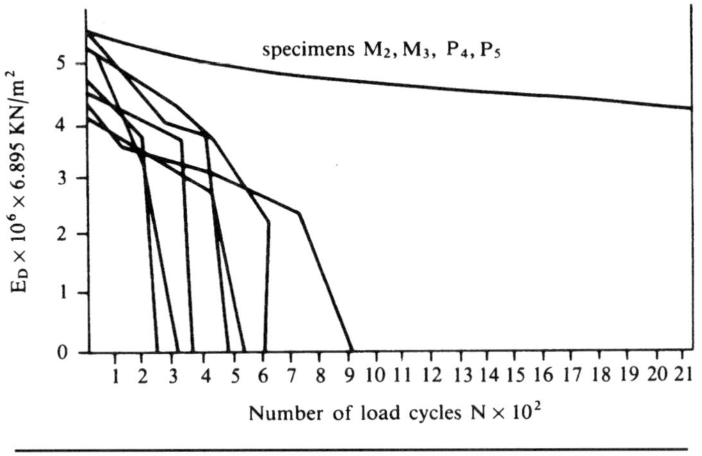

Test no: M_2, M_3, P_5	Cyclic load (% of ultimate)	Curing period (90, 28 days)
A_1	70	7 years
A_2	70	7 years
A_3	70	7 years
A_4	70	7 years
A_5	70	7 years
A_6	70	7 years
A_7	70	7 years
A_8	70	7 years

Fig. V.1. Dynamic modulus of elasticity versus number of load cycles (after Wilson [285]).

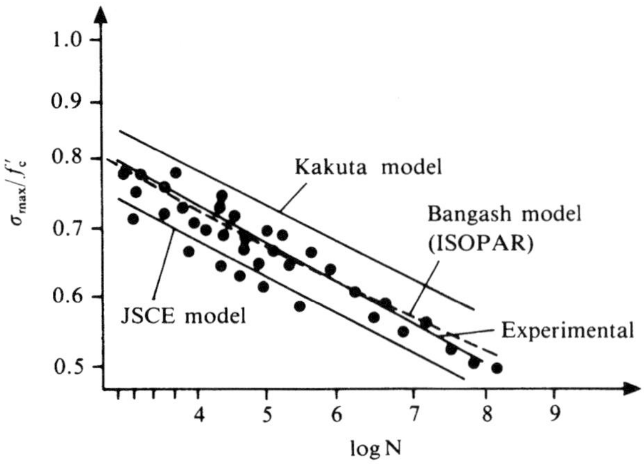

Fig. V.2. Comparison of two fatigue models with experimental tests.

where N_{fi} = the specified number of cycles at which the dynamic load is terminated.

Several tests have been carried out for N_i and N_{fi} less or greater than 1800. The average durability factor for early freezing was 15.2, compared with 19.2 in a standard case without freezing. The average cycles to failure for freezing and non-freezing concrete were 851 and 1079. A slight improvement in the compressive strength was noted in early freezing concrete.

Figure V.1 shows some of these results. Figure V.2 shows a comparison of the Kakuta and JSCE models with the finite element experimental tests carried out on a prestressed concrete beam by the author.

PART II

This section contains a step-by-step finite element analysis for concrete structures in which provision is made for elastic–elasto-plastic static, dynamic, impact and blast effects. An attempt is made to integrate the elements of Part I and to present the simulation technique of concrete numerical models in a global finite element analysis.

VI. Finite element modelling of concrete structures

VI.1 Introduction

A step-by-step finite element analysis is developed in which a provision is made for elasto-plasticity, thermal creep, creep recovery, and cracking under static and dynamic conditions. Various concrete strength theories described in earlier chapters are included in the finite element analysis. These give facilities for the comparative study of results necessary for some sensitive structures. The analysis presented here also includes facilities for loading and unloading phenomena. Although greater emphasis is given to the use of isoparametric elements in the text, tables are provided which will assist in the automatic inclusion of other types of finite elements. Equations are set up in such a way that they can easily be replaced or modified to include other case studies. The initial stress or modified Newton–Raphson method is used for the major solution procedures. They, together with acceleration and convergence procedures can easily solve nonlinear, plasticity and cracking problems. Where dynamic problems exist, equations of motion are introduced in both linear and nonlinear cases. Expressions for displacements, velocities and accelerations are then obtained. Direct integration and Wilson-θ methods have been used throughout for the time dependent solutions of dynamically loaded structures. The main analysis of this chapter is supported by analyses given in various tables and appendices. For further study and additional information, the references cited in this text may be found useful.

VI.2 Solid isoparametric element representing concrete

Three-dimensional isoparametric elements are used and the functions relating to co-ordinate systems are expressed as follows:

$$\begin{aligned}
X &= \sum_{i=1}^{n} N_i(\xi,\eta,\zeta)X_i = N_1x_1 + N_2x_2 + \cdots = \{N\}^T\{X_n\} \\
Y &= \sum_{i=1}^{n} N_i(\xi,\eta,\zeta)Y_i = N_1y_1 + N_2y_2 + \cdots = \{N\}^T\{Y_n\} \qquad \text{(VI.1)} \\
Z &= \sum_{i=1}^{n} N_i(\xi,\eta,\zeta)Z_i = N_1z_1 + N_2z_2 + \cdots = \{N\}^T\{Z_n\}
\end{aligned}$$

where $N_i(\xi, \eta, \zeta)$, $i = 1$ to n, are the interpolation functions in the curvilinear co-ordinates ξ, η, ζ, and X_i, Y_i and Z_i are the global X, Y, Z co-ordinates of mode i. The interpolation function N is also known as the shape function.

The terms $\{X_n\}$, $\{Y_n\}$ and $\{Z_n\}$ represent the nodal co-ordinates and $\{N\}^T$ is dependent on ξ, η, and ζ. Reference is made to Fig. VI.1 for important aspects of the mapping procedures. The most important aspect is to establish a one-to-one relationship between the derived and the parent element (Fig. VI.1). The necessary condition for a one-to-one relationship is the Jacobian determinant given in eqn (VI.2).

$$\mathrm{d}[J] = \frac{\partial(X, Y, Z)}{\partial(\xi, \eta, \zeta)} = \begin{bmatrix} \frac{\partial X}{\partial \xi} & \frac{\partial X}{\partial \eta} & \frac{\partial X}{\partial \zeta} \\ \frac{\partial Y}{\partial \xi} & \frac{\partial Y}{\partial \eta} & \frac{\partial Y}{\partial \zeta} \\ \frac{\partial Z}{\partial \xi} & \frac{\partial Z}{\partial \eta} & \frac{\partial Z}{\partial \zeta} \end{bmatrix} \tag{VI.2}$$

The Jacobian can also be written in a transposed form, i.e. columns replace rows and vice versa.

VI.2.1 The shape function

The relationship between displacements at any point in the local system (ξ, η, ζ) within the element and the nodal displacements is expressed in the most general form with the aid of a shape or interpolation function as

$$\begin{aligned} U(\xi, \eta, \zeta) &= N_1 u_1 + N_2 u_2 + \cdots + (N)^T (u_n) = \sum_{i=1}^{n} N_i(\xi, \eta, \zeta) u_i \\ V(\xi, \eta, \zeta) &= N_1 v_1 + N_2 v_2 + \cdots + (N)^T (v_n) = \sum_{i=1}^{n} N_i(\xi, \eta, \zeta) v_i \\ W(\xi, \eta, \zeta) &= N_1 w_1 + N_2 w_2 + \cdots + (N)^T (w_n) = \sum_{i=1}^{n} N_i(\xi, \eta, \zeta) w_i \end{aligned} \tag{VI.3}$$

where u_i, v_i and w_i are the nodal displacements in the x, y, z directions at node i.

The interpolation function N_i can also be expressed in terms of a local dimensionless co-ordinate system (ξ, η, ζ). Equation (VI.3) is written in a more generalised form as

$$\bar{u} = \sum_{i=1}^{n} N_i \bar{u}_i \tag{VI.4}$$

where $\bar{u}$ is in any co-ordinate and $\bar{u}_i$ is the current value of $\bar{u}$ at node i and N_i is the shape function for node i.

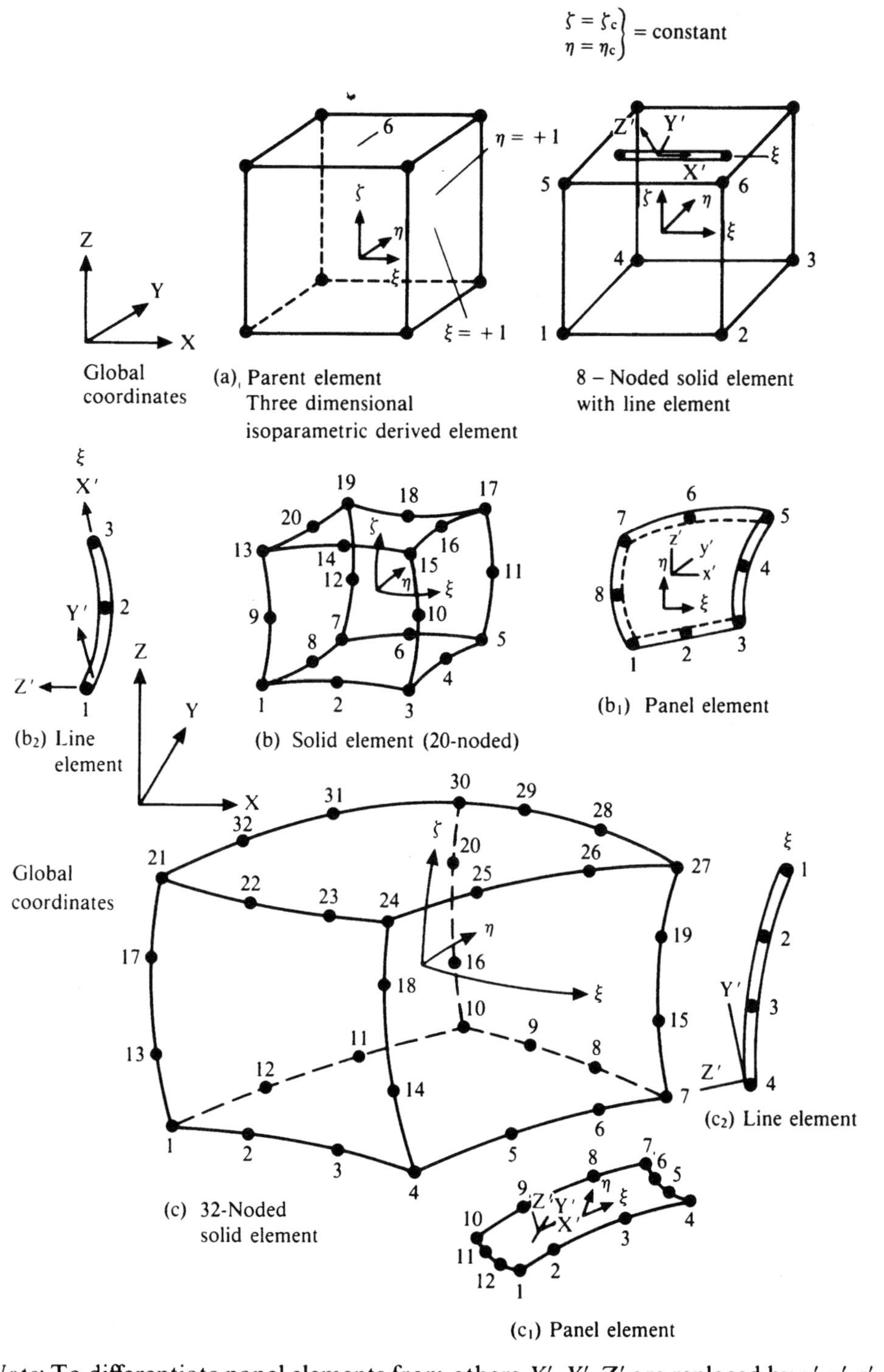

Note: To differentiate panel elements from others X', Y', Z' are replaced by x', y', z'.

Fig. VI.1. Finite elements.

VI.2.2 Derivatives and the Jacobian matrix

With the shape functions known, the global co-ordinates and displacements at any point within the element are expressed in terms of the nodal values (eqns (VI.1) and (VI.3))

$$X = \sum_{i=1}^{n} N_i X_i \qquad Y = \sum_{i=1}^{n} N_i Y_i \qquad Z = \sum_{i=1}^{n} N_i Z_i \tag{VI.5}$$

$$U = \sum_{i=1}^{n} N_i u_i \qquad V = \sum_{i=1}^{n} N_i v_i \qquad W = \sum_{i=1}^{n} N_i w_i \tag{VI.6}$$

where n is the number of nodes on element and X_i, Y_i, Z_i and u_i, v_i, w_i are the nodal co-ordinates and nodal displacements respectively.

The derivatives of shape functions with respect to global co-ordinates require the following transformation.

$$\begin{Bmatrix} \dfrac{\partial N_i}{\partial \xi} \\ \dfrac{\partial N_i}{\partial \eta} \\ \dfrac{\partial N_i}{\partial \zeta} \end{Bmatrix} = [J] \begin{Bmatrix} \dfrac{\partial N_i}{\partial X} \\ \dfrac{\partial N_i}{\partial Y} \\ \dfrac{\partial N_i}{\partial Z} \end{Bmatrix} \tag{VI.7}$$

where i is the current node number

$$\begin{Bmatrix} \dfrac{\partial N_i}{\partial X} \\ \dfrac{\partial N_i}{\partial Y} \\ \dfrac{\partial N_i}{\partial Z} \end{Bmatrix} = [J] \begin{Bmatrix} \dfrac{\partial N_i}{\partial \xi} \\ \dfrac{\partial N_i}{\partial \eta} \\ \dfrac{\partial N_i}{\partial \zeta} \end{Bmatrix} \tag{VI.7a}$$

where

$$[J] = \begin{bmatrix} \dfrac{\partial X}{\partial \xi} & \dfrac{\partial Y}{\partial \xi} & \dfrac{\partial Z}{\partial \xi} \\ \dfrac{\partial X}{\partial \eta} & \dfrac{\partial Y}{\partial \eta} & \dfrac{\partial Z}{\partial \eta} \\ \dfrac{\partial X}{\partial \zeta} & \dfrac{\partial Y}{\partial \zeta} & \dfrac{\partial Z}{\partial \zeta} \end{bmatrix} \tag{VI.7b}$$

where $[J]$ is a 3×3 Jacobian matrix. This matrix eventually plays a major role in the equilibrium equations:

$$\{F\} = [K]\{U\} \tag{VI.8}$$

where

$\{U\}$ = the nodal displacement vector

$$[K] = \sum_{i=1}^{n} [K_e] = \text{the total stiffness matrix}$$

$[K_e]$ = the element stiffness matrix

n = the number of elements

The determinant d[J] as represented in eqn (VI.2) can be evaluated for a one-to-one relationship. Such derivatives of shape functions are given in Appendix I.

Displacements at nodes can now be expressed in the form of element displacement vectors

$$\{U\} = \begin{Bmatrix} \{u_1\}_1 \\ \{u_2\}_2 \\ \vdots \\ \{u_n\}_n \end{Bmatrix} \tag{VI.9}$$

The displacement field within each element can be expressed as

$$\{u\} = [N]\{u^e\} = \sum_{i=1}^{n} (N_i[I]\{u\}_i) \tag{VI.10}$$

where $\{u^e\}$ = the element nodal displacement vector; $\{u\}_i$ = the displacements at node i; $[N]$ = the element shape function matrix; N_i = the shape functions of node i.

VI.2.3 Determination of strains

With the displacement known at all points within each element, the strains $\{\varepsilon\}$ can be expressed in the following form:

$$\{\varepsilon\} = \sum_{i=1}^{n} [B_i]\{u\}_i = [B]\{u\} \tag{VI.11}$$

For the three-dimensional element the [B] matrix of node i and $\{\varepsilon\}_{6\times 1}$ are given below:

$$\underset{6\times 3}{[B_i]} = \begin{bmatrix} \frac{\partial N_i}{\partial X} & 0 & 0 \\ 0 & \frac{\partial N_i}{\partial Y} & 0 \\ 0 & 0 & \frac{\partial N_i}{\partial Z} \\ \frac{\partial N_i}{\partial Y} & \frac{\partial N_i}{\partial X} & 0 \\ 0 & \frac{\partial N_i}{\partial Z} & \frac{\partial N_i}{\partial Y} \\ \frac{\partial N_i}{\partial Z} & 0 & \frac{\partial N_i}{\partial X} \end{bmatrix} \tag{VI.12}$$

$$\underset{6\times 1}{\{\varepsilon\}} = [\varepsilon_X, \varepsilon_Y, \varepsilon_Z, \gamma_{XY}, \gamma_{YZ}, \gamma_{ZX}]^{T''} \tag{VI.12a}$$

Appendix II gives full details of the co-ordinate transformations between Cartesian and curvilinear axes. The dimension of $[B]$ for 8-, 20- and 32-noded elements is 6×24, 6×60 and 6×96 respectively.

VI.2.4 Determination of stresses

The stresses $\{\sigma\}$ can be determined within each element from the strains as

$$\{\sigma\} = [D](\{\varepsilon\} - \{\varepsilon_0\}) + \{\sigma_0\} \tag{VI.13}$$

where $\{\sigma_0\}$ = initial stresses; $\{\varepsilon_0\}$ = initial strains; and $[D]$ = the material compliance matrix.

VI.2.5 Load vectors and material stiffness matrix

Appendix III gives the material matrices $[D]$ for both concrete and steel. Now virtual displacement $\{du^e\}$ is applied at the nodes; and the sum of work done, dW, by the stresses, and distributed body and surface forces over the element volume, vol, and surface, S, is given by

$$dW = [\{du^e\}^{T''}]\left(\int_{\text{vol}} [B]^{T''}\{\sigma\}\, dV - \int_S [N]^{T''}\{p\}\, dS - \int_{\text{vol}} [N]^{T''}\{G\}\, d\,\text{vol}\right) \tag{VI.14}$$

In order to maintain equilibrium within the element, a system of external nodal forces $\{p\}^e$ must be applied which will reduce virtual work (dW) to zero. Equation (VI.14) will take the form

$$\{du^e\}^{T''}\{p\}^e = \{du^e\}^{T''}\left(\int_{\text{vol}} [B]^{T''}\{\sigma\}\, d\,\text{vol} - \int_S [N]^{T''}\{p\}\, dS - \int_{\text{vol}} [N]^{T''}\{G\}\, d\,\text{vol}\right) \tag{VI.14a}$$

where $\{p\}$ = the surface force per unit surface area; $\{G\}$ = the force per unit volume; and T'' = transpose of the matrix.

When the external work dW is related to the internal work dU

$$dW = dU \tag{VI.15}$$

The element stiffness matrix $[K]$

$$[K] = \int_{\text{vol}} B^{T''} DB\, d\,\text{vol} = \int_{-1}^{+1}\int_{-1}^{+1}\int_{-1}^{+1} B^{T''} DB \det J\, d\xi\, d\eta\, d\zeta \tag{VI.16}$$

The right-hand side of eqn (VI.16) is based on the Gaussian integration rule described in Appendix IV.

Equation (VI.14a) is valid for any virtual displacement $\{du^e\}$, thus it can be eliminated from both sides of eqns (VI.14a). Substituting eqns (VI.11) and (VI.13) into eqn (VI.14a) one obtains

$$\{P\}^{\mathrm{e}} = \left(\int_{\mathrm{vol}} [B]^{T''} [D][B]\,\mathrm{d}V\right)\{u^{\mathrm{e}}\} - \int_{\mathrm{vol}} [B]^{T''} [D]\{\varepsilon_0\}\,\mathrm{d}V$$
$$+ \int_{\mathrm{vol}} [B]^{T''}\{\sigma_0\}\,\mathrm{d}V - \int_S [N]^T\{p\}\,\mathrm{d}S - \int_{\mathrm{vol}} [N]^{T''}\{G\}\,\mathrm{d}V \qquad \text{(VI.17)}$$

Table (VI.1) gives miscellaneous loads and forces. Equation (VI.17) is the force–displacement relation with stiffness transformation. The terms in the eqn (VI.17) are defined as:

The element stillness matrix

$$[K]^{\mathrm{e}} = \int_{\mathrm{vol}} [B]^{T''}[D][B]\,\mathrm{d}V = \int_{-1}^{+1}\int_{-1}^{+1}\int_{-1}^{+1} B^{T''} DB \det J\,\mathrm{d}\xi\,\mathrm{d}\eta\,\mathrm{d}\zeta \qquad \text{(VI.17a)}$$

The nodal force due to body force

$$\{P_{\mathrm{b}}\}^{\mathrm{e}} = -\int_{\mathrm{vol}} [N]^{T''}\{G\}\,\mathrm{d}V = -\int_{-1}^{+1}\int_{-1}^{+1}\int_{-1}^{+1} N^{T''} G \det J\,\mathrm{d}\xi\,\mathrm{d}\eta\,\mathrm{d}\zeta \qquad \text{(VI.17b)}$$

The nodal force due to surface force

$$\{P_{\mathrm{s}}\}^{\mathrm{e}} = -\int_S [N]^{T''}\{p\}\,\mathrm{d}S \qquad \text{(VI.17c)}$$

The nodal force due to initial stress

$$\{P_{\sigma_0}\}^{\mathrm{e}} = \int_{\mathrm{vol}} [B]^{T''}\{\sigma_0\}\,\mathrm{d}V = \int_{-1}^{+1}\int_{-1}^{+1}\int_{-1}^{+1} B^{T''} \sigma_0 \det J\,\mathrm{d}\xi\,\mathrm{d}\eta\,\mathrm{d}\zeta \qquad \text{(VI.17d)}$$

The nodal force due to initial strain

$$\{P_{\varepsilon_0}\}^{\mathrm{e}} - \int_{\mathrm{vol}} [B]^{T''}[D]\{\varepsilon_0\}\,\mathrm{d}V = -\int_{-1}^{+1}\int_{-1}^{+1}\int_{-1}^{+1} B^{T''} D\varepsilon_0 \det J\,\mathrm{d}\xi\,\mathrm{d}\eta\,\mathrm{d}\zeta \qquad \text{(VI.17e)}$$

Equation (VI.17) can be rewritten as

$$\{F\}^{\mathrm{e}} = [K]^{\mathrm{e}}\{u\}^{\mathrm{e}} + \{P_{\mathrm{b}}\}^{\mathrm{e}} + \{P_{\mathrm{s}}\}^{\mathrm{e}} + \{P_{\sigma_0}\}^{\mathrm{e}} + \{P_{\varepsilon_0}\}^{\mathrm{e}} \qquad \text{(VI.17f)}$$

The initial strain vector can be replaced, or the strains due to creep and shrinkage, $\{P^{\mathrm{a}}\}^{\mathrm{e}}$ or $\{P^{\mathrm{s}}\}^{\mathrm{e}}$ can be added to it. Hence these expressions can be added to eqn (VI.17f). In addition, the strain in eqn (VI.17e) can be the element swelling strain vector $\{P^{\mathrm{s}'}\}^{\mathrm{e}}$. In the case of dynamic loadings applied to structures, an acceleration load vector $\{P^{\mathrm{a}}\}^{\mathrm{e}}$ can be introduced into eqn (VI.17f) such that

$$\{P^{\mathrm{a}}\}^{\mathrm{e}} = [M]\{A_{\mathrm{n}}\} = \sum_{i=1}^{n} [M_{\mathrm{e}}] = P\sum_{i=1}^{n} [A_{\mathrm{n}}]^{T''}[A_{\mathrm{n}}]\,\mathrm{d}[J]\,\mathrm{d}\xi\,\mathrm{d}\eta\,\mathrm{d}\zeta \qquad \text{(VI.17g)}$$

where $[M]$ = the total mass matrix; $[M_{\mathrm{e}}]$ = the element mass matrix; and $\{A_{\mathrm{n}}\}$ = the nodal acceleration vector. If large displacements are considered, the nodal vector $\{P^*\}^{\mathrm{e}}$ can also be incorporated into eqn (VI.17f).

Table VI.1. Miscellaneous loads and forces [216]

Gravitational forces (surface, forces)
Equivalent nodal force in the line of gravity Z-direction

$$\{P_\mathrm{s}\}_i = \int_v [N^T]_i \begin{Bmatrix} 0 \\ 0 \\ -pg \end{Bmatrix} \mathrm{d\ vol} = \sum_{i=1}^{n}\sum_{i=1}^{n}\sum_{i=1}^{n} [N]_{i,jK}^T \begin{Bmatrix} 0 \\ 0 \\ -pg \end{Bmatrix} |J|_{i,j,K} W_i W_j W_K$$

Body forces
Body force component per unit volume at (X, Y) point is

$$\begin{Bmatrix} f_x \\ f_y \\ f_z \end{Bmatrix} = \{\bar{f}\} = P_\mathrm{b}\omega^2 \begin{Bmatrix} X \\ Y \\ 0 \end{Bmatrix} \qquad 0 \neq \int_v \{N\}^T \begin{Bmatrix} f_x \\ f_y \\ f_z \end{Bmatrix} \mathrm{d\ vol}$$

in the case of isoparametric elements.

Concentrated loads
Concentrated loads away from the point

$$\{P_{\sigma_0}\} = N_i(\xi_1, \eta_1, \zeta_1)P \qquad \xi_1 = \xi \qquad \eta_1 = -\eta \qquad \zeta = +1$$

Distributed loads

$$\{P_S\} = \int_{-1}^{+1}\int_{-1}^{+1}\int_{-1}^{+1} N_i^{T''}[p_x, p_y, p_z]^{T''} \begin{Bmatrix} \dfrac{\partial Y}{\partial \xi}\dfrac{\partial Z}{\partial \eta} - \dfrac{\partial Z}{\partial \xi}\dfrac{\partial Y}{\partial \eta} \\ \dfrac{\partial Z}{\partial \xi}\dfrac{\partial X}{\partial \eta} - \dfrac{\partial Z}{\partial \eta}\dfrac{\partial X}{\partial \xi} \\ \dfrac{\partial X}{\partial \xi}\dfrac{\partial Y}{\partial \eta} - \dfrac{\partial Y}{\partial \xi}\dfrac{\partial X}{\partial \eta} \end{Bmatrix} \mathrm{d}\xi\,\mathrm{d}\eta \quad \text{for } \zeta = \pm 1$$

similarly for $\xi = \pm 1$; $\eta = \pm 1$.

Thermal loads

$$\{P\}_\mathrm{T} = \int B^T D \varepsilon_\mathrm{T}\,\mathrm{d}v = \sum_{i=1}^{n}\sum_{j=1}^{n}\sum_{k=1}^{n} [B]_{i,jK}[D]\{\varepsilon_\mathrm{T}\}|J|_{i,j,K} W_i W_j W_K$$

$$\{\varepsilon_\mathrm{T}\} = [\alpha_\mathrm{T}T, \alpha_\mathrm{T}T, \alpha_\mathrm{T}T \quad 0 \quad 0 \quad 0]^T \qquad T = \sum_{i=1}^{n} N_i T_i$$

Creep loads

$$\{P\}_{ev} = \int B^T D \varepsilon_\mathrm{c}\,\mathrm{d}v = \sum_{i=1}^{n}\sum_{j=1}^{n}\sum_{k=1}^{n} [B]_{i,jK}[D]\{\varepsilon_\mathrm{c}\}_{i,jk} W_i W_j W_K \,\mathrm{d}\xi\,\mathrm{d}\eta\,\mathrm{d}\zeta$$

Equation (VI.17f) is the force–displacement relation for each element. For the whole structure the stiffness matrix and load vectors are assembled according to the nodal incidences; overall equilibrium equations can be written in the following form, as before,

$$[K]\{U\} = \{F\} \qquad \text{(VI.18)}$$

Equation (VI.18) is solved for unknown displacements. The element strains can be obtained from nodal displacements. Linear material stresses are obtained from eqn (VI.13).

The same procedure is adopted for choosing other element types. Appendix I gives data for some well-known elements.

If boundary conditions are specified on $\{U\}$ to guarantee a unique solution, eqn (VI.17f) can be solved to obtain nodal point displacements at any node in the given structure. The equations with all degrees of freedom can be written as:

$$\left[\begin{array}{c|c} K & K_{\mathrm{R}} \\ \hline K_{\mathrm{R}}^{T''} & K_{\mathrm{RR}} \end{array}\right]\left\{\begin{array}{c} U \\ \hline U_{\mathrm{R}} \end{array}\right\}=\left\{\begin{array}{c} F \\ \hline F_{\mathrm{R}} \end{array}\right\} \tag{VI.19}$$

The subscript R represents reaction forces. The top half of the eqn (VI.19) is used to solve for $\{U\}$.

$$\{U\} = -[K]^{-1}[K_{\mathrm{R}}]\{U_{\mathrm{R}}\} + [K]^{-1}\{F\} \tag{VI.20}$$

The reaction forces $\{F_{\mathrm{R}}\}$ are computed from the bottom half of the equation as

$$\{F_{\mathrm{R}}\} = [K_{\mathrm{R}}]^{T''}\{U\} + \{K_{\mathrm{RR}}\}\{U_{\mathrm{R}}\} \tag{VI.21}$$

Equation (VI.20) must be in equilibrium with eqn (VI.21).

VI.2.6 The superelement and substructuring

For large structures with complicated features, a substructure (superelement) may be adopted on the lines suggested in eqn (VI.19). This superelement may then be used as a reduced element from the collection of elements. If subscripts γ and γ' represent the retained and removed degrees of freedom of the equations partitioned into two groups, then the expressions in eqn (VI.19) can be written as

$$\left[\begin{array}{c|c} K_{\gamma\gamma} & K_{\gamma\gamma'} \\ \hline K_{\gamma'\gamma} & K_{\gamma'\gamma'} \end{array}\right]\left\{\begin{array}{c} U_{\gamma} \\ \hline U_{\gamma'} \end{array}\right\}=\left\{\begin{array}{c} F_{\gamma} \\ \hline F_{\gamma'} \end{array}\right\} \tag{VI.22}$$

Equation (VI.22) when expanded assumes the following form:

$$\{F_{\gamma}\} = [K_{\gamma\gamma}]\{U_{\gamma}\} + [K_{\gamma\gamma'}]\{U_{\gamma'}\} \tag{VI.22a}$$

$$\{F_{\gamma'}\} = [K_{\gamma'\gamma}]\{U_{\gamma}\} + [K_{\gamma'\gamma}]\{U_{\gamma'}\} \tag{VI.22b}$$

When a dynamic analysis is carried out, the subscript γ (retained) represents the dynamic degrees of freedom.

When eqn (VI.22b) is solved, the value of $U_{\gamma'}$ is then written, similarly to eqn (VI.20),

$$\{U_{\gamma'}\} = [K_{\gamma'\gamma}]^{-1}\{F_{\gamma'}\} - [K_{\gamma'\gamma'}]^{-1}[K_{\gamma'\gamma}]\{U_{\gamma}\} \tag{VI.23}$$

Substituting $\{U_{\gamma'}\}$ into eqn (VI.22a)

$$[\![[K_{\gamma\gamma}] - [K_{\gamma'\gamma}][K_{\gamma\gamma'}]^{-1}[K_{\gamma'\gamma}]]\!]\{U_{\gamma}\} = [\{F_{\gamma}\} - [K_{\gamma\gamma'}][K_{\gamma\gamma'}]^{-1}\{F_{\gamma'}\}] \tag{VI.24}$$

or

$$[\bar{K}]\{\bar{U}\} = \{\bar{F}\} \tag{VI.24a}$$

where

$$[\bar{K}] = [K_{\gamma\gamma}] - [K_{\gamma\gamma'}][K_{\gamma'\gamma'}]^{-1}[K_{\gamma'\gamma}] \tag{VI.24b}$$

$$\{\bar{F}\} = \{F_{\gamma}\} - [K_{\gamma\gamma'}][K_{\gamma'\gamma'}]^{-1}\{F_{\gamma'}\} \tag{VI.24c}$$

$$\{\bar{U}\} = \{U_{\gamma}\} \tag{VI.24d}$$

and $[\bar{K}]$ and $\{\bar{F}\}$ are generally known as the substructure stiffness matrix and load vector, respectively.

In the above equations, the load vector for the substructure is taken as a total load vector. The same derivation may be applied to any number of independent load vectors. For example, one may wish to apply thermal, pressure, gravity and other loading conditions in varying proportions. Expanding the right-hand sides of eqns (VI.22a) and (VI.22b)

$$\{F_{\gamma}\} = \sum_{i=1}^{n} \{F_{\gamma i}\} \tag{VI.25}$$

$$\{F_{\gamma'}\} = \sum_{i=1}^{n} \{F_{\gamma' i}\} \tag{VI.26}$$

where $n =$ the number of independent load vectors.

Substituting into eqn (VI.24c)

$$\{\bar{F}\} = \sum_{i=1}^{n} \{F_{\gamma\gamma'}\} - [K_{\gamma\gamma'}][K_{\gamma'\gamma'}]^{-1} \sum_{i=1}^{n} \{F_{\gamma' i}\} \tag{VI.27}$$

The left-hand side of eqn (VI.27) is written as

$$\{\bar{F}\} = \sum_{i=1}^{n} \{\bar{F}_i\} \tag{VI.28}$$

Substituting eqn (VI.28) into eqn (VI.27), the following equation is achieved:

$$\{\bar{F}_i\} = \{F_{\gamma i}\} - [K_{\gamma\gamma'}][K_{\gamma\gamma'}]^{-1}\{F_{\gamma' i}\} \tag{VI.29}$$

VI.3 The membrane isoparametric elements

The membrane steel plane element acting with a concrete element to form a composite section is adopted for many concrete structures, such as bridge decks, concrete pressure and containment vessels, offshore gravity platforms, etc. They are treated as thin shell elements. These elements are capable of transmitting only the stresses in plane while assuming a constant strain along the thickness. These elements are compatible with the one face of the solid isoparametric elements representing concrete. The element local, global and curvilinear co-ordinate systems are shown in Figs VI.1a–c. Appendix II gives details of shape functions. Knowing these functions, and using eqns (VI.5) and (VI.6), the strain–displacement relation can be obtained. The thickness d of the element is interpolated as

$$d = \sum_{i=1}^{n} N_i d_i \tag{VI.30}$$

where n = the number of nodes, and d_i = the nodal thickness at node i.

As shown in Fig. VI.1, the local co-ordinate system x', y', z' with corresponding nodal freedom u', v', w' can be used to obtain the local strain field at that point.

$$\varepsilon_{x'} = \frac{\partial u'}{\partial x'} \qquad \varepsilon_{y'} = \frac{\partial v'}{\partial y'} \qquad \gamma_{x',y'} = \frac{\partial u'}{\partial y'} + \frac{\partial v'}{\partial x'} \tag{VI.31}$$

The x'-axis is tangential to the x_i-axis, the z'-axis is normal to the plane of the element, and the y'-axis is determined from the right-hand co-ordinate system. If $\hat{F}_\xi$ is the vector tangential to the ξ-axis on the η-axes, the vector is written as

$$\hat{F}_\xi = \begin{Bmatrix} \dfrac{\partial x'}{\partial \xi} \\ \dfrac{\partial y'}{\partial \xi} \\ \dfrac{\partial z'}{\partial \xi} \end{Bmatrix} \tag{VI.32}$$

Similarly the $\hat{F}_\eta$ vector, in matrix form, is written as

$$\hat{F}_\eta = \begin{Bmatrix} \dfrac{\partial x'}{\partial \eta} \\ \dfrac{\partial y'}{\partial \eta} \\ \dfrac{\partial z'}{\partial \eta} \end{Bmatrix} \tag{VI.33}$$

The vector normal to these vectors will be shown as:

$$\hat{F}_{z'} = \hat{F}_\xi \times \hat{F}_\eta = \begin{bmatrix} \dfrac{\partial y'}{\partial \xi} & \dfrac{\partial z'}{\partial \eta} & -\dfrac{\partial z'}{\partial \xi} & \dfrac{\partial y'}{\partial \eta} \\ \dfrac{\partial z'}{\partial \xi} & \dfrac{\partial x'}{\partial \eta} & -\dfrac{\partial x'}{\partial \xi} & \dfrac{\partial z'}{\partial \eta} \\ \dfrac{\partial x'}{\partial \xi} & \dfrac{\partial y'}{\partial \eta} & -\dfrac{\partial y'}{\partial \xi} & \dfrac{\partial x'}{\partial \eta} \end{bmatrix} \tag{VI.34}$$

The above vectors are normalised in order to obtain direction cosines

$$\hat{\hat{F}}_z = \frac{\hat{F}_z}{|\hat{F}_z|} \tag{VI.35}$$

For the x'-axis

$$\hat{F}_{x'} = \frac{\hat{F}_\xi}{|\hat{F}_\xi|}$$

and for the y'-axis

$$\hat{\hat{F}}_y = \hat{\hat{F}}_z \times \hat{\hat{F}}_x$$

where

$$|\hat{F}_z| = \sqrt{\left(\frac{\partial x'}{\partial \xi}\frac{\partial z'}{\partial \eta} - \frac{\partial z'}{\partial \xi}\frac{\partial y'}{\partial \eta}\right)^2 + \left(\frac{\partial z'}{\partial \xi}\frac{\partial x'}{\partial \eta} - \frac{\partial x'}{\partial \xi}\frac{\partial z'}{\partial \eta}\right)^2 + \left(\frac{\partial x'}{\partial \xi}\frac{\partial y'}{\partial \eta} - \frac{\partial y'}{\partial \xi}\frac{\partial x'}{\partial \eta}\right)^2} \tag{VI.35a}$$

$$|\hat{F}_\xi| = \sqrt{\left(\frac{\partial x'}{\partial \xi}\right)^2 + \left(\frac{\partial y'}{\partial \xi}\right)^2 + \left(\frac{\partial z'}{\partial \xi}\right)^2}$$

The direction cosines in a matrix form for the local orthogonal Cartesian system assume the following form:

$$\underset{3\times 3}{\{\hat{F}\}} = \begin{Bmatrix} \hat{\hat{F}}_{x'} \\ \hat{\hat{F}}_{y'} \\ \hat{\hat{F}}_{z'} \end{Bmatrix} = \begin{bmatrix} \hat{F}_{x',x} & \hat{F}_{y',x} & \hat{F}_{z',x} \\ \hat{F}_{x',y} & \hat{F}_{y',y} & \hat{F}_{z',y} \\ \hat{F}_{x',z} & \hat{F}_{z',z} & \hat{F}_{z',z} \end{bmatrix} \tag{VI.36}$$

VI.3.1 Derivatives

The local derivative of eqn (VI.31) must be obtained. To start with, sets of transformations are considered and the global derivatives are obtained using the method given in Appendix I.

$$\partial U_G = \begin{bmatrix} \frac{\partial u}{\partial x'} & \frac{\partial v}{\partial x'} & \frac{\partial w}{\partial x'} \\ \frac{\partial u}{\partial y'} & \frac{\partial v}{\partial y'} & \frac{\partial w}{\partial y'} \\ \frac{\partial u}{\partial z'} & \frac{\partial v}{\partial z'} & \frac{\partial w}{\partial z'} \end{bmatrix} = J^{-1} \begin{bmatrix} \frac{\partial u}{\partial \xi} & \frac{\partial v}{\partial \xi} & \frac{\partial w}{\partial \xi} \\ \frac{\partial u}{\partial \eta} & \frac{\partial v}{\partial \eta} & \frac{\partial w}{\partial \eta} \\ 0 & 0 & 0 \end{bmatrix} \tag{VI.37}$$

where

$$[J] = \begin{bmatrix} \frac{\partial x'}{\partial \xi} & \frac{\partial y'}{\partial \xi} & \frac{\partial z'}{\partial \xi} \\ \frac{\partial x'}{\partial \eta} & \frac{\partial y'}{\partial \eta} & \frac{\partial z'}{\partial \eta} \\ \hat{F}_{z',x}d & \hat{F}_{z',y}d & \hat{F}_{z',z}d \end{bmatrix} \tag{VI.37a}$$

The next step is to obtain the local derivatives and they are summarised below:

$$\partial U_{\mathrm{L}} = \hat{F}^{T''}\partial\hat{U}_{\mathrm{G}}\hat{F} = \sum_{i=1}^{n}\begin{bmatrix} \partial\hat{F}_1^i\,\partial u_1^i & \partial\hat{F}_1^i\,\partial u_2^i & \partial\hat{F}_1^i\,\partial u_3^i \\ \partial\hat{F}_2^i\,\partial u_1^i & \partial\hat{F}_2^i\,\partial u_2^i & \partial\hat{F}_2^i\,\partial u_3^i \\ \partial\hat{F}_3^i\,\partial u_1^i & \partial\hat{F}_3^i\,\partial u_2^i & \partial\hat{F}_3^i\,\partial u_3^i \end{bmatrix} \tag{VI.38}$$

where

$$\begin{aligned}
\partial F_1^i &= \hat{F}_{x',x}\frac{\partial N_i}{\partial x} + \hat{F}_{x',y}\frac{\partial N_i}{\partial y} + \hat{F}_{x',z}\frac{\partial N_i}{\partial z} \\
\partial F_2^i &= \hat{F}_{y',x}\frac{\partial N_i}{\partial x} + \hat{F}_{y',y}\frac{\partial N_i}{\partial y} + \hat{F}_{y',z}\frac{\partial N_i}{\partial z} \\
\partial F_3^i &= \hat{F}_{z',x}\frac{\partial N_i}{\partial z} + \hat{F}_{z',y}\frac{\partial N_i}{\partial y} + \hat{F}_{z',z}\frac{\partial N_i}{\partial z} \\
\partial u_1^i &= \hat{F}_{x',x}u_i + \hat{F}_{x',y}v_i + \hat{F}_{x',z}w_i \\
\partial u_2^i &= \hat{F}_{y',y}u_i + \hat{F}_{y',y}v_i + \hat{F}_{y',z}w_i \\
\partial u_3^i &= \hat{F}_{z',z}u_i + \hat{F}_{z',y}v_i + \hat{F}_{z',z}w_i
\end{aligned} \tag{VI.38a}$$

Using eqns (VI.31) and (VI.38), the strain value is computed as

$$\begin{aligned}
\{\varepsilon\} &= \sum_{i=1}^{n}\left\{\begin{matrix} \partial\hat{F}_1^i\,\partial u_1^i \\ \partial\hat{F}_2^i\,\partial u_2^i \\ \partial\hat{F}_2^i\,\partial u_1^i + \partial\hat{F}_1^i\,\partial u_2^i \end{matrix}\right\} \\
&= [B]\{u^{\mathrm{e}}\} \\
&= \left\{\begin{matrix} [B_1] \\ [B_2] \\ \vdots \\ [B_i] \\ \vdots \\ [B_n] \end{matrix}\right\} \times \left\{\begin{matrix} \{u_1\} \\ \{u_2\} \\ \\ \{u_i\} \\ \\ \{u_n\} \end{matrix}\right\}
\end{aligned} \tag{VI.39}$$

in which $[B_i]$ and $\{u_i\}$ are given as

$$\underset{3\times 3}{[B]} = \begin{bmatrix} \bar{\hat{F}}_{x',x}\,\partial\hat{F}_1^i & \hat{F}_{x',y}\,\partial\hat{F}_1^i & \hat{F}_{x',z}\,\partial\hat{F}_1^i \\ F_{y',x}\,\partial\hat{F}_2^i & \hat{F}_{y',y}\,\partial\hat{F}_2^i & \hat{F}_{y',z}\,\partial\hat{F}_2^i \\ (\hat{F}_{x',x}\,\partial\hat{F}_2^i + \hat{F}_{y',x}\,\partial\hat{F}_1^i) & (\hat{F}_{x',y}\,\partial\hat{F}_2^i + \hat{F}_{y',y}\,\partial\hat{F}_1^i) & (\hat{F}_{x',z}\,\partial\hat{F}_2^i + \hat{F}_{y',z}\,\partial F_1) \end{bmatrix} \tag{VI.40}$$

$$\underset{3\times 1}{\{u_i\}} = \begin{Bmatrix} u_i \\ v_i \\ w_i \end{Bmatrix} \tag{VI.40a}$$

The local stresses at any point are written in usual manner in form of eqn (VI.13)

$$\{\sigma\} = [D](\{\varepsilon\} - \{\varepsilon_0\}) \tag{VI.41}$$

in which

$$\{\sigma'\} = [\sigma_{x'}, \sigma_{y'}, \tau_{x'y'}]^{T''} \qquad \{\varepsilon'_0\} = [\varepsilon_{x'0}, \varepsilon_{y'0}, 0]^{T''} \tag{VI.41a}$$

For the plane stress case, the elastic material matrix is given by

$$[D] = \frac{E_s}{1-\nu_s^2}\begin{bmatrix} 1 & \nu_s & 0 \\ \nu_s & 1 & 0 \\ 0 & 0 & \frac{1-\nu_s}{2} \end{bmatrix} \tag{VI.42}$$

where E_s and ν_s are the modulus of elasticity and Poisson's ratio of steel. For $[D]$ in three-dimensional situations please refer to Appendix III. The element stiffness matrix for this element is given by

$$[K] = \int_{-1}^{+1}\int_{-1}^{+1} B^{T''} D B \det J \, d\xi \, d\eta \tag{VI.43}$$

Again (as before), the three-dimensional situation can occur; in which case eqn VI.32 is modified. For further details about other elements, reference should be made to Appendix I.

VI.4 Isoparametric line elements

The method of derivation here for line elements is the same as for solid or membrane elements, the difference is only in the choice of displacement polynomials and the shape functions.

The line elements represent the conventional reinforcing bars or the prestressing tendons in concrete. The nodes of these elements depend upon the type of solid elements used for concrete. They are given below:

(a) A two-noded line element corresponds to an 8-noded solid element.
(b) A three-noded line element corresponds to a 20-noded solid element.
(c) A four-noded line element corresponds to a 30-noded solid element.

These line element nodes can be matched by placing them on top of the solid elements. Sometimes it is difficult to idealise them in this manner owing to the reinforcement layout. Such line elements can be placed in the body of the solid element. Where bond–slip analysis is carried out, these line elements can be

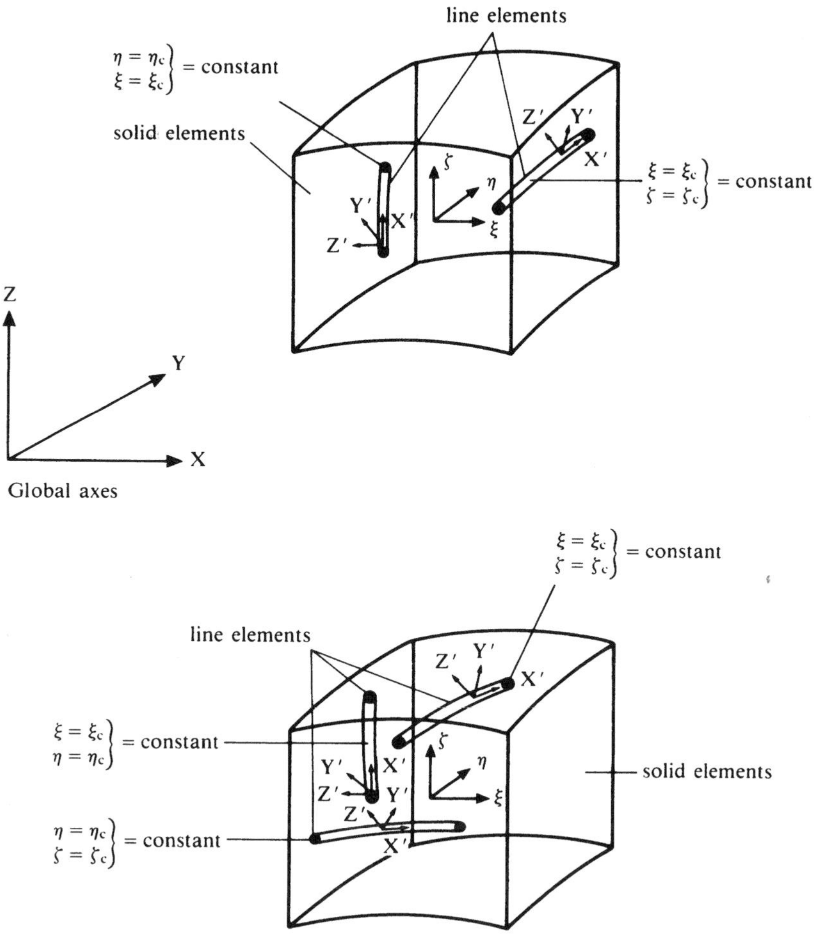

Fig. VI.2. Line elements in the body of a solid element.

placed on the nodes of the solid element or can be attached to spring elements as shown in Fig. VI.2.

VI.4.1 Two-, three- and four-noded elements

The shape functions and derivatives for the isoparametric line elements are given below:

(a) Two-noded line element (Fig. VI.1a)

Shape functions	*Derivatives*	
$N_1 = \frac{1}{2}(1-\xi)$	$\dfrac{\partial N_i}{\partial \xi} = -\frac{1}{2}$	(VI.44)
$N_2 = \frac{1}{2}(1+\xi)$	$\dfrac{\partial N_2}{\partial \xi} = \frac{1}{2}$	

(b) Three-noded line element (Figs VI.1 b_1 and b_2)

Shape functions *Derivatives*

$$N_1 = \tfrac{1}{2}(\xi - 1)\xi \qquad \frac{\partial N_1}{\partial \xi} = \xi - \tfrac{1}{2}$$

$$N_2 = 1 - \xi^2 \qquad \frac{\partial N_2}{\partial \xi} = 2\xi \tag{VI.45}$$

$$N_3 = \tfrac{1}{2}(\xi + 1)\xi \qquad \frac{\partial N_3}{\partial \xi} = \xi + \tfrac{1}{2}$$

(c) Four-noded line element (Figs VI.1 c_1 and c_2)

Shape functions *Derivatives*

$$N_1 = \tfrac{1}{3}(1 - \xi)(2\xi^2 - \tfrac{1}{2}) \qquad \frac{\partial N_1}{\partial \xi} = \tfrac{1}{3}(4\xi - 6\xi^2 + \tfrac{1}{2})$$

$$N_2 = \tfrac{4}{3}(\xi^2 - 1)(\xi - \tfrac{1}{2}) \qquad \frac{\partial N_2}{\partial \xi} = \tfrac{4}{3}(3\xi^2 - \xi - 1) \tag{VI.46}$$

$$N_3 = \tfrac{4}{3}(1 - \xi^2)(\xi + \tfrac{1}{2}) \qquad \frac{\partial N_3}{\partial \xi} = \tfrac{4}{3}(1 - 3\xi^2 - \xi)$$

$$N_4 = \tfrac{1}{3}(1 + \xi)(2\xi^2 - \tfrac{1}{2}) \qquad \frac{\partial N_4}{\partial \xi} = \tfrac{1}{3}(4\xi + 6\xi^2 - \tfrac{1}{2})$$

VI.4.2 The strain–displacement relation

At any point in the line element, the local Cartesian axis X' is tangential to the curvilinear axis. The local strain in the axial direction at any point can be written as

$$\varepsilon_{X'} = \partial U'/\partial X' \tag{VI.47}$$

Using the displacement transformation, eqn (VI.47) is written as

$$\varepsilon_{X'} = \frac{1}{L}\left(l_1 \frac{\partial U}{\partial \xi} + m_1 \frac{\partial V}{\partial \xi} + n_1 \frac{\partial W}{\partial \xi}\right) \tag{VI.47a}$$

where l_1, m_1, n_1 are the direction cosines of the X' axis and are written as

$$l_1 = \frac{\partial X}{\partial \xi} \Big/ L$$

$$m_1 = \frac{\partial Y}{\partial \xi} \Big/ L \tag{VI.47b}$$

$$n_1 = \frac{\partial Z}{\partial \xi} \Big/ L$$

$$L = \sqrt{(\partial X/\partial \xi)^2 + (\partial Y/\partial \xi)^2 + (\partial Z/\partial \xi)^2} \tag{VI.47c}$$

U, V, and W are the global nodal freedoms at any node, and U' is the local freedom in the X' direction. It is related according to

$$U' = l_1 U + m_1 V + n_1 W \tag{VI.48}$$

Equation (VI.47) in terms of the shape function derivatives is now written as

$$\varepsilon_{x'} = \frac{1}{L}\sum_{i=1}^{n}\left[l_1 \frac{\partial N_i}{\partial \xi}, m_1 \frac{\partial N_i}{\partial \xi}, n_1 \frac{\partial N_i}{\partial \xi}\right]\begin{Bmatrix} U_i \\ V_i \\ W_i \end{Bmatrix} \tag{VI.49}$$

where n is the number of nodes on the element.

Equation VI.49 is written also as

$$\{\varepsilon_{x'}\} = [B]\{U^{\mathrm{e}}\} \tag{VI.50}$$

where

$$[B] = [[B_1], [B_2], [B_3] \ldots [B_i] \ldots \{B_n\}]^T \tag{VI.51}$$

$$\{U^{\mathrm{e}}\} = [[U_1], [U_2], [U_3] \ldots [U_i] \ldots \{U_n\}]^T \tag{VI.52}$$

The stiffness matrix assumes the form

$$[K] = \int_{-1}^{+1} B^T E_{\mathrm{s}} B A(\xi) L \,\mathrm{d}\xi = \sum_{j=1}^{N} B_j^T E_{\mathrm{s}} B_j L_j W_j A(\xi_j) \tag{VI.53}$$

where

$$A(\varepsilon)_j = \sum_{i=1}^{n} N_i A_j \tag{VI.53a}$$

where A = the cross-sectional area at node j; n = the number of nodes on the element; N = the number of integration points; and N_i = the shape function at node i.

The strain and stress are calculated as

$$\{\varepsilon_{x'}\} = [B]\{U^{\mathrm{e}}\} \qquad \{\sigma_{x'}\} = [E_{\mathrm{s}}]\{\varepsilon_{x'}\} \tag{VI.54}$$

where $\{U^{\mathrm{e}}\}$ is the global nodal displacement of the element.

VI.4.3 Line element in the body of a solid element

The procedure for the strain–displacement and the stiffness matrix is the same as for the other line elements. The element must lie parallel to one of the curvilinear axes (ξ, η, ζ) of the solid element. The element may be anywhere in the solid element with maximum curvilinear co-ordinates $\xi = \pm 1$, $\eta = \pm 1$ and $\zeta = \pm 1$. The displacement U inside the element is written (Fig. VI.2) as

$$\{U\} = [\bar{N}]\{U^{\mathrm{e}}\} = \sum_{i=1}^{n} [N_i][I]\{U_i\} \tag{VI.55}$$

$$[\bar{N}] = [N(\xi, \eta_{\mathrm{c}}, \zeta_{\mathrm{c}})] \text{ (solid element)} \tag{VI.55a}$$

such that

$$\zeta = \zeta_c \qquad \eta = \eta_c \text{ (constant)}$$

For the tangential directions of ξ and η

$$\hat{X}_\xi = \begin{Bmatrix} \dfrac{\partial X}{\partial \xi} \\ \dfrac{\partial Y}{\partial \xi} \\ \dfrac{\partial Z}{\partial \xi} \end{Bmatrix} \text{ (at } \eta = \eta_c, \zeta = \zeta_c) \qquad \hat{X}_\eta = \begin{Bmatrix} \dfrac{\partial X}{\partial \eta} \\ \dfrac{\partial Y}{\partial \eta} \\ \dfrac{\partial Z}{\partial \eta} \end{Bmatrix} \text{ (at } \eta = \eta_c, \zeta = \zeta_c) \tag{VI.56}$$

A normal $\hat{Z}'$-axis vector, and $\hat{X}'$-axis and $\hat{Y}'$-axis vectors are defined as

$$\hat{Z}' = (\hat{X}_\xi \times \hat{X}_\eta)/|\hat{X}_\xi \times \hat{X}_\eta|$$
$$\hat{X}' = \hat{X}_\xi/|\hat{X}_\xi|$$
$$\hat{Y}' = \hat{Z}' \times X'$$
$$\hat{F} = [X', Y', Z']^{T''} = \begin{bmatrix} l_1 & l_2 & l_3 \\ m_1 & m_2 & m_3 \\ n_1 & n_2 & n_3 \end{bmatrix} \tag{VI.57}$$

Following eqn (VI.35) onward, the strains can be evaluated as

$$\varepsilon_{x'} = \frac{\partial U'}{\partial X'} = \sum_{i=1}^{n} \bar{a}(l_1 U_i + m_1 V_i + n_1 W_i) = [B]\{U^e\} \tag{VI.58}$$

$$\bar{a} = l_1 \frac{\partial N_i}{\partial x} + m_1 \frac{\partial N_i}{\partial y} + n_1 \frac{\partial N_i}{\partial z} \tag{VI.59}$$

Stresses are computed as

$$\{\sigma'_x\} = \{E_s\}\{\varepsilon'_x\} \tag{VI.60}$$

This analysis is needed when the prestressing tendons or conventional reinforcement cannot be modelled as line elements lying at the nodes of the solid elements. They can then be located at the body of the concrete element. This analysis will replace the ones adopted for the other line elements. For additional information, reference should be made to Appendix I.

VI.5 Plastic flow rule and stresses during elastoplastic straining

Many materials have been examined, including concrete. They behave elastically up to a certain stage of the loading beyond which plastic deformation takes place. During this plastic deformation the state of strain is not uniquely determined by the state of stress, as stated previously. In a uniaxial state of stress a simple rule is required to initiate yielding at any Gaussian point of the isoparametric element. In the multiaxial state of stress, there are an infinite number of possible combinations of stresses at which yielding starts. These can be examined by the flow rule. Moreover, the flow rule supplements the elastic constitutive relationship; and the plastic strain increments are related to the plastic stress increments during the occurrence of plastic flow.

The yield criterion described by a failure surface in a multi-dimensional stress space is given by Bangash [350, 351]

$$L_F(\sigma_{ij}^s, \varepsilon^p, S_H) = 0 \tag{VI.61}$$

where

L_F = the yield function

σ_{ij}^s = the multi-dimensional stress state

ε^p = accumulated plastic strain

S_H = strain hardening or softening parameter

The general form of the yield surface given by eqn (VI.61) allows either isotropic or kinematic hardening of the material. For a given previous history, $L_F(\sigma_{ij})$ is always considered as a function of the current state of stress for which S_H is variable.

To give added generality, the plastic potential to which the normality principle is applicable is assumed as

$$L_Q(\sigma_{ij}^s, \varepsilon^p, SH_0) = 0 \tag{VI.62}$$

This allows non-associated plasticity to be dealt with and associated rules to be obtained as a special case by making

$$L_F = L_Q \tag{VI.63}$$

For a perfectly plastic material the yield surface of eqn (VI.61) remains constant. For a strain hardening material the yield surface must change with continued straining beyond the initial yield.

This phenomenon is included in eqn (VI.61) by allowing both L_F and S_H to be functions of the state of stress and the plastic deformation history. This means S_H will have a new value for every time dependent yielding. Further, if the material is unloaded and then loaded again, additional yielding cannot take place, unless the current value of S_H has been exceeded.

A unified approach for arriving at the incremental stress–strain equation based on eqn (VI.62) can be written in a combined tensor form in three dimensions as

$$f(\sigma, \varepsilon, \dot{\sigma}, \dot{\varepsilon}, S_H) = 0 \tag{VI.64}$$

where f is a definite representation of L_F and is a stress function. The only change is in the value of S_H, accounting for isotropic and anisotropic hardening, and allowing the function f to be dependent not only on the present state of stress or strain, but also on the hardening history according to previous states of stress and strain. The value of $(\sigma, \dot{\sigma})$ and $(\varepsilon, \dot{\varepsilon})$ must be in the plastic range, having total values of σ_{ij}^{p} and $\varepsilon_{ij}^{\mathrm{p}}$ respectively.

When eqn (VI.64) is satisfied then the total differential of f is written as

$$\mathrm{d}f = \frac{\partial f}{\partial \sigma_{ij}^{\mathrm{p}}}\, \mathrm{d}\sigma_{ij}^{\mathrm{p}} + \frac{\partial f}{\partial \varepsilon_{ij}^{\mathrm{p}}}\, \mathrm{d}\varepsilon^{\mathrm{p}} + \frac{\partial f}{\partial S_{\mathrm{H}}}\, \mathrm{d}S_{\mathrm{H}} \tag{VI.65}$$

The yield condition with $\varepsilon_{ij}^{\mathrm{p}}$ and S_{H} held constant can be interpreted as a yield surface in the multidimensional stress space, and is in conformity with eqn (VI.61). When $f < 0$ the condition indicates a purely elastic change towards the inside of the yield surface. In cases where plastic flow does not occur, the increments of plastic strain $\mathrm{d}\varepsilon_{ij}^{\mathrm{p}}$ and the change of hardening parameter $\mathrm{d}S_{\mathrm{H}}$ will be automatically zero, and hence, in the case of the unloading, eqn (VI.65) is reduced to

$$\mathrm{d}f = \frac{\partial f}{\partial \sigma_{ij}^{\mathrm{p}}}\, \mathrm{d}\{\sigma_{ij}^{\mathrm{p}}\} < 0 \tag{VI.66}$$

When $\mathrm{d}f = 0$, which is the case for neutral loading, no plastic strain changes occur and the hardening factor remains unchanged, then

$$\mathrm{d}f = \frac{\partial f}{\partial \sigma_{ij}}\, \mathrm{d}\{\sigma_{ij}^{\mathrm{p}}\} = 0 \qquad \mathrm{d}\{\varepsilon_{ij}^{\mathrm{p}}\} = 0 \tag{VI.66a}$$

The quantity $\mathrm{d}\sigma_{ij}$ is tangent to the surface for neutral loading. For the vector products to be zero, $\partial f/\partial \sigma_{ij}$ will be normal to the surface. When $\mathrm{d}\sigma_{ij}$ is pointing to the outside of the surface the vector product will be positive; and this constitutes loading, with plastic flow taking place such that

$$\mathrm{d}f = \frac{\partial f}{\partial \sigma_{ij}}\, \mathrm{d}\{\sigma_{ij}^{\mathrm{p}}\} > 0 \tag{VI.67}$$

The definition for the structural material stability postulates that during a load cycle that includes loading and unloading, the work performed has to be greater than zero, i.e.

$$\mathrm{d}\varepsilon_{ij}^{\mathrm{p}}\, \mathrm{d}\sigma_{ij}^{\mathrm{p}} \geq 0 \tag{VI.68}$$

From eqns (VI.67) and (VI.68) for ideally plastic material the plastic strain increment $\mathrm{d}\varepsilon_{ij}^{\mathrm{p}}$ is proportional to the stress gradient of the yield surface

$$\mathrm{d}\varepsilon_{ij}^{\mathrm{p}} = \frac{\partial f}{\partial \sigma_{ij}^{\mathrm{p}}}\, \mathrm{d}\lambda \tag{VI.69}$$

where $\mathrm{d}\lambda$ = a constant of proportionality.

The normality rule given by eqn (VI.69) shows that the plastic strain increment has the same direction as the normal surface f in the stress space. At this stage both isotropic and kinematic hardening cases can be included in eqn (VI.69).

In order to perform the plastic analysis, the loading path is discretised into several linear load steps. The increments in stresses and strains for each step are then related by the rate constitutive laws mentioned earlier. The assumption is that the extension in the linearised step can be regarded as infinitesimal, and eqn (VI.69), which is true for all infinitesimal increments of stresses and strains, can be employed for small increments. Concrete structures for nuclear and offshore installations require small increments. The total strain increment $\mathrm{d}\varepsilon'_{ii}$ is the algebraic sum of the increments in elastic and plastic strains $\mathrm{d}\bar{\varepsilon}_{ij}$ and $\mathrm{d}\varepsilon^{\mathrm{p}}_{ij}$ respectively. The equation for $\mathrm{d}\varepsilon'_{ij}$ can be written as

$$\mathrm{d}\varepsilon_{ij} = \mathrm{d}\bar{\varepsilon}_{ij} + \mathrm{d}\varepsilon^{\mathrm{p}}_{ij} = \mathrm{d}\varepsilon^{\mathrm{e}}_{ij} + \mathrm{d}\varepsilon^{T}_{ij} + \mathrm{d}\varepsilon^{\mathrm{p}}_{ij} \tag{VI.70}$$

The stresses can be written as

$$\mathrm{d}\sigma'_{ij} = [D]\,\mathrm{d}\varepsilon_{kl} \tag{VI.71}$$

$$\mathrm{d}\sigma_{ij} = [D]\,\mathrm{d}\varepsilon_{kl} - \mathrm{d}\lambda[D]^{-1}\frac{\partial f}{\partial \sigma_{kl}} \tag{VI.72}$$

The plastic strain relationship is given as

$$\mathrm{d}\varepsilon = [D^*]^{-1}\,\mathrm{d}\{\sigma\} \tag{VI.73}$$

The total strain increment $\{\Delta\varepsilon\}_{\mathrm{TOT}}$ becomes

$$\{\Delta\varepsilon\}_{\mathrm{TOT}} = [D^*]^{-1}\{\Delta\sigma\} + \frac{\partial f}{\partial \sigma}\,\mathrm{d}\lambda \tag{VI.74}$$

When eqn (VI.65) is written in a compressed form as

$$\mathrm{d}f = \frac{\partial f}{\partial\{\sigma\}}\,\mathrm{d}\{\sigma\} + \frac{\partial f}{\partial S_{\mathrm{H}}}\,\mathrm{d}S_{\mathrm{H}} = 0 \tag{VI.75}$$

Replacing the second term for the hardening characteristics by $\mathrm{d}\lambda$, the following equations are obtained

$$\left\{\frac{\partial f}{\partial \sigma}\right\}^T \mathrm{d}\{\sigma\} - C\,\mathrm{d}\lambda = 0 \tag{VI.76}$$

$$\mathrm{d}\varepsilon_{\mathrm{TOT}} = [D^*]^{-1}\,\mathrm{d}\{\sigma\} + \left\{\frac{\partial f}{\partial \sigma}\right\}\mathrm{d}\lambda \tag{VI.77}$$

$$0 = \left\{\frac{\partial f}{\partial \sigma}\right\}^T \mathrm{d}\{\sigma\} - C\,\mathrm{d}\lambda \tag{VI.78}$$

Multiplying eqn (VI.77) with $[D^*]$ throughout, eqn (VI.77) can be expressed as

$$\mathrm{d}\{\sigma\} = [D^*]\{\mathrm{d}\varepsilon\}_{\mathrm{TOT}} - [D^*]\left\{\frac{\partial f}{\partial \sigma}\right\}\mathrm{d}\lambda \tag{VI.79}$$

Equation (VI.79) is then substituted into eqn (VI.78), and eqn (VI.80) is obtained

$$0 = \frac{\partial f^T}{\partial \sigma}[D^*]\{\mathrm{d}\varepsilon\}_{\mathrm{TOT}} - \left\{\frac{\partial f}{\partial \sigma}\right\}^T [D^*]\left\{\frac{\partial f}{\partial \sigma}\right\}\mathrm{d}\lambda - C\,\mathrm{d}\lambda \tag{VI.80}$$

Equation (VI.70) is then combined with eqns (VI.69), (VI.77) and (VI.65) to give the final value of $\mathrm{d}\lambda$

$$\mathrm{d}\lambda = \frac{\left\{\frac{\partial f}{\partial \sigma}\right\}^T [D^*]\{\mathrm{d}\varepsilon\}_{\mathrm{TOT}}}{C + \left\{\frac{\partial f}{\partial \sigma}\right\}^T [D^*]\left\{\frac{\partial f}{\partial \sigma}\right\}} \tag{VI.81}$$

The factor C in eqn (VI.81) is an unknown factor defining the state of elasto-plasticity. When $C = 0$, eqn (VI.81) gives a value of $\mathrm{d}\lambda$ for a perfectly plastic situation with no hardening. The stress increments are given by:

$$\{\mathrm{d}\sigma\}_{\mathrm{TOT}} = [D^*] - \frac{[D^*]\left\{\frac{\partial f}{\partial \sigma}\right\}\left\{\frac{\partial f}{\partial \sigma}\right\}^T [D^*]}{C + \left\{\frac{\partial f}{\partial \sigma}\right\}^T [D^*]\left\{\frac{\partial f}{\partial \sigma}\right\}}\{\mathrm{d}\varepsilon\}_{\mathrm{TOT}} \tag{VI.82}$$

for the elasto-plastic case,

$$\{\mathrm{d}\sigma\}_{\mathrm{TOT}} = [D^*] - [D_{\mathrm{P}}]\{\mathrm{d}_{\varepsilon}\}_{\mathrm{TOT}} \tag{VI.83}$$

$$[D_{\mathrm{P}}] = \frac{[D^*]\left\{\frac{\partial f}{\partial \sigma}\right\}\left\{\frac{\partial f}{\partial \sigma}\right\}^T [D]}{C + \left\{\frac{\partial f}{\partial \sigma}\right\}^T [D]\left\{\frac{\partial f}{\partial \sigma}\right\}} \tag{VI.84}$$

The true stress increment in eqn (VI.84) is the difference between the stress increment $[D_{\mathrm{p}}]\{\mathrm{d}\bar{\varepsilon}\}_{\mathrm{TOT}}$ and the algebraic sum of stresses $[D]\{\mathrm{d}\varepsilon\}$ due to elastic, creep, shrinkage, temperature, fatigue and other effects given in Part I.

The matrix $[D]$ has the flexibility to include any concrete failure models described elsewhere in this book.

The parameter C is given by Hill [336] as

$$C = 4/9\sigma_{\mathrm{eq}}^2 S_{\mathrm{H}} \tag{VI.85}$$

where S_{H} is the slope of the curve and represents hardening.

Where the influence of the studs, lugs and any other type of anchorages is to be included in the finite element analysis, the steps given in Table VI.2 are considered together with solid, panel and line elements.

VI.5.1 Sample cases

VI.5.1.1 Plastic potential of the same form as the yield surface

$$f(J_2) = f(\sigma_{ij}) = 3I_2 = \tfrac{3}{2}S_{ij}S_{ij} \tag{VI.86}$$

where

$$S_{ij} = \sigma_{ij} - \tfrac{1}{3}\sigma_{kk}\delta_{ij} \qquad \delta_{ij} = \text{the Kronecker delta}$$

Table VI.2. Analytical formulation of the steel anchors/studs

$$[K_{\mathrm{TOT}}]\{\delta\}^* + \{F_{\mathrm{T}}\} - \{R_{\mathrm{T}}\} = 0$$

where

$$[K_{\mathrm{TOT}}] = [K_l] + [K_{\mathrm{a}}]$$

$$\{\delta\}^* = \begin{Bmatrix} \delta_{\mathrm{un}} \\ \delta_{\mathrm{b}} \end{Bmatrix} \qquad \{F_{\mathrm{T}}\} = \begin{Bmatrix} F_{\mathrm{un}} \\ F_{\mathrm{b}} \end{Bmatrix} \qquad \{R_{\mathrm{T}}\} = \begin{Bmatrix} R_{\mathrm{un}} \\ R_{\mathrm{b}} \end{Bmatrix}$$

$[K_{\mathrm{TOT}}]$ = total stiffness matrix

$[K_{\mathrm{e}}]$ = liner stiffness matrix

$[K_{\mathrm{a}}]$ = a stud stiffness matrix

$\{F_{\mathrm{T}}\}$ = total initial load vector

$\{R_{\mathrm{T}}\}$ = total external load vector

Subscript

un = quantities corresponding to unknown displacement

b = quantities corresponding to restrained boundaries

$$[k_l]\{\delta_{\mathrm{un}}\} + \{F_{\mathrm{un}}\} = 0$$

$$\{\varepsilon\} = [B]\{\delta\}$$

$$\{\sigma\} = [D](\{\varepsilon\} - \{\varepsilon_0\})$$

$$\{\bar{S}\} = \text{anchor shear forces} = [K_{\mathrm{a}}]\{\delta_{\mathrm{un}}\}$$

$$\{F_{\mathrm{un}}\} = \int_\nu [B]^T[D]\{\varepsilon_0\}\,\mathrm{d}\nu = \int_\nu [B]^T[D]\{\varepsilon_0\}\det[J]\,\mathrm{d}\xi\,\mathrm{d}\eta\,\mathrm{d}\zeta$$

The plastic buckling matrix is given by

$$(K + \lambda K_{\mathrm{G}})F_{\mathrm{T}} = 0$$

where K = the elasto-plastic stiffness matrix as a function of the current state of plastic deformation; and K_{G} = the initial stress geometric stiffness matrix.

The determinant

$$|K + \lambda_{\mathrm{c}} K_{\mathrm{G}}| = 0$$

The essential equation is characteristically triangularised for the ith loading step as

$$(K^i + \lambda_{\mathrm{c}} K^i_{\mathrm{G}})F^i_{\mathrm{T}} = 0 \qquad \lambda_{\mathrm{c}} = 1 + E_{\mathrm{ps}}$$

E_{ps} is an accuracy parameter

Differentiating

$$\frac{\partial f}{\partial \sigma_{ij}} = \tfrac{3}{2}\partial(S_{kl}S_{kl})/\partial\sigma_{ij} = \tfrac{3}{2}S_{kl}\partial S_{mn}/\partial\sigma_{ij} = 3S_{ij} \tag{VI.87}$$

The plastic strain increment is stated as

$$d\lambda\varepsilon_{ij}^{p} = \lambda S_{ij} \tag{VI.88}$$

The equivalent plastic strain $\lambda\bar{\varepsilon}^{p}$ can be obtained as

$$d\lambda\bar{\varepsilon}^{p} = (\tfrac{2}{3}\, d\varepsilon_{ij}^{p}\, d\varepsilon_{ij}^{p})^{1/2} \tag{VI.89}$$

where

$$d\lambda = \frac{3}{2}\frac{d\bar{\varepsilon}^{p}}{\sigma_{eq}}$$

VI.5.1.2 Von Mises yield surface associated with isotropic hardening

The yield function f is written as

$$f = \frac{T}{2}(\sigma_{ij}) - \tfrac{1}{3}\sigma_{eq}^{2} \tag{VI.90}$$

By differentiating, eqn (VI.90) becomes

$$df = (S_{ij}\, d\sigma_{ij}) - \tfrac{2}{3}\sigma_{eq}(\partial\sigma_{eq})/(\partial\bar{\varepsilon}^{p})\{d\bar{\varepsilon}^{p}\} \tag{VI.91}$$

Using eqn (VI.66) onwards, the values of $d\lambda$ and $d\sigma$ given in eqns (VI.81) and (VI.82) are given by

$$d\lambda = \left([D^{*}]_{i,j,k,l} - \frac{[D^{*}]S_{mn}S_{pq}[D^{*}]_{pqkl}}{\sigma_{ij}[D^{*}]_{i,j,k,l}S_{kl} + (\tfrac{2}{3}\sigma_{eq})^{2}S_{H}}\right)\{d\varepsilon\}_{kl} \tag{VI.92}$$

VI.6 Dynamic analysis

Dynamic analysis is a subject by itself. In this text a specific approach to the dynamic analysis of concrete structures is discussed. The equations of motion are discretised in time. The direct integration technique [285–363] and the Wilson-θ method are summarised. Where dynamic problems are tackled, the nonlinear equations of motion (coupled or uncoupled), have been solved using these methods.

VI.6.1 Nonlinear transient dynamic analysis

The dynamic equilibrium at the nodes of a system of structural elements is formulated at a given time t as

$$[M]\{\ddot{U}_t\} + [C_t]\{\dot{U}_t\} + K_t'\{U_t\} = \{R_t\} \tag{VI.93}$$

where $\{U_t\}$ and $\{R_t\}$ are the vectors of displacement and specified load, respectively. $[M]$ represents the mass matrix which is regarded as constant; $[C_t]$ and $[K_t']$ are the damping and stiffness matrices, respectively. The subscript t is used for quantities at time t and a dot denotes a derivative with respect to time.

To formulate eqn (VI.93), discretisation with respect to time, using isoparametric finite elements, is performed. The simple applicable method is the numerical step-by-step integration of the coupled equations of motion, such as eqn VI.93. The response history is divided into time increments Δt, which are of equal length. The system is calculated for each Δt, with properties determined at the beginning of the interval. Only one matrix based on M, is excited. In addition, the direct integration technique allows a general damping matrix $[C_t]$ (which has to be specified explicitly), to be used without resorting to complex eigenvalues.

VI.6.1.2 Discretisation in the time domain

The equation of motion formulated at time $t = 0$, is written in the form

$$[M]\{\ddot{U}_0\} + [C_0]\{\dot{U}_0\} + [K_0]\{U_0\} = \{R_0\} \quad \text{(VI.94)}$$

where the subscript zero has been introduced. At time t, all quantities are known. Equation (VI.93) is specified as

$$[M]\{\ddot{U}_t\} + ([C_0] + [\Delta C_{0\to t}])\{\dot{U}_t\} + ([K_0'] + [\Delta K_{0\to t}])\{U_t\} = \{R_t\} \quad \text{(VI.95)}$$

or as

$$[M]\{\ddot{U}_t\} + [C_0]\{\dot{U}_t\} + [K_0]\{U_t\} = \{R_t\} + \{P_t\} = \{F\}_t \quad \text{(VI.96)}$$

where the initial load $\{P_t\}$ is specified by

$$\{P_t\} = -[\Delta C_{0\to t}]\{\dot{U}_t\} - \{\Delta K_{0\to t}\}\{\delta_t\} \quad \text{(VI.97)}$$

To obtain the solution at time $t + \Delta t$, the equation is stated as

$$[M]\{\ddot{U}_{t+\Delta t}\} + [C_0]\{\dot{U}_{t+\Delta t}\} + [K_0]\{U_{t+\Delta t}\} = \{R_{t+\Delta t}\} + \{P_t\} + \{\Delta P_{t\to t+\Delta t}\} \quad \text{(VI.98)}$$

$\{\Delta P_{t\to t+\Delta t}\}$ represents the influence of the nonlinearity during the time increment t and is determined by iteration and satisfied for $t + \tau$, where $\tau = \theta\Delta t$ ($\theta > 1.37$ for an unconditionally stable method) applied to a linear problem. $[\Delta C_{0\to t}]$ and $[\Delta K_{0\to t}]$ represent the change of $[C]$ and $[K]$, respectively, from $t = 0$ to t.

To obtain the solution at time $t + \Delta t$, eqn (VI.98) can be written as

$$[M]\{\ddot{U}_{t+\Delta t}\} + [C_0]\{\dot{U}_{t+\Delta t}\} + [K_0]\{U_{t+\Delta t}\} = \{R_{t+\Delta t}\} + \{F_t\} + \{\Delta F_{t\to t+\Delta t}\} \quad \text{(VI.99)}$$

$\{\Delta P_{t\to t+\Delta t}\}$ represents the influence of the nonlinearity during the time increment t and is determined by iteration

$$\begin{aligned}\{\Delta P_{t\to t+\Delta t}\} = &- [\Delta C_{0\to t}]\{\Delta\dot{U}_{t\to t+\Delta t}\} - [\Delta C_{t\to t+\Delta t}](\{\dot{U}\} + \{\Delta\dot{U}_{t\to t+\Delta t}\})\\ &- [\Delta K_{0\to t}]\{\Delta U_{t\to t+\Delta t}\} - [\Delta K_{t\to t+\Delta t}](\{U_t\} + \{\Delta U_{t\to t+\Delta t}\})\end{aligned} \quad \text{(VI.100)}$$

$(\Delta P_{t\to t+\Delta t})$ is calculated using the initial stress approach.

A modified Newton–Raphson or initial stress approach is adopted for solving these nonlinear equations. A step-by-step integration method is given in Table VI.3. Using these methods along with acceleration and convergence procedures described in this chapter and Appendix V can allow successful solution of finite element based problems.

VI.6.2 Reduced linear transient dynamic analysis

This is a reduced form of nonlinear transient dynamic analysis. This analysis is carried out faster than the nonlinear analysis since the matrix in eqn (IV.93) requires

Table VI.3. Summary of step-by-step integration method

Initialisation
(1) Effective stiffness matrix $[K_0^*] = (6/\tau^2)[M] + (3/\tau)[C_0] + [K_0]$ (A)
(2) Triangularise $[K_0^*]$

For each time step:
Calculation of displacement $\{U_{t+\tau}\}$
(1) Constant part of the effective load vector

$$\{R_{t+\tau}^*\} = \{R_t\} + \theta(\{R_{t+\Delta t}\} - \{R_t\}) + \{F_t\} = [M]$$
$$\times \left(\left(\frac{6}{\tau^2}\right)\{U_t\} + \frac{6}{\tau}\{\dot{U}_t\} + 2\{\ddot{U}_t\}\right) + [C_0]\left(\frac{3}{\tau}\{U_t\} + 2\{\dot{U}_t\} + \frac{\tau}{2}\{U_t\}\right) \quad \text{(B)}$$

(2) Initialisation $i = 0$, $\{\Delta P_{t \to t+\tau}^i\} = 0$
(3) Iteration
 (a) $i \to i + 1$
 (b) Effective load vector $\{R_{t+\tau_{\text{tot}}}^*\} = \{R_{t+\tau}^*\} + \{\Delta P_{t \to t+\tau}^{i-1}\}$ (C)
 (c) Displacement $\{U_{t+\tau}^i\}[K_0^*]\{U_{t+\tau}^i\} = \{R_{t+\tau_{\text{tot}}}^{*i}\}$ (D)
 (d) Velocity $\{\dot{U}_{t+\tau}^i\} + (3/\tau)(\{U_{t+\tau}^i\} - \{U_t\}) - 2\{\dot{U}_t\} - (\tau/2)\{\ddot{U}_t\}$
 (e) Change of initial load vector caused by the nonlinear behaviour of the material $\{\Delta P_{t \to t+\tau}^i\}$

$$\{\Delta P_{t \to t+\tau}^i\} = -[\Delta C_{0 \to t}](\{\dot{U}_{t+\tau}^i\} - \{\dot{U}_t\}) - [\Delta C_{t \to t+\tau}^i]\{\dot{U}_{t+\tau}^i\}$$
$$\times [\Delta K_{0 \to t}](\{U_{t+\tau}^i\} - \{U_t\}) - [\Delta K_{t \to t+\Delta t}^i]\{U_{t+\tau}^i\} \quad \text{(E)}$$

 In fact, $\{\Delta P_{t \to t+\tau}^i\}$ is calculated using the initial-stress method.
 (f) Iteration convergence

$$\|\{\Delta P_{t \to t+\tau}^i\} - \{\Delta P_{t \to t+\tau}^{i-1}\}\| / \|\{\Delta P_{t \to t+\tau}^i\}\| < \text{tol} \quad \text{(F)}$$

 or analogously, on stresses.
Note: $\{P\}$ could be any value of $\{F\}$.

Calculation of velocity, acceleration
Calculate new acceleration $\{\ddot{U}_{t+\Delta t}\}$, velocity $\{\dot{U}_{t+\Delta t}\}$, displacement $\{U_{t+\Delta t}\}$ and initial load $\{P_{t+\Delta t}\}$:

$$\{\ddot{U}_{t+\Delta t}\} = (6/\theta\tau^2)(\{U_{t+\tau}\} - \{U_t\}) - (6/\tau\theta)\{\dot{U}_t\} + \left(1 - \frac{3}{\theta}\right)\{\ddot{U}_t\}$$

$$\{\dot{U}_{t+\Delta t}\} = \{\dot{U}_t\} + \frac{\tau}{2\theta}\{\ddot{U}_t\} + \{\ddot{U}_{t+\Delta t}\} \quad \text{(G)}$$

$$\{U_{t+\Delta t}\} = \{U_t\} + \frac{\tau}{\theta}\{\dot{U}_t\} + (\tau^2/6\theta^2)(2\{\ddot{U}_t\} + \{\ddot{U}_{t+\Delta t}\})$$

$$\{P_{t+\Delta t}\} = \{P_t\} + \{\Delta P_{t \to t+\tau}^i\}$$

Calculation by quadratic integration
When the velocity varies linearly and the acceleration is constant across the time interval, appropriate substitutions are made into eqn (VI.93) giving

$$[f_1[M] + f_2[C_t] + [K_t']]\{U_t\} = \{F(t)\} + \{f_3([C_t], [M], U_t, U_{t2}, \ldots)\} \quad \text{(H)}$$

where f_1, f_2, f_3 = functions of time.

This results in an implicit time integration procedure. The only unknown is $\{U_t\}$ at each time point and this is calculated in the same way as in static analysis. Equation (H) is then written as:

$$\left(\frac{2}{\Delta t_0\,\Delta t_{01}}[M]+\frac{2\Delta t_0+\Delta t_1}{\Delta t_0\,\Delta t_{01}}[C]+[K_t']\right)\{U_t\}$$
$$=\{F(t)\}+[M]\left(-\frac{2}{\Delta t_0\,\Delta t_1}\{U_{t-1}\}-\frac{2}{\Delta t_1\,\Delta t_{01}}\{U_{t-2}\}\right)$$
$$+[C_t]\left(\frac{\Delta t_{01}}{\Delta t_0\,\Delta t_1}\{U_{t-1}\}-\frac{\Delta t_0}{\Delta t_{01}\,\Delta t_1}\{U_{t-2}\}\right) \qquad \text{(I)}$$

where

$\Delta t_0 = t_0 - t_1$ $\qquad t_0$ = time of current iteration

$\Delta t_1 = t_1 - t_2$ $\qquad t_1$ = time of previous iteration

$\Delta t_2 = t_2 - t_3$ $\qquad t_2$ = time before previous iteration

$\qquad t_3$ = time before t_2

$\Delta t_2 = \Delta t_0 + \Delta t_1 = t_0 - t_2$

Calculation by cubic integration
Equation (H) becomes cubic and hence is written as

$$(a_1[M]+a_2[C_t]+[K_t'])\{U_t\}=\{F(t)\}+[M](a_3\{U_{t-1}\}-a_4\{U_{t-2}\}+a_5\{U_{t-3}\})$$
$$+[C](a_6\{U_{t-1}\}-a_7\{U_{t-2}\}+a_8\{U_{t-3}\}) \qquad \text{(J)}$$

where a_1 to a_8 are functions of the time increments; these functions are derived by inverting a four by four matrix.

For clear-cut solutions, the size of the time step between adjacent iterations should not be more than a factor of 10 in nonlinear cases and should not be reduced by more than a factor of 2 where plasticity exists.

to be inverted once, and the analysis is reduced to a series of matrix multiplications and essential degrees of freedom (dynamic or master of freedoms) to characterise the response of the system. The analysis generally has restrictions such as constant $[M]$, $[C_t]$, $[K_t]$ and time interval for all iterations and nodal forces applied at dynamic or master degrees of freedom.

Equations (I) and (J) of Table VI.3 are written in a different form by considering the above-mentioned restraints.

VI.6.2.1 Quadratic integration

$$\left(\frac{1}{\Delta t^2}[M]_R+\frac{3}{2\Delta t}[\hat{C}_t]_R+[K_t]_R\right)\{U_t\}_R=\{F(t)\}_R$$
$$+[M]_R\frac{1}{\Delta t^2}(2\{U_{t-1}\}_R-\{U_{t-2}\}_R)$$
$$+\frac{1}{\Delta t}(2\{U_{t-1}\}_R-\tfrac{1}{2}\{U_{t-2}\}_R) \qquad \text{(VI.101)}$$

The symbol R represents reduced matrices and vectors.

VI.6.2.2 Cubic integration

$$\left(\frac{2}{\Delta t^2}[M]_{\mathrm{R}} + \frac{11}{6\Delta t}[C_t]_{\mathrm{R}} + [K_t]_{\mathrm{R}}\right)\{U_t\}_{\mathrm{R}} = \{F(t)\}$$

$$+ [M]_{\mathrm{R}}\frac{1}{\Delta t^2}(5\{U_{t-1}\}_{\mathrm{R}} - 4\{U_{t-2}\}_{\mathrm{R}} + \{U_{t-3}\})$$

$$+ [C_t]_{\mathrm{R}}\frac{1}{\Delta t^2}(3\{U_{t-1}\}_{\mathrm{R}} - \tfrac{3}{2}\{U_{t-2}\}_{\mathrm{R}} + \tfrac{1}{3}\{U_{t-3}\}_{\mathrm{R}}) \tag{VI.102}$$

VI.6.3 Mode frequency analysis

The equation of motion for an undamped structure with no applied forces is written as

$$[M]\{\ddot{U}_t\} + [K_t']\{U_t\} = \{0\} \tag{VI.103}$$

$[K_t']$, the structure stiffness matrix, may include stress-stiffening effects.

The system of equations is initially condensed down to those involved with the master (dynamic) degrees of freedom.

The number of dynamic degrees of freedom should at least be equal to two times the selected frequencies. The reduced form of eqn (VI.103) can be written as

$$[M]_{\mathrm{R}}\{\ddot{U}_t\}_{\mathrm{R}} + [K_t']_{\mathrm{R}}\{U\}_{\mathrm{R}} = \{0\} \tag{VI.104}$$

For a linear system, free vibrations of harmonic type are written as

$$\{U_t\}_{\mathrm{R}} = \{\psi_i\}_{\mathrm{R}}\cos\omega_i t \tag{VI.105}$$

where $\{\psi_i\}_{\mathrm{R}}$ = the eigenvector representing the shape of the ith frequency; ω_i = the ith frequency (radians/unit time); and t = time.

Equation (VI.103) assumes the form

$$(-\omega_i^2[M]_{\mathrm{R}} + [K_t']_{\mathrm{R}}\{\psi_i\}_{\mathrm{R}} = \{0\} \tag{VI.106}$$

which is an eigenvalue problem with n values of ω^2 and n eigenvectors $\{\psi_i\}_{\mathrm{R}}$ which satisfy eqn (VI.106), where n is the number of dynamic degrees of freedom. Using standard iteration procedures, eqn (IV.106) will yield a complete set of eigenvalues and eigenvectors.

Each eigenvector $\{\psi_i\}_{\mathrm{R}}$ is then normalised such that:

$$\{\psi_i\}_{\mathrm{R}}^{T''}[M]_{\mathrm{R}}\{\psi_i\}_{\mathrm{R}} = 1 \tag{VI.107}$$

These n eigenvectors are now expanded to the full set of structure modal displacement degrees of freedom. Using Section VI.2.6

$$\{\psi_{\gamma' i}\}_{\mathrm{R}} = [K_{\gamma'\gamma'}]^{-1}[K_{\gamma'\gamma}]\{\psi_i\}_{\mathrm{R}} \tag{VI.108}$$

where $\{\psi_i\}_{\mathrm{R}}$ = the slave degree of freedom vector of mode i; and $[K_{\gamma'\gamma'}]$, $[K_{\gamma'\gamma}]$ = submatrix parts as shown in eqn (VI.22) onwards.

The above dynamic analysis approach is generally adopted for structures subjected to normal dynamic loads, wind, wave and seismic loads. The above analysis, with modifications, is also applied to missile and aircraft explosions/impact problems.

VI.6.4 Spectrum analysis

Spectrum analysis is an extension of the mode frequency analysis, with both base and force excitation options. The base excitation option is generally suitable for seismic and wave applications. A direction vector and a response spectrum table will be needed in addition to the data and parameters required for the reduced modal analysis. The response spectrum table generally includes displacements, velocities and accelerations. The force excitation is, in general, used for wind and space structures and missile/aircraft impact. It requires a force distribution and an amplitude multiplier table in addition to the data and parameters needed for the reduced modal analysis. A study of the mass distribution is made. Generally the masses are kept close to the reaction points on the finite element mesh rather than the (master) degrees of freedom. It is important to calculate the participation factors in relation to a given excitation direction. The base and forced excitations are given below:

$$\tilde{\gamma}_i = \{\psi_i\}_{\mathrm{R}}^{T''}[M]\{\tilde{\mathrm{b}}\} \quad \text{for the base excitation} \tag{VI.109}$$

$$\tilde{\gamma}_i = \{\psi_i\}_{\mathrm{R}}^{T''}\{F_t\} \quad \text{for the force excitation} \tag{VI.110}$$

where $\{\tilde{\mathrm{b}}\}$ = the unit vector of the excitation direction; and $\{F_t\}$ = an input force vector, such as eqn (VI.93).

The values of $\{\psi\}_{\mathrm{R}}$ are normalised, and the reduced displacement vector is calculated from the eigenvector by using a mode coefficient $\tilde{M}$.

$$\{\tilde{U}\}_i = [\tilde{M}_i]\{\psi\}_i \tag{VI.111}$$

where $\{\tilde{U}\}_i$ = the reduced displacement vector; and $[\tilde{M}_i]$ = the mode coefficient and where (a) for velocity spectra

$$[\tilde{M}_i] = \frac{[V_{\mathrm{s}i}]\{\tilde{\gamma}_i\}}{\omega_i} \tag{VI.112}$$

($V_{\mathrm{s}i}$ = spectral velocity for the ith mode); (b) for force spectra

$$[\tilde{M}_i] = \frac{[\bar{f}_{\mathrm{s}i}]\{\tilde{\gamma}_i\}}{\omega_i^2} \tag{VI.113}$$

($\bar{f}_{\mathrm{s}i}$ = spectral force for the ith mode); (c)

$$[\tilde{M}_i] = \frac{[a_{\mathrm{s}i}]\{\tilde{\gamma}_i\}}{\omega_i^2} \tag{VI.114}$$

($a_{\mathrm{s}i}$ = spectral acceleration for the ith mode); (d)

$$[\tilde{M}_i] = \frac{[U_{\mathrm{s}i}]\{\tilde{\gamma}_i\}}{\omega_i^2} \tag{VI.115}$$

($U_{\mathrm{s}i}$ = spectral displacement for the ith mode).

$\{U\}_i$ may be expanded to compute all the displacements, as was done in eqns (VI.19) onwards

$$\{U_{\gamma'}\}_i = [K_{\gamma'\gamma'}]^{-1}[K_{\gamma'\gamma}]\{U_i\}_{\mathrm{R}} \tag{VI.116}$$

where $\{U_{\gamma'}\}_i$ = the slave degree of freedom vector of mode i; and $[K_{\gamma'\gamma'}]$, $[K_{\gamma'\gamma}]$ = submatrix parts.

Sometimes an equivalent mass M_i^e is needed for the ith mode since it may not be a function of excitation direction. This M_i^e is computed as

$$[M_i^e] = 1/\{\psi_i\}_R^{T''}\{\psi_i\}_R \quad \text{(VI.117)}$$

This is derived from the definition of the diagonal matrix of equivalent masses $[M^e]$

$$[\psi]_R^{T''}[M_e][\psi]_R = [I] \quad \text{(VI.117a)}$$

where $[I]$ = the identity matrix; and $[\psi]_R$ = a square matrix containing all mode shape vectors.

Where damping is included, the damping ratio D_{Ri} for the data input, including damping C_e, is given for a matrix of coupling coefficients as

$$D_{Ri} = C_e\omega_{i/2} \quad \text{(VI.118)}$$

where ω_i is the undamped natural frequency of the ith mode.

In between the modes i and j, a modified damping ratio D'_{Ri} is needed to take into account the concrete structures subjected to wave and seismic effects.

$$D'_{Ri} = D_{Ri} + 2/t_e\omega_i \quad \text{(VI.119)}$$

where t_e is the duration.

VI.6.5 Impact/explosion

Structural response of concrete structures subjected to relatively fast loading rates, such as those from missile and aircraft impact/explosion, bombs and nuclear detonations, etc., can be influenced by the effect of strain rate on the material properties. These material changes lead to an instantaneous change in the strength of materials such as concrete. Section VI.5 is invoked along with the above equations of motion. The normality rules and the proportionality factor $d\lambda$ are used as given in Section VI.5.

The $d\lambda$ values are substituted into eqn (VI.72) to give an expression for the stress state of the form

$$d\dot{\sigma}_{ij} = [D]\{d\dot{\varepsilon}_{kl}\} + \tilde{\gamma}_{ij}\{\ddot{\varepsilon}\}^p \quad \text{(VI.120)}$$

where, using Von Mises criterion,

$$\tilde{\gamma}_{ij} = \frac{[D^*]\{S_{mn}\}\{S_{kl}\}\{\bar{D}\}}{\alpha^*} \quad \text{(VI.121)}$$

$$\begin{aligned} \alpha^* &= \tfrac{4}{9}\sigma_{eq}^2\frac{\partial\sigma_{eq}}{\partial\varepsilon^p} + S_{ij}[D]\{S_{kl}\} \\ [D^*] &= D_{ijmn} \\ [\bar{D}] &= D_{pqkl} \end{aligned} \quad \text{(VI.122)}$$

the term $\gamma_{ij}\dot{\varepsilon}^{\mathrm{p}}$ can be implemented. Where deformation rates do not change too rapidly the term $\gamma_{ij}\varepsilon^{\mathrm{p}}$ is neglected. The rest of the procedure is the same as for general dynamic analysis.

VI.7 Buckling analysis

The eigenvalue buckling (bifurcation) analysis is vital where a steel liner is fully or partially anchored to concrete, such as in concrete bridges, tunnels, pressure and containment vessels, cells of offshore platforms, underground shelters, structures for hydroelectricity and irrigation, etc.

Buckling of the liner or its embedded anchors in concrete is possible. The buckling matrix is developed so that at appropriate stages the liner and its anchor system is checked against buckling. The equation below gives the eigenvalue buckling by bifurcation:

$$([K]^{\mathrm{e}} + \lambda_{\mathrm{e}i}[K^{\mathrm{s}}])\{\psi\}_i = \{0\} \tag{VI.123}$$

where

$[K]^{\mathrm{e}}$ = the elastic stiffness matrix

$[K^{\mathrm{s}}]$ = the stress stiffness matrix

$\lambda_{\mathrm{e}i}$ = the ith eigenvalue (used to multiply the loads which generated $[K^{\mathrm{s}}]$

$\{\psi\}_i$ = the ith eigenvector of displacements

The first step in the solution is to reduce eqn (VI.123) to its static or dynamic buckling (master) degrees of freedom. The elastic stiffness matrix $[K]^{\mathrm{e}}$ is reduced and the stress stiffness matrix $[K^{\mathrm{s}}]$ is reduced in a manner identical to that by which the mass matrix is reduced. Hence eqn (VI.123) becomes:

$$([K]_{\mathrm{R}} + \lambda_{\mathrm{e}i}[K^{\mathrm{s}}])\{\psi_i\}_{\mathrm{R}} = 0 \tag{VI.124}$$

where $[K]_{\mathrm{R}}$ can be replaced by $[K_t']_{\mathrm{R}}$ when dynamic problems are involved.

When a geometric stiffness matrix is included, the plastic buckling matrix is given as

$$([K^{\mathrm{p}}]_{\mathrm{R}} + \lambda_{\mathrm{c}i}[K]_{\mathrm{G}})\{\psi_i\}_{\mathrm{R}} = 0 \tag{VI.125}$$

where $\lambda_{\mathrm{c}} = 1 + E_{\mathrm{ps}}$, and E_{ps} = accuracy parameters.

Table VI.2 gives a brief breakdown of the buckling phenomenon. $\{\psi_i\}_{\mathrm{R}}$ is related to the total initial load vector $\{F_{\mathrm{T}}\}$ of the liner and the studs. Equations (VI.124) and (VI.125) are further condensed by again reducing out those rows and columns of the matrices that have a positive value on the main diagonal of the stress or geometric matrix. The matrices are inverted such that an accuracy is achieved having the lowest eigenvalue. A standard iteration procedure is adopted for the solution of these matrices. The eigenvectors are then normalised such that each has the largest value of 1.0.

VII. Material modelling simulation for finite element formulation

VII.1 Introduction

This chapter deals with the step-by-step numerical modelling of various concrete strengths and failure theories as applied to finite elements. Theoretical expressions are arranged in such a way that they can easily be incorporated into the finite element analysis given in Chapter VI. The simulation of the material modelling gives a unified analysis for the strength and failure conditions of concrete structures. The theoretical models given in this chapter cover One- to Five-Parameter Models: Bulk and Shear Moduli Models, Endochronic Theory, Temperature, Heat Conduction, Shrinkage, Creep, Bond, Fatigue and Cracking.

VII.2 Uniaxial stress–strain curve

Figure VII.1 shows an idealised stress–strain curve in compression and in tension. Curve B is obtained in a number of ways, as described in Chapter 1; from A to B$'$ eqns (I.1), (I.2), (I.5), (I.7) and (I.8) can be chosen depending upon the choice of the individual concrete strength theory. The stress–strain values are incorporated into eqn (VI.71). Using incremental techniques described in Chapter VI, a series of points can be chosen on the curve from A to B$'$ and then, invoking eqn (VI.71), they can be included in the finite element cycles. Where softening is included (B$'$ to B^{IV}), eqn (I.5) is used. If B$'$ to B^{IV} is treated as a straight line, then the value of σ^0 shall not be less than $0.15\sigma_{cu}$. In the case of cyclic loads, the value of σ^0 shall be limited to $0.2f_c'$. The B$'$–B^{IV} curve can also be evaluated from the following equation:

$$\frac{\sigma}{\sigma_{cu}} = \frac{2(\varepsilon/\varepsilon_{cu})}{1 + (\varepsilon/\varepsilon_{cu})^2} \qquad \text{(VII.1)}$$

Curve A represents the tension cut-off scheme.

Bicanic and Zienkiewicz [297] define the descending curve for concrete softening, i.e. B$'$B^{IV}:

$$\sigma^0 = c'/\gamma^* \qquad \text{(VII.2)}$$

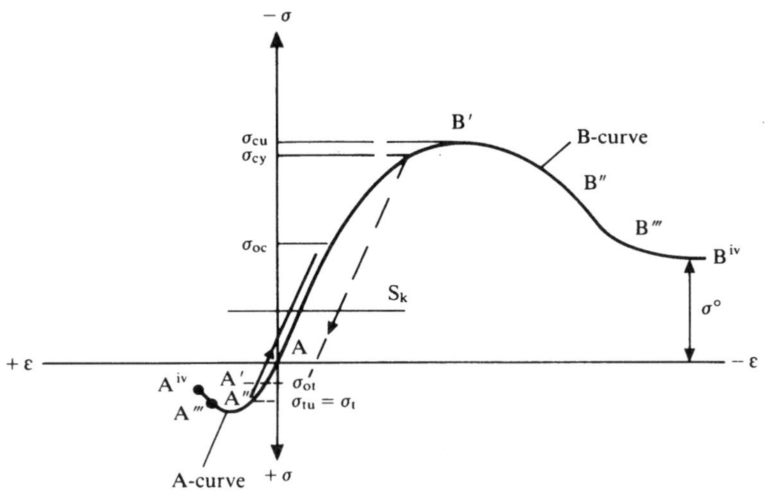

Fig. VII.1(a). Typical stress–strain curve.

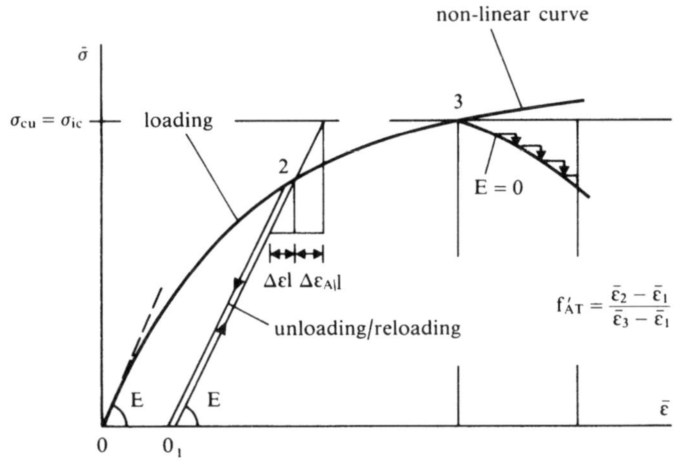

Fig. VII.1(b). Solution technique for stress–strain curve (loading–unloading scheme).

where

$$c' = \frac{\mathrm{d}\sigma'}{\mathrm{d}t'}\frac{1}{E_0} + \gamma^* \sigma'$$

$$\log \gamma = a_0 + a_1 \log \dot{\varepsilon}_c \quad \text{or} \quad \gamma^* = 10^{a_0}(\dot{\varepsilon}_c)^{a_1}$$

$$\dot{\varepsilon}_c = \text{strain rate} \qquad \sigma' = \sigma - \sigma_{0c} \qquad \sigma_{0c} = \text{initial stress of concrete}$$

$$a_0 = -2.9299 \qquad a_1 = 0.76 \qquad 0 \leq \dot{\varepsilon}_c \leq 10^{-5}$$

Figure VII.1 shows an idealisation of the tension part denoted by curve A. The maximum value is not greater than $0.55\sqrt{f'_c}$, and beyond this tensile failure is assumed to occur. In finite element analysis, this limitation is easily achieved in standard computer output, giving normal and principal stresses and cracks.

VII.2.1 Solution technique for descending stress–strain curve

Figure VII.1 shows the descending portion of the stress–strain curve. The tangent modulus of elasticity for this portion of the known curve is negative. Using a negative value of E in the constitutive matrix may lead to a non-positive definite stiffness matrix. In order to overcome this problem, the tangent modulus on the descending portion of this known curve is set to zero. For the purposes of numerical solution this is essential and permissible so long as the stress is corrected to the proper value at the end of each iteration.

Unloading of concrete in compression is treated differently from that in tension. For compression, the unloading of a point is calculated using the maximum equivalent strain in the finite element analysis. If the equivalent strain ($\sqrt{\frac{2}{3}\varepsilon_{ij}\varepsilon_{ij}}$) at any point is less than that of the equivalent strain in the previous load increment, the point is maintained. The unloading is treated as elastic, as shown in Fig. VII.2. The equivalent strains of all points from where unloading started can now be stored. For example, unloading in Fig. VII.2 starts at point 2 and follows the linear line 2–0_1, and the equivalent strain of point 2 is stored. Reloading of this point will follow the unloading line 0_1–2 until the equivalent strain of point 2 is reached. After that, the nonlinear curve is followed. Now upon reloading at point

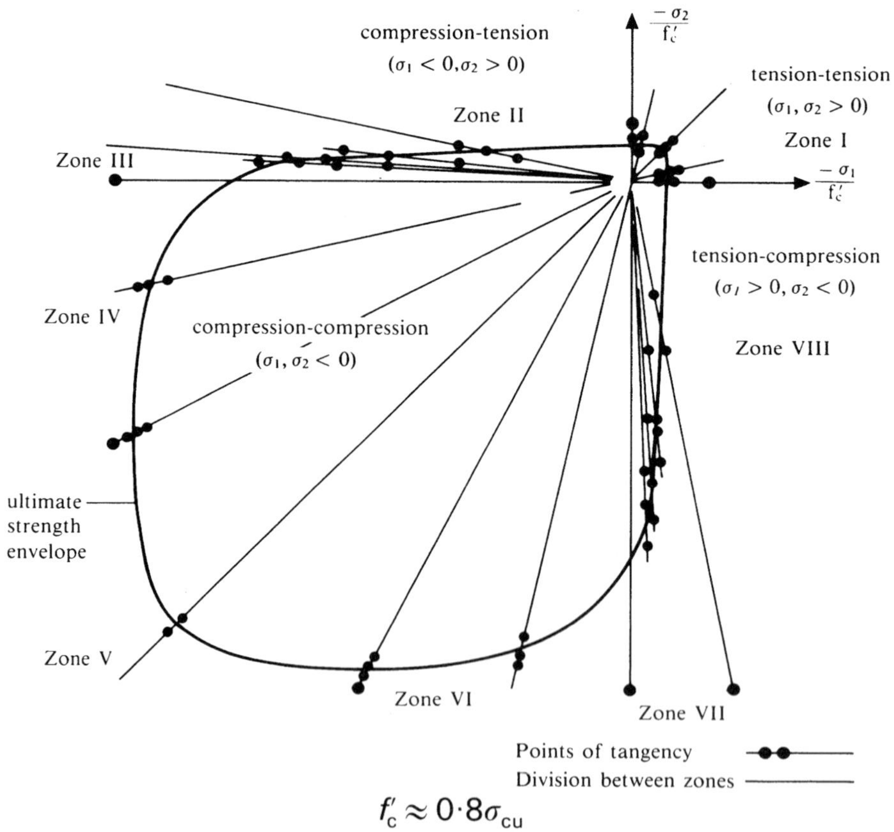

Fig. VII.2. Simulation of a typical biaxial state curve.

1 (on the same curve) the path 1–2–3 is followed. Path 1–2–3 of the strain increment covers both linear and nonlinear parts. The linear and nonlinear strains are calculated as described below.

If F_{AT} is a fraction of a linear strain increment, its value is computed as

$$F_{AT} = (\bar{\varepsilon}_2 - \bar{\varepsilon}_1)/(\bar{\varepsilon}_3 - \bar{\varepsilon}_1)$$

where $\bar{\varepsilon}_1$, $\bar{\varepsilon}_2$ and $\bar{\varepsilon}_3$ are the equivalent strains at points 1, 2 and 3, respectively:

$$\bar{\varepsilon}_1 = \sqrt{\tfrac{2}{3}\varepsilon_1^{T''}\varepsilon_1} \qquad \bar{\varepsilon}_2 = \sqrt{\tfrac{2}{3}\varepsilon_2^{T''}\varepsilon_2} \qquad \bar{\varepsilon}_3 = \sqrt{\tfrac{2}{3}\varepsilon_3^{T''}\varepsilon_3} \tag{VII.3}$$

The linear part of the strain increment is given as

$$\{\Delta\varepsilon_l\} = F_{AT}\{\Delta\varepsilon\}_{TOT} = F_{AT}\{\Delta\varepsilon_l + \Delta\varepsilon_{nl}\} \tag{VII.4}$$

The nonlinear part of the strain increment is then computed as

$$\{\Delta\varepsilon_{nl}\} = (1 - F_{AT})\{\Delta\varepsilon\}_{TOT} \tag{VII.5}$$

where

$$\{\Delta\varepsilon\}_{TOT} = \text{the total strain increment} \tag{VII.6}$$

and

$$\Delta\varepsilon_{nl} = \text{nonlinear strain} \qquad \Delta\varepsilon_l = \text{linear strain}$$

Unloading of a point in tension is treated as elastic such that the point is uncracked. For a cracked point, the strain across the crack is checked, and if it is positive then the crack is assumed to be open and the material matrix of this point is adjusted. If the strain across the crack is negative, it is then assumed that the compressive strain in that direction has been developed and the material matrix in compression is evaluated. On loading, this point may develop tensile strain and the crack is assumed to be open if the strain across the crack becomes positive. For details of cracking of concrete reference should be made to Sections II.2.3, II.2.4 and Chapter VI.

VII.3 Biaxial stress envelope

A number of methods of modelling the biaxial stress envelope can be incorporated into the finite element analysis. Equation (I.13) is generally used. The curve can be defined in the biaxial principal stress space as shown in Fig. VII.3 in terms of ultimate strength in tension σ_{tu} and compression σ_{cu}, respectively, as obtained from the uniaxial stress–strain curve shown in Fig. VII.1. If f is the yield function, various regions of the stress space termed 'zones' can then be given analytical expressions. The peak values and limitations given in Section I.3 are marked. In the finite element analysis, curves for tension–tension regions, compression–tension regions and compression–compression regions are represented by their respective equations in Section I.3. In the compression–compression region under directions 1 and 2, the mathematical relations are given for the biaxial stress envelope given in Fig. VII.3. The initial yield curve and current yield curve can easily be plotted within the ultimate strength envelope, as indicated in Fig. VII.3. Three different

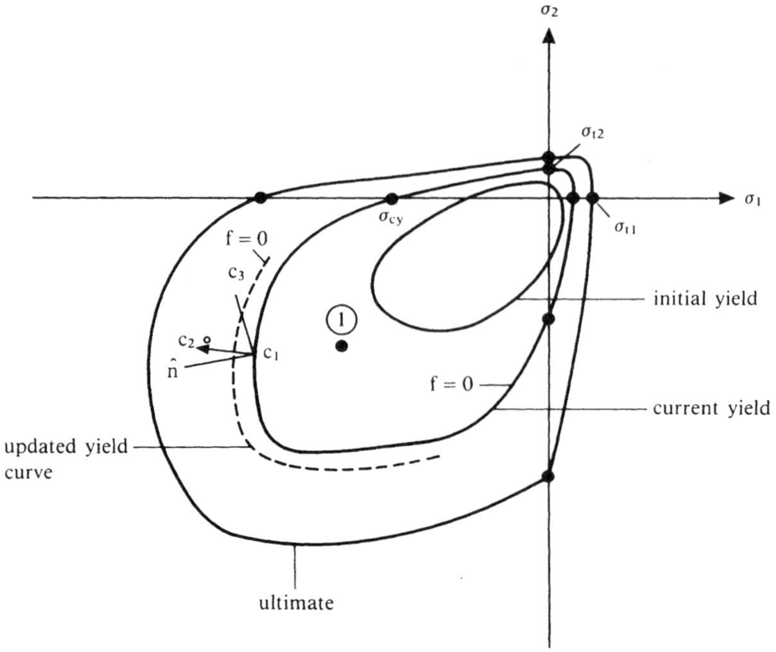

Fig. VII.3. Yield curves and points in stress space.

yield strengths are required in order to define the current yield curve which, in turn, requires three different strain hardening measurements. Assuming that increments in plastic strain are normal to the yield curve and the stress point is located on the current yield curve, then

$$\frac{\partial \varepsilon^{\mathrm{p}}}{\partial t} = \text{vector of the principal plastic strain rate}$$

$$= \dot{\varepsilon}^{\mathrm{p}}$$

$$= \left[\frac{\partial f'}{\partial \sigma_1}, \frac{\partial f'}{\partial \sigma_2}\right]^t \mathrm{d}\lambda \qquad \text{(VII.7)}$$

where f' is $f(\sigma_1, \sigma_2, \sigma_{\mathrm{cy}}, \sigma_{t1}, \sigma_{t2})$; $\mathrm{d}\lambda$ is constant of proportionality; and t is time. For any plastic strain, the rate of equivalent plastic strain $\partial \varepsilon^{\mathrm{p}}_{\mathrm{eq}}/\partial t$ can be determined on the lines suggested in eqn (VI.71).

This equivalent plastic strain rate is now decomposed arbitrarily into proportions which contribute to hardening of any yield strength. If three decomposition parameters are A_1, A_2 and A_3, the equivalent plastic strains are written as

$$\varepsilon^{\mathrm{p}}_{\mathrm{c}} = \sum_0^t A_1 \frac{\partial \varepsilon^{\mathrm{p}}_{\mathrm{eq}}}{\partial t}\,\mathrm{d}t \qquad \varepsilon^{\mathrm{p}}_{t1} = \sum_0^t A_2 \frac{\partial \varepsilon^{\mathrm{p}}_{\mathrm{eq}}}{\partial t}\,\mathrm{d}t \qquad \varepsilon^{\mathrm{p}}_{t2} = \sum_0^t A_3 \frac{\partial \varepsilon^{\mathrm{p}}_{\mathrm{eq}}}{\partial t}\,\mathrm{d}t \qquad \text{(VII.8)}$$

where $\varepsilon^{\mathrm{p}}_{\mathrm{c}}$, $\varepsilon^{\mathrm{p}}_{t1}$ and $\varepsilon^{\mathrm{p}}_{t2}$ are equivalent compressive plastic strain and tensile strains. Table VII.1 gives the decomposition parameters adopted by Elwi and Murray [5].

Table VII.1. Yield function and decomposition parameters (Elwi & Murray [5])

Zone	f	Limits		A_1	A_2	A_3
		$\frac{\sigma_1}{f'_c} = \frac{\sigma}{\sigma_{cy}}$	$\frac{\sigma_2}{f'_c} = \frac{\sigma}{\sigma_{cy}}$			
1	$\sqrt{\sigma_{t1}\sigma_{t2}} f_1\left(\frac{\sigma_1}{\sigma_{t1}}, \frac{\sigma_2}{\sigma_{t2}}\right)$	≥ 0	≥ 0	0	$^a\left(\frac{b_1}{b}\right)^2$	$^a\left(\frac{b_2}{b}\right)^2$
2	$\sigma_2 + \sigma_{t2} f_2\left(\frac{\sigma_1}{\sigma_{cy}}\right)$	$\left(-\frac{\sigma_2}{2\sigma_{t2}}, 0\right)$	> 0	$-\frac{\sigma_1}{\sigma_{cy}}$	0	$1 - A_1$
3	$\sigma_2 + \sigma_{t2} f_3\left(\frac{\sigma_1}{\sigma_{cy}}\right)$	$< -\frac{\sigma_2}{2\sigma_{t2}}$	> 0	$-\frac{\sigma_1}{\sigma_{cy}}$	0	$1 - A_1$
4	$\sigma_{cy} f_4\left(\frac{\sigma_2}{\sigma_{cy}}\right) - \sigma_1$	$< \frac{1.58491\sigma_2}{\sigma_{cy}}$	≤ 0	1	0	0
5	$\sigma_{cy} f_5\left(\frac{\sigma_1}{\sigma_{cy}}, \frac{\sigma_2}{\sigma_{cy}}\right) - \sigma_{cy}$	$\left(\frac{1.58491\sigma_2}{\sigma_{cy}}, \frac{\sigma_2}{1.58491\sigma_{cy}}\right)$	≤ 0	1	0	0
6	$\sigma_{cy} f_4\left(\frac{\sigma_1}{\sigma_{cy}}\right) - \sigma_2$	$\left(\frac{\sigma_2}{1.58491\sigma_{cy}}, 0\right)$	≤ 0	1	0	0
7	$\sigma_1 + \sigma_{t1} f_3\left(\frac{\sigma_2}{\sigma_{cy}}\right)$	> 0	$< -\frac{\sigma_1}{2\sigma_{t1}}$	$-\frac{\sigma_2}{\sigma_{cy}}$	$1 - A_1$	0
8	$\sigma_1 + \sigma_{t1} f_2\left(\frac{\sigma_2}{\sigma_{cy}}\right)$	> 0	$\left(-\frac{\sigma_1}{2\sigma_{t1}}, 0\right)$	$-\frac{\sigma_2}{\sigma_{cy}}$	$1 - A_1$	0

[a] $b = \left|\frac{\partial f}{\partial \sigma}\right|$ = the magnitude of $\frac{\partial f}{\partial \sigma}$; b_1, b_2 are the components of b.

Table VII.2. Step-by-step simulation of the biaxial envelope using the Kupfer model

The Kupfer model is chosen such that I_1 and I_2 are the first and second stress invariants. The yield criterion is written as

$$f(I_1, I_2) = \sqrt{K_1' I_1 + 3K_2' I_2} = \sigma_{eq} \tag{A}$$

where K_1' and K_2' are material parameters.

Kupfer's results for

$$\sigma_1/\sigma_2 = \infty \quad \text{(uniaxial compression)}$$

and

$$\sigma_1/\sigma_2 = 1 \quad \text{(equal biaxial compression)}$$

are

$$K_1' = 0.356\sigma_{eq} \qquad K_2' = 1.356\sigma_{eq} \tag{B}$$

For other values of σ_1/σ_2 reference should be made to Section I.3.

The strain increment $\Delta\varepsilon^p$ for the flow rule is given by

$$\Delta\varepsilon^p = d\lambda \frac{\partial f(\sigma)}{\partial \sigma} \tag{C}$$

where $d\lambda$ is a factor for the plastic strain increment from eqn (VI.81) and $\partial f/\partial\sigma$ is a vector normal to the current load surface.

The parameter or stress function, in this case from eqn (C), is written as

$$f(\sigma) = \left[2\frac{K_1'}{2\sigma_{eq}}\sigma_{eq} I_1 + 3K_2' I_2\right]^{1/2} = \frac{K_1}{2\sigma_{eq}} I_1 + \left[\left(\frac{K_1}{2\sigma_{eq}}\right)^2 + 3K_2 I_2\right]^{1/2} \tag{D}$$

Flow vectors

$$\frac{\partial f(\sigma)}{\partial \sigma} = \left[\underset{①}{\frac{\partial f}{\partial \sigma_x}}, \underset{②}{\frac{\partial f}{\partial \sigma_y}}, \underset{③}{\frac{\partial f}{\partial \tau_{xy}}}, \underset{④}{\frac{\partial f}{\partial \sigma_z}}, \underset{⑤}{\frac{\partial f}{\partial \tau_{yz}}}, \underset{⑥}{\frac{\partial f}{\partial \tau_{zx}}}\right] \tag{E}$$

where

$$\frac{\partial f(\sigma_1)}{\partial \sigma_1} = \frac{K_1'}{2\sigma_{eq}} + \left[2\left\{\left(\frac{K_1'}{2\sigma_{eq}}\right)^2 + K_2\right\}\sigma_x + \left\{2\left(\frac{K_1'}{2\sigma_{eq}}\right)^2 - K_2'(\sigma_y + \sigma_z)\right]\Big/ K^*$$

$$= L_1 + [L_2\sigma_x + L_3(\sigma_y + \sigma_z)]/K^*$$

$$\frac{\partial f(\sigma_2)}{\partial \sigma_2} = L_1' + [L_2'\sigma_y + L_3'(\sigma_z + \sigma_x)]/K^*$$

$$\frac{\partial f(\sigma_3)}{\partial \sigma_3} = L_1' + [L_2'\sigma_z + L_3'(\sigma_x + \sigma_y)]/K^*$$

$$\frac{\partial f(\sigma_4)}{\partial \sigma_4} = 6K_2'\tau_{xy}/K^*$$

$$\frac{\partial f(\sigma_5)}{\partial \sigma_5} = 6K_2'\tau_{yz}/K^*$$

$$\frac{\partial f(\sigma_6)}{\partial \sigma_6} = 6K_2'\tau_{zx}/K^*$$

$$K^* = 2[L_2'(\sigma_x^2 + \sigma_y^2 + \sigma_z^2) + L_3'(\sigma_x\sigma_y + \sigma_y\sigma_z + \sigma_z\sigma_x) + 3K_2'(\tau_{xy}^2 + \tau_{yz}^2 + \tau_{zx}^2)]^{1/2}$$

$$\{d\sigma\} = \text{the elasto-plastic incremental stress} = \{D^*\}\{d\varepsilon\} \tag{F}$$

The elasto-plastic material matrix $\{D^*\}$ is given in Chapter VI.

The rest of the procedure for crushing, compression–tension combinations, and cracking are given in earlier chapters and Section VII.11. For example, for crushing

$$3J_2 = \varepsilon_{cu}^2 \tag{G}$$

where J_2 is the deviatoric strain invariant.

For compressive behaviour of cracked concrete and where hardening S_H is involved, the value of S_H given below is substituted in the elasto-plastic material matrix $\{D^*\}_{ep}$:

$$S_H = E\left[\left\{1 - 20 \text{ to } 50 \times \frac{\varepsilon_t}{\varepsilon_p}\right\}^2 - 1\right]$$

where E = Young's modulus; ε_t = tensile strain; and ε_p = plastic strain.

Assuming $\{\sigma\}_1$, $\{\varepsilon\}_1$, $\{\varepsilon^p\}_1$, A_1, A_2 and A_3, a new value of $\{\varepsilon\}$ is imposed such that

$$\{\Delta\varepsilon\} = \{\varepsilon\} - \{\varepsilon\}_1 \tag{VII.9}$$

Next $\sigma_{cy}(A_1)$, $\sigma_{t1}(A_2)$ and $\sigma_{t2}(A_3)$ are computed. It is then easy to compute (Fig. VII.4) the corresponding values of $\{\Delta\sigma\}$ and $\{\sigma\}_2$:

$$\{\Delta\sigma\} = [D]\{\Delta\varepsilon\} \quad \text{and} \quad \{\sigma\}_2 = \{\sigma\}_1 + \{\Delta\sigma\} \tag{VII.10}$$

If

$$f[\{\sigma\}_2, \{\sigma_{cy}\}, \{\sigma_{t1}\}, \{\sigma_{t2}\}] \leq 0 \tag{VII.11}$$

then

$$\{\sigma\}_1 = \{\sigma\}_2 \qquad \{\varepsilon\}_1 = \{\varepsilon\}$$

When $f[\{\sigma\}_1, \{\sigma_{cy}\}, \{\sigma_{t1}\}, \{\sigma_{t2}\}] \geq 0$ it is necessary to take into consideration a factor which can bring the stress point to the yield surface. If that factor is F_{AT} as shown in Fig. VII.1, then

$$\{\sigma\} = \{\sigma\}_1 + F_{AT}[D]\{\Delta\varepsilon\} \qquad \{\Delta\varepsilon\} = (1 - F_{AT})\{\Delta\varepsilon\} \tag{VII.12}$$

The third case is when $\{\sigma\} = \{\sigma\}_1$ hardening parameters and decomposition parameters are updated, and the following value of $\{\Delta\sigma\}$ is computed:

$$\{\Delta\sigma\} = [D]_{ep}\{\Delta\varepsilon\} \tag{VII.13}$$

The rest of the solution procedure given in Sections II.2.3, II.2.4 and Appendix V is adopted for strength and crack evaluations. Table VII.2 gives a second version of the biaxial stress envelope and the tension curve in the finite element analysis. Loading surfaces for concrete constitutive models are given in Chapters I and II.

VII.4 Simulation procedures for one- to five-parameter models

VII.4.1 One-parameter model

Equations (II.1)–(II.4) give various cases of the one-parameter model. During loading it is necessary that the yield function of any one of these equations must be satisfied such that the stress state is on the yield surface. The simplest form of the isotropic-hardening Von Mises model is $f = J_2 = \frac{1}{2}S_{ij}S_{ij}$, where f is the loading function. If λ_s is a scalar function such that $\lambda_s = \lambda_s(J_2)$, then

$$d\varepsilon_{ij}^p = \lambda(J_2)\frac{\partial f}{\partial \sigma_{ij}}\,dJ_2 = \lambda_s(J_2)S_{ij}\,dJ_2 \tag{VII.14}$$

When both sides of eqn (VII.14) are squared, it yields

$$d\varepsilon_{ij}^p\,d\varepsilon_{ij}^p = \lambda_s^2(J_2)2J_2(dJ_2)^2 \tag{VII.15}$$

The value of $\int \sqrt{d\varepsilon_{ij}^p\,d\varepsilon_{ij}^p}$ is a function of $J_2 = \frac{1}{3}\sigma_{eq}^2$.

For a one-parameter model

$$\sigma_{eq} = \sqrt{3J_2} \tag{VII.16}$$

where

$$J_2 = \tfrac{1}{2}(S_x^2 + S_y^2 + S_z^2) + \tau_{xy}^2 + \tau_{yz}^2 + \tau_{zx}^2 \tag{VII.17}$$

$$S_x = \sigma_x - \sigma_0 \qquad S_y = \sigma_y - \sigma_0 \qquad S_z = \sigma_z - \sigma_0 \tag{VII.17a}$$

$$\sigma_0 = \text{mean stress} = \tfrac{1}{3}(\sigma_x + \sigma_y + \sigma_z)$$

The equivalent plastic strain ε_p is defined as

$$\varepsilon_p = \int \sqrt{\{d\varepsilon_{ij}^p\}^{T''}\{d\varepsilon_{ij}^p\}} \tag{VII.18}$$

$$d\varepsilon_p = d\lambda\sqrt{\left\{\frac{\partial f}{\partial \sigma_{ij}}\right\}^{T''}\left\{\frac{\partial f}{\partial \sigma_{ij}}\right\}} \tag{VII.18a}$$

where $d\lambda$ is a proportionality factor and f is a stress function of the given flow rule.

Knowing $d\lambda$, and upon solving for $d\lambda$, the following equation yields

$$\{d\varepsilon_{ij}^p\} = \{d\varepsilon_p\}\frac{\{\partial f/\partial \sigma_0\}}{\sqrt{\{\partial f/\partial \sigma_{ij}\}^{T''}\{\partial f/\partial \sigma_{ij}\}}} \tag{VII.19}$$

The stress increment can now be written as

$$\{d\sigma\} = [D](\{d\varepsilon\} - \{d\varepsilon^p\}) \tag{VII.20}$$

where $\{d\varepsilon\}$ and $\{d\varepsilon^p\}$ are the strain vectors and are related as

$$\{d\varepsilon^p\} = [C]\{d\varepsilon\} \tag{VII.21}$$

The matrix $[C]$ transforms the total strain increments into plastic strain increments. Substituting eqn (VII.21) into eqn (VII.20), the well-known equation expressed in the finite element analysis is derived as

$$\{d\sigma\} = [D]_{ep}\{d\varepsilon\} \tag{VII.22}$$

The plastic material matrix $[D_p]$ is derived in the way shown in Chapter VI.

VII.4.2 Two-parameter model

Equations (II.5)–(II.10) give various expressions for the two-parameter model. All these equations can easily be linked to the associated flow rule. For example, eqn

(II.10) is written as

$$f(\sigma) = f(I, \sqrt{J_2}) = K \tag{VII.23}$$

$$\frac{\partial f}{\partial \sigma_{ij}} = \frac{\partial f}{\partial I_1}\frac{\partial I_1}{\partial \sigma_{ij}} + \frac{\partial f}{\partial \sqrt{J_2}}\frac{\partial \sqrt{J_2}}{\partial \sigma_{ij}} \tag{VII.24}$$

$$= \frac{\partial f}{\partial I}\delta_{ij} + \tfrac{1}{2}\sqrt{J_2}\frac{\partial f}{\partial \sqrt{J_2}}S_{ij} \tag{VII.24a}$$

where δ_{ij} is the Kronecker delta and S_{ij} are the principal stress deviators.

This value is substituted in an equivalent plastic strain equation, such as is given in eqn (VII.18a). The procedure given from eqn (VII.19) onwards is now adopted for the stress and strain increments.

When eqn (II.10) is written in its original form

$$f = \alpha I_1 + \sqrt{J_2} - K = 0 \tag{VII.25}$$

the plastic stress–strain relation is written after computing $\partial f/\partial \sigma_{ij}$. The final expression is given as

$$\frac{\partial f}{\partial \sigma_{ij}} = \alpha\delta_{ij} + \tfrac{1}{2}\sqrt{J_2}S_{ij} \tag{VII.25a}$$

The plastic strain $\mathrm{d}\varepsilon_{ij}^{\mathrm{p}} = \mathrm{d}\lambda(\partial f/\partial \sigma_{ij})$ is then evaluated, and this is followed by the rest of the procedure given earlier.

Now the expression given in eqn (II.5) for the Mohr–Coulomb criterion can be rewritten for the elasto-plastic formulation to suit the finite element analysis. Chen [151] proposes the form of yield function f for the Mohr–Coulomb criterion by rewriting eqn (II.5) as

$$f(\sigma_{ij}) = I_1 \sin\phi + \frac{3\sin\theta(1-\sin\phi) + \sqrt{3}\cos\theta(3+\sin\phi)}{2}\sqrt{J_2} - 3c\cos\phi = 0 \tag{VII.26}$$

The chain rule given in eqn (VII.24) is extended to cater for the third parameter θ. The gradient $\partial f/\partial \sigma_{ij}$ is written as

$$\frac{\partial f}{\partial \sigma_{ij}} = \frac{\partial f}{\partial I_1}\frac{\partial I_1}{\partial \sigma_{ij}} + \frac{\partial f}{\partial I_2}\frac{\partial J_2}{\partial \sigma_{ij}} + \frac{\partial f}{\partial J_3}\frac{\partial J_3}{\partial \sigma_{ij}} \tag{VII.27}$$

Partial differentiation of the above terms is given below:

$$\frac{\partial f}{\partial I_1} = \sin\phi \qquad \frac{\partial \theta}{\partial J_2} = \frac{3\sqrt{3}}{4\sin 3\theta}\frac{J_3}{J_2^{5/2}} \qquad \frac{\partial \theta}{\partial J_3} = \frac{-\sqrt{3}}{2\sin\theta J_2^{3/2}} \tag{VII.28}$$

$$\frac{\partial f}{\partial J_2} = \frac{3\sin\theta(1-\sin\phi) + \sqrt{3}(3+\sin\phi)\cos\theta}{4J_2}$$
$$+ \frac{3\sqrt{3}J_3}{8J_2^2\sin 3\theta}[3(1-\sin\phi)\cos\theta - \sqrt{3}(3+\sin\phi)\sin\theta] \tag{VII.29}$$

$$\frac{\partial f}{\partial J_3} = -\frac{\sqrt{3}[3(1-\sin\phi)\cos\theta - \sqrt{3}(3+\sin\phi)\sin\theta]}{4J_2\sin 3\theta} \tag{VII.30}$$

The value of $\partial f/\partial\sigma_{ij}$ is calculated and the procedure given earlier in Section VII.4.1 is followed for the stress–strain calculation. Crushing and cracking can then be linked to this model as stated in other cases elsewhere in the text.

VII.4.3 Three-parameter model

The three-parameter model of William and Wanke has already been discussed in Section II.1.3. In the case of elasto-plastic formulation of the three-parameter model, eqn (II.13) is chosen; this predicts failure or yielding provided the average stresses σ_a, τ_a and the angle of similarity θ satisfy this equation. The gradient direction is again evaluated using the chain rule of differentiation:

$$\frac{\partial f}{\partial\sigma_{ij}} = \frac{\partial f}{\partial\sigma_a}\frac{\partial\sigma_a}{\partial\sigma_{ij}} + \frac{\partial f}{\partial\tau_a}\frac{\partial\tau_a}{\partial\sigma_{ij}} + \frac{\partial f}{\partial\theta}\frac{\partial\theta}{\partial\sigma_{ij}} \tag{VII.31}$$

The terms for σ_a and τ_a are given in eqn (II.13a). Various terms on the right-hand side of eqn (VII.31) are evaluated on the lines suggested in eqn (VII.27):

$$\frac{\partial f}{\partial\sigma_a}\frac{\partial\sigma_a}{\partial\sigma_{ij}} = \frac{1}{zf'_c}\tfrac{1}{3}\delta_{ij} \tag{VII.31a}$$

$$\frac{\partial f}{\partial\tau_a}\frac{\partial\tau_a}{\partial\sigma_{ij}} = \frac{1}{\rho(\theta)f'_c}\frac{1}{5\tau_a}S_{ij} \tag{VII.31b}$$

$$\frac{\partial f}{\partial\theta}\frac{\partial\theta}{\partial\sigma_{ij}} = \frac{\partial f}{\partial\rho}\frac{\partial\rho}{\partial\theta}\frac{\partial\theta}{\partial\sigma_{ij}} \tag{VII.31c}$$

The values of $\rho(\theta)$ are given in eqn (II.12a). In eqn (VII.31c), the terms are calculated as

$$\frac{\partial f}{\partial\theta} = -\frac{1}{\rho^2}\frac{\tau_a}{f'_c} \tag{VII.31d}$$

From eqn (II.12a)

$$\frac{\partial\rho}{\partial\theta} = \frac{V\dfrac{dU}{d\theta} - U\dfrac{dV}{d\theta}}{V^2} \tag{VII.31e}$$

where

$$\begin{aligned} U &= 2\rho_c(\rho_c^2 - \rho_t^2)\cos\theta + \rho_c(2\rho_t - \rho_c)[\alpha''_1 + 5\rho_t^2 - 4\rho_t\rho_c]^{1/2} \\ \alpha''_1 &= 4(\rho_c^2 - \rho_t^2)\cos^2\theta \end{aligned} \tag{VII.31f}$$

$$\frac{dU}{d\theta} = -2\rho_c(\rho_c^2 - \rho_t^2)\sin\theta\frac{4\rho_c(\rho_c - 2\rho_t)(\rho_c^2 - \rho_t^2)\sin\theta\cos\theta}{[\alpha' + 5\rho_t^2 - 4\rho_t\rho_c]^{1/2}} \tag{VII.31g}$$

$$V = \alpha''_1 + (\rho_c - 2\rho_t)^2 \qquad \frac{dV}{d\theta} = -8(\rho_c^2 - \rho_t^2)\sin\theta\cos\theta \tag{VII.31h}$$

The term $\partial\theta/\partial\sigma_{ij}$ is computed by differentiating the angle θ given in eqn (II.12b).

Again the angle $\cos\theta$ is denoted by similar terms given in eqn (VII.31e). Equation (II.12b) is written as

$$\theta = \cos^{-1}\left(\frac{U}{V}\right) \qquad \frac{\partial\theta}{\partial\sigma_{ij}} = \frac{\partial\cos^{-1}}{\partial\sigma_{ij}}\left(\frac{U}{V}\right) = \frac{1}{\sin\theta}\left[\frac{V\dfrac{\partial U}{\partial\sigma_{ij}} - U\dfrac{\partial V}{\partial\sigma_{ij}}}{V^2}\right] \tag{VII.31i}$$

$$U = 2\sigma_1 - \sigma_2 - \sigma_3 + \tau_{xy} = 0 + \tau_{yz} = 0 + \tau_{xz} = 0 \tag{VII.31j}$$

$$\frac{\partial U}{\partial\sigma_{ij}} = [2, -1, -1, 0, 0, 0] \tag{VII.31k}$$

$$V = \sqrt{2}[(\sigma_1 - \sigma_2)^2 + (\sigma_2 - \sigma_3)^2 + (\sigma_3 - \sigma_1)^2]^{1/2}$$

$$\frac{\partial V}{\partial\sigma_{ij}} = \frac{\sqrt{6}}{\sqrt{5}}\,\frac{\sqrt{15}\frac{1}{3}[2\sigma_1 - \sigma_2 - \sigma_3, 2\sigma_2 - \sigma_3 - \sigma_1, 2\sigma_3 - \sigma_1 - \sigma_2, 0, 0, 0]}{\sqrt{(\sigma_1 - \sigma_2)^2 + (\sigma_2 - \sigma_3)^2 + (\sigma_3 - \sigma_1)^2}}$$

$$= \frac{5[2\sigma_1 - \sigma_2 - \sigma_3, 2\sigma_2 - \sigma_3 - \sigma_1, 2\sigma_3 - \sigma_1 - \sigma_2, 0, 0, 0]}{\sqrt{(\sigma_1 - \sigma_2)^2 + (\sigma_2 - \sigma_3)^2 + (\sigma_3 - \sigma_1)^2}} \tag{VII.31l}$$

Equations (VII.31a)–(VII.31l) are substituted into eqn (VII.31) for evaluation of the gradient $\partial f/\partial\sigma_i$. The procedure given in Section VII.4.1 onward is adopted together with the finite element technique given in Chapter VI for crushing and cracking.

VII.4.4 Four-parameter model

An analytical failure model in which four parameters are included has been developed by Ottoson [24, 25]. This failure surface is described in Section II.1.4 and is defined by eqn (II.14). The same chain rule of partial differentiation is applied. The gradient $\{\partial f/\partial\sigma\}$ is given by Bangash [351, 749]:

$$C = \frac{\partial f}{\partial(\sigma)} = \frac{\partial f}{\partial I_1}\frac{\partial I_1}{\partial\sigma} + \frac{\partial f}{\partial J_2}\frac{\partial J_2}{\partial\sigma} + \frac{\partial f}{\partial\cos 3\theta}\frac{\partial\cos 3\theta}{\partial\sigma}$$

$$= a_1c_1 + a_2c_2 + a_3c_3$$

$$a_1 = \frac{\partial f}{\partial I_1} = \frac{B}{f_c'} \tag{VII.32}$$

$$a_2 = \frac{\partial f}{\partial J_2} = \frac{\partial}{\partial J_2}\frac{AJ_2}{f_c'^2} + \frac{\partial}{\partial J_2}\frac{\lambda J_2^{1/2}}{f_c'^2} = \frac{A}{f_c'^2} + \frac{\partial}{\partial J_2}\frac{\lambda J_2^{1/2}}{f_c'^2}$$

$$\frac{\partial}{\partial J_2}\frac{\lambda J_2^{1/2}}{f_c'^2} = \frac{\lambda}{f_c'}\frac{1}{2}J^{-1/2} + \frac{J^{1/2}}{f_c'}\frac{\partial(\lambda)}{\partial J_2} \tag{VII.33}$$

$$\frac{\partial(\lambda)}{\partial J_2} = \frac{\partial}{\partial J_2}\left\{K_1\cos\left[\tfrac{1}{3}\cos^{-1}\left(K\frac{3\sqrt{3}}{2}\frac{\sqrt{3}}{J_2^{3/2}}\right)\right]\right\} \tag{VII.34}$$

where

$$\cos 3\theta \geq 0 = -K_1 \sin(P)\frac{1}{3}\frac{(-1)}{\sqrt{1-t^2}}K_2\left(\frac{3}{2}\right)\sqrt{3}J_3\left(-\frac{3}{2}\right)J_2^{-6/2}$$

$$\equiv -K_1 \sin(P)\frac{1}{3J}\left(K_2\frac{3\sqrt{3}}{2}\frac{\sqrt{3}}{J_2^{3/2}}\right)\frac{1}{\sqrt{1-t^2}}\left(+\frac{3}{2}\right)$$

$$\equiv -K_1\frac{t}{\sqrt{1-t^2}}\sin(P)\frac{1}{2J} = -\frac{K_1}{2J}\frac{t}{\sqrt{1-t^2}}\sin(P)$$

$$t = K_2\frac{3\sqrt{3}}{2}\frac{J_3}{J_2^{3/2}}$$

$$\frac{\partial(\lambda)}{\partial J_2} = \frac{\partial}{\partial J_2}\left\{K_1\cos\left[\frac{\pi}{3}-\frac{1}{3}\cos^{-1}\left(-K_2\frac{3\sqrt{3}}{2}\frac{J_3}{J_2^{3/2}}\right)\right]\right\} \qquad \cos\theta \leq 0$$

$$= K_1(-\sin P)\left[-\frac{1}{3}\frac{(-1)}{\sqrt{1-t^2}}\left(-K_2\frac{3\sqrt{3}}{2}J_3\right)\left(-\frac{3}{2}\right)J^{-6/2}\right]$$

$$= \frac{K_1}{T}(-\sin P)\left[-\frac{1}{3}\frac{t}{\sqrt{1-t^2}}\left(-\frac{3}{2}\right)\right] = \frac{-K}{2J}(\sin P)\frac{t}{\sqrt{1-t^2}} \tag{VII.35}$$

From this equation

$$a_2 = \frac{A}{f_c'^2} + \frac{1}{2J_2^{1/2}f_c'}\left\{\lambda - K_1\sin\left[\frac{1}{3}\cos^{-1}\left(K_2\frac{3\sqrt{3}}{2}\frac{J_3}{J_2^{3/2}}\right)\right]\frac{t}{\sqrt{1-t^2}}\right\}; \ \cos 3\theta \geq 0$$

$$= \frac{A}{f_c'^2} - \frac{1}{2J_2^{1/2}f_c'}\left\{\lambda - K_1\sin\left[\frac{\pi}{3}-\frac{1}{3}\cos^{-1}\left(-K_2\frac{3\sqrt{3}}{2}\frac{\sqrt{3}}{J_2^{3/2}}\right)\frac{t}{\sqrt{1-t^2}}\right]\right\} \tag{VII.36}$$

$$a_3 = \frac{\partial}{\partial\cos 3\theta}\left(\lambda\frac{J_2^{1/2}}{f_c'}\right) \qquad \cos 3\theta \geq 0 \tag{VII.37}$$

$$= \frac{J_2^{1/2}}{f_c'}\frac{1}{3}K_1\sin\left[\frac{1}{3}\cos^{-1}(K_2\cos 3\theta)\right]\frac{K_2}{\sqrt{1-t^2}}$$

$$= \frac{1}{3}\frac{K_1K_2J_2^{1/2}}{3f_c'\sqrt{1-t^2}}\sin\left[\frac{1}{3}\cos^{-1}(K_2\cos 3\theta)\right] \qquad \cos\theta \geq 0 \tag{VII.38}$$

$$a_3 = \frac{K_1K_2J_2^{1/2}}{3\sqrt{1-t^2}f_c'}\sin\left[\frac{\pi}{3}-\frac{1}{3}\cos^{-1}(-K_2\cos 3\theta)\right] \qquad \cos 3\theta \leq 0$$

Where $t = K_2 \cos 3\theta$

$$J_2 = [\tfrac{1}{2}(S_x^2 + S_y^2 + S_z^2) + \tau_{xy}^2 + \tau_{yz}^2 + \tau_{zx}^2]$$

VII.4.4.1 The values of C_1, C_2, C_3

$$C_1 = \frac{\partial I_1}{\partial\{\sigma\}} = \begin{Bmatrix} 1 \\ 1 \\ 1 \\ 0 \\ 0 \\ 0 \end{Bmatrix}$$

$$C_2 = \frac{\partial J_2}{\partial\{\sigma\}} = \frac{\partial J_2}{\partial S_x}\frac{\partial S_x}{\partial\{\sigma\}} + \frac{\partial J_2}{\partial S_y}\frac{\partial S_y}{\partial\{\sigma\}} + \frac{\partial J_2}{\partial S_z}\frac{\partial S_z}{\partial\{\sigma\}}$$
$$+ \frac{\partial J_2}{\partial \tau_{xy}}\frac{\partial \tau_{xy}}{\partial\{\sigma\}} + \frac{\partial J_2}{\partial \tau_{yz}}\frac{\partial \tau_{yz}}{\partial\{\sigma\}} + \frac{\partial J_2}{\partial \tau_{zx}}\frac{\partial \tau_{zx}}{\partial\{\sigma\}}$$
$$= S_x\begin{Bmatrix} \frac{2}{3} \\ -\frac{1}{3} \\ \frac{1}{3} \\ 0 \\ 0 \\ 0 \end{Bmatrix} + S_y\begin{Bmatrix} -\frac{1}{3} \\ \frac{2}{3} \\ -\frac{1}{3} \\ 0 \\ 0 \\ 0 \end{Bmatrix} + S_z\begin{Bmatrix} -\frac{1}{3} \\ -\frac{1}{3} \\ \frac{2}{3} \\ 0 \\ 0 \\ 0 \end{Bmatrix} + 2\begin{Bmatrix} 0 \\ 0 \\ 0 \\ \tau_{xy} \\ \tau_{yz} \\ \tau_{zx} \end{Bmatrix} \quad \text{(VII.39)}$$

or

$$C_2 = \begin{Bmatrix} \frac{1}{3}(2S_x - S_y - S_z) \\ \frac{1}{3}(2S_y - S_x - S_z) \\ \frac{1}{3}(2S_z - S_x - S_y) \\ 2\tau_{xy} \\ 2\tau_{yz} \\ 2\tau_{zx} \end{Bmatrix} = \begin{Bmatrix} S_x \\ S_y \\ S_z \\ 2\tau_{xy} \\ 2\tau_{yz} \\ 2\tau_{zx} \end{Bmatrix} \quad \text{(VII.40)}$$

$$\cos 3\theta = \frac{3\sqrt{3}}{2} \times \frac{J_3}{J_2^{3/2}} \quad \text{(VII.41)}$$

$$C_3 = \frac{\partial \cos 3\theta}{\partial J_3} = \frac{\partial \cos 3\theta}{\partial J_3}\frac{\partial J_3}{\partial\{\sigma\}} + \frac{\partial \cos 3\theta}{\partial J_2}\frac{\partial J_2}{\partial\{\sigma\}} \quad \text{(VII.42)}$$

From eqn (VII.42)

$$\frac{\partial \cos 3\theta}{\partial J_3} = \frac{3\sqrt{3}}{2J_2^{3/2}} \qquad \frac{\partial \cos 3\theta}{2J_2} = \left(\frac{3\sqrt{3}}{2}J_3\right)\left(\frac{-\frac{3}{2}}{J_2^{5/2}}\right) = -\frac{9}{4}\sqrt{3}\frac{J_3}{J_2^{5/2}} \tag{VII.43}$$

Hence

$$C_3 = \frac{3}{2}\frac{\sqrt{3}}{J_2^{3/2}}\frac{\partial J_3}{\partial\{\sigma\}} - \frac{9}{4}\sqrt{3}\frac{J_3}{J_2^{5/2}}\frac{\partial J_2}{\partial\{\sigma\}} = \frac{3}{2}\frac{\sqrt{3}}{J_2^{3/2}}\left[\frac{\partial J_3}{\partial\{\sigma\}} - \frac{3}{2}\frac{J_3}{J_2}\frac{\partial J_2}{\partial\{\sigma\}}\right] \tag{VII.44}$$

Now

$$J_3 = [S_x S_y S_z + 2\tau_{xy}\tau_{yz}\tau_{zx} - S_x\tau_{yz}^2 - S_y\tau_{xz}^2 - S_z\tau_{xy}^2]$$

$$\frac{\partial J_3}{\partial\{\sigma\}} = \frac{\partial J_3}{\partial S_x}\frac{\partial S_x}{\partial\{\sigma\}} + \frac{\partial J_3}{\partial S_y}\frac{\partial S_y}{\partial\{\sigma\}} + \frac{\partial J_3}{\partial S_z}\frac{\partial S_z}{\partial\{\sigma\}}$$

$$+ \frac{\partial J_3}{\partial \tau_{xy}}\frac{\partial \tau_{xy}}{\partial\{\sigma\}} + \frac{\partial J_3}{\partial \tau_{yz}}\frac{\partial \tau_{yz}}{\partial\{\sigma\}} + \frac{\partial \sqrt{3}}{\partial \tau_{zx}}\frac{\partial \tau_{zx}}{\partial\{\sigma\}}$$

$$\frac{\partial J_3}{\partial S_x} = S_y S_z - \tau_{yz}^2 \qquad \frac{\partial J_3}{\partial S_y} = S_x S_z - \tau_{xz}^2 \qquad \frac{\partial J_3}{\partial S_z} = S_x S_y - \tau_{xy}^2 \tag{VII.45}$$

$$S_x = \tfrac{1}{3}(2\sigma_x - \sigma_y - \sigma_z) \tag{VII.46}$$

$$\frac{\partial S_x}{\partial\{\sigma\}} = \tfrac{1}{3}\begin{Bmatrix} 2 \\ -1 \\ -1 \\ 0 \\ 0 \\ 0 \end{Bmatrix} \qquad \frac{\partial S_y}{\partial\{\sigma\}} = \tfrac{1}{3}\begin{Bmatrix} -1 \\ 2 \\ -1 \\ 0 \\ 0 \\ 0 \end{Bmatrix} \qquad \frac{\partial S_y}{\partial\{\sigma\}} = \tfrac{1}{3}\begin{Bmatrix} -1 \\ -1 \\ 2 \\ 0 \\ 0 \\ 0 \end{Bmatrix} \tag{VII.47}$$

$$\frac{\partial J_3}{\partial \tau_{xy}} = 2\tau_{yz}\tau_{zx} - 2S_z\tau_{xy} \qquad \frac{\partial J_3}{\partial \tau_{yz}} = 2\tau_{xy}\tau_{zx} - 2S_x\tau_{yz}$$

$$\frac{\partial J_3}{\partial \tau_{xz}} = 2\tau_{xy}\tau_{yz} - 2S_y\tau_{xz} \tag{VII.48}$$

$$\frac{\partial \tau_{xy}}{\partial\{\sigma\}} = \begin{Bmatrix} 0 \\ 0 \\ 0 \\ 1 \\ 0 \\ 0 \end{Bmatrix} \qquad \frac{\partial \tau_{yz}}{\partial\{\sigma\}} = \begin{Bmatrix} 0 \\ 0 \\ 0 \\ 0 \\ 1 \\ 0 \end{Bmatrix} \qquad \frac{\partial \tau_{zx}}{\partial\{\sigma\}} = \begin{Bmatrix} 0 \\ 0 \\ 0 \\ 0 \\ 0 \\ 1 \end{Bmatrix} \tag{VII.49}$$

$$\frac{\partial J_3}{\partial\{\sigma\}} = \begin{Bmatrix} \frac{1}{3}[2(S_yS_z - \tau_{yz}^2) - (S_xS_z - \tau_{xz}^2) - (S_xS_y - \tau_{xy}^2)] \\ \frac{1}{3}[-(S_yS_z - \tau_{yz}^2) + 2(S_xS_z - \tau_{xz}^2) - (S_xS_y - \tau_{xy}^2)] \\ \frac{1}{3}[-(S_yS_z - \tau_{yz}^2) - 2(S_xS_z - \tau_{xz}^2) + 2(S_xS_y - \tau_{xy}^2)] \\ 2(\tau_{yz}\tau_{zx} - S_z\tau_{xy}) \\ 2(\tau_{xy}\tau_{zx} - S_x\tau_{yz}) \\ 2(\tau_{xy}\tau_{yz} - S_y\tau_{xz}) \end{Bmatrix} \tag{VII.50}$$

Equation (VII.50) is further simplified as

$$\frac{\partial J_3}{\partial\{\sigma\}} = \begin{Bmatrix} \frac{1}{3}[2S_yS_z - S_xS_z - S_xS_y - 2\tau_{yz}^2 + \tau_{xz}^2 + \tau_{xy}^2] \\ \frac{1}{3}[2S_xS_z - S_yS_z - S_xS_y - 2\tau_{xz}^2 + \tau_{yz}^2 + \tau_{xy}^2] \\ \frac{1}{3}[2S_xS_y - S_yS_z - S_xS_z - 2\tau_{xy}^2 + \tau_{yz}^2 + \tau_{zx}^2] \\ 2(\tau_{yz}\tau_{zx} - S_z\tau_{xy}) \\ 2(\tau_{xy}\tau_{zx} - S_x\tau_{yz}) \\ 2(\tau_{xy}\tau_{yz} - S_y\tau_{xz}) \end{Bmatrix} \tag{VII.50a}$$

From the flow rule of the normality principle, the following relationship exists between the plastic strain increment and the plastic stress increment:

$$d\{\varepsilon\} = \lambda \frac{\partial f}{\partial\{\sigma\}} \tag{VII.51}$$

This equation can be interpreted as requiring the normality of the plastic strain increment vector to yield the surface in the hyper space of n stress dimensions. As before $d\lambda$ is the proportionality constant.

For stress increments of infinitesimal size, the change of strain can be divided into elastic and plastic parts, thus (as before)

$$d\{\varepsilon\} = d\{\varepsilon\}_e + d\{\varepsilon\}_p \tag{VII.52}$$

The elastic increment of stress and strain is related to an isotropic material property matrix $[D]$ by

$$d\{\varepsilon\}_e = [D]^{-1}\, d\{\sigma\} \tag{VII.53}$$

From eqns (VII.51)–(VII.53) the following equation is established:

$$d\{\varepsilon\} = [D]^{-1}\, d\{\sigma\} + d\lambda \left\{ \frac{\partial f}{\partial \sigma} \right\} \tag{VII.54}$$

The function stresses, on differentiation, can be written as

$$df = \frac{\partial f}{\partial \sigma_1} d\sigma_1 + \frac{\partial f}{\partial \sigma_2} d\sigma_2 + \cdots \qquad 0 = \left\{ \frac{\partial f}{\partial \sigma} \right\}^{T''} d\{\sigma\} \tag{VII.55}$$

Equation (VII.55) together with eqn (VII.56) can be written in matrix form as

$$d\{\varepsilon\} = [D]_{ep}^{-1}\, d\{\sigma\} \tag{VII.56}$$

or

$$\begin{Bmatrix} d\varepsilon_x \\ d\varepsilon_y \\ d\varepsilon_z \\ d\gamma_{xy} \\ d\gamma_{yz} \\ d\gamma_{zx} \\ \hdashline 0 \end{Bmatrix} = \left[\begin{array}{cccccc:c} & & & & & & \dfrac{\partial f}{\partial \sigma_x} \\ & & & & & & \dfrac{\partial f}{\partial \sigma_y} \\ & & & & & & \dfrac{\partial f}{\partial \sigma_z} \\ & & & [D]^{-1} & & & \dfrac{\partial f}{\partial \tau_{xy}} \\ & & & & & & \dfrac{\partial f}{\partial \tau_{yz}} \\ & & & & & & \dfrac{\partial f}{\partial \tau_{zx}} \\ \hdashline \dfrac{\partial f}{\partial \sigma_x} & \dfrac{\partial f}{\partial \sigma_y} & \dfrac{\partial f}{\partial \sigma_z} & \dfrac{\partial f}{\partial \tau_{xy}} & \dfrac{\partial f}{\partial \tau_{yz}} & \dfrac{\partial f}{\partial \tau_{zx}} & 0 \end{array}\right] \begin{Bmatrix} d\sigma_x \\ d\sigma_y \\ d\sigma_z \\ d\tau_{xy} \\ d\tau_{yz} \\ d\tau_{zx} \\ \hdashline d\lambda \end{Bmatrix} \tag{VII.57}$$

Inversion of the above matrix $[D]^{-1}$ will give stresses

$$d\{\sigma\} = [D]_{ep}\, d\{\varepsilon\} \tag{VII.58}$$

The explicit form of the elasto-plastic material matrix $[D]_{ep}$ is given by

$$[D]_{ep} = [D] - [D]\left\{\frac{\partial f}{\partial \sigma}\right\}\left\{\frac{\partial f}{\partial \sigma}\right\}^T [D]\left[\left\{\frac{\partial f}{\partial \sigma}\right\}^T [D]\left\{\frac{\partial f}{\partial \sigma}\right\}\right]^{-1} \tag{VII.59}$$

The value of $\{\partial f/\partial \sigma\}$ has been evaluated above.

The rest of the procedure using finite elements is described in Chapter VI.

VII.4.5 Five-parameter model

A full description of this model is given in Section II.1.5. Equation (II.17) predicts failure. The same failure surface in terms of tensile and compressive meridians is given by eqn (II.18). The average normal stress σ_a is given by eqn (II.19) and in brief is

$$\rho(\sigma_a, \theta) = \frac{(c+t)}{v} \tag{VII.60}$$

$$c = 2(\rho_c^3 - \rho_c \rho_t^2)\cos\theta$$

$$t = (2\rho_t\rho_c - \rho_c^2)[\alpha_1'' + 5\rho_t^2 - 4\rho_t\rho_c]^{1/2}$$

$$v = \alpha_1'' + (\rho_c - 2\rho_t)^2$$

The chain rule of differentiation is used in order to evaluate the gradient $\partial f/\partial\sigma_{ij}$:

$$\frac{\partial f}{\partial\sigma_{ij}}=\frac{\partial f}{\partial\rho}\frac{\partial\rho}{\partial\sigma_{\mathrm{a}}}\frac{\partial\sigma_{\mathrm{a}}}{\partial\sigma_{ij}}+\frac{\partial f}{\partial\rho}\frac{\partial\rho}{\partial\theta}\frac{\partial\theta}{\partial\sigma_{ij}}+\frac{\partial f}{\partial\tau_{\mathrm{a}}}\frac{\partial\tau_{\mathrm{a}}}{\partial\sigma_{ij}} \tag{VII.61}$$

Values for the terms in eqn (VII.61) are given below:

$$\frac{\partial f}{\partial\rho}=\frac{-\tau_{\mathrm{a}}}{f_{\mathrm{c}}'}\frac{1}{\rho^2} \tag{VII.61a}$$

where

$$\frac{\partial\rho}{\partial\sigma_{\mathrm{a}}}=\frac{1}{v}\left[\left(\frac{\partial c}{\partial\sigma_{\mathrm{a}}}\right)+\left(\frac{\partial t}{\partial\sigma_{\mathrm{a}}}\right)\right]-(c+t)\frac{\partial v}{\partial\sigma_{\mathrm{a}}} \tag{VII.61b}$$

$$\frac{\partial c}{\partial\sigma_{\mathrm{a}}}=2\cos\theta\left[(3\rho_{\mathrm{c}}^2-\rho_{\mathrm{t}}^2)\frac{\mathrm{d}\rho_{\mathrm{c}}}{\mathrm{d}\sigma_{\mathrm{a}}}-2\rho_{\mathrm{c}}\rho_{\mathrm{t}}\frac{\mathrm{d}\rho_{\mathrm{t}}}{\mathrm{d}\sigma_{\mathrm{a}}}\right] \tag{VII.61c}$$

$$\frac{\partial t}{\partial\sigma_{\mathrm{a}}}=\left[2\rho_{\mathrm{c}}\frac{\mathrm{d}\rho_{\mathrm{t}}}{\mathrm{d}\sigma_{\mathrm{a}}}+2(\rho_{\mathrm{t}}-\rho_{\mathrm{c}})\frac{\mathrm{d}\rho_{\mathrm{c}}}{\mathrm{d}\sigma_{\mathrm{a}}}\right][\alpha_1''+5\rho_{\mathrm{t}}^2-4\rho_{\mathrm{c}}\rho_{\mathrm{t}}]^{1/2}$$
$$+\frac{(2\rho_{\mathrm{c}}\rho_{\mathrm{t}}-\rho_{\mathrm{c}}^2)\left[(+10\rho_{\mathrm{t}}-8\rho_{\mathrm{t}}\cos^2\theta-4\rho_{\mathrm{c}})\frac{\mathrm{d}\rho_{\mathrm{t}}}{\mathrm{d}\sigma_{\mathrm{a}}}+(8\rho_{\mathrm{c}}\cos^2\theta-4\rho_{\mathrm{t}})\frac{\mathrm{d}\rho_{\mathrm{c}}}{\mathrm{d}\sigma_{\mathrm{a}}}\right]}{2[\alpha_1''+5\rho_{\mathrm{t}}^2-4\rho_{\mathrm{c}}\rho_{\mathrm{t}}]^{1/2}} \tag{VII.61d}$$

$$\frac{\partial v}{\partial\sigma_{\mathrm{a}}}=(8\rho_{\mathrm{t}}\sin^2\theta-4\rho_{\mathrm{c}})\frac{\mathrm{d}\rho_{\mathrm{t}}}{\mathrm{d}\sigma_{\mathrm{a}}}+(8\rho_{\mathrm{c}}\cos^2\theta+2\rho_{\mathrm{c}}-4\rho_{\mathrm{t}})\frac{\mathrm{d}\rho_{\mathrm{c}}}{\mathrm{d}\sigma_{\mathrm{a}}} \tag{VII.61e}$$

From eqn (II.20) the vectors ρ_{c} and ρ_{t} are evaluated.

In eqn (VII.61d) $\mathrm{d}\rho_{\mathrm{c}}/\mathrm{d}\sigma_{\mathrm{a}}$ is taken from eqn (II.20) as

$$\frac{\mathrm{d}\rho_{\mathrm{c}}}{\mathrm{d}\sigma_{\mathrm{a}}}=\sqrt{5}b_1+\frac{2\sqrt{5}b_2\sigma_{\mathrm{a}}}{f_{\mathrm{c}}'} \tag{VII.61f}$$

Similarly

$$\frac{\mathrm{d}\rho_{\mathrm{t}}}{\mathrm{d}\sigma_{\mathrm{a}}}=\sqrt{5}a_1+\frac{2\sqrt{5}a_2\sigma_{\mathrm{a}}}{f_{\mathrm{c}}'}$$

The value of the term $\partial\sigma_{\mathrm{a}}/\partial\sigma_{ij}$ is already known and is equal to $\frac{1}{3}S_{ij}$.

Once the gradient is known the elasto-plastic formulation described in this chapter and in Chapter VI can be adopted.

VII.5 Creep model

Experiments have been carried out under various stress combinations for volumetric and deviatoric creep strain behaviour. The stress–strain relationship is given in Part I for creep under multiaxial conditions. The normal modulus of elasticity, E, is replaced by inverse of the uniaxial creep compliance, $J(t'-\tau')$. The elastic Poisson ratio is replaced by the creep Poisson ratio. Hence the multiaxial creep strain as given in Part I may be written as

$$\varepsilon_{ij}^{c} = J(t'' - \tau')\{(1 + \nu_c)\sigma_{ij} - \nu_c^{*}\delta_{ij}\sigma_{kk}\} \tag{VII.62}$$

where ν_c^* is the creep Poisson ratio along any principal axis. In incremental form over a pseudo-time $\Delta t''$, this may be written as

$$\Delta\varepsilon_{ij}^{c} = \Delta J(t'' - \tau')\{(1 + \nu_c)\sigma_{ij} - \nu_c^{*}\delta_{ij}\sigma_{kk}\} \tag{VII.63}$$

where $\Delta J(t'' - \tau')$ is the uniaxial creep compliance over an increment of $\Delta t''$; ν_c^* is the creep Poisson ratio; and δ_{ij} is the Kronecker delta. Equation (VII.63) can be written by rearranging it in matrix form as

$$\begin{Bmatrix} \Delta\varepsilon_x^c \\ \Delta\varepsilon_y^c \\ \Delta\varepsilon_z^c \\ \Delta\gamma_{xy}^c \\ \Delta\gamma_{yz}^c \\ \Delta\gamma_{zx}^c \end{Bmatrix} = \Delta J(t'' - \tau') \begin{bmatrix} 1 & -\nu_c^* & -\nu_c^* & 0 & 0 & 0 \\ -\nu_c^* & 1 & -\nu_c^* & 0 & 0 & 0 \\ -\nu_c^* & -\nu_c^* & 1 & 0 & 0 & 0 \\ 0 & 0 & 0 & 2(1+\nu_c^*) & 0 & 0 \\ 0 & 0 & 0 & 0 & 2(1+\nu_c^*) & 0 \\ 0 & 0 & 0 & 0 & 0 & 2(1+\nu_c^*) \end{bmatrix} \times \begin{Bmatrix} \sigma_x \\ \sigma_y \\ \sigma_z \\ \tau_{xy} \\ \tau_{yz} \\ \tau_{zx} \end{Bmatrix} \tag{VII.64}$$

$$\Delta\varepsilon^{c} = \Delta J(t'' - \tau')[D]\{\sigma\} \tag{VII.64a}$$

where $[D]$ is the material compliance matrix with a constant value of ν_c^*. For variable Poisson ratio value and Young's modulus along different axes, reference should be made to matrices $[D]$ in Appendix I with the relevant constituents modified.

The creep strain (or increment of strain) in finite element analysis is treated as 'initial strain'. The kinematically equivalent loads arising from creep strains have been discussed elsewhere. This equivalent load may be written using the creep strain increment as

$$\Delta F^{c} = \int_{\text{vol}} B^{T''} D\,\Delta\varepsilon^{c}\,\mathrm{d\,vol} \tag{VII.65}$$

or

$$= \int_{-1}^{+1}\int_{-1}^{+1}\int_{-1}^{+1} B^{T''} D\,\Delta\varepsilon^{c} \det J\,\mathrm{d}\xi\,\mathrm{d}\eta\,\mathrm{d}\zeta \tag{VII.65a}$$

or

$$= \sum_{i=1}^{nx}\sum_{j=1}^{ny}\sum_{k=1}^{nz} W_i W_j W_k \bar{g} \det J$$

where nx, ny, nz are Gauss points; W_i, W_j, W_k are the weighting coefficients; and $\det J$ is the determinant of the Jacobian:

$$\bar{g} = B^{T''} D \, \Delta\varepsilon_{\mathrm{c}} \tag{VII.65b}$$

where $B =$ the strain–displacement matrix; $D =$ the elastic or nonlinear material compliance matrix; and $\Delta\varepsilon_{\mathrm{c}} =$ the creep strain increment.

The following steps are performed to include the effects of creep:

1. At time $t' = 0$, elastic or linear analysis is carried out. If thermal loads are considered, calculate the kinematically equivalent load as

$$\{\Delta F\}_i = \int_V B^{T''} D \, \Delta\varepsilon_{\mathrm{c}} \, \mathrm{d\,vol} \tag{VII.65c}$$

   Calculate stresses, strains and other parameters as before.
2. Apply a small time increment, $\Delta t''$, during which the stresses are assumed to remain constant. The creep strain increment for concrete is calculated using eqn (VII.64) of the form

$$\Delta\varepsilon_{\mathrm{c}} = f(\sigma, \Delta t'', t', T) \tag{VII.65d}$$

3. Convert the creep strain increment into the kinematically equivalent load using eqn (VII.65).
4. The rest of the procedure is given in this chapter and in Chapters II and VI for crushing, cracking, plasticity and solution procedures.

VII.6 Thermal stress–strain model

The thermal strain in finite element analysis is treated as 'initial strain' similar to the creep strain. The kinematically equivalent load due to temperature rise, $\Delta T'$, has a similar form:

$$\begin{aligned}\{\Delta F\}^{T''} &= \int [B]^{T''} [D] \{\Delta\varepsilon_{\mathrm{T}}\} \, \mathrm{d\,vol} \\ &= \int_{-1}^{+1}\int_{-1}^{+1}\int_{-1}^{+1} [B]^{T''} [D] \{\Delta\varepsilon_{\mathrm{T}}\} \, \mathrm{d}|J| \, \mathrm{d}\xi \, \mathrm{d}\eta \, \mathrm{d}\zeta\end{aligned} \tag{VII.66}$$

where $\{\Delta F\}$ is the thermal strain increment of the material. For concrete, as a three-dimensional material, the thermal strain increment is written as

$$\begin{Bmatrix} \Delta\varepsilon_x \\ \Delta\varepsilon_y \\ \Delta\varepsilon_z \\ \Delta\gamma_{xy} \\ \Delta\gamma_{yz} \\ \Delta\gamma_{zx} \end{Bmatrix}^{T''} = \begin{Bmatrix} \alpha_{\mathrm{T}}\Delta T' \\ \alpha_{\mathrm{T}}\Delta T' \\ \alpha_{\mathrm{T}}\Delta T' \\ 0 \\ 0 \\ 0 \end{Bmatrix} \tag{VII.67}$$

For a plane stress condition

$$\left\{\begin{array}{c}\Delta\varepsilon_x \\ \Delta\varepsilon_y \\ \Delta\gamma_{xy}\end{array}\right\} = \left\{\begin{array}{c}\alpha_T\Delta T' \\ \alpha_T\Delta T' \\ 0\end{array}\right\} \tag{VII.68}$$

where α_T = the coefficient of thermal expansion for the materials.

Tendons and reinforcing bars are treated as uniaxial material, for which the thermal strains are

$$\{\Delta\varepsilon_x\} = \{\alpha_T\Delta T'\} \tag{VII.69}$$

The rest of the procedure given in Chapter VI is followed, as mentioned in Section VII.4.4.

VII.7 Nonlinear bond linkage element [262]

The Ahmlink element procedure described in Part I can easily be included in the finite element analysis by adopting the following steps.

(a) Calculate the incremental slip from the nodal displacements:

$$\{\Delta S_i\} = [T'']\{\Delta U_i\} \tag{VII.70}$$

where $[T'']$ is the transformation matrix and $\{\Delta U_i\}$ are the element nodal displacements.

The total slip at iteration i is calculated as

$$\{S_i\} = \{S_{i-1}\} + \{\Delta S_i\} \tag{VII.71}$$

(b) Calculate the incremental stress based on the bond stress at iteration $i - 1$:

$$\{\Delta\sigma_{b_i}\} = [E_b^t]\{\sigma_{b_{i-1}}\}\{\Delta S_i\} \tag{VII.72}$$

Total stress is equal to

$$\{\sigma_{b_i}\}_{TOT} = \{\sigma_{b_{i-1}}\} + \{\Delta\sigma_{b_i}\} \tag{VII.73}$$

(c) Check the state of the bond, i.e. whether the bond is broken or not, and calculate the stress accordingly.

— If $|S_i| > S_{max}$, set flag $I_{flat} = 1$, i.e. bond is broken. At this point, the bond stress is instantaneously dropped to zero, i.e. $\{\sigma_{b_i}\} = 0.0$, where $\{S_{max}\}$ is the maximum slip allowed.

— If $|S_i| < S_{max}$, calculate the bond stress which is compatible with the slip, S_i. This is obtained by linear interpolation of a nonlinear bond–slip curve. The curve is simulated by multi-linear lines. Figure VII.4 gives the scheme for the linear interpolation. Let $\{\sigma_b^r\}$ be the bond stress compatible with the slip S_i. The difference between $\{\sigma_{b_i}\}$ and $\{\sigma_b^r\}$ is treated as initial stress and this may be converted into nodal loads on the structure under

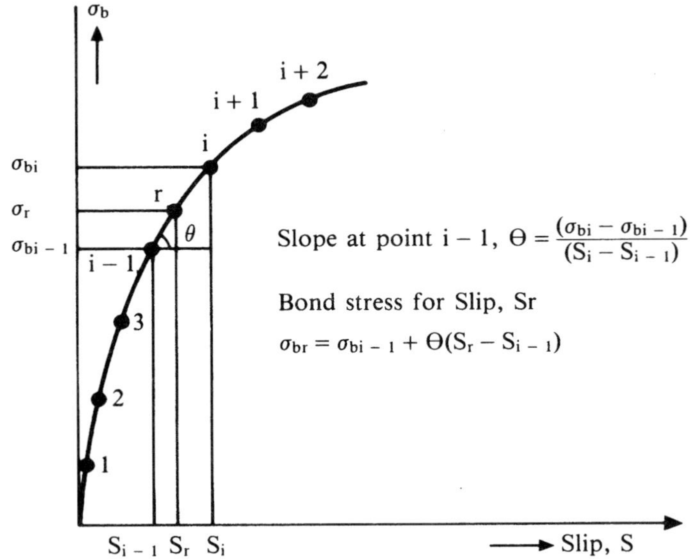

Fig. VII.4. Linear interpretation of nonlinear bond–slip curve.

consideration:

$$\{\Delta\sigma_{\mathrm{D}}\} = \{\sigma_{\mathrm{b}_i}\} - \{\sigma_{\mathrm{b}}^{\mathrm{r}}\} \tag{VII.74}$$

The correct stress is written as

$$\{\sigma_{\mathrm{b}_i}\}_{\mathrm{cor}} = \{\sigma_{\mathrm{b}_i}\} - \{\Delta\sigma_{\mathrm{D}}\} \tag{VII.75}$$

(d) Total internal equivalent loads and residuals are calculated as

$$\{F\}_{\mathrm{internal}} = \pi\,\mathrm{d}L[T]^{T''}\{\sigma_{\mathrm{b}_i}\} \qquad \{R\} = \{F\}_{\mathrm{external}} - \{F\}_{\mathrm{internal}} \tag{VII.76}$$

The rest of the procedure is the same as for the other above cases.

VII.8 Bulk and shear moduli model

This has been given in detail in Section II.1.7. The material compliance matrix $[D]$ is given in eqn (II.55). Using various parameters given in eqns (II.56)–(II.62), the stress–strain equation is written as

$$\{\Delta\sigma\} = [D]\{\Delta\varepsilon\} \tag{VII.77}$$

The rest of the procedure for the finite element analysis is given in detail in this chapter as well as in Chapter VI.

VII.9 Endochronic cracking model

The basic constitutive equations of the endochronic model are given in Section II.1.8, in which deviatoric stress–strain increments are formulated together with stress invariants. The inelastic dilatancy using shear and bulk moduli is included in the derivation of various expressions. Based on the cracking criteria given in Chapters II and VI, the endochronic constitutive relations can be extended for cracking. Based on the material matrix given in eqn (II.55), the left-hand side of the same equation is modified as

$$\begin{Bmatrix} \Delta\sigma_x + \Delta\sigma_x^{\mathrm{p}} \\ \Delta\sigma_y + \Delta\sigma_y^{\mathrm{p}} \\ \Delta\sigma_z + \Delta\sigma_z^{\mathrm{p}} \\ \Delta\tau_{xy} + \Delta\tau_{xy}^{\mathrm{p}} \\ \Delta\tau_{yz} + \Delta\tau_{yz}^{\mathrm{p}} \\ \Delta\tau_{zx} + \Delta\tau_{zx}^{\mathrm{p}} \end{Bmatrix} = \begin{bmatrix} D_{11} & D_{12} & D_{13} & 0 & 0 & 0 \\ & D_{22} & D_{23} & 0 & 0 & 0 \\ & & D_{33} & 0 & 0 & 0 \\ & & & D_{44} & 0 & 0 \\ \text{sym.} & & & & D_{55} & 0 \\ & & & & & D_{66} \end{bmatrix} \begin{Bmatrix} \Delta\varepsilon_x \\ \Delta\varepsilon_y \\ \Delta\varepsilon_z \\ \Delta\gamma_{xy} \\ \Delta\gamma_{yz} \\ \Delta\gamma_{zx} \end{Bmatrix} \tag{VII.78}$$

where

$$\begin{aligned} &D_{11} = D_{22} = D_{33} = K + 4/3G \\ &D_{12} = D_{13} = D_{23} = K - 2/3G \\ &D_{44} = D_{55} = D_{66} = \beta' G \end{aligned} \tag{VII.79}$$

(β' = a factor for concrete aggregate interlocking)

or

$$\{\Delta\sigma\} + \{\Delta\sigma^{\mathrm{p}}\} = [D]\{\Delta\varepsilon\} \tag{VII.80}$$

When concrete cracks, a sudden drop of tensile strength across the crack occurs. This creates a non-equilibrium state in the concrete structure. The residual stress from the non-equilibrium state is redistributed to another part of the structure. The material matrix $[D]$ is modified to include the reduction of the stiffness across the crack. The material constitutive relations in the crack co-ordinate system are written as

$$\{\Delta\sigma^*\} = \{\Delta\sigma^{\mathrm{p}*}\} = [D^*]\{\Delta\varepsilon^*\} \tag{VII.81}$$

Assuming concrete has a crack normal to any axis, x, then

$$\Delta\sigma_x^* = 0$$

The procedure described in Section II.2 is adopted. The value of $\Delta\sigma_x^{\mathrm{p}}$ is given by

$$\begin{aligned} &\Delta\sigma_x^{\mathrm{p}*} = D_{11}\Delta\varepsilon_x^* + D_{12}\,\Delta\varepsilon_y^* + D_{13}\,\Delta\varepsilon_z^* = \Delta\sigma_x^{\mathrm{p}*} \\ &\Delta\varepsilon_x^* = \frac{\Delta\sigma_x^{\mathrm{p}*}}{D_{11}} - \frac{D_{12}}{D_{11}}\Delta\varepsilon_y^* - \frac{D_{13}}{D_{11}}\Delta\varepsilon_y^* \end{aligned} \tag{VII.82}$$

Using eqn (VII.82), the stresses represented by eqn (VII.81) are written in a local crack co-ordinate system as

$$\Delta\sigma_y^* + \Delta\sigma_y^{p*} - \frac{D_{12}}{D_{11}}\Delta\sigma_x^{p*} = \left(D_{22} - \frac{D_{12}D_{21}}{D_{11}}\right)\Delta\varepsilon_y^* + \left(D_{23} - \frac{D_{12}D_{13}}{D_{11}}\right)\Delta\varepsilon_z^*$$

$$\Delta\sigma_z^* + \Delta\sigma_z^{p*} - \frac{D_{13}}{D_{11}}\Delta\sigma_x^{p*} = \left(D_{23} - \frac{D_{12}D_{13}}{D_{11}}\right)\Delta\varepsilon_y^* + \left(D_{33} - \frac{D_{13}D_{31}}{D_{11}}\right)\Delta\varepsilon_z^*$$

$$\Delta\tau_{xy}^* + \Delta\tau_{xy}^{p*} = \beta' D_{44}\gamma_{xy}^*$$

$$\Delta\tau_{yz}^* + \Delta\tau_{yz}^{p*} = D_{55}\Delta\gamma_{yz}^*$$

$$\Delta\tau_{zx}^* + \Delta\tau_{zx}^{p*} = \beta' D_{66}\Delta\gamma_{zx}^* \tag{VII.83}$$

Cracks in other directions are dealt with by modifying the matrix $[D]$ using the same principles. Where cracks are open in all three directions, the value of $[D]$ has all elements equal to zero such that

$$[D] = [0] \tag{VII.84}$$

Other details of crack initiation, closure and reopening, with and without the inclusion of concrete aggregate interlocking, are given in Section II.2 and in Appendix II.

VII.10 Heat-conduction model

Steady state and transient heat-conduction models are developed and problems are solved using finite elements. The equations of heat conduction generally include the effects of surface areas with convection and radiation boundary conditions.

VII.10.1 Steady state

The basic thermal equilibrium equation is written as

$$[\bar{K}]\{T\} = \{Q\} \tag{VII.85}$$

where $[K]$ = the thermal conductivity matrix; $\{Q\}$ = the heat flow vector; and $\{T\}$ = the vector of the nodal point temperatures.

VII.10.2 Transient

The basic thermal diffusion equation is written as

$$[C]_s\{T\} + [K]\{T\} = \{Q\} \tag{VII.86}$$

where $[C]_s$ is the specific heat matrix.

For large and complicated structures a substructuring system on the lines suggested in Section VI.2.6 is adopted. In order to distinguish the dynamic transient equations (eqns VI.23 and VI.24) from these, the governing equation for the thermal substructure is written as

$$[C]_s\{T\} + [K]_c\{T\} = \{Q\} \tag{VII.86a}$$

where

$$[K]_c = [K_{\gamma\gamma}] - [K_{\gamma\gamma'}][K_{\gamma'\gamma'}]^{-1}[K_{\gamma'\gamma}]$$

$$[C]_s = [C_{\gamma\gamma}]_s - [K_{\gamma\gamma'}][K_{\gamma'\gamma'}]^{-1}[C_{\gamma'\gamma}]_s - [C_{\gamma'\gamma}]_s[K_{\gamma'\gamma'}]^{-1}[K_{\gamma'\gamma}] + [K_{\gamma\gamma'}][K_{\gamma'\gamma'}]^{-1}[C_{\gamma'\gamma'}]_s[K_{\gamma'\gamma'}]^{-1}[K_{\gamma'\gamma}]$$

$$\{Q\} = \{Q_\gamma\} - [K_{\gamma\gamma'}][K_{\gamma'\gamma'}]^{-1}\{Q_{\gamma'}\}$$

$$\{T\} = \{T_\gamma\}$$

Hence

$$\{T_{\gamma'}\} = [K_{\gamma'\gamma'}]^{-1}\{Q_{\gamma'}\} - [K_{\gamma'\gamma'}]^{-1}[K_{\gamma'\gamma}]\{T_\gamma\} \tag{VII.86b}$$

In the above equations the convection surfaces are not included.

The temperature at point X, Y, Z is written as

$$T(X, Y, Z) = \{N\}^{T''}\{T\} \tag{VII.87}$$

where $\{N\}$ = the vector of shape functions and $\{T\}$ = the nodal temperature vector.

$\{S\}^T$, the vector of temperature, is related to $\{T\}$ using the definition of $\{S\}^T$ and eqn (VII.87) to give

$$\{S\}^T = [B]\{T\} \tag{VII.88}$$

where

$$[B] = \begin{bmatrix} \left\{\dfrac{\partial N}{\partial X}\right\}^{T''} \\ \left\{\dfrac{\partial N}{\partial Y}\right\}^{T''} \\ \left\{\dfrac{\partial N}{\partial Z}\right\}^{T''} \end{bmatrix} \qquad \{S\}^T = \begin{bmatrix} \left(\dfrac{\partial T}{\partial X}\right) \\ \left(\dfrac{\partial T}{\partial Y}\right) \\ \left(\dfrac{\partial T}{\partial Z}\right) \end{bmatrix}_{x,y,z}$$

In thermal terms, the virtual work within one element, i.e. internal work, is written as

$$\Delta U = \int \{\Delta S^T\}^{T''}\{Q\}_{X,Y,Z}\,\mathrm{d\,vol} \tag{VII.89}$$

where

$$\{Q\} = \begin{Bmatrix} Q_X \\ Q_Y \\ Q_Z \end{Bmatrix} = \text{the vector of heat flow}$$

Q_X = heat flow in the x direction per unit area

Q_Y = heat flow in the y direction per unit area

Q_Z = heat flow in the z direction per unit area

The heat flows are related to the temperature gradients using the following relation:

$$\{Q\} = [K^{c}]\{S\}^{T} \tag{VII.90}$$

where the conductivity matrix $[K^{c}]$ is given by

$$[K^{c}] = \begin{bmatrix} K_{XX} & 0 & 0 \\ 0 & K_{YY} & 0 \\ 0 & 0 & K_{ZZ} \end{bmatrix}_{c}$$

The internal work is written again as

$$\Delta U = \{\partial T\}^{T''} \int_{vol} [B]^{T''} [D][B] \, d\,vol \{T\} \tag{VII.91}$$

The virtual internal work associated with convection surfaces is given by

$$\Delta U = \int_{A} \partial \Delta T \hat{Q} \, dA \tag{VII.92}$$

where $\hat{}$ = the direction normal to the surface:

$$\partial \Delta T = \partial T(X, Y, Z)|_{s} \tag{VII.92a}$$

The heat flow $\hat{Q}$ over the unit area is given by

$$\hat{Q} = e_{f} \, \Delta T \tag{VII.93}$$

where e_{f} = the film coefficient for heat transfer of the surface and $\partial T(X, Y, Z)|_{s}$ is the temperature function evaluated at the convection surface.

In finite element form the above equations are then written as

$$\Delta U = \{\partial T_{e}\}^{T''} c_{f} \int_{area} \{N_{s}\}^{T''} d(area)\{T_{e}\} - \{\partial T_{e}\}^{T''} c_{f} T_{B} \int_{area} \{N_{s}\} \, d(area) \tag{VII.93a}$$

where $\{N_{s}\}$ are the shape functions evaluated at the convection surface and T_{B} is the temperature of the coolant.

Similarly the virtual work associated with specific heat, i.e. thermal damping, is written as

$$\Delta U = \int_{V} \partial T(X, Y, Z) H \, d\,vol \tag{VII.94}$$

where

$H = D_{c} C_{p} \dfrac{\partial T(X, Y, Z)}{\partial t}$ = total heat change per unit volume per unit time

D_{c} = density

C_{p} = specific heat

When D_c, C_p and T do not vary over the element, eqn (VII.94) is written

$$\partial U = \{\partial T\}^{T''} D_c C_p \int_V \{N\}\{N\}^{T''} \,\mathrm{d\,vol} \frac{\partial}{\partial t}\{T\} \tag{VII.94a}$$

If H_r is the heat generation rate per unit volume, the expression for ∂U is written as

$$\partial U = \{\partial T\}^{T''} H_r \int_V \{N\} \,\mathrm{d\,vol} \tag{VII.94b}$$

When all the above effects are combined, the following expression is obtained:

$$[[K] + [K^c]]\{T\} + [C]_s \frac{\partial}{\partial t}\{T_e\} = \{Q^c\} + \{Q^h\} + \{Q\} \tag{VII.95}$$

where

$[K]$ is the material matrix $= \int_V B^{T''} DB \,\mathrm{d\,vol}$

$[K^c]$ is the conductivity matrix $= e_f \int_A \{N_s\}\{N_s\}^{T''} \,\mathrm{d}A$

$[C]_s$ is the specific heat matrix $= \rho C_p \int_V \{N\}\{N\}^{T''} \,\mathrm{d}V$

$\{Q^c\}$ is a total element matrix heat flow matrix per unit area

$= e_f T_B \int_A \{N_s\} \,\mathrm{d}A$

$\{Q^h\}$ is a total emeent matrix heat generated matrix per unit volume

$= H_r \int_V \{N\} \,\mathrm{d}V$

t is time

For the nonlinear transient analysis a time integration scheme is adopted on the lines suggested in the dynamic analysis given in Chapter VI.

The usual heat-conduction matrices and thermal load vectors are given below:

$$[K_e] = \begin{bmatrix} K_T + S_1 & -K_T & 0 & 0 & 0 \\ -K_T & 2K_T & -K_T & 0 & 0 \\ 0 & -K_T & 2K_T & -K_T & 0 \\ 0 & 0 & -K_T & 2K_T & -K_T \\ 0 & 0 & 0 & -K_T & K_T + S_2 \end{bmatrix} \tag{VII.96}$$

where

$$K_T = \frac{4(A)K^c_{XX}}{d}$$

$$S_i = C_{i,f}(A)$$

K^{c}_{XX} = thermal conductivity

$\bar{t}$ = thickness

f = film coefficient

A = area

The thermal damping (specific heat) matrix is

$$[C_e]_s = \begin{bmatrix} 2B_t & B_t & 0 & 0 & 0 \\ B_t & 4B_t & B_t & 0 & 0 \\ 0 & B_t & 4B_t & B_t & 0 \\ 0 & 0 & B_t & 4B_t & B_t \\ 0 & 0 & 0 & B_t & 2B_t \end{bmatrix} \qquad \text{(VII.96a)}$$

where

$$B = \frac{(A)C_d dD_c}{24}$$

D_c = density

C_p = specific heat (°C)

A = area

The thermal load vector is

$$\{Q\} = \begin{Bmatrix} k_t + \bar{g}_1 \\ 2k_t \\ 2k_t \\ 2k_t \\ k_t + \bar{g}_2 \end{Bmatrix} \qquad \text{(VII.97)}$$

where

$$k_t = \frac{(A)H_r t}{8}$$

H_r = heat generation rate

$\bar{g}_i = S_i T_b$

T_b = bulk temperature

A = area

VII.11 Miscellaneous orthotropic constitutive models

VII.11.1 Liu, Nilson and Slate [70]

An introduction is given in Section I.3 to this orthotropic constitutive model for biaxial compression of concrete subjected to proportional loading. Liu *et al.* [70]

developed an expression similar to that of Saenz [4] (eqn (I.8)) using specific boundary conditions for biaxial loading and this is given as

$$\sigma_{cu} = \sigma_i = E_0\varepsilon_{ie}/(1-\nu)\bar{\alpha}\left[1+\left(\frac{1}{1-\nu\bar{\alpha}}\frac{E_0}{E_{sec}}-2\right)\frac{\varepsilon_{ie}}{\varepsilon_1}+\left(\frac{\varepsilon_{ie}}{\varepsilon_1}\right)^2\right] \quad \text{(VII.98)}$$

where $E_{sec} = \sigma_p/\varepsilon_p$; σ_p and ε_p = the maximum principal stress and corresponding strain; and $\bar{\alpha}$ = the principal stress ratio (eqn (I.18), $\bar{\alpha} = \sigma_2/\sigma_1$).

If $\bar{\alpha}$ is kept at zero, eqn (VII.98) reduces to the uniaxial stress–strain curve suggested by Saenz [4]. This model is extended to include compression–tension and tension–tension cases. They are given below in conjunction with eqns (I.15)–(I.19).

Compression–tension

$$\sigma_1/f_c' = \frac{1}{\bar{\alpha}f_c'/\sigma_t} \quad \text{(VII.99)}$$

Tension–tension

$$\sigma_1/f_c' = \frac{1}{f_c/\sigma_t} \quad \text{for } 1 \geq \bar{\alpha} \geq 0 \quad \text{(VII.100)}$$

Using the experimental data of Tasuji and co-workers [73, 74], the above equations are extended to cover all states of stress. The values of ε_1 in the biaxial state at maximum stress in the major and minor principal directions are given below:

	Major direction	*Minor direction*
eqn (VII.99):	$\varepsilon_1 = (360 + 0.59\sigma_1) \times 10^{-6}$	$\varepsilon_1 = (1180 - 2.58\sigma_1) \times 10^{-6}$
eqn (VII.100):	$\varepsilon_1 = 150 \times 10^{-6}$	$\varepsilon_1 = (-26 + 52.2\sigma_1) \times 10^{-6}$

Using the finite element method for the plain stress conditions, the incremental stress–strain curve with the above boundary limits is written as

$$\begin{Bmatrix} \Delta\sigma_1 \\ \Delta\sigma_2 \\ \Delta\tau_{12} \end{Bmatrix} = \begin{bmatrix} \dfrac{E_{01}'}{L_{11}} & \nu E_{02}' & 0 \\ \dfrac{\nu E_{02}'}{L_{11}} & \dfrac{E_{02}'}{L_{11}} & 0 \\ 0 & 0 & \dfrac{E_{01}'E_{02}'}{E_{01}'E_{02}'+L_{22}} \end{bmatrix} \times \begin{Bmatrix} \Delta\varepsilon_1 \\ \Delta\varepsilon_2 \\ \Delta\gamma_{12} \end{Bmatrix} \quad \text{(VII.101)}$$

where

$$L_{11} = 1 - \nu^2\frac{E_{02}}{E_{01}} \qquad L_{22} = \frac{2\nu E_{02}'}{1-\nu^2 E_{02}'/E_{01}'}$$

where E_{01}' and E_{02}' are the tangent moduli of elasticity as a function of the state of stress and strain in the principal stress directions 1 and 2, which are also the principal axes of orthotropy. Equation (VII.98) is considered for the evaluation of the E_0' values. These values are given in Chapter I for various conditions. Liu *et al.* [70]

obtain the value of E_0' by differentiating eqn (VII.98) with respect to strain and taking into consideration the microcrack confining effect:

$$E_{01}' = \frac{E_0\left[1 - \left(\frac{\varepsilon_{ie}}{\varepsilon_1}\right)^2\right]}{\left[1 + \left(\frac{1}{1 - \nu\bar{\alpha}}\frac{E_0}{E_{\text{sec}}} - 2\right)\frac{\varepsilon_{ie}}{\varepsilon_1} + \left(\frac{\varepsilon_{ie}}{\varepsilon_1}\right)^2\right]} \tag{VII.102}$$

The value of $i = 1, 2, 3$ for the respective orthotropic axes. The shear modulus G (eqn (VII.101)) has been obtained using the assumption that no interaction exists between the principal normal stress and shear strains caused by the stress-induced anisotropy. In the finite element analysis the value of G is taken from the data given in Chapter I and the normal stresses are obtained from eqn (VII.101). The rest of the method for computing stresses, strains and cracks is given in Chapter VI. The program ISOPAR (described elsewhere) or any other finite element package can be extended to include the above model for the stress–strain relationships of a concrete structure subjected to biaxial loading.

VII.11.2 Darwin and Pecknold [7a,b]

The constitutive model proposed by Darwin and Pecknold [7a,b] has already been introduced in Chapter I. As mentioned earlier, concrete is treated as an orthotropic nonlinear elastic material. The stress–strain relations for use in the finite element analysis are written in incremental form as

$$\begin{Bmatrix} \Delta\sigma_1 \\ \Delta\sigma_2 \\ \Delta\tau_{12} \end{Bmatrix} = \frac{1}{1 - \nu^2}\begin{bmatrix} E_{01}' & \nu\sqrt{E_{01}'E_{02}'} & 0 \\ & E_{02}' & 0 \\ \text{sym.} & & \frac{1}{4}(E_{01}' + E_{02}' - 2\nu E_{01}'E_{02}') \end{bmatrix}\begin{Bmatrix} \Delta\varepsilon_1 \\ \Delta\varepsilon_2 \\ \Delta\gamma_{12} \end{Bmatrix} \tag{VII.103}$$

The tangent moduli E_{01}' and E_{02}' are evaluated using Saenz's equation (I.8). The failure envelope given by Kupfer and Gerstle [31] is used as described in Section I.3 for computing $\sigma_i = \sigma_{\text{cu}}$ and ε_{i0}'. The Poisson ratio given in eqn (I.57) is used for the uniaxial compression and tension–compression states. For various limiting cases, namely compression–compression and compression–tension, reference should be made to Section I.3. In this model the biaxial tension stress state is given by choosing the strain as

$$\varepsilon_{it} = \sigma_{\text{t}}/E_0 \tag{VII.104}$$

It is estimated that a constant Poisson ratio of 0.2 gives a good account of this model for the stress levels between 75 and 80% of the value of f_{c}'. Beyond this stress range, the Poisson ratio increases sharply. As explained in the finite element formulation in Chapter VI, the equivalent uniaxial strains are accumulated during a typical loading history in order to determine the tangent moduli E_{01}' and E_{02}' from the stress–strain curves. This work has been extended as described in Chapter I to include cyclic loading capability. The stress–strain relationship given in eqn (VII.103) replaces other relations in the finite element analysis and the rest of Chapter VI is then adhered to for the analysis of concrete structures.

VII.11.3 Elwi and Murray [5]

Elwi and Murray [5] have developed a three-dimensional constitutive model for concrete using the hypoelastic concept given in Chapter II, Section II.1.6.1. Here they have extended it to include orthotropicity. In addition, the model incorporates the equivalent uniaxial strain concept of Darwin and Pecknold [7a,b], the non-linearity of Saenz [4], and the failure surface of Willam and Wanke [130], which is described in Chapter II, Section II.1.3. The incremental stress–strain relations referred to the orthotropic principal axes 1, 2 and 3 are written in the following form:

$$\{\mathrm{d}\sigma\} = [D]\{\mathrm{d}\varepsilon\} \tag{VII.105}$$

in which the tangent material stiffness matrix $[D]$ is given by

$$[D] = \frac{1}{\nu}\begin{bmatrix} D_{11} & D_{12} & D_{13} & 0 \\ & D_{22} & D_{23} & 0 \\ & & D_{33} & 0 \\ & & & D_{44} \end{bmatrix} \tag{VII.106}$$

The tangent material stiffness matrix given in eqn (VII. 106) can easily be compared with similar compliance matrices for concrete given in Chapters II and VI and in Appendix III. Where other cases are included, they can replace eqn (VII.106) in the finite element analysis. The elements of the matrix $[D]$ in eqn (VII.106) are defined below:

$$\begin{aligned}
&D_{11} = E'_{01}(1-\nu_{32}^2) \qquad D_{12} = \sqrt{E'_{01}E'_{02}}[\nu_{13}\nu_{32}+\nu_{12}]\\
&D_{13} = \sqrt{E'_{01}E'_{03}}[\nu_{12}\nu_{32}+\nu_{13}] \qquad D_{22} = E'_{20}(1-\nu_{13}^2)\\
&D_{23} = \sqrt{E'_{02}E'_{03}}[\nu_{12}\nu_{13}+\nu_{32}] \qquad D_{33} = E'_{03}(1-\nu_{12}^2)\\
&D_{44} = \bar{\nu}G_{12}\\
&\quad \bar{\nu} = 1-\nu_{12}^2-\nu_{23}^2-\nu_{13}^2-2\nu_{12}\nu_{23}\nu_{13}\\
&\nu_{12}^2 = \nu_{12}\nu_{21} \qquad \nu_{23}^2 = \nu_{23}\nu_{32} \qquad \nu_{13}^2 = \nu_{13}\nu_{31}\\
&G_{12} = \frac{1}{4\bar{\nu}}[E'_{01}+E'_{02}-2\nu_{12}\sqrt{E'_{01}E'_{02}}-(E'_{01}\nu_{23}+E'_{02}\nu_{31})^2]
\end{aligned} \tag{VII.106a}$$

Equation (VII.105) is then written in the form

$$\begin{Bmatrix} \mathrm{d}\sigma_1 \\ \mathrm{d}\sigma_2 \\ \mathrm{d}\sigma_3 \\ \mathrm{d}\tau_{12} \end{Bmatrix} = \begin{bmatrix} E'_{01}B_{11} & E'_{02}B_{12} & E'_{01}B_{13} & 0 \\ E'_{02}B_{21} & E'_{02}B_{22} & E'_{02}B_{23} & 0 \\ E'_{03}B_{31} & E'_{03}B_{32} & E'_{03}B_{33} & 0 \\ 0 & 0 & 0 & G_{12} \end{bmatrix} \begin{Bmatrix} \mathrm{d}\varepsilon_1 \\ \mathrm{d}\varepsilon_2 \\ \mathrm{d}\varepsilon_3 \\ \mathrm{d}\gamma_{12} \end{Bmatrix} \tag{VII.107}$$

Using the equivalent uniaxial strain ε_{ie} ($i = 1, 2, 3$), the right-hand side strain matrix is then

$$d\varepsilon = B_{i1}\, d\varepsilon_1 + B_{i2}\, d\varepsilon_2 + B_{i3}\, d\varepsilon_3 \tag{VII.108}$$

In order to be truly three-dimensional, additional values of G_{23} and G_{31} are included in $[D]$ of eqns (VII.106) and (VII.107) as given in Appendix III. Throughout the finite element analysis the above material matrix $[D]$ is adopted.

It is interesting to note that the biaxial stress–strain relationship (eqn (I.8)) suggested by Saenz [4] is generalised for assessing tensile and compressive response. Equation (I.8a) represents the equivalent uniaxial strain. The incremental moduli in eqn (VII.107) can then be determined by differentiating eqn (I.8a) which is given in eqn (I.66). The Poisson ratio ν_i used is given in eqn (I.66). The maximum value of ν_i is 0.5 corresponding to zero incremental volume change. The five-parameter model can be used in conjunction with the above equations to obtain the three values of σ_i for each stress ratio. The values of ε_{ie} corresponding to σ_i values are determined by assuming a surface in the equivalent uniaxial strain space that has the same form as the ultimate strength surface. The rest of the finite element procedure given in Chapter VI can then easily be adopted for analysing concrete structures. In this regard reference should be made to the finite element formulation given in Chapter VI and the associated appendices.

VII.11.4 Strain hardening models

VII.11.4.1 Chen and Chen [148]

This constitutive model for concrete has been developed using the principle of work hardening plasticity; concrete is modelled as a continuous, isotropic and linearly elasto-plastic strain hardening fraculure material. The transition from the elastic to plastic stage is represented by a discontinuous point on an initially discontinuous surface. Unloading and loading in the plastic range are assumed to follow the initial elastic modulus. At a limiting tensile stress or strain, a crack is assumed to occur in a plane normal to the direction of the stress or strain; under compression the concrete becomes perfectly plastic at the ultimate stress. The surface fails or fractures when crushing occurs at a limiting crushing strain or stress. The failure surface is simulated using biaxial test data for two different conditions:

Compression–compression

$$F_u(\sigma_{ij}) = J_2 + \tfrac{1}{3}A_u I_1 - \tau_u^2 = 0 \tag{VII.109}$$

Tension–tension or tension–compression

$$F_u(\sigma_{ij}) = J_2 - \tfrac{1}{6}I_1^2 + \tfrac{1}{3}A_u I_1 - \tau_u^2 = 0 \tag{VII.110}$$

where A_u and τ_u are material constants generally determined from tests.

Using a combination of isotropic and kinematic hardening rules in which the loading surface experiences translation and uniform expansion, eqns (VI.109) and (VII.110) are rewritten as

Compression–compression

$$F(\sigma_{ij}) = \frac{J_2 + (\beta''/3)I_1}{1 - (\alpha/3)I_1} - \tau^2(\varepsilon_p) = 0 \qquad \text{(VII.111)}$$

Tension–tension and tension–compression

$$F(\sigma_{ij}) = \frac{J_2 - (1/6)I_1^2 + (\beta''/3)I_1}{1 - (\alpha''/3)I_1} - \tau^2(\varepsilon_p) = 0 \qquad \text{(VII.112)}$$

where α'' and β'' are constants; $\tau = \tau_u$ for the failure surface; $\tau = \tau_0$ for the yield surface; and $\tau = f(\varepsilon_p)$ = a function of plastic strain.

The finite element method described in Chapter VI is used for the stress–strain relations with a particular associated flow rule.

VII.11.4.2 Buyukozturk [71]

The strain hardening model proposed by Buyukozturk [71] predicts yield and failure of concrete under multiaxial stress states using the Mohr–Coulomb internal friction theory. The failure law is based on biaxial experimental data. For biaxial compression and compression–tension states, the following expression is proposed:

$$f(I_1, J_2) = 3J_2 + \sqrt{3}\beta''\tau I_1 + \alpha'' I_1^2 - \tau^2 = 0 \qquad \text{(VII.113)}$$

where α'', β'' and τ are experimental constants. The experimental constants are evaluated, and upon insertion eqn (VII.113) is rewritten as

$$3J_2 + f_c' I_1 + \frac{I_1^2}{5} = \frac{f_c'^2}{9} \qquad \text{(VII.114)}$$

The initial yield surface to a stress of one-third of the ultimate strength ($f_c'/3$) is given by

$$3J_2 + f_c'/3I_1 + \frac{I_1^2}{5} = \frac{(f_c'/3)^2}{9} \qquad \text{(VII.115)}$$

This now defines the motion of the subsequent yield surfaces during plastic deformation. The flow rules described in Chapter VI can now be invoked to obtain the incremental stress–strain relations. The rest of the procedure in the finite element formulation is then followed for the analysis of concrete structures.

VII.12 Strain effects

From studies, concrete is considered a rate-dependent material. At high rate loading, concrete possesses a higher yield stress than its quasi-static value, thus having a higher elastic modulus. Various investigations have accounted for strain rate by enhancing the static values to represent its dynamic response. One presented by Bischoff and Perry [1225] takes the form

$$\frac{\sigma_d}{\sigma_s} A + B \log\left(\frac{\sigma_d}{\sigma_s}\right) + C \log\left(\frac{\sigma_d}{\sigma_s}\right)^{a} \qquad \text{(VII.116)}$$

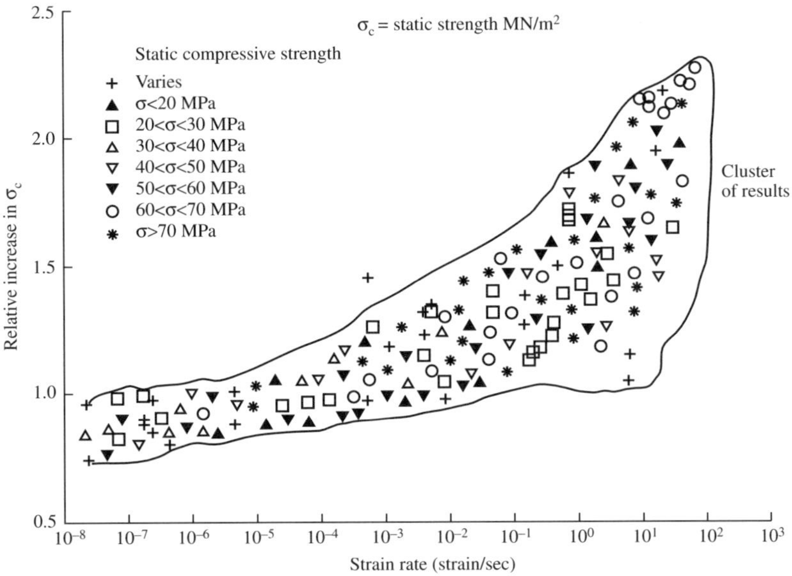

Fig. VII.5. Relative increase in stress upon increase in strain rate.

Another by Mishasshi and Wittmann gives

$$\frac{\sigma_d}{\sigma_s} = \left(\frac{\sigma_d}{\sigma_s}\right)^{1/(\beta+1)} \quad \text{(VII.117)}$$

where

σ_d = dynamic compressive stress of rate

σ_s = quasi-static compressive stress of rate $\sigma_s \approx 0.25\,\text{MPa s}^{-1}$

A, B, C, α, β are all material parameters obtained empirically recommended for the dynamic to static enhancement given by

$$\frac{\sigma_d}{\sigma_s} = \left(\frac{\sigma_d}{\sigma_s}\right)^{\alpha} \qquad \sigma_d \leq 10^6\,\text{MPa s}^{-1}$$
$$\sigma = \beta(\sigma_d)^{1/3} \qquad \sigma_d \leq 10^6\,\text{MPa s}^{-1} \quad \text{(VII.118)}$$

$\sigma_s = 1\,\text{MPa s}^{-1}$; $\alpha = (5 + 3f_c/4)^{-1}$; f_c = static compressive cube strength (MPa) and $\log \beta = 6\alpha - 2$.

Work carried out on the influence of strain rate is given in a review of Bischoff and Perry [1225] summarised in Fig. VII.5.

VII.12.1 Rate-dependent constitutive law

An empirical rate-dependent cap model is proposed, assuming that the effects of strain rate on the plastic loading surface and damage surface are separable and isolated from the static effect by considering the strength enhancement at high strain rates. Using the proposed rate-dependent constitutive model for concrete, the numerical simulations capture the overall qualitative behaviour of concrete response to planar impact at different velocities.

Assume that the damage D is a scalar which represents randomly oriented micro-cracks. The Helmholtz free energy of the material is written:

$$\rho\psi = 1/2(1 - D)K_i\varepsilon^e{:}\varepsilon^e \qquad \text{(VII.119)}$$

where K_i = initial stiffness matrix, ρ = mass density, ε^e = elastic strain tensor

$$\sigma = \frac{\delta\rho\psi}{\delta\varepsilon^e} = (1 - D)K_i{:}\varepsilon^e \qquad \text{(VII.120)}$$

The constitutive equations are derived from thermodynamics of irreversible processes where σ = stress.

The rate constitutive relation is obtained

$$\sigma = (1 - D)K_i{:}\varepsilon^e - DK_i{:}\varepsilon^e \qquad \text{(VII.121)}$$

In plasticity formulation, the strain increments $d\varepsilon$ are assumed separable into elastic $d\varepsilon^e$ and inelastic $d\varepsilon^p$

$$d\varepsilon = d\varepsilon^e + d\varepsilon^e \qquad \text{(VII.122)}$$

The plastic strain increment is given by the flow rule

$$\varepsilon^p = \lambda\frac{\delta f}{\delta\sigma} \qquad \text{(VII.123)}$$

Where f is the yield, λ denotes a scalar coefficient and is determined by requiring that $f = 0$ for loading

$$\sigma = (1 - D)K_i{:}\left[\varepsilon - \lambda\frac{\delta f}{\delta\sigma}\right] - DK_i{:}\varepsilon^e \qquad \text{(VII.124)}$$

$$E_c = C_r f_{ct} \text{ or } f_{cu} \qquad \text{(VII.125)}$$

where E_c and f_{ct}, f_{cu} are elastic modulus or tensile and compressive strength of concrete when subjected to dynamic load or static load, respectively. C_r is an empirical function of strain rate and is approximated as the function

$$C_r = C_0 + C_1 \ln\left(\frac{\dot{\varepsilon}_{ef}}{\dot{\varepsilon}_N}\right) + C_2\left[\ln\left(\frac{\dot{\varepsilon}_{ef}}{\dot{\varepsilon}_N}\right)\right]^2 \qquad \text{(VII.126)}$$

where C_0, C_1 and C_2 are obtained by least squares fit of C_r to exponential data, and εN represents the unit strain rate in 1/sec. The effective strain rate is defined as

$$\dot{\varepsilon}_{ef}[(\dot{\varepsilon}_0)^2 + \tfrac{1}{2}(\dot{\gamma}_0)^2]^{1/2} \qquad \text{(VII.127)}$$

where ε_0 and γ_0 denote octahedral normal strain rate and octahedral shear strain rate, respectively.

Therefore, we assume that the effects of strain rate on the plastic loading surface and damage surface are separable and isolated from the static effect by considering the strength enhancement at high strain rates. The yield function $f(\sigma, h, \varepsilon_{ef})$ of the two-invariant cap model is assumed to be expressed as

$$f(\sigma, h, \varepsilon_{\mathrm{ef}}) = \begin{cases} J_2' - g_1(\varepsilon_{\mathrm{ef}})F_{\mathrm{e}}(J_1) \leq 0 & \text{(failure envelope)} \\ J_2' - g_1(\varepsilon_{\mathrm{ef}})F_{\mathrm{c}}(J_1, h) \leq 0 & \text{(cap surface)} \end{cases} \tag{VII.128}$$

where h = hardening parameter, $g_1(\varepsilon_{\mathrm{ef}})$ = function of $\varepsilon_{\mathrm{ef}}$

$$D = 1 - \bar{W} = 1 - \frac{\bar{\rho}}{\bar{\rho}^*} \tag{VII.129}$$

thus, in the case of spallation, the damage D is given by eqn (VII.129). This is based on eqns (VII.121) and (VII.124). The stress σ in spalling process is given by

$$\sigma = \sigma_{\mathrm{f}}\left(1 + \frac{\bar{\rho}}{\bar{\rho}^*}\right) \tag{VII.130}$$

where ρ and ρ^* are the gross density of damaged material at the times t and t^* respectively. t^* is the time when the tensile stress reaches the critical value σ_{f}

$$\sigma_{\mathrm{f}} = \sigma_{\mathrm{f}}^0 g_1(\varepsilon_{\mathrm{ef}}) \tag{VII.131}$$

the rate-dependent bulk and shear moduli are assumed as

$$K(J_1, \varepsilon_{\mathrm{ef}}) = K(J_1)g_2(\varepsilon_{\mathrm{ef}})$$

$$G(J_2', \varepsilon_{\mathrm{ef}}) = G(J_2')g_2(\varepsilon_{\mathrm{ef}})$$

where K_0, K_2, G_0, G_2 are parameters and $g_2(\varepsilon_{\mathrm{ef}})$ is a function of $\varepsilon_{\mathrm{ef}}$

$$K(J_1) = K_0 + K_2 J_1 \qquad \text{bulk modulus} \tag{VII.132}$$

$$G(J_2') = G_0 + G_2 \qquad \text{shear modulus} \tag{VII.133}$$

When the gross damage D reaches a critical value D_{c}, the rapid coalescence of cracks results in complete separation and the stress drops to zero. This method accounts explicitly for individual cracks and adaptively create new surface by duplicating the finite element nets. It is reported that the critical tensile strain at high strain rates is expected to increase by 250%. Then the critical damage D_{c} is also assumed as

$$D_{\mathrm{c}} = D_{\mathrm{c}}^0 g_3(\varepsilon_{\mathrm{ef}})$$

Parametric values: generally, in the absence of appropriate values, the following values are adopted: $J_{10} = 61.0\,\mathrm{MPa}$, $n = 1.0$, $K_0 = 6.3\,\mathrm{GPa}$, $G_0 = 4.7\,\mathrm{GPa}$, $K_2 = -60.0$ and $G_2 = 180$. $P_{\mathrm{H}}(\varepsilon_v^{\mathrm{P}})$.

PART III

This part considers the application of the analyses to a variety of concrete structures, ranging from beams, slabs and concrete-filled columns to containment vessels, offshore platforms, LNG tanks, etc. It is supported by relevant appendices (I–X).

VIII. Concrete structures — seismic criteria, numerical modelling of soil–concrete structure interaction and isolators

VIII.1 Structural design requirements for concrete structures

VIII.1.1 Introduction to the design basis

The seismic analysis and design procedures to be used in the analysis and design of concrete structures and their components shall be as prescribed in various codes. The design ground motions can occur along any direction of a structure. The design seimic forces, and their distribution over the height of the structure, shall be established by a specific code and the corresponding internal forces in the members of the concrete structure shall be determined using a linearly elastic model. An approved alternative procedure may be used to establish the seismic forces and their distribution, in which case the corresponding internal forces and deformations in the members shall be determined by using a theoretical model.

Individual members shall be designed and sized for the shears, axial forces and moments determined in accordance with these provisions, and connections shall develop the strength of the connected members or the forces. The foundation shall be designed to accommodate the forces developed or the movement imparted to the structure by the design ground motions. In the determination of the foundation design criteria, special recognition shall be given to the dynamic nature of the forces, the expected ground motions, and the design basis for strength and energy dissipation capacity of the structure. The foundations can be flexible or rigid. In the layout the foundations shall have the isolators carefully arranged to offset damaging seismic forces.

VIII.2 Structure framing systems

The basic structural framing systems to be used are indicated in Table VIII.1. Each type is subdivided by the types of vertical element used to resist lateral seismic forces. The structural system used shall be in accordance with the seismic performance category and height limitations indicated in Table VIII.1. The appropriate response modification coefficient, R, and the deflection amplification factor, C, indicated in Table VIII.1 shall be used in determining the base shear and design storey drift.

Table VIII.1. Structural systems

Basic structural system and seismic force resisting system	Response modification coefficient R[a]	Deflection amplification factor C_d[b]	Structural system limitations and building height (ft) limitations Seismic performance category			
			A & B	C	D[d]	E[e]
Bearing wall system						
Light frame walls with shear panels	$6\frac{1}{2}$	4	NL[c]	NL	160	100
Reinforced concrete shear walls	$4\frac{1}{2}$	4	NL	NL	160	100
Reinforced masonry shear walls	$3\frac{1}{2}$	3	NL	NL	160	100
Concentrically braced frames	4	$3\frac{1}{2}$	NL	NL	160	100
Plain concrete shear walls	$1\frac{1}{2}$	$1\frac{1}{2}$	NL	[g]	NP[c]	100
Building frame system						
Eccentrically braced frames, moment-resisting connections at columns away from link	8	4	NL	NL	160	100
Eccentrically braced frames, non-moment resisting connections at columns away from link	7	4	NL	NL	160	100
Composite eccentrically braced frames (C-EBF)	8	4	NL	NL	160	100
Light frame walls with shear panels	7	$4\frac{1}{2}$	NL	NL	160	100
Concentrically braced frames	5	$4\frac{1}{2}$	NL	NL	160	100
Composite concentrically braced frames (C-CBF)	5	$4\frac{1}{2}$	NL	NL	160	100
Reinforced concrete shear walls	$5\frac{1}{2}$	5	NL	NL	160	100
RC shear walls composite with steel elements	$5\frac{1}{2}$	5	NL	NL	160	100
Plain (unreinforced) masonry shear walls	$1\frac{1}{2}$	$1\frac{1}{2}$	NL	[f]	NP	NP
Plain concrete shear walls	2	2	NL	[g]	NP	NP
Dual system with an intermediate moment frame of reinforced concrete or an ordinary moment frame of steel capable of resisting at least 25% of prescribed seismic forces						
Concentrically braced frames	5	$4\frac{1}{2}$	NL	NL	160	100

Composite concentrically braced frames (C-CBF)	5	4½	NL	NL	160	100
Reinforced concrete shear walls	6	5	NL	NL	160	100
RC shear walls composite with steel elements	6	5	NL	NL	160	100
Special moment frames of reinforced concrete	2½	2½	NL	NL	NL	NL
Moment-resisting frames system						
Special moment frames of reinforced concrete	8	5	NL	NL	NL	NL
Composite special moment frames (C-SMF)	8	5½	NL	NL	NL	NL
Intermediate moment frames of reinforced concrete	5	4½	NL	NL	NP	NP
Composite ordinary moment frames (C-OMF)	4½	4	NL	NL	160	100
Composite partially restrained frames (C-PRF)	6	5½	160	160	100	NP
Ordinary moment frames of reinforced concrete	3	2½	NL[h]	NP	NP	NP
Dual system with a special moment frame capable of resisting at least 25% of prescribed seismic forces						
Eccentrically braced frames, moment-resisting connections at columns away from link	8	4	NL	NL	NL	NL
Eccentrically braced frames, non-moment resisting connections at columns away from link	7	4	NL	NL	NL	NL
Composite eccentrically braced frames (C-EBF)	8	4	NL	NL	NL	NL
Concentrically braced frames	6	5	NL	NL	NL	NL
Special concentrically braced frames — steel	8	6½	NL	NL	NL	NL
Composite concentrically braced frames (C-CBF)	6	5	NL	NL	NL	NL
Reinforced concrete shear walls	8	6½	NL	NL	NL	NL
RC shear walls composite with steel elements	8	6½	NL	NL	NL	NL
Steel plate reinforced composite shear walls	8	6½	NL	NL	NL	NL

[a]Response modification coefficient R for use through the provisions.
[b]Deflection amplification factor C_d.
[c]NL = not limited and NP = not permitted. If using metric units, 100 ft approximately equals 30 m and 160 ft approximately equals 50 m.
[d]For descriptions of building systems limited to buildings with a height of 240 ft (70 m) or less.
[e]For building systems limited to buildings with a height of 160 ft (50 m) or less.
[f]Structures excluded.
[g]Plain concrete shear walls shall have nominal reinforcement.
[h]Ordinary moment frames of reinforced concrete are not permitted as part of the seismic force resisting system in seismic performance category B buildings founded on soil profile type E or F.

VIII.2.1 Dual system

For a dual system, the moment frame shall be capable of resisting at least 25% of the design forces. The total seismic force resistance is to be provided by the combination of the moment frame and the sheet walls or braced frames in proportion to their rigidities.

VIII.2.2 Combinations of framing systems

Different structural framing systems are permitted along the two orthogonal axes of the building. Combinations of framing systems shall comply with the requirements of this section.

VIII.2.3 Combination factor

The response modification coefficient R in the direction under consideration at any storey shall not exceed the lowest response modification factor R for the seismic force resisting system in the same direction considered above that storey.

VIII.2.4 Combination framing detailing requirements

The detailing requirements of the higher response modification coefficient R shall be used for structural components common to systems having different response modification coefficients.

VIII.2.5 Seismic performance categories A, B and C

The structural framing system for buildings assigned to seismic performance categories A, B and C shall comply with the building height and structural limitations in Table VIII.1.

VIII.2.6 Seismic performance category D

The structural framing system for a building assigned to seismic performance category D shall comply with the above provisions and the additional provisions of this section.

VIII.2.7 Limited building height

The height limit in Table VIII.2 may be increased to 240 ft (70 m) in buildings that have concrete cast-in-place shear walls. Such buildings shall have braced frames or shear walls arranged in one plane such that they resist no more than the following portion of the seismic forces in each direction, including torsional effects.

(a) 60% where the braced frame or shear walls are arranged only on the perimeter.
(b) 40% where some of the braced frames or shear walls are arranged on the perimeter.
(c) 30% for other arrangements.

Table VIII.2. Plan structural irregularities

Irregularity type and description	Seismic performance category application
1. *Torsional irregularity* To be considered when diaphragms are rigid in relation to the vertical structural elements that resist the lateral seismic forces	D and E
Torsional irregularity Shall be considered to exist when the maximum storey drift, computed including accidental torsion, at one end of the structure transverse to an axis is more than 1.2 times the average of the storey drifts at the two ends of the structure	C, D and E
2. *Re-entrant corners* Plan configurations of a structure and its lateral force-resisting system contain re-entrant corners, where both projections of the structure beyond a re-entrant corner are greater than 15% of the plan dimension of the structure in the given direction	D and E
3. *Diaphragm discontinuity* Diaphragms with abrupt discontinuities or variations in stiffness, including those having cut out or open areas greater than 50% of the gross enclosed diaphragm area, or changes in effective diaphragm stiffness of more than 50% from one storey to the next	D and E
4. *Out-of-plane offsets* Discontinuities in a lateral force resistance path, such as out-of-plane offsets of the vertical elements	D and E
5. *Non-parallel systems* The vertical lateral force-resisting elements are not parallel to or symmetric about the major orthogonal axes of the lateral force-resisting system	C, D and E

VIII.2.8 Interaction effects

Moment-resisting frames that are enclosed or adjoined by more rigid elements not considered to be part of the seismic force resisting system shall be analysed and designed so that the action or failure of those elements will not impair the vertical load and seismic force resisting capability of the frame. The analysis and design shall consider and provide for the effect of these rigid elements on the structural system at building deformations corresponding to the design storey drift.

VIII.2.9 Deformational compatibility

Every structural component not included in the seismic force resisting system in the direction under consideration shall be designed to be adequate for the vertical load-carrying capacity and the induced moments resulting from the design storey drift.

VIII.2.10 Special moment frames

A special moment frame that is used but not required by Table VIII.1 is permitted to be discontinued and supported by a more rigid system with a lower response modification coefficient. Where a special moment frame is required by Table VIII.1 the frame shall be continuous to the foundation.

VIII.2.11 Seismic performance category E

The framing systems of buildings assigned to category E shall conform to the code requirements for category D and to the additional requirements and limitations of this section. The height limitiation shall be reduced from 240 ft to 160 ft (from 70 m to 50 m).

VIII.2.12 Building configuration

Buildings shall be classified as regular or irregular based on the criteria in this section. Such classification shall be based on the plan and vertical configuration.

VIII.2.13 Plan irregularity

Buildings having one or more of the features given in Table VIII.2 shall be designated as having plan structural irregularity and shall comply with the requirements in the codes.

VIII.2.14 Vertical irregularities

Buildings having one or more of the features given in Table VIII.3 shall be designated as having vertical irregularity and shall comply with the requirements in the relevant sections of the code.

Table VIII.3. Vertical structural irregularities

Irregularity type and description	Seismic performance category application
1. *Stiffness irregularity — soft storey* A soft storey is one in which the lateral stiffness is less than 70% of that in the storey above or less than 80% of the average stiffness of the three storeys above	D and E
2. *Weight (mass) irregularity* Mass irregularity shall be considered to exist where the effective mass of any storey is more than 150% of the effective mass of an adjacent storey. A roof that is lighter than the floor below need not be considered	D and E
3. *Vertical geometric irregularity* Vertical geometric irregularity shall be considered to exist where the horizontal dimension of the lateral force-resisting system in any storey is more than 130% of that in an adjacent storey	D and E
4. *In-plane discontinuity in vertical lateral force-resisting elements* An in-plane offset of the lateral force-resisting elements greater than the length of those elements or a reduction in stiffness of the resisting element in the storey below	D and E
5. *Discontinuity in capacity — weak storey* A weak storey is one in which the storey lateral strength is less than 80% of that in the storey above. The storey strength is the total strength of all seismic-resisting elements sharing the storey shear for the direction under consideration	B, C, D and E

VIII.3 Combination of load effects

The effects on the building and its components due to gravity loads and seismic forces shall be combined in accordance with the factored load combinations as presented in ANSI and ASCE 7 (Ref. 2-1) except that the effect of seismic loads E shall be as defined herein. For other codes a reference is made [1375].

The effect of seismic load E shall be defined as follows for load combinations in which the effects of gravity and seismic loads are additive.

$$E = Q_{\mathrm{E}} + 0.5C_{\mathrm{a}}D \qquad \text{(VIII.1)}$$

where E is the effect of horizontal and vertical earthquake-induced forces; C_{a} is the seismic coefficient based on the soil profile type and the values of A_{a} as determined; D is the effect of dead load, and Q_{E} is the effect of horizontal seismic forces.

The effect of seismic load E shall be defined by eqn (VIII.2) as follows for load combinations in which the effects of gravity counteract seismic load:

$$E = Q_{\mathrm{E}} - 0.5C_{\mathrm{a}}D \qquad \text{(VIII.2)}$$

where E, Q_{E}, C_{a} and D are as defined above.

The load factor on E need not be taken as greater than 1.0 in factored load combinations when using the seismic loads as defined in these provisions.

For columns supporting discontinuous lateral force-resisting elements, the axial force in the columns shall be computed using the most critical load combinations. Load combinations including seismic loads shall also be investigated, except that the effect of seismic load E shall be defined by eqn (VIII.3) as follows:

$$E = \left(\frac{2R}{5}\right)Q_{\mathrm{E}} + 0.5C_{\mathrm{a}}D \qquad \text{(VIII.3)}$$

where E, Q_{E}, C_{a} and D are as defined above, and R is the response modification coefficient as given in Table VIII.1. The axial forces in such columns need not exceed the capacity of other elements of the structure to transfer such loads to the column.

Brittle materials, systems, and connections shall be designed using the most critical load combinations. Load combinations including seismic loads shall also be investigated, except that the effect of seismic load E shall be defined by eqn (VIII.4) as follows:

$$E = \left(\frac{2R}{5}\right)Q_{\mathrm{E}} - 0.5C_{\mathrm{a}}D \qquad \text{(VIII.4)}$$

where E, R, Q_{E}, C_{a} and D are as defined above.

The factor $(2R/5)$ shall be equal to or greater than 1.0. The term $0.5C_{\mathrm{a}}$ is permitted.

VIII.4 Deflection and drift limits

The design storey drift Δ as determined shall not exceed the allowable storey drift Δ_{a} as obtained for any torsional effects. All portions of the building shall be designed

Table VIII.4. Allowable storey drift Δ_a (in. or mm)

Building	Seismic hazard	Exposure group	
	I	II	III
Buildings, other than masonry shear wall or masonry wall frame buildings, four storeys or less in height with interior walls, partitions, ceilings and exterior wall systems that have been designed to accommodate the storey drifts	$0.025h_{sx}$[a,b]	$0.020h_{sx}$	$0.015h_{sx}$
Masonry cantilever shear wall buildings	$0.010h_{sx}$	$0.010h_{sx}$[c]	$0.010h_{sx}$[c]
Other masonry shear wall buildings	$0.007h_{sx}$[c]	$0.007h_{sx}$[c]	$0.007h_{sx}$[c]
Masonry wall frame buildings	$0.013h_{sx}$[c]	$0.013h_{sx}$	$0.010h_{sx}$
All other buildings	$0.020h_{sx}$	$0.015h_{sx}$	$0.010h_{sx}$

[a] h_{sx} is the storey height below level x.
[b] There shall be no drift limit for single-storey buildings with interior walls, partitions, ceilings and exterior wall systems that have been designed to accommodate the storey drifts.
[c] Buildings in which the basic structural system consists of masonry shear walls designed as vertical elements cantilevered from their base foundation support which are so constructed that moment transfer between shear walls (coupling) is negligible.

and constructed to act as an integral unit in resisting seismic forces unless separated structurally by a distance sufficient to avoid damaging contact under total deflection. A reference is made to ductility analysis of concrete beams in Appendix XVI.

VIII.5 Equivalent lateral force procedure

VIII.5.1 General

This section provides the required minimum standards for the equivalent lateral force procedure of seismic analysis of buildings. For purposes of analysis, the building is considered to be fixed at the base. See Section VIII.3 for limitations on the use of this procedure.

VIII.5.2 Seismic base shear

The seismic base shear V in a given direction shall be determined in accordance with the following equation:

$$V = C_3 W \tag{VIII.5}$$

where C_3 is the seismic response coefficient and W is the total dead load and applicable portions of other loads listed below.

(a) In areas used for storage, a minimum of 25% of the floor live load shall be applicable. Floor live load in public garages and open parking structures is not applicable.
(b) Where an allowance for partition load is included in the floor load design, the actual partition weight or a minimum weight of 10 psf (0.5 kN/m^2) of floor area, whichever is greater, shall be applicable.
(c) Total operating weight of permanent equipment.

VIII.5.2.1 Calculation of seismic response coefficient

When the fundamental period of the building is computed, the seismic response coefficient C_s shall be determined in accordance with the following equation:

$$C_s = (1.2C_v)/(RT^{2/3}) \tag{VIII.6}$$

where

C_v = the seismic coefficient based on the soil profile type and the value of A_v

R = the response modification factor

T = the fundamental period of the building (sec.).

A soil–structure interaction reduction is permitted when determined using the generally accepted procedures approved by the regulatory agency.

Alternatively, the seismic response coefficient need not be greater than the following:

$$C_s = (2.5C_v)/R \tag{VIII.7}$$

where R is as above and C_s is the seismic coefficient based on the soil profile type and the value of A_a.

VIII.5.3 Period determination

The fundamental period T of the building in the direction under consideration shall be established using the structural properties and deformational characteristics of the resisting elements in a properly substantiated analysis. The fundamental period T shall not exceed the product of the coefficient for upper limit on calculated period C_u from Table VIII.5 and the approximate fundamental period T_a determined from the appropriate requirements.

VIII.5.3.1 Approximate fundamental period for concrete and steel moment-resisting frame buildings

The approximate fundamental period, T_a in seconds, shall be determined from the following equation:

$$T_a = C_T h_n^{3/4} \tag{VIII.8}$$

Table VIII.5. Coefficient for upper limit on calculated period

Seismic coefficient C_v	Coefficient C_u
≥ 0.4	1.2
0.3	1.3
0.2	1.4
0.15	1.5
0.1	1.7
0.05	1.7

where $C_T = 0.039$ for moment-resisting frame systems of reinforced concrete in which the frames resist 100% of the required seismic force and are not enclosed or adjoined by more rigid components that will prevent the frames from deflecting when subjected to seismic forces (the metric coefficient is 0.0731); $C_T = 0.020$ for all other building systems (the metric coefficient is 0.0488), and h_n = the height (ft or m) above the base to the highest level of the concrete structures.

Alternatively, the approximate fundamental period, T_a in seconds, shall be determined from the following equation for concrete structures and moment-resisting frame buildings not exceeding twelve storeys in height and having a minimum storey height of 10 ft (3 m).

$$T_a = 0.1N$$

where N is the number of storeys.

VIII.5.4 Vertical distribution of seismic forces

The lateral forces F_x (kip or kN) induced at any level shall be determined from the following equations:

$$F_x = C_{vx} V \qquad \text{(VIII.9)}$$

and

$$C_{vx} = \frac{W_x h_x^k}{\sum_{i=1}^{n} w_i h_i^k} \qquad \text{(VIII.10)}$$

where

C_{vx} = vertical distribution factor

V = total design lateral force or shear at the base of the building (kip or kN)

w_i and w_x = the portion of the total gravity load of the building W located or assigned to level i or x

h_i and h_x = the height (ft or m) from the base to level i or x, and

k = an exponent related to the building period as follows:

for buildings having a period of 0.5 s or less, $k = 1$
for buildings having a period of 2.5 s or less, $k = 1$
for buildings having a period between 0.5 and 2.5 s, k shall be 2 or shall be determined by linear interpolation between 1 and 2.

VIII.6 Horizontal shear distribution

The seismic design storey shear in any storey V_x (kip or kN), shall be determined from the following equation:

$$V_x = \sum_{i=1}^{x} F_i \qquad \text{(VIII.11)}$$

where F_i is the portion of the seismic base shear V (kip or kN), induced at level i.

The seismic design storey, V_x (kip or kN), shall be distributed to the various vertical elements of the seismic force-resisting system in the storey under consideration, based on the relative lateral stiffness of the vertical resisting elements and the diaphragm.

VIII.6.1 Torsion

The design shall include the torsional moment, M_i (kip-ft or kNm), resulting from the location of the structure (masses plus the accidental torsional moments, M_{ta} (kip-ft or kNm), caused by assumed displacement of the base each way from its actual location by a distance equal to 5% of the dimension of the building perpendicular to the direction of the applied forces.

Buildings of seismic performance categories C, D and E, where type 1 torsional irregularity exists, shall have the effects accounted for by increasing the accidental torsion at each level by a torsional amplification factor A_x determined from the following equation

$$A_x = \left(\frac{\delta_{max}}{1.2\delta_{avg}} \right)^2 \qquad \text{(VIII.12)}$$

where δ_{max} is the maximum displacement at level x (in. or mm) and δ_{avg} is the average of the displacements at the extreme points of the structure at level x (in. or mm).

The torsional amplification factor A_x is not required to exceed 3.0. The more severe loading for each element shall be considered for design.

VIII.6.2 Overturning

The building shall be designed to resist overturning effects caused by the seismic forces. At any storey, the increment of overturning moment in the storey under consideration shall be distributed to the various vertical force-resisting elements in the same proportion as the distribution of the horizontal shears to those elements.

The overturning moments at level x, M_x (kip-ft or kNm), shall be determined from the following equation:

$$M_x = \tau \sum_{i=x}^{n} F_i(h_i = h_x) \qquad \text{(VIII.13)}$$

where

F_i = the position of the seismic base shear V included at level i,

h_i and h_x = the height (ft or m) from the base to level i or x,

τ = 1.0 for the top 10 storeys,

τ = 0.8 for the 20th storey from the top and below, and

τ = a value between 1.0 and 0.8 determined by a straight line interpolation for storeys between the 20th and 10th storeys below the top.

The foundations of buildings except inverted type structures shall be designed for the foundation overturning design moment, M_f (kip-ft or kNm), at the foundation–soil interface determined using the equation for the overturning moment at level x, M (kip-ft or kNm) with an overturning moment reduction factor τ of 0.75 for all building heights.

VIII.7 Drift determination and P–Δ effects

Storey drifts and, where required, member forces and moments due to P–Δ effects shall be determined in accordance with this section.

VIII.7.1 Storey drift determination

The design storey drift shall be computed as the difference between the deflections at the top and bottom of the storey under consideration. The deflections of level x at the centre of the mass, δx (in. or mm) shall be determined in accordance with the following equation:

$$\delta_x = C_d \delta_{xc} \tag{VIII.14}$$

where

C_d = the deflection amplification factor

δ_{xc} = the deflections determined by an elastic analysis (in. or mm).

Where applicable, the design storey drift, Δ (in. or mm) shall be increased by the incremental factor relating to the P–Δ effects.

VIII.7.2 P–Δ effects

P–Δ effects on storey shears and moments, the resulting member forces and moments, and the storey drifts induced by these effects are not required to be considered when the stability coefficient θ, as determined by the following equation, is equal to or less than 0.10:

$$\theta = (P_x \Delta)/(V_x h_{sx} C_d) \tag{VIII.14a}$$

where

P_x = the total vertical design load at and above level x (kip or kN); when calculating the vertical design load for purposes of determining P–Δ, the individual load factors need not exceed 1.0

Δ = the design storey drift occurring simultaneously with V_x (in. or mm)

V_x = the seismic shear force acting between level x and $x - 1$ (kip or kN)

h_{sx} = the storey height below level x (ft or m)

C_d = the deflection amplification factor.

The stability coefficient θ shall not exceed θ_{max}, determined as follows:

$$\theta_{max} = 0.5/(\beta C_d) \leq 0.25 \tag{VIII.14b}$$

where β is the ratio of shear demand to shear capacity for the storey between level x and $x-1$. The ratio may be conservatively taken as 1.0.

When the stability coefficient θ is greater than 0.10 but less than or equal to θ_{max} the incremental factor related to P–Δ effects a_d shall be determined by rational analysis.

To obtain the storey drift for including the P–Δ effect, the design storey drift determined shall be multiplied by $1.0/(1-\theta)$. When θ is greater than θ_{max} the structure is potentially unstable and shall be redesigned.

VIII.8 Modal analysis procedure

VIII.8.1 Introduction

This section provides required standards for the model analysis procedures of seismic analysis of structures. The symbols used in this method of analysis have the same meaning as those for similar terms with the subscript m denoting quantities in the *m*th mode. The structure shall be modelled as a system of masses lumped at various levels with each mass having one degree of freedom — that of lateral displacement in the direction under consideration.

The analysis shall include, for each of two mutually perpendicular axes, at least the lowest three modes of vibration or all modes of vibration with periods greater than 0.4 s. The number of modes shall equal the number of storeys for buildings less than three storeys in height.

The required periods and mode shapes of the building in the direction under consideration shall be calculated by established methods of structural analysis for the fixed base condition using the masses and elastic stiffnesses of the seismic forces resisting system.

VIII.8.2 Modal base shear

The portion of the base shear contributed by the *m*th mode, V_m shall be determined from the following equations:

$$V_m = C_{sm}\bar{W}_m \qquad \text{(VIII.15)}$$

$$\bar{W}_m = \frac{\left(\sum_{i=1}^{n} w_i\phi_{im}\right)}{\sum_{i=1}^{n} w_i\phi_{im}^2} \qquad \text{(VIII.16)}$$

where

C_{sm} = the modal seismic response coefficient determined below

W_m = the effective modal gravity

w_i = the portion of the total gravity load of the building at level i and

ϕ_{im} = the displacement amplitude at the ith level of the building when vibrating in its mth mode.

The modal seismic response coefficient C_{sm} shall be determined in accordance with the following equation:

$$C_{sm} = (1.2C_v)/(RT_m^{2/3}) \tag{VIII.17}$$

where

C_v = the seismic coefficient based on the soil profile type and the value A_v

R = the response modification factor determined from Table VIII.1 and

T_m = the modal period of vibration (in seconds) of the mth mode of the building.

The modal sesimic design coefficient C_{sm} is not required to exceed 2.5 times the seismic coefficient C_a divided by the response modification factor R.

The following are the exceptions.

(a) The limiting value is not applicable to seismic performance category D and E buildings with a period of 0.7 s or greater located on soil profile type E or F sites.

(b) For buildings on sites with soil profile type D, E or F, the modal seismic design coefficient C_{sm} for modes other than the fundamental mode that have periods less than 0.3 s is permitted to be determined by the following equation:

$$C_{sm} = \frac{C_a}{R}(1.0 + 5.0T_m) \tag{VIII.18}$$

where R and T_m are as defined above and C_a is the seismic coefficient based on the soil profile type and the value of A_a.

(c) For buildings where any modal period of vibration T_m exceeds 4.0 s, the modal seismic design coefficient C_{sm} for that mode is permitted to be determined by the following equation:

$$C_{sm} = \frac{3C_a}{RT_m^{4/3}} \tag{VIII.19}$$

where R and T_m are as defined above and C_a is the seismic coefficient based on the soil type and the value of A_v.

The reduction due to soil structure–concrete interaction may be used.

VIII.8.3 Modal forces, deflection and drifts

The modal force F_{xm} at each level shall be determined by the following equations:

$$F_{xm} = C_{vxm}V_m \tag{VIII.20}$$

and

$$C_{vxm} = \frac{w_x\phi_{xm}}{\sum_{i=1}^{n} w_i\phi_{im}} \tag{VIII.21}$$

where C_{vsm} = the vertical factor in the mth mode except that for buildings where the gravity load is concentrated at a single level, it shall be taken as equal to W.

The reduced based shear V shall in no case be taken as less than $0.7V$.

V_m = the total design lateral force or shear at the base in the mth mode

w_i, w_x = the portion of the total gravity load of the building W located or assigned to level i or x

ϕ_{xm} = the displacement amplitude at the xth level of the building when vibrating in its mth mode and

ϕ_{im} = the displacement amplitude at the δ_{xm} level of the building when vibrating in its mth mode.

$$\delta_{xm} = C_d \delta_{xem} \tag{VIII.22}$$

and

$$\delta_{xem} = \left(\frac{g}{4\pi^2}\right)\left(\frac{T_m^2 F_{xm}}{w_x}\right) \quad \text{VIII.23)}$$

where

C_d = the deflection amplification factor determined from Table VIII.1

δ_{xem} = the deflection level x in the mth mode at the centre of the mass at level x determined by an elastic analysis

g = the acceleration due to gravity (ft/s^2 or m/s^2)

T_m = the model period of vibration in seconds of the mth mode of the building

F_{xm} = the portion of the seismic base shear in the mth mode induced at level x and

w_x = the portion of the total gravity load of the building W located or assigned to level x.

The modal drift in a storey Δ_m shall be computed as the difference between the deflections δ_{xm} at the top and bottom of the storey under consideration.

VIII.8.4 Soil–concrete structure interaction effects

VIII.8.4.1 General

The provisions set forth in this section may be used to incorporate the effects of soil–concrete structure interaction in the determination of the design earthquake forces and the corresponding displacement of the building. The use of these provisions will decrease the design values of the base shear lateral forces and overturning moments but may increase the computed values of the lateral displacements and the secondary forces associated with the P–Δ effects.

VIII.8.4.2 Equivalent lateral forces procedure

The following provisions are supplementary to those presented above.

VIII.8.4.3 Base shear

To account for the effects of soil–concrete structure interaction, the base shear V determined may be reduced to

$$V = V - \Delta V \tag{VIII.24}$$

The reduction ΔV shall be computed as follows:

$$\Delta V = \left[C_\mathrm{s} - C_\mathrm{s}\left(\frac{0.05}{\beta}\right)^{0.4}\right]\bar{W} \tag{VIII.25}$$

where

C_s = the seismic response coefficient computed using the fundamental natural period of fixed base structure T or T_a, or the seismic response coefficient computed using the fundamental natural period of flexibly supported structure (T)

β = the fraction of critical damping for the structure–foundation system

W = the effective gravity load of the building which shall be taken as $0.7W$, except that for buildings where the gravity load is concentrated at a single level, it shall be taken as equal to W.

The reduced base shear V shall in no case be taken as less than $0.7V$.

VIII.8.4.4 Effective structural period

The effective period T shall be determined as follows:

$$\bar{T} = T\sqrt{\left(1 + \frac{\bar{k}}{K_\mathrm{y}}\right)\left(1 + \frac{K_\mathrm{y}\bar{h}^2}{K_\theta}\right)} \tag{VIII.26}$$

where

T = the fundamental period of the concrete structures

$\bar{k}$ = the stiffness of the concrete structure when fixed at the base

$\bar{h}$ = the effective height of the structure which shall be taken as 0.7 times the total height

$$\bar{k} = 4\pi^2\left(\frac{\bar{W}}{gT^2}\right) \tag{VIII.27}$$

$\bar{h}$ except that for buildings where the gravity load is effectively concentrated at a single level, it shall be taken as the height to that level; K_y is the lateral stiffness of the foundation defined as the static horizontal force at the level of the foundation necessary to produce a unit deflection at that level, the force and the deflection

Table VIII.6. Values of G/G_0 and v_s/v_{so}

	Ground ≤ 0.10	Acceleration ≤ 0.15	Coefficient ≤ 0.20	A_v ≥ 0.30
Value of G/G_0	0.81	0.64	0.49	0.42
Value of V_s/v_{so}	0.90	0.80	0.70	0.65

being measured in the direction in which the structure is analysed; K_θ is the rocking stiffness of the foundation defined as the static moment necessary to produce a unit average rotation of the foundation, the moment and rotation being measured in the direction in which the structure is analysed, and g is the acceleration of gravity.

The foundation stiffnesses K and K_θ shall be computed by established methods using soil properties that are compatible with the soil strain levels associated with the design earthquake motion. The average shear modulus G for the soils beneath the foundation at large strain levels and the associated shear wave velocity v_s needed in these computations shall be determined from Table VIII.6.

V_{so} = the average shear wave velocity for the soils beneath the foundation at small strain levels (10^{-3}% or less)

$G_0 = \gamma v_{so}^2/g$ = the average shear modulus for the soils beneath the foundation at small strain levels and

γ = the average unit weight of the soils.

For concrete structures supported on mat foundations that rest at or near the ground surface or are embedded in such a way that the side wall makes contact with the soil, cannot be considered to remain effective during the design ground motion; the effective period of the building may be determined as follows:

$$\bar{T} = T\sqrt{1 + \frac{25\alpha r_a \bar{h}}{v_s^2 T^2}\left(1 + \frac{1.12 r_a \bar{h}^2}{r_m^3}\right)} \quad \text{(VIII.28)}$$

where α = the relative weight density of the structure and the soil defined by

$$\alpha = \frac{\bar{W}}{\gamma A_0 \bar{h}} \quad \text{(VIII.29)}$$

r_a and r_m = characteristic foundation lengths defined by

$$r_a = \sqrt{\frac{A_0}{\pi}} \quad \text{(VIII.30)}$$

and

$$r_m = \sqrt[4]{\frac{4I_0}{\pi}} \quad \text{(VIII.31)}$$

where

A_0 = the area of the foundation

I_0 = the static moment of the foundation about a horizontal centroidal axis normal to the direction in which the structure is analysed.

VIII.8.4.5 *Effective damping*

The effective damping factor for the structure–foundation system β shall be computed as follows:

$$\bar{\beta} = \beta_0 + \frac{0.05}{\left(\frac{\bar{T}}{T}\right)^3} \qquad \text{(VIII.32)}$$

where β_0 = the foundation damping factor as specified in Fig. VIII.1.

The value corresponding to $A_v = 0.15$ in Fig. VIII.1 shall be determined by averaging the results obtained from the solid lines and the dashed lines.

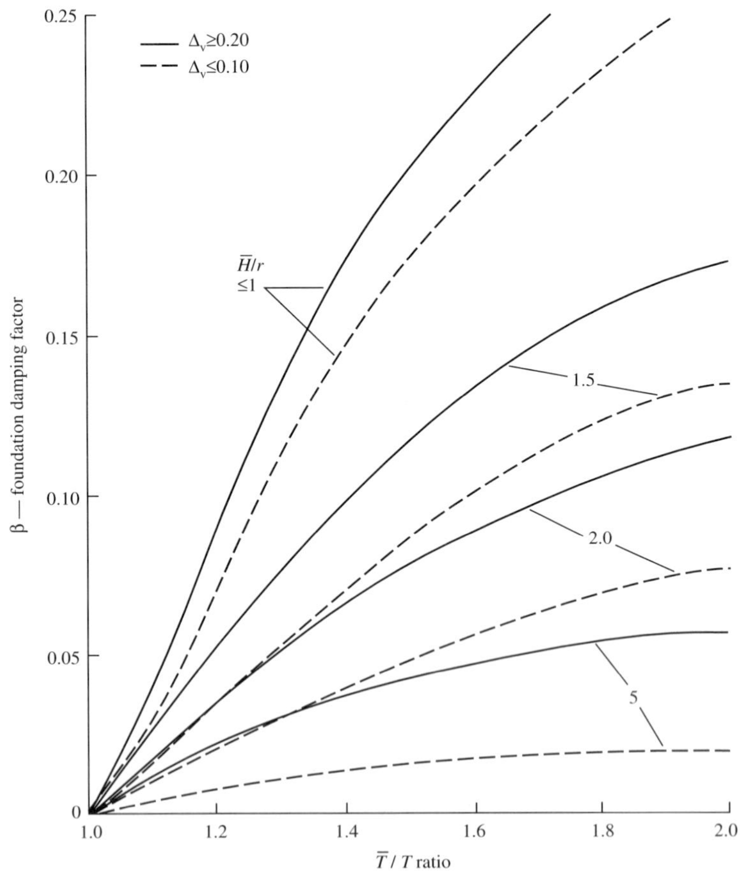

Fig. VIII.1. Foundation damping factors.

The quantity r in Fig. VIII.1 is a characteristic foundation length that shall be determined as follows.
For $\bar{h}/L_0 \leq 0.5$

$$r = r_a = \sqrt{\frac{A_0}{\pi}} \tag{VIII.33}$$

For $\bar{h}/L_0 \geq 1$

$$r = r_m = \sqrt[4]{\frac{4I_0}{\pi}} \tag{VIII.34}$$

where

L_0 = the overall length of the side of the foundation in the direction being analysed

A_0 = the area of the load-carrying foundation, and

I_0 = the static moment of inertia of the load-carrying foundation.

For intermediate values of h/L_0 the value of r shall be determined by linear interpolation.

VIII.9 Methods of analysis

VIII.9.1 General introduction

Dynamic analysis may be used for the design of any concrete structures interacting with soils. Some analyses are given below:

(a) response-spectrum analysis with and without seismically isolated structures
(b) time-history analysis with and without seismically isolated structures.

VIII.9.2 Response-spectrum analysis

Response-spectrum analysis shall be performed using a damping value equal to the effective damping of the isolation system or 30% of critical, whichever is less.

Response-spectrum analysis is used to determine the total design displacement and the total maximum displacement shall include simultaneous excitation of the model by 100% of the most critical direction of ground motion and 30% of the ground motion on the orthogonal axis. The maximum displacement of the isolation system shall be calculated as the vectorial sum of the two orthogonal displacements.

The design shear at any storey shall not be less than the storey shear obtained using the value of V_s taken as equal to the base shear obtained from the response-spectrum analysis of the direction of interest.

VIII.9.3 Time-history analysis

Time-history analysis shall be performed with at least three appropriate pairs of horizontal time-history components as defined.

Each pair of time histories shall be applied simultaneously to the model, considering the most disadvantageous location of mass eccentricity. The maximum displacement of the isolation system shall be calculated from the vectorial sum of the two orthogonal components at each time step.

The parameter of interest shall be calculated for each time-history analysis. If three time-history analyses are performed, the maximum response of the parameter of interest shall be used for design. If seven or more time-history analyses are performed, the average value of the response parameter of interest shall be used for design. For each pair of horizontal ground motion components, the square root sum of the squares (SRSS) of the 5% damped spectrum of the scaled, horizontal components shall be constructed. The motions shall be scaled such that the average value of the SRSS spectra does not fall below 1.3 times the 5% damped spectrum of the design earthquake (or maximum capable earthquake) by more than 10% in the period range of *TI* from minus 1.0 s to *TI* plus 1.0 s.

Time histories developed for sites within 15 km of a major active fault shall incorporate near fault phenomena. In both cases, the analyses shall be performed using specific earthquake input.

The design earthquake shall be used to calculate the total design displacement of the isolation system and the lateral forces and displacements of the isolated structure. The maximum capable earthquake shall be used to calculate the total maximum displacement of the isolation system.

VIII.9.4 Characteristics of interaction

The interaction effects in the approach used here are expressed by an increase in the fundamental natural period of the structure and a change (usually an increase) in its effective damping. The increase in period results from the flexibility of the foundation soil, whereas the change in damping results mainly from the effects of energy

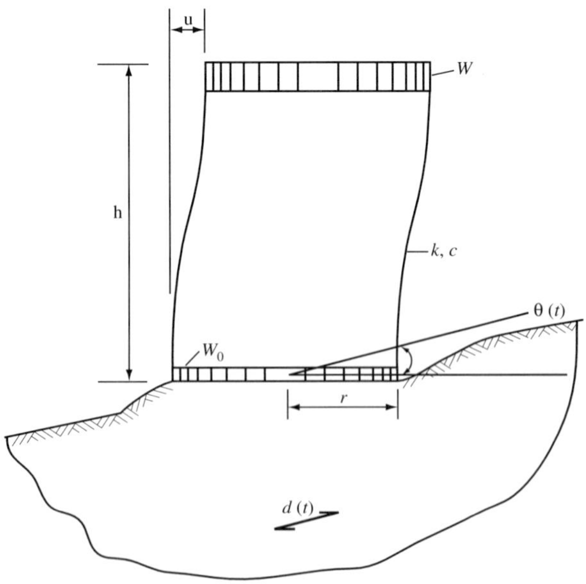

Fig. VIII.2. Concrete structure–soil interaction.

dissipation in the soil due to radiation and material damping. These statements can be clarified by comparing the responses of rigidity and elastically supported systems subjected to a harmonic excitation of the base. Consider a linear structure of weight W, lateral stiffness k, and coefficient of viscous damping c (shown in Fig. VIII.2) and assume that it is supported by a foundation of weight W_0 at the surface of a homogeneous elastic halfspace.

As an example, the foundation weight and the weight of the structure are assumed for a circular footing to be uniformly distributed over a circular area of radius r. The base excitation is generally specified by the free-field motion of the ground surface. This is taken as horizontally directed motion with a period T_0 and an acceleration of amplitude a_m. The rotation of the foundation is θ and the displacement to the base of the top of the structure designated as u.

The response spectra for $h/r = 1$ and $h/r = 5$ for the structure can be plotted as shown in Figs VIII.3 and VIII.4. The solid lines represent response spectra for the steady state amplitude of the total shear for different values of ϕ_0 — the relative flexibility parameter for the soil and the structure concerned. The results are displayed in a dimensionless form, the abscissa representing the ratio of the period of the excitation T_0 to the fixed base natural period of the system T and

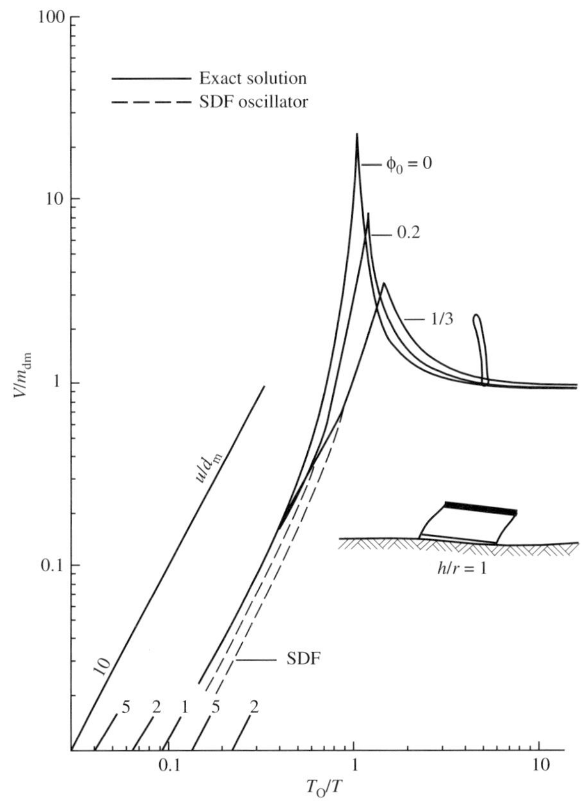

Fig. VIII.3. A comparative study of exact and SDF oscillator results ($h/r = 1$).

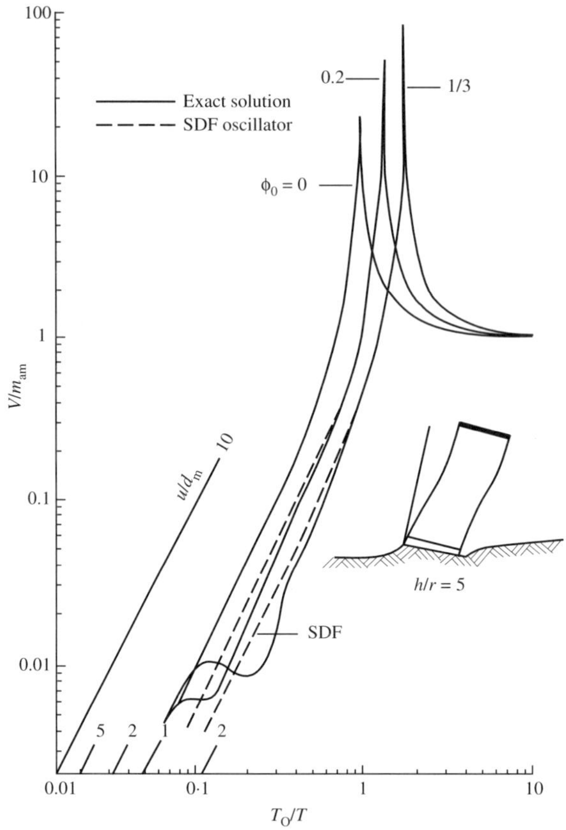

Fig. VIII.4. A comparative study of exact and SDF oscillator results ($h/r = 5$).

the ordinate showing the ratio of the amplitude of the actual base shear V to the amplitude of the base shear. The value of $m = W/g$ in a product m_{am} where a_{m} is the acceleration amplitude of the free-field ground motion. The inclined scales represent deformation amplitude of the superstructure. They are the plots produced by Veletsos and Meek [1421].

Having given the background to the seismic criteria and the soil–concrete–structure interaction, it now becomes essential to present mathematical data on the site seismic response. They are given in Figs VIII.1 to VIII.3. A typical example of a multi-storey building using recent data on the Kobe earthquake is examined in Section IX.14.

VIII.10 Site response — a critical problem in soil–structure interaction analyses for embedded structures

VIII.10.1 Vertical earthquake response and P–Δ effect

The methodology for vertical and horizontal response analysis is the same in principle. However, the difference in vibrational properties of the structure has strong influences in response analysis. At least the following three influences exist.

First, the amplitude and spectural contents of earthquake motion vary with the magnitude and distance of the earthquake, but differently in the horizontal and vertical directions; the relative amplitude of the vertical motion is higher for sites in the epicentral region.

Secondly, the possible variation for both amplitude and frequency content is to be considered when analysing the response in the vertical direction.

Thirdly, the vertical vibration response of vertical members is mainly from axial forces while the horizontal response is from the shear and bending forces. The hysteretic property for different loads may be different, especially in the nonlinear range; the P–Δ effect will produce a descending branch in the force–deformation curve.

VIII.10.2 P–Δ effects

In an elastic system, P–Δ may also introduce nonlinearly. This effect is very important for high rise structures and inelastic structures.

For a single mass system as shown in Fig. VIII.5 the P–Δ effect is the secondary moment produced by the axial force $F(t)$ and the vertical inertial force $M(v + v_g)$ on the deflection u, which is equivalent to an additional horizontal force.

$$[F(t) + M(\ddot{v} + \ddot{v}_g)]u/H \qquad \text{(VIII.35)}$$

where $\ddot{v}_g$ is the vertical ground acceleration and $\ddot{v}$ is the relative vertical acceleration of the mass. Because vertical vibration is also considered, there are now two degrees of freedom and two equations of motion

$$M\ddot{u} + C_u\ddot{u} + k_u u = P(t) - Mu_g - [F(t) + M(\ddot{v} + \ddot{v}_g)]u/H \qquad \text{(VIII.36)}$$

$$M\ddot{v} + C_v\dot{v} + k_v v = F(t) - M\ddot{v}_g \qquad \text{(VIII.37)}$$

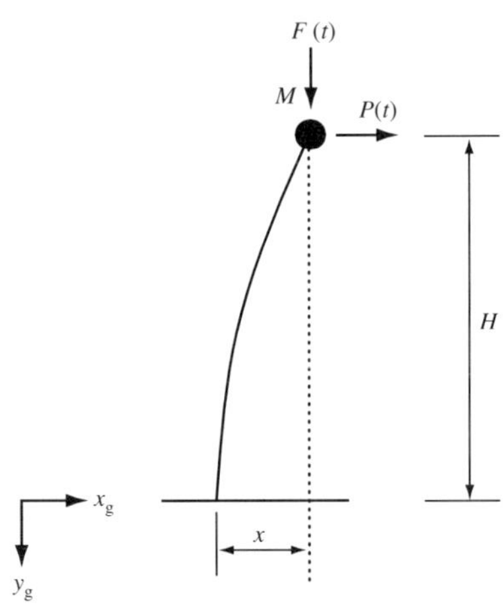

Fig. VIII.5. P–Δ effects.

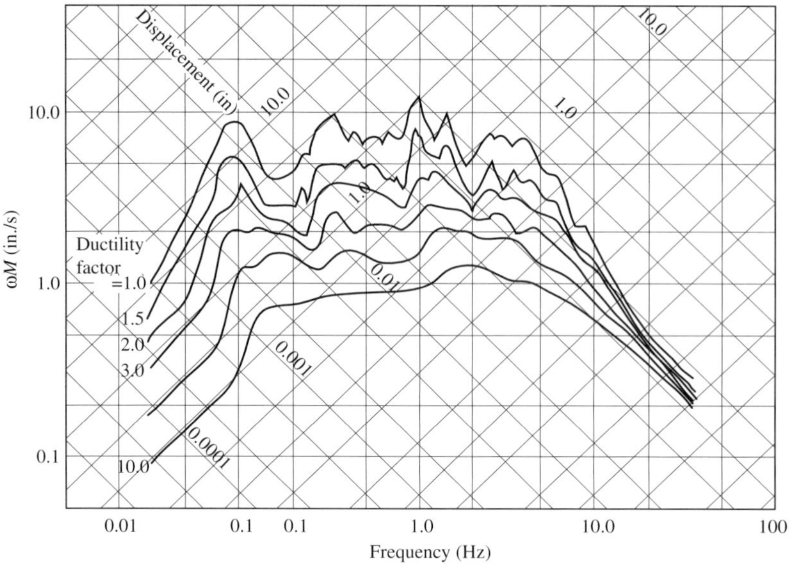

Fig. VIII.6. Frequency spectra.

where C_u, C_v and k_u, k_v are respectively the damping coefficient and stiffness of the system in the horizontal and vertical direction. It is seen here that the last term in eqn (VIII.38) makes the problem nonlinear.

$$\begin{aligned} M\ddot{u} + C\dot{u} + ku &= P - m\ddot{u}_g = p(t) \\ &= a_0 + \sum [a_j \cos \omega_j t + b_j \sin \omega_j t] \quad (j = 1, \infty) \\ &= \sum C_j \exp(i\omega_j t) \qquad (j = -\infty, \infty) \end{aligned} \tag{VIII.38}$$

where coefficient $C_j = (a_j - ib_j)/2$ for $j > 0$ or $C_j = (a_j = ib_j)/2$ for $j < 0$ is complex and $C_0 = a_0$; $i = (-1)^{1/2}$; $\omega_{-j} = \omega_j$; $C_j = C_{-j}$, the Fourier development gives the following result:

$$\begin{aligned} p(t) &= \sum C_j \exp(i\omega_j t) \\ C_j &= \frac{1}{T_p} \int_0^{T_p} p(t) \exp(-i\omega_t t)\, dt \end{aligned} \tag{VIII.39}$$

where T_p is the total duration of motion, as shown in Fig. VIII.7, including a dynamic part and a quiescent part, and the latter is added for the convenience of Fast Fourier Transform (FFT) computation. In FFT, the total duration T_p is divided into N intervals of i, and $N = 2M$, where M and N are integers. The most commonly used N values are 1024, 2048 or 4096. The selection of $\Delta t = T_p/N < T_{min}/z$ is controlled by high frequency components and T_p should be several times the greater period considered.

The transfer function plays an important role in frequency domain analysis. It is the ratio of the stationary response to the simple harmonic input. Where

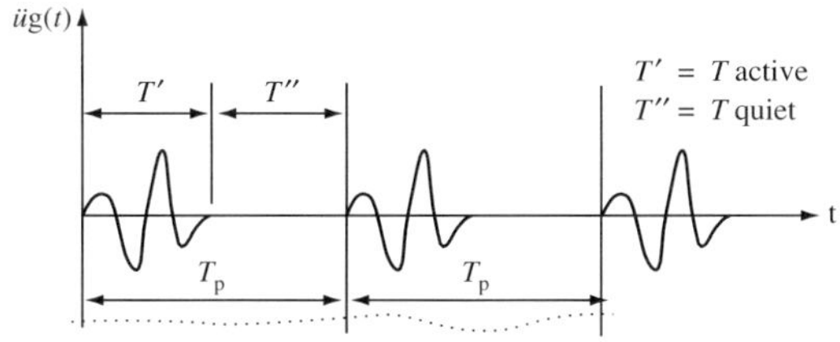

Fig. VIII.7. Acceleration–time relationship.

$p(t) = C\exp(i\omega t)$, the stationary response of an SDF system is

$$u(t) = H(i\omega)C\exp(i\omega t) \tag{VIII.40}$$

and the transfer function is then

$$H(i\omega) = \frac{u(t)}{p(t)} = \frac{1}{-m\omega^2 + iC\omega + k} \tag{VIII.41}$$

$$= \frac{(1/k)}{1 - \beta^2 + 2i\xi\beta} \tag{VIII.42}$$

where $\beta = \omega_0$ is the ratio of the input motion frequency ω and the natural frequency $\omega_0 = (k/m)^{1/2}$ of the system. The transfer function is complex; its modulus $[H(iw)]$ is the ratio of the amplitudes of the response and the input, and the phase angle of $H(iw)$ is the difference between the phase angles of the response and the input. There are many transfer functions. For earthquake motion $u_g(t)$,

$$\ddot{u} + 2\xi\omega_0\dot{u} + \omega^2 u_0 = \ddot{u}_g \tag{VIII.42a}$$

and the transfer function of the relative displacement $u(i)$ of the system to the input $\ddot{u}_g(t)$ is from eqn (VIII.42)

$$Hu(i\omega) = (\omega_0^2 - \omega^2 + 2i\xi\omega_0\omega)^{-1} = [(\omega_0^2 - \omega^2)^2 + 4\xi^2\omega_0^2\omega^2]^{-1/2}\exp(i\phi)$$

$$\phi = \tan^{-1}[(2\xi\omega_0\omega)/(\omega_0^2 - \omega^2)] \tag{VIII.42b}$$

From relations between the spectra of a function and its derivative, it is easy to find the transfer functions of relative velocity $u(t)$ and the absolute acceleration $\ddot{u}(t) + \ddot{u}_g(t)$ to $\ddot{u}_g(t)$. With the transfer function found, the corresponding response in the time domain can be obtained from eqn (VIII.42)

$$u(t) = \sum H_u(i\omega_j)C_j(\omega)\exp(i\omega_j t) \tag{VIII.43}$$

VIII.10.3 Frequency domain analysis of an MDF system

The procedure is exactly the same as for the SDF system. Take earthquake motion input as an example. The equation of motion is

$$M\ddot{u} + C\dot{u} + Ku = -MI\ddot{u}_g(t) \tag{VIII.44}$$

The Fourier transformation of this equation is

$$(-\omega^2 M + i\omega C + K)U(i\omega) = -MI\ddot{U}_g(i\omega) \tag{VIII.44a}$$

or, in terms of the transfer function matrix,

$$H(i\omega) = U^0(i\omega) \tag{VIII.44b}$$

which is the ratio of the stationary system response matrix $u(t) = U^0(i\omega)\exp(i\omega t)$ to input in the frequency domain and hence

$$U(i\omega) = H(i\omega)^* I\ddot{U}_g(i\omega) \tag{VIII.44c}$$

The computation of transfer function matrix $H(i\omega)$ is time consuming. The response matrix $u(t)$ can then be found from $U(i\omega)$ by a Fourier transformation. To avoid the computation of $H(i\omega)$, the model response in frequency domain can be found

$$Q_j(i\omega) = H_j(i\omega)\ddot{U}_g(i\omega) \tag{VIII.44d}$$

After finding the model response $Q_j(t)$ in the time domain by Fourier transformation, the total response can be found.

As mentioned before, any other response can be found in the same way if the corresponding transfer function is used.

Instead of finding the structural response for a given ground motion $U_g(t)$ in the step-by-step integration method, the response spectrum method finds the maximum structural response from the given response spectrum $S_a(\omega)$ of the ground motion.

VIII.10.4 Some nonlinear response spectra

In the linear case, because of the approximate relation,

$$S_a = \omega S_v = \omega S_d \tag{VIII.45}$$

All three spectra can be plotted on one tripartite grid diagram. These approximations do not hold in the nonlinear case; there are then three possible nonlinear response spectra.

For the nonlinear yield response spectrum, the ordinate is not displacement response but the yield deformation which produces the displacement response $u_{max} = \mu u_y$ where $\mu = [u_{max}/u_y]$ is the ductility factor. The corresponding approximation in this case is

$$S_a = \omega S_v = \omega^2 u_y \tag{VIII.46}$$

and the yield strength of the concrete structure is

$$P_y = P_{max} = ku_y = m\omega^2 u_y \tag{VIII.47}$$

In an ideal elasto-plastic system, P_y is the maximum internal force elastic limit of the deformation capacity u_y which the system should have in order to limit the spectral acceleration, velocity and displacement to not over $\omega^2 u_y$, ωu_y and u_y respectively. Here u_y and ductility factor μ are requirements of the capacity for the system. Finally, four structural factors (natural period, damping ratio, type of hysteretic curve and the ductility factor μ) and three ground motion parameters (peak acceleration A, velocity V and displacement D) are obtained for analysis.

VIII.10.5 Soil–structure interaction numerical technique

Soil–structure interaction (SSI) was first studied a long time ago in analysis of the vibration of machinery foundations but has developed greatly in the past two decades because of the rapid advances in computer technology. It may be classified into time and frequency domains in methodology or total structure substructure and hybrid structure in structural decomposition. Most investigators have adopted frequency-domain analysis which has the following advantages.

(a) It is convenient to consider the frequency-dependent foundation impedance when the whole system is divided into structure and ground.
(b) The transmitting boundary for a finite element model or wave propagation theory is frequency dependent.

Generally speaking, SSI analysis consists of four fundamental problems as follows.

(i) The free-field problem: the rock (ground) motion is found from other given motions.
(ii) The scattering problem: the response of the ground is studied with the structure removed at the contacts (Fig. VIII.7) of the structure and ground. The different contact points of the hybrid structure are studied, including the foundation soil and the remaining half space.
(iii) Impedance: the displacements $U_c(\omega)\exp(i\omega t)$ at the contact points are found for forces applied at the contact point $F_c(\omega)\exp(i\omega t)$. The impedance matrix $Is(\omega) = C^{-1}(\omega)$ or the flexibility matrix $C(\omega)$ of the ground is defined in the following ratio:

$$U_c(\omega) = C(\omega)F_c(\omega) \tag{VIII.48}$$

If the foundation of the structure is on the ground surface, the ground substructure is a half space; for the case with only one horizontal layer on top of homogeneous space, there is an analytical solution for $C(\omega)$.
(iv) Finally, there is the soil–structure interaction (SSI) system.

VIII.10.6 Substructure method in the frequency domain

Figure VIII.8(a) shows the FF system of the total structure or the frequency domain analysis. This method is also called the method of impedance, which divides the system into two substructures (substructures 1 and 2) at the contact surface as shown in Fig. VIII.8(b) and (c), studies their responses separately first and combines them later to satisfy the continuity condition. It has been used both in time and frequency domains.

Let the known motion be a general three-dimensional time history $u_g(t)$, whose frequency domain expression is $U_g(\omega)$ at some depth below the surface. The four primary problems are encountered:

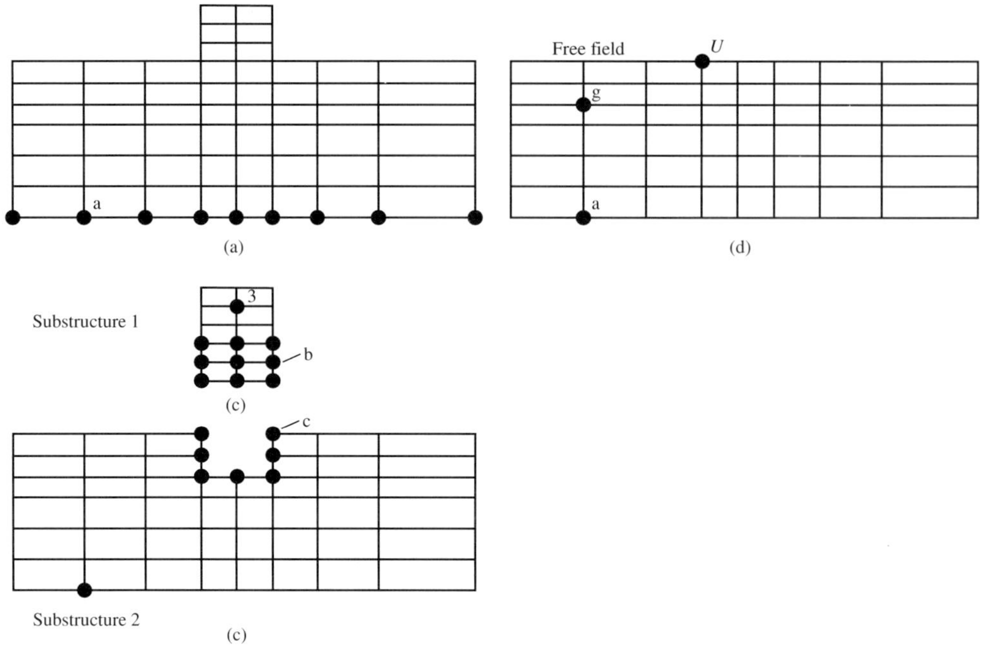

Fig. VIII.8. A total structure and substructures.

(a) Free-field response: the input motion $u_a(t)$ or $U_a(\omega)$ is evaluated as bedrock point 'a' in Fig. VIII.8(d), the lower boundary of the ground substructure for given motion at point g. Let the analysis use an operator L_1^*:

$$L_1^*: u_a(t), U_a(\omega) = u_g(t), U_g^0(\omega) \tag{VIII.49}$$

(b) Scattering problem: the response of substructure 2 is studied (Fig. VIII.8(c)); let there be an operator L_2^*:

$$L_2^*: u_g^0(t), U_g^0(\omega) = u_a(t), U_a(\omega) \tag{VIII.49a}$$

(c) Impedance problem: the response of substructure 2 is studied (Fig. VIII.8(c)) under an external excitation $f_c(t) = F_c(\omega)\exp(i\omega t)$ acting along one direction at any point c. The response at point c, $u_c(t) = U_c(\omega)\exp(i\omega t)$ for a given force $f_b(t) = F_b(\omega)\exp(i\omega t)$ at point c is

$$U_c(\omega) = C(\omega)F_c(\omega) \text{ or } F_c(\omega) = I_s(\omega)U_c(\omega)c(\omega) \tag{VIII.50}$$

(d) The response $u_b(t)$ or $U_b(\omega)$ of substructure 1 under external forces $f_b(t) = F_b(\omega)\exp(i\omega t)$ is studied at contact point b:

$$F_b(\omega) = M_b(\omega)\omega^2 U_b(\omega) \tag{VIII.51}$$

where $M_b(\omega)$ is the equivalent, frequency-dependent mass matrix of the superstructure.

The four primary problems are four steps to find the operators L_1^*, L_2^*, $C(\omega)$ or $I_s(\omega)$ and $M_b(\omega)$. The final solution is only an assemblage of the four to satisfy the compatibility requirement at the boundary points b and c.

The continuity condition at boundary points b and c is that the displacement $u_b(t)$ or $U_b(\omega)$ is equal to the sum of displacements $u_c^0(t)$ or $U_c^0(\omega)$ of substructure 2 and the displacements $u_c(t)$ or $U_c(\omega)$ due to the SSI effect of substructure 1 on substructure 2:

$$u_b(t) = u_c^0(t) + u_c(t) \text{ or } U_b(\omega) = U_c^0(\omega) + U_c(\omega) \tag{VIII.52}$$

The equilibrium condition requires that the sum of forces $f_c(t)$ or $F_c(\omega)$ between the ground and foundation and the forces $f_b(t)$ or $F_b(\omega)$ between the structure and foundation resist the inertial force $f_i(t)$ or $F_i(\omega)$ of the rigid foundation, i.e.

$$f_c(t) = f_i(t) + f_b(t) \text{ or } F_c(\omega) = M_0\omega^2 U_b(\omega) + F_b(\omega) \tag{VIII.53}$$

where $M_0\omega^2 U_b(\omega)$ is the Fourier transform of the inertial force $f_i(t)$ and M_0 the mass matrix of the rigid foundation.

From these Fourier equations it is finally found that

$$U_b(\omega) = \{I - \omega^2 C(\omega)[M_0 + M_b(\omega)]\}^{-1} U_c^0(\omega) \tag{VIII.54}$$

or

$$U_b(\omega) = \{I_s(\omega) - \omega^2[M_0 + M_b(\omega)]\}^{-1} F_c^0(\omega) \tag{VIII.55}$$

$$F_c^0(\omega) = I_s(\omega) U_c^0(\omega) \tag{VIII.56}$$

where I is the identity matrix. If three-dimensional ground motion is considered, a rigid foundation has six components; three translational and three rotational, and I is 6×6. Equation (VIII.54) clearly shows the interactions. The effect of input motion $u_g(t)$ is represented by $U_c^0(\omega)$, the foundation input motion obtained by operators L_1 and L_2; the interaction of the superstructure foundation and ground by the term $C(\omega)[M_0 + M_b(\omega)]$ and $U_b(\omega)$ is the result of the input motion on the soil–structure system, from which the structural response is obtained.

In eqn (VIII.55), $F_c^0(\omega) = I_s(\omega)U_c^0(\omega)$ is the driving force vector which includes the forces and moments from the rigid foundation on the ground when points c of the substructure 2 tend to deform and the foundation resists the deformation. Equations (VIII.55) and (VIII.56) are easier to use than eqn (VIII.54).

VIII.11 Mode superposition method — numerical modelling

Mode superposition method is a method of using the natural frequencies and mode shapes from the modal analysis to characterise the dynamic response of a structure to transient or steady harmonic excitations.

The equations of motion may be expressed as

$$[M]\{\ddot{u}\} + [C]\{\dot{u}\} + [K]\{u\} = \{F\} \tag{VIII.57}$$

$\{F\}$ is the time-varying load vector given by

$$\{F\} = \{F^{nd}\} + s\{F^s\} \tag{VIII.58}$$

where $\{F^{nd}\}$ = time-varying nodal forces and s = load vector scale factor. $\{F^s\}$ = load vector from the modal analysis.

The load vector $\{F^s\}$ is computed when doing a modal analysis and its generation is the same as for a substructure load vector.

Define a set of modal co-ordinates y_i such that

$$\{u\} = \sum_{i=1}^{n} \{\phi_i\} y_i \tag{VIII.59}$$

where $[\phi_i]$ = the nodal shape of mode i and n = the number of modes to be used.

$$[M]\sum_{i=1}^{n}\{\phi_i\}\bar{y}_i + [C]\sum_{i=1}^{n}\{\phi_i\}\dot{y}_i + [K]\sum_{i=1}^{n}\{\phi_i\}y_i = \{F\} \tag{VIII.60}$$

premultiply by a typical mode shape $\{\phi_j\}^{\mathrm{T}}$.

$$\{\phi_j\}^{\mathrm{T}}[M]\sum_{i=1}^{n}\{\phi_i\} + \{\phi_j\}^{\mathrm{T}}[C]\sum_{i=1}^{n}\{\phi_i\}\dot{y}_i \tag{VIII.61}$$

The orthogonality condition of the natural modes states that

$$\begin{aligned} \{\{\phi_j\}^{\mathrm{T}}\}[M]\{\phi_i\} &= 0 \qquad i \neq j \\ \{\phi_j\}^{\mathrm{T}}[K]\{\phi_i\} &= 0 \qquad i \neq j \end{aligned} \tag{VIII.62}$$

In the mode superposition method, only Rayleigh or constant damping is allowed so that

$$\{\{\phi_j\}^{\mathrm{T}}\}[C]\{\phi_j\} = 0 \qquad i \neq j \tag{VIII.63}$$

Applying these conditions to eqn (VIII.57), only the $i = j$ terms remain:

$$\{\{\phi_j\}^{\mathrm{T}}\}[M]\{\phi_j\}\ddot{y}_j + \{\phi_j\}^{\mathrm{T}}[C]\{\phi_j\}\dot{y}_j + \{\phi_j\}^{\mathrm{T}}[K]\{\phi_j\}y_j = \{\phi_j\}^{\mathrm{T}}[F] \tag{VIII.64}$$

The coefficients of $\ddot{y}_j$, $\dot{y}_j$ and y_j are derived as follows:

(a) *Coefficient of* $\ddot{y}_j$: by the normality condition

$$\{\phi_j\}^{\mathrm{T}}[M]\{\phi_j\} = 1 \tag{VIII.65}$$

(b) *Coefficient of* $\dot{y}_j$: the damping term is based on treating the modal co-ordinate for which

$$\{\phi_j\}^{\mathrm{T}}[C]\{\phi_j\} = C_j\phi_j^2 \tag{VIII.66}$$

and

$$\{\phi_j\}^{\mathrm{T}}[M]\{\phi_j\} = M_j\phi_j^2 = 1 \tag{VIII.67}$$

$$\phi_j = 1/\sqrt{M_j}$$

$$C_j = 2\xi_j\sqrt{(K_jM_j)} \tag{VIII.68}$$

where ξ_j = fraction of critical damping for mode j.

And

$$\omega_j = \sqrt{(K_j/M_j)} \tag{VIII.69}$$

where ω_j = natural circular frequency of mode j.

VIII.12 Modal analysis of concrete cyclic structures

Combining equations (VIII.59) to (VIII.64) through equations (VIII.66)

$$\{\phi_j\}^{\mathrm{T}}[C]\{\phi_j\} = 2\xi_j\sqrt{K_jM_j}\left\{\frac{1}{\sqrt{M_j}}\right\}^2 \qquad \text{(VIII.70)}$$

Coefficient of y_j:

$$[K]\{\phi_j\} = \omega_j^2[M]\{\phi_j\} \qquad \text{(VIII.71)}$$

premultiply by $\{\phi_j\}^{\mathrm{T}}$

$$\{\phi_j\}^{\mathrm{T}}[K]\{\phi_j\} = \omega_j^2\{\phi_j\}^{\mathrm{T}}[M]\{\phi_j\} \qquad \text{(VIII.72)}$$

substituting equation for the mass term

$$\{\phi_j\}^{\mathrm{T}}[K]\{\phi_j\} = \omega_j^2 \qquad \text{(VIII.73)}$$

for convenient notation, let $f_j = \{\phi_j\}^{\mathrm{T}}[F]$, the equation of motion of the modal co-ordinates is obtained:

$$\ddot{y}_j + 2\omega_j\xi_j\dot{y}_j + \omega_j^2 y_j = f_j \qquad \text{(VIII.74)}$$

Since j represents any mode, eqn (VIII.74) represents n uncoupled equations in the n unknown y_j. The advantage of the uncoupled system is that all the computationally expensive matrix algebra has been in the eigensolver and long transients may be analysed inexpensively.

The y_j are converted back into geometric displacements $\{u\}$ (the system response to the loading) by using eqn (VIII.74). That is, the individual modal responses y_j are superimposed to obtain the actual response and hence the name 'mode superposition'.

VIII.12.1 Modal damping

The modal damping ξ_j is the combination of several damping inputs:

$$\xi_j = (\alpha/2\omega_j) + (\beta\omega_j/2) + \xi + \xi_{\mathrm{m}j} \qquad \text{(VIII.75)}$$

where

α = uniform mass damping multiplier

β = uniform stiffness damping multiplier

ξ = constant damping ratio

$\xi_{\mathrm{m}j}$ = modal damping ratio.

VIII.12.2 Complete mode shape derivation

The mode shape in each sector is determined from the modal displacements. The displacement component (x, y or z) at any point i in the full structure is

$$u_{\mathrm{i}} = A\cos(\alpha(I-1)+\phi) \qquad \text{(VIII.76)}$$

where

$$I = \text{sector number}$$

$$A = u_{\mathrm{A}}/\cos\phi$$

$$\phi = \tan^{-1}(u_{\mathrm{B,i}}/u_{\mathrm{A,i}})$$

$u_{\mathrm{B,i}}, u_{\mathrm{A,i}}$ = modal analysis component displacement values at point i.

VIII.12.3 Mass moments of inertia

The computation of the mass moments and products of inertia as well as the model centroids is described in this section. The model centroids are computed as

$$X_{\mathrm{c}} = A_x/M$$
$$Y_{\mathrm{c}} = A_y/M \qquad \text{(VIII.77)}$$
$$Z_{\mathrm{c}} = A_z/M$$

where typical terms are

$X_{\mathrm{c}} = X$ co-ordinate of model centroid

$$A_x = \sum_{i=1}^{N} m_i X_i$$

N = number of elements

m_i = mass of element i

$X_i = X$ co-ordinate of the centroid of element i

$$M = \sum_{i=1}^{N} m_i = \text{mass of model centroid}$$

The moments and products of inertia with respect to the origin are

$$I_{xx} = \sum_{i=1}^{N} m_i[(Y_i)^2 + (Z_i)^2] \qquad I_{zz} = \sum_{i=1}^{N} m_i[(X_i)^2 + (Y_i)^2] \qquad \text{(VIII.78)}$$

$$I_{yy} = \sum_{i=1}^{N} m_i[(X_i)^2 + (Z_i)^2] \qquad I_{xy} = -\sum_{i=1}^{N} m_i[(X_i)(Y_i)] \qquad \text{(VIII.79)}$$

$$I_{yz} = -\sum_{i=1}^{N} m_i[(Y_i)(Z_i)] \qquad I_{xz} = -\sum_{i=1}^{N} m_i[(X_i)(Z_i)] \qquad \text{(VIII.80)}$$

where typical terms are I_{xx} = mass moment of inertia about the X axis through the model centroid and I_{xy} = mass product of inertia with respect to the X and Y axes through the model centroid.

The moment and products of inertia with respect to the model centroid (the components of the inertia tensor) are

$$
\begin{aligned}
I'_{xx} &= I_{xx} - M[(Y_c)^2 + (Z_c)^2] \\
I'_{yy} &= I_{yy} - M[(X_c)^2 + (Z_c)^2] \\
I'_{zz} &= I_{zz} - M[(X_c)^2 + (Y_c)^2] \\
I'_{xy} &= I_{xy} + MX_c Y_c \\
I'_{yz} &= I_{yz} + MY_c Z_c \\
I'_{xz} &= I_{xz} + MX_c Z_c
\end{aligned}
\qquad \text{(VIII.81)}
$$

where typical terms are I'_{xx} = mass moment of inertia about the X axis through the model centroid and I'_{xy} = mass product of inertia with respect to the X and Y axes through the model centroid.

It may be seen from the above development that only the mass (m_i) and the centroid (X_i, Y_i and Z_c) of each element are included. Effects which are not considered are

(a) the mass being different in different directions
(b) the presence of rotational inertia terms.

VIII.12.4 Energies

Energies are available by setting

$$
E_e^{po} = \begin{cases} \frac{1}{2}\sum_{i=1}^{NINT} \{\sigma\}^T\{\varepsilon^{el}\}\,\text{vol}_i + E_e^{pl} & \text{if element allows only displacement and rotational degree of freedom (DOF), and either is nonlinear or uses integration points} \\ \frac{1}{2}\{u\}^T[K_e]\{u\} & \text{all other cases} \end{cases}
$$

$$
= \text{potential energy} \qquad \text{(VIII.82)}
$$

$$
E_e^{ki} = \frac{1}{2}\{\dot{u}\}^T[M_e]\{\dot{u}\} \qquad \text{(VIII.83)}
$$

$$
= \text{kinetic energy}
$$

$$
E_e^{pl} = \sum_{i=1}^{NINT}\sum_{j=1}^{NCS}\{\sigma\}^T\{\Delta\varepsilon^{pl}\}\,\text{vol}_i
$$

$$
= \text{plastic energy} \qquad \text{(VIII.84)}
$$

where

$\{u\}$ = element DOF vector

$\{\dot{u}\}$ = time derivative of element DOF vector

$[K_e]$ = element stiffness/conductivity matrix

$[M_e]$ = element mass matrix

NINT = number of integration points

NCS = total number of converged substeps

$\{\sigma\}$ = stress vector

$\{\varepsilon^{\mathrm{el}}\}$ = elastic strain vector

$\{\Delta\varepsilon^{\mathrm{pl}}\}$ = plastic strain increment

vol_i = volume of integration points.

VIII.13 Modelling of isolators

VIII.13.1 Introduction

The design process for an isolation system will generally begin with a preliminary design using parameters from a previous project or from data from a manufacturer to estimate the possible maximum displacement of the system and minimum values of various controlling quantities (such as shear strain) and also for estimating the structural base shear, stability of the isolators, and possibility of uplift. After this preliminary design process is completed, examples of the final design of the isolators will be ordered and subjected to the code-mandated prototype test programme.

An innovative design strategy called 'seismic isolation' provides an economic practical alternative for the design of new structures and the seismic rehabilitation of existing buildings, bridges and industrial equipment. Rather than resisting the large forces generated by earthquakes, seismic isolation decouples the structure from the ground motion and thus reduces earthquake forces by factors of 5 to 10. This level of force reduction is very significant because it may eliminate the ductility demand on a structural system.

The principle of seismic isolation is to introduce flexibility at the base of a structure in the horizontal plane, while at the same time introducing damping elements to restrict the amplitude or extent of the motion caused by the earthquake. The essential feature is to ensure that the period of structure is well above that of the predominant earthquake input.

The advantages of seismic isolation include:

(a) the ability to eliminate ductility demand
(b) significant reduction of structural and non-structural damage
(c) enhancement of the safety of the building contents, occupants and architectural facades
(d) reduction of seismic design forces.

Seismic isolation should be considered if any of the following situations apply:

(i) increased building safety, post-earthquake operability and business survivability are desired
(ii) reduced lateral design forces are desired

(iii) alternative forms of construction with limited ductility capacity (such as precast and prestressed concrete) are desired in an earthquake region
(iv) an existing structure is not currently safe for earthquake loads.

Seismic isolation gives the option of providing a building with better performance characteristics than all current codes give. It therefore represents a major step forward in the seismic design of civil engineering structures. In the case of building retrofit, the need for isolation may be more obvious: the structure may simply not be safe in its present condition. In such cases, seismic isolation should be compared with alternative solutions, such as strengthening, to determine the cost effectiveness.

Structures are generally suitable for seismic isolation if the following conditions exist:

(1) height of structure is two storeys or greater (unless unusually heavy)
(2) site clearance permits horizontal displacements at the base of about 6 in. (152 mm)
(3) geometry of structure is not slender in elevation
(4) lateral wind loads or other non-seismic loads are less than approximately 10% of the weight of the structure
(5) site ground motion is not dominated by long period components.

New impetus was given to the concept of seismic isolation by the successful development of mechanical energy dissipators and elastomers with high damping properties. Mechanical energy dissipators, when used in combination with a flexible isolation device, can control the response of the structure by limiting displacements and forces, thereby significantly improving seismic performance.

VIII.13.2 Isolation system components

Base isolation is now a mature technology and is used in many countries, and there are a number of acceptable isolation systems, the construction of which is well understood. Most systems used today incorporate either elastomer bearings with the elastomer being either natural rubber or neoprene, or sliding bearings, with the sliding surface being Teflon and stainless steel (although other sliding surfaces have been used). Systems that combine elastomeric bearings and sliding bearings have also been proposed and implemented. Some systems of base isolation are given below.

VIII.13.2.1 Elastomeric-based systems

The bearings are large rubber blocks without the steel reinforcing plates used today and compress by about 25% under the weight of the building. The bearings have a vertical stiffness that is only a few times the horizontal stiffness and the rubber is relatively undamped.

(a) Low-damping natural and synthetic rubber bearings. Low-damping natural rubber bearings and synthetic rubber bearings have been widely used in Japan in

conjunction with supplementary damping devices, such as viscous dampers, steel bars, lead bars, frictional devices, and so on. The elastomer used in Japan comprises natural rubber, while in France neoprene has been used in several projects. The isolators have two thick steel endplates and many thin steel shims.

(b) Lead-plug bearings. The lead-plug bearing was invented in New Zealand in 1975 and has been used extensively in New Zealand, Japan and the United States. Lead plug bearings are laminated rubber bearings similar to low damping rubber bearings but contain one or more lead plugs that are inserted into holes, as shown in Fig. VIII.11. The lead in the bearing deforms physically at a flow stress of around 10 MPa (1500 psi), providing the bearing with a bilinear response. The lead must fit tightly in the elastomeric bearing and this is achieved by making the lead plug slightly larger than the hole and forcing it in.

(c) High-damping natural rubber (HDNR) systems. The development of a natural rubber compound with enough inherent damping to eliminate the need for supplementary damping elements was achieved. Adding extra fine carbon block, oils or resins, and other proprietary fillers increases the damping. The damping is increased to levels between 10 and 20% at 100% shear strains, with the lower levels corresponding to low hardness (50–55 durometer) and a shear modulus around 0.34 MPa (50 psi) and the high levels to high hardness (70–75 durometer) and a high shear modulus (1.40 MPa (200 psi)). The methods of vulcanisation, bonding, and construction of the isolators are unchanged.

The material is nonlinear at shear strain less than 20%, see Fig. VIII.9, and is characterised by higher stiffness and damping, which tends to minimise response under wind load and low-level seismic load. Over the range of 20–120% shear

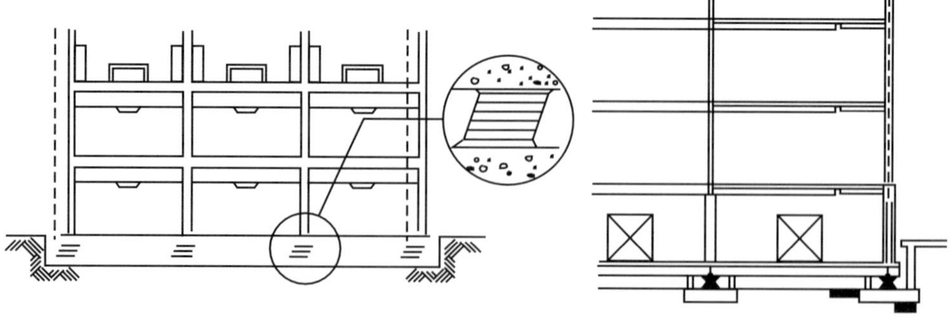

Fig. VIII.9. High damping systems and their locations.

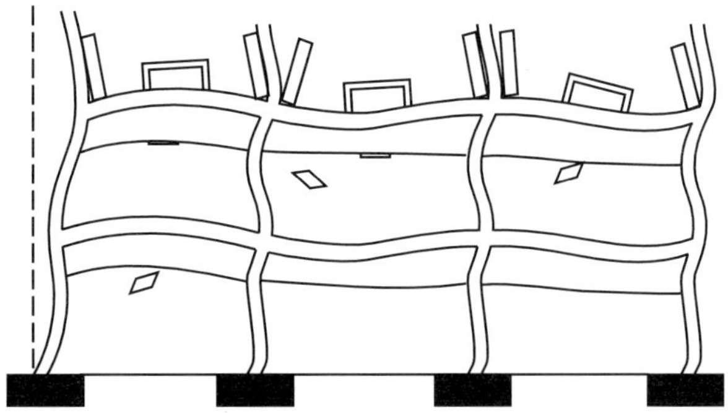

Fig. VIII.10. Performance under seismic effects.

strain, the modulus is low and constant. The damping in the isolator is neither viscous nor hysteretic but somewhere in between. As a linear viscous element, the energy dissipation is quadratic.

VIII.13.2.2 Isolation systems based on sliding

A purely sliding system is the earliest and simplest isolation system to be proposed. A system using pure sliding was proposed in 1909 by Johannes Avetican Calantarients, a medical doctor in England. He suggested separating the structure from the foundation by a layer of talc (shaded portion in Fig. VIII.9). As is evident in his diagrams (Figs VIII.12–14), Calantarients clearly understood that the isolation system reduced accelerations in the isolated building at the expense of large relative displacements between the building and the foundation, for he designed a set of ingenious connections for utilities — in those days restricted to gas lines and sewage pipes (Fig. VIII.10) — to accommodate these displacements. In fact, Calantarients' system incorporates all the elements now considered necessary in a base isolation system: a method of decoupling the building and the foundation and a method whereby utility lines can withstand large relative displacements. The system for a lead-plug isolator is shown in Fig. VIII.11).

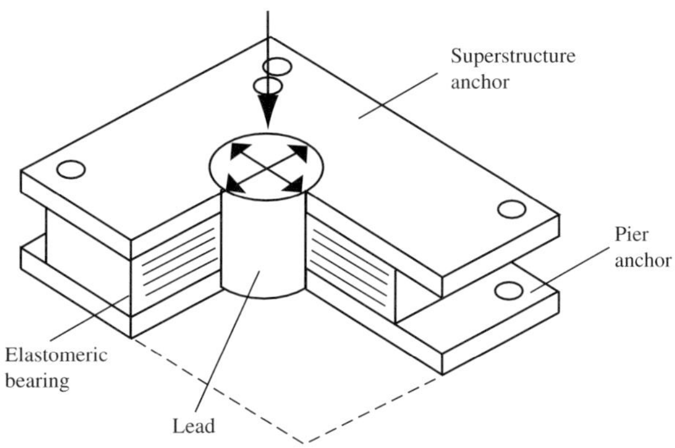

Fig. VIII.11. Typical isolation base using lead-plug.

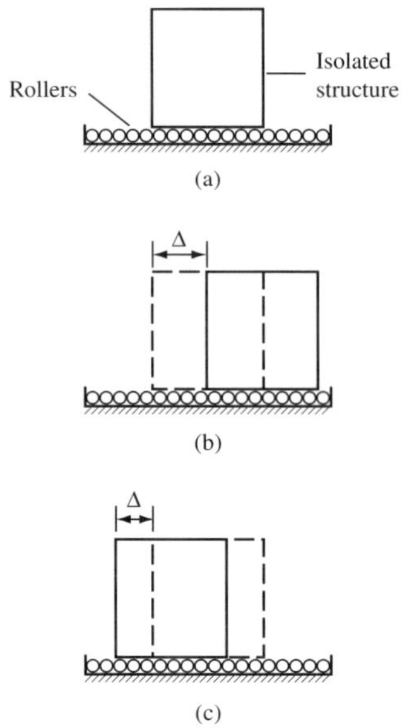

Fig. VIII.12. Concrete structures–displacement phenomena.

(a) Resilient-friction base isolation system. The resilient-friction base isolation (R-FBI) bearing attempts to overcome the problem of the high friction coefficient of Teflon on stainless steel at high velocities by using many sliding interfaces in a single bearing. Thus, the number of layers divides the velocity between the top

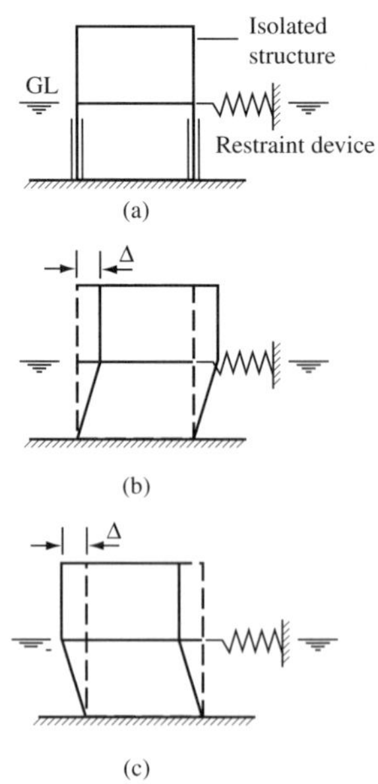

Fig. VIII.13. Displacement of isolated structures.

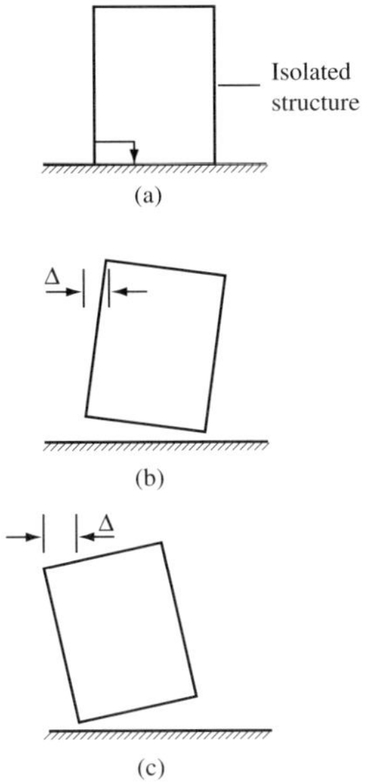

Fig. VIII.14. Displacement and rotation of concrete structures.

and bottom of the bearing so that the velocity at each face is small, maintaining a low friction coefficient (Fig. VIII.15). In addition to the sliding elements, there is a central core of rubber that carries no vertical load but provides a restoring force. This is shown in Fig. VIII.15.

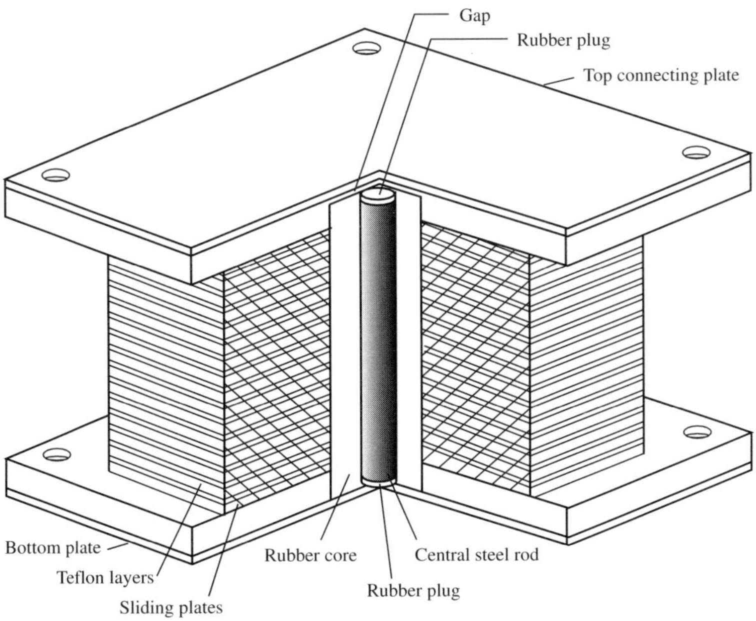

Fig. VIII.15. Isolators with low friction.

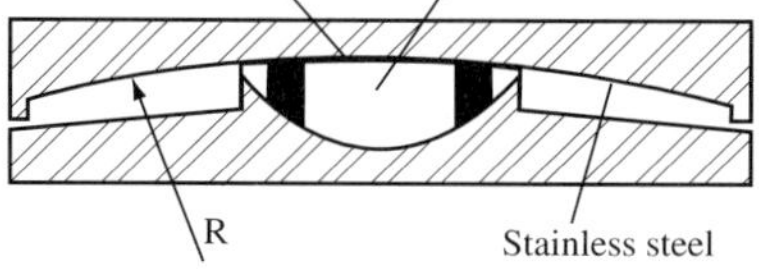

Fig. VIII.16. Isolators using friction pendulum systems.

(b) Friction pendulum system. The friction pendulum system (FPS) is a frictional isolation system that combines a sliding action and a restoring force by geometry. The FPS isolator, shown schematically in Fig. VIII.16, has an articulated slider that moves on a stainless steel spherical surface. The side of the articulated slider in contact with the spherical surface is coated with a low-friction composite material. The other side of the slider is also spherical, coated with stainless steel, and sits in a spherical cavity, also coated with the low-friction composite material. As the slider moves over the spherical surface, it causes the supported mass to rise and provides the restoring force for the system. Friction between the articulated slider and the spherical surface generates damping in the isolators. The effective stiffness of the isolator and the isolation period of the structure is controlled by the radius of curvature of the concave surface.

VIII.13.2.3 Spring-type systems

Elastomeric and sliding isolation systems are usually configured to provide only horizontal isolation. When full three-dimensional isolation is required, it is possible, but not common, to use elastomeric bearings. Generally, spring-type systems have been used in these cases.

GERB system. The GERB system for seismic isolation was developed originally for the vibration isolation of power plant turbine generating equipment. It uses large helical steel springs, that are flexible both horizontal and vertically. The vertical frequency is around 3–5 times the horizontal frequency. These steel springs are completely without damping and the system is always used in conjunction with GERB viscodamper. As in all three-dimensional systems, there is very strong coupling between horizontal motion and rocking motion because the centre of gravity of the isolated structure is above the centre of stiffness of the isolation system. This type of system is practical in situations where the centre of gravity and the centre of stiffness are at the same level — in reactor vessels in a nuclear power plant, for example.

VIII.14 Various systems in constructed facilities

Tables VIII.7 to VIII.10 summarise the various systems adopted in buildings and bridges in Japan and the USA respectively.

Table VIII.7. Seismically isolated buildings in Japan

Licence date	Type	Building name	Storey	Total floor area (m^2)	Isolation system
1982	Dwelling	Yachiyodai	2	114	EB + F
1985	Institute	Research Lab.	4	1330	EB + S
1986	Institute	High-tech. Research Lab.	5	1623	EB + S
1986	Laboratory	Oiles Tech. Centre	5	4765	LRB + E
1986	Dormitory	Tikuyu-Ryo	3	1530	EB + V
1986	Institute	Acoustic Lab.	2	656	EB + S
1986	Museum	Elizabeth Sanders (re-desi)	2	293	EB + S
1986	Test Model	Tohhoku University	3	208	EB
1986	Apartment	Apt. Hukumiya	4	681	EB + S
1987	Office	Sibuya Simizu Building	5 + B1	3385	EB + S
1987	Institute	Research Lab. No. 6	3	306	LRB
1987	Institute	Tsukuba Muki-Zaiken	1	616	EB + S
1987	Office	Tsuchiura branch	4	636	LRB
1987	Institute	Lab. J Building	4	1173	SL + R
1987	Apartment	Kousinzuka	3	476	EB + S
1987	Office	Toranomon Building	8	3373	EB + S
1988	Apartment	Itoh mansion	10	3583	LRB
1988	Dormitory	Itinoe dormitory	3	770	EB + S
1988	Institute	Clean room lab.	2	405	EB + V
1988	Rest house	Atagawa Hoyojo	1	140	SL + S
1988	Apartment	Ogawa Mansion	4	1186	HDR
1988	Office	Asano Building	7	3255	LRB
1988	Store	Kusada Building	4 + B1	1047	HDR
1988	Dwelling	Ichikawa residence	2	297	EB
1988	Computer	Computer centre	6	10032	HDR
1988	Office	Sagamihara Centre	3	255	HDR
1988	Clinic	Gerontology Res. Lab.	2 + B1	1615	EB + S
1989	Dwelling	M-300 Hoyosyo	2	309	LRB
1989	Apartment	Harvest Hills	6	2065	EB + S
1989	Institute	Acoustic Lab.	2	656	EB + S
1989	Office	Toshin Building	9 + B1	7573	EB + S
1989	Laboratory	Dwell. Test Lab.	3	680	EB + S
1989	Office	MSB-21 Ootuka	12 + B	5962	LRB
1989	Institute	Wind Laboratory	3	555	HDR
1989	Office	CP Fukuzumi	5	4406	EB + F
1989	Apartment	Employees' Buildings	4	652	LRB + HD
1989	Office	Toho-Gas Centre	3	1799	SL + RS
1989	Dormitory	Tudanuma dormitory	2	202	EB + S
1989	Dwelling	M-300 Yamada's	2	214	LRB
1989	Apartment	Koganei apartment	3	714	LRB + EB
1989	Computer	Operation Centre	2	10463	LRB
1989	Factory	Urawa-Kogyo	5	1525	HDR
1990	Office	Kanritou	3	955	EB + V
1990	Computer	Noukyou Centre	3	5423	LRB
1990	Office	C-1 building	7 + B1	37846	LRB
1990	Office	Keisan Kenkyusyo	3	627	EB + V
1990	Office	Kasiwa Kojyo	4	2186	HDR
1990	Institute	Acoustic Laboratory	2	908	EB + F
1990	Dormitory	Yamato-ryo	8	1921	EB + S
1990	Dormitory	Kawaguchi-ryo	4	659	LRB
1991	Computer	Dounen Computer Centre	4	3310	EB + LD

EB = elastic bearing
LRB = lead-rubber bearing
HDR = high damping rubber bearing
SL = sliding system (PTFE)
S = steel damper
V = viscous damper
F = friction damper
RS = rubber spring
LD = lead damper
B = basement
B1, B2 = basements

Table VIII.8. Seismically isolated bridges in Japan

Completion (scheduled)	Bridge name	Site	Superstructure type	Bridge length (m)	Isolation system
1991	On-netoh Oh-hashi Bridge	Hokkaida	4-span continuous steel girder	102	RB (12) LRB (18)
1991	Nagakigawa Bridge	Akita	3-span continuous steel girder	99	LRB (20)
1991	Maruki Bridge	Iwate	3-span continuous PC girder	122	LRB (8)
1991	Miyagawa Bridge	Shizuoka	3-span continuous steel girder	104	LRB (10)
1991	Metropolitan Highway Bridge No. 12	Tokyo	6-span continuous PC slab	138	LRB (10)
1990	Hokuso Line Viaduct	Chiba	2-span continuous steel girder	80	LRB (8)
1991	Kanko Bridge	Tochigi	6-span continuous PC girder	296	LRB (10)
1991	Matsunohama Bridge	Osaka	4-span continuous steel girder	211	LRB (12)
1991	Uehara Bridge	Aichi	2-span continuous steel girder	65	LRB (18)
1994	Karasuyama No. 1 Bridge	Tochigi	6-span continuous PC girder	245	HDR (14)
1996	Shirasuji Viaduct (Railway)	Chiba	2-span continuous steel girder	76	RB (4) LRB (4)
1998	Trans-Tokyo Bay Highway Bridge	Tokyo Bay	10-span continuous steel girder	800	LRB (18)

RB = rubber bearing
LRB = lead-rubber bearing
F = friction damper

VIII.15 Mechanical characteristics and numerical modelling with isolations

VIII.15.1 Mechanical characteristics

The mechanical characteristics of the isolator systems have been summarised in Tables VIII.9 and VIII.10. Various parameters and associated equations for these systems are needed as inputs for the numerical modelling of the above isolator systems which are linked to the global seismic analysis of the structures.

VIII.15.2 Numerical modelling with the isolators

VIII.15.2.1 Introduction

The only engineering solution that mitigates both drift and high floor accelerations is through the use of base isolation. When a building is built on an isolation system, it should have a fundamental frequency that is lower than both its fixed-base frequency and the dominant frequencies of the ground motion. These high modes have very low participation factors, so that if there is high energy in the ground motion at the high mode frequencies, this energy cannot be imparted to the structure. In this

Table VIII.9. Seismically isolated buildings in the United States

Date	Building	Height/ storeys	Floor area (m^2)	Isolation system
1985/6	Foothill Communities Law and Justice Centre	4	17000	10% damped elastometric bearings (EB-HDR)
1987/8	Salt Lake City and County Building (retrofit)	5	16000	Rubber and lead-rubber bearing (EB-LRB)
1987/8	Salt Lake City Manufacturing Facility (Evans and Sutherland Building)	4	9300	Lead-rubber bearing (LRB)
1989	USC University Hospital	8	33000	Rubber and LRB
1989	Fire Command and Control Facility	2	3000	10% damped EB + HDR
1989	Rockwell Building (retrofit)	8	28000	LRB
1991	Kaiser Computer Centre	2	10900	LRB
1991	Mackay School of Mines (retrofit)	3	4700	10% damped EB + HDR
1991	Hawley Apartments (retrofit)	4	1900	Friction-pendulum/slider (FSL)
1991	Channing House Retirement Home (retrofit)	11	19600	LRB
1991	Long Beach VA Hospital (retrofit)	12	33000	LRB

EB = elastic bearing
LRB = lead-rubber bearing
HDR = high damping rubber bearing

way, the demand on the structural system is reduced, and the accelerations transmitted to the internal non-structural components and contents are reduced. Figure VIII.17 represents the idealised force–deformation relations that may occur in isolation elements.

(a) Horizontal stiffness K_H

$$K_{\mathrm{H}} = \frac{GA}{t_{\mathrm{r}}} \tag{VIII.85}$$

where G is the shear modulus of the elastomer, A is the full cross-sectional area (which may differ from the area of the reinforcing shims), and t_{r} is the total thickness of the rubber. The maximum horizontal displacement D is related to the maximum shear strain γ by

$$\gamma = \frac{D}{t_{\mathrm{r}}} \tag{VIII.86}$$

The vertical stiffness K_{v} and the bending stiffness (expressed as EI by analogy with beam theory) are also given by a simple linear elastic theory and are needed for designing a bearing.

The vertical frequency of an isolated structure, often an important design criterion, is controlled by the vertical stiffness of the bearings that comprise the system. In order to predict this vertical frequency, the designer need only compute the vertical stiffness of the bearings under a specified dead load, and for this a

Table VIII.10. Seismically isolated bridges in the United States

Completion date	Bridge	Superstructure type	Bridge length (m)	Isolation system
1984/5	Sierra Point Bridge, California (US 101) (retrofit)	Longitudinal steel plate girder	190	LRB
1986/7	Santa Ana River Bridge, California (retrofit)	Steel trusses	310	LRB
1987	Main Yard Vehicle Access Bridge, California	Steel plate girders	80	LRB
1987	Eel River Bridge, California (US 101) (retrofit)	Steel through truss simple span	185	LRB
1988	All American Canal Bridge, California (retrofit)	Continuous steel plate girders	125	LRB
1990	Sexton Creek Bridge, Illinois	Cont. steel plate	120	LRB
1990	Toll Plaza Road Bridge, Pennsylvania	Simple span steel plate girder	55	LRB
1991	Lacey V. Murrow Bridge West Approach, Washington (retrofit)	Cont. concrete box girder	340	LRB
1991	Cache River Bridge, Illinois (retrofit)	Cont. steel plate girder	85	LRB
1991	Route 161 over Dutch Hollow Road, Illinois	Steel plate girder	110	LRB
1991	West Street Overpass, New York (retrofit)	Steel beam	50	LRB
1991	US 40 Wabash River Bridge, Indiana	Cont. steel plate girder	270	LRB
1991	Metrolink Light Rail, St. Louis (7 dual bridge)	Concrete box girder	65–280	LRB
1991	Pequannock River Bridge, New Jersey	Steel plate girder	260	LRB
1992	Blackstone River Bridge, Rhode Island	Steel plate girder	305	LRB
1992	Bridges B764 E & W, Nevada (retrofit)	Steel plate girder	135	LRB
1992	Squamscott River Bridge, New Hampshire	Steel plate girder	270	LRB
1992	Olympic Blvd. Separation, California	Steel plate girder	210	LRB
1992	Carlson Blvd. Bridge, California	Concrete box girder	45	LRB
1992	Clackamas Connector, Oregon	Concrete box girder	305	LRB
1992	Cedar River Bridge, Washington	Steel plate girders	160	LRB

LRB = lead-rubber bearing

linear analysis is adequate. The initial response of a bearing under vertical load is very nonlinear and depends on several factors. Normally, bearings have a substantial run-in before the full vertical stiffness is developed.

The vertical stiffness of a rubber bearing is given by the formula

$$K_v = \frac{E_c A}{t_r} \qquad \text{(VIII.87)}$$

Table VIII.11. Seismic isolation base shear forces and displacements

Soil type	Isolation period	Elastic base shear	Isolator displacement
S1	2.0 s	0.11W	4.5 in.
S1	2.5 s	0.10W	5.9 in.
S2	2.0 s	0.15W	6.0 in.
S2	2.5 s	0.12W	7.5 in.
S3	2.0 s	0.20W	8.0 in.
S3	2.0 s	0.16W	10.0 in.

where A is the cross-sectional area of the bearing (in this case it is usually taken as the area of the shim plates), t_r is the total thickness of rubber in the bearing, and E_c is the instantaneous compression modulus of the rubber–steel composite under the specified level of vertical load. The value of E_c for a single rubber layer is controlled by the shape factor S.

For example, in an infinite strip of width $2b$ and with a single-layer thickness t

$$S = \frac{b}{t} \tag{VIII.88}$$

For a circular pad of diameter ϕ or radius R and thickness t

$$S = \frac{\phi}{4t} \quad \text{or} \quad S = \frac{R}{2t} \tag{VIII.89}$$

And for a square pad of side dimension a and thickness t

$$S = \frac{a}{4t} \tag{VIII.90}$$

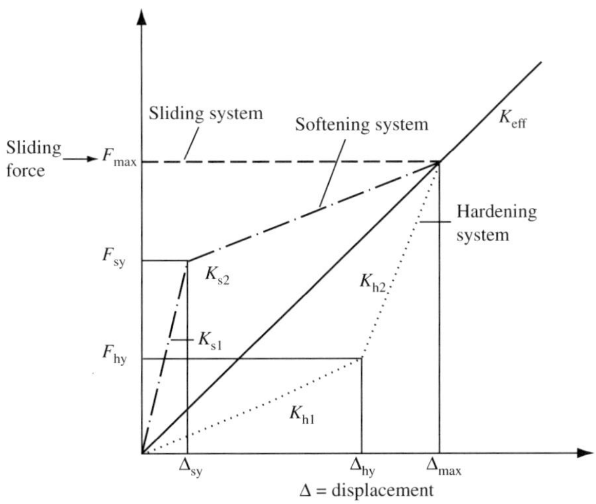

Fig. VIII.17. Load displacement for lead-plug bearing.

For a single pad in the form of a complete circle; the compression modulus E_c is given by

$$E_c = 6GS^2 \quad \text{(VIII.91)}$$

And for a square pad, the result is

$$E_c = 6.73GS^2 \quad \text{(VIII.92)}$$

In some cases bearings are designed with unfilled central holes. The result for a bearing with an inside radius a and outside radius b is

$$E_c = 6\lambda GS^2 \quad \text{(VIII.93)}$$

where

$$\lambda = \frac{b^2 + a^2 - [(b^2 - a^2)/(\ln b/a)]}{(b-a)^2} \quad \text{(VIII.94)}$$

If $a/b \to 0$, then $\lambda \to 1$; hence, $E_c = 6GS^2$, which is the result for the full circular pad. If $a/b \to 1$, by writing $a/b = 1 - \varepsilon$ and letting $\varepsilon \to 0$, we find that $\lambda \to 2/3$ and $E_c \to 4GS^2$, which is the result for the infinite strip.

Under direct compression, the constraint provided by the bonded steel shims produces a shear strain in the rubber, which is denoted by γ_c. If the nominal compression strain ε_c is given by

$$\varepsilon_c = \frac{\Delta}{t_r} \quad \text{(VIII.95)}$$

where Δ is the vertical displacement, then

$$\gamma_c = 6S\varepsilon_c \quad \text{(VIII.96)}$$

(b) Bending stiffness. The pad is assumed loaded by a simple moment M and a beam theory is adopted

$$M = (EI)_{\text{eff}} \frac{\alpha}{t} \quad \text{(VIII.97)}$$

The relative angle between the top and bottom plates is denoted by α; for a circular pad of radius R the result is

$$M = \frac{3\alpha G}{2t^3} \frac{\pi R^6}{12} \quad \text{(VIII.98)}$$

and for a beam of circular section

$$I = \frac{\pi R^4}{4} \quad \text{(VIII.99)}$$

$$(EI)_{\text{eff}} = E_c(0.3291) \quad \text{(VIII.100)}$$

For the circular pad with a central hole, we have

$$(EI)_{\rm eff} = 2GS^2 I \frac{(b+a)^2}{b^2 - a^2} \qquad \text{(VIII.101)}$$

The shear stresses induced by the vending are given by

$$\gamma_{\rm b} = 6S\varepsilon_{\rm b} \qquad \text{(VIII.102)}$$

where $\varepsilon_{\rm b} = R(\alpha/t)$ is the edge compression strain produced by bending.

When rewritten for $E_{\rm c}$, the ad hoc equation becomes

$$E_{\rm c} = \frac{E'_{\rm c}K}{E'_{\rm c} + K} \qquad \text{(VIII.103)}$$

And for the circular pad, we have $E'_{\rm c} = 6GS^2$; thus

$$E_{\rm c} = \frac{6GS^2K}{6GS^2 + K} = 6GS^2\left[\frac{1}{1 + 6GS^2/K}\right] \qquad \text{(VIII.104)}$$

The value of K varies quite widely in reference material, ranging from a low 1000 MPa (145 000 psi) to 2500 MPa (363 000 psi). The value most commonly used, and which appears to fit laboratory test results on bearings, is 2000 MPa (290 000 psi).

When S is small, we have

$$E_{\rm c} = 6GS^2\left(1 - \frac{6GS^2}{K}\right) \qquad \text{(VIII.105)}$$

VIII.15.2.2 Mechanical characteristics of lead-plug bearing under concrete structures

In practice, all isolation bearings are modelled by a bilinear model based on the three parameters K_1, K_2 and Q. The elastic stiffness K_1 is either estimated from available hysteresis loops from elastomeric bearing tests or as a multiple of K_2 for lead-plug bearings and friction pendulum bearings. The characteristic strength Q is estimated from the hysteresis loops for the elastomeric bearings. For lead-plug bearings, Q is given by the yield stress in the lead and the area of the lead, while in the friction pendulum bearings it is given by the friction coefficient of the sliding surface and the load carried by the bearing. The post-yield stiffness can be accurately estimated or predicted for all three types of bearing.

The effective stiffness, defined as the secant slope of the peak-to-peak values in a hysteresis loop, is given by

$$K_{\rm eff} = K_2 + \frac{Q}{D} \qquad D \geq D_{\rm y} \qquad \text{(VIII.106)}$$

where $D_{\rm y}$ is the yield displacement. The natural frequency ω is given by

$$\omega = \sqrt{\frac{K_{\rm eff}g}{W}} \qquad \sqrt{\omega_0^2 + \mu\frac{g}{D}} \qquad \text{(VIII.107)}$$

$$T = \frac{2\pi}{\omega} = \frac{2\pi}{\sqrt{\omega_0^2 + \mu \dfrac{g}{D}}} \tag{VIII.108}$$

The effective damping β_{eff} for $D \geq D_y$ is defined to be

$$\beta_{\text{eff}} = \frac{\text{area of hysteresis loop}}{2\pi K_{\text{eff}} D^2} \tag{VIII.109}$$

The area of the hysteresis loop is given by $4Q(D - D_y)$; to put β_{eff} in terms of these basic parameters

$$D_y = \frac{F_y}{K_1} \qquad F_y = Q + K_2 D_y \tag{VIII.110}$$

so that

$$D_y = \frac{Q}{K_1 - K_2} \tag{VIII.111}$$

Using the definition of β_{eff} and the result for K_{eff}, we have

$$\beta_{\text{eff}} = \frac{4q(D - D_y)}{2\pi(K_2 D + Q)D} \tag{VIII.112}$$

As a general rule of thumb, elastic stiffness K_1 is taken as $10K_2$, so that $D = Q/(9K_2)$, giving

$$\beta_{\text{eff}} = \frac{4q(D - Q/9K_2)}{2\pi(K_2 D + Q)D} \tag{VIII.113}$$

Data:

Bearing loads 4 No. 50 t = 200 t
8 No. 100 t = 800 t total = 1600 t
3 No. 200 t = 600 t

Column 3 rows 10 m apart each direction (40 m × 20 m)

$G = 0.4\,\text{MN/m}^2$

Target period = 2.5 s; $C_{vd} = 0.56$; $\beta_D = 1.2$; $K_1 = 10K_2$

Υ = max. shear strain = 1.5, $\beta_{\text{eff}} = 0.15$

$$\text{Design displacement } D_D = g/4\pi^2 \frac{CVDTD}{\beta_D} = 0.258\,\text{m}$$

$$K_H = 1600 \times 1000\left(\frac{2\pi}{1.5}\right) = 10.1\,\text{MN/m}$$

W_D = global energy dissipated = $2\pi K_{\text{eff}} D^2 \beta_{\text{eff}} = 0.634\,\text{MN/m}$

$Q = W_D/4D = 0.611\,\text{MN}$

$K_2 = 10.1 - Q/D = 7.72\,\text{MN/m} = 0.009\,\text{m}$; area of $LP = A_{pb} = 0.637\,\text{m}^2$

$D_y = 0.611/9(7.72)$

K_Υ = total rubber stiffness = $K_h(Q/D_D) = 7.64\,\text{MN/m}$

VIII.15.2.3 Numerical modelling of equations of motion with isolators

As mentioned in the previous sections, the elements used for base isolating structures (high damping rubber bearings, lead-rubber bearings, hysteretic and viscous dampers) behave in the nonlinear range under horizontal forces. Therefore, the nonlinear behaviour of these isolation systems themselves has to be included in the discrete mathematical mode of an isolated structure. The dynamic equilibrium equations of an elastic system with nonlinear isolators and energy-dissipating elements are given by the following expression:

$$\boldsymbol{M}\ddot{U}(t) + \boldsymbol{C}\dot{U}(t) + \boldsymbol{K}U(t) + \boldsymbol{R}_{\mathrm{N}}(t) = R(t) \qquad \text{(VIII.114)}$$

where M, K and C are the mass matrix, the stiffness matrix and the damping matrix, respectively, for the whole elastic system. $R(t)$ is a time varying load and $\boldsymbol{R}(t)_{\mathrm{N}}$ presents a vector of nodal forces in the nonlinear isolators and the dissipating elements (dampers).

The more economical way of solving eqn (VIII.114), through transformation of eqn (VIII.114) into generalised modal co-ordinates, is as follows:

$$IX(t) + \Lambda X(t) + \Omega^2 X(t) = F(t) - F_{\mathrm{N}}(t) \qquad \text{(VIII.115)}$$

where

$$\phi^{\mathrm{T}}\boldsymbol{M}\phi = I \qquad \text{(VIII.116)}$$

$$\phi^{\mathrm{T}}\boldsymbol{K}\phi = \mathrm{diag}[\omega^2] = \Omega^2 \qquad \text{(VIII.117)}$$

$$F_{\mathrm{N}}(t) = \phi^{\mathrm{T}}\boldsymbol{R}(t)_{\mathrm{N}} \qquad \text{(VIII.118)}$$

where $F_{\mathrm{N}}(t)$ versus $\Delta(t)$ can be realised from Fig. VIII.17.

If C is assumed to be proportional damping, then $\Lambda = \phi^{\mathrm{T}} C\phi = \mathrm{diag}[2\xi_i\omega_i]$, where ξ represents a modal damping ratio. The transformation matrix ϕ can represent either the eigen or the Ritz vectors for the whole model without the nonlinear elements. Further direct integration of the system of equations can be done by means of the modal superposition method where only the first few vectors that participate in the total response of the system need to be included in the direct integration.

During the integration of eqn (VIII.115), the loads in the nonlinear elements are calculated at the end of each time step. Then, the loads are transformed into modal loads and are added on the right side of eqn (VIII.115). Iteration is performed within each time step until the loads converge.

If linear isolators and viscous dampers are used to isolate the structure, matrix C will not be proportional to matrices $\boldsymbol{M}$ and $\boldsymbol{K}$, and will be given as follows:

$$\boldsymbol{C} = C_{\mathrm{d}} + C = \mathrm{diag}[2\xi_i\omega_i] + C \qquad \text{(VIII.119)}$$

By substituting matrix $\boldsymbol{C}$, the system eqn (VIII.115) assumes the following form:

$$x_i + 2\xi_i\omega_i x + \omega^2 x_i = f_i - \sum_{j=1}^{m} c_{ij}x_j, \qquad i = 1, 2, \ldots, m \qquad \text{(VIII.120)}$$

Further solving of eqn (VIII.114) is the same when no linear elements are introduced, i.e. the system is solved through iteration.

If the isolation systems are represented as equivalent linear springs and equivalent proportional damping, then terms $R_N(t)$ of eqn (VIII.114) and $F_N(t)$ of eqn (VIII.115) are eliminated. In this case, it is necessary to conduct time-history analyses of such linearised equations of dynamic equilibrium by means of the mode superposition method, because the time needed to obtain the dynamic response of the system by using the mode superposition method is much shorter than that needed when performing direct integration of eqn (VIII.114).

Discrete energy expressions. Apart from the need for computing the time-history acceleration, displacement and forces for both the structure and the isolation system, it is also necessary to determine the distribution of input energy into the structure. In order to obtain the energetic time-history in the structure, it is necessary to compute the distribution of energy at each time step. To define the energy expressions, it is necessary that eqn (VIII.114) be pre-multiplied by $U(t)^T$ and then integrated in the entire time domain. After these operations, the following

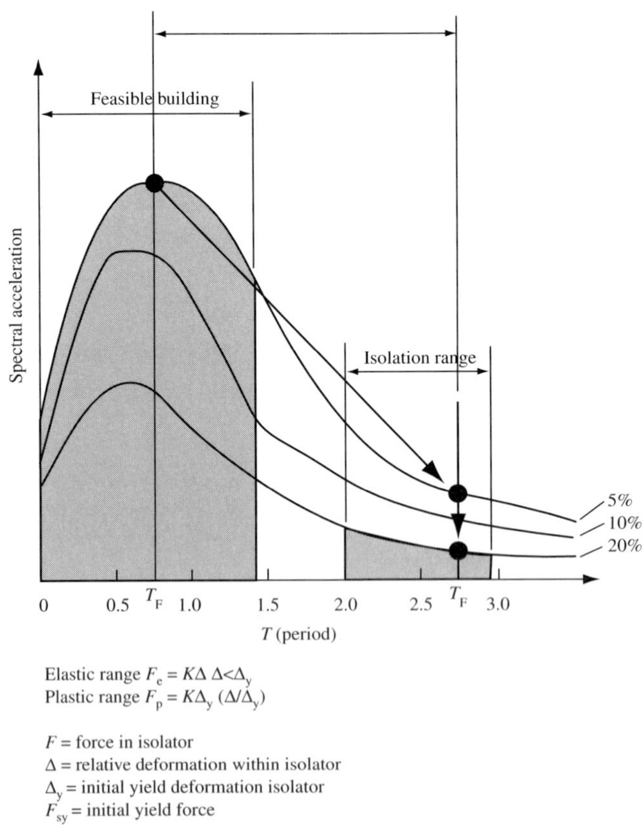

Fig. VIII.18. Feasible building and isolated range–response spectra.

expressions are obtained:

$$K(t) + D(t) + U(t) + BI(t) = I(t) \tag{VIII.121}$$

where

$K(t)$ (kinetic energy) $= \frac{1}{2}[U(t)^{\mathrm{T}} M U(t)]$

$D(t)$ (modal damping energy)

$$= D(t) = D(t - \Delta t) + \tfrac{1}{2}[U(t - \Delta t) + U(t)^{\mathrm{T}} C[U(t) - U(t - \Delta t)]]$$

$U(t)$ (strain energy stored in the superstructure) $= \frac{1}{2}[U(t)^{\mathrm{T}} K_{\mathrm{L}} U(t)]^{\mathrm{T}}$

$BI(t)$ (ene.abs. by isolators)

$$= BI(t) = BI(t - \Delta t) + \tfrac{1}{2}[U(t) - U(t - \Delta t)^{\mathrm{T}}[R_{\mathrm{N}}(t - \Delta t) + R_{\mathrm{N}}(t)]]$$

$I(t)$ (input energy) $= I(t - \Delta t) - \frac{1}{2}[U(t) - U(t - \Delta t)^{\mathrm{T}}[R(t - \Delta t) + R(t)]]$

Computing the above time-histories gives insight into that part of the input energy dissipated through dampers and that dissipated through viscous damping in the structure. The typical response spectra with and without isolation are given in Fig. VIII.18.

IX. Application to engineering problems

IX.1 Introduction

Various analyses and numerical models given in earlier chapters are now applied to selected engineering problems. As far as possible the analytical results produced for various case studies have been compared with those available from experimental results and other published data. For all experiments concrete cubes and cylinders were tested under uniaxial, biaxial and triaxial loading conditions. Appendix VI gives a summary of the experimental and theoretical results.

IX.2 Case Study No. 1 — reinforced and prestressed concrete beams

IX.2.1 Reinforced concrete beams

A model reinforced concrete beam with reinforcement details is shown in Fig. IX.1. Cube crushing tests showed that the concrete mix is of grade 30 ($f_{cu} = \sigma_{cu} = 30\,\text{N/mm}^2$). Separate tensile tests were carried out on the reinforcements. The characteristic strength of a steel bar was determined as $410\,\text{N/mm}^2$. The beam was placed on simple supports as shown in Fig. IX.2. Dimensions and the positions of studs on strain gauges were recorded. The load was applied in increments to failure. The first load recorded was due to self-weight of the beam and the equipment. The second was of 10 kN, and all other loads were in increments of 5 kN. At the end of each load increment, the inclinometer, demec and strain gauge readings were recorded. Figures IX.3–IX.6 show the values of strains for various depths at all four stud columns shown in Fig. IX.2. Figure IX.7 shows the load–displacement relationship. At a load of 55 kN, the beam was finally crushed.

Figures IX.8a and b show the final cracking stages.

Figure IX.9 shows the finite element mesh of the same concrete beam with 20-noded isoparametric elements. The thickness of the beam was divided into half, such that the depth of each element was 95 mm. The total number of solid isoparametric elements chosen was 20. The top and bottom reinforcements were treated as three-noded elements placed in the body of the solid elements.

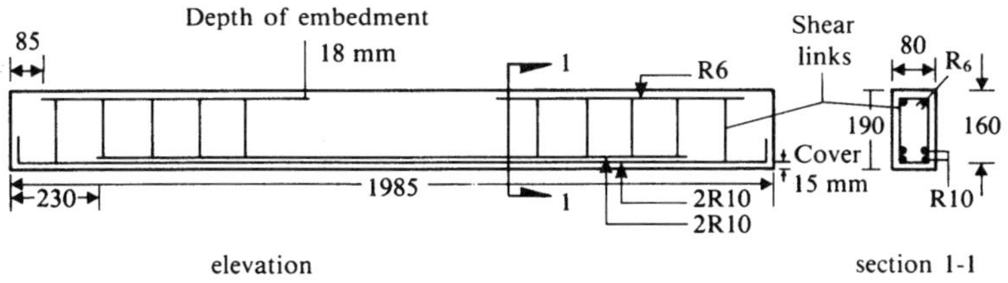

Fig. IX.1. Reinforcement details.

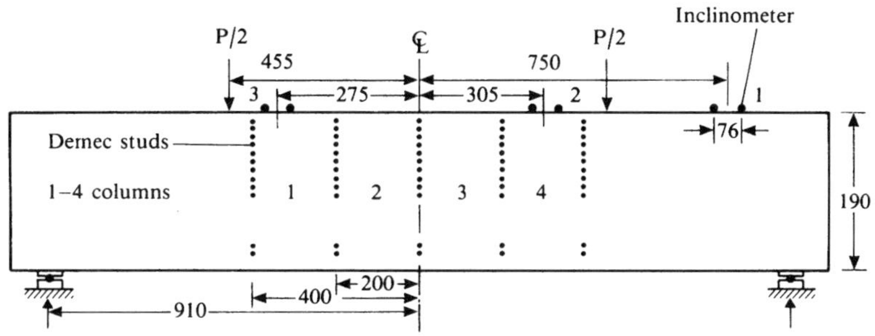

Fig. IX.2. Position of supports, loads, and demec and inclinometer studs.

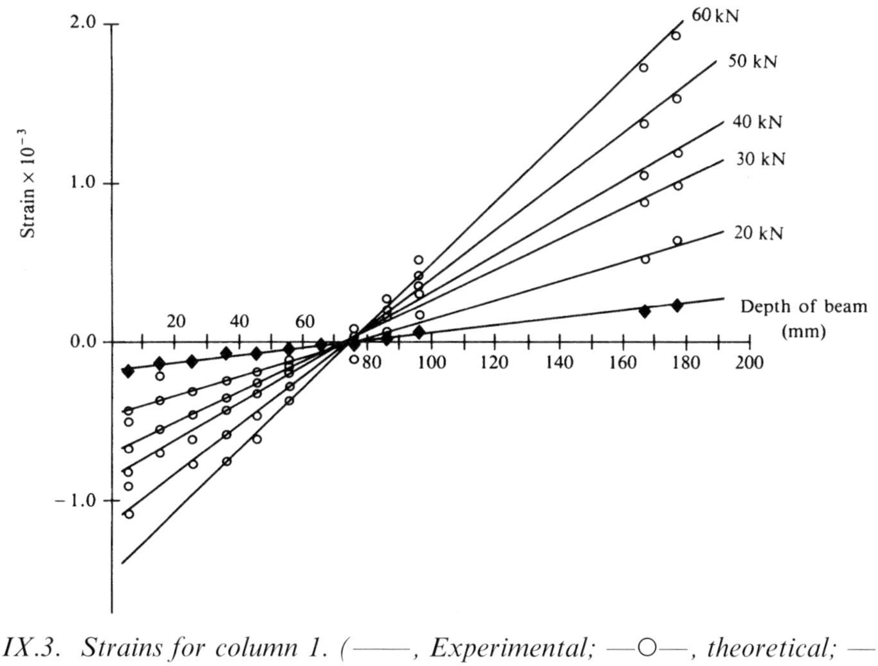

Fig. IX.3. Strains for column 1. (——, Experimental; —○—, theoretical; —◆—, additional theoretical, 10 kN.)

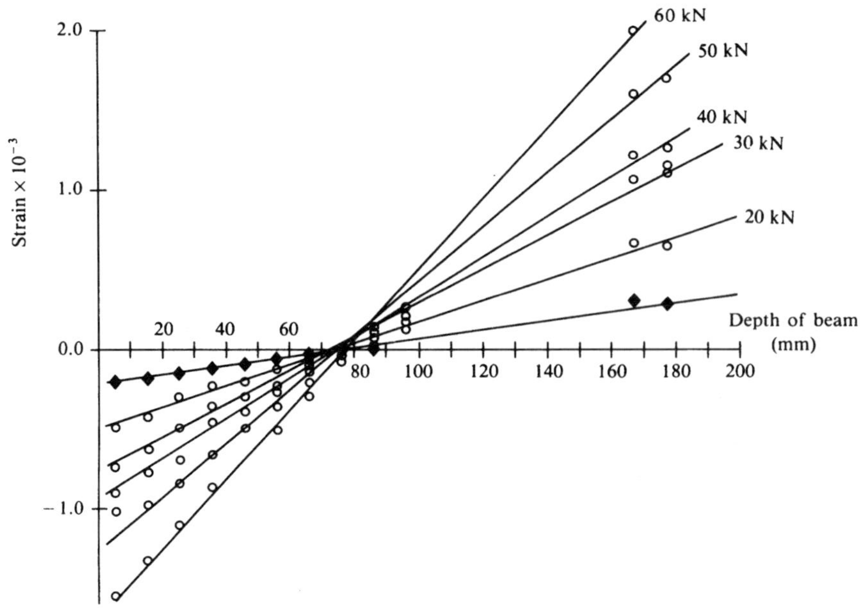

Fig. IX.4. Strains for column 2. (——, Experimental; —O—, theoretical; —◆—, additional theoretical, 10 kN.)

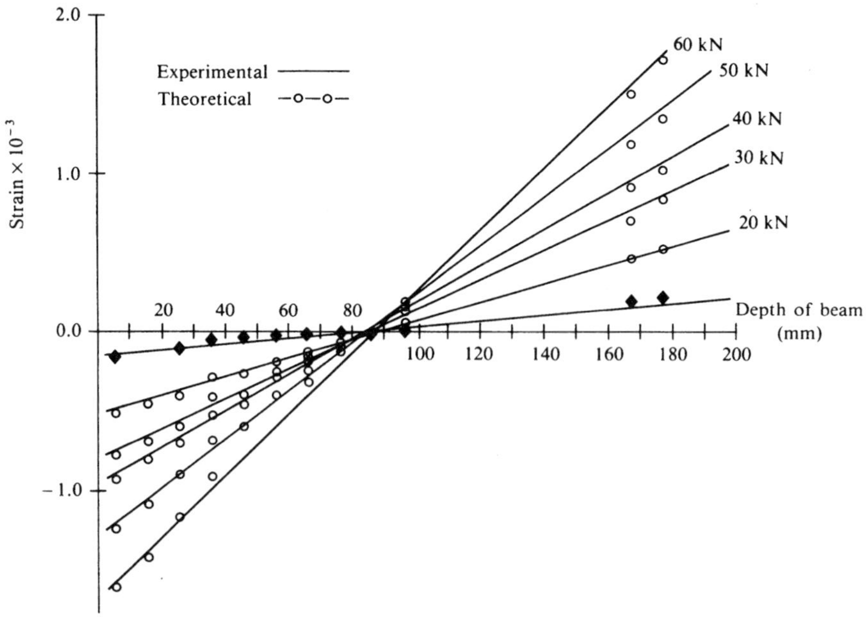

Fig. IX.5. Strains for column 3. (——, Experimental; —O—, theoretical; —◆—, additional theoretical, 10 kN.)

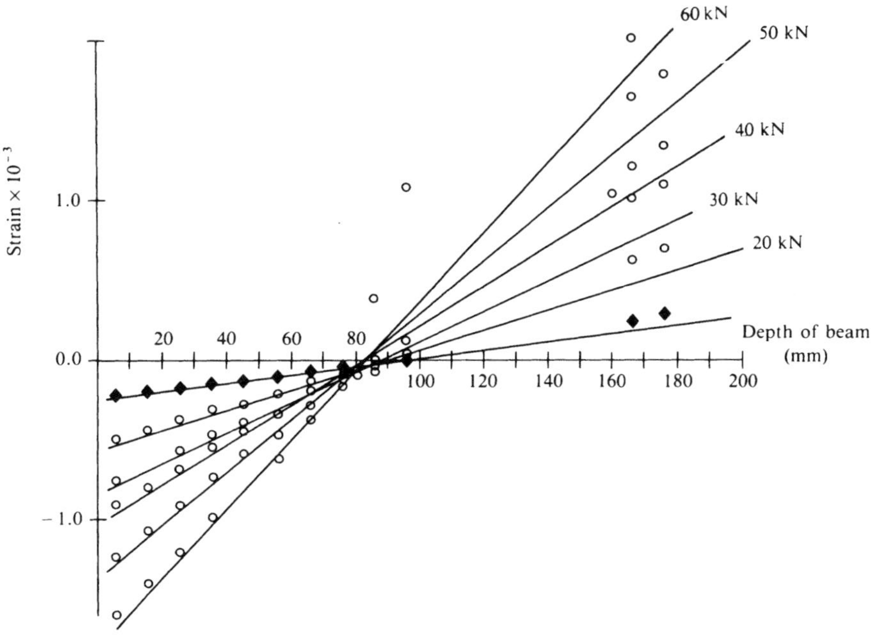

Fig. IX.6. Strains for column 4. (——, Experimental; —○—, theoretical; —◆—, additional theoretical, 10 kN.)

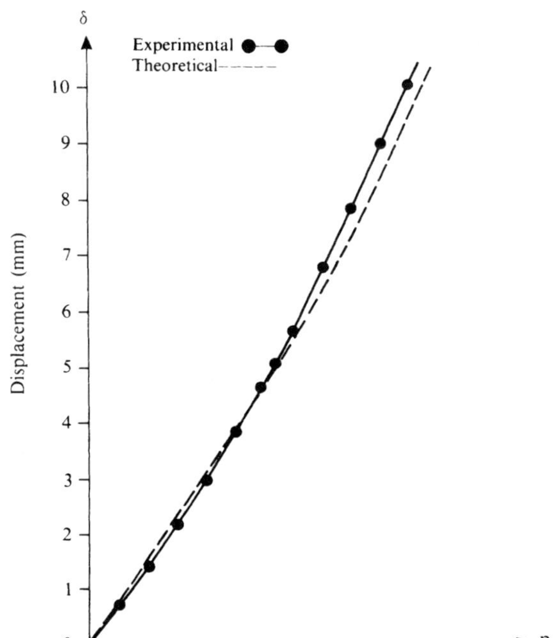

Fig. IX.7. Load–displacement relation.

Fig. IX.8a. Cracking of a beam (at 40 kN).

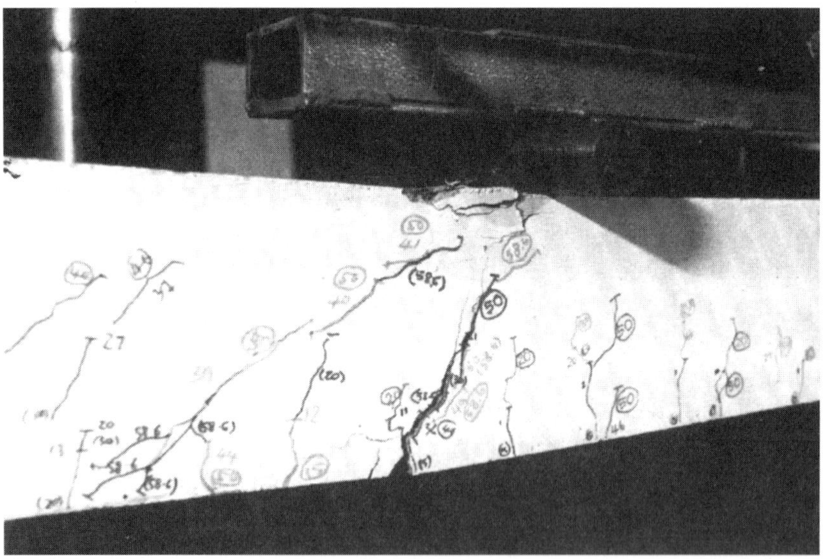

Fig. IX.8b. Cracking and crushing (at 55 kN).

A three-dimensional finite element analysis, as given in Chapters VI and VII, was carried out using a five-parameter model and the cracking criteria given in Sections II.1.5 and VII.4.5. Both E and ν values were assumed to change along the three principal axes with initial values of 200 GN/m^2 and 0.15, respectively. Concrete

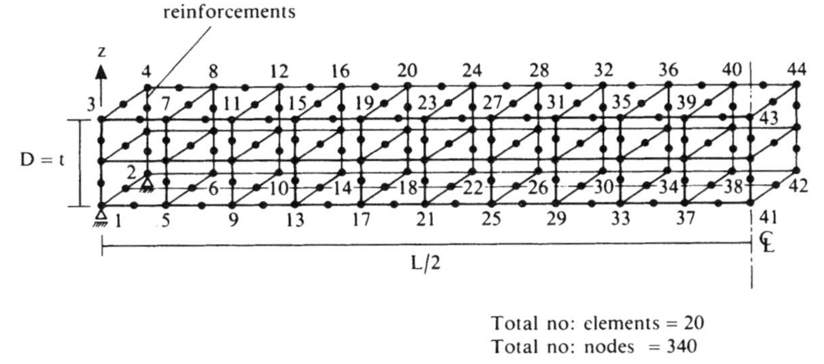

Fig. IX.9. Finite element mesh.

aggregate interlocking factor $\beta' = 0.5$ was considered in the final evaluation of the material matrix $[D]$. In the final element analysis using the program ISOPAR, the load was applied in increments of 5 kN.

Figures IX.3–IX.6 show the comparative values of strain for various depths. These results are in good agreement with those from the experiment.

The load–displacement relationship obtained from the finite element analysis again gives a good correlation with that given by the experiment. These results are plotted in Fig. IX.7. The failure load predicted by the finite element analysis was about 7% higher than that given by the experiment. Figure IX.10 shows a comparative study of crack patterns from both theory and experiment. The finite element analysis predicted the maximum stress in the reinforcement as 395 N/mm^2 in the bottom bars. One of the bottom bars in the experiment was severely damaged with a V-notch. The safety margins against the design load predicted by the analysis and the experiment were found to be 2.95 and 2.81, respectively.

IX.2.2 Prestressed concrete beams

In this case the size of the beam is the same as in the case of the reinforced concrete beam given in Fig. IX.1. This beam was recast with certain minor changes, as shown

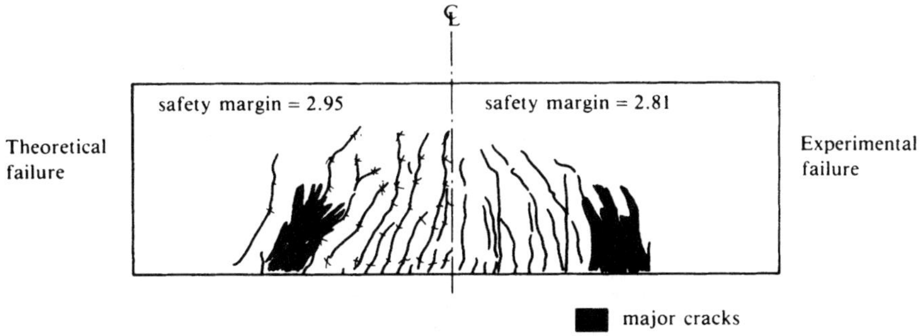

Fig. IX.10. Comparative results of theoretical and experimental failure.

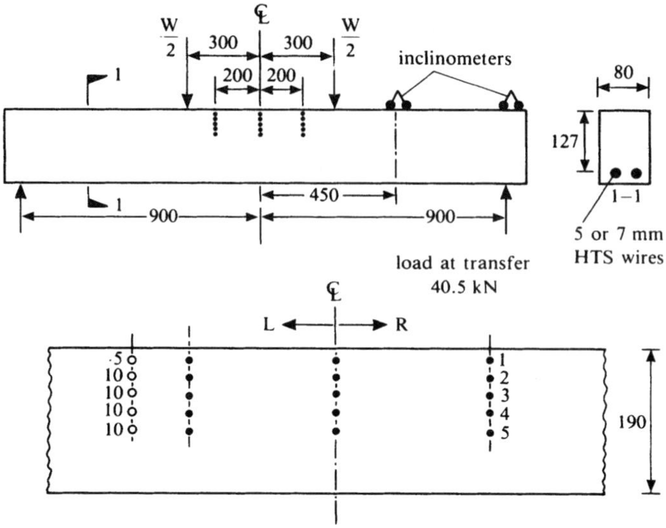

Fig. IX.11. Details of the prestressed concrete beam.

in Fig. IX.11. These changes are the span of 1800 mm between the supports and the effective depth of 127 mm due to two straight ducts for prestressing tendons.

The beam was pretensioned by means of a hydraulic jack. The prestressing force at transfer per tendon was 40.5 kN. Grade 30 concrete was selected for this beam. First, ignoring applied load, displacements, rotations and strains were measured due to dead loads of the equipment and the self-weight of the beam. Loads at two intervals of 5 kN were applied. Afterwards the beam was subjected to loads of five increments

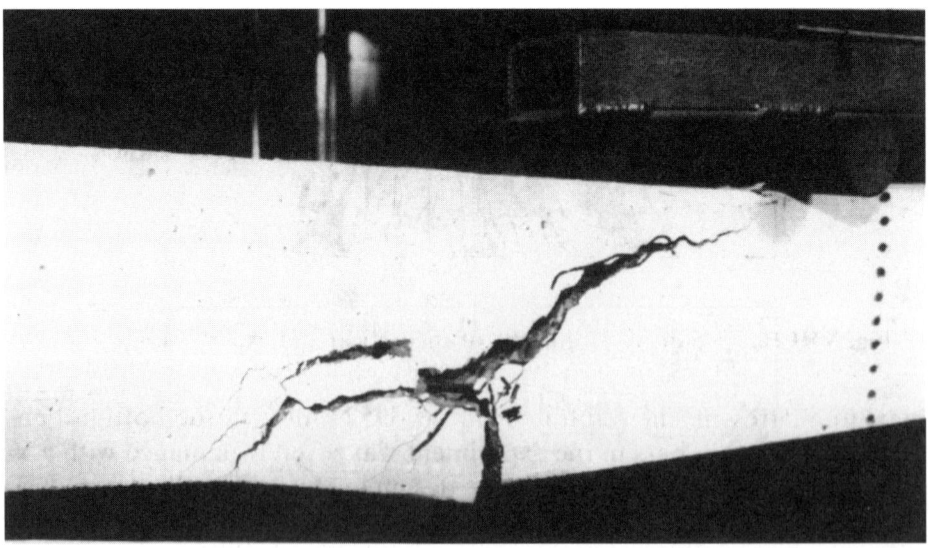

Fig. IX.12. Collapse mode of the beam.

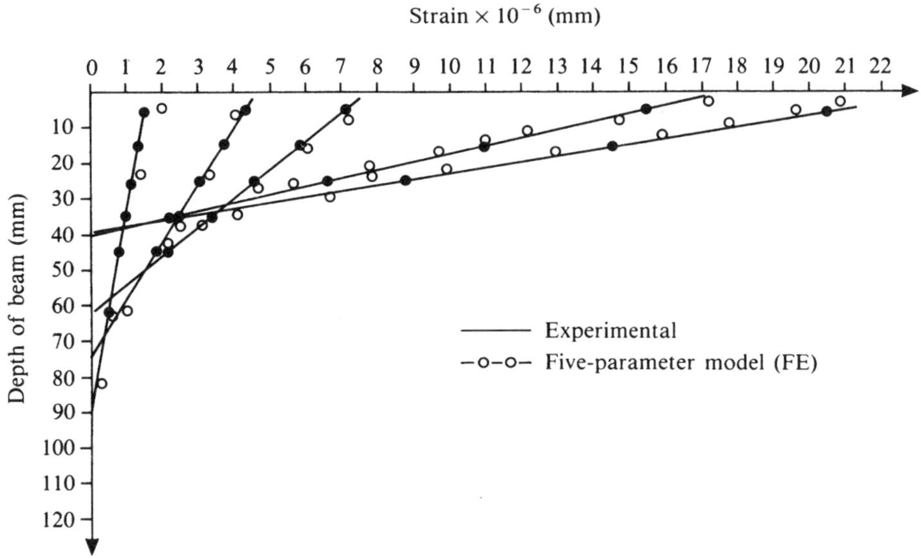

Fig. IX.13. Strain distribution along the depth of the beam.

each of 1 kN. The last four loads prior to failure of the beam were in increments of 0.5 kN. The beam finally failed at about 17.5 kN. Figure IX.12 shows the collapse mode of the beam. Figure IX.13 shows the strain distribution along the depth of the beam plotted for 5, 10, 12, 15 and 17.5 kN loads. Figure IX.14a shows the final load–displacement curve, given by a solid line.

In the finite element analysis, the same mesh as given in Fig. IX.10 was adopted. This time the line elements were placed at 127 mm from the top of the beam. As the tendons were grouped, the Ahmlink element (representing bond) was used along with 20-noded solid and three-noded line elements. The material properties were the same except for the stress–strain relation of the prestressing tendons. Only the strain distribution results using the five-parameter model are plotted along with the experimental results in Fig. IX.13. The results are in good agreement. The load–displacement relation using endochronic, four- and five-parameter models are plotted in Fig. IX.14a. The endochronic theory gave close results above the 12 kN load, otherwise it showed a lower bound effect. The five-parameter model bore close resemblance to the experimental behaviour of this beam. The four-parameter model, as expected, gave results which indicated a lower bound phenomenon as soon as the beam deformation becomes nonlinear. In all cases, as is evident from Fig. IX.14a, the collapse load was slightly lower than that obtained from the experiment. The difference ranges between 1 and 5%. This difference, although acceptable, could have been due to non-optimisation of the mesh, assumptions made in failure models, and the assumptions for the boundary conditions of the experimental model. Material properties at the time of experiment might be slightly different from those considered in the finite element analysis.

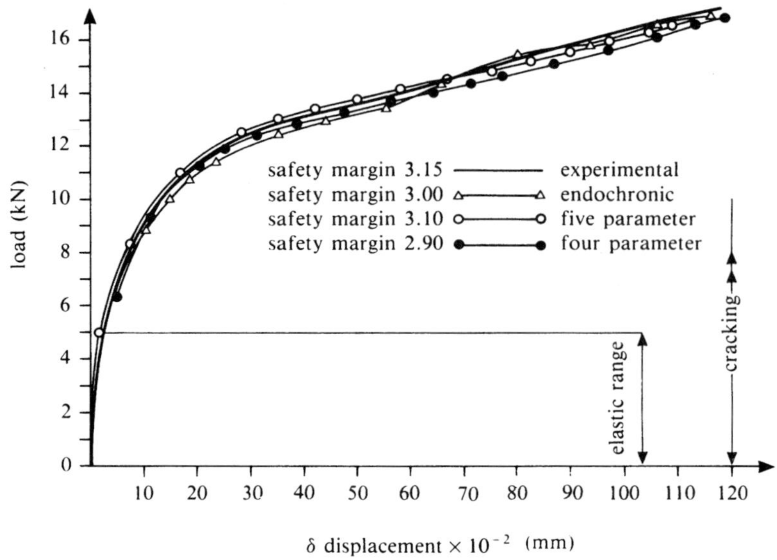

Fig. IX.14a. Load–displacement relation.

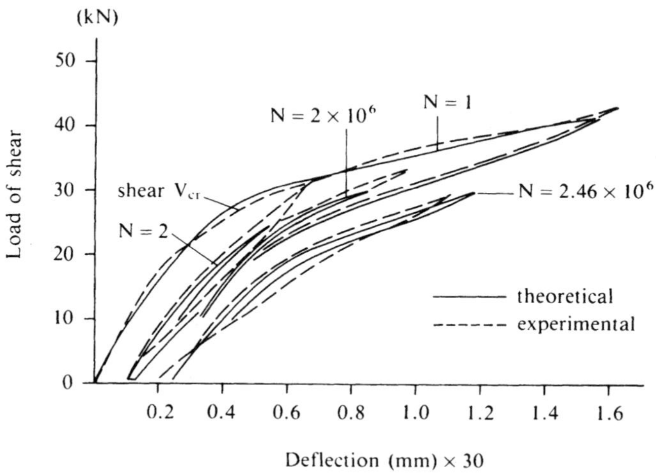

Fig. IX.14b. Shear deflection.

The idealised stress–strain curve for tendons given in Appendix IX had been adopted for this particular problem. The program ISOPAR has the flexibility to include other types of stress–strain curves and material properties. For this particular problem concrete aggregate interlocking was ignored. The crack pattern given in Fig. IX.11 is identical for this beam as well; only the major crack is obtained at a much lower load.

Next the same beam was studied in order to evaluate the repeated-load shear strength. The beam was subjected to a symmetrical two-point loading and was loaded statically in increments to approximately 80% of the flexural capacity

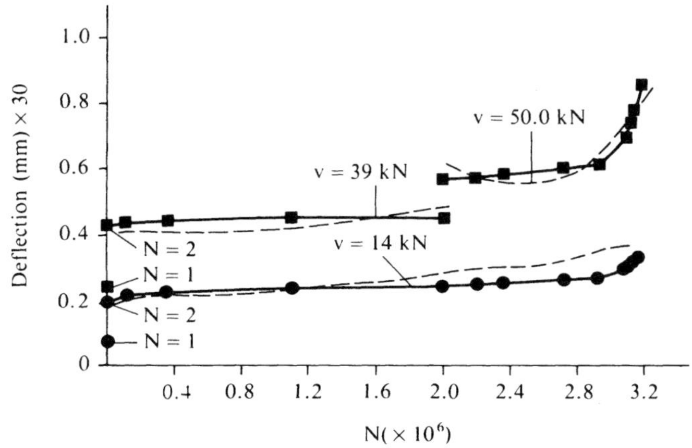

Fig. IX.14c. Deflection–N curve. (●—●, ■—■ Experimental; – – –, theoretical (ISOPAR).)

determined above. This was sufficient to cause significant inclined cracking. After unloading (the procedure described in Chapters VI and VII) the beam was then subjected to additional cycles of statically applied loads (4 cycles/s). The repeated load was stopped at 2×10^6 cycles of its design repeated-loading. The load increments adopted were equal to 5% of the flexural capacity of the beam. The maximum crack width computed was 10 mm. Figure IX.14b shows the load–deflection curve. Figure IX.14c shows the deflection–cycle curve.

IX.3 Case Study No. 2 — reinforced concrete slabs

IX.3.1 Reinforced concrete rectangular slabs subject to concentrated loads

The dimensions of a reinforced concrete slab made of grade 30 concrete are shown in Fig. IX.15a. The slab is 610 mm × 910 mm × 52 mm thick. It is reinforced at the bottom only with a light steel mesh reinforcement (6 mm diameter at 80 mm c/c). Based on several tensile tests, the yield strength of the mesh is around 413 N/mm^2. This reinforced concrete slab, as shown in Fig. IX.15b, was simply supported at the corners. The increments of the point-load were applied centrally. Four deflection gauges were located 150 mm from the corners. Table IX.1 shows the proofing ring calibration. The slab was subjected to incremental loadings. At a certain stage of loading between 10 and 11 kN, the tensile stress in the concrete reached the modulus of rupture and cracking started in the zones of maximum tensile stress. The deflections recorded immediately after this stage were nonlinear. Excessive cracks were formed around 16 kN. As shown in Fig. IX.16 the major cracks were formed in the central area of the slab and the corners of the newly developed crack zone of a square shape (240 mm square) were linked to the corners of the slab by the radial cracks. The preferential cracks around these major cracks started developing and propagating in large numbers at a load of 14.94 kN. The slab failed completely at a load of 18.59 kN. It was a plug type of failure. The reinforcement mesh was

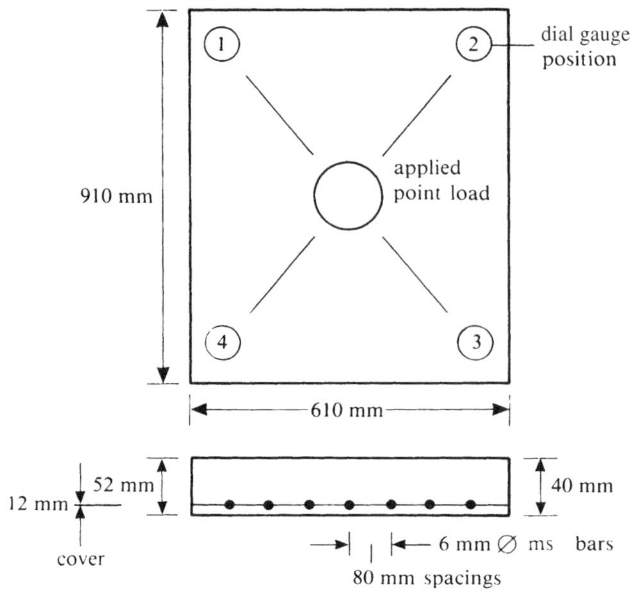

Fig. IX.15a. Reinforced concrete slab model.

Fig. IX.15b. A reinforced concrete slab on simple supports with a proofing ring and dial gauges.

Table IX.1. Proofing ring calibration

Load (kN)	Division	Proofing ring calibration
10	91	
20	182	
30	270	
40	360	
50	450	0.110 67 kN
60	540	per division
70	635	
80	720	
90	820	
100	910	

examined. At the level of major cracks the mesh was either yielded or completely ruptured. The load–displacement relationship for this slab is given in Fig. IX.17a and IX.17b together with the relationship obtained using finite elements. The concrete modelling was based on the five-parameter concept.

Three-dimensional isoparametric 20-noded solid elements together with three-noded line elements placed in the body of the element were considered for this slab. The finite element mesh scheme for this slab is given in Fig. IX.18. Three failure models, namely endochronic, five-parameter and hypoelastic, were used. The load–displacement relation was plotted for each of these failure models. The collapse loads obtained from the finite element analysis and with the use of the four- and five-parameter models were 17.5 and 16.80 kN, respectively. The collapse load from the endochronic theory was 19.5 kN. The three failure models associated with the finite element analysis gave identical crack patterns. There were changes in the direction of such cracks but these were confined to only 10 small peripheral cracks. In all cases, the crack patterns obtained from the finite element analysis were close to the ones shown in Fig. IX.19. Comparative study of the theoretical and experimental crack patterns given in Figs IX.16 and IX.19 shows marked similarities in the initiation of major cracks. No attempts have been made gradually to remove the load and to assess the closure of some of the cracks. The finite element analysis showed stresses in the reinforcement well above 20% of the yield stress at the time of collapse. The loads were applied in increments of 2 kN up to 10 kN, and afterwards small increments of 0.5 kN were considered in order to assess the final stress trajectory and collapse conditions. The modified Newton–Raphson method was adopted as a solution procedure. The convergence criterion described in Chapters VI and VII has been adopted so as to converge the solutions of the non-linear equations.

IX.3.2 Reinforced micro-concrete square slabs subject to patch loads

An identical test was carried out with a micro-concrete slab subjected to four patched loads. The slab was 510 mm square and 25 mm thick reinforced with

Fig. IX.16. Major cracks in the model slab.

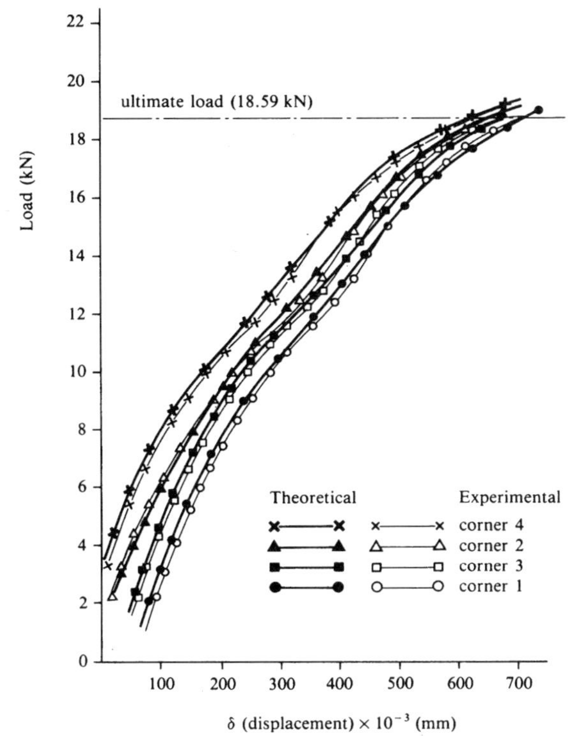

Fig. IX.17a. Load versus deflection at corners. (Note: The slab has four corners where displacements are evaluated.)

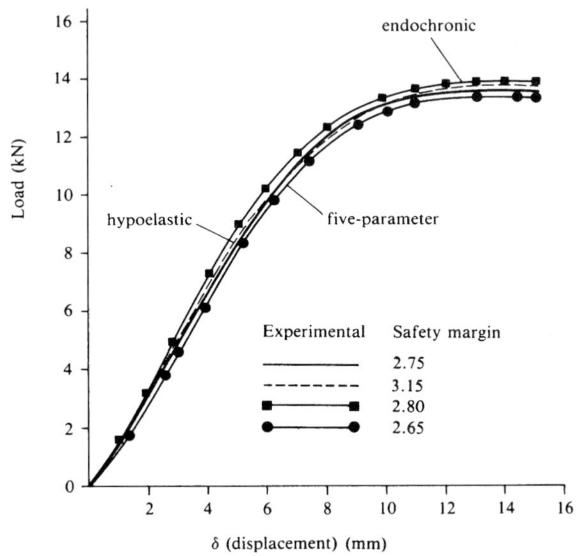

Fig. IX.17b. Load–displacement relationship for a square slab under patch loads.

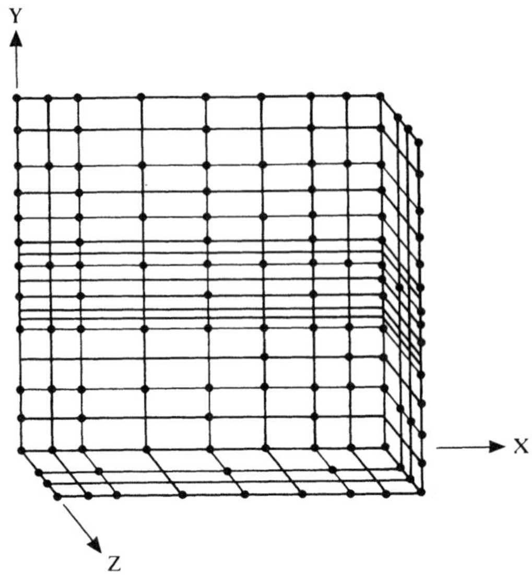

Fig. IX.18. Finite element mesh. (Number of solid elements = 384; number of nodes = 7400; Gauss points 2 × 2 × 2; tol > 0.001.)

1.55 mm mesh with 13 mm spacings. Figure IX.17b shows a comparative study of the load–displacement relationship using the finite element mesh scheme of Fig. IX.18.

IX.3.3 Reinforced concrete slabs under impact loads

Two other reinforced concrete slabs were tested under impact loads by the UK Atomic Energy Authority at Winfrith [769–772]. Figures IX.20 and IX.21 show the reinforcement layout for the slabs. Table IX.2 shows the data from the impact test. Figure IX.22 shows a typical soft missile used as an impactor, and Fig. IX.23 shows a load–time function.

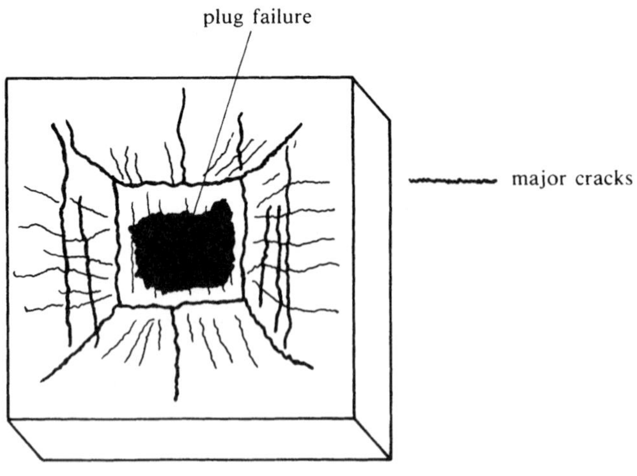

Fig. IX.19. Crack pattern from finite element analysis.

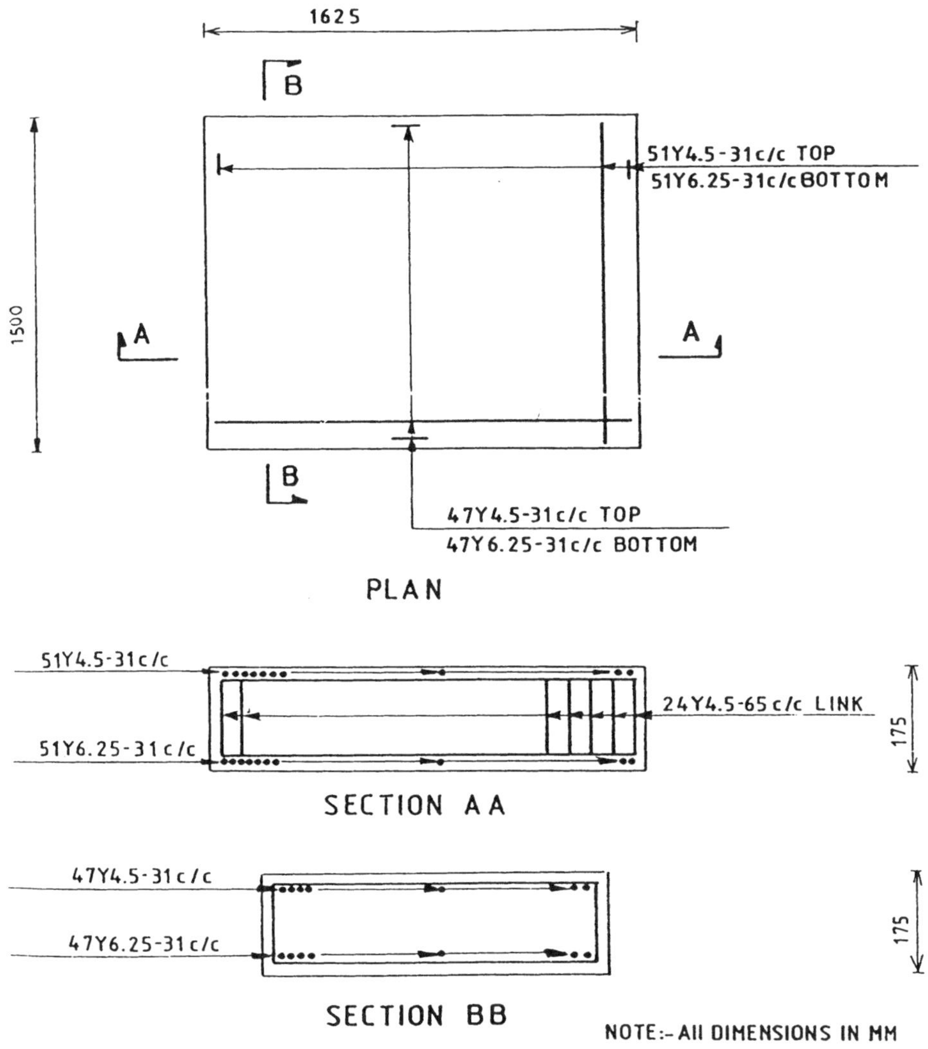

Fig. IX.20. Detail of target slab B16 (courtesy of UK Atomic Energy Authority, Winfrith).

A three-dimensional finite element analysis has been carried out using the program ISOPAR.

Three-dimensional 20-noded isoparametric elements have been used to represent the concrete. Each edge of this isoparametric element forms a parabola, so that eight nodes define the corner of the element and a further 12 nodes define the position of the 'mid-point' of each edge. This element has 27 Gaussian integration points located in three layers within the element. There are three global co-ordinates in the x-, y- and z-directions, and there are three global degrees of freedom, U, V and W. There are six components of stress and strain.

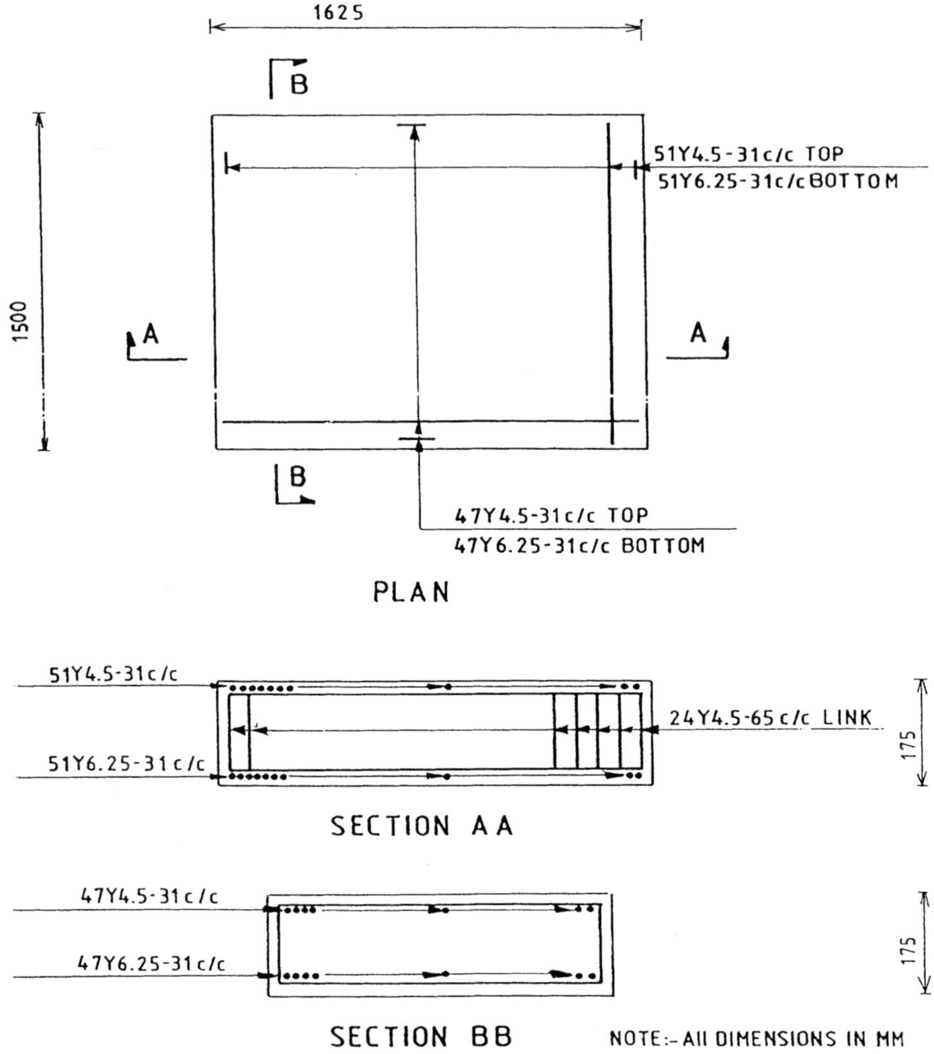

Fig. IX.21. Detail of target slab B16 (courtesy of UK Atomic Energy Authority, Winfrith).

The reinforcement is represented by the three-dimensional line element. This element is an isoparametric, three-dimensional empty block which contains reinforcing bars running in patterns designed by the user through the user subroutine, LINE. These elements are used in conjunction with 20-noded elements, which represent the concrete. The combination of these two elements approximates the reinforced concrete behaviour, since, in the isoparametric elements, each edge of this element forms a parabola. The line elements are integrated using a numerical scheme based on Gauss quadrature. Each layer contains nine integration points. In each of these elements there are five layers and a total of 45 integration points.

Table IX.2. Data for slabs subject to impact loads

Test No. M126	
Missile: collapsing steel tube type $11\frac{1}{4}$	
Impact velocity: 237m/s	
Target: type $16\frac{1}{4}$	1.625 m × 1.5 m × 0.175 m
bending reinforcement	∅4.5 mm impact face, ∅6.25 mm rear face on 31-mm pitch
shear reinforcement	∅4.5 mm 0.37% of plan area in central region
Concrete strength measurements:	
Age (days)	28
152 mm cubes	34.4 MPa std dev. 2.0 MPa
75 mm cubes	32.0 2.6 MPa
75 mm × 75 mm × 300 mm beams	3.0 0.4 MPa
Brazilian	2.5 0.3 MPa
∅152 mm × 300 mm cylinders	33.7 2.0 MPa
Panel age at date of test	28 days
Test No. M289	
Missile: type $11\frac{1}{4}$, mass 15.6 kg	
Impact velocity: 240 m/s	
Target: B26($\frac{1}{4}$)	1500 mm × 1625 mm × 150 mm
bending reinforcement	∅4.5/∅6.25 mm at 37/28mm front/rear
shear reinforcement	∅5.0 mm at 62 mm
Concrete strength measurements:	
Age (days)	35
152 mm cubes	45.1 + 1.6 MPa
75 mm cubes	61.3 + 2.9 MPa
75 mm × 75 mm × 300 mm beams	3.8 + 0.4 MPa
Brazilian	3.8 + 0.9 MPa
∅152 mm × 300 mm cylinders	37.0 + 4.7 MPa
Panel age at date of test	36 days

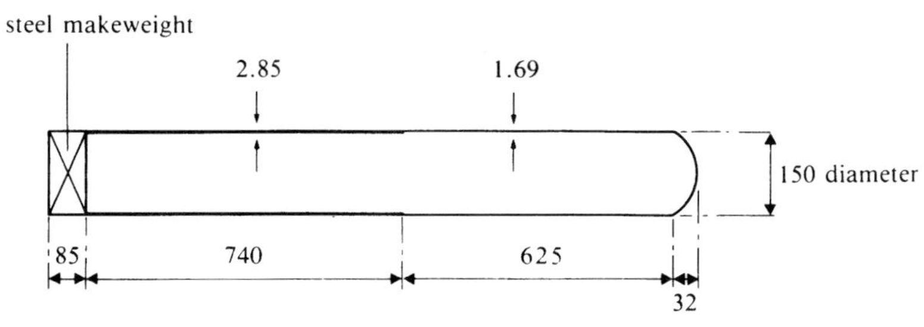

Fig. IX.22. Dimensions of soft missiles, type 12, used for tests M126 and M289 (all dimensions in mm) (courtesy of UK Atomic Energy Authority, Winfrith).

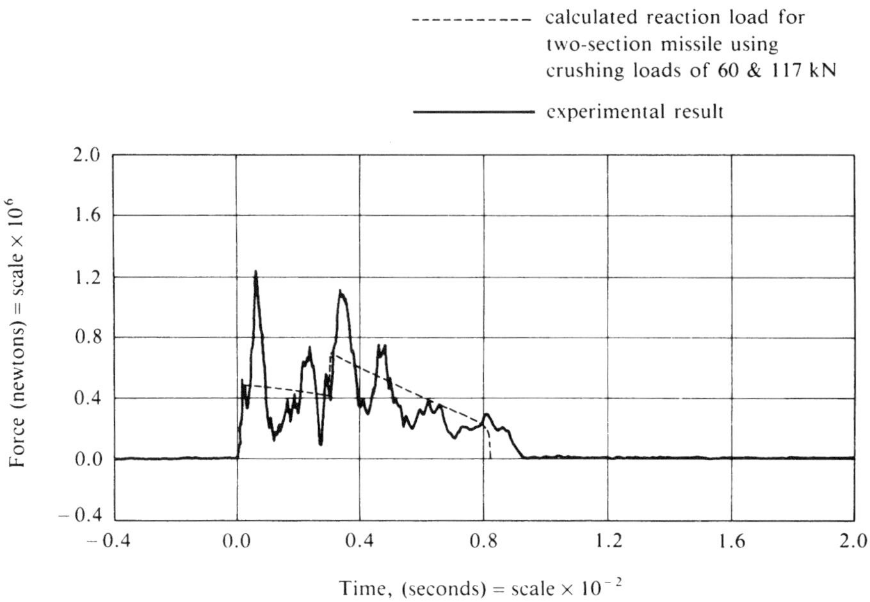

Fig. IX.23. Load–time function for type 11 missile load at 236 m/s impact (courtesy of UK Atomic Energy Authority, Winfrith).

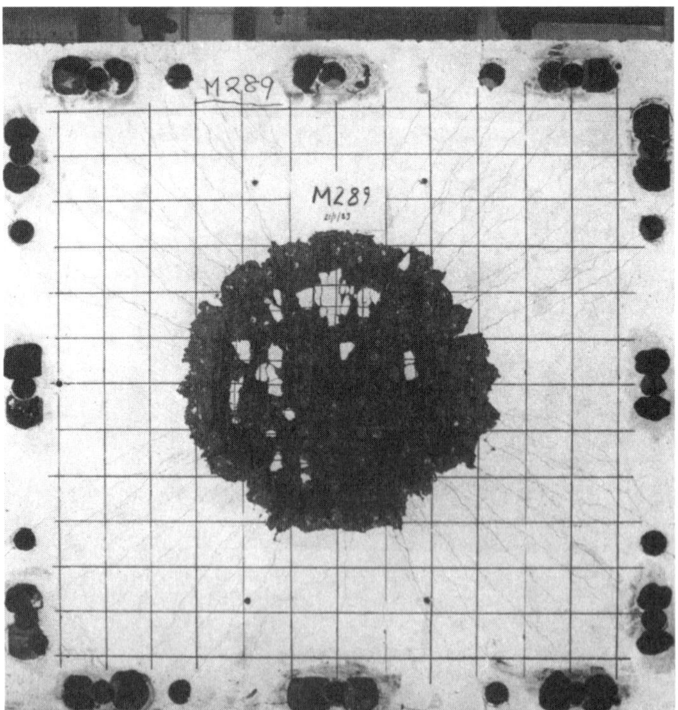

Fig. IX.24. Bottom face of target B26 after impact test.

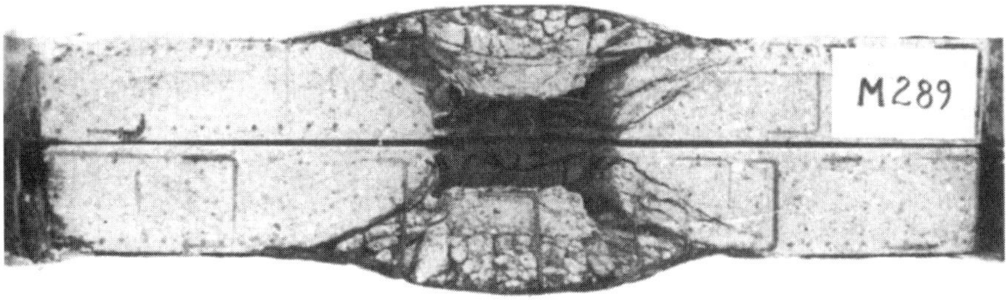

Fig. IX.25. Cross-section through target B26 after impact test.

The Newmark direct integration method ($\alpha = \frac{1}{2}$; $\beta = \frac{1}{4}$) was applied. The force was applied keeping in mind the size of the time step, the numerical damping factor and the total transient time. Nodal displacements, velocities, accelerations, principal stresses and strains, plastic strains and crack strains were computed. Concrete modelling adopted was based on Ottoson's four-parameter model and the cracking criteria adopted in this book. The total number of elements, 509, was chosen using a close mesh at the centre. Figures IX.24 and IX.25 show the results from the experiment. Figures IX.26a, IX.26b and IX.26c summarise the results obtained from the program ISOPAR. The maximum penetration found on the face at which the impact occurred was 17.5 mm compared with the experimental value of 20 mm. On the rear face the diameter of the concrete cover scabbed was 700 mm

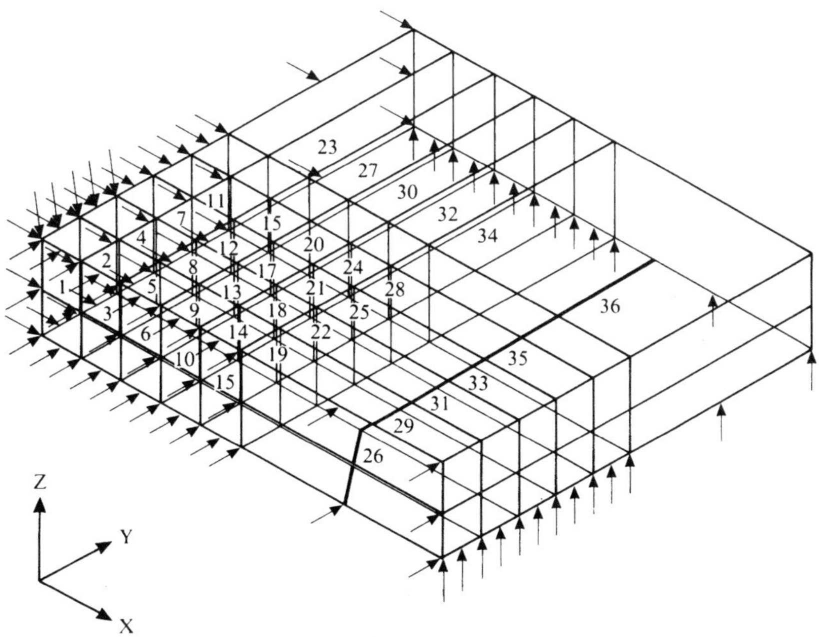

Fig. IX.26a. Model of one-quarter of the slab showing support condition load.

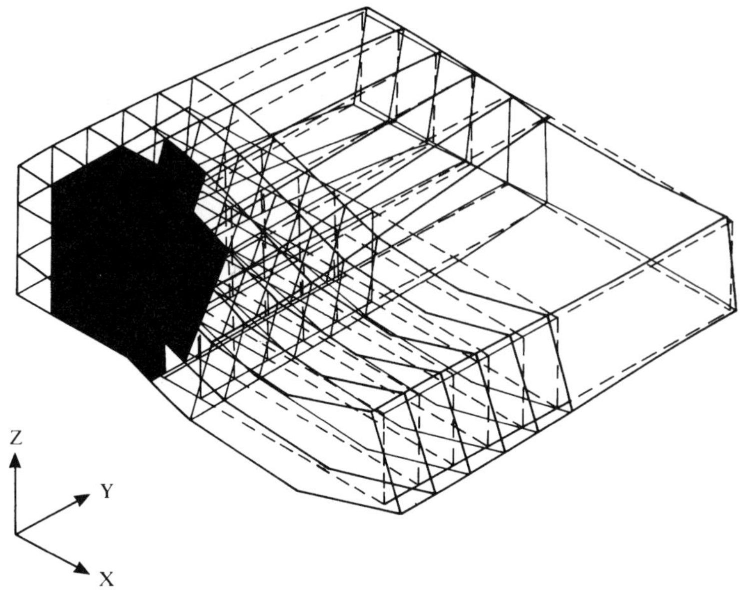

Fig. IX.26b. Mode 2 of target B16 — plugged-out zone in three dimensions.

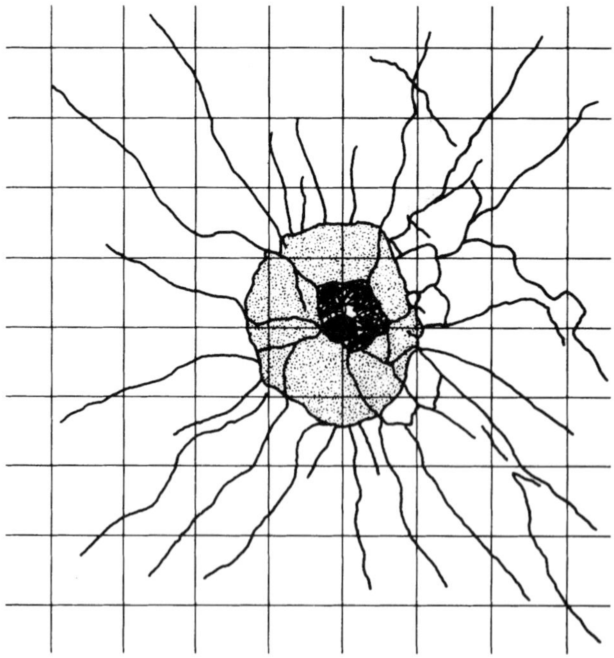

Fig. IX.26c. Final element results for shear cone and cracking.

compared with the experimental value of 650. The spread of cracking was 370 mm compared with the experimental value of 300 mm.

IX.4 Case Study No. 3 — bonded prestressed concrete specimens and slabs

Studies of the local bond strength and bonded post-tensioned slab specimens detailed below were carried out as described by Ahmad and Bangash [873], together with modifications proposed by Bangash [749].

IX.4.1 Bond–slip test specimen

The test specimen chosen was based on the 'flexural only case'. A rectangular beam (cross-section 152 mm × 152 mm) of one metre length was chosen (Fig. IX.27a) with two rectangular holes on each side of a 38 mm bonded length (L_b). A circular duct running along the beam through the centre of the cross-section was cast void for later insertion of the prestressing wire. This duct between rectangular holes in the central part of the specimen was grouted using cement grout after post-tensioning.

Two strain gauges (gauge factor 2.5) were mounted onto the wires 50 mm from the centre of the wires prior to the insertion of the wires in the ducts. The strains on the wires were recorded using a Peekel strain indicator (Fig. IX.27b). Four high-precision displacement transducers were placed at both ends of the bonded length

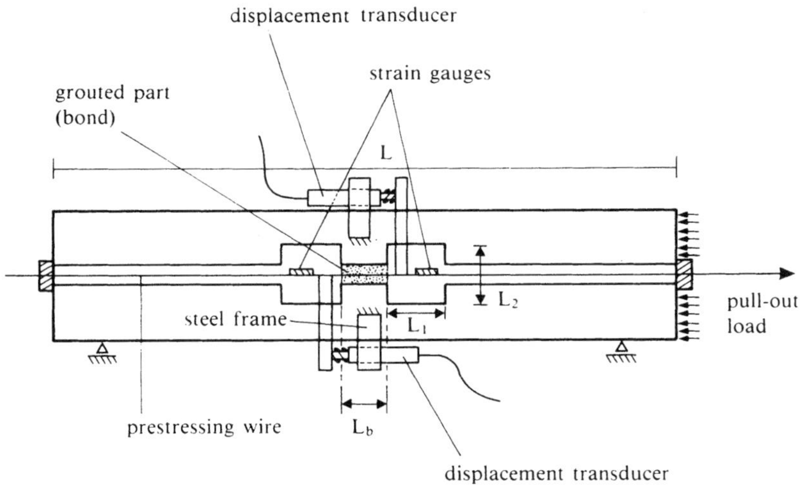

dimensions:	Material data
L = 1000.0 mm	$E_s = 200 \times 10^3$ N/mm^2
L_b = 38.0 mm	$E_c = 27.5 \times 10^3$ N/mm^2
L_1 = 76.0 mm	$f_{cu} = 45$ N/mm^2
L_2 = 38.0 mm	$v = 0.17$
	σ_y (conventional steel) = 297 N/mm^2
	σ_y (high tensile wire) = 1340 N/mm^2

Fig. IX.27a. Details of prestressed concrete pull-out specimen [262].

Fig. IX.27b. General views of the test specimen with transducers and load cells (after [873] and [262], with permission).

(with two at each end). The slip of the wires was measured using these transducers with a measurement accuracy of 10^{-5} mm. Slip was directly recorded from the transducer measuring unit. The load was applied using a hydraulic jack and the load increment was recorded using a load cell. The load was applied in increments of 0.25 kN. To study the influence of various parameters on the bond between the prestressing wire and the grout, two main parameters were investigated: the magnitude of the prestress force and the wire diameter. Wires of 5 and 7 mm diameter were prestressed to 65 and 75% of the guaranteed ultimate tensile strength (GUTS). Ten

Table IX.3. Parameters of bond–slip specimen

Beam number	Diameters (mm)	Amount of prestress (% GUTS)
1	5	75
2	5	75
3	5	75
4	7	75
5	7	75
6	7	75
7	7	65
8	7	65
9	5	65
10	5	65

beam specimens were tested. Table IX.3 shows the prestressing loads and wire diameters for these beams. The slip was obtained by taking an average value of the transducer readings. Table IX.4 gives the results of bond test.

The finite element analysis was carried out using three-dimensional isoparametric elements representing concrete and line elements for tendons with three nodes in conjunction with an Ahmlink element. The mesh elements were similar to those shown in Fig. IX.9 but with reduced sizes. The load was applied incrementally; stresses, strains and cracks were produced at the onset of the nonlinear behaviour of the specimen. Bond stress–slip relationships were computed, for all the specimens, using the program N-SARVE [262]. The analytical results are plotted in Figs IX.28–IX.31. In some cases the results obtained from the analysis disagreed with those from the experiments. This is because of the complicated nature of the steel–concrete interface factors, such as local shrinkage and settlement of grout, which were not taken into the finite element analysis carried out by the program ISOPAR.

Table IX.4. Bond test results

Beam number	Diameter (mm)	Prestress force (% GUTS)	σ_{bmax} (N/mm^2)	Average σ_{bmax} (N/mm^2)	Slip (mm)	Average slip (mm)	E_h (N/mm^2/mm)	Average E_h (N/mm^2/mm)
1		75	1.6		0.044		380	
2		75	1.64	1.64	0.052	0.048	540	473.3
3	5	75	1.68		0.05		500	
9		65	2.2	2.21	0.019	0.0205	509	526.0
10		65	2.22		0.022		543	
4		75	1.84		0.055		510	
5		75	1.81	1.8	0.065	0.057	520	523.3
6	7	75	1.75		0.052		540	
7		65	2.1	2.05	0.026	0.025	466	483.0
8		65	2.0		0.024		500	

σ_{bmax} = ultimate (maximum) bond stress.
E_h = initial slope or slip modulus (average value of E_h = 501.4 N/mm^2/mm).

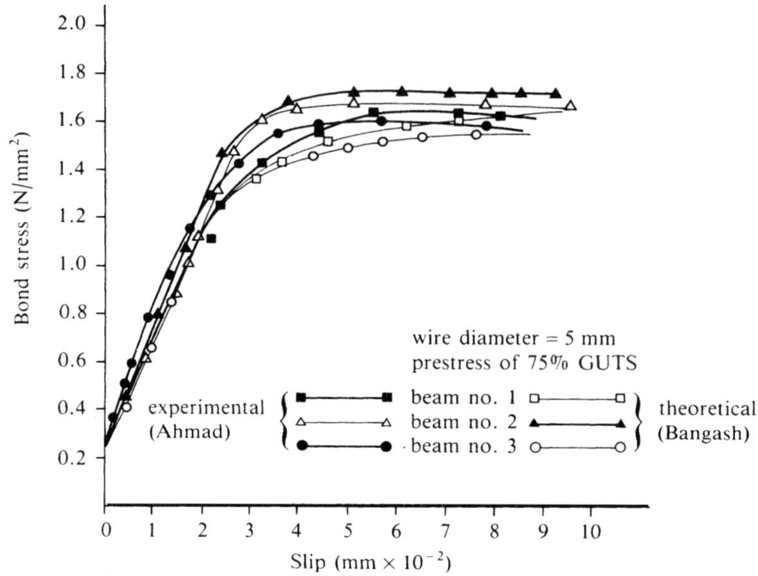

Fig. IX.28. Experimental bond stress–slip curve of 5 mm diameter prestressing wire.

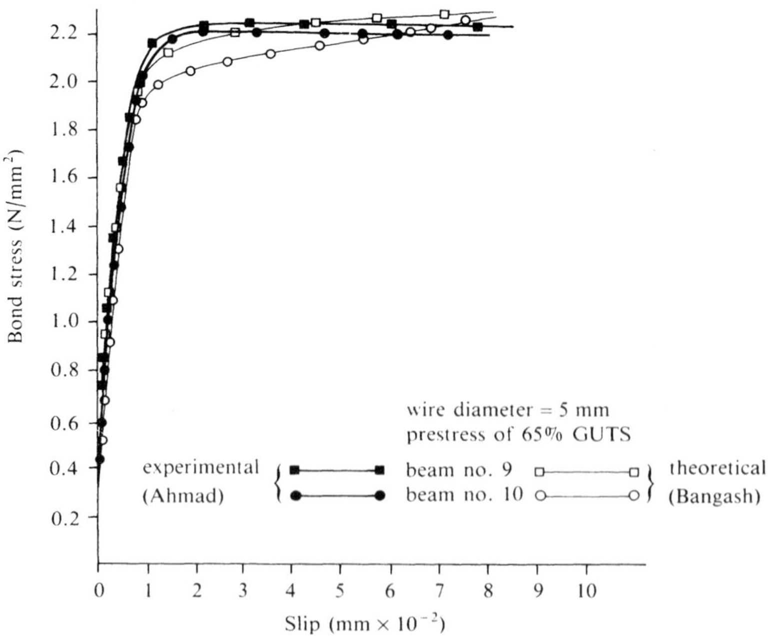

Fig. IX.29. Experimental bond stress–slip curve of 5 mm diameter prestressing wire.

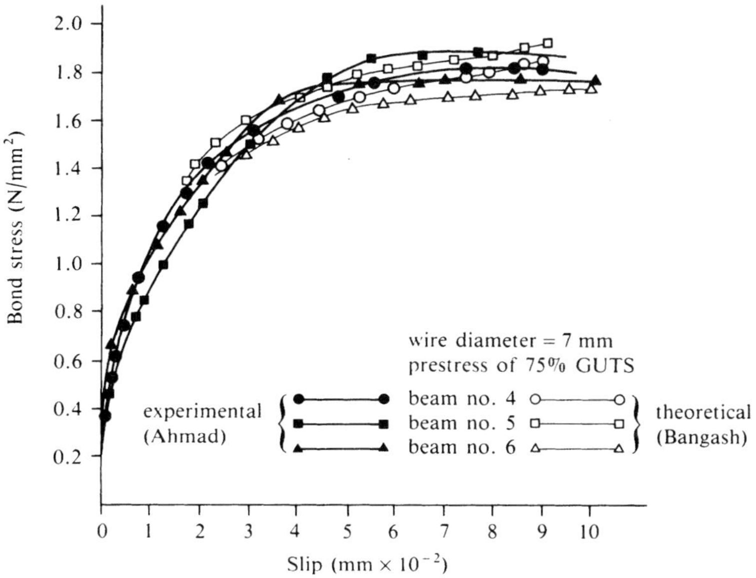

Fig. IX.30. Experimental bond stress–slip curve of 7 mm diameter prestressing wire.

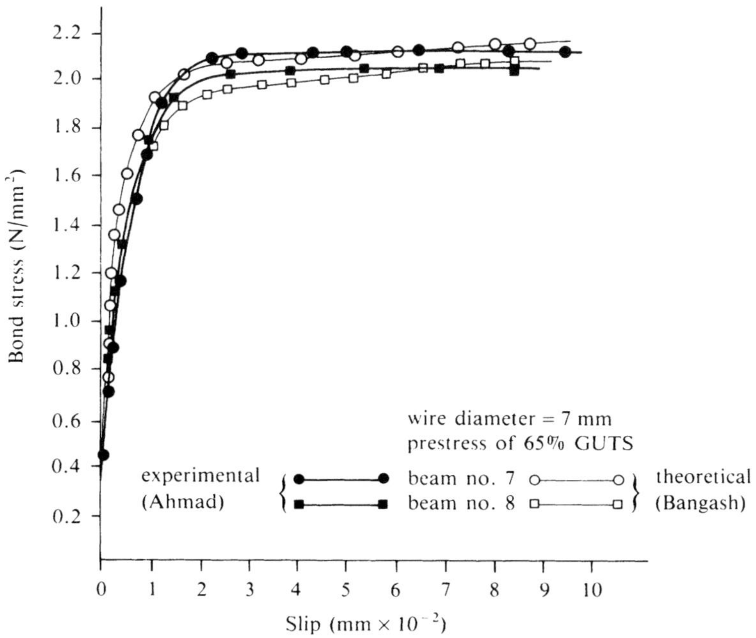

Fig. IX.31. Experimental bond stress–slip curve of 5 mm diameter prestressing wire.

Figure IX.31 shows the results for beams 7 and 8 in which 7 mm diameter wire was used with a prestress of 65% GUTS. In this case, as shown in Fig. IX.31, a relatively large scatter of results was obtained for maximum bond stress. An average value of maximum bond stress was found to be 2.30 N/mm^2 (2.05 N/mm^2 from experiment) at an average slip of 0.025 mm when bond stress stops increasing. In this case curves indicate less nonlinearity, and maximum bond stress is relatively larger than that for beams 3, 4 and 5. The maximum slip at which the bond stress stops increasing is less here than for beams 3, 4 and 5. Figure IX.29 shows the results for beams 9 and 10 in which 5 mm diameter wire with a prestress of 65% GUTS was used. The average maximum bond stress is found to be 2.40 N/mm^2 (test 2.21 N/mm^2). The slip at which this maximum bond stress was obtained is 0.0205 mm. Comparing these results with those of beams 1, 2 and 3 it can be seen that the maximum bond stress is larger with a smaller slip.

From the above results it may be concluded that the maximum bond stress decreases with the increase of prestressing force in the wire and at a lower slip value. The effect on maximum bond stress due to the variation of steel wire diameters is negligible. For any definite conclusions about large tendons, much more testing with variations of prestress and tendon diameter needs to be performed.

IX.4.2 Details of the bonded slab

A scale model of a prestressed concrete slab (Fig. IX.32) representing the top cap of a typical prestressed concrete reactor vessel was tested [873]. The objects of the test were: (a) to study the physical behaviour of a prestressed concrete slab with bonded tendons under increasing load up to the point of failure, thereby providing load–displacement relationships, and information about the distribution of cracking and the failure mechanism; and (b) to compare these results with those obtained from the analytical models.

An octagonal prestressed concrete slab representing the top cap of a vessel was prestressed with 5 and 7 mm diameter wires in two orthogonal directions. A conventional reinforcement of 12 mm diameter mild steel bars was provided, as shown in Fig. IX.32. Four 7 mm diameter wires were placed along the centre line BB, two on either side of the centre line at distances of 115 mm (c_1) and 350 mm (c_2). These wires (shown by dashed lines) were at a height of 36.5 mm from the bottom surface of the slab. Eight 5 mm diameter wires were placed along the centre line AA, four on either side of the centre line at distances of 115 and 350 mm, and at heights of 18 and 55 mm from the bottom surface of the slab. Eight 12 mm diameter bars spaced at 65 mm were provided in one quadrant of the slab. A total of 32 bars were provided in the slab, as shown in Fig. IX.33. Each wire was stressed up to 75% GUTS using the CCL prestressing system. Dial gauges were placed on one-quarter part of the top surface of the slab (Fig. IX.34(a)) to measure the transverse deflection. Rosette strain gauges were also placed (Fig. IX.34(b)) on the top surface of the slab.

The slab was loaded with the loading rig producing four load points on the main steel frame and 12 load points on the slab. The loading rig system is shown in

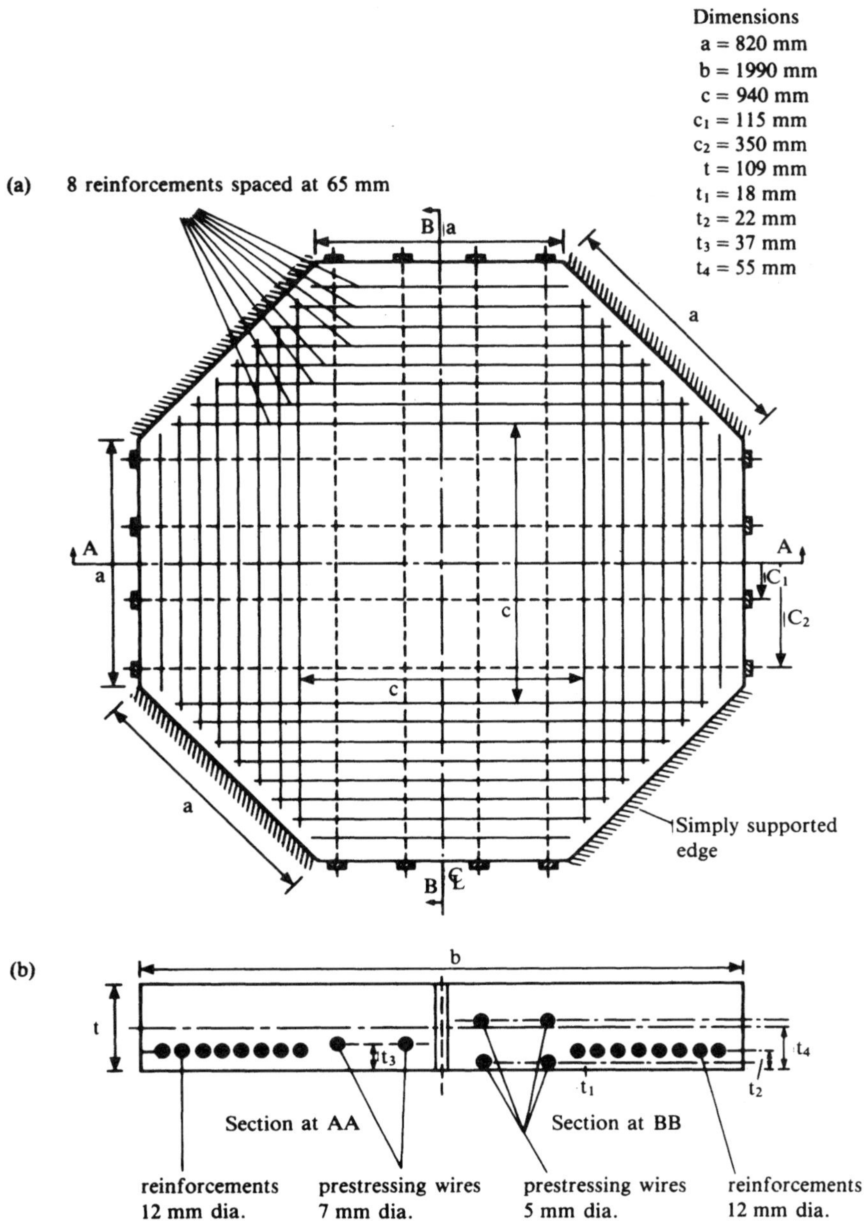

Fig. IX.32. (a) Prestressed concrete slab and (b) a detailed section (see (a) for dimensions) [262].

Fig. IX.35. Four deep steel I sections were welded together at 90° to make a main rigid steel frame. This frame was supported by two strong steel beams running across the slab, which were supported by the columns. Each rigid frame was connected to a hydraulic jack. Hydraulic jacks were connected to a common hydraulic pump. Each hydraulic ram was lowered onto a three-legged spreader frame, with the

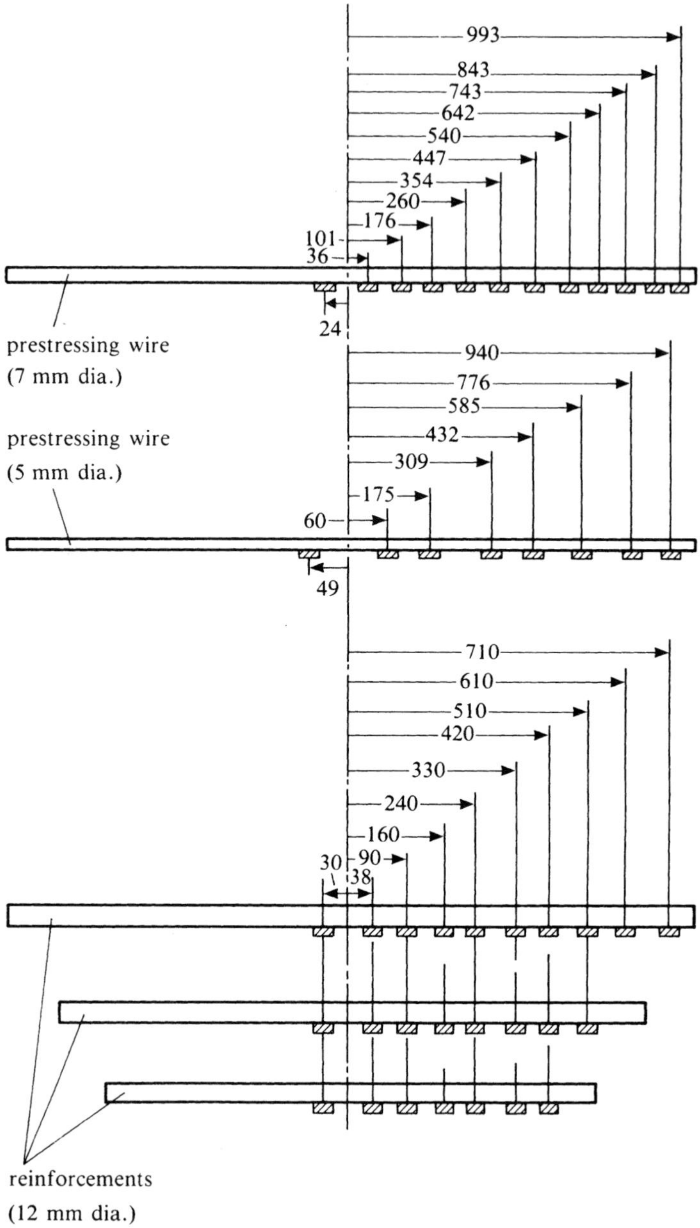

Fig. IX.33. Strain gauge positions on prestressing wires and reinforcements [262] (all dimensions in mm).

load being transferred between each leg and the slab by a steel plate. This steel plate had a rubber pad underneath it to distribute the load. The steel plates and rubber pads were spread over a diameter of approximately 1780 mm. Load cells (100 kN) were used on each hydraulic jack for accurate measurement of load. The surface strains of the concrete were measured using rosette strain gauges (Fig. IX.34). The

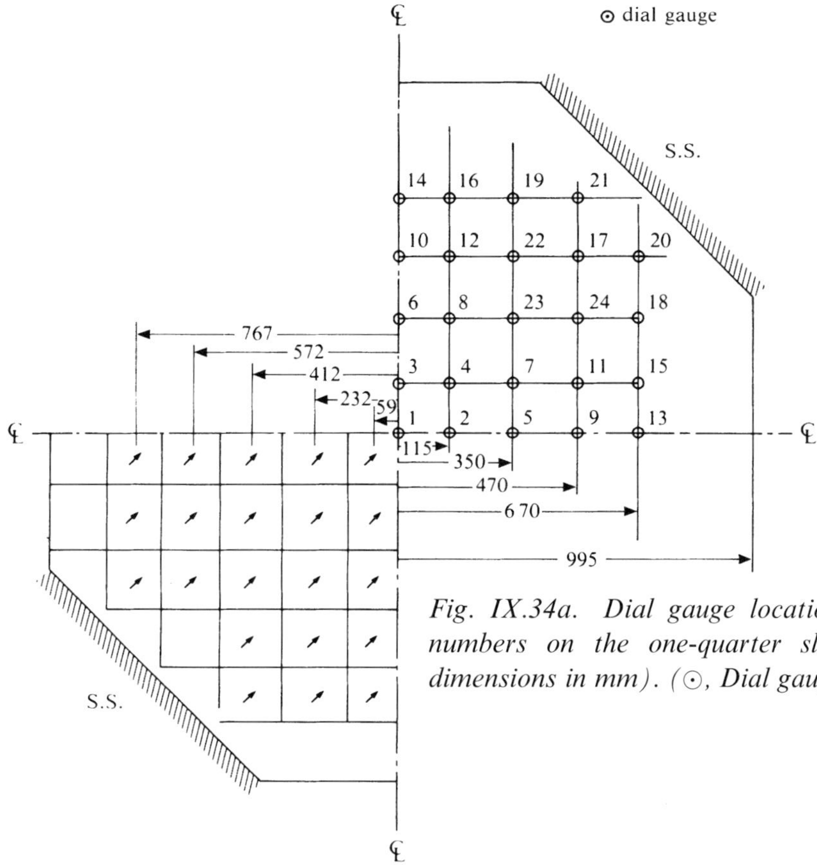

Fig. IX.34a. Dial gauge locations and numbers on the one-quarter slab (all dimensions in mm). (⊙, Dial gauge.)

Fig. IX.34b. Rosette strain gauge locations on the one-quarter slab [873, 262] (all dimensions in mm). (↗, Rosette strain gauge.)

strain gauges and the rosette strain gauges were connected to a data logger (compulog) to process the results. For each load increment, the strains were recorded by compulog, and these were also printed on a teletype.

The vertical deflections were measured on the top surface of the slab using dial gauges which were capable of measuring with an accuracy of 2×10^{-3} mm. The dial gauges were placed on only one-quarter of the slab due to symmetry and were clamped using the rigid steel beam running across the slab. Figure IX.36 shows the model slab ready for testing.

The load was applied in small increments using a hydraulic pump which had four hydraulic jacks connected to it as stated earlier. At each load increment, the readings of the dial gauges and the strain gauges were recorded. In the early stages of the loading larger load increments of 20 kN (0.032 N/mm^2) were applied.

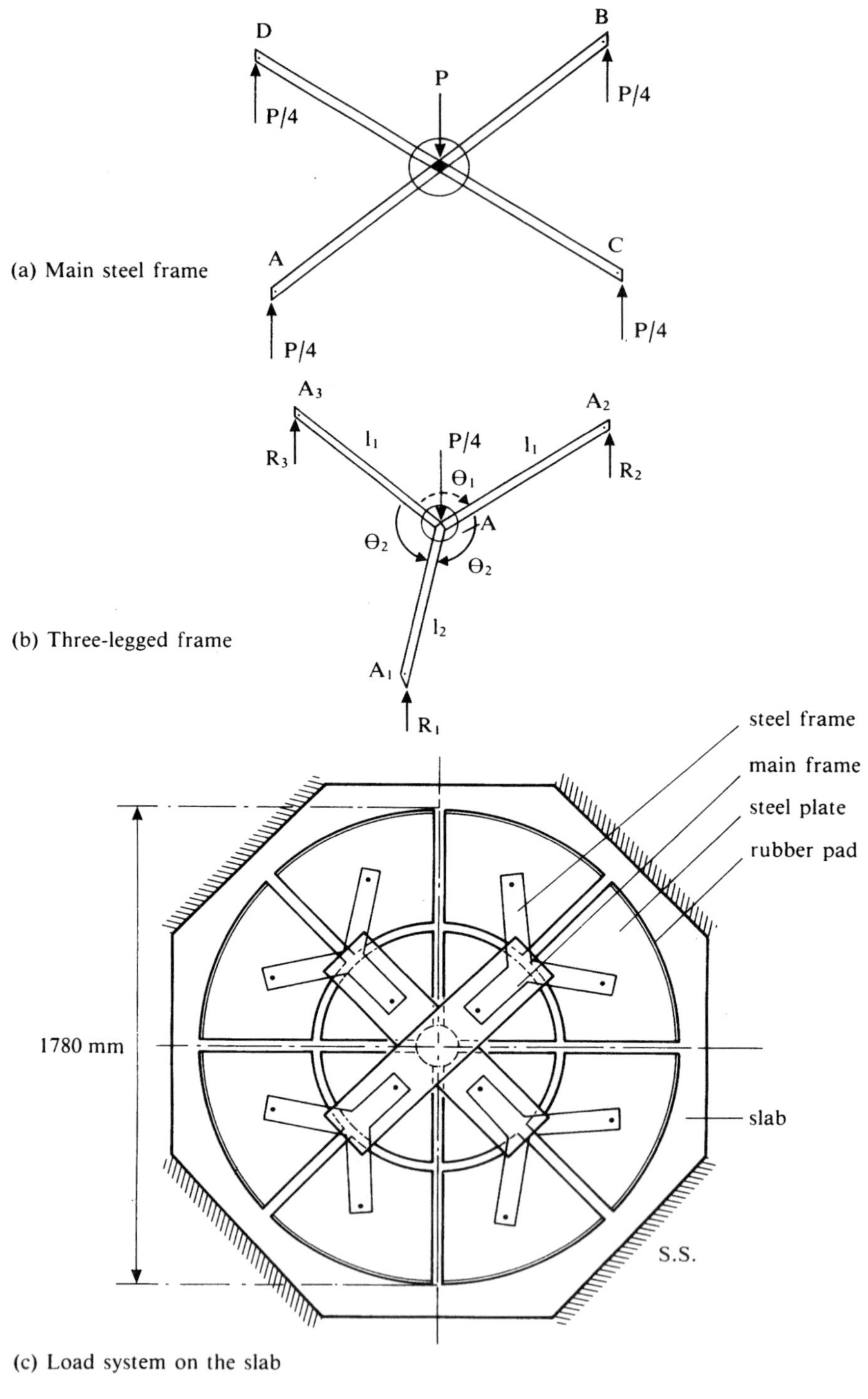

Fig. IX.35. Load system of the rig, [873] and after Ahmad [262]. ($l_1 = 284$ mm; $l_2 = 204$ mm; $\theta_1 = 141.5$; $\theta_2 = 109.25$.)

Fig. IX.36. Model slab of octagonal shape with instrumentation, [873] and after Ahmad [262].

In the later stages of loading, smaller load increments of 8 kN (0.012 86 N/mm^2) were applied. The bottom part of the slab was painted with whitewash in order to show the cracks. The part of the slab which was vulnerable to cracking was divided into small, 50 mm square, divisions in order to accurately estimate the positions and propagation of cracks. The slab was examined for cracks at each load increment. As cracks appeared, they were marked with a dark pencil so that they were visible in a mirror placed underneath the slab. A camera was mounted at the corner of the slab to take photographs of the cracks reflected in the mirror.

The total load up to the point of failure of the slab was applied in 53 increments. The results obtained are plotted in the form of load–displacement curves. The experimental load–displacement curve at the centre of the slab for the entire load history is shown in Fig. IX.37. The deflected shapes of the slab along two centre lines (i.e. AA and BB) for various loads are shown in Figs IX.38 and IX.39. Photographs of cracks at the bottom surface of the slab were taken at various stages of loading. The collapse mode of Fig. IX.40 showed crack patterns at failure. Just before failure the cracks at the bottom surface had developed, as shown in Fig. IX.41.

The slab behaved linearly up to a load of 160 kN. The initiation of the first set of cracks occurred at a load of 180 kN. The vulnerable load stages are 200–250 kN. Figure IX.41 indicates the position and distribution of cracking at the surface of the slab. The slab finally failed at a load of 544 kN, when deflections could not be

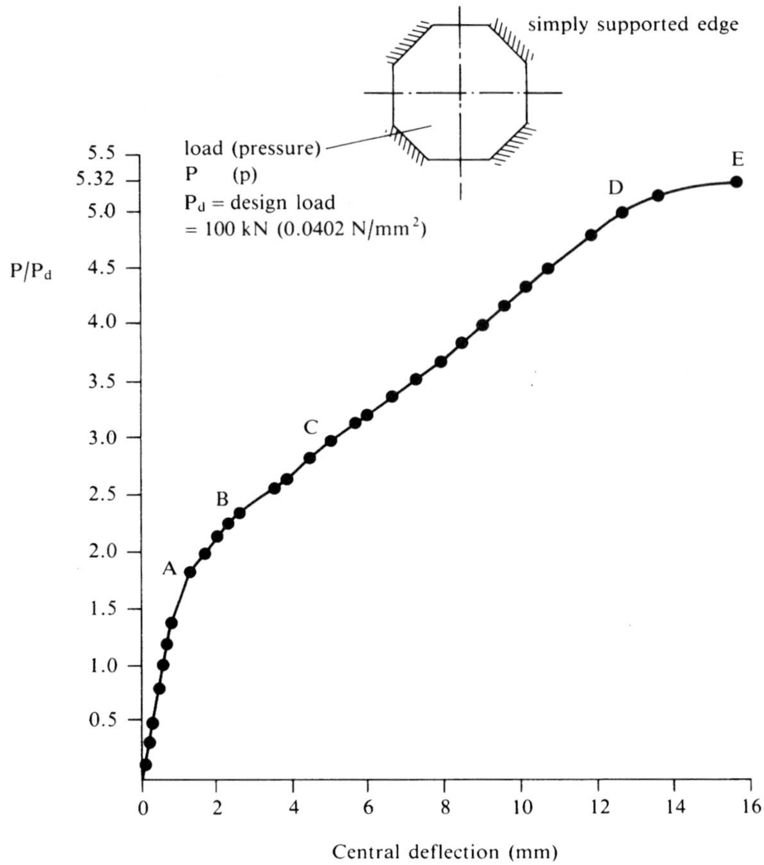

Fig. IX.37. Experimental load–displacement curve of the octagonal prestressed concrete slab, [873] and after Ahmad [262].

measured as they were beyond the limit of the measuring gauges. The safety margin against the design load was 3.95.

Ahmad and Bangash [873, 749] carried out finite element analyses for the following three cases:

Case I Bonded Prestressed Concrete Slab — the interface behaviour between the steel and concrete was modelled using the linkage (Ahmlink) element.

Case II Perfectly Bonded Prestressed Concrete Slab — the steel element was placed on one side of the solid element and a perfect (rigid) bond assumed between the two.

Case III Unbonded Prestressed Concrete Slab — the steel elements were not included in the analysis but their prestress forces were included in the analysis of the slab.

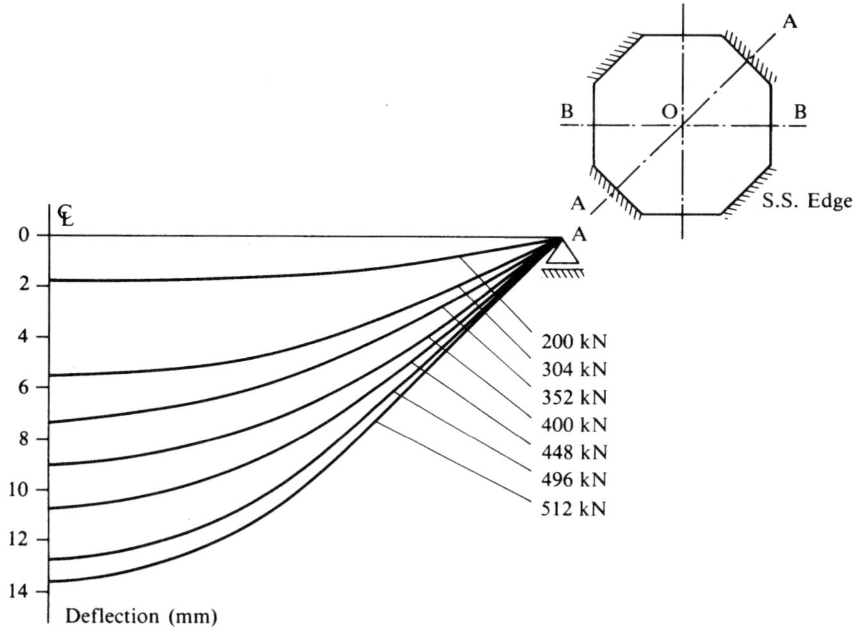

Fig. IX.38. Experimental deflected shapes along AA.

Due to symmetry, only one-quarter of the slab was analysed (Fig. IX.42). This slab was restrained from moving in the y direction along AB (i.e. $v = 0$) and in the x direction along AC (i.e. $u = 0$). The centre point A was restrained in both the x and y directions (i.e. $u = 0$, $v = 0$). The edge DE was supported in the z direction (i.e. $w = 0$). Figure IX.36 shows these boundary conditions. The finite element mesh of one-quarter of the slab was dimensioned, as shown in Fig. IX.43. The choice

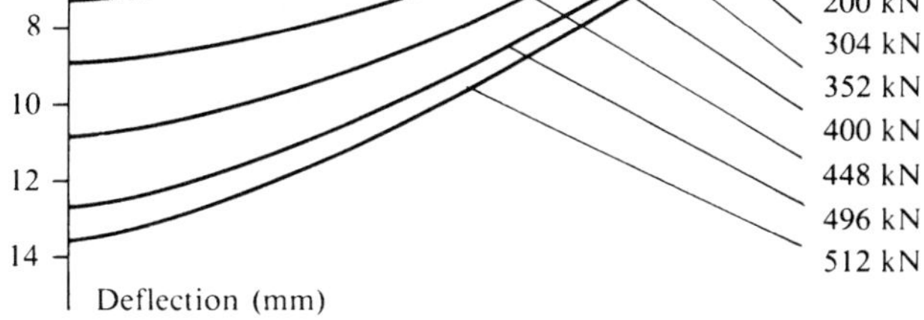

Fig. IX.39. Experimental deflected shapes along BB, [873] and after Ahmad [262].

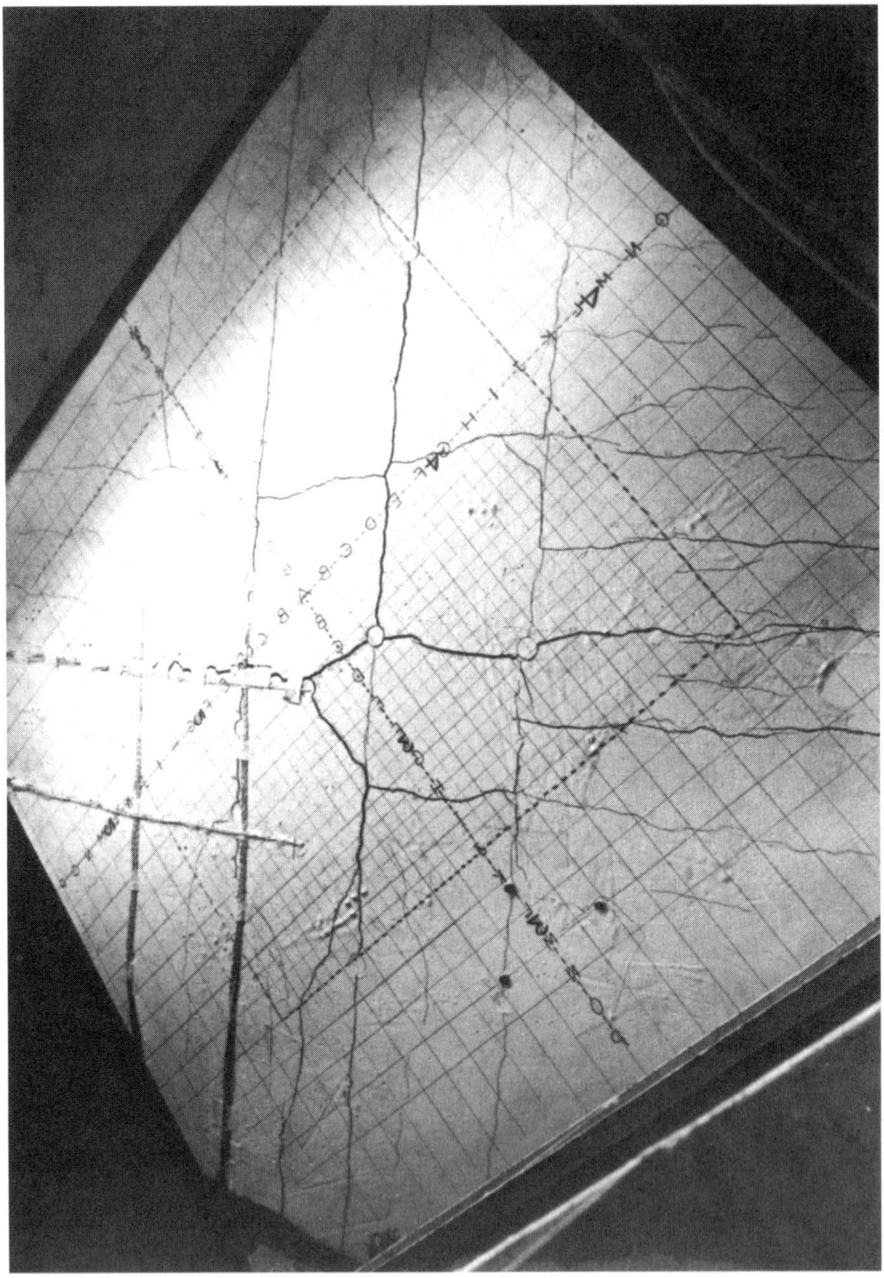

Fig. IX.40. A 'post-mortem' of the octagonal slab.

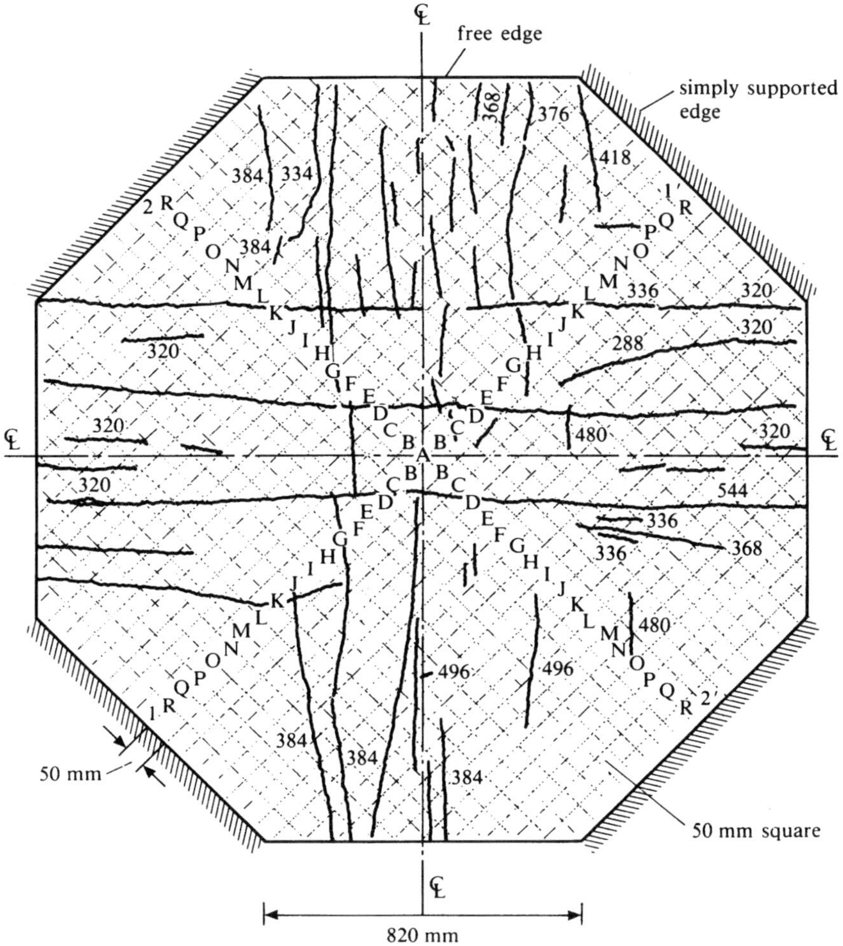

Fig. IX.41. Experimental cracks at the bottom surface of the slab just before failure.

of this mesh was made so that the prestressing wires and reinforcements lay on the sides of the solid elements. The nodes of solid elements and line elements were connected either by Ahmlink elements for the bonded slab or rigidly interconnected for the perfectly bonded slab. Where a large amount of ordinary reinforcement is closely spaced in the slab, they were modelled as line elements in the body of the solid elements. Such elements are shown as dashed lines in Fig. IX.43. Two solid elements were chosen through the thickness of the slab. The line elements were placed at an effective depth of 72.5 mm (Fig. IX.43). The details of finite element meshes for the three different cases are given below.

IX.4.2.1 Case 1: Bonded slab (Ahmlink)

The slab concrete was modelled using eight-noded isoparametric solid elements, the reinforcements and prestressing wires were modelled using two-noded axial line

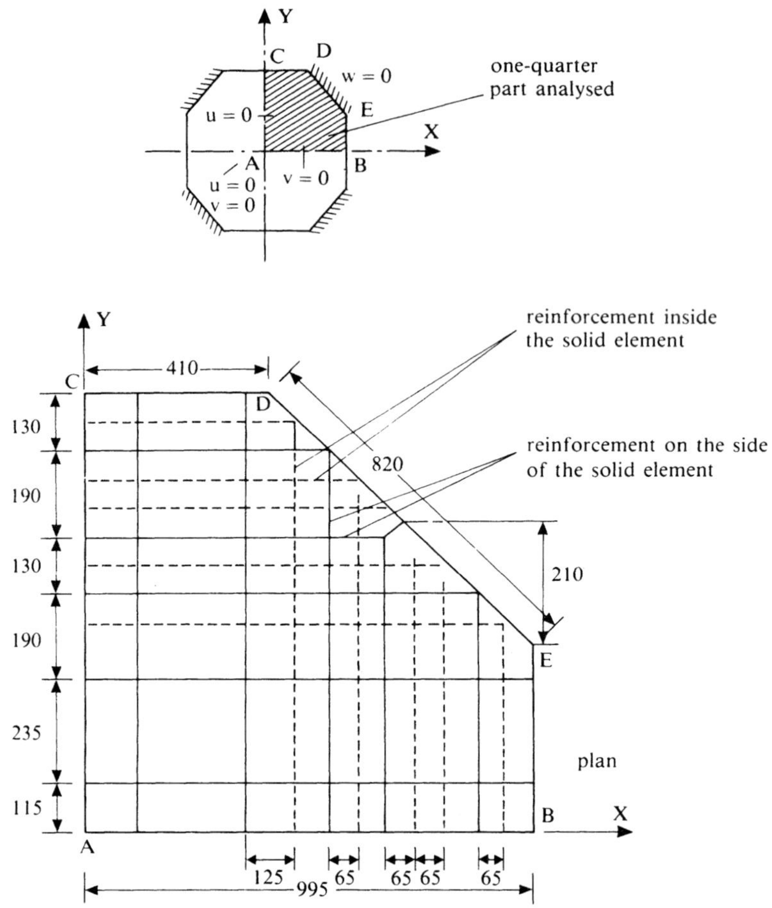

Fig. IX.42. Octagonal prestressed concrete slab (all dimensions in mm).

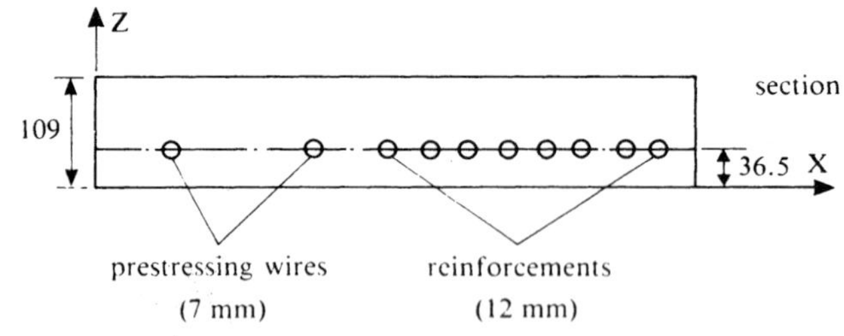

Fig. IX.43. Main dimensions of finite element mesh of the slab [873] (all dimensions in mm).

elements and the bond between the steel and concrete modelled using two-noded Ahmlink elements. Figure IX.44 shows the finite element mesh for this case only. The mesh comprises 60 solid isoparametric elements, 48 line elements, 28 Ahmlink elements and 40 body axial line elements, which are shown as dashed lines in

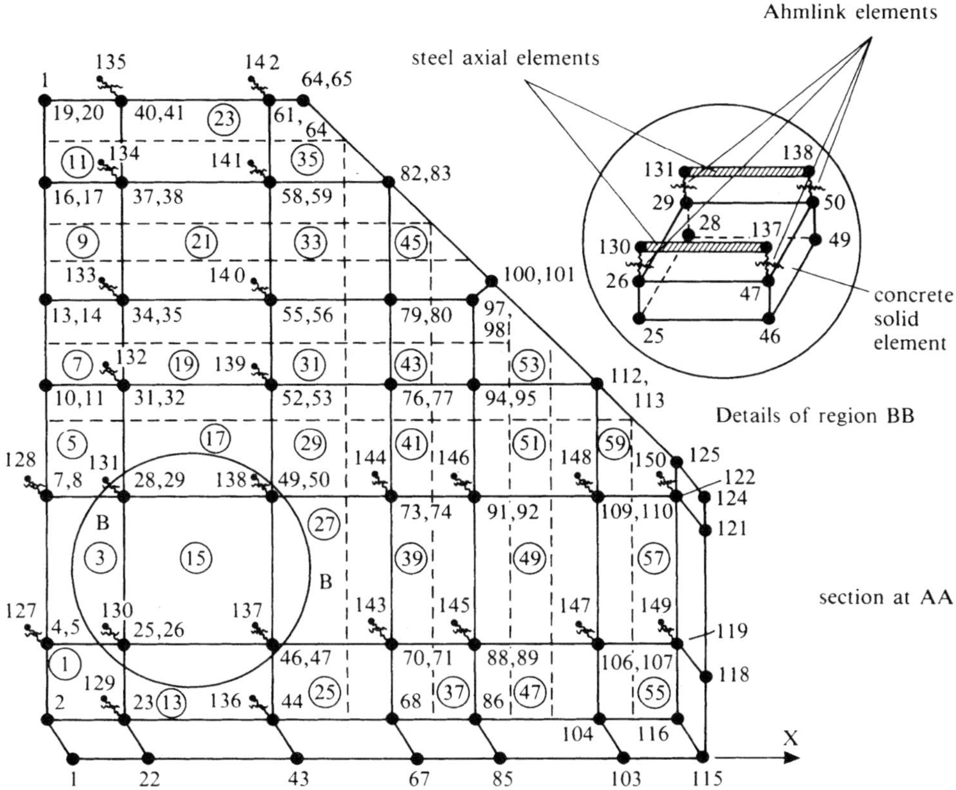

Fig. IX.44. Finite element mesh of a bonded slab [873].

Fig. IX.44. The nodes for steel and concrete are represented by different numbers, although they occupy the same points in space. The Ahmlink elements (joining steel and concrete) are located at nodal points along the steel–concrete interfaces. Figure IX.44 shows the bond of Ahmlink elements connected to solid elements. Here solid element nodes 26, 29, 47 and 50 are connected to the line element nodes 130, 131, 137 and 138 through the Ahmlink elements. Node numbers 26 and 130, 29 and 131, 47 and 137, and 50 and 138 have the same co-ordinates in space. There are a total of 150 nodes and 450 degrees of freedom for this case.

IX.4.2.2 Case II: Perfectly bonded slab

The finite element mesh for this case is shown in Fig. IX.45. The mesh is similar to the bonded slab case except that the Ahmlink elements are replaced by traditional line elements rigidly placed on solid elements. It is assumed to have a perfect bond between the solid element and the line element. The steel line elements are assumed to lie along the sides of the solid element sharing the same nodes (Fig. IX.45). The finite element mesh for this case comprises 60 isoparametric

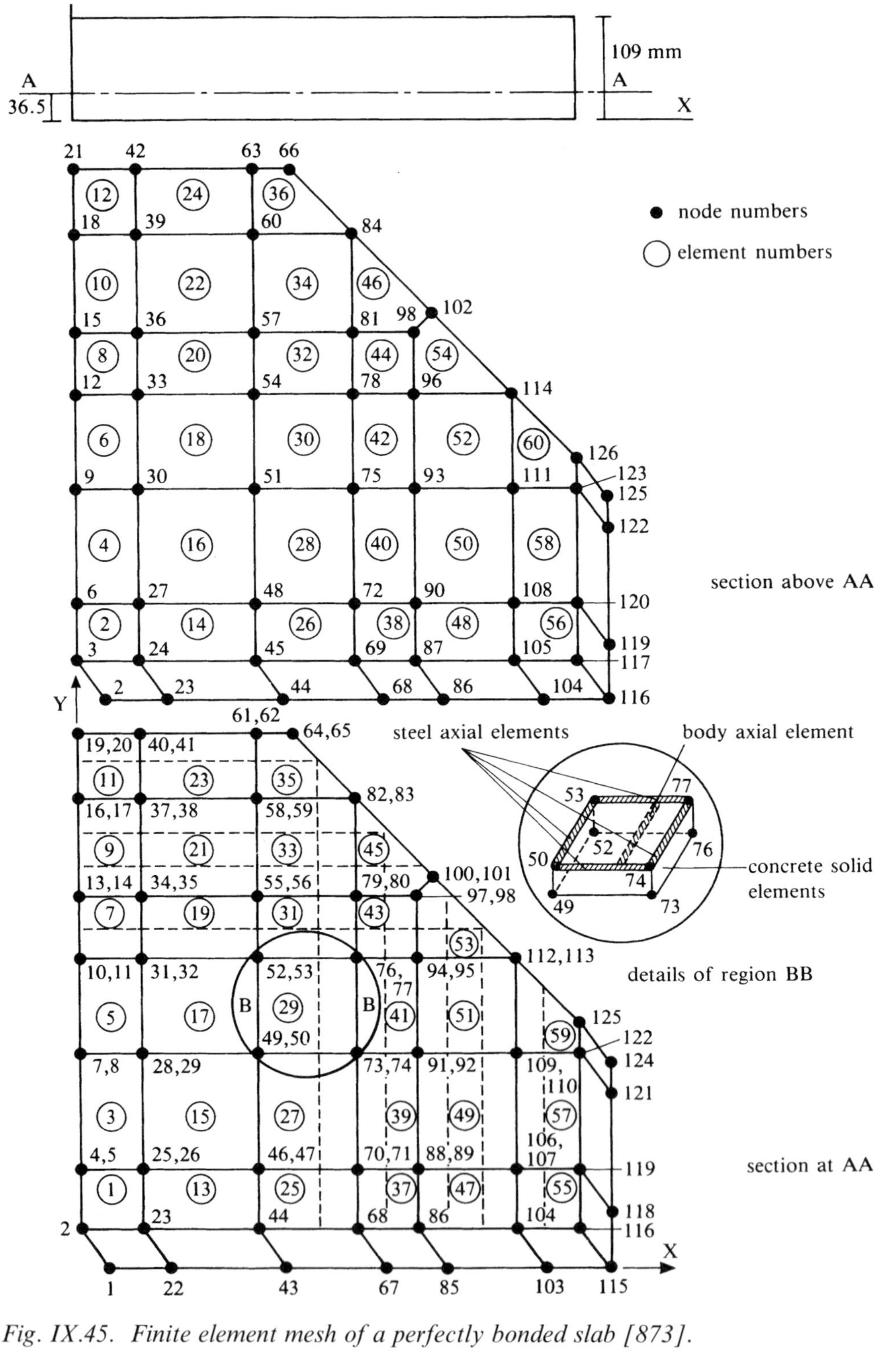

Fig. IX.45. Finite element mesh of a perfectly bonded slab [873].

solid elements, 48 line elements and 40 body axial line elements. There are a total of 126 nodes and 378 degrees of freedom.

IX.4.2.3 Case III: Unbonded slab

In this case the line elements representing prestressing steel are not included in this analysis since they are unbonded. The finite element mesh (Fig. IX.45) in this case is identical to the case of perfectly bonded slab except that here the line elements representing prestressing wires are excluded. The finite element mesh comprises 60 isoparametric solid elements representing concrete, 24 axial line elements and 40 body axial elements representing ordinary reinforcements. There are a total of 126 nodes and 378 degrees of freedom.

The total load (pressure) on the stab was applied in 11 increments. The first load increment was due to prestress applied on the edges of the slab. The second and following increments were due to transverse pressure applied on the top surface of the slab. These pressures were:

prestress (N/mm^2): 1.89
pressures (N/mm^2): 0.0603, 0.0804, 0.1005, 0.1206, 0.1407, 0.1608, 0.1809, 0.201, 0.2074, 0.2138

The bond–slip curve used is given for each load increment (Fig. IX.46); iterations were performed to correct the equilibrium. A maximum of 12 iterations per increment were allowed and iterations were generally terminated when the Euclidian

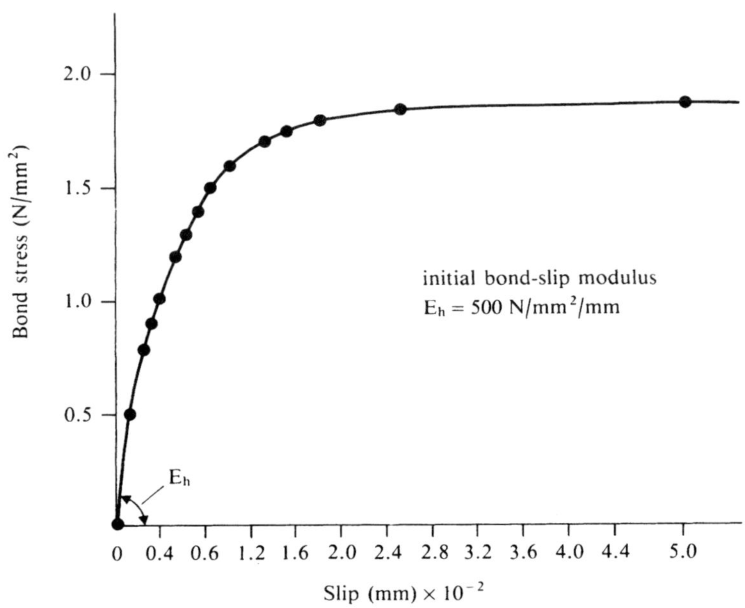

Fig. IX.46. Bond stress–slip curve used in the finite element analysis of a prestressed concrete slab.

norm of the residual loads became less than the specified tolerance (Tol $= 3 \times 10^{-2}$). The constant stiffness option (initial stress method) was employed for all cases. The orthotropic concrete model in compression, the tension cut-off model for cracking of concrete and the elasto-plastic model with strain hardening for steel were adopted throughout.

The results obtained from these analyses are plotted in the form of load–displacement curves, crack patterns and deflected shapes, as shown in Figs IX.47–IX.52. Cracking of concrete has a particular strong influence on the behaviour of the slab. Figure IX.47 shows the load–central deflection history of the slab. The

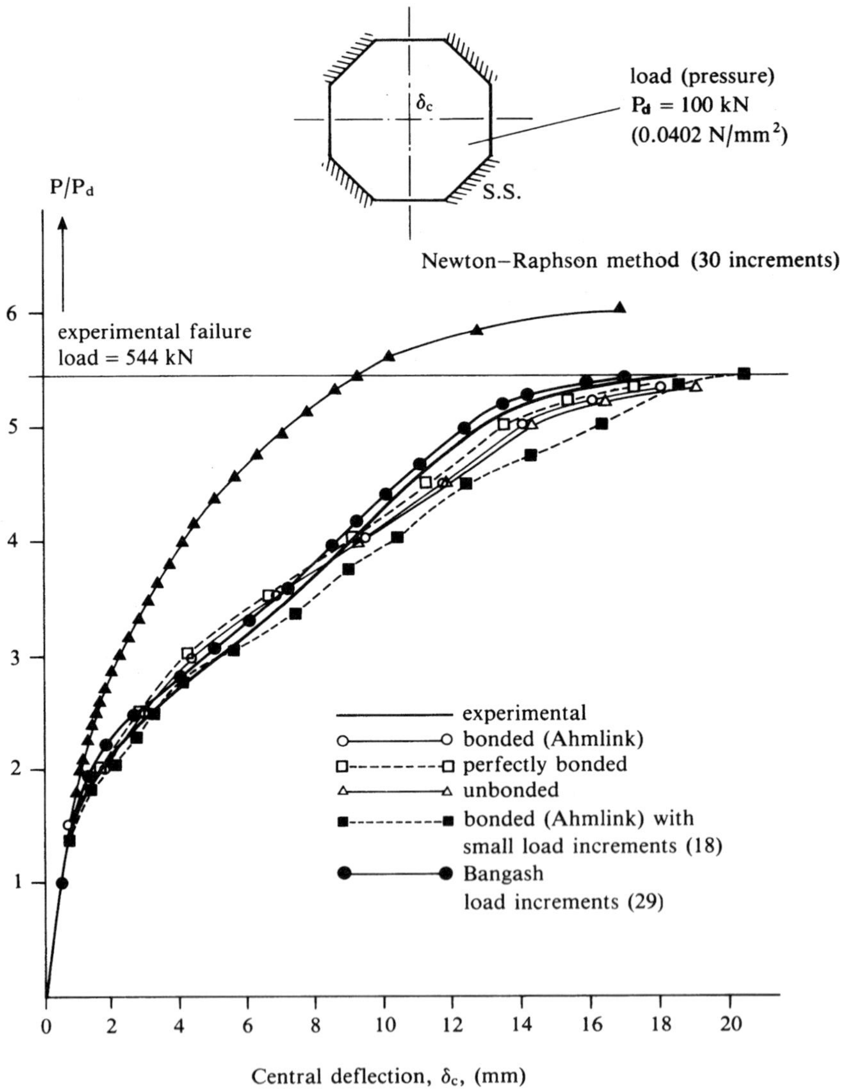

Fig. IX.47. Load–displacement curves of prestressed concrete slabs.

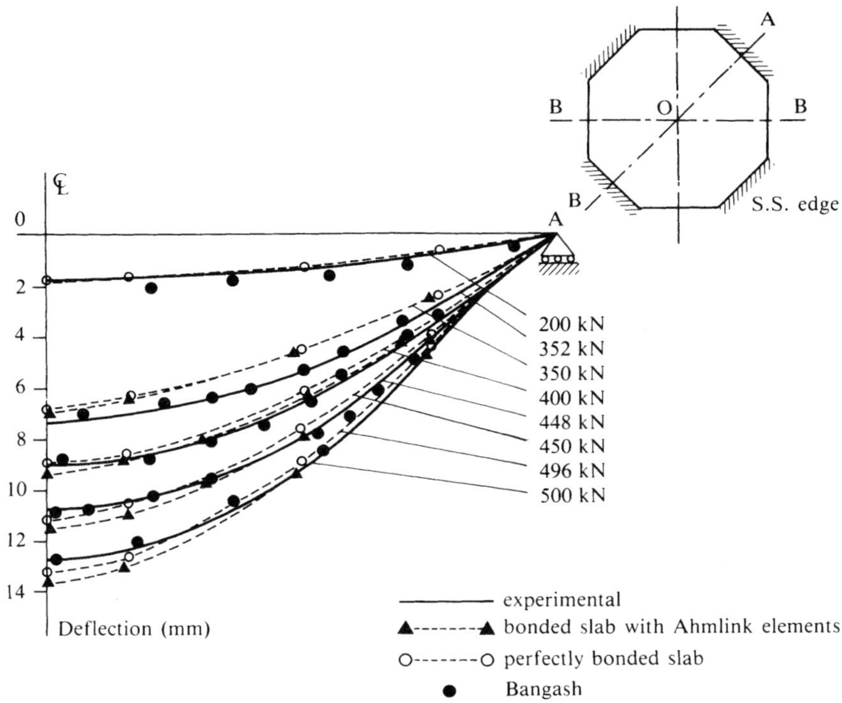

Fig. IX.48. Deflected shapes at various loads along AA.

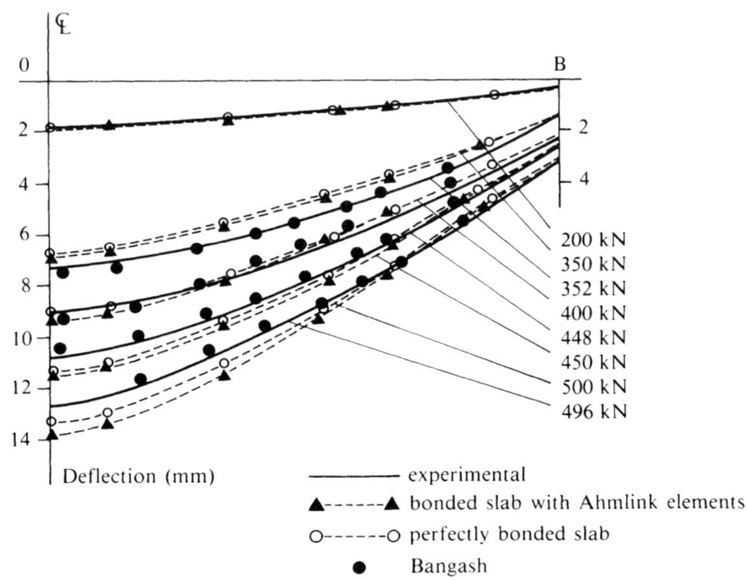

Fig. IX.49. Deflected shapes at various loads along BB [749, 873].

experimental curve is also shown for comparison. It is seen that the predicted results for three cases compare favourably with the experimental results. These curves show that they are in very good agreement in the early stages of loading (between 180 kN (0.0723 N/mm^2) and 280 kN (0.112 N/mm^2)). At this stage most of the cracking takes place.

The difference in the ultimate strength of the two cases is about 1.4%. In addition, the third case of the unbonded slab has also been analysed for comparison. The computed ultimate load is reasonably close to the other two analyses, but the computed displacements at the ultimate load are much too large. The reason is explained earlier. Figures IX.48 and IX.49 illustrate the deflected shapes along the two centre lines of the slab. The computed and the experimental deflected shapes are in close agreement. Figure IX.50 shows the maximum steel stress (reinforcement

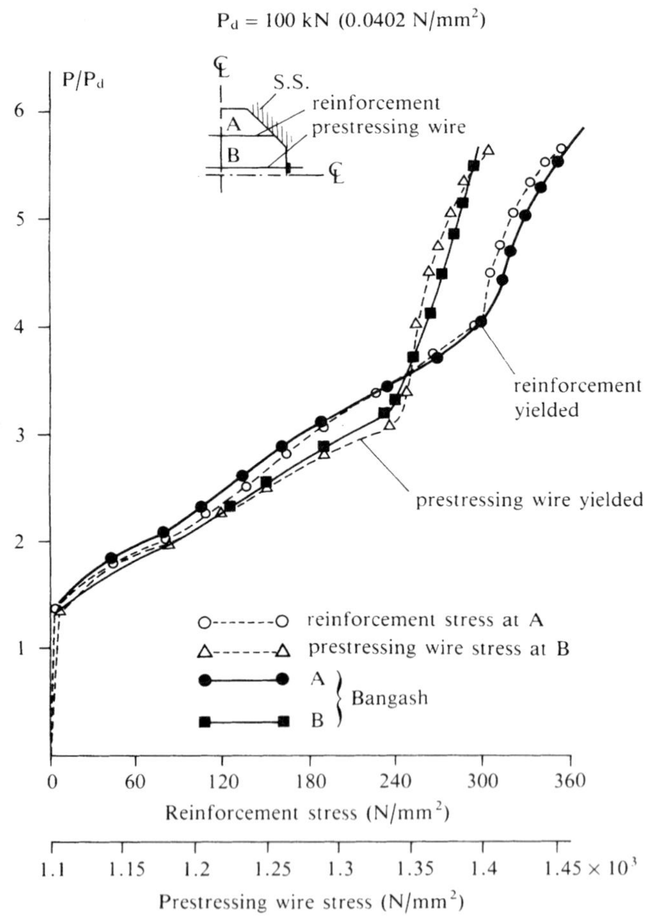

Fig. IX.50. Variation of steel stress with load in a prestressed concrete slab [749, 873].

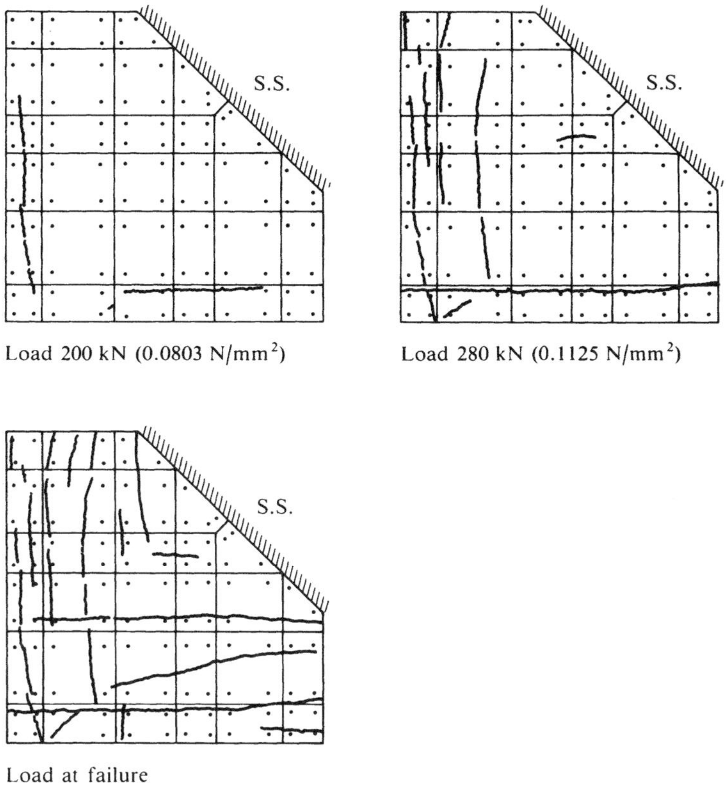

Fig. IX.51. Experimental cracks at the bottom surface of the slab [873].

and prestressing tendon) as a function of applied load as computed for the analysis.

A re-analysis has been carried out for the above two cases, I and II. The bonded slab of case I was analysed this time with smaller load increments. The total number of 18 increments was adopted using the initial stress method and equilibrium iterations. A further degree of sophistication in results was achieved and the results are in good agreement with those from experiment.

Bangash [749] carried out a further sophisticated analysis with a fine mesh using the program ISOPAR. Twenty-noded solid elements and three-noded line elements replaced the previous ones. A perfect bond was assumed. All three cases, I, II and III, were re-analysed. Material properties, the layout of the slab and the iterations were kept the same. The number of increments was raised to 29. Instead of the four-parameter model used by Ahmad, Bangash [749] used a five-parameter model in conjunction with the finite element analysis. Cracks developed using ISOPAR match well in size(s) and positions. The number of increments was 53, which gave comparative results on an increment-by-increment basis. The results are plotted in Figs IX.47–IX.54. They compare much better than in the work carried out by Ahmad and Bangash [873].

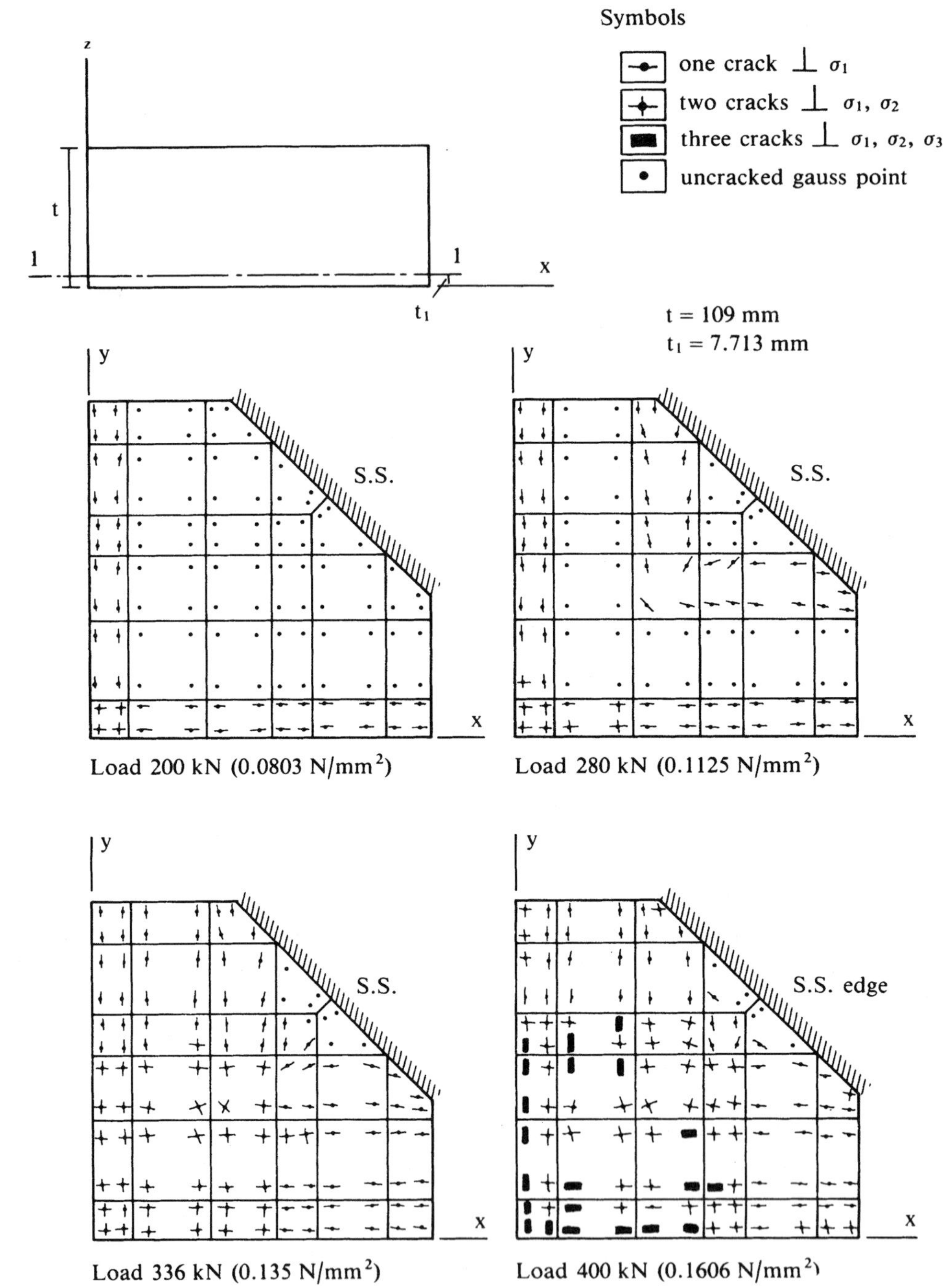

Fig. IX.52. Crack patterns of the slab at 1-1; slab type—perfectly bonded [873].

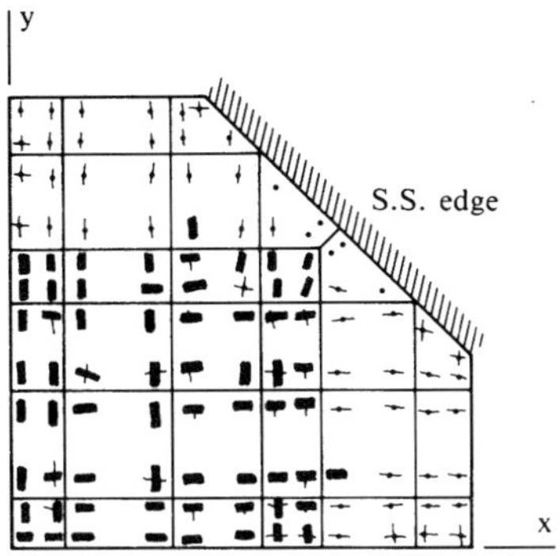

Fig. IX.53. Load of 544 kN. (Symbols as for Fig. IX.51.)

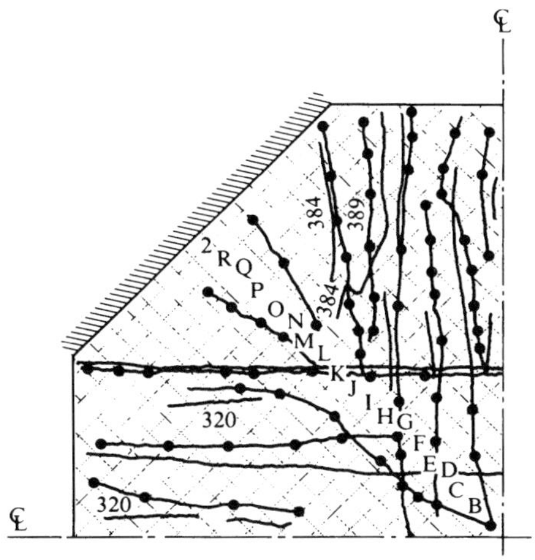

Fig. IX.54. Cracks imposed on Fig. IX.52. (—●—, Bangash 644 kN; ——, experiment 400 kN [849, 873].)

IX.5 Case Study No. 4 — reinforced concrete deep beams

IX.5.1 Introduction

A deep beam can be defined as a beam having a ratio of span to depth of about 5 or less, or having a shear span less than about twice the depth, and which is loaded at the top or compression face only. They are encountered in transfer girders, pile caps, foundation walls, raft beams, offshore gravity type structures, walls of rectangular tanks, hoppers, floor diaphragms and shear walls. Because of their proportions, deep beams are likely to have strength controlled by shear rather than flexure. On the other hand, their shear strength is expected to be significantly greater than predicted by the usual equations. They differ because of a special capacity to redistribute internal forces before failure and to develop a mechanism of force transfer quite different from beams of common proportions.

The aggregates used in these beams are stone chips, coarse aggregate and brick aggregates. In this section, finite element analysis is carried out to correlate the results obtained from various theories [1207 to 1271].

This section also covers the behaviour of reinforced concrete flanged deep beams. Flanged beams are usually deep with a thin web. The application of flanged deep beams may not normally be apparent in ordinary reinforced concrete structures but they are, for instance, a major structural component in the foundations of offshore gravity type structures and in the horizontal and vertical diaphragms used to transmit wind forces in tall buildings.

IX.5.2 History and development

Classic literature reviews have been compiled by Albritton [1268], the Cement and Concrete Association (C&CA) [1270] and Construction Industry Research and

Information Association (CIRIA) [1274], which have been supplemented by the reviews of Tang [1255] and Wong [1256]. These show that the early investigations were mostly on the elastic behaviour. Of course, elastic studies can easily be carried out today using the standard finite difference and finite element techniques. It was not until the 1960s that ultimate load analyses were carried out by Leonhardt and Walther [1243]. The tests were a major step forward in deep beam research. They revealed a concern for empirical evidence which reflected the philosophy of the Comité Européen du Béton (CEB). A number of authors then carried out tests [1220 to 1232].

IX.5.3 A review of current literature

Considerable literature is available on the elastic behaviour of ordinary (not flanged) deep beams (Albritton [1268]; Cement and Concrete Association (C&CA) [1270]). Dischinger [1275] used trigonometric series to determine the stresses in continuous deep girders. Chow *et al.* [1272] used finite-difference equations to solve simple-span deep beams. Cheng and Pei [1271] contributed much to the theory of deep beams by solving the case in which displacement was permitted. Leonhardt and Walther [1243] undertook extensive investigation of deep beams under all loading and support conditions and presented comprehensive findings for directly and indirectly loaded beams. Gogate [1278, 1279] developed a finite element approach for deep beams with progressive cracking.

Around 1965, an extensive long-term programme was initiated in the UK which is still continuing at the University of Newcastle upon Tyne under Ong and his co-workers. Their published work does not include investigations of different loading conditions of deep beams. El-Behairy [1276] included these types of loading condition in the study.

Nylander [1281–1285] undertook an extensive investigation of deep beams under various loadings and support conditions. He also studied 28 flanged deep beams. He observed that the variation in the amount of transverse reinforcement designed to prevent bursting did not produce any significant effect on the strength of the beam. He also observed that provision of heavy flexural reinforcement at the inner edge of flange increased the strength substantially.

Paul [1286] studied 18 wall panels with variable tensile and web reinforcement and different loading patterns. He concluded that the panels loaded below the compression zone (i.e. indirectly loaded) are weaker than the panels which were directly loaded at the compression face under similar conditions. He also concluded that adequate suspension or hanger reinforcement at the location of the load and good anchoring in the compression zone increased the load-carrying capacity of indirectly loaded beams.

Regan and Hamadi [1287] studied six beams using a simple point load from the top. They concluded that web strength is limited by crushing of the point. The web strength does not appear to be influenced by instability for height/thickness ratios up to 50.

Subedi [1290] studied two micro-concrete models. He concluded that the diagonal splitting force depends on the limiting tensile strength of concrete in a biaxial-compression–tension state of stress. In general, failure modes of deep beams with loads and boundary conditions are summarised in Tables IX.5 and IX.6.

IX.5.4 Current design practice

The subsequent sections under IX.5.6 of this chapter will summarise the main design recommendations of the CIRIA Guide 2, the (draft) Eurocode and the CEB-FIP Model Code, the ACI Code 318-83 (revised 1986) and the Canadian Code CAN-A23.3-M84.

IX.5.5 Experimental tests on modes of failure

The structural behaviour of flanged deep beams could be analysed with those of the conventional deep or panel beams of rectangular cross-sections. The methods of design for conventional deep beams are available in several documents. The main differences between a conventional deep beam and a flanged deep beam can be categorised as follows.

(a) Overall depth/web thickness ratio h/t

Conventional deep beams $h/t = 10$ or less

$$\text{Flanged beams } h/t = 60 \tag{IX.5.1}$$

(b) For conventional deep beams — main tensile reinforcements are within the plane of the web.

For flanged beams, wider flanges are accommodated. The usual modes of failure for flanged beams are given below.

(i) *Flexural shear failure.* Cracks develop from the bottom flange at or near mid-span. Flexural failures are accompanied by diagonal web cracking. As the load increases, excessive strains in the main reinforcement and large deflections occur.
(ii) *Flexural-shear compression failure.* As for (i) above but with the crushing of the compression flange before full tensile capacity is realised.
(iii) *Diagonal splitting failure.* The splitting is generally located at mid-depth of the beam with excessive cracking in the web with a moderate amount of reinforcement in the top and bottom flanges.
(iv) *Splitting with compression failure.* This is similar to mode (iii) except that diagonal cracks eventually penetrate into the compression flange, causing the crushing of the flange.

Subedi [1287] has finally summarised these failures and translated them into equations, as detailed below.

IX.5.5.1 Subedi summary of deep-flanged beams

Flexural-shear failure. Flexural-shear failure occurs in under-reinforced beams, i.e. beams with a very small amount of main tensile reinforcement A_u.

Table IX.5. Deep beam failure cases I

I *Case (IA)*

(Ia) In a single span deep beam, when loaded at the top with a uniform load, the stress pattern along the depth of the beam shall be as shown. Assuming h_a is the depth of the beam, all zones of the stress pattern are defined by x_1 to x_3. The direction of the principal stress is clearly indicated.

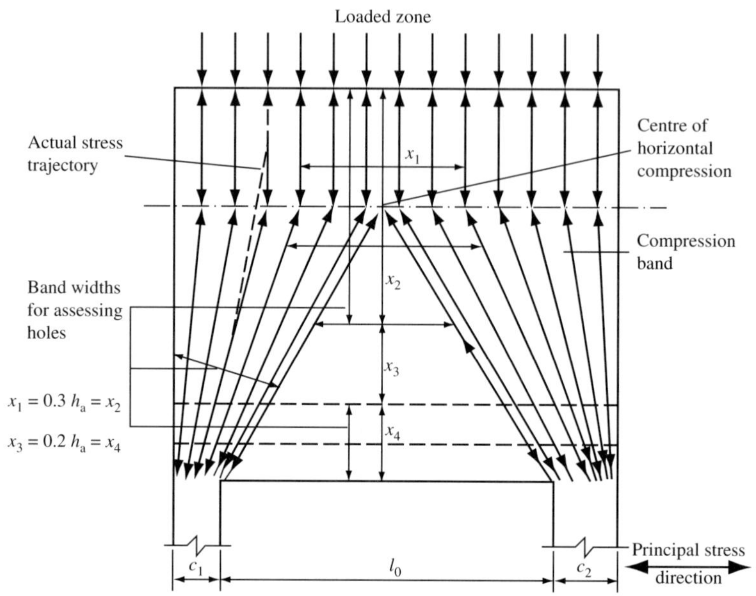

(Ia) Top loaded

(Ib) When the bottom of this beam is uniformly loaded, the action is represented by a tied analogy. This is true when the concrete in tension has cracked. The centre of compression force rises as an arch approximately half the span of the tie.

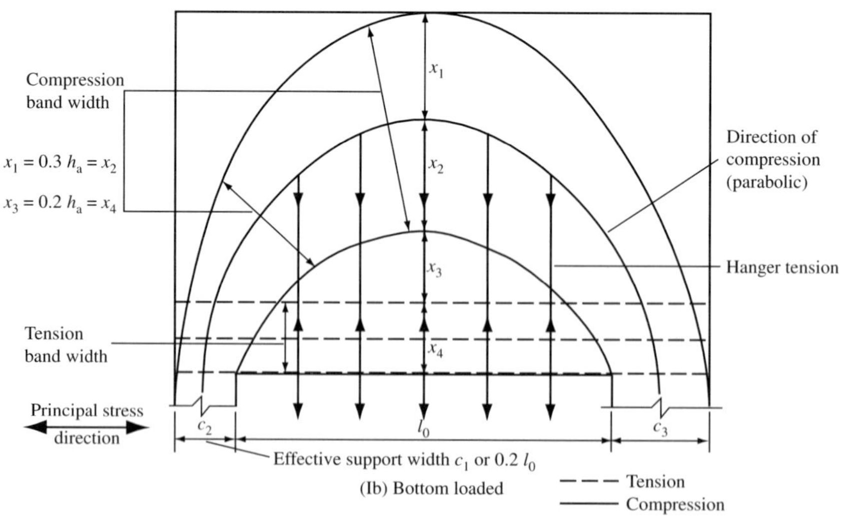

(Ib) Bottom loaded

Table IX.5. — contd.

(Ic) *For multi-span deep beams*
In this case of a multi-span deep beam, the load is transferred to the supports by a combination of tied arch and suspension system analogies. The thickness of this beam should not be less than the sum of the six dimensions given below:

(1) 2 × minimum concrete cover to the outer layers of bars,
(2) a dimension allowing 4 layers of web steel, one vertical and one horizontal at each face of the beam including bar deformation.

In all cases the bearing stresses shall not exceed 0.4 × the cubic strength of concrete. For bearing stresses, the reactions must be uniformly distributed over

(i) the wall thickness
(ii) actual support length
(iii) the length $0.2l_0$ if less.

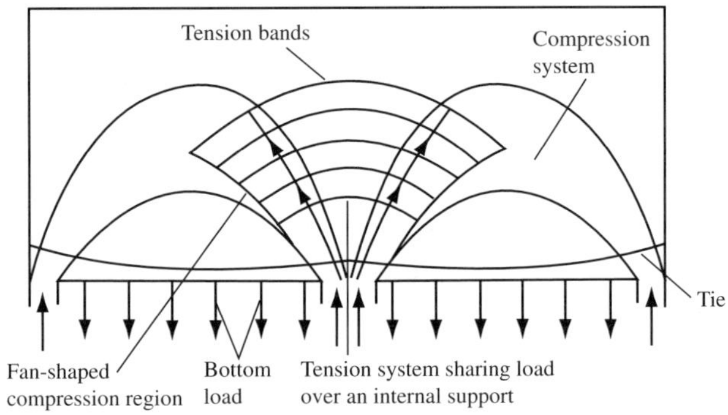

(Ic) Bottom loaded

The shear failure which occurs in a beam when the horizontal component of the diagonal splitting force, $f_{tc}t_w h_w + A_h f_s$

(a) > the capacity of the tensile reinforcement $A_{st} f_y$
(b) > the compression capacity, $A_{sc} f_y + (A_c - A_{sc}) f_{cu}$

Flexural-shear compression failure. This mode of failure is described as flexural-shear compression failure. The criteria for this type of failure are when the horizontal component of the diagonal splitting force $f_{tc}t_w h_w + A_h f_d$

(a) > the capacity of the tensile reinforcement, $A_{st} f_y$
(b) > the compression capacity, $A_{sc} f_y + (A_c - A_{sc}) f_{cu}$; $A_{st} f_y > A_{sc} f_y + (A_c - A_{sc}) f_{cu}$

Diagonal splitting failure. Therefore, diagonal splitting mode failure will occur when the horizontal component of the diagonal splitting force $f_{tc}t_w h_w + A_h f_s$

(a) < the capacity of the tensile reinforcement $A_{st} f_y$

Table IX.6. Deep beam failure cases II

Case (IIa). Single span beam partially loaded
Here the single span deep beam is loaded at the top. The case is similar to (Ia) but the load is partially placed along the centre. The stress pattern is identical.

$$x_1 = 0.3h_a = x_2$$

$$x_3 = 0.2h_a = x_4$$

The symmetry exists along the centre of compression. Because of the load application on a portion of this beam, the stress trajectory is reduced. All other conditions are similar.

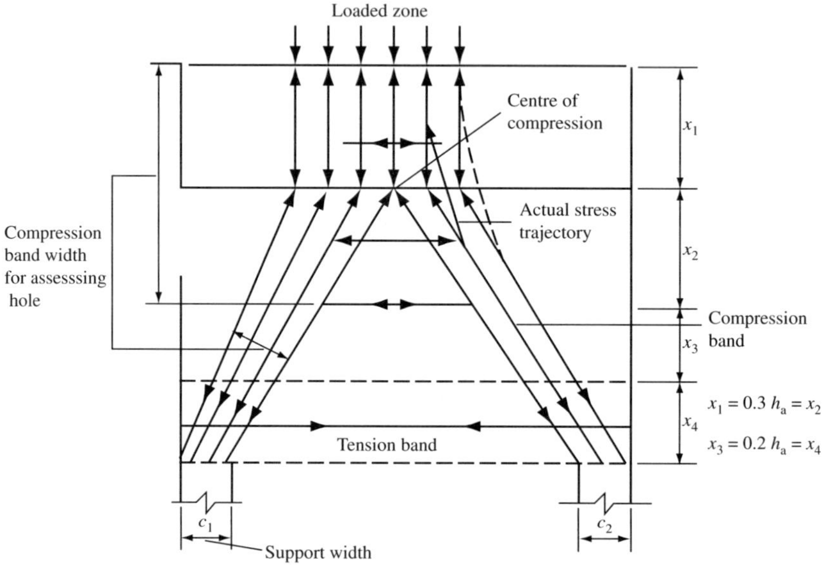

(IIa) Loaded partially at top

(b) $<$ the compression capacity $A_{sc}f_y + (A_c - A_{sc})f_{cu}$

If the tensile capacity is less than the compression capacity, the diagonal crack may eventually extend into the tensile flange.

Diagonal splitting with compression failure. This mode of failure is similar to diagonal splitting failure except that, in this case, the diagonal crack may eventually penetrate into the compression flange.

Here the criteria are that the horizontal component of the diagonal splitting force, $f_{tc}t_wh_w + A_hf_s$

(a) $<$ the capacity of the tensile reinforcement $A_{st}f_y$
(b) $<$ the compression capacity $A_{sc}f_y + (A_c - A_{sc})f_w$

the compression capacity $<$ the tensile capacity.

Table IX.6.— contd.

Case (IIb). Stress pattern over the internal support

The deep stress pattern forms a fan shape when the internal support carries a uniform reaction. The horizontal tension is represented by straight horizontal lines. The compression region is fan shaped, spreading out of the internal. The two lengths $l_{1/4}$ and $l_{2/4}$ can be different. The overall span for the tension area is

$$\frac{(l_1 + l_2)}{4} + C_1$$

The region is extended vertically by $x_1 = 0.6h_a$ and $x = 0.2h_a$.

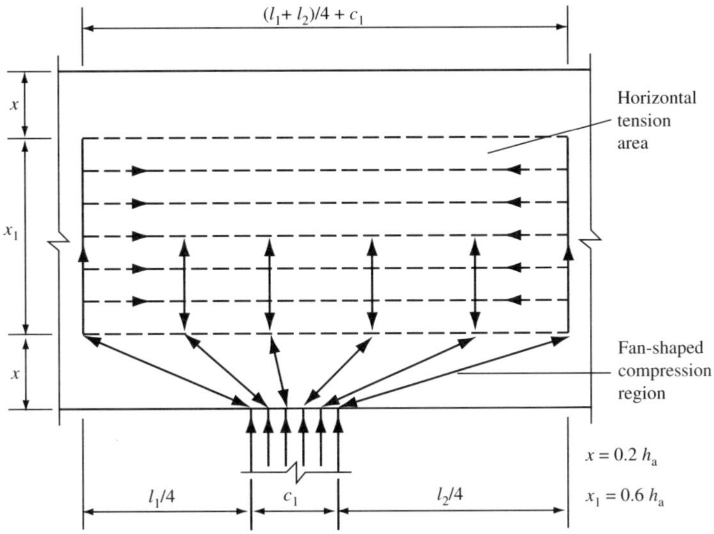

(IIb) Loaded at internal support as reaction

The modes of failure are summarised in Table IX.6. In this table it is assumed that the amount of horizontal web reinforcement is small and that

$$f_{tc}t_w h_w > A_h.f_{s,y}$$

IX.5.6 Main design recommendations on deep beams

IX.5.6.1 Introduction

A number of researchers have produced elastic and ultimate methods loads [1226–1290] for evaluating the behaviour of deep beams.

In designing normal (shallow) reinforced concrete beams of the proportions more commonly used in construction, it is assumed that plane sections remain plane after loading. This assumption is not strictly true but the errors resulting from it do not become significant until the depth of the beam becomes equal to, or more than, about half the span. Hence, special methods are taken into account in order to evaluate the overall applied moments and shears, but also the stress patterns and internal deformations within these deep beams. Tables IX.5 and IX.6 show the deep beams under different conditions. Table IX.6 shows the failure modes of deep beams.

In this section, various codified methods for the analysis and design of deep beams are given. Some are compared among themselves and with the finite element method.

IX.5.6.2 Design based on Eurocode and CEB-FIP Model Code

The Eurocode 2 (1984): Common Unified Rules for Concrete Structures does not directly provide guidelines for the design of deep beams. It refers instead to clause 18.1.8 of the CEB-FIB Model Code (1978). The CEB-FEP Model Code applies to simply supported beams of span/depth ratio e/h less than 2 and to continuous beams of e/h less than 2.5.

Flexural strength: simply supported deep beams. The area of the longitudinal reinforcement is calculated from the equation in which the lever arm z is to be taken as follows:

$$\begin{aligned} z &= 0.2l(l+2h) \quad \text{for } 1 < (l/h) < 2 \\ z &= 0.6l \quad \text{for } (l/h) < 1 \end{aligned} \qquad \text{(IX.5.2)}$$

For continuous deep beams, the lever arm z is taken as

$$\begin{aligned} z &= 0.2l(l+1.5h) \quad \text{for } 1 < l/h < 2.5 \\ z &= 0.6l \quad \text{for } 1/h < 1 \end{aligned} \qquad \text{(IX.5.3)}$$

The main longitudinal steel in the span should be detailed as for simply supported beams. Over the support, half the steel should extend across the full length of the adjacent span; the remaining half is stopped at $0.4l$ or $0.4h$, whichever is smaller, from the face of the support.

The two equations (IX.5.2) and (IX.5.3) show that in deep beams the lever arm varies at a lower rate with the depth h. When the depth exceeds the span, the lever arm becomes independent when $h > l$. The calculated main reinforcement should extend from one support to the other, be fully anchored at both ends, and no curtailment can be adopted. Vertical hooks causing cracks at the anchorage level shall be avoided. The required steel from the soffit of the beam shall be uniformly distributed over a depth of $0.25h$ to $0.05l$ using smaller diameter bars. Such an action will limit the crack width under service loads.

These small bars will facilitate the anchorage at the support and will produce an efficient design.

Shear strength and web reinforcement. The design shear strength should not exceed the lesser of

$$0.10bh(f_c'/\Upsilon_m) \quad \text{and} \quad 0.10bl(f_c/\Upsilon_m) \qquad \text{(IX.5.4)}$$

where

b = width

h = beam depth

f_c' = characteristic cylinder strength of concrete

Υ_m = material partial safety factor

The web reinforcement is provided in the form of a light mesh of orthogonal reinforcement consisting of vertical stirrups and horizontal bars placed near each face and surrounding the extreme vertical bars. The web steel ratio should be about 0.20% in each diameter, 0.20% for high bond bars.

IX.5.6.3 Design based on ACI Building Code 318-83 (revised 1986)

Flexural strength. For flexural design, ACI Code 318-83 (revised 1986) defines a deep beam as a beam in which the ratio of the clear span l_0 to the overall depth h is less than the limits given below:

simple spans: $l_0/h < 1.25$

continuous spans: $l_0 < 2.5$

Minimum tension reinforcement. The main steel ratio ρ shall not be less than ρ_{min}

$$\rho_{min} = 200/f_y \tag{IX.5.5}$$

where $\rho_{min} = A_s/bd$, is the main tension reinforcement, b is the beam width, d is the effective depth and f_y is the steel strength (lb/in.^{2}). For $f_y = 460\,\mathrm{N/mm^2}$ (about 66 500 lb/in.^{2}), $\rho_{min}$is about 0.3%.

$$A_s = \frac{M_u}{\phi \mathrm{j}\, \mathrm{d} f_y} \tag{IX.5.6}$$

Web reinforcement. An orthogonal mesh of web reinforcement is required. The minimum areas of the vertical and horizontal bars shall satisfy

$$A_v/bs_v > 0.15\%$$

$$A_h/bs_v > 0.25\% \tag{IX.5.7}$$

where A_v is the area of the vertical bars with the spacing s_v and A_h is the area of the horizontal bars within the spacing s_v.

Flexural design. Apart from the above requirement, the ACI Code does not give further detailed guidelines. It merely states that account shall be taken of the non-linear distribution of strain and lateral buckling.

The shear provisions of ACI Code 318-83 (revised 1986) apply to top-loaded simple or continuous beams having a (clear span) (effective depth) ratio l_0/d less than 5.

Shear strength: simply supported deep beams. Calculations are carried out using a concentrated load, taken as half-way between the load and the face of the support. The shear reinforcement required at the critical section shall be used throughout the span. The design is based on

$$V_u < 0\phi V_n$$

$$V_n = V_c + V_s \tag{IX.5.8}$$

where V_u is the design shear force at the critical section (lb), V_n is the nominal shear strength (lb) and ϕ is the capacity reduction factor for shear, taken as 0.85, V_c is the shear strength provided by concrete (lb) and V_s is the shear strength provided by steel (lb). The nominal shear strength V_n should not exceed the following:

$$\begin{aligned} &V_n < 8\sqrt{f'_c}bd && \text{for } l_0/d < 2 \\ &V_n < (2/3)(10 + l_0 d)\sqrt{f'_c}bd && \text{for } 2 < l_0/d < 5 \end{aligned} \qquad \text{(IX.5.9)}$$

where f'_c is the concrete cylinder compressive strength (lb/in.^{2}), h is the beam width (in.) and d is the effective depth (in.).

The shear provided by concrete is calculated from

$$V_c = (3.5 - 2.5M_u/V_u d)(1.9\sqrt{f'_c} + 2500\rho V_u d/M_u)bd$$

Shear force V_u at factored load exceeds the design shear strength of the concrete ϕV_c; the contribution from the web steel is

$$V_s = \left[A_v/S\left(\frac{1 + l_n/d}{12}\right) + \frac{A_{vh}}{S_2}\left(\frac{11 - l_n/d}{12}\right)\right]f_y d$$

where

A_v = shear reinforcement area 1 to flexural steel within a distance S

A_{vh} = shear reinforcement area 11 to flexural steel within a distance S_2

Shear criteria. ACI (318-83) revised 1986. Restriction

$$K = [3.5\text{–}2.5(M_u/V_u d] < 2.5$$

$$\text{Shear provided by concrete} = K(1.9\sqrt{f'_c} + 2500\rho V_u d/M_u)bd \qquad \text{(IX.5.10)}$$

where M_u = design bending moment (in. lb); V_u = shear strength at critical section.

$$V_c < 8\sqrt{f'_c}bd;\ V_c = \text{allowable shear or nominal shear} \qquad \text{(IX.5.11)}$$

where $V_u > \phi V_c$, a system of orthogonal shear reinforcement to carry the excess shear.

The contribution of V_s to shear reinforcement is then given by

$$\frac{V_s}{f_y d} = \frac{A_v}{S_v}\left[\frac{1 + l_0/d}{12}\right]^* + \frac{A_h}{S_h}\left[\frac{11 - l_0/d}{12}\right]^* \quad (^* = \text{weighting factor}) \qquad \text{(IX.5.12)}$$

Combining between equations

$$A_v[1 + l_0/d] \qquad A_h[11 - l_0/d] \qquad (V_u/\phi) - V_c \qquad \text{(IX.5.13)}$$

where A_v is the area (in.^{2}) of vertical web reinforcement within a spacing S_y, A_h is the area (in.^{2}) of horizontal web reinforcement within a spacing S_h, f_y is the strength of the web steel which should not be taken as more than 60 000 lb/in.^{2} (410 N/mm^2). S_v is the spacing (in.) of the vertical web bars — which must exceed neither $d/5$ nor 18 in. — and S_h is the spacing (in.) of the horizontal web bars — which must exceed neither $d/3$ nor 18 in. The l_0/d is generally 5.

IX.5.6.4 Continuous deep beams: shear strength

The design is generally based on

$$V_u < \phi V_n \qquad V_n = V_c + V_s \tag{IX.5.14}$$

However, for continuous deep beams, the concrete nominal shear strength V_c is to be taken as the least value given by

$$V_c = 3.5\sqrt{f_c'}bd$$
$$V_c = [1.9\sqrt{f_c'} + 2500\rho]bd \tag{IX.5.15}$$
$$V_c = [1.9\sqrt{f_c'} + 2500\rho(V_u d/M_u)bd$$

where M_u is the design moment occurring simultaneously with V_u at the section considered: ρ is the main steel ratio A_s/bd and f_c' is the concrete cylinder strength (lb/in.^{2}).

Where V_u exceeds $0.5\phi V_c$, vertical shear reinforcement should be provided to satisfy the condition. Where the design shear force V_u exceeds ϕV_c, vertical shear reinforcement shall be provided to carry the excess shear. The contribution V_s of this shear reinforcement is given by

$$V_s = (A_v f_y/S_v) \tag{IX.5.16}$$

where the symbols are

$$(A_v/bs_v) > (V_u/\phi - V_c)/bdf_y \tag{IX.5.17}$$

such that

$$A_s/bs_v \not< 0.0015; \quad S_v \not< d/5 \text{ nor } 18'' \text{ (450 mm)}$$
$$A_h/bs_h \not< 0.0025; \quad s_h \text{ of horizontal steel} \not> d/3 \text{ nor } 18'' \text{ (450 mm)}$$
$$F_y \not> 60\,000 \text{ lb ft/in.}^2 \text{ (410 N/m}^2\text{m)}$$

IX.5.6.5 CIRIA Guide 2

The CIRIA Guide applies to beams having an effective span/depth ratio l/h of less than 2 for single-span beams and less than 2.5 for continuous beams. Figure IX.55 shows the meaning of symbols.

The CIRIA Guide considers that the active height h_a of a deep beam is limited to a depth equal to the span; that part of the beam above this height is taken merely as a load-bearing wall between supports.

CIRIA Guide 'simple rules'. CIRIA's 'simple rules' are intended primarily for uniformly loaded deep beams. They can be applied to both single-span and continuous beams.

Flexural strength.

Step 1: Calculate the capacity of the concrete section

$$M_u = 0.12 f_{cu} b h_a^2 \tag{IX.5.18}$$

where f_{cu} is the concrete characteristic strength and h the beam thickness.

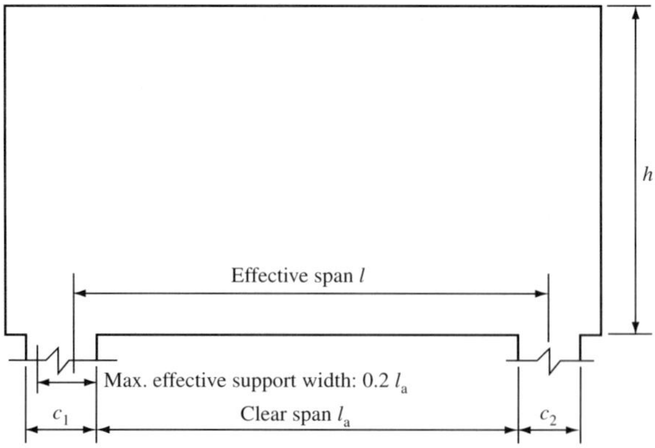

Fig. IX.55. CIRIA Guide 2 — meanings of symbols c_1, c_2, h, l and l_a

Step 2: If $l/h_a < 1.5$ go to Step 3. If $l/h_a > 1.5$ check that the applied moment M does not exceed M_u.

Step 3: Calculate the area A_s of the main longitudinal reinforcement

$$A_s > M/0.87f_y z, \tag{IX.5.19}$$

where M is the applied moment, f_y the steel characteristic and z the lever arm, which is to be taken as follows:

$z = 0.21 + 0.4h_a$ for single-span beams

$z = 0.21 + 0.3h_a$ for continuous beams

Step 4: Distribute the reinforcement A over a depth of $0.2h_a$

Shear strength: bottom-loaded beams

Step 1: Calculate the concrete shear capacity:

$$V_u = 0.75bh_a v_u \tag{IX.5.20}$$

where V_u is the maximum shear stress.

Step 2: Check that the applied shear force V does not exceed V_u.

Step 3: Provide hanger bars in both faces to support the bottom loads, using a design stress of $0.87f_y$. The hanger bars should be anchored by a full bond length above the active height h_a or, alternatively, anchored as links around longitudinal bars at the top.

Step 4: Provide nominal horizontal web reinforcement over the lower half of the active height h_a and over a length of the span equal to $0.4h_a$ measured from each support. The area of this web reinforcement should not be less than 80% of the uniformly distributed hanger steel, per unit length. The bar spacing and reinforcement percentage should also meet the requirements.

Shear strength: top-loaded beams. The proven concept of the clear shear span x_e as used by Kong *et al.* has been adopted by the CIRIA Guide. The CIRIA Guide has also accepted that for uniformly distributed loading, the effective clear shear span x_e may be taken as $l/4$.

Step 1: Calculate the effective clear shear span x_e which is to be taken as the least of:

(a) the clear shear span for a load which contributes more than 50% of the total shear force at the support

(b) $l/4$ for a load uniformly distributed over the whole span

(c) the weighted average of the clear shear spans where more than one load acts and none contributes more than 50% of the shear force at the support. The weighted average will be calculated as $\sum(V_r X_r)/\sum V_r$, where $\sum V_r = V$ is the total shear force at the face of the support, $V_r =$ an individual shear force and $x_r =$ clear shear span of V_r.

Step 2: Calculate the shear capacity V_u to be taken as the value given by

$$V_u = 1.2bh_a^2 v_c/x_e \text{ for } h_a/h \geq 4$$

$$V_u = bh_a v_u \tag{IX.5.21}$$

where the x_e is the effective clear shear span calculated from Step 1.

Step 3: Check that the applied shear force V does not exceed the shear capacity V_u calculated in Step 2.

Step 4: Provide nominal web reinforcement in the form of a rectangular mesh in each face. The amount of this nominal reinforcement should not be less than that required for a wall which is 0.25%.

Bearing strength. For deeper beams ($l/h < 1.5$), the bearing capacity may well be the governing design criterion, particularly for those beams with shorter shear spans. To estimate the bearing stress at the support, the reaction may be considered uniformly distributed over an equal area to (the beam width b) $\times$ (the effective support length) where the effective support length is to be taken as the actual support length c or $0.2l_0$ whichever is the lesser. The bearing stress so calculated should not exceed $0.4f_{cu}$.

Crack control. The minimum percentage of reinforcement, in the horizontal or vertical direction, should comply with the requirements for a wall, as given BS 8110: Part 1: clauses 3.12.5.3 and 3.12.11.2.9. The maximum bar spacing should not exceed 250 mm. In a tension zone, the steel ratio ρ, calculated as the ratio of the total steel area to the local area of the concrete in which it is embedded, should satisfy the condition

$$\rho > (0.25\sqrt{f_{cu}})/(0.87f_y) \tag{IX.5.22}$$

The maximum crack width should not be allowed to exceed 0.3 mm in a normal environment; in a more aggressive environment, the maximum crack width may have to be limited to 0.1 mm, to control maximum crack widths to within 0.3 and 0.1 mm.

CIRIA Guide 'supplementary rules'. *Flexural strength*. The 'simple rules' can be used without modification.

Shear strength: top-loaded beams.

Step 1: If the beam is under uniformly distributed loading, go to Step 3 or use the more conservative 'simple rules'.

Step 2: If the beam is under concentrated loading go to Step 3.

Step 3: Check that the applied shear force V does not exceed the limit imposed for a beam with a system of orthogonal web reinforcement which can be expressed in the more convenient form over the range 0.23–0.70 for X_e/h. In using these equations, ignore any web reinforcement which is above the active height h_a.

$$\frac{V}{bh_a} < \lambda_1\left[1 - 0.35\frac{x_e}{h_a}\right]\sqrt{f_{cu}} + \lambda_2\sum\frac{100A_r v_r \sin 2\theta_r}{bh_a^2} \quad \text{(IX.5.23)}$$

where λ_1 is 0.44 for normal weight concrete and 0.32 for lightweight concrete, λ_2 is 1.95 N/mm^2 for deformed bars and 0.85 N/mm^2 for plain bars, h is the beam thickness, h_a is the active height of the beam. A_r is the area of a typical web (the main longitudinal bars are considered also to be web bars), v_r is the depth at which the typical web bar intersects the critical diagonal crack, which can be represented by a line Y–Y, θ_r is the angle between the bar being considered and the line Y–Y $(0 < \pi/2)$ and x_e is effective clear shear span.

On the right-hand side of the term, $\lambda_1[1 - 0.35x_e/h_a]\sqrt{f_{cu}}$ is the concrete contribution to the shear orthogonal reinforcement and can be expressed in a convenient form

$$V/bh_a < \lambda_1 v_x + \beta(v_{ms} + v_{wh} + v_{wv}) \quad \text{(IX.5.24)}$$

where $\lambda_1 v_x$ = concrete shear stress parameter

β = 1.0 deformed bars

v_{ms} = 0.4 plain bars

v_{wh} = horizontal steel shear stress parameter for web

v_{wv} = vertical steel shear stress parameter for web

Step 4: From the calculations in Step 3, check the total contribution of the (main web) reinforcement to the shear capacity. If this is less than $0.2v$, increase the web reinforcement to bring the total steel contribution up to at least $0.2v$.

Step 5: Check that the applied shear force v is less than the shear capacity of the concrete section.

IX.5.6.6 Design example using ACI Code

Data

$f_c^1 = 5000$ psi (34.5 N/mm^2); $f_y = 78\,500$ psi (541 N/mm^2)

Gauge 6 wire mesh of 50 000 psi (345 N/mm^2) for #10 (32 mm dia. bars)

$$h = h_n = 44'' \text{ (1120 mm)}; \; b = 1.5'' \text{ (38 mm)}$$

$$d = 42.875'' \text{ (1089 mm) } l_n = 89 - 2 \times 10 = 69'' \text{ (1750 mm)} = l_0$$

$$W = \text{total load concentrated reaction} = V_u = 30\,923 \text{ lbft (137 kN)}$$

Solution
Check l_n/d, evaluate factored shear force, V_u, $d = h - \text{cover} - 1/2$ dia. of bar $= 44 - 0.5 - 0.625 = 42.875$ in. (1089 mm), $l_n/d = 69/42.875 = 1.609 < 5$: hence the beam is treated as a deep beam.
Factored shear force V_u and resisting capacity V_c

$$V_u \leq \phi(8\sqrt{f_c'}b_w d) = 0.85(8\sqrt{5000}(1.5)(42.875) = 30\,923 \text{ lb (137 kN)}$$

$$M_u = Wl_u/4 = 30\,923 \times 69/4 = 533\,421 \text{ in.-lb (60.25 kNm) [assume } W = V_u]$$

$$M_u/V_u d = 533\,421/(30\,923 \times 42.875) = 0.4023$$

$$3.5 - 2.5M_u/V_u d = 3.5 - 2.5 \times 0.4023 = 2.5$$

$$\rho_w = 2 \times 1.27/1.5 \times 42.875 = 0.0395 V_u V_u d/M_u = 2.5$$

$$\begin{aligned} V_c &= 2.5(1.9(f_c' + 2500\rho_w(V_u d/M_u))/b_u d \\ &= 2.5(1.9\sqrt{5000} + 2500 \times 0.0395 \times 2.5)1.5 \times 42.875 \\ &= 60\,234 \text{ lbs (268 kN)} \end{aligned}$$

$$6\sqrt{f_c'}b_w d = 28\,030 \text{ lb (124.7 kN)}$$

The contribution from the web steel is given by

$$V_s = \left[\frac{A_v}{S}\left\{\frac{1 + l_n/d}{12}\right\} + \frac{A_{vh}}{S_2}\left\{\frac{11 - l_n/d}{12}\right\}\right] f_y d$$

$= 15\,710$ lb (69.9 kN) initially $S = S_2 = 6''$ (150 mm) where

$$S = d/5 = 42.875/5 = 8.6 \text{ in. (218 mm)} < 18 \text{ in. (457 mm)}$$

$$S_2 = d/3 = 42.875/3 = 14.3 \text{ in. (363 mm), 18 in. (457 mm)}$$

$$A_v = A_{vh} : l_n/d = 1.609$$

Minimum $A_v = 0.0015bS = 0.0015(1.5)(8.6) = 0.019$ in.^{2} (12.3 mm^2)
Provided $A_v = 0.04$ in.^{2} (19.3 mm^2)
Minimum $A_{vh} = 0.0025bS_2 = 0.0025(1.5)(14.3) = 0.054$ in.^{2} (34.6 mm^2)
Provided $A_{vh} = 0.067$ in.^{2} (43.0 mm^2)
Total shear force: $V_u = V_c + V_s = 60\,234 + 15\,710 = 75\,944$ (387 kN)

Provide 6T12 two-legged horizontal stirrups (1356 mm^2) such that 4 No. bars at 100 mm c/c and 2 bars at 100 mm c/c above and below the opening if it exists in

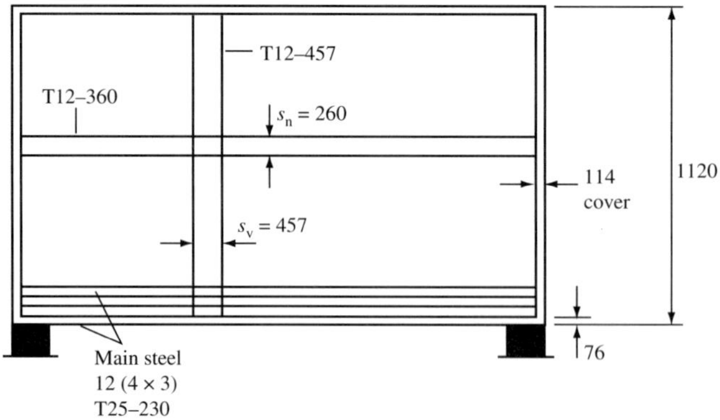

Fig. IX.56. Reinforcement details of the deep beam.

the beam. In addition, nominal two-legged stirrups of R6 or T8, whichever is available, may be provided.

$$M_u = \frac{75\,944 \times 69}{4} = 1\,310\,034 \text{ in lb}$$

$$\frac{M_u}{V_u d} = \frac{0.4 V_u d}{M_u} = 2.5$$

$$V_c = 60\,234\,\text{lb}\,(268\,\text{kN}); \quad V_s = 15\,690\,\text{lb}\,(69.9\,\text{kN}); \quad V_u = 75\,944\,\text{lb}\,(337.8\,\text{kN})$$

Flexural reinforcement

$$M_u = 1\,310\,034 \text{ in lb } (148\,\text{kNm})$$

$$l/h = 1.568 < 2; \qquad A_s = 0.783\,\text{in.}^2\,(505\,\text{mm}^2)$$

Adopt 2.54 in.^{2} (1639 mm^2)
Minimum reinforcement is 0.166 in.^{2} (166 mm^2) O.K.
Final reinforcement detail is given in Fig. IX.56

IX.5.6.7 Beam with web openings

$$l_n = 1750\,\text{mm}; \quad d = 750\,\text{mm}; \quad X_N = 250\,\text{mm}$$

Bearding width = 75 mm; $W = P = 330\,\text{kN}$; $f_c' = 30\,\text{N/mm}^2$;

$$f_{sy} = 250\,\text{N/mm}^2; \quad f_{cu} = 34.5\,\text{N/mm}^2; \quad f_c' = 3.48\,\text{N/mm}^2$$

Co-ordinates for opening centre

$$K_1 X_N = 137.5\,\text{mm}; \quad K_2 D = 475\,\text{mm}$$

two-point loading each 330 kN

$$M = 6025 \times 10^4\,\text{Nmm}$$

$$0.455 A_s \frac{f_{sy} d}{\gamma_m} = 107.72 \times 10^3 A_s\,\text{Nmm}$$

$$A_s = \frac{6025 \times 10^4}{107.72 \times 10^3} = 554.32\,\text{mm}^2$$

2T20 ($626\,\text{mm}^2$)

Shear resistance = 0.65 × 50% of the solid web

$$\tfrac{1}{2} \times 0.65 \times 268 \times 10^3 = 87.1 \times 10^3\,\text{N}$$

$b = 143.76$; adopt 150 mm value of b

Web steel

$$A_{wv} = A_v b = 69.9\,\text{kN}$$

Horizontal web steel

$$0.75\left[0.01 \times 0.5 \times 0.85 \frac{A_{wh} \times 100 \times 250}{150 \times 1089}\right] \frac{150 \times 1089}{1000} = 0.0693 A_{wh}\,\text{kN}$$

$$A_{wh} = 1009\,\text{mm}^2$$

Reinforcement details are shown in Fig. IX.57.

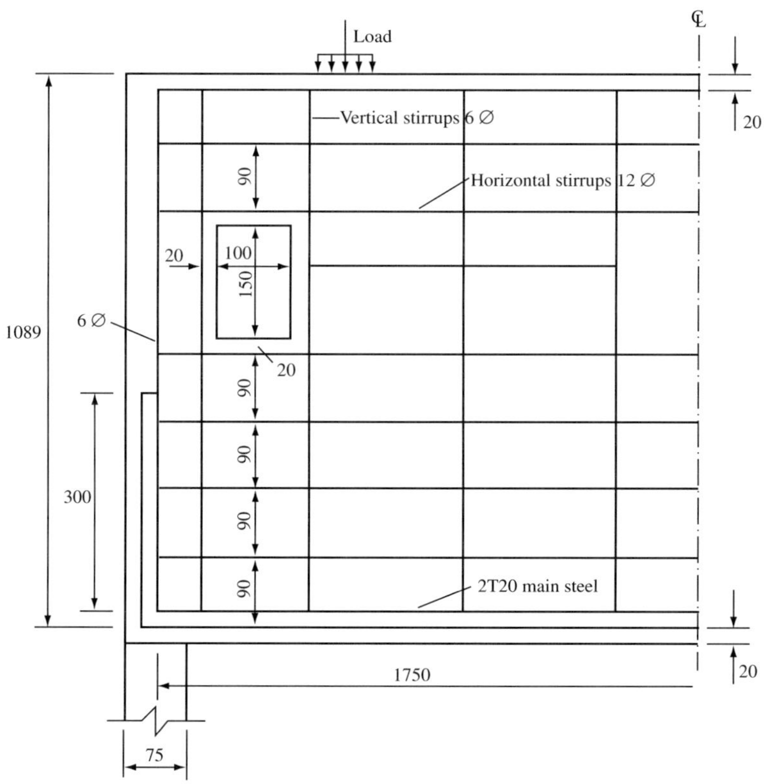

Fig. IX.57. Reinforcement details of a deep beam with a hole.

IX.5.6.8 Finite element analysis of deep beams

Introduction. It is first necessary to discretise the continuum into specific series of elements. The analysis of deep beams is affected by the choice of elements. Finite element modelling is covered in detail in Chapter VI. The equivalent load vector, stiffness matrix evaluation on local and global axes, discrete cracking model or sheared cracking model, solution and convergence procedures etc. have been given in the text. They are still applicable to deep beams. The design model given in Fig. IX.56 is implemented with the finite element method and the analytical results are compared with those obtained from the experiments [1238–1287]. The strain softening characteristics of the model have allowed satisfactory predictions of the response up to failure. The areas where cracks developed and the propagation of the crack zones up to the final collapse of the beam are predicted and compared with experimental data.

Finite element model. A three-dimensional constitutive model based on incrementally linear elastic (hypoelastic) behaviour of material has been formulated recently. The material is assumed initially isotropic but it becomes anisotropic at later stages of loading. In order to express the nonlinear stress–strain relations, the path dependency, the elastic moduli as a function of total strain history are to be considered. The strain can be decomposed into two parts, the octahedral part associated with change in volume and the deviatoric part associated with change in shape. The motivation of the decomposition in volumetric (octahedral) and deviatoric components comes from a space truss concept. The advantage of this modelling is that the triaxial behaviour of concrete can be modelled by a combination of uniaxial models and stiffness of the members can be calculated at each strain increment from the uniaxial and stress–strain relationship. The stiffnesses of the truss members is equivalent to the hydrostatic and deviatoric stiffness of the material. It is noted that aggregate interlock, multiple cracking and crack rotation are taken into account in this model.

Finite element data.

(a) *Beam 1 — mesh scheme (Fig. IX.58)*

No element	400
Solid (isoparametric)	
Number of nodes	10 000
Bar elements (4 noded)	1800
Number of loads	2 concentrated loads spaced
Number of iterations prior to failure	21

(b) *Beam 2 — mesh scheme (Fig. IX.59)*

No element: main type	4200
Solid isoparametric (20 noded)	4200
Bar element (4 noded)	750
Number of loads	2 concentrated loads
Number of iterations prior to failure	22

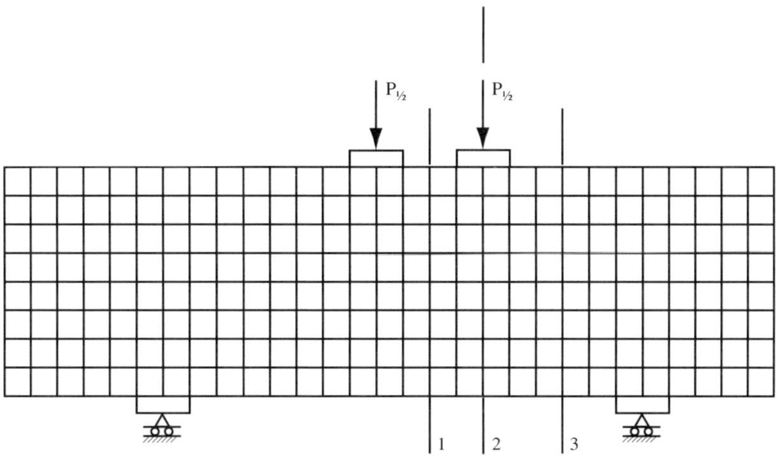

Fig. IX.58. Finite element mesh scheme for a deep beam no. 1.

(c) *Common data — beams 1 and 2*

Material properties:

Main reinforcement	T25
Secondary reinforcement	T12
L_n = length (Beam 1): 1750 mm	(beam 2) 750 mm
$d = 1089$ mm (Beam 1)	(beam 2) 750 mm
$D = 1120$ mm	$D = 965$ mm
$f_{cu} = 34.514\,\text{mm}^2$	
$f'_c = 0.78 f_{cu}$	
$\nu = 0.2$	

Analysis of results. The two RC beams are subjected to two-point loading and they are both simply supported. The widths of the beams is between 150 and 200 mm. The

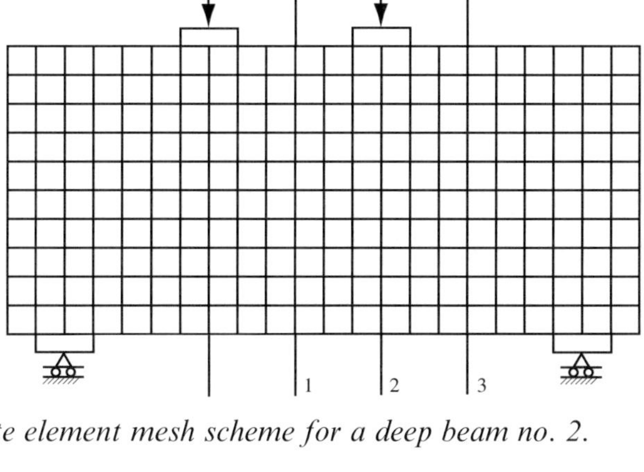

Fig. IX.59. Finite element mesh scheme for a deep beam no. 2.

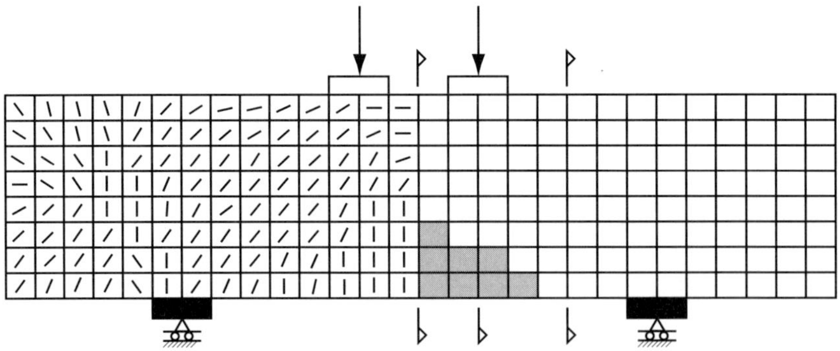

Fig. IX.60. Beam 1—compressive stresses with cracks at 40 kN.

length of the bearing is not less than 76 mm. As stated earlier, the failure of reinforced concrete deep beams is classified in four modes.

(a) flexure failure
(b) flexure-shear failure
(c) diagonal splitting
(d) bearing crush.

The most common are (b) and (c) but (d) is not uncommon. Some load zones and magnitudes such as 40 kN to 114 kN are noted for cracks and load–deflection curves. In some cases, stresses and strains are also identified.

(i) *Beam 1.* At 40 kN load level, heavy compressive principal stresses and resulting cracks are shown in Fig. IX.60. At an incremental load up to 65 kN, additional stresses and cracks have appeared as shown in Fig. XI.61. It was necessary to carry out experimental tests on this beam. Both theoretical and experimental results are plotted in Fig. IX.63 as a comparative study. The experimental loads are further applied incrementally. The beam failed at 105 kN load and the corresponding crack patterns are shown in Fig. IX.62. The analytical predicted

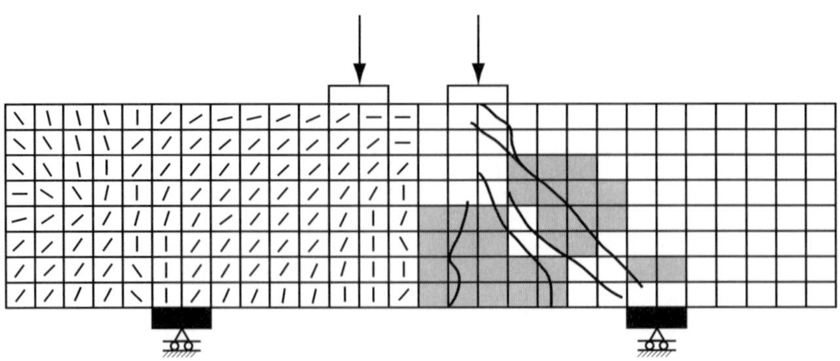

Fig. IX.61. Stress and crack patterns at a load of 65 kN (arch effects visible).

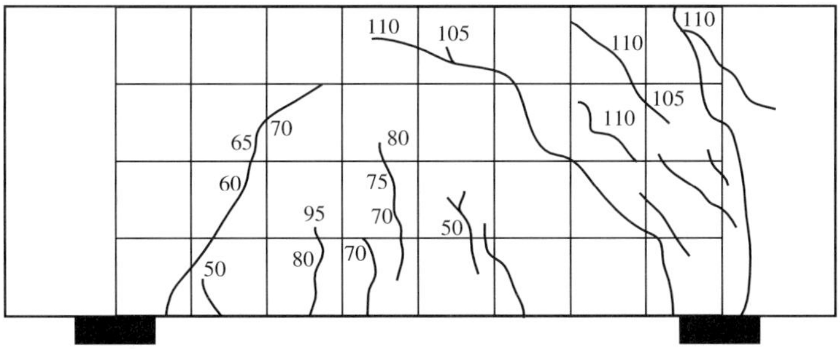

Fig. IX.62. An experimental test result on beams at failure (arch effects visible).

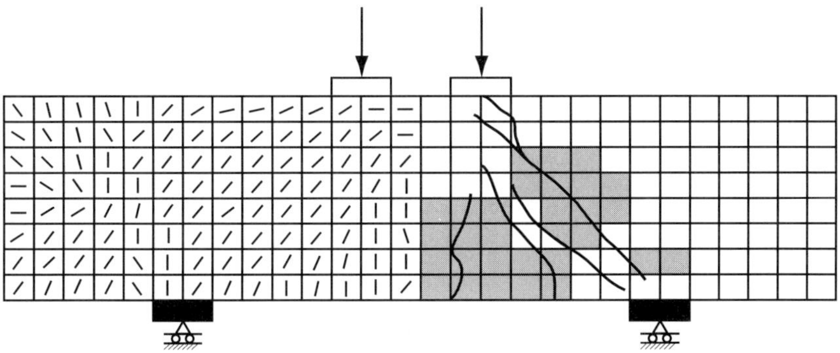

Fig. IX.63. Arch effects with cracks.

failure load was at 118 kN, as shown on the right-hand side (hatched and thick lines) on Fig. IX.63.

In Fig. IX.64 are shown the predicted load–deflection curves and they are compared with the experimentally obtained ones. For beams with larger span/depth ratio, the predicted deflections are smaller than the experimental. This is probably

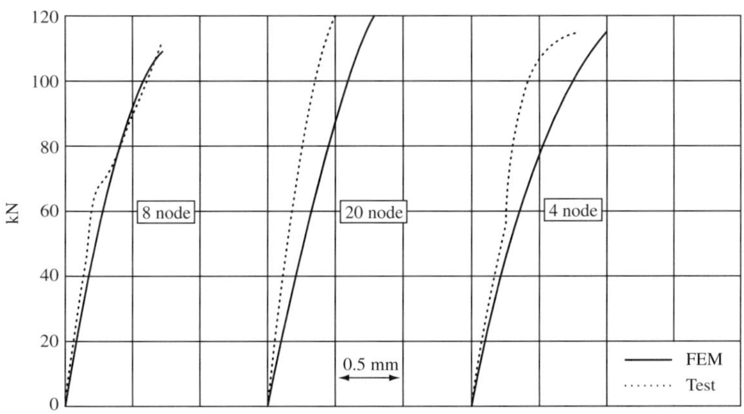

Fig. IX.64. Deflection results for 4 node, 8 node and 20 node—a comparative study with experimental results.

due to the arch effect which was apparent in Figs IX.61 and IX.63. In this mode, the main steel is acting as a tie of the tied arch. However, the fact that in the finite element analysis it is allowed that a perfect bond exists between steel and concrete, a stiffer structure with smaller deflections resulted. Another reason that may have contributed to some degree to the discrepancies is the crushing of the concrete at the interface of the bearing plates, since the bearing plates in the experiments were not embedded. There is a considerable deflection at the beginning of loading which may be attributed to compaction and crushing of the concrete under the plates.

The experimental tests on Beam 1 showed that most areas of various elements had principal compressive stresses at various stages of loading. Using the finite element, the same behaviour was observed at various stages of loading. The beam is shown with the cracked zones at relevant elements that developed zero or negative stiffness.

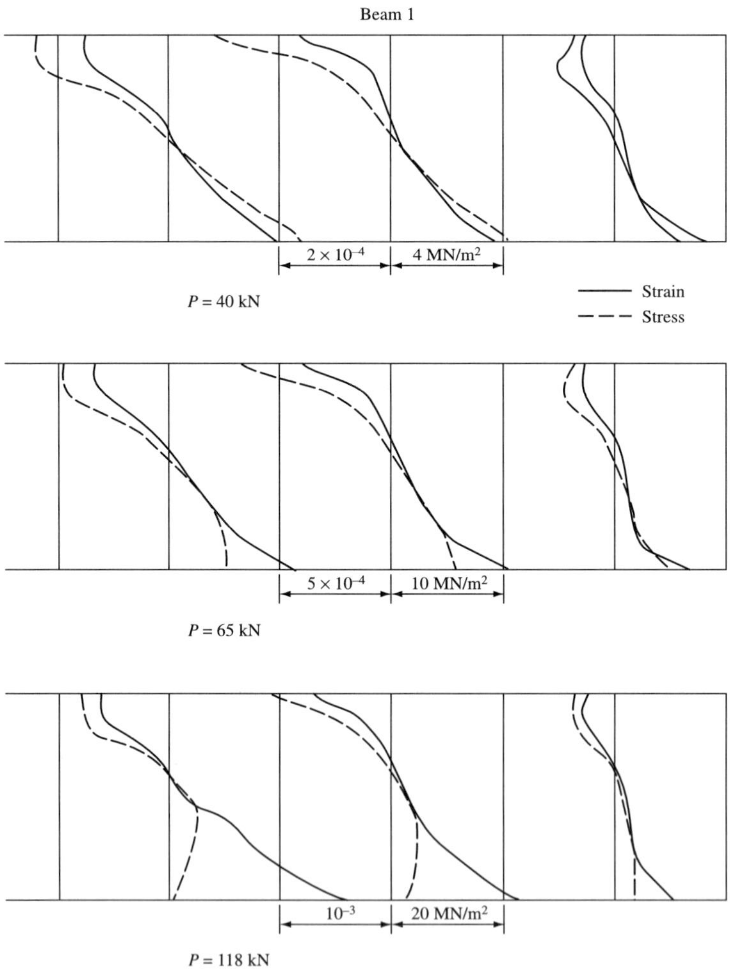

Fig. IX.65. Deflection results for stresses and strains at $P = 40\,kN$, $P = 65\,kN$ and $P = 118\,kN$.

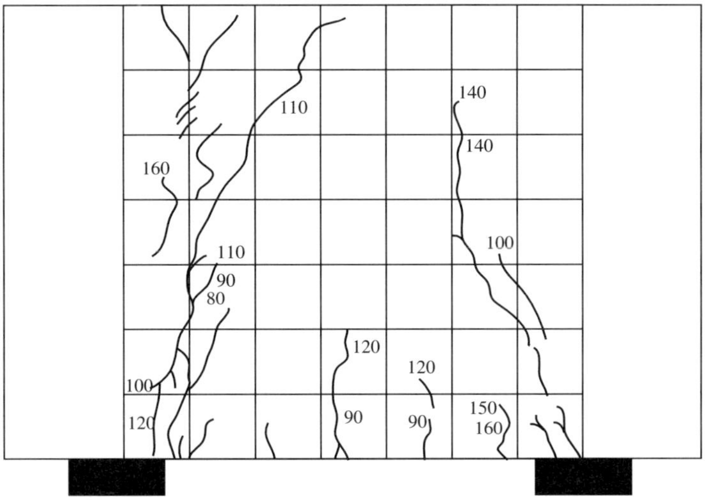

Fig. IX.66. Experiment test results for beam no. 2.

From the comparison with the experimentally observed cracking patterns, it is seen that predictions of cracks are still satisfactory. The principal stresses indicate an almost direct transfer of the compressive force from the loading point onto the support. The arching effect is apparent. The stress–strains are plotted for three cases, namely, loads of 40 kN, 65 kN and 118 kN.

(ii) *Beam 2.* The finite element mesh scheme is given in Fig. IX.59. The experimental test results are shown in Fig. IX.66. The procedure for testing this beam was identical to Beam 1, except that the loading interval in the computer analysis was different. Figure IX.67 gives the two-case study of Beam 2. Crack zones and stress–strain distribution at Sections 1–1, 2–2 and 3–3 at 40 kN and 180 kN (cracking failure mode) are shown in Figs IX.65 and IX.68. These can be compared with those of Beam 1 in Figs IX.61 and IX.63. The results are in good agreement with those predicted by the experiment carried out at Middlesex University by the author.

From these two beams, it can clearly be seen that principal directions do not change when cracks develop at failure load conditions. In both cases, the first cracks appear at mid-span and they are flexural cracks in a vertical direction. They then propagate as shear cracks at certain positions. At these stages, they are turned into shear cracks and they deviate from the previous vertical directions. In Beam 1, the flexural cracks appear. In Beam 2, comparatively few flexural cracks appear since the span/depth ratio is almost 1. After a few load increments, in Beam 1 and 2 diagonal cracks appear, which in Beam 2 are sudden out.

Conclusions. The finite element models given for these two beams compared favourably with the experimental results. The failure modes of these two beams were predominantly flexural-shear and diagonal splitting and they are apparent from

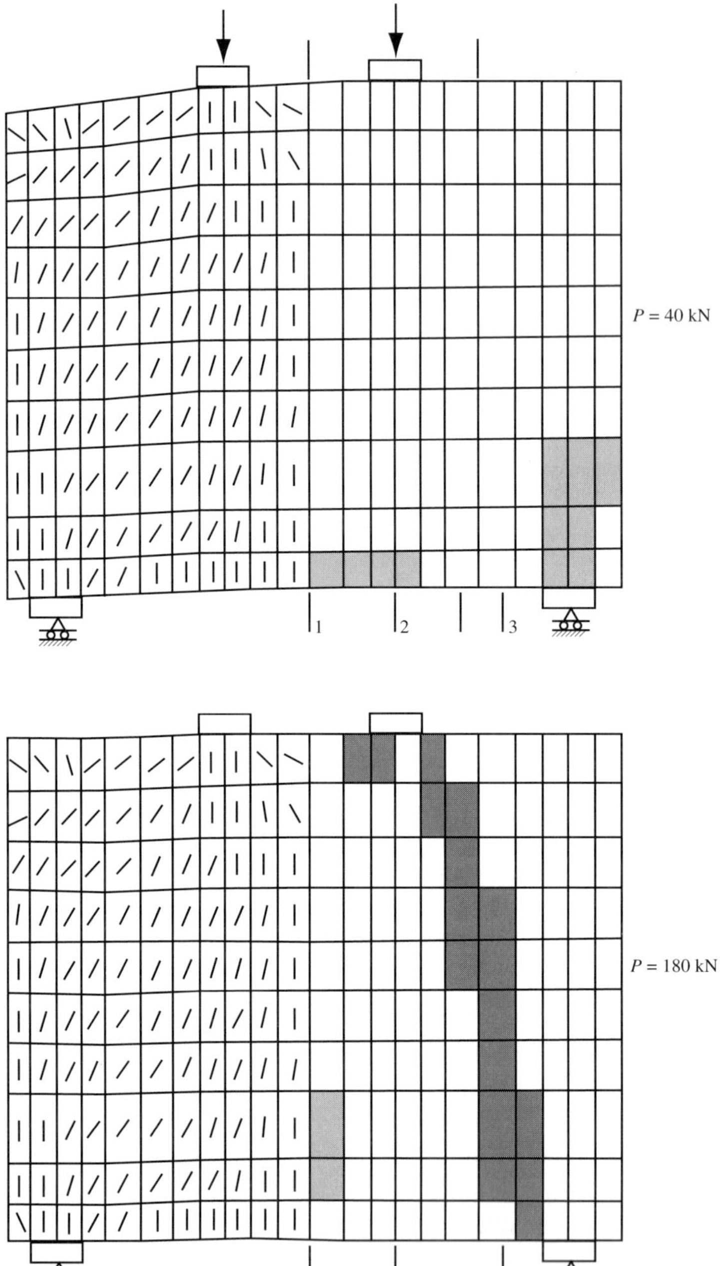

Fig. IX.67. Finite element stress and crack zones for the beam no. 2 for two loads.

the directions of principal stresses. Analytical predicted crack zones with directions and ultimate strength of the deep beams are in good agreement with the experimental data. It is interesting to note that Beam 1 with a large span/depth ratio is less ductile. This is due to the perfect bond assumed between the steel and concrete

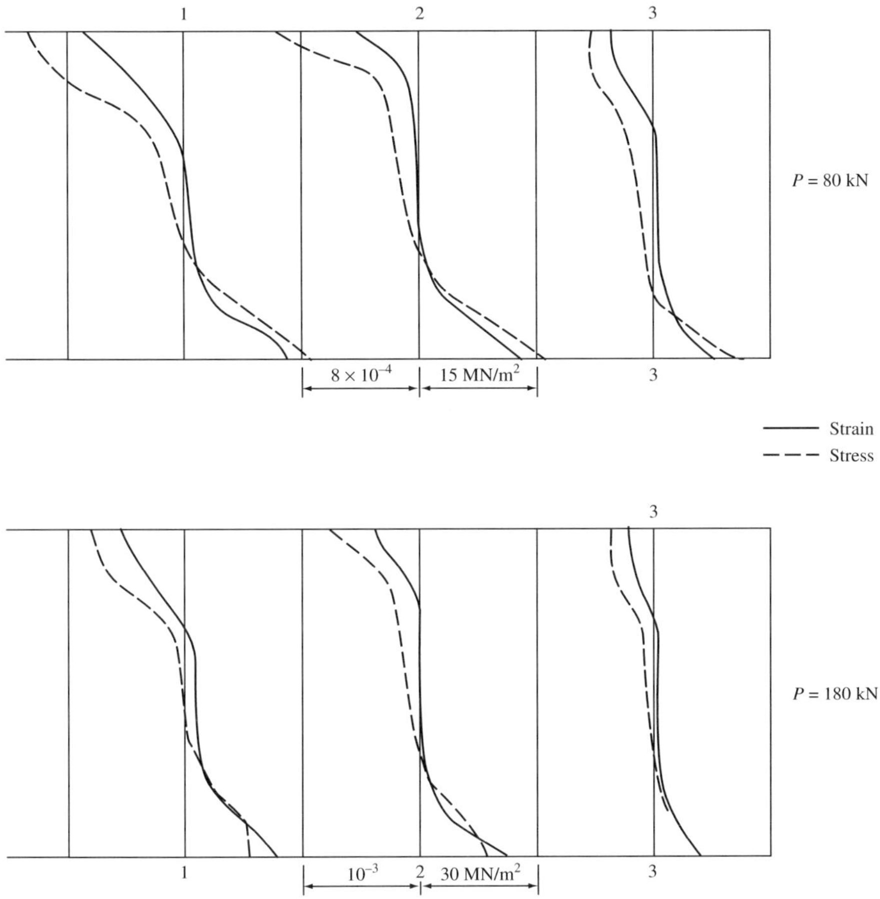

Fig. IX.68. Beam no. 2 results for stresses and strains.

causing the arch effect. Assuming no bond and 50% bond, this effect was reduced by at least 50–70%. Everywhere, the cracks and stress–strains were within 5–8% of those predicted in the experiments.

IX.6 Case Study No. 5 — strengthening of damaged concrete structural components

IX.6.1 Introduction

The application of the plate bonding technique is required to strengthen structurally damaged reinforced concrete beams. Epoxy-bonded steel plates are now [1292–1325] being accepted as an efficient and convenient way of improving the serviceability performance and/or ultimate strength of concrete structures. This section presents comprehensive test data on the effect of glued steel plate on the first crack load, cracking behaviour, deformations and ultimate/serviceability load of concrete members strengthened with such plates on the tension face. An analytical procedure is given for the deformation of this composite structure using an age-adjusted effective modulus method and independent composite material incorporating individual material properties. The numerical model is tested against the experimental tests carried out by various researchers [1292–1325]. The results obtained from the finite element analysis are plotted along with the experimental test results. Both types of result are in good agreement thus making the finite element analysis a powerful tool to obtain results without subjecting such structures to experimental tests, whether static or time dependent.

IX.6.2 The limit state and serviceability analyses

Prior to comprehensive analytical modelling in three dimensions, it is essential to check the concrete structure or its components using the limit state and serviceability methods. Table IX.7 gives a brief description of these methods.

Recent studies [1293–1325] have shown that the plate bonding technique results in considerable reduction of deflection and flexural crack widths and that, by suitable choice of epoxy adhesive, structural integrity and composite action between the concrete member and steel plate can be ensured at all stages of loading. However, the interface shear stresses can be critical at the ends of a plate, leading to plate separation and premature failure. The provision of additional glued anchor plates has been shown to overcome this problem of anchorage failure and enables the full theoretical flexural strength of the structural member to be achieved with considerable ductility.

IX.6.3 Experimental tests

IX.6.3.1 Introduction

Several experimental tests have been carried out [1293–1325] on externally bonded beams. Two experimental tests and their data have been selected to test the given numerical model based on three-dimensional finite element analysis. The analytical results are then checked with those obtained from the tests on these two beams. In the finite element work, the bonded plate material properties for steel and CFRP plates with glue are part of the input to programs ISOPAR and ANSYS, whichever is applicable.

IX.6.3.2 A comparative study of experimental data

Figure IX.69 shows the key diagram for the dimensional and load parameters. Figures IX.70 and IX.71 give the two programmes and tests which are under investigation.

Basic material properties

CFRP	composite fibre reinforced plastic (carbon)
GFRP	glass fibre reinforced plastic
AFRP	aramid fibre reinforced plastic
CON	concrete
STE	steel

Table IX.7. Limit state and serviceability methods

Based on BS 8110 E Code 2

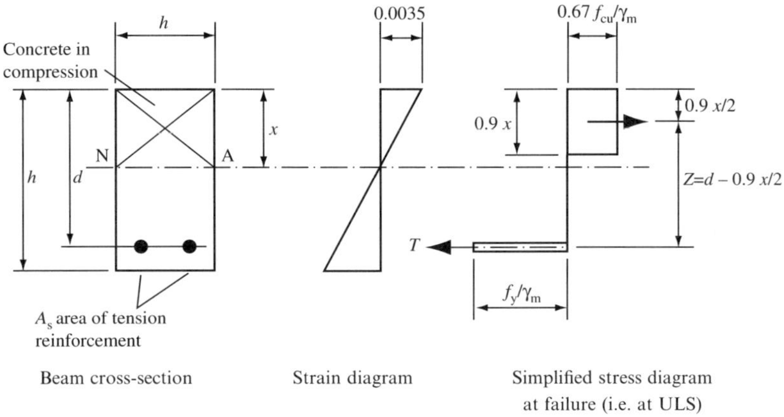

To avoid lateral buckling failure, the distance between lateral restraints should be the lesser of $60b_c$ or $250b_c^2/d$. The relevant BS 8110 Part 1 formulae for simply supported singly reinforced beams are as follows:

$$K = \frac{M}{bd^2 f_{cu}} \tag{a}$$

where for singly reinforced beams $K \leq K' = 0.156$ (b)

$$z = d[0.5 + \sqrt{(0.25 - K/0.9)}] \not> 0.95d \tag{c}$$

$$x = \frac{d - z}{0.45} \tag{d}$$

$$A_s = \frac{M}{0.87 f_y z} \tag{e}$$

where

A_s = area of tension reinforcement

b = compression width of section

d = effective depth of tension reinforcement

f_{cu} = characteristic strength of concrete

f_y = characteristic strength of steel

M = design ultimate resistance moment: equals design ultimate moment M_u

x = depth to neutral axis

z = lever arm

Minimum spacing of reinforcement

During concreting, the aggregate must be allowed to move between the bars in order to achieve adequate compaction. For this reason BS 8110 Part 1 recomments a minimum bar spacing of 5 mm greater than the maximum coarse aggregate size h_{agg}. That is,

Minimum distance between bars or groups of bars $= h_{agg} + 5\,\text{mm}$ (f)

Table IX.7. — contd.

Maximum spacing of reinforcement
When the limitation of crack widths to 0.3 mm is acceptable and the cover to reinforcement does not exceed 50 mm, the maximum bar spacing rules given in BS 8110 Part 1 may be adopted.

Deflection SLS
In all normal situations the deflection of beams will be satisfactory if the basic span to effective depth ratios is equal to those given in BS 8110 Part 1 Table 3.10.

Basic span to effective depth ratios for rectangular or flanged beams (BS 8110 Part 1 1985 Table 3.10)

Supporting conditions	Rectangular sections	Flanged beams with $b_w/b \leq 0.3$
Cantilever	7	5.6
Simply supported	20	16.0
Continuous	26	20.8

It should be understood that the span to depth ratios given above are based on the following:

(a) the span does not exceed 10 m
(b) the total deflection is not greater than span/250, and the deflection after application of finishes and erection of partitions (that is, due to imposed loads) is limited to 20 mm or span/500
(c) the loading pattern is uniformly distributed.

If the span exceeds 10 mm,

$$\text{Revised ratio} = \text{basic ratio} \times 10/\text{span}$$

If the total deflection must not exceed span/β,

$$\text{Revised ratio} = \text{basic ratio} \times 250/\beta$$

If deflection after the application of finishes must be less than the 20 mm limit,

$$\text{Revised ratio} = \text{basic ratio} \times \alpha/20$$

where α is the revised limit.

Deflection of beams is also influenced by the amount of tension steel present in the beam. This is expressed in terms of the design service stress in the steel and the ratio M/bd^2. To allow for this, the basic span to effective depth ratio is modified by a factor from BS 8110 Part 1 Table 3.11.

To determine the design service stress in the steel, the following expression, from the second footnote to Table 3.11, is used:

$$f_s = \frac{5}{8} f_y \frac{A_{s,req}}{A_{s,prov}} \times \frac{1}{\beta_b} \quad \text{(g)}$$

where

$A_{s,prov}$ = area of tension reinforcement provided at mid-span or at the support of a cantilever

$A_{s,req}$ = area of tension reinforcement required

f_s = estimated design service stress in the tension reinforcement

f_y = characteristic strength of steel

β_b = ratio to take account of any bending moment redistribution that has taken place

Table IX.7. — contd.

Modification factor for tension reinforcement (BS 8110 Part 1 1985 Table 3.11)

Service stress	M/bd^2								
	0.50	0.75	1.00	1.50	2.00	3.00	4.00	5.00	6.00
100	2.00	2.00	2.00	1.86	1.63	1.36	1.19	1.08	1.01
150	2.00	2.00	1.98	1.69	1.49	1.25	1.11	1.01	0.94
156 ($f_y = 250$)	2.00	2.00	1.96	1.66	1.47	1.24	1.10	1.00	0.94
200	2.00	1.95	1.76	1.51	1.35	1.14	1.02	0.94	0.88
250	1.90	1.70	1.55	1.34	1.20	1.04	0.94	0.87	0.82
288 ($f_y = 460$)	1.68	1.50	1.38	1.21	1.09	0.95	0.87	0.82	0.78
300	1.60	1.44	1.33	1.16	1.06	0.93	0.85	0.80	0.76

Note 1: The values given above derive from the equation

$$\text{Modification factor} = 0.55 + \frac{477 - f_s}{120(0.9 + M/bd^2)} \leq 2.0 \qquad \text{(h)}$$

where M is the design ultimate moment at the centre of the span or, for a cantilever, at the support.

Note 2: The design service stress in the tension reinforcement in a member may be estimated from the equation

$$f_s = \frac{5}{8} f_y \frac{A_{s,req}}{A_{s,prov}} \times \frac{1}{\beta_b} \qquad \text{(i)}$$

Note 3: For a continuous beam, if the percentage of redistribution is not known but the design ultimate moment at mid-span is obviously the same as or greater than the elastic ultimate moment, the stress f_s given above may be taken as $5/8 f_y$.

The ratio β_b does not apply to the simply supported beams dealt with in this manual. Hence the expression for simply supported beams becomes

$$f_s = \frac{5}{8} f_y \frac{A_{s,req}}{A_{s,prov}} \qquad \text{(j)}$$

Using the tables, the minimum effective depth d for a singly reinforced beam to satisfy deflection requirements may be written as follows:

$$\text{Minimum effective depth } d = \frac{\text{span}}{\text{Table (factor)} \times \text{Table (factor)}} \qquad \text{(k)}$$

Stresses in the anchorage zone

When relatively thick plates are used to strengthen reinforced concrete beams, high interface shear stresses may occur near the ends of the plates. Associated with these stresses are normal forces tending to peel off the plates. The peeling forces are relatively small, and begin to have a significant effect only after interface shear failure has been initiated. The development and magnitudes of these anchorage forces, together with methods of assessment and techniques of avoiding anchorage failure have not been included.

The beams designed with thick plates have high interface shear stresses and hence illustrate the problems that may occur in the anchorage zones.

An elastic analysis of the plated section may be carried out to determine the neutral axis depth and the second moment of area of the transformed section. In doing this, the modular ratio should be based on the long-term elastic modulus of the concrete.

Use the conventional elastic theory to determine the horizontal shear stress in beams to obtain an interface shear stress at the ends of the plate:

$$\tau = \frac{\nu A \bar{y}}{Ib} \qquad \text{(l)}$$

Table IX.7. — contd.

where

ν is the maximum shear due to ultimate beam loading
A is the transformed area of the plate
$\bar{y}$ is the distance from the neutral axis to the centroid of the plate
I is the transformed second moment of area of the section
b is the width of plate

To avoid diagonal compression failure of the concrete when shear reinforcement is present, ν must never exceed the lesser of $0.8\sqrt{f_{cu}}$ or $5\,N/mm^2$. If the calculated value of ν did exceed these limits then the size of beam would need to be increased.

The next step is to determine the form and area of the shear reinforcement needed by comparing the calculated value of ν with the guidance given in BS 8110 Part 1 Table 3.8, and the design shear stress capacity of concrete ν_c given in BS 8110 Table 3.9.

Form and area of shear reinforcement in beams (BS 8110 Part 1 1985 Table 3.8)

Value of ν (N/mm^2)	Form of shear reinforcement to be provided	Area of shear reinforcement to be provided
Less than $0.5\nu_c$ throughout the beam	See Note 1	
$0.5\nu_c < \nu < (\nu_c + 0.4)$	Minimum links for whole length of beam	$A_{sv} \geq 0.4 b_v s_v / 0.87 f_{yv}$ (see Note 2)
$(\nu_c + 0.4) \leq \nu < 0.8\sqrt{f_{cu}}$ or $5\,N/mm^2$	Links or links combined with bent-up bars. Not more than 50% of the shear resistance provided by the steel may be in the form of bent-up bars (see Note 3)	Where links only provided: $A_{sv} \geq b_v s_v (\nu - \nu_c)/0.87 f_{yv}$ Where links and bent-up bars provided: see clause 3.4.5.6 of BS 8110

Note 1: While minimum links should be provided in all beams of structural importance, it will be satisfactory to omit them in members of minor structural importance such as lintels or where the maximum design shear stress is less than $0.5\nu_c$.
Note 2: Minimum links provide a design shear resistance of $0.4\,N/mm^2$.
Note 3: See clause 3.4.5.5 of BS 8110 for guidance on spacing of links and bent-up bars.

Values of design concrete shear stress ν_c (N/mm^2) (BS 8110 Part 1 1985 Table 3.9)

$100A_s/b_v d$	Effective depth (mm)							
	125	150	175	200	225	250	300	≥400
≤0.15	0.45	0.43	0.41	0.40	0.39	0.38	0.36	0.34
0.25	0.53	0.51	0.49	0.47	0.46	0.45	0.43	0.40
0.50	0.67	0.64	0.62	0.60	0.58	0.56	0.54	0.50
0.75	0.77	0.73	0.71	0.68	0.66	0.65	0.62	0.57
1.00	0.84	0.81	0.78	0.75	0.73	0.71	0.68	0.63
1.50	0.97	0.92	0.89	0.86	0.83	0.81	0.78	0.72
2.00	1.06	1.02	0.98	0.95	0.92	0.89	0.86	0.80
≥300	1.22	1.16	1.12	1.08	1.05	1.02	0.98	0.91

Note 1: Allowance has been made in these figures for a γ_m or 1.25.
Note 2: The values in the table are derived from the expression

$$0.79[100A_s/(b_v d)]^{1/3}(400/d)^{1/4}/\gamma_m \qquad \text{(m)}$$

where $100A_s/b_v d$ should not be taken as greater than 3, and $400/d$ should not be taken as less than 1. For characteristic concrete strengths greater than $25\,N/mm^2$, the values in the table may be multiplied by $(f_{cu}/25)^{1/3}$. The value of f_{cu} should not be taken as greater than 40.

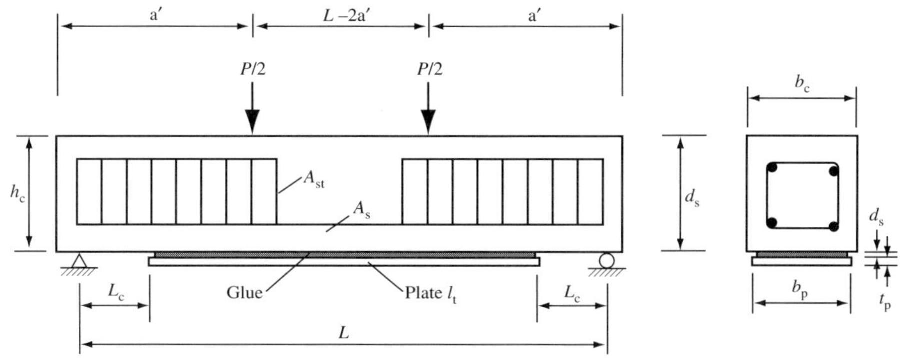

Fig. IX.69. A glue laminated reinforced concrete beam.

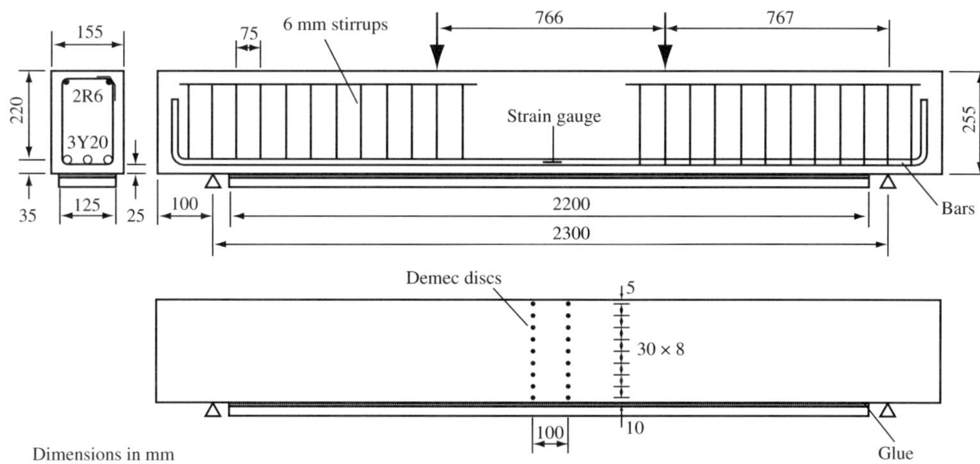

Fig. IX.70. Swamy et al. bonded steel plate beams.

Data

Swamy et al. beam (Fig. IX.70) — bonded steel plate (STE)

Basic dimensions:

$L = 2300\,\text{mm}$ t_p = steel plate thickness = 1.5–6 mm with No. layers 1 to 2

$H = 255\,\text{mm}$ t_g = glue thickness = 1.5–3 mm

$B = 125\,\text{mm}$ A_p = plate area = 187–750 mm^2

$S_p = 766\,\text{mm}$ l_t = bond length = 2200 mm

Material strength:

$$f_{cu} = 63.9\text{–}72\,\text{MN/mm}^2$$

Loads: $P = 35\text{–}165\,\text{kN}$ @ $S_p = 766\,\text{mm}$

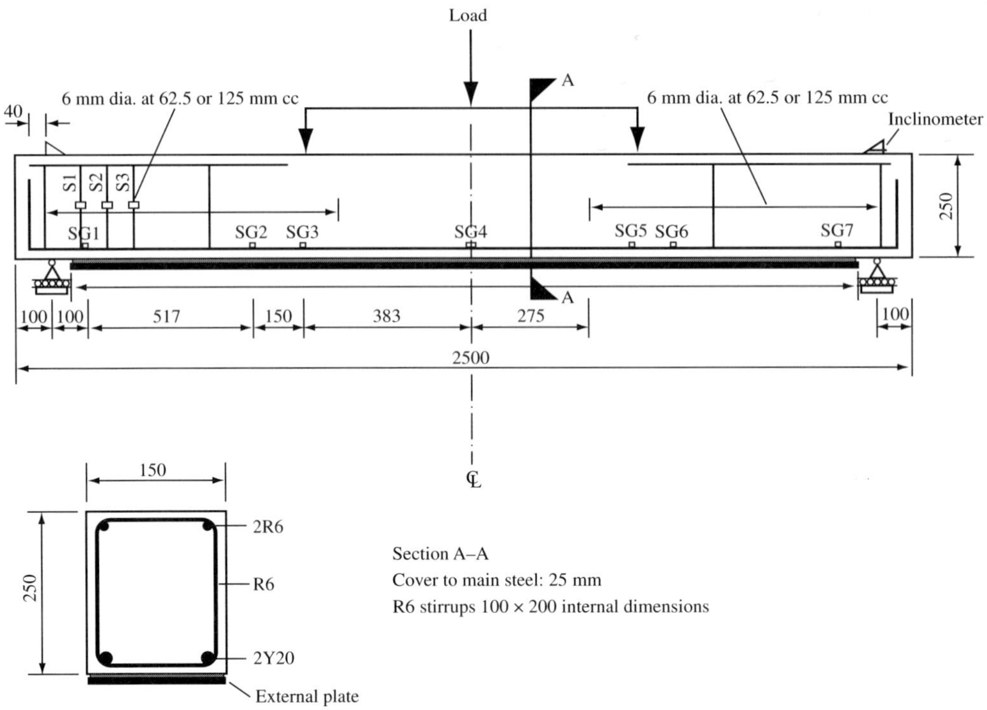

Fig. IX.71. Hong et al. beams with external plates with FRP.

Hong He et al. beam (Fig. IX.71) — CFRP plate
Basic dimensions:

$$L = 2500\,\text{mm} \qquad t_p = \text{plate thickness} = 1.00\text{–}1.5\,\text{mm}$$

$$H = 250\,\text{mm} \qquad t_g = \text{glue thickness} = 1.5\text{–}3\,\text{mm}$$

$$B = 150\,\text{mm} \qquad A_p = \text{plate area} = 187\text{–}750\,\text{mm}^2$$

$$S_p = 658\,\text{mm} \qquad l_t = \text{bond length} = 2200\,\text{mm}$$

Material strength:

$$f_{cu} = 63.9\text{–}72\,\text{MN/m}^2 \qquad \nu_c = 0.16 \qquad \sigma_{tc} = 5.66\,\text{N/mm}^2$$

$$E_{CFRP} = 186\text{–}900\,\text{GN/m}^2 \qquad \sigma_t = 1.450\text{–}3\,\text{MN/m}^2$$

$$f_{yt} = 425\,\text{N/mm}^2 \qquad E_{AFRP} = 79\,\text{GN/m}^2 \qquad \sigma_t = 610\,\text{MN/m}^2$$

$$E_{GFRP} = 44\,\text{GN/m}^2 \qquad \sigma_t = 370\,\text{MN/m}^2$$

Loads: $P = 50\text{–}116.5\,\text{kN}$ @ $S_p = 658\,\text{mm}$

IX.6.3.3. Finite element analysis of externally bonded RC beams

Initially, ISOPAR main finite element analysis was carried out. The analytical results were checked by the program ANSYS. In the three-dimensional analysis, two material options were utilised to model cracking of concrete, namely Drucker-Prager failure criteria and bilinear kinematic hardening, adopted on many occasions by other researchers using ANSYS. For the ISOPAR program the concrete material properties are the same as for ANSYS, as advocated in the text and in Section VIII.14.3.2. The FE model in both cases and for both types of beam had to be fine tuned by adjusting material properties for these two individual beams. The steel plate and the CFRP plate materials were first treated as bilinear kinematic hardening by using Von Mises yield criterion with the associated flow rule.

The multilinear criterion was also examined with these two experimental results and gave closer correlation than the bilinear one. The bilinear hardening was eventually ruled out with two-dimensional analysis. The elements in both cases were 8-noded isoparametric elements. The cracking criteria which cover the three orthogonal directions at each integration point as advocated in crushing and plastic deformation were included. The cracking in both computer programs was a smeared band of cracks rather than discrete cracks. In the three-dimensional ANSYS program, tension elements were used to model the reinforcement while the reinforcement was assumed smeared in concrete in ISOPAR. Multilinear material criteria were used for the CFRP plated one. The finite element mesh adopted for Section VIII.14.2 is repeated here with plates attached. The glue between the concrete beam and the plate in both cases is repeated as the gap element whose properties are defined in Appendix IB.

In many cases, the results plotted are based on the following case studies:

(a) unplated experiment
(b) unplated finite element
(c) plated experiment
(d) plated finite element.

They are mutually compared on the basis of the experiment and also on the type of finite element computer program. In this case, both ISOPAR and ANSYS computer packages have been used, using the same finite element mesh scheme shown in Fig. IX.72. Since both programs have the flexibility to change parameters in no time, it was decided to alter certain dimensions in order to see the effects of such changes.

Swamy et al. beam with steel plate. In the finite element analysis, based on a nonlinear approach, the incremental loads established 20 kN with 12 increments. During the 13th increment, the beam started to lose stiffness and cracks started to develop. The first crack appeared at 70 kN, against 60 kN based on the experiments. Figures IX.73 and IX.74 give a summary of the load deflection curves based on both ISOPAR and ANSYS computer programs with various steel plates thickness and 1.5 mm glue. The glue thickness now is changed to 2 mm and the results are examined for their relation with deflection at the centre of the beam and they are plotted on the same Fig. IX.73 for 3 mm thick plate. The results obtained are in close agreement, thus qualifying the numerical methods adopted for this case. Figure IX.75 shows the

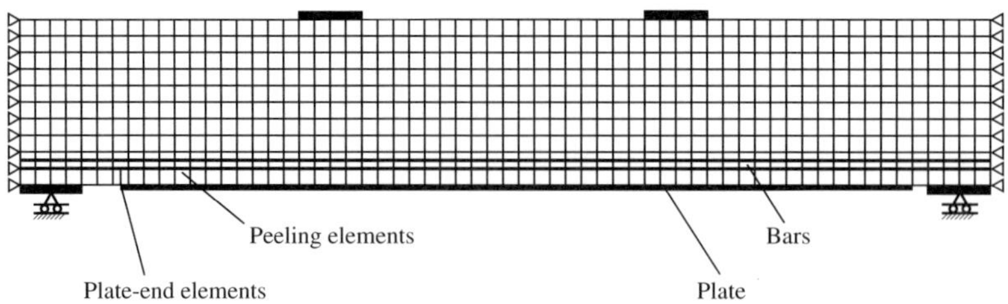

Fig. IX.72. Finite element mesh scheme of retrofit beams.

steel bar strains obtained by the numerical models and these are compared with those from the experiment. Figure IX.76 shows the results obtained by the experiments for concrete strains which were computed under incremented loads for both unplated and plated RC beams. In this case, the plate thicknesses of 1.5 mm with 1.5 mm glue as used in the experiment were adopted. A comparative study of these results is shown in Fig. IX.77.

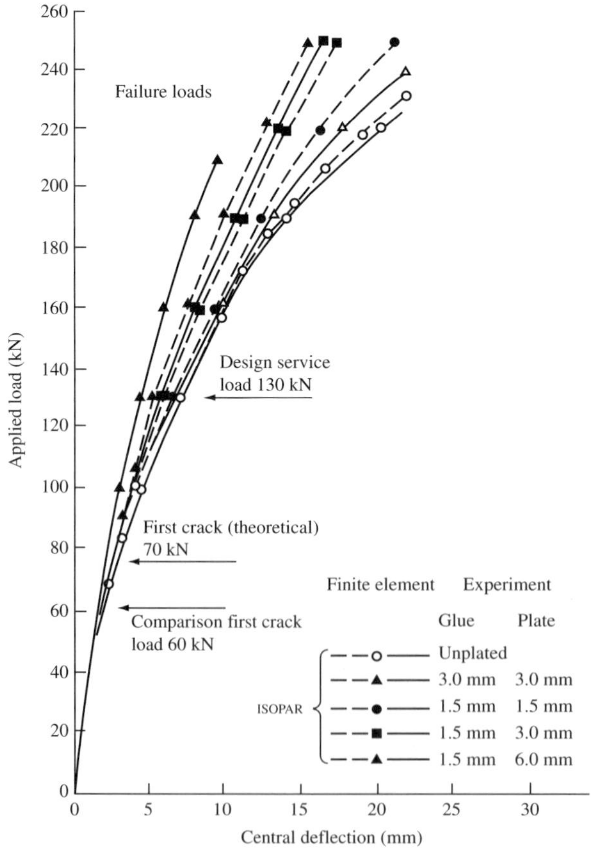

Fig. IX.73. A comparative study of load–deflection (experimental versus finite element results).

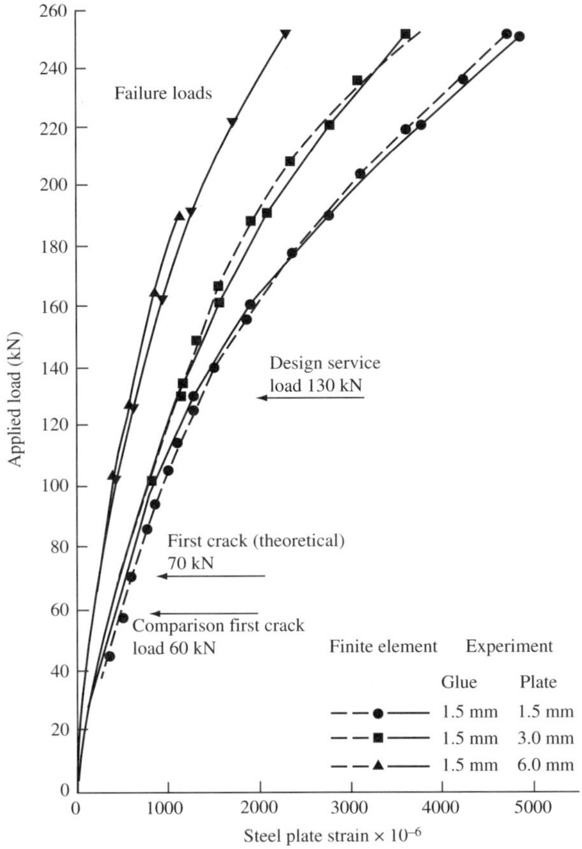

Fig. IX.74. A comparative study of load–strains (experimental versus finite element results).

Hong He et al. beam with CFRP plate (Fig. IX.71). The analysis in this case was identical to that for the externally bonded steel plate as described above. Similar results were produced for plate thicknesses of 1.0–1.5 mm, using 1.5–3 mm of glue (Fig. IX.78). Geometric and material properties were different. In order to save time and space within the text, they are not repeated. A summary of the comparative study of deflections and strains for various incremental loads is given in Fig. IX.77 and Fig. IX.78 respectively. The strains in the CFRP plate computed for various Load F developed at various positions are plotted in Fig. IX.79. The bond–slip relation is given in Fig. IX.80. The relationship of the F_{pu}/b (load/plate width) is given for various bond lengths l_a (Fig. IX.81). Cracks were developed at values beyond $l_a = 1000$ mm. The final post-mortem of this beam is shown in Figs IX.82, using both ISOPAR and ANSYS computer programs.

IX.6.4. Conclusion

Finite element analysis has a key role to play in developing design rules and guidelines for plate bonded solutions using advanced composite materials and steel.

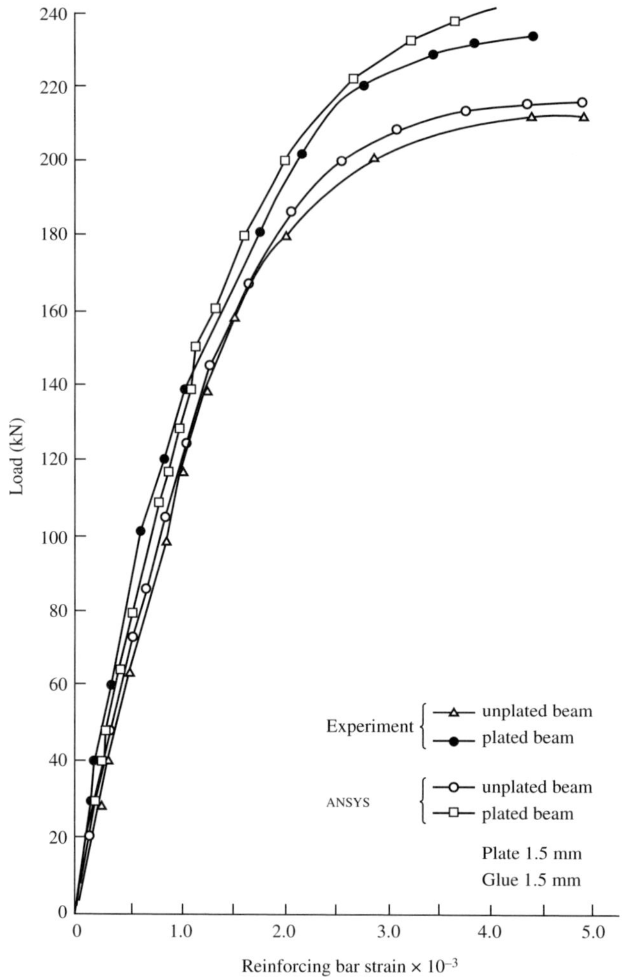

Fig. IX.75. Load–strain relationship for plated and unplated beams.

Accurate prediction of cracking of the concrete and yielding of steel rebars in a reinforced prestressed concrete beam using the finite element method is not a simple task, but has been achieved satisfactorily under both the ISOPAR and ANSYS programs. Both ISOPAR and ANSYS finite element codes using three-dimensional brick elements with a modified concrete material criterion have been used successfully.

Experimental work was undertaken on RC beams strengthened with steel and equivalent in stiffness to CFRP plates. The plates used are substantially thicker. Results are presented for beams strengthened with steel and CFRP plates bonded in a conventional manner. The following conclusions were derived.

(a) When using plates of equivalent axial stiffness of steel, there is no overall saving in volume of material, the weight of CFRP plates is less than one-third of the equivalent steel plate.

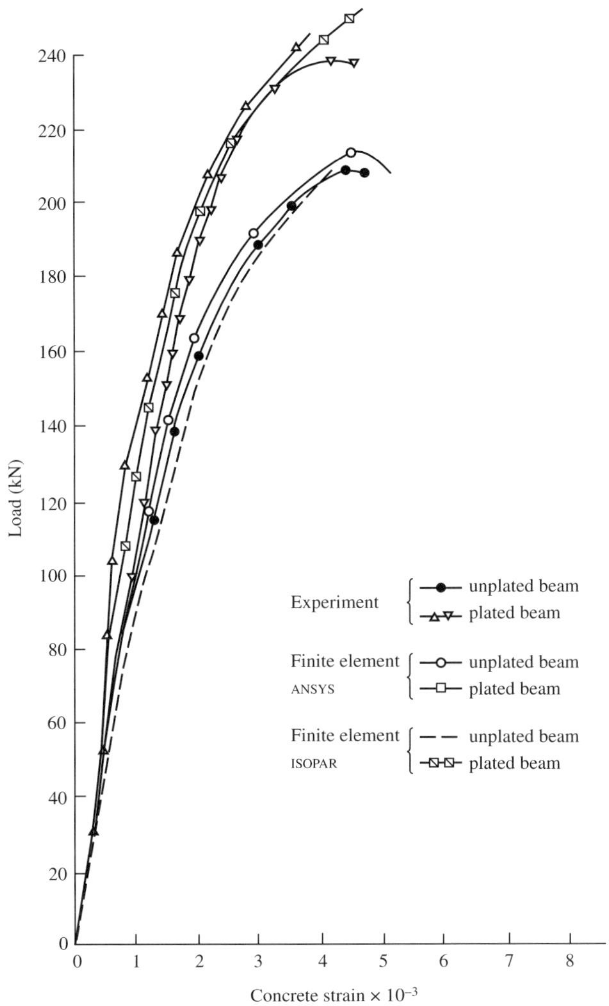

Fig. IX.76. Load–strain relations (a comparative study of experimental results from two sets of finite element program results).

(b) By using the external plate bonding technique, a significant strength enhancement can be achieved at the serviceability limit state of RC beams cracked prior to strengthening. A reduction of strains on the original reinforcement and a resulting reduction in deflections and crack widths can be obtained at this limit state.

(c) The peeling off of the external CFRP plate from the strengthened beam is the typical failure mode for such thick plates. Peeling off occurs at the end of the plates and causes the separation from the beam of the plate, together with the concrete cover, and results in the abrupt and premature failure of beams. This failure mode is mainly associated with the stiffness and thickness of the plate used. The steel plate performs better.

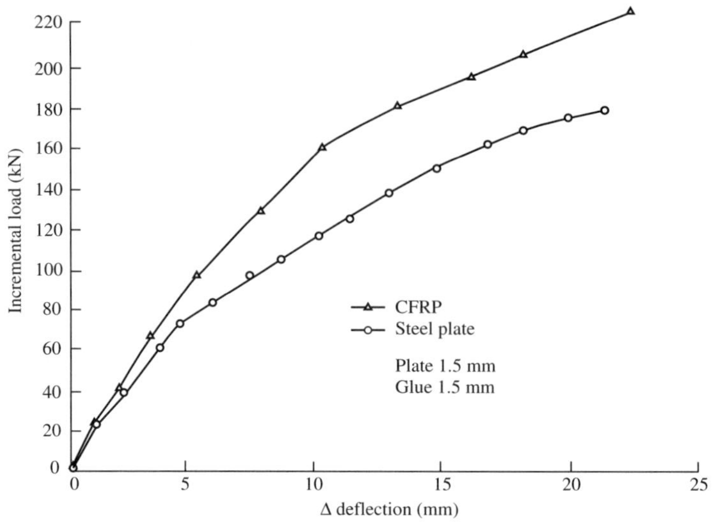

Fig. IX.77. Load–deflection for CFRP and steel-plated beams—a comparative study.

(d) The peeling off type of failure is associated with high shear stress concentration at the end of the plate. This stress concentration causes the yielding of shear links in this critical region.
(e) If serviceability is the main criterion for strengthening, then the high stress and strain potential of the CFRP will not be utilised unless the material is somehow prestressed.

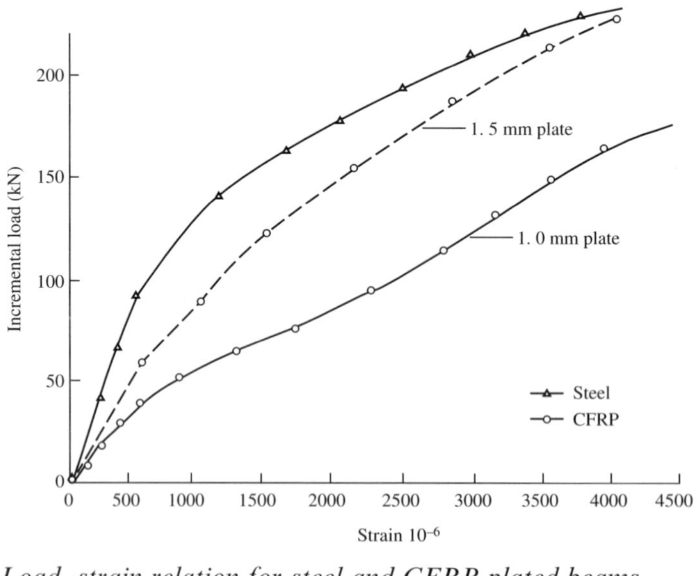

Fig. IX.78. Load–strain relation for steel and CFRP plated beams.

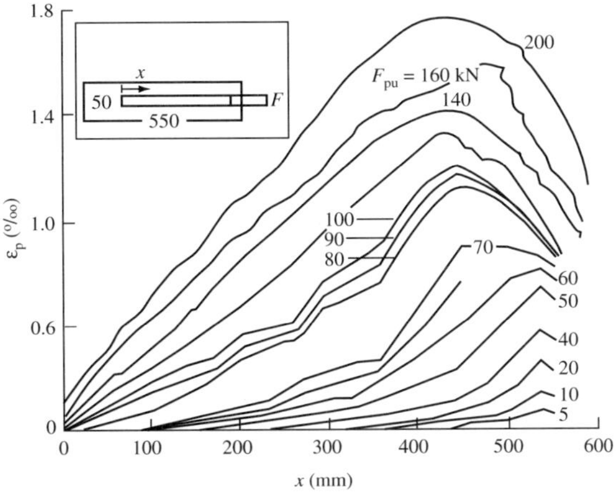

Fig. IX.79. Load–strain developed for bond-slip of the CFRP plated beam.

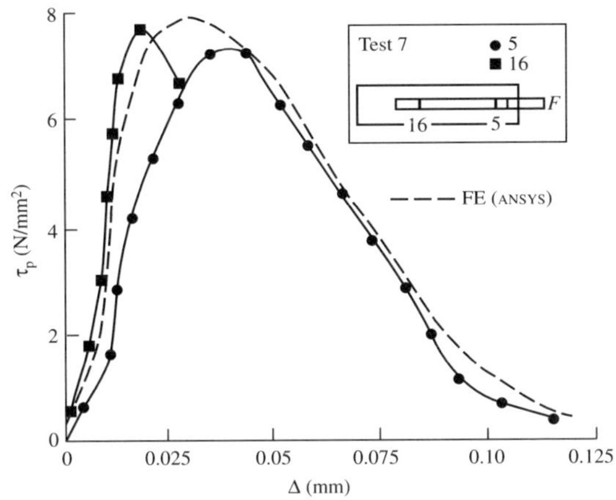

Fig. IX.80. Bond–slip relations—experimental versus finite element isopar results.

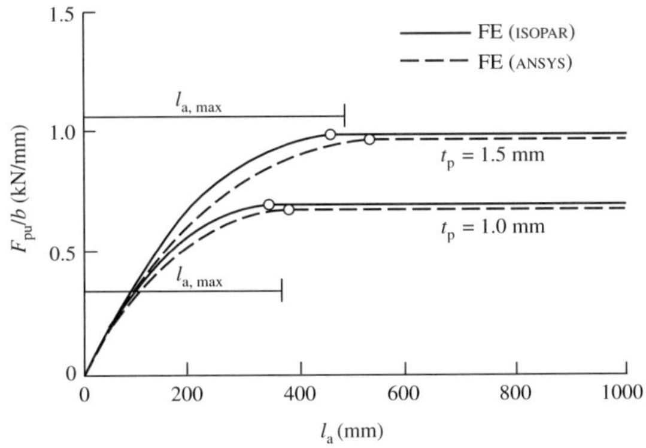

Fig. IX.81. F_{pu}/b versus l_a for various values of thickness—a comparative study of ISOPAR *and* ANSYS *finite element analyses.*

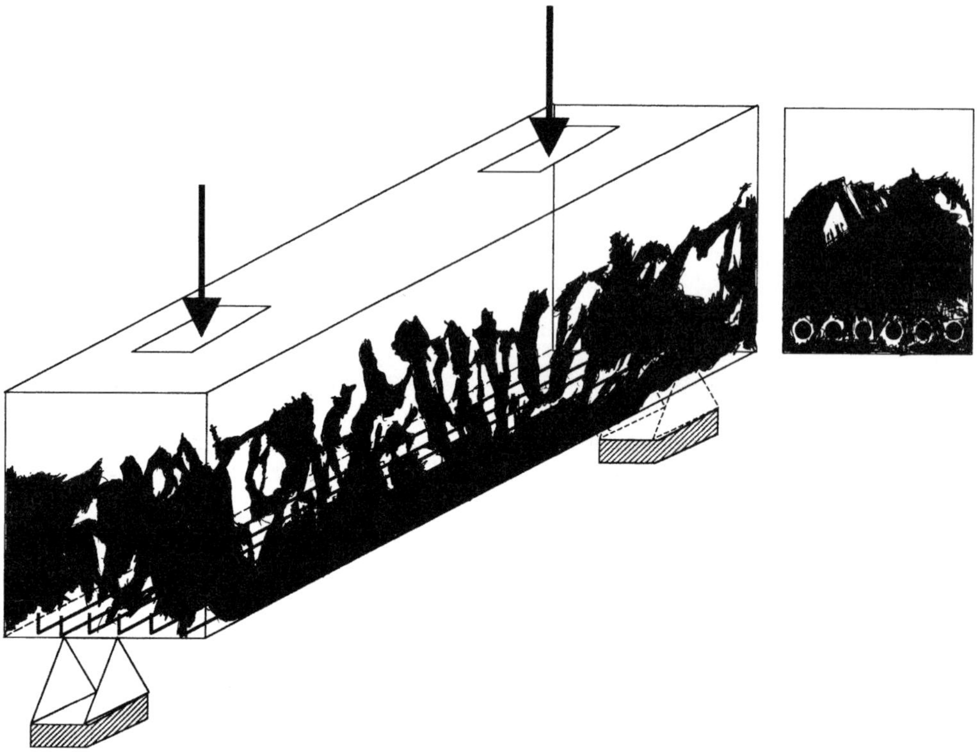

Fig. IX.82. Final post-mortem of the beam.

(f) Reinforced concrete beams, structurally damaged by loading up to 70% of their ultimate flexural strength, were strengthened by bonding steel plates under two conditions:

(i) while under no load, and

(ii) while under load;

the following conclusions are drawn from the tests of these two types of strengthened beams.

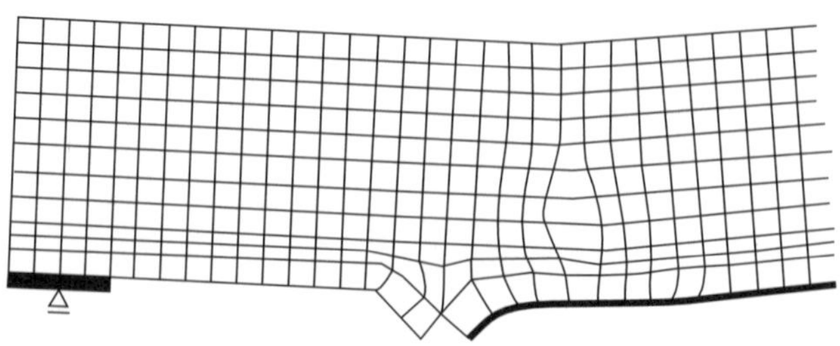

Fig. IX.83. Final post-mortem of the plated beams.

Table IX.8. Strength reduction

Based on ACI strength reduction method

The load corresponding to flexural mode of failure P_{flex} is given by

$$P_{flex} = 2M_u/a \quad \text{(a)}$$

$$M_u = A_s f_{ys}[d_s - (\bar{a}/2)] + t_p b_p f_{yp}[h_p - (\bar{a}/2)] \quad \text{(b)}$$

$$\bar{a} = \frac{A_s f_{ys} + b_p t_p f_{yp}}{0.85 f'_c b_c} \quad \text{(c)}$$

$$h_p = h_c + d_a + (t_p/2) \quad \text{(d)}$$

where M_u is the ultimate moment capacity, A_s is the area of tension steel, d_s, h_p are the distances to the centroid of tension steel and plate from the top fibre; t_p, b_p are thickness and width of the plate; d_a is glue thickness; b_c, h_c are width and depth of the concrete section; f'_c is concrete compressive strength; and f_{ys}, f_{yp} are the yield strengths of internal steel and plate respectively.

In order to check the shear capacity of the plated RC beams, the following quantities were computed:

$$V_c = \frac{1}{6}(\sqrt{f'_c} + 100\rho_w)b_c d_s \quad \text{(e)}$$

$$V_s = A_{st} f_{yst} d_s/s \quad \text{(f)}$$

$$V_u = 0.5P_{expt} \quad \text{(g)}$$

where V_c, V_s are shear capacities of the plain concrete and stirrups; V_u is the experimentally measured ultimate shear capacity of the section; A_{st} is the area of the stirrups; f_{yst} is the yield strength of the stirrups; s is the stirrup spacing; and $\rho_w = A_s/b_c d_s$.

Notation

a	Distance from support to loaded point
A_s	Area of tension steel
A_{st}	Area of stirrups
b_a	Width of adhesive
b_c	Width of concrete section
b_p	Width of plate
d_a	Adhesive thickness
d_s	Distance to centroid of tension steel from top
E_a	Elastic modulus of adhesive
E_p	Elastic modulus of plate
f'_c	Concrete compressive strength
f_{yp}	Yield strength of plate
f_{ys}	Yield strength of internal steel
f_{yst}	Yield strength of stirrups
G_a	Shear modulus of adhesive
h	Distance from top of beam to neutral axis of equivalent transformed steel section
h'_c	Depth of concrete section
h_p	Distance to centroid of plate from top fibre
I_p	Second moment of area of equivalent transformed steel section about neutral axis for a cracked section
K_n	Normal stiffness of interface layer
K_s	Shear stiffness of interface layer
L_a	M_0/V_0
M_u	Ultimate moment capacity
P_{expt}	Failure load
P_{flex}	Load corresponding to flexural mode of failure
s	Stirrup spacing
t_p	Thickness of plate
V_c	Shear capacity of plain concrete
V_s	Shear capacity of stirrups
V_u	Ultimate shear capacity of section
	$A_s/(b_c d_s)$

Table IX.9. Bonded plated beam shear

Shear strength of the repaired beam

$$V_u = V_c + V_s + V_p \qquad \text{(a)}$$

where

V_c = concrete shear resistance

V_p = shear resistance of the plate

V_s = shear resistance of the internal shear reinforcement

$$V_p \text{ (strips employed)} = 2F_p = 2[\tau_{ult}(dh_{j/2})] \qquad \text{(b)}$$

where

h_j is the depth of the jacket $\not> 120$ mm

d is the distance from extreme compression fibre to the centroid of steel reinforcement

F_p is the force in the plate

$$V_p \text{ (strips employed)} = \frac{2F_p d}{s_p} = \frac{2[\tau_{ave}(t_s h_s/2)]d}{s_p} \qquad \text{(c)}$$

where

s_p = spacing of strips $\not> 60$ mm

τ_{ult} = interface shear resistance $\approx 3.5\,MN/m^2$

τ_{ave} = average shear stress $\approx 1.2\,MN/m^2$

$$V_p \text{ (wings employed)} = 2F_p = 2[\tau_{ave}(dh_w/2)] \qquad \text{(d)}$$

where

h_w = depth of the wing $\not> 120$ mm

$$l_t = \text{the maximum bonded length (mm)} = \frac{\sqrt{E_p t_p}}{4f_{ctm}} \qquad \text{(e)}$$

where

E_p = Young's modulus

b_p = plate width

t_p = plate thickness

f_{ctm} = tensile strength of concrete

$$F = T_{max} = 0.4k_b k_c b_p \sqrt{E_p t_p f} \qquad \text{(f)}$$

G_F = fracture energy = $C_F f_{ctm}$ ranges from 0 to 2.0 (0.59 for 30 N/mm)

k_b, k_c = factors ≈ 1.0 in general

c_F = 0 to 0.70 (mm)

$$s_u = \text{ultimate deformation} = \frac{18}{5}\frac{G_F}{\sigma_{tu}}$$

Table IX.9. — contd.

where

$$\sigma_t = 0.9\sigma_{tu}$$

σ_{tu} = ultimate tensile splitting strength

When the plate is stopped before the support, the plate and shear capacity is to be evaluated.

$$\tau_{cu} = V_{cu} = \text{plate end shear capacity} = \tau_{cu} bd \tag{h}$$

where

$$\tau_{cu} = \text{nominal shear stress} = \xi \sqrt[3]{\frac{3d}{a_L}} \left[lt\sqrt{\frac{200}{d}} \right]^3 \sqrt{\rho_s} f_c \tag{i}$$

where $\xi = 0.18$ for mean stress, $\xi = 0.15$ for design stress

$$a_L = \text{fictitious shear span} = \sqrt[4]{\frac{(1-\sqrt{\rho_s})^2}{\rho_s} dL^3} \tag{j}$$

where

a = distance from end to the first load

$a > L + d$

$a_L < a$

$$a_c = 3.3\left[\frac{\rho(d/a)^2}{(1-\sqrt{\rho})^2}\right]^{1/3} a \xrightarrow{\text{modelling analogy}} L = 3.3\left[\frac{\rho_s(d_s/a_L)^2}{(1-\sqrt{\rho_s})^2}\right]^{1/3} a_L \tag{k}$$

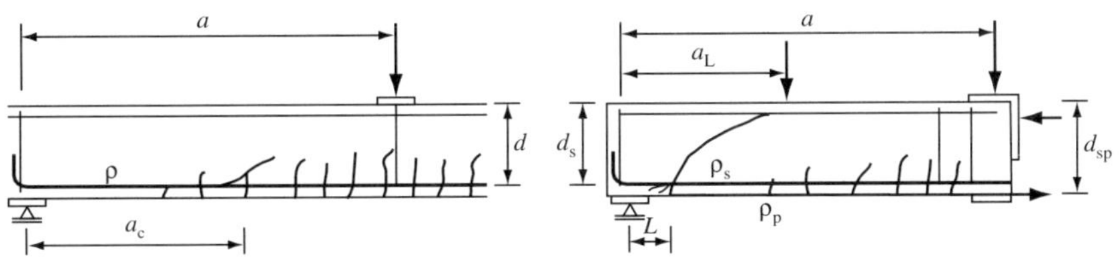

Anchorage length
Anchorage length (Betonkalender [1996]) based on Holzenkämpfer

$$F_{pu} = 0.35\left\{1 + \frac{\sigma_n}{f_{ctm}}\right\} b_p k_b k_c \sqrt{f_{ctm} E_p t_p} \tag{l}$$

$$l_{a,max} + \sqrt{\frac{E_p t_p}{4 f_{ctm}}} \quad \text{with } f_{ctm} \text{ of surface} \tag{m}$$

Plate–epoxy–concrete interface at the plate end
The approximate solutions for shear and normal stress at the plate end are assumed to be τ_{max} and σ^N_{max}. They are expressed as (based on ACI-SP-165, 1997)

$$\tau_{max} = \left(V_0 + \left\{\frac{K_s}{E_p b_p t_p}\right\}^{1/2} M^*\right) \frac{b_p t_p}{I b_g}(d_p - x) \tag{n}$$

$$\sigma^N_{max} = \tau_{max} t_p \left\{\frac{K_n}{4 E_p I_p}\right\}^{1/4} \tag{o}$$

Table IX.9. — contd.

in which

$$K_s = G_g \frac{b_g}{t_g} \quad \text{and} \quad K_n = E_g \frac{b_g}{t_g} \qquad \text{(p)}$$

$$V_0 = F/2$$

$$M^* = \text{bending moment at location } L + \frac{h + t_p}{2} \qquad \text{(q)}$$

Epoxy resin adhesives can ensure full composite action of structurally damaged RC beams strengthened by bonded steel plates, and result in increased stiffness and strength for the plated beams.

(g) Damaged beams, strengthened while under very high loads, show deformations and crack widths slightly higher than those of the continuous beam plated when in an undamaged condition. Thus, plates with severely damaged beams under load are structurally efficient and can restore a structural member to stiffness and strength conditions which are better than those of the original undamaged beam.

The yield strain and plastic properties of the plates should be chosen so as to enable full exploitation of the added plates.

Apart from the predictable increase in stiffness and strength, external steel plates are able to restrain the free widening of the tensile cracks and allow a high stress transfer between the cracks, resulting in an improved performance of concrete in the tension zone.

IX.7 Case Study No. 6 — concrete-filled tubular columns [364–386]

Chan [364] has carried out tests on concrete-filled tubular columns and compared their results with those from the known empirical formulae summarised in Table IX.10 [364–376]. For the concrete mix design a coarse aggregate of uncrushed irregular gravel of 38 mm maximum size was suggested. The characteristic strength obtained of the concrete was 61.47 N/mm^2 with an average value for E_c of 57.14 kN/mm^2. A tensile test specimen of the steel tube was taken by cutting the tube longitudinally into pieces of about 20 mm width. A universal testing machine was used and the extension of the coupon was measured using an extensometer over a gauge length of 50 mm. A complete stress–strain curve was obtained, from which the yield stress of 305 kN/mm^2 corresponded to the design yield strain of 2000×10^{-6}. A short length of tube about 250 mm long was tested in compression to failure. A 2000 kN Losenhausen (Fig. IX.84) compression machine was used. The longitudinal strain at the centre was measured using two electrical resistance gauges positioned opposite to each other. Strain readings were taken at suitable increments until the load reached 100 kN. The procedure was repeated three times to get the average

Table IX.10. Empirical column formulae

1. *ACI–NBC allowable load formulae* [374]

$$p = 0.25f_c\left(1 - 0.000\,025\frac{L^2}{r_c^2}\right)A_c + f_s A_s$$

where the allowable steel stress f_s is given by

$$f_s = \left[20\,000 - 70\frac{KL}{r_s}\right]\frac{\sigma_y}{33\,000} \qquad f_s \nleq \frac{145\,000\,000}{(\bar{K}L/r_s)^2}$$

where $\bar{K}$ = the length fixity factor for columns, and r_c and r_s are radii of gyration. Safety margins are 2.18–4.32.

2. *Knowles and Park formulae* [365, 366]
The ultimate load of concrete-filled tubular columns is given by the recommended equation (all values in imperial units):

$$P_u = 2qf_c'A_c(\sqrt{q^2+1} - q) + f_y A_s\left[1 - \frac{f_y(\bar{K}L/T_s)^2}{4\pi^2 E_s}\right]$$

when

$$\frac{\bar{K}S}{r_s} < \sqrt{\frac{2\pi^2 E_s}{f_y}}$$

$$P_u = 2qf_c'A_c(\sqrt{q^2+1} - q) + \frac{A_s\pi^2 E_s}{(\bar{k}L/r_s)^2}$$

when

$$\frac{\bar{K}S}{r_s} < \sqrt{\frac{2\pi^2 E_s}{f_y}}$$

where

P_u = ultimate load of composite column = $162.8D_c^{1.5}/\sqrt{f_c}\bar{K}L/r_c$

$f_{cu} = 0.85f_c$ (lb/in^2)

f_y = yield strength of steel

D_c = unit weight of concrete (lb/ft^3)

$\bar{K}$ = effective length factor

f_c = compressive strength of concrete (lb/in^2)

For SI units the following conversion is adopted:

$$1\,\text{lb/in}^2 = 6.845\,\text{kN/m}^2 \qquad 1\,\text{lb/ft}^3 = 0.27\,\text{kN/m}^3$$

Safety margins are 1.40–3.91.

3. *Furlong, Kloppel and Godr formulae* [371–374]
The strength interaction formula is (all values are in imperial units)

$$\left(\frac{P}{P_0}\right)^2 + \left(\frac{M}{M_0}\right)^2 \leq 1.0$$

Table IX.10. — contd.

in which for circular tubes

$$P_0 = A_s f_y + A_c f_c \sqrt{\frac{f_y}{0.0018E_s}}$$

$$M_0 = \frac{f_y}{6}(D_o^3 - D_i^3)$$

$$M_0 = \frac{f_y}{6} D_o^3\{1 - (1 - p)^{3/2}\}$$

and for square tubes

$$P_0 = A_g p f_y + (1 - p) f_c \sqrt{\frac{f_y}{0.0018E_s}}$$

$$M_0 = \frac{f_y}{4}(D_o^3 - D_i^3)$$

$$M_0 = \frac{f_y}{4} D_o^3\{1 - (1 - p)^{3/2)}\}$$

where

P = ultimate axial load when a finite moment also exists
P_0 = axial load capacity in the absence of moment
M = ultimate flexural load when a finite axial load also exists
M_0 = flexural capacity in the absence of axial load
D_o = outside diameter of steel tubes
D_i = inside diameter of steel tubes
p = ratio between steel area and gross area

Safety margins are 1.16–4.41.

elastic modulus of steel. The column lengths chosen were 0.25, 0.75, 1.3, 2.1, 2.3, 2.5, 2.7 and 2.9 metres. The load–strain curves were produced and the possible failure mode was established.

A three-dimensional finite element scheme devised by the author for tubular columns is shown in Figs IX.85a and IX.85b. A Von Mises failure criterion was chosen for the steel tube together with the bulk and shear moduli model for the concrete described in Chapters VI and VII. In order to assess the frictional effect between the concrete and the steel, two cases were examined, namely the application at nodes and loads caused by friction and the introduction of a special element known as the gap element, the details of which are given in Appendix I. Owing to low frictional forces, the results in both cases differed by between 4 and 6%. Integration order (3 × 3 × 3) was used for computing stresses, strains, plasticity of the steel and cracking of the concrete. After 100 kN of load, an incremental load of 10 kN was adopted. At the end of each iteration the results were examined.

Fig. IX.84. Top: arrangement of the apparatus in the Losenhausen compressive machine for strain and deflection measurements. Bottom: failure of 2.3 m lengths of the column [364].

Acceleration and convergence procedures discussed in Chapter VI were adopted. All long columns failed in flexure and the short columns failed by yielding of steel and cracking of concrete. Figures IX.86–IX.93 give a comparative study of analytical and experimental results. These results indicate that the parabolic

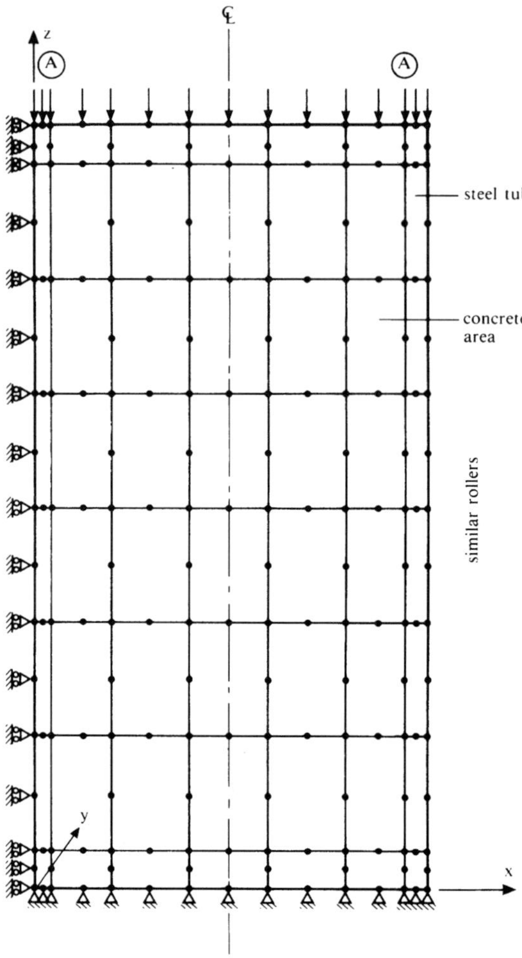

Fig. IX.85a. Finite element mesh scheme for a concrete-filled tubular elevation (20-noded solid isoparametric column element).

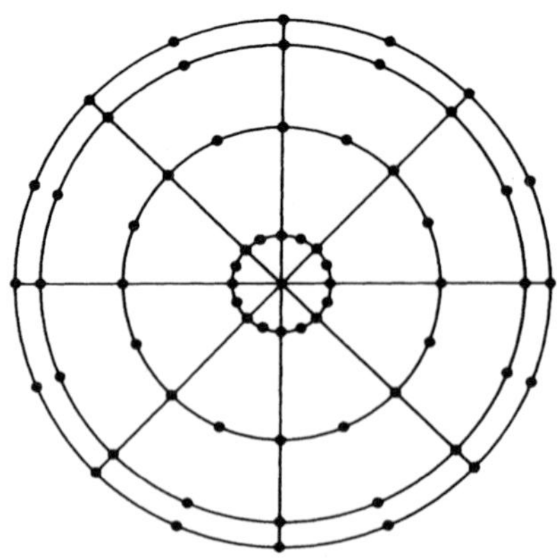

Fig. IX.85b. Finite element mesh scheme for a concrete tubular column plan.

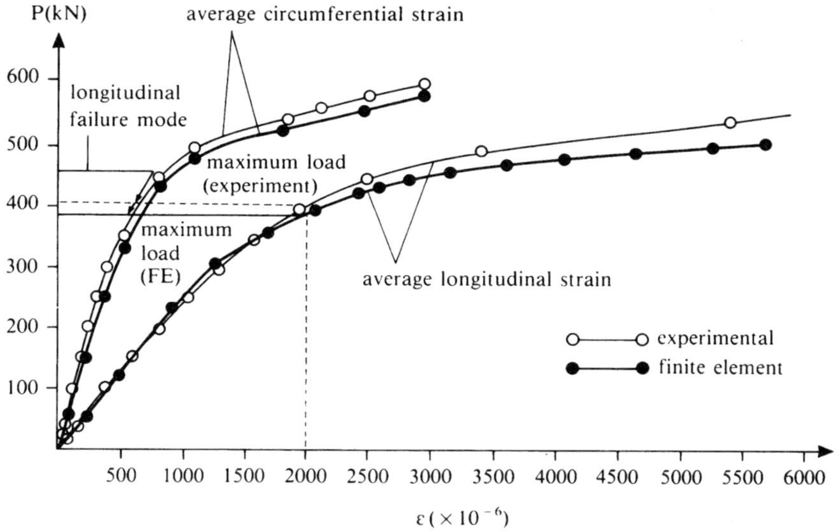

Fig. IX.86. Load–strain curves for specimens of 0.25 m length.

decrement hypothesis gives a good representation of the redistribution of longitudinal and circumferential stresses in the column in the nonlinear range. In the finite element analysis the boundary conditions (Figs IX.85a and IX.85b) chosen for the column were very close to those imposed in the experiment. Results from the finite element analysis are around 10–12% lower than those obtained from the experiment. The difference may be ascribed to the slight end fixity not considered in the analysis; and the exact boundary conditions at the bottom could

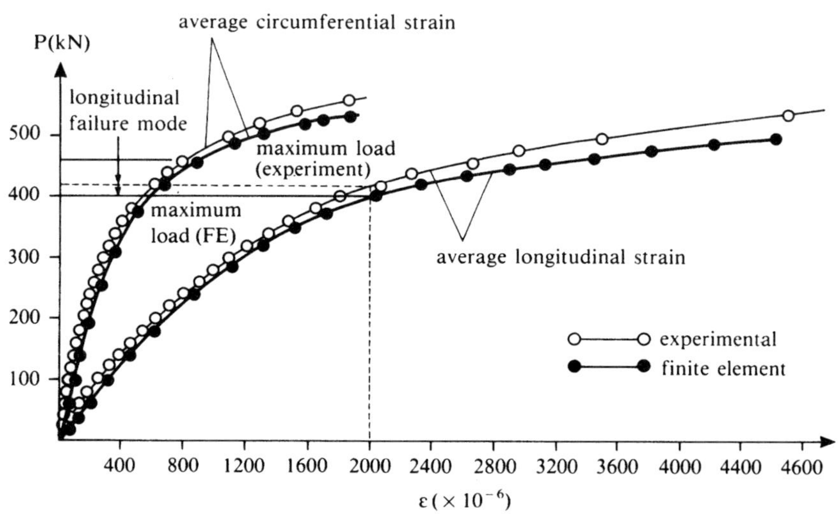

Fig. IX.87. Load–strain curves for specimens of 0–75 m length.

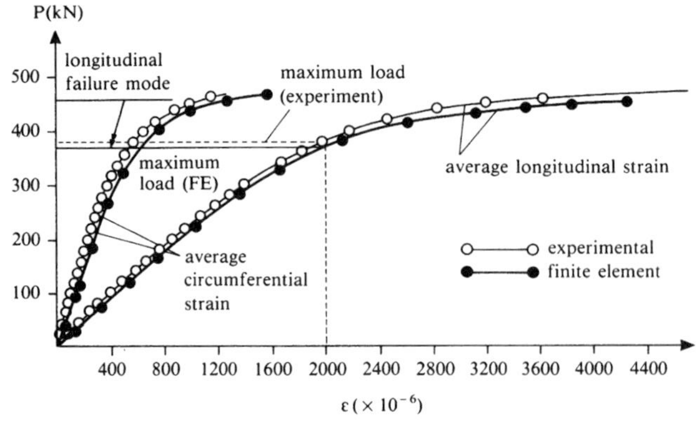

Fig. IX.88. Load–strain curves for specimens of 1.3 m length.

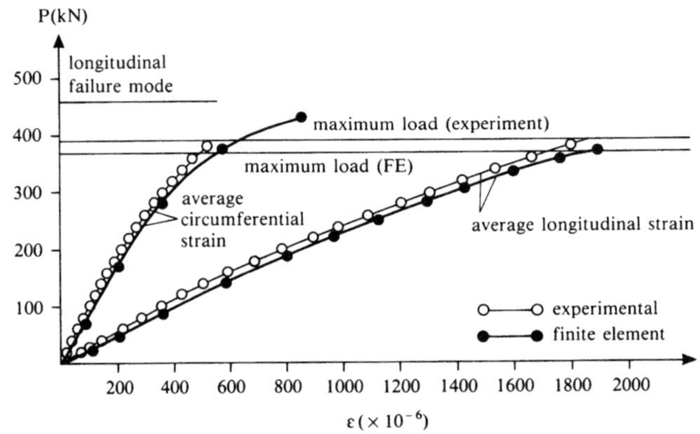

Fig. IX.89. Load–strain curves for specimens of 3.1 m length.

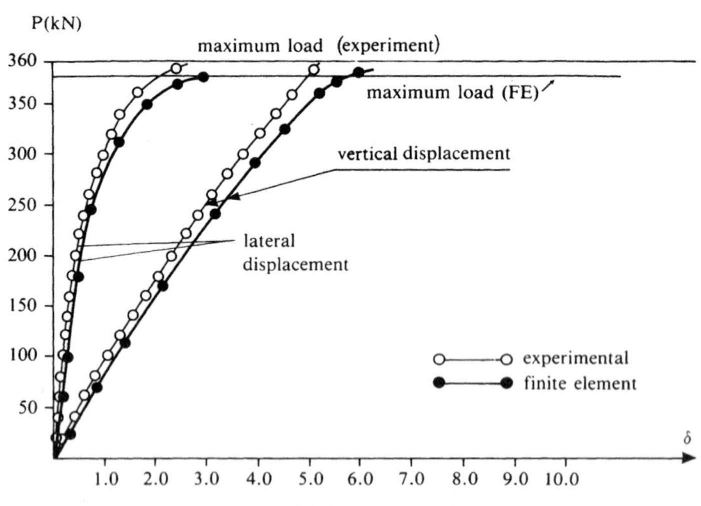

Fig. IX.90. Load versus displacement curves for specimens of 2.1 m length.

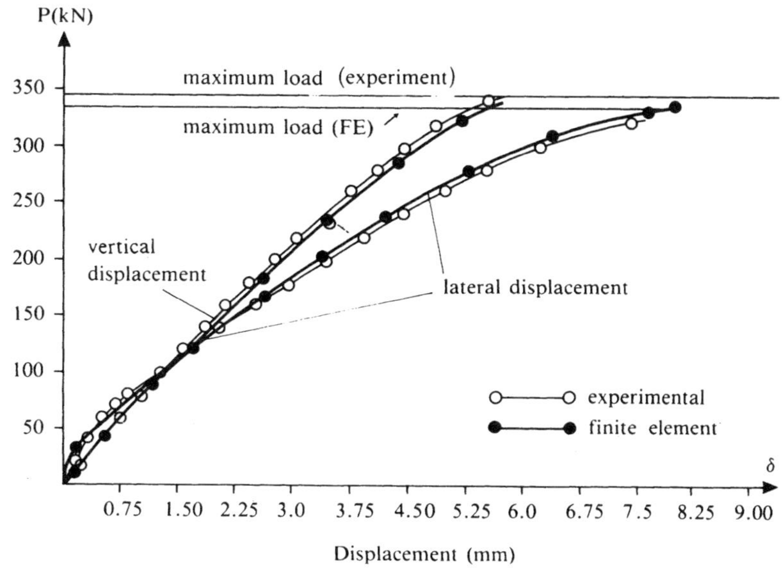

Fig. IX.91. Load versus displacement curves for specimens of 2.5 m length.

not be established from the report produced by Chan [364]. It is interesting to note from the experiment that no slip between the concrete and steel had occurred. Hence it was not necessary to include the bond stress–slip analysis discussed earlier. Using the Von Mises criterion for the steel tube, the results from the finite element analysis showed that during the elastic stage the deviation of strains from the average strain was generally in the range of 15%, corresponding to 20%

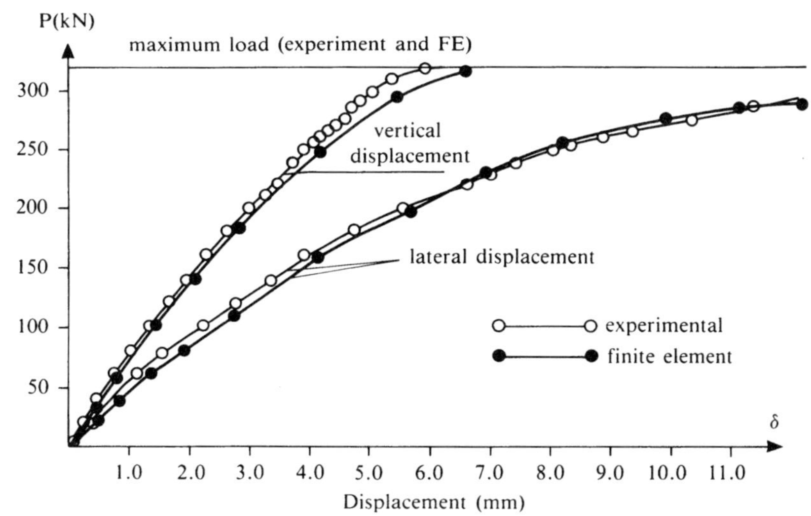

Fig. IX.92. Load versus displacement curves for specimens of 2.7 m length.

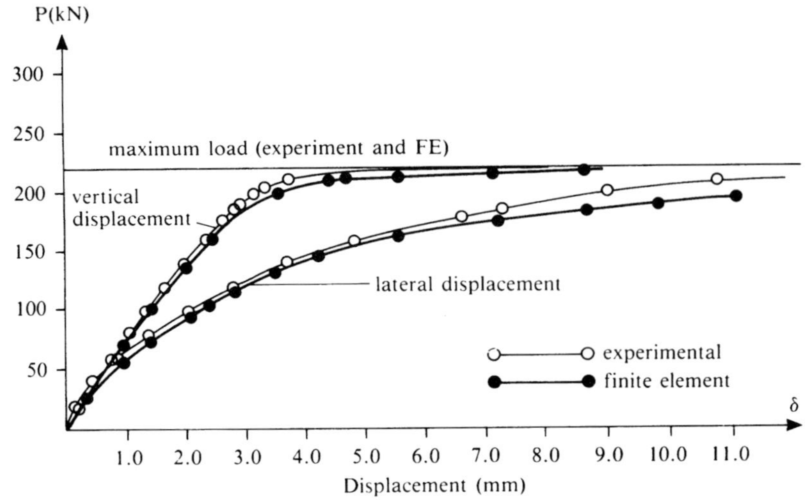

Fig. IX.93. Load versus displacement curves for specimens of 2.9 m length.

evaluated from the experiments. The finite element analysis also indicated the early initiation of plastic zones along the longitudinal axis. The concrete corresponding to such zones had crushed when reaching a strain value of 0.0035. The safety margins from both analyses differ by 12–15%.

IX.8 Case Study No. 7 — concrete power and diversion tunnels for hydro-electric schemes

IX.8.1 Introduction

Tunnels vary in their shapes, and their analysis and design depend on a number of factors [389–398] including the site conditions and loadings. Such details are beyond the scope of this book. However, an attempt has been made to demonstrate the capability of the finite element technique and the concrete failure models in the area of hydro-electric structures.

IX.8.2 Diversion, power and pressure tunnels

Bangash [1005] has developed a systematic approach for the design of tunnels used in hydro-electric schemes.

IX.8.2.1 General classification

The tunnels used in hydro-electric schemes are generally classified as:

(a) tunnels to divert flow during construction which are eventually closed using a combination of gates and concrete plugs
(b) tunnels to divert water from one scheme to another
(c) tunnels to be converted to power or pressure tunnels.

IX.8.2.2 Components of tunnels

The following are the general components of a tunnel:

(a) approach channel
(b) tunnel intake structure
(c) tunnel proper
(d) gate shaft
(e) tunnel transition and plug
(f) slide gate chamber
(g) portal structure and stop logs
(h) discharge channel.

This section only deals with the portal structure and the tunnel proper. Loadings pertaining to other components are not discussed; they can be included in certain cases.

IX.8.2.3 Shape and thickness of lining

Many shapes based on hydraulic and structural criteria have been investigated and compared. Only circular tunnels that have transitioned from rectangular openings at the intake level have become popular. The thicknesses of single concrete linings of tunnels and also of steel linings for concrete tunnels depend on the following well-known factors:

(a) size and shape of the tunnel and the character of the rock formation around it
(b) external and internal loading conditions, including stability of the steel lining
(c) stress distribution before and after rock mass excavation
(d) boundary conditions, including types of supports and spaces for concreting requirements.

Prior to an accurate estimate of a designed thickness of lining, a rule of thumb is available to give an initial thickness for the analysis. Both the US Bureau of Reclamation and the US Corps of Engineers generally adopt thicknesses of the order of 19–25.4 mm of concrete lining per $\frac{1}{3}$ m of finished diameter of tunnel. Regarding steel liners, the thickness based on an unstiffened shell is generally 6 mm per 344.75 kN/m^2 internal pressure based on a 3 m internal diameter. Table IX.11 gives a detailed description of various parameters for recent designs of diversion, power and pressure tunnels.

IX.8.2.4 Hydraulic and structural criteria

Generally the design of a tunnel is governed by two independent criteria, namely hydraulic criteria and structural criteria. The final design is always a compromise between the salient features of both these criteria. Again both these criteria are specifically geared to the individual project or scheme and the problems surrounding each one of them. The following criteria based on existing available statistics give reasonable guidance for tunnel design.

Table XI.11. Existing tunnel parameters [1005]

Type[a]	Scheme	Rock type	E (tons/ft^2) Young's modulus	f_c (tons/ft^2) compressive strength	Average inside diameter (ft/in)	Average lining (concrete) thickness (in)	Steel lining (in)	Pressure inside tunnel (psi)	Steel liner stress (tons/in^2)	Concrete strength (lb/in^2)	Safety factor
D	Fort Peck (USA)	Clay-shale	750–10 000	5–50	24–8	36	—	—	—	3000	3.21
P	Garrison (USA)	Clay-shale containing lignite coal beds	800–4000	5–75–150	29–0	36	—	—	—		
P			800–4000	5–75–150	26–0	33	—	—	—	3000	3.95
D			800–4000	5–75–150	22–0	30	—	—	—		
P	Oahe (USA)	Shale	2000–10 000	5–150	24–0	30	—	—	—	3000	4.5
D			2000–10 000	5–150	19–9	28	—	—	—		
D	Akosombo (Ghana)	Quartzite	3000–5000	720–1370	30–0	24 u/s[b] 12 d/s	—	—	—	3000	2.75
P	Bandama (Ivory Coast)	Greenstone	2850 570	1000–600 (500 design)	23–0 (7 m)	24 and 23	$\frac{9}{16}$–$1\frac{7}{8}$ $1\frac{1}{2}$	410	20 000 permissible 60 000 yield	3000	2.75
D	Rama-ganga (India)	Clay-shale interbedded with sand rock	3000–7000 7850–520 000	1.83–6.44	30–0	30	—	—	—	3000	3.3
P			3000–7000 7850–520 000		31–0	30	—	—	—	3000	
D	Mangla (Pakistan)	Clay and sandstone bedrock	6200–14 400	26.5–41.5	26 and 30 u/s 30 and 31 d/s	36	—	—	—	3750	3.85
P	Burrendong (Australia)	Weathered greywacke and slate	3000–4000	1200	27–0	24	—	—	—	3000	6.9

P	Yamuna (India)	Quartzite and slate and phyllites	Average 3000	1000	18–0	18	$\frac{3}{8}$–$2\frac{1}{8}$	+74	—	3000	5.1
D	Roselend (France)	—	—	—	12–6	20 (average)	$\frac{1}{2}$ with $4.7 \times 4.7 \times \frac{1}{2}$ angle iron stiffeners	230	—	3000	3.93
P	Bersimis No. 1				5 No. 10–0		$1\frac{1}{16}$–$2\frac{1}{16}$	381	—	3000	5.2
	No. 2				5 No. 17–0 reduced to 12–0	20 (average)	$1\frac{1}{16}$–$2\frac{1}{16}$	162			
	No. 2 (Canada)	—	—	—							
P	Warsak (Pakistan)	Metamorphosis schists and gneisses with phyllite	700 (average)	80	39–0 and 6 No. 18–0	18 (average)	$\frac{5}{8}$–$\frac{15}{16}$	100 (average)	—	3000	4.0
P	Tumut No. 1	Granite gneiss with biotite granite	700–1425	12	12–0	12	$\frac{1}{2}$–$1\frac{5}{8}$	325	Ultimate tensile strength 26–32 yield 15	3000	3.5
	No. 2				11–6	12	$\frac{1}{2}$–$1\frac{5}{8}$	320	Ultimate tensile strength 26–32 yield 15		

[a] D = diversion tunnel; P = power tunnel or pressure tunnels.
[b] u/s = upstream; d/s = downstream.
(1 ton/m^2 ≈ 10 N/m^2, 1 inch = 25.4 mm.)

(a) Hydraulic criteria. It is essential to have a complete record of the following major items.

(i) A complete history of the catchment area, and specifically the hydrology of the site and operational conditions.
(ii) Establishment of upstream and downstream elevation at the design discharge and normal water surface and an estimate of hydraulic pressures.
(iii) Design discharge through the tunnels, the size of tunnels, etc.
(iv) The closure water level of the bulkhead and plug completion if it is a diversion tunnel; the minimum velocity during the diversion period.
(v) For power and pressure tunnels an accurate estimate of the surge tank diameter and the area, including load demand and rejection.

(b) Structural criteria (external loading or pressure). For design computation, empirical values are given by Bangash, based on Terzaghi [389], assuming the rock load acts as a uniformly distributed load over a tunnel diameter. These loads are based on assumptions that the rock is associated with some roof yielding due to failing of the interlocking action. In loose and fracture rock the sides also yield, thereby developing an active pressure state along the shear planes at an angle of $45^\circ + \alpha$, where α denotes the angle of internal friction. If B and H represent the width and height of the tunnel, the following values from Terzaghi give a fair estimate of the extent of rock load on tunnels.

Rock load acting at depth of more than $1.5(B+H)$

(i) Moderately blocky and seamy rock: $0.25B$–$0.35(B+H)$
(ii) Very blocky and seamy: 0.35–$1.1(B+H)$
(iii) Completely crushed but chemically intact: $1.1(B+H)$
(iv) Squeezing rock, moderate depth: 1.1–$2.1(B+H)$
(v) Squeezing rock, great depth: 2.1–$4.5(B+H)$
(vi) Swelling rock: up to 85 m.

(c) Load combination. Lateral restraint offered by the surrounding rock is extremely important and must be investigated while designing the tunnel lining. Various theories exist to take into account the pre-existing stress in rock and the redistribution of stresses in the surrounding rocks after tunnelling.

In addition, tunnels and their components are designed for the following loading cases, separately and in combination:

(1) Transition and the portion of the tunnel upstream from the plug: External pressure from rock + hydrostatic pressure at normal water surface.
(2) Tunnel downstream from the plug:
External pressure from rock + weight of lining + zero hydrostatic pressure.
(3) Rock load + weight of lining:
Rock load with submerged weight of the rock below water table + weight of lining + minimum hydrostatic head + empty tunnel.

(4) Rock load + submerged weight of the rock below water table + weight of lining + contained water with internal pressure + zero external hydrostatic load.
(5) Stop logs:
Closure at specific level + hydrostatic pressure at normal water surface + any external loading.
(6) Tunnel plug:
Hydrostatic pressure at design elevation.
(7) Pier walls:
Stop log + wheelgate closed + maximum external pressure.

In the absence of suitable data, where applicable the tunnel components are designed to take into account the following average loads:

(i) Gate shaft house: 1.676 kN/m^2
(ii) Hoist deck: 9.576 kN/m^2
(iii) Inspection deck: 12.768 kN/m^2
(iv) Gate support deck: 12.768 kN/m^2
(v) Cut and cover section: 12.768 kN/m^2
(vi) Fixed wheel gate storage area: surcharge equivalent of 1 m minimum of fill
(vii) Wind pressure: 0.96 kN/m^2 on the vertical projection of exposed area
(viii) 10% seismic value of dead and permanent live loads applied at the centre of the mass in any horizontal direction
(ix) Hydrostatic + full uplift pressures arising from upstream or tail water conditions

(d) Materials and stresses and other elastic constants. Materials such as concrete and steel have been discussed in this text in detail.

IX.8.2.5 Data for diversion tunnels

A typical diversion scheme, similar to the Akosombo scheme [390] with altered parameters, is presented as a design example. It is considered to be the most economic layout of a diversion tunnel in rock. An interesting feature is the introduction of a single pier housing slide gate chambers and a fixed wheel gate shaft with stop logs covering rectangular openings. The pier is placed symmetrically about the centre line of the tunnel. Figures IX.94 and IX.95 show a plan and sectional elevation of this tunnel. Figures IX.96–IX.100 give a full account of the way the 12.192 m transition is made from a rectangular portal to the beginning of a circular tunnel. It is assumed that after the diversion of flow the tunnel is finally closed with a permanent concrete plug.

Tunnel diameter = 7.62 m
Area of tunnel section = 45.54 m^2
For reinforcement (plan and sections) see Figs IX.98 and IX.99
External loading intensity and water elevation
Maximum water surface elevation = 206.00
All other elevations are shown in Fig. IX.100

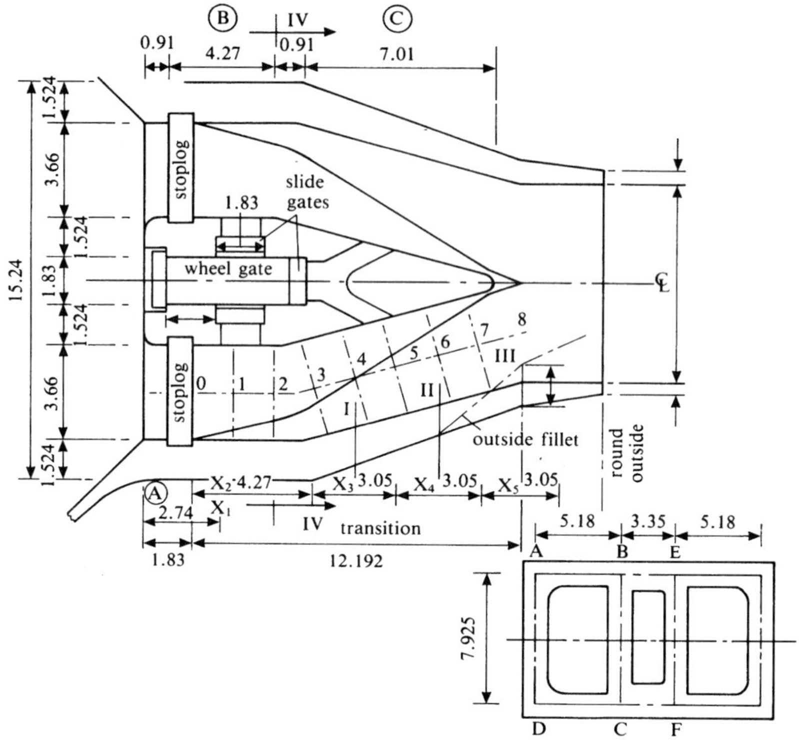

Fig. IX.94. Plan and section IV-IV of the tunnel. (Note: All dimensions in metres.)

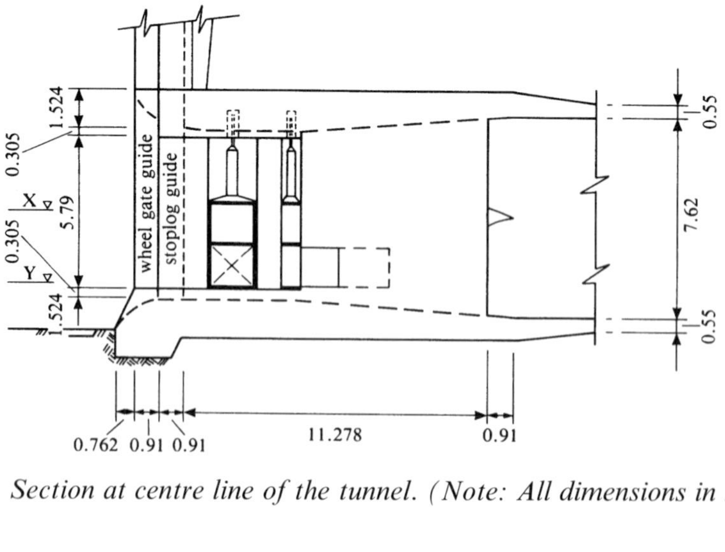

Fig. IX.95. Section at centre line of the tunnel. (Note: All dimensions in metres.)

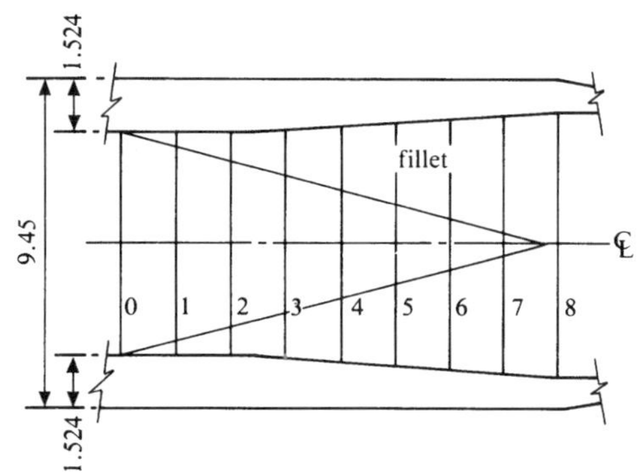

Fig. IX.96. Profile at centre line of the tunnel. (Note: All dimensions in metres.)

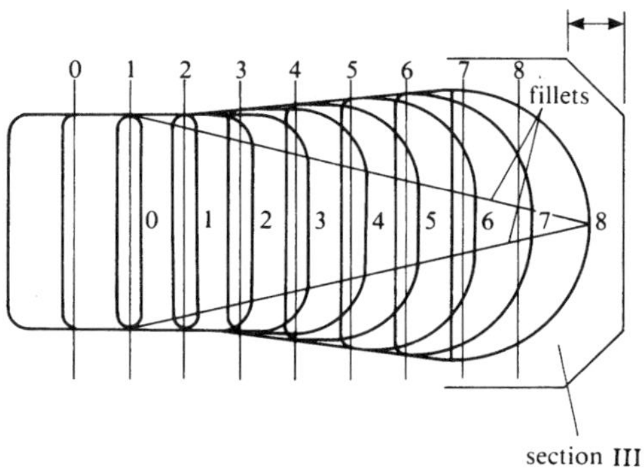

Fig. IX.97. Sections along centre line of the tunnel.

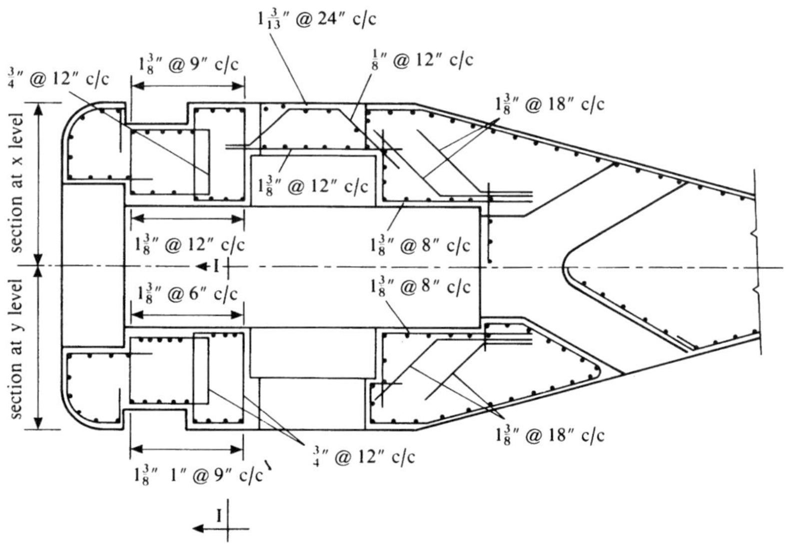

Fig. IX.98. Reinforcement plan. All vertical bars = $\frac{7}{8}$ in. × 12 in. (c/c) and all horizontal bars = $\frac{3}{4}$ in. × 12 in. (c/c), unless otherwise noted. (Note: All dimensions in imperial units; 1 in = 25.4 mm.)

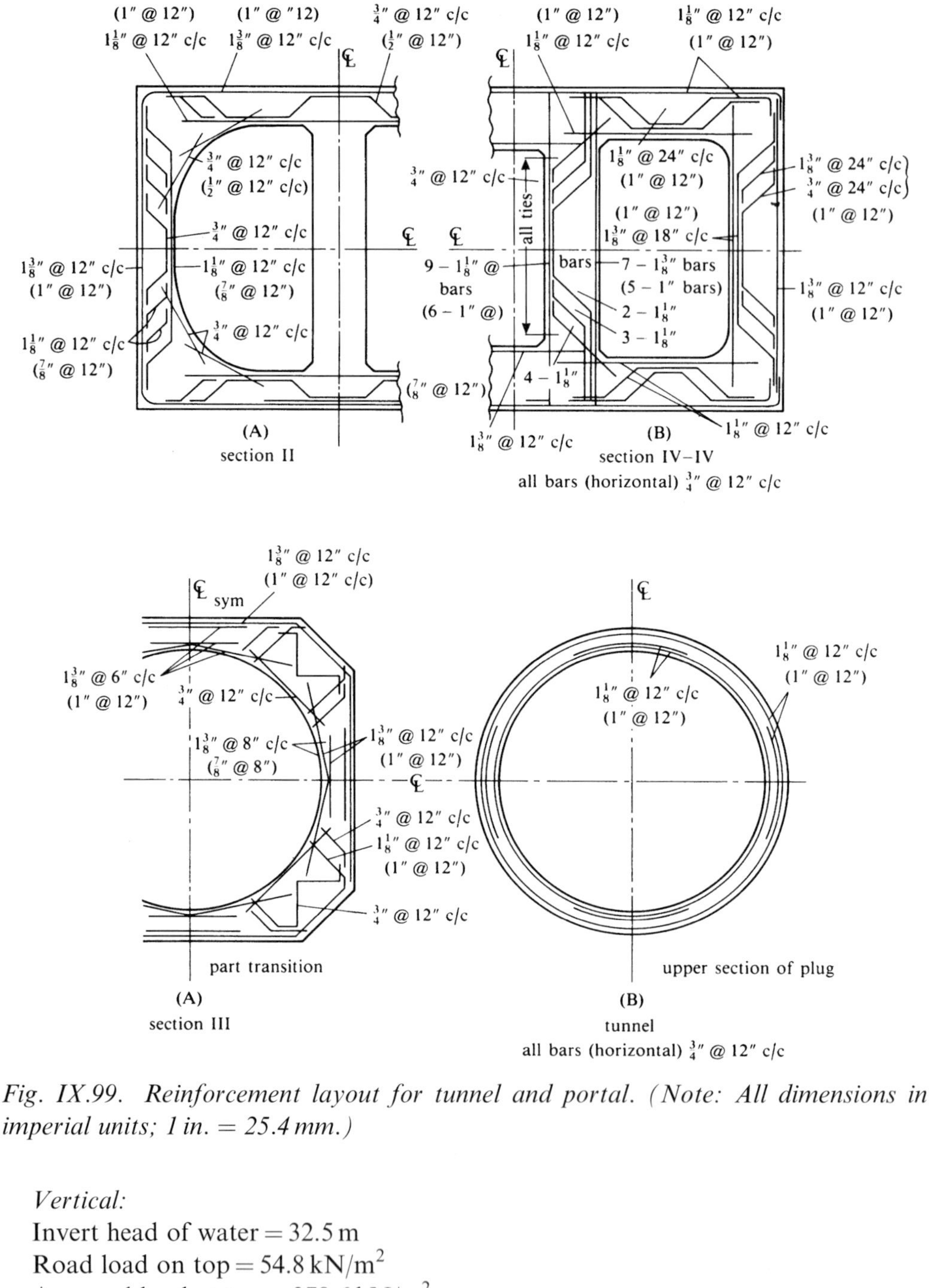

Fig. IX.99. Reinforcement layout for tunnel and portal. (Note: All dimensions in imperial units; 1 in. = 25.4 mm.)

Vertical:

Invert head of water = 32.5 m

Road load on top = 54.8 kN/m^2

Assumed load at top = 378.6 kN/m^2

Self load of section = 196 kN/m^2

Intensity of loading due to self load = 21.22 kN/m^2

therefore load at bottom = 590.86 kN/m^2

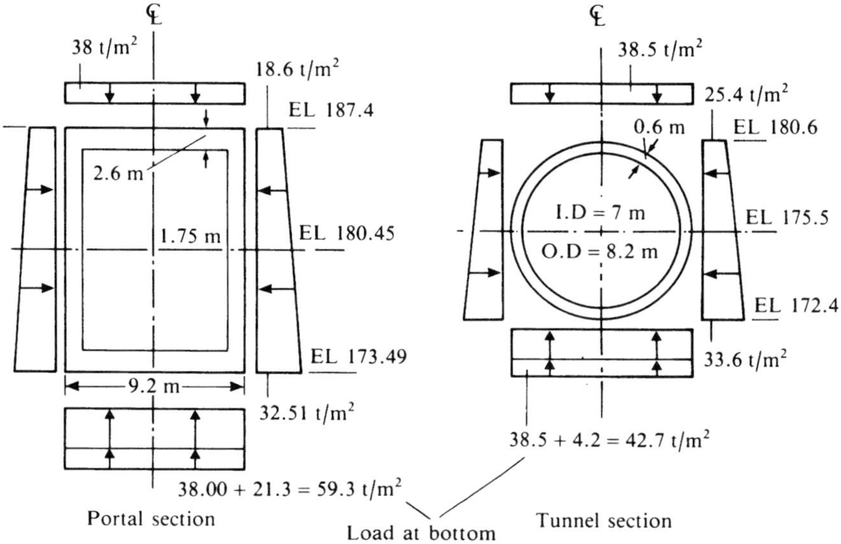

Fig. IX.100. Bandama tunnel: portal and tunnel sections. (Note: Loads in original MKS units; 1 ton = 9.964 kN.)

Lateral:
At top = 185.33 kN/m^2
At bottom = 323.93 kN/m^2

Internal loading:
254.58 kN/ m^2
Tunnel section (Fig. IX.99)
External loading: maximum water surface elevation 206.00

Vertical:
Invert head of water = 334.79 kN/m^2
Rock load on top = 488.2 kN/m^2
therefore load at top = 383.6 kN/m^2
Self load of section = 340.77 kN/m^2
Intensity of loading due to self load = 41.85 kN/m^2
therefore load at bottom = 425.46 kN/m^2

Lateral:
Hydrostatic head at top = 253.086 kN/m^2
Hydrostatic head at bottom = 334.79 kN/m^2

Internal loading:
Water head = 293.94 kN/m^2
Shear in rock as for plug key τ = 206.85 kN/m^2

IX.8.2.6 Finite element analysis

Three-dimensional finite element analyses using eight-noded solid isoparametric elements and two-noded line elements were used. Where the steel lining interfaced with concrete, displacements at the nodes of the two meshes were matched. When lugs were provided the finite element analysis covered the lugs by invoking Table VI.2. External and internal loads were transformed into nodal loads. The tunnel concrete was analysed using the finite element mesh scheme given in Fig. IX.101. The existing layout of the reinforcement designed by conventional analysis was used. The line elements

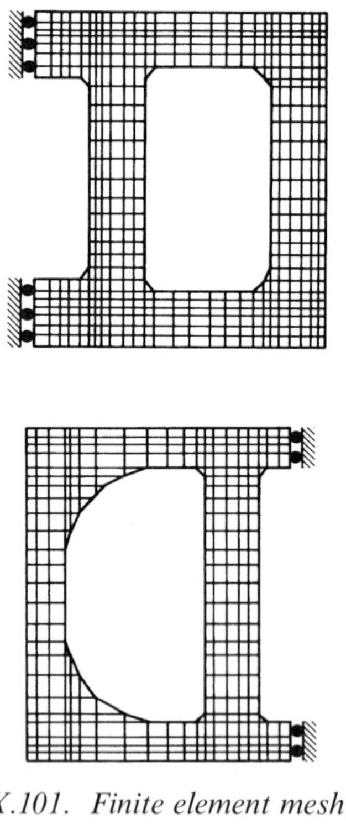

Fig. IX.101. Finite element mesh.

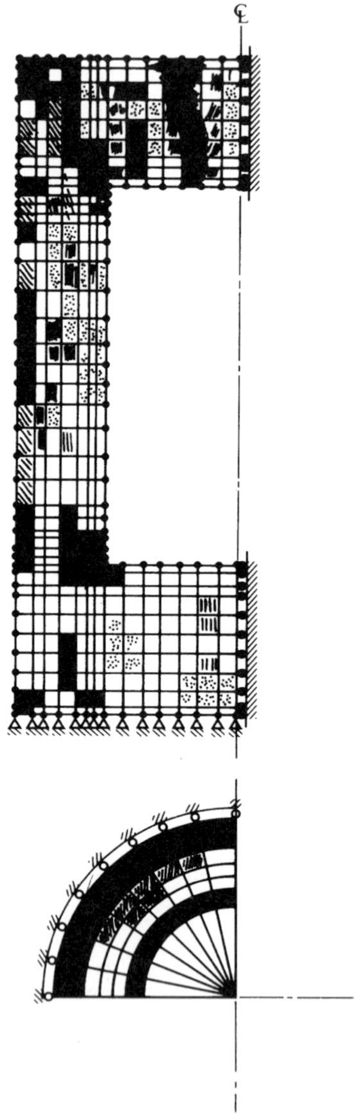

Fig. IX.102. Finite element mesh and cracking. (■, Cracks in three dimensions; ▥, radial cracks; ▧, circumferential cracks; ▩; heavy principal stresses.)

and their positions in the solid elements were as in the original layout. Stresses, strains and principal stresses were computed. Afterwards inelastic analysis was carried out as described in Chapter VI using the endochronic cracking models formulated in Chapters VI and VII. Twenty increments each of 10 kN were considered. The early cracking in concrete started at about 1.65 times the load. The concrete section of the tunnel crushed at about 2.75 times the design load.

Figure IX.102 shows a 'post mortem' of the tunnel at about 2.75 times the indicated design loads shown in Fig. IX.100. Figure IX.103 shows reinforcement for maximum principal stresses at 2.5 times the design loads.

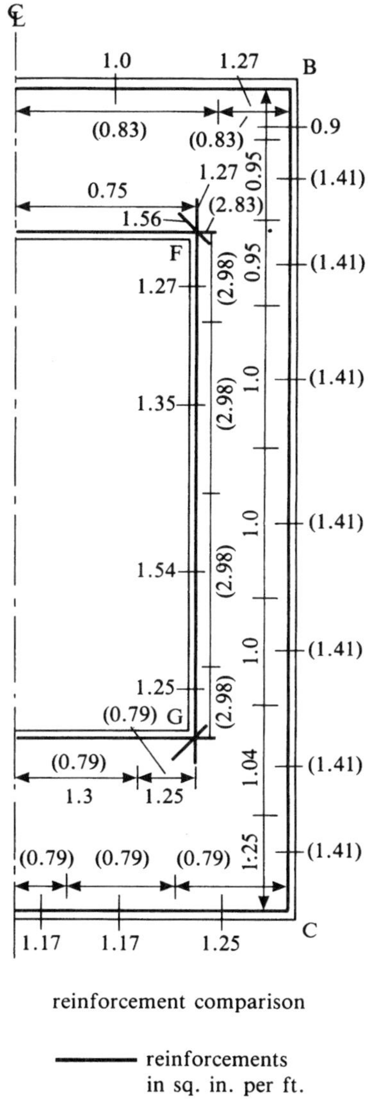

Fig. IX.103. Tunnel reinforcement. (Note: All dimensions in imperial units; 1 in. = 25.4 mm.)

Table IX.11 shows the parameters and the safety margins for other types of tunnels using finite element analysis inclusive of the concrete failure models (endochronic) and the cracking equations given in Chapters VI and VII.

IX.9 Case Study No. 8 — prestressed concrete nuclear reactor vessels

IX.9.1 General

The use of nuclear energy in the production of electric power involves substantial structural systems comprising pressure vessels to house the reactor. Most of the light water and pressurised water reactors in operation at present are sheltered by two major structural units: the primary container, which is a steel pressure vessel for holding the coolant systems under pressure, and a secondary container providing a second protective shield. An alternative to the use of liquid coolants is a pressurised gas system. The coolant in this case is either carbon dioxide or helium. The current British and the French reactors, using these coolants, are called Advanced Gas-Cooled Reactors (AGR) and High-Temperature Gas-Cooled Reactors (HTR or HTGCR). Efficient operation of these nuclear power plants requires a large electrical output. This requirement calls for a large nuclear core space and much larger primary pressure vessels and supporting structures of sizeable dimensions. The inherent difficulties that would be involved in the fabrication and transportation of the pressure vessel steel units of the size necessary for such a reactor have led to the need for a more versatile type of structure. The vessel must, undoubtedly, be built on site and meet all the serviceability and safety requirements. The solution adopted has been the use of prestressed concrete reactor vessels (PCRV).

One incentive for adopting PCRV is the economic advantages to be expected. These advantages arise, for example, from:

(a) the ability to contain large reactors with high pressures and temperatures;
(b) simplification of plant layout;
(c) the fact that a highly developed steel fabrication industry is not necessary.

Attractive features from a safety point of view include:

(a) physical isolation of the steel prestressing tendons and reinforcement from sources of heat and radiation and from the primary coolant;
(b) the high degree of redundancy in the prestressing systems.

The shapes used, so far, for PCRVs have varied from a cylinder bounded by two inverted, non-prestressed, hemispherical domes, torospherical domes to spherical shells. The current trend in PCRV configuration is the use of thick-walled cylinders, the ends of which are closed by flat slabs known as caps. The boilers and circulators are either housed within the main cavity or within the thickness of the walls — the latter is known as a *multi-cavity type vessel*. This study is concerned with such vessels.

IX.9.2 Historical development

The first application of prestressed concrete as a reactor vessel was in the Marcoule G-2 and G-3 installations in France. These are horizontally placed concrete cylinders with concave domes as caps. The vessel main cylinder is wrapped with prestressing cables describing an arc of 270° which are finally anchored to prestressed concrete foundations under the vessel. The domes are not prestressed, resulting in a substantial loss of vessel volume. The stand pipe and the control rod areas are badly situated for the charge machine. The French have recognised these limitations of the Marcoule vessels. However, in the meantime the requirements for high-quality steel and more rigorous inspections for both British and French reactor vessels has reawakened interest in concrete vessels. Consequently, in planning the construction of the EDF-3 reactor, the French designers decided to re-examine the application of prestressed concrete to reactor vessels. The design selected provides a vessel with a cylindrical shape and with prestressing cables placed in layers parallel and perpendicular to the long axis of the vessel. Supplemental prestressing steel is also provided at the corners where the stress concentrations are high.

The British interest in concrete pressure vessels began with some experimental work done by the General Electric Company and the Simon-Carves group on cylindrical vessels prestressed circumferentially by wire winding. Based on these tests and other tests carried out by the United Kingdom Atomic Energy Authority (UKAEA), Waters and Barrett [399] published design studies for cylindrical and spherical vessels containing advanced gas-cooled reactors. They carried out a comparative study of steel and concrete pressure vessels and concluded that there were significant economic and technical attractions to adopting such vessels. They then presented a detailed design philosophy in which particular attention was paid to the short- and long-term effects of tendon loads on concrete, the premature failure of the liner, and the optimum choice of the conventional reinforcement as an anti-crack steel. Apart from highlighting design and constructional problems, they suggested a list of research programmes. At a time when no suitable code of practice and no completely reliable mathematical tools existed for analysing and designing such structures, their recommendations prompted many UK firms with interests in nuclear technology to carry out research and development in this field. The first order placed by the Central Electricity Generating Board (CEGB) was for the Oldbury Power Station. The Nuclear Power Group (TNPG) proposed a vessel design that was a hybrid of the French Marcoule and EDF-3 concepts. The vessel is of cylindrical shape with flat top and bottom caps, and it utilises alternate layers of helical cables which wrap the vessel in reverse directions. These cables are anchored in groups in specially provided top and bottom galleries. Both longitudinal and lateral prestressing loads are produced from this single prestressing system.

By the time of completion of this vessel a great deal of knowledge had been gained from both the British and the French experience in this field. Another concept was under consideration at the CEGB, namely a spherical vessel containing the entire reactor internals. Owing to immediate requirements for more electrical output, the CEGB gave the go-ahead to a group member of the British Nuclear Design Company [412, 456] to design and construct such a vessel for the Wylfa power plant. This vessel has a spherical inner surface and the prestressing is provided by a six-sided polyhedron made up of cables anchored to buttresses along the vessel circumference. Each tendon is laid perpendicular to the preceding one. Since the prestressing tendons in the Wylfa Nuclear Power Station are nearly straight, the possibility of large frictional effects is remote. The prestressing loads are uniform and this is partly due to its spherical shape. Compared with cylindrical vessels, the Wylfa spherical vessel effectively utilises the entire prestressing system, and the peak stressing is considerably less.

In the meantime, a number of variations on the French EDF-4 (now Bugey) reactor were under consideration. It was decided to have a vertical cylindrical pressure envelope divided into two chambers: the upper one supports the reactor core whilst the lower one contains the boilers or heat exchangers. These affect the size of the internal cavity and the prestressing system. Complete data for these vessels was published by Marsh and Melese [400], based on detailed literature surveys and the preliminary designs carried out by the Frankline Institute [401].

Both the Frankline Institute [401] and the Oak Ridge Laboratory [886] in the USA made a start on the research and development required for the Fort St Vrain high temperature gas-cooled reactor. The choice was made of a cylinder with flat top and bottom caps. Both in the wall and in the cap the prestressing was by means of cables of variable curvature.

In the meantime, in the UK a great deal of experience was gained from the Oldbury and Wylfa vessels. Problems associated with the plant layout and the construction of the Wylfa vessel were such that the British designers had to rethink the use of cylindrical vessels. The CEGB finally approved the cylindrical vessel of the Dungeness B Power Station using large tendons in the vertical and in the circumferential directions. The circumferential tendons are anchored on specially formed buttresses. This time, using a minimum number of large tendons, a greater concentration of forces has been achieved for the limited available space in the vessel.

However, the Oldbury contractors carried out optimisation studies on the Hinkley B and Hunterston B power stations. Slight changes in the vessel parameters were inevitable. This time true helices were adopted for the prestressing tendons rather than the 'barrelised' type adopted for the Oldbury vessel.

At this stage, the concept of multi-cavity vessels was under active consideration. Boilers and circulators occupied a large space within the main cavity. The dimensions of the main cavity were reduced by sheltering the boilers and circulators within the thickness of the wall. Burrow reported [456] on providing the wire-winding system for the circumferential prestressing loads on the vessel, thus leaving the main vessel thicknesses for the large capacity longitudinal tendons only. Such a vessel was accepted for the AGR system adopted for both the Hartlepool and Heysham power stations. In the last few years substantial progress has been made in the development of such vessels and much experimental data has been produced from both model tests and on-the-spot measurement of the prototype vessel.

IX.9.3 Problems associated with vessels

Although the multi-cavity vessel has much to offer, it is by no means an easy structure to model. Many designers and structural analysts have tried to tackle the various problems associated with these structures. In many cases they have claimed some achievements but for some reason they have not made them public. However, most are united on the design criteria for such vessels and these are:

(a) the vessel should be designed for elastic response to all possible combinations of loads during operation;
(b) it must show a progressive mode of failure under increasing gas pressures with large deformation to warn against impending failure; and
(c) the vessel must have an acceptable safety margin against failure.

In order to meet the above criteria satisfactorily, one has first to validate the performance of the individual components forming such structures. The major components are:

(a) concrete,
(b) prestressing tendons,
(c) the liner and other penetrations, and
(d) bonded primary and secondary reinforcements.

The problems associated with each one of these are enormous. For example, a certain amount of evidence must be available to demonstrate that for short- and long-term loadings (under multiaxial compressive stresses) concrete can safely withstand higher compressive stresses than are generally acceptable under uniaxial loading. For all service load conditions, including start-up and shut-down, both initially and at the end of the vessel's life, the stress–strain characteristics for the concrete should take account of the age, temperature and time under load. It is considered also that limited cracking may be accepted provided due regard is paid to any significant redistribution of the stresses which may arise, due to lack of integrity and leak tightness of the liner. Where local concentrations of stress occur, due to the presence of embedments or other discontinuities in the vessel geometry, these should be assessed individually. In such cases, due regard should be paid to the effects of increased creep rates or tensile cracking on the distribution of stresses in the vessel concrete and the influence which they may have on the strains in the vessel liner.

Stresses and strains, deflections and cracking in the vessel should be analysed for all relevant combinations of mechanical and thermal loads which can arise under normal service and ultimate conditions. In such cases the prestressing forces play a great part. The tendon forces adopted in each analysis should include allowances for the most severe effects of friction and loss of prestress. A proper method of analysis together with short- and long-term experimental tests are required to design tendon systems for cyclic loads produced by gas pressures. These requirements are translated in terms of range of stress or strain cycles in the steel liner, the concrete, or the other components in the nuclear islands. Beyond the elastic range, the method of assessing the ultimate behaviour of prestressing tendons and their anchorages is very important for predicting a safety margin for the vessel. This, of course, depends on the time at which the incident occurs to the vessel and the rate of increased gas pressure and whether or not the tendons are grouted.

The liner undoubtedly represents a vital safety element of the vessel. Indeed, it serves the fundamental purpose of forming an impervious barrier to the cooling gas. The reactor's operability depends on its integrity. The vessel will collapse, not at the ultimate collapse pressure coincident with the failure of the prestressing tendons but at the liner failure pressure which might be lower. This would require knowledge of the topology and magnitude of the cracking on the internal face of the concrete in contact with the liner. It would be necessary, therefore, to have a clear idea of the cracking mode of the vessel for increasing pressures up to the collapse pressure. Moreover, the liner integrity is also dependent upon the performance of the liner anchorages and cooling pipes, buckling stability and fatigue in general areas, the insulation and temperature distribution.

The choice and the distribution of the conventional bonded steel reinforcement in the vessel main areas and around the penetrations are extremely important should initial cracking occur due to moisture migration and unpredictable shrinkage and creep in concrete. The bonded steel reinforcement is needed in areas where extreme stresses under serviceability conditions cannot be avoided. Above all, the progressive failure of vessels depends on the amount and the distribution of such reinforcements.

Now turning to the vessel behaviour under extreme loads, it is necessary to provide some means of limiting the effect on the vessel of an excessive rise in internal operating pressure. The magnitude and rate of a postulated pressure rise can be determined only by reference to the characteristics of the particular reactor contained. It is common practice to provide automatic venting devices, such as safety valves, for this purpose. An alternative is to design the vessel so that it is self-venting by partial structural failure. It is generally recognised that this alternative cannot be relied upon at present.

It is customary and advantageous to make provision to verify the state of the vessel. This may be done by installed instrumentation and/or by periodic in-service inspection. Such measures, which are taken in a manner appropriate to the particular situation, serve to verify the vessel integrity and to confirm the design criteria. However, this does not rule out the main problem of numerically assessing such design criteria. Under any circumstances the design philosophy is based on the recognition of two or more modes in the vessel response to increasing pressure, and this cannot be met by simple experimental models.

The design objective is to ensure that a particular response to the imposed load can be achieved in each mode and that this behaviour is consistent with the appropriate operational and predetermined fault conditions.

Over a range of pressures and temperatures, including normal operating conditions, the vessel will respond to short-term variations in pressure in an elastic manner. This facilitates machine analysis of stress and strain in the vessel. Long-term stresses and strains are affected by shrinkage and creep of the concrete, relaxation of tendons and possibly fatigue. In this range of response the effects of short- and long-term behaviour can be combined to demonstrate that stresses and strains are limited to acceptable values.

Beyond the elastic range the response becomes increasingly inelastic and non-linear. The vessel would not be expected to enter this phase except under the most severe overpressure fault conditions. In this phase the vessel is stable but may experience permanent damage. It is in this phase of vessel response that some limit states occur.

The ultimate load condition, in which the vessel is incapable of sustaining any further increase of internal pressure, is a further limit state. Evaluation of the ultimate load provides a measure of the factor of safety above design conditions.

IX.9.4 Vessel layouts and finite element mesh schemes

Typical layouts of the Dungeness B, Oldbury and HTGCR vessels are given in Figs IX.104–IX.106 respectively. Figures IX.107–IX.110 show the finite element mesh layout of the Dungeness B, Oldbury, Hartlepool and HTGCR vessels respectively.

The mesh layouts take into consideration the vessel penetrations, prestressing configurations and other features such as the existence of buttresses, galleries and the liner lugs and cooling pipes.

IX.9.5 Design analysis

The objective of this analysis is to demonstrate that reactor vessels will meet both the serviceability and ultimate limit state conditions, taking into consideration the gas increasing pressure.

IX.9.5.1 Service conditions

Under service conditions it is important to know the technique and precision with which the vessel geometry can be reproduced by analytical means. A number of key loading cases are given below.

(a) Initial prestressing loads: Analysis includes losses due to friction and elastic shortening and creep; this is to see that no abnormal stresses develop in the vessel concrete.
(b) Proof test: In this analysis the vessel is tested at the commissioning stage for changes in strains and deflection due to proof pressure loading.

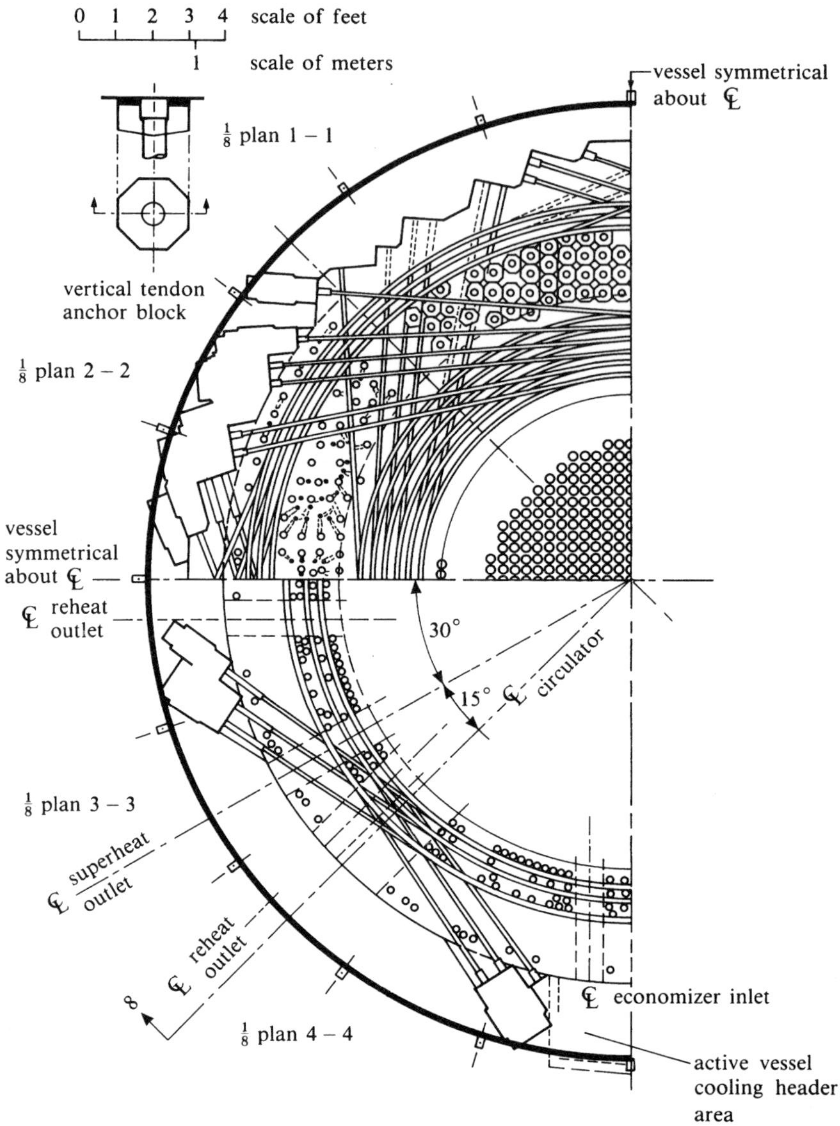

Fig. IX.104. Dungeness B vessel prestressing layout (plan).

(c) Normal operating conditions: This is intended to observe the vessel operating normally under extreme loads caused by the combination of prestress, design pressure and temperature gradient together with the effects of applicable environmental loadings. This can be at an early-life or late-life operating condition.

(d) Start-up and shut-down conditions: These are transient and quasi-transient conditions normally imposed on the steady conditions of normal operation. Together they reproduce the extremes of the early-life shut-down/start-up

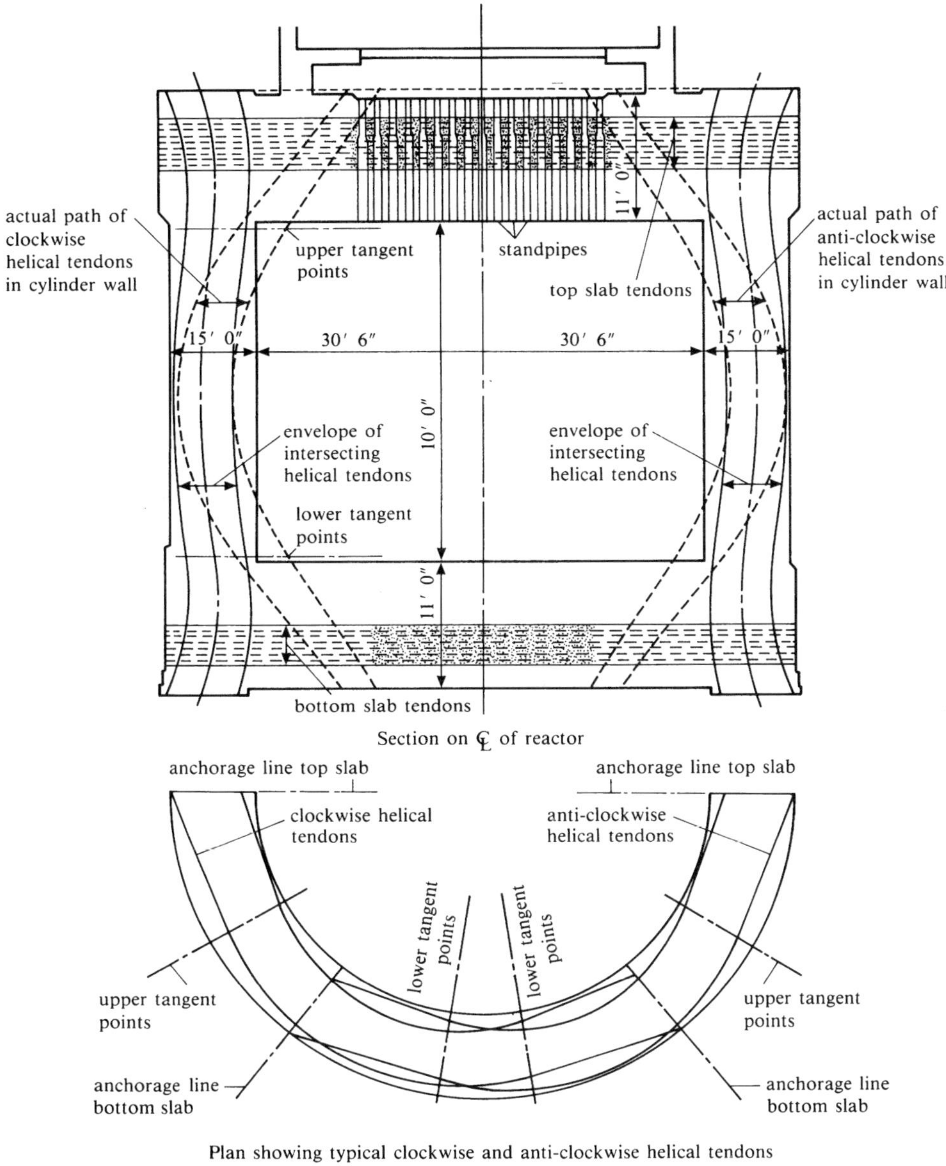

Fig. IX.105a. Principal prestressing systems (patented). (Note: All dimensions in imperial units; 1 ft = 0.3048 m, 1 in. = 25.4 mm.) [14, 415, 418]

cycle. The late-life conditions are also examined by changing the material properties and temperature transient conditions.

(e) Deviations from operating conditions: A number of inaccuracies arise from the simple treatment of loading cases; these mainly concern the effect of concrete creep on stresses, unloading and cooling effects on the vessel, and the allowance for low-probability loadings arising from accident or fault conditions. Analyses are required of these uncertainties.

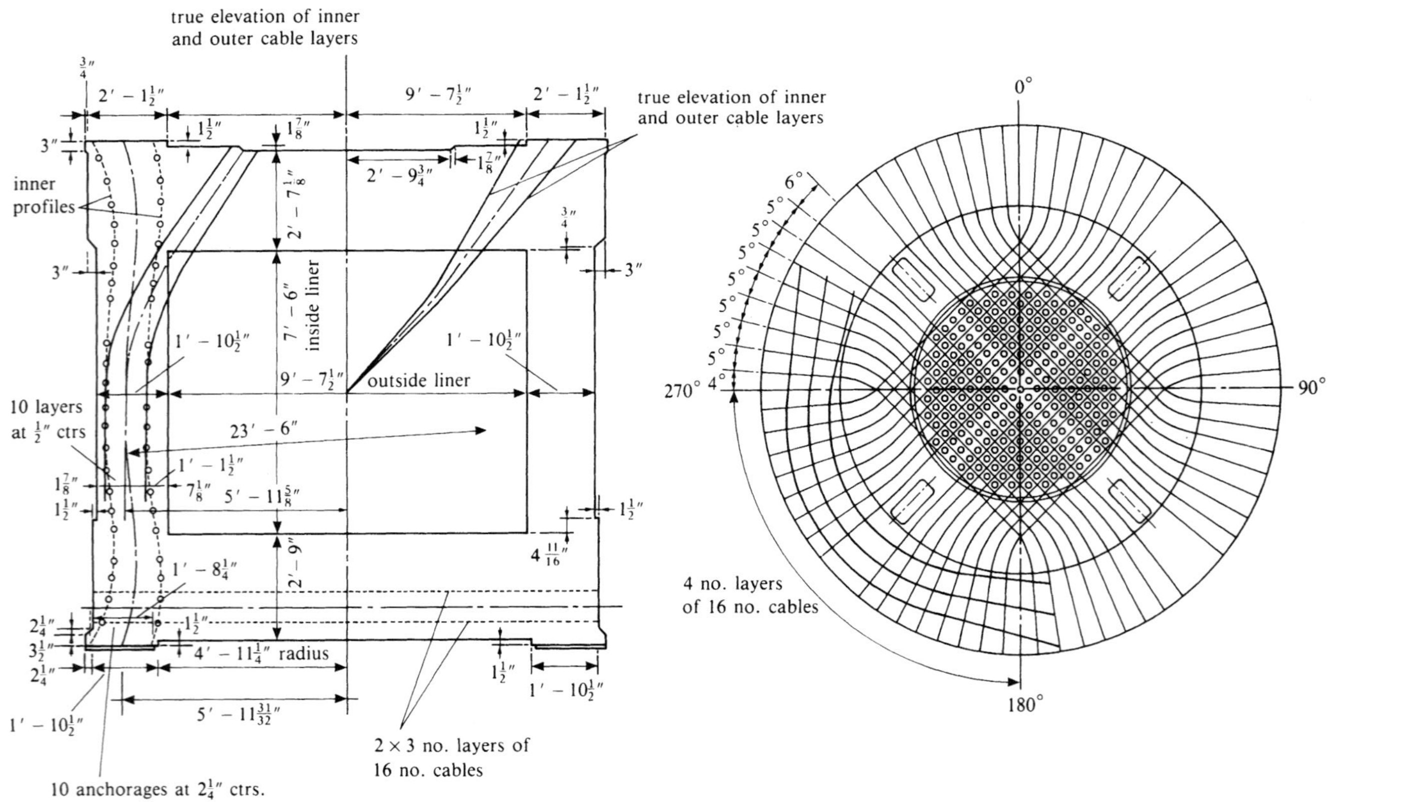

Vertical section through vessel centre line

Plan view of top slab

Fig. IX.105b. Model for Oldbury nuclear power station — layout of prestressing cables and cable profiles. (Note: All dimensions in imperial units; 1 ft = 0.3048 m, 1 in. = 25.4 mm.) [14, 415, 418]

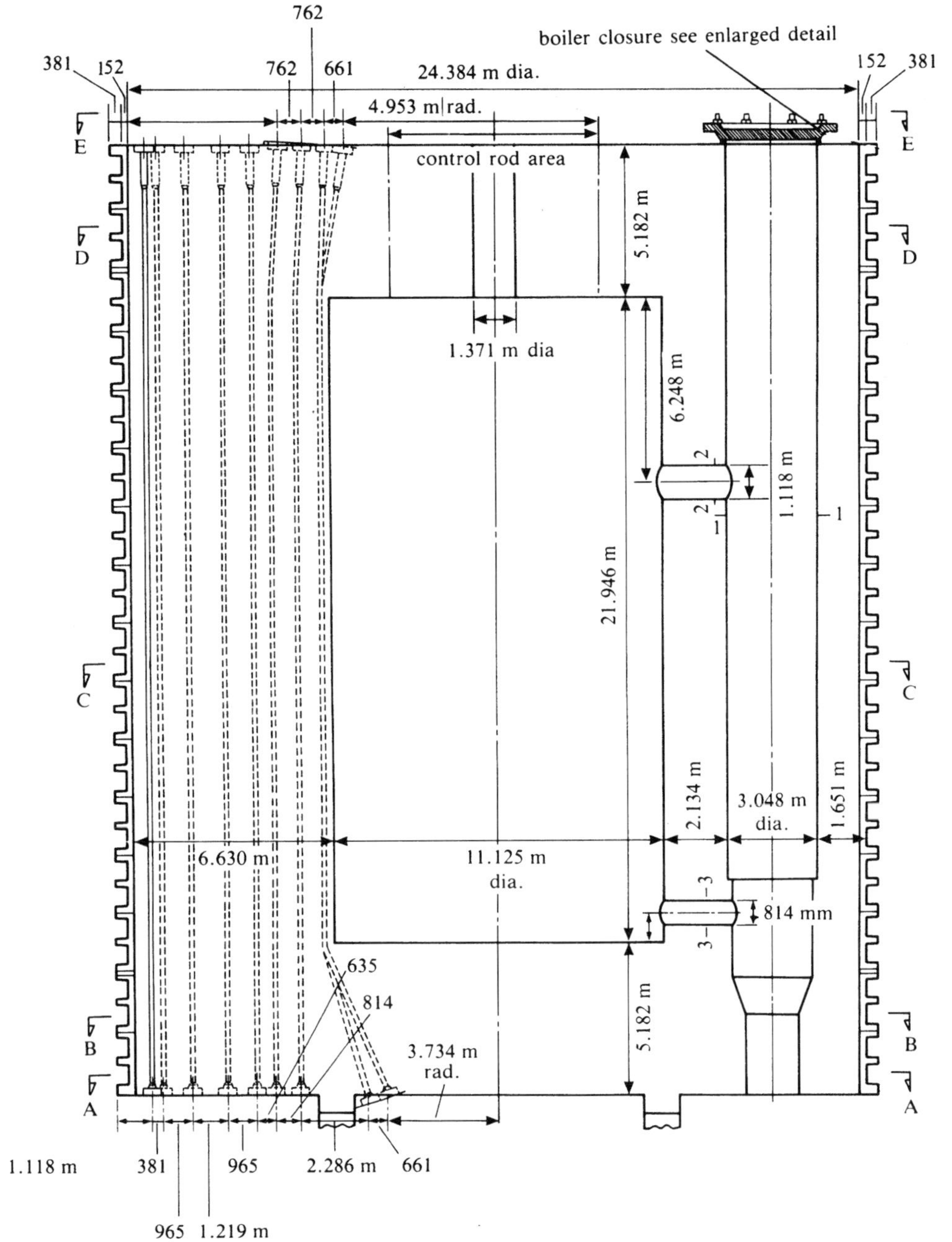

Fig. IX.106. Prestressed concrete cylindrical multi-cavity reactor vessel, HTR, section F–F.

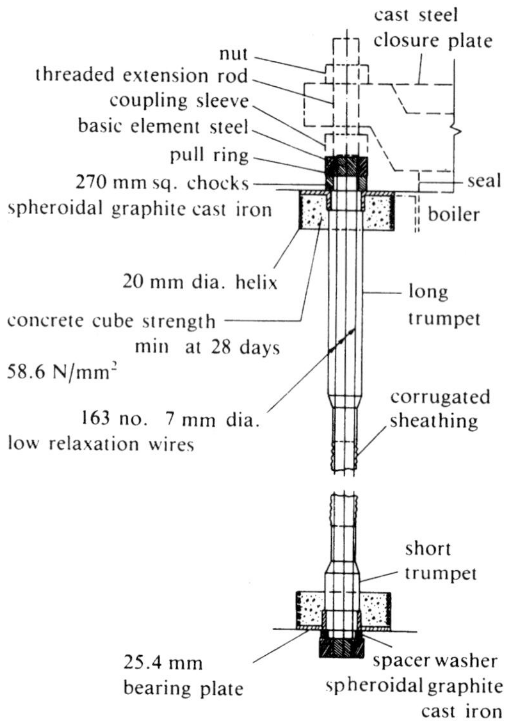

Details of prestressing tendon.

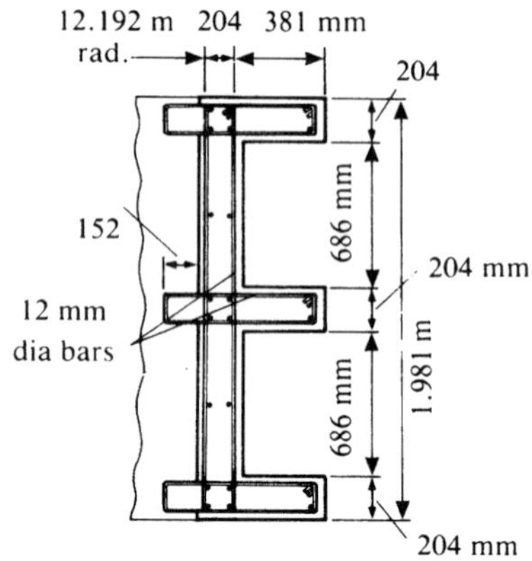

Section through precast channel shutter.

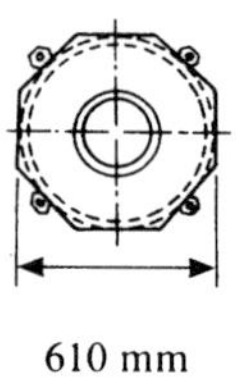

Plan of bearing plate assembly.

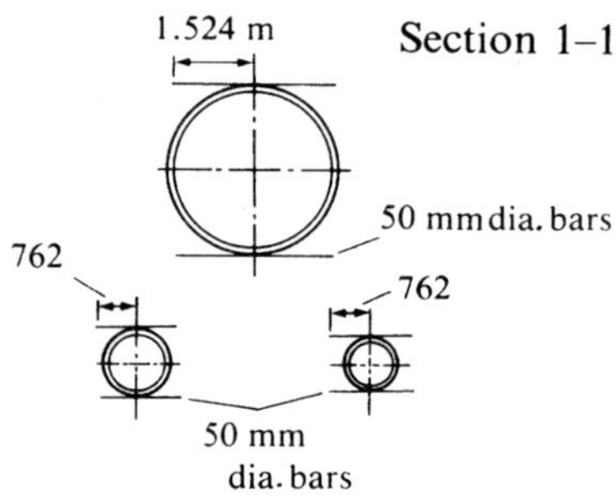

Section 1–1.

Section 2–2. Section 3–3.

Fig. IX.106 — contd.

IX.9.5.2 Ultimate conditions

A validity analysis is required to check the mode of failure assumed or calculated or derived from the experimental tests. This analysis should cover cases such as the intact and ruptured liner, failure of local areas, the possibility of shear failure in top caps or at cavity/wall ligaments and cracks in hot spots (areas around cooling pipes).

IX.9.5.3 Methods of analysis

Many methods are available to designers. The most common [399–429, 750, 751] are based on the finite element, finite difference or dynamic relaxation, lumped parameter, and limit state methods. This book gives earlier the step-by-step approach

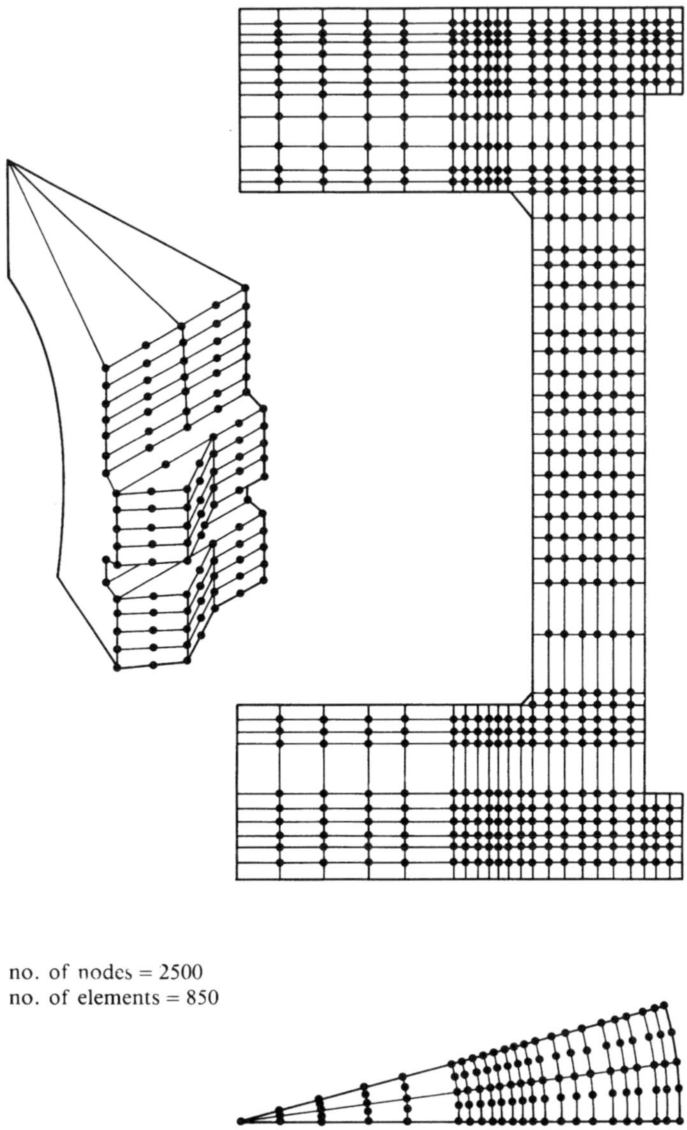

Fig. IX.107. Finite element idealisation of Dungeness B vessel. (Note: For clarity certain nodes are not shown.)

of the finite element method. In some service and fault conditions it is necessary to consider the influence of external hazards and environmental conditions. Major external hazards are seismic disturbances, wind, missiles and aircraft crashes. These hazards are considered for the case of containment vessels.

IX.9.5.4 Model testing

The purpose of a model test is to verify the methods of analysis and to provide a visible physical demonstration of the adequacy of the design requirements. As

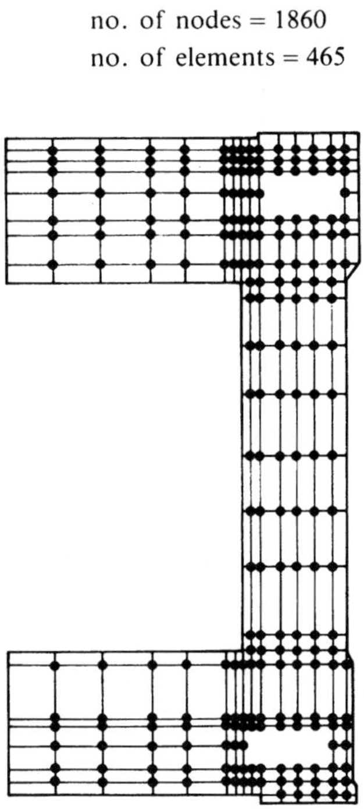

Fig. IX.108. Finite element idealisation of Oldbury vessel.

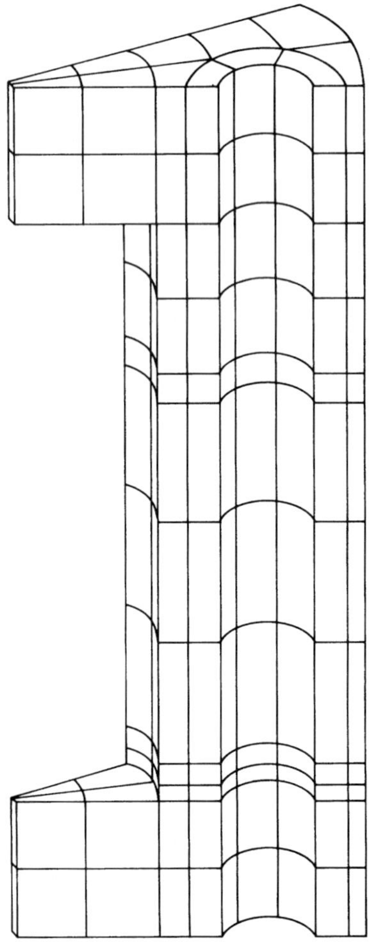

Fig. IX.109. Prestressed concrete cylindrical multi-cavity reactor vessel (HTR) — three-dimensional finite element.

described earlier, model tests have been carried out to verify the service and ultimate state behaviour of vessels [418–429], and to assess prestressing, reinforcement and liner requirements [418–429]. The scale chosen for a specific model depends on the objective of the test and on the reduction in size of important vessel components. For a complete vessel a suitable scale varies from 1 : 25 to 1 : 10, and scales of 1 : 25 and 1 : 50 have been used for the investigation of the top slab behaviour. A large number of models for shear modes of failure have been tested [418–423] with varying geometries, materials, prestressing loads, reinforcement and penetrations. Some pneumatic tests have been carried out using gas as the pressurising medium [418–423]. Most of these tests produced progressive failures With sufficient means

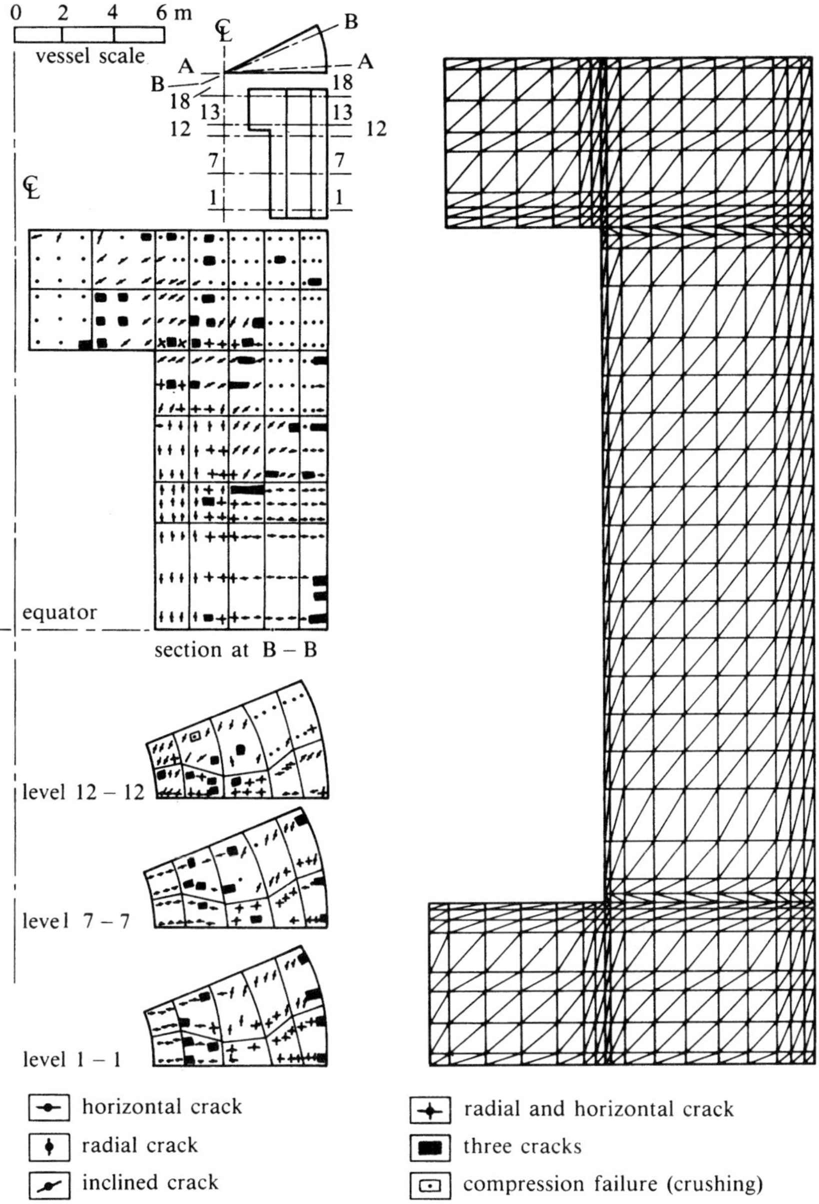

Fig. IX.110. Prestressed concrete cylindrical 2-D finite element pressure vessel.

available for detecting the elastic limit, the onset of cracking and the failure mode. Many individual tests have been carried out to assess the integrity of the liners, penetrations, closures and thermal protection systems [461–483]. The finite element analysis and constitutive model discussed earlier are applied in Section IX.9.5.5 to validate the proposed analysis.

IX.9.5.5 Analysis of results

(a) Dungeness B vessel. Figures IX.111 and IX.112 show gas pressure versus vessel deflection for the top cap. Here only 12 increments of pressure are plotted. For various pressures the trend in the incremental deflection is fairly regular and no abrupt change in the deflections between any two increments has been found. Finite element analysis using three concrete numerical models, namely the four-parameter, hypoelastic and endochronic models, has been carried out on the model as well. Figure IX.113 shows the pressure–deflection relation for two areas of the barrel wall. These results from the analysis are summarised in Figs IX.111 to IX.113b. The final curves are the idealised curves for pressure versus deflection of the wall and the top cap. The shapes of these curves are similar to those produced by Brading and McKillen [418, paper 3].

Figures IX.113a and IX.114 show the various principal stresses under normal operation of the vessel. Figure IX.115 shows the maximum and minimum principal stresses at 2.5 times the design pressure. Based on these stress contours, assessment of various cracks is made. The results are in good agreement with those provided by Brading and McKillen [418, paper 3] and as shown in Fig. IX.116a. Figure IX.116b shows a failure mode of the top cap along the thickness; this was provided by ENL of Italy [423].

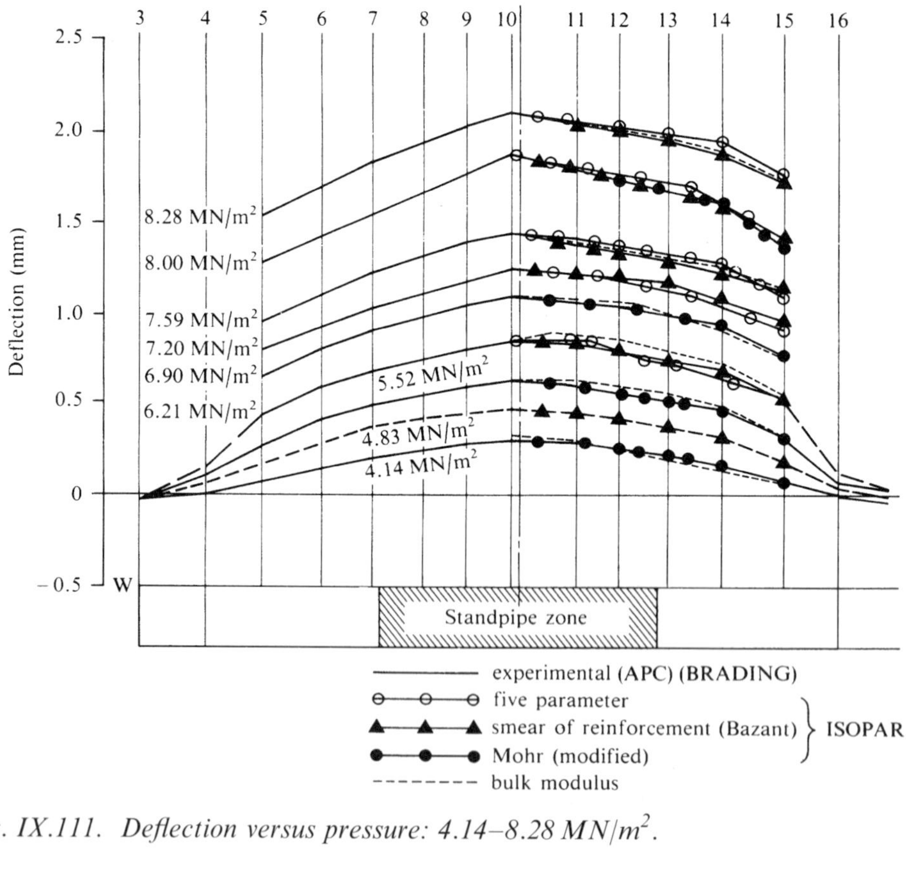

Fig. IX.111. Deflection versus pressure: 4.14–8.28 MN/m².

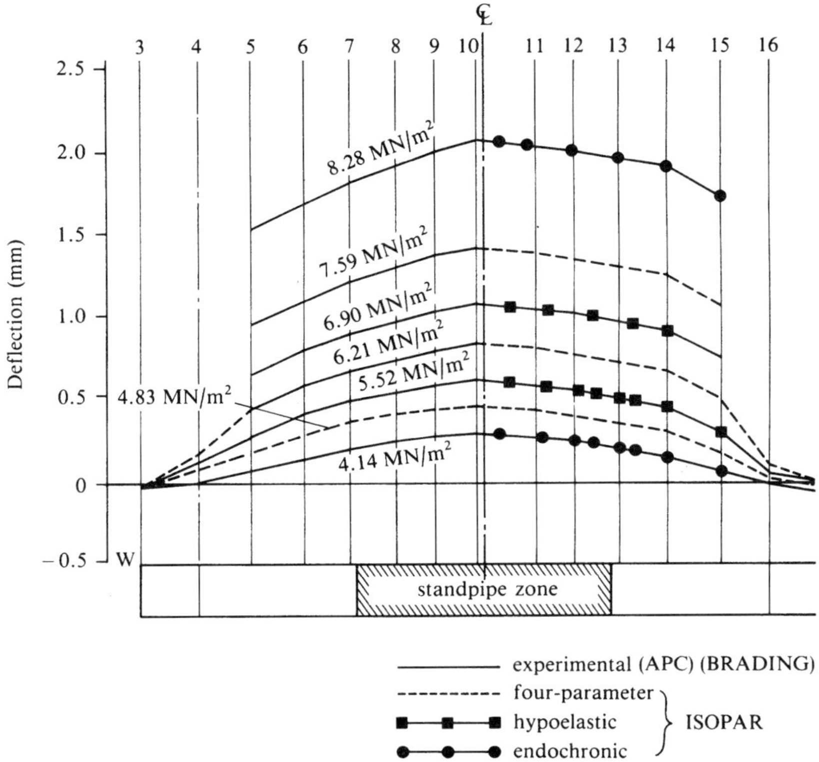

Fig. IX.112. Deflection versus pressure: 4.14–8.28 MN/m².

Since the Dungeness B vessel is stressed by circumferential tendons as well, it is essential to give the stress pattern under the anchorages. Figures IX.117a and IX.117b show maximum and minimum principal stresses, for normal operation, in zones where buttresses are located.

For ease of discussion in the following paragraphs M denotes cracks derived from the published model test results and FE denotes cracks in various zones predicted by the finite element and the limit state analyses.

Various stress trajectories at the design gas pressure ($P_{GD} = 3.3\,\mathrm{MN/m^2}$) and at 2.5 times the design gas pressures ($8.25\,\mathrm{MN/m^2}$) have already been discussed. At pressure $P_G = 4\,\mathrm{MN/m^2}$ ($1.21P_{GD}$) both FE and M indicate minor local cracking in the haunch zone initiated in both the radial and circumferential directions. In the top cap radial cracks appear from FE analysis. Model tests [418–423] indicate that initial local cracking at the top cap in the radial direction has occurred at a pressure of $3.46\,\mathrm{MN/m^2}$. This value has been raised to $3.8\,\mathrm{MN/m^2}$ in an isolated top cap test [418–423].

When the pressure is raised to $4.7\,\mathrm{MN/m^2}$ ($1.42P_{GD}$), no change is found in the radial crack, but in both the radial and circumferential directions a narrow band indicating a plastic zone has appeared. A slight increment in the length of crack is noticed in the haunch zone. Here both FE and M agree with each other. At the

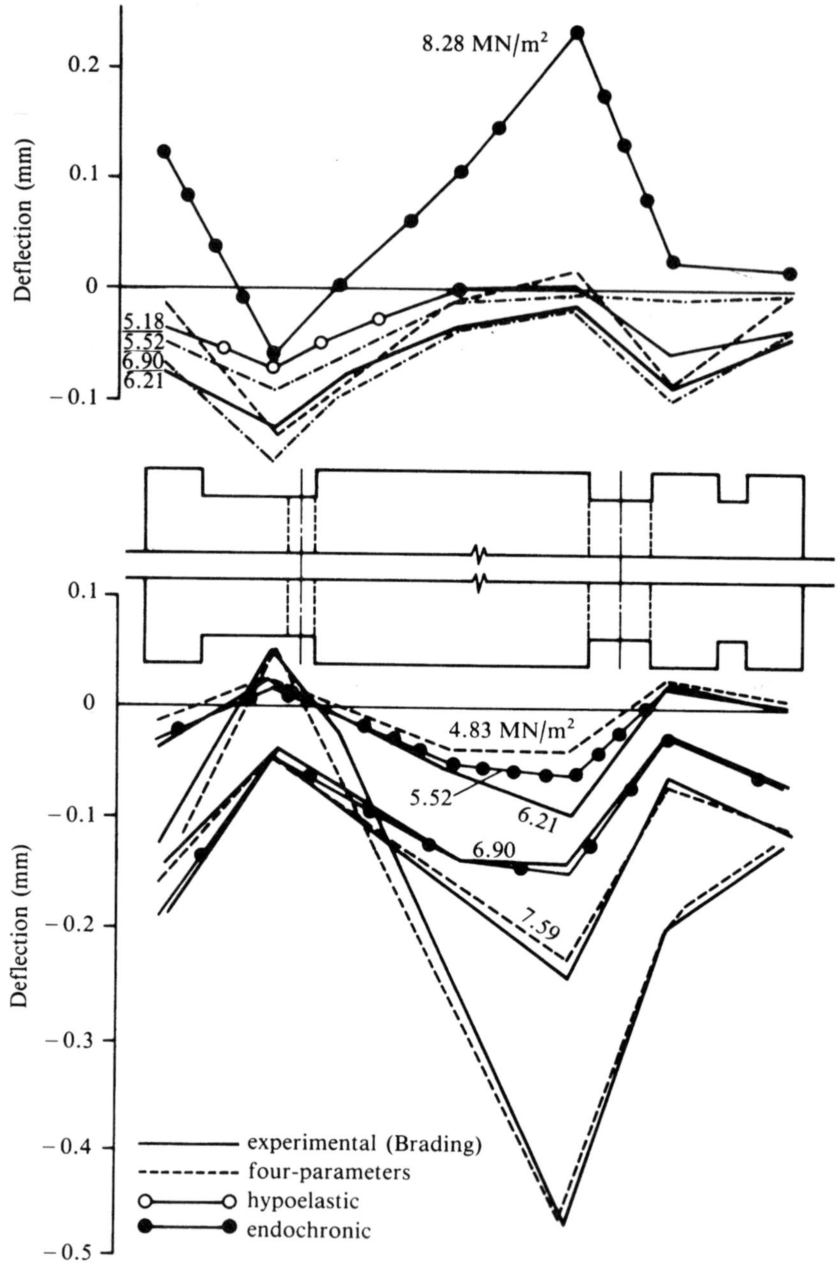

Fig. IX.113a. Deflected shape — sides of model, ultimate condition.

same pressure at the haunch near the bottom cap, a localised plastic zone has developed. However, at the same pressure the M analysis indicates a plastic hinge together with a crack at the bottom haunch. The 1/25th scale model test [418–423] has not indicated this feature, probably because of the difficulty of seeing and

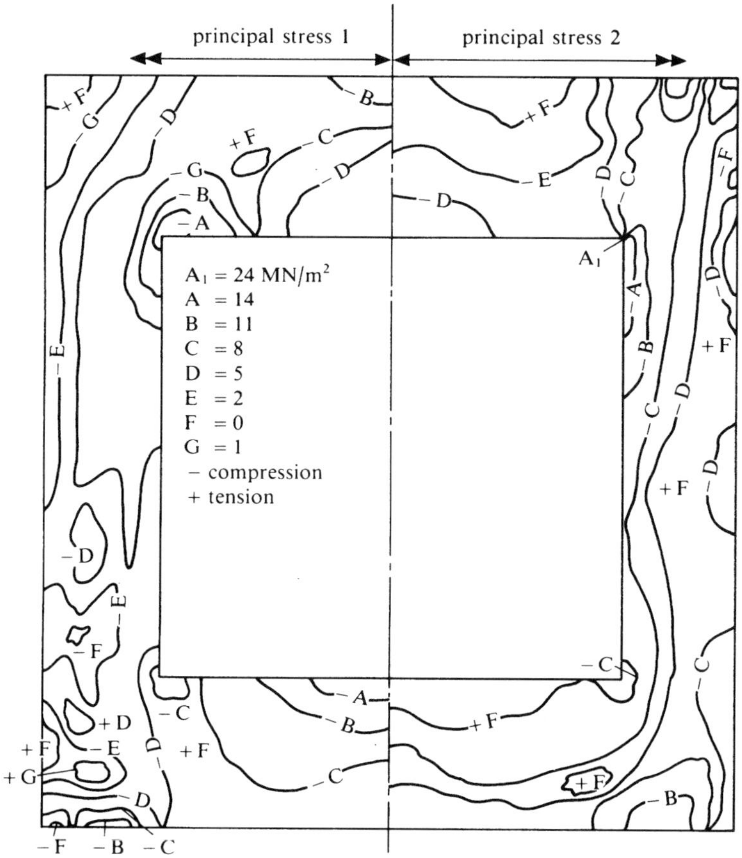

Fig. IX.113b. Principal stresses 1 and 2 — normal operation.

measuring such a crack. However, the tests show extreme stress concentration around this pressure in these local areas.

Raising the pressure to 5.1 MN/m^2 (1.545P_{GD}) the plastic zone has been increased only in depth. FE suggests that plasticity is followed by minor flexural cracks and some premature shear cracks initiating within this zone. Radial-cum-circumferential cracks have also been predicted by model tests.

At this pressure the top haunch crack indicated has been extended and so has the bottom haunch crack under this loading. These crack positions are in full agreement with those from the model tests M. Due to high stress concentrations produced in the top haunch by the model tests, there might be evidence of some kind of cracking but this has not been confirmed. Both FE and M predict surface cracking at the equator of the vessel. Clearly this again has not been revealed in published test data. At a pressure of 5.8 MN/m^2 (1.76P_{GD}), FE data indicate no change in the top cap plastic zone. It is clearly established that the localised shear-compression cracks predicted in the top cap by the model tests M have not changed much. There is an overall improvement in the plastic zones in the haunch areas.

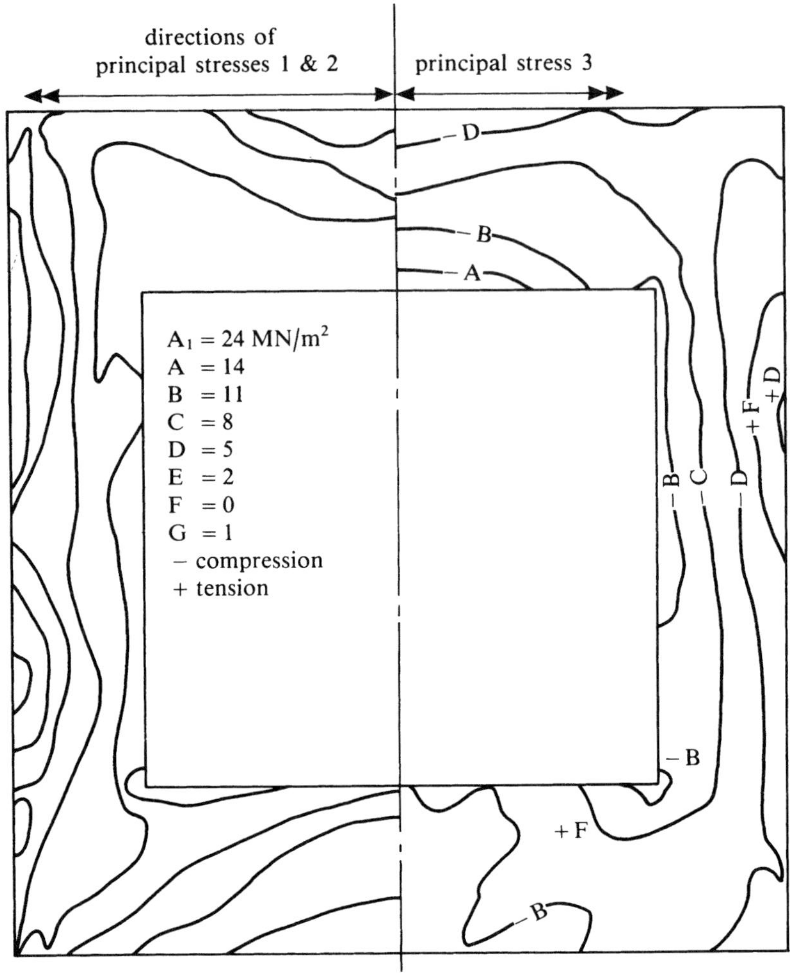

Fig. IX.114. Principal stress 3 and directions of the major principal stresses — normal operation.

However, there is no change of the loading in the surface crack at the equator except that FE indicates a slight extension of the plasticity zone at and around the equator with a possibility of isolated flexural cracks. Just opposite to the bottom haunch crack at the outside face of the vessel, a plastic zone has developed. Simultaneously, another plastic zone appears in areas along the centre line of the bottom cap. They have both been predicted by FE. No evidence is available from model tests.

When the pressure is raised to 6.5 MN/m^2 (1.97P_{GD}) it becomes evident that the top cap plastic zone has now been slightly modified by the FE data. In addition, both flexural and shear cracks in isolated pockets within this plasticity zone are visible.

The plastic zones at the haunches and areas close to them predicted by FE are now suddenly increased. These zones have practically the same size as predicted by M earlier.

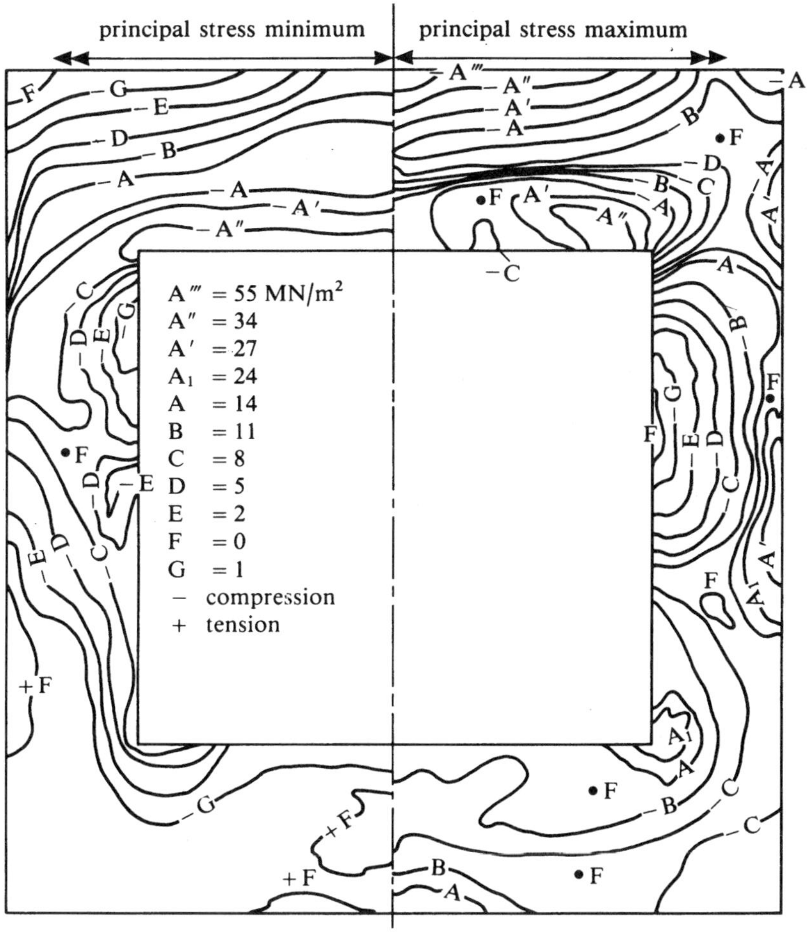

Fig. IX.115. Principal stresses at 2.5P_{GD}.

However, there is a change in the plastic zone in the bottom cap; but at the equator FE and M show flexural cracks, although the positions indicated by both vary. The experimental tests [418–423] show only surface flexural cracks half-way between those predicted by the finite element analysis.

When the pressure is increased to 7.2 MN/m^2 (2.14P_{GD}) the plastic zone has been extended far beyond that predicted in the earlier results. Nevertheless, within this zone the crack sizes predicted by FE have been stabilised and no further change has been observed. The shear compression crack and the inclined crack predicted by FE have been enlarged. This has been observed using M and the results are superimposed on the emerging inclined crack. There is a further extension of plasticity zones in the top and bottom haunches predicted by FE. From FE results and change in the wall cracks in all directions this is negligible, as there is no substantial improvement on the previous results. However, at the haunch in the bottom cap a two-pronged crack has been observed. This is due to the fact that results from M indicate crack positions both in the cap and in the barrel wall. The bottom cap

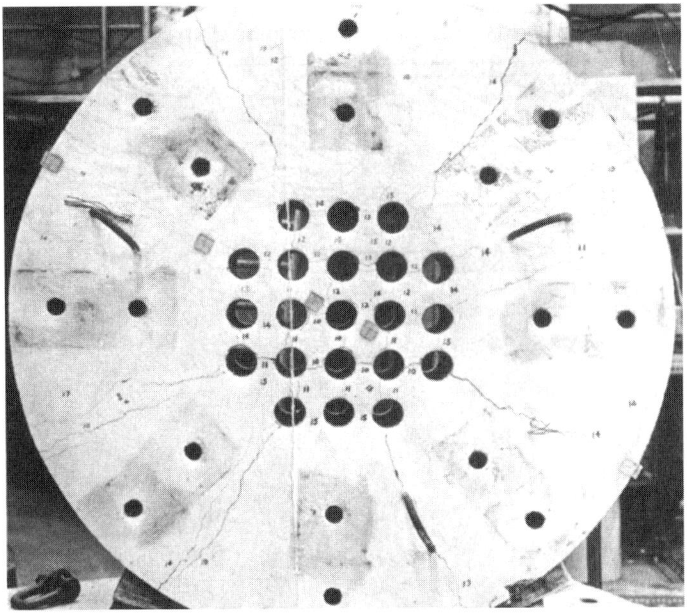

Fig. IX.116a. Flexural failure of caps (courtesy of Atomic Power Construction and ICE [418]).

Fig. IX116b. Plug-cum-flexure failure (courtesy of Professor Scotto, ENL, Italy).

plastic zone near the centre line predicted by FE has been extended radially and circumferentially with a possibility of a minor flexural crack near the centre line. There is a marked improvement in the plastic zone at the equator and near the outside face of the wall. Additional superficial cracks at the outside face have also

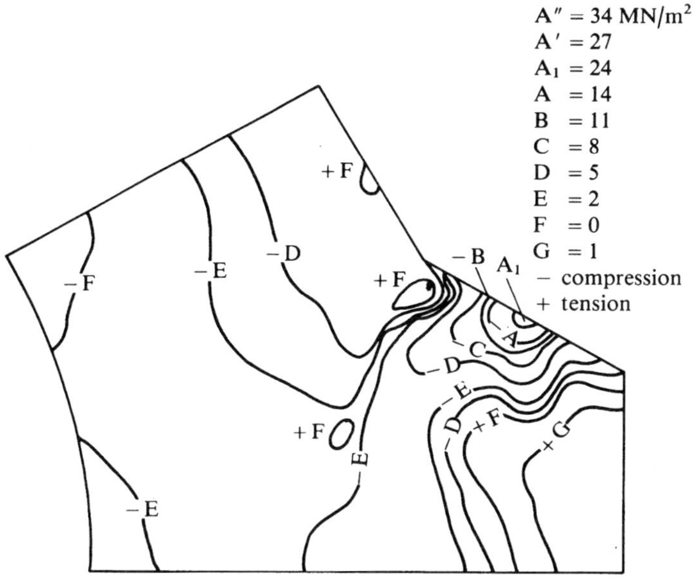

Fig. IX.117a. Maximum principal stresses — normal operation.

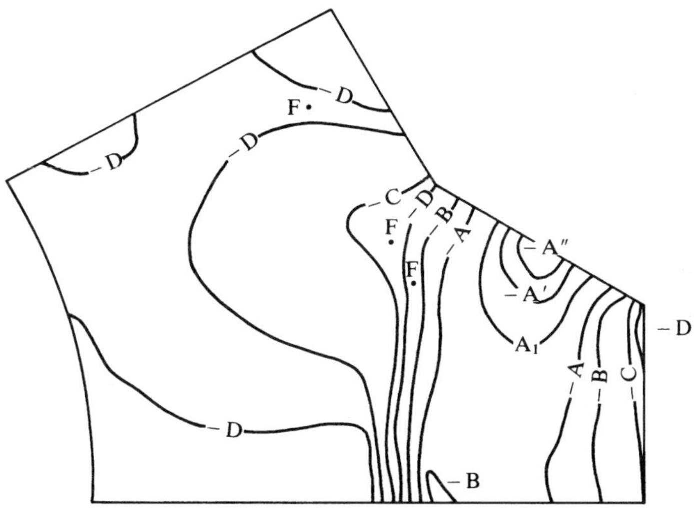

Fig. IX.117b. Minimum principal stresses — normal operation.

been predicted within this zone. It is interesting to note that the plastic zone under FE is almost the same as that predicted by M.

Now, with a further increase in pressure reaching 7.9 MN/m^2 (2.39P_{GD}), additional flexural cracks have been observed using FE. Some cracks have disappeared in minor areas and new cracks have emerged. The crack at the top haunch is two-pronged and jointed together at the corner. From M cracks have been predicted in the top cap and the inside zone in the wall. FE gives a slight

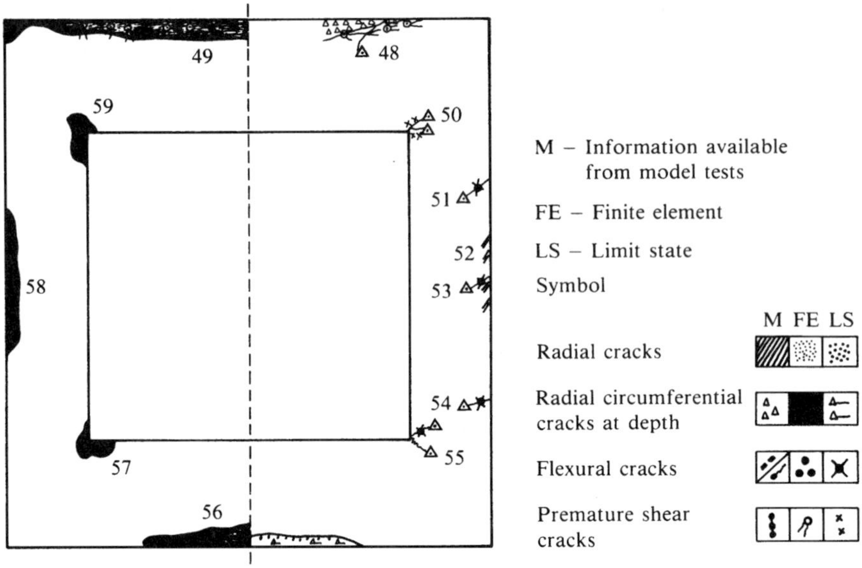

Fig. IX.118a. Dungeness B prestressed concrete reactor vessel ($P_G = 7.9\ MN/m^2$).

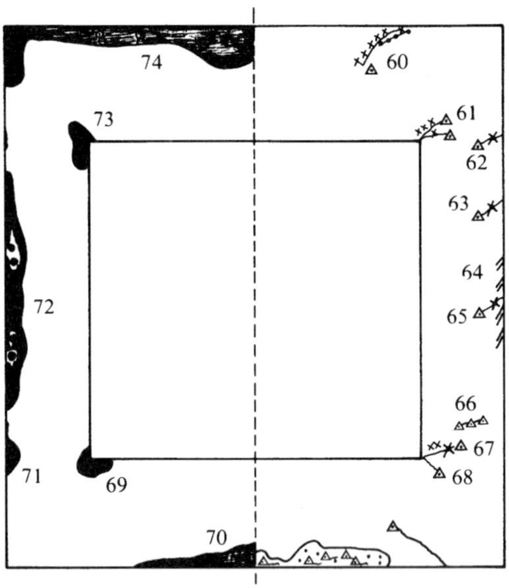

Fig. IX.118b. Dungeness B prestressed concrete reactor vessel ($P_G = 8.6\ MN/m^2$). (See Fig. IX.118a for key.)

improvement in the plastic zones in the top and bottom haunches. The plastic zone has been further extended with additional surface cracks in the circumferential directions. The depth of these cracks and the plastic zones have not been modified. In the top cap M predicts a complete initiation of the inclined crack which forms the initial surface for a plug failure. Nevertheless, FE results show extensions in the existing plastic zones, and development in the supplementary flexural cracks in radial and circumferential directions. It is interesting to observe that experimental tests [418, paper 3] indicate that the haunch crack has been observed at 7.7 MN/m^2 which has joined up with the first crack detected at 5.18 MN/m^2 in the circumferential direction. At this stage the central plug has been formed. In the bottom cap, the plastic zone predicted by the FE has been extended further, and an additional circumferential crack slightly away from the centre line of the vessel is formed. Figure IX.118a shows the stage reached at the pressure level $P_G = 7.9\,MN/m^2$.

At a pressure of 8.6 MN/m^2 ($2.6P_{GD}$) in the top cap (Fig. IX.118b) FE predicts a much enlarged plastic zone under mark 74 together with widespread cracks initiating radially and circumferentially. Corresponding to this limited state analysis (LS) shows cracking (mark 60) with a visible line for a plug failure. Along the barrel wall an additional crack (mark 62) has been predicted by M. Cracks under marks 61, 63, 65, 67 and 68 have been slightly modified and extended. The results from the experimental tests [418–423] show a crack above the duct (mark 66). This is in line with the plastic zone and cracks developed (mark 72) by FE. Results from LS do not predict a crack at this stage. No change has been envisaged by FE in the top and bottom haunch plastic zones (mark 73 and 69) and the bottom cap plastic zone, and cracking radially and circumferentially under mark 70.

Now when the pressure has reached 10.7 MN/m^2 ($\approx 3.25P_{GD}$) the top cap (Fig. IX.119) (by both FE and M) shows deep cracks for plug failure (marks 75 and 76) and the plastic zone in the top cap (mark 75) almost reaches the plastic zone coupled with cracks at the top haunch (mark 88). The plug failure line (mark 76) is close to that predicted by model tests [418–423]. Apart from that, at this stage, the circumferential cracks in the top cap are still discontinuous and there is

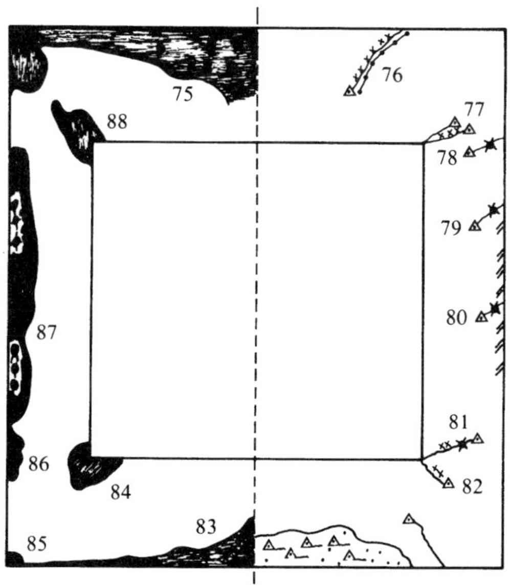

Fig. IX.119. Dungeness B prestressed concrete reactor vessel ($P_G = 10.7\,MN/m^2$). (See Fig. IX.118a for key.)

no further damage apparent to the top cap. The plastic zone in the barrel wall (mark 87) and cracks contained therein have enlarged their boundaries. The trend is identical to that exhibited by LS over the area shown under marks 77 to 82. In both cases cracks have not gone so deep despite excessive deflections. The bottom haunch (mark 84) and the bottom cap (marks 83 and 85) have their plastic zones extended and visible cracks have been predicted by FE. Corresponding to this the M shows an inclined crack, indicating the first complete initiation of the plug failure. The corner crack 71 (Fig. IX.119) discussed earlier has been stabilised but an additional plastic zone has been discovered under mark 86.

At this stage it is fairly evident that both caps are more vulnerable to failure than the barrel wall.

When the pressure is raised to 12.8 MN/m^2 (3.88P_{GD}) new cracks have been predicted by LS and now the total number of cracks is shown under marks 90 to 98 in Fig. IX.120. Both M and FE have predicted the extended failure zone in the top cap (marks 89 and 105). It is interesting to observe that both LS and FE crack zones (marks 102 to 104) have been extended further along the height of the vessel but have not made any further advance within the thickness of the wall. On the other hand, both top and bottom haunch cracks (marks 101 and 104) have made a significant advance. The bottom cap plastic zone (mark 101) has reached the plastic zone of the wall (mark 102). The continuity has not yet been broken. On the other hand, the top haunch plastic zone (mark 104) has reached the wall under its own influence. However, there is still a continuity. The bottom cap plastic zone (mark 100) has been enlarged with a few additional cracks. Here LS gives two closed cracks in the bottom cap but the depth of these cracks remains practically the same.

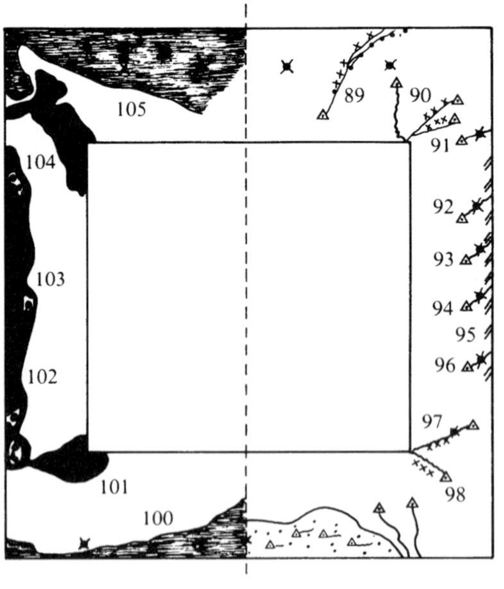

Fig. IX.120. Dungeness B prestressed concrete reactor vessel (P_G = 128 MN/m^2). (See Fig. IX.118a for key.)

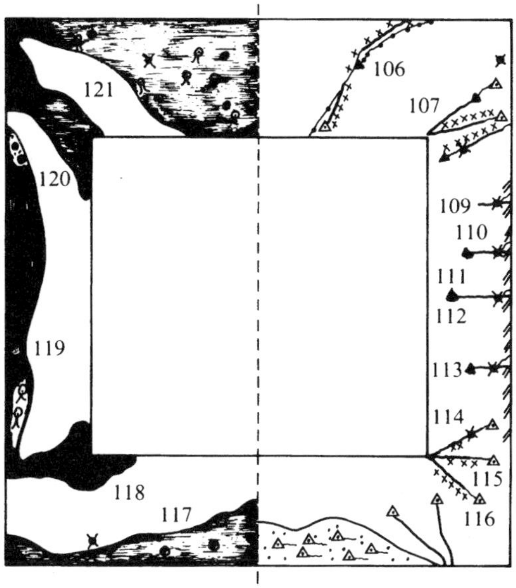

Fig. IX.121. Dungeness B prestressed concrete reactor vessel ($P_G = 13.5\,MN/m^2$). (See Fig. IX.118a for key.)

Now, raising the pressure to 13.5 MN/m^2 (4.1P_{GD}) there is a complete plug failure in the top cap, and a shear compression failure at the junction of the top cap and the wall, as are shown in Fig. IX.121 under marks 106, 107, 120 and 121. The study of the LS results indicates that haunch cracks (mark 107) have made a significant advance within the thickness of the wall causing a shear–compression failure. The remaining cracks in the wall (marks 107 to 115) have, likewise, predicted by FE under mark 119, not made much progress. This is very much true for the bottom cap, except that a shear–compression failure exists near the bottom haunch under mark 118.

With an assumed 2% ultimate strain in the tendons, the possibility of failure of some longitudinal tendons and a group of circumferential tendons lying close to marks 106, 107, 120 and 121 cannot be ruled out. From the experimental tests [418–423] it is evident that the central core (plug) has come out at a pressure of 13.8 MN/m^2 (1850 lb/in^2) and two tendons in the model have failed at a pressure of 10.2 MN/m^2.

It is concluded that the failure of this vessel has occurred owing to the complete removal of the central core or plug in the top cap at a pressure around four times the design pressure. The FE and the M analyses are in good agreement with each other and with most of the published data from the experimental tests. Throughout the vessel there is no premature failure of the liner.

(b) Oldbury vessel. For the model test of the Oldbury vessel, gas pressure versus deflections is plotted in Figs IX.122 and IX.123. The results from finite element analysis have been plotted for comparison in the same figures. Because of great variations in tendon layout, the pressure versus deflection curves are given for

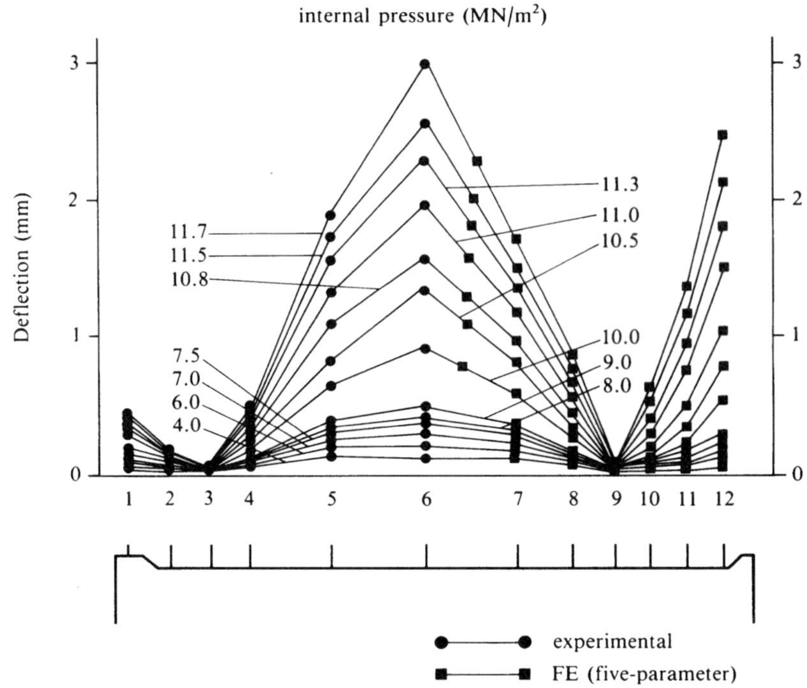

Fig. IX.122. Horizontal deflection of wall (ultimate pressure) — Oldbury vessel. [418; experimental]

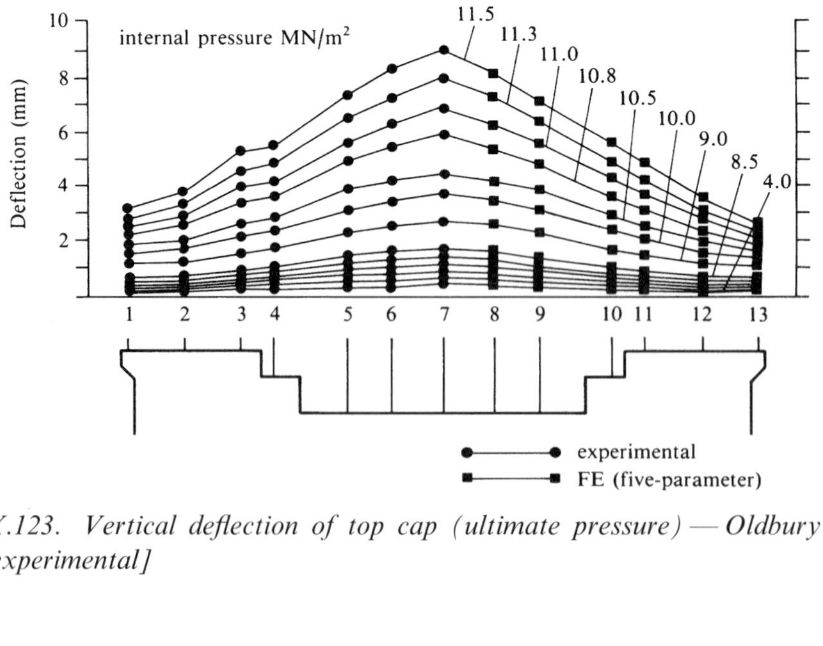

Fig. IX.123. Vertical deflection of top cap (ultimate pressure) — Oldbury vessel. [418; experimental]

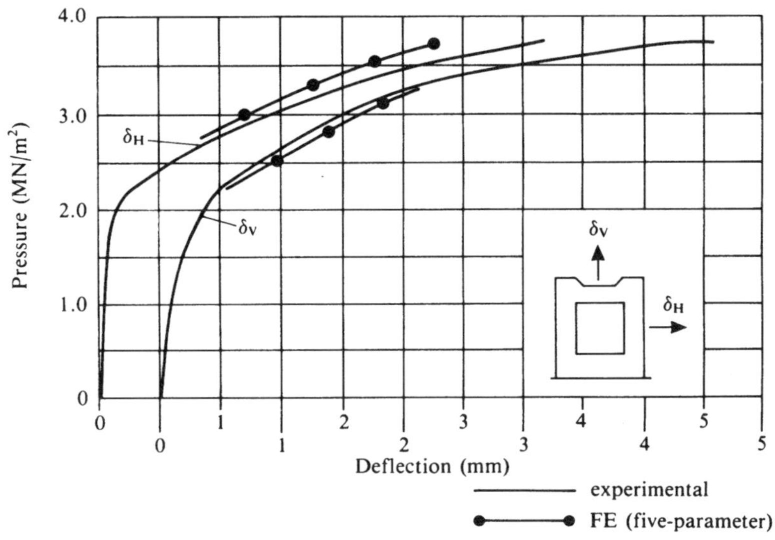

Fig. IX.124. Pressure–deflection curve of Oldbury vessel. [418; experimental]

various positions or locations. Therefore each curve takes into account the contribution it has received from the exact position of tendons, conventional steel and the liner in the original layout. In order to test these results, some of them are plotted in Fig. IX.124 along with those from the model tests (four tests) carried out by Eddie *et al.* and others [418–423]. Both the horizontal and vertical deflections are in good agreement. However, in Fig. IX.125 although the computed axial strains are in good agreement with those measured on the prototype [418–423], the hoop or circumferential strains are wide apart. The only explanation that can be offered is the difference between the exact chosen tendon layout of the vessel parameters and that of the prototype. The percentage difference is still within 5–8%. A further comparison is given in Fig. IX.125(a–c) which is based on the monitoring of the vessel's performance for a period of 10 years.

Figures IX.126 and IX.127 show plots for various principal stresses. It is interesting to note that in Fig. IX.128 the effect of removal of cracked concrete at 1.81 times the design pressure is shown. Figures IX.129 and IX.130 show the principal stresses at 2.5 times design pressure. No cracks have been observed for a pressure range of 2.66–3.75 MN/m^2. At a pressure of 3.99 MN/m^2 (1.455P_{GD}) a plastic zone in the top cap centre has been predicted by FE. Both FE and M show a plastic zone and a minor crack at the top haunch. Radial hair cracks together with a plastic zone have also been predicted by FE. However, M assumes a crack width at the equator and finally proves that it is the only crack that exists in the wall at this pressure.

At a pressure of 4.665 MN/m^2 (1.75P_{GD}), a wider and deeper plastic zone appeared (mark 6) in the top cap. It is interesting to note that the top haunch plastic zone has also been increased. In addition, a minor plastic zone has been predicted by FE at the bottom cap haunch. Moreover, at the upper part of the side wall a wider radial crack has occurred and this crack has also been confirmed by the model tests,

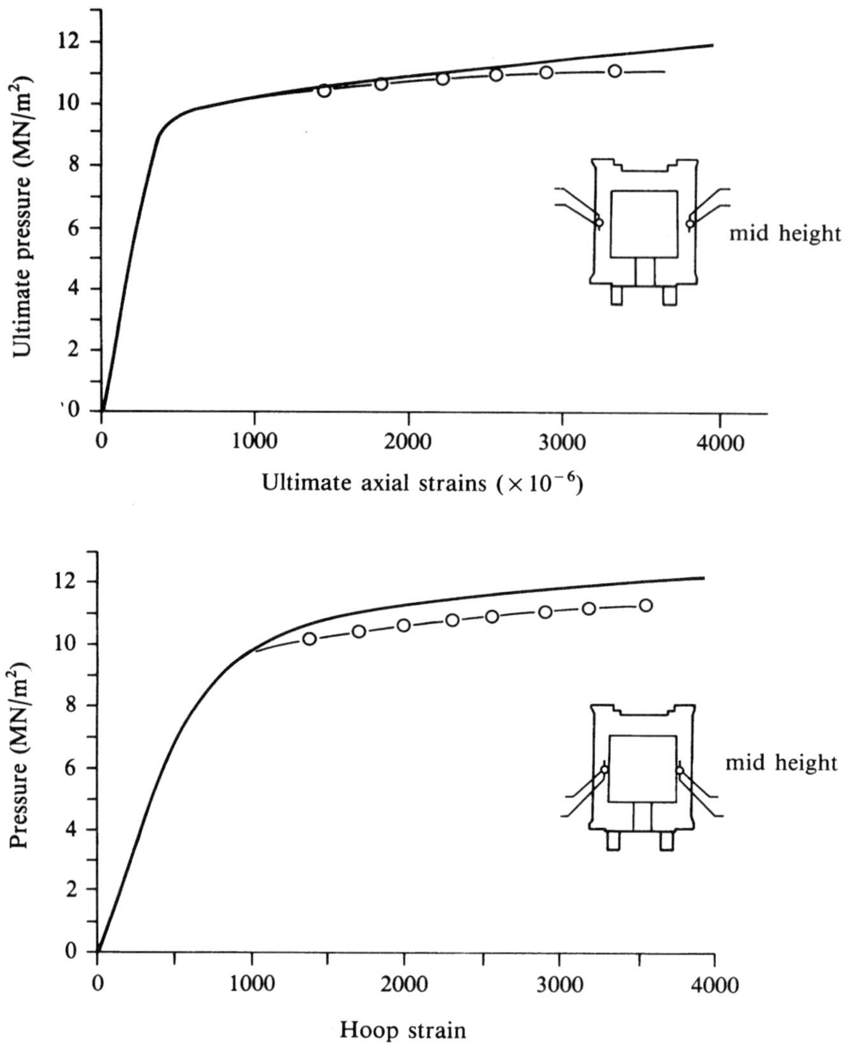

Fig. IX.125. Hoop and axial strains at wall mid-height — Oldbury reactor vessel (conditions at ultimate pressure). (——, Finite element (five-parameter); —O—, experimental.)

but at a pressure of 4.825 MN/m^2 (1.815P_{GD}). This must have been produced even earlier in the model, since such a crack cannot suddenly appear at such a marginal pressure difference of 0.065P_{GD}. The results from FE clearly indicate that there is a plastic zone appearing at the equator.

Raising the pressure to 5.32 MN/m^2 (1.995P_{GD}) FE indicates that the vessel top cap plastic zone has been abruptly increased and additional flexural cracks have emerged. The plastic zone in this case has reached the areas surrounding the central hole in the stand pipe area and the prestressing galleries. Model tests [418–423] show similar effects in the top cap. The results from M give two surface flexural cracks and

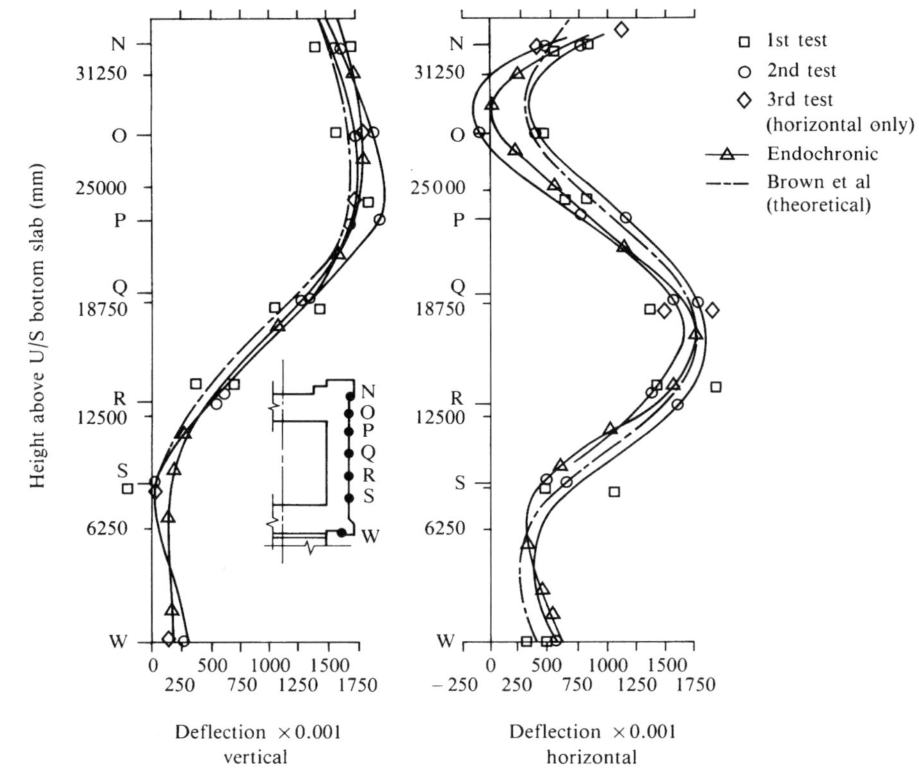

Fig. IX.125a. Wall horizontal and vertical deflections.

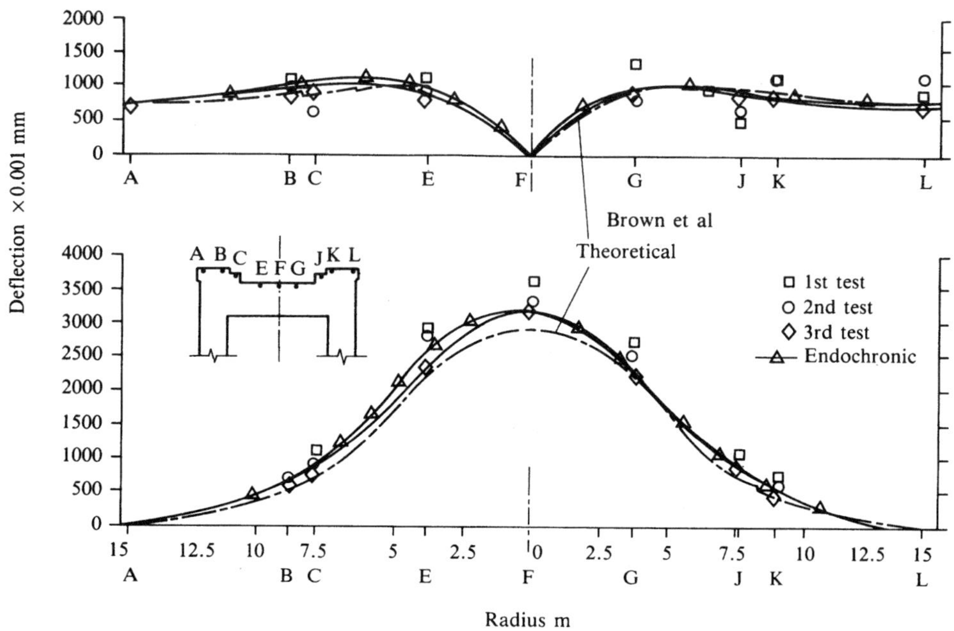

Fig. IX.125b. Top slab horizontal and vertical deflections.

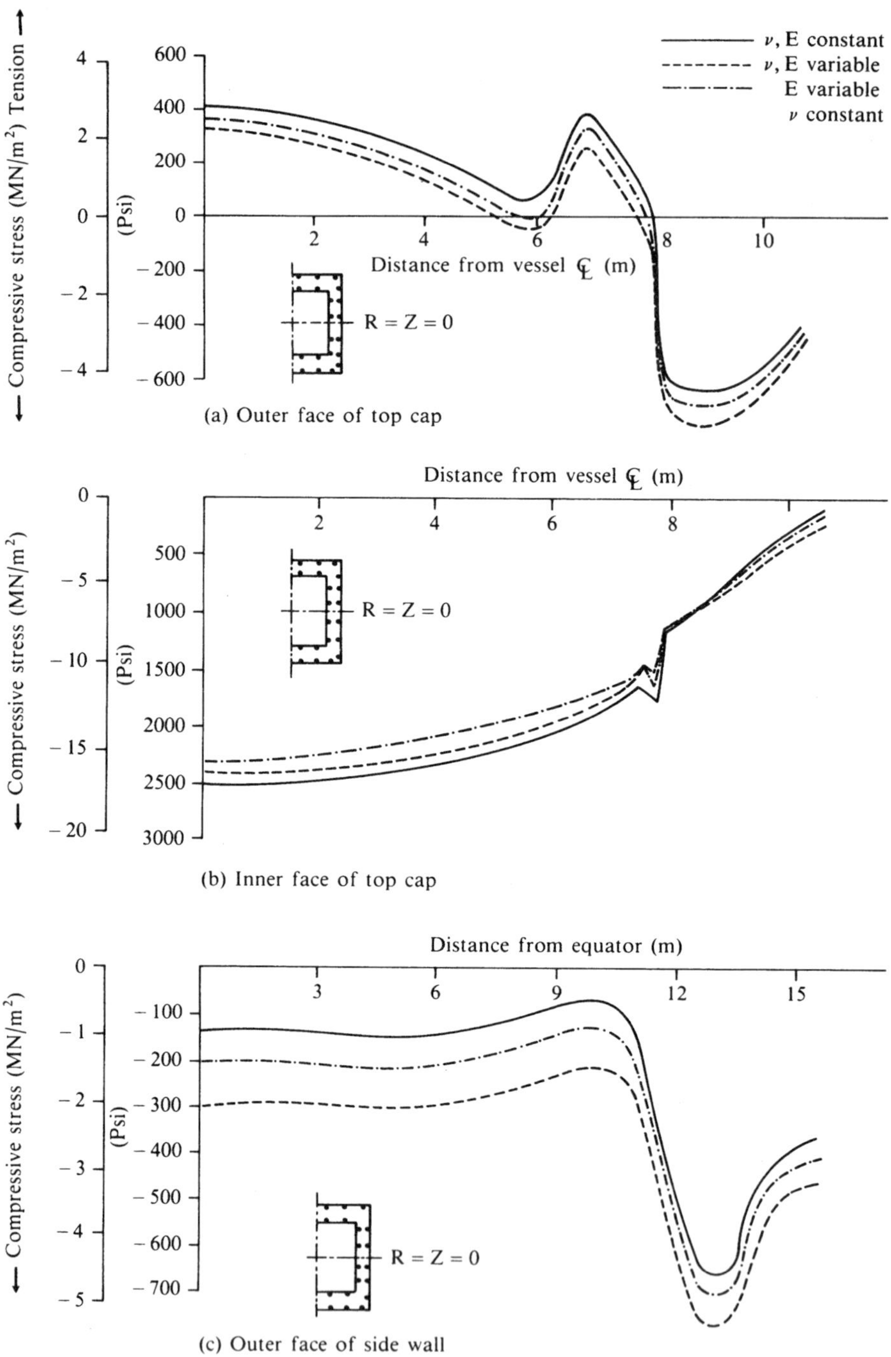

Fig. IX.125c. Circumferential stresses — operating conditions only.

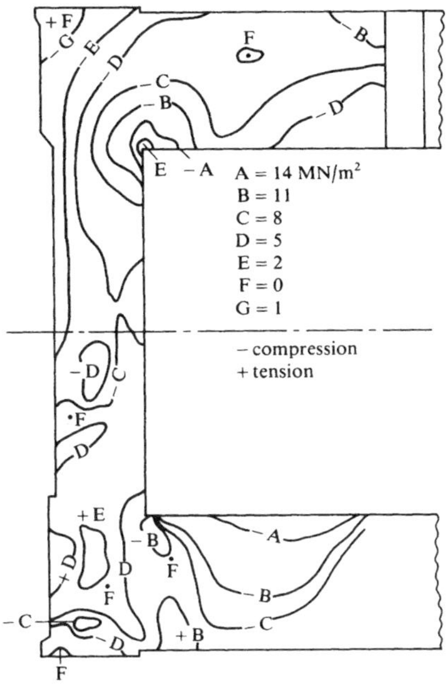

Fig. IX.126. Principal stress 1 — normal operation.

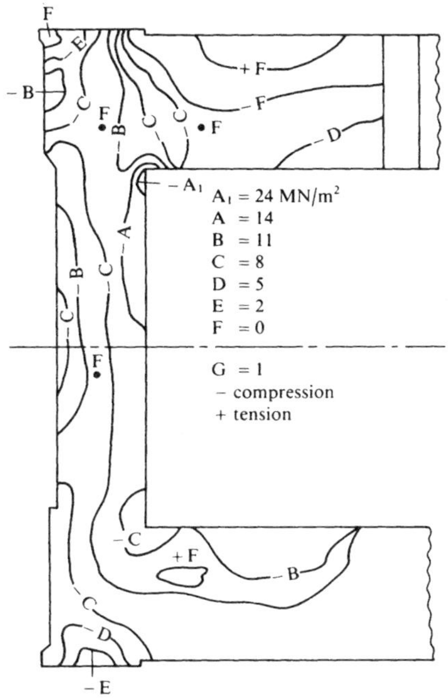

Fig. IX.127. Principal stress 2 — normal operation.

a shear compression crack initiating from the gallery zone. Both top and bottom haunch plastic zones of this vessel predicted by FE are fairly large. Here there is some disagreement between FE and M on the type of crack. The top haunch M results show both flexural and premature shear cracks. On the other hand, the results from FE clearly indicate the possibility of cracks caused by the principal tensile stresses caused by shear. However, there is an agreement between the two analyses on the initiation of flexural cracks in the bottom cap haunch. The magnitude of these cracks is not greater than 0.007 mm. Practically there is no change in the flexural crack. The bottom cap has an enlarged plastic zone.

Reports [418–424] suggest that a horizontal-cum-radial crack has appeared above the circulator penetration, but the published data [418428] have not confirmed it. In the barrel wall M shows flexural cracks in both radial and circumferential directions. Examination shows that cracks in radial directions do exist at pressures between 4.664 and 5.32 MN/m^2. They have been superimposed for comparison. The results obtained from FE show plastic zones and surface cracks. Obviously there is a slight discrepancy. It seems that FE does not agree with M up to a pressure of 5.32 MN/m^2 for the creation of plastic zones or surface cracks. The difference is in the modelling of the layouts. No test data are available for this vessel or any other cylindrical vessels [418-428] for such cracks appearing at 1.995 times the design pressure. However, there seems to be some agreement in this area at pressures above 5.32 MN/m^2.

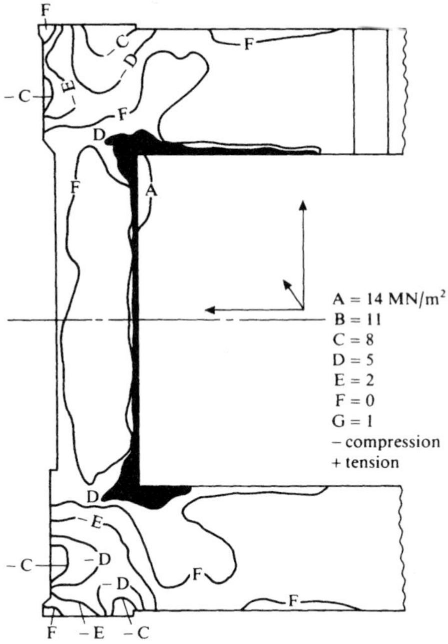

Fig. IX.128. *Minimum principal stress — pressurisation after long-term shut-down showing effect of removal of cracked concrete (A).*

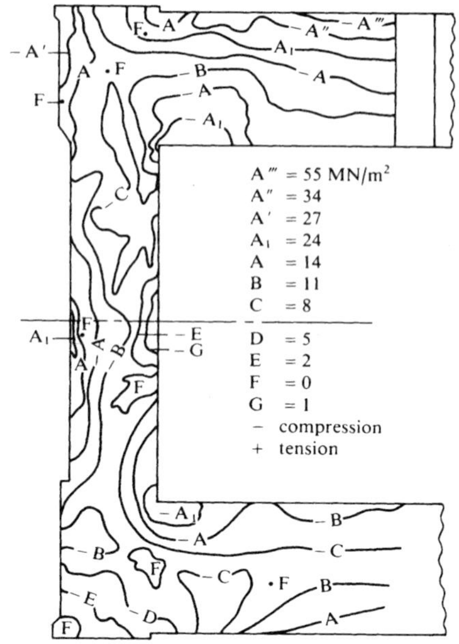

Fig. IX.129. Maximum principal stresses at 2.5P_{GD}.

For a pressure of 5.985 MN/m^2 (2.26P_{GD}) additional radial and circumferential cracks have appeared in the top cap of this vessel and the plastic zone predicted in the top cap has been extended within its thickness. There is virtually no change in the overall length of this plastic zone. Results from M predict two flexural cracks, one within and the other just outside the stand pipe perimeter. In addition, the shear–compression failure line has been initiated. At the outside of the top prestressing gallery, a circumferential crack has been predicted in a localised zone. However, in the top cap haunch (mark 27) FE predicts a much wider plastic zone. Here both FE and M are in agreement with the model test results. At the upper part of the vessel side wall, both FE and the model tests are now in agreement.

In the bottom cap haunch the plastic zone predicted by FE is now slightly changed in shape but still no crack has appeared. In the bottom cap itself, the plastic zone predicted by FE now extends from the centre line of the vessel to the inside edge of the lower prestressing gallery. Some flexural cracks are visible in the bottom cap of a magnitude not greater than 0.009 mm. Here both M and FE partially agree with each other. No information from model tests is available to support this. However, in critical regions of the wall cracks have appeared and here both FE and M are in full agreement. Again within this area model cracks measured from test data [418–430] have been shown side-by-side with the computed cracks. Finite element analysis predicts radial cracks, and model tests [418–430] show

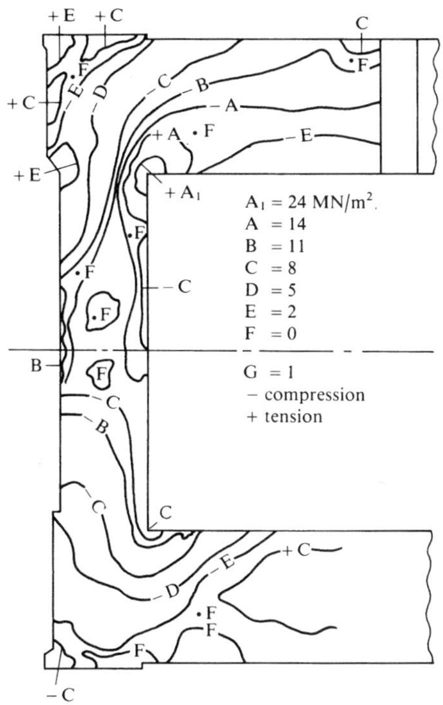

Fig. IX.130. Minimum principal stresses at 2.5P_{GD}.

that these cracks are justifiable, although the FE 'spread zone' is twice that of the model test data.

At a pressure of 6.65 MN/m^2 (2.5P_{GD}) additional cracks have appeared and the plastic zones have been increased along the thickness of the vessels. Here only lengths and widths of cracks in the radial and circumferential directions have been extended.

It is interesting to note that in the top cap there is a straight competition between the flexural and premature shear cracks. Superimposed on these are the reproduction of cracks obtained from the cracked model [418–431]. Both crack pattern analyses from the model in the top cap are very similar. One novel feature is the re-emergence of the radial crack predicted by the model test [418–431] reaching the circular penetration. This has also been predicted by FE.

When the pressure is raised to 7.98 MN/m^2 (3.0P_{GD}) model test results in the top cap of the vessel have shown zones of considerable cracking. In the actual prototype vessel, FE predicts a crack zone, but not as much as is given by the model test. On the other hand, in the top cap the flexural failure zone advances beyond the plug failure. At this stage it is necessary to show a cracked (before failure model) model, which is given in Fig. IX.131. Apart from a slight modification of the plastic zones, the cracking behaviour is fairly consistent.

As shown in Fig. IX.131 for a pressure of 8.645 MN/m^2 (3.25P_{GD}), the top cap plasticity zone predicted by FE comes very close to the model prediction, as is

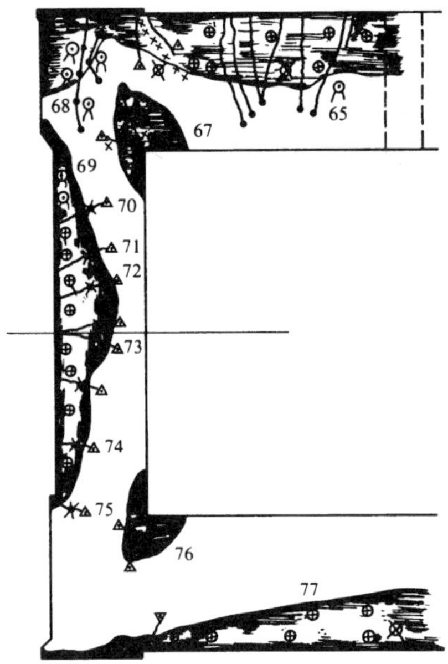

Fig. IX.131. Oldbury prestressed concrete reactor pressure vessel ($P_{GD} = 8.645\ MN/m^2$). (See Fig. IX.118a for key.)

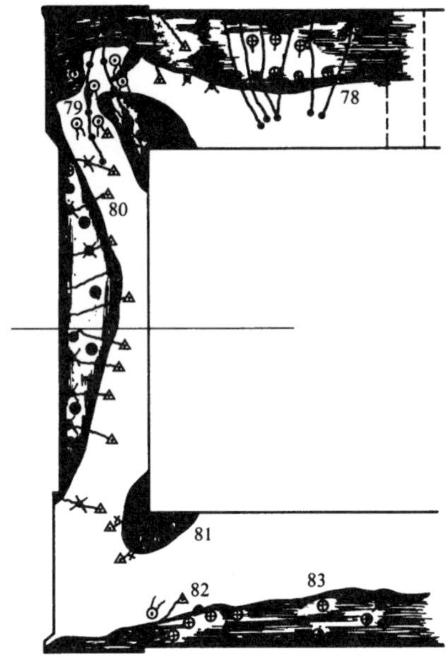

Fig. IX.132. Oldbury prestressed concrete reactor pressure vessel ($P_{GD} = 9.975\ MN/m^2$). (See Fig. IX.118a for key.)

evident from marks 65 and 68. In the top cap, the plug failure line is catching up with the flexural cracks, and at one place it is a pure shear failure line proceeding beyond the flexural zone. Both haunches (marks 67 and 76) have not gained much. The bottom cap (mark 77) plastic zone and cracks have increased in size and in number. The sizes of these cracks is no more than 0.9 mm, but the tendons have also reached 3% strain, which is very high and almost reaches the yield point. The barrel wall is developing cracks and plastic zones under pure flexure. The biggest crack in the barrel wall is at the equator. Most of the top gallery zone in the top cap has cracked. Model tests predicted a portion of gallery cracking at a pressure of 7.98 MN/m^2.

When the pressure is raised to 9.975 MN/m^2 (3.75P_{GD}), cracks have changed their positions (Fig. IX.132), but the failure patterns given by FE and M are consistent. The plastic zones near the top gallery (mark 79) have been extended. It seems that the top haunch plastic zone is very close to the one approaching from the top gallery. Both bottom cap (marks 82 and 83) and the barrel wall plastic zones predicted by FE have been very slightly modified. Tendons passing through these cracks have almost reached 3.8% strain, and their failure has been predicted. The top cap is broken into two sectors. The longitudinal wall cracks have joined the circumferential cracks of the wall and the caps.

At a pressure of 11.305 MN/m^2 (4.24P_{GD}) the plastic zone at the top haunch has reached the gallery zone (mark 85), as shown in Fig. IX.133. Both in the wall and in

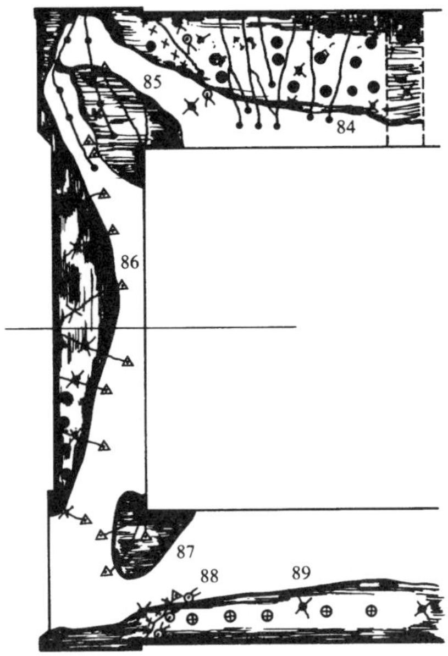

Fig. IX.133. Oldbury prestressed concrete reactor pressure vessel ($P_{GD} = 11.505\ MN/m^2$). (See Fig. IX.118a for key.)

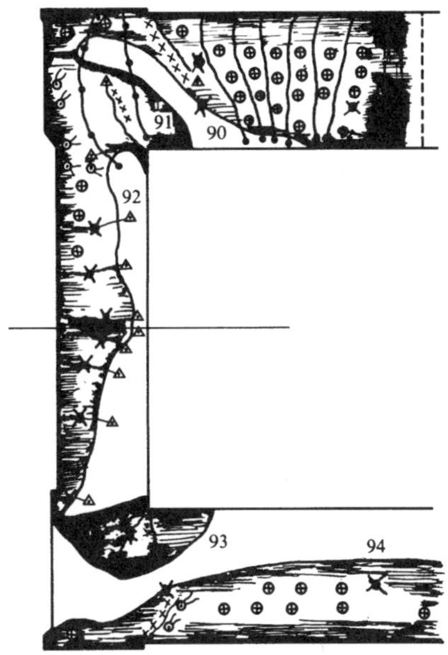

Fig. IX.134. Oldbury prestressed concrete reactor pressure vessel ($P_{GD} = 11.97\ MN/m^2$). (See Fig. IX.118a for key.)

the caps the cracks have advanced further. Model test results are in agreement with those predicted by FE and M. However, at a pressure of 11.97 MN/m^2 (4.5P_{GD}) the vessel top cap and top corner (Fig. IX.133) have been separated. The bottom cap haunch plastic zone has reached the wall plastic zone but the cracks (mark 93) have not joined. The wall and the bottom cap are broken into several pieces but have not been separated yet. The top cap failure is predominantly flexural, although shear–compression failure causing the removal of the plug is just one step away. The tendons in most places have reached 4.25% strain. The crack sizes are not exceeding permissible values, and premature liner failure is not possible provided no initial defects exist. The conventional steel at the haunches and at equator has yielded.

It is concluded that the vessel predominant failure is governed by the failure of the top cap in flexure and the tendons at a pressure of 9.975 MN/m^2 (3.75P_{GD}). The remaining failure mechanisms (Figs IX.133 and IX.134) are purely an academic exercise for investigating any reserve strength left in the vessel.

This failure mechanism is different from that predicted for the Dungeness B vessel, although they both have an identical shape and reactor system layout but with different prestressing system and gas design pressure.

(c) High-temperature gas-cooled reactor vessel (HTGCR). Figures IX.135 and IX.136 show gas pressure versus deflection, and Figs IX.137 and IX.138 show principal stresses at normal operation. Figures IX.139 and IX.140 show principal stresses at 2.5 times the

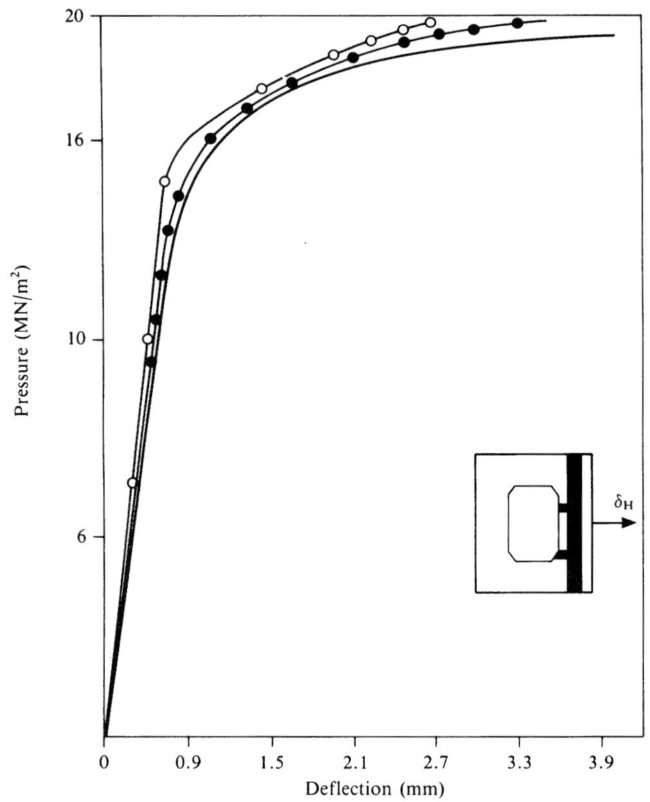

Fig. IX.135. Deflection curve of wall of HTGCR reactor vessel. (——, Experimental; —○—, FE (four-parameter); —●—, FE (five-parameter).)

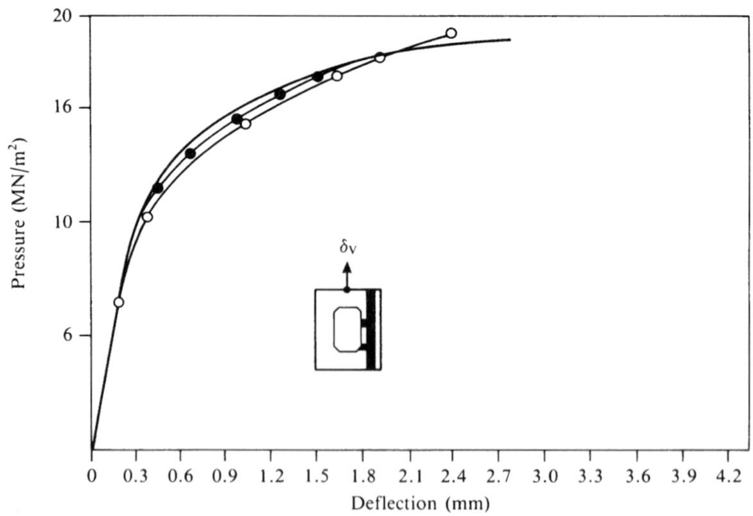

Fig. IX.136. Pressure–deflection curve for top cap of HTGCR reactor vessel. (——, Experimental; —○—, FE (four-parameter); —●—, FE (five-parameter).)

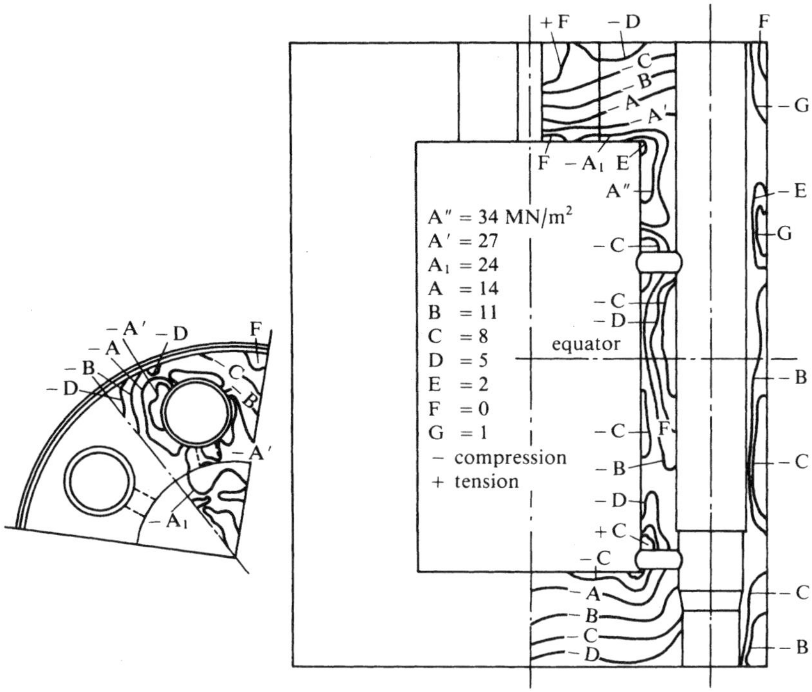

Fig. IX.137. Maximum principal stresses at normal operation.

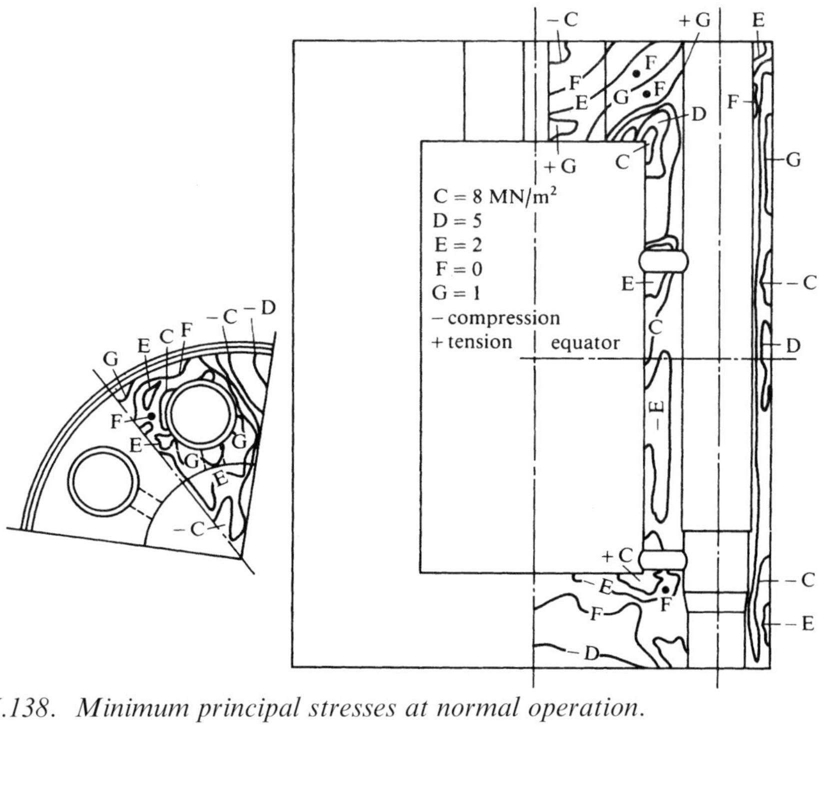

Fig. IX.138. Minimum principal stresses at normal operation.

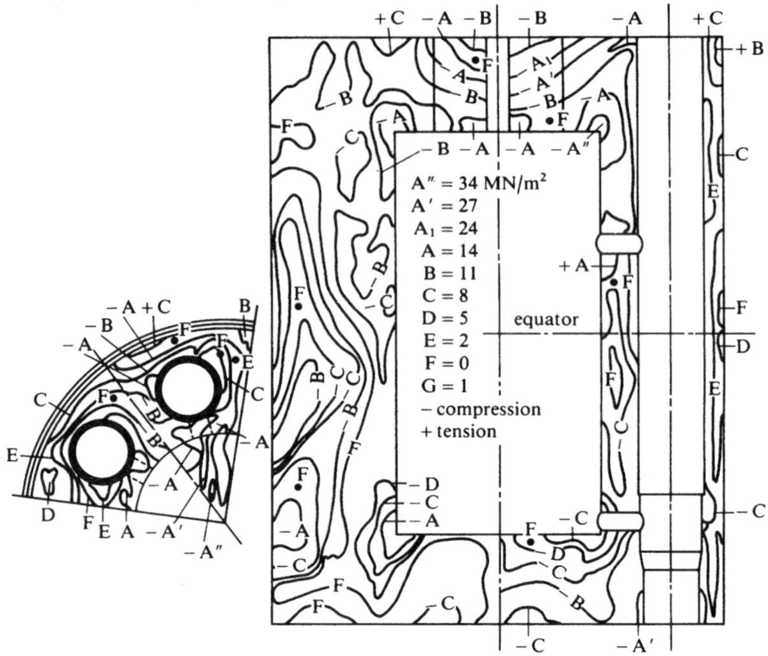

Fig. IX.139. Maximum principal stresses at $2P_{GD}$.

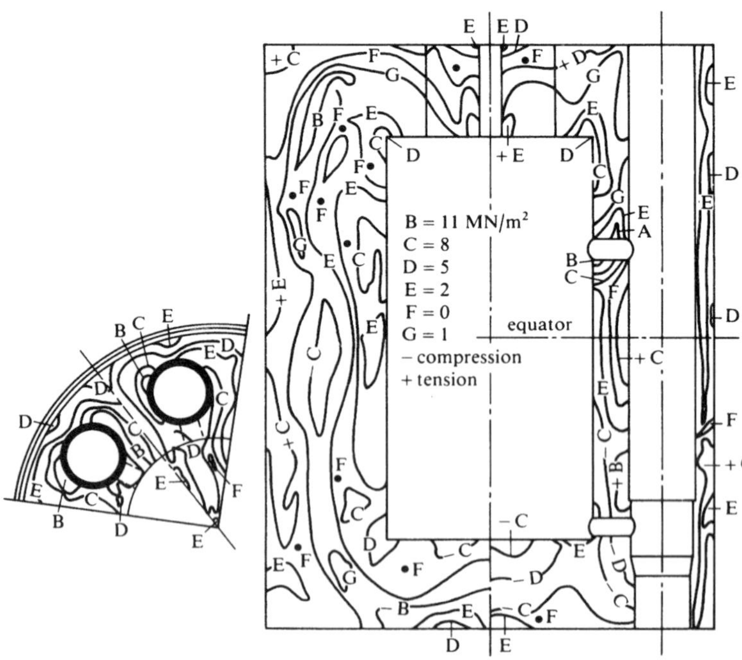

Fig. IX.140. Minimum principal stresses at $2P_{GD}$.

design pressure. Although this vessel layout is different in many respects to that of the Hartlepool reactor, the stress flow in many areas is identical. The stress concentrations around the boilers and at the junction between the caps and the walls are similar in both cases. However, the quantities do differ. It is because of these and some dissimilar stress contours that the final failure modes in both vessels differ.

Figures IX.109 and IX.110 show three-dimensional and two-dimensional finite element mesh generation schemes of a prestressed concrete vessel designed for a high-temperature gas-cooled reactor. Both internal and external loads have been computed in accordance with the method given in Chapter VI. The program CREEP, which is a part of the program ISOPAR, gives results for when the vessel undergoes the effect of creep. First a normal operation condition (prestress + pressure + temperature) is considered. This vessel is analysed with and without creep. The axial, radial and circumferential stresses have been computed. The influence of creep when the Young's modulus is constant or variable is considerable. In some local areas, because of creep, the stresses change significantly with passage of time. Figure IX.141 shows

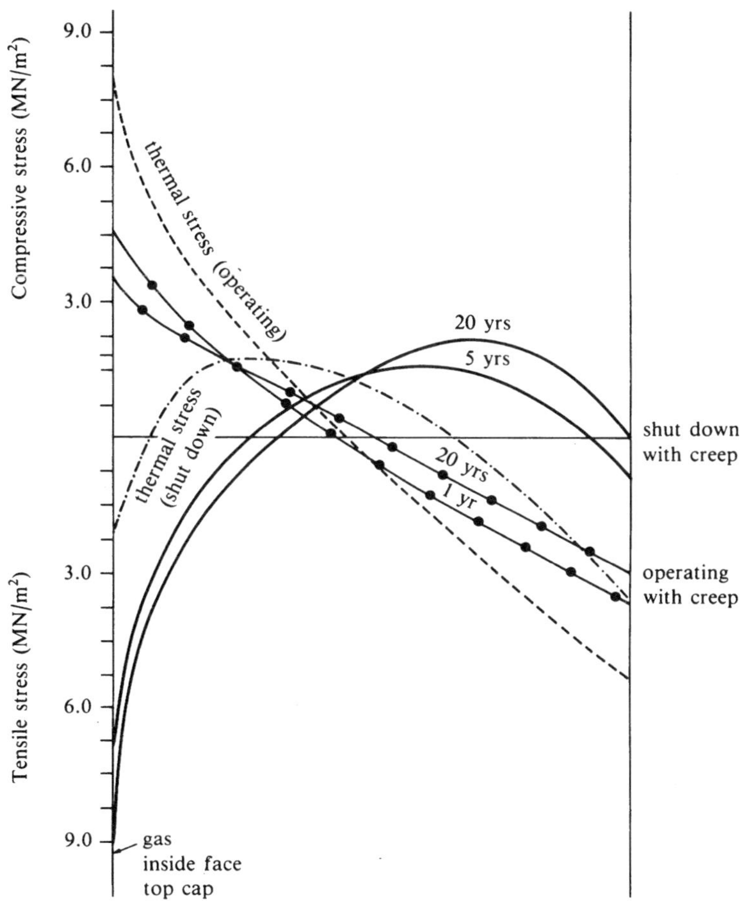

Fig. IX.141. Stresses through top cap — at operating temperature and shut-down.

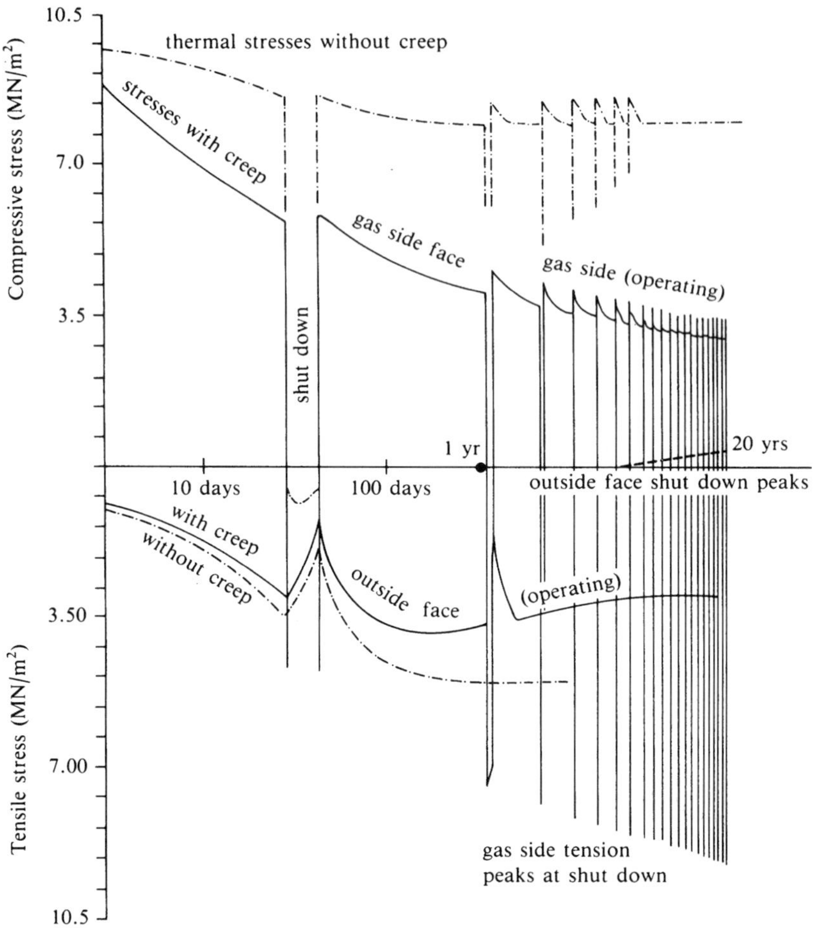

Fig. IX.142. Stresses in top cap showing influence of shut-down.

stresses at the top cap with and without creep for up to 20 years. Figure IX.142 shows the behaviour of the vessel under operating and thermal shut-down conditions with and without the influence of creep.

The vessel is then taken to elasto-plastic and cracking conditions. As discussed later on, the relationships between gas load, prestressing strains, and deflections and cracks with and without the influence of creep have been established. These results show that where the influence of creep is not considered, the load-carrying capacity is overestimated by as much as 25%. With creep, considerable changes were discovered in the cracking pattern of the vessel and the overstress conditions in local areas (stand pipes and boiler/circulator penetrations).

Ahmad [262] has idealised the same vessel, as shown in Fig. IX.143. He has used the vessel bonded and unbonded, and produced deformations (Figs IX.143–IX.146) for normal operation and for 40 years of creep. He has also produced a graph

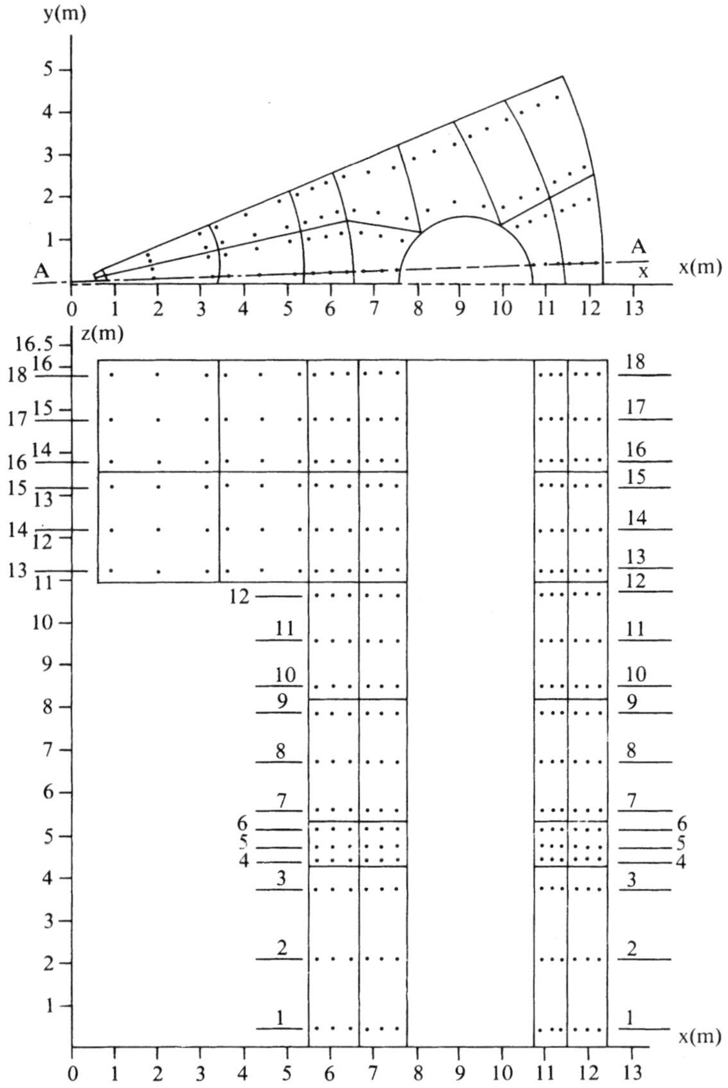

Fig. IX.143. Gauss point locations in the reactor vessel [262] (all dimensions in mm).

(Fig. IX.147) showing deformation for this vessel at various pressures. He has included creep in his analysis. This vessel has also been considered for elasto-plastic and cracking effects by Bangash [749]. The deformation of the vessel at the cracking stage under the influence of creep and with the steel liner anchored to concrete is shown in Fig. IX.147. The analyses for these cases are given in Chapters VI and VII.

The stress trajectories at normal operation and at 2.5 times design pressures can be obtained in the same manner as for vessels discussed previously. The vessel design pressure P_{GD} here is 5.68 MN/m^2. At a pressure of 7.48 MN/m^2 (1.315P_{GD}), this

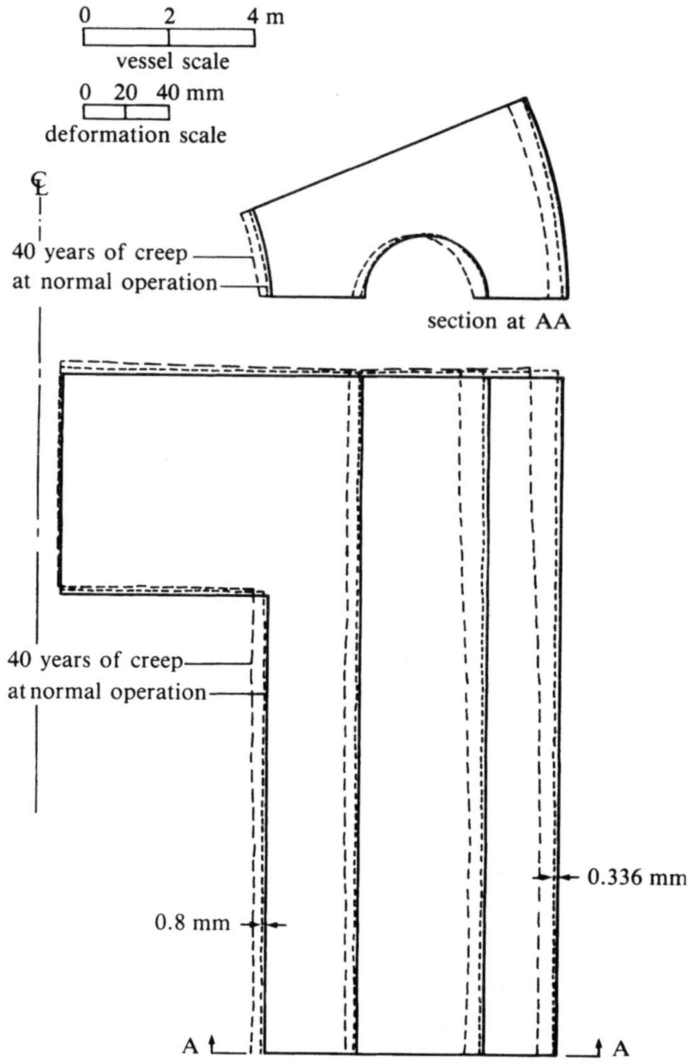

Fig. IX.144. Deformations under normal operation and after 40 years of creep [262] — vessel type, bonded (with Ahmlink element).

vessel's behaviour (Fig. IX.148) is identical to that of the Hartlepool vessel [418–428]. The only change is the plastic zone developed around the gas inlet duct under mark 6.

At a pressure of 9.28 MN/m^2 ($1.635P_{GD}$) the behaviour of this vessel (Fig. IX.149) in most zones is identical to that of the Hartlepool vessel [423, papers H3/5, H5/3] and others [423, 424]. When the pressure was raised to 6 MN/m^2 the only difference is in the position of cracks (marks 11, 12 and 13) at the top haunch and the extent of the plastic zones (marks 7, 14, 20, 21 and 22). The crack sizes have not been predicted in the top cap. The bottom cap has been affected only at the bottom haunch. Both

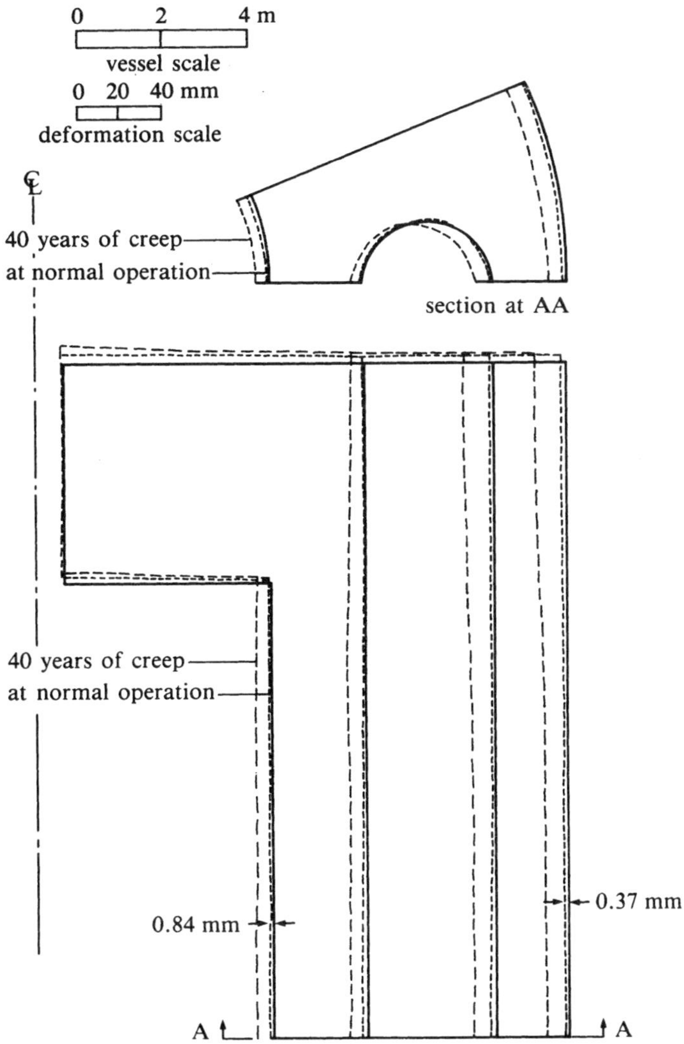

Fig. IX.145. Deformations under normal operation and after 40 years of creep [262] — vessel type, perfectly bonded.

FE and M are in agreement in most areas, as shown in Fig. IX.149. Raising the pressure to 11.08 MN/m^2 ($1.95P_{GD}$) there is a significant difference in the behaviour of this vessel (Fig. IX.150).

The HTGCR vessel failure mode is moving more towards flexural criteria than shear criteria. The initiation and propagation of cracks and their positions is very different. The reason is purely the dominant role that the barrel wall/cap slenderness ratio has played in producing flexural cracks. In plan around the boiler and circulator penetrations and in the stand pipe area, most of the plastic zones are identical.

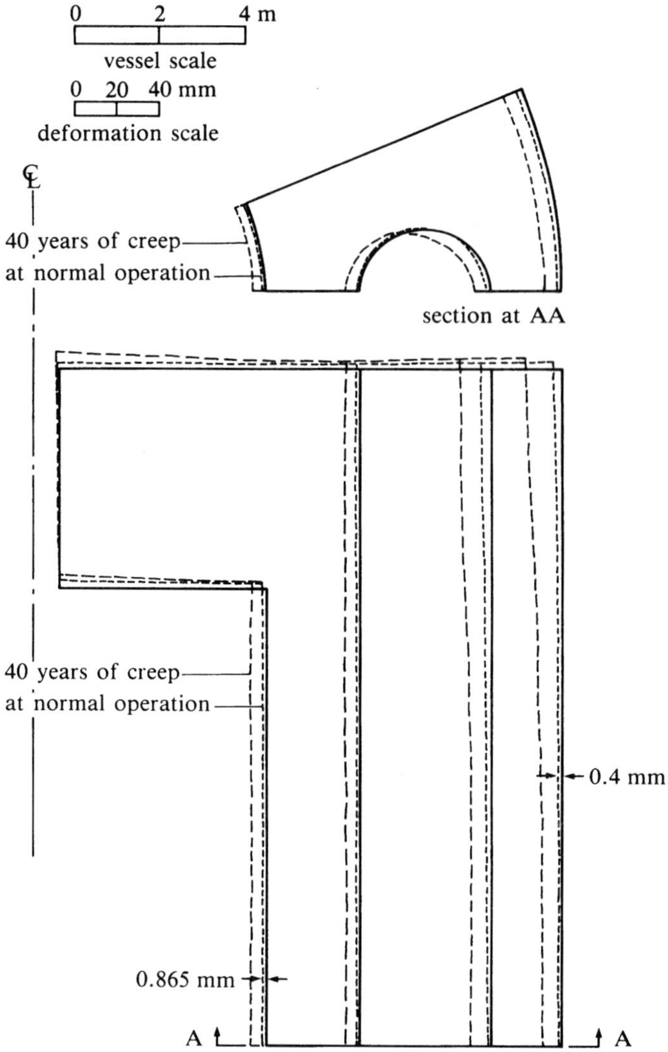

Fig. IX.146. Deformations under normal operation and after 40 years of creep [262] — vessel type, unbonded.

At a pressure of 12.88 MN/m^2 (2.24P_{GD}) the top cap mode (marks 46 to 53) is given in Fig. IX.151.

There is a similarity in the development of plastic zones between the two vessels in the bottom caps. In plan, the plastic patch zones are very similar between these two vessels. The only difference is the extent of cracks developed at the equator (mark 49) by FE. At this pressure in the top cap the plug failure (mark 60) is imminent. It is interesting to note that the crack sizes are identical in these zones. Another difference is in the sudden appearance of a plastic zone just above the gas inlet (mark 55) and a flexural crack in the circumferential direction in the top corner (mark 54) at the

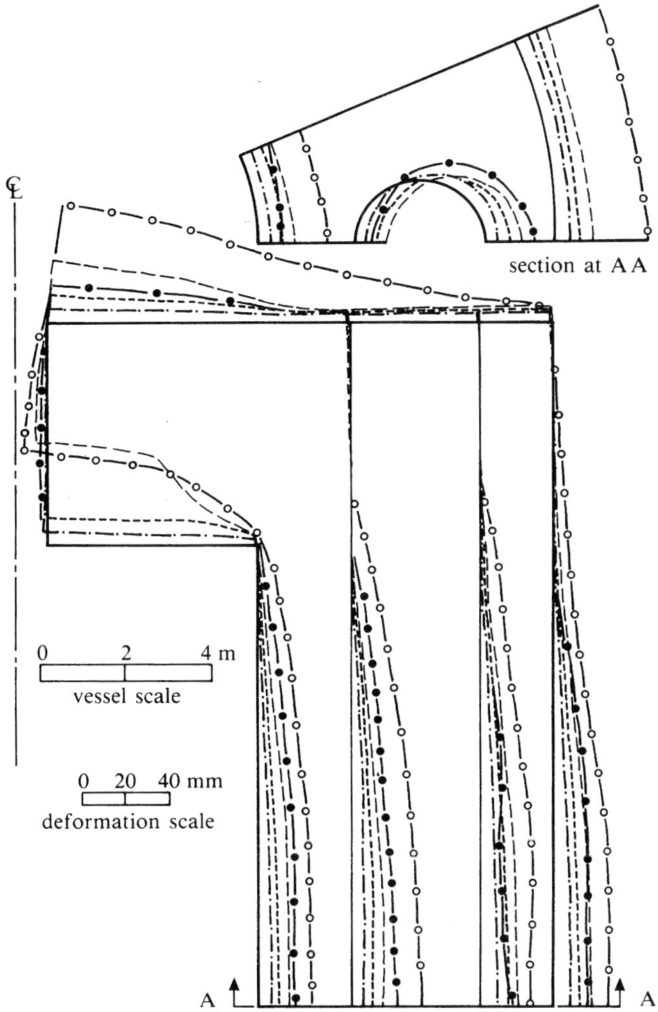

Fig. IX.147. Deformed shapes at various pressures [262, 749] — vessel type, bonded vessel with Ahmlink element. (—·—, deformed shape at 3.0; – – –, deformed shape at 3.3; — — —, deformed shape at 3.5; —O—, FE (four-parameter, Bangash); —●—, FE (five-parameter, Bangash).)

outside of the boiler penetration. There is also a slight shift in the position of the crack (mark 57) around the circulator. The crack widths have reached about 0.2 mm in most places. In many places strains in the tendons and in the bands do not exceed 0.009.

Raising the pressure to 14.68 MN/m^2 (2.58P_{GD}), sufficient flexural cracking in the barrel wall (Fig. IX.152) has occurred.

The nearest case that can be compared is the behaviour of the Hartlepool vessel [418–431] under 2.5P_{GD}. In the top cap, plastic zones have identical locations.

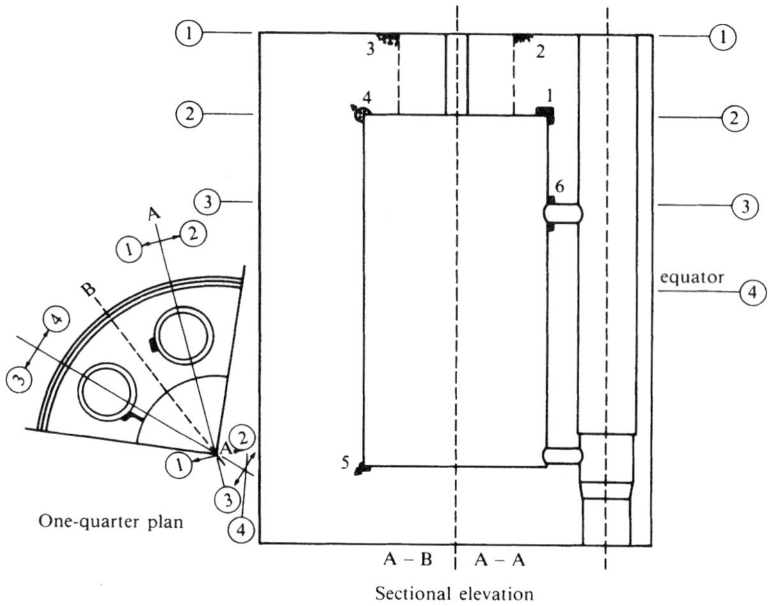

Fig. IX.148. High-temperature gas-cooled reactor pressure vessel ($P_G = 7.48\ MN/m^2$). *(See Fig. IX.118a for key.)*

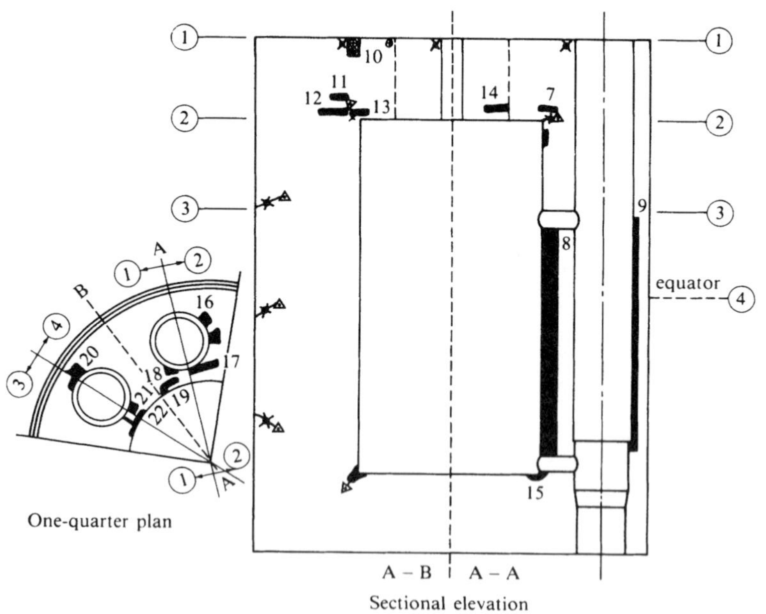

Fig. IX.149. High-temperature gas-cooled reactor pressure vessel ($P_G = 9.28\ MN/m^2$). *(See Fig. IX.118a for key.)*

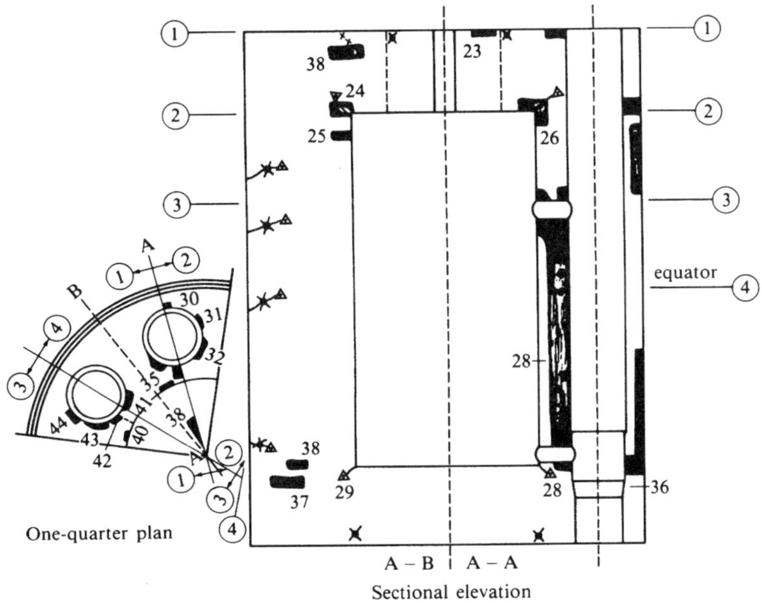

Fig. IX.150. High-temperature gas-cooled reactor pressure vessel ($P_G = 11.08\ MN/m^2$). (See Fig. IX.118a for key.)

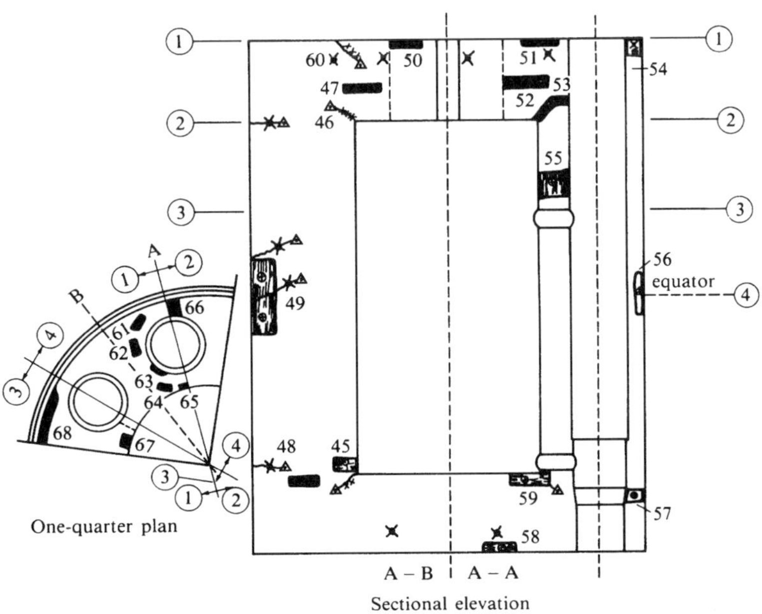

Fig. IX.151. High-temperature gas-cooled reactor pressure vessel ($P_G = 12.88\ MN/m^2$). (See Fig. IX.118a for key.)

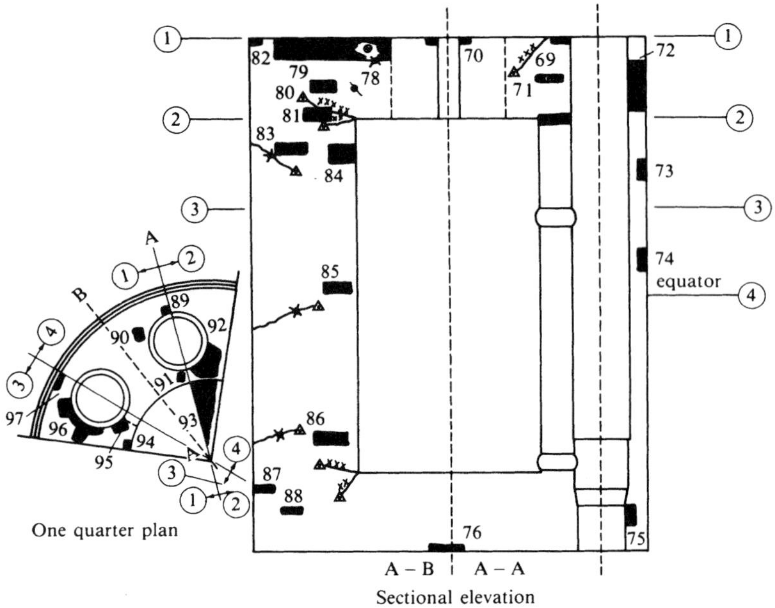

Fig. IX.152. High-temperature gas-cooled reactor pressure vessel ($P_G = 14.68$ MN/m^2). (See Fig. IX.118a for key.)

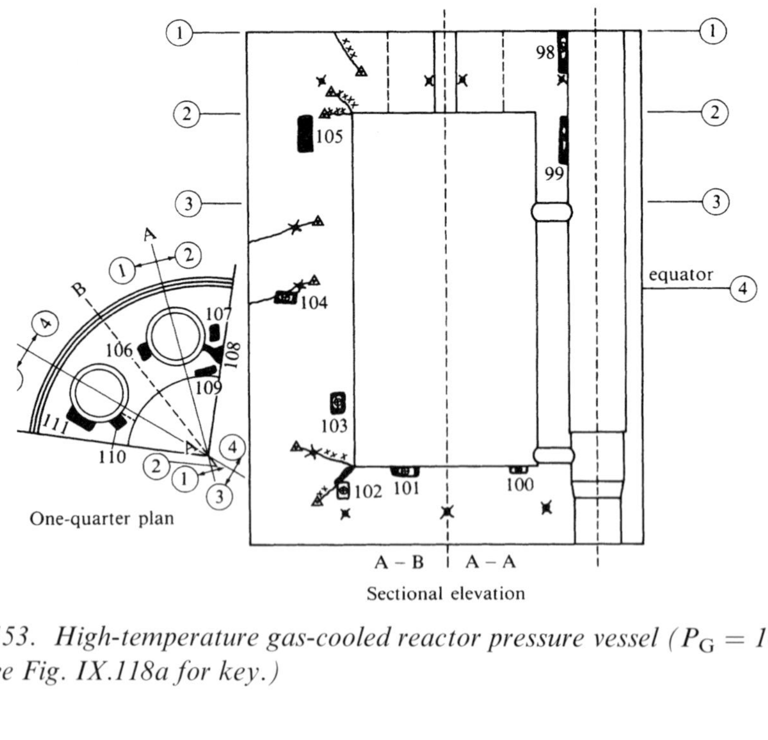

Fig. IX.153. High-temperature gas-cooled reactor pressure vessel ($P_G = 16.48$ MN/m^2). (See Fig. IX.118a for key.)

However, in the HTGCR vessel no internal cracking has been predicted by LS around the boiler and the circulator penetrations.

In plan, the plastic zones are quite similar. There is a change in the plastic zone around the boiler penetration between sections (3)-(3) and (4)-(4). In the top cap a plug failure under mark 69 has been initiated. The strain and crack widths in most places do not exceed 0.01% and 30 mm.

At a pressure of 16.48 MN/m^2 (2.81P_{GD}) the behaviour of this vessel (Fig. IX.153) is identical in many respects to the behaviour of the Hartlepool vessel [418–431] for 2.82P_{GD}. In plan, the cracks and plastic zones are very similar in both cases. In both the top and bottom caps flexural cracks have been extended in depth. Not enough change is visible of the plug failure line in the top cap. In addition, the barrel (Figs IX.152 and IX.153) wall cracks and the haunch cracks have advanced considerably. These flexural cracks are more pronounced than in the case of the Hartlepool vessel. The circumferential prestressing bands have reached the ultimate strain of 1.5% and the cracks in most places are around 50 mm. When the pressure is raised to 18.28 MN/m^2 (3.21P_{GD}), Fig. IX.154a shows the positions of plastic zones between sections (1)-(1) and (2)-(2).

Additional flexural cracks have appeared in the top cap. It seems that the influence of cracks due to shear compression and plug failure have been minimised. At the equator (marks 115 and 116) a larger plastic zone has been created. The bottom haunch crack (mark 117) position is another difference in the behaviour

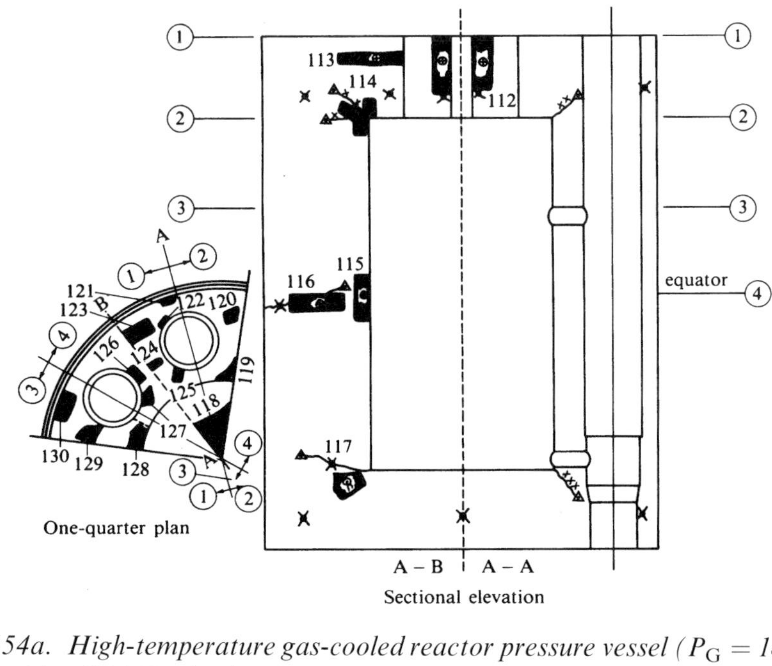

Fig. IX.154a. High-temperature gas-cooled reactor pressure vessel (P_G = 18.28 MN/m^2). (See Fig. IX.118a for key.)

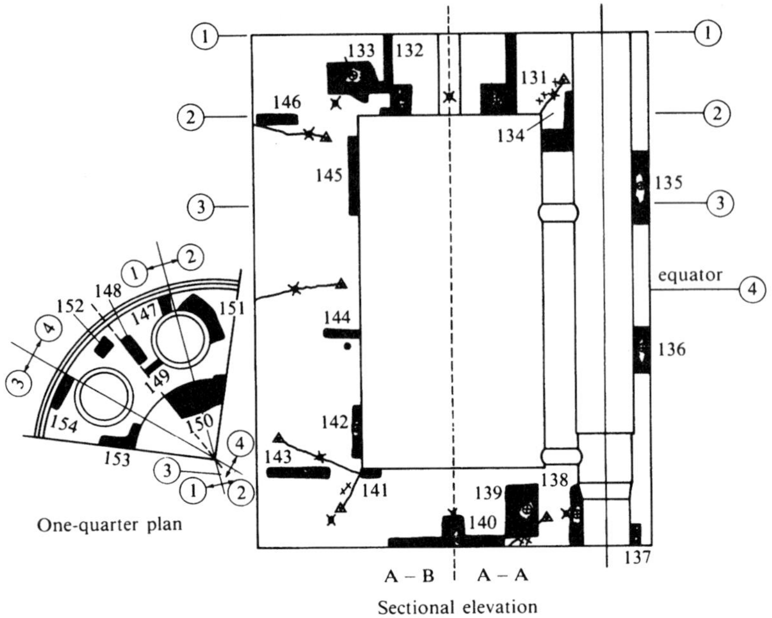

Fig. IX.154b. High-temperature gas-cooled reactor pressure vessel ($P_G = 20.08$ MN/m^2). (See Fig. IX.118a for key.)

of these two vessels. The plastic zones also differ from those in the Hartlepool vessel. Their shapes and positions and crack sizes do not match. This again is due to the difference in the dominant role being played by the span/depth ratios. Longitudinal prestressing has failed and the crack sizes are well over 90 mm. Bonded reinforcement has yielded. Although there is sufficient yielding of the liner, it still has not failed.

At a pressure of 20.08 MN/m^2 (3.55P_{GD}) both the top cap and the barrel wall have failed in flexure one after another. All cracks around the boilers and circulators have widened.

Circumferential cracks around the top haunch have joined the longitudinal cracks. The bottom cap has cracked but the various pieces are held together.

It is concluded that the vessel has failed due to the failure of a strand band at a pressure of 3.21P_{GD} and enough cracks have been produced in the top cap to define a premature flexural failure intervening in the shear failure in the top cap. Flexural failure has also occurred at several places in the barrel wall. Figures IX.154a and b give additional cases for failure modes but these are not governing cases. Figure IX.155 gives the deflection and cracking history of this vessel. Figures IX.156a and IX.156b show cutaway sections of the HTGCR vessel of ENL (Italy), the cracking zones of which correspond with the final collapse mode of this vessel.

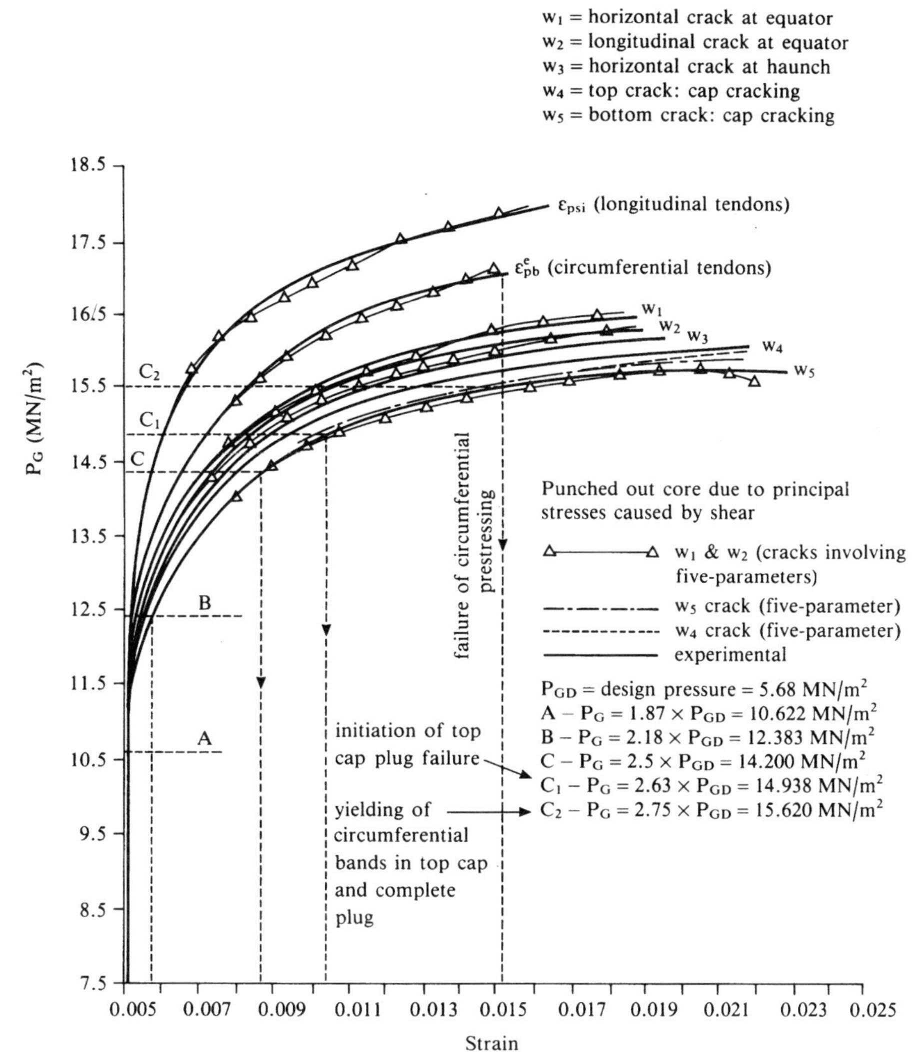

crack width top cap & wall (mm)	5	10	15	20	25	30	35	40	45	50	55	60	65	70	75	80	85	90	95	100
crack width bottom cap (mm)		8		16		24		32		40		48		56		64		70		78
top cap central deflection (mm) × 10		0.9		1.2		1.5		1.8		2.1		2.4		2.7		3.0		3.3		3.6
bottom cap central deflection (mm) × 10		0.6		0.9		1.2		1.5		1.8		2.1		2.4		2.7		3.0		3.3
wall horizontal deflection (mm) × 10		0.9		1.2		1.5		1.8		2.1		2.4		2.7		3.0		3.3		3.6

Fig. IX.155. High-temperature gas-cooled reactor pressure vessel versus cracks and deflections.

Fig. IX.156a. Ultimate load tests on HTGCR—top view (courtesy of Professor Scotto, ENL, Italy [423]).

Fig. IX.156b. Ultimate load tests on HTGCR — cutaway section (courtesy of Professor Scotto, ENL, Italy [423]).

(d) Thermal analysis of vessels. Thermal or temperature analysis has been carried out for all vessels on the lines suggested in Chapters II, VI and VII. The analytical results include the effects of temperature. Figures IX.157–IX.160 give the results for the temperature distributions for the vessels. Thermal loads are computed and are put at the nodes of the elements as concentrated loads or patch loads. Alternatively

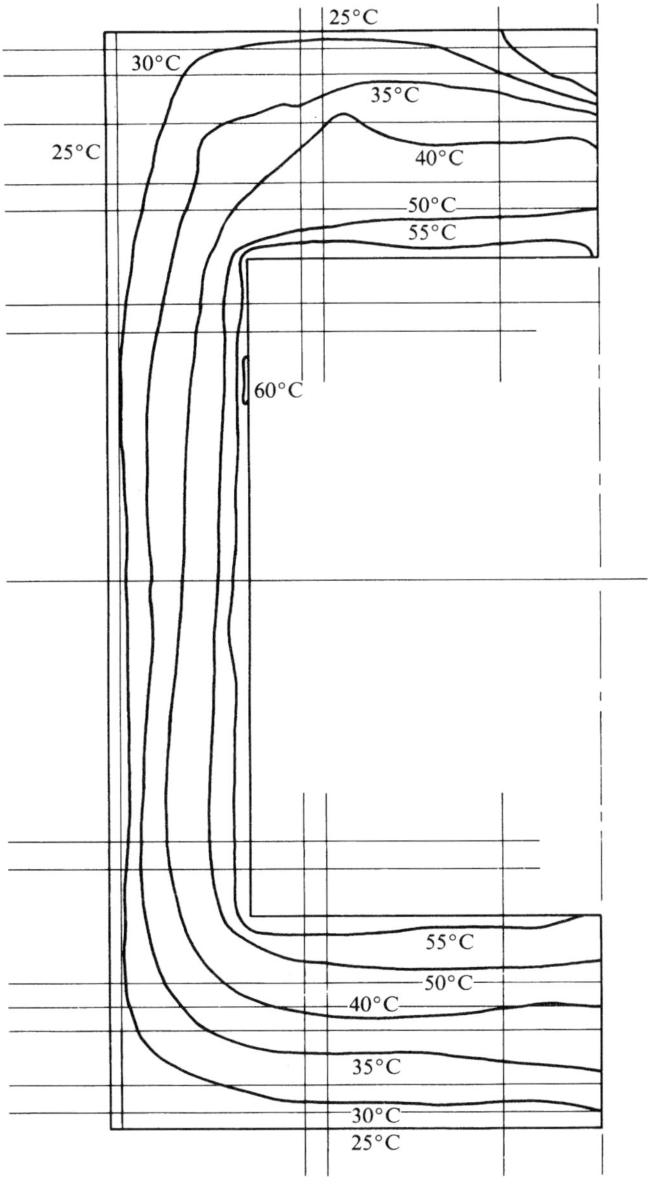

Fig. IX.157. Dungeness B type vessel — temperature distribution.

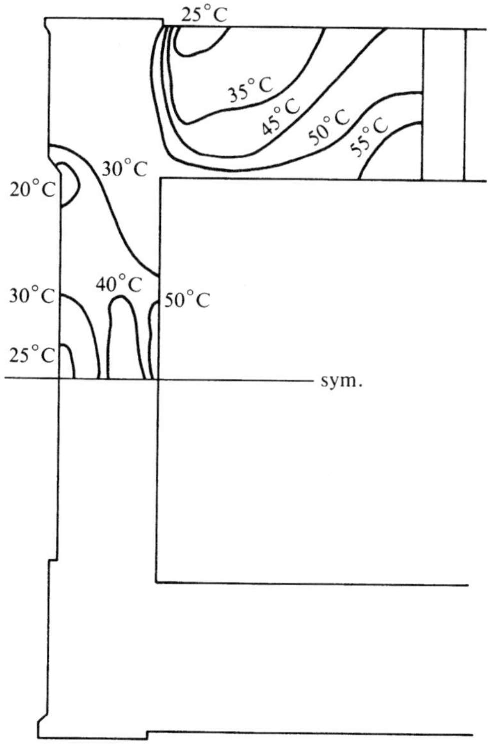

Fig. IX.158. Oldbury vessel — temperature distribution.

strains are evaluated from these temperatures and they are included in the finite element formulations as initial strains. Stress trajectories can be drawn for the temperature-only case on the lines given for vessels in normal operation — cases which in fact include temperature effects.

(e) General conclusions. Much fuller details are given here at various stages of the deformation and cracking of vessels under operational and overload conditions. In the following pages concerning circular concrete structures, such detailed analysis, although avoided, has been used to predict the behaviour of such structures. This analysis, therefore, should form the basis of investigation of all circular concrete structures under operational and extreme loads.

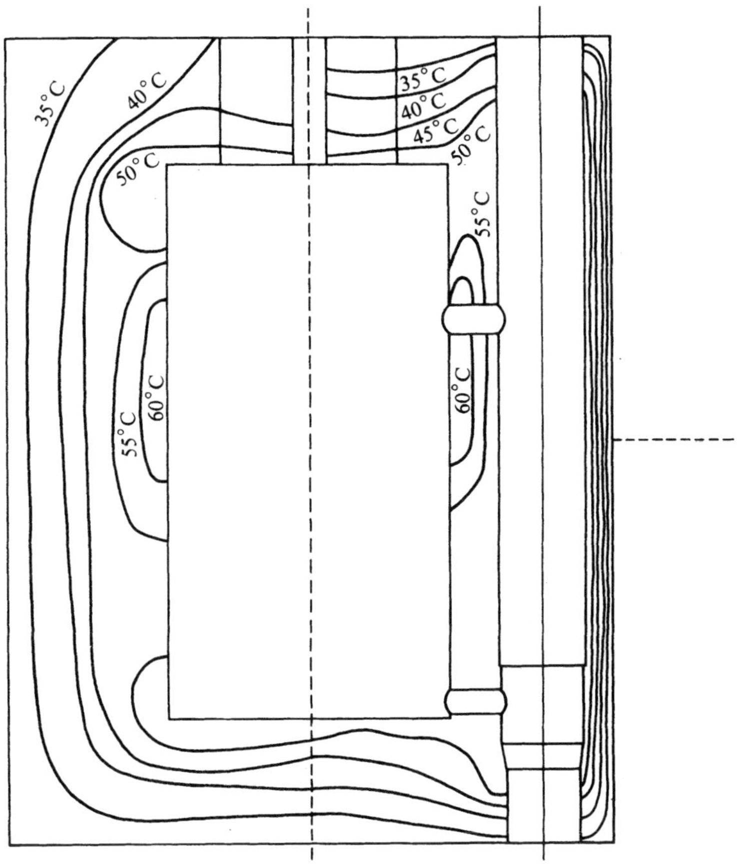

Fig. IX.159. HTGCR vessel — temperature distribution.

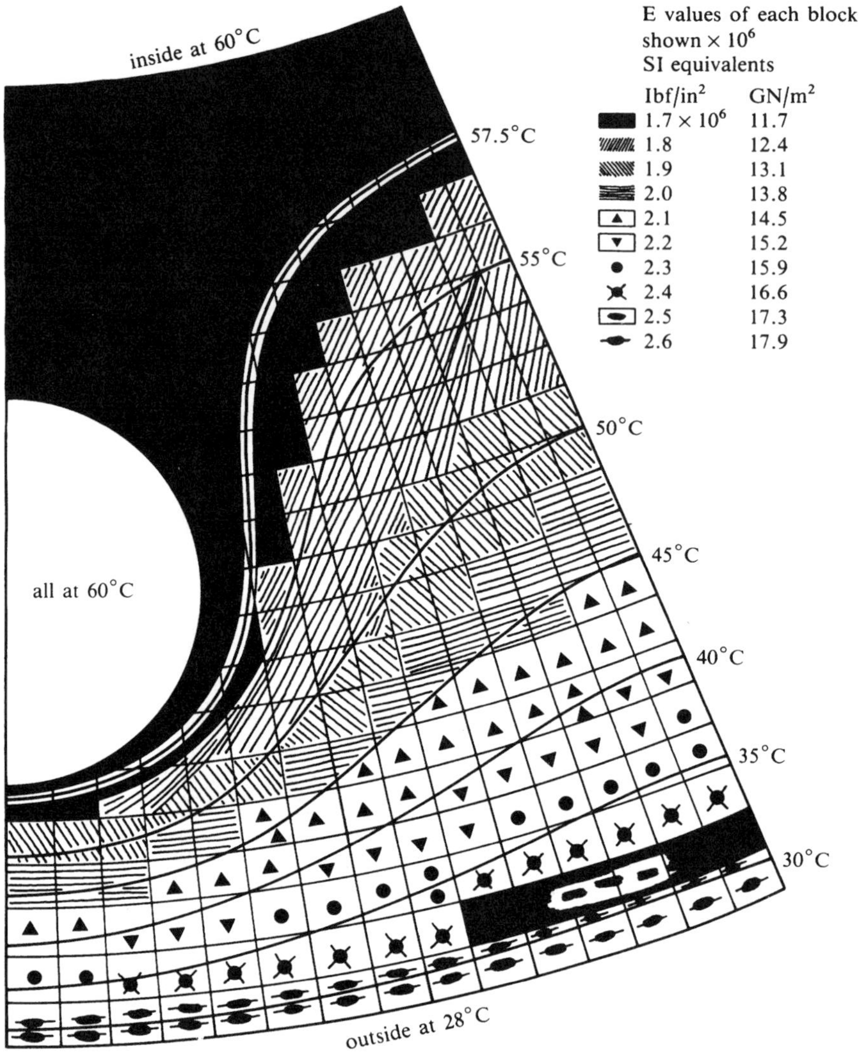

Fig. IX.160. Temperature and Young's moduli at equator — HTGCR vessel.

IX.10 Case Study No. 9 — concrete containment vessels

IX.10.1 Introduction

Bangash [843] has produced a comprehensive analysis of containment vessels. The purpose of the containment is to:

(a) prevent the escape of radioactive materials into the environment;
(b) protect the reactor vessel from external hazards; and
(c) provide biological shielding against nuclear radiation.

Concrete vessels are divided into two classes, namely reinforced and prestressed vessels. The walls are generally of cylindrical shape supporting spherical, elliptical and torospherical domes. Figure IX.161 shows a typical containment with spherical

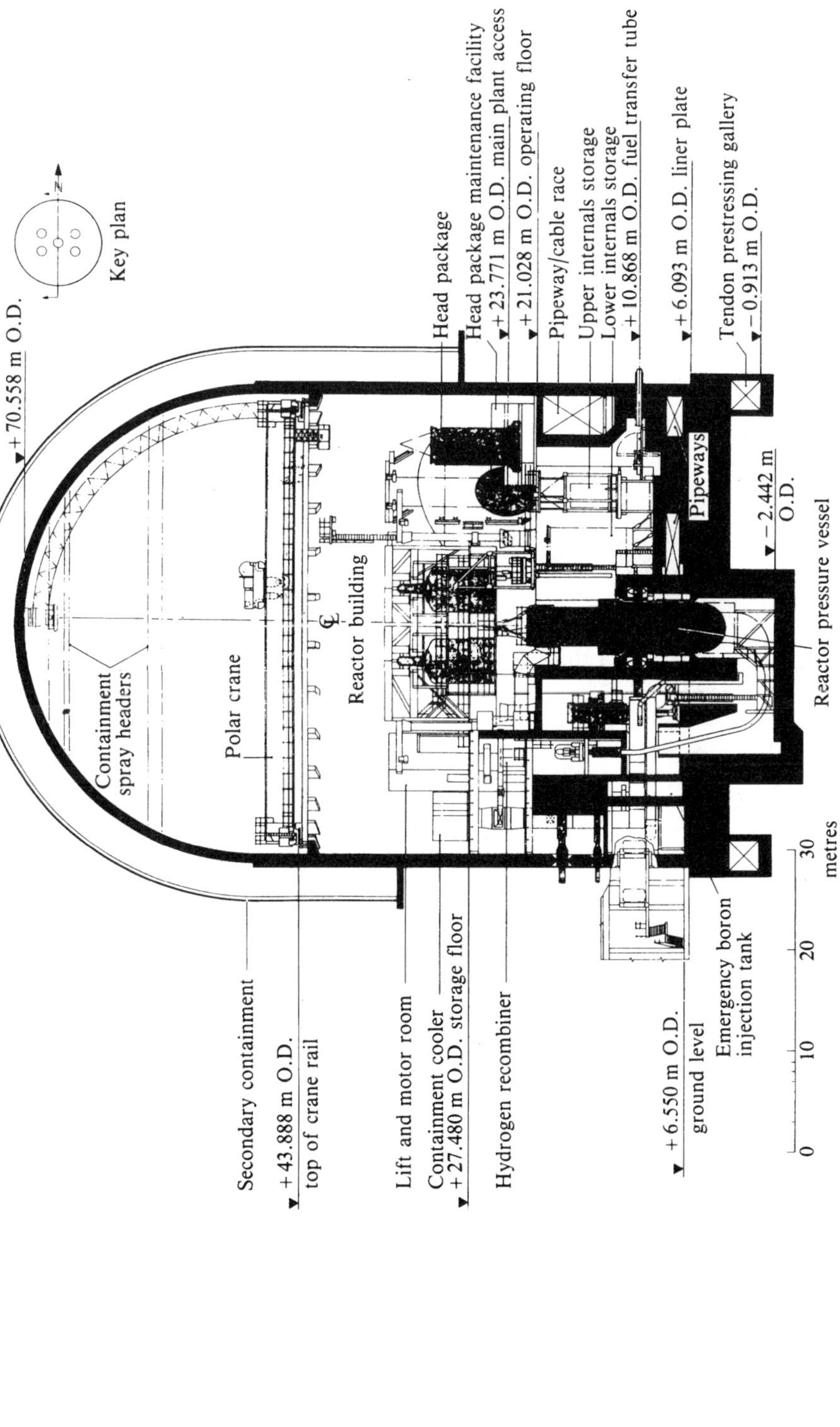

Fig. IX.161. Sizewell B power station reactor building — sectional elevation on north/south centre line. (With compliments of CEGB, UK)

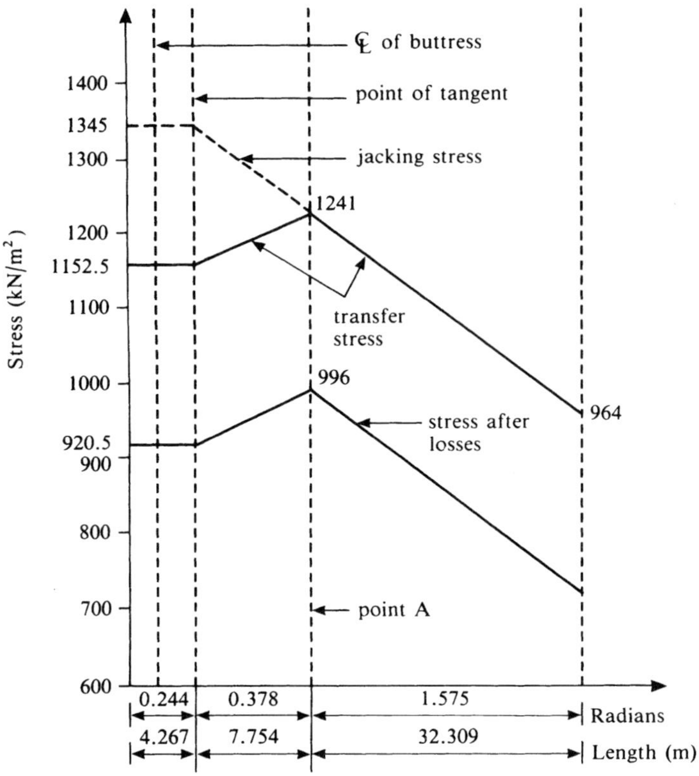

Fig. IX.162. Post-tensioning loads — circumferential tendons [843].

dome (Sizewell B). Figures IX.162–IX.165 show a typical reinforcement system. In many cases a secondary dome (Fig. IX.161) is provided to protect the primary containment from external hazards and the public from the internal hazards. Some well-known vessel parameters are given in Table IX.12. Various loading criteria are given in Table IX.13 [843].

The type of such a vessel depends on the nuclear steam supply system (NSSS), site, environmental and economic conditions. The following are two types of containment vessel.

(a) *Pressurised Water Reactor (PWR)*
These are based on the pressure containment concept. Apart from over-pressure conditions, these vessels are checked for high energy steam released during a loss-of-coolant accident (LOCA). In some cases pressure suppression capability is provided (ICE-condenser containments) which allows reduction in containment volume and design pressure. Where double barrier containments are provided they have been devised for when the dose rate at the site boundary cannot be maintained during stagnant atmospheric conditions or where danger exists internally or externally.

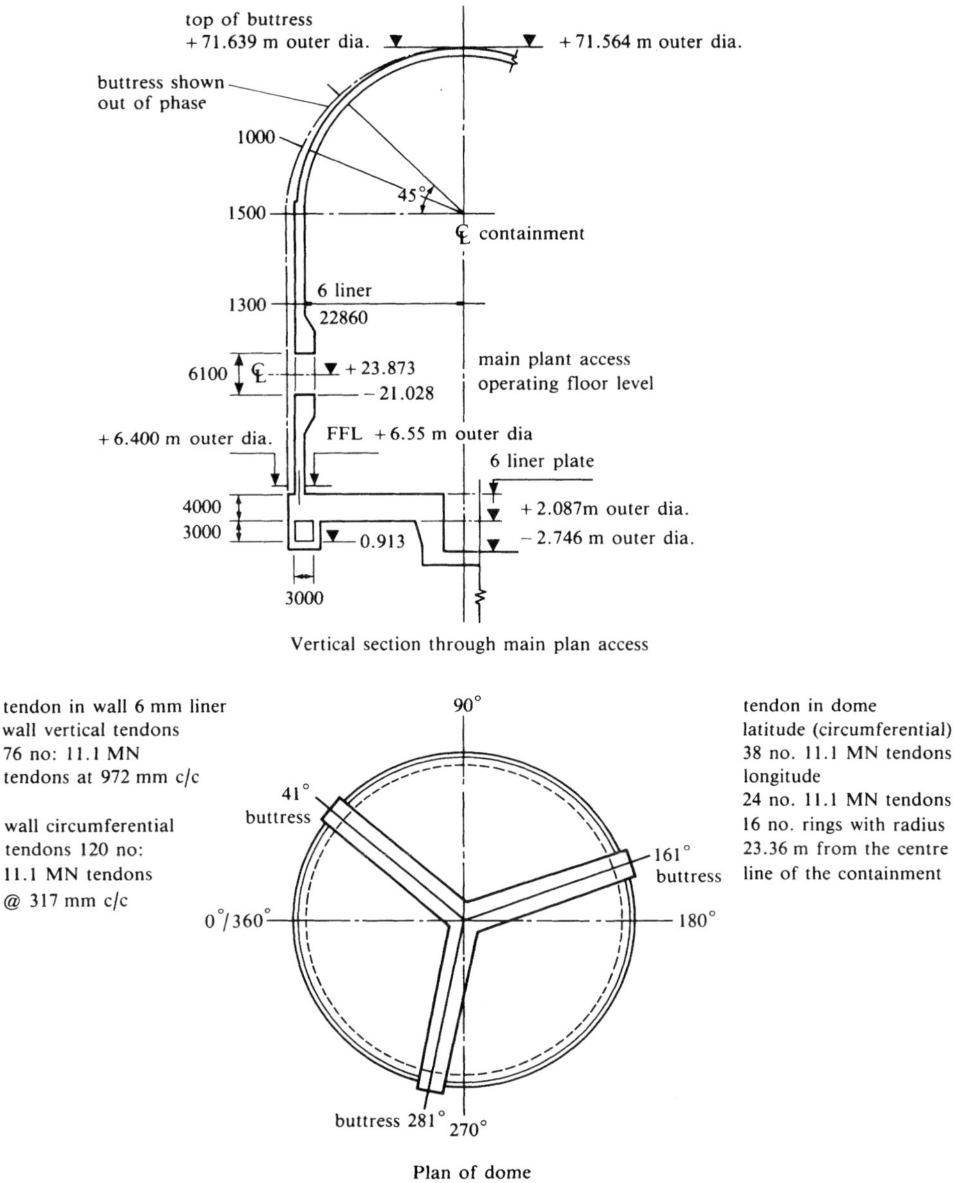

Fig. IX.163. Sizewell B dimensional layout [599].

(b) *Boiling Water Reactors (BWR)*

The containment vessels for BWRs are designed using the pressure suppression concept. The layout contains both primary and secondary containments. The primary containment has a dry well which encloses the reactor and its cooling system and a wet well which channels the steam released during a LOCA into the pool of water where a quenching process is carried out.

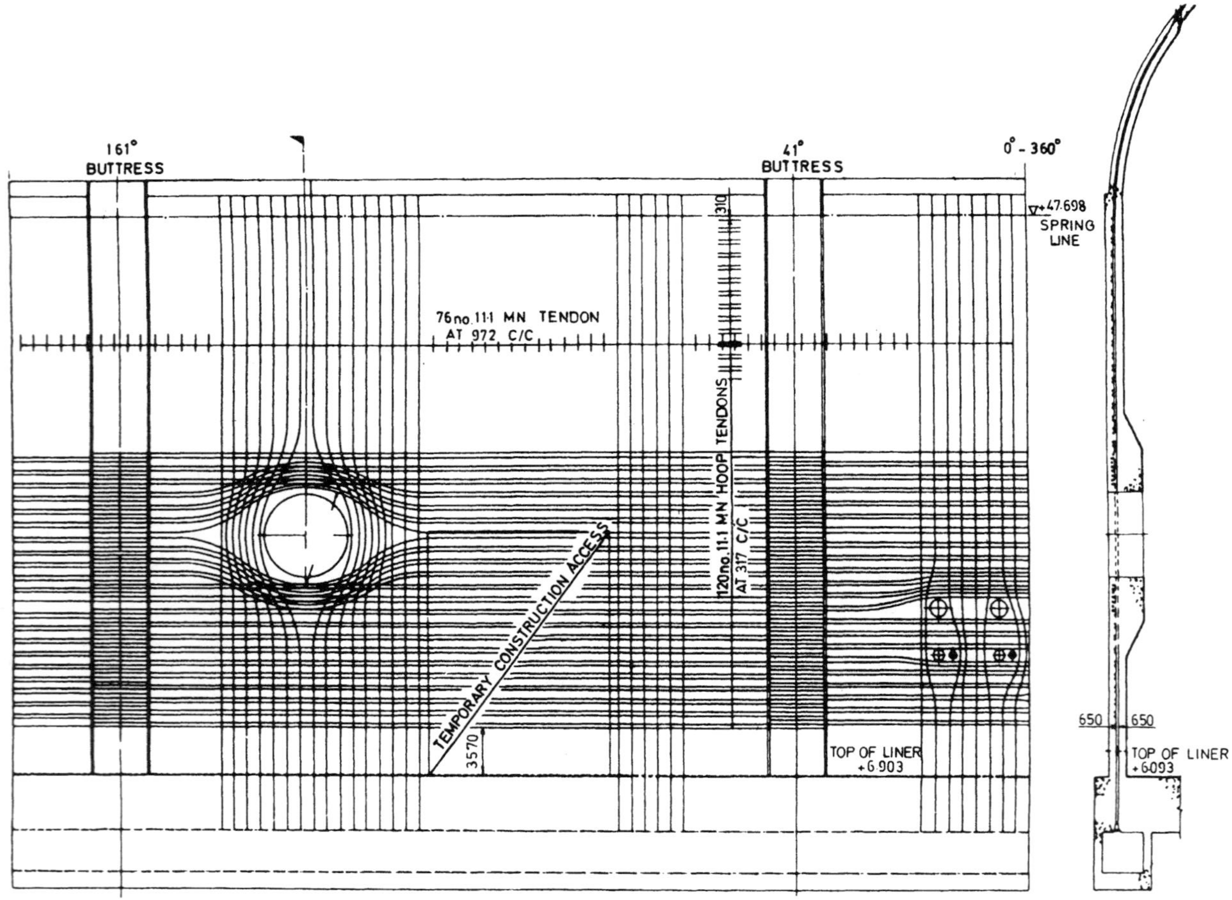

Fig. IX.164. Developed elevation (section — right) showing tendon layout 0°–180° (all dimensions in mm except reduced level) [599] (courtesy of CEGB and Sizewell Task Force).

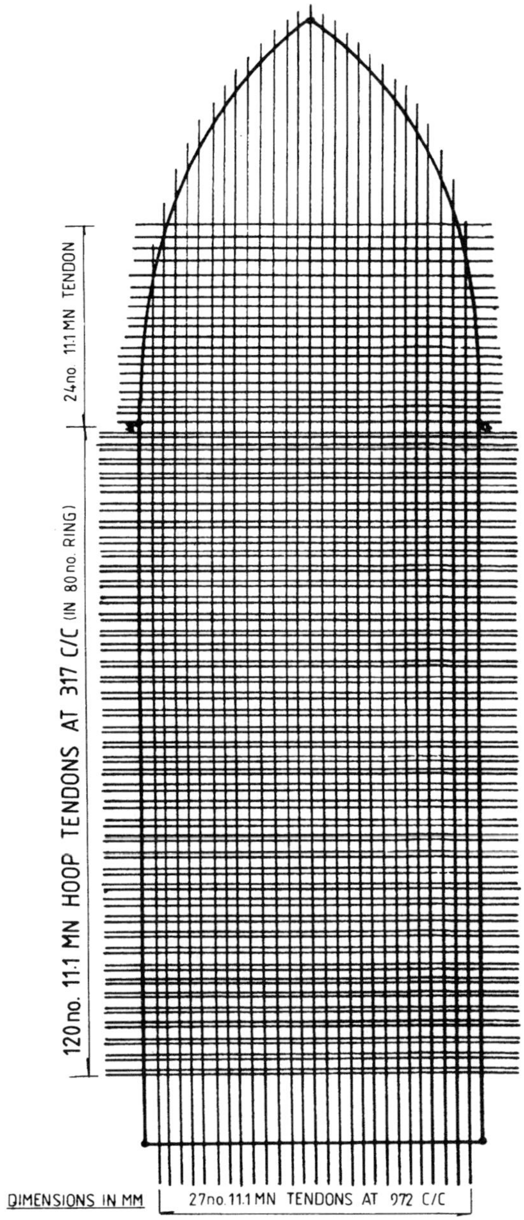

Fig. IX.165. Arrangement of longitudinal and circumferential tendons (all dimensions in mm).

IX.10.2 Loads and stresses

IX.10.2.1 Dead and service loads

(a) Normal loads.

(i) Dead load (D') — weight of all permanent structural and non-structural components.

Table IX.12. Available data for major vessels [843]

Reactor	Containment dimensions	Available data
CVTR	Concrete containment 34.565 high 17.632 m diameter	Liner 6.4–12.7 mm Wall thickness 1.19 m Factored load of 1.5 × design and accident pressure
Enricu Fermi	Steel containment 30.4 m high 21.9 m diameter	ANS and ASME Codes
San Onofre	Steel containment 42.56 m diameter sphere	ANS and ASME Codes
EGCR	Concrete containment 60.8 mm high 30.4 m diameter	Liner 6.5 mm Wall thickness 1.19 m ACI/ASME Codes
Stone and Webster	Concrete containment 56.39 m high 38.40 mm inside diameter	Liner 6.4–12.7 mm
Westinghouse 900 MWe	Concrete containment 53.10 m high 37.0 mm inside diameter	0.90 m thick wall 0.75 m dome thickness Prestress 1000 tonne/m of cylinder
1350 MWe	69.0 m high 42.0 m inside diameter	Factored load of 1.5 × design and accident pressure Vertical prestress 50 bar
Rancho Seco	56.43 m high inside 39.65 m inside diameter	1.19 m wall thickness; pressure 414 kN/m^2. Four groups of dome tendons 81 tendons total divided into three groups, 126 vertical tendons, 117 hoop tendons, 6mm liner thickness
Palisades	58.825 m high 35.264 m inside diameter	1.18 m wall thickness Six buttresses, 165 dome tendons divided into three groups, 180 vertical tendons, 6 mm liner thickness
Trojan	63.232 m high 37.696 m inside diameter	1.18 m wall thickness 3 buttresses, 150 hoop tendons, 70 vertical tendons, 6 mm liner thickness
Doel (Belgium)	Concrete containment 63.3 m high 42.5 m inside diameter	1.20 m wall thickness 2.9 m annulus for outer containment 2 buttresses
Ringhals II	Concrete containment 63.4 m high 35.40 m inside diameter	4 buttresses 1.10 m wall thickness
Gentilly	Concrete containment 56.0 m high 36.48 m inside diameter	1.216 m thick wall 0.508 m dome thickness 4 buttresses
Ginna	Concrete containment 46.056 m high 31.92 m inside diameter	1.18 m wall thickness 0.976 m dome thickness 3 mm liner thickness 19 mm wire tendons vertical 3 No.18 S bars hoop steel
Mol (Belgium)	Steel containment 32.3 m high 16.5 m diameter	30 mm thickness
Seebrook	57.456 m high 21.28 m inside diameter	1.368 m wall thickness No.18 at 300 mm hoop and cylinder vertical bars No.18 at 550 mm at 45° angles
Bellefonte	67.285 m high 41.14 m inside diameter	1.067 m wall thickness
Fessenheim and Bugey	Concrete containment 56.6 m high 37.0 m inside diameter	0.9 m wall thickness 6.0 mm liner
Creys-Malville	Concrete containment 58.5 m high 42.5 m inside diameter	0.85 m wall thickness 6.0 mm liner
Three Mile Island Sizewell B	54.465 m high 45.0 m inside diameter	2.5 m thick wall cylinder 1.5 m thick dome 6.0 mm liner

Table IX.13. Partial load factors

Category	D'	L	F	P_t	P_0	T_t	T_0	T_a	E_0	SSE	W	W_t	R_0	R_a	R_r	P_r
Service																
Test	1.0	1.0	1.0	1.0	—	1.0	—	—	—	—	—	—	—	—	—	—
Construction	1.0	1.0	1.0	—	—	—	1.0	—	—	—	—	—	—	—	—	—
Normal	1.0	1.0	1.0	—	—	—	1.0	—	—	—	—	—	1.0	—	—	1.0
Severe environmental	1.0	1.0	1.0	—	—	—	1.0	—	—	—	1.0	—	1.0	—	—	1.0
	1.0	1.0	1.0	—	—	—	1.0	—	—	—	1.0	—	1.0	—	—	1.0
Factored																
Severe environmental	1.0	1.3	1.0	—	—	—	1.0	—	1.5	—	—	—	1.0	—	—	1.0
	1.0	1.3	1.0	—	—	—	1.0	—	—	—	1.5	—	1.0	—	—	1.0
Extreme environmental	1.0	1.0	1.0	—	—	—	1.0	—	—	1.0	—	—	—	1.—	—	1.0
	1.0	1.0	1.0	—	1.—	—	1.0	—	—	—	—	1.0	—	—	—	1.0
Abnormal	1.0	1.0	1.0	—	1.5	—	—	1.0	—	—	—	—	—	1.0	—	—
	1.0	1.0	1.0	—	1.0	—	—	1.0	—	—	—	—	—	1.25	—	—
Abnormal/severe environmental	1.0	1.0	1.0	—	1.25	—	—	1.0	1.25	—	—	—	—	1.0	—	—
	1.0	1.0	1.0	—	1.25	—	—	1.0	—	—	1.25	—	—	1.0	—	—
	1.0	1.0	1.0	—	—	—	1.0	—	1.0	—	1.0	—	—	—	—	—
Abnormal/extreme environmental	1.0	1.0	1.0	—	1.0	—	—	1.0	—	1.0	—	—	—	1.0	1.0	—

D' Dead loads, including hydrostatic and permanent equipment loads
E_0 Loads generated by the operating basis earthquake
SSE Loads generated by the safe shutdown earthquake (weights considered will be the same as for E_0)
f Loads due to prestress
L Live loads
P_0 Design pressure load within the containment generated by the design basis accident (DBA)
P_t Pressure during the structural integrity and leak rate tests
P_r External pressure loads (resulting from pressure variation either inside or outside the containment)
R_a Pipe reaction from thermal conditions due to the DBA, including R_0
R_r The local effects on the containment due to DBA
R_0 Pipe reactions during normal operating or shutdown conditions
T_a Thermal loads generated by the DBA, including T_0
T_0 Thermal loads during normal operating or shutdown conditions
T_t Thermal effects and loads during the test
W_t Tornado loading, including the effects of missile impact
W Design wind load

(ii) Operating live load — live load is the load superimposed (L, LL) on the vessel and its components.
(iii) Construction loads — construction loads are large heavy loads due to erection of equipment such as those from cranes, fork lift forms, and transport and temporary storage of equipment.
(iv) Concentrated loads (L_C) — rail and truck supports (C_F, H_{20}), craneways (25% of wheel load), and ordinary machinery impact loads.

(b) Environmental loads.

(i) Wind loads (W) (excluding loads induced by tornadoes) — wind loads are the same as those used in conventional structural design; these are based on statistical data for the vicinity of the structure, ground topography and general climate.

The pressure (P) on any part of the vessel surface exposed to wind varies from point to point over the surface, depending on the direction of the wind and flow pattern. The pressure (P) can be expressed by $P = C_p q$, where C_p is the pressure coefficient and q the dynamic pressure. In the calculation of the wind it is essential to take into account the pressure difference between opposite faces. It is convenient to use distinguishing pressure coefficients:

$$F = (C_{pe} - C_{pi})qA$$

where F = resultant force, C_{pe} = external pressure coefficient, C_{pi} = pressure coefficient, and A = surface area. A positive value for F would mean the pressure is inwards, and a negative value that pressure is outwards. In addition to the existing formula, the total wind load on any section of the vessel may be found using

$$F = C_f q A_e$$

where C_f = force coefficient and A_e = effective frontal area of the section.

(ii) Snow loads (S) — complete data on snow load requirements are given by Bangash [843, 105] and by many other researchers [853, 1014–1020].

(c) Soil and hydrostatic pressure (E_p) and buoyancy (B).

(i) Lateral pressures of adjacent soil including active, passive and at-rest pressures.
(ii) Dynamic pressures due to earthquakes.
(iii) At the free water surface the weight of the soil diminished by buoyancy plus hydrostatic pressure.

(d) Piping equipment reaction load (R_0). Piping systems attached through hangers, struts, lugs, restraints and anchors.

(i) Thermal expansion and dynamic effects; together with
(ii) imposed restraints such as changes in momentum, water and steam hammer in the equipment.

(e) Operating pressure and temperature loads (P_0, T_0).

(i) Differential pressure ranges applied to the structures.
(ii) Loads from differential thermal expansion or stresses from thermal gradients within the structure.

(f) Operating basis earthquake (E_0). The minimum acceptable acceleration for E_0 is at least one half of the safe shut-down earthquake.

(g) Extreme loads. Nuclear structures are designed against a much broader range of loads such as major earthquakes, tornadoes, environmental hazards and military missiles. Nuclear structures are also designed for a third level of load known as the extreme load. Details of extreme loads are:

(i) The maximum earthquake potential for a site, taking into consideration regional and local geology, seismology, local foundation conditions, tornadoes, winds and associated airborne missile loads.
(ii) Postulated design basis accidental loads due to high-energy system ruptures resulting in pipe break action, impingement loads, pipe whip and associated accidentally generated missiles, flooding and high thermal transients.

(h) Safe shut-down earthquake (SSE). The items listed briefly below should be considered to establish the ground motion input associated with the SSE:

(i) Geological conditions at the site under consideration.
(ii) Tectonic structure evaluation.
(iii) Identification of effects of prior earthquakes.
(iv) Determination of static and dynamic characteristics of the underlying materials of containment vessels.
(v) Determination of fault locations:
    (1) The vibratory ground motion associated with the SSE shall be considered to be acting at the ground surface in that portion of the site not affected by adjacent topographical structures.
    (2) For a vessel founded on competent rock or other equally competent material, the soil–structure interaction shall preferably be considered.
    (3) The minimum acceleration for the vibratory ground motion associated with the SSE shall be $0.10g$ within a 200-mile radius of the site.
    (4) The values determined for the vertical motion associated with the SSE should not be less than two-thirds of the maximum horizontal ground acceleration for the SSE.
    (5) The strength of each horizontal component shall be equal to the SSE intensity defined for the site.

(i) Tornado loads (W_t). Structures should be designed to resist maximum tornado conditions for a given plant site.

The effects of a tornado are generated by three separate phenomena:

(i) Wind.
(ii) Differential atmospheric pressure.
(iii) Missiles interacting with the vessel structure causing damage through three principal failure modes:
   (1) pressure forces created by drag and lift as air flows around and over the vessel;
   (2) pressure forces due to rapid changes in atmospheric pressure resulting in differential pressure between the interior and exterior of the building; and
   (3) penetration, spalling and impact forces created by environmental and other missiles.

The criteria suggest that, at the point in a tornado where the wind velocity is a maximum, the atmospheric pressure change is taken as one-half of its maximum value, if a Rankine vortex is assumed. Combining the effects, the velocity pressures shall be multiplied by the appropriate pressure coefficients, and the design loads are determined for:

(i) maximum atmospheric pressure change acting alone; and
(ii) maximum wind velocity plus one-half total atmospheric pressure change combined with tornado missile loads.

(j) Hurricane loads. A hurricane is a cyclonic storm having rotational wind velocities in excess of 119 km/h. The basic characteristics of hurricanes are:

(i) Wind distributions — maximum wind velocities:
   (1) The vectorial addition of the translational and rotational components of the wind.
   (2) The direction of the wind inclined 20–30° towards the centre of the hurricane.
   (3) Wind gusts within the hurricane exceeding sustained winds by 30–50%.
(ii) Atmospheric pressure change.
(iii) Size — hurricanes vary in diameter from 24 km to more than 160 km. Gale force winds (64 km/h) exist within a 560–640 km diameter.

(k) Tsunami. Tsunami are long ocean water-waves generated by earthquakes or underwater explosions, which impinge on coastal areas and structures located therein. Factors for consideration are:

(i) Location and records of tsunami and their mechanism.
(ii) Size of an offshore earthquake or underwater explosion.
(iii) Tsunami wave height.
(iv) Influence of hydrography, harbours and breakwaters.

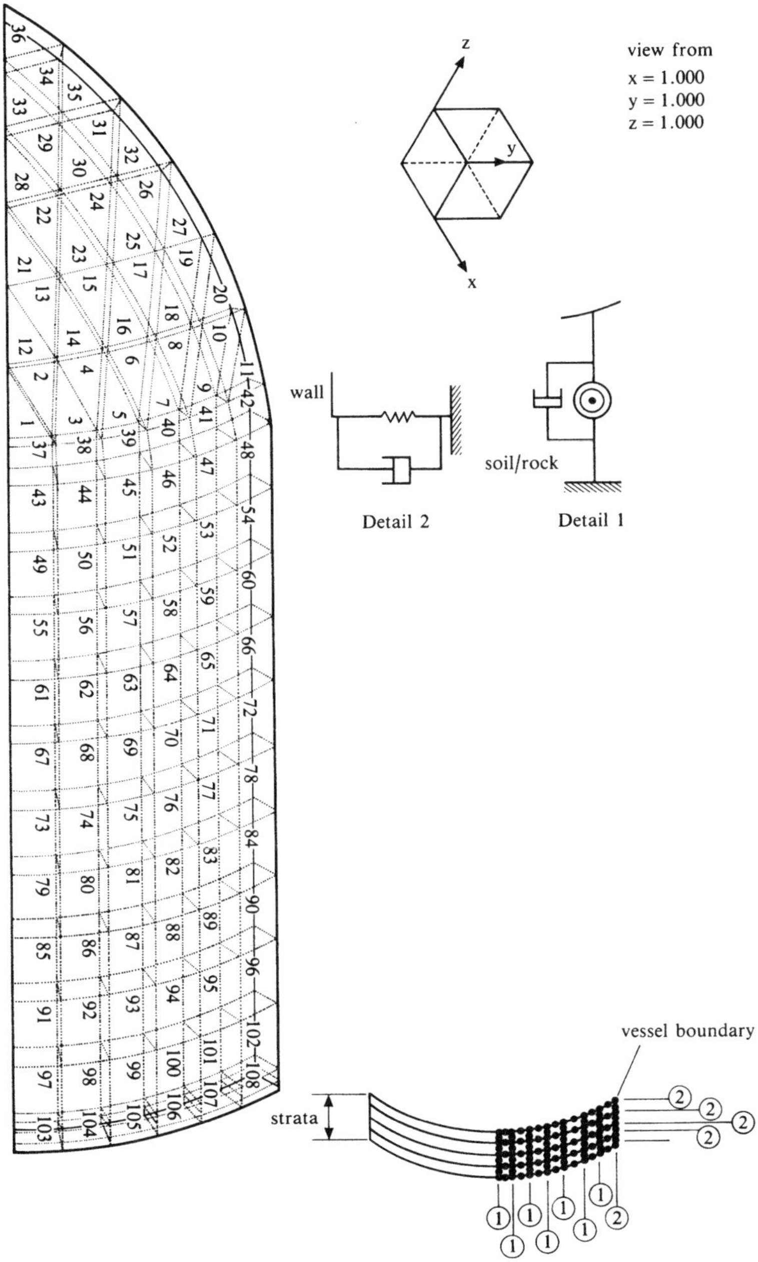

Fig. IX.166. Nonlinear model — Sizewell B vessel.

(l) Missile loads (Y_m). In nuclear plant design, components and equipment should be protected against loss of function due to plant-generated and extreme environmental missiles. Where a containment building is close to a major airport, the damage caused by a potential missile such as an aircraft crash should be considered.

The effect of a missile impact on a target depends on the material and the geometrical properties of the missile and its target. For a rapid impact, local effects must also be considered. These include local structural failures, such as penetration, perforation, scabbing and spalling. In other cases, for long-duration impacts in which momentum exchange takes place, the global response of the containment should be considered. The energy absorption capacity of the containment should be included in the load effects.

The generated missiles may vary widely in size, weight and impact velocity. Extreme environmental missiles include tornado-generated missiles and aircraft.

Factors to consider for tornado-generated missiles are:

(i) Designed tornado wind field.
(ii) Identification of typical objects located at or near a plant site.
(iii) Analytical approaches for the motion of an object should derive the motion of the object until the end of its flight (the point where such missiles are ejected from the tornado wind field).

In the study of aircraft impacts the distance of an airport from a particular power plant site must be considered. This is normally done for the analysis and design on the basis of studies of probability of occurrence.

IX.10.2.2 Stresses

The allowable stresses are adopted according to the relevant codes of practice, including those given in Section IX.9 for pressure vessels.

IX.10.3 Containment vessel analysis

A finite element analysis has been carried out using solid, panel and line elements as explained in Chapter VI. Figure IX.166 shows a typical finite element mesh for a seismic case. Where seismic analysis and aircraft impact analysis are needed, the soil strata are also idealised as layered elements together with a spring and dashpot system, given in Details 1 and 2 of Fig. IX.166. Table IX.14 shows safety margins

Table IX.14. Safety margin (finite element) [843]

Containment vessel	F_s (safety margin)	Containment vessel	F_s (safety margin)
CVTR	3.1	Ringhals	3.10
EGCR	3.25	Gentilly	3.30
Stone/Webster	3.13	Ginna	2.95
Westinghouse (400)	3.21	Mol	4.00
(1350)	3.18	Seebrook	3.75
Rancho Seco	3.30	Bellefonte	3.75
Palisade	3.4	Fessenheim	3.00
Trojan	3.23	Creys-Malville	2.90
Doel	3.00	Sizewell B	2.5

$$F_s = \frac{\text{final pressure at complete cracking}}{\text{operating pressure}}$$

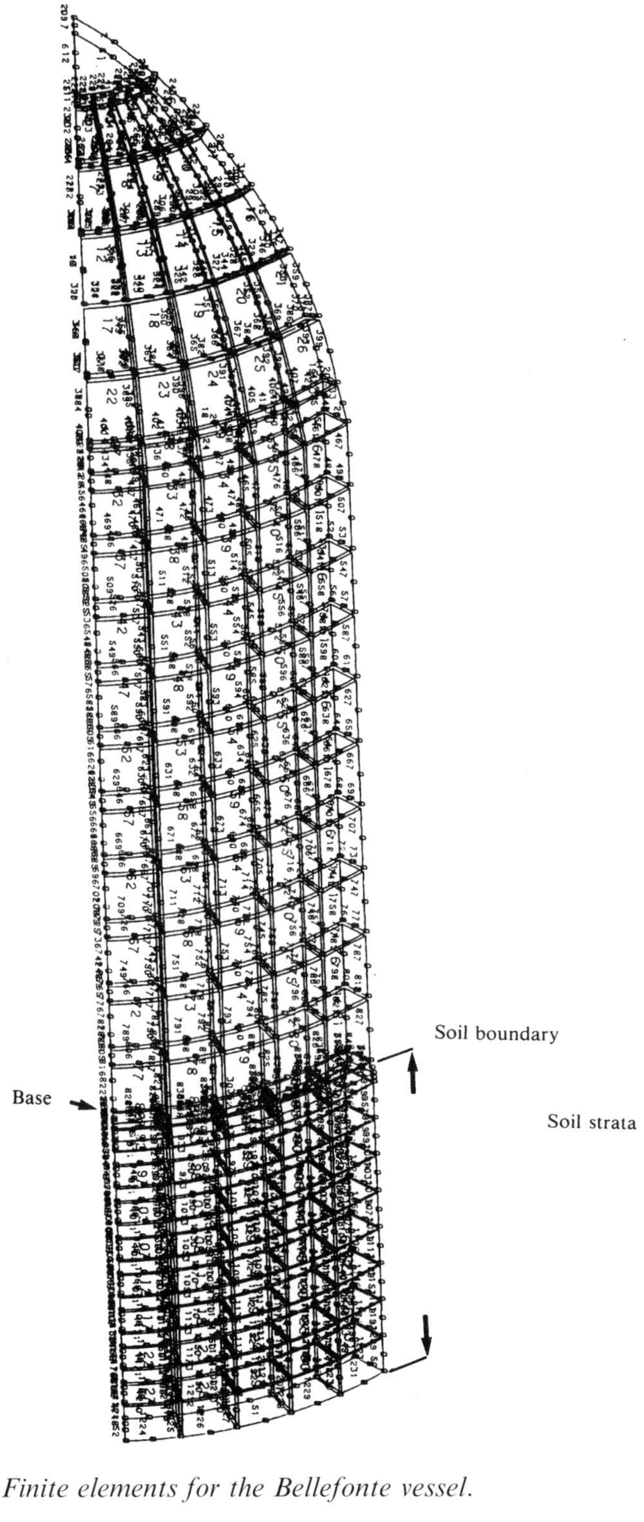

Fig. IX.167a. Finite elements for the Bellefonte vessel.

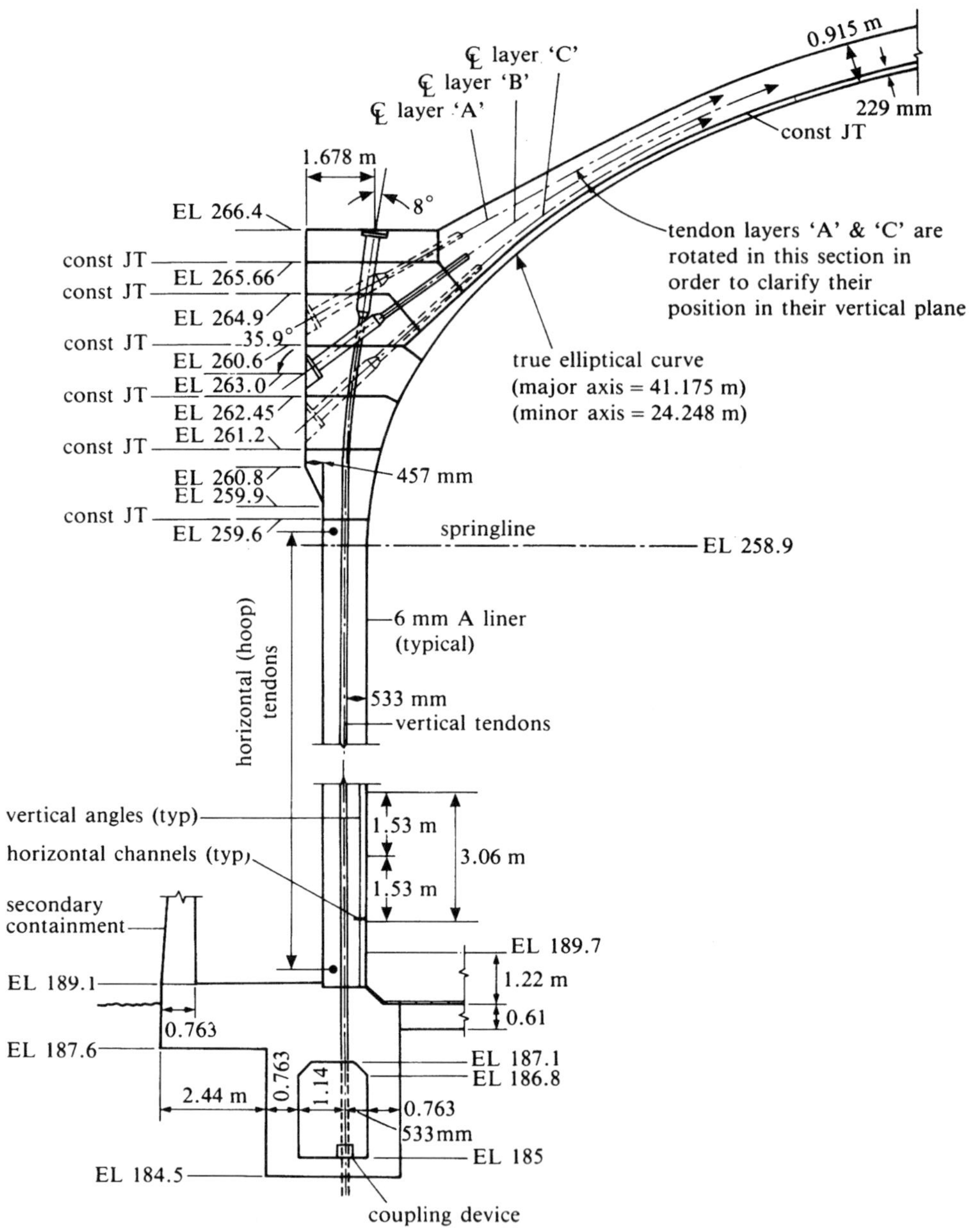

Fig. IX.167b. Typical section — wall and dome (courtesy of TVA and Bellefonte NPS).

for containment vessels subjected to internal pressures up to 344.75 kN/m^2 along with dead, live and prestressing loads. The idealised strata in case of the Bellefonte vessel are given in Fig. IX.167a for the vessel parameters shown in Figs IX.167b and IX.167c.

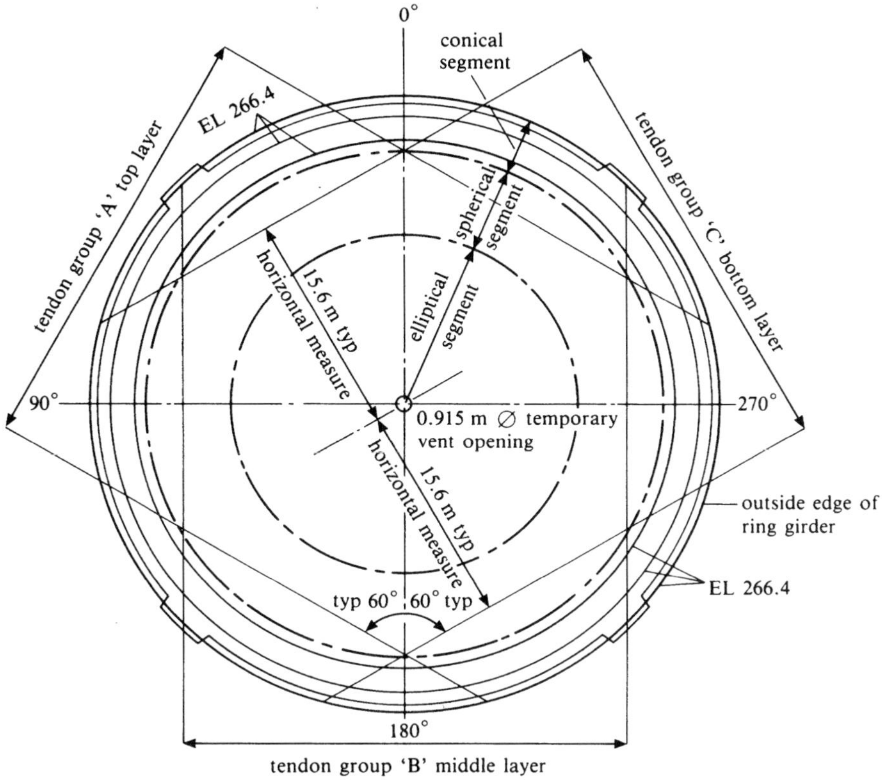

Fig. IX.167c. Plan of the dome. (Note: The typical dome tendon spacing is 978 mm, measured horizontally. The final prestressing force for each tendon group is 5453.25 kN/m.)

IX.10.3.1 Aircraft impact

The design and analysis of a containment vessel to withstand the effect of an aircraft crash or impact involves many complex parameters, including loadings and durations. Bangash [843] has produced modified force–time impact functions for various aircraft, and these are shown in Fig. IX.168. First, the Bellefonte vessel parameters were chosen for the analysis (Figs IX.167b and IX.167c). The impact point is chosen to be the apex of the vessel and at the junction of the dome and the cylinder. Results are produced in the form of acceleration–time histories, displacements and stress states for selected times and acceleration response spectra. Newmark and Wilson θ direct integration methods (discussed in Chapter VI) are used for the analysis. Crack propagation, and linear and nonlinear displacements for four aircraft are shown in Figs IX.169 to IX.171a. The impact zones are shown in Fig. IX.171b. Table IX.15 shows comparative results for penetrations and perforations based on both empirical (Section II.2.3) and finite element analysis for a number of aircraft impact loads. The endochronic model is used throughout the analysis. A similar analysis was carried out for the Sizewell B vessel using load–time functions for Phantom and Tornado aircraft. The 'post-mortem' of the vessel is shown in Fig. IX.172. The load–displacement as a function of time is given in Fig. IX.173.

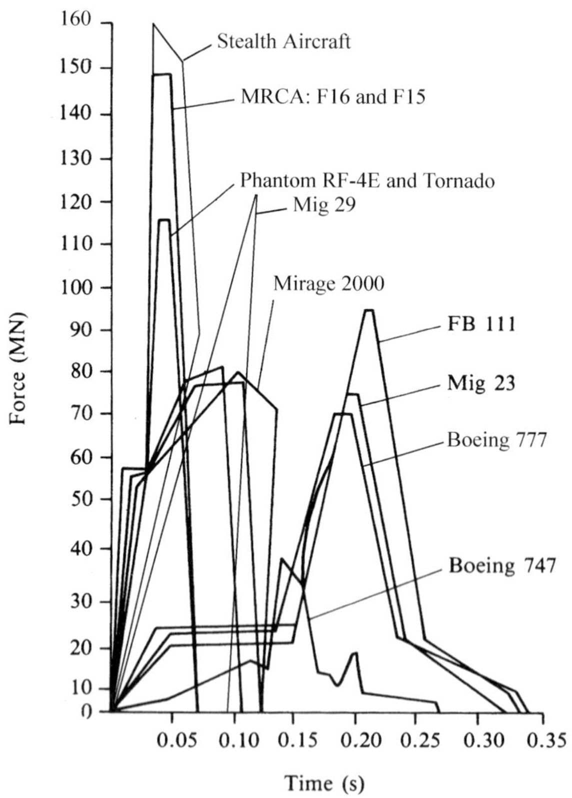

Fig. IX.168. Time–impact function.

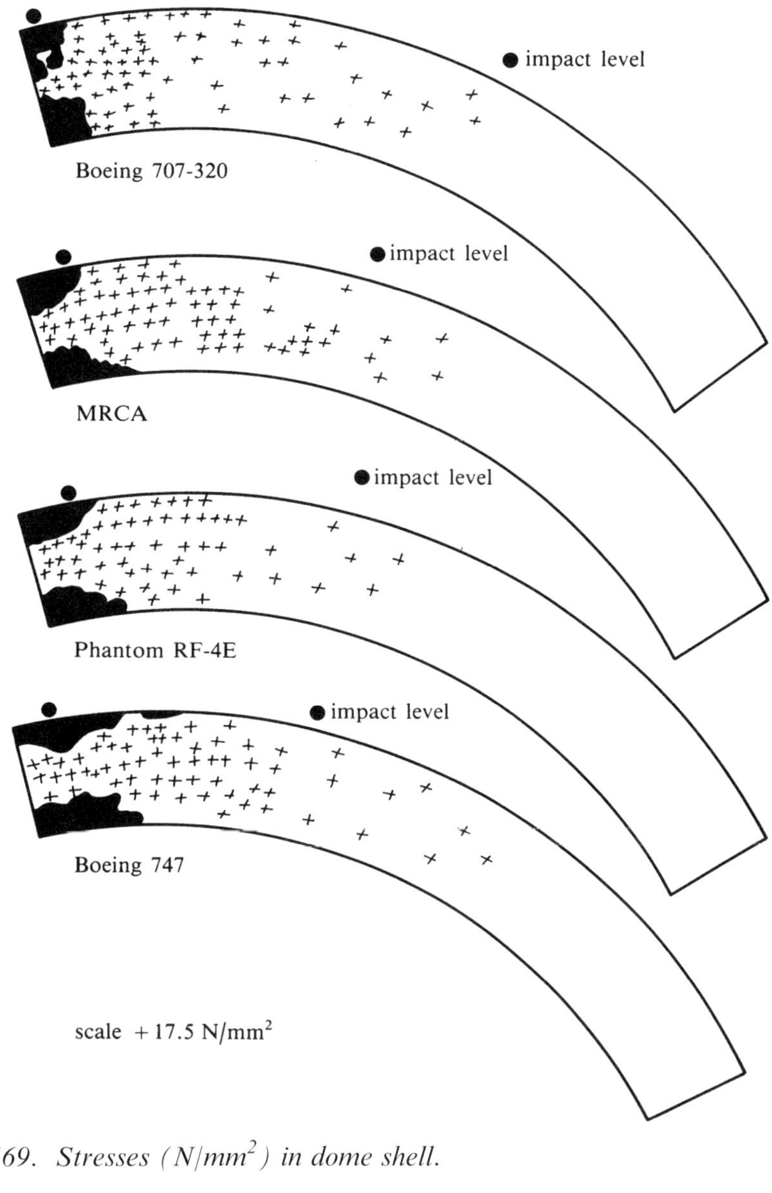

Fig. IX.169. Stresses (N/mm²) in dome shell.

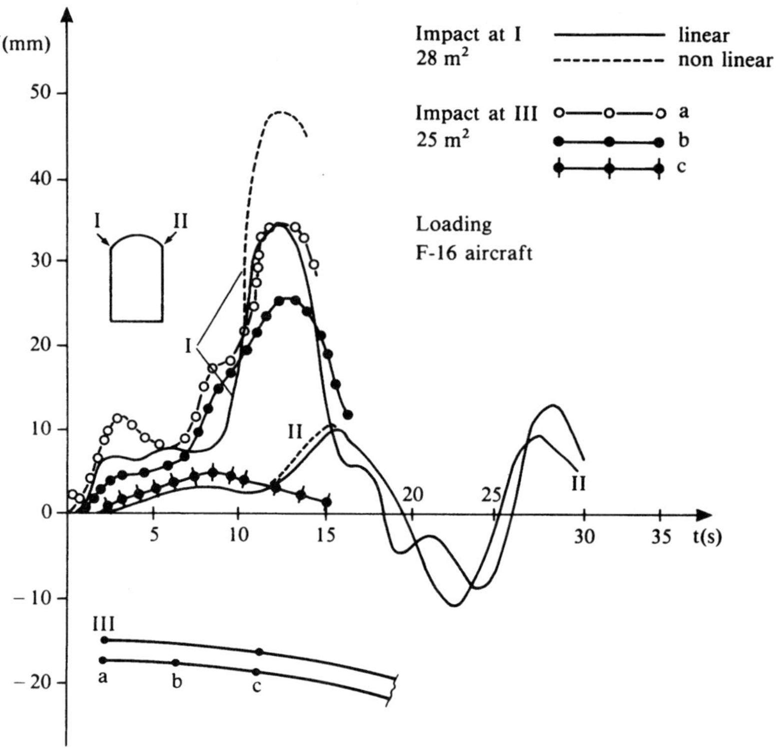

Fig. IX.170. Linear and nonlinear displacement as a function of time.

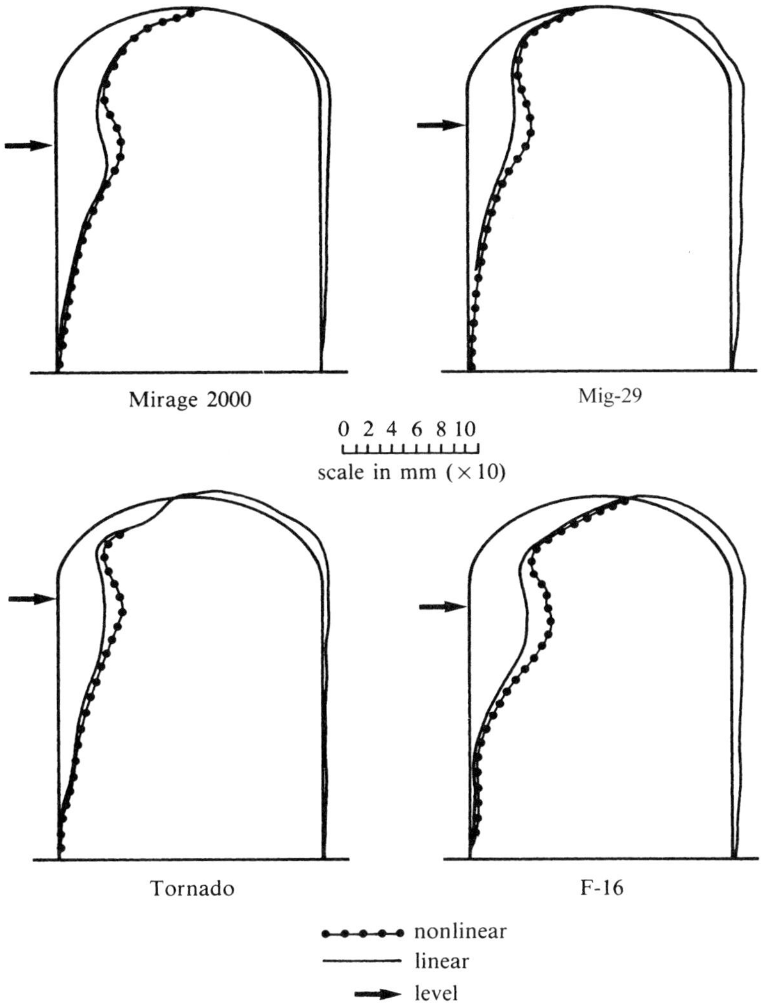

Fig. IX.171a. Linear and nonlinear displacements for impacts due to four aircraft.

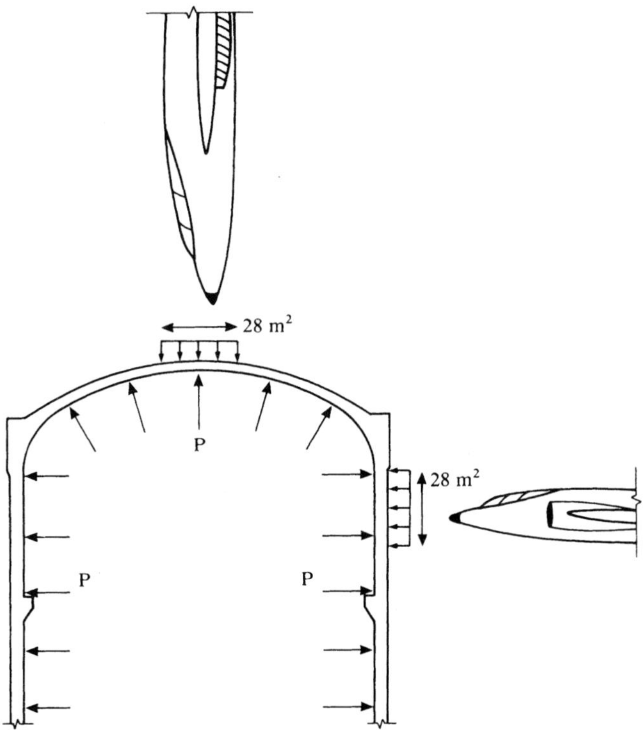

Fig. IX.171b. Extreme loading condition.

Table IX.15a. Aircraft impact load versus comparative vessel thicknesses

Aircraft [Speed (mph)]	Depth of penetration (m) [Thickness for no perforation (m)]						Finite element thickness[a]/Displacement (mm)	
	IRS	HN–NDRC	ACE	DF–BRL	BRL	DF–ACE	Depth of penetration (m)	Thickness[b]
MRCA	2.0 [2.59]	3.1 [3.75]	2.5 [2.85]	1.57 [2.30]	1.59 [2.75]	2.5 [2.95]	1.80 (55 max)	3.0 —
Phantom RF-4E [482]	1.85 [2.50]	2.90 [3.70]	2.25 [3.00]	1.25 [2.00]	1.35 [1.75]	2.10 [2.75]	1.55 (55 max)	2.35 —
Phantom (C) F482]	1.75 [2.50]	2.80 [3.50]	2.15 [2.85]	1.10 [1.75]	1.35 [1.75]	2.0 [2.75]	1.55 (60 max)	2.35 —
Boeing 707-320 (E) [730]	1.80 [2.50]	2.75 [3.25]	2.00 [2.75]	1.20 [1.75]	1.50 [2.0]	2.0 [2.75]	1.10 (55 max)	2.00 —
Boeing 747 [250]	2.0 [2.50]	3.0 [3.75]	2.30 [2.85]	1.55 [2.30]	1.85 [2.50]	2.30 [2.85]	1.90 (65 max)	2.75 —
FB-111 (CH) [200]	2.0 [2.50]	2.50 [3.75]	2.00 [2.75]	1.05 [2.00]	1.35 [2.00]	1.75 [2.25]	1.25 (45 max)	2.30 —
Mirage [470]	1.75 [2.75]	2.85 [3.75]	2.20 [3.00]	1.25 [2.00]	1.50 [2.20]	2.50 [3.00]	1.45 (40 max)	2.35 —
Mig-23 [480]	1.2 [1.6]	1.8 [5.0]	2.2 [6.0]	1.0 [3.0]	1.0 [1.15]	2.1 [3.10]	1.80 (45 max)	2.85 —

[a] All thicknesses mentioned under finite element are computed thicknesses.
[b] For wide perforation.

Table IX.15b. Foundation parameters (major data)

Foundation radius $R = 24.86\,\mathrm{m}$

Foundation thickness $t_f = 3.76\,\mathrm{m}$

Seismic shear force $= 1822.52\,\mathrm{kN/m}$

Uplift from vertical earthquake component $= 470\,\mathrm{kN/m}$

Tangential shear from earthquake $= 1272.96\,\mathrm{kN/m}$

$$K_u = \text{translational spring constant} = \frac{32(1-\nu)GR_0}{(7-8\nu)}$$

$$K_z = \text{vertical constant} = \frac{4GR}{(1-\nu)}$$

$$K_\phi = \text{rotational spring constant} = \frac{8GR_0^3}{3(1-\nu)}$$

$$K_t = \text{torsional spring constant} = \frac{16GR^3}{3}$$

$\nu = 0.14$

$\beta' = 0.5$

Material damping: 8% for the soil and 2% for the vessel

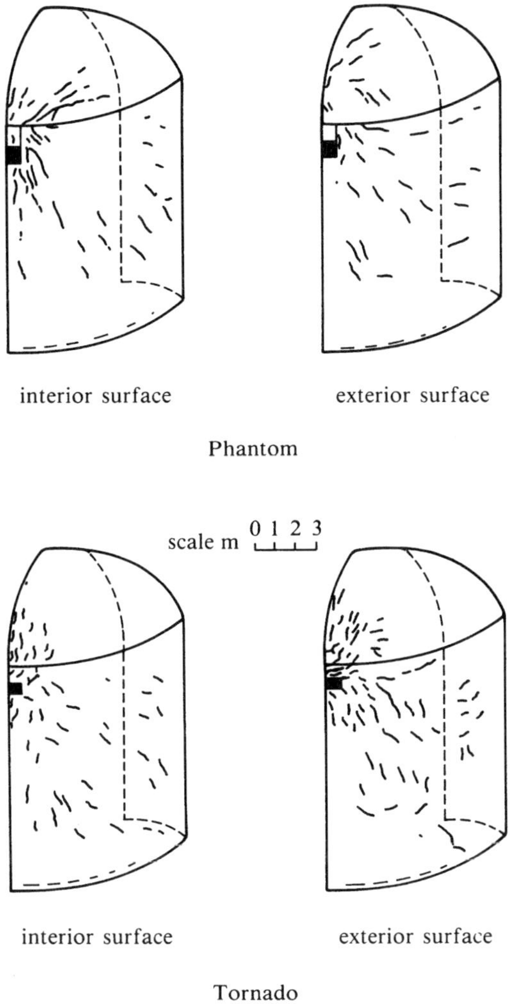

Fig. IX.172. Phantom and Tornado aircraft impact 'post-mortem' [1007].

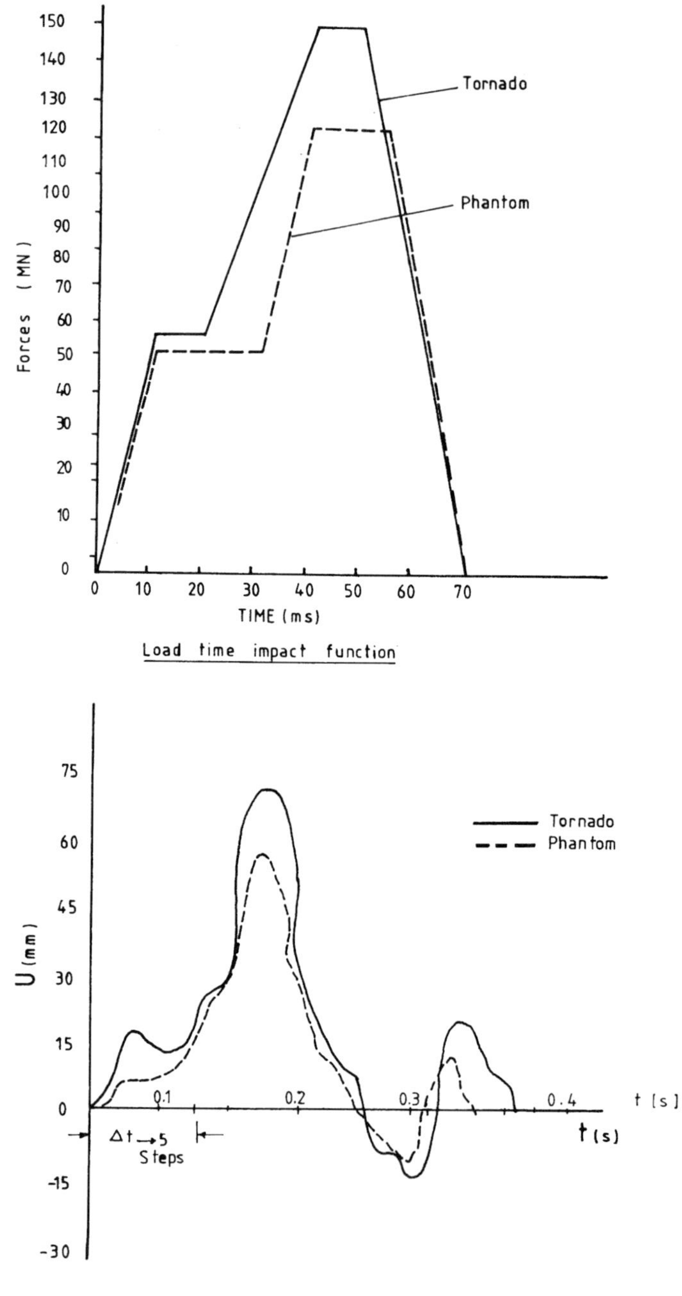

Fig. IX.173. Load–displacement as a function of time [1007].

IX.10.3.2 Seismic analysis

A nonlinear dynamic analysis was carried out for the containment vessel. A direct approach was adopted in which the whole system (containment–soil/bedrock) was modelled and analysed in the time domain using finite elements. Inertia forces were reproduced by distributed masses by involving consistent mass matrices. In general the equation of motion contains a mass matrix (M), a damping matrix (C) and a stiffness matrix (K). The excitation is taken in the form of a base motion. As shown in Figs IX.166 and IX.167, a 20-noded isoparametric element together with line and panel elements representing concrete, prestressing tendons or reinforcement and steel liner respectively are used to model the structure and the supporting medium. This method allows treatment of nonlinear behaviour and material properties which are adjusted at each step. Special care is exercised in using the numerical integration procedure together with time steps in order to achieve stability in the solution. In the stiffness matrix evaluation of the vessel, the four-parameter model is considered. The cracking cases are evaluated using the endochronic cracking model given in Chapters VI and VII and associated appendices. Table IX.15b shows the foundation parameters for impact and earthquake problems.

The program ISOPAR has been written to perform seismic analysis incorporating the effects of circumferential cracking. The program uses numerical integration rather than the method of normal modes. The cracks were obtained when the overall stiffness of the vessel was no longer constant but was a function of the shearing stresses in the vessel.

The equations of motion were solved using the Wilson and Newmark θ methods. The convergence criterion for ending the iterations was 0.0001.

The size of the time step (t) is critical for obtaining accuracy and rapid iteration convergence (optimum $t = 0.0025$ s). A method given by Elwi and Murray [5] was considered along with that presented by Bangash [843].

Stiffness changes will occur during some time steps. A typical subroutine that keeps track of where each crack is on the hysteresis loop is shown in Figs IX.174 and IX.175, and makes changes in the crack stiffness when necessary. A time step is repeated only when stiffness change from line to line occurs. This is because the high velocities which occur during the unloading from certain lines can cause the crack tip to go far below the specified slips at certain points, which causes the loop to grow much wider than originally specified. Shear crack relations for a typical vessel are given in Fig. IX.176.

At time t, one of the cracks is at the position marked on a line by the hysteresis loop. At time ($t + \Delta t$), a certain point can be missed by a significant amount. The program at this point goes back to time t and refines the time step and computes the shear stress and crack displacement with the new refined time step. If the shear stress is not within a specified limit then the time step is refined further and the process is repeated until the specified limits are met. The same process is performed for the stiffness change from line to line.

The changing of hysteresis loops due to cycling is shown in Figs IX.174 and IX.175. All cracks start on the cycle 1 line. Once a shear stress is exceeded, unloading

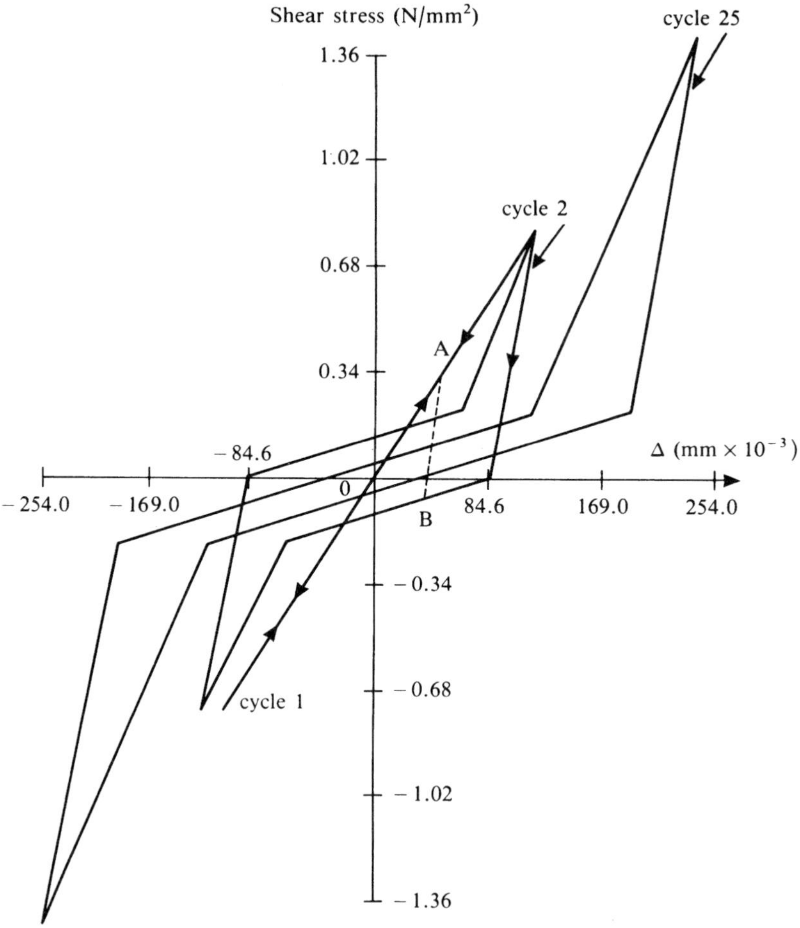

Fig. IX.174. Shear stress versus slip.

starts and proceeds along a line parallel to the dotted line A–B. The second cycle is reached when A–B intercepts a line. This process continues.

The circumferential (horizontal) and longitudinal (vertical) cracks in the vessel have a significant effect on the dynamic response of the vessel due to the SSE. The important design parameters which are affected are the liner distortion and the maximum shear stress in the concrete vessel.

Current design criteria specify that the nuclear containment vessel must be able to withstand the simultaneous occurrence of a LOCA with strong (SSE) earthquake motions. The internal pressure creates tensile stresses in both the longitudinal and circumferential directions, while the earthquake causes inertia forces which in turn cause shearing stresses and bearing stresses in the vessels. These stresses cause forces which must be transferred across horizontal and vertical cracks. The crack patterns, crack widths and spacing are computed. Crack widths vary between 0.25 and 0.38 mm.

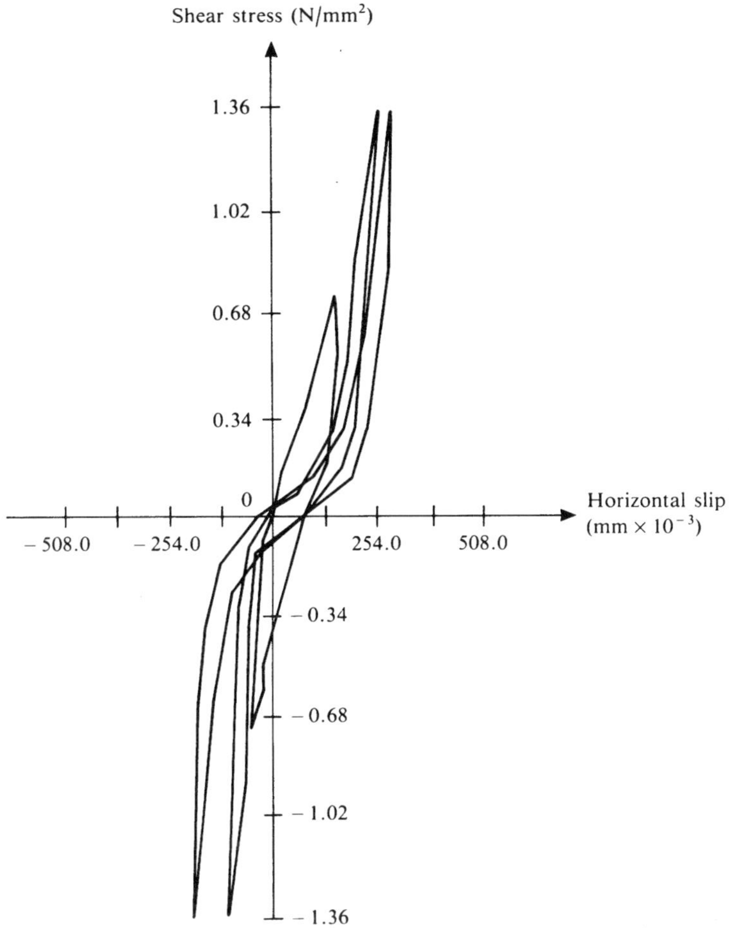

Fig. IX.175. Shear stress versus horizontal slip.

The shearing stiffness of the cracks due to aggregate interlock affect the shear stress distribution in the containment vessel after immediate short-term internal pressurisation.

During an earthquake, the stresses caused by inertia forces cause the crack width to change. Where the crack was too narrow over the whole of some portion of its circumference, it results in alteration in the shear stress distributions. Computer results show that for a total unbonded length at the crack of 64 mm the change in crack width is small compared to the initial crack width. Therefore the shear stress distribution is not significantly altered from the sinusoidal distribution. The vertical cracks present may effectively decrease the flexural and shear stiffness of the containment vessel.

Figure IX.177 shows a mass displacement–time relation for the vessel. Figure IX.178 shows response spectra under various conditions and at the top and bottom of the vessel. The control motion is specified at the free surface of the soil

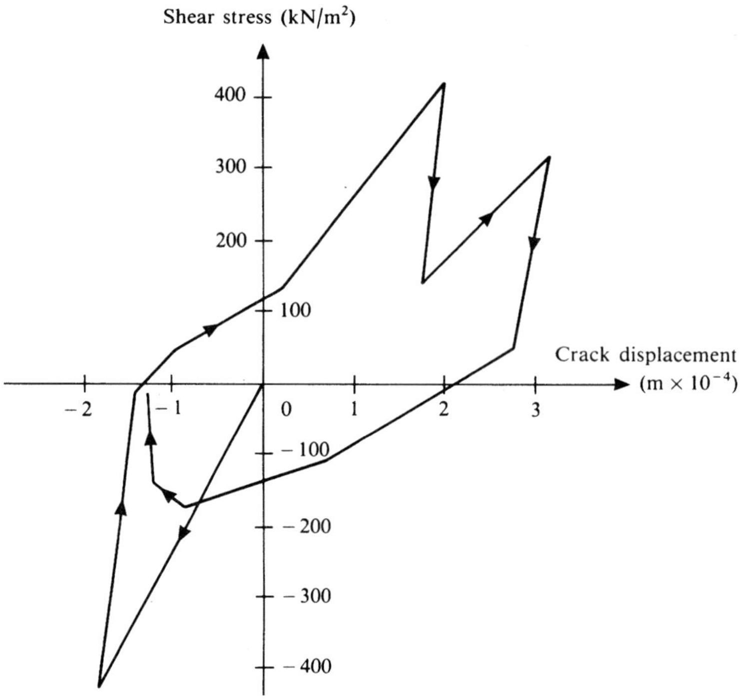

Fig. IX.176. Shear–crack relation.

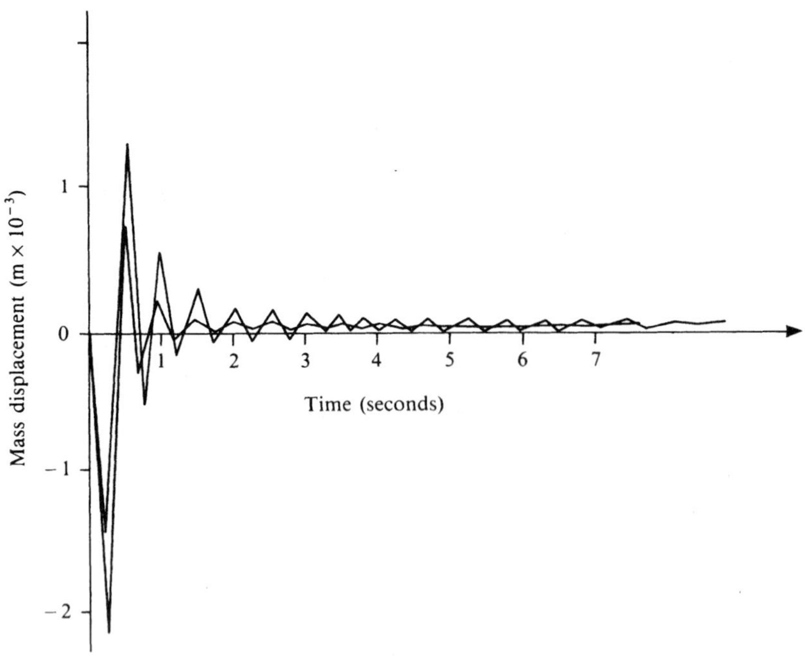

Fig. IX.177. Mass displacement–time relation for a typical containment vessel.

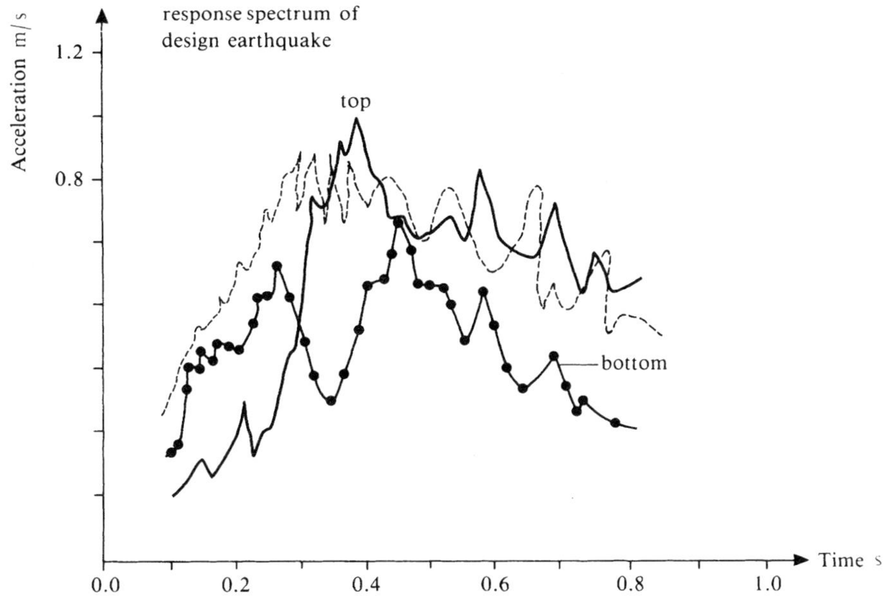

Fig. IX.178. Containment vessel response spectrum. (– – –, Measured; ——, non-linear; —●—, linear.)

deposit. The curves show the motion specified at bedrock and at the foundation level, respectively. Figure IX.179 shows a 'post-mortem' of the Sizewell B vessel at 0.3*g*. Throughout the vessel excitation the damping for the vessel was 2% and that of the soil was 8%.

IX.10.3.3 Containment overpressurisation analysis

A 1/6th scale reinforced concrete model[1006] (as shown in Figs IX.180a and IX.180b) has been designed for a design pressure of 317 kPa. The reinforcement layouts are shown in Figs IX.181 and IX.182. A pre-test analysis has been carried out using the program ISOPAR. The endochronic model along with cracking criteria proposed in this text have been adopted.

The results are compared with others in Figs IX.183–IX.186 using the finite element mesh of Fig. IX.187. The final mode is shown in Fig. IX.188 and some crack zones are indicated in a portion of the wall in Fig. IX.189.

The Sizewell B model was overpressurised. Figure IX.190a shows the displacement pattern. The vessel failed at 3.15 by rupturing of tendons and excessive cracking. The mode is of the type indicated in Figs IX.189 and IX.190b.

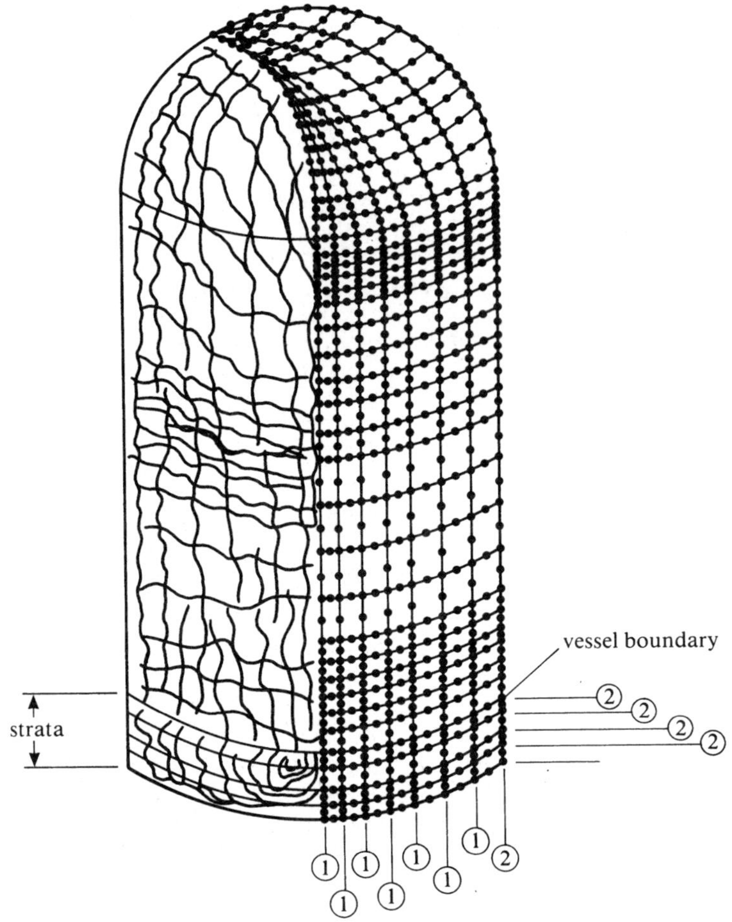

Fig. IX.179. Cracked containment.

Fig. IX.180a. One-sixth scale RC model containment (courtesy of Lawrence Livermore Laboratory [1006]).

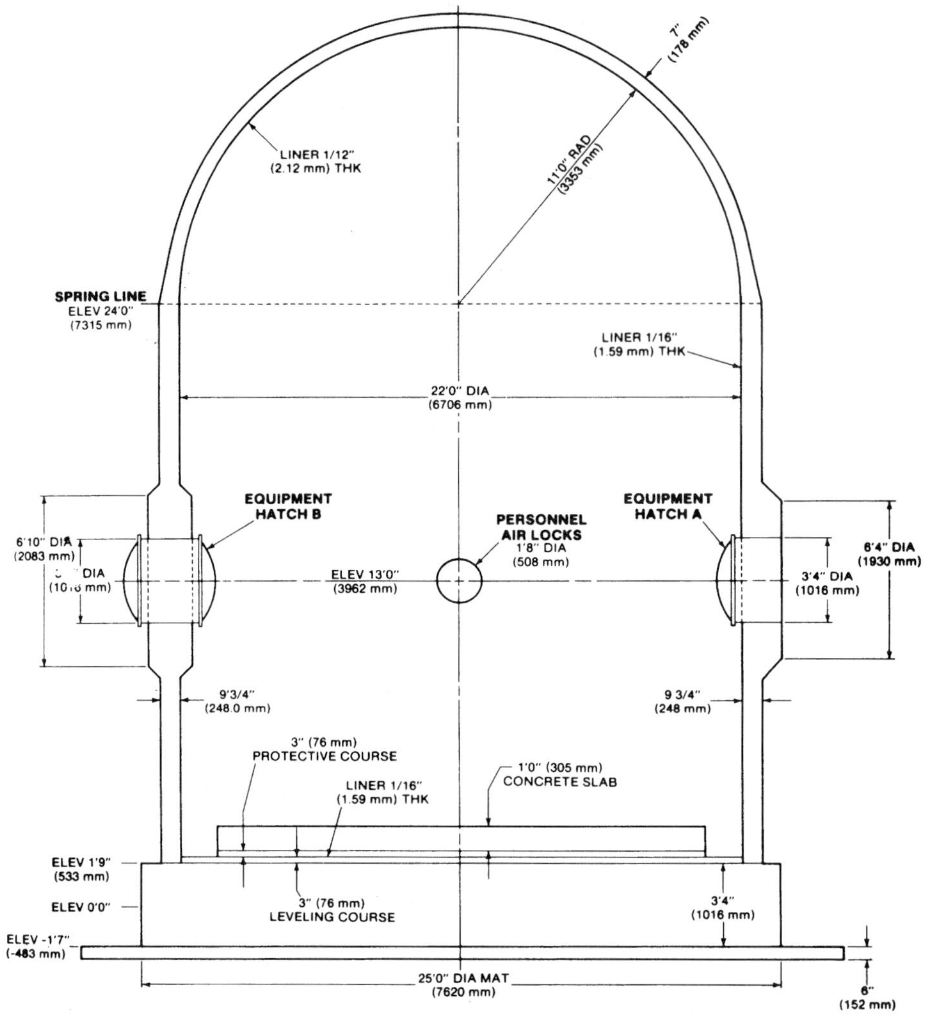

Fig. IX.180b. Schematic of the 1/6 scale reinforced concrete containment model elevation view (courtesy of Lawrence Livermore, USA).

REINFORCING BAR DETAILS

LAYER	TYPE	SIZE
1	HOOP	#4
2	MERIDIONAL	#4
3	HOOP	#4
4	HOOP	#4
5	MERIDIONAL	#4
6	HOOP	#4
7	SEISMIC (45° DIAG)	#4
8	SEISMIC (45° DIAG)	#4

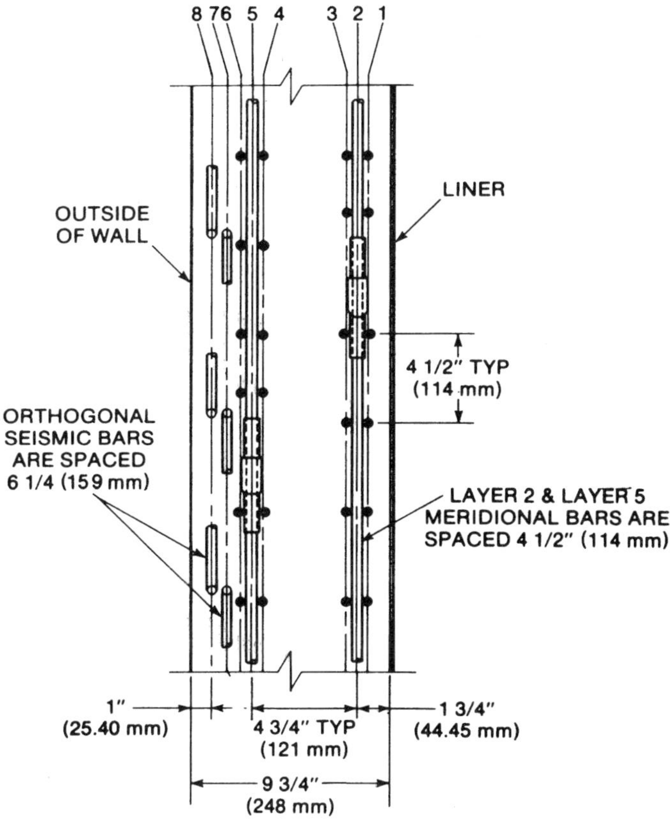

Fig. IX.181. Reinforcement in the cylinder (free field) (courtesy of Lawrence Livermore, USA).

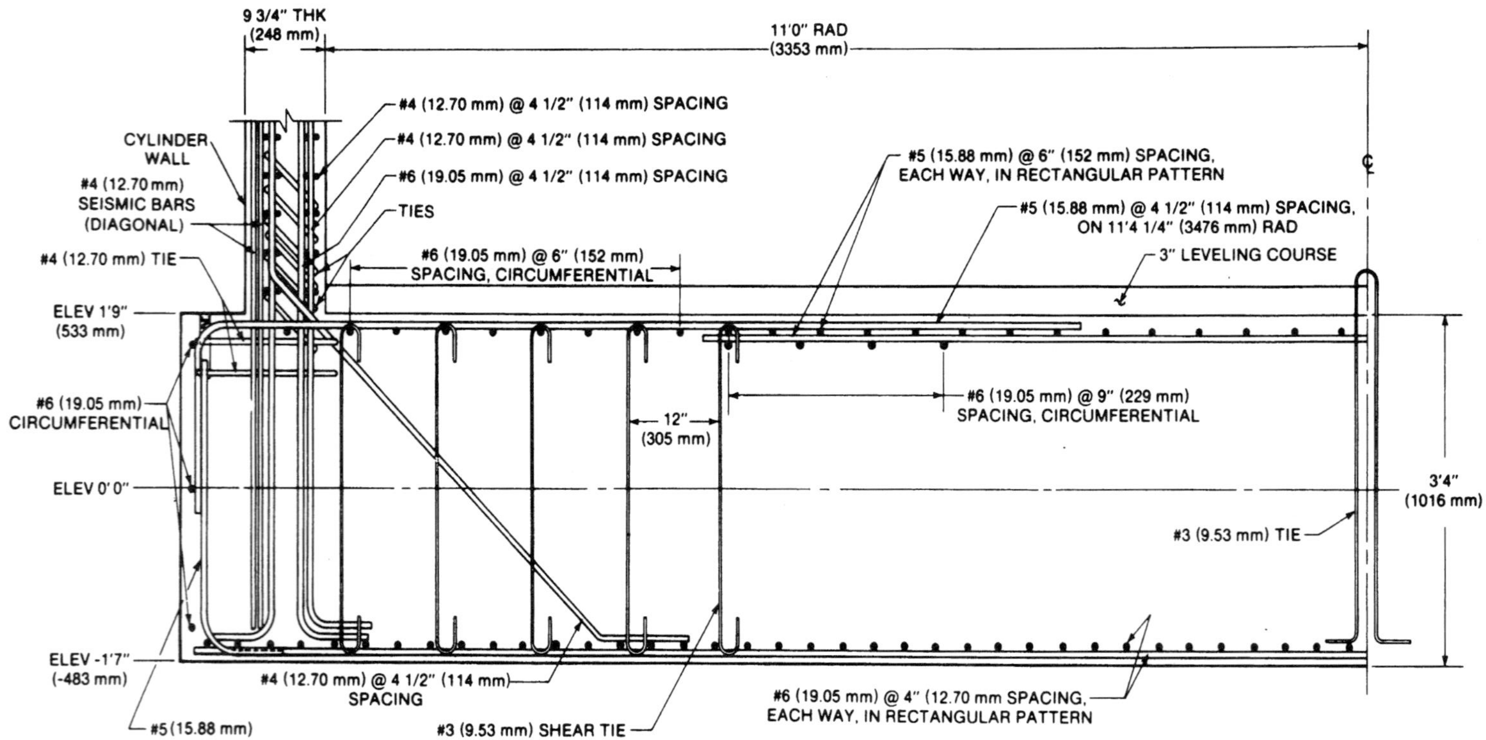

Fig. IX.182. Reinforcement in the basement and cylinder–basement junction (courtesy of Lawrence Livermore, USA).

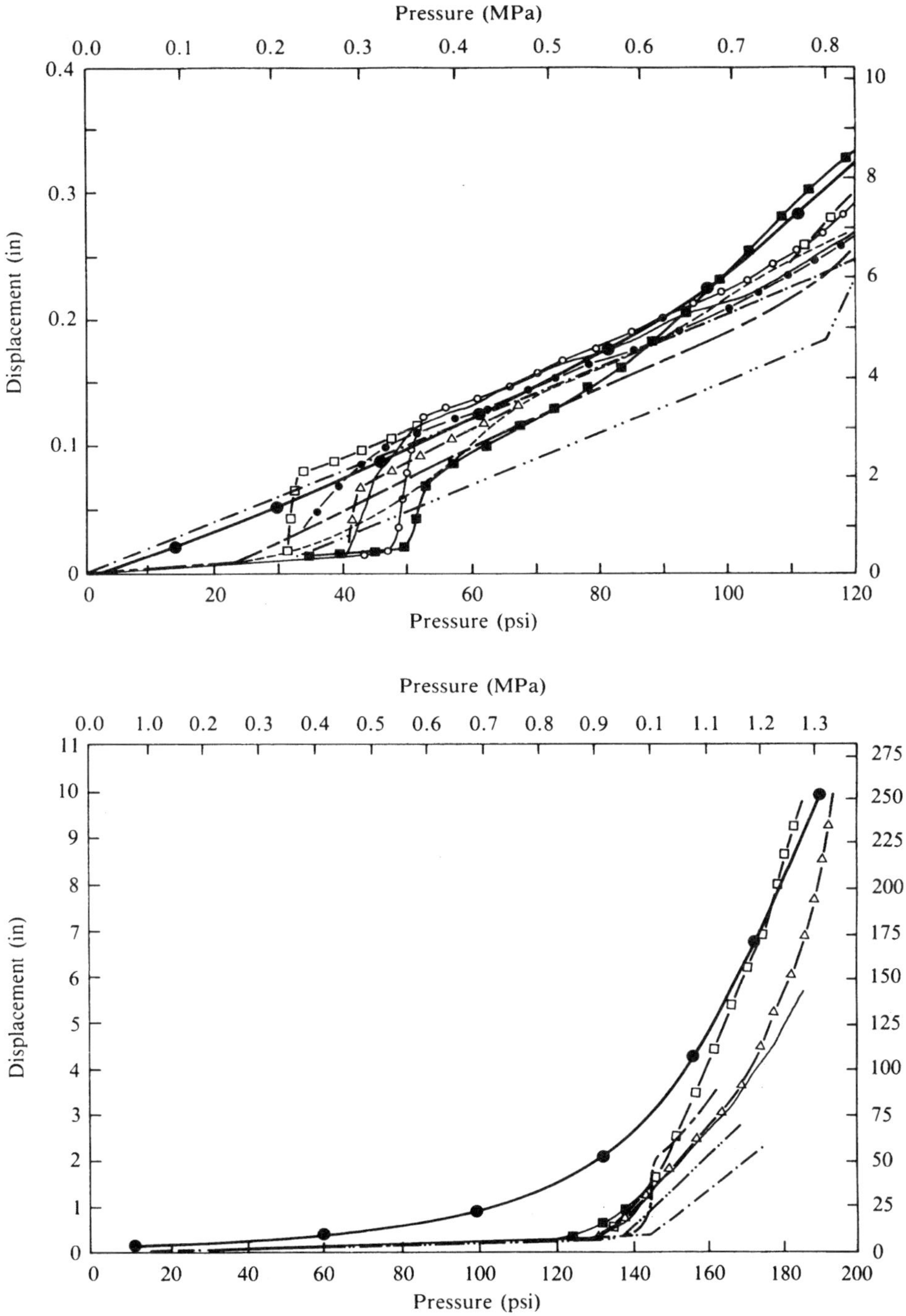

Fig. IX.183. Comparative Study—1. (—··—, SNL; —·—, NIL; — — —, SERD; —■—, CEGB; —○—, EPRI; - - -, BNL; —□—, ENEA; ——, ANL; •—•, CEA; —△—, GRS; —●—, Bangash.) (Courtesy of Lawrence Livermore, USA.)

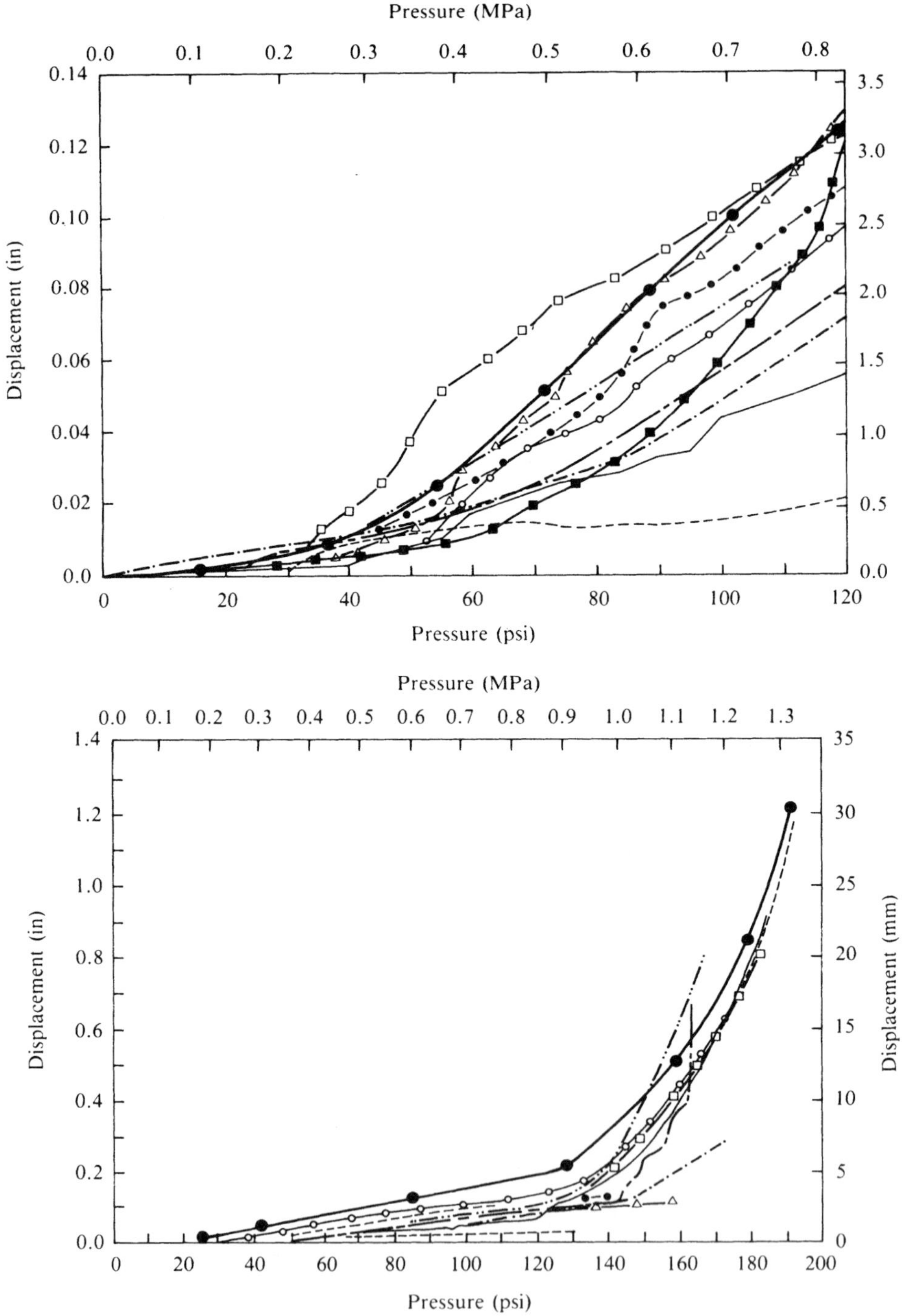

Fig. IX.184. Comparative Study — 2. (See Fig. IX.183 for key.) (Courtesy of Lawrence Livermore, USA.)

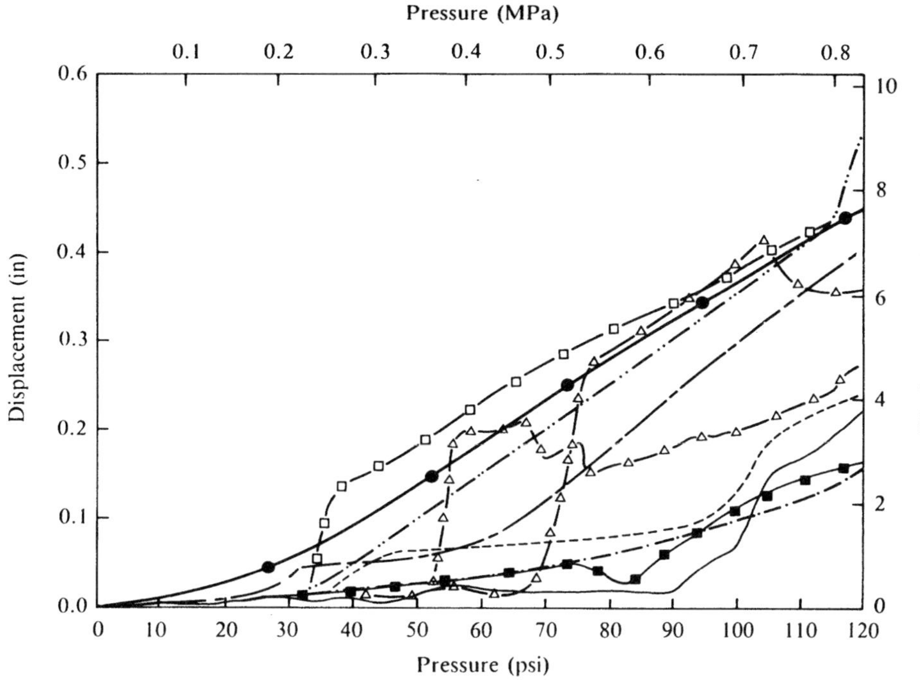

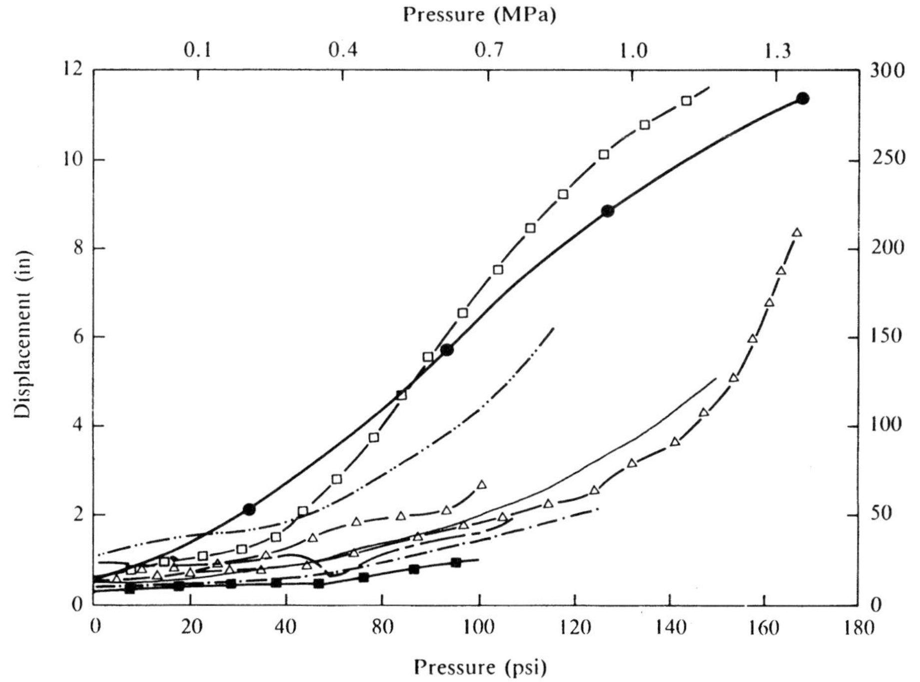

Fig. IX.185. Comparative Study — 3. (See Fig. IX.183 for key.) (Courtesy of Lawrence Livermore, USA.)

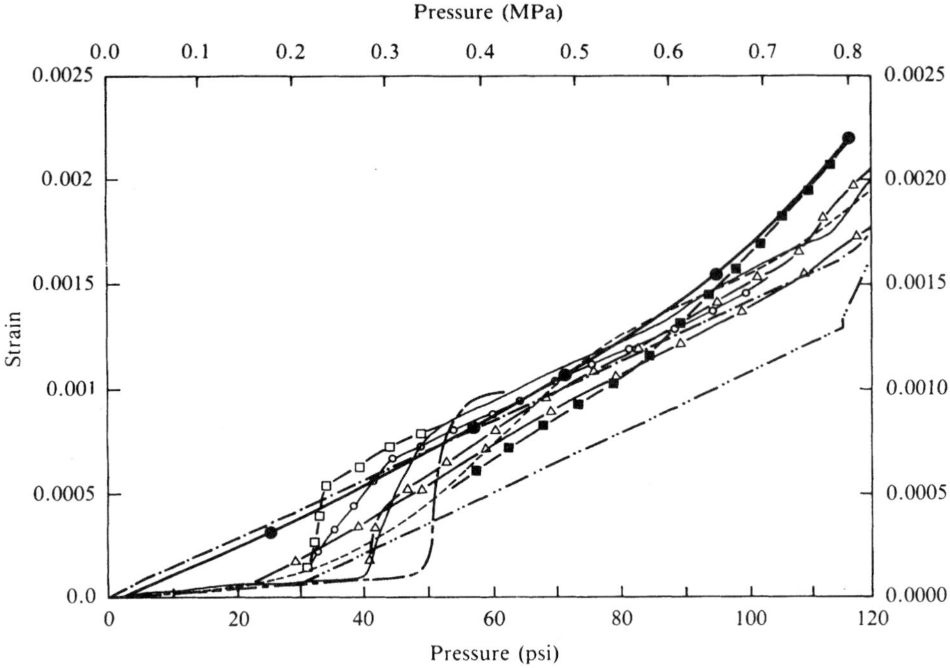

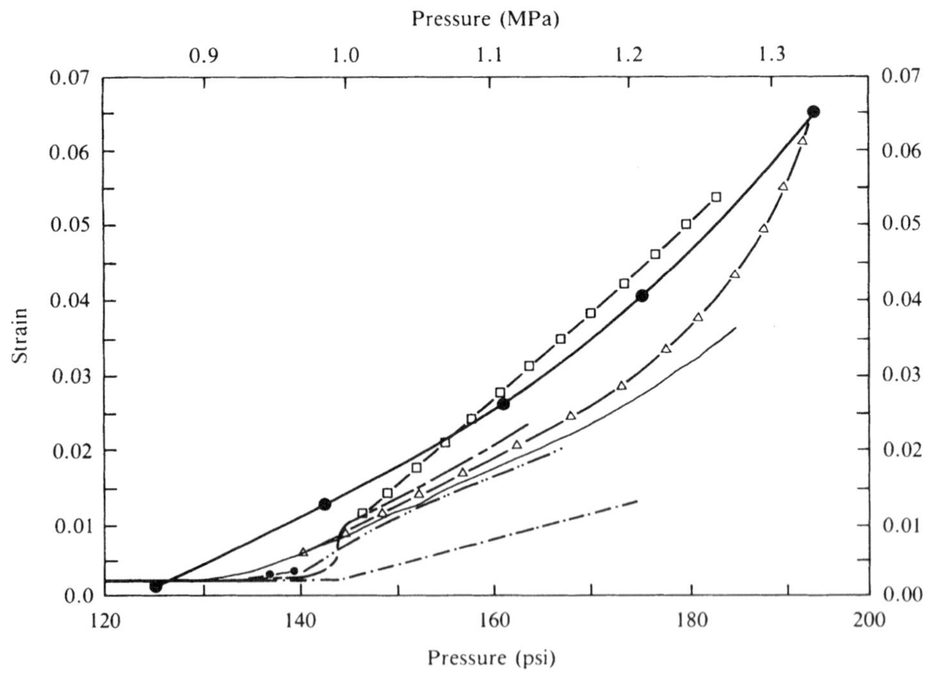

Fig. IX.186. Comparative Study — 4. (See Fig. IX.183 for key.)

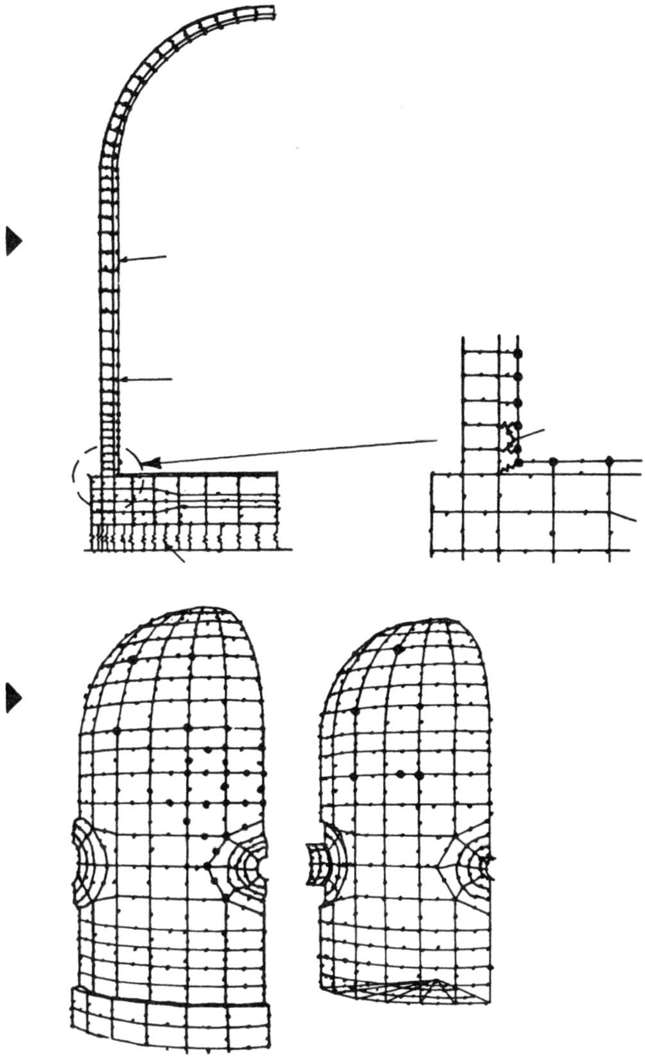

Fig. IX.187. One-sixth scale RC containment — finite element mesh.

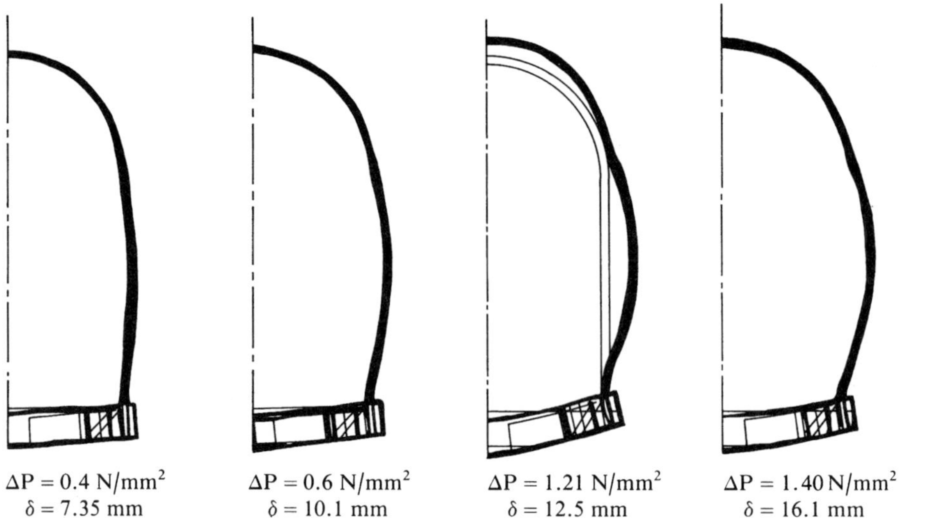

Fig. IX.188. Containment vessel final modes [1006; ISOPAR].

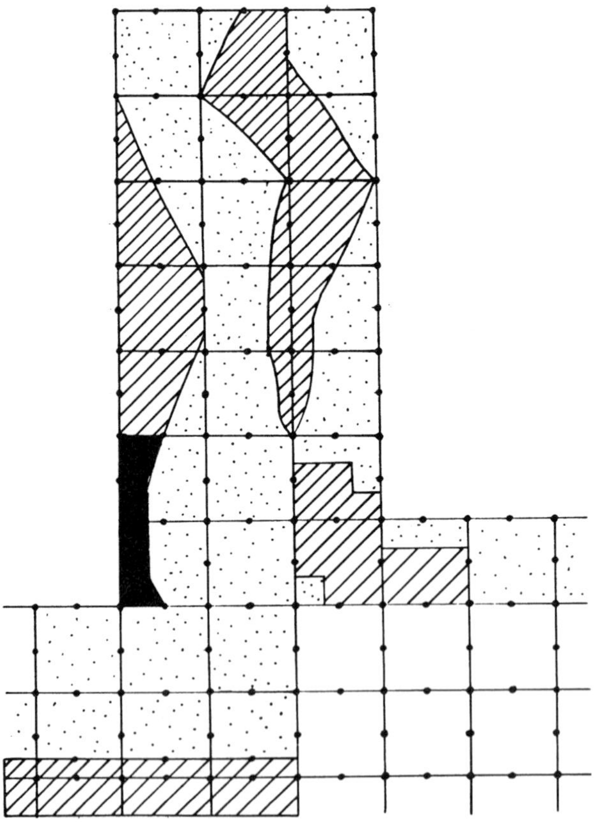

Fig. IX.189. Major crack zones [1006]. (▧, denotes zero tensile stress; ■, denotes crushing; ▨ denotes radial vertical cracking.)

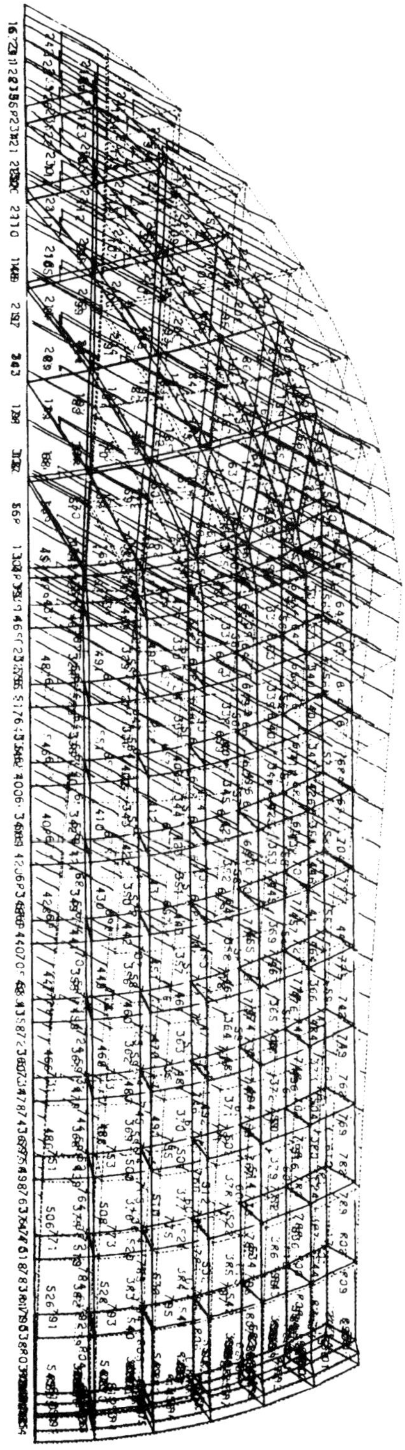

Fig. IX.190a. Sizewell B overpressurisation analysis — final mode.

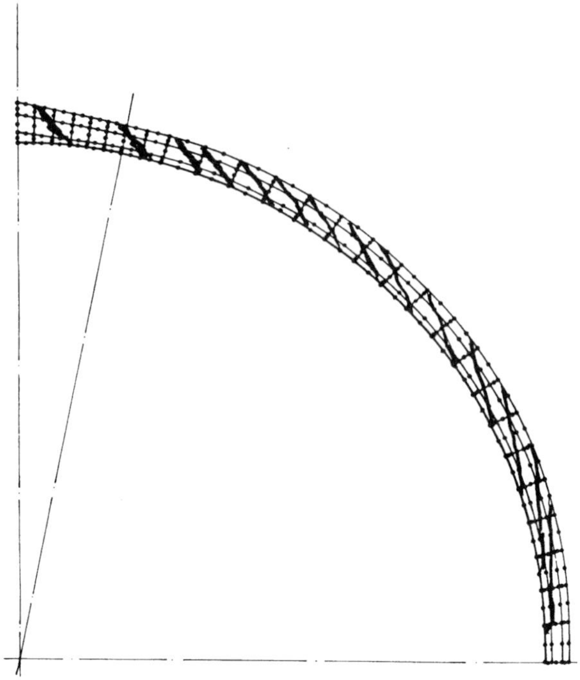

Fig. IX.190b. Cracking of Sizewell B dome.

IX.11 Case Study No. 10 — concrete nuclear shelters

IX.11.1 Introduction

There is increasing current concern about safety from nuclear hazards, including nuclear blasts and radiation. There will be greater involvement in protecting people against such hazards. A nuclear shelter is just one of many ideas to protect and shield a person from the effects of nuclear explosions. These structures can range from a deep buried rigid structure to a concrete framed box covered with soil.

The vast quantities of nuclear energy released by detonation are distributed approximately in the following way [617]:

Blast and shock wave: 45%
Light and heat radiation: 35%
Initial nuclear radiation: 5%
Residual radiation: 15%

Secondary effects such as flooding and soil liquefaction are also considered in the overall safety analysis of shelters. Factors which affect the response of shelters to air blast are:

(a) strength and mass
(b) structural layout
(c) ductility.

The airburst from an atom bomb propagates through the atmosphere to great distances with minimum energy losses. On the other hand, in an underground burst much of the energy is absorbed in cratering and melting of the ground. A burst occurring above 30 000 m is known as an airburst provided the fireball at maximum brilliance does not touch the surface of the earth. If it touches, it becomes a 'surface burst'. These cases together with their loading conditions are described in this section. A typical concrete shelter design for a six-person family is chosen for finite element analysis using the five-parameter constitutive law. Analytical results show agreement with those provided by the various codes [639, 640, 667].

IX.11.2 Characteristics of the blast wave in air

Most of the material damage from an airburst (nuclear weapon) is caused mainly by the shock (or blast) wave which accompanies the explosion. The majority of structures will suffer some damage from air blast when the overpressure in the blast wave, i.e. the excess over atmospheric pressure (101.3 kN/m^2 at sea level), is about 3.5 kN/m^2 or more. The distance to which this overpressure level will extend depends on the yield of the explosion and on the height of the burst.

A difference in air pressure acting on separate surfaces of a structure produces a force on that structure; the size of these forces is dependent on the difference between ambient air pressure and the overpressure. The maximum value of overpressure, known as the peak overpressure, occurs at the front of the blast wave.

As the blast wave travels in the air away from its source, the overpressure at the front steadily decreases, and the pressure behind the front falls off in a regular manner. After a short time, when the shock front has travelled a certain distance from the fireball, the pressure behind the front drops below that of the surrounding atmosphere and a so-called 'negative phase' of the blast wave forms, as given in Fig. IX.191.

For the curves marked t_1 to t_4 the pressure in the blast wave has not fallen below atmospheric pressure, but on the curve t_5 it is seen that at some distance behind the

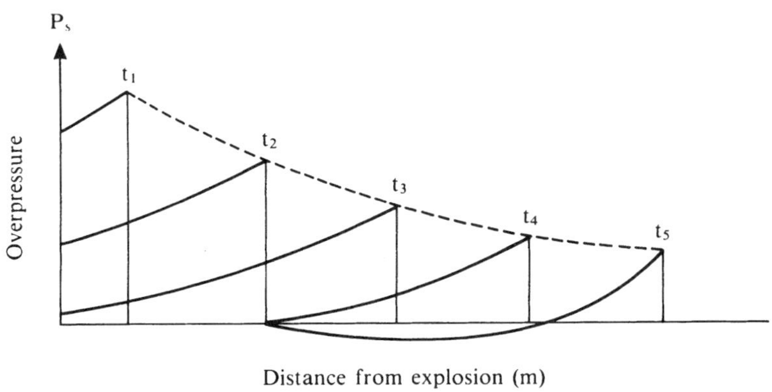

Fig. IX.191. Variation of overpressure with distance at successive times.

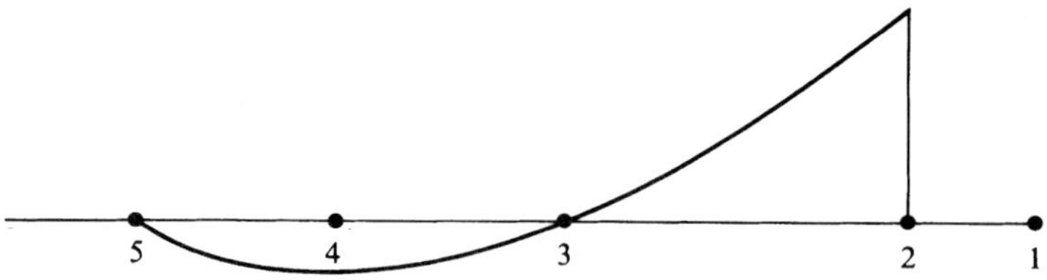

Fig. IX.192. Variation of pressure with time at a fixed location.

shock front the overpressure is below that of the original atmosphere so that an 'underpressure' exists.

During the negative (rarefaction or suction) phase a partial vacuum is produced and the air is sucked in, instead of being pushed away, as it is when the overpressure is positive.

The peak values of the underpressure are generally small compared with the peak overpressures, the former having a maximum value of 27.5 kN/m^2 below the ambient pressure. With increasing distance from the explosion, both peak values decrease, the positive more rapidly than the negative, but they do not approach equality until the peak pressures have decayed to a very low level (Fig. IX.192).

For a short interval after the detonation, point 1, there will be no increase in pressure since it takes the blast wave some time to travel the distance from the point of explosion to the given location. Point 2 indicates the time of arrival of the shock front; a strong wind commences to blow away from the explosion. This is often referred to as a 'transient' wind, as its velocity decreases fairly rapidly with time.

Following the arrival of the shock front the pressure falls rapidly, and at the time corresponding to point 3 it is the same as that of the original atmosphere. Although the overpressure is now zero, the wind will continue in the same direction for a short time. The interval from points 2 to 3 (roughly 2–4 s for a 1 megaton explosion) represents the passage of the positive (or compression) phase of the blast wave. It is during this interval that most of the destructive action of the airburst will be experienced.

As the pressure in the blast wave continues to decrease, it sinks below that of the surrounding atmosphere. In the time interval from point 3 to point 5, which may be several seconds, the negative (or suction) phase of the blast wave passes the given location. For most of this period the transient wind blows towards the explosion. As the negative phase passes, the pressure at first decreases below ambient and then increases towards that of the ambient atmosphere which is reached at the time represented by point 5. The blast wind has then effectively ceased and the direct destructive action of the airblast is over.

IX.11.2.1 Dynamic pressure

Although the destructive effects of the blast wave are usually related to values of the peak overpressure, there is another quantity of equivalent importance called the 'dynamic pressure'.

Table IX.16. Overpressure, dynamic pressure and wind velocities in air at sea level for an ideal shock front

Peak overpressure, kN/m^2 (atmos.)	Peak dynamic pressure, kN/m^2 (atmos.)	Maximum wind velocity, km/h
1 378 (13.78)	2 274 (22.74)	3 347
1 034 (10.34)	1 537 (15.37)	2 860
689 (6.89)	848 (8.48)	2 275
496 (4.96)	551 (5.51)	1 883
345 (3.45)	275 (2.75)	1 512
207 (2.07)	110 (1.10)	1 078
138 (1.38)	55 (0.55)	756
69 (0.69)	15 (0.14)	467
34 (0.34)	5 (0.05)	257
14 (0.14)	0.7 (0.007)	97

The dynamic pressure is proportional to the square of the wind velocity and to the density of the air behind the shock front. For very strong shocks the dynamic pressure is larger than the overpressure but below 480 kN/m^2 (4.7 atmospheres) overpressure at sea level; the dynamic pressure is smaller. The peak overpressure dynamic pressure decreases with increasing distance from the explosion centre, although at a different rate.

Some indication of the corresponding values of peak overpressure, peak dynamic pressure and maximum blast wind velocity for an ideal shock front in air at sea level are given in Table IX.16.

When the shock front reaches a given point, both the overpressure and the dynamic pressure increase almost immediately from zero to their maximum values and then decrease. The dynamic pressure (and wind velocity) will fall to zero somewhat later than the overpressure because of the momentum of the air in motion behind the shock front, but for purposes of estimating damage the difference is not significant. During the negative (suction) phase of the blast wave the dynamic pressure is very small and acts in the opposite direction.

IX.11.2.2 Arrival time and duration

There is a finite time interval required for the blast wave to move out from the explosion centre to any particular location. This time interval is dependent upon the energy yield of the explosion and the distance involved. At 1.6 km from a 1 megaton burst, the arrival time would be about 4 s. Initially, the velocity of the shock front is quite high, many times the speed of sound, but as the blast wave progresses outward, so it slows down as the pressure at the front weakens. Finally, at long ranges, the blast wave becomes essentially a sound wave and its velocity approaches ambient sound velocity.

The positive phase duration is shortest at close ranges and increases as the blast wave moves outward. At 1.6 km from a 1 megaton explosion, the duration of the positive phase of the blast is about 2 s. The period of time over which the positive

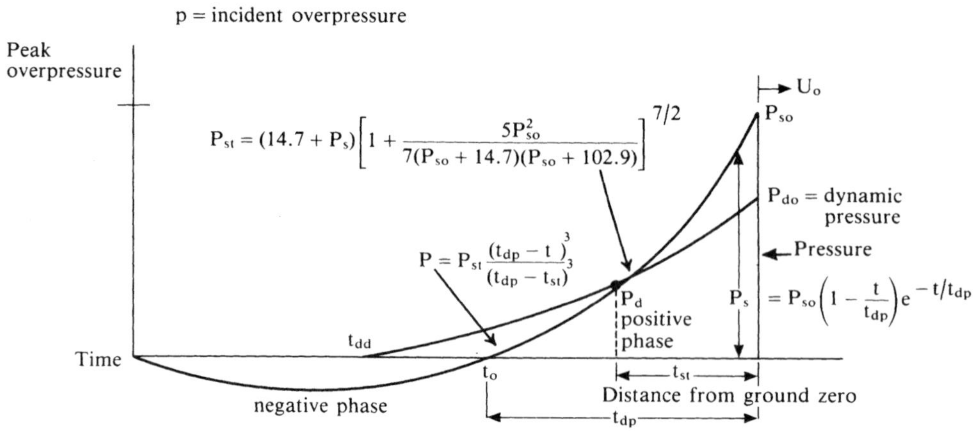

Fig. IX.193. Variation of overpressure with time at a fixed location.

dynamic pressure is effective may be taken as essentially the positive phase duration of the overpressure.

IX.11.2.3 Reflection of the blast wave at a surface

When the incident blast wave from an explosion in air (Fig. IX.193) strikes a more dense medium, such as the Earth's surface, either land or water, it is reflected.

When such reflection occurs, an object precisely at the surface will experience a single pressure increase, since the reflected wave is formed instantaneously. The value of the overpressure thus experienced at the surface is generally considered to be entirely a reflected pressure. In the region near ground zero, this total reflected overpressure will be more than twice the value of the peak overpressure of the incident blast wave.

There are two important destructive aspects of the blast wave involved with reflection:

(i) Only a single pressure increase is experienced in the Mach region below the triple point as compared to the separate incident and reflected waves in the region of regular reflection.

(ii) (a) Since the Mach stem is nearly vertical, the accompanying blast wave is travelling in a horizontal direction at the surface.

(b) The transient winds are approximately parallel to the ground, thus, in the Mach region, the blast forces on aboveground structures and other objects are directed nearly horizontally, so that vertical surfaces are loaded more intensely than horizontal surfaces.

IX.11.2.4 Blast from a surface burst

In a surface explosion, the front of the blast wave in air is hemispherical in form (Fig. IX.194), there is no region of regular reflection, and all objects and structures on the surface, even close to ground zero, are subjected to air blast similar to that in the Mach region below the triple point of an airburst.

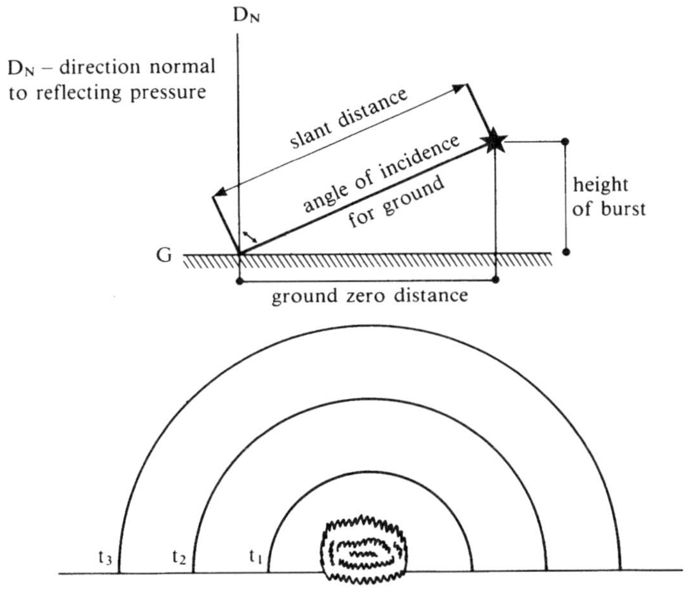

Fig. IX.194. Blast wave from a contact burst.

The diameter of the rupture zone D_r is $1.5D_a$; the overall diameter, including the lip D_t, is $2D_a$; and the height of the lip H_t is $0.25H_a$.

The dependence of crater radius and crater depth upon the depth of burst for a 1 kiloton explosion in dry soil is shown in Fig. IX.195. Also shown are the range of crater dimensions possible from a surface burst to the approximate maximum for

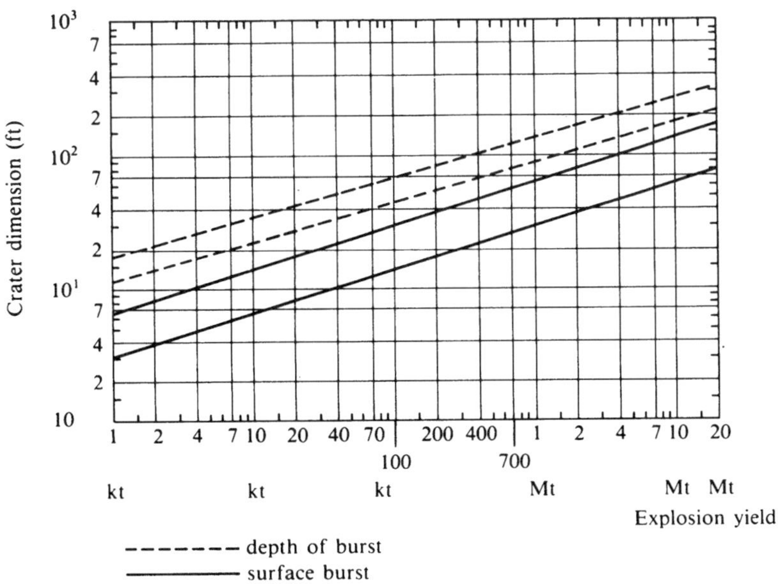

Fig. IX.195. Crater versus yield (courtesy of US Airforce),

an underground burst, for any explosion energy yield from 1 kiloton to 10 megatons. Therefore a 1 megaton blast wave front may be assumed to be vertical for most structures, with both overpressure and dynamic pressure decaying at different rates behind the blast wave as previously stated.

All the above descriptions are based on an ideal blast wave in ideal atmospheric conditions; unfortunately, nothing is perfect, especially meteorological conditions, which can greatly alter the expected performances of explosions, either increasing them or decreasing them.

IX.11.2.5 Ground shock from air blast

Another aspect of the blast wave problem is the possible effect of an airburst on underground structures as a result of the transfer of some of the blast wave energy into the ground. A minor oscillation of the surface is experienced and a ground shock is produced. The strength of the shock at any point is determined by the overpressure in the blast wave immediately above it. For large overpressures with long positive phase duration, the shock will penetrate some distance into the ground; but blast waves which are weaker and of shorter duration are attenuated more rapidly. The major principal stress in the soil will be nearly vertical and about equal in magnitude to the air blast overpressure.

Table IX.17. Blast loadings[a]

p_{d0} = peak dynamic pressure	U_0 = velocity of the shock front
p_s = overpressure at time	t = time after detonation of yield 'y'
t_{dd} = delayed time (dynamic)	t_{dp} = duration of dynamic overpressure
$U_0 = [U_s = 1117 \times \sqrt{1 + 6p_{s0}/7p_0}]$	p_0 = atmospheric pressure = 14.7

$$p_r = \text{reflected pressure} = 2p_0\left(\frac{102.9 + 4p_{s0}}{102.9 + p_{s0}}\right)$$

$$t_c = \text{clearing time of the reflection effect}$$
$$= 3 \times \text{height of the reflecting surface above ground}/U_0$$

Scaling laws (for two bombs)

$R_2 = R_1(y_2/y_1)^{1/3}$ $\quad$ R, R_1, R_2 = radii or distances from explosions

$t_{dd} = t_{dp}(y_2/y_1)^{1/3}$

Reference should be made to Fig. IX.163 for other details

Additional loadings

$$\frac{p_{s0}}{p_0} = 3.2 \times 10^6 R^{-3}\sqrt{1 - \left(\frac{R}{87}\right)^2} \times \left(\frac{800 + R}{800}\right)$$

$$\text{time in seconds} = t_{dp} = y^{1/3}\left\{\frac{180[1 + (0.01R)^3]^{1/2}}{1 + (R/40)\sqrt[6]{1 + (R/285)^5}\ \sqrt[6]{(1 + 0.0002R}}\right\}$$

Table IX.18. Characteristics of blast waves from 1- and 10-megaton nuclear burst optimum height air blasts (UK Home Office Data)

Distance from ground zero km (miles)	Peak overpressure, kPa (psi)		Windspeed, m/s (mph)		Arrival time, s		Duration, s	
	1 Mt	10 Mt	1 Mt	10 Mt	1 Mt	10 Mt	1 Mt	10 Mt
1 (0.6)	980 (140)	4200 (600)	980 (2200)	—	0.9	—	0.9	1
2 (1.25)	200 (29)	1050 (150)	400 (900)	980 (2200)	4.0	1.8	1.8	2
3 (1.9)	130 (18)	380 (54)	180 (400)	805 (1800)	6.0	4.7	2.3	2.4
4 (2.5)	63 (12)	220 (31)	128 (285)	490 (1100)	9.0	8	2.6	3.7
5 (3.1)	56 (8)	170 (24)	106 (238)	277 (620)	12	11	2.8	4.2
6 (3.8)	42 (6)	135 (19)	81 (182)	200 (450)	14	12.5	3.0	4.7
7 (4.4)	33 (4.7)	110 (16)	69 (155)	170 (380)	18	16	3.2	5.0
8 (5)	27 (3.8)	90 (13)	56 (125)	143 (320)	22	18	3.4	5.3
9 (5.6)	24 (3.4)	77 (11)	48 (108)	125 (280)	26	20	3.5	5.5
10 (6.2)	21 (3)	63 (9)	40 (90)	114 (255)	28	22	3.6	5.9
12 (7.5)	15 (2.2)	46 (6.5)	31 (70)	96 (215)	31	30	3.7	6.3
15 (9.4)	10 (1.5)	34 (4.8)	24 (53)	72 (160)	45	38	3.8	6.7
20 (12.5)	7 (1)	22 (3.3)	17 (37)	46 (103)	59	53	4.0	7.5
25 (16)	5 (0.75)	15 (2.2)	9 (20)	32 (72)	70	67	4.2	7.8

Table IX.19. Blast loads on shallow buried and surface shelters

Member	Shallow buried		Surface
	In dry ground	High water level	
Roof and floors	p_{s0}	p_{s0}	p_{s0}
Walls	$0.5p_{s0}$	p_{s0}	$2.3p_{s0}$

p_{s0} = overpressure.
Dead loads, soil and water loads should be added to the blast loads given above.

For high airbursts, where relatively large blast pressures are not expected at ground level, the effects of ground shock induced by the air blast will be negligible, but if the overpressure at the surface is large there may be damage to buried structures. However, even if the structure is strong enough to withstand the effect of the ground shock, the sharp jolt resulting from the impact of the shock front can cause injury to occupants and damage to loose equipment.

IX.11.2.6 Technical aspects of a blast wave

The basic relationships among the properties of a blast wave having a sharp front at which there is a sudden pressure discontinuity, i.e. a true shock front, are given in Tables IX.17–IX.20 and Fig. IX.196.

IX.11.3 Air blast loading and target response

The behaviour of an object or structure exposed to the blast wave from a nuclear explosion may be considered under two main headings.

(1) Loading, the forces which result from the action of that pressure.
(2) Response, or distortion of the structure due to the particular loading.

Table IX.20. Generalised data in relation to loads

1-megaton burst at a distance of 1.6 km (1 mile) from ground zero
Velocity of shock front 500 m
Ductility ratio $\mu = 5$
Maximum allowable time = 0.15 s
Where applicable, drag coefficients: roof = −0.4, wall = +0.9
Yield strength of steel reinforcement = 410 or 425 N/mm^2
f_{cu} (dynamic) = $1.25f_{cu}$ (static)
f_y (dynamic) = $1.10f_y$ (static)
Main reinforcement $\not< 0.25\%bd$ (b = width of section, d = effective depth of section)
Secondary reinforcement $\not< 0.15bd$
The ultimate shear stress $\not> 0.04f_{cu}$
The dynamic shear stress for mild steel $\not>$ 1.172 N/mm^2
Bolts should be black bolts to BS 4190
Allowable dynamic stresses:

Bolts:	tension	275 N/mm^2	Welds:	tension/compression	275 N/mm^2
	shear	170 N/mm^2		shear	170 N/mm^2
	bearing	410 N/mm^2			

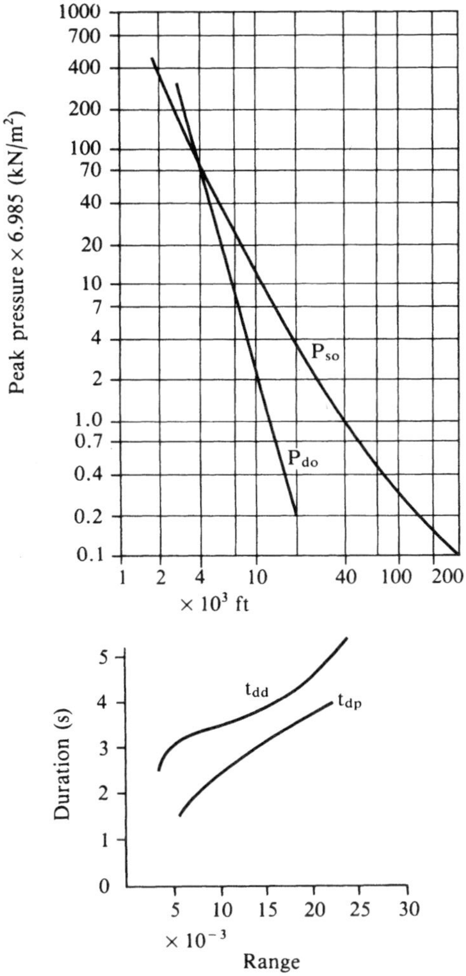

Fig. IX.196. Megaton weapon curves (US Department of Defense and Atomic Energy Commission). For all other weapon yields Table IX.16 should be used. For example:

$$1\,Mt: \quad P_{s0} = 20\,lbf/in^2 \rightarrow 7100\,ft$$

$$\tfrac{1}{2}\,Mt: \quad \frac{R_1}{7100} = \left(\frac{0.5}{1}\right)^2$$

$$\text{therefore} \quad R_1 = 5650\,ft$$

(Note: 1 ft = 0.3048 m.)

As a general rule, response may be taken to be synonymous with damage, since permanent distortion of a sufficient amount will impair the usefulness of a structure.

Direct damage to structures attributable to air blast can take various forms; the blast may deflect structural steel frames, collapse roofs, dish-in walls, shatter panels and break windows.

IX.11.3.1 Loading

(a) Diffraction loading. When the front of an air blast wave strikes the face of a building, reflection occurs. As a result the overpressure builds up rapidly to at least twice that in the incident wave front; the actual pressure attained is determined by various factors, such as the peak overpressure of the incident wave and the angle between the direction of motion of the wave and the face of the building. As the wave front moves forward, the reflected overpressure on the face drops rapidly to that produced by the blast wave without reflection, plus an added drag force due to the wind (dynamic) pressure. At the same time, the air pressure wave bends or 'diffracts' around the structure so that the structure is eventually engulfed by the blast (Fig. IX.197), and approximately the same pressure is exerted on the side walls and the roof.

The front wall, however, is still subject to wind pressure, although the back wall is shielded from it.

The pressure differential between the front and back faces will have its maximum value when the blast wave has not yet completely surrounded the structure. In this case, such a pressure differential will produce a lateral (or translational) force, tending to cause the structure to deflect and thus move bodily, usually in the same direction as the blast wave. This force is known as the 'diffraction loading' because it operates while the blast wave is being diffracted around the structure. The extent and nature of the actual motion will depend upon the size, shape and weight of the structure, and how firmly it is attached to the ground.

When the blast wave has engulfed the structure, the pressure differential is small, and the loading is almost entirely due to the drag pressure exerted on the front face. The actual pressures on all faces of the structure are in excess of the ambient atmospheric pressure and will remain so, although decreasing steadily, until the positive

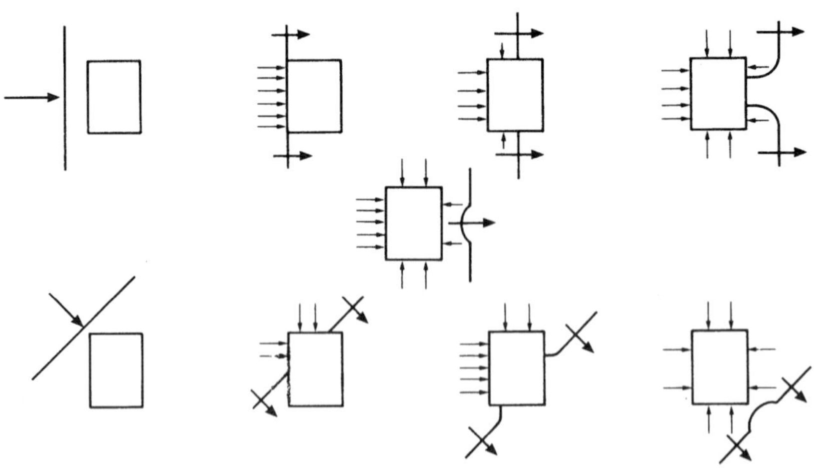

Fig. IX.197. Stages in diffraction (plan view) without openings (courtesy of US Department of Defense and Atomic Energy Commission).

phase of the blast wave has ended, thus the diffraction loading on a structure without openings is eventually replaced by an inwardly directed pressure.

The damage caused during the diffraction stage will be determined by the magnitude of the loading and by its duration. The loading is related to the peak overpressure in the blast wave and this consequently is an important factor. If the structure under consideration has no openings, as has been assumed for this discussion, the duration of the diffraction loading will be very roughly the time required by the wave front to move from the front to the back of the building, although wind loading will continue for a longer period.

If the building exposed to the blast wave has openings, or if it has windows, panels, light sidings or doors which fail in a very short space of time, there will be a rapid equalisation of pressure between the inside and outside of the structure. This will tend to reduce the pressure differential while diffraction is taking place. The diffraction loading on the structure as a whole will thus be decreased, although the loading on interior walls and partitions will be greater than for essentially closed structures. Furthermore, if the building has many openings after the diffraction stage, the subsequent squeezing (crushing) action, due to the pressure being higher outside than inside, will not occur.

IX.11.3.2 Loading due to drag

During the whole of the overpressure positive phase, a structure will be subjected to the dynamic pressure loading or drag loading caused by transient winds behind the blast wave front. Like the diffraction loading, the drag loading, especially in the Mach region, is equivalent to a lateral (air translational) force acting upon the structure.

IX.11.4 Loading versus structural characteristics

There are basically two types of building when considering blast wave loading:

(1) Diffraction type
(2) Drag type

As these names imply, in a nuclear explosion, the former would be affected mainly by diffraction loading and the latter by drag loading. While it is true that some structures will respond mainly to diffraction forces and others mainly to drag forces, all such buildings will respond to both types of loading.

A diffraction-type building is one that is primarily sensitive to the peak overpressure in the shock wave, e.g. reinforced concrete buildings with small window area and large wall-bearing structures.

When the pressures on different areas of a structure are quickly equalised, because of its small size, the characteristics of the structure or the rapid formation of numerous openings by the action of blast, the diffraction forces operate for a very short time. The response of the structure is then mainly due to the dynamic pressures (or drag forces) of the blast wind, e.g. for telephone poles, radio and television transmitter towers, and tall chimneys.

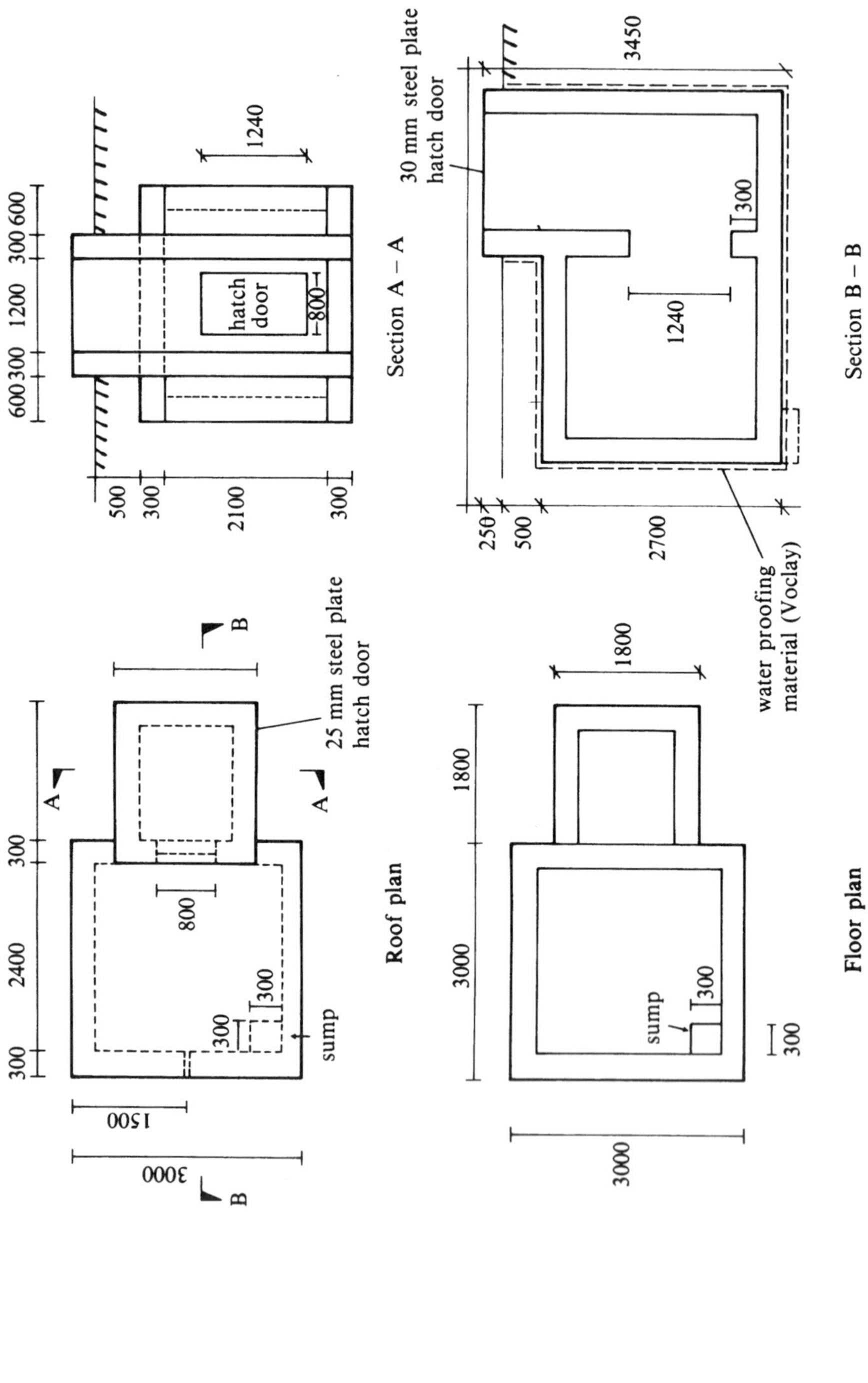

Fig. IX.198a. Domestic nuclear shelter — general arrangement.

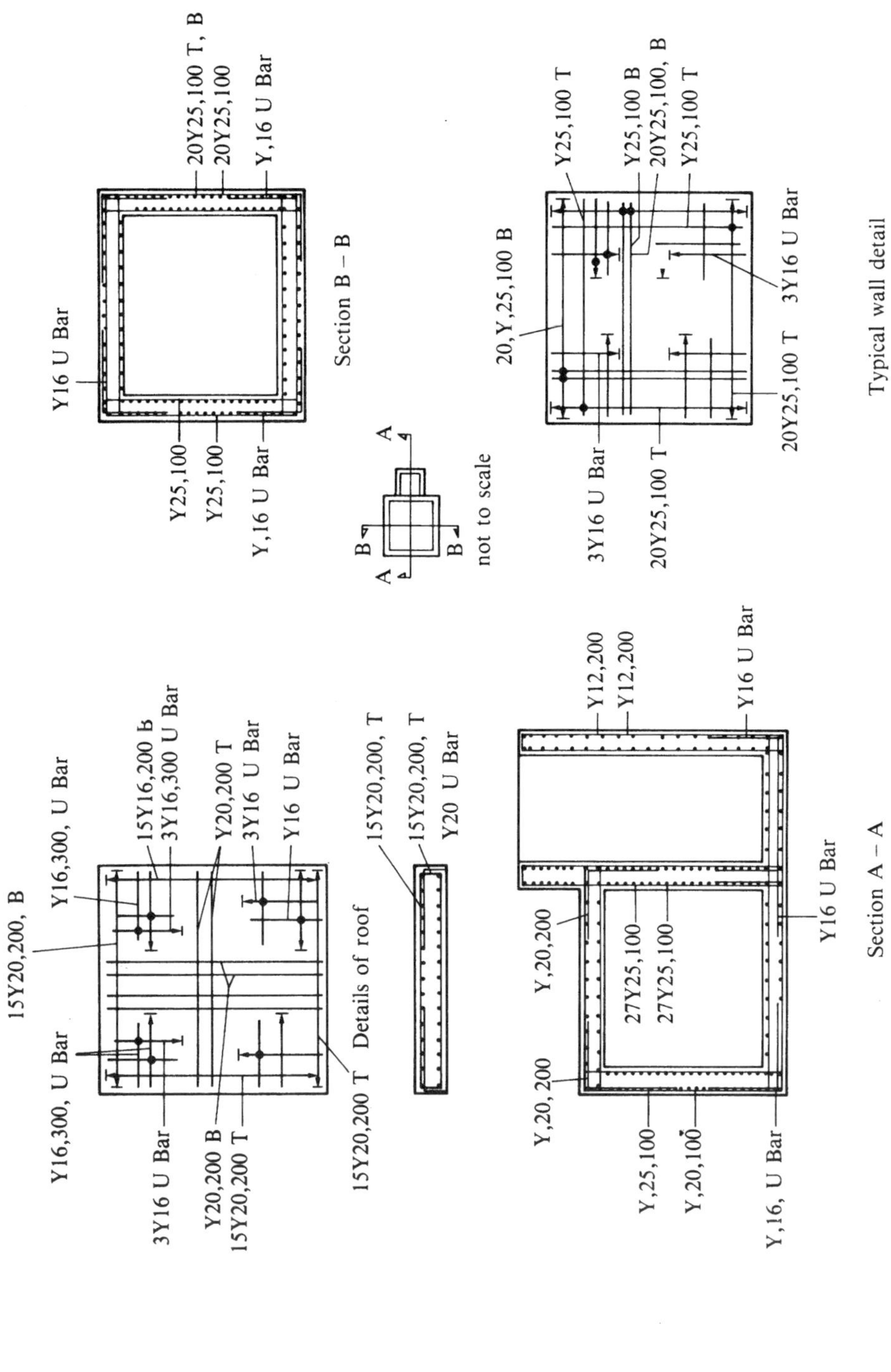

Fig. IX.198b. Domestic nuclear shelter (reinforced concrete) — detail.

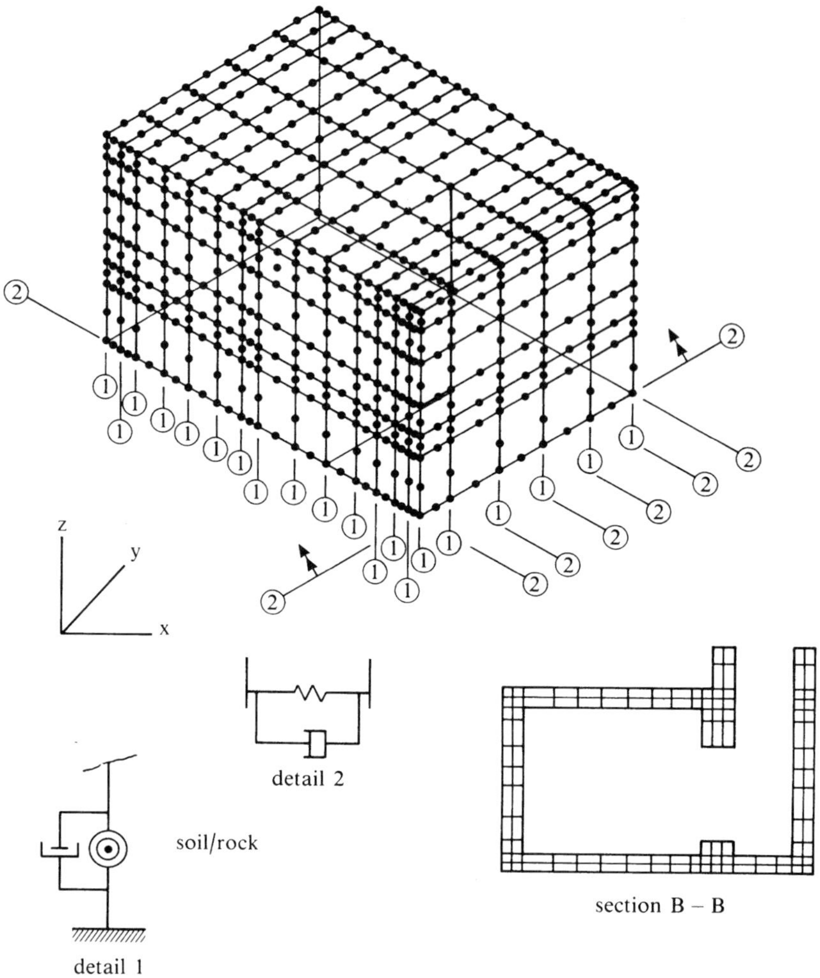

Fig. IX.199. Dynamic model for nuclear shelter.

IX.11.5 Damage classification

Damage to structures or objects above ground can be divided into three categories, as follows.

Severe damage: A degree of damage that precludes further use of the structure or object for its intended purpose, without essentially complete reconstruction. For a structure or building, collapse is generally implied.

Moderate damage: A degree of damage to principal members that precludes effective use of the structure or object for its intended purpose, unless major repairs are made.

Slight damage: A degree of damage to buildings resulting in broken windows, slight damage to roofing and cladding, blowing down of light interior partitions, and slight cracking of certain walls in buildings.

IX.11.6 Blast loads and stresses

Table IX.17 and Fig. IX.196 give a summary of the blast loads generated from a bomb burst. Table IX.20 gives generalised data, including allowable stresses used later on in the analysis.

Based on HMSO reports [639, 640], the American Defence Agency [642–646] carried out limit state analysis for a six-person domestic shelter, as given in Figs IX.198a and IX.198b. This was designed for a 1 megaton weapon yield, for which the data are given in Table IX.18.

IX.11.7 Finite element analysis of a domestic nuclear shelter

Figure IX.199 shows a typical finite element mesh for the dynamic analysis of this shelter. Again 20-noded elements representing concrete in the walls and slabs with smeared reinforcements are used to analyse the shelter resistance against a 1 megaton weapon yield. The dynamic pressure p_{d0} is applied in the form of nodal loads. Time duration is included corresponding to the specific values of p_{d0}. The dynamic finite element analysis together with cracking is given in Chapter VI, and the numerical model for the material law developed for the five-parameter case has been adopted. The program ISOPAR is modified to include the blast loads. The total number of increments adopted is 12. The structure is yielded, cracked and crushed in vulnerable areas when assumed to be placed above ground. The same structure is placed underground with 0.5 m overburden. The stress trajectories are plotted. Figures IX.200a and IX.200b show the cracking of the structure when placed above ground. Figure IX.200c shows the stresses and partial cracking when placed underground. The safety factor for the structure is 3.1 when placed underground and the protection factor based on the HMSO reports [639, 640] is 4000. The shelter received no damage above ground for $\frac{1}{2}$Mt yield at a ground-zero distance of 20 000 m.

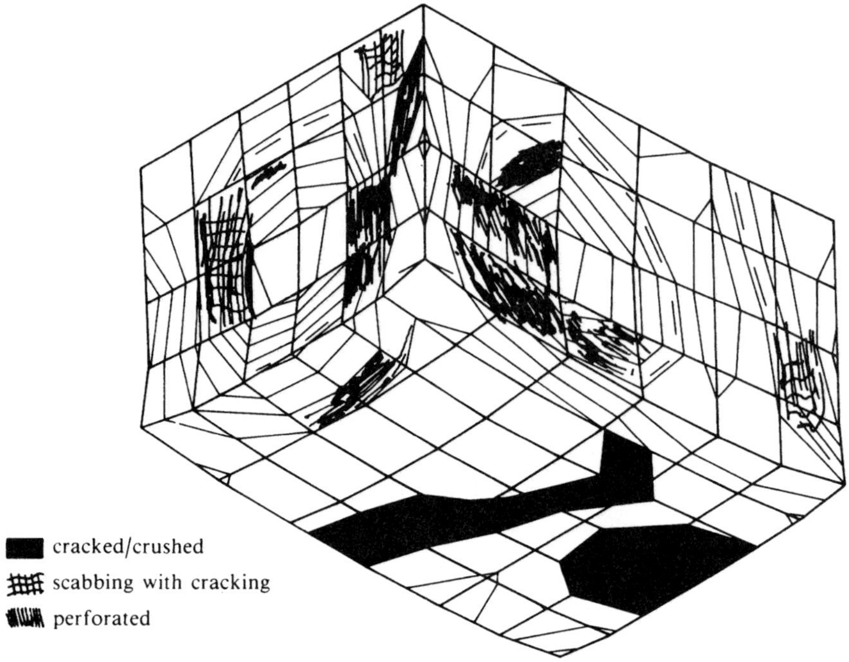

Fig. IX.200a. Stress trajectories/cracking.

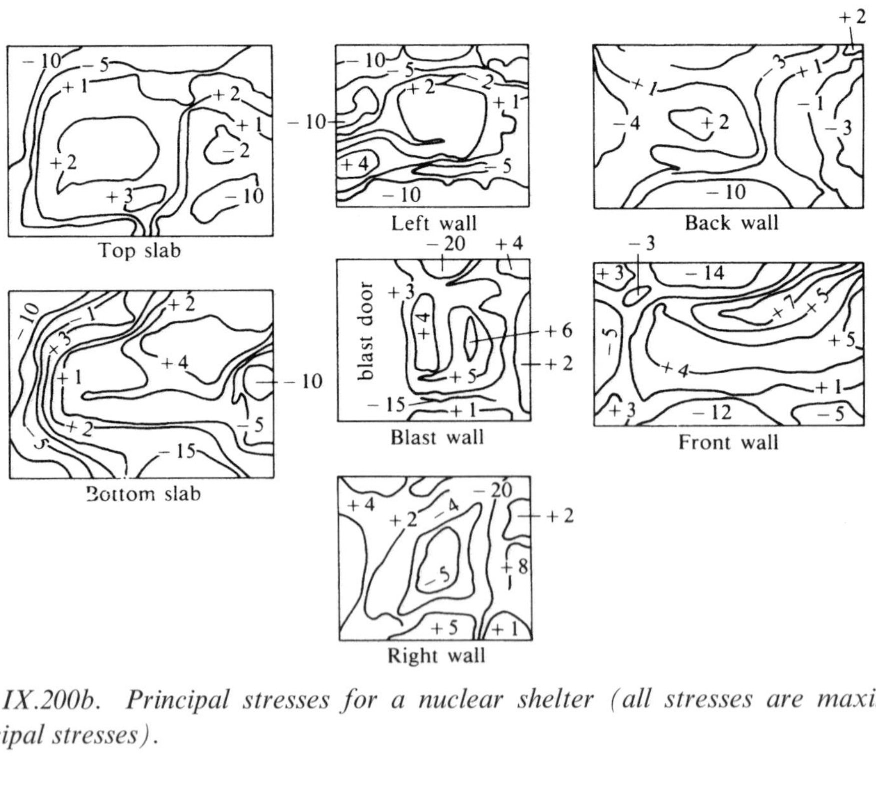

Fig. IX.200b. Principal stresses for a nuclear shelter (all stresses are maximum principal stresses).

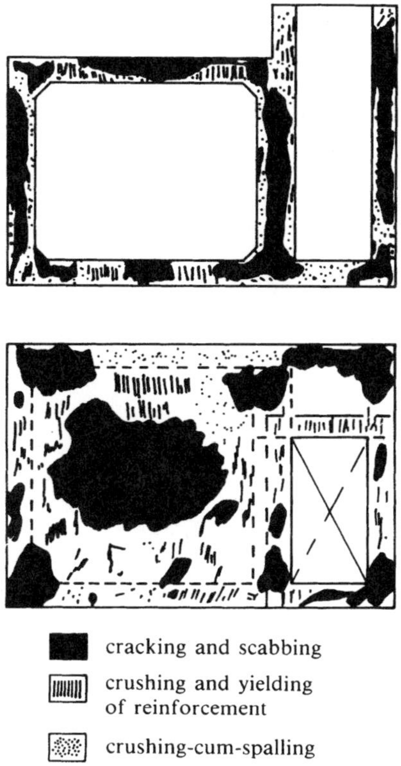

Fig. IX.200c. Shelter 'post-mortem'.

IX.12 Case Study No. 11 — silos and bunkers

IX.12.1 Introduction

Analysts and designers [677–702] have assumed in the past that granular material behaves quasi-statically, and that the pressures exerted by such a material could be calculated similarly to hydrostatic pressure. Such assumptions are only useful for static horizontal loads and do not consider the material friction in the vertical direction along the wall. Janssen [cited in Ref. 692] derived formulae for granular material pressures on silo walls and bottoms. Many others came forward [677–702] using modified Janssen's formulae. These formulae are summarised in Table IX.21. Janssen's and Airy's formulae have been derived on the lines of Rankine and Coulomb's theories of earth pressure. Figure IX.201 gives a summary of silo wall pressures based on various codes of practice. These loads are converted to nodal and patch loads and are applied at appropriate positions on the finite element mesh chosen for a silo. The load is applied incrementally using the material constitutive law based on bulk and shear moduli given in Chapters II and VII. A time dependent analysis is also carried out in which the interaction between the materials and the silo's wall and floor is included. The details of the time dependent dynamic analysis are fully described in Chapter VI. A typical flow pattern is given in Fig. IX.202a.

Table IX.21. Formulae for silos and bunkers (reference is made to Fig. IX.105)

(A) Janssen's theory

δPV

Total pressure

$$P = \sum_{0}^{z} P_{\mathrm{H}}\,\delta z = \frac{D_{\mathrm{c}}A/p}{\mu}\left[z - \left(\frac{A/p}{c\mu}\right)\{1 - e^{-c\mu z p/A}\}\right]$$

(B) Airy's theory

$\mu R = \tan\bar{\theta}_1 W \qquad P_{\mathrm{F}} = \tan\bar{\theta}_2 W = \mu p$

$\bar{\theta}_1, \bar{\theta}_2$ = angles of friction

A = sectional area

p = perimeter

P_{F} = frictional force

P_{V} = vertical pressure/unit area on top

c = constant = 0.5 during filling

= 1.0 during emptying

(i) For shallow bins/silos

Depth = $D\tan\theta'$

$$P = \tfrac{1}{2}WH_{\mathrm{T}}^2\left\{1\Big/\left[\mu R\left(\mu R + \frac{P_{\mathrm{F}}}{P_{\mathrm{V}}}\right) + \sqrt{1 + (\mu R)^2}\right]^2\right\}$$

where maximum depth $\not> D\left[\mu R + \left(\mu R + \sqrt{\mu R\dfrac{1 + (\mu R)^2}{\mu R + P_{\mathrm{F}}/P_{\mathrm{V}}}}\right)\right]$

$$P_{\mathrm{H}} = WH_{\mathrm{T}}\Big/\left[\mu R\left(\mu R + \frac{P_{\mathrm{F}}}{P_{\mathrm{V}}}\right) + \sqrt{1 + (\mu R)^2}\right]^2$$

at depth H_{T} below.

(ii) For deep silos

$$P = \frac{\tfrac{1}{2}WD^2}{(\mu R + P_{\mathrm{F}})}\left[\sqrt{\left\{\frac{2H_{\mathrm{T}}}{D}(\mu R + P_{\mathrm{F}}) + 1 - \mu R\frac{P_{\mathrm{F}}}{P_{\mathrm{V}}}\right\}} - \sqrt{1 + (\mu R)^2}\right]^2$$

$$P_{\mathrm{H}} = \frac{WD}{(\mu R + P_{\mathrm{F}})}\left[1 - \frac{\sqrt{1} + (\mu R)^2}{\left\{\dfrac{2H_{\mathrm{T}}}{D}(\mu R + P_{\mathrm{F}}) + 1 - \mu R\dfrac{P_{\mathrm{F}}}{P_{\mathrm{V}}}\right\}}\right]$$

at depth H_{T} below.

(C) British code (material handling board)

$$P_{\mathrm{zf}} = P_{\mathrm{zfV}} = P_{\mathrm{V(at\,z)}} = 0.6gD_{\mathrm{c}}\left(z + \frac{H_{\mathrm{s}}}{2 + m}\right) \quad \text{(in Pa)}$$

$$P_{\mathrm{zf}} = \frac{D_{\mathrm{c}}gR}{\mu_1} \times \left[1 - \exp\left\{-\frac{\mu_1 K_{\mathrm{u}}}{R}\left(z + \frac{H_{\mathrm{s}}}{2 + m}\right)\right\}\right] \quad \text{(in Pa)}$$

(parallel section wall pressure)

Table IX.21 — contd.

P_{zf} = the pressure normal to the wall surface at a depth z (Pa) or
P_H = horizontal pressure/unit area on a vertical surface
P_{zfV} = the vertical pressure at height z (Pa)
D_c = the bulk density (kg/m^2)
g = acceleration due to gravity (m/s^2)
z = the distance from the lowest point of the top surface of the material in the parallel section (m)
μ_l = lower bound wall friction coefficient
H_s = the height of the surcharge (m)
K_u = the upper bound pressure ratio
m = 0 for rectangular silo bins ($L = 3D$)
= 1 for axisymmetric silo bins
R = hydraulic radius of the parallel section

$$R = \frac{\text{internal cross-sectional area}}{\text{internal perimeter}} \text{ (m)}$$

D = width or diameter

Material	Unit weight (kg)	$\bar{\theta}$ (degrees)	μR	$P_F/P_V = \mu p$
Wheat	800	25	0.466	0.444
Cement	1442	17.5	0.316	0.554
Coal	800	35	0.716	0.700

IX.12.2 Loads on silos

The design loads include silo wall loadings caused by stored material, thermal loads, wind loads, prestressing loads, wall-base-restraint loads and seismic loads. The dominating load for the silo wall design is the lateral load from the stored material. Figure IX.201 shows several curves. Janssen's formula is modified to take various factors into consideration. The most important of these factors is the pressure increase, in relation to the static load, experienced during overloading of a silo and caused by dynamic effects associated with the flow of materials. These modifications are carried out in numerous ways including overpressure factors prescribed by ACI 313-77 for the formula. Adjustments to the basic parameters, such as density and angle of friction, in the formula have caused the difficult curves given in Fig. IX.201. These details are given in Refs 678 and 679.

Dead, live, prestressing, thermal and wind loads have already been discussed in greater length in earlier chapters. Tables IX.21 and IX.22 give a summary of load computations for silos and bunkers with reference to Fig. IX.202b.

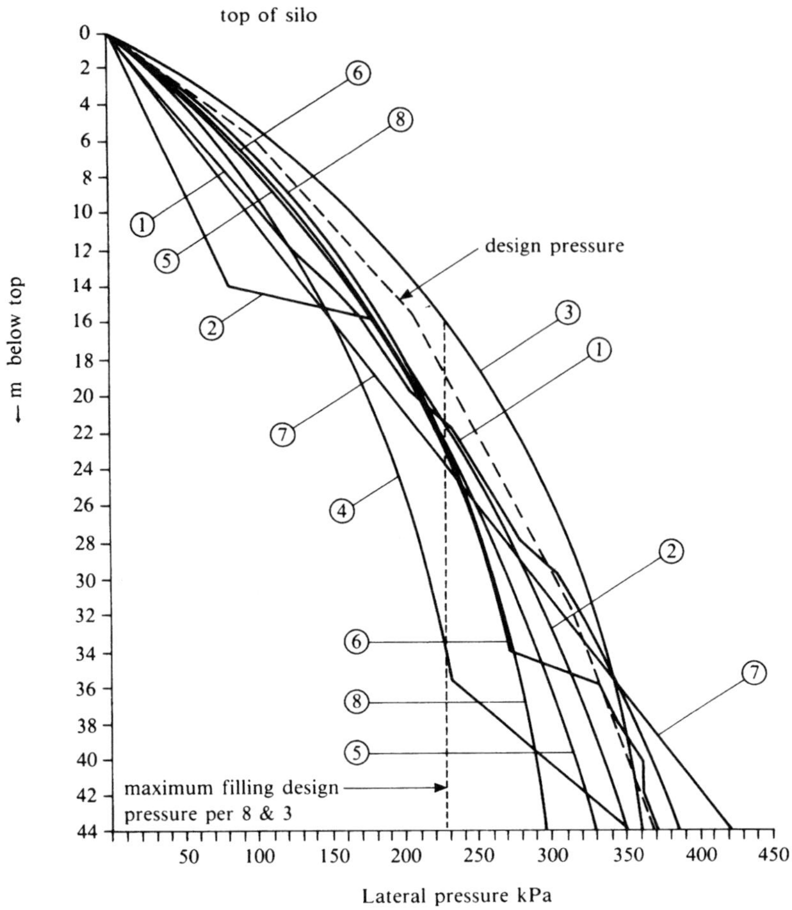

NA • 1 ACI 313-77 (US Code)
NA •• 2 CH-302-65 (Soviet Code)
NA • 3 DIN 1055-Blatt 6, 1977 (German Code)
A • 4 Peter Martens, Braunschweig
A 5 R. T. Jenkyn
A • 6 Regles de Conception et de Calcul des Silos en Beton (French Code)
A 7 .6 γ Z ($\gamma = 1.6$)
NA •• 8 DIN 1055-Blatt 6, 1964

• incl. effect of eccentric discharge
•• excl. effect of eccentric discharge
NA = no aeration
A = aeration

Fig. IX.201. Silo wall pressures [699].

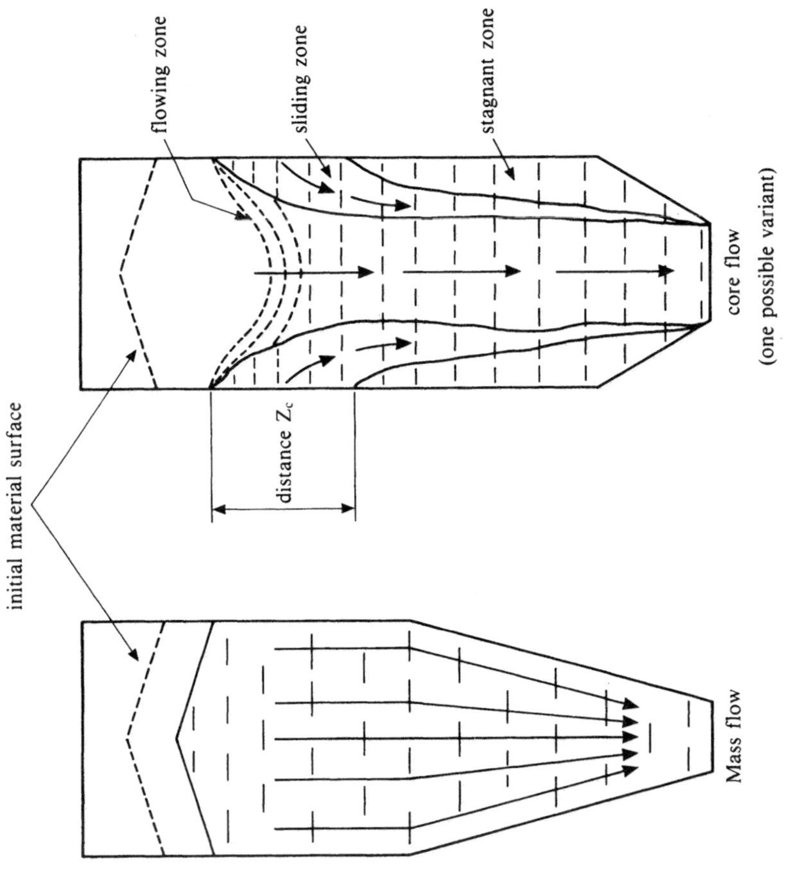

Fig. IX.202a. Flow patterns.

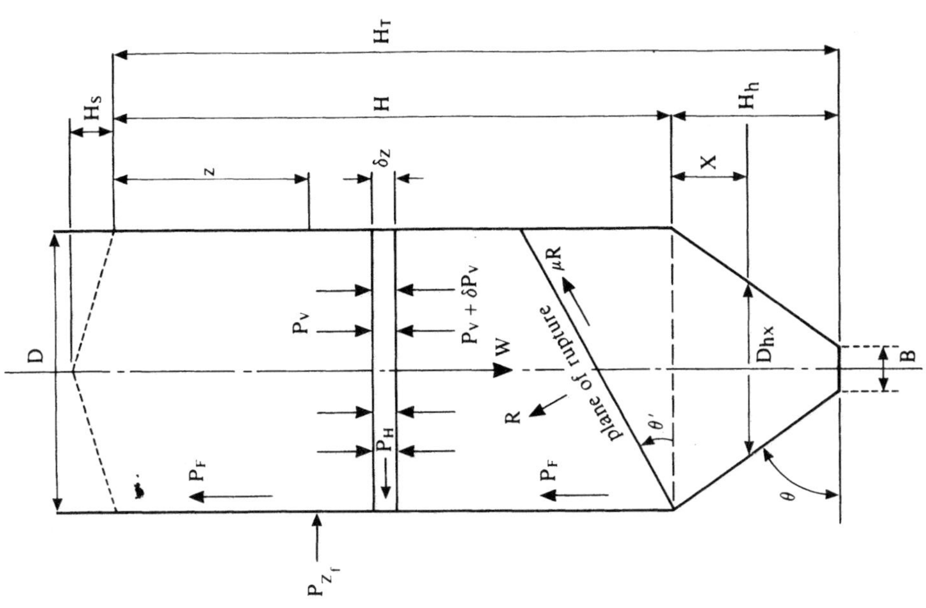

Fig. IX.202b. Silo bin dimensions.

Table IX.22. Design loads

The design requirements for offshore gravity structures are covered by several codes. In the light of their recommendations, the following load combinations are recommended for the design of gravity platforms:

Case I	$G + Q + H + C_1 + BO$	100% unit stress*
Case II	$G + Q + BO + WF + W$	125% unit stress
Case III	Case I + 50% W + WL	130% unit stress
Case IV	Case I + SM + 50% W	125% unit stress
Case V	Case II + SM + 50% W	130% unit stress
Case VI	Case III + SM + 50% W	140% unit stress
Case VII	$G + BO + H + WF + EQ$	150% unit stress
Case VIII	Case II + C_1 + H	140% unit stress
Case IX	Case I + C_1 + 50% W	130% unit stress
Case X	Case I + W + SM	150% unit stress
		(*stresses for service load)

where

G = dead load

Q = live load

H = hydrodynamic load due to wave

C_1 = load due to ice and other debris

BO = buoyancy

SM = seabed movement

WF = wave force steady state or slamming forces

W = wind forces

EQ = earthquake effects

IX.12.3 Material properties

The properties of concrete, conventional and prestressing steels are given full treatment in Chapters I, II and VII, together with the relevant appendices.

IX.12.4 Application to silos

A typical prestressed concrete silo [699], 28 m internal diameter and 44 m wall height, is shown in Fig. IX.203. Figures IX.204a, IX.204b and IX.204c show a typical reinforcement plan and elevation. The strand is 12.5 mm of low relaxation wire with a guaranteed minimum breaking force of 184 kN that was adopted for the VSL multi-strand system having 19 prestressing anchorages. The total losses covering elastic shortening friction, creep, shrinkage, wobble effects and relation of steel are of the order of 40%. This cement silo capacity is currently located on a wharf in Birkinhead, South Australia [699]. Figures IX.205 and IX.206 show the finite element mesh for this silo. The ACI 313-77 curve given in Fig. IX.201 is adopted for the finite element analysis. A 20-noded isoparametric element is adopted for concrete. A three-noded line element represents the prestressing tendons. As shown in Fig. IX.206, the mesh layout is identical to that adopted for the Dungeness

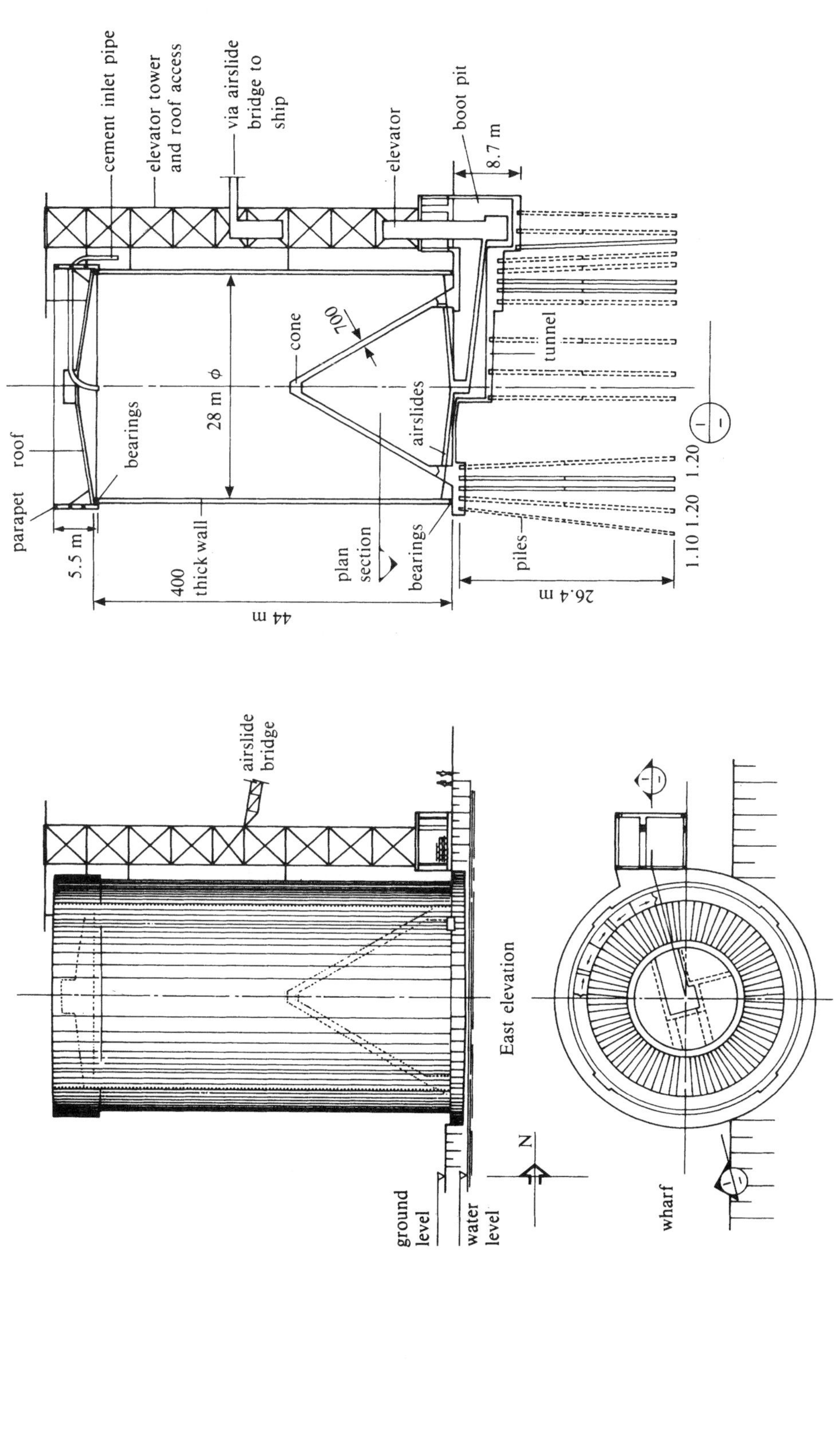

Fig. IX.203. A general layout [699].

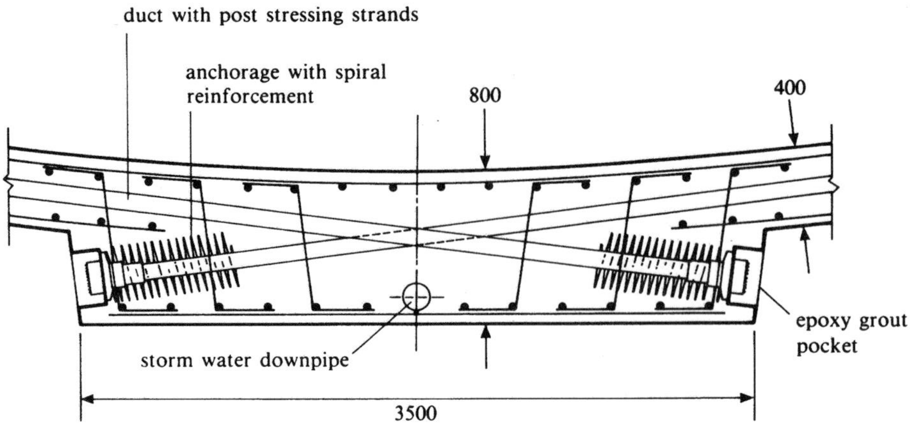

(i)

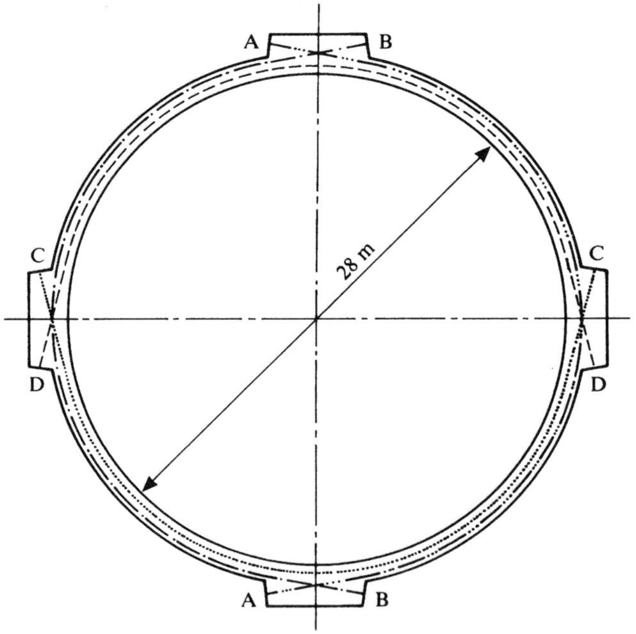

(ii)

Fig. IX.204a. (i) Plan of typical buttress; (ii) diagram of typical duct layout [699].

B vessel. Figures IX.207–IX.209 show the stresses and corresponding failure mode of this silo using 14 load increments. Assuming a 3 mm crack width as a permissible value, the safety margin computed prior to collapse for the design load is 3.65. The detailed cracking analyses given in Chapters VI and VII together with the relevant appendices have been adopted for both static and dynamic investigations for this silo.

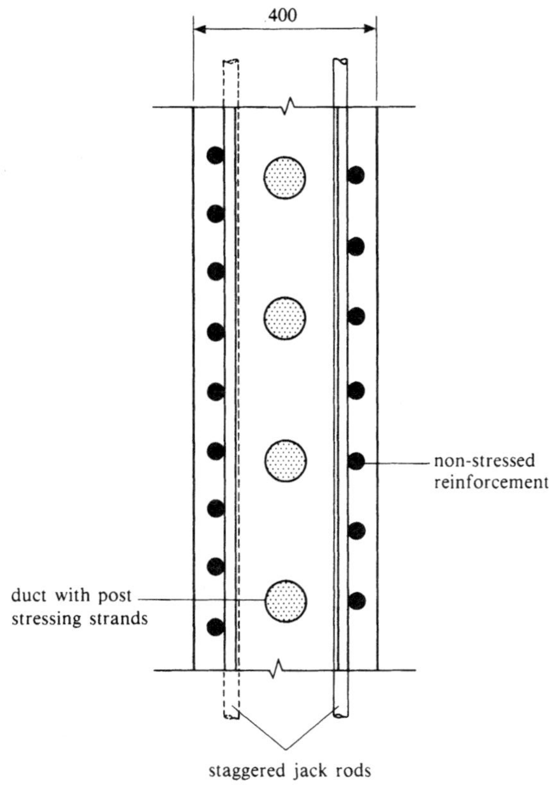

Fig. IX.204b. *Typical wall section [699].*

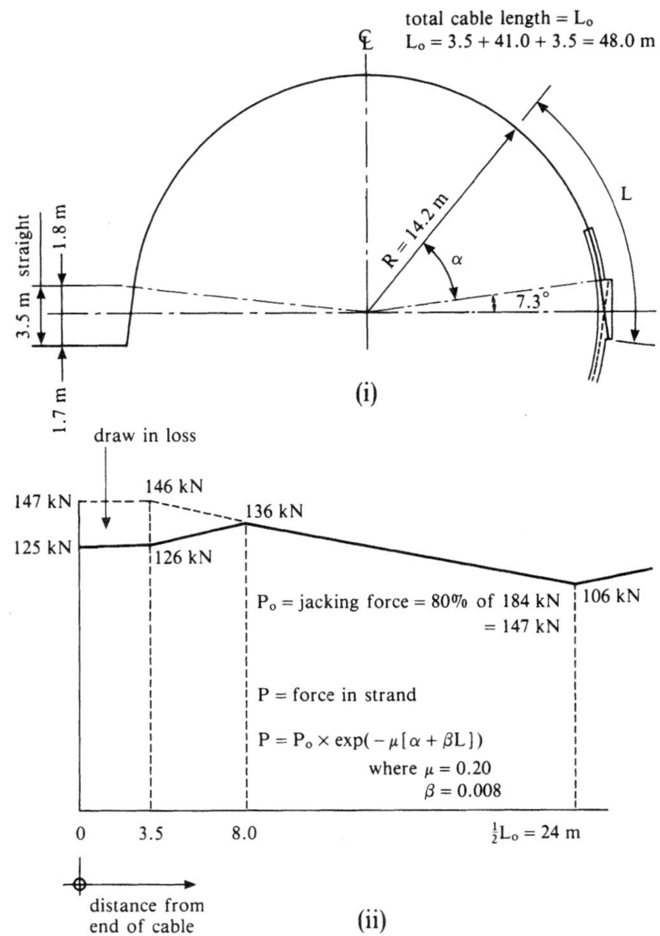

Fig. IX.204c. *(i) Typical plan of strand; (ii) typical force diagram for one strand.*

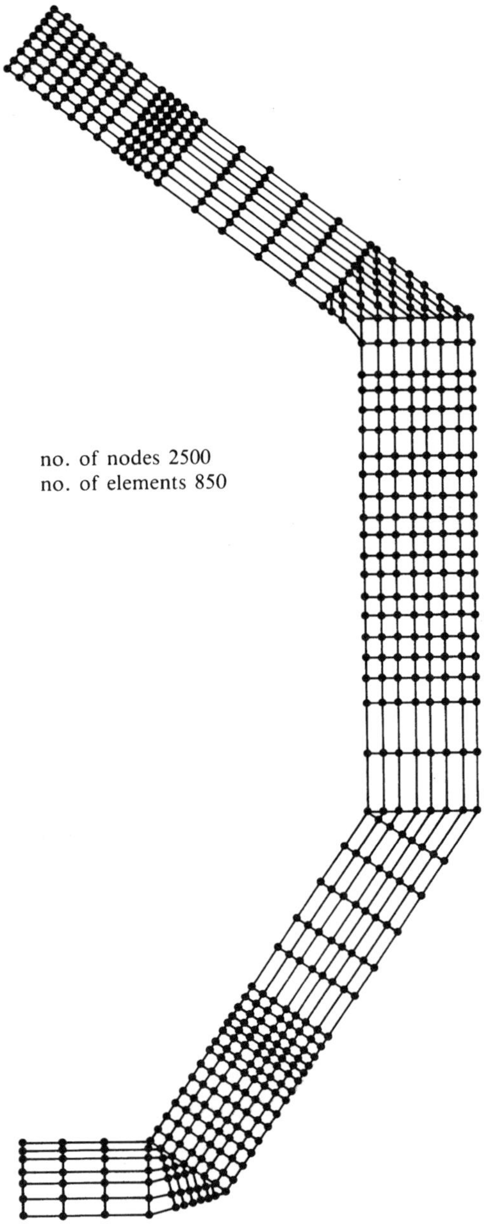

Fig. IX.205. Finite mesh scheme for silos (exaggerated).

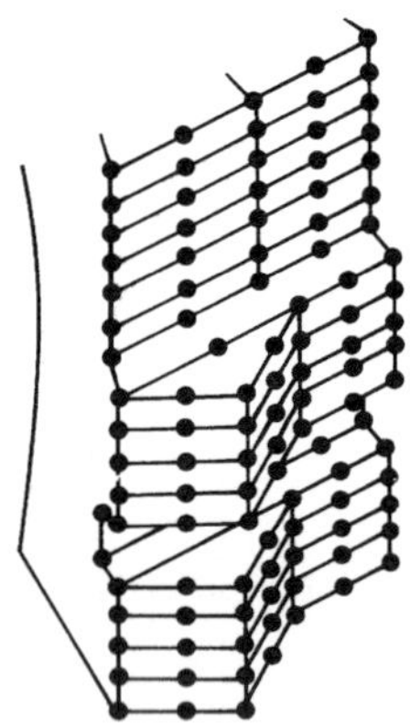

Fig. IX.206. Finite element mesh for buttresses.

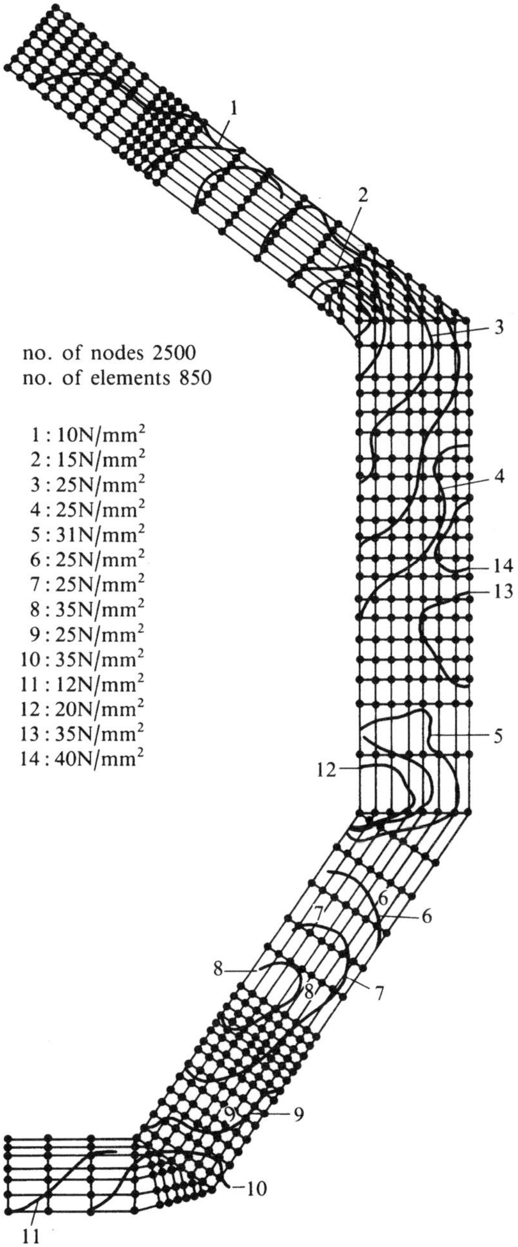

Fig. IX.207. Silo-stress trajectory at 3.5 × design pressure (exaggerated section).

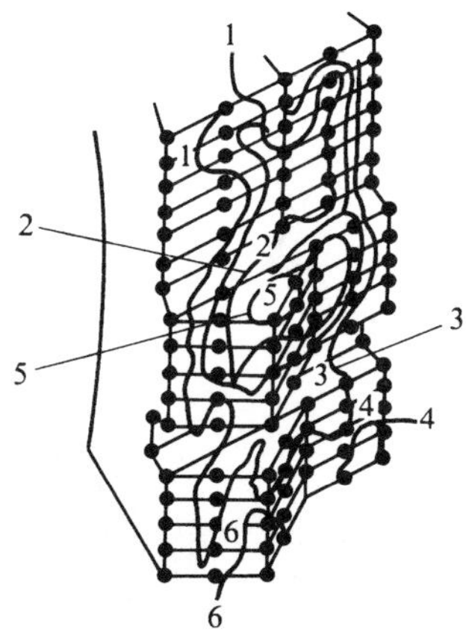

Fig. IX.208. Stress trajectory buttresses at 3.5 × design pressure.

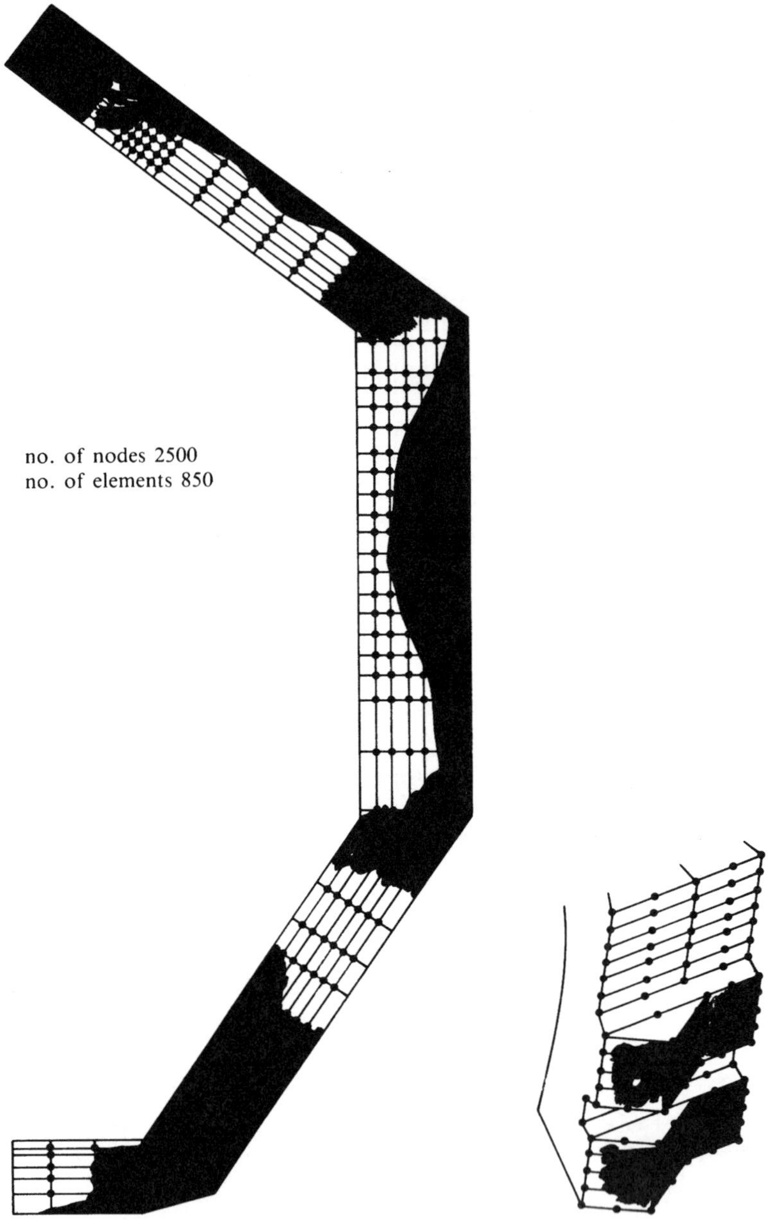

Fig. IX.209. Silo 'post-mortem' (exaggerated sections).

IX.13 Case Study No. 12 — offshore structures

IX.13.1 Concrete gravity platforms

The continuing search for petroleum has led to exploration and production of oil and gas in deeper waters and in areas subject to severe storm conditions, such as in the North Sea and the northwestern Atlantic. Since the periods of good weather that permit heavy construction at sea in these hostile offshore environments are relatively short, there is considerable interest in production of structures that can be constructed onshore or in sheltered waters, towed semi-submerged to location on completion and then installed on location in a short time by ballasting or flooding. Concrete gravity platforms meet this requirement [703–735, 1008–1029].

These structures are supported by relatively large foundation elements bearing on the near-seafloor soils, and depend principally on their weight and shape to resist vertical and horizontal forces and overturning moments. Concrete gravity platforms require less time and have reduced demands for costly marine equipment during installation than those for conventional pile-supported steel jacket platforms. In addition, oil storage facilities can easily be incorporated in the cellular configuration of many of the proposed gravity platforms. Unlike steel structures, concrete structures present no severe maintenance and corrosion problems, and are virtually free from problems of fatigue.

Traditionally, offshore platforms are designed on a basis of a static application of a design wave. Hostile sea environments require that the dynamic behaviour of structures in deep water be studied. In addition to the dynamic nature of the environmental loads on the platform, these loads are random in nature. Therefore methods of analysis and design which reflect the dynamic and random nature of the environmental loads on deep water platforms are needed. However, such methods are usually very complex and a large amount of computer time is involved. Nevertheless, the dynamic analysis given in this book can easily be applied for random loads using the ISOPAR routines.

Many concrete gravity platforms have been developed during the past few years. Some examples of these platforms are given in Appendix IX. The first concrete gravity structure installed in the North Sea was the Ekofisk Doris tank for Phillips Petroleum, placed in the summer of 1973. The next two were the Condeep Production Platform Beryl A for Mobil and Brent B for Shell/Esso, both installed in the summer of 1975. A typical case study of the Condeep Platform is examined in this section. A similar procedure is adopted for the other concrete platforms.

IX.13.2 Design of the Condeep platform

The concrete part of the Condeep structure (Figs IX.210a and IX.210b) consists of 19 cylindrical cells, of which three have been extended upwards to form towers to support the deck. Two of the towers are used for drilling production wells, while the third is filled with pipework and equipment. When the platform is in production the cells will always be filled either with oil or with seawater.

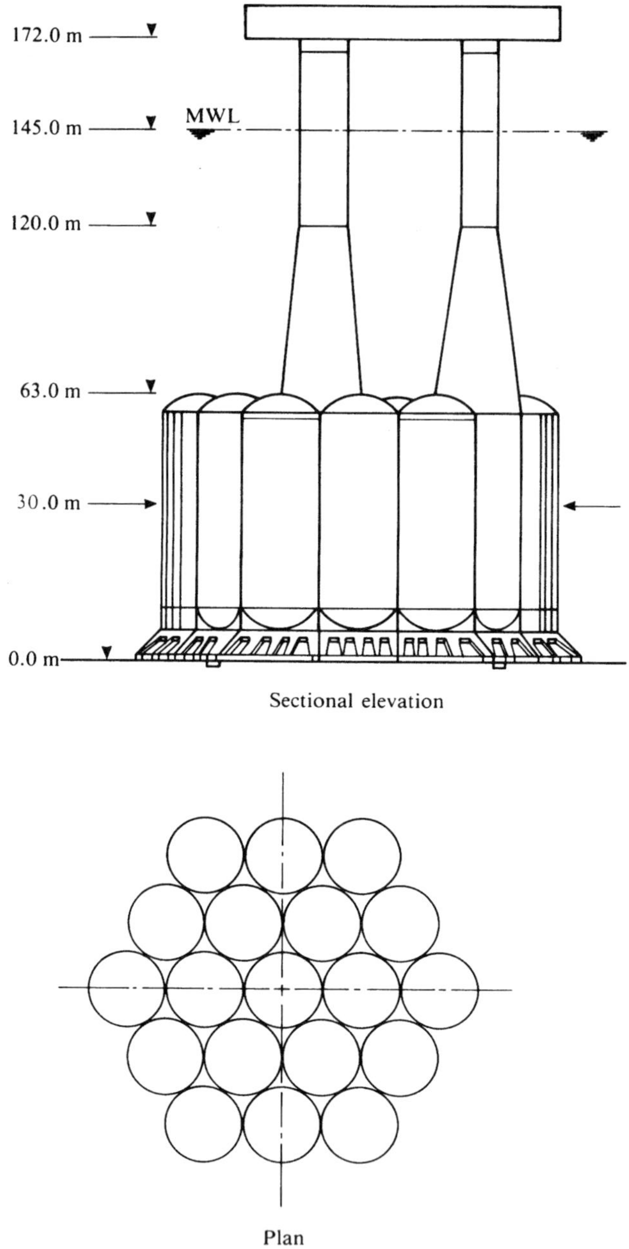

Fig. IX.210a. Condeep platform.

IX.13.2.1 Environmental loads

The design environmental loads used for offshore structures are normally based on environmental conditions, with a recurrence period of 50–100 years while the design lifetime is usually 20–30 years. For a concrete gravity platform, the wave load

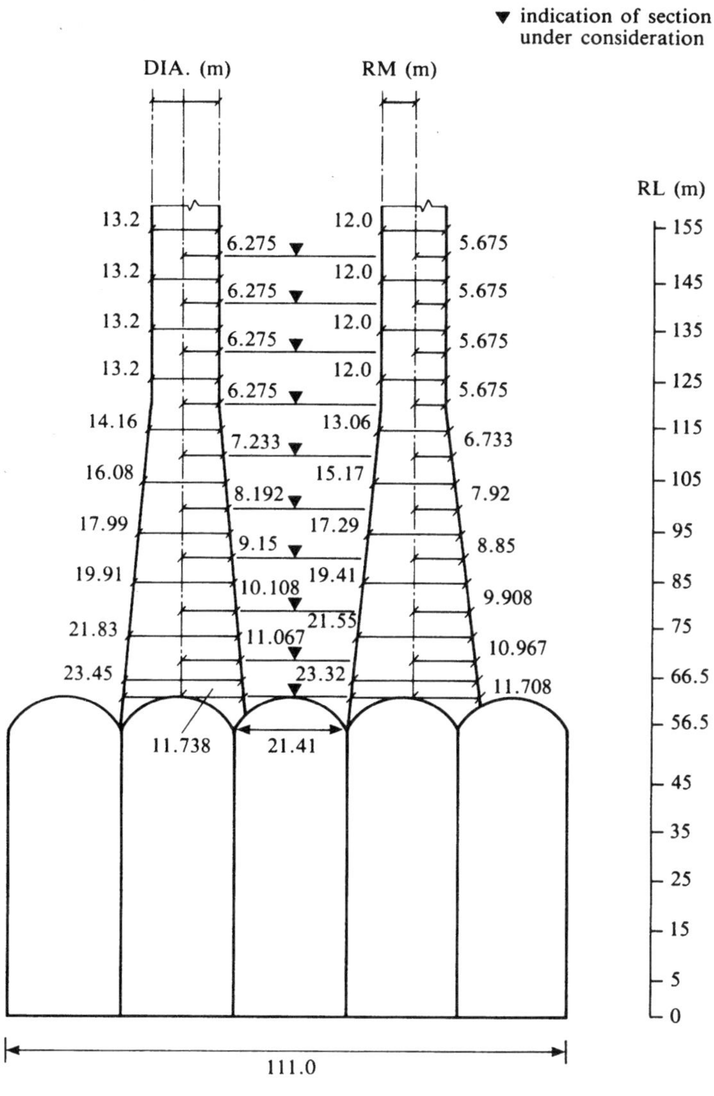

Fig. IX.210b. Details of Condeep input data.

accounts for the major part of the environmental loads. The design wave loads are usually determined on the basis of regular waves by means of analyses and model tests. Loads due to wind and current are usually a relatively small fraction of the total forces. They are, however, important and must be accounted for. For further details refer to Appendix IX.

IX.13.2.2 Requirements of concrete structures

The methods of design of offshore concrete structures will in principle be the same as those used for structures on land. Therefore recognised standards for design and construction of reinforced concrete structures such as BS 8110 are used. Reference is also made to the CEB–FIB and other recommendations [703–735].

The proposed requirements for design of offshore concrete structures may be placed in two categories:

(a) requirements for material quality, and
(b) requirements for strength and serviceability.

(a) Materials. The main requirements for materials are strength and durability. Only high-strength concrete with a 28-day characteristic minimum strength of 45 N/mm^2 measured on a 100 mm cube should be used for structural concrete exposed to marine or crude oil environments. For details reference should be made to Chapter I.

The characteristic strength of the reinforcement may vary depending on its purpose. The qualities most used are:

—ordinary reinforcement $f_y = 410\ \text{N/mm}^2$
—prestress reinforcement
  bars $f_y = 800–1250\ \text{N/mm}^2$
  tendons $f_y = 1200–1800\ \text{N/mm}^2$

In the splash zone a minimum cover over the reinforcement of 70 mm is recommended for ordinary reinforcement, and 100 mm for ducts for prestressing tendons.

(b) Structural requirements. The structural design of reinforced concrete should normally be based on the limit state method and the limit states are:

(i) The ultimate limit states.
(ii) The serviceability limit states.

(i) Ultimate limit state requirements:
  design imposed load $= 1.6Q_k$
  design dead load $= 1.4G_k$
  design environmental load $= 1.0V_k$
  the factor of safety for material $\gamma_m = 1.5$ for concrete and I.15 for steel
(ii) Serviceability limit state requirements:
  design imposed load $= 1.0Q_k$
  design dead load $= 1.0G_k$
  design environmental load $= 1.0V_k$
  the factor of safety for material $\gamma_m = 1.0$ for both concrete and steel

These factors vary from code to code. For other conditions reference should be made to load combinations based on data collected for other cases [703–735]. Table IX.22 gives the best possible load combination for the design of offshore facilities.

(c) Cracking. Cracking of concrete should not adversely affect the durability of the structure. The engineer should satisfy himself that any cracking will not be excessive. For reinforced concrete, the surface crack width should not exceed 0.2 mm, or 0.004 times the nominal coverage of the main reinforcement.

The concrete strength and cracking theories and numerical models are fully described in Chapters I, II, VI and VII. Any one of these can be examined in relation to the behaviour of such structures.

IX.13.2.3 Implosion

Implosion phenomena in concrete gravity platforms are extremely important. They can occur very rapidly and result in catastrophic failure. The cells of the caissons can be vulnerable to implosion. A particular check is required against implosion.

Several analysis methods will be presented here, because different design approaches were found appropriate for cylindrical structures of different geometries. The following four cases are treated separately: (a) short, (b) moderately long and (c) long thin-walled cylinders, and (d) thick-walled cylinders of all lengths.

The separation point between thin-walled and thick-walled cylinders is selected at $t/D_0 = 0.063$, where t is the thickness and D_0 is the outer diameter.

Initial implosion pressure formulae are given in Appendix IX.

IX.13.3 Data for Condeep

Steel stress $f_y = 410\,\text{N/mm}^2$
Concrete stress $f_{cu} = 45\,\text{N/mm}^2$ or $f_c = 0.87 f_{cu}$
Modulus of elasticity of steel reinforcement $E_s = 200\,\text{kN/Mm}^2$
Minimum cover to reinforcement = 75 mm
Main reinforcement bar diameter = 20 mm
Steel ratio $p = 0.01$
Thickness of tower = 600 mm
Height of tower = 100 m
Diameter of dome on caisson cell = 28 m
Diameter of caisson cylindrical wall = 24 m
Thickness of dome = 600 mm
Height of caisson cell = 60 m
Area of foundation base = 6400 m^2
Equivalent side of base = 80 m
Weight of superstructure = 175 000 t

A section and the diameters of the towers and caisson cell are given in Figs IX.210a and IX.210b. The depths of the sections below mean seawater level are also given in the figures.

Wave forces and wave moments are obtained using various theories [703–720, 1008–1029] including the Hogber and Standing diffraction wave theories. Appendix IX gives some basic wave theories based on these, for maximum forces and wave moments at phase angle 90°.

Maximum wave force $= 0.865 \times 10^6$ kN
Maximum wave moment $= 0.396 \times 10^8$ kN m

IX.13.3.1 Soil information

The soil conditions in the North Sea have the following approximate properties:

Stiff clay:
Modulus of elasticity of soil $= 500\,\text{t/m}^2$ ($5 \times 10^5\,\text{kN/m}^2$)
Coefficient of consolidation $(M_v) = 0$

Soft clay:
Modulus of elasticity of clay $= 1 \times 10^5\,\text{kN/m}^2$
Coefficient of consolidation $(M_v) = 1.7 \times 10^{-5}\,\text{N/m}^2$

The minimum factors of safety against failure of the foundations are 1.5 and 2.0 for extreme and operating conditions, respectively.

IX.13.3.2 Waves

(a) Extreme conditions. Extreme sea conditions are the highest sea state having a duration of 6 h during a once-in-a-hundred-years storm. This sea state is represented by a sea spectrum with a significant wave height (H_s) of 15.00 m and a significant period (T_s) of 15 s. The spectrum is defined by the spectral density function:

$$S(w) = H_s^2/4w \times A(w) \times e^{-A(w)}$$

where

H_s = significant wave height from crest to trough

w = circular wave frequency in radians per second

$A(w) = 690/(wT)$ ($T = 0.88 \times$ significant wave period (T_s))

(b) Hydrodynamic coefficients. For the calculation of wave forces using Morison's theory (Appendix XII), a drag coefficient of 0.6 and an inertia coefficient of 2.0 is used.

IX.13.3.3 Wind velocities

(a) Storm wind. The maximum one minute sustained wind velocity of 125 mph is used in conjunction with the storm wave.

(b) Operating wind. A wind velocity of 60 mph is used in conjunction with the operating wave.

(c) Instantaneous gust. A maximum instantaneous gust velocity of 160 mph is used throughout.

IX.13.3.4 Tides

(a) Storm tide. The maximum tide of 2.15 m is assumed and finally combined with a storm surge of 0.6 m to produce a total storm tide rise of 2.75 m.

(b) Operating tide. The maximum tide of 2.15 m is combined with a storm surge of 0.12 m to produce a total operating tide rise of 2.27 m.

IX.13.3.5 Currents

(a) Storm current. The maximum storm current velocity shall vary with water depth according to the following profile:

(a) At the water surface = 1.38 m/s
(b) Above bottom = 0.54 m/s at 15 m
(c) At bottom = 0.24 m/s

(b) Operating current. The maximum current velocity under operating conditions shall vary with water depth according to the following profile:

(a) At water surface = 1.17 m/s
(b) Above bottom = 0.45 m/s
(c) At bottom = 0.24 m/s

IX.13.3.6 Temperatures

Minimum water temperature is assumed as +7.0°C. Crude oil temperature in storage is taken as 40°C and in production wells it is taken as 75°C.

IX.13.4 Ship impact on concrete gravity platforms

IX.13.4.1 Introduction

There is growing public interest in the effects of collisions involving ships [1030–1038] and tankers carrying particularly hazardous material (including the impact of nuclear vessels on vital installations such as the gravity platforms in the sea). Table IX.23 gives useful formulae and data for the ship impact phenomenon. A theoretical model based on finite elements is fully described in earlier chapters. The level at which the kinetic energy of impact is lost depends on the ship/tanker/vessel angle of encounter, mass and speed, location of the platform and of the impacter-surrounding water, shock effects, etc. The analysis carried out assumes throughout that the collision force, owing to its quasi-static nature, is constant with the depth of penetration.

IX.13.4.2 Present methods of assessment

Frequent impacts against offshore platforms may come from vessels operating close to the platform. There is growing concern that an impact may occur due to navigational errors by heavy vessels. The DnV rules for mobile offshore units specify a ship of 5000 tons displacement with impact speed of 2 m/s. This is treated as the design

Table IX.23. Ship impact data

$$F_I = \text{impact force} = \frac{E_D}{e} = \tfrac{1}{2} m_s v_s^2 \left[\frac{m_s + m_t}{m_t} \bigg/ 1 + \frac{m_s + m_t}{m_t} \right] \bigg/ e$$

m_s, v_s = mass and velocity of the striking ship

m_t = equivalent added mass of the target

e_t = penetration

E_D = energy of deformation $= \tfrac{1}{2} m_s v_s^2 \phi_1 \phi_2 \phi_4 + \tfrac{1}{2} m_t v_t^2 \phi_5 - v_s v_t m_t \phi_6$

ϕ represents the non-dimensional functions representing the colliding structure/vessel, the impact location, the angle of encounter, the speed of the target, and the product of both speeds.

(a) Duration for constant collision force:

$$T = \phi_1 \phi_2 \frac{m_t}{F_I} v_s$$

(b) Linear collision force — duration:

$$T = \frac{\pi}{2} \sqrt{\frac{\delta(2-\delta)}{\gamma \eta^2 \tan \theta} \phi_1 \phi_2 m_t}$$

Acceleration at the time of collision

$$\ddot{\delta}'_{\max(x)} = \frac{F_{I_{(x)}}}{m_t} \qquad \ddot{\delta}_{(y)} + \frac{F_{I_{(y)}}}{m_t + m_s} + \frac{\delta x (F_I)_y e}{I'}$$

where

x = the distance between the CG of the target and the point at which acceleration is computed

$F_{I_{(x)}}, F_{I_{(y)}}$ = the maximum components of the collision force in the transverse and longitudinal directions

I' = the mass moment of inertia of the target

limit state. Another method adopted recently is based on the energy absorption capability rather than the design limit state, and this DnV impact energy is given below.

(a) 14 megajoules (MJ) for a side collision equal to 40% of added mass.
(b) 11 megajoules (MJ) for a bow or stern collision equal to 10% of added mass.

IX.13.4.3 Impact mechanics

The ship/platform impact mechanics are based on two criteria, namely conservation of momentum and conservation of energy. For a short impact duration (Fig. IX.211)

$$E_s + E_p = \tfrac{1}{2}(m_s = \Delta m_s) v_t^2 \frac{(1 - v_t/v_s)^2}{1 + \dfrac{(m_s + \Delta m_s)}{(m_t + \Delta m_t)}}$$

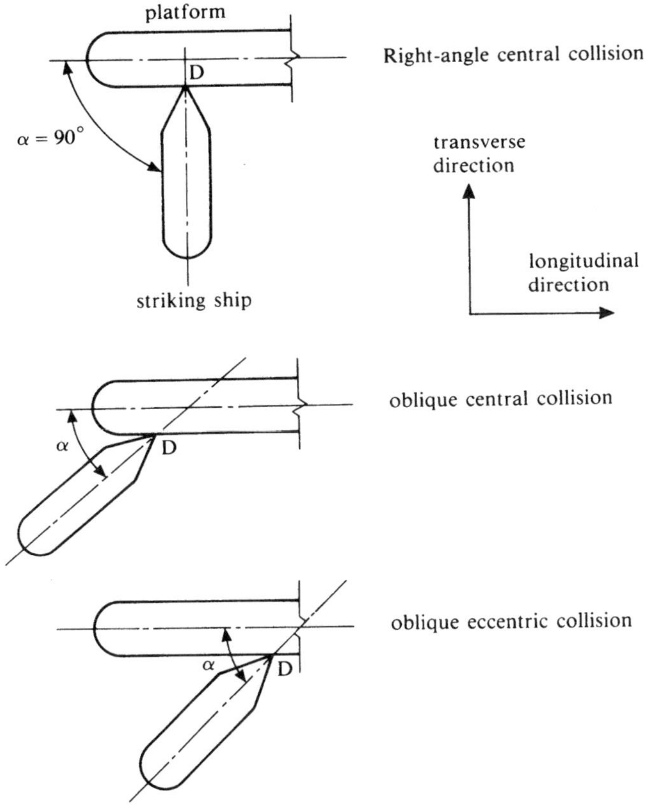

Fig. IX.211. Ship impact data — typical collision cases.

where

m_s = mass of striking ship (Δm_s = added mass (40% of vessel displacement for sideways collision and 10% for bow or stern collision))
m_t = mass of the platform including added mass
v_s = velocity of the striking ship immediately before collision
v_t = velocity of the semisubmersible platform immediately before collision
E_s = energy absorbed by the ship
E_p = energy absorbed by the platform

For a fixed platform, the corresponding energy expression is obtained using $(m_t + \Delta m_t) = \infty$ and $v_t = 0$.

For a supply ship m_t is 5000 t and $\Delta m_t = 0.1$ and 0.4; v_s is 2 m/s or $0.5H_s$, where H_s is the maximum significant wave height.

Table IX.23 and Figs IX.212–IX.214 can be used for this analysis.

IX.13.5 Dynamic finite element analysis

An offshore platform oscillating in the ocean has been discretised by lumping masses and hydrodynamic forces at the specified heights of the towers. The hydrodynamic

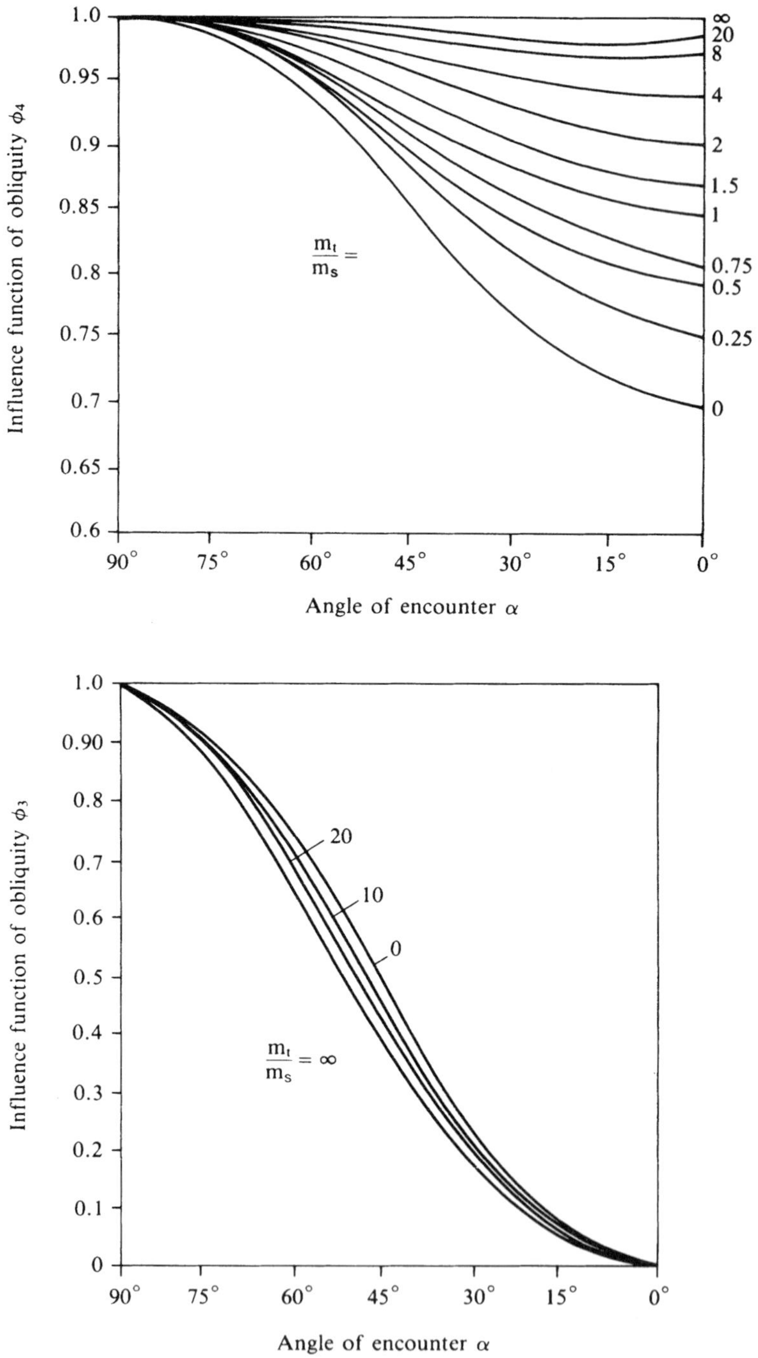

Fig. IX.212. Ship impact data — ϕ_3, ϕ_4 versus α.

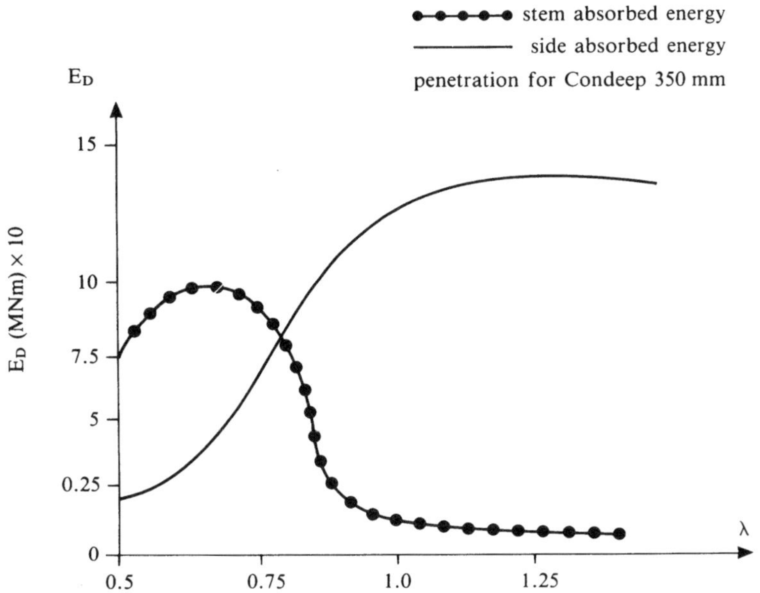

Fig. IX.213. Ship impact data — relations between strength ratio (λ) of stern to side and absorbed energy (E_D).

forces at these nodal points include inertia and drag terms resulting from the acceleration and velocity of the waves relative to the platform oscillations.

IX.13.5.1 Main platform analysis

A finite element program, ISOPAR, has been developed in which both soil and structure are considered. A mesh node of six degrees of freedom [375, 749] is assumed to have a compatible variation of displacements alongside of each element. Each of the points is forced to have the same displacement as the corresponding points of the adjacent element. The unknown displacements associated with the mid nodal points are then expressed in terms of corner nodal points and are thus eliminated from the systematic procedure. Matrix [*K*] is formed from the remaining matrix and a direct stiffness matrix. The global matrix [*M*] is based on lumped mass approximation in which one-third of the mass of each element is assumed to be concentrated at the corner nodal points. A *soft boundary technique* based on one-dimensional wave propagation is first introduced into the program in order initially to reduce unwanted defractions and reflections to minimum amplitudes. The program ISOPAR now picks up elastic stress–strain properties by elastic–perfectly plastic relations when expressing the stiffness matrix and the stress–strain relations. The incremental theory of plasticity is invoked in which the strain increment is expressed in terms of current stresses and the stress increments, as described in Chapter VI. Soil structure is considered in the analysis. The platform foundation structure is interfaced with seabed strata using an interface element. Dynamic

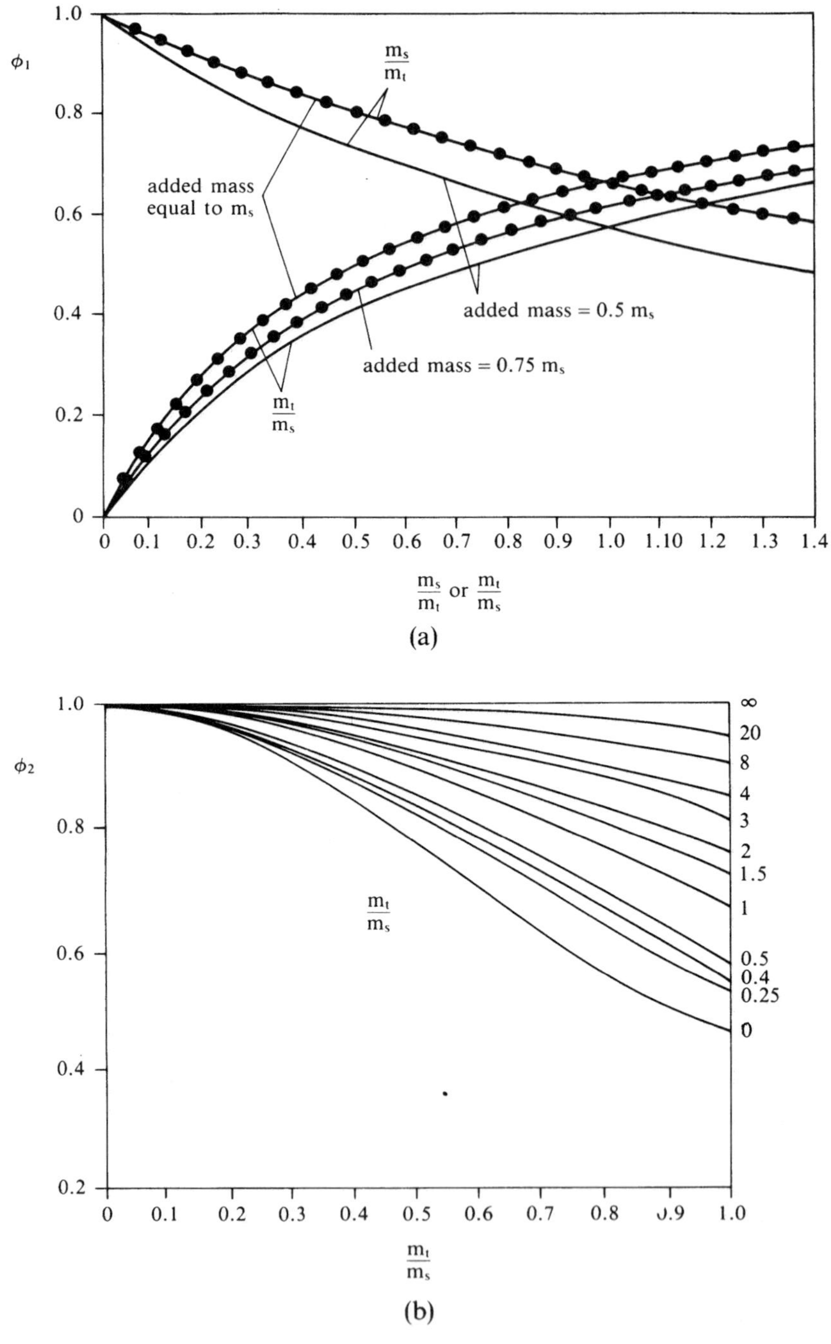

Fig. IX.214. (a) Ship impact data — influence of mass ratio ϕ_1; (b) influence function of eccentricity ϕ_2.

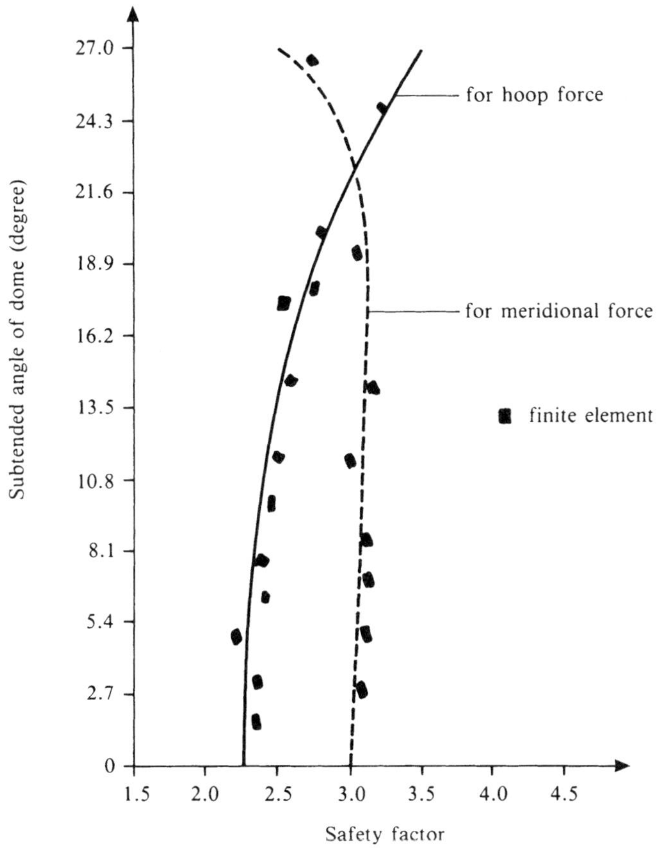

Fig. IX.215. Dome safety factors.

analysis is performed in conjunction with the specified sea spectrum using the time-domain technique. At every time step a test is included to determine the element elastic, plastic and cracked conditions. A five-parameter model is included for the concrete failure surface together with cracking conditions. Figures IX.215 and IX.216 show a comparative study for safety factors at various levels.

IX.13.5.2 Implosion analysis

Implosion cases described in Section IX.13.2.3 and in Appendix IX are considered. One typical cell is treated in the finite element program. Figure IX.217 shows a typical mesh scheme using 20-noded isoparametric elements for the concrete and three-dimensional line elements for the reinforcement in the vertical, radial and horizontal directions. Based on implosion pressures, the damaged part is shown in Figs IX.218 and IX.219 when the final incremental failure pressure is 3.5 times the wave pressure based on extreme load cases. For Condeep the safety factor is thus 3.5 against these types of pressure. For further details see Appendix IX.

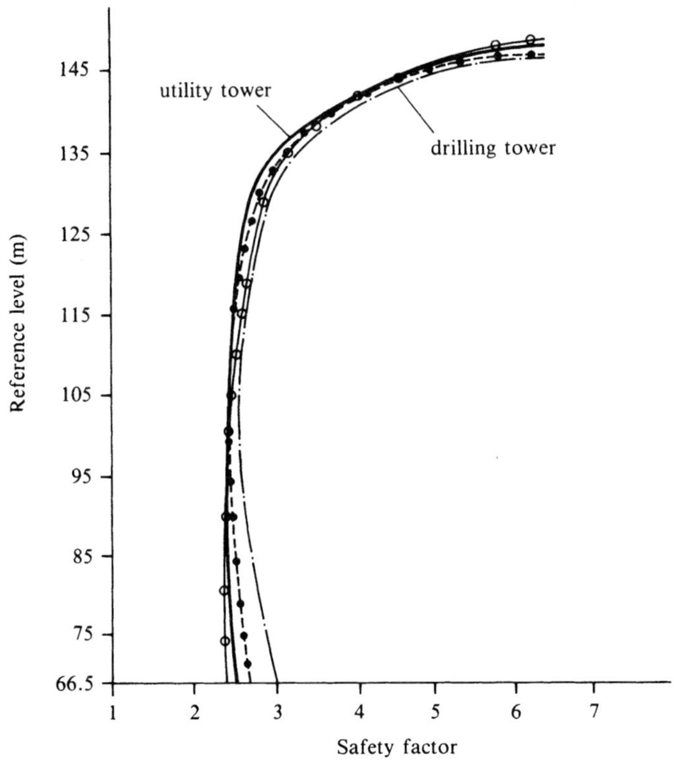

Fig. IX.216. Comparative study of safety factors.

drilling —— · —— · —— } *current analysis* AL-NOURY
utility ————————
drilling ● – – – ● – – – ● *finite element* ISOPAR
utility ⊖———⊖———⊖ ISOPAR

IX.13.5.3 Ship impact analysis

The ship impact case given in Section IX.13.4.3 is examined using the loading cases and material criteria for the Condeep platform. The platform is analysed using a given supply ship. The load–displacement curve is given in Fig. IX.220. Three failure modes are given in Figs IX.221 and IX.222, and the cracking pattern of the shafts is given in Figs IX.223 and IX.224 for the cell. The major results associated with final mode of failure are:

	Type	*Maximum displacement (m)*	*Time (s)*
(a)	Base of the shaft	0.02	3.50
(b)	Top of the cell	0.018	3.10
(c)	Major part of the cell	0.0159	3.03

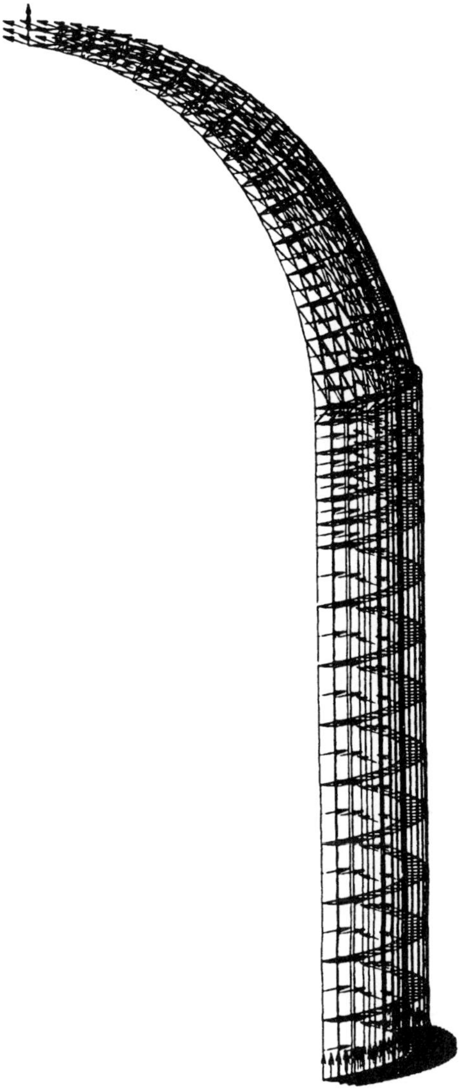

Fig. IX.217. Finite element mesh scheme for a cell.

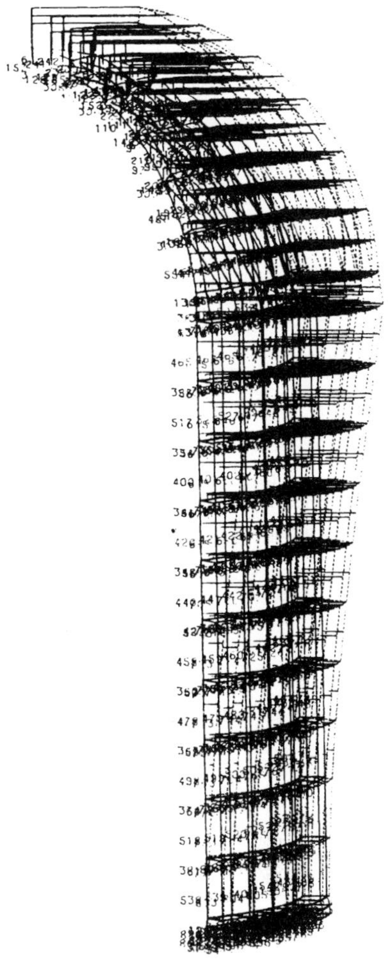

Fig. IX.218. Damage caused by implosion at a level 2.75 times extreme loads (minimum safety factor 2.5).

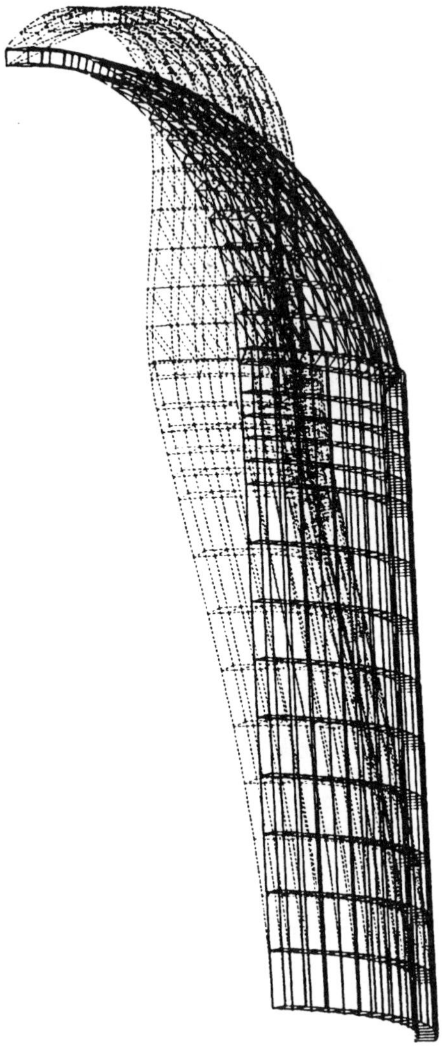

Fig. IX.219. Damage caused by implosion at a level 3.5 times extreme loads (concrete cracked/scabbed; steel yielded/plastic).

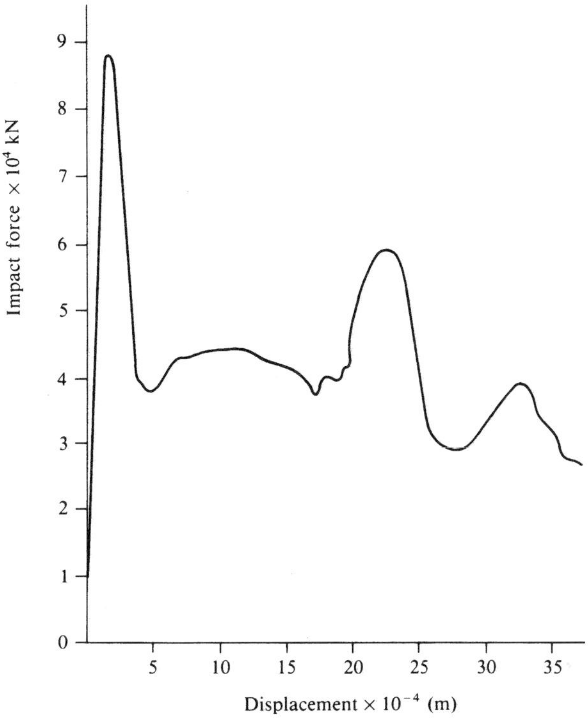

Fig. IX.220. Impact force–displacement relation.

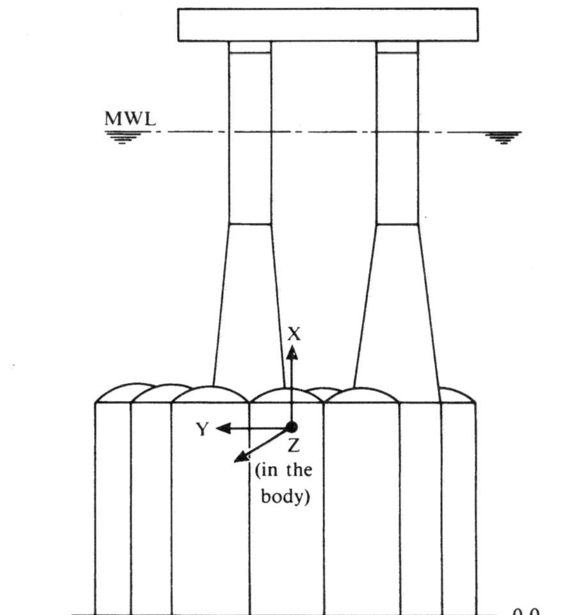

Fig. IX.221. Concrete gravity platform.

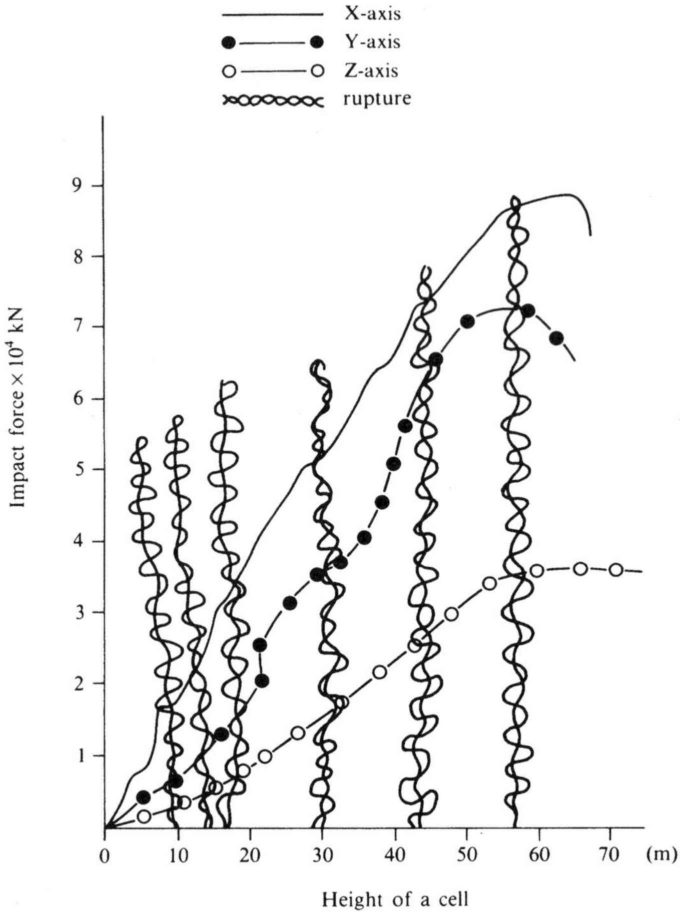

Fig. IX.222. Failure modes of a cell (see Fig. IX.221).

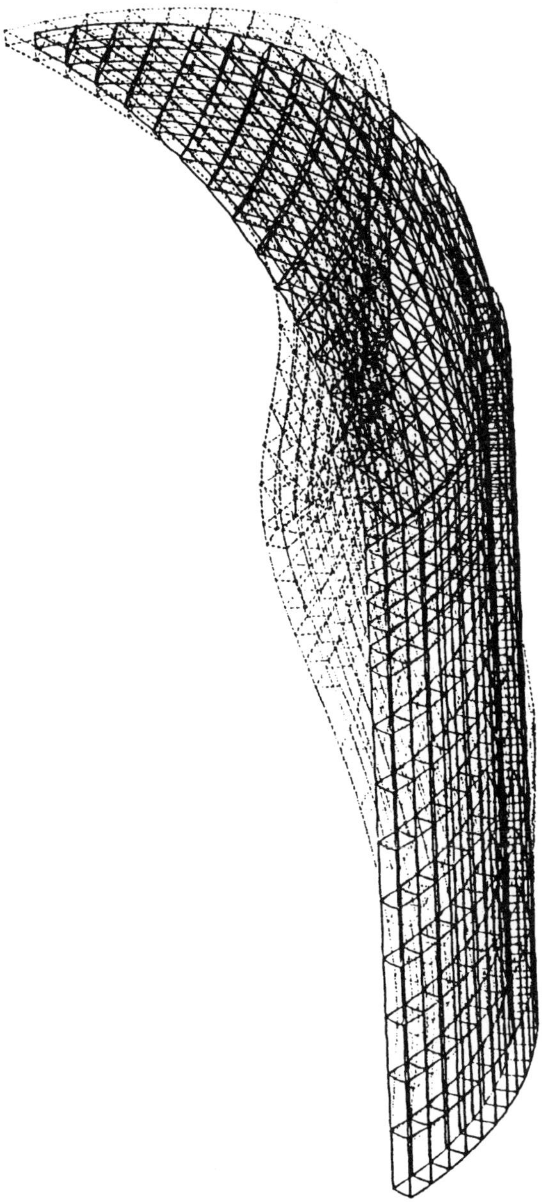

Fig. IX.223. Dark areas representing damage zones.

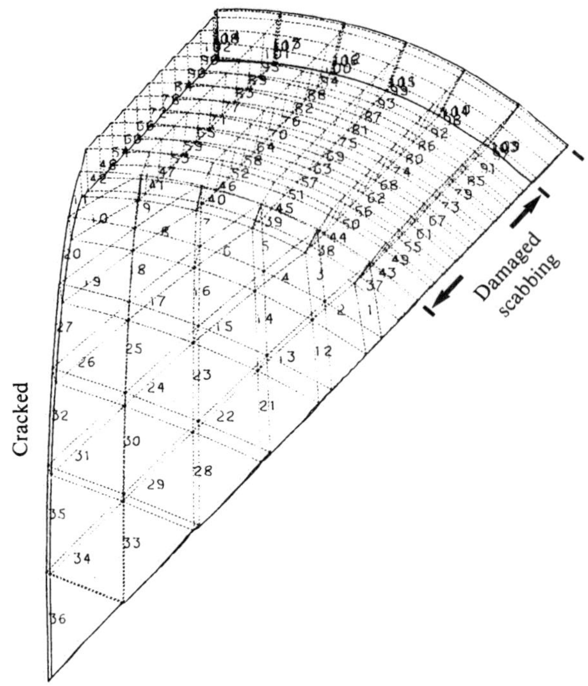

Fig. IX.224. Junction between the dome and the wall.

IX.14 Case Study No. 13 — liquefied natural gas (LNG) tanks

IX.14.1 Introduction

Due to the growth and expansion of industrial natural gas facilities the storage of large volumes of gas is needed to meet winter demands. In many areas there exist underground reservoirs which provide a satisfactory and economic solution. When these do not exist, or have already been used, alternative storage facilities must be provided to store LNG at a temperature of $-165°C$.

In natural gas storage systems the container is usually the main expense. Storage of the gas in liquid form is preferable since it requires less storage space. Conventional tanks are flat-bottomed cylindrical vessels with an external mild steel shell and internal tank of aluminium or steel, with insulation in the spaces between the walls. These tanks can be expensive, but recent developments in alloy tanks have made storage more economical.

Prestressed concrete tanks have also been developed for the industry; in 1953 in the USA they were used to store liquid oxygen [737]. The running costs and construction cost of prestressed concrete vessels were, in 1961, substantially less than for conventional steel tanks. Following this, a demonstration tank was built underground for the storage of liquefied natural gas and lengthy studies carried out on its behaviour; these studies included temperature changes, freezing of the soil, costs and evaporation rates. Using the experience gained from the demonstration tanks, larger tanks were designed.

General design features for a 140 000 cubic metre tank, prestressed and situated below ground, include the following features:

(a) The base of the tank can be a reinforced concrete slab covered with 300 mm of foamglass insulation within a steel liner.
(b) The concrete walls on the outside may have a steel liner wrapped with prestressed wire and protected with a special mortar.
(c) The roof is a dome supported by a prestressed concrete ring beam covered with a thin steel liner and foamglass insulation.
(d) Heating coils are installed 600 mm below the tank base and halfway up the side of the underground walls in order to counter freezing and frost heaving in the winter months.

Properties of the structural components of such tanks were studied and found to be adequate for use at cryogenic temperatures, both the compressive strength of moist concrete and the tensile strength of high carbon steel prestressing wire increased considerably at temperatures below room temperature.

In a cryogenic tank there are two tension-producing effects: thermal and hydrostatic. Since concrete has good compressive strength, it is vital that the concrete is kept in compression so that no tension-producing force can ever quite overcome the initial compression.

The tank can be designed either as a rigid structure, that is with the base and the walls fixed, or as a free-sliding [736–748] wall-to-base structure, the latter design being more feasible for the pressures exerted on the walls. This design using independent movement for the side walls relative to the base is made possible by the introduction of an expansion joint in the steel liner at the bottom of the tank wall.

The data obtained were adequate to be scaled up from the demonstration model to the prototype. The data also suggested a change in the insulation of the base. In order to improve the performance of tanks subject to such temperatures it is necessary to design the base ring for varying temperatures.

Extensive tests on the expansion joint using liquid nitrogen proved its adequacy and showed a large safety factor. Checks were made for leaks. Tests showed that there were 220 leaks in the structure of the tank.

Experimental tests showed no extensive vertical movement of the tank during cool-down contraction of the circumference. The maximum movement that occurred was around 25 mm. When the tank was filled again no noticeable movement was observed.

Tests were also conducted on the effectiveness of the foamglass insulation used. The calculated values compared with measured values during use of the tank showed about 83% efficiency. Tests were also carried out on various soils and their freezing zones. It was observed that this zone moved further away when the inside temperature was brought up to room temperature.

Conclusions from all tests showed that prestressed concrete tanks were a viable proposition for the storage of liquid gases. A typical concrete tank prestressed by Dykman (BBR Tank) is shown in Fig. IX.225.

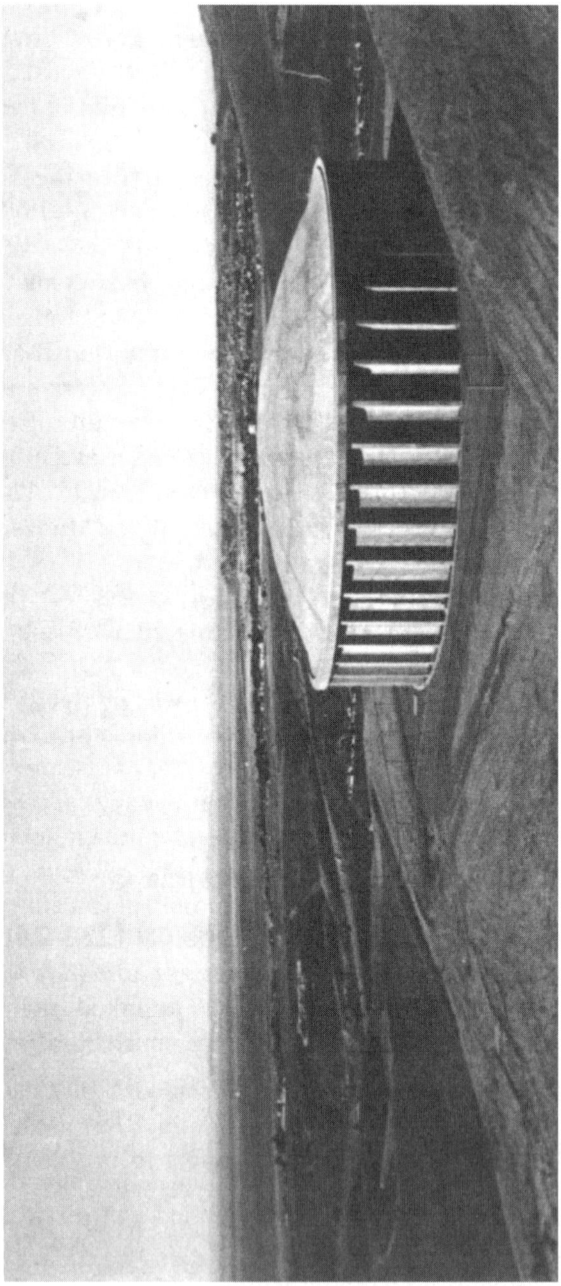

Fig. IX.225. A typical natural gas reservoir (San Clemente, California). Tank diameter 35.4 m (116 ft), wall height 7.93 m (26 ft). (Courtesy of BBRV Prestress Tanks Inc., California, USA.)

IX.14.2 Methods of analysis and design [736–748, 1039]

IX.14.2.1 Methods of analysis and design considerations

There are many considerations which have to be taken into account even before design can be carried out. The four basic methods of design are:

(a) *Empirical*—Promoters of this concept do not go into deep theoretical considerations for their design. They rather try to make corrections in future designs, responding to problems from earlier structures.
(b) *Crack control*—This theory is largely based on the idea of eliminating all straight tensions through prestressing and reinforcing the concrete with mild steel against bending stresses. It is then assumed that the crack developments in concrete will not be necessary provided that the crack size can be controlled using smaller diameter reinforcing bars with particular bar spacings and concrete cover. Under this concept, stresses are permitted close to the yield point of the steel.
(c) *Crack prevention*—Most 50-year-rated designs are now prepared with this concept in mind. A truly non-cracked tank condition can most easily be accomplished by prestressing the concrete in two directions. To minimise leakage the crack prevention concept seems more reliable for the design of prestressed concrete tanks. In order to prevent leakage, the strength of linings may be introduced into the design. For lighter fluids such as kerosene, gasoline and jet fuels, it is necessary to provide a steel, plastic or rubber lining.
(d) *Three-dimensional numerical modelling with well-disposed cracks*—Analyses include the prediction of cracks under increasing pressures and cryogenic temperatures for both short- and long-term conditions.

IX.14.2.2 Design consideration

(a) *Life span*—The demand for fully prestressed concrete structures has increased rapidly in the last decade. Around 5000 circular prestressed concrete structures are now in service on a global basis, some of which have already given over 30 years service. On the basis of what is known today, it is possible to build concrete storage structures with substantial trouble-free performance for periods of at least 50 years. It is expected that fully prestressed systems will outlast ordinary reinforced structures by a considerable margin.
(b) *Corrosion*—Special precautions are needed for the storage of acids, brine and salts. It is generally not sufficient to rely on the concrete to protect the steel. Thus it is necessary to resort to steel, plastic or rubber linings, or use suitable stainless steel reinforcement throughout. Particular attention must be given to preventing sulphates and chlorides from getting to carbon steel. Provision for cathodic protection is generally ignored when proper design and construction techniques are followed. The use of galvanised wire for prestressed reinforcement is particularly advantageous for a long life of the tank.

(c) *Leakage* — Concrete storage structures have a negligible loss factor when such liquids as water and oils are stored. For lighter fluids such as kerosene, gasoline and jet fuels, it is necessary to provide a steel, plastic or rubber lining.

(d) *Safety against failure* — Concrete storage structures have a good operational record and a relatively small number of failures are on record, all of which are

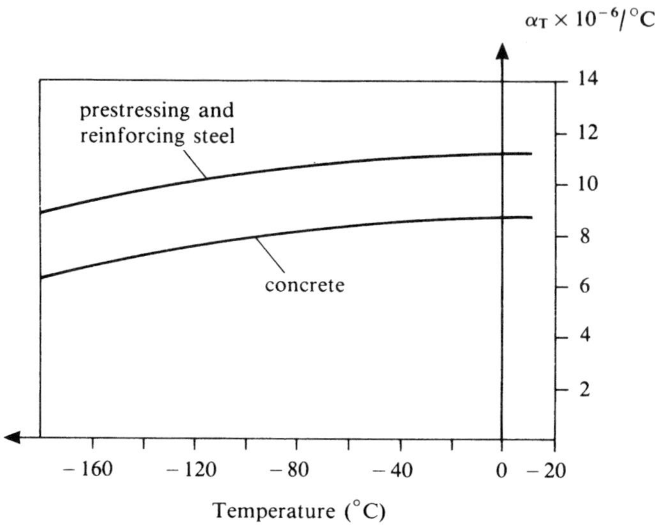

Fig. IX.226a. Coefficient of thermal expansion (α_T) for dry to moderately moist concrete and steel (data collected from [743–746]).

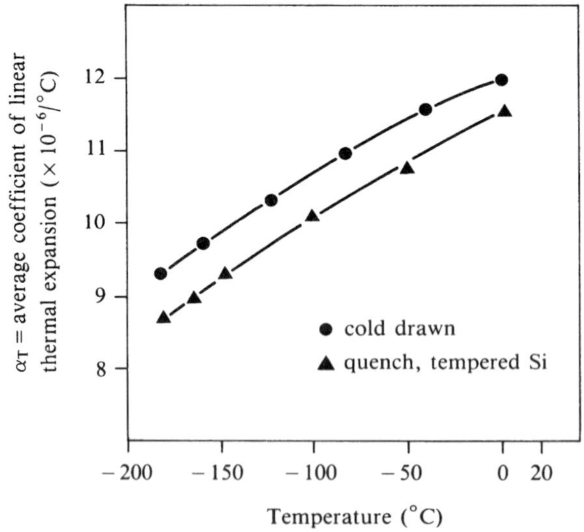

Fig. IX.226b. Relation between temperature and coefficient of thermal expansion for several steels for prestressing (data collected from [743–746]).

the result of poor detailing and inadequate design and specification. The minimum factor of safety for design shall not be less than 2.5.

(e) *Formulae for preliminary sizes* — For dome roofs having diameters up to 60 m and live loads to 195 kg/m^2 the following empirical formulae are suggested:

Shell thickness d	$0.00417R_d$ (m)
Fillet thickness t	$d + 0.000\,78W$ (m)
Fillet length F	$0.2023R_d$ (m)
Dome radius R_d	$2.125R_d$ (m)
Dome shell weight	$8025.657R_d \times d$ (kg)
Dome fillet weight	$1.216R_d^2 w$ (kg)
Dead weight (metric) of circumference	$0.1592W/R_d$ (kg)
Dead weight of projected dome area	$2553.48d + 0.377W$ (kg m)
Shell reinforcing A_s	$625d$ (mm^2/m)
Fillet reinforcing A_s	$0.221R_d$ (m)
Top and bottom reinforcement At	$635t$ (mm^2/m)

The wall has a uniform thickness:

$$M/bt^2 = 0.35\,\text{N/mm for grade 30 where } d/t = 0.75$$
$$M/bt^2 = 0.32\,\text{N/mm for grade 20 where } d/t = 0.75$$

The assumptions give the moment at the bottom of the wall.

IX.14.2.3 LNG tank data

Chapter I gives the material properties of concrete under cryogenic conditions. In addition, some data are given here for both concrete and tendons. Figures IX.226a and IX.226b give the thermal coefficient relationships for steel.

Design calculations for a prestressed concrete cylindrical tank (Figs IX.227 and IX.228) with a spherical dome are carried out using the numerical models described in this book. The main dimensions are shown in the figures.

(a) Design parameters for prestressing steel

$$P = 413.7\,\text{N/m}^2$$
$$f_c' = 41.37\,\text{N/m}^2$$
$$f_y = 413.7\,\text{N/m}^2$$
$$f_{pu} = 16.548 \times 10^5\,\text{kN/m}^2$$
$$E_c = 28.27\,\text{N/m}^2$$
$$E_s = 200\,\text{GN/m}^2$$
$$\mu = 0.14$$
$$k = 0.0003$$

and the assumed prestress level $= 1.2PR$.

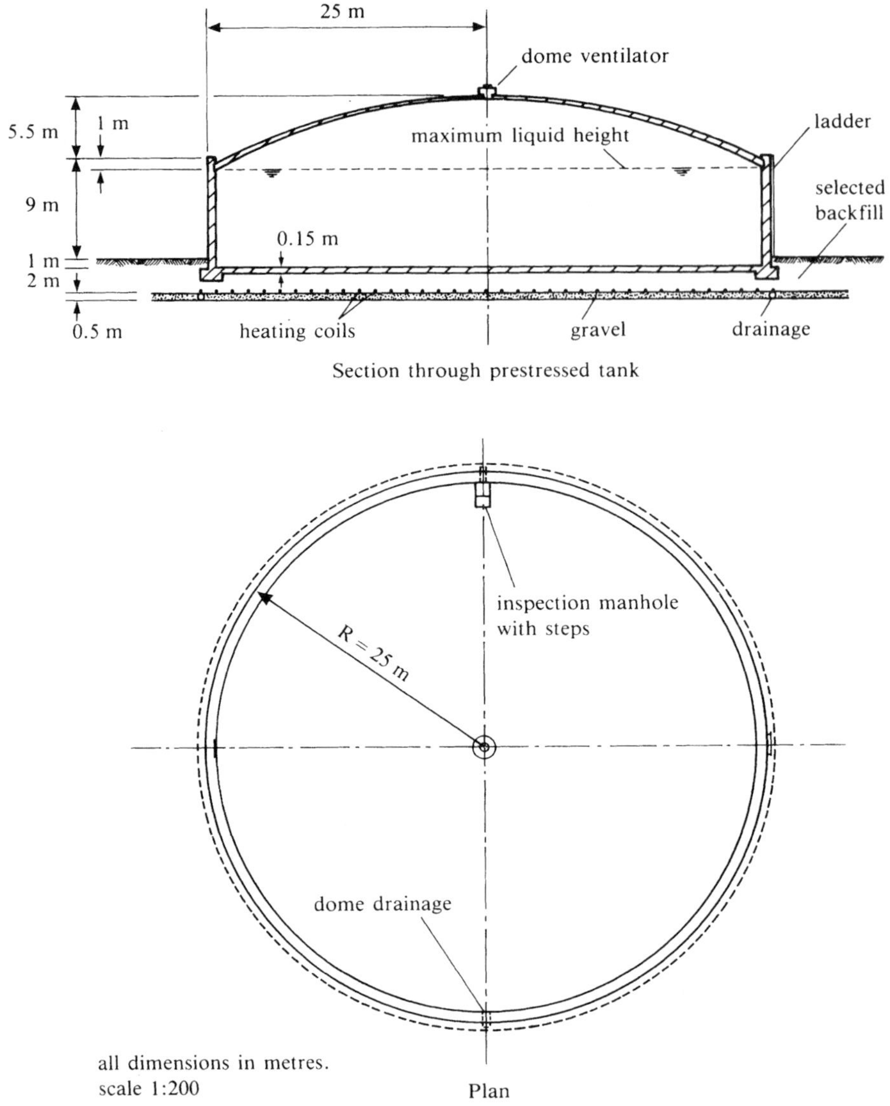

Fig. IX.227. LNG tank parameters (all dimensions in metres).

In addition, long-term losses are taken as:

Creep and shrinkage = 500×10^{-6}
Tendon relaxation = 8% at $0.7f_{pu}$
Tendon sheathing diameter = 150 mm
Wall (hoop) prestressing $FHW = 1.2PR$

(a) Tendon stress after friction loss:

$$f_{pu} = f_{px}\, e^{(kl+\mu\theta)} \qquad f_{ps} = 0.8 f_{pu} \qquad f_{px} = f_{ps}\, e^{-(kl+\mu\theta)}$$

(b) Tendon stresses at stress transfer after friction loss:

$$f_{pA} = 0.75f_{pu} \qquad f_{ps} = f_{pA}\, e^{(kl+\mu\theta)}$$

(c) Average tendon stress and stress transfer due to friction loss: Since the hoop tendons are anchored at the buttresses alternately, it may be assumed that the stresses at the wall are uniformly distributed. The average stress, f_{avg}, is written as

$$f_{avg} = \frac{\text{area under the stress curve at transfer}}{\text{length of tendon}}$$

(d) Loss of tendon stress due to elastic shortening of concrete:

$$f_{ces} = \tfrac{1}{2} f_c \frac{E_s}{E_c}$$

assuming $f_c = 0.3f_c'$

$$f_{ces} = \frac{3}{20} f_c' \frac{E_s}{E_c}$$

(e) Loss of tendon stress due to creep and shrinkage of concrete:

$$f_{ces} = \varepsilon_{cs} E_s$$

(f) Loss of tendon stress due to steel relaxation:

$$f_{sr} = 0.08 \times 0.7f_{pu}$$

(g) Average effective tendon stress after all losses:

$$f_{se} = f_{avg} - f_{sr}$$

(h) Tendon requirements:

$$N = \frac{Fl}{\tfrac{2}{3} A_{ps} f_{se}}$$

(b) Tank dome (meridional) and wall (vertical) prestressing. The losses due to elastic shortening and concrete creep are assumed to be the same in the vertical and dome tendons as in the wall hoop tendons. At the apex of the dome $FM = 1.2(PR/2)$.

(a) Tendon stresses after a friction loss:

apex of the dome $f_{ps} = f_{pxe}(kl + \mu\theta)$

(b) Tendon stress at stress transfer after friction loss:

$$f_{pA} = 0.7f_{pu}$$

(c) Loss of tendon stress due to elastic shortening of concrete:

$$f_{ces} = \tfrac{1}{2} f_c \frac{E_s}{E_c}$$

assuming

$$f_c = 0.3f_c'$$

(d) Loss of tendon stress due to creep and shrinkage of concrete:

$$f_{cses} = \varepsilon_{cs}E_s$$

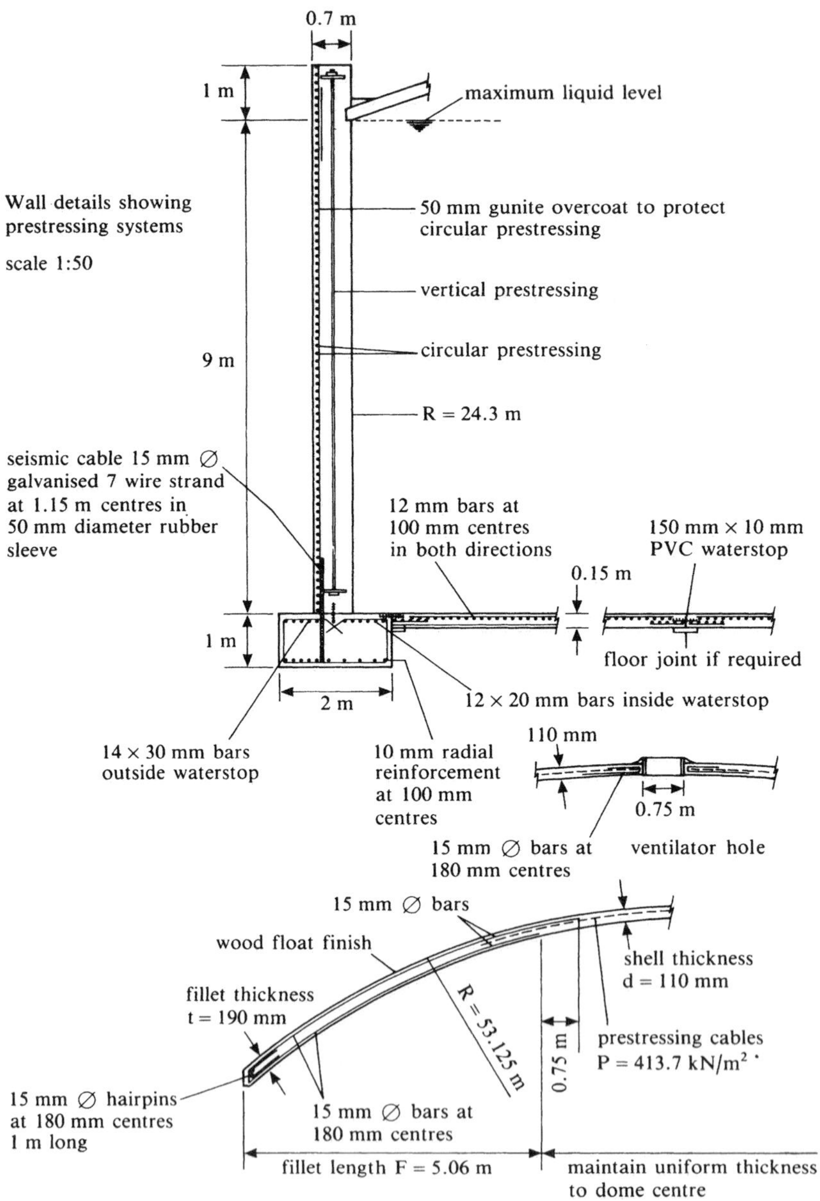

Section through dome Scale 1:50

(e) Loss of tendon stress due to steel relaxation:

$$f_{sr} = (0.08)(0.7f_{pu})$$

(f) Effective tendon stress after all losses:

$$\text{losses other than friction} = (f_{ces} + f_{cses} + f_{sr})$$

(g) Tendon requirements:

$$N = \frac{Fl}{A_{ps}(F_{se})}$$

(h) Tank dome (hoop) prestressing:

$$FHD = \frac{1.2PR}{2}$$

Assuming the hoop tendon stresses in the dome are the same as the hoop tendon stresses in the cylindrical wall.

Tendon requirement — hoop tendons are provided for the hemispherical dome to a line which forms the 90° solid angle of the dome

$$N = \frac{Fl}{\frac{1}{2}A_{ps}f_{se}}$$

Tendon spacing $= 1/N$.

Fig. IX.228. LNG tank — data and structural details. General design parameters:

1. Capacity			*15 000 m*^{3}
2. Dimensions	*(a)*	*Inside diameter*	*50 m*
	(b)	*Liquid height*	*8 m*
	(c)	*Free height*	*1 m*
3. Loadings	*(a)*	*Live load on roof*	*100 kg/m*^{2}
	(b)	*Specific gravity*	*1.5*
	(c)	*Liquid temperature:*	
		Summer average	*60° F*
		Winter average	*50° F*
	(d)	*Air temperature:*	
		High summer average	*70° F*
		Low winter average	*40° F*
		Temperature peaks	*±15° F*
	(e)	*Seismic acceleration*	*10%g*
	(f)	*Backfill height*	*1 m*
	(g)	*Backfill density*	*1700 kg/m*^{3}
	(h)	*Allowable soil pressure*	2.5×10^4 *kN/m*^{2}
4. Roof type			*Dome*

(c) Concrete material characteristics. Reference is made to material properties of concrete in Chapter I. Recent tests in different countries, to measure the compressive strength of concrete at temperatures down to -196°C, indicate that compressive strength increases with reducing temperature. The following are concrete material properties used in the design of a number of tanks.

(a) Coefficient of contraction:
Below $-70^\circ\text{C} = 53 \times 10^{-6}/^\circ\text{C}$.

(b) Compressive strength:
Strength increment at temperatures below $-120^\circ\text{C} = T^\circ\text{C}$

$$= \frac{-T \times m_{\text{m}}}{12} \quad \text{N/mm}^2$$

(m_{m} = moisture content, 5%). The compressive strength is independent of the strength grade in this case.

(c) Tensile strength:
Tensile strength increases as the temperature is reduced down to about -70°C. At temperatures lower than -70°C, there is a small reduction. At -165°C there is an increase in residual strength up to 4 N/mm^2 for moist concrete and 2.5 N/mm^2 for dry concrete.

(d) Elastic moduli:
Reference is made to the relevant values in Chapter I.

(e) Creep:
At a temperature of -100°C, 60% reduction of creep is adopted.

(f) Poisson ratio:
At temperatures of -165°C and below a value of 0.22 is adopted. For details refer to Chapter I.

(g) Thermal contraction:
The coefficient of contraction is $5\text{–}6 \times 10^{-6}/^\circ$C at -165°C. A value of 0.13 % is chosen for the range 20 to -165°C.

(h) Thermal conductivity:
An increased value of thermal conductivity: moist concrete, 60% at -165°C; partially dried concrete, 20%.

(i) Impact resistance:
Reference should be made to earlier chapters for impact on concrete.

IX.14.3 Finite element analysis of LNG tanks

Finite element analysis is carried out on this tank using three-dimensional eight-noded isoparametric elements for concrete with two-noded line elements, representing steel, placed within the body and on top of the solid concrete elements. The choice of eight-noded elements was made on economic grounds. Shape functions and stiffness matrices for these elements are fully discussed in Chapter VI and associated appendices. Figure IX.229 shows a typical $22\frac{1}{2}^\circ$ mesh for the tank. The total numbers of solid elements and line elements were 230 and 375, respectively. A 2×2

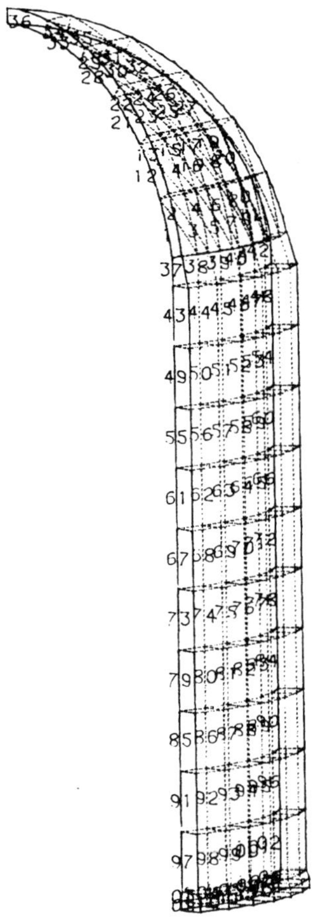

Fig. IX.229. Finite element mesh scheme for an LNG tank.

Gaussian quadrature is adopted. The prestressing and the LNG pressure (Fig. IX.230) loads are computed. The prestressing loads were transferred to the nodal points while LNG pressures are assumed to act on the solid element surfaces. The total number of load increments chosen was 15. Figures IX.231–IX.233 show principal stress trajectories and crack positions in the walls and dome of the tank. The procedure for analysing crack sizes under incremental loads is the same as described for pressure vessels in this chapter and in Chapter VI. The concrete of the tank is assumed to behave as a three-parameter model as described in Chapter VII. The safety margin computed over the design loads is 2.91.

Seismic analysis was carried out on this tank. For soils a layered finite element scheme with springs (Table IX.15b) was chosen. A seismic acceleration of 0.25g was considered for modes such as translation, rocking and bending given in Fig. IX.234; recorded acceleration was plotted in real time. The forced vibration spectral data is plotted in Fig. IX.235. The computed frequencies are given

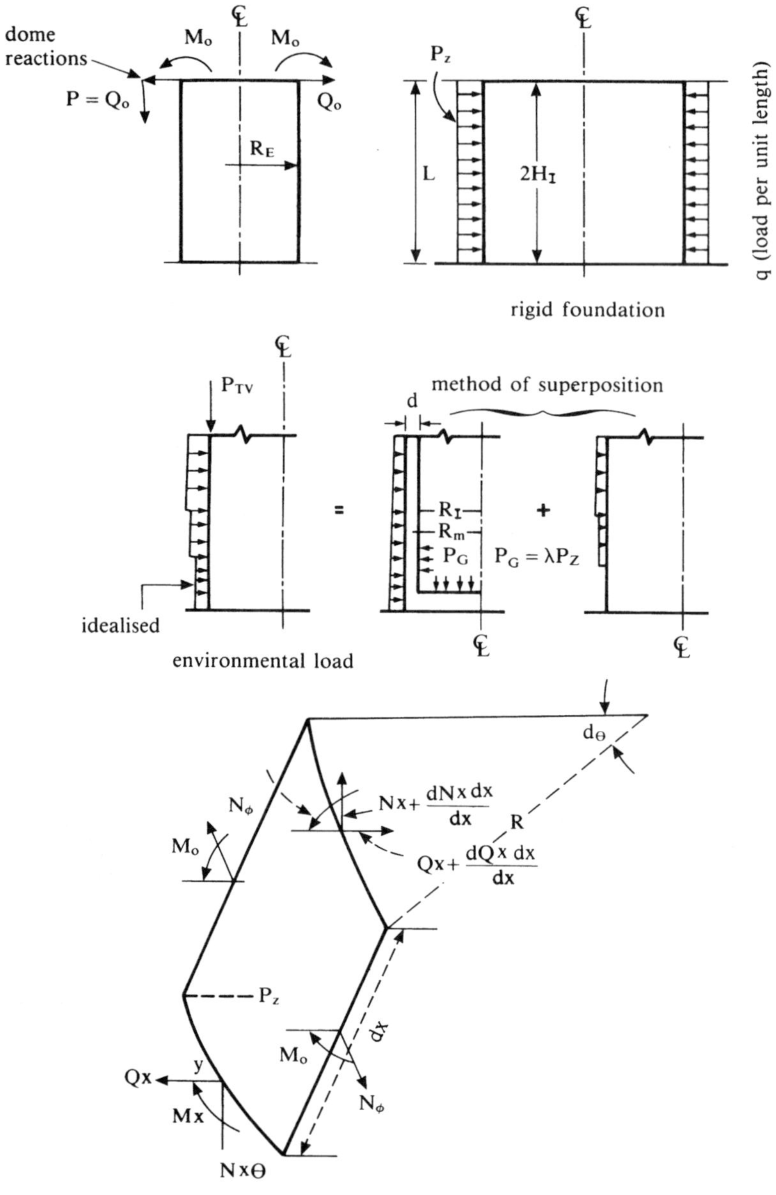

Fig. IX.230. Forces on barrel wall.

below:

Vertical	8.8 Hz
Horizontal bending	2.5 Hz
Rocking	3.5 Hz
Translation	5.5 Hz

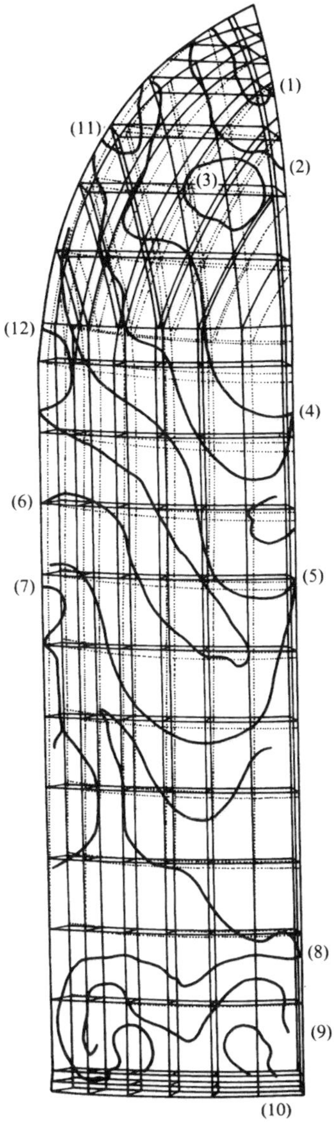

Fig. IX.231. Principal tensile stress at 3.5 times design pressure. (1, 30 N/mm^2; 2, 24 N/mm^2; 3, 5 N/mm^2; 4, 10 N/mm^2; 5, 5 N/mm^2; 6, 8 N/mm^2; 7, 11 N/mm^2; 8, 10 N/mm^2; 9, 11 N/mm^2; 10, 12.0 N/mm^2; 11, 16 N/mm^2; 12, 23 N/mm^2.

Fig. IX.232. Cracking (3.5 times design pressure).

The data are useful for an earthquake safety review of the tank. The program ISOPAR can use two levels of exciting force in order to make it possible to distinguish between steady-state motions from nearby mechanical equipment and the transient response together with associated mode shapes.

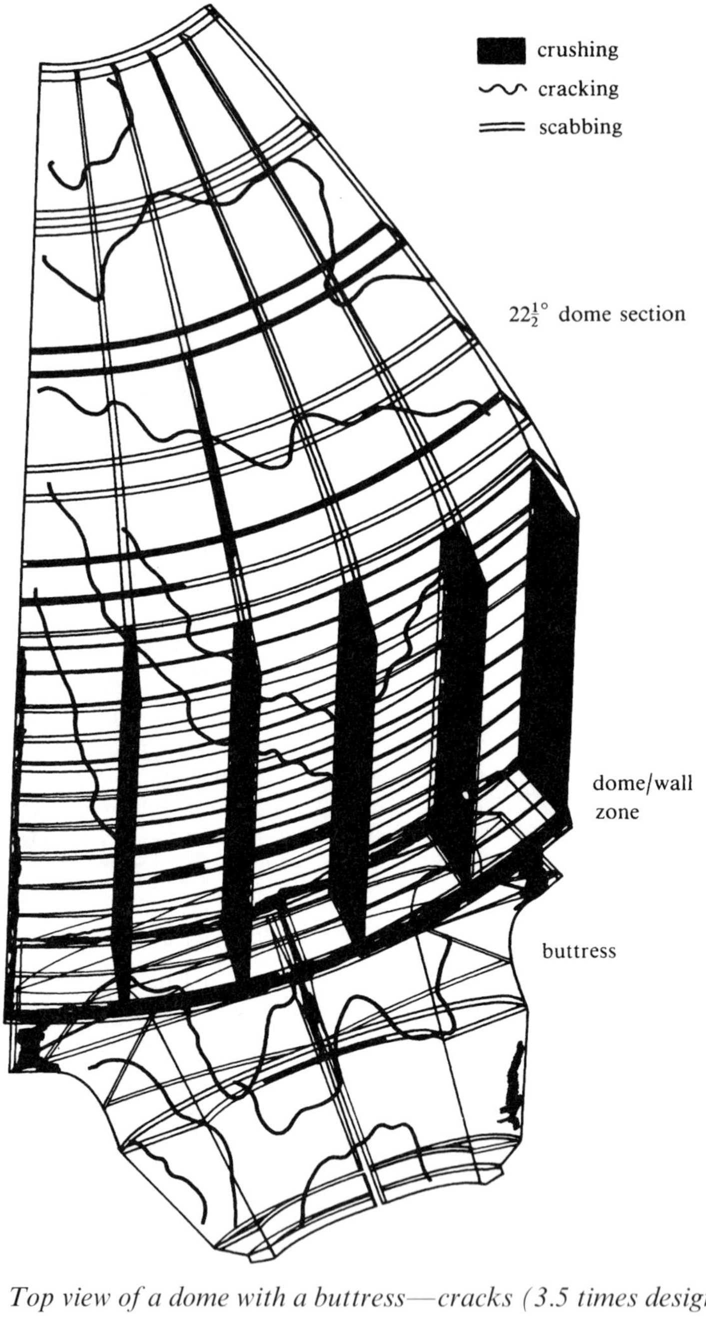

Fig. IX.233. Top view of a dome with a buttress—cracks (3.5 times design pressure).

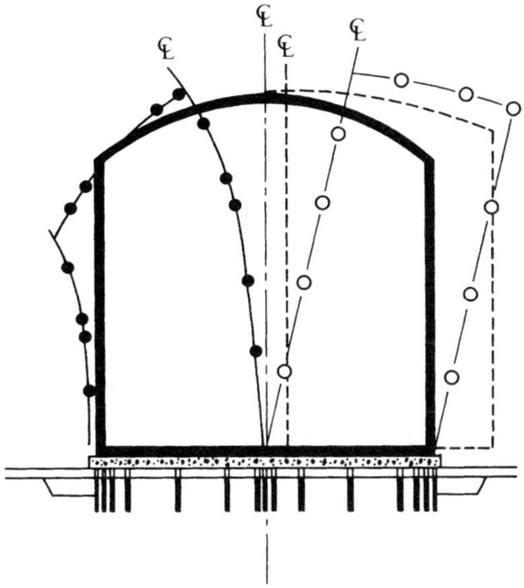

Fig. IX.234. Translation, rocking and bending during forced vibration under seismic loads. (– – –, Translation; —○—, rocking; —●—, bending.

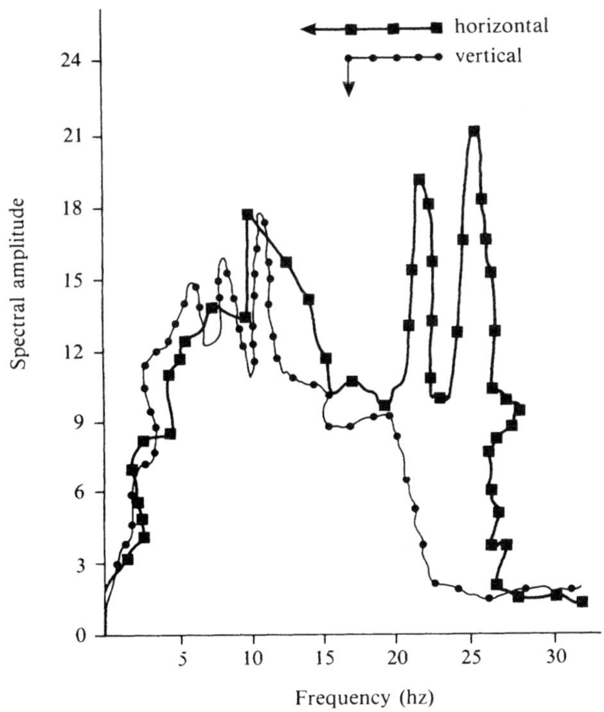

Fig. IX.235. Forced vibration spectral data for an LNG tank.

IX.15 Case Study No. 14 — fire analysis of concrete elements

IX.15.1 Introduction

The fire resistance of concrete structures has been achieved [1326–1338] using a number of techniques. A nonlinear finite element analysis described in the text is used here to predict the behaviour of reinforced concrete and composite concrete/steel members subject to fire. Complex features such as thermal expansion, shrinkage, creep, cracking or crushing etc. are included. Concrete and steel material properties affected by fire are included in the data input. The splitting and scabbing effects are excluded. Predictions from the analysis are compared with the experimental results from other researchers.

IX.15.2 Proposed analytical procedure

The members are modelled as an assemblage of concrete elements and reinforced bar elements. The concrete elements are divided into several layers and each layer is assumed to have a uniform temperature. The degree of sophistication depends on the number of layers. Throughout this analysis, the main reinforcement is modelled as 3-node line elements so that they can match up with 8-noded isoparametric elements representing solid concrete elements. Where shear reinforcement or any unusual reinforcement exists, these are modelled in the smear manner as described in detail in Section II.2.4. The examples included are those for those structural components which are in service conditions, i.e. they are loaded and are functioning. The structural components are described in series of elements and nodes. The deformation history at various nodes is developed, together with temperature histories within members. Hence, all parameters in the analysis are time dependent. In general, the failure model for concrete is based on the four-parameter model developed by Ottoson [24, 25]. The smear reinforcement model as stated above, is devised using Von Mises yield surface is assumed. A three-dimensional cracking criterion given in Appendix III has been adopted where crack investigations are needed or necessary. The sections are first checked according to the two methods given below.

Wickstrom's method. The temperature rise in a normal-weight concrete element T is given by the following:

for uniaxial heat flow

$$T = (n_x n_w)\,\Delta T_g \tag{IX.15.1}$$

for biaxial heat flow

$$T = (n_w(n_x + n_y - 2n_x n_y) + n_x n_y)\,\Delta T_g \tag{IX.15.2}$$

where n_w is given by

$$n_w = 1 - 0.0616t - 0.88 \tag{IX.15.3}$$

and n_x (or n_y with y substituted for x) by

$$n_x = 0.18\ln(t/x^2) - 0.81 \tag{IX.15.4}$$

with x (or y) subject to the limit

$$x > 2h - 3.6\sqrt{0.0015t} \tag{IX.15.5}$$

where t is the time (hours), x and y are the depths into the slab (m), and h is the overall depth of the section (m). ΔT_g is given by the rise in furnace temperature above ambient at time t. Note that the temperature rise above ambient on the surface of the element is given by $n_w \Delta T_g$.

Hertz's method. The unidimensional time-dependent temperature $T(x, t)$ is given by

$$T(x, t) = f_1(x, t) + f_2(x, t) + f_3(x, t) \tag{IX.15.6}$$

where the functions $f_1(x, t)$, $f_2(x, t)$ and $f_3(x, t)$ are solutions to the heat transfer equation for specific boundary conditions. These functions are given by

$$f_1(x, t) = E'\left[1 - \frac{x}{3.363\sqrt{at}}\right]^2 \tag{IX.15.7}$$

$$f_2(x, t) = D' \exp\left[-x\left(\frac{\pi}{2C'a}\right)^{1/2}\right] \sin\left[\frac{\pi t}{C'} - x\left(\frac{\pi}{2C'a}\right)^{1/2}\right] \tag{IX.15.8}$$

and

$$f_3(x, t) = \frac{D' + E'}{2(e^{LC'} - 1)}\left\{1 - \exp\left[(L(t - C))x\left(\frac{L}{a}\right)^{1/2}\right]\right\} \tag{IX.15.9}$$

where a is the thermal conductivity.

Note that f_1 is set equal to zero if

$$1 - \frac{x}{3.363\sqrt{at}} \leq 0 \tag{IX.15.10}$$

f_2 is set equal to zero if

$$\frac{\pi t}{C'} - x\left(\frac{\pi}{2C'a}\right)^{1/2} \leq 0 \tag{IX.15.11}$$

f_3 is set equal to zero if

$$L(t - C') - x\left(\frac{L}{a}\right)^{1/2} \leq 0 \tag{IX.15.12}$$

The parameters E', D' and C' are dependent on the heating regime and L is dependent on the temperature curve during cooling and is given by

$$L = \frac{2}{C} In\left(\frac{3D'}{E' - 2D'}\right) \tag{IX.15.13}$$

Note that if $E' - 2D'$ is negative then $E' - 2D'$ is set equal to 0.02. The temperature on the surface during heating is given by $D' + E'$.

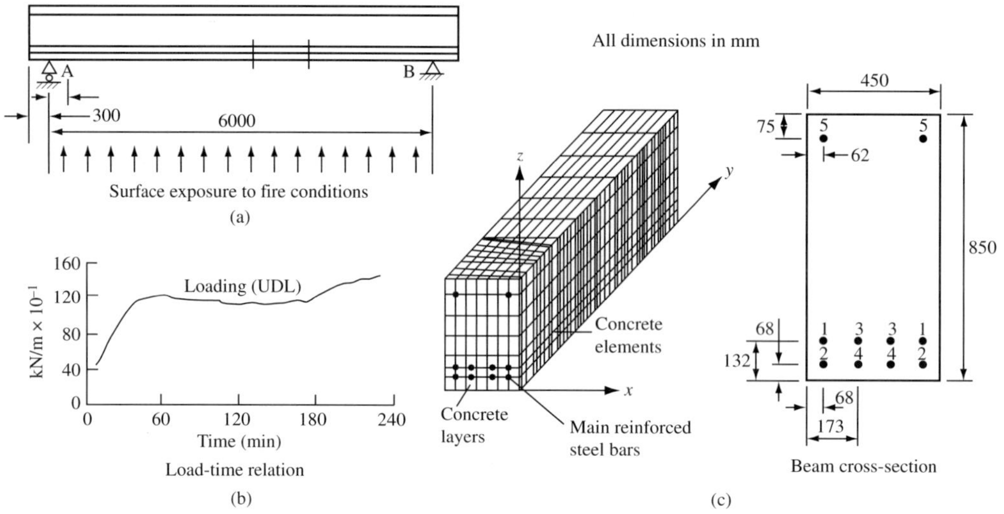

Fig. IX.236. Reinforced concrete beam with load–time function and finite element mesh scheme.

Based on the European Code ENV 1991-1-2 for the half depth of the beam; the values of $\xi_{\theta x}$ for T_x at different levels for the reinforcing bars under temperature are given below:

$$\xi_{\theta x} \text{ or } \xi_{\theta y} = (T_x \text{ or } T_y)/(D' + E') \qquad \text{(IX15.14)}$$

Shielded parameters. No change in the values of ξ_s. $T_s = 0.9 \times$ the above values.

IX.15.3 Geometrical and numerical data

IX.15.3.1 Reinforced concrete beams

A typical reinforced concrete beam is shown, along with its optimum finite element mesh scheme, in Fig. IX.236. The mesh identifies the positions of the main reinforcement and helicals or links.

IX.15.3.2 Numerical data

Figure IX.237 gives the load–time relation during the fire test. Table IX.24 gives the values of various fire parameters at room temperature. The following material properties form the basic input data for the finite element analysis.

IX.15.3.3 Analysis of results of the RC beam tests

The beam is heated from the bottom and the sides. The concrete is assumed to be affected throughout. The flow of heat is in the compression zone, i.e. the top half of the beam, while the beam is subjected to the serviceable loads as shown in Fig. IX.237. The temperature distribution for various parameters is given in Figs IX.238 and IX.239.

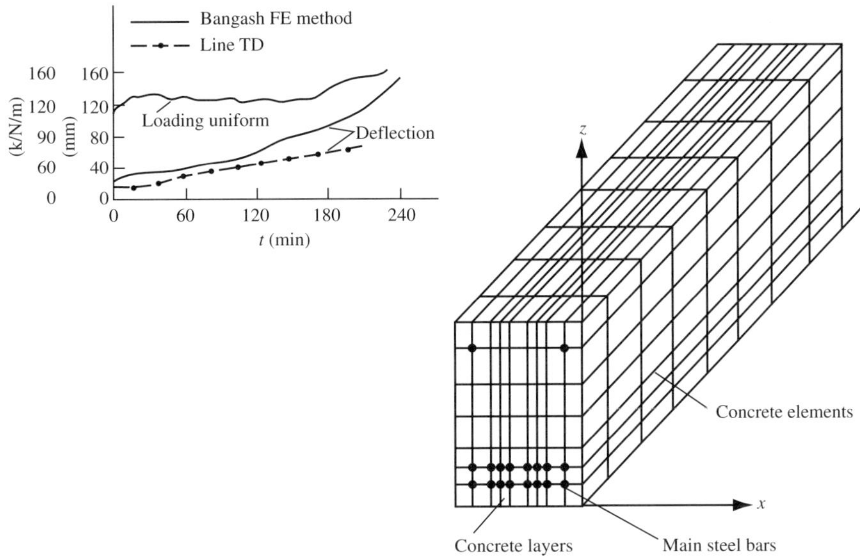

Fig. IX.237. Load–time function and time-dependent deflection.

Figure IX.240 shows the comparative study of deflection–time relation under loads specified in Fig. IX.236. Table IX.25 specifies the basis of the analysis given by Wickstrom and Hertz and the European Code ENV 1992-1-2. The idealised stress–strain curve for reinforced steel combined with the nonlinear elastic model proposed by Ottoson [24, 25] for concrete and room temperature are adopted at initial stages. The model is capable of representing, in a simplistic manner, most of the characteristics of concrete behaviour for the general state of stress. The features include all three stress invariants, taking dilation into consideration. The prediction of realistic failure stresses and the applicability of all stress states are included in those where tensile stresses occur. The cracking in this criterion is well defined in the text. Time-dependent temperature is defined in this section. The load is varied

Table IX.24. Fire parameters at room temperature

$\sigma_t = 2\,\mathrm{MN/m^2}$; $V_C = 0.18\varepsilon_{co} = 2.4588 \times 10^{-3} E_C = 20\,\mathrm{kN/mm^2}$; $E_{sp} = 1034\,\mathrm{MN/m^2}$

$F_{yk} = 460\,\mathrm{MN/m^2}$; $\sigma_{yt} = 350\,\mathrm{MN/m^2}$; $T = 560°\mathrm{C}$; $r_w = 23.4\,\mathrm{kN/m^3}$; $f_{cu} = 30\,\mathrm{MN/m^2}$

Bars	f_y ($\mathrm{MN/m^2}$)	E_y (mm/mm)
10	483.4	0.0028
20	481.3	0.0025
25	510.0	0.0028

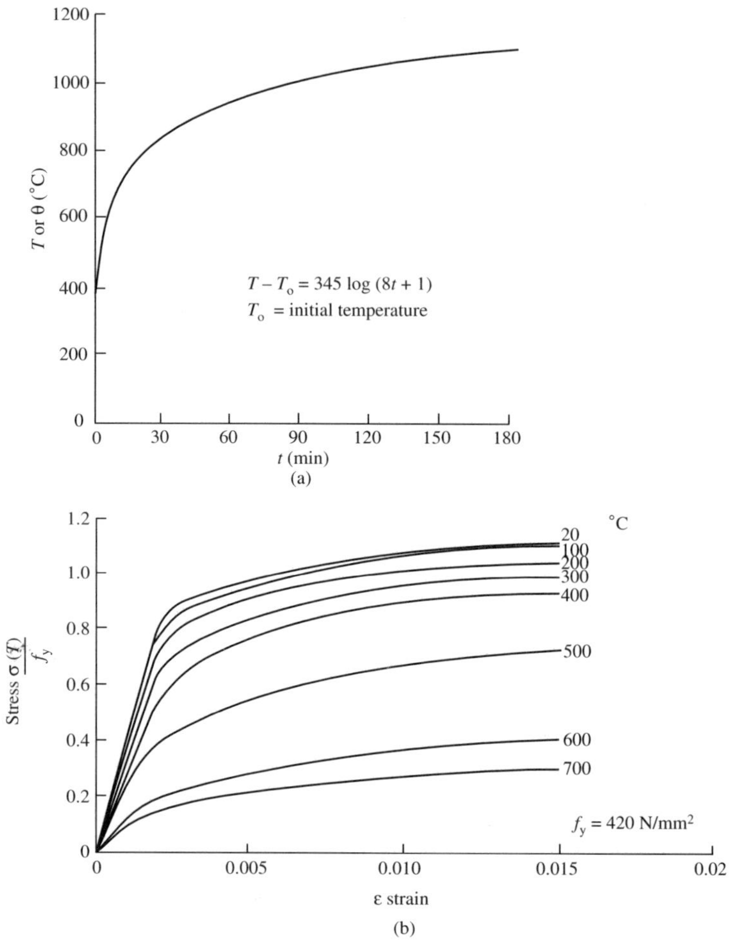

Fig. IX.238. Temperature distribution and results: (a) temperature–time relation; (b) $\sigma(T)/f_y$ versus strains.

with the temperature for a specific time. Figure IX.241 shows the damaged beam when the temperature reaches 600°C, 2.5 h maximum time. Cracks closing and re-opening are not considered in the period of 2.5 h. Tension–stiffness is considered. For the finite element analysis, the shear reduction factor ($0 < \beta < 1.0$) chosen is 0.25. Free thermal strains have been ignored in this particular analysis. In order to start the first iteration, the Gauss point of each layer of the concrete elements and each steel bar element is considered. Crushing occurs at these points before or after cracking.

IX.15.3.4 Reinforced concrete slabs subjected to fire

Introduction and limit state design excluding fire load. A simply supported concrete slab supported on edge beams is shown in Fig. IX.242. BS 8110 is adopted initially for evaluating and checking and checking the thickness and the reinforcement. The

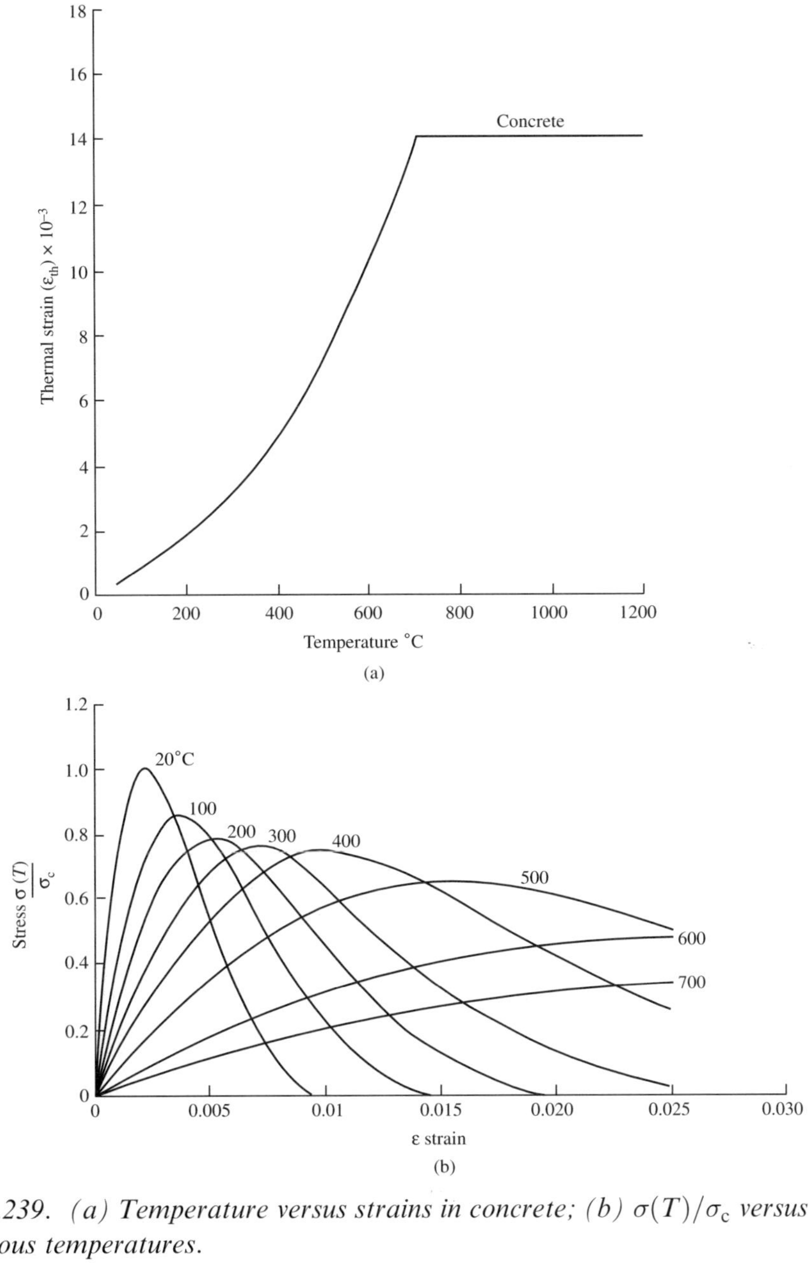

Fig. IX.239. (a) Temperature versus strains in concrete; (b) $\sigma(T)/\sigma_c$ versus ε strain for various temperatures.

slab has finishes. The slab has a thickness of 125 mm with an effective span of 2.5 m. Based on BS 8110, the following formula is adopted.

$$W = r_f G_k + r_L Q_k$$

$$1.4 \times \left[\frac{125 \times 23.6 + 1.27}{1000}\right] + 1.6 \times 3.5 = 5.9 + 5.60 = 11.50\,\text{kN/m}$$

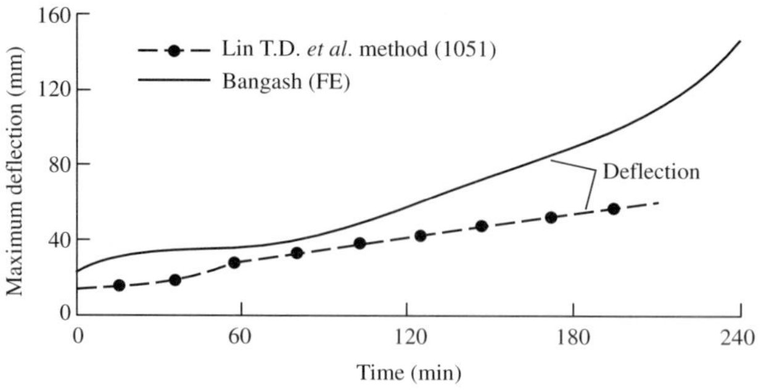

Fig. IX.240. Deflection–time relations.

Normal data:

ρ = density of concrete = 23.6 kN/m^3

finishes = 1.27 kN/m; $T = 670°C$

imposed load = 3.5 kN/m

$f_{cu} = 30\,N/mm^2$; $f_y = 460\,N/mm^2$

depth of embedment = 25 mm

$$BM = Wl^2/8 = \frac{11.5 \times (2.5)^2}{8} \qquad d = 125 - 25 = 100\,mm$$

$$K = \frac{M}{bd^2 f_{cu}} = \frac{9 \times 10^6}{1000(100)^2 30} = 0.03 \prec K' = 0.156$$

$= 9\,kNm$ No compression steel is required

$$Z = d(0.5 + \sqrt{0.25 - K/0.9}) \approx 0.95d$$

$$A_s = \frac{M}{0.87 f_{yz}} = \frac{9 \times 10^6}{0.87 \times 460 \times 0.95 \times 100} = 237\,mm^2/m \text{ width}$$

Use $T10 - 250(314\,mm^2/m = A_{sprov}) = 162.5\,mm^2/m$ OK

$$\text{Minimum steel} = \frac{0.13}{100} \times 1000 \times 25 = (A_s = 1625\,mm^2/m \prec 314\,mm^2/m)$$

$h > 127\,mm > 200\,mm$, the clear distance between bars $> 3d$ or 750 mm or 300

actual = 250 mm c/c OK

Load during fire $= G_k \gamma f_g + Q_k \gamma f_q = 4.22 \times 1.05 + 3.5 \times 1.0 = 7.931\,kN/m$

$$\text{Applied moment} = \frac{4.22 \times 1.05 \times 2.52}{8} = 3.462\,kNm$$

A reference is made to the IStructE/Concrete Society report *Design and detailing of concrete structures for fire resistance. Interim guidance* (April, 1978, p. 59).

Table IX.25a. Parameters for temperature analysis of concrete beams

Time (h)	C' (h)	D' (°C)	E' (°C)
0.5	1.0	150	600
1.0	2.0	220	600
1.5	3.0	310	600
2.0	4.0	360	600
3.0	6.0	410	600
4.0	8.0	460	600

Data from Hertz, 1981.

Table IX.25b. Unshielded parameters: bar temperatures and strength reductions

Time (h)	Bar number													
	Bar 1 $x = 68$ mm, $y = 132$ mm				Bar 2 $x = y = 68$ mm			Bar 3 $x = 132$ mm, $y = 173$ mm				Bar 4 $x = 68$ mm, $y = 173$ mm		
	$\xi_{\theta x}$	$\xi_{\theta y}$	T_s	ξ_s	$\xi_{\theta x} = \xi_{\theta y}$	T_s	ξ_s	$\xi_{\theta x}$	$\xi_{\theta y}$	T_s	ξ_s	$\xi_{\theta x}$	$\xi_{\theta y}$	ξ_s
0.5	0.05	0	41	0.98	0.05	80	0.96	0	0	0	1.0	0.05	0	0.98
1.0	0.17	0	138	0.93	0.17	251	0.84	0	0	0	1.0	0.17	0	0.93
1.5	0.25	0.03	243	0.84	0.25	391	0.67	0.03	0.01	30	0.99	0.25	0.01	0.85
2.0	0.31	0.07	340	0.73	0.31	498	0.50	0.07	0.02	80	0.97	0.31	0.01	0.76
3.0	0.39	0.15	486	0.50	0.39	639	0.21	0.15	0.06	199	0.88	0.39	0.06	0.60
4.0	0.45	0.21	598	0.27	0.45	740	0.10	0.21	0.11	307	0.77	0.45	0.11	0.40

Fig. IX.241. Fire damage scenario of the proposed reinforced concrete beam at 600°C at 2.5 h.

At depth	T (°)	ϕ_{CT}	ϕ_{ST}	Temperature at centroid of steel
d_c	<300	1.0	1.0	$= 670°C \rightarrow \phi_{ST} = 0.32$
T	670	0.44	0.32	

$$F_t = \phi_{st} f_y A_s = 0.32 \times 460 \times 314 = 46\,220\,\text{N}$$

$$F_C = \frac{0.67\phi_{CT} f_{cu} d_c \times T}{\gamma_{mc}} \qquad F_t = F_c$$

$$d_c = \frac{46\,220}{0.67 \times \dfrac{36}{1.3} \times 1000} = 3.00\,\text{mm}$$

$$Z = 0.94 \times 100$$

$$M_u = F_t Z = \frac{46\,220 \times 95}{10^6} = 4.39\,\text{kNm}$$

Reducing moment = 9.00 − 4.39 = 4.61 kNm

Strength reduction % age = 41

This reduction is taken in the finite element analysis. Alternatively,

$\phi_{c,T}$ = strength reduction factor in concrete

$= 1.623 - 0.00178T$

$= 1.023 - 0.00178 \times 670 = 0.43$ (identical to above)

$\phi_{s,0}$ = strength reduction factor in steel reinforcement

$= 1.57 - 0.0019T_s$

$= 1.57 - 0.0019 \times 670$

$= 0.297$ or 29.7%

This stress reduction factor is taken into the finite element analysis.

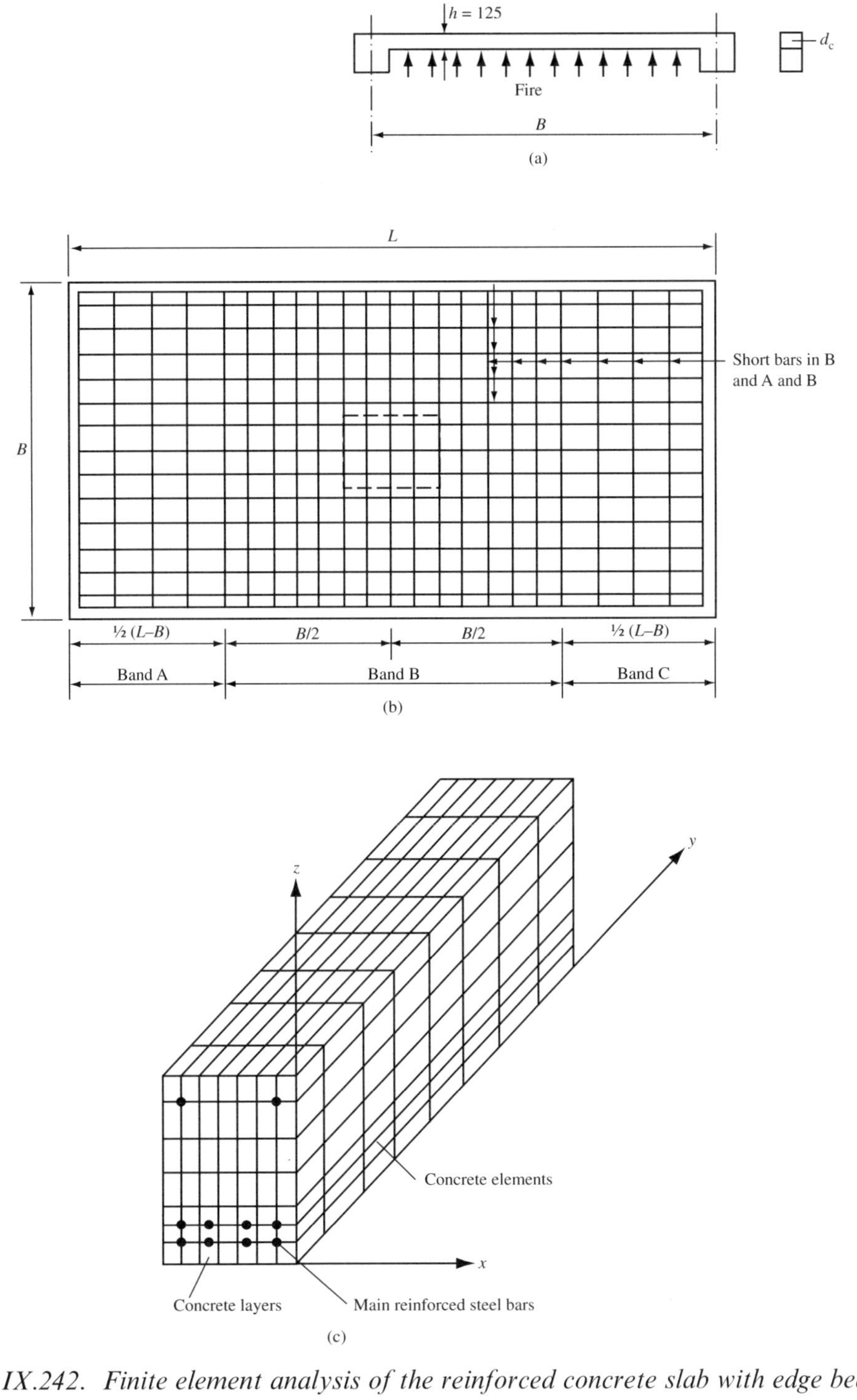

Fig. IX.242. Finite element analysis of the reinforced concrete slab with edge beams: (a) reinforced concrete slab with edge beams; (b) finite element mesh coinciding with reinforcement layout in plan; (c) edge beam finite element mesh scheme.

IX.15.4 Loading criteria and material properties for FE analysis

A uniformly distributed load given below is adopted.

Material properties:

finishes $0.75\,kN/m^2$

imposed load $3.5\,kN/m^2$

load intensity $23.6\,kN/m^2$

fire load as defined in Section IX.13.3

slab width varies from 1500 mm to 3000 mm

IX.15.4.1 Analysis of results

A three-dimensional finite element analysis has been carried out using the ISOPAR program. Concrete was represented by solid 20-noded isoparametric elements throughout in the slab and in the supporting beams. The reinforcement is represented by 3-noded elements placed at nodal points. The slab has three layers and the supporting beams also have three layers of isoparametric elements. The bars are in line with nodal points. Each element has 27 Gaussian integration points. The subroutine LINE is invoked to analyse the slab by assuming three-dimensional line elements acting in conjunction with the solid elements. Elasto-plastic and cracking analyses were carried out on a time increment and load increment basis. Solution procedures and convergence procedures described in the text have been adopted. A total number of 21 iterations were considered for the solution procedure for a period of two-hour fire loads. Figures IX.243 and IX.244 show the deflection–time history of constant thickness length and variable width with certain areas damaged in the final stages. Figure IX.245 shows the final post-mortem of the slab under consideration. Sufficient cracks are produced which are defined by black or dark areas. The bars yielded at the centre of the slab.

IX.15.5 Fire protection and numerical analysis of a composite beam

IX.15.5.1 Introduction

The development of an appropriate numerical model to predict the response of steel–concrete composite sections to fire has been developed using the analyses and solution procedures given in this text. It is essential to provide adequate fire protection for a composite beam. This analysis is based on BS 5950 and Eurocode ENV 1994-1-2 practices. The insulation thickness is obtained initially prior to finite element modelling. The three-dimensional analytical model for the beam connected to an RC slab is developed using the finite element technique. The main components are represented by the three-dimensional isoparametric solid with eight nodal points, having three degrees of freedom at each node on the x, y and z directions. The steel reinforcement mesh inside the concrete slab is represented by a three-dimensional bar element (in tension or compression) with three degrees of

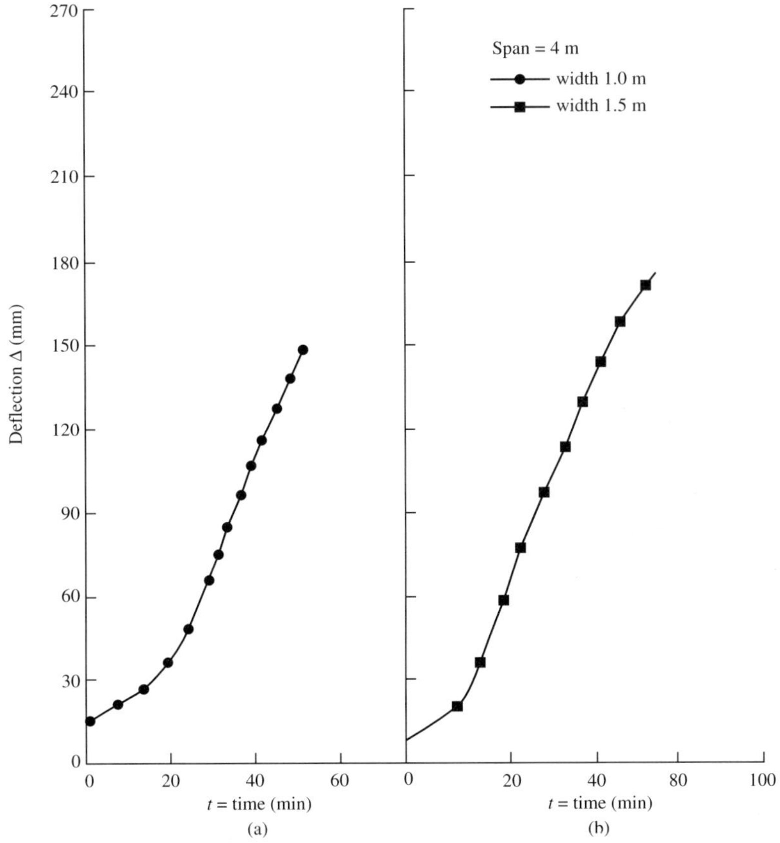

Fig. IX.243. Deflection–time history for 4 m span slab with two different widths of (a) 1 m and (b) 1.5 m.

freedom at each node, without involving bending of these elements. Owing to symmetry, only one-quarter of the beam under investigation need be considered. The temperature distribution in the section is carried out using the time-step integration and having small time steps of $\Delta T = 5$ minutes. Various stress–strain–temperature and deflection graphs are drawn for times of 30, 60 and 90 minutes. These results are discussed later in this section. The final post-mortem of the section is established and those results are also discussed later in this section.

IX.15.5.2 The use of codes

Initially, two codes, BS 5950 and ENV 1994-1-2, are used for the steel beam section and BS 8110 for concrete slabs with or without a steel section embedded in RC work. Figure IX.246 shows a steel–concrete composite section without steel section embedded into concrete and Fig. IX.247 shows the same cross-section without embedment. The reinforcement details in the slab and the embedded steel section are given in Fig. IX.248.

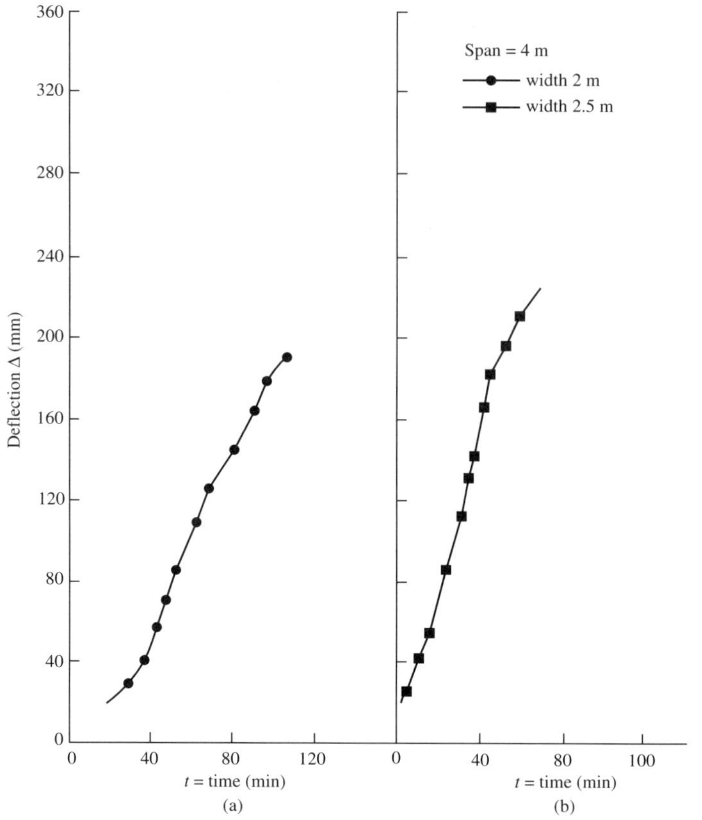

Fig. IX.244. Deflection–time history for a 4 m span slab with two different widths of (a) 2 m and (b) 2.5 m.

Fig. IX.245. The final post-mortem of the reinforced concrete slab under fire at 600°C (2.5 h duration).

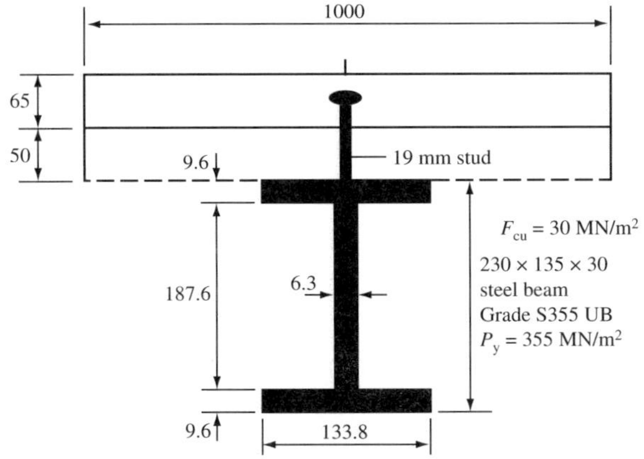

Fig. IX.246. Steel–concrete composite beam.

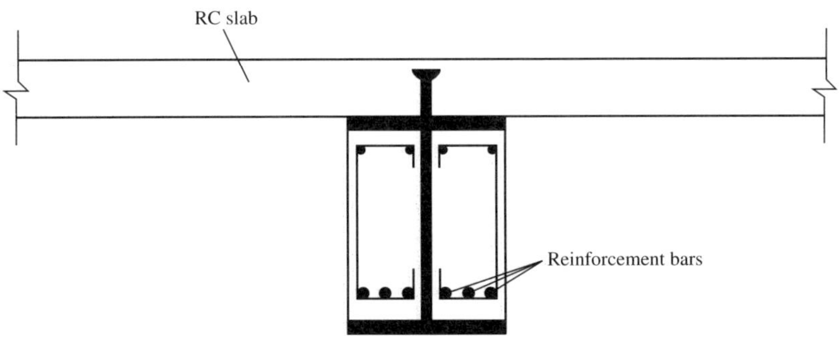

Fig. IX.247. Encased steel beam in concrete with reinforced concrete slab.

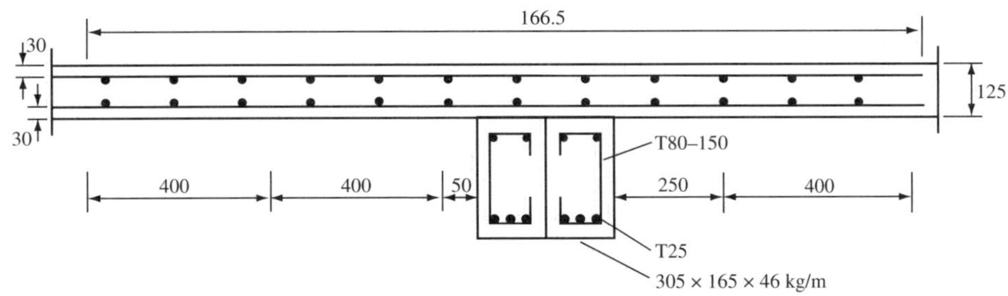

Fig. IX.248. Reinforcement layout for composite section of Fig. IX.247 (embedded steel not shown).

IX.15.5.3 Section without embedment (Fig. IX.216)

Steel section analysis based on BS 5950

1. Plastic moment of resistance, M_p;
   limit of elastic moment of resistance, M_e;
   elastic moment, M (1)

2. Plastic and compact section $1.0d/t < 79\varepsilon$ (2)
3. Shear capacity $P_V = 0.6p_y A_v$
   where p_y = yield stress in shear;
   A_V = area (3)
4. Buckling resistance moment $M_b = S_{xx} p_b$
   where S_{xx} = plastic modulus;
   p_b = bending strength (4)
5. $\bar{M} \le M_b$
   where $\bar{M}$ = equivalent uniform moment on any portion of a member between adjacent lateral restraints as mM_A;
   m = equivalent uniform factor, generally 1 or less than 1;
   M_A = maximum moment or a portion of a member (5)
6. The factor $\beta = \dfrac{\text{smaller end moment}}{\text{large end moment}}$ such that $m = 0.57 + 0.33\beta + 0.1\beta^2$ (6)
7. Equivalent slenderness ratio $\lambda_{LT} = nuv\lambda$
   where n = the slenderness correction factor;
   u = buckling parameter;
   v = uniform factor for flanged members with at least one axis of symmetry (7)
8. Web buckling and web bearing
   $P_w = (b_1 + n_1)tp_C$—web buckling
   $P_{crip} = (b_1 + n_2)tp_{yw}$—web bearing
   where b_1 = stiff bearing length; t = web thickness;
   n_1 = length obtained by dispersion at 45° through half the depth of flange;
   n_2 = length obtained by dispersion through the flange to top of the root fillet sloped 1:2.5 to the plane of the flange;
   p_c, p_{yw} = compressive strength and design web strength respectively (8)
9. Deflection: $\delta_{actual} = \dfrac{\text{span}}{360}$ (9)
10. Shear value $A_\nu = \dfrac{D}{D+B}$ (10)
11. $P_V > F_V$ (11)

Steel section analysis using ENV 1994-1-2 method

1. Shear resistance, $V_{Pl.Rd}$;
   moment resistance, $M_{c.Rd}$;
   buckling resistance, $R_{b.Rd}$ (1)
2. Permanent load $\gamma_G = 1.35$;
   variable load $\gamma_G = 1.5$ (2)
3. Compute maximum moment (design) M_{Sd} and coincident shear V_{Sd} (design)
4. Section classification

5. Flange subject to compression c/t_f ratio; c/t_w ratio (3)

6. $V_{Pl.Rd}$ = design plastic resistance $= A_v(f_y/\sqrt{3})/g_{MD}$ (4)

7. $M_{c.Rd} = W_{pl}f_y/\gamma_{MD}$ with coincident shear force $\ll 0.5V_{Pl.Rd}$ (5)

8. Shear buckling requirement: d/t_w ratio to check (6)

9. Web crushing (i.e. bearing)
   Design crushing resistance $R_{y.Rd} = [(S_s + S_y)t_w f_{yw}]/\gamma_{Ml}$ (7)
   where S_s = length of the stiff bearing;
   t_f = flange thickness;
   f_{yf} = yield stress in flange;
   b_f = flange breadth $b \leq 25t_f$;
   t_w = web thickness;
   f_{yw} = yield stress in web;
   $\sigma_{f.Ed}$ = longitudinal stress in flange;
   $\gamma_{Mo} = 1.05$; $\gamma_{Ml} = 1.05$;
   S_y (for end bearing point loads)
   $$= Xt_f(b_f/t_w)^{0.5}[f_{yf}/f_{yw}]^{0.5}[1 - (\gamma_{Mo}\sigma_{f.Ed}/f_{yf})^2]^{0.5} \quad (8)$$
   where $X = 1$
   S_y (bearing with the length or within span) (9)
   Equation (8) where $X = 2$

10. Design web buckling
    $R_{b.Rd} = (\beta_A f_C A)/\gamma_{Ml}$ (10)
    $A = b_{eff} x t_w$
    f_c = compressive strength (11)
    $$b_{eff} = \text{effective breadth } [h^2 + S_s^2]^{0.5} \not> \left(\frac{h}{2} + a\right) \quad (12)$$
    a = from the centre line to end—a dimension in m

11. Deflection
    $$\delta_{max} = \frac{0.104 \times B\max L^2}{EI} \quad (13)$$

12. Lateral torsional buckling
    $$\lambda_{LT} = \frac{[K/C_1]^{0.5} L/i_{LT}}{\left[1 + \frac{(L/a_{LT})^2}{25.66}\right]^{0.25}} \quad (14)$$
    C_1 = shape factor
    K = effective length factor
    i_{LT} = 35 generally
    a_{LT}/i_{LT} = slenderness ratio λ values

Numerical modelling of a composite beam and analysis of results. As stated earlier, composite beams consist of steel sections that are connected to reinforced concrete slabs or composite sections with profiled steel sheets connected to concrete slabs by anchors embedded in concrete. The anchors in this case are welded to flanges of the steel profiles. The analytical method using the finite element technique offers a better optimisation of this composite section subjected to fire. Figures IX.246 and IX.247 show two such cases as described earlier. The steel beam and slab are initially connected by studs and the steel beam is initially not embedded. The second case is of beams consisting of steel cross-section profiles concreted between the flanges and reinforced with 2T25 with stirrups T8-150 c/c along the beam span. The beam is connected to a 125 mm thick RC slab by shear connectors: T19 × 100 mm at 300 mm spacings. As indicated, the total effective span is 4.0 m and the steel section is shown in Fig. IX.248.

The slab reinforcement is shown in Fig. IX.248. The finite element mesh scheme is shown in Fig. IX.249. Programs ISOPAR and ANSYS are used throughout for the analysis of this beam. ANSYS is used to validate ISOPAR for this type of analysis and for double checking.

IX.15.5.4 Analysis of results

The composite section is statically loaded at six points, as shown in Fig. IX.249, by six hot positions under the ambient temperature of 20°C. Stresses and strains under

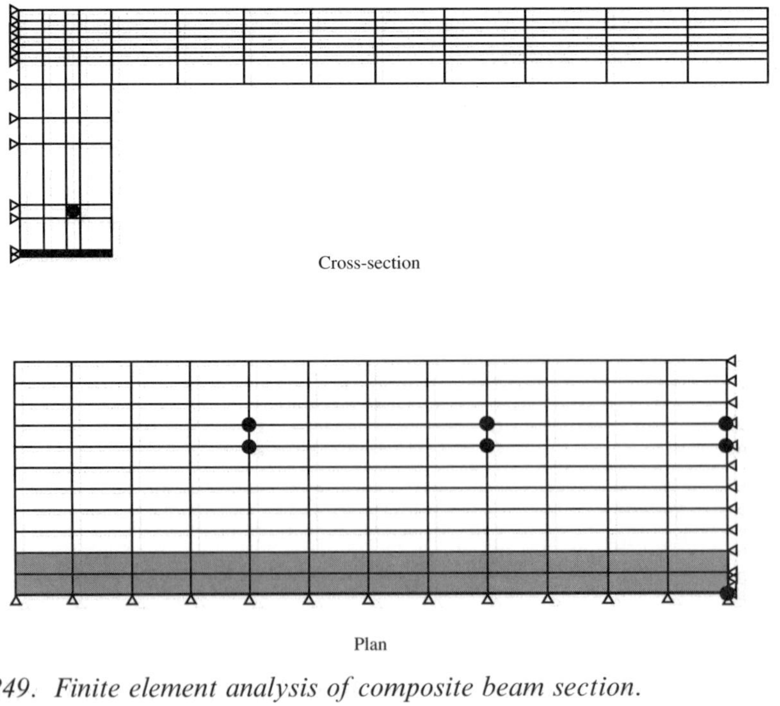

Fig. IX.249. Finite element analysis of composite beam section.

Table IX.26. Temperature quantities versus time t

t (min)	T_f (°C)	$\alpha = \alpha_c + \alpha_r$ (W/m^2) (°C)	ΔT_s (°C)	$T_f - T_s$ (°C)	T_s (°C)
0.00					20
5.00	113.20	29.92	16.80	76.40	36.80
10.00	163.54	31.17	23.55	126.74	60.35
15.00	224.71	33.60	32.92	164.36	93.27
25.00	312.69	38.00	49.75	219.48	143.02
35.00	402.61	44.83	69.36	259.59	212.38
55.00	529.73	57.70	109.14	317.36	321.52
75.00	680.46	81.38	174.10	358.94	495.62
100.00	894.21	134.08	318.52	398.59	814.14

Steel section 230 × 135 × 30 grade S355UB
Connectors: 19 mm studs
Data: Insulation

$$d_i = \text{thickness} = 0.025\,\text{m}$$

$$k_1 = 0.15\,W/\text{m}\,(^\circ\text{C})$$

$$c_i = 1200\,\text{J/kg}\,(^\circ\text{C})$$

$$\Delta t \ngtr \frac{2500}{D \text{ or } P_s/A_z}$$

$$T = \text{ambient temperature } 20^\circ\text{C at zero time}$$

$$\Delta T_s = \frac{P_i}{A_s}\frac{T_f - T_s}{C_s P_s} \times \Delta t(60)\frac{C_s P_s}{C_s P_s + C_1 \dfrac{P_i d_i P_i}{2A_s}} \times \frac{k_1}{d_i} = 10^\circ\text{C}$$

$$C_s = 250\,\text{J/kg}\,(^\circ\text{C})$$

$$P_s = 7850\,\text{kg/m}^3$$

$$T_s = 20 + 10 = 30^\circ\text{C}$$

elastic conditions have been computed. Thermal loading is then raised on the lines suggested in ISO 843 (time–temperature curve) given in Fig. IX.238. Temperature is raised from 0°C to 1200°C in increments of 200°C for a time interval of between 30 min. and a maximum of 180 min. Throughout, the slab is treated for serviceability conditions of $1.4G_k + 1.6Q_k$, where $Q_k = 5.4\,\text{kN/m}^2$. Nonlinear finite element analysis with a time-step integration has been carried out for the calculation of temperature distribution. As stated earlier, the time step was taken as $D_t = 5$ min.

Since these values are temperature dependent, the following values of the above quantities taken for various temperatures are tabulated in Table IX.26.

Figure IX.250 shows the temperature field of a composite structure exposed to fire after 30, 60 and 90 min. The thermal strain–temperature and temperature-dependent stress–strain relations given in Chapter I are considered for the embedded case. A

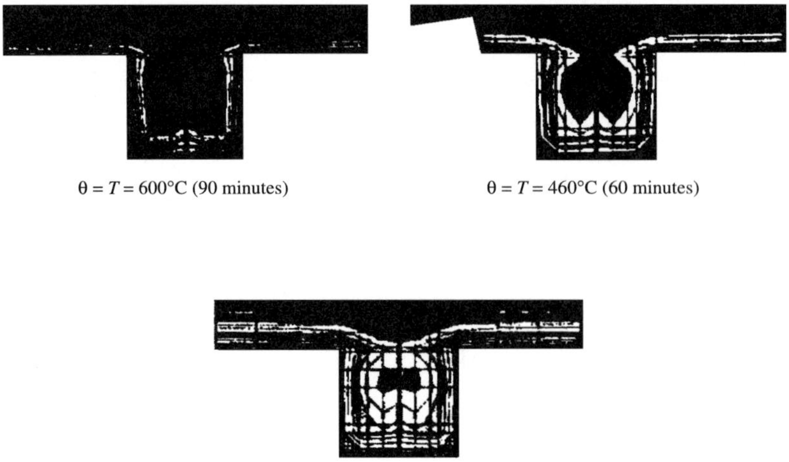

θ = *T* = 600°C (90 minutes)   θ = *T* = 460°C (60 minutes)

θ = *T* = 260°C (30 minutes)

Fig. IX.250. Temperature field of a composite structure—damage scenario.

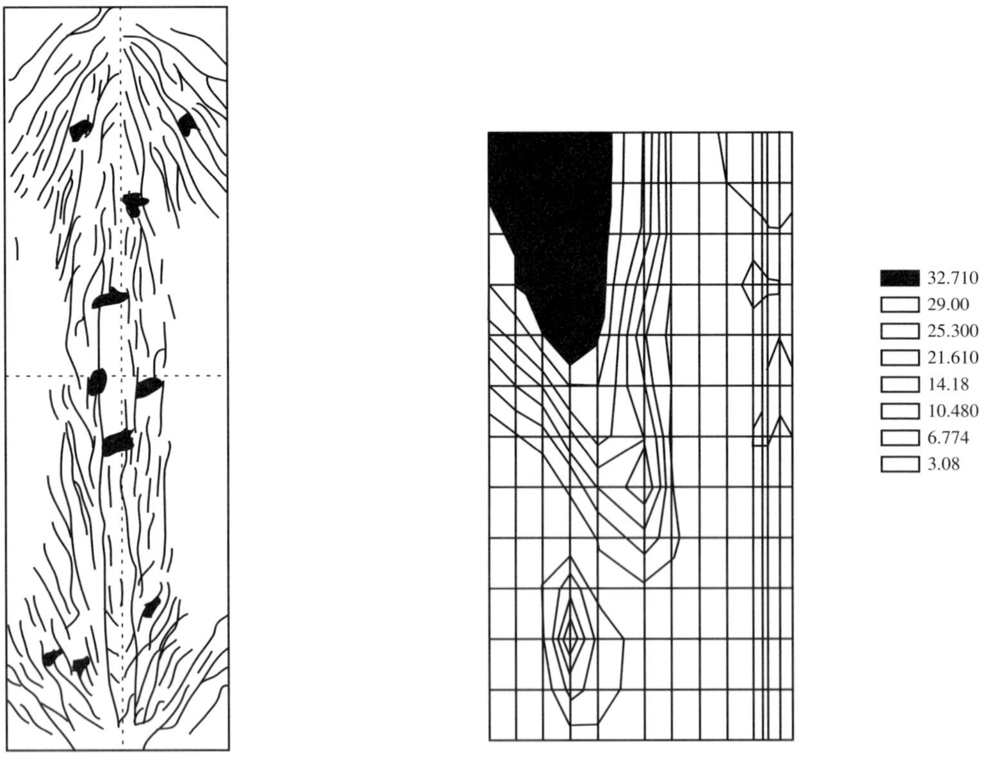

(a) Cracked concrete   (b) Stress distribution

Fig. IX.251. Cracks and stresses.

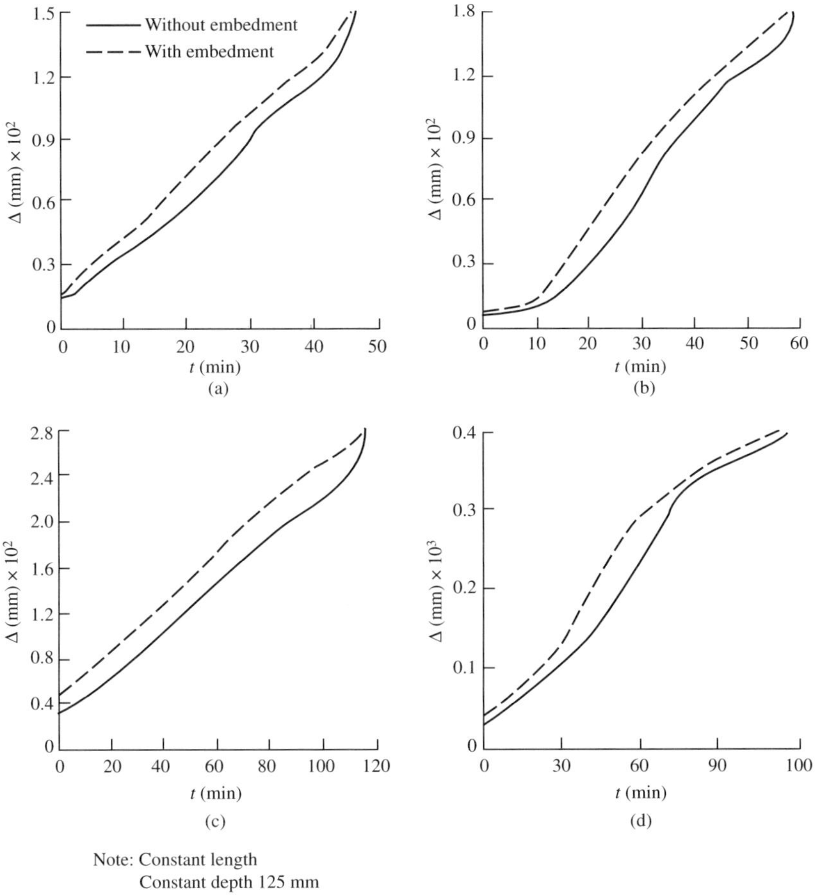

Fig. IX.252. A comparative study of time-dependent deflections with and without embedment.

comparison is given for the non-embedded case for the same period in Fig. IX.251. The corresponding stress distributions together with cracked zones are shown for the embedded case. Figure IX.252 shows the comparative study deflections. For a typical composite steel–concrete section (Fig. IX.253(a)), the load–deflection curve is shown in Fig. IX.253(b). Figure IX.254 shows the damage scenario of the beam.

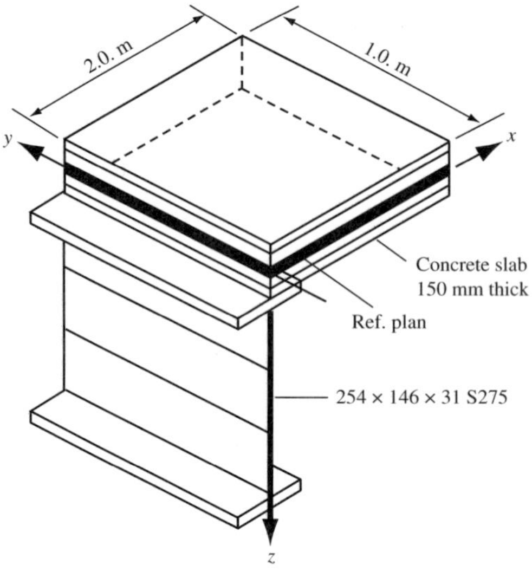

(a)

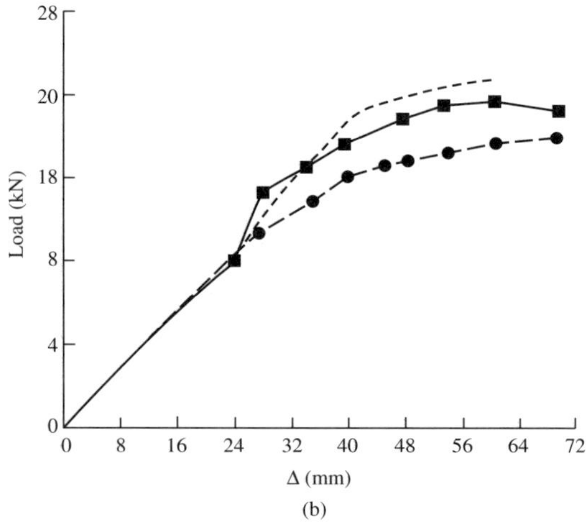

(b)

Fig. IX.253. The load–deflection curve.

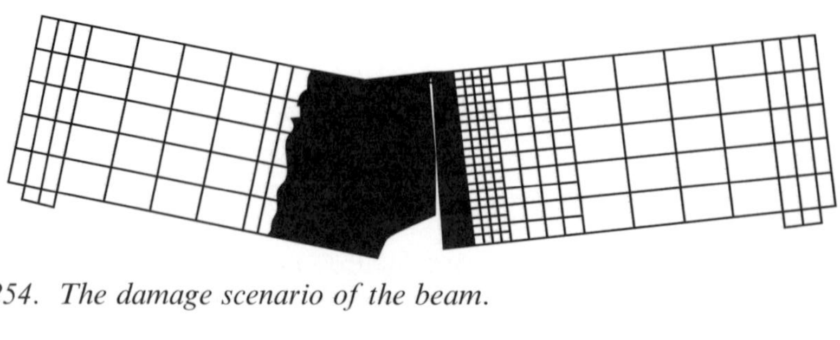

Fig. IX.254. The damage scenario of the beam.

IX.16 Case Study No. 15 — concrete structures–soil interaction

IX.16.1 Introduction

For detailed analysis and background criteria, a reference is made to Chapter VIII of this book and to Section A3 'Buildings and earthquakes' in *Prototype building structures* [1602].

Fundamental to the design provisions presented in Chapter VIII is the assumption that the motion experienced by the base of a concrete structure during an earthquake is the same as the free-field ground motion, a term that refers to the motion that would occur at the level of the foundation if no concrete structure was present. This assumption is true only for structures supported on essentially rigid ground. For structures supported on soft soil, the foundation motion generally is different from the free-ground motion and may include an important rocking component in addition to a lateral or translational component. The rocking component may be particularly significant for tall concrete structures.

A flexibly supported concrete structure also differs from a rigidly supported concrete structure in that a substantial part of its vibrational energy may be dissipated into the supporting medium by radiation of waves and by hysteretic action in the soil. The importance of the latter increases with increasing intensity of ground-shaking. There is, of course, no counterpart of this effect of energy dissipation in a rigidly supported concrete structure.

The effect of soil–structure interaction detailed in the literature [1340–1602] represents the difference in the response of the structure computed by assuming the motion of the foundation to be the same as the free-field ground motion and considering the modified or actual motion of the foundation. This difference depends on the characteristics of the free-field ground motion as well as on the properties of the concrete structure and the supporting medium [1421–1452].

The interaction effects accounted for should not be confused with 'site effects', which refer to the fact that the characteristics of the free-field ground motion induced by a dynamic event at a given site are functions of the properties and layout of the subsurface soil and rock. The interaction effects, on the other hand, refer to the fact that the dynamic response of a concrete structure built on that site depends, in addition, on the interrelationship of the structural characteristics and the properties of the local underlying soil deposits. The site effects are reflected in the values of the seismic coefficients employed.

Two different approaches may be used to assess the effects of soil–structure interaction. The first involves modifying the stipulated free-field design ground motion and evaluating the response of the given structure to the modified motion of the foundation, whereas the second involves modifying the dynamic properties of the structure and evaluating the response of the modified structure to the prescribed free-field ground motion. When properly implemented, both approaches lead to equivalent results.

In the same study of various concrete structures interacting with soils, the following cases have been chosen using the finite element technique for specific structures:

(a) shallow foundations
(b) deep foundations including pole supporting structures
(c) retaining walls
(d) buildings.

Wherever possible, the results obtained from the finite element method are compared with those obtained from the experimental and/or site monitoring data.

IX.16.2 Individual case studies for the concrete–soil interaction analysis

IX.16.2.1 General statement

Individual case studies of concrete structures–soil interactions have been examined. These have been properly analysed and designed under normal load conditions using specific codes. Concrete and reinforcement parameters and quantities are determined prior to analysing individual cases for concrete structure–soil interaction in the seismic environment.

IX.16.2.2 Shallow foundations supporting superstructures

RC pad footing design. A reinforced concrete pad footing supporting a concrete column is shown in Fig. IX.255. It is necessary to design the footing using Eurocode 2. The following data are chosen for the analysis:

Geometry: 3 m × 3 m square pad, 0.60 m deep
0.4 m × 0.4 m square column
Loads: permanent axial (load) = 1500 kN including footing 50 kN load
variable load = 400 kN
γ_u = bearing pressure = $200\,\text{kN/m}^2$
Material: $F_{ck} = 30\,\text{N/mm}^2$; $f_{yk} = 460\,\text{kN/mm}^2$

(a) Serviceability limit state
design axial load $= 1.0G_k + 1.0Q_k = 1550\,\text{kN}$
base area $= 1550/200 = 7.75\,\text{m}^2 < 3 \times 3 = 9\,\text{m}^2$ OK
(b) Ultimate limit state
N_{sd} = column axial design load $= 1.35G_k + 1.5Q_K$
earth pressure $= 1950/(3)^2 = 217\,\text{kN/m}^2$

Overall depth = 600 mm

Effective depth, $d = 600 - 50 - 30$

Depth of embedment = 50 mm

Cover = 30 mm

At the column face,

$$\text{maximum shear resistance } V_{RD2} = 0.9u_d\sqrt{f_{ck}}$$

$$= 0.9 \times 1600 \times 520\sqrt{30} \times 10^3$$

$$= 4100\,\text{kN} > N_{sd} = 1950\,\text{kN}$$

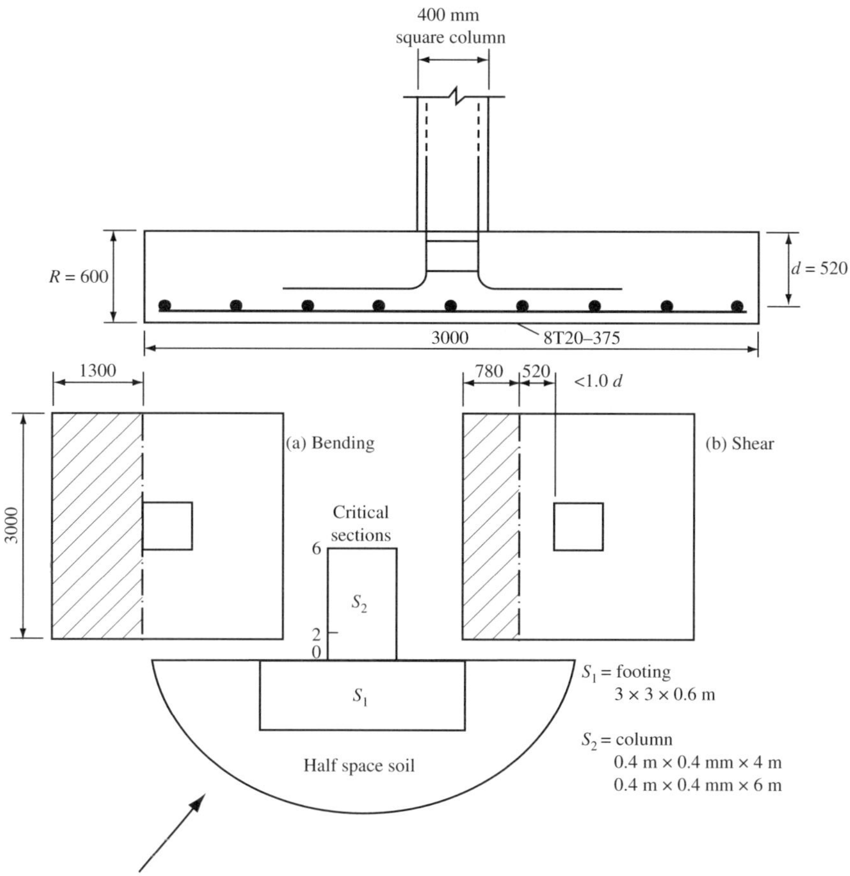

Fig. IX.255. Concrete pad footing soil interaction.

Punching shear:

$$\text{critical perimeter} = \text{column perimeter} = 3\Pi D$$

$$= (400 + 3d)^2 + 3\Pi \times 520 = 650\,\text{mm}$$

$$\text{area within the perimeter} = (400 + 3d)^2 - [4 - \Pi(1.5d)^2]$$

$$= -3.32 \times 10^6\,\text{mm}^2$$

S_1 = factoring $3 \times 3 \times 0.6\,\text{m}$

S_2 = columns: $0.4\,\text{m} \times 0.4\,\text{m} \times 4\,\text{m}$; $0.4\,\text{m} \times 0.4\,\text{m} \times 6\,\text{m}$

V_{sd} = punching shear force $= 217(3^2 - 3.32) = 1233\,\text{kN}$

v_{sd} = punching shear stress $= V_{sd}/\text{perimeter} \times d = 0.365\,\text{N/mm}^2$

hence $h = 600\,\text{mm}$ OK

Bending reinforcement

At the column face: $M_{sd} = (217 \times 3 \times 1.3) \times 1.3/2 = 500\,\text{kNm}$

For concrete $M_{bal} = 0.107 f_{ck} bd^2 = 4046\,\text{kNm} > 550\,\text{kNm}$

$$A_s = \frac{M_{sd}}{0.87 f_{ykz}} = \frac{550 \times 10^6}{0.87 \times 460(0.98 \times 520)} = 2.696\,\text{mm}^2$$

8 T20 bars — 100 mm c/c $\quad A_s = 3142\,\text{mm}^2$
$100A_s/bd = 0.20 > 0.15$ $\quad$ Minimum steel requirement is satisfied

Crack control

Minimum steel area $A_s = k_e f_{ct.ef} A_{ct}/f_{yk}$

$$A_{ct} = bh/2$$

$$k_e = 0.4; k = 1.0$$

$$A_s = k_e \times 1.0 \times 3 \times (3000 \times 600/2)/460 = 2348\,\text{mm}^2$$

8 T20 bars — 100 mm c/c $\quad A_s = 314.2\,\text{mm}^2$ OK

Maximum bar size

Steel stress under the action of the quasi-permanent loading F_s will be

$$f_s = \frac{f_{yk}(G_k + 0.3Q_k)}{1.15(1.35G_k + 1.5Q_k)} = 230\,\text{N/mm}^2 \qquad G_k = 1000\,\text{kN} \quad Q_k = 400\,\text{kN}$$

Maximum bar size allowable by the Eurocode is 20 mm.

Check on punching shear

$F_{ck} = 30$; $\tau_{Rd} = 0.34\,\text{N/mm}^2$
V_{Rdl} = shear resistance of concrete $= [\tau_{Rd} k(1.2 + 40\rho)]u_d$
$k = 1.6 - d = 1.08$
$\rho = A_s/bd = 0.00201$
$V_{Rdl} = 1572\,\text{kN} > V_{sd} = 1233\,\text{kN}$

Maximum shear force

At critical sections for shear, 1.0d from the column face
V_{sd} design shear $= 217 \times 3.0 \times 0.78 = 508\,\text{kN}$
$V_{Rdl} = [\tau_{Rd}(1.2 + 40\rho)]b_w d = 679\,\text{kN} > V_{sd} = 508\,kN \qquad b_w = 3000$

Soil–footing interaction. To analyse the soil–structure interaction, it is sufficient to take the input motion as being derived from an observed record at the site, or at least a similar site. Starting from this control motion in one point, the earthquake motion throughout the free field, characterised by its spatial and temporal variation, is calculated. The structure will interact with the soil along the embedment, thus modifying the seismic motion of the free field. Therefore, the seismic motion acting on the structure is not known before the soil–structure interaction analysis is performed. In summary, for seismic excitation, the analysis of soil–structure interaction consists of two distinct parts. First, determining the free-field response of the site. Second, calculating the scatter response when the structure is inserted into the seismic environment of the free field.

When infinite domain problems are encountered, not only the efficiency of this method is called into question, but the accuracy of analysis solution as well. In this case either the infinite domain problems should be truncated by some kind of artificial boundary, a non-reflecting boundary, or special elements should be introduced.

In the analytical model, as shown in Fig. IX.256, the time domain boundary element model, used for the soil medium which is modelled as a homogeneous, isotropic, elastic semi-halfspace, is combined with the finite element model for the superstructure and footing.

A 3 m $\times$ 3 m and 0.6 m square, rigid footing and superstructure is chosen to examine this earthquake response analysis. The material properties of the structure part, included in superstructure and footing, and the halfspace soil which is assumed for this work, are as shown in Table IX.27.

In this study, a parametric study is conducted for the heights of the superstructure as columns. The distribution of contact pressure at the concrete footing–soil interface is treated as a function of the interactive behaviour of the system described above. The longitudinal direction of the footing has been discretised as beam elements and the soil is modelled as a Winkler bed or linear elastic homogeneous halfspace. The solution is obtained by using the displacement compatibility at the footing–soil interface. Cheung and Nag [1439] used this procedure for the analysis of a footing on elastic halfspace. The contact pressure obtained is significantly different from that calculated without interaction.

In this model, the time step of the analysis was taken as $\Delta t = 0.01$ s. The analysis is used to obtain the response of the footing and structure subjected to the harmonic seismic excitation with magnitude A to be 300 gal and 500 gal [1439] in the frequency domain. Although these results are plotted versus frequency domain, which is a common practice for frequency domain techniques, they are obtained through the time domain methodology. The response of the footing and the superstructure due to harmonic excitation (the horizontal and the vertical displacement amplitude at two characteristic points) are the base of footing and the top of superstructure in the non-dimensional frequency range. The time-history displacement is shown in Fig. IX.256 for the Kobe earthquake.

IX.16.3 Deep foundations including pile supported structures

IX.16.3.1 Introduction

In this category, pile foundations are considered. The behaviour of a pole when it is axially loaded can be analysed using FEM. Any complex behaviour of soil medium and the interface friction can be considered. A group of piles with pile caps can also be studied with FEM.

The response of a pole subjected to lateral and vertical loading is very much a function of the interaction of the pile–soil system. The pile can be treated as

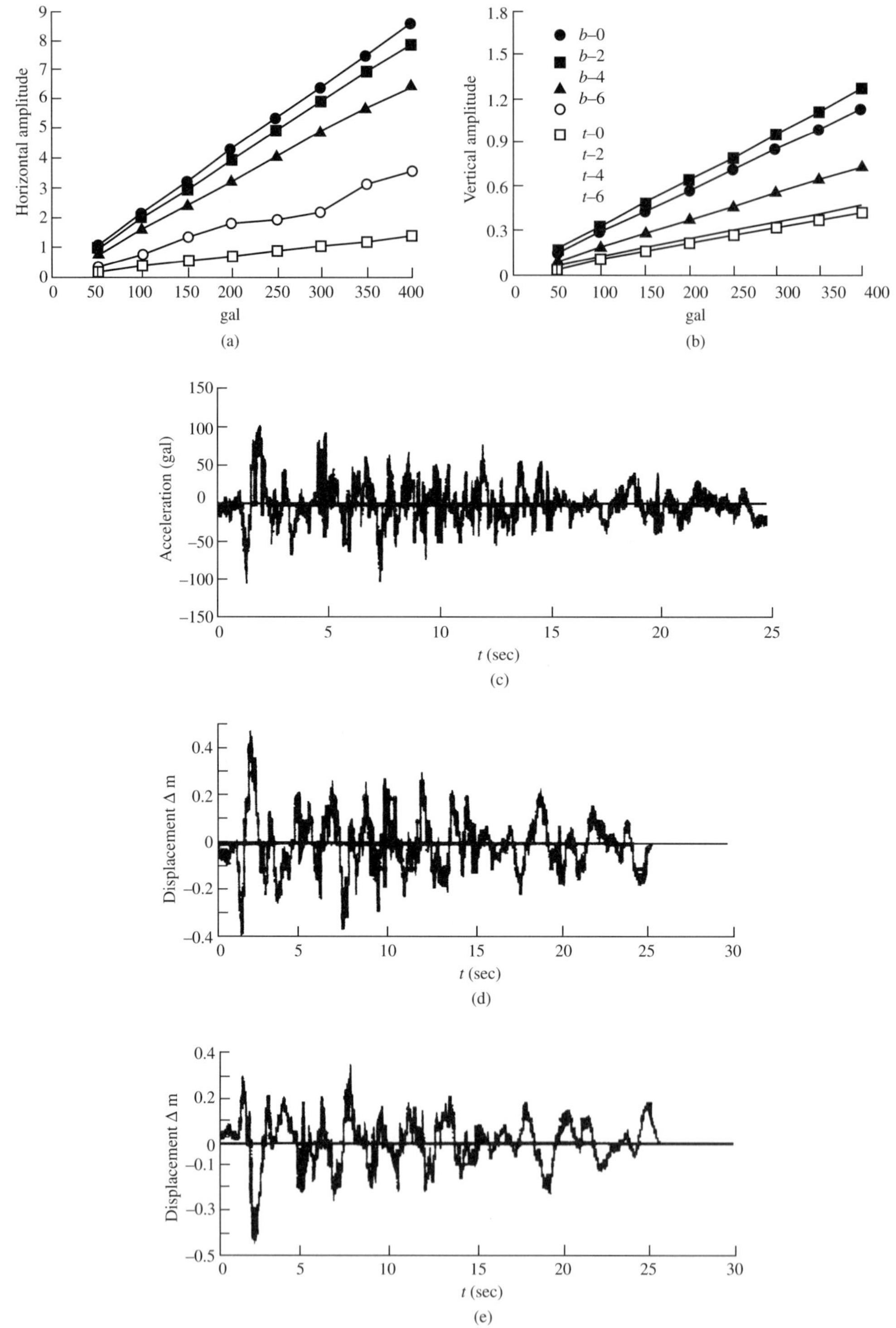

Fig. IX.256. Amplitude, acceleration and displacements.

Table IX.27. Material properties of soil–footing (S1)–superstructure (S2)

	Dilatational wave C_1	Shear wave velocity C_2	Mass density ρ	Poisson's ratio ν	Young's Modulus E	Parameters
Soil	345.033 m/s	140.859 m/s	1.8 t/m^3	0.4	10.5×10^6 kN/m^2	$\phi = 40$; $c = 1$ MN/m^2
S_1, S_2	1013.464 m/s	640.778 m/s	2.4 t/m^3	0.167	21×18 kN/m^2	

$S_1 = 3\,\text{m} \times 3\,\text{m} \times 0.6\,\text{m}$; $S_2 = 0.4\,\text{m} \times 0.4\,\text{m}$ and $0.4\,\text{m} \times 0.4\,\text{m} \times 6\,\text{m}$ etc.

behaving like a beam and the soil can be modelled as the Winkler bed. A more realistic method is to consider the problem as a geometrically symmetric axis with asymmetric loading.

An alternative procedure is to consider the pile section at different depths in the soil medium. The problem is thus transformed into one of plane strain. In this case, the interface behaviour as well as the modified stress condition in the soil due to pile driving can be incorporated. Since the building is assumed to have pile supports, a comprehensive analysis is needed.

The response of a pile-supported ten-storey building to earthquake disturbance has been studied by the finite element method. The soil has been modelled as a nonlinear hysteretic material whose stress–strain law is described by a curved description model. The piles are connected to a rigid mat foundation that supports the building. The rigid bedrock supporting the piles and the soil is excited simultaneously in all three directions to simulate the Kobe earthquake. Results computed and presented graphically include the shear forces and twisting moments induced in different storeys of the building, one component of the horizontal displacement of the rigid mat foundation, roof and the fifth floor. These values of displacement are compared with the corresponding values for the same building resting on a rigid base.

IX.16.3.2 Major data

Figure IX.257 shows the finite element model for a ten-storey frame supported on a rigid foundation block, with soil medium beneath to bedrock. External loads are as shown in Fig. IX.257.

Soil zone −3–0.5 high + top 2 m = 17 m total
Base dimensions 5–7.5 m = 37.5 m square
Weight of foundation block = 1300 kN
Intensity with depth = 2.5 m
Soil modulus of elasticity (E_{soil}) = 70 MN/m^2
Poisson's ratio $\nu = 0.45$
Weight density (γ_s) = 1923 kg/m^3
Damping coefficients $\xi(\alpha) = 0.556$, $\xi(\beta) = 0.018$

Per pile data: pile centres = 3 × diameter
Force F on each pile = 1454 kN with pile reinforcement 8T25 (3930 mm^2)
Pile size 450 mm at 1.35 m; pile cap depth = 750 mm

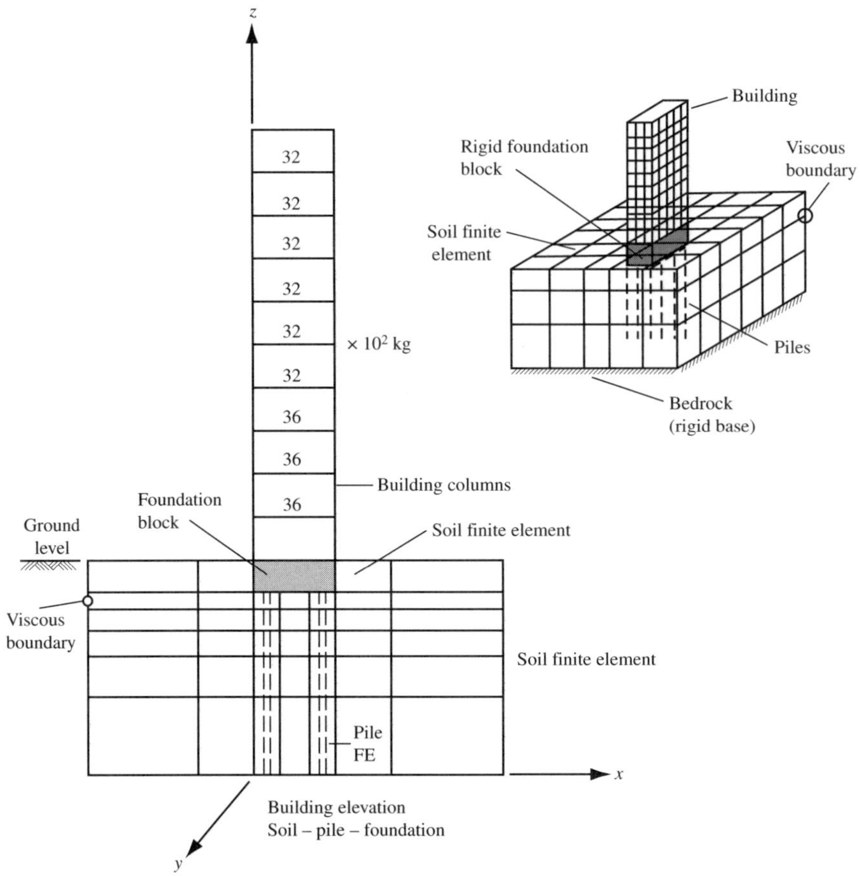

Fig. IX.257. Finite element model for a ten-storey framed building on piles.

Column data:
column size: 500 mm square columns and beams
torsional constant 20
$E_s = 200\,\text{GN/m}^2$; Poisson's ratio $\nu_c = 0.15$
beam sizes from 105 cm × 90 cm

Finite element data:

$$\left.\begin{array}{ll}\text{no elements foundation} & 360\\ \text{no elements piles} & 180\end{array}\right\}\text{4-noded elements}$$

total finite element increments 24

A soil supported structure. After establishing the basis of pile elements, it is advisable to determine the response of pile supported structures to seismic disturbances using three-dimensional finite element method. Major data in IX.16.3.2 shows a ten-storey building in concrete supported by reinforced concrete piles and mat foundation. The multistorey building has a concrete frame of beams and columns with reinforced

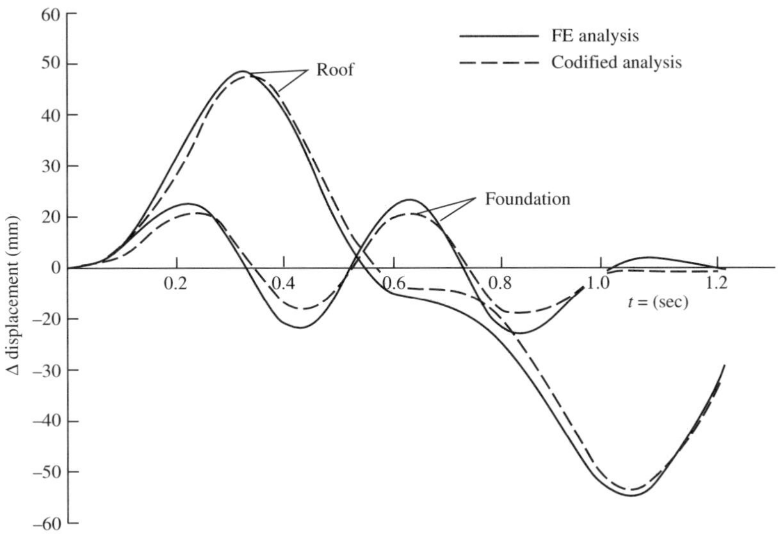

Fig. IX.258. Time–dependent displacements—a comparison between finite element and codified analysis.

concrete floors and a foundation block. The soil has been modelled as a nonlinear hysteretic material. The piles are connected to a rigid mat foundation. The two horizontal components (N–S and E–W) and the vertical component (0.5 × horizontal) of acceleration of the Kobe earthquake are applied at the bedrock. The soil is excited in all three directions to simulate this earthquake. The program ISOPAR checked by ANSYS condenses out the translational degrees of freedom perpendicular to the bedrock. The soil properties, as input to both these programs, are the loading bulk modulus K_L, the loading shear modulus G_L and the unloading bulk modulus k_u for various discrete values of the volumetric strain. A silty clay is considered. The tensile stress limit $f_t = \sigma_L$, tensile bulk modulus K_T and the corresponding shear modulus G_T for these programs have been adapted from the uniaxial strain tests. Owing to the rapidly varying values of acceleration with respect to time, the time step had to be adjusted.

Figures IX.258 and IX.259 show displacement–time histories in the lateral direction and accelerations in the horizontal and vertical directions.

Cracks appearing in the foundation block and pile cap range from 10 mm to 30 mm. Columns were displaced by 150 mm to 280 mm. Pile cap reinforcements were burst.

IX.16.4 Retaining walls

IX.16.4.1 Introduction

Cantilever retaining walls subjected to seismic forces are examined. It is essential first to establish a design permitting appropriate wall and footing thicknesses to be evaluated. It is also necessary to detail reinforcement for wall stem and footing prior to subjecting the retaining wall to seismic actions and to examine retaining

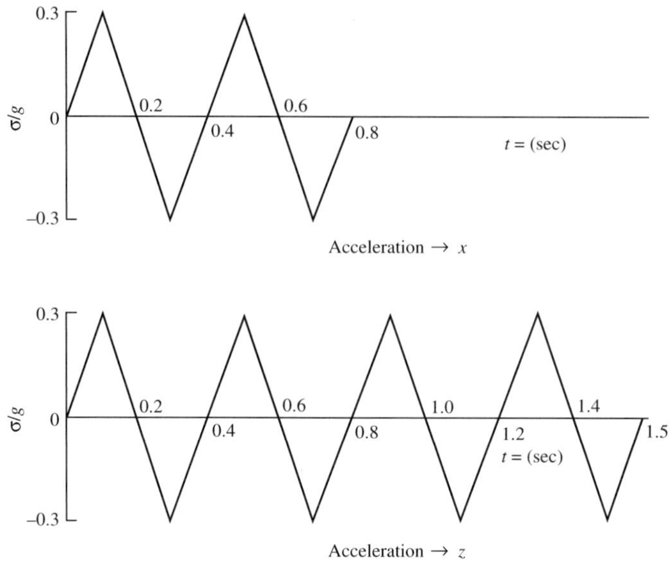

Fig. IX.259. Stresses versus accelerations.

wall–soil interaction. Moreover, a facility must be constructed which will enable a realistic picture to be presented for a real structure.

IX.16.4.2 Design of a retaining wall data input

The cantilever retaining wall shown in Fig. IX.260 supports a granular material of saturated density 2000 kg/m^3 and the allowable bearing pressure is 110 kN/m^2. Use BS 8110, concrete Grade 30 and $f_{yk} = 460\,\text{N/mm}^2$.

$P_H = ka\rho gh$ is the horizontal force $= 31.7\,\text{kN/m}^2$

$ka = 0.33$ coefficient for active pressure

$\rho =$ density of the back-fill $= 2000\,\text{kg/m}^3$

$h =$ depth under consideration

$P_T =$ total vertical load $= 264.6\,\text{kN}$ (walls, base, earth)

$\mu =$ coefficient of friction 0.33 to 0.40

$F_f =$ friction force $= 105.8\,\text{kN} = \gamma_f H_k$

$F_{sl} =$ sliding force $= \mu(1.0G_k + 1.0V_k) \geq \gamma_f H_k = 116.6\,\text{kN}$

$H_k = 0.5\rho h = 77.7\,\text{kN}$

$H_p =$ passive earth pressure $= \gamma_f(0.5kp\rho g a^2) = 14.4\,\text{kN}$

$P =$ bearing pressure static $= 110\,\text{kN/m}^2$

wall $A_s =$ T20–250 main steel base slab included

$A_{sd} =$ T12–200 distribution steel base slab bottom

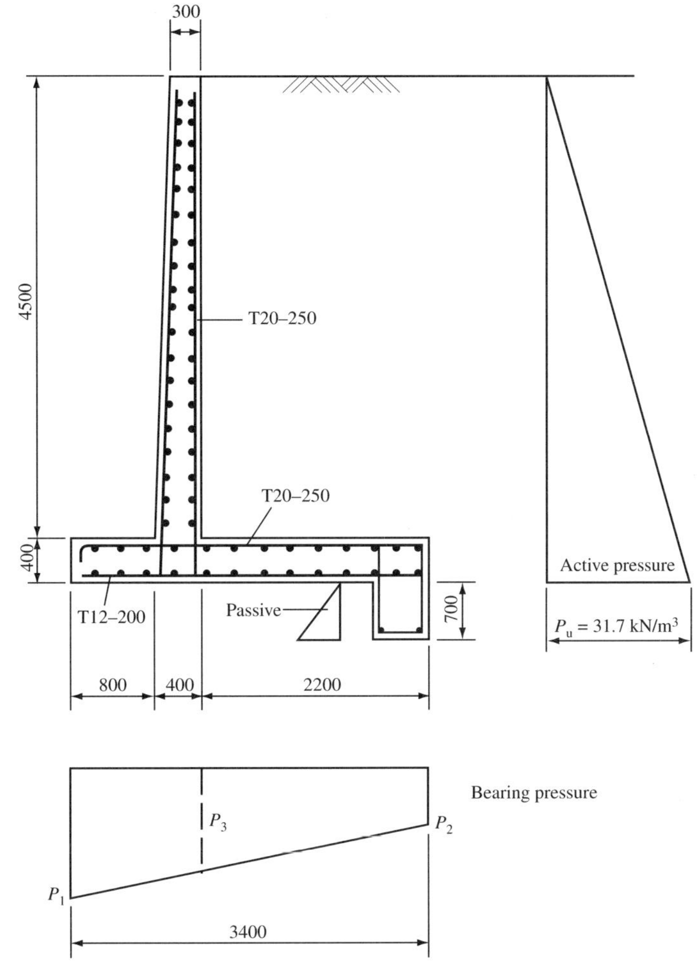

Fig. IX.260. Reinforced concrete retaining wall.

Finite element mesh scheme (see Fig. IX.261) is adopted for this study. Additional data can be seen in Table IX.28.

RC retaining walls–soil interaction. Retaining structures are of two types: rigid and flexible. One of the important forces to be considered in the design of retaining structures is due to the soil pressure. The distribution of soil pressure is influenced by the flexibility of the retaining structure and nature of its movement.

The effects of the deformation parameters of the back-fill and foundation subsoil, construction sequence, wall flexibility and other complex initial conditions have not so far taken into consideration. In this case, the finite element method is employed to investigate the behaviour of concrete cantilever retaining walls. The analyses reveal that the wall and soil movements are of considerable importance in predicting the lateral earth pressure.

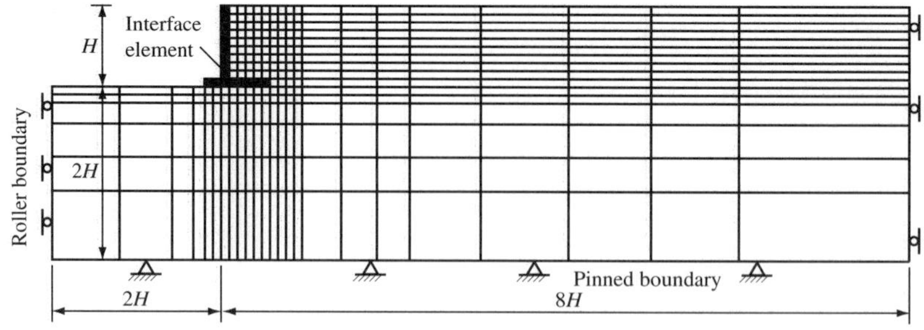

Fig. IX.261. Finite element scheme for retaining wall—soil interaction.

Table IX.28 shows the properties employed in the dynamic finite element analysis, with the program ISOPAR essentially following the 'initial stress' solution procedure for the finite element formulations. Eight-noded isoparametric elements with reduced integration were used to model the soil and wall elements. The soil was modelled as linear elastic perfectly plastic with a shear bulk moduli model for the

Table IX.28. Properties employed in finite element analyses

Property (1)	Symbol (2)	Value (3)
(a) Wall properties		
Wall-base width (m)	B	3.0
Wall-base thickness (m)	t	0.2
Wall-stem thickness (m)	s	0.4
Wall-heel thickness (m)	h	0.4
Wall height (m)	H	4.5
Wall-stem height (m)	D	4.1
Young's ratio	E_c	25.0
Poisson's ratio	ν_c	0.2
Total unit weight (kN/m^2)	γ_c	22.0
(b) Back-fill soil		
Elastic modulus (MN/m^2)	E_b	40
Poisson's ratio	ν_b	0.33
Friction angle	ϕ_b	30°
Cohesion (kN/m^2)	c	0
Total unit weight (kN/m^3)	γ_b	15.7
Coefficient of earth pressure at rest	K_0	0.5
(c) Foundation subsoil		
Elastic modulus (MN/m^2)	E_f	70
Poisson's ratio	ν_f	0.3
Friction angle	ϕ_f	35°
Cohesion (kN/m^2)	c	0
Total unit weight (kN/m^3)	γ_f	18.0
Coefficient of earth pressure at rest	K_0	0.426
(d) Wall base interface		
Initial shear stiffness (MN/m^2/m)	K_s	490
Failure ratio	R_f	0.9
Friction angle	δ_f	30°

interaction between the soil and the retaining wall. The wall thickness varies from 0.2 to 0.4 m. The wall height is 4.5 m with maximum influence boundary as 36 m on one side and 9.0 m on the other side. The interface stiffness is 1×10^6 MN/m. The interface shear displacement is based on

$$K_{st} = K_s(1 - n_f\sigma_L)$$

where

n_f = failure ratio (50 MN/m interval)

K_s = initial shear stiffness = 500 MN/m

σ_L = interface stress level

The smallest value at the initial stage taken for shear value $K_s = 0.001$ MN/m. The Kobe (Kinshan) earthquake data was employed. Fig. IX.262 shows the

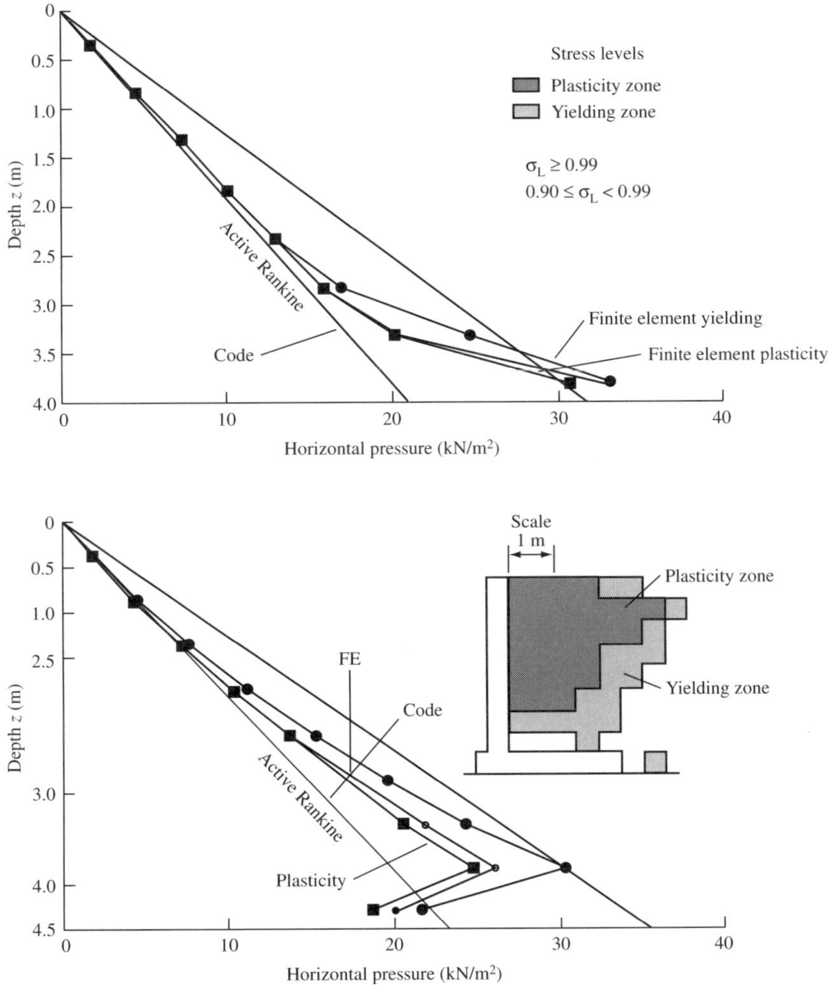

Fig. IX.262. Depth versus horizontal pressure, zero earthquake to Kobe's level.

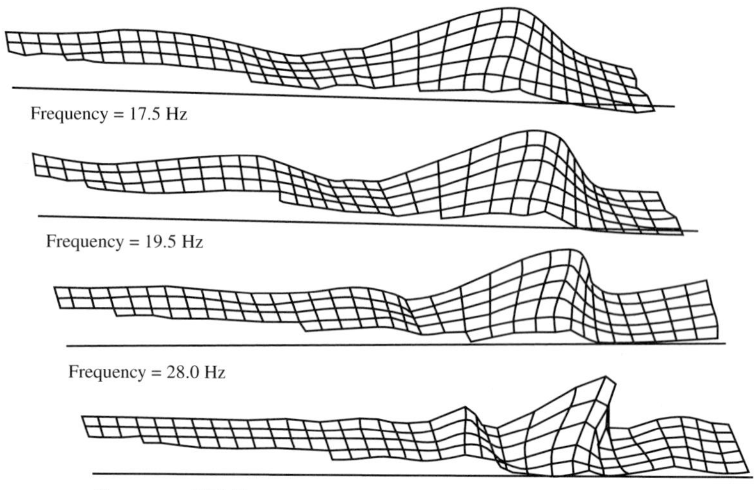

Fig. IX.263. Retaining wall mode shapes and frequencies.

relationship between the wall depth z with horizontal pressure varying from zero earthquake to the level of the Kobe earthquake. The stress level $\sigma_L$was greater than 1.0. The maximum displacement at 1/3 height from the wall was 40 mm. The predicted displacement in the vertical direction was 1.9 mm. The shift in the eccentricity was 0.20 m. The wall heel and the wall foot settlement due to the Kobe earthquake was 16.7 and 19.8 mm. The wall cracking exceeded 3.0 mm along the 2/3 height from the base. Sufficient cracks exceeding 3.0 mm existed in the base. The deformation of soil below the base showed inward ballooning. The reinforcing bars at the stem show maximum yield/plasticity with 25 mm cracking below and above. Through the analysis, it is assumed that the mass was rough, having some friction between soil and stem which existed until it was broken due to cracking in the overburden or excessive deformation below the footing. Figure IX.263 shows typical mode shapes of the wall with frequencies ranging from 17.5 to 85.0 Hz.

IX.16.5 Buildings under seismic actions

IX.16.5.1 Introduction

High rise buildings have a number of storeys with underground basements and top floors for auxiliary equipment and services. For the purpose of analysis, Bürohaus Trianon situated in Mainzer landstraße in Frankfurt (Germany) is chosen. Figure IX.264 shows a typical floor plan and details. A sectional elevation from one typical view is shown in Fig. IX.265. The nonlinear response spectrum is adopted with vertical and horizontal acceleration effects for the building. The three-dimensional finite element mesh scheme shown in Fig. IX.266 represents the true picture of the building frame in three dimensions without a core structure. Structural details shown for a typical floor in Fig. IX.265 is treated as constant for all floors,

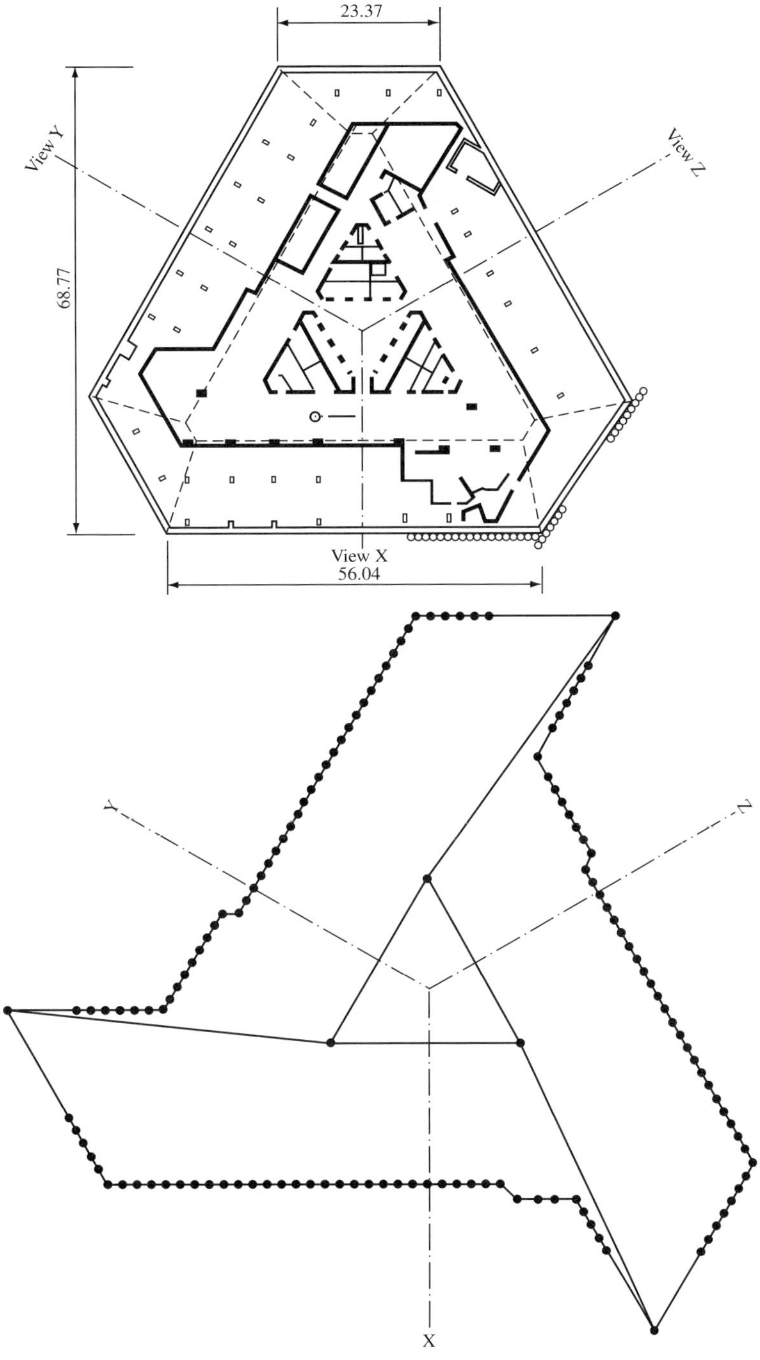

Fig. IX.264. Trianon Building floor plan.

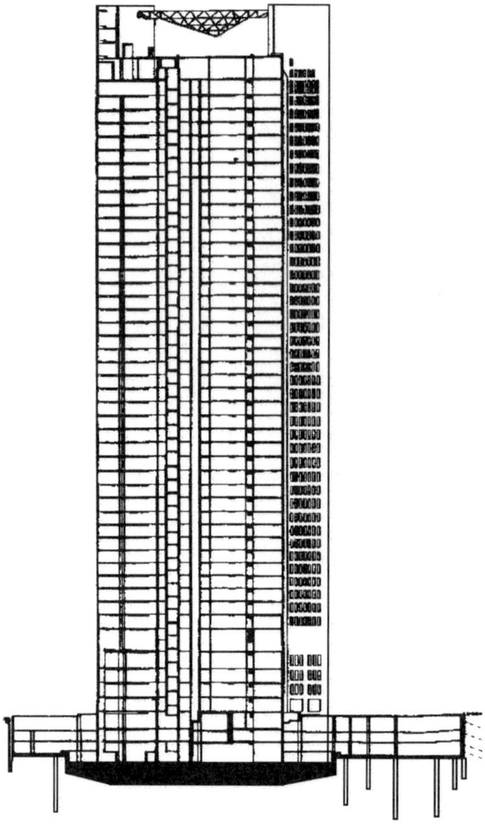

Fig. IX.265. Sectional elevation of the Trianon Building at Frankfurt (55 storeys).

although some variations do exist in the layouts and structural details for certain floors.

The following four linear elastic models are employed for the investigation of the dynamic behaviour of the building under these earthquake conditions.

(a) Model 1: the complete three-dimensional building on a fixed support.
(b) Model 2: the complete three-dimensional building with soil–structure interaction effects.
(c) Model 3: Model 2 plus P–Δ secondary effects.
(d) Model 4: Model 3 plus finite rigid connection zones.

The building is analysed, and the results compared with each other. They are as follows.

(i) Model 1: The structure is defined in complete three-dimensional form, and its foundation is assumed to be fixed.
(ii) Model 2: The soil–structure interaction effects are considered.
(iii) Model 3: The effects of secondary moments P–Δ have been added to Model 2. These moments are produced by the axial forces and the displace-

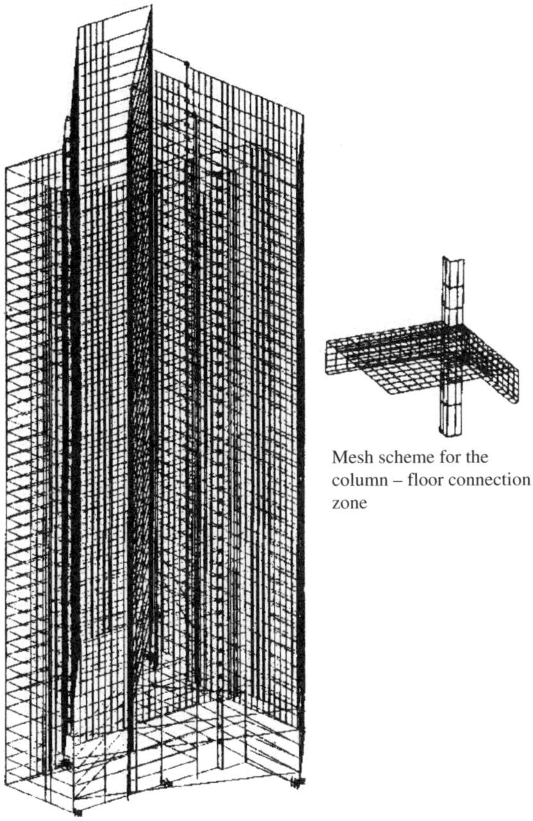

Fig. IX.266. A three-dimensional finite element mesh scheme for the seismic analysis of the Trianon building.

ments perpendicular to the axis of the element. They increase the internal forces and displacements.

(iv) Model 4: The rigidity effects of connection zones have been added to Model 3. This action reduces the effective length of the elements and thus increases their stiffness.

The data for the input of the ANSYS program, including geometry, material specifications, constructional conditions, masses, and degrees of freedom have been precisely and accurately provided.

To calculate the precise values of the masses, the building has been inspected and the structural maps have been used. The details relating to the partitions and movable equipment have also been considered.

To consider the soil–structure interaction effects by ANSYS software, an extra massless basement is assumed such that its vertical elements, like the walls and columns, have the same stiffness as the soil. For these basements, horizontal springs are also assumed in order to consider the horizontal resistance of the soil. Since the stiffness of these springs is too high, only the rocking phenomenon is considered. It is assumed that there is a long distance between the building and the epicentre of the earthquake so it is subjected to the vertical components of *P*, *S* and *R* waves and begins to rock. Simultaneously, the horizontal component of the earthquake is considered, the Kobe earthquake criteria being the main focus of this work. The situation of the building is assumed to have been inspected before the occurrence of the earthquake.

The live load for earthquake forces is 20% of the total live load for official buildings according to Iran's regulations. For the Kobe earthquake, only the live load of stairs and corridors has been considered for computational purposes.

Since the program itself computes the translational and rotational mass moment of inertia from the mean distributed loads, so the mass per unit area of each floor is given as the input. Notice that the dead and live load of the floors are almost the same. The difference between the total mass of them, therefore, is due to the thickness of their shear walls.

Some cracks occur in concrete sections. It is therefore impossible to obtain precise flexural rigidity. On the other hand, the presence of steel bars in concrete sections increases this rigidity. So the flexural rigidity of uncracked concrete sections is an appropriate average value and is most commonly used. Since the whole area of the foundation is covered by the underlying soil, and the area of load-carrying walls is defined as a substitute for the soil, these walls must have a modulus of elasticity proportional to their area. This means that the modulus of elasticity of the soil must be multiplied to the inverse of the ratio of the foundation. This ratio is approximately equal to 1/9. Enormous results have been obtained which are beyond the scope of this book. However, important results are given later in this section for the Trianon Building.

IX.16.5.2 Structural model and method of analysis

Three-dimensional finite element analysis using mixed hybrid elements of 20-node isoparametric and 3-node beam elements are adopted for the concrete/reinforcement respectively. Smeared cracking is simulated. A conventional bar–isoparametric combination is abandoned owing to numerical problems encountered in a preliminary simple frame. However, the interface elements with discrete cracking were added to the concrete cover at the vicinity of the internal reinforcement to enable the simulation of peeling. These elements were necessary to allow the concrete cover to peel away without any restraints due to element locking. As a consequence of this modelling approach, no multidimensional smeared cracks were allowed in the integration points of the three-dimensional elements when joint separation was simulated. This was to avoid crack localisation in the three-dimensional elements connected to the interface elements, which are made up of layered soil, base isolations and viscous dampers. The base isolators are used throughout along the three dimensions of the building. The finite element mesh scheme shown in Fig. IX.266 has been adopted using the Kobe earthquake data given in Appendix XIV. The basement of the building is assumed to have four levels. The building has been examined for the following effects:

(a) characteristics of building vibration and response spectrum of input motions
(b) eccentricity of buildings and torsional vibration
(c) multidirectional input of ground motion
(d) fracture
(e) non-structural walls
(f) base isolation devices.

Earthquake response analysis

(a) *Hyper-dynamics.* A three-dimensional nonlinear frame analysis system developed by Bangash [1337] was used in static and dynamic analyses of buildings.
(b) *Analytical models.* A building is modelled in three dimensions using 6-degrees-of-freedom beam elements. Each member, such as a column, beam, brace and structural wall (including non-structural walls which may be regarded as structural walls) is idealised as an isoparametric element. Lumped masses are assumed to concentrate at nodal points. The base of the analytical model is assumed to be fixed in three translational directions at the ends of the foundation. The basement of the buildings is linked to the assumed Kobe soil strata as shown in Fig. IX.267 using the interface elements here called gap elements, the details of which are given in Appendix XIII.
(c) *Materials.* The basic values of nodules of elasticity, shear modulus, and material strengths for concrete and steel materials were obtained from field tests such as core sampling tests and, if test data were not available, design standard values were adopted.
(d) *Input earthquake motions.* The incident waves on the seismic bedrock at the Trianon building site were estimated using the three components (NS, EW, UD) of the recorded motions at Kobe University. Taking into account the ground conditions, seismic waves at foundation level were converted into X, Y and Z coordinates of a structure.
(e) *Analytical methods for vibration.* The Newmark-β method is employed to perform time–history response analysis. The increment period for the analysis is set at 1/100 seconds, with a duration of 10 to 12 seconds including the principal vibration. Damping is assumed to be proportional to instantaneous stiffness with a damping coefficient of 3.0% for RC and SRC members, and 2% for S and SC members.

The ANSYS software can calculate the time–history and dynamic response spectrum but, in the output, for lack of space only the maximum displacements and the corresponding times have been reported. So, with the use of the model superposition method (MSM) program, and with the application of few mode shapes and mode periods, the displacement of the floors were obtained.

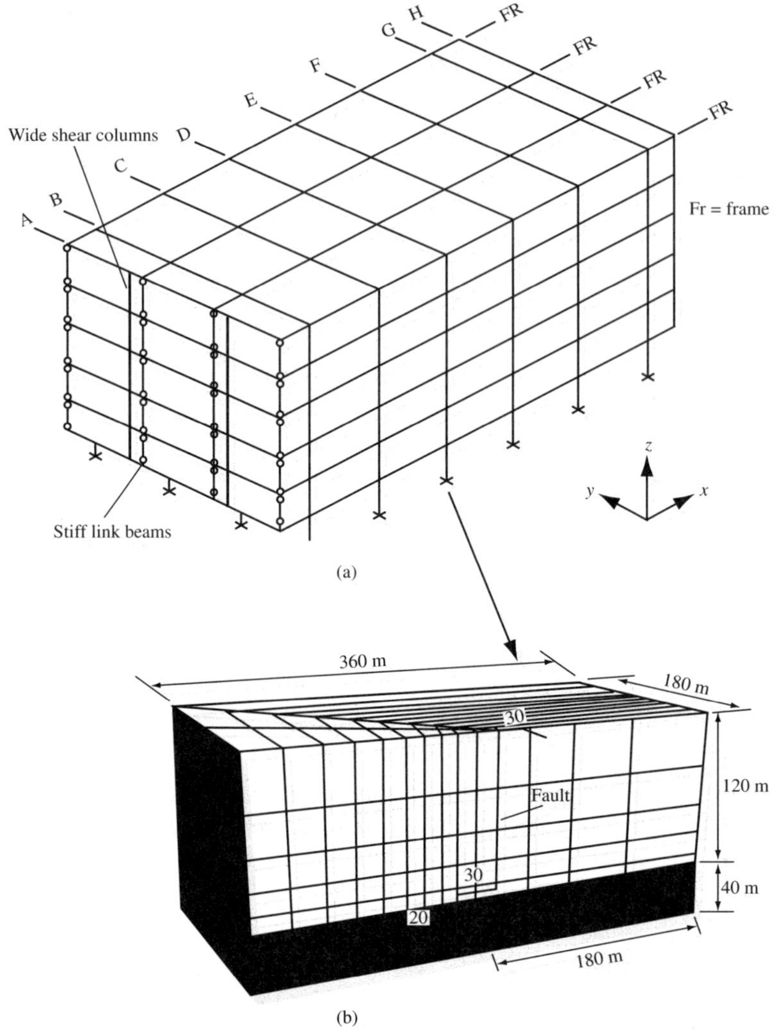

Fig. IX.267. Trianon building podium—soil structure interaction: (a) shear columns, links and frames for finite element analysis; (b) three-dimensional shaped podium with soil medium all around finite element mesh scheme.

(i) Maximum displacement of the floors, in both longitudinal (X) and transverse (Y) directions, is relatively small. This is due to the high rigidity of the building in both directions and it also indicates the linear elastic behaviour of the building.

(ii) In spite of the fact that the ratio of the width to length is 1/3, the model periods of Model 1 in both directions are very close to each other. This is because the flexural rigidity of the building in both longitudinal and transverse directions is almost the same. Note that the periods of the second mode of vibration in the Y direction is less than in the X direction.

(iii) The third mode of vibration is due to torsion. This indicates the importance of torsion in these computations. Unfortunately, there were insufficient instruments in the building to evaluate this problem. Hence, no such effects in the FE analysis are included.

(iv) The observed periods in two horizontal directions are very close to the computed fundamental periods in all four different models, and it represents the considerable participation of the fundamental modes in earthquake excitations. This phenomenon is confirmed by the comparison between the computed mode shapes and observed displacements of the floors.

(v) The twelfth and twenty-second floors show that both of these floors pass through the point of zero displacement at the same time. It means that the fundamental mode dominates the motion. Furthermore, the response of the fundamental mode and the first three modes are very close to each other. This indicates that the effect of the fundamental mode is much more significant than all other modes.

(vi) Comparison between the modal periods for different models shows the following.
- The soil–structure interaction effects are negligible in the longitudinal (X) direction, but they increase to more than 30% of the period of the first mode in the transverse (Y) and (Z) directions.
- The consideration of secondary moments of P–Δ and also finite rigid connection zones have little effect on the dynamic behaviour of the building.
- The observed periods are very similar to the periods of Model 2, and are representative of soil–structure interaction effects.
- The values of the periods rapidly decrease in all three directions for second and third modes. This, again, indicates the high participation of the fundamental mode in earthquake excitations.

Figures IX.264 and IX.265 show the actual building assumed to be located in the Kobe area. Figure IX.268 and IX.269 show the mode shapes as predicated by the finite element analysis using FEMGEN software. Figure IX.270 shows how the entire building moved under Kobe's earthquake. A typical damage scenario of the core wall with some idea of the damage to the opening is given in Fig. IX.271. The lift shaft of the building received cracks during the third mode. The displacement and acceleration of the building plotted against time is shown in Fig. IX.269. It seems that despite the severity of this earthquake, the majority of the building components are intact, albeit with certain bruises as may be expected. Although this building was originally designed to withstand the severest wind effects, it seems that, had this building been located within the Kobe area, it would have survived the intensity of the Kobe earthquake. Sufficient ductility exists to resist three-dimensional forces against Kobe seismic excitation. From this analysis it can be concluded that the building is somewhat stiffer than anticipated in the design, and for low level response included in the curtain wall system has no noticeable effect on the structural behaviour of this building.

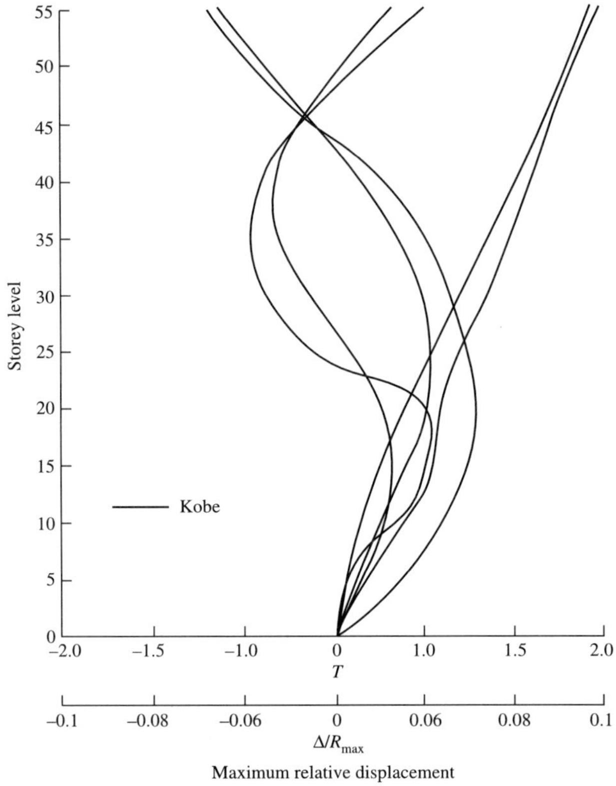

Fig. IX.268. Mode shapes at various storey levels.

IX.16.5.3 Data on finite element analysis isolators

The finite element mesh scheme with boundary conditions is shown in Fig. IX.266. A CRAY-3 computer is employed and 55 storeys (frames and floors) are considered.

Main structure.
Floor: 4-noded solid elements (55 500 total).
Frame: 4-noded line elements (220 000 total).
The entire structure is divided into seven No. 8 substructures

Soil–structure zone.
Ground floor: 2400 nodes.
Ground floor: 2200 nodes.
Interface elements (10 000) are used.
Loading reference to Kobe earthquake magnitude $M = 7.2$ with peak horizontal and vertical accelerations. 1000 mc/s/s respectively with actual values taken for the finite element analysis as m_a. Horizontal acceleration = 447 gal maximum. Period $T = 5$ seconds.
Maximum horizontal velocity = 90 kine (m/sec).
Maximum vertical velocity = 40 kine (m/sec).

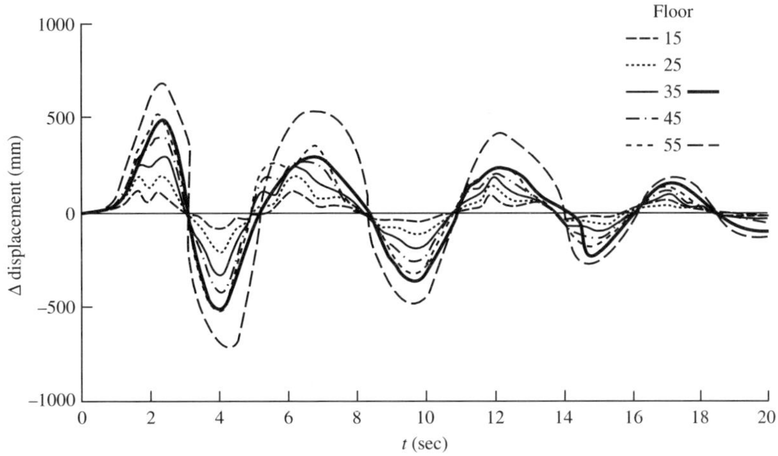

Fig. IX.269. Displacement–time relation for floor levels.

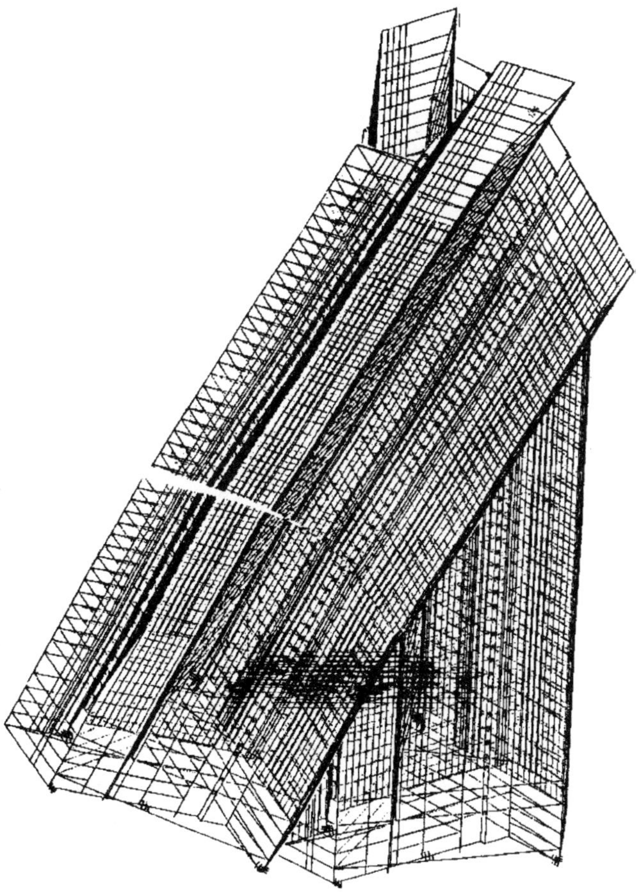

Fig. IX.270. A typical damage scenario for the Trianon building.

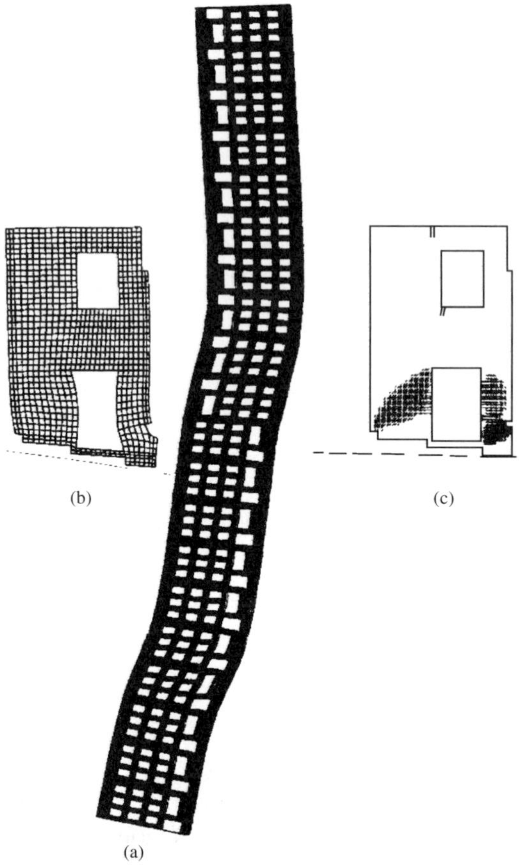

Fig. IX.271. Damage scenario of the typical core wall and its openings.

All columns are 500 mm × 500 mm RC with 3% steel rigidity connected with beams. Transverse framing consists of wall frames, open frames and braced frames. Nonlinear earthquake $\nless 0.5g$.

Elastoplastic spectra, with 5% damping.

Isolators: lead-plug type 3500 total with viscous dampers. They are mentioned under isolators.
h_x = total horizontal stiffness of the isolators = 130 kN/m for $T = 2$ s.

IX.16.5.4 Influence of isolators and dampers

The design philosophy, analysis and performance of the isolators have been dealt with in detail in Chapter VIII. The building structure given in Section IX.16.5 is now re-analysed using isolators and dampers. Again, the Kobe earthquake and its parameters are included because the deformation–force relationship of most isolators can be modelled using a bilinear diagram. Analysis of the same structure with incorporated bilinearly behaving isolators is presented. The discretisation of the superstructure was performed using the same number of hybrid elements as in the previous analysis. Because the number of nonlinear elements was predefined, a nonlinear analysis was performed by means of the nodal superposition method. Ritz's vectors were used for transforming equations into modal coordinates, and 2% of critical damping was adopted for all modes. The integration step was $\Delta t = 0.005\,\mathrm{s}$. For convergence of the results within each time interval, an average of four iterations were necessary. The N–S component of the Kobe earthquake was the only seismic output used with $A_{\max} = 0.3g$. A total number of 500 isolators with uniform characteristics were installed at the base of the building.

The last group of isolation systems is represented by the Sliding Isolation System. Under normal load, the fundamental period of this isolator is the same as that of the non-isolated structure on which this system is mounted. When the friction force is exceeded during large earthquake loads, the system tends to infinity. The disadvantage of this system is that large accelerations occur in the isolated structure due to the initial short period. A combination of a sliding bearing mounted on top of rubber bearings (a material is used that has a low friction coefficient (0.05–0.2) dependent on the sliding velocity) is used to overcome this disadvantage.

To determine the effect of viscous dampers on the response of the isolated building structure an analysis was performed applying a two-component isolator, i.e. natural rubber bearings and a linear viscous damper. Because viscous damping, that was not proportionate to either the mass of the system or the stiffness, was considered, the dynamic response of the building was obtained by computing the complex eigenvectors that transform the system into a set of complex equations. Because the complex eigenvalue problem is twice as big as the undamped eigenvalue problem,

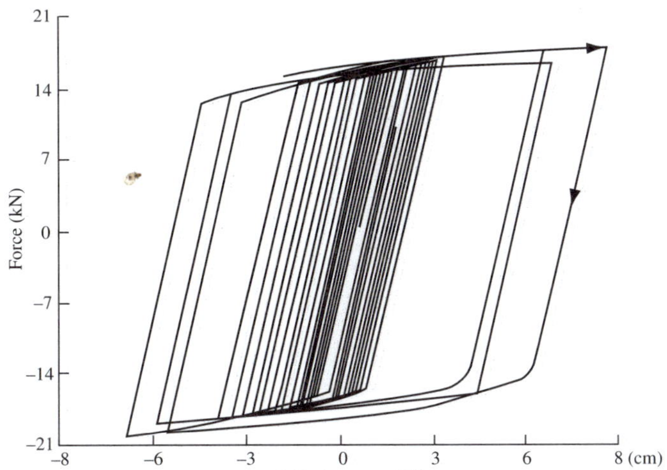

Fig. IX.272. Force–displacement relation for an isolator.

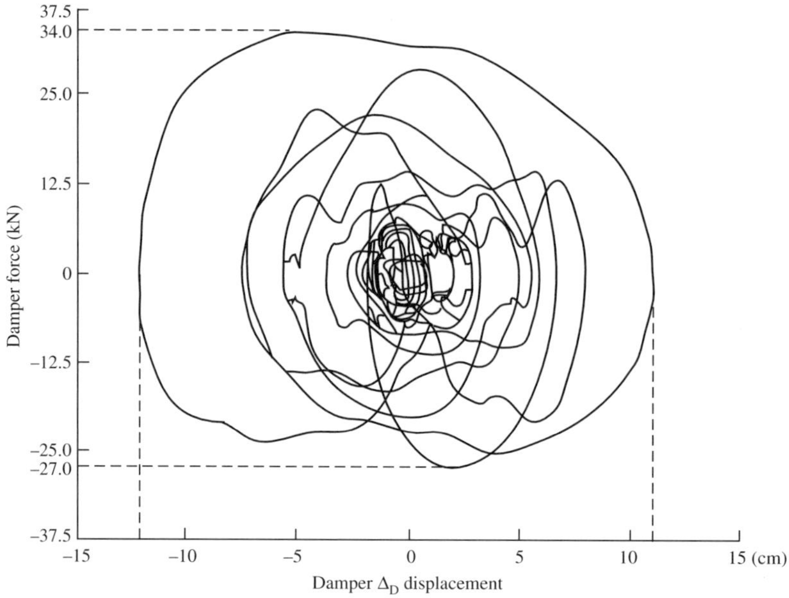

Fig. IX.273. Force–displacement relation for viscous damper.

the force in the damper in this analysis was computed as a modal force. Therefore, Ritz vectors were again used to transform equations into results. The discretisation of the superstructure was performed using the same number of isoparametric elements as in the previous analysis.

Figures IX.272 and IX.273 show the force–displacement relations for a typical isolator and a viscous damper. Figure IX.274 shows displacement–time relations for the building using fixed base and isolated base.

Along the height of the isolated building Fig. IX.275 presents the energy–time history for the isolated building with a viscous damper. The dashed line in

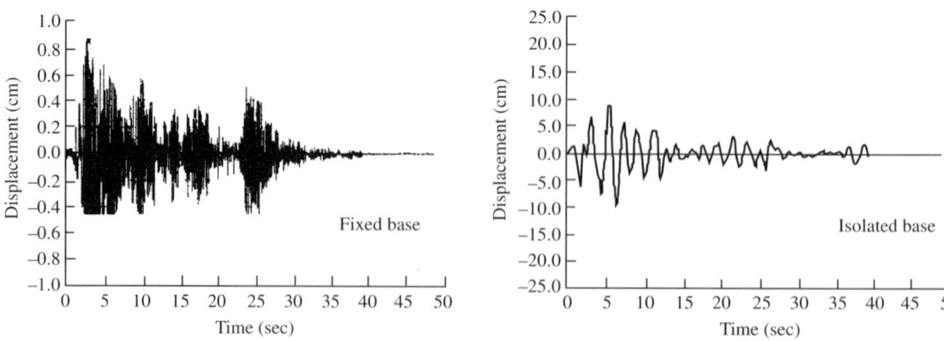

Fig. IX.274. A comparative study of displacement–time relation using fixed and isolated bases (inclusive of dampers).

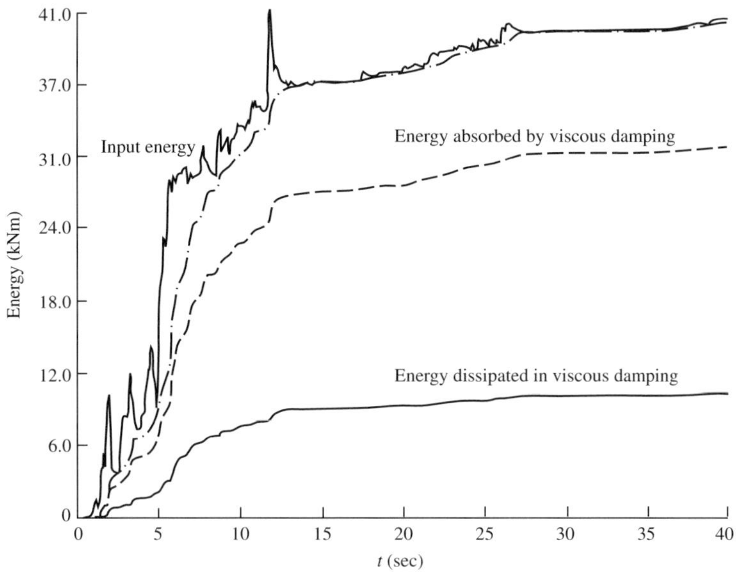

Fig. IX.275. Energy–time history with a viscous damper for the isolated building.

Fig. IX.276 represents the time history of the base moment of the building with a viscous damper. For the purpose of comparison, the same diagram shows the time history of the base moment obtained (solid line) by linear modal analysis and equipment proportional damping (15% damping ratio for the first mode and 2% for the remaining four modes). Figure IX.277 shows the time history of base shears with fixed base and with isolated base.

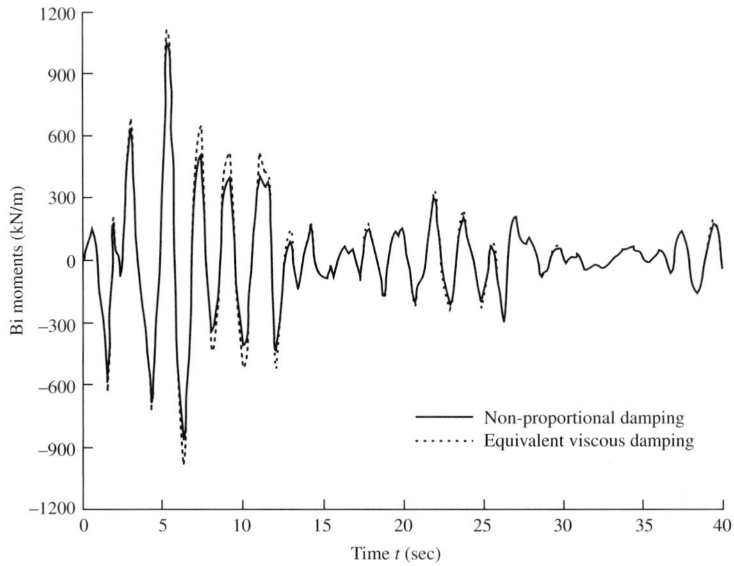

Fig. IX.276. Time history of the building with and without viscous damping.

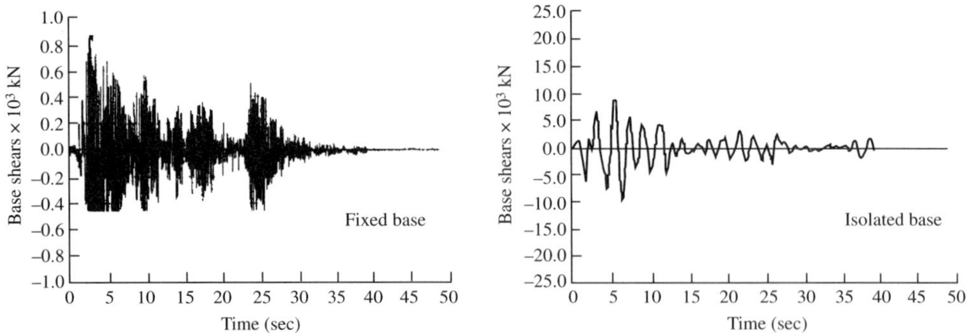

Fig. IX.277. Base shears with fixed and isolated bases.

IX.16.5.5 Conclusions

The two analyses were examined with fixed and isolated bases. The overall displacement and the deformation of the building were by 61.5% with isolated bases. Less damage was predicted by about 71% compared with those of isolated bases. The distribution of cracks in the podium in three dimensions was reduced to 51.5% with 90% crack depth were below 0.3 mm along three dimensions. The upper part of the building received only minor damage of no significance. With isolators and dampers, the structure in question withstood Kobe earthquake conditions. Most of the elements received minimum spalling effects in the upper region. In general, the predicted scenario in Fig. IX.270 simply does not exist. The core wall received damage with less intensity than was shown in Fig. IX.271. The maximum of fixed base moments were between 3600 and 4070 kNm within 5 seconds, assuming fixed bases. Similarly, with isolated bases the base moments were between 570 and 890 kNm within 5 seconds. They die away in approximately 40 seconds with intermediate peak values for fixed bases (2000 kNm, 20 seconds) and for isolated bases (400 kNm, 20 seconds).

IX.17 Case Study No. 16 — incremental collapse analysis of bridge deck

IX.17.1 Incremental collapse analysis

IX.17.1.1 Introduction

The scope of this short report encompasses a theoretical analysis based on incremental collapse mechanism as a numerical technique. The grids of composite bridges are assumed to be continuous across supports of fixed elevation. The equations developed do not include the effects due to torsion and dynamic plasticity. However, these equations can be adjusted later on to include such effects.

The present analysis is flexible and can take spans with any boundary conditions or rigidities. Apart from dead load, slow but heavy vehicular loading moving from point to point across the bridge is also included. Again, the vehicular loading can be replaced by u.d.l. (uniformly distributed loads) or patched loading representing some kind of dynamic loads. A numerical example is included in order to explain the use of these equations.

IX.17.1.2 Problem description and analytical equations

Composite bridge. The slab is divided into rectangular meshes and the longitudinal grid lines must be in line with longitudinal beams as shown in Fig. IX.278. Several plastic hinge mechanisms for spans must be examined, the one that gives the lowest load is chosen — the value of $\bar{K}\bar{M}_p/C_3L_3$. This kind of optimisation can easily be carried out by computer. $\bar{K} = 2$ for the sample grid/line. Figure IX.279 shows a continuous span ABC loaded with dead and imposed vehicular loadings. The grid points are assumed to be supported on springs K or K_1. If $L_1 = L$, then $K_1 = K = \bar{K}\bar{M}_p/C_3L_3$ (Fig. IX.280). For sample grid $K = 4$.

The analysis involves the algebraic sum of residual moments to maximum elastic moments at several nodal or grid points in simple or continuous structures. These moments are then equated to full plastic moments. The resulting equilibrium equations will give several possible collapse loads. The minimum value of such loads will then be considered as the final collapse load.

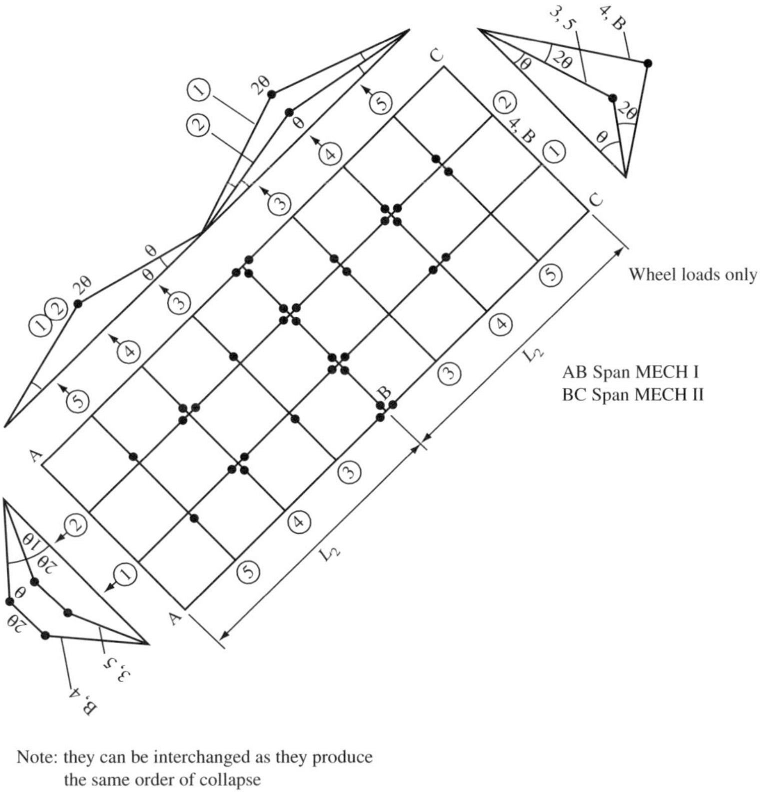

Fig. IX.278. Plane grid for the bridge deck.

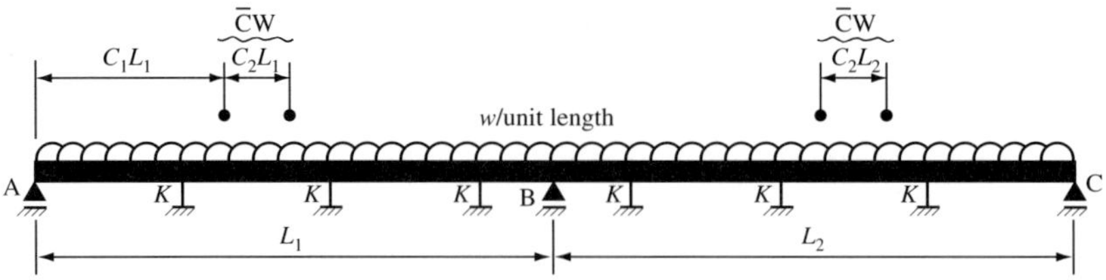

Fig. IX.279. Beam grid.

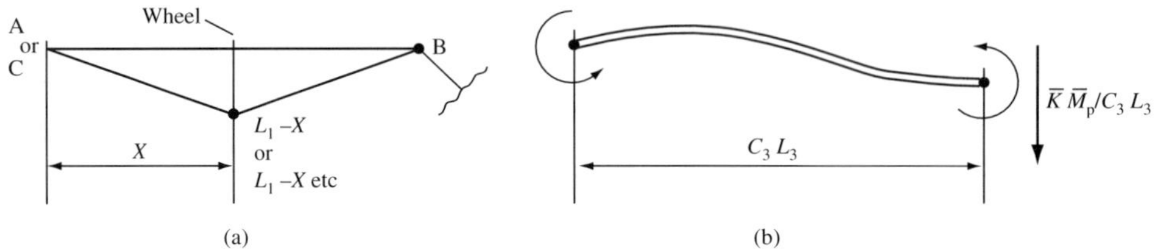

Fig. IX.280. Typical mechanisms.

Symbols

$L_{1,2}$ = spans of main members

C_3L_3 = spacings of main members across

M_p = full plastic moment

$\bar{M}_p$ = plastic moment from grid across

$\bar{C}W$ = vehicular loading

w = dead load/unit length

C_1 = distance from the origin to the nearest load

C_2 = spacings between any two loads

$C_2 = 0$ for single concentrated

K, K_1 = spring stiffness

M_{ej} = maximum elastic moment at any zone of failure

M_R = residual moment at any zone of failure

IX.17.1.3 Main analysis

$$M_p = M_e + M_R \tag{IX.17.1}$$

Case A — dead load alone

$$M_x = \frac{wL_{1,2}^2}{8}\bar{s} - \frac{4.125\bar{M}p}{C_3L_3}x \quad \text{for } 0 \le x \le \frac{L_{1,2}}{4} \tag{IX.17.2}$$

$$M_x = \frac{wL_{1,2}^2}{8}\bar{s} - \frac{\bar{M}pL_{1,2}}{8}\left[\frac{x}{L_{1,2}} + \frac{8}{C_3L_3}\right] \quad \text{for } \frac{L_{1,2}}{4} \leq x \leq \frac{L_{1,2}}{2} \qquad \text{(IX.17.3)}$$

$$M_x = \frac{wL_{1,2}^2}{8}\bar{s} + \frac{3\bar{M}p}{C_3L_3}[1.291x - L_{1,2}] \quad \text{for } \frac{L_{1,2}}{2} \leq x \leq \frac{3L_{1,2}}{4} \qquad \text{(IX.17.4)}$$

$$M_x = \frac{wL_{1,2}^2}{8}\bar{s} + \frac{6\bar{M}p}{C_3L_3}[1.3125x - L_{1,2}] \quad \text{for } \frac{3L_{1,2}}{4} \leq x \leq L_{1,2} \qquad \text{(IX.17.5)}$$

Case B — vehicular loading

$$M_x = \frac{Px}{8}\bar{s}_1 \quad \text{for } 0 \leq x \leq C_1L_{1,2} \qquad \text{(IX.17.6)}$$

$$M_x = \frac{Px}{8}\bar{s}_1 + \frac{C_1}{2}WL_{1,2} \quad \text{for } C_1L_{1,2} \leq x \leq (C_1 + C_2)L_{1,2} \qquad \text{(IX.17.7)}$$

$$M_x = \frac{Px}{8}\bar{s}_1 + \frac{W}{2}L_{1,2}(2C_1 + C_2) \quad \text{for } (C_1 + C_2)L_{1,2} \leq x \leq L_{1,2} \qquad \text{(IX.17.8)}$$

where

$$\bar{s} = 3\left(\frac{x}{L}\right) + 4\left(\frac{x}{L}\right)^2$$

$$\bar{s}_1 = 3C_1^2C_2 + 3C_1C_2^2 + 2C_1^3 + C_2^3 - 5(2C_1 + C_2) \qquad \text{(IX.17.9)}$$

$$C_1 = \frac{-C_2 + \sqrt{1.334 - C_2^2}}{2}$$

Plastic moment formula (non-dimensional)

$$\frac{wL_{1,2}^2}{2Mp}[C_1 - C_3^2] - \frac{\bar{M}pL_{1,2}}{MpC_3L_3}(1 + 2C_1) + \frac{wL_{1,2}}{8Mp}\bar{s}_1 + \frac{C_1}{2}\left(\frac{4}{3} - \frac{C}{2}\right) = 1 + C_1 \qquad \text{(IX.17.10)}$$

$$M_p = \frac{wL_{1,2}^2}{8}[3C_1 - 4C_1^2] - \frac{\bar{M}p \cdot L_{1,2}}{C_3L_3}\left[\frac{C_1}{8} + 1\right] \qquad \text{(IX.17.11)}$$

Numerical example

$w = 15\,\text{kN/m}$ from beam and slab

$C_2 = 0.1$ for $K_1 = K_2 \ldots \bar{K} = 4$

$C_3 = \frac{1}{8}$

$\bar{M}_p = 410\,\text{kN/m}$ (computed)

$M_p = 5100\,\text{kN/m}$

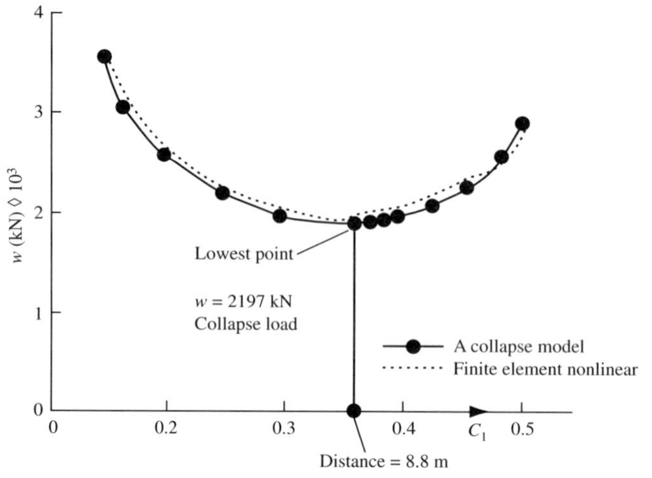

Fig. IX.281. W–C_1 graph. A comparative study of results from plastic and nonlinear finite element analyses.

W is obtained from eqn (IX.17.10) by substituting the above values. Results are plotted and $W = 2197$ kN at 8.8 m is the collapse load and its point of application, i.e. failure zone, is 8.8 m from the left support as shown in Fig. IX.281. Results from the nonlinear finite analysis proposed in the text indicate the collapse load of 2200 kN at 9.1 m from the same left support.

Thus it will determine what type of vehicles may cross a typical bridge or what sizes of bridge components are necessary for typical vehicular loadings.

Bibliography

1. Hognestad, E., A study of combined bending and axial load in reinforced concrete members. Bulletin No. 399, Engineering Experiment Station, University of Illinois, Urbana, Illinois, Vol. 49, No. 22, November 1951.
2. Desayi, P. & Krishnan, S., Equation for stress strain curve of concrete. *J. ACI Proc.*, **3** (1964).
3. British Standard CP110, Code of practice for the structural use of concrete, Part 1, design materials and workmanship. BSI London, November, 1972.
4. Saenz, L. P., Equation for stress strain curve of concrete in uniaxial and biaxial compression of concrete. *J. ACI*, **61** (1965) 1229–35.
5. Elwi, A. & Murray, D. W., A three-dimensional hypo-elastic concrete constitutive relationship. *J. Eng. Mech. Div., ASCE*, **105** (1976) 623–41 (No. EM4, Proc. paper 14746).
6. Bangash, Y., Discussion of Ref. 20. Paper 7865, Institution Civil Engineers, 1971.
7. Darwin, D. & Pecknold, D. A. (a) Analysis of cyclic loading of plane reinforced concrete structures. *J. Computers and Structures*, **7** (1977) 137–47. (b) Non-linear biaxial stress–strain law for concrete. *J. Eng. Mech. Div., ASCE*, **103** (1977) 229–49 (No. EM2, Proc. paper 12839).
8. Karson, I. D. & Jirsa, J. O., Behaviour of concrete under compressive loadings. *J. Struct. Div., ASCE*, **95** (1969) 2543–63.
9. Welch, P. F., Fracture of plain concrete. *Indian Concrete Journal*, **46** (1972) 468–79.
10. Evans, R. H., Effect of rate of loading on the mechanical properties of some materials. *J. ICE*, **18** (1962) 296.
11. Hughes, B. P. & Chapman, G. P. The deformation of concrete and micro-concrete and tension with particular reference to aggregate size. *Magazine of Concrete Research*, **18**(54) (1966) 19–24.
12. Popovics, S., A numerical approach to the complete stress–strain curve of concrete. *Cem. Concr. Res.*, **3** (1973) 583–99.
13. Johnson, R. D., *Structural Concrete*. McGraw-Hill, London, 1969.
14. Richart, F. E., Brandzaeg, A. & Brown, R. L., A study of the failure of concrete under combined compressive stresses. Bulletin No. 185, Engineering Experiment Station, University of Illinois, Vol. 26, 1928.
15. Glucklich, J., On the compression failure of plain concrete. T & A M, Report No. 125, University of Illinois, March 1962.
16. Balmer, G. G., Shearing strength of concrete under high triaxial stress — computation of Mohr's envelope as a curve. Struct. Research Lab., Report No. SP-23, United States Bureau of Reclamation (USBR), 1949.
17. Westergaard, H. M., On the resistance of ductile materials to combined stresses in two or three directions perpendicular to one another. *J. Frankline Inst.*, **189** (1920) 627–40.

18. Nadai, A., *Theory of Flow and Fracture of Solids*, Vol. 1. McGraw-Hill, New York, 1950.
19. Merkle, J. G., An engineering approach to multiaxial plasticity. ORNL-4138, 1968.
20. Sandbye, P., A plastic theory for plain concrete. In: *Bygnings-Statiske Meddelelser*, Vol. 36. Teknisk Forlag, Copenhagen, 1965, pp. 41–62.
21. Newman, K. & Newman, J. B., Failure theories and design criteria for plain concrete structure. In: *Solids, Solid Mechanics*, Wiley Interscience, London, 1971, pp. 963–95.
22. Wastlung, G., Nye ron angaende betongens grundlaggende hallfasthetsegenskabar. Betong Haft 3, Stockholm, 1937.
23. Bresler, B. & Pister, K. S., Failure of plain concrete under combined stresses. *Trans. ASCE*, **122** (1957) 1049–68.
24. Ottoson, N. S., A failure criteria for concrete. *J. Eng. Mech. Div., ASCE*, **103** (1977) 527–35 (No. EM4, Proc. paper 13111).
25. Ottoson, N. S., Constitutive model for short-time loading of concrete. *J. Eng. Mech. Div., ASCE*, **105** (1979) 127–41 (No. EM 1, Proc. paper 14375).
26. Ottoson, N. S., Structural failure of thick walled concrete elements. 4th International Conference on Structural Mechanics in Reactor Technology (SMiRT), Paper H4/3, San Francisco, August 1977.
27. Suidan, M. & Schnobrich, W., Finite element analysis of reinforced concrete. *J. Struct. Div., ASCE*, **99** (1973) 2109–22.
28. Cervenka, V., Inelastic finite element analysis of reinforced concrete panels under in-plane loads. PhD Thesis, University of Colorado, Boulder, Colorado, 1970.
29. Kupfer, H. B., Hilsdorf, H. K. & Rüch, H., Behaviour of concrete under biaxial stresses. *ACI Journal*, **66** (1969) 556–66.
30. Liu, T. C., Nilson, A. H. & Slate, F. O., Stress strain response and fractures of concrete in uniaxial and biaxial compression. *J. ACI Proc.*, **69** (1972) 291–5.
31. Kupfer, H. B. & Gerstle, K. H., Behaviour of concrete under biaxial stresses. *J. Eng. Mech. Div., ASCE*, **99** (1973) 552–866 (EM4, Proc. paper 9917).
32. 1st International Conference on Structural Mechanics in Reactor Technology (SMiRT), Berlin, September 1971.
33. 2nd International Conference on Structural Mechanics in Reactor Technology (SMiRT), Berlin, 1973.
34. 3rd International Conference on Structural Mechanics in Reactor Technology (SMiRT), London, September 1975.
35. Von Mises, R., Mechanik der festen Koerper im plastisch deformablen Zustant, *Getlinger Nachr., Math. Phys.*, **K1** (1913) 582–92.
36. Prandtl, L. & Reuss, A., (a) *The Mathematical Theory of Plasticity*, ed. R. Hill. Oxford University Press, London, 1950. (b) *Plasticity, Theory and Application*, ed. A. Mendelson. Macmillan, London, 1968.
37. Hencky, H. Z. & Nadai, A., Zur Theorie Plastischer Deformationen und der hierdurch im Material hervorgerufenen Nachspannungen, Report, 1924, pp. 323–34. (Available National Lending Library.)
38. Sturman, G. M., Shah, S. P. & Winter, G., Effect of flexural strain gradients on micro-cracking and stress–strain behaviour of concrete. *J. ACI*, **62** (1965) 805–22.
39. Schleicher, F., Der Spannungszustand an der Fliessgrenze (Plasticitatsbedonging). *Zietschrift für Angewandte Mathematik und Mechanik*, Band 6 Heft 3. Berlin, June 1926.
40. Vile, G. W. D., The strength of concrete under short term static biaxial stress. In: *The Structure of Concrete and its Behaviour under Load*, Proc. Int. Conf., London, 1968, pp. 275–88.
41. Smith, G. M., Failure of concrete under combined tensile and compressive stresses. *J. ACI*, **50** (1953) 137–40.
42. Bresler, B. & Pister, K. S., Strength of concrete under combined stresses. *J. ACI*, **55** (1958) 321–45.

43. Bellamy, C. J., Strength of concrete under combined stresses. *J. ACI*, **58** (1961) 367–80.
44. Bazant, Z. P., Comment on orthotropic models for concrete and geomaterials, *J. Eng. Mech. Div., ASCE*, **109** (1983) 849–65.
45. Bazant, Z. P. & Cedolin, L., Fracture mechanics of reinforced concrete. *J. Eng. Mech. Div., ASCE*, **106** (1980) 1287–307.
46. Bazant, Z. P. & Kim, S. S., Plastic-fracturing theory for concrete. *J. Eng. Mech. Div., ASCE*, **105** (1979) 407–28.
47. Bazant, Z. P. & Shieh, C., Hysteretic fracturing endochronic theory for concrete. *J. Eng. Mech. Div., ASCE*, **106** (1980) 929–50.
48. Akroyd, T. N. W. Concrete under triaxial stress. *Magazine of Concrete Research*, **13**(39) (1961) 111–18.
49. Nashizawa, N., Strength of concrete under combined tensile and compressive loads. *Japan Cement Engineering Association Review*, May 1961, 126–31.
50. Campbell Allen, D., Strength of concrete under combined tensile and compressive loads. *Constructional Review*, **35** (1962) 29–37.
51. Sigvaldason, O. T., Failure characteristics of concrete. PhD Thesis, University of London, June 1965.
52. Baker, A. L. L. (a) An analysis of deformation and failure characteristics of concrete. *Magazine of Concrete Research*, **11**(33) (1959) 119–28. (b) A criterion of concrete failure. *Proc. Inst. Civ. Eng.* London, (1970) 269–78.
53. Anson, M. An investigation into a hypothetical deformation and failure mechanism for concrete. PhD Thesis, University of London, November 1962.
54. Robinson, G. S., The failure mechanism of concrete under particular reference to the biaxial compressive strength. PhD Thesis, University of London, November 1964.
55. Rüsch, H., Der Zuzammenhang Zwischen Risbilding und Haftfestigkeit unter besonderer Berucksichtingung der Anwendunghober Stahlspannungen, 5. Konferenz der Int. Vereiningung für Brucken und Hochbau, Lissabon, 1956.
56. Fumagalli, E., *Strength Characteristics of Concrete under Conditions of Multi-Axial Compression*. Cement and Concrete Association, London (Translation No. CU 128), 1975.
57. Weigher, H. & Becker, G., Untersuchungen uber das Bruch und verformungsverhalten von Beton bei Zweiachsiger Beausprachung. *Deutscher Ausschuss fur Stahlbeton*, Heft 157. Berlin, 1963.
58. Hannant, D. J. & Frederick, C. D., Failure for concrete in compression. *Magazine of Concrete Research*, **20** (1968) 137–44.
59. Link, J., Numerical analysis oriented biaxial stress–strain relation and failure criterion of plain concrete. 3rd International Conference on Structural Mechanics in Reactor Technology (SMiRT), Paper H1/2, London, September, 1975.
60. Baker, A. L. L., Safety of pressure vessels, group B. International Conference on PCPV, Paper 8, Inst. Civ. Eng., London, 1967.
61. Chawalla, E., *Ein Fuhrung in die Baustatik*, Stahlbauverlag, Koln, 1954.
62. Chinn, J. & Zimmerman, R. M., Behaviour of plain concrete under various high triaxial compression loading conditions. Technical Report No. WL TR 64-163 (AD468460), Air Force Weapon Laboratory, New Mexico, August 1965.
63. Reimann, H., Kritische spannungszustande der Betons bei mehrachsiger, ruhender Kurzzeitbelastung. *Deutscher Ausshuss fur Stahlbeton*, Heft 175, Berlin, 1965.
64. Argyris, J. H., Recent developments in the finite element analysis of prestressed concrete reactor vessels. IABSE Seminar, ISMES, Bergamo, Italy, May 1974.
65. Willam, K. J., Constitutive model for the triaxial behaviour of concrete. IABSE Seminar, ISMES, Bergamo, Italy, May 1974.
66. Launay, P. & Gachon, H. (a) Strain and ultimate strength of concrete under triaxial stress. 1st International Conference on Structural Mechanics in Reactor Technology (SMiRT), Paper H1/3, Berlin, September 1971. (b) Strain and ultimate strength of concrete under

triaxial stress. ACI SP-34, Paper 13, 1970. (c) Deformation et resistance ultime du beton sous etreinte triaxiale. Annales de l'Institut Technique du Batiment et du Travaux Publics, No. 269, 1970.

67. York, G. P., Kennedy, T. W. & Perry, E. S., An experimental approach to the study of the creep behaviour of plain concrete subjected to triaxial stresses and elevated temperatures. Union Carbide, Report No. 28642, The University of Texas at Austin, August 1971.
68. Garas, F. K., Contributions to paper 6, pp. 100–1. International Conference on Model Techniques for Prestressed Concrete Pressure Vessels, BNES, London, 1969.
69. Liu, T. C. Y., Stress–strain response and fracture of concrete in biaxial compression. Research Report No. 339, Department of Structural Engineering, Cornell University, February 1971.
70. Liu, T. C. Y., Nilson, A. H. & Slate, F. O., Biaxial stress–strain relation of concrete. *J. Struct. Div., ASCE*, **98** (1972) 1025–34.
71. Buyukozturk, O., Nonlinear analysis of reinforced concrete structures. *J. Computers and Structures*, **7** (1977) 149–56.
72. Popovics, S., A numerical approach to the complete stress–strain curve of concrete. *Cement and Concrete Research*, **3** (1973) 538–99.
73. Tasuji, M. E., Nilson, A. H. & Slate, F. O., Biaxial stress–strain relationships for concrete. *Magazine of Concrete Research*, **31**(109) (1979) 217–24.
74. Tasuji, M. E., The behaviour of plain concrete subject to biaxial stress. Research Report No. 360, Department of Structural Engineering, Cornell University, March 1976.
75. Tassoulas, J. L., Inelastic behaviour of concrete in compression. MS Thesis, Department of Civil Engineering, MIT, Cambridge, Mass., 1979.
76. Hussain, M. & Saugy, S., Elevation of the influence of some concrete characteristics on non-linear behaviour of a prestressed concrete reactor vessel. Paper SP-34-8, Concrete for Nuclear Reactors, American Concrete Institute, Detroit, Michigan, October 1970.
77. Linse, D., Aschl, H. & Stockl, S., Concrete for PCRV's: strength of concrete under triaxial loading and creep at elevated temperatures. 3rd International Conference on Structural Mechanics in Reactor Technology (SMiRT), Paper H1/3, London, September 1975.
78. Gardner, J. J., Triaxial behaviour of concrete. *Proc. Am. Concr. Inst.* No. 2, (1969) 136–47.
79. Ross, A. D., Shrinkage in concrete pressure vessels. *Nucl. Eng. Des.*, **5** (1967) 150–60.
80. Kasami, H., Properties of concrete exposed to sustained elevated temperature. 3rd International Conference on Structural Mechanics in Reactor Technology (SMiRT), Vol. 3, Paper H1/5, London, September 1975.
81. Nasser, K. W. & Neville, A. M., Creep of concrete at elevated temperatures. *Proc. Am. Concr. Inst.*, **12** (1965) 1567–79.
82. Irving, J. & Carmichael, G. D. T., The influence of creep on the behaviour of concrete structures subjected to cyclic heating. Conference Solid Mechanics and Engineering Design in Civil Engineering Materials, Paper No. 107, Southampton University, April 1969.
83. Karni, J. & McHenry, D., Strength of concrete under combined tensile and compressive stresses, *J. ACI* (April 1958).
84. Browne, R. D., Properties of concrete in reactor vessels. ICE Conference Prestressed Concrete Pressure Vessels, Institution of Civil Engineers, London, 1958, pp. 131–51.
85. Taylor Woodrow Construction, Tests on Wylfa models. Private communications, 1968.
86. England, G. L. & Phok, M., Time-dependent stresses in a long thick cylindrical prestressed concrete vessel subjected to sustained temperature crossfall. *Nucl. Eng. Des.*, **9** (1969) 488–95.
87. Zienkiewicz, O. & Watson, M., Some creep effects in stress analysis with particular reference to concrete pressure vessels. *Nucl. Eng. Des.*, No. 4 (1966).
88. Hannant, D. J., Creep and creep recovery of concrete subjected to a multiaxial compressive stress. *Proc. Am. Concr. Inst.*, **66** (1969) 391–4.
89. British Standards CP110, Parts 1 and 2, 1972, and CP8110, 1985.

90. Truesdell, C., Hypo-elasticity. *J. Ration., Mech. Analysis*, **13** (1955) 1019–20.
91. Rivlin, A. & Erickson, J. L., Stress–deformation relations for isotropic materials. Report, Graduate Institute for Maths and Mechanics, Indiana University, December 1954.
92. Ross, A. D., Creep of concrete under variable stress. *ACI Journal*, **54** (1958) 739–58.
93. England, G. L. & Jardaan, I. J., Time-dependent and steady state stresses in concrete structures with steel reinforcement at normal and raised temperatures. *Magazine of Concrete Research*, **27** (1975) 131–42.
94. Bangash, Y. & England, G. L., The influence of thermal creep on the operational behaviour of complex structures. International Conference on Fundamental Creep and Shrinkage, Lausanne, Switzerland, 1980.
95. 4th International Conference on Structural Mechanics in Reactor Technology (SMiRT), Berlin, 1977.
96. 5th International Conference on Structural Mechanics in Reactor Technology (SMiRT), Berlin, 1979.
97. 6th International Conference on Structural Mechanics in Reactor Technology (SMiRT), Paris, 1981.
98. Nayak, G. C. & Zienkiewicz, O. C., Note on the 'alpha' constant stiffness method for analysis of non-linear problems. *Int. J. Num. Meth. Eng.* (1972) 579–82.
99. Vallapan, S. & Doolan, T. F., Non-linear stress analysis of reinforced concrete. *J. Struct. Div., ASCE*, **98** (1972) 885–98.
100. Schrobrich, W. C. *et al.*, Discussion on Ref. 134. *J. Struct. Div., ASCE*, **98** (1972) 2419–32.
101. Timoshenko, S., *Theory of Elasticity*. McGraw-Hill, New York, 1951.
102. Marin, J., *Mechanical Behaviour of Engineering Materials*. Prentice Hall, New York, 1962, p. 117.
103. Lame, A., In: *Theory of Flow and Fracture of Solids*, Vol. 1, ed. A. Nadai, McGraw-Hill, New York, 1950.
104. Clapeyron, Z., *Theory of Flow and Fracture of Solids*. (2nd edn) ed. A. Nadai, McGraw-Hill, New York, 1950.
105. Rankine, T., Soil mechanics and plastic analysis or limit design. *Q. Appl. Math.*, **10** (1952).
106. Poncelot, F., *Foundations of the Non-Linear Theory of Elasticity*, ed. V. V. Novozhilov, Greylock Press, USA, 1953.
107. St Venant, *Mem. Savants Etrangers*, **14** (1855). (A copy of this reference is held at the British Museum Library, London).
108. Tresca, H., Sur l'ecoulement des corps solids soumis a de fort pression. *Comp. Rend.*, **59** (1864) 754.
109. Coulomb, C. A., Essai sur une application des regles des maximis et minimis a quelques problems de statique. Royale Academie des Sciences, Paris, Vol. 7, 1776.
110. Navier, C. F., In: *A Theory of Deformation and Failure of Concrete*, ed. Reinius, Stockholm, Vol. 40, No. 1, 1955.
111. Mohr, O., *Abhandlungen aus dem Gebiet der technischen Mechanik* (2nd edn), Wilhelm Ernst und Sohn, Berlin, 1914, p. 192.
112. Sell, R., Investigation into the strength of concrete under sustained load. RILEM Bulletin No. 5, December 1959, pp. 5–13.
113. Kupfer, H., *Das Verhalten des Betons Unter Mehrachsigen Kurzzeitbelastung det Zweiachsigen Beanspruchung Deutscher Ausshuss fur Stahlbeton*, Heft 229, Berlin, 1973.
114. Evans, R. J. & Pister, K. S., Constitutive equations for a class of non-linear elastic solids. *Int. J. Solids Structures*, **2** (1966) 427–45.
115. Chiorino, M. A. & Levi, R., Influence de l'Elasticite Differee sur le Regime des Contraintes des Constructions en Beton. Cahiers de la Recherche No. 24, Institut Technique du Batiment et des Travaux Publics, Eyrolles, Paris, France, 1967 (see also *Giornale del Genio Civile*, 1967; and Academia Nazionale dei Lincei, Fasc. 5, Series 8, Vol. 38, May 1965).

116. Chinn, J. & Zimmerman, R. M., Behaviour of plain concrete under various high triaxial compression loading conditions. Technical Report No. WL TR 64-163 (AD4684460), Air Force Weapon Laboratory, New Mexico, 1965.
117. Mills, L. L. & Zimmerman, R. M., Compressive strength of plain concrete under multiaxial loading conditions. *ACI Journal*, **67** (1970) 802–7.
118. Sozen, M. A. & Corley, W. G., Time-dependent deflection of reinforced concrete beams. *ACI Journal*, March 1966, 374–85.
119. Conference on Prestressed Concrete Pressure Vessels, Inst. Civ. Eng., London, 1967.
120. Bangash, Y., The automated three-dimensional cracking analysis of prestressed concrete vessels. Sixth International Conference on Structural Mechanics in Reactor Technology (SMiRT), Paper H3/2, Paris, August 1981.
121. Chen, E. S. & Buyukozturk, O., Damage model for concrete in multiaxial cyclic stress. Department of Civil Engineering, MIT, Cambridge, Massachusetts, 1983.
122. Bakker, W. T., Properties of refractory concretes. Publication SP-57, American Concrete Institute, Detroit, Michigan, 1978, pp. 11–52.
123. Bakker, W. T., Recent advances in refractory concrete technology, Publication SP-74. American Concrete Institute, Detroit, Michigan, 1982, pp. 1–16.
124. Calvo, J. J., Design of reinforced concrete containment walls under combined axial and shear forces. MS Thesis, Department of Civil Engineering, MIT, Cambridge, Massachusetts, 1982.
125. Langan, D. & Garas, F. K., Behaviour of end slabs in cylindrical prestressed concrete pressure vessels. 2nd International Conference on Structural Mechanics in Reactor Technology (SMiRT), Paper H3/4, Berlin, 1973.
126. Bangash, Y., Reactor pressure vessel design and practice. *Progress in Nuclear Energy*, **10** (1982) 69–124.
127. Bangash, Y., PWR steel pressure vessel design and practice. *Progress in Nuclear Energy*, **16** (1985) 1–40.
128. Saugy, B., Three-dimensional rupture analysis of a prestressed concrete pressure vessel including creep effects. 2nd International Conference on Structural Mechanics in Reactor Technology (SMiRT), Berlin, 1973.
129. Saouma, V., Automated nonlinear finite element analysis of reinforced concrete: a fracture mechanics approach. Thesis presented to Cornell University, Ithaca, NY, in partial fulfilment of the requirements for the degree of Doctor of Philosophy, 1981.
130. Willam, K. J. & Wanke, E. P., Constitutive nodal for triaxial behaviour of concrete. IABSE Seminar on Concrete Structures Subjected to Triaxial Stresses, Part III-I, Bergamo, Italy, May 1974.
131. Hobbs, D. W., Design stresses for concrete structures subjected to multiaxial stresses. *Struct. Eng.*, **55** (1977) 157–64.
132. Kotsovos, M., Failure criteria for concrete under generalised stress states. PhD Thesis, University of London, 1974, p. 284.
133. Langan, D. & Garas, F. K., Capacity of prestressed vessels to resist shear. *ACI Journal* (*Seminar on Concrete for Nuclear Reactors*), (October 1970) SP34–12.
134. Langan, D. & Garas, F. K., The design of pod boiler pressure vessels with particular reference to nuclear power stations. 2nd International Conference on Structural Mechanics in Reactor Technology (SMiRT), Paper H4/4, Berlin, 1973.
135. Elwi, A. A. & Murray, D. W., A three-dimensional hypo-elastic concrete constitutive relationship. Private communication, 1980.
136. Turner, F. H., *Concrete and Cryogenics*. View Point, New York, 1978.
137. Indian Standard IS 456, Delhi, 1979.
138. British Standard BS8110. The structural use of concrete, 1985.
139. ACI 318. Building code requirements for reinforced concrete, 1971.
140. DIN 4227. Standard specification for reinforced concrete, Germany, 1978.

141. Ross, A. D., Private communications, Kings College, London, 1968.
142. Buyukozturk, O., A constitutive model for concrete in compression. Proc. Third ASCE Engineering Division Specialty Conference, Austin, September 1979.
143. Kasami, H. *et al.*, Private communications.
144. USBR (United States Bureau of Reclamation), Private communications.
145. Green, S. J. & Swanson, S. R., Static constitutive relations for concrete. Tech. Report AFWC-TR-72-2, 1973.
146. Zienkiewicz, O. C., Watson, M. & King, I. P., A direct method of visco-elastic creep. *Int. J. Mech. Sci.*, **10** (1968) 807–27.
147. Dibsi, J., PhD Thesis, Imperial College, London, 1979.
148. Chen, A. C. T. & Chen, W. F., Constitutive relations for concrete. *J. Eng. Mech. Div. ASCE*, **101** (1975) 465–81.
149. Kojic, M. & Cheathem, J. B., Theory of plasticity of porous media with fluid flow. *Soc. Pet. Eng. J.*, **257** (1974) 263.
150. Hsu, T. T. C., Slate, F. O., Sturman, G. M. & Winter, G., Microcracking of plain concrete and the shape of the stress–strain curve. *J. Am. Concr. Inst.*, **60** (1963) 209–24.
151. Chen, W. F. *et al.*, *Constitutive Equations for Engineering Materials*, Wiley, New York, 1983.
152. Suh, N. P., A yield criterion for plastic frictional, work hardening granular materials, *Int. J. Powder Materials* (1969) 69.
153. Zienkiewicz, O. C. & Pande, G. C., Private communication, 1983.
154. Drucker, D. C. & Prager, W., A more fundamental approach to plastic stress–strain relation. Proc. Natl. Cong. US Inst. Appl. Mech., Chicago, 1951, pp. 487–91.
155. England, G. L., The direct calculation of stresses in creeping concrete and concrete structures. *Proc. Conference on Structure, Solid Mechanics and Design, Southampton*, Part 2, ed. M. Te'eni, 1969.
156. England, G. L. & Illston, J. M., Method of computing stress in concrete from history to measured strain. *Civil Engineering and Public Review, London*, **60** (1965).
157. Parrot, L. J., Some observations on the components of creep in concrete. *Magazine of Concrete Research*, **22** (1970).
158. Bazant, Z. P., Endochronic inelasticity and incremental plasticity. *Int. J. Solids and Structures*, **14** (1978) 691–714.
159. Coons, M. D. & Evans, R. J., On the recoverable deformation of granular media. Private report, University of Washington, Seattle, 1972.
160. Love, A. E. H., *Mathematical Theory of Elasticity*, McGraw-Hill, New York, 1939.
161. Lekhnitskii, S. G., *Theory of Elasticity of an Anisotropic Elastic Body*, Holden-Day, San Francisco, California, 1963.
162. Nye, J. F., *Physical Properties of Crystals*, Clarendon Press, Oxford, 1957.
163. Green, A. E., Hypo-elasticity and plasticity. *Proc. Roy. Soc. London*, **A234** (1956) 46–57.
164. Hughes & Chapman, J. C., Private communications, University of Birmingham, 1970.
165. Charlton, S. *et al.*, Prestressing and proof pressure testing of the Wylfa prestressed concrete pressure vessels and comparison of measured data with predicted performance. 1st International Conference on Structural Mechanics in Reactor Technology (SMiRT), Berlin, September 1971.
166. Noll, W. J., On the continuity of solids and fluid states rational mechanics. *Anal.*, **4** (1955) 3.
167. Phillips, D. V. *et al.*, Finite element methods in the analysis of reactor vessels. *Nucl. Eng. and Design*, **20** (1972).
168. Phillips, D. V. & Zienkiewicz, O. C., Finite element non-linear analysis of concrete structures. *Proc. Inst. Civ. Eng. Research and Theory*, **61** (1976).
169. Cedolin, L., Crutzen, Y. R. J. & Dei Foli, S., Triaxial stress–strain relationships for concrete. *J. Eng. Mech. Div., ASCE*, **103** (1977) 423–39. No. EM3, Proc. paper 12969.
170. Saugy, B. *et al.*, Private communication, 1979.

171. Valanis, K. C., A theory of visco-plasticity without a yield surface. *Archiwum Mechaniki Stosowanej* (Warsaw), **3** (1971) 517–51.
172. Bazant, Z. P. & Bhat, P. D., Endochronic theory of inelasticity and failure of concrete, *J. Eng. Mech. Div., ASCE,* (1976) 701–21.
173. Bazant, Z. P. & Shieh, H., Endoelectronic model for non-linear triaxial behaviour of concrete. *Nucl. Eng. and Des.*, **47** (1978) 305–15.
174. Bangash, Y., A report on the progress of G/H Division, SMiRT 7. BNES, 1984.
175. Pfiffer, P. A. *et al.*, Blunt crack propagation in finite element analysis for concrete structures. 7th International Conference on Structural Mechanics in Reactor Technology (SMiRT), North-Holland, Amsterdam, Paper H/2, 1983.
176. Gopalakrishnan, K. S., Neville, A. M. & Ghali, A., Creep Poisson's ratio of concrete under multiaxial compression. *J. Am. Conc. Inst.*, **66** (1969) 1008–19.
177. Gambarora, P. G. *et al.*, Rough cracks in reinforced concrete. *J. Struct. Div., ASCE,* **106** (1980) 819–42.
178. Rebora, B., Zimmermann, T. H. & Wolf, J. P., Dynamic rupture analysis of reinforced concrete shells. *Nucl. Eng. Des.*, **37** (1976) 269–97.
179. Ree, H. V. & Hoek, M. J. V. D., Analysis of the behaviour of concrete structure due to an airplane. Private communication, 1979.
180. Zerna, W. *et al.*, Optimized reinforcement of nuclear power plant structures for aircraft impact forces. *Nucl. Eng. Des.*, **37** (1976) 313–20.
181. Davies, I. L., Damaging effects from the impact of missiles against reinforced concrete structures. *Nuclear Energy*, **19** (1980) 199–205.
182. Zukas, J. A., *Impact Dynamics*. Wiley, New York, 1981.
183. Wolf, J. P., Bucher, K. M. & Skirikerud, P. E., Response of equipment to aircraft impact. *Nucl. Eng. Des.*, **47** (1978) 169–93.
184. Rice, J. S. & Bahar, L. Y., Reaction-time relationship and structural design of reinforced concrete slabs and shells for aircraft impact. 3rd International Conference on Structural Mechanics in Reactor Technology (SMiRT). Paper J5/3, London, 1975.
185. Riera, J. D., On the stress analysis of structures subjected to aircraft impact forces. *Nucl. Eng. Des.*, **8** (1968) 415–26.
186. Rock, T. A., Dynamic analysis of civil engineering structures. PhD Thesis, University of Wales, Swansea, 1974.
187. Ronstad, K. M., Taylor, M. A. & Harmann, L. R., Numerical biaxial characterization for concrete. *J. Eng. Mech. Div., ASCE,* **100** (1974) 935–48.
188. Rynen, J., Villafane, E. & Crutzen, Y., Impulsive loading on concrete structures, 6th International Conference on Structural Mechanics in Reactor Technology (SMiRT), Paper J10/1, Paris, 1981.
189. Smith, G. M., Failure of concrete under combined tensile and compressive stresses. *J. Am. Concr. Inst.*, **50** (1953) 137–47.
190. Stephenson, A. E., Full-scale tornado-missile impacts tests. EMPRI NP-440, Sandia Laboratories, 1977.
191. Suidan, M. & Scnobrich, W. C., Finite element analysis of reinforced concrete. *J. Struct. Div., ASCE,* **99** (1973) 2109–22.
192. Varpasuo, P. & Kenttala, J., The analysis of the containment building for global effects of an aircraft crash. International Conference on Structural Mechanics in Reactor Technology (SMiRT), 1977. Paper J9/10.
193. Kar, A. E., Impactive effects of tornado missiles and aircraft, *Trans. ASCE,* **105** (1979) 2243–60.
194. Kar, A. E., Impact load for tornado generated missiles. *Nucl. Eng. Des.*, **47** (1978) 107–14.
195. Kar, A. E., Loading time history for tornado generated missiles. *Nucl. Eng. Des.*, **51** (1979) 487–93.

196. Zapp, F. C., Testing of containment systems used in light-water-cooled power reactors. Report ORNL-NSIC-26, United States Atomic Energy Commission, 1968.
197. Kennedy, R. P., Effects of an aircraft crash into a concrete reactor containment building. Holmes and Narver Inc., Oakland, California, July 1966.
198. Yeh, G. C. K., Probability and containment of turbine missiles. *Nucl. Eng. Des.*, **37** (1976) 307–12.
199. Zienkiewicz, O. C., *The Finite Element Method*. McGraw-Hill, New York, 1980.
200. General design criteria for nuclear power plant construction permit. *Federal Register*, 33 FR 19213, 11 July 1967.
201. Gwaltney, R. C., Missile generation and protection in light-water-cooled power reactor plants. Report ORNL-NSIC-22, United States Atomic Energy Commission, 1968.
202. Burns, W. J., Kennedy, W. J. L. & Weinzimmer, F., Shoreham nuclear power station with concrete containment. *Proc. Am. Power Conf.*, **30** (1968) 365–70.
203. Owen, D. R., Implicit finite element methods for the dynamic transient analysis of solids with particular reference to non-linear situations. In: *Advanced Structural Dynamics*, ed. J. Donea. Applied Science Publishers, London, 1978, pp. 123–52.
204. Nasser, K. W. & Neville, A. M., Creep of concrete at elevated temperatures. *ACI Journal*, **62** (1965) 1567–79; and Private communication, 1965.
205. Hannant, D. J., The strain behaviour of concrete under compressive stresses at elevated temperatures. CERI, Note. No. RD/LIN 67/66, 1967.
206. Arthanari, S., Influence of temperature and biaxial stresses on creep in concrete. PhD Thesis, University of London, 1966.
207. Ross, A. D., Concrete creep data. *Proc. Inst. Struct. Eng.*, **15** (1957) 314–25.
208. Ross, A. D., Creep and shrinkage in plain, reinforced and prestressed concrete. *Proc. Inst. Civ. Eng.*, **21** (1943).
209. Zienkiewicz, O. C., Watson, M. & King, I. P., A numerical method of viscoelastic stress analysis. *Int. J. Mech. Sci.*, **10** (1968) 807–27.
210. Zienkiewicz, O. C., Gago, J. P. S. E. & Kelly, D. W., The hierarchical concept of finite element analysis. *Computers and Structures*, **16** (1983).
211. Thorpe, D. E., Elastic analysis of the Dungeness B vessel. Conference on PCPV, Inst. Civ. Eng., London, Paper 36, March 1967.
212. Indian Standard Code of Practice for Prestressed Concrete, IS:1345, October 1975.
213. Hilsdorf, H. K. & Muller, H. S., Comparison of prediction methods for creep coefficient of structural concrete with experimental data. In: *Fundamental Research on Creep and Shrinkage*, Sidhoff and Nishoff, The Hague, 1982.
214. ACI prediction of creep shrinkage and temperature effects. Draft, ACI 209, October 1978.
215. CEB–FIP Model Code, 1978.
216. Hansen, T. C., Creep of concrete. A discussion of some fundamental problems. Bulletin 33, Swedish Cement and Concrete Institute, 1975.
217. Zienkiewicz, O. C. & Cormeau, I. C., Viscoplasticity–plasticity and creep in elastic solids. *Int. J. Num. Meth.*, (1974) 821–45.
218. Hilsdorf, H. K., *Drying and Shrinkage of Concrete and Reinforced Concrete Structures*. Ernst and Sohn, Verlag, 1969.
219. ACI prediction of creep, shrinkage and temperature effects. ACI 209, 1970.
220. CEB–FIP 70, method for shrinkage. FIP, 1978.
221. Bangash, Y., Movements in prestressed concrete reactor vessels. *Nucl. Eng. Des.*, **50** (1978) 462–73.
222. Ross, A. D., Illston, J. M. & England, G. L., Short and long-term deformations of concrete as influenced by its physical structure and state. International Conference on the Structure of Concrete, London, 1965.
223. Sharp, T. J., The influence of elevated temperature on the physical behaviour of water in concrete. PhD Thesis, University of London, August 1971.

224. Schimmelwitz, P. & Hundt, J., A note on moisture condition of concrete. SP-34, Vol. 2, American Concrete Institute, Detroit, Michigan, 1972.
225. Davis, R. E., A summary of the results of investigations having to do with volumetric changes in cements, mortars and concretes due to causes other than stress. *ACI Journal, Proc.*, **26** (1930) 407–43.
226. Mitchell, L. J., Thermal expansion tests on aggregates, neat cement, and concretes. *Proc. ASTM*, **53** (1953) 963–77.
227. Meyers, S. L., Thermal expansion characteristics of hardened cement paste and of concrete. *Proc. Highway Research Board*, **30** (1950) 193–203.
228. Bonnel, D. G. R. & Harper, F. C., Thermal expansion of concrete. *J. Inst. Civ. Eng.*, **33** (1950) 320–30.
229. Hatt, W. K., The effect of moisture on concrete. *Trans. ASCE*, **89** (1926) 270–370.
230. Walker, K., Stanton, L., Bloem, D. L. & Mullen, W. G., Effects of temperature changes on concrete as influenced by aggregates. *ACI Journal, Proc.*, **48** (1952) 661–80.
231. Berwanger, C., The modulus of concrete and the coefficient of expansion of concrete and reinforced concrete at below normal temperatures. *Temperature and Concrete*, SP-25, American Concrete Institute, Detroit, Michigan, 1971, pp. 191–233.
232. Emanuel, H. J. & Hulsey, L. J., Prediction of the thermal expansion of concrete. *ACI Journal*, (1977) 149–55.
233. Philleo, R., Some physical properties of concretes at high temperatures. *ACI Journal, Proc.*, **54** (1958) 857–64.
234. Malhotra, H. L., The effect of temperature on the compressive strength of concrete. *Magazine of Concrete Research*, **8** (1956) 63–8.
235. Sullivan, P. J. & Poucher, M. P., The influence of temperature on the physical properties of concrete and mortar in the range 20°C to 400°C. *Temperature and Concrete*, SP-25, American Concrete Institute, Detroit, Michigan, 1971, pp. 103–35.
236. Ali, I. & Kesler, Å., Rheology of concrete. A review of research. The Engineering Experiment Station, University of Illinois Bulletin, Vol. 62, No. 68, 1965.
237. Newman, B. J., Deformational behaviour, failure mechanisms and design criteria for concrete under combinations of stress. PhD Thesis, University of London, 1973.
238. Buettner, D. R. & Hollrah, R. L., Creep recovery of plain concrete. *J. Am. Concr. Inst.*, **64** (1968) 452–61.
239. Brice, L. P., Etude des conditions de formation des fissures de glissement et de decohesion dans les solides, *Travaux*, June (1964).
240. Saliger, I., *Der Stahlbetonbau*, Vol. 7, Auglage, 1956.
241. Dressel, J., Square mesh reinforcement of plane-supporting membrane structures. Results of Wastlund Saliger Technical Report No. 71, Technical University, Dresden, 1964.
242. Somayaji, S. & Shah, S. P., Bond stress versus slip relationship and cracking response of tension members. *J. ACI* (May/June, 1981) 217–25.
243. Somayaji, S., Composite response, bond stress–slip relationships and cracking in ferrocement and reinforced concrete. PhD Thesis, University of Illinois, 1979.
244. Marshall, W. T. & Krishnamurthy, P., Design of end zone reinforcement for pretensioned concrete beam. Proc. 6th FIP, Prague, June 1970.
245. Hoyer, E. *et al.*, Beitrag zur frage der Haftspannung in Eisten beton bauteilen. *Beton und Eisen, Berlin*, **38** (1939) 107–10.
246. Marshall, G. G., End anchorage and bond stresses in prestressed concrete. *Magazine of Concrete Research*, **1** (1949) 123–7.
247. Marshall, W. T., A theory for end zone stresses in pretensioned concrete beams. *J. ACI*, **11** (1966) 45–51.
248. Bangash, Y., The riddle of bonded and unbonded tendons. TPI Report, 1978; also *Concrete*, **8**(6) (1974).

249. Edward, A. D. & Picard, D., Bonding properties of $\frac{1}{2}$ inch diameter strand. *ACI Journal* (November 1972) 684–9.
250. Yannopoulos, P. J., Fatigue, bond and cracking characteristics of reinforced concrete tension members. PhD Thesis, University of London, 1976.
251. Nilson, A. H., Bond stress–slip relations in reinforced concrete. Report No. 345, Department of Structural Engineering, Cornell University, December 1970.
252. Rehm, G., The basic principles of bond between steel and concrete. Translation No. 134, Cement and Concrete Association, London, 1968, p. 66.
253. Evans, R. H. & Robinson, G. W., Bond stresses in prestressed concrete from X-ray photographs. *Proc. ICE, Part 1*, **4** (1955).
254. Hanson, N. W. & Kaar, P. H., Flexural bond tests of pretensioned prestressed beams, *ACI Journal*, **55** (1959).
255. Dably, L. J., Static tests on prestressed concrete beams using 7/16 inch strands. Progress Report No. 13 on Prestressed Concrete Bridge Members, Leigh University, Fritz Engineering Laboratory, November 1956.
256. Base, G. D., An investigation of transmission length in pretensioned concrete. Congress of Federation Internationale de la Precontrainte, Session 3, Paper 9, Berlin, 1958, pp. 603–23.
257. Ratz, E. H. *et al.*, The transmission of prestressed concrete by bond. Congress of Federation Internationale de la Precontrainte, Session 3. Paper 10, Berlin, 1958, pp. 624–40.
258. Dinshore, G. A. *et al.*, Anchorage and bond in pretensioned prestressed concrete members. Report No. 223-19, Leigh University, Fritz Engineering Laboratory, December 1953.
259. Ruch, H. & Rehm, G., Tests for determination of transfer lengths of prestressing steel. *Deutscher Aussuchuss fur Stahlbeton*, Heft 147, Berlin, 1963, pp. 138.
260. Kaar, P. H. *et al.*, Influence of concrete strength and strand transfer length. *J. Prestressed Concrete Institute* (*PCI*), (October 1963).
261. Preston, H. K., Characteristics of 15 per cent stronger seven wire strand. *PCI J.*, **8** (1963) 39–45.
262. Ahmed, M., Bond strength history in prestressed concrete reactor vessels. R-84, Thames Polytechnic, 1984; PhD Thesis, (CNAA), 1983.
263. Hanson, J. M. & Hulsbos, C. L., Overload behaviour of pretensioned prestressed concrete I-beams with web reinforcement. *Highway Record*, No. 76, (1965) 1–33.
264. Over, R. S. & Au, T., Prestressed transfer bond of pretensioned strands in concrete. *ACI Journal*, **62** (1965).
265. Stocker, M. F. & Sozen, M. A., Investigation of prestressed reinforced concrete for highway bridges, Part 5: bond characteristics of prestressing strand. University of Illinois, College of Engineering, Bulletin No. 503, 1970.
266. Lutz, L. A. & Gergely, P., Mechanics of bond and slip for deformed bars in concrete. *ACI Journal*, **64** (1967) 711–21.
267. Wahl, A. *et al.*, Direct measurement of bond–slip in reinforced concrete. *Proc. Am. Soc. Agr. Eng.*, (1969).
268. Tanner, J. A., An experimental investigation of bond–slip in reinforced concrete. MS Thesis, Cornell University, November 1971.
269. Morris, S. & David, W. J., Bond development length tests of grouted 54 strand post-tensioned tendons. *ACI Journal* (October, 1974) 522–5.
270. Morris, S., Large post-tensioned tendons, *ACI Journal* (May/June, 1972).
271. Morris, S., Grouting tests on large post-tensioned tendons for secondary nuclear containment structures. *ACI Journal* (March/April, 1971).
272. Ngo, D. & Scorcelis, A. S., Finite element analysis of reinforced concrete beams. *ACI Journal*, **64** (1967).
273. Mirza, M. S. & Houde, J., Study of bond stress–slip relationships in reinforced concrete. *ACI Journal, Proc.*, **76** (1979) 19–146.

274. Naus, W., A new method for the measurement of local slip between steel and concrete. PhD Dissertation, Technical University, Darmstadt, 1979.
275. Shah, S. P., Workshop on high strength concrete, Department of Materials Engineering, University of Illinois at Chicago Circle, December 1979.
276. Holman, J. O., Fatigue of concrete by constant and variable amplitude loading. Recent research in fatigue of concrete structures, ACJ SP-75, 1982, pp. 72–110.
277. Kakuta, Y. *et al.*, New concepts for concrete fatigue design procedures in Japan. *Proc. IABSE, Lausanne*, **37** (1982) 51–8.
278. Aas-Jacobson, K., Fatigue of concrete beams and columns. Norwegian Institute of Technology, Bulletin No. 70-1, Trondheim, 1979, p. 148.
279. Miner, M. A., Cumulative damage in fatigue. *Trans. Am. Soc. Mech. Eng.*, **67** (1945) 159–64.
280. Waagaard, K., Design recommendations for offshore concrete structures. *Proc. IABSE, Lausanne*, **37** (1982) 59–67.
281. Japan Society of Civil Engineers, Trans. 122, October 1965.
282. Proc. CEB–FIP, Recommendation on fatigue, 1982.
283. American Association of State Highway and Transportation Officials, Report No. 4, Publication No. 953, 1962, p. 217.
284. NCHR (The National Cooperative Highway Research). Private communication, Transportation Research Board, Washington, 1984.
285. Wilson, A., Durability of concrete and concrete structures by slow cycle fatigue. SP-47, No. 12, American Concrete Institute, Detroit, Michigan, 1975, pp. 259–88.
286. Cook, R. D., *Concepts and Applications of Finite Element Analysis* (2nd edn), John Wiley and Sons, New York, 1981.
287. Zienkiewicz, O. C., *The Finite Element in Engineering Science*. McGraw-Hill, London, 1977.
288. Martin, H. C. & Carrey, G. F., *Introduction to Finite Element Analysis*. McGraw-Hill, New York, 1973.
289. Bathe, K. J. & Wilson, E. L., *Numerical Methods in Finite Element Analysis*. Prentice-Hall, Englewood Cliffs, New Jersey, 1972.
290. Desai, C. S. & Abel, J. F., *Introduction to the Finite Element Method.* Van Nostrand Reinhold, New York, 1972.
291. Ahmad, S., Curved finite element in the analysis of solid shell and plate structures. PhD Thesis, University of Wales, Swansea, 1969.
292. Clough, R. W., Comparison of three-dimensional finite elements. Proc. Symposium on Application of Finite Element Methods in Civil Engineering, Vanderbilt University, 1969.
293. Ngo, D., Scordelis, A. C. & Franklin, H. A., Finite element study of reinforced concrete beams with diagonal tension cracks. University of California, Berkeley, UC-SESM Report No. 70-19, December 1970.
294. Marcal, P. V., Finite element analysis with material nonlinearities — theory and practice. Conference on Finite Element Methods in Civil Engineering, Montreal, 1972, pp. 71–113.
295. ASCE Committee on Concrete and Masonry Structures. State-of-the-Art Report on Finite Element Analysis of Reinforced Concrete, Task Committee on Finite Element Analysis of Reinforced Concrete Structures, ASCE Special Publication, 1982.
296. Argyris, J. H. *et al.*, Recent development in the finite element analysis of prestressed concrete reactor vessels. *Nucl. Eng. Des.*, **28** (1974).
297. Bicanic, N. & Zienkiewicz, O. C., Constitutive models for concrete under dynamic loading. *Earthquake Engineering and Structural Dynamics*, **11** (1983) 689–710.
298. Argyris, J. & Faust, W., Limit load analysis of thick-walled concrete structures — a finite element approach to fracture. *Computer Methods in Applied Mechanics and Engineering*, **8** (1976) 215–43.
299. Mohraz, B. *et al.*, Crack development in a prestressed concrete reactor as determined by lumped parameter method. *Nucl. Eng. and Des.* (1970).

300. Sozen, M. A. & Paul, S. L., Structural behaviour of small scale prestressed concrete reactor vessels. *Nucl. Eng. and Des.* (1969) 403–14.
301. Campbell-Allen, D. & Low, E. W. E., Pressure tests on end slabs for prestressed concrete pressure vessels. *Nucl. Eng. and Des.*, **6** (1967) 345–59.
302. Rashid, Y. R., Ultimate strength analysis of prestressed concrete pressure vessels. *Nucl. Eng. and Des.*, **7** (1968).
303. Rashid, Y. R., Analysis of concrete composite structures by the finite element method. *Nucl. Eng. and Des.*, **3** (1966).
304. Rashid, Y. R. & Rockenhauser, W., Pressure vessel analysis by finite element techniques. *Conference on Prestressed Concrete Pressure Vessels*, Inst. Civ. Eng., London, 1968.
305. Connor, J. J. & Sarne, Y., Non-linear analysis of prestressed concrete reactor pressure vessels. 3rd International Conference on Structural Mechanics in Reactor Technology (SMiRT), Paper H2/2, London, 1975.
306. Takeda, T., Yamaguchi, T. & Imoto, K., Inelastic analysis of multi-cavity PCRV under internal pressure. 3rd International Conference on Structural Mechanics in Reactor Technology (SMiRT), Paper H2/2, London, 1975.
307. Akyuz, F. A. & Merwin, J. E., Solution of non-linear problems of elastoplasticity by finite element method. *AIAA J.*, **6** (1968).
308. Alexander, J. M. & Gunasekera, J. S., On the geometrically similar expansion of a thin infinite plate. *Proc. Roy. Soc. London*, **A326** (1972) 361–72.
309. Anand, S. C., Lee, S. L. & Rossow, E. C., Finite element analysis of elastic–plastic plane stress problems based on Tresca's yield criterion. *Ingenieur Archiv*, **39** (1970).
310. Argyris, J. H., *Energy Theorems and Structural Analysis*. Butterworths, London, 1960.
311. Argyris, J. H., Three-dimensional anisotropic and inhomogenous elastic media, matrix analysis for small and large displacements. *Ingenieur Archiv*, **34** (1965).
312. Argyris, J. H., Elasto-plastic matrix displacement analysis of three-dimensional continua. *J. RAeS*, **69** (1965).
313. Argyris, J. H., Continua and discontinua. Proc. Conference Matrix Methods in Structural Mechanics, Ohio, 1965.
314. Argyris, J. H., Application of the matrix displacement method to the analysis of pressure vessels. *ASME, J. Eng. Industry*, (1970).
315. Argyris, J. H. *et al.*, *Matrix Methods of Structural Analysis*. Pergamon Press, Oxford, 1964.
316. Argyris, J. H. *et al.*, Some general considerations of the natural mode technique, Part 1. *J. RAeS*, **73** (1969).
317. Argyris, J. H. *et al.*, Some general considerations of the natural mode technique, Part 2. *J. RAeS*, **73** (1969).
318. Argyris, J. H. *et al.*, Method of elasto-plastic analysis. Symposium of Finite Element Techniques at the Institute fur Statik und Dynamik der Luft und Raumfahrtstrucktionen, University of Stuttgart, 1969.
319. Armen, H. *et al.*, Discrete element methods for the plastic analysis of structures subjected to cyclic loading. *Int. J. Num. Meth. Eng.*, **2** (1970).
320. Bogner, F. K. *et al.*, A cylindrical shell discrete element. *AIAA J.*, **5** (1967).
321. Brebbia, C. *et al.*, Geometrically non-linear finite element analysis. *J. Eng. Mech. Div., ASCE*, **95** (1969).
322. Bushnell, D., Non-linear axisymmetric behaviour of shells of revolution. *AIAA J.*, **5** (1967).
323. Chan, A. S. L. *et al.*, The analysis of cooling towers by the matrix finite element method. *J. RAeS*, **74** (1970).
324. Dupuis, G. A., Incremental finite element analysis of large elastic deformation problems. Brown University, May 1971.
325. Ford, H. & Alexander, J. M., *Advanced Mechanics of Materials*. Longman, London, 1963.
326. Gallagher, R. H. *et al.*, Discrete element procedure for thin shell instability analysis. *AIAA J.*, **5** (1967).

327. Gallagher, R. H. *et al.*, Discrete element approach to structural instability analysis. *AIAA J.*, **1** (1963).
328. Gunasekera, J. S., A preliminary study of closed die forging. MSc Thesis, University of London, 1969.
329. Gunasekera, J. S. *et al.*, Matrix analysis of the large deformation of an elastic–plastic axially symmetric continuum. Proc. Symposium on Foundations of Plasticity, Warsaw, September 1972.
330. Green, A. E. *et al.*, A general theory of an elastic–plastic continuum. *Arch. Ration. Mech. Anal.*, **18** (1965).
331. Green, A. E. & Zerna, W., *Theoretical Elasticity*, Clarendon Press, Oxford, 1954.
332. Hartley, H. O., The modified Gauss–Newton method for the fitting of nonlinear regression functions by least squares. *Technometrics*, **3** (1961).
333. Hartz, B. J. *et al.*, Finite element formulation of geometrically non-linear problems of elasticity. Proc. Japan–US Seminar on Matrix Methods of Structural Analysis and Design, Tokyo, 1969.
334. Hibbit, H. D. *et al.*, Hybrid finite element analysis with particular reference to axisymmetric structures. *AIAA J.* (8th Aerospace Sciences Meeting, New York, January 1970).
335. Hibbit, H. D. *et al.*, A finite element formulation for problems of large strain and large displacement. *Int. J. Solids Struct.*, **6** (1970).
336. Hill, R., *The Mathematical Theory of Plasticity*. Clarendon Press, Oxford, 1950.
337. Mallet, R. H. *et al.*, Finite element analysis of non-linear structures. *J. Struct. Div., ASCE*, **94** (1968).
338. Marcel, P. V., A comparative study of numerical method of elastic–plastic analysis. *AIAA J.*, **6** (1968).
339. Marcel, P. V., Finite element analysis of combined problems of non-linear material and geometric behaviour. Brown University, March 1969.
340. Marcel, P. V. *et al.*, Elastic–plastic analysis of two-dimensional stress systems by the finite element method. *Int. J. Mech. Sci.*, **9** (1967).
341. Martin, H. C., Derivation of stiffness matrices for the analysis of large deflection and stability problems. Proc. 1st Conf. on Matrix Methods in Structural Mechanics, Ohio, October 1965.
342. Washizu, K., *Variational Methods in Elasticity and Plasticity*. Pergamon Press, Oxford, 1968.
343. Webster, G. A., Iterative procedures for elastic, plastic and creep deformation in beams. *J. Mech. Eng. Sci.*, **9** (1967).
344. Wilson, E. L., Structural analysis of axisymmetric solids. *AIAA J.*, **3** (1965).
345. Wissman, J. W., Non-linear structural analysis; tensor formulation. Proc. 1st Conf. on Matrix Methods in Structural Mechanics, Ohio, October, 1965.
346. Yamada, Y. *et al.*, Plastic stress–strain matrix and its application for the solution of elastic–plastic problems by the finite element method. *Int. J. Mech. Sci.*, **10** (1968).
347. Zienkewicz, O. C., Valliappan, S. & King, I. P., Stress analysis of rock as a no-tension material. *Géotechnique* (1968).
348. Murray, D. W. *et al.*, An approximate non-linear analysis of thin-plates. Proc. 2nd Conf. on Matrix Methods in Structural Mechanics, December 1969.
349. Argyris, J. H. & Sharpe, D. W., Methods of elasto-plastic analysis. Proc. ISA–ISSC Symposium on Finite Element Technology, Stuttgart, 1969, pp. 381–416.
350. Bangash, Y., The structural integrity of concrete containment vessels under external impacts. 6th International Conference on Structural Mechanics in Reactor Technology (SMiRT), J7/6, Paris, August 1981.
351. Bangash, Y., The automated three-dimensional cracking analysis of prestressed concrete vessels. 6th International Conference on Structural Mechanics in Reactor Technology (SMiRT), H Paper 3/2, Paris, August 1981.

352. Bathe, K. J. & Wilson, E. L., Stability and accuracy analysis of direct integration method. *Int. J. of Earthquake Engineering and Structural Dynamics*, **1** (1977) 283–91.
353. Armen, H. *et al.*, Discrete element methods for the plastic analysis of structures subjected to cyclic loading. *Int. J. Num. Meth. Eng.*, **2** (1970).
354. Bogner, F. K. *et al.*, A cylindrical shell discrete element. *AIAA J.*, **5** (1967).
355. Brebbia, C. *et al.*, Geometrically non-linear finite element analysis. *J. Eng. Mech. Div., ASCE*, **95** (1969).
356. Bushnell, D., Non-linear axisymmetric behaviour of shells of revolution. *AIAA J.*, **5** (1967).
357. Chan, A. S. L. *et al.*, The analysis of cooling towers by the matrix finite element method. *J. RAeS*, **74** (1970).
358. Dupuis, G. A., Incremental finite element analysis of large elastic deformation problems. Brown University, May 1971.
359. Newmark, N. M., A method for computation of structural dynamics. *Proc. Am. Soc. Civ. Eng.*, **85** (1959) 67–94.
360. Akyuz, F. A. & Merwin, J. E., Solution of non-linear problems of elastoplasticity by finite element method. *AIAA J.*, **6** (1968).
361. Alexander, J. M. & Gunasekera, J. S., Private communication.
362. Anand, S. C. *et al.*, Finite element analysis of elastic–plastic plane stress problems based upon Tresca's yield criterion. *Ingenieur Archiv*, **39** (1970).
363. Bathe, K. J. *et al.*, Finite element formulation and solution of non-linear heat transfer. *Nucl. Eng. Des.*, **51** (1979) 389–405.
364. Chan, R. C. Y., Concrete filled tubular columns. Project Report Civil Engineering, Middlesex Polytechnic, May 1984.
365. Knowles, R. B. & Park, R., Strength of concrete filled steel tubular columns. *J. Struct. Div., ASCE*, **95** (1969).
366. Knowles, R. B. & Park, R., Axial load design for concrete filled steel tubes, *J. Struct. Div., ASCE*, **96** (1970).
367. Neogi, P. K., Sen, H. K. & Chapman, I. C., Concrete filled tubular columns under eccentric loading. *ACI Journal*, **47** (1969) 187–95.
368. Russel, W. A., Structural properties of light-gage tubular columns. Housing Research Paper No. 21, Housing and Home Finance Agency, October 1953.
369. Salani, H. J. & Sims, J. R., Behaviour of mortar filled steel tubes in compression. *ACI Journal Proc.*, **61** (1964) 1271–83.
370. Sen, H. K. & Chapman, J. C., Ultimate load tables for concrete filled steel tubular. CIRIA Technical Note 13, August 1970.
371. Furlong, R. W., Strength of steel-encased concrete beam columns. *J. Struct. Div., ASCE*, **93** (1967) 113–24.
372. Furlong, R. W., Design of steel-encased concrete beam columns. *J. Struct. Div., ASCE*, **94** (1968) 267–81.
373. Furlong, R. W., Concrete-encased steel columns — design tables. *J. Struct. Div., ASCE*, **100** (1964).
374. Gardner, N. J. & Jacobson, R., Structural behaviour of concrete filled steel tubes. *ACI Journal* (1967) 404–13.
375. Bangash, Y., The finite element analysis of concrete filled tubes under complex load — Program CONFT. Middlesex Polytechnic, 1984.
376. Sen, H. K. & Chapman, J. C., Ultimate load tables for concrete filled tubular steel columns. CIRIA Tech. Note 13, August 1970.
377. BS 5950, The structural use of steel, 1985.
378. Basu, A. & Sommerville, J., Derivation of formulae for the design of rectangular composite columns. ICI Paper No. 7206S, May 1969.
379. BS 4360, Weldable structural steels, 1968.
380. BS 4: Part 2, Structural steel sections — hot rolled hollow sections, 1969.

381. Pink, A., *Winter Concreting*. Cement and Concrete Association, London, 1967.
382. BS 938. General requirements for metal — arc welding of structural steel tubes to BS 1775, 1962.
383. *ACI Journal*, January 1971 to February 1983.
384. BS 449: Part 1. The use of structural steel in building (Imperial Units), 1970.
385. Portland Cement Association. Private communication, Concrete Filled Tubes, Skokie, USA, 1981.
386. British Steel Corporation. Private communication, Literature on concrete filled tubes, 1981.
387. Timoshenko, S. P. & Gere, J. M., *Theory of Elastic Stability*. McGraw-Hill, New York, 1958.
388. Bangash, Y., The ultimate load analysis of containment vessels under accidental pressures. International Conference on Structural Mechanics in Reactor Technology (SMiRT), August 1987, pp. 275–80.
389. Terzaghi, K., *Theoretical Soil Mechanics*. John Wiley, New York, 1943.
390. Ghana Department of Irrigation. Private communication, Akra, Ghana, 1973.
391. Brink, A., Recent developments in the design of submerged tunnels. *Struct. Engr.*, **44** (1966).
392. Bickel, J. O., Trench type subaqueous tunnels: design and construction. *Struct. Engr*, **44**, **45** (1966, 1967).
393. Pequignot, C. A., Selective bibliography on immersed tubes. *Tunnels and Tunnelling*, **1** (1969).
394. Palmer, W. F. *et al.*, Developments in trench-type tunnel construction. *J. Const. Div., ASCE*, **101** (1975).
395. Glerum, A. *et al.*, Motorway tunnels built by the immersed tube method. *Rijkswaterstaat Comm.*, **25** (1976).
396. Brakel, J., Submerged tunnelling. Technische Hogeschool, Delft, 1978.
397. United States Bureau of Reclamation. Hoover Dam Tunnels & Spillways, Washington DC, 1955.
398. WAPDA. Reports on tunnels for Mangla and World Bank. Tarbela dams. West Pakistan Water and Power Development Authority, Reports 2, 3, 1975.
399. Waters, T. C. & Barrett, N. T., Prestressed concrete pressure vessels for nuclear reactors. *J. BNES* (July 1963).
400. Marsh, R. O. & Melese, G. B., Prestressed concrete pressure vessels. *Nucleonics*, **23** (1975) 63–74.
401. Frankline Institute Laboratories. State of Art of Prestressed Concrete Pressure Vessels for Nuclear Power Reactors — A Critical Review of Literature. USEAC Report ORNL-TM-812, Oak Ridge National Laboratory, June 1964.
402. Bender, M., A status report on prestressed concrete reactor vessels, pressure vessel technology, Parts 1 and 2. *Nuclear Structural Engineering*, (1965) 83–90; 206–23.
403. Jaeger, T. A., Note on stress analysis of prestressed concrete reactor vessels. *Nuclear Structural Engineering*, **1** (1965) 133–6.
404. Rashid, Y. R., Analysis of axisymmetric composite structures by the finite element. *Nucl. Eng. Des.*, **3** (1966) 163–82.
405. Tottenham, H. & Kanchi, M. B., Structural characteristics of cylindrical pressure vessels of medium thickness. *Nucl. Eng. Des.* (1966) 177–92.
406. Kanchi, M. B., Elastic theories of thick plates and shells. CE Dept. Report No. CE11-65, University of Southampton, 1965.
407. Reisnner, E., Stress–strain relations in the theory of thin elastic shells. *J. Math. Physics*, **31** (1952) 109.
408. Orr, R. S. & Holland, D. A., Theoretical analysis of model spherical vessels. Conf. on PCPV, London, Paper 33, March 1967.
409. UKAEA, Private communication, 1967.

410. Common, D. K. & Hannah, I. W., Specifications of concrete vessels for gas cooled reactors. Conference on PCPV, London, Paper 7, March 1967.
411. Finigen, A., Ultimate analysis of Dungeness B vessels. Conference on PCPV, London, Paper 31, Group F, March 1967.
412. Zbirohowski-Kuscia, K. & Carlton, D., Analysis of vessel structures with particular reference to Wylfa. Conference on PCPV, London, Paper 30, March 1967.
413. Morice, P. B., Discussion on group F papers on design and analysis of vessel structures. Conference on PCPV, London, March 1967, pp. 397–434.
414. Zudnas, Z. & Tans, C. P., Feasibility study of prestressed concrete pressure vessel design. USAEC Report CRNL-TM-813, Oak Ridge National Laboratory, June 1967.
415. (a) Brown, A. H., The Oldbury vessels. Conference on PCPV, London, Paper 1, Group A, March 1967. (b) Harris, A. J. & Hay, J. D., Rupture design of the Oldbury vessels. Conference on PCPV, London, Paper 29, Group F, March 1967.
416. Rashid, Y. R., Ultimate strength analysis of prestressed concrete pressure vessels. *Nucl. Eng. Des.*, **7** (1968) 334–44.
417. Gomez, A. E. *et al.*, Lumped parameter analysis of cylindrical prestressed concrete reactor vessels. University of Illinois, Vols. 1 & 2, 1968.
418. Papers on Ultimate Load Tests, Conference on PCPV Model Techniques, British Nuclear Energy Society, The Institution of Civil Engineers, London 1968. (a) Paper 3 — Deflection measurement technique, Brading, K. F. & McKillen, R. R. (b) Paper 4 — Techniques for rupture testing of prestressed concrete vessel models, Scotto, F. (c) Discussion on papers 2 and 4, Bangash, Y. (d) Paper 6 — The use of simplified models for the design of end slabs, Campbell-Allen, D. & Low, E. W. E. (e) Paper 7 — small scale model prestressed concrete pressure vessel tests at Foulness, Davidson, I. (f) Paper 8 — An experimental investigation into the load behaviour of perforated end slabs for concrete pressure vessels under temperature and external load, Stefanou, G. D. *et al.* (g) Paper 9 — Model philosophy in relation to prestressed concrete pressure vessel design problems, Langan, D. (h) Discussion on Papers 7 and 9, Bangash, Y. (i) Paper 11 — The behaviour of the Oldbury model vessel with time under thermal and pressure loadings, Hornby, T. W. (j) Paper 12 — The two Bugey 1/5 models, Launay, P. (k) Paper 13 — Elastic and ultimate pressure tests on a one-tenth scale model of the Dungeness B concrete pressure vessel, Brading, K. F. & Finigin, A. (l) Paper 14 — Testing the one-tenth scale model of the Hinkley B and Hunterston B power station prestressed concrete pressure vessels, Eadie, D. McD. & Bell, D. J. (m) Discussion Session C including P on Paper 13, Bangash, Y.
419. Sozen, M. A. *et al.*, Strength and behaviour of prestressed concrete vessels for nuclear reactors. Vols 1, 2 & 3, University of Illinois, 1969.
420. Koerner, R. J., Over pressure analysis of prestressed concrete pressure vessels. *Nucl. Eng. Des.* (December 1970).
421. ACI-BAM Concrete for Nuclear Reactors. Proc. International Seminar at Bundesantalt fur Materialsprufungin, Berlin, October 1970, ACI Special Publication SP-34, Vols 1, 2 & 3. (a) Paper SP34-12, Capacity of prestressed concrete pressure vessels to resist shear, Garas, F. K. & Langan, D. (h) Paper SP34-67, Concrete behaviour under continued stresses up to failure: test results on small dimension prestressed concrete pressure vessel models, Scotto, F. L. (c) Paper SP34-68, Testing of a perforated prestressed concrete pressure vessel top slab model, Meerwald, K. & Schwiers, G. (d) Paper SP34-69, Apparatus, instrumentation and concrete models of Bugey 1 prestressed concrete pressure vessel.
422. ACI-BAM Concrete for Nuclear Reactors. Proc. International Seminar at Bundesanstalt fur Materialsprufungin, Berlin, October 1970. ACI Special Publication SP-34, Vols 1, 2 & 3. (a) Paper SP34-7, PCRV problems of calculation, material properties and evaluation of results, Hofmann, B. H. H. (b) Paper SP34-9, Discussions on finite element analysis of prestressed concrete reactor vessels, Argyris, J. H. *et al.* (c) Paper SP34-10, Analysis of prestressed concrete pressure vessels for nuclear reactors in Czechoslovakia, David, M. (d)

Paper SP34-70, Strain behaviour of the Oldbury concrete pressure vessel under operating pressure and temperature, Carmichael, G. D. T. *et al.* (e) Paper SP34-39, An investigation of the time dependent behaviour of prestressed concrete pressure vessels, Whitman, G. D. *et al.*

423. CEC & BAM Proc. 1st International Conference on Structural Mechanics in Reactor Technology, 20–24 September 1971, Berlin, Vol. 4, Reactor Pressure Vessels, Part H, Prestressed Concrete Pressure Vessels. (a) Paper H3/4, Behaviour of eng. slabs in cylindrical prestressed concrete pressure vessels, Langan, D. & Garas, F. K. (b) Paper H3/5, Analyse tridimensionelle du comportement non lineaire d'un caisson de reacteur nucleaire en beton precontraint, Hussain, M. K. *et al.* (c) Paper H3/7, Allegeineine Berechnung Von Spannbeton-Reaktordruck behaltern unter Berucksichtigung von Nichtlinearen Spannungs-Dehnungsgesetzen Nach Der Methode Der Dynamischen Relaxation, Schnellenbach, G. (d) Paper H4/3, Der Spannberunbehalter des THTR-300 — MWe Prototyp-Kernkraftwerkes, Lotz, H. *et al.* (e) Paper H4/4, The design of pod boiler pressure vessels with particular reference to Hartlepool nuclear power station, Langan, D. *et al.* (f) Paper H5/3, The Scandinavian PCRV model project stress calculation and experimental verification, Andersen, S. I. *et al.* (g) Paper H5/4, die Dreidimensionalen Elastischen Spannungszustande im Bereich der Offnungen des THTR-Spannbeton behalters Ein Vergleich Zwischen Numerischer Berechnung und Messergebnissen Eines Elastischen Modells, Hansson, V. (h) Paper H5/5, Modeliversuche zur Vertiefung der Kenntnisse Uber das Tragverhalten des Spannberonbehalters fur das THTR-300-MWe Prototyp-Kernkraftwerk, Lotz, H. & Meerwald, K. (i) Paper H5/6, Thin-walled 1 : 20 prestressed concrete pressure vessel model for THTR reactor type, Scotto, F. L. (j) Paper H5/7, An investigation on a prestressed concrete reactor vessel under internal gas pressure mode of rupture, Lenschow, R. (k) Paper H5/8, Theoretische und Experimentale Untersuchungen des Spannungs-Dehnungs Verhaltens der Druckbehalter aus Vorgespanntem Beton, Servit, R.

424. CEC & BAM Proc. 2nd International Conference on Structural Mechanics in Reactor Technology, September 1973. (a) Paper H2, Simple approximate method for the analysis of the ultimate carrying capacity of prestressed concrete reactor pressure vessels, Huber, A. E. & Hoffman, H. H. (b) Paper H2/2, On the behaviour of PCPV under ultimate load conditions and its influence on calculations and design especially applied to the THTR vessel, Binseil, P. (c) Paper H2/3, Investigation on ultimate load design of prestressed concrete reactor pressure vessels, Schimmelpfennig, K. & Hansen, V. (d) Paper 2/4, Inelastic behaviour and failure modes of prestressed concrete vessels, Fluge, F., Gausel, E. & Lenschow, R. (e) Paper 2/5, Three-dimensional rupture analysis of a prestressed concrete pressure vessel including creep effects, Sang, B. *et al.*

425. CEC & BNES Trans. 3rd International Conference on Structural Mechanics in Reactor Technology, 15 September 1975, London, Vol. 3, Reactor Vessels, Part H, Structural Analysis of Prestressed Concrete Reactor Pressure Vessels. (a) Paper H2/2, Non-linear analysis of prestressed concrete reactor pressure vessels, Connor, J. J. & Sarne, Y. (b) Paper H2/3, Experience in the application of a finite element system to the analysis of complex prestressed concrete pressure vessels, Wade, M. J. & Henrywood, R. K. (c) Paper H2/4, Analysis of prestressed concrete reactor vessel under high thermal gradient, Gulkan, P. & Akay, H. U. (d) Paper H2/5, Inelastic analysis of a multicavity PCRV under internal pressure, Takeda, T. *et al.* (e) Paper H2/6, The method of slice substructures in the analysis of boiler-podded PCPV, Kawamata, S. *et al.* (f) Paper H2/9, Probabilistic determination of partial safety factor for the design of prestressed concrete high pressure vessels, Rackwitz, R. (g) Paper H3/6, Theoretical and experimental studies for optimization of PCRV top closures, Ottasen, N. S. & Andersen, S. I. (h) Paper H4/1, The analysis of crack structures, Davidson, I. (i) Paper H4/2, Crack analysis of multicavity prestressed concrete reactor vessels, Gallix, R. *et al.* (j) Paper H4/3, Failure mode analysis of PCRV influence of

some hypothesis, Zimmermann, Th. *et al.* (k) Paper H4/4, Comparison of two methods of analysing three-dimensional cracking in the upper section of the THTR prestressed concrete pressure vessel, Bindseil, P. & Hansson, V. (l) Paper H4/5, Three-dimensional analysis of cracked concrete in prestressed concrete pressure vessels and comparison with experimental results, Hansson, V. & Stover, R. (m) Paper H4/6, Ultimate load analysis of prestressed concrete reactor pressure vessels considering a general material law, Schimmelpfenning, K. (n) Paper H4/7, Inelastic behaviour, failure modes and ultimate load design of prestressed concrete pressure vessels, Berquam, T. *et al.* (o) Paper H4/8, Ultimate strength design of prestressed concrete reactor vessels, Meyer, C. & Goldman, B. I.

426. Irving, J. *et al.*, A full-scale model test of hot spots in the prestressed concrete pressure vessels of Oldbury nuclear power station. *Proc. Inst. Civ. Eng. Part 2*, **57** (1974) 331–51 & 795–8 (Discussion).
427. Brunton, J. D. *et al.*, Wire-winding of Hartlepool and Heysham reactor pressure vessels. International Conference in Design, Construction and Operation PCPV's and Containments for Nuclear Reactors, Paper 147/75, University of York, September 1975.
428. Cahill, T. & Branch, G. D., Long term relaxation behaviour of stabilized prestressing wires and strands. Inst. Civ. Eng., March 1967.
429. Bangash, Y., A circumferential prestressing load analysis due to wire/strand winding systems on prestressed concrete reactor vessels. *Proc. Inst. Civ. Eng.*, **157** (1974) 437–50 (Discussion).
430. Hruska, F. H., Radial forces in wire ropes. *Wire and Wire Products* (May 1952).
431. Leissa, A. W., Contact stresses in wire ropes. *Wire and Wire Products*, **34** (1959) 307–14 & 372–73.
432. Luskin, J. L., *Contact problems*. In: *Handbook of Engineering Mechanics*, W. Flugge, ed. McGraw-Hill, New York, 1962.
433. Starkey, W. L. & Cress, H. A., Analysis of critical stress and mode of failure of a wire rope. *J. Eng. for Industry, ASME*, **81,** (1959) 307–16.
434. Durelli, A. J. & Machida, S., Response to oversized epoxy strand models to axial, torsional and bending loads. Catholic University of America, Washington DC. Technical Report 72-2, 1972, p. 17.
435. Machida, S. & Durelli, A. J., Response of a strand in twisted wire cables. Catholic University of America, Washington DC, Technical Report 72-1, 1972, p. 22.
436. Hertz, H., *Uber die Behruhrung Fester Elastische Korper*, Leipzig, 1965, pp. 174–96 & 155–73.
437. Prandtl, L., Spannungs Verteilung in Plastischen Koerpern. Proc. 1st International Congress Applied Mechanics, Delft, Technische Boekhandel en Druckerij, J. Waltman Jr., 1925, pp. 43–54.
438. Reusse, E., Beruecksichtigung der Elastischen Formaenderungen in der Plastizitaetstheorie. *Z. Angew. Math. Mech.*, **10** (1930) 226–74.
439. Bland, D. R., The associated flow rule of plasticity, *J. Mech. Phys. Solids*, **5** (1957) 71–8.
440. Budiansky, B., A reassessment of deformation theories of plasticity, *J. Appl. Mech.*, **26** (1959) 259–64.
441. Johnson, W. & Mellor, P. B., *Plasticity of Mechanical Engineers*. Van Nostrand, Princeton, NJ, 1962.
442. Mendelson, A. & Manson, S. S., Practical solution of plastic deformation problems in elastic–plastic range, NASA Tech. Report R-28, 1959.
443. Berg, O. Y., The problem of strength and plasticity of concrete. *Doklady Akademii Nauk*, **70** (1950) 617–20.
444. Hill, R., *The Mathematical Theory of Plasticity*. Oxford University Press, Oxford, 1956, pp. 34–45.
445. Green, A. E. & Adkins, J. E., *Large Elastic Deformations*. Oxford University Press, Oxford, 1960.

446. Marcal, P. V. & King, I. P., Elastic–plastic analysis of two-dimensional stress systems by the finite element method. *Int. J. Mech. Sci.*, **9** (1967) 143–55.
447. Eringen, A. C., Linear theory of micropolar elasticity, *J. Math. and Mech.*, **15** (1966).
448. Yamada, Y. *et al.*, Plastic stress–strain matrix and its application for the evolution of elastic–plastic problems by the finite element method. *Int. J. Mech. Sci.*, **10** (1969) 343–54.
449. Zienkiewicz, O. C. *et al.*, Finite element methods in the reactor vessels. *Nucl. Eng. Des.*, **20** (1972) 507–41; Private communication, 1972.
450. Schalkwijk, R. & Konter, A., Dynamic analysis with MARC. *Zoetermeer* (1986).
451. Simon Carves and Atomic Power Construction. BBRV tests on axial and circumferential tendons for short and long term conditions. Private communication, 1973.
452. Warner, P. C., The Dungeness B vessels. Conference on PCPV. Inst. Civ. Eng., Paper 3, Group A, London, March 1967.
453. Thorpe, D. E., Elastic analysis of the Dungeness B vessel. Conference on PCPV, Inst. Civ. Eng., Paper 36, London, March 1967.
454. Cable Covers & Taylor Woodrow. Cable covers multiforce tendons, Private communication, 1973.
455. Dawson, P., Development of the prestressing system for the Wylfa vessel, Conference on PCPV, Paper 23, March 1967.
456. Burrow, R. E. D., Prestressing tendon systems. Conference on PCPV, Inst. Civ. Eng., Group E, Paper 22, London, 1967.
457. Sokolikoff, I. S. & Redheffer, R. M., *Mathematics of Physics and Modern Engineering*. McGraw-Hill, New York, 1958.
458. Taylor, S. J. *et al.*, Development, testing and installation of prestressing of the PCPV's at Hinkley B and Hunterston B. International Conference on Experience in Design, Construction and Operation of PCPV's and Containments for Nuclear Reactors, Inst. Mech. Eng. & University of York, Paper C152/75, September 1975.
459. Thorpe, W., BBRV post-tensioning systems as applied to reactor containments and prestressed concrete pressure vessels. International Conference on Experience in Design, Construction and Operation of PCPV's and Containments for Nuclear Reactors, Inst. Mech. Eng. & University of York, Paper C145/75, September 1975.
460. Ross, A. D., The elasticity, creep and shrinkage of concrete. Proc. Conference Mechanical Properties of Non-Metallic Brittle Materials, London, April 1958.
461. Hardingham, R. R., Parker, J. V. & Spruce, T. W., Liner design and development for the Oldbury vessel. Conference on PCPV, Inst. Civ. Eng., Paper 56, London, March 1967.
462. Young, A. G. & Tate, L. A., Design of liners for reactor vessels. Conference on PCPV, Inst. Civ. Eng., London, Paper 57, March 1967.
463. Chapman, J. C. & Carter, A., Interaction between a pressure vessel and its liner. Conference on PCPV, Inst. Civ. Eng., Paper 58, London, March 1967.
464. Bishop, R. F. *et al.*, Liner design and construction. Conference on PCPV, Inst. Civ. Eng., Paper 59, London, March 1967.
465. Parker, J. V., Stress analysis of liners for prestressed concrete pressure vessels. Proc. 1st International Conference on Structural Mechanics in Reactor Technology, Paper H6/1, Berlin, September 1971.
466. White, C. M. *et al.*, Basis of design of liners for prestressed concrete pressure vessels and practical examples of the application of the design basis. Proc. 1st International Conference on Structural Mechanics in Reactor Technology, Paper 6/8, Berlin, September 1971.
467. Chan, H. C. & McMinn, S. J., The stabilisation of the steel liner for a prestressed concrete pressure vessel. *Nucl. Eng. Des.*, **3** (1966) 66–73.
468. Peinado, C. O., Private communications, 1977.
469. Kitcher, T. P., Buckling of a stud supported thin cylindrical shell encased in concrete. Report prepared for Nelson Stud Welding Co., Lorain, Ohio, August 1969.

470. Tate, L. A. *et al.*, The design of penetration — liner junctions for prestressed concrete pressure vessels. 2nd International Conference on Structural Mechanics in Reactor Technology, Paper H5/4, Berlin, September 1973.
471. NSW Design Data, Nelson concrete anchor studs. Nelson Stud Welding, Division of Gregory Industries, Ohio, Manual No. 21, 1961.
472. Nemet, J., Reaktordruck Behalter aus Spannbeton mit Heiber Dichthaut. 1st International Conference on Structural Mechanics in Reactor Technology, Paper H4/6, Berlin, 1971.
473. Nemet, J., The Austrian PCRV project with a hot liner — A status Report. 2nd International Conference on Structural Mechanics in Reactor Technology, Paper 3/8, Berlin, September 1973.
474. Howarth, H., Inelastic buckling of the lining units of a prestressed concrete pressure vessel. MSc Thesis, The Victoria University of Manchester, May 1969.
475. Oberpichler, R. *et al.*, Stress analysis of liners for prestressed concrete reactor pressure vessels with regard to non-linear behaviour of liner material and of anchor characteristics. 3rd International Conference on Structural Mechanics in Reactor Technology, Paper H3/8, London, September 1975.
476. Yang, T. Y., A matrix displacement matrix on pre and post buckling analysis of liners for reactor vessels. 1st International Conference on Structural Mechanics in Reactor Technology, Paper H6/2, Berlin, 1971.
477. Schnelenbach, G., Beitrag zur Numerischen Berechung des Raumlichen Spannungs-Zustandes in Hohlzylindern mit Ausschnitten. PhD University of Bochum, 1969.
478. BS 4975. Specification for prestressed concrete pressure vessels for nuclear reactors, 1973.
479. Doyle, J. M. & Chu, S. L., Liner plate buckling and behaviour of stud and rip type anchors. 1st International Conference on Structural Mechanics in Reactor Technology, Paper H6/3, Berlin, September 1971.
480. Proposed standard code for concrete reactor vessels and containments. ACI-HSME Technical Committee on Concrete Pressure Components for Nuclear Service, August 1972.
481. Structural analysis of critical areas of cavity liner. Fort St. Vrain, Gulf General Atomic Report GADR-17, May 1972.
482. Containment Building Liner Plate Design Report. Bechtel Corporation, San Francisco, December 1972.
483. Tan, C. P., A study of the design and construction of particles of prestressed concrete and reinforced concrete containment vessels. Clearing House for Federal Scientific and Technical Information, Report No. TID-25176, 1969.
484. Burdett, E. G. & Rogers, L. W., Liner anchorage tests. *J. Struct. Div., ASCE*, (July 1975).
485. Mutzl, J. *et al.*, Buckling analysis of the hot liner of the Austrian PCRV Concept. 3rd International Conference on Structural Mechanics in Reactor Technology, Paper H3/9, London, 1975.
486. Levy, S., Bending of rectangular plates with large deflections. NACA TR-737, 1942.
487. Preliminary safety analysis report. Bellefonte Nuclear Plant, Docket Nos. 50-438 and 50-439, Tennessee Valley Authority, Knoxville, Tennessee, 1974.
488. Onat, E. T. & Drucker, D. C., Inelastic stability and incremental theories of plasticity. *J. Aeronautical Sci.*, **20** (1953) 181–6.
489. Murphy, L. M. & Lee, L. H. N., Inelastic buckling process of axially compressed cylindrical shells subject to edge constraints. *Int. J. Solids and Structures*, **7** (1971) 1153–70.
490. Shanley, F. R., Inelastic column theory. *J. Aeronautical Sci.*, **14** (1947) 261–7.
491. Hill, R., The essential structure of constitutive laws for metal composites and polycrystals. *J. Mech. Phys. Solids*, **15** (1967) 79–95.
492. Haghdi, P. M. Stress–strain relations in plasticity and thermoplasticity. Proc. 2nd Symposium Naval Structural Mechanics, Brown University, 1960, pp. 121–67.
493. Sewell, M. J., A yield surface corner lowers the buckling stress of an elastic/plastic compressed plate. MRCTR 1226, University of Wisconsin, 1973.

494. Gerard, G., Plastic stability theory of stiffened cylinders under hydrostatic pressure. *J. Ship Research*, **6** (1962) 1–7.
495. Koiter, W. T., The effect of axisymmetric imperfections on the buckling of cylindrical shells under axial compression. *Proc. Koninkl. Nederl. Akademic van Wenschappen*, **66** (Series B) (1963).
496. Odquist, F. K. G., *Mathematical Theory of Creep and Creep Rupture*. Oxford University Press, Oxford, 1966.
497. Alnajafi, A. M. J., Non-linear behaviour of cylindrical shells under lateral pressure loading. Report TT7003, Loughborough University of Technology, October 1970.
498. Crose, J. G. & Ang, A. H. S., A large deflection analysis method for elastic–perfectly plastic circular plates. Civ. Eng. Studies, Struct. Research Series No. 323, University of Illinois, 1967.
499. Yaghmai, S., Incremental analysis of large deformations in mechanics of solids with applications to axisymmetric shells of revolution. PhD Thesis, University of California, Berkeley, 1969.
500. Irons, B. M., Quadrature rule of brick based finite-elements. *Int. J. Num. Methods Engineering*, **3** (1971) 293–4.
501. Hellen, T. K., Effective quadrature rules for quadratic solid isoparametric finite elements. *Int. J. Num. Methods Engineering*, **4** (1972) 597–600.
502. Yamada, Y., Yoshimura, N. & Sakurai, T., Plastic stress–strain matrix and its application for the solution of elastic–plastic problems by the finite-element method. *Int. J. Mech. Sci.*, **10** (1968).
503. Owen, D. & Hinton, E., *Finite Elements in Plasticity Theory and Practice*. Pineridge Press, 1980.
504. Nayak, G. C. & Zienkiewicz, O. C., A generalization of various constitutive relations including strain softening. *Int. J. Num. Methods Engineering*, **5** (1972) 113–35.
505. Yagawa, G. & Ando, Y., Three-dimensional finite-element method of thermoelasto-plasticity with creep effect. 2nd International Conference on Structural Mechanics in Reactor Technology (SMiRT), Berlin, September 1973.
506. Drucker, D. C., A more fundamental approach to plastic stress–strain relations. 1st US Congress of Applied Mechanics, ASME, New York, 1952, pp. 487–91.
507. Reckling, K. A., *Plastizitaetstheorie und ihre Anwendung auf Festigkeitsprobleme*. Springer Verlag, Berlin/Heidelberg/New York, 1967.
508. Foye, R. L., Theoretical post-yielding behaviour of composite laminates. Part 1 — inelastic micromechanics. *J. Comp. Mats*, **7** (1973) 178–93.
509. Lin, T. H., Salinas, D. & Ito, Y. M., Elastic–plastic analysis of unidirectional composites. *J. Comp. Mats*, **6** (1972) 48–60.
510. Heide & Zumsteg, Mentat: A System for Finite Element Pre- and Post-processing. MARC, Zentech, London, 1968.
511. Davidson, I. & Barrett, N. T., Design philosophy and safety. International Conference on PCPV, Inst. Civ. Eng., London, Paper 6, 1967.
512. Freudenthal, A. M., Safety and probability of structural failure. *Trans. ASCE*, **121** (1956) 1337.
513. Freudenthal, A. M. *et al.*, The analysis of structural safety. *J. Struct. Div., ASCE*, **92** (1966) 267–325.
514. Pugsley, A., Report on structural safety. *The Structural Engineer*, **34** (1955) 141.
515. Baker, M. J., The evaluation of safety factors in structures. CIRIA, Project 72, Final Report, Imperial College, London, 1970.
516. Baker, A. L. L., Safety pressure vessels. International Conference on PCPV, Inst. Civ. Eng., London, Group B, Paper 8, 1967.
517. Bangash, Y., Riddle of bonded and unbonded tendons in prestressed concrete reactor vessels. *Concrete*, **8** (1974) 46–8.

518. Paul, S. L., Structural behaviour of prestressed concrete reactor vessels. *J. Struct. Div., ASCE* (July 1971) 1897–6.
519. Ravindra, M. K. *et al.*, Illustration of reliability-based design. *J. Struct. Div., ASCE* (September 1974) 1789–811.
520. Lind, N. C., Deterministic formats for the probabilistic design of structures. University of Waterloo, Report 1, 1969.
521. Rosenblueth, E. *et al.*, Reliability basis for some Mexico codes. Probabilistic design of reinforced concrete buildings, Publication SP31, ACI, 1972.
522. GKN, Tests on GKN reinforcements. Private communication, 1970.
523. Baker, A. L. L., *The Ultimate Load Theory Applied to the Design of Reinforced and Prestressed Concrete Frames*. Concrete Publications Ltd, London, 1956.
524. Benjamin, J. R. & Cornell, A., *Probability, Statistics and Decisions for Civil Engineers*. McGraw-Hill, New York, 1970.
525. European Code Bulletin d'Information CEB. (a) No. 50, 201–226 (July 1965); (b) No. 50, 174–200 (July 1965); (c) No. 56 (August 1966).
526. Corso, J. M., Discussions on earthquake codes. *Proc. ASCE*, **80** (1954).
527. Cave, M., Atomic Power Construction, Sutton, Surrey. Private communication on reactor safety, 1969.
528. Mischke, C. R., A method of relating factors of safety and reliability. *J. Engineering for Industry* (August 1970) 537–42.
529. Mischke, C. R., Implementing mechanical design to a reliability specification. Des. Eng. Tech. Conf. New York, CASME Publications, October 1974.
530. Mischke, M. R., Rationale for mechanical design to a reliability specification. Iowa State University, Ames, October 1974.
531. Drucker, D. C., Limit analysis of cylindrical shells under axially symmetric loadings. Proc. First Mid-Western Conf. on Solid Mechanics, USA, 1953.
532. Onat, E. T., Plastic collapse of cylindrical shells under axially symmetrical loading. *Quart. App. Math.*, **13** (1959) 63–72.
533. Hodge, P. G., The rigid–plastic analysis of symmetrically loaded cylindrical shell. *J. Appl. Mech.*, **21** (1954) 336–42.
534. Hopkins, J. & Prager, W., The load carrying capacities of circular plates. *J. Mech. Phys. Solids*, **2** (1954).
535. Bangash, Y., Prestressed concrete reactor vessel. Time-saving ultimate load analysis. *J. Inst. Nucl. Eng.*, **13** (1972).
536. Bangash, Y., A basis for the design of bonded reinforcement in the prestressed concrete reactor vessels. Paper 7478S, Supplement (8), Inst. Civ. Eng., 1972.
537. Smee, D. J., The effect of aggregate size and concrete strength on the failure of concrete under multiaxial compression. Civil Engineering Transaction, Inst. of Engineers, Australia, Vol. CE9, No. 2, October 1967, pp. 339–44.
538. The Nuclear Power Group Company, Risley, Lancashire. Private communication, 1968.
539. Corum, J. M. & Smith, J. E., Use of small models in design and analysis of prestressed concrete reactor vessels. ORNL-4346, USA, May 1970.
540. Smith, J. R., Problems of assessing the correlation between the observed and predicted behaviour of models. Model Techniques for Prestressed Concrete Pressure Vessels, BNES, London, Paper 10, 1969.
541. Treharne & Davies, Tests on samples and calculation of standard deviations for bonded reinforcement. Private communication, 1969.
542. Paul, S. L. *et al.*, Mortar models of prestressed concrete reactor vessels. *J. Struct. Div.*, ASCE (1969).
543. Brown, G., Advances in the gas-cooled reactor. Lecture Joint Session of IEE, I MechE and BNES, University of Strathclyde, Glasgow, 16 October 1968 (available from British Nuclear Energy Society).

544. Cluckman, A. L., Reactor containment structures abroad. *J. Power Div., ASCE* (1969) 61–76.
545. Lorenz, H., Design of the concrete containment vessel for the R.E. GINNA nuclear power plant. *Nucl. Eng. Des.*, **6** (1967) 360–6.
546. Wahl, W. W. *et al.*, Design and construction aspects of large prestressed concrete (PWR) containment structures. *ACI Journal* (May 1969) 400–12.
547. Halligan, D. W., Structural design criteria for secondary containment structures. *Nucl. Eng. Des.*, **8** (1968) 427–34.
548. Kulka, F., Prestressing systems for secondary and primary containment structures. *Nucl. Eng. Des.*, **8** (1968) 435–9.
549. Hofman, H., Spannbeton Reactordruckbehalter mit Stutzbeton. *Nucl. Eng. Des.*, **8** (1968) 467–70.
550. Aymot, P., Construction of the containment building for the Gentilly nuclear power reactor. *Nucl. Eng. Des.*, **9** (1969) 479–87.
551. Harstead, G. A. *et al.*, Grouted tendons for nuclear containments. *J. Power Div., ASCE* (October 1969) 277–92.
552. Chauvin, G. A., Post-Tensioned Concrete Nuclear Structures. Private Report No. 615, Fressynet, 1971.
553. Attala, I. & Nowatony, B., Missile impact on a reinforced concrete structure. *Nucl. Eng. Des.*, **37** (1976) 321–32.
554. Bannerjee, A. K. *et al.*, Design of reinforced concrete containments. 3rd International Conference on Structural Mechanics in Reactor Technology (SMiRT), Paper J3/1, London, 1975, pp. 1–12.
555. Green, D. E. *et al.*, Design of concrete containments for tangential shear loads. 3rd International Conference on Structural Mechanics in Reactor Technology (SMiRT), Paper J3/4, London, 1975, pp. 1–12.
556. Buchart, K. P. *et al.*, Analysis of reinforced concrete containment vessels considering concrete cracking, 3rd International Conference on Structural Mechanics in Reactor Technology (SMiRT), Paper J4/1, London, 1975, pp. 101–10.
557. Singh, M. P. *et al.*, Statistical aspects of containments tendon surveillance. Am. Soc. Civ. Eng., PO1, 1975, pp. 23–33.
558. Chu, K. Y., Prestress system for nuclear concrete containment domes. Am. Soc. Civ. Eng., PO1, 1974, pp. 111–23.
559. Bahar, L. Y. & Rice, J. S., Simplified derivation of the reaction time history in aircraft impact on a nuclear power plant. *Nucl. Eng. Des.*, **49** (1978) 263–8.
560. Bangash, Y., The structural integrity analysis of concrete containment vessels under external impacts. 6th International Conference on Structural Mechanics in Reactor Technology (SMiRT), Paper J7/6, Paris, 1981.
561. Bangash, Y., The automated three-dimensional cracking analysis of prestressed concrete containment vessels. 6th International Conference on Structural Mechanics in Reactor Technology (SMiRT), Paper H3/2, Paris, 1981.
562. Barr, P., Brown, M. L., Carter, P. G., Howe, W. P., Jowet, J., Neilson, A. J. & Young, R. L. D., Studies of missiles impact with reinforced concrete structures. BNES Symposium, 1980.
563. Bartlett, W. R. & Davies, I., Aircraft impact design for SGHWR containment. International Conference on Experience in the Design, Construction and Operation of PCPV, York, England, Paper 128/75, 1975.
564. Bathe, K. J. & Wilson, E. L., *Numerical Methods in Finite Element Analysis*. Prentice Hall, New York, 1976.
565. Brais, A. & Godbout, P., Modelling and simulation of local failures for reinforced concrete structures under concentrated impulsive loads. Private communication, 1978.
566. Bresler, B. & Pister, K. S., Strength of concrete under combined stresses. *J. Am. Concr. Inst.*, **55** (1957) 321–45.

567. Bresler, B. & Pister, K. S., Failure of plain concrete under combined stresses. *Trans. ASCE*, **122** (1957).
568. Brown, M. K., Curtsen, N. & Jowett, J., Local failure of reinforced concrete under missile impact loading. 5th International Conference on Structural Mechanics in Reactor Technology (SMiRT), Berlin, 1979.
569. Carlton, D. & Bedi, A., Theoretical study of aircraft impact on reactor containment structures. *Nucl. Eng. Des.*, **45** (1978) 197–206.
570. Cedolin, L., Crutzen, R. J. & Poli, J. D., Triaxial stress–strain relationship for concrete. *J. Eng. Mech. Div., ASCE*, **103** (1977) 423–39.
571. Kameswara, S. & Prasad, S., Impact loads on beams on elastic foundations. 3rd International Conference on Structural Mechanics in Reactor Technology (SMiRT), Paper J5/8, London, 1975.
572. Davis, I., Studies of the response of reinforced concrete structures to short duration dynamic loads. In: *Design for Dynamic Loading*, Construction Press, UK, 1982.
573. Chinn, J. & Zimmermann, R. M., Behaviour of plain concrete under various high triaxial compression loading conditions. Technical Report No. WLTR 64-163 (AD468460), Air Force Weapon Laboratory, New Mexico, 1965.
574. Constantopoulos, I. V., Vardanega, C. & Attalla, I., Dynamic response to aircraft impact of reactor building with protective shell on independent foundation. 7th International Conference on Structural Mechanics in Reactor Technology (SMiRT), Paper J9/6, 1981.
575. Degen, P., Furrer, H. & Jemielewski, J., Structural analysis and design of a nuclear power plant building for aircraft crash effects. *Nucl. Eng. Des.*, **37** (1976) 241–68.
576. Dritler, K. & Guner, P., Calculation of the total force acting upon a rigid wall by projectiles. *Nucl. Eng. Des.*, **47** (1976) 231–44.
577. AEC takes long hard look at its regulation policies, *Nuclear News* (August 1969) 21–2.
578. Amirikian, A., Design of protective structures. Bureau of Yards and Docks, Publication No. NavDocks P-51, Department of the Navy, Washington DC, 1950.
579. A Seismic Design and Testing of Nuclear Facilities. Technical Reports, Series No. 88, International Atomic Energy Agency, June 1968.
580. Berg, G. V. & Stratta, J. L., Anchorage and the Alaska earthquake of March 27 1968. American Iron and Steel Institute, 1968.
581. Brown, T., Large capacity unbonded tendons. Presented at ACSE National Meetings on Water Resources Engineering, New Orleans, Louisiana, Preprint No. 841, 3–8 February 1969.
582. Burns, W. J., Kennedy, W. J. L. & Weinzimmer, F., Shoreham nuclear power station with concrete containment. *Proc. Am. Power Conf.*, **30** (1968) 365–70.
583. Cornell, C. A., Design seismic inputs. Seminar on Seismic Design for Nuclear Power Plants, MIT, 31 March–3 April 1969.
584. Fundamentals of protective design. Report AT 1297821, Army Corps of Engineers, Office of the Chief of Engineers, 1966.
585. General design criteria for nuclear power plant construction permit. Published in the *Federal Register*, 33 F.R. 19213, July 11 1967.
586. Gwaltney, R. C., Missile generation and protection in light-water-cooled power reactor plants. United States Atomic Energy Commission Report, ORNL-NSIC-22, 1968.
587. Kennedy, R. P., Effects of an aircraft crash into a concrete reactor containment building. Holmes and Narver Inc., Oakland, California, June 1966.
588. National Defence Research Committee. Effects of impact and explosion. Summary Technical Report of Division 2, Vol. 1, Washington DC, 1946.
589. Newmark, N. M., Design criteria for nuclear reactors subjected to earthquake hazards. Unpublished paper, 25 May 1967.
590. Report of pressure testing of reactor containment for Connecticut Yankee atomic power plant. Connecticut Yankee Atomic Power Company, Stone and Webster Engineering Corporation, October 1967.

591. Review of methods of mitigating spread of radioactivity from a failed containment system. United States Atomic Energy Commission, Report ORNL-NSIC-27, September 1968.
592. Riera, J. D., On the stress analysis of structures subjected to aircraft impact forces. *Nucl. Eng. Des.*, **8** (1968) 415–26.
593. Sharpe, R. L., Earthquake engineering for nuclear power plants. *Civil Engineering, ASCE* (March 1969) 39–43.
594. Spohn, H. R. & Waite, P. J., Iowa tornadoes. *Monthly Weather Review*, (September 1962) 398–406.
595. Tan, C. P., A study of the design and construction practices of prestressed concrete and reinforced concrete containment vessels. United States Atomic Energy Commission, Report No. TID-24176, 1969.
596. Transcript of nuclear containment. Seminar co-sponsored by Power Engineering Magazine and Graver Tank and Manufacturing Co., Washington DC, 21 March 1969, Unpublished report.
597. Marti, J., Kalsi, G. & Attalla, I., Three-dimensional aircraft impact analysis. 7th International Conference on Structural Mechanics in Reactor Technology (SMiRT), Paper J9/7, 1983.
598. Labra, J., Protective structures response to vehicle impact. *J. Struct. Div., ASCE* (1979).
599. CEGB, A case for Sizewell B, Vol. 1. Sizewell Inquiry, 1981.
600. Zapp, F. C., Testing of containment systems used in light-water-cooled power reactors. United States Atomic Energy Commission, Report ORNL-NSIC-26, 1968.
601. Zudans, Z., Dynamic response to shell type structures subjected to impulsive mechanical and thermal loadings. *Nucl. Eng. Des.*, **3** (1966) 117–37.
602. Halligan, D. W., Structural design criteria for secondary containment structures. *Nucl. Eng. Des.*, **8** (1968) 427–34.
603. Richardson, K. *et al.*, Dynamic analysis of footing on layered media. *J. Eng. Mech. Div., ASCE*, **101** (1975) 167–88.
604. Newmark, N. M. & Rosenblueth, E., *Fundamentals of Earthquake Engineering*, Prentice Hall, New York, 1971.
605. Hudson, D. E., Response spectrum techniques in engineering seismology. Proceedings of the World Conference on earthquakes engineering, Earthquake Engineering Research Institute, June 1956.
606. Housner, G. W., Behaviour of structures during earthquakes. *Proc. ASCE*, **85** (1959).
607. Newmark, N. M., Hall, W. J. & Noraz, B., Seismic design spectra for nuclear power plants. *J. Power Div., ASCE*, **99** (1973).
608. Carmichael, G. D. T. & Irving, J., Calculation of PCPV prestressing forces. *Nucl. Eng. Des.* (1968).
609. Elwi, A. A. & Murray, D. W., A three-dimensional hypo-elastic concrete constitutive relationship. *J. Eng. Mech. Div., ASCE*, **105** (1979) 623–41.
610. Filho, F. V., Coombi, R. F. & Beerreto, L. C., Design of the reinforced concrete containment shell of a nuclear reactor for aircraft impact. 7th International Conference on Structural Mechanics in Reactor Technology (SMiRT), Paper J9/11, 1979.
611. Gardner, N. J., Triaxial behaviour of concrete. *Proc. Am. Concr. Inst.*, **2** (1969) 136–47.
612. Gupta, Y. M. & Seaman, L., Local response of reinforced concrete to missile impacts. *Nucl. Eng. Des.*, **45** (1978) 507–14.
613. Halder, A., Turbine missile — A critical review. *Nucl. Eng. Des.*, **55** (1979) 293–304.
614. Hammel, J., Aircraft impact on a spherical shell. *Nucl. Eng. Des.*, **37** (1976) 206–23.
615. Hammel, J., Impact loading on a spherical shell. 5th International Conference on Structural Mechanics in Reactor Technology (SMiRT), 1979, Berlin.
616. Hannant, D. J. & Frederick, C. O., Failure criteria for concrete in compression. *Magazine of Concrete Research*, **20** (1969) 134–44.

617. Barber, E. M. *et al.*, A study of air distribution in survival shelters using a small-scale modelling technique. USA National Bureau of Standards, Report 10689, 1972.
618. Barneby, H. L. *et al.*, Toxic gas protection. Paper to ASHRAE Symposium on Survival Shelters, 1962; ASHRAE, 1963.
619. Barthel, R., Research on the climate in an underground shelter. *Ingenieur*, **77** (1965) 143–7 (in Dutch).
620. Barthel, R., Theoretical and experimental research regarding the indoor climate in an underground shelter. *Ingenieur*, **77** (1965) 6143–7 (in Dutch).
621. Baschiere, R. J. *et al.*, Analysis of above-ground fallout shelter ventilation requirements. *ASHRAE Journal*, 7 (1965) 88–90.
622. Baschiere, R. J. *et al.*, Control of fallout shelter environments. ASHRAE Symposium on Survival Shelter Problems, 1970; ASHRAE, 1971.
623. Baschiere, R. J. *et al.*, Development of manually powered pedal ventilator for shelter use. ASHRAE Symposium on Survival Shelter Problems, 1970; ASHRAE, 1971.
624. Breistein, M., Climate conditions in a survival shelter — ventilation and filtration. *Norsk VVS*, **10** (1967) 353–64 (in Norwegian).
625. Broido, A., Environmental hazards of fire. ASHRAE Symposium on Survival Shelters, 1962; ASHRAE, 1963.
626. Coal mine rescue survival system. *Heating and Ventilating Engineer*, **45** (1972) 353–4.
627. Cooper, J., After the bomb. *Journal of the Chartered Institution of Building Services*, **2** (1980) 48–9.
628. Dasler, A. R. & Minrad, D., Environmental physiology of shelter habitation. *ASHRAE Transactions*, **71** (1965) 115–24.
629. Drucker, E. E. & Cheng, H. S., Analogue study of heating in survival shelters. ASHRAE Symposium on Survival Shelters, 1962; ASHRAE, 1963.
630. Bigg, J. M., *Introduction to Structural Dynamics*. McGraw-Hill, New York, 1964.
631. Ducar, G. J. & Engholm, G., Natural ventilation of underground fallout shelters. *ASHRAE Journal*, **7** (1965) 91–2.
632. Engle, P. M., Air filtration — biological and radiological. ASHRAE Symposium on Survival Shelters, 1962; ASHRAE, 1963.
633. Everetts, J. *et al.*, Conditional air for improving shelter habitability. ASHRAE Symposium on Survival Shelter Problems, 1970.
634. Flannigan, F. M. & Morrison, C. A., Moisture storage in survival shelter structures. ASHRAE Symposium on Survival Shelter Problems, 1971.
635. Gates, A. S., Air revitalization in sealed shelters. ASHRAE Symposium on Survival Shelters, 1962; ASHRAE, 1963.
636. Gessner, H., The ventilation of air-raid shelters. *Schweizerische Blatter fur Heizung und Luftung*, **28** (1961) 1–12 (in German).
637. Hanna, G. M., Ventilation design for fallout and blast shelters. *Air Engineering*, **4** (1962) 19–21.
638. Heating and ventilation of nuclear fallout shelters. *Heating and Ventilating Engineer*, **36** (1962) 221–9.
639. Home Office, Domestic nuclear shelters (DNS) — Technical Guidance. HMSO, 1981.
640. Home Office, Domestic Nuclear Shelters. HMSO, 1981.
641. Hummell, J. D. & Beck, W. D., The study of a shelter cooling system using methanol as the heat sink and as the fuel. Battelle Memorial Institute, AD 637953,1966.
642. Hummell, J. D. *et al.*, Survival shelter cooling: conventional and novel systems. *ASHRAE Journal*, **7** (1965) 35–42, 87.
643. Humphreyes, C. M. *et al.*, Sensible and latent heat losses from occupants of survival shelters. *ASHRAE Transactions*, **72** (1966) 255–63.
644. Improving the conditions in fallout shelters. *Engineering* (26 January 1962) 158–9.

645. Kusuda, T. & Achenbach, P. R., Outdoor air psychrometric criteria for summer ventilation of protective shelters. *ASHRAE Transactions*, **71** (1965) 76–87.
646. Laine, P., Air conditioning system for a new type bomb shelter. *Rakennustekniikka*, **9128** (1972) 205–7 (in Finnish).
647. Moore, M., Building for the bomb. *Building*, **239** (1980) 39–42.
648. Morrison, C. A. & Flanigan, F. M., A graphical method for determining ventilation requirements for underground shelters. ASHRAE Symposium on Survival Shelter Problems, 1971.
649. Murphy, H. C., Ventilation on survival areas for civil defence. *Heating, Piping and Air Conditioning*, **23** (1951) 97–9; **23** (1951) 85–6, 90.
650. Nevins, R. G., Physiological aspects of survival. ASHRAE Symposium on Survival Shelters, 1962; ASHRAE, 1963.
651. Nykanen, J., Air and water services in air-raid shelters. *LVI*, **23** (1971) 8–12 (in Finnish).
652. Parry, T. H., Mechanical requirements of bomb shelters. *Air Conditioning, Heating and Ventilating*, **65** (1980) 108–10.
653. Rathman, C. E., Air distribution in shelters due to Kearney pumps and pedal ventilators. General American Research Division, Report No. 1476-1, 1970.
654. Rehm, F. R., Radiological instruments for shelters. *Air Engineering*, **4** (1962) 32–5.
655. Rickenbach, H., Heating, ventilation, cooling and sanitary engineering for large air-raid shelters. *Schweizerische Blatter fur Heizung und Luftung*, **28** (1961) 13–22 (in German).
656. Rosell, A., How underground workshops are made livable. *Heating, Piping and Air Conditioning*, **30** (1958) 118–20.
657. Rosell, A., Swedish air-raid shelter ventilation designed for atomic war emergency. *Heating, Piping and Air Conditioning*, **30** (1958) 136–9.
658. Rossetti, G., Technical installations in civil defence shelters. *Schweizerische Blatter fur Heizung und Luftung*, **37** (1970) 119–29 (in French).
659. Roubinet, M., Air conditioned anti-atomic shelters. *Industries Thermiques et Aerauliques*, **12** (1966) 301–31 (in French).
660. Scripture, D. G., How to engineer air for a municipal emergency center. *Air Engineering*, **4** (1962) 25–31.
661. Sisson, G. N., Underground for nuclear protection. *Underground Space*, **4** (1980) 341–8.
662. Small, S., Tests disclose serious defects in air-raid shelters. *Norsk VVS*, **15** (1972) 539–47 (in Norwegian).
663. Taylor, D. W. & Gonzales, J. O., Air distribution in multi-room shelter using a package ventilation kit. US Army Florida Engineering Industry Experimental Station, AD 633916, 1965.
664. Trayser, D. A. & Flanigan, L. J., Shelter power systems: their effect on environmental control. *ASHRAE Transactions*, **72** (1966) 246–54.
665. Urdahl, T. H., Success of underground structures depends on air conditioning. *Heating, Piping and Air Conditioning*, **30** (1958) 108–11.
666. Van Biesen, J. A. H., Calculation of the climate in an underground shelter when the period of occupancy is limited. *Ingenieur*, **77** (1965) 6133–41 (in Dutch).
667. Ventilation for the piping in private air-raid shelters. *Sanitar Heizungs Technik*, **36** (1971) 732–3 (in German).
668. Viessman, W., Environment protection in subterranean shelters. *Refrigerating Engineering*, **59** (1951) 1175–8.
669. Viessman, W., How to plan air conditioning for protective shelters. *Heating Piping and Air Conditioning*, **26** (1954) 84–8, 122–7.
670. Viessman, W., Protective shelters in modern warfare. *Heating, Piping and Air Conditioning*, **23** (1951) 77–82, 111–16.
671. Weise, E., The time of residence inside ventilated shelters. *Sanitar Heizungs Technik*, **35** (1970) 143–4 (in German).

672. Wood, R., Blast protection valves shield building's Achilles heel. *Process Engineering* (October 1977) 88–9.
673. Wright, M. D. & Hill, E. L., Ventilation kit application studies. ASHRAE Symposium on Survival Shelter Problems, 1970; ASHRAE, 1971.
674. Yaglou, C. P., Limits for cold, heat and humidity in underground shelters. *Archives of Environmental Health*, **2** (1961) 110–15.
675. Moazami, M., Course work on nuclear shelters. Thames Polytechnic, London, 1982/83.
676. American Concrete Institute, ACI 313: recommended practice for design and construction of concrete bins, bunkers and silos for storing granular material, 1984.
677. Arnold, P. C., McLean, A. & Roberts, A., Bulk solids: storage, flow and handling. TUNRA Bulk Solids Handling Associates, 1982.
678. ACI Standard 313-77. Recommended practice for design and construction of concrete bins and bunkers for storing granular materials. Second Printing 1982.
679. Brown, R. L. & Hawskley, P. G. W., The flow of granular material. *Fuel*, No. 26-159 (1947).
680. Butters, G., *Plastics Pneumatic Conveying and Bulk Storage*. Applied Science Publishers, 1981.
681. Durand, R. & Condolios, E., The flow of granular material through a circular orifice. *Revue Travaux* (*m*) (1980).
682. Enstad, G., On the theory of arching in mass flow hoppers. *Chem. Eng. Science*, **30** (1975) 1273–83.
683. Faber, J. & Alsop, D. J. A., Economics of reinforced concrete multibin grain silo configurations. *The Structural Engineer*, **60A** (1982) 121.
684. ISO 8456, Storage equipment for loose bulk materials (safety code), 1980.
685. Jenike, A. W., Johanson, J. R. & Carson, J. W., Bin loads — Part 2; concepts. *ASME J. Eng. for Ind.*, **95** (1973) 15.
686. Jenike, A. W., Johanson, J. R. & Carson, J. W., Bin loads — Part 3; mass flow bins. *ASME J. Eng. for Ind.*, **95** (1973) 6–12.
687. Jenike, A. W., Effect of solid flow properties and hopper configurations on silo loads. In: *Unit and Bulk Material Handling*. J. F. Loefler and C. R. Proctor, eds. ASME, New York, 1980.
688. Leonhardt, F., Boll, K. & Spiedel, E., The safe design of cement silos. Translation by Amerongen, C & CA, Translation No. 94, London, 1961.
689. McLean, A. G., Arnold, A. G. & Martin, A. G., Draft code for the evaluation of bin wall design. Association of Consulting Structural Engineers, Australia, September 1983.
690. Nguyen, T. V., Brenner, C. E. & Sabersky, R. H., Funnel flow in hoppers. *ASME J. Appl. Mech.*, **47** (1980) 729–35.
691. Nedderman, R. M. & Thorpe, R. B., The rate of discharge of powders. Seminar on Bunker/Feeder Interface Problems, Institution of Mechanical Engineers, September 1983.
692. Reimbert, M. & Reimbert, A., *Silos — Theory and Practice*. Trans. Tech. Publications, 1976.
693. Jenike, A. W., Johnson, J. R. & Carson, J. W., Bin loads, Part 1, concepts. ASME Paper No. 72-MH-1, 1975.
694. Jenike, A. W., Johnson, J. R. & Carson, J. W., Bin loads, Part 3, Mass Flow Bins. ASME Paper No. 72-HM-2, 1975.
695. Emery, R. B., Aeration apparatus converts bin from funnel flow to mass flow characteristics. ASME Paper No. 72-MH-5, ACI Standard 318 and 313, 1975.
696. API-620, Recommended rules for design and construction of large, welded, low pressure storage tanks, 1980.
697. IS-4996, Part 1 and Part 2, Criteria for design of reinforced concrete bins for the storage of granular and powdery materials. Indian Code, 1980.
698. Proposed ACI Standard (P 529), Recommended practice for design and construction of concrete bins and bunkers for storing granular materials and its commentary, October 1975.

699. Janson, Birkenhead Project, Private communication, 1983.
700. Janson, Analysis. Private communication, 1985.
701. DIN 1055, German codes on loads on bins and storage, 1964.
702. Soviet Code 313-77, Soviet code on loads on bins and storage, 1980.
703. Myers, J. J., Holm, C. H. & McAllister, R. F., *Design Considerations, Handbook of Ocean and Underwater Engineering*, McGraw-Hill, New York, 1970.
704. Browne, R. D. & Domone, P. L. J., The long term performance of concrete in the marine environment. Off-shore Structures, Inst. Civ. Eng., Paper 5, 1949.
705. Bury, M. R. C. & Domone, P. L. J., The role of research in the design of concrete off-shore structures. Off-shore Technology Conference, Houston, Texas, Paper No. OTC 1949, 1974.
706. Netherlands Committee for Concrete Research, Cases of damage to corrosion of prestressing steel. CUR, July 1971.
707. Hove, K. & Foss, I., Quality assurance for off-shore concrete gravity structures. Off-shore Technology Conference, Houston, Texas, Paper No. OTC 2113, 1974.
708. API, Planning, designing and constructing fixed off-shore platforms, 1976.
709. Bishop, R. E. D. & Hassan, Y., The lift and drag forces on a circular cylinder in a flowing fluid. *Proc. Roy. Soc.*, **A277** (1963) 32–50.
710. Det Norske Veritas, Rules for the design, construction and inspection of fixed offshore structures. Oslo, Norway, 1974.
711. La Fond, E. C., Deep current measurements with the bathyscaphe 'Trieste'. *Deep Sea Res.*, **9** (1962) 115–16.
712. Nowroozi, A. *et al.*, Deep ocean current and its correlation with the ocean tide off the coast of Northern California. *J. Geophys. Res.*, **73** (1968) 1921–32.
713. Condeep safely on the bottom. *NCE* (July 1975) 11–13.
714. Cracks damage Condeep pride but not design integrity. *NCE* (July 1974) 10.
715. Special feature: North Sea oil. *NCE* (1973).
716. Special review: North Sea oil. *NCE* (May 1974).
717. Hirayama, K. I., Schwarz, J. & Wu, H. C., Effects of the ice thickness on ice forces. Off-Shore Technology Conference, Houston, Texas, Paper No. OTC 1974, 1974.
718. McDowell, D. N. & Holmes, P., General research problems in the design of off-shore structures. Off-Shore Structures, ICE, Paper No. 13, 1975.
719. Department of Energy, Guidance on the design and construction of off-shore installations. HMSO, 1974.
720. Shell Exploration, Design of floating concrete platform for deep water oil production. Private communication, London, 1974.
721. Skjelbreia, L. & Hendrickson, J., Fifth order gravity wave theory. Proc. 7th Conf. on Coastal Engineering, Vol. 1, Chapter 10, 1961.
722. Garrison, C. J. & Rao, V. S., Interaction of waves with submerged objects. *Journal of the Waterways, Harbours and Coastal Engineering Division, Proc. ASCE* (May 1961).
723. Boreel, L. J., Wave action on large off-shore structures. Off-Shore Structures, ICE, Paper No. 2, 1975.
724. Zienkiewicz, O. C. *et al.*, *Numerical Methods in Offshore Engineering*. Wiley, New York, 1982.
725. BBR Tanks Inc., BBR tanks in operation. In: *Prestressed Concrete Tanks*. Vols. 1 and 2. BBR Tanks Inc., USA.
726. Morrison, J. R., O'Brien, M. P., Johnson, J. W. & Scnaaf, S. A., The force exerted by surface wave on piles. *Petroleum Trans., AIME*, **189** (1950).
727. Hogben, N. & Standing, R. G., Wave loads on large bodies. Int. Symp. on the Dynamics of Marine Vehicles and Structures in Waves, London, April 1974.
728. Garrison, C. J., Torum, A., Iversen, C. & Leivseth, S., Wave forces on large volume structures — a comparison between theory and model tests. Off-shore Technology Conference, Houston, Texas, Paper No. OTC 2137, 1974.

729. Valenta, O., Kinetics of water penetration into concrete as an important factor of its deterioration and of reinforcement corrosion. Proc. International Symposium Durability of Concrete, Prague, A177–A193, 1969.
730. Taylor, R. E., Paper No. 5: Structural dynamics of off-shore platforms. Off-Shore Structures, Inst. Civ. Eng., Paper No. 5, 1975.
731. Horie, F. & Odaka, T., Consideration on effect of hysteresis damping for composite systems. Proc. 5th World Conference on Earthquake Engineering, Vol. II, Rome, 25–29 June 1973, pp. 1734–43.
732. Clausen, C. J. F., Stability problems related to off-shore gravity structures. Norwegian Geotechnical Institute, 1983.
733. Mansour, A. E. & Millman, D. N., Dynamic random analysis of fixed off-shore platforms. Off-Shore Technology Conference, Houston, Texas, Paper No. OTC 2049, 1974.
734. Biggs, M. J., *Introduction to Structural Dynamics*, McGraw-Hill, New York, 1964.
735. Fertis, D. G., *Dynamics and Vibration of Structures*, John Wiley & Sons, New York, 1973.
736. Preload Co. London, Design of LNG tanks. Private communications, 1971.
737. First, Second and Third Conferences on LNG Available, American Petroleum Institute, 1975, 1976, 1979.
738. J. Zarafa, The design of concrete LNG tanks. Partial fulfilment for the degree of BSc, CNAA, 1973.
739. Adesola, O. T., A feasibility study of non-conventional storage of liquefied natural gas. Final Year BSc (Hons) Project, Thames Polytechnic, London, 1983.
740. Bruggeling, A. S. G., *Prestressed Concrete for the Storage of Liquified Gases.* Viewpoint Publication No. 12.083, Eyre & Spottiswoode, London, 1981.
741. Trotter, H. G., Turner, F. H., Sullivan, P. J. E. & Brooks, W. T., Behaviour of prestressed concrete materials at very low temperatures. *Build International*, **8** (1975) 85–91.
742. Anchor, R. O., *Handbook on BS 5337; The Structural Use of Concrete for Retaining Aqueous Liquids.* Viewpoint Publications, 1976.
743. Bridon Wire, Cryogenic temperatures; the behaviour of carbon steel prestressing wire and strand. Technical Note No. 13, Bridon Wire, Doncaster, May 1970.
744. FIP, *Cryogenic Behaviour of Material for Prestressed Concrete.* Viewpoint Publications, 1983.
745. FIP, *Recommendations for the Design of Prestressed Concrete Oil Storage Tanks.* Viewpoint Publications, 1983.
746. *Prestressed Concrete for Storage of Liquified Gases.* Viewpoint Publications, 1982.
747. Turner, F. H., *Concrete and Cryogenics.* Viewpoint Publications, 1979.
748. Marchel, J. C., Variations in modulus of elasticity and Poisson's ratio with temperature. ACI Publication, Sp. 35–27, 1972.
749. Bangash, Y., A three-dimensional rupture analysis of steel liners anchored to concrete pressure and containment vessels. *Eng. Fract. Mech.*, **28** (1987) 157–85.
750. Bangash, Y., The simulation of endochronic model in the cracking analysis of PCPV. 9th International Conference on Structural Mechanics in Reactor Technology (SMiRT), Lausanne, 1987, Vol. 4, pp. 333–40.
751. Al-Noury, S. I. & Bangash, Y., Prestressed concrete containment structures—circumferential hoop tendon calculation. 9th International Conference on Structural Mechanics in Reactor Technology (SMiRT), Lausanne, 1987, Vol. 4, pp. 395–401.
752. Petersson, J., Crack growth and development of fracture zones in plain concrete and similar materials. PhD Thesis, University of Lund, 1981.
753. Eibl, J., Behaviour of critical regions under soft missile impact and impulsive loading. IBAM, June 1982.
754. Hughes, G. & Speirs, D., An investigation of the beam impact problem. Technical Report 546, C and CA, 1982.
755. Bate, S., The effect of impact loading on prestressed and ordinary reinforced concrete beams. National Building Studies Research Paper 35, 1961.

756. Billing, I., *Structure Concrete*. Macmillan, London, 1960.
757. Watson, A. & Ang, T., Impact resistance of reinforced concrete structures. In: *Design of Dynamic Loading*, Construction Press, 1982.
758. Watson, A. & Ang, T., Impact response and post-impact residual strength of reinforced concrete structures. International Conference and Explosion of Structural Impact and Crashworthiness, July 1984.
759. Perry, S. & Brown, I., Model prestressed slabs subjected to hard missile loading. In: *Design for Dynamic Loading*, Construction Press, 1982.
760. Perry, S., Brown, I. & Dinic, G., Factors influencing the response of concrete slabs to impact. International Conference and Exposition of Structural Impact and Crashworthiness, July 1984.
761. Kufuor, K. & Perry, S., Hard impact of shallow reinforced concrete domes. International Conference and Exposition of Structural Impact and Crashworthiness, July 1984.
762. Burgess, W. & Campbell-Allen, D., Impact resistance of reinforced concrete as a problem of containment. School of Civil Engineering, The University of Sydney, Research Report No. R251, 1974.
763. Stephenson, A., Tornado-generated missile full-scale testing. Proc. Symposium on Tornadoes, Assessment of Knowledge and Implications for Man, Texas University, June 1976.
764. Jankov, Z., Turnahan, J. & White, M., Missile tests of quarter-scale reinforced concrete barriers. Proc. Symposium on Tornadoes, Assessment of Knowledge and Implications for Man, June 1976.
765. Stephen, A. & Silter, G., Full-scale tornado-missile impact tests. 4th International Conference on Structural Mechanics in Reactor Technology (SMiRT). Paper J10/1, Berlin, 1977.
766. Jonas, W. & Rudiger, E., Experimental and analytical research on the behaviour of reinforced concrete slabs subjected to impact loads. 4th International Conference on Structural Mechanics in Reactor Technology (SMiRT), Paper J7/6, Berlin, 1977.
767. Beriaud, C. *et al.*, Local behaviour of reinforced concrete walls under hard missile impact. 4th International Conference on Structural Mechanics in Reactor Technology (SMiRT), Paper J7/9, Berlin, 1977.
768. Gupta, Y. & Seaman, L., Local response of reinforced concrete to missile impacts. 4th International Conference on Structural Mechanics in Reactor Technology (SMiRT), Paper J10/4, Berlin, 1977.
769. Barr, P. *et al.*, An experimental investigation of scaling of reinforced concrete structures under impact loading. In: *Design for Dynamic Loading*, Construction Press, 1982.
770. Barr, P. *et al.*, Experimental studies of the impact resistance of steel faced concrete composition. 7th International Conference on Structural Mechanics in Reactor Technology (SMiRT), Paper J8/4, 1983.
771. Barr, P. *et al.*, Studies of missile impact with reinforced concrete structures. *Nuclear Energy*, **19** (1980).
772. Neilson, A. Missile impact on metal structures. *Nuclear Energy*, **19** (1980).
773. Anderson, W., Watson, A. & Armstrong, P., Fibre reinforced concrete for the protection of structures against high velocity impact. International Conference and Exposition of Structural Impact and Crashworthiness, July 1984.
774. Berriaud, C. *et al.*, Test and calculation of the local behaviour of concrete structures under missile impact. 5th International Conference on Structural Mechanics in Reactor Technology (SMiRT), Paper J7/1, Berlin, 1979.
775. Brandes, K., Limberger, E. & Hertes, J., Experimental investigation of reinforced concrete behaviour due to impact load. 5th International Conference on Structural Mechanics in Reactor Technology (SMiRT), Paper J7/3, Berlin, 1979.
776. Sage, E. & Pfeiffer, A., Response of reinforced concrete targets to impacting soft missile. 5th International Conference on Structural Mechanics in Reactor Technology (SMiRT), Paper J8/4, Berlin, 1979.

777. Jonas, W. *et al.*, Experimental investigations to determine the kinetic ultimate capacity of reinforced concrete slabs subject to deformable missiles. 5th International Conference on Structural Mechanics in Reactor Technology (SMiRT), Paper J8/4, Berlin, 1979.
778. Dulac, J. & Giraud, J., Impact testing of reinforced concrete slabs. 6th International Conference on Structural Mechanics in Reactor Technology (SMiRT), Paper J7/1, Paris, 1981.
779. Chiba, N. *et al.*, Nonlinear dynamic analysis of steel plates subjected to missile impact. 6th International Conference on Structural Mechanics in Reactor Technology (SMiRT), Paper J7/4, Paris, 1981.
780. Ohte, S. *et al.*, The strength of steel plates subjected to missile impact. 6th International Conference on Structural Mechanics in Reactor Technology (SMiRT), Paper J7/10, Paris, 1981.
781. Yamamoto, S. *et al.*, Nonlinear analysis of missile impact on steel plates. 7th International Conference on Structural Mechanics in Reactor Technology (SMiRT), Paper J7/3, 1983.
782. Woodfin, R. & Sliter, G., Modelling and contact of turbine missile concrete impact experiments. 6th International Conference on Structural Mechanics in Reactor Technology (SMiRT), Paper J8/1, Paris, 1981.
783. Woodfin, R. & Sliter, G., Results of full-scale turbine missile concrete impact experiments. 6th International Conference on Structural Mechanics in Reactor Technology (SMiRT), Paper J8/2, Paris, 1981.
784. Rudiger, E. & Riech, H., Experimental and theoretical investigations on the impact of deformable missiles onto reinforced concrete slabs. 7th International Conference on Structural Mechanics in Reactor Technology (SMiRT), Paper J8/3, 1983.
785. Brandes, K., Limberger, E. & Herter, J., Strain rate dependent energy absorption capacity of reinforced concrete members under aircraft impact. 7th International Conference on Structural Mechanics in Reactor Technology (SMiRT), Paper J9/5, 1983.
786. Gueraud, R. *et al.*, Study of the perforation of reinforced concrete slabs by rigid missiles — general introduction and experimental study, Part I. *Nucl. Eng. Des.*, **41** (1977).
787. Figuet, G. & Dacquet, S., Study of the perforation of reinforced concrete slabs by rigid missiles — experimental study, Part II. *Nucl. Eng. Des.*, **41** (1977).
788. Goldstein, S., Berriaud, C. & Labrot, R., Study of the perforation of reinforced concrete slabs by rigid missiles — experimental study, Part III. *Nucl. Eng. Des.*, **41** (1977).
789. Davidson, I. & Bradbury, J., The analysis of dynamically loaded non-linear structures. *Nucl. Eng. Des.*, **45** (1978).
790. Gupta, Y. & Seamae, L., Local response of reinforced concrete to missile impact. *Nucl. Eng. Des.*, **45** (1978).
791. Limberger, E., A simple mode for predicting energy dissipation of thin plates being perforated by hard missiles. *Nucl. Eng. Des.*, **51** (1979).
792. Stefanou, G., An investigation into the behaviour of perforated slabs for concrete reactor vessels under temperature and external load. *Nucl. Eng. Des.*, **52** (1979).
793. Bignon, P. & Riera, J., Verification of methods of analysis of soft missile impact problems. *Nucl. Eng. Des.*, **60** (1980).
794. Albertini, C. & Montagnani, M., Constitutive laws of materials in dynamics — outline of a program of testing on small and large specimens for containment of extreme dynamic loading conditions. *Nucl. Eng. Des.*, **68** (1981).
795. Nachtsheim, W. & Stangenberg, F., Investigation of results of Meppen slab tests — comparison with parametric investigation. *Nucl. Eng. Des.*, **75** (1982).
796. Kussmaul, K., The investigation of the tensile and notch impact bend test into an experimentally validated fracture mechanics concept. *Nucl. Eng. Des.*, **72** (1982).
797. Riera, J., Basic concept and load characteristic in impact problem. IBAM, June 1982.
798. Degen, P., Perforation of reinforced concrete slabs by rigid missiles. *J. Struct. Div., ASCE*, **106** (1980).

799. Ghaboussi, J., Millavec, W. & Jsenberg, J., Reinforced concrete structures under impulsive loading. *J. Struct. Div., ASCE* (March 1984).
800. Zerna, W. & Stangenberg, F., On the shock behaviour of reinforced concrete structural systems. IBAM, June 1982.
801. Douglas, R. & Bingham, W., Strains and wave velocities in high velocity impact. Proc. Second Annual Engineering Mechanics Division Speciality Conference, May 1977.
802. Broman, R. *et al.*, Report on the ASCE committee on impactive and impulsive loads. *Civil Engineering and Nuclear Power*, **V** (1980).
803. Broman, R. *et al.*, Analysis and design for impact loads. *Civil Engineering and Nuclear Power*, **V** (1980).
804. Rotz, J., Evaluation of tornado missile impact effects on structures. Proc. Symposium on Tornadoes, Assessment of Knowledge and Implications for Man, June 1976.
805. Baker, A., Partition of impact energy between local deformation and target response. *Nuclear Energy*, **19** (1980).
806. Jonker, J., Dynamic response of a clamped/free hollow circular cylinder under travelling torsional impact loads. *Nucl. Eng. Des.*, **67** (1981).
807. Broman, R. *et al.*, Analysis and design for impulsive loads. *Civil Engineering and Nuclear Power*, **V** (1980).
808. Chang, W., Burdette, E. & Barnett, R., Missile penetration. *J. Struct. Div., ASCE* (1976).
809. Chang, W., Impact of solid missiles on concrete barriers. *J. Struct. Div., ASCE* (February 1981).
810. Kar, A., Barrier design for tornado generated missiles. 4th International Conference on Structural Mechanics in Reactor Technology (SMiRT), Paper J10/3, Berlin, 1977.
811. Kar, A., Local effects of tornado generated missiles. *J. Struct. Div. ASCE* (May 1978).
812. Kar, A., Projectile penetration into steel. *J. Struct. Div., ASCE* (October 1979).
813. Kar, A., Impactive effects of tornado missiles and aircrafts. *J. Struct. Div., ASCE* (November 1979).
814. Kar, A., Projectile penetration into buried structures. *J. Struct. Div., ASCE* (January 1978).
815. Kar, A., Impact load for tornado-generated missiles. *Nucl. Eng. Des.*, **47** (1978).
816. Kar, A., Loading time history for tornado-generated missiles. *Nucl. Eng. Des.*, **51** (1979).
817. Sliter, G., Assessment of empirical concrete impact formulae. *J. Struct. Div., ASCE* (May 1980).
818. Halder, A. & Miller, F., Penetration depth in concrete for non-deformable missiles. *Nucl. Eng. Des.*, **71** (1982).
819. Haldar, A., Hatami, M. & Miller, F., Concrete structures: penetration depth estimation for concrete structures. *J. Struct. Div., ASCE* (January 1983).
820. Haldar, A., Hatami, M. & Miller, F., Penetration and spalling depth estimation for concrete structures. 7th International Conference on Structural Mechanics in Reactor Technology (SMiRT), Paper J7/2, 1983.
821. Berriaud, C. *et al.*, Local behaviour of reinforced concrete walls under missile impact. *Nucl. Eng. Des.*, **45** (1978).
822. Dubois, J., Chedmail, J. & Bianchini, J., Numerical analysis of impact-penetration problems for nuclear reactor safety. 4th International Conference on Structural Mechanics in Reactor Technology (SMiRT), Paper J7/4, Berlin, 1977.
823. Drittler, K., Gruner, P. & Krivy, J., Calculation of forces arising from impacting projectiles upon yielding structures. 4th International Conference on Structural Mechanics in Reactor Technology (SMiRT), Paper J7/4, Berlin, 1977.
824. Alderson, M. *et al.*, Reinforced concrete behaviour due to missile impact. 4th International Conference on Structural Mechanics in Reactor Technology (SMiRT), Paper J7/7, Berlin, 1977.

825. McMahon, P., Sen, S. & Meyers, B., Behaviour of reinforced concrete barriers subject to the impact of turbine missiles. 5th International Conference on Structural Mechanics in Reactor Technology (SMiRT), Paper J7/6, Berlin, 1979.
826. Stangenberg, F. & Buttmann, P., Impact testing of steel fibre reinforced concrete slabs with liner. 5th International Conference on Structural Mechanics in Reactor Technology (SMiRT), Paper J7/5, Berlin, 1979.
827. Haldar, A., Turbine missile — a critical review. *Nucl. Eng. Des.*, **55** (1979).
828. Haldar, A., Impact loading — damage predicting equations. 6th International Conference on Structural Mechanics in Reactor Technology (SMiRT), Paper J8/4, Paris, 1981.
829. Davis, I., Design and analysis of concrete structures under impact and impulsive loading. IBAM, June 1982.
830. Healing, J., Local effects of missile impact on protective barriers. Proc. Symposium on Tornadoes, Assessment of Knowledge and Implications for Man, June 1976.
831. Yakeda, J., Crashworthiness of concrete structures subject to impact or explosion. International Conference and Exposition of Structural Impact and Crashworthiness, July 1984.
832. Zielinski, A., Fracture of concrete under impact loading. International Conference and Exposition of Structural Impact and Crashworthiness, July 1984.
833. Zorn, N. & Reinhardt, H., Concrete structures under high intensity tensile waves. International Conference and Exposition of Structural Impact and Crashworthiness, July 1984.
834. Tulacz, J. & Smirth, R., Assessment of missile generated by pressure component failure and its application to recent gas-cooled nuclear plant design. *Nuclear Energy*, **19** (1980).
835. Nashtsheim, W. & Stangenberg, F., Impact of deformable missiles on reinforced concrete plates — comparisonal calculation of Meppens tests. 6th International Conference on Structural Mechanics in Reactor Technology (SMiRT), Paper J7/3, Paris, 1981.
836. Nashtshein, W. & Stangenberg, F., Selected result of Meppen slab tests — state of interpretation, comparison with computation investigation. 7th International Conference on Structural Mechanics in Reactor Technology (SMiRT), Paper J8/1, 1983.
837. Davis, I., Damaging effects from the impact of missiles against reinforced concrete structures. *Nuclear Energy*, **19** (1980).
838. Kennedy, R., A review of procedures for the analysis and design of concrete structures to resist missile impact effects. *Nucl. Eng. Des.*, **37** (1976).
839. Attalla, I., Missile impact on reinforced concrete structures. *Nucl. Eng. Des.*, **37** (1976).
840. Hughes, G., Hard missile impact on reinforced concrete. *Nucl. Eng. Des.*, **77** (1984).
841. Reyen, J., Villafune, E. & Crutzen, Y., Impulsive loading on concrete structures. 6th International Conference on Structural Mechanics in Reactor Technology (SMiRT), Paper J10/1, Paris, 1981.
842. Bangash, Y., The limit state analysis of a prestressed concrete containment vessel for PWR. 7th International Conference on Structural Mechanics in Reactor Technology (SMiRT), Paper J3/8, 1983.
843. Bangash, Y., Containment vessel design and practice. *Nuclear Energy*, **11** (1982).
844. Krulzik, N., Analysis of aircraft impact problem. In: *Advanced Structural Dynamics*, Applied Science Publishers, London, 1980.
845. Glasstone, S. & Dolan, P. J., The effect of nuclear weapons (3rd edn). US Govt Printing Office, Washington DC 1977.
846. Rice, J. & Bahar, L., Reaction-time relationship and structural design of reinforced concrete slabs and shells for aircraft impact. 3rd International Conference on Structural Mechanics in Reactor Technology (SMiRT), Paper J5/3, London, 1975.
847. Bahar, Y. & Rice, S., Simplified derivation of the reaction-time history in aircraft impact on a nuclear power plant. *Nucl. Eng. Des.*, **49** (1978).

848. Sharpe, R., Ramil, H. & Scunlan, R., Analysis of aircraft impact on reactor building. 3rd International Conference on Structural Mechanics in Reactor Technology (SMiRT), Paper J5/4, London, 1975.
849. Zerna, W., Schnellenbach, G. & Stangenberg, F., Optimized reinforcement of nuclear power plant structures for aircraft impact forces. *Nucl. Eng. Des.*, **37** (1976).
850. Meder, G., Dynamic response of a soft elastic–plastic systems subjected to aircraft impact pulses. *Nucl. Eng. Des.*, **47** (1982).
851. Schnellenbach, G. & Stangenberg, F., Design of concrete containments for aircraft impact. 4th International Conference on Structural Mechanics in Reactor Technology (SMiRT), Paper J8/4, Berlin, 1977.
852. Riera, J., A critical reappraisal of nuclear power plant safety against accidental aircraft impact. *Nucl. Eng. Des.*, **57** (1980).
853. Steveson, J., Current summary of international extreme load design requirements for nuclear power plant facilities. *Nucl. Eng. Des.*, **60** (1980).
854. Hornyk, K., Analytical modelling of the impact of soft missiles on protective walls. 4th International Conference on Structural Mechanics in Reactor Technology (SMiRT), Paper J7/3, Berlin, 1977.
855. Stoykovich, M., Impact load time histories for viscoelastic missiles. 4th International Conference on Structural Mechanics in Reactor Technology (SMiRT), Paper J7/5, Berlin, 1977.
856. Chiapetta, R. & Costello, J., Automobile impact forces on concrete walls. 6th International Conference on Structural Mechanics in Reactor Technology (SMiRT), Paper J8/8, Paris, 1981.
857. Porter, W., Generation of missiles and destructive shock fronts and their consequences. *Nuclear Energy*, **19** (1980).
858. McMahon, P., Meyers, B. & Buchert, K., The behaviour of reinforced concrete barriers to the impact of tornado generated deformable missiles. 4th International Conference on Structural Mechanics in Reactor Technology (SMiRT), Paper J10/2, Berlin, 1977.
859. Jones, W. *et al.*, Approximate calculation of the impact of missiles onto reinforced concrete structures and comparison of test results. 5th International Conference on Structural Mechanics in Reactor Technology (SMiRT), Paper J8/6, Berlin, 1979.
860. Hanagud, S. & Bernard, R., A penetration theory for axisymmetric projectiles. 4th International Conference on Structural Mechanics in Reactor Technology (SMiRT), Paper J7/10, Berlin, 1977.
861. Brown, M., Curtress, N. & Jewett, J., Local failure of reinforced concrete under missile impact loading. 5th International Conference on Structural Mechanics in Reactor Technology (SMiRT), Paper J8/7, Berlin, 1979.
862. Jamet, P. *et al.*, Perforation of a concrete slab by a missile: finite element approach. 7th International Conference on Structural Mechanics in Reactor Technology (SMiRT), Paper J7/6, 1983.
863. Wittmann, F. & Boulahdour, T., Variability of resistance of concrete slabs under impact load. 6th International Conference on Structural Mechanics in Reactor Technology (SMiRT), Paper J7/2, Paris, 1981.
864. Stangenberg, F. & Zerna, W., Extreme load resistance design of nuclear power plant structures. 4th International Conference on Structural Mechanics in Reactor Technology (SMiRT), Paper J8/5, Berlin, 1977.
865. Munday, G., Initial velocities attained by plant generated missile. *Nuclear Energy*, **19** (1980).
866. Yen, G., Probability of a containment of turbine missile. *Nucl. Eng. Des.*, **37** (1976).
867. Twisdale, L., Dunn, W. & Davies, T., Tornado missile transport analysis. *Nucl. Eng. Des.*, **51** (1979).
868. Goodman, J. & Koch, J., The probability of a tornado missile hitting a target. *Nucl. Eng. Des.*, **75** (1983).

869. Haldar, A. & Hamieh, H., Local effects of solid missiles on concrete structures. *J. Struct. Div., ASCE* (May 1984).
870. Costello, J., Simiu, E. & Cordes, M., Tornado borne missile speed. *J. Struct. Div., ASCE* (June 1977).
871. Watwood, V., Jr., The finite element method for prediction of crack behaviour. *Nucl. Eng. Des.*, **11** (1969).
872. Hopkirk, R., Lympany, S. & Marti, J., Three-dimensional numerical predictions of impact effects. *Nuclear Energy*, **19** (1980).
873. Ahmad, M. & Bangash, Y., A three-dimensional bond analysis using finite element. *Computers and Structures*, **25** (1987) 281–96.
874. Hodge, P. G., The theory of piece-wise linear isotropic plasticity. IUTAM Colloquium, Madrid, 1955.
875. Martin, H. C. & Carrey, G. F., *Introduction to Finite Element Analysis*. McGraw-Hill, New York, 1973.
876. Bathe, K. J. & Wilson, E. L., *Numerical Methods in Finite Element Analysis*. Prentice-Hall, Englewood Cliffs, New Jersey, 1972.
877. Desai, C. S. & Abel, J. F., *Introduction to the Finite Element Method*. Van Nostrand Reinhold, New York, 1972.
878. Ahmad, S., Curved finite element in the analysis of solid shell and plate structures. PhD Thesis, University of Wales, Swansea, 1969.
879. Clough, R. W., Comparison of three-dimensional finite elements. Proc. Symposium on Application of Finite Element Methods in Civil Engineering, Vanderbilt University, 1969.
880. Scotto, F. L., Thin walled 1:20 prestressed concrete pressure vessel model for THTR reactor type. 2nd International Conference on Structural Mechanics in Reactor Technology (SMiRT), Paper H5/6, Berlin, 1973.
881. Ottosen, N. S. & Andersen, S. I., Theoretical and experimental studies for optimisation of PCRV top closures. 3rd International Conference on Structural Mechanics in Reactor Technology (SMiRT), Paper H3/6, London, 1975.
882. Langan, D. & Smith, J. R., 1/10th scale model of Hartlepool pressure vessel — introduction statement. Structural Research Laboratory, Taylor Woodrow Construction Ltd, November 1968.
883. Houde, J., Study of force–displacement relationships for the finite element analysis of reinforced concrete. Report No. 73-2, Department of Civil Engineering and Applied Mechanics, McGill University, Montreal, December 1973.
884. Houde, J. & Mirza, M. S., A finite element analysis of shear strength of reinforced concrete beams. In: *Shear in Reinforced Concrete*, Vol. 1, Special Publication SP-42, American Concrete Institute, Detroit, Michigan, 1974.
885. Houde, J. & Mirza, M. S., Investigation of shear transfer across cracks by aggregate interlock. Research Report No. 72-06, Ecole Polytechnique de Montreal, Department of Genie Civil, Division de Structures, 1972.
886. Fort St. Vrain Nuclear Generating Station, Final Safety Analysis Report, Appendix E. PCRV Data, Public Service Co. of Colorado Report, 1979.
887. Freudenthal, A. M. & Roll, F., Creep and creep recovery of concrete under high compressive stress. *J. Am. Concr. Inst.*, **54** (1958) 1111–42.
888. Hungspreug, S., The local interaction between reinforcing bars and concrete under high-level reversed cyclic loading. Thesis presented to Cornell University in partial fulfilment of the requirements for the degree of Doctor of Philosophy, 1981.
889. Herrmann, L. R., Finite element analysis of contact problems. *J. Eng. Mech. Division, ASCE*, **104** (1978).
890. Hatt, W. K., Notes on the effect of time element in loading reinforced concrete beams. *Proc. ASTM*, **7** (1907) 421–33.

891. Van der Houwen, P. J., *Construction of Integration Formulas for Initial Value Problems.* North-Holland, Amsterdam, 1977.
892. Huet, C., Application of Bazant's algorithm to the analysis of viscoelastic composite structures. *Materials and Structures* (RILEM, Paris), **13** (1980) 91–8.
893. Gilbert, R. I. & Warner, R. F., Nonlinear analysis of reinforced concrete slabs with tension stiffening. Uniciv Report R-167, University of New South Wales, January 1977.
894. Gilbert, R. I. & Warner, R. F., Time dependent behaviour of reinforced concrete slabs. Uniciv Report No. R-173, University of New South Wales, October 1977.
895. Grootenboer, H. J., Finite element analysis of two dimensional reinforced concrete structures, taking account of nonlinear physical behaviour and the development of discrete cracks. Doctoral Dissertation, Technical University of Delft, March 1979.
896. Fanning, D. N. & Dodge, W. G., Methods of calculating inelastic response of prestressed concrete pressure vessels. 5th International Conference on Structural Mechanics in Reactor Technology (SMiRT), Berlin, August 1979.
897. Echeverria, A., Mohraz, B. & Schnobrich, W. C., Crack development in a prestressed concrete reactor vessel as determined by a lumped parameter method. *Nucl. Eng. Des.*, **11** (1970) 286–94.
898. Bertero, V. V., Bresler, B. & Liao, H., Stiffness degradation of reinforced concrete members subjected to cyclic flexural loads. Earthquake Engineering Research Center, Report No. EERC 69-12, University of California, Berkeley, 1969.
899. Bertero, V. V., Experimental studies concerning reinforced, prestressed and partially prestressed concrete structures and their elements. Symposium on Resistance and Ultimate Deformability of Structures Acted on by Well Defined Repeated Loads, IABSE, Lisbon, 1973.
900. Bertero, V. V., Popov, E. P. & Wang, T. Y., Hysteretic behaviour of reinforced concrete flexural members with special web reinforcement. Earthquake Engineering Research Center, Report No. EERC 74-9, University of California, Berkeley, 1974.
901. Bertero, V. V. & Popov, E. P., Hysteretic behaviour of ductile moment-resisting reinforced concrete frame components. Earthquake Engineering Research Center, Report No. EERC 75-16, University of California, Berkeley, 1975.
902. Bertero, V. V. & Popov, E. P., Seismic behaviour of moment-resisting reinforced concrete frames. In: *Reinforced Concrete Structures in Seismic Zones.* SP-53, American Concrete Institute, Detroit, Michigan, 1977.
903. Backlund, J., Finite element analysis of nonlinear structures. Doctoral Thesis, Chalmers Technical University, Goteborg, Sweden, November 1973.
904. Becker, J. M. & Bresler, B., FIRES-RC — A computer program for the fire response of structures — Reinforced concrete frames. Report No. UCB FRG 74-3, Fire Research Group, Structural Engineering and Structural Mechanics, Department of Civil Engineering, University of California, Berkeley, July 1974.
905. Bell, J. C., A complete analysis of reinforced concrete slabs and shells. PhD Dissertation, Department of Civil Engineering, University of Canterbury, Christchurch, New Zealand, 1970.
906. Bell, J. C. & Elms, D. G., A finite element post elastic analysis of reinforced concrete shells. *Bulletin of the International Association for Shell and Spatial Structures*, No. 54 (April 1974).
907. Bell, J. C. & Elms, D., Partially cracked finite elements, *J. Struct. Div., ASCE*, **97** (1971).
908. Browne, R. D. & Bamforth, P. P., The long term creep of the Wylfa P. V. concrete for loading ages up to $12\frac{1}{2}$ years. 3rd International Conference on Structural Mechanics in Reactor Technology, London, Paper H1/8, September 1975.
909. Browne, R. D. & Blundell, R., The influence of loading age and temperature on the long term creep behaviour of concrete in a scaled moisture stable state. *Materials and Structures* (RILEM, Paris), 2 (1969) 133–44.

910. Sharma, N. K., Splitting failures in reinforced concrete members. PhD Thesis, Department of Structural Engineering, Cornell University, 1969.
911. Taylor, H. P. J., The fundamental behaviour of reinforced concrete beams in bending and shear. In: *Shear in Reinforced Concrete*, Vol. 1, SP-42, American Concrete Institute, Detroit, Michigan, 1974.
912. Taylor, H. P. S., Investigation of the forces carried across cracks in reinforced concrete beams in shear by interlock of aggregate. Cement and Concrete Association, London, Report No. 42-447, November 1970.
913. Taylor, R., A note on the mechanism of diagonal cracking in reinforcement. *Magazine of Concrete Research*, **11** (1959) 151–8.
914. Teller, L. W. & Cashell, H. D., Performance of doweled joints under repetitive loading. Bureau of Public Roads, April 1958.
915. Rostasy, F. S., Teichen, K. Th. & Engleke, H., Beitrag zur Klarung der Zusammenhanges von Kriechen und Relaxation bei Normal-beton. *Amtliche Forschungs- und Material-prufungsanstalt fur das Bauwesen*. Otto-Graf-Institut, Universitat Stuttgart, Strassenbau und Strassenverkehrstechnick, Heft 139, 1972.
916. Ruetz, W., An hypothesis for the creep of hardened cement paste and the influence of simultaneous shrinkage. International Conference on the Structure of Concrete, London, 1965; Cement and Concrete Association, 1968, pp. 365–87. (See also Deutscher Ausschuss fur Stahlbeton Heft 183, 1966.)
917. Oesterle, R. G. & Russel, H. G., Shear transfer in large scale reinforced concrete containment elements. Report for US Nuclear Regulatory Commission, NUREG/CR-1374, Construction Technology Laboratories, Portland Cement Association, April 1980.
918. Pauley, T., Design aspects of shear walls for seismic areas. Department of Civil Engineering, Research Report No. 71-11, University of Canterbury, Christchurch, New Zealand, October 1974.
919. Pauley, T. & Loeber, P. S., Shear transfer by aggregate interlock. In: *Shear in Reinforced Concrete*, Vol. 1, SP-42, American Concrete Institute, Detroit, Michigan, 1974.
920. Pauley, T., Park, R. & Phillips, M. H., Horizontal construction joints in cast-in-place reinforced concrete. In: *Shear in Reinforced Concrete*, Vol. 2, American Concrete Institute, Detroit, Michigan, 1974.
921. Mahaidi, A. L., Nonlinear finite element analysis of reinforced concrete deep members. Thesis presented to Cornell University, in partial fulfilment of the requirements for the degree of Doctor of Philosophy, 1979.
922. Aktan, H. M. & Hanson, R. D., Nonlinear cyclic analysis of reinforced concrete plane stress members. *J. Am. Concr. Inst.*, SP-63, No. 6 (1980).
923. Arutyunian, N. Kh., *Some Problems in the Theory of Creep*. Technteorizdat, Moscow, 1952 (in Russian); Engl. trans. Pergamon Press, 1966.
924. Agrawal, A. B., Jaeger, L. G. & Mufti, A. A., Crack propagation and plasticity of reinforced concrete shear-wall under monotonic and cyclic loading. Conference on Finite Element Methods in Engineering, Adelaide, Australia, December 1976.
925. Agrawal, A. B., Nonlinear analysis of reinforced concrete planar structures subjected to monotonic, reversed cyclic and dynamic loads. PhD Thesis, University of New Brunswick, Fredericton, Canada, March 1977.
926. Agrawal, A. B., Mufti, A. A. & Jaeger, L. G., A program for the nonlinear analysis of reinforced concrete planar structures subject to monotonic and reversed cyclic loads. School of Computer Science, Acadia University, Wolfsville, N.S., Canada, Report No. 1, April 1977.
927. Agrawal, A. B., Mufti, A. A. & Jaeger, L. G., A program for the nonlinear analysis of reinforced concrete structures underground accelerations. School of Computer Science, Acadia University, Wolfsville, N.S., Canada, Report No. 2, April 1977.

928. Mikkola, M. J. & Schnobrich, W. C., Material behaviour characteristics of reinforced concrete shells stressed beyond the elastic range. Civil Engineering Studies, SRS No. 367, University of Illinois at Urbana-Champaign, Urbana, Illinois, August 1970.

929. Marcus, H., Load carrying capacity of dowels at transverse pavements joints. *J. Am. Conc. Inst., Proc.*, **48** (1951).

930. Mattock, A. H., Effect of moment and tension across the shear plane on single direction shear transfer strength in monolithic concrete. Report No. SM-743, Department of Civil Engineering, University of Washington, Seattle, Washington, October 1974.

931. Mattock, A. H., Shear transfer in concrete having reinforcement at an angle to the shear plane. In: *Shear in Reinforced Concrete*, Vol. 1, SP-42, American Concrete Institute, Detroit, Michigan, 1974.

932. Mattock, A. H., Shear transfer under cyclically reversing loading across an interface between concretes cast at different times. Report No. SM-77-1, Department of Civil Engineering, University of Washington, Seattle, Washington, November 1974.

933. Mattock, A. H., Shear transfer under monotonic loadings across an interface between concrete cast at different times. Report No. SM-76-3, Department of Civil Engineering, University of Washington, Seattle, Washington, September 1976.

934. Mattock, A. H., The shear transfer behaviour of cracked monolithic concrete subject to cyclically reversing shear. Report No. SM-74-4, Department of Civil Engineering, University of Washington, Seattle, Washington, November 1974.

935. Mattock, A. H. & Hawkins, N. M., Shear transfer in reinforced concrete — recent research. *J. Prestressed Concrete Institute*, **17** (1972) 55–75.

936. McDonald, J. E., An experimental study of multiaxial creep in concrete. In: *Concrete for Nuclear Reactors*, SP-34, American Concrete Institute, Detroit, Michigan (1972) pp. 732–68.

937. McHenry, D., A new aspect of creep in concrete and its application to design. *Proc. ASTM*, **43** (1943) 1069–86.

938. Macmillan, M., Evolution de fluage et des proprietes de beton. *Annales, Inst. Tech. du Batiment et des Travaux Publics*, **21** (1969) 1033; and **13** (1960) 1017–52.

939. Macmillan, M. & Lelan, M., Le fluage de beton. *Annales, Inst. Tech. du Batiment et des Travaux Publics* (Supplement) **23** (1970) 7–13; and **21** (1968) 847–50.

940. Mandel, J., Sur les corps viscoelastiques lineaires dont les proprietes dependent de l'age. *Comptes Rendus des Seances de l'Academie des Sciences*, **247** (1958) 175–8.

941. Marechal, J. C., Contribution a l'etude des proprietes thermiques et mecaniques du beton en function de la temperature. *Annales, Inst. Tech. du Batiment et des Travaux Publics*, **23** (1970) 123–45.

942. Marechal, J. C., Fluage de beton en fonction de la temperature. *Materials and Structures* (RILEM, Paris), **2** (1969) 111–15; see also *Materials and Structures*, **3** (1970) 395–405.

943. Marechal, J. C., Fluage du beton en fonction de la temperature. *Annales, Inst. Tech. du Batiment et des Travaux Publics*, **23** (1970) 13–24.

944. Maslov, G. N., Thermal stress states in concrete masses, with account of concrete creep. *Izvestia Nauchno-Issledovatelskogo Instituta Gidrotekhniki, Gosenergoizdat*, **28** (1940) 175–88 (in Russian).

945. McNeice, G. M., Elastic–plastic bending of plates and slabs by finite element method. Thesis presented to the University of London in partial fulfilment of the requirements for the degree of Doctor of Philosophy, 1967.

946. Saouma, V. E., Interactive finite element analysis of reinforced concrete: a fracture mechanics approach. PhD Thesis, Cornell University, January 1981.

947. Sandler, I. S., On the uniqueness and stability of endochronic theories of material behaviour. *J. Appl. Mech.*, **45** (1978) 263–6.

948. Schapery, R. A., A thermodynamic constitutive theory and its applications to various nonlinear materials. *Proceedings IUTAM Symposium, Kilbride, June 1968*, B. A. Boley, ed. Springer Verlag, New York, 1968.

949. Sinha, B. P., Gerstle, K. H. & Tulin, L. G., Stress–strain relations for concrete under cyclic loading. *J. Am. Concr. Inst.*, **61** (1964) 195–211.
950. Valanis, K. C. & Fan, J., Endochronic analysis of cyclic elastoplastic strain fields in a notched plate. *ASME, J. Appl. Mech.*, **50** (1983) 789–94.
951. Valanis, K. C. & Fan, J., A numerical algorithm for endochronic plasticity and comparison with experiment. *Computers and Structures*, **19** (1984) 717–24.
952. Van der Houwen, P. J., *Construction of Integration Formulas for Initial Value Problems.* North-Holland, Amsterdam, 1977.
953. Van Greunen, J., Nonlinear geometric, material and time dependent analysis of reinforced and prestressed concrete slabs and panels. PhD Thesis, Division of Structural Engineering and Structural Mechanics, University of California, Berkeley, Report No. UC-SESM 79-3, October 1979.
954. Van Zyl, S. F., Analysis of curved segmentally erected prestressed concrete box girder bridges. PhD Dissertation, Division of Structural Engineering and Structural Mechanics, University of California, Berkeley, Report No. UC-SESM 78-2, January 1978.
955. Van Zyl, F. S. & Scordelis, A. C., Analysis of curved prestressed segmental bridges. *J. Struct. Div., ASCE*, **105** (1979).
956. Volterra, V., *Lecons sur les Fonctions de Ligne.* Gauthier-Villars, Paris, 1913; and *Theory of Functionals and of Integral and Integrodifferential Equations.* Dover, New York, 1959.
957. Wegner, R. & Duddeck, H., Der Gerissenen Zustand Zweiseitig Gelagerter Platten unter Einzellasten-Nichtlineare Berechnung mit Finiten Elementen. *Beton und Stahlbetonbau*, **70** (1975).
958. Wegner, R., Harbord, R. & Duddeck, H., Flach- und Pilzgecken im Ungerissenen und gerissenen Zustand. *Bauingenieur*, **50** (1975).
959. Wanchoo, M. K. & May, G. W., Cracking analysis of reinforced concrete plates. *J. Struct. Div., ASCE*, **101** (1975).
960. Wittmann, F. H., Vergleich einiger Kriechfunktionen mit Versuchsergebnissen. *Cement and Concrete Research*, **1** (1971) 679–90.
961. Yuzugullu, O. & Schnobrich, W. C., A numerical procedure for the determination of the behaviour of a shear wall frame system. *J. Am. Conc. Inst.*, **70** (1973) 474–9.
962. Wittmann, F., Einfluss des Feuchtigkeitsgehaltes auf das Kriechen des Zementsteines. *Rheol. Acta*, **9** (1970) 282–7.
963. Wittmann, F., Kriechen bei Gleichzeitigem Schwinden des Zementsteins. *Rheol. Acta*, **5** (1966) 198–204.
964. Wittmann, F., Kreichverformung des Betons unter Statischer und unter Dynamischer Belastung. *Rheol. Acta*, **10** (1971) 422–8.
965. Wittmann, F., Surface tension, shrinkage and strength of hardened cement paste. *Materials and Structures* (RILEM, Paris), **1** (1968) 547–52.
966. Wittmann, F., Bestimmung Physikalischer Eigenschauften des Zementsteins. *Deutscher Aussehuss fur Stahlbeton*, Heft 232, W. Ernst, Berlin, 1974.
967. Isenberg, J., Wojcik, G. L. & Nikooyeh, H., Dynamic analysis of buried structures including catastrophic failure. Weidlinger Associates, Final Report under Contract DNA001-79-C-0227 to Defense Nuclear Agency, 1 December 1979.
968. Isenberg, J., Vaughan, D. K. & Wong, F. S., Effects of scale in the response of MX vertical shelters. Weidlinger Associates, Final Report under Contract DNA001-79-C-0253 to Defense Nuclear Agency, 15 April 1980.
969. Ingraffea, A. R., On discrete fracture propagation in rock loaded in compression. Proc. First International Conference on Numerical Methods in Fracture Mechanics, Swansea, 9–13 January, 1978, pp. 235–48.
970. Jimenez, R., Gergely, P. & White, R. N., Shear transfer across cracks in reinforced concrete. Report 78-4, Department of Structural Engineering, Cornell University, August 1978, p. 357.

971. Jimenez, R., White, R. N. & Gergely, P., Bond and dowel capacities of reinforced concrete. *J. Am. Concr. Inst.*, **76** (1979) 73–92.
972. Jones, R., The ultimate strength of reinforced concrete beams in shear. *Magazine of Concrete Research*, **8** (1956).
973. Kesler, C. E., Wallo, E. M., Yuan, R. L. & Lorr, J. L., Sixth Progress Report, Prediction of Creep in Structural Concrete from Short Time Tests. Department of Theoretical and Applied Mechanics, University of Illinois, Urbana, Illinois, August 1965.
974. Kimishima, H. & Kitahara, H., Creep and creep recovery of mass concrete. Technical Report C-64001, Central Research Institute of Electric Power Industry, Tokyo, Japan, September 1964.
975. Klug, P. & Wittmann, T., The correlation between creep deformation and stress relaxation in concrete. *Materials and Structures* (RILEM, Paris), **3** (1970) 75–80.
976. Komendant, G. J., Polivka, M. & Pirtz, D., Study of concrete properties for prestressed concrete reactor vessels. Final report Part II, Creep and strength characteristics of concrete at elevated temperatures. Report No. UCSEMS76-3, Structures and Materials Research, Department of Civil Engineering, Report to General Atomic Company, San Diego, California, Berkeley, California, April 1976.
977. Lin, C., Hseih, B. J. & Valentin, R. A., The application of endochronic plasticity theory in modelling and dynamic inelastic response of structural systems. *Nucl. Eng. Des.*, **66** (1981) 213–21.
978. Lauer, R. K. & Slate, F. J., Autogeneous healing of cement paste. *J. Am. Concr. Inst.*, **52** (1956) 1083–98.
979. LeCamus, B., Recherches experimentales sur le deformation du beton et du beton arme, Part II. *Annales Inst. Tech. du Batiment et des Travaux Publics*, (1947).
980. Lambotte, H. & Mommens, A., L'evolution du fluage du beton en fonction de sa composition, du taux de contrainte et de l'age. Groupe de travail GT 22, Centre National de Recherches Scientifiques et Techniques pour l'Industrie Cimentiere, Bruxelles, July 1976.
981. Leombruni, P., Buyukozturk, O. & Connor, J. J., Analysis of shear transfer in reinforced concrete with application to containment wall specimens. Report MIT-CER-79-26, Department of Civil Engineering, Massachusetts Institute of Technology, June 1979.
982. Mufti, A. A., Mirza, M. S., McCutcheon, J. O. & Houde, J., A study of the behaviour of reinforced concrete elements using finite elements. Structural Concrete Series No. 70-5, McGill University, Montreal, September 1970.
983. Mufti, A. A., Mirza, M. S., McCutcheon, J. O. & Spowski, R. W., A finite element study of reinforced concrete structures. Structural Concrete Series No. 71-8, McGill University, Montreal, October 1971.
984. Mikkola, M. J. & Schnobrich, W. C., Material behaviour characteristics for reinforced concrete shells stressed beyond the elastic range. Civil Engineering Studies, SRS No. 367, University of Illinois at Urbana-Champaign, Urbana, Illinois, August 1970.
985. Mroz, Z., Mathematical models of inelastic concrete behaviour. In: *Inelasticity and Non-Linearity in Structural Concrete*, M. Z. Cohn, ed. University of Waterloo Press, Waterloo, Ontario, Study No. 8, 1972, pp. 47–72.
986. Mueller, G., Numerical problems in nonlinear analysis of reinforced concrete. US-SESM Report No. 77-5, University of California, Berkeley, September 1977.
987. Murray, D. W., Octahedral based incremental stress–strain matrices. *J. Eng. Mech. Division, ASCE*, **105** (1979) 501–13.
988. Murray, D. W., Chitnuyanondh, L., Agha, K. Y. & Wong, C., A concrete plasticity theory for biaxial stress analysis. *J. Eng. Mech. Division, ASCE*, **105** (1979) 989–1006.
989. Ma, S.-Y.M., Bertero, V. V. & Popov, E. P., Experimental and analytical studies on the hysteretic behaviour of reinforced concrete rectangular and T-beams. Earthquake Engineering Research Center, Report No. EERC 76-2, University of California, Berkeley, 1976.

990. Mahin, S. A. & Bertero, V. V., Rate of loading effects on uncracked and repaired reinforced concrete members. Earthquake Engineering Research Center, Report No. EERC 72-9, University of California, Berkeley, 1972.
991. Marchertas, A. H., Fistedis, S. H., Bazant, Z. P. & Belytschko, T. B., Analysis and application of prestressed concrete reactor vessels for LMFBR containment. *Nucl. Eng. Des.*, **49** (1978) 155–73.
992. Marchertas, A. H., Belytschko, T. B. & Bazant, Z. P., Transient analysis of LMFBR reinforced/prestressed concrete containment. 5th International Conference on Structural Mechanics in Reactor Technology (SMiRT), Paper H8/1, Berlin, August 1979.
993. dePaiva, H. A. R. & Siess, C. P., Strength and behaviour of deep beams in shear. *J. Struct. Div., ASCE*, **91** (1965) 19–41.
994. Dulacska, H., Dowel action of reinforcement crossing cracks in concrete. *J. Am. Concr. Inst.*, **69** (1972).
995. Bathe, K. J. & Wilson, E. L., *Numerical Methods in Finite Element Analysis*. Prentice-Hall, Englewood Cliffs, New Jersey, 1976.
996. Bathe, K. J. *et al.*, Some computational capabilities for nonlinear finite element analysis. *Nucl. Eng. Des.*, **46** (1978) 429–55.
997. Colville, J. & Abbasi, J., Plane stress reinforced concrete finite elements. *J. Struct. Div., ASCE*, **100** (1974).
998. Crisfield, M. A., A faster modified Newton–Raphson integration. *Computer Methods in Applied Mechanics and Engineering*, **20** (1979) 267–78.
999. Glemberg, R., Dynamic analysis of concrete structures. PhD Thesis Department of Structural Mechanics, Chalmers University of Technology, Stockholm, 1984.
1000. Belytschko, T. & Mullen, R., Stability of explicit–implicit time mesh partitions in time integration. *Int. J. Num. Meth. Eng.*, **12** (1978) 1575–86.
1001. Berwanger, C., The modulus of concrete and the coefficient of thermal expansion below normal temperatures. In: *Temperature and Concrete*, SP-25, American Concrete Institute, Detroit, Michigan, 1971, pp. 181–234.
1002. Ergataudis, J. G., Isoparametric finite elements in two and three dimensional analysis. PhD Thesis, University of Wales, Swansea, 1968.
1003. Fahmi, H. M., Polivka, M. & Bresler, B., Effect of sustained and cyclic elevated temperature on creep and concrete. *Cement and Concrete Research*, **2** (1972) 591–606.
1004. Bangash, Y. & England, G. L., The influence of thermal creep on the operational behaviour of complex structures. *Proc. International Conference on Fundamental Research on Creep and Shrinkage, Lausanne*, Martinus Nijhoff, The Hague, 1982.
1005. Bangash, Y., The structural design of circular diversion, power and pressure tunnels. *Tunnels and Tunnelling*, Part 1, May/June (1971); Part 2, July/August (1971); Part 3, September/October (1971).
1006. Sandia Laboratories, 1 : 6 scaled model of R.C. containment — a pre-test analysis. Report: N/CR 4913 Sand 87-0891, Sandia Laboratories, 1987.
1007. Bangash, Y., Aircraft crash analysis of the proposed Sizewell B containment vessel. 9th International Conference on Structural Mechanics in Reactor Technology (SMiRT), Vol. J, Lausanne, 1987, pp. 307–14.
1008. Bangash, Y., Loads and moments due to currents and wind on offshore gravity platforms. *Ocean Engineering*, **6** (1979) 285–96.
1009. Nataraja, R. & Kirk, C. L., Dynamic response of a gravity platform under random wave forces. Proc. OTC, Paper OTC-2904, 1977.
1010. J. Mowlem Co., London. Private communications on CONDEEP, 1983.
1011. BS 5337: The structural use of concrete for retaining aqueous liquids, London, 1976.
1012. Dean, R. G., Relative validities of water wave theories. Civil Engineering the Oceans, ASCE Conference, 6–8 September 1967.

1013. FIP, *Recommendations for the design and Construction of Concrete Sea Structures*. Federation Internationale de Constrainte, Paris, France (2nd edn), 1973.
1014. Frimann Clausen, C. J., *The Condeep Story*. John Mowlam & Co., London, 1976.
1015. Hogben, N. & Standing, R. G., Experience in computing wave loads on large bodies. NPL Offshore Technology Conference, Paper No. OTC 2189, Vol. 1, May 1975, & private notes.
1016. Maccamy, R. C. & Fuchs, R. A., Wave forces on piles — a diffraction theory. Beach Erosion Board Technical Memorandum, No. 69, 1954.
1017. Morison, J. R., O'Brien, M. P., Johnson, J. W. & Schaof, S. A., The force exerted by surface waves on piles. *Petroleum Transactions, AIME*, **189** (1950).
1018. Myers, J. J., Holm, C. H. & McCallister, R. F., *Handbook of Ocean and Underwater Engineering*. McGraw-Hill, New York, 1969.
1019. Pinfold, G. M., *Reinforced Concrete and Towers*. McGraw-Hill, New York, 1969.
1020. Skjelbreia, L. & Hendrickson, J., Fifth order gravity wave theory. Proc. Seventh Conference Coastal Engineering, 1961.
1021. Gibson, J. E., *Thin Shells*. Pergamon Press, Oxford, 1980.
1022. Stephen, P., Timoshenko, S. & Woinowsky-Krieger, S., *Theory of Plates and Shells*. McGraw-Hill, New York, 1981.
1023. Ursell, F., Mass transport in gravity waves. *Proc. Cambridge Phil. Soc.*, 1953.
1024. Werenskiold, S. Private communication, 1976.
1025. Al-Noury, S., Three-dimensional analysis of the TLP platform. *Computers and Structures*, **23** (1986) 699–714.
1026. Sachs, P., *Wind Forces in Engineering*. Pergamon Press, Oxford, 1972.
1027. *ABAQUS User's Manual*. University of Manchester, 1986.
1028. Newman, K., The structure and engineering properties of concrete. International Symposium Theory Arch. Dams, Southampton, England, 1965.
1029. Olsen, O., Implosion analysis of concrete cylinders under hydrostatic pressure. *ACI Journal*, March 1978, 83–85.
1030. Det Norske Veritas, Rules for classification of mobile offshore units, 1981.
1031. Minorsky, V. V., An analysis of ship collisions with reference to nuclear power plants. *Journal of Ship Research*, **3** (1959) 1–4.
1032. Søreide, T. H., *Ultimate Load Analysis of Marine Structures*. Tapir Publishing Company, Trondheim, Norway, 1981.
1033. JABSF, Ship collision with an offshore structure. IABSE Colloquium, Copenhagen, 1983.
1034. Woisin, G., Conclusion from collision examinations for nuclear merchant ships in the FRG. Proc. of Symposium on Naval Submarine, Hamburg, 1977.
1035. Reckling, K. A., On the collision protection of ships. PRADS Symposium, Tokyo, 1977.
1036. Nagasawa, H., Arita, K., Tani, M. & Oka, S., A study on the collapse of ship structure in collision with bridge piers. *J. Soc. of Nav. Arch. of Japan*, **142** (1977).
1037. Faulkner, D., A review of effective plating for use in the analysis of stiffened plating in bending and compression. *J. of Ship Research*, **19** (1975).
1038. Ohnishi, T., Kawakami, H., Yasukawa, W. & Nagasawa, H., On the ultimate strength of bow construction. *J. Soc. of Nav. Arch. of Japan*, **151** (1982).
1039. BBRV Tanks Inc., California, Design of prestressed concrete tanks, Vols 1, 2 and 3, 1977.
1040. Lew, H. S., West Virginia cooling tower collapse caused by inadequate concrete strength. *Civ. Engng (New York)*, **50**(2) (1980) 62–67.
1041. Hobbs, D. W., Expansion of concrete due to alkali–silica reaction — an explanation. *Mag. Concr. Res.*, **30**(105) (1978) 215–20.
1042. Hobbs, D. W., *Influence of mix proportions and cement alkali content upon expansion due to the alkali–silica reaction*. Cement and Concrete Association Technical Report 534, 1980, 31 pp.
Shrinkage of natural aggregates in concrete. Building Research Station Digest 35 (2nd Series), HM Stationery Office, London, 1963.

1043. McKaig, T. H., *Building failures — case studies in construction and design*. McGraw-Hill, New York, Toronto and London, 1963, 261 pp.
1044. Feld, J., *Lessons from failures of concrete structures*. American Concrete Institute, Monograph No. 1, 1964, 179 pp.
1045. Smolira, M., *Analysis of defects in concrete and brick structures during construction and in service*. Department of the Environment, London, 1969, 142 pp.
1046. Smith, W. D., Bridge failures. *Proc. Instn Civ. Engrs*, **60** (1976) 367–82.
1047. Melchers, R. E., *Studies of civil engineering failures — a review and classification*. Monash University, Civil Engineering Research Reports, No. 6/1976, 1976, 34 pp.
1048. Blockley, D. I., Analysis of structural failures. *Proc. Instn Civ. Engrs*, **62**, Feb. 1977, pp. 51–74.
1049. Sibly, P. G. & Walker, A. C., Structural accidents and their causes. *Proc. Instn Civ. Engrs*, **62** (1977) 191–208.
1050. Wearne, S. H., A review of reports of failures. *Proc. Instn Mech. Engrs*, **193**(20) (1979) 125–36.
1051. Hauser, R., Lessons from European failures. *Concr. Int.*, **1**, No. 12 (1979) 21–25.
1052. Walker, A. C., Study and analysis of the first 120 failure cases. *Proc. symp. struct. failures in buildings*. Institution of Structural Engineers, London, 30 April 1980, 78 pp.
1053. Masterman, O. J., High alumina cement concrete with data concerning conversion. *Civ. Eng. Publ. Wks Rev.*, **56**, No. 657 (1961) 483–86.
1054. Bate, S. C. C., High alumina cement concrete in existing building superstructure, 1982.
Building Regulations Advisory Committee. *Report by Sub-Committee P (High alumina cement concrete)*. Department of the Environment, London, 1975, 12 pp.
Report of Working Party on high alumina cement concrete. *Struct. Eng.*, **54**, No. 9 (1976) 352–61.
1055. Bate, S. C. C., High alumina cement concrete — an assessment from laboratory and field studies. *Struct. Eng.*, **58A**, No. 12 (1980).
1056. British Standards Institution, CP 110: 1972 (as amended 1977). *The structural use of concrete*, London, 1980, Parts 1–3, 156 pp., 104 pp., and 93 pp.
1057. Blenkinsop, J. C., Effect on normal $\frac{3}{8}$ in. reinforcement of calcium chloride additions to dense and porous concretes. *Mag. Concr. Res.*, **15**, No. 43 (1963) 33–38.
1058. Roberts, M. H., Effect of calcium chloride on the durability of pre-tensioned wire in prestressed concrete. *Mag. Concr. Res.*, **14**, No. 42 (1962) 143–54.
1059. Braestrup, M. W., Shear capacity of reinforced concrete beams. *Arch. Inzyn. Lad.*, **26**, No. 2 (1980) 295–317. (See also: *Effect of main steel strength on shear capacity of reinforced concrete beams with stirrups*. Technical University of Denmark, Structural Research Laboratory, Copenhagen, Report R 110, 1979, 49 pp.)
1060. Jensen, J. F., Plastic solutions for reinforced concrete beams in shear. *IABSE* (21), Aug. 1979, 71–78.
1061. Nielsen, M. P. & Braestrup, M. W., Shear strength of prestressed concrete beams without web reinforcement. *Mag. Concr. Res.*, **30**, No. 104 (1978) 119–27.
1062. Collins, M. P. & Mitchell, D., Shear and torsion design of prestressed and non-prestressed concrete beams. *J. Prestressed Concr. Inst.*, **25**, No. 5 (1980) 32–100.
1063. Gvozdev, A. A., Obosnovanie § 33 norm proektirovaniya zhelobetonnykh konstruktsii [Comments to par. 33 of the design standard for reinforced concrete structures]. *Stroit. Prom.*, **17**, No. 3 (1939) 51–58.
1064. Janas, M., Limit analysis of non-symmetric plastic shells by a generalized yield line method. Olszak, W. and Sawczuk, A. (eds), *Non-classical shell problems*. North-Holland, Amsterdam, 1964, pp. 997–1010.
1065. Janas, M. & Sawczuk, A., Influence of position of lateral restraints on carrying capacity of plates. *Bull. Inf. Com. Eur. Béton*, No. 58 (1966) 164–89.
1066. Wood, R. H., *Plastic and elastic design of slabs and plates*. Thames & Hudson, London, 1961, 344 pp.

1067. Janas, M., Large plastic deformations of reinforced concrete slabs. *Int. J. Solids & Struct.*, **4**, No. 1 (1968) 61–74.

1068. Braestrup, M. W., *Yield lines in discs, plates, and shells*. Technical University of Denmark, Structural Research Laboratory, Copenhagen, Report R 14, 1970, 54 pp.

1069. Braestrup, M. W., Dome effects in RC slabs: Rigid-plastic analysis. *J. Struct. Div. Am. Soc. Civ. Eng.*, **106**, No. ST6 (1980) 1237–53.

1070. Nielsen, M. P., Yield conditions for reinforced concrete shells in the membrane state. Olszak, W. and Sawczuk, A. (eds), *Non-classical shell problems*. North-Holland, Amsterdam, 1964, pp. 1030–40.

1071. Nielsen, M. P., On the strength of reinforced concrete discs. *Acta Polytech. Scand., Civ. Eng. Bldg Constr. Ser.*, No. 70, 1971, 261 pp.

1072. Behaviour of reinforced concrete barriers subject to the impact of turbine missiles. 5th International Conference on Structural Mechanics in Reactor Technology (SMiRT), Paper J7/5, Berlin, 1979.

1073. NDRC Effects of impact and explosion. Summary Technical Report of Division 2, Vol. 1, National Defence Research Committee, Washington DC, 1946.

1074. Jenkov, Z. D. *et al.*, Missile tests of quarter-scale reinforced concrete barriers. Prepared for Stone & Webster Inc. by Texas Technical University, 1976.

1075. BPC Design of structures for missile impact. Topical Report BC-TOP-9.A, Bechtel Power Corporation, 1974.

1076. Haldar, A., Hatami, M. & Miller, F., Penetration and spalling depth estimation for concrete structures. 7th International Conference on Structural Mechanics in Reactor Technology (SMiRT), Paper J7/2, 1983 (also Turbine missile — a critical review, *Nucl. Eng. Des.*, **55**, 1979).

1077. Takeda, J. *et al.*, Proc. Inst. Assoc. Symp. Concrete Structures Under Impact and Impulsive Loading, pp. 289–95, BAM, Berlin, 1982.

1078. Kar, A. K., Impactive effects of tornado missiles and aircraft. Trans. ASCE, **105**, ST2, 2243–60 (1979) (also Loading time-history for tornado generated missiles. *Nucl. Eng. Des.*, **51** (1979) 487–93).

1079. Rotz, J., Evaluation of tornado missile impact effects on structures, in Proc. Symp. on Tornadoes, Assessment of Knowledge and Implications for Man, USA, 1976.

1080. Bangash, M. Y. H., *Prototype Building Structures — Analysis and Design*. Thomas Telford, London, 1999, Section A4.

1081. Bangash, M. Y. H., *Prototype Bridge Structures — Analysis and Design*. Thomas Telford, London, 1999.

1082. Barson, J. M. & Rolfe, S. T., *Fracture and Fatigue Control in Structures: Applications of fracture mechanics*. 2nd edn, Prentice-Hall, Englewood Cliffs, NJ, USA, 1987.

1083. Bazant, Z. P. & Oh, B. H., Crack band theory for fracture of concrete. *Materiaux et Constructions*, Paris, France, **16**(93) (1983) 155–77.

1084. Cornelissen, H. A. W., Hordijk, D. A. & Reinhardt, H. W., Experimental determination of crack softening characteristics of normal weight and lightweight concrete. *Heron*, Delft, The Netherlands, **31**(2), 1986.

1085. Dodds, R. H., Darwin, D. & Leibengood, L. D., Stress controlled smeared cracking in R/C beams. *J. Struct. Engng, ASCE*, **110**(9) (1984) 1959–76.

1086. Frank, K. H., Test on a continuous composite beam. Ferguson Struct. Engrg Lab., University of Texas, Austin, Texas, 1991.

1087. Gopalaratnam, V. S. & Shah, S. P., Softening response of plain concrete in direct tension. *ACI Journal*, **82**(3) (1985) 310–23.

1088. Hawkins, N. M., Wyss, A. N. & Mattock, A. H., Fracture analysis of cracking in concrete beams. *J. Struct. Div. ASCE*, **103**(5) (1977) 1015–30.

1089. Hillerborg, A., Modeer, M. & Peterson, P. E., Analysis of crack formation and crack growth in concrete by means of fracture mechanics and finite elements. *Cem. and Concr. Res.*, **6**(4) (1976) 773–82.

1090. Hillerborg, A., Numerical methods to simulate softening and fracture of concrete. *Fracture mechanics of concrete*, G. C. Sih and A. Ditommaso (eds), Martinus Nijhoff Publishers, Dordrecht, The Netherlands, 1984, pp. 141–46.

1091. Hordijk, D. A., Van Mier, J. G. M. & Reinhardt, H. W., Material properties. *Fracture mechanics and concrete*, L. Elfgren (ed), Chapman & Hall, London, 1989, pp. 67–80.

1092. Ingrafles, A. R., Gerstle, W. H., Gergely, P. & Souma, V., Fracture mechanics of bond in reinforced concrete. *J. Struct. Engng, ASCE*, **110**(4) (1984) 871–90.

1093. Kupfer, H. & Hilsodorf, H. K., Behaviour of concrete under biaxial stresses. *ACI Journal*, (1969) 656–66.

1094. Kupfrer, H. & Gerstle, K. H., Behaviour of concrete under biaxial stresses. *J. Struct. Div., ASCE*, **101**(3) (1973) 853–66.

1095. Lin, C. S. & Scordelis, A. C., Nonlinear analysis of RC shells of general form. *J. Struct. Div., ASCE*, **101**(3) (1975) 523–28.

1096. McNeice, G. M., Elastic-plastic bending of plates and slabs by the finite element method. PhD dissertation, University of London, England, 1967.

1097. Morcos, S. S., Fracture in continuous composite beams. PhD dissertation, University of Pittsburgh, PA, USA, 1991.

1098. Morcos, S. S. & Bjorhovde, R., Fracture in continuous composite girders. Proc. 9th Intl. Bridge Conf., Engineers Society of Western Pennsylvania, Pittsburgh, PA, USA, 1992(a) 262–76.

1099. Morcos, S. S. & Bjorhovde, R., Fracture criteria of concrete. Fracture mechanics of concrete structures (Proc. 1st Int. Conf. on Fracture Mech. of Concrete Struct.), S. P. Bazant (ed.), Elsevier Applied Science Publishers, New York, NY, 1992(b).

1100. Petersson, P. E., Fracture energy of concrete: method of determination. *Cem. and Concr. Res.*, **10** (1982(a)) 78–89.

1101. Petersson, P. E., Fracture energy of concrete: practical performance and experimental results. *Cem. and Concr. Res.*, **10** (1980(b)) 91–101.

1102. Petersson, P. E., Crack growth and development of fracture zones in plain concrete and similar materials. Report No. TVB.M. 1006, University of Lund, Lund, Sweden, 1981.

1103. Reinhardt, H. W., Fracture mechanics of an elastic softening material like concrete. *Heron*, Delft, The Netherlands, **29**(2) (1984).

1104. Reinhardt, H. W. & Cornelissen, H. A. W., Strain-softening analysis of concrete fracture specimens. *Fracture toughness and fracture energy of concrete*, F. H. Wittmann (ed.), Elsevier Applied Science Publishers, Amsterdam, The Netherlands, (1985) 137–48.

1105. Rots, J. G. & Blaauwendraad, J., Crack models for concrete — discrete or smeared? Fixed, multi-directional or rotating? *Heron*, Delft, The Netherlands, **34**(1) (1989).

1106. Rots, J. G., Nauta, P., Kusters, G. M. A. & Blaauwendraad, J., Smeared crack approach and fracture localization in concrete. *Heron*, Delft, The Netherlands, **31**(1) (1985).

1107. Testa, T. B. & Stubbs, N., Bond failure and inelastic response of concrete. *J. Engrg Mech. Div., ASCE*, **103**(2) 295–310.

1108. Walsh, P. F., Crack initiation in plain concrete. *Mag. Concr. Res.*, **28** (1976).

1109. Shadbolt, P. J., Impact loading of plates. Dphil thesis, Oxford University, 1981.

1110. Salvatorelli, F. & Rollins, M. A., Some Applications of ABAQUS, ABAQUS User's Conference, pp. 377–96. HKS Inc., Newport, Rhode Island, 1988.

1111. Xia, Y. R. & Ruiz, C., Reponse of layered plates to projectile impact. In Proc. 4th Int. Conf. on Mechanical Properties of Materials at High Rates of Strain. Institute of Physics, Oxford, 1989.

1112. Xia, Y. R. & Ruiz, C., Shear effects between the layers of laminated beams under impact loading. *Proc. Inst. Mech. Engrs, London* (1989) 87–92.

1113. Winter, R. & Rifko, A. B., Finite element crash analysis of automobiles and helicopters. In *Structural impact and crash worthiness*, J. Morton (ed), Vol. II, pp. 278–309. Conference papers, Elsevier, 1988.

1114. Chan, A. S. L. & Li, F.-N., Impact of thin-walled shells using a simplified finite element model. Private communication, 1989.
1115. Chan, A. S. L., Reinforcements in the impact analysis of thin-walled shells using a simplified finite element model. Private communication, 1989.
1116. Varpason, P. & Kentalla, J., The analysis of the containment building for global effects of an aircraft crash. 6th International Conference on Structural Mechanics in Reactor Technology (SMiRT), Paper J9/10, Paris, 1981.
1117. Ohnuma, H., Dynamic response and local rupture of reinforced concrete beam and slab under impact loading. 8th International Conference on Structural Mechanics in Reactor Technology (SMiRT), Paper J5/3, Brussels, 1985.
1118. Crutzen, Y., Non-linear transient dynamic analysis of think shells using the SEMILOOF finite element, PhD thesis, University of Brussels, 1979.
1119. NDRC Effects of Impact and Explosion. Summary Technical Report of Division 2, Vol. 1, National Defence Research Committee, Washington DC, 1946.
1120. Jenkov, A. D. *et al.*, Missile tests of quarter-scale reinforced concrete barriers, prepared for Stone & Webster Inc. by Texas Technical University, 1976.
1121. BPC Design of structures for missile impact. Topical Report BC-TOP-9-A, Bechtel Power Corporation, 1974.
1122. Haldar, A., Hatami, M. & Miller, F., Penetration and spalling depth estimation for concrete structures. 7th International Conference on Structural Mechanics in Reactor Technology (SMiRT), Paper J7/2, 1983. Also Turbine missile — a critical review. *Nucl. Eng. Des.*, **55** (1979).
1123. Takeda, J. *et al.*, Proc. Inst. Assoc. Symp. Concrete Structures Under Impact and Impulsive Loading, pp. 289–95, BAM, Berlin, 1982.
1124. Kar, A. K., Impactive effects of tornado missiles and aircraft. *Trans. ASCE*, **105**, ST2, 2243–60, 1979. (Also Loading time-history for tornado generated missiles. *Nucl. Eng. Des.*, **51** (1979) 487–93.)
1125. Rotz, J., Evaluation of tornado missile impact effects on structures, in Proc. Symp. on Tornadoes, Assessment of Knowledge and Implications for Man, USA, 1976.
1126. PLA, *Design Manual for Concrete Pavements*. Port of London Authority, 1989.
1127. Kar, A., Projectile penetration into steel. *J. Struct. Div., ASCE* (1979).
1128. ACE Fundamentals of protective design. Report AT 1207821. Army Corps of Engineers, USA, 1946.
1129. Rinaldi, L. J., *Containerization: the new method of intermodal transport*. Stirling Publishing, 1972.
1130. D'Arcangelo, A. M. (ed), *Ship design and construction*. The Society of Naval Architects and Marine Engineers, New York, 1975.
1131. Petry, S. & Brown, I., Model prestressed concrete slab subjected to hard missile loading. *Design for Dynamic Loading*, Construction Press, 1982.
1132. International Conference: Structural Impact and Crash Worthiness. Imperial College, London, Elsevier, Vols 1, 2, 3, 1984.
1133. Stevenson, J., *Design for nuclear power plant facilities*. ASCE, New York, 1976.
1134. Behaviour of reinforced concrete barriers subject to the impact of turbine missiles. 5th International Conference on Structural Mechanics in Reactor Technology (SMiRT), Berlin, Paper J7/5, 1979.
1135. SRI Report No. 361. Stanford Research Institute, California, USA, 1960.
1136. Nelson, A. J., Winfrith perforation energy model. UK Atomic Energy Report No. 31, Winfrith, Dorset, UK, 1985.
1137. Neilson, A. J., Missile impact on metal structures. *Nucl. Energy*, **19**, No. 3 (1980) 191–98.
1138. Chellis, R. D., *Pile foundations*. McGraw-Hill, New York, 1961.
1139. Chandrasekaran, V., Analysis of pile foundations under static and dynamic loads. PhD thesis, University of Roorkee, India, 1974.

1140. Van der Meer, J. W. & Pilarczyk, K. W., Stability of rubble mound slopes under random wave attack. In Proc. 19th ICCE, Chapter 176, Houston, USA, 1984.
1141. Chiesa, M. L. & Callabressi, M. L., Non-linear analysis of a mitigating steel nose cone. In *Computational methods in non-linear structural and solid mechanics*, A. K. Noor and H. G. McComb Jr. (ed), 295–300, Pergamon, New York, 1981.
1142. Shular, W., Propagation of study shockwaves in polymethyl methocrylate. *J. Mech. Phys. Solids*, **18** (1970) 277–93.
1143. Jernstrom, C., Computer simulation of a motor vehicle crash dummy and use of simulation in the design/analysis process. *Int. J. Vehicle Design*, **4**, No. 2 (1983) 136–49.
1144. Kormeling, H. A., Impact tensile strength of steel fibre concrete. Report 1.5–84–8, Department of Civil Engineering, The Delft University of Technology, 1984.
1145. Suaris, W. & Shah, S. P., Properties of concrete subjected to impact. *ASCE St. Div.*, **109**(7) (1982) 1727–41.
1146. Wirzbicki, T., Chryssostomidis, C. & Wiernicki, C., Rupture analysis of ship plating due to hydrodynamic wave impact. Society of Naval Architects and Mechanical Engineers Ship Structure Symposium '84, 1984.
1147. Zayas, V. A., Dao, B. V. & Mammett, D. S., Experimental and analytical comparisons of semi-submersible offshore rig damage resulting from a ship collision. Offshore Technology Conference, Houston, Texas, 1985.
1148. Banthia, N., Impact resistance of concrete. PhD thesis, University of British Columbia, Vancouver, 1987.
1149. Banthia, N., Mindess, S. & Bentur, A., Impact behaviour of concrete beams. *Materials and Structures*, **20** (1987) 293–303.
1150. Ravid, M. & Bodner, S. R., Dynamic perforation of viscoplastic plates by rigid projectiles. *Int. J. Eng. Sci.*, **21** (1983) 577–91.
1151. Curran, D. R., Seaman, L. & Shockey, D. A., Dynamic failure of solids. *Physics Report*, 147, 254–387.
1152. Anderson, C. E. & Bodner, S. R., Ballastic impact: the status of analytical and numerical modelling. *Int. J. Impact Eng.*, **7** (1988) 9–35.
1153. Goldsmith, W. & Finnegan, S. A., Normal and oblique impact of cylindroconical and cylindrical projectiles on metallic plates. *Int. J. Impact Eng.*, **4** (1986) 83–105.
1154. Lambert, J. P., ARBRL-TR-02033. Ballistic Research Laboratory, Alabama, USA, 1978.
1155. Meunier, Y., Sangoy, L. & Pont, G., Steels for ballistic protection. In Proc. 4th Israel Materials Engineering Conference (ed. J. Baran and D. Eliezer), Beer-Sheva, Israel, 7–8 December 1988, 1989.
1156. Wilkins, M. L. & Guinan, M. W., Impact of cylinders on a rigid boundary. *J. Appl. Phys.*, **44** (1973) 1200–6.
1157. Yatteau, J. & Recht, R., High speed penetration of speed plates by computer fragments. In Proc. 9th International Symposium on Ballistics 2, pp. 365–74, Shrivenham, 1970.
1158. Naz, P., Spaced plated penetration by spherical high density fragments at high velocity. In Proc. 10th International Symposium on Ballistics 2, San Diego, 1987.
1159. Richards, A. & Bowden, A., The development of a projectile velocity measuring system for the RARDE No. 2 Hypervelocity Launcher. *Int. J. Impact Eng.*, **5** (1987) 519–32.
1160. Zhong, Z. H., On contact-impact problems. Dissertation No. 178, Linkoping University, Sweden, 1988.
1161. Zhong, Z. H. & Nilson, L. A., A contact searching algorithm for general contact problems. *Computers & Structures* (1988).
1162. Zhong, Z. H., New algorithms for numerical treatments of general contact — impact interfaces. In Proc. 4th SAS — World conference FEMCAD, Vol. 1, China, 1988.
1163. Chaudhary, A. B. & Bathe, K. J., A solution method for static and dynamic analysis of three-dimensional contact problems with friction. *Computers and Structures*, **24** (1986).

1164. Perry, S. & Brown, I., Model prestressed slabs subjected to hard missile loading. In *Design for Dynamic Loading*, Construction Press, London, UK, 1982.
1165. Perry, S., Brown, I. & Dinic, G., Factors influencing the response of concrete slabs to impact. International Conference and Exposition of Structural Impact and Crashworthiness, Imperial College, London, UK, 1984.
1166. Barr Petal, An experimental investigation of scaling of reinforced concrete structures under impact loading, in *Design for Dynamic Loading*. Construction Press, London, UK, 1982.
1167. Barr Petal, Experimental studies of the impact resistance of steel faced concrete composition. 7th International Conference on Structural Mechanics in Reactor Technology (SMiRT), Paper J8/4, Chicago, USA, 1983.
1168. Barr Petal, An experimental investigation of scaling of reinforced concrete structures under impact loading. In *Design for Dynamic Loading*, Construction Press, London, UK, 1982.
1169. Johns Hopkins University, The resistance of various metallic materials to perforation by steel fragments: empirical relationships for fragment residual velocity and residual weight. Ballistic Res. Lab. Proj. Thor TR No. 47, Johns Hopkins University, Baltimore, MD, USA, 1961.
1170. Johns Hopkins University, The resistance of various non-metallic materials to perforation by steel fragments: empirical relationships for fragment residual velocity and residual weight. Ballistic Res. Lab. Proj. Thor TR No. 47, Johns Hopkins University, Baltimore, MD, USA, 1961, a special report.
1171. Stevenson, J, (ed.), *Design for nuclear power plant facilities*. ASCE, New York, 1976.
1172. Carino, N. J., Sansalone, M. & Hsu, N. N., Flaw detection in concrete by frequency spectrum analysis of impact-echo waveforms. *International Advances in Nondestructive Testing*, 12th edn, W. J. McGonnagle (ed.), Gordon & Breach Science Publishers, New York, pp. 117–46, 1986.
1173. Malhotra, M., Testing hardened concrete: nondestructive methods, ACI Monograph No. 9, 1976.
1174. Maser, K. R. & Roddis, W. M. K., Principles of thermography and radar for bridge deck assessment. *ASCE J. Transp. Engng*, **116**, No. 5 (1990) 583–601.
1175. Naik, T. R. & Malhotra, V. M., The ultrasonic pulse velocity method. Chapter 7 in *Handbook on nondestructive testing of concrete*, V. M. Malhotra and N. J. Carino (eds), CRC Press, Boca Raton, FL, 1991, pp. 169–88.
1176. Pratt, D. & Sansalone, M., Impact-echo signal interpretation using artificial intelligence. *ACI Materials J.*, **89**, No. 2, Mar.–Apr. 1992, pp. 178–87.
1177. Rutenbeck, T., Shotcrete strength testing — comparing results of various specimens, shotcrete for underground support, ACI SP-54, Oct. 1976.
1178. Sansalone, M. & Carino, N. J., Stress wave propagation methods. Chapter 12 in *Handbook on nondestructive testing of concrete*, V. M. Malhotra and N. J. Carino (eds), CRC Press, Boca Raton, FL, 1991.
1179. Sansalone, M. & Carino, N. J., Laboratory and field study of the impact-echo method for flaw detection in concrete. In *Nondestructive testing of concrete*. ACI-SP 112, American Concrete Institute, Detroit, pp. 1–20, 1988b.
1180. Sansalone, M., Lin, Y., Pratt, D. & Cheng, C., Advancements and new applications in impact-echo testing. Proc. ACI Int. Conf. on Evaluation and Rehabilitation of Concrete Structures and Innovations in Design, Hong Kong, V. M. Malhotra (ed.), ACI SP-128, 1991, pp. 135–50.
1181. Maidi, B. *et al.*, Bericht des III. Kolloquiums für Bauverfahrenstecheit Spritzbeton für Tunnelbau und Bergbau, Technische-wissenschaftliche Mitteilungen des Institutes für Konstruktiven Ingenieurbau der Ruhr-Universität Bochum, Heft 85-6, Bochum, 1985.
1182. Maak, H., Maidi, B., Maidi, R., Springenschmidt, R., Bewehrte und nichtbewehrte Innenschale im Felstunnelbau, TT 1986, S. 373–402.

1183. Maidi, B. *et al.*, Arbeitsplatzverbesserung bei Spritzbetonarbeiten im Tunnelbau und im Bergbau, Schriftenreihe zur Humanisierung des Arbeitslebens im Bergbau, Band 13, Verlag Gluckauf, Essen, 1987.
1184. Maidi, B. & Handke, D., Stahlfaserspritzbeton im Tunnelbau. *Bauingenieur*, **62** (1987).
1185. Maidi, B. & Guthoff, K., Stand der Entwicklung in der Spritzbetonforschung an der Ruhr-Universität Bochum. *Felsbau*, **6**, Nr. 2, (1988) 78–82.
1186. Koenning, R., Stahlfaserbeton im Tunnelbau, Vortrag HdT, Essen 1988 (unveröffentlicht), 1989.
1187. Schreyer, J., Forschungsbericht der STUVA, Konstruktive und wirtschaftliche Möglichkeiten zum Bau von Tunneln in einschaliger Bauweise (unveröffentlicht), 1988.
1188. Swoboda, G., Moussa, A. & Hafez, N., Two- and three-dimensional modelling of layered shotcrete tunnel lining. Proc. Int. Conf. on Computational Methods in Struct. and Geotech. Engng, Hong Kong, **4** (1994) 1077–84.
1189. Swoboda, G. & Moussa, A., Numerical modelling of shotcrete in tunnelling. Proc. 4th Int. Symp. on Num. Models in Geomech., Swansea, UK, Pande, G. & Pietruszczak, S. (eds), Vol. 2, Balkema, Rotterdam, 1992, 717–27.
1190. Hafez, N., Post-failure modelling of three-dimensional shotcrete lining for tunneling. PhD thesis, Innsbruck University, Austria (in preparation), 1995, 1996.
1191. Swoboda, G., Numerical modelling of tunnels. In *Numerical Methods and Constitutive Modelling in Geomechanics, Courses and Lectures*, C. S. Desai and G. Gioda (eds), No. 311, Springer-Verlag, Wien, 1990, 277–318.
1192. Swoboda, G., Finite element analysis of the new Austrian tunneling method. 3rd Int. Conf. on Numerical Methods in Geomechanics, Aachen, 1979, 581–86.
1193. Schmid, A., Beitrag zur Berechnung geschichteter Schalentragwerke mittels der Methode der finiten Elemente. PhD thesis, Innsbruck University, Austria.
1194. Ahmad, S., Irons, B. M. & Zienkiewicz, O. C., Analysis of thick and thin structures by curved finite elements. *Int. J. for Num. Meth. in Engng*, **2** (1970) 419–51.
1195. Chen, A. C. T. & Chen, W., Constitutive equations and punch-indentation of concrete. *Proc. ASCE*, **101**, No. EM6, 889–906.
1196. Swoboda, G., Programmsystem FINAL — finite element analysis program for linear and nonlinear structures. Version 6.7, University of Innsbruck, 1994.
1197. Pottler, R., Time-dependent rock–shotcrete interaction — a numerical shotcrete. *Computer and Geotechnics*, **9** (1990) 149–69.
1198. Sezaki, M., Kibe, T. & Kawamoto, T., An experimental study on mechanical properties of shotcrete. *J. Soc. Material Science*, Japan, **38**, No. 434 (1989).
1199. Panet, M. & Guenota, Analysis of convergence behind the face of a tunnel. *Tunnelling '82*, 3rd Int. Symp., Inst. Min. and Metallurgy, England, 1982, 197–204.
1200. Nguyen-Minh, D. & Corbetta, D., New methods for rock-support analysis of tunnels in elasto-plastic media. Proc. of Rock Support in Mining and Underground Construction, Rotterdam, Canada, 1992, pp. 83–90.
1201. Chang, Y., Influence of early-age properties of shotcrete on tunnel construction sequences. Licentiate Thesis, Dept. of Soil and Rock Mechanics, Royal Institute of Technology, Stockholm, Sweden.
1202. Hungerford, H. E., Mantey, R. E. & Maele, L. P. V., New shielding material for high-temperature application. *Nuclear Science & Engineering*, **6** (1959) 396–408.
1203. ASTM Designation C 637-69T, Aggregates for radiation-shielding concrete, ASTM Standards Part 10. *Concrete and Mineral Aggregates* (1969) 394–98.
1204. Clarke, F. W., Data of geochemistry. U.S.G.S. Bulletin 770, Dept. of Interior, U.S. Government Printing Office, 1977, p. 617.
1205. Woodrooffe, H. M., Pyrolysis of chrysotile asbestos fibre. The Canadian Mining and Metallurgical Bulletin, Transactions V. LIX, 1956, 1965, pp. 363–68.
1206. Davis, H. S., N-reactor shielding. Paper presented at the ACI Seminar on Concrete for Nuclear Reactors, West Berlin, Oct. 1970.

1207. Rots, J. G., Computational modelling of concrete fracture. Dissertation, Delft University of Technology, Faculty of Civil Engng, 1988.
1208. Schellekens, J. C. J., Interface elements in finite element analysis. Report, Delft University of Technology, Faculty of Civil Engng, in preparation 1990 (1991).
1209. Kufuor, K. & Perry, S., Hard impact of shallow reinforced concrete domes. International Conference and Exposition of Structural Impact and Crashworthiness, Imperial College, London, UK, 1984.
1210. Burgess, W. & Campbell-Allen, D., Impact resistance of reinforced concrete as a problem of containment. *Research Report No. R251*, School of Civil Engineering, University of Sydney, 1974.
1211. Stephenson, A., Tornado-generated missile full-scale testing, in *Proceedings of the Symposium on Tornadoes, Assessment of Knowledge and Implications for Man*. Texas University, 1976.
1212. Jankov, Z., Turnham, J. & White, M., Missile tests of quarter-scale reinforced concrete barriers, in *Proceedings of the Symposium on Tornadoes, Assessment of Knowledge and Implications for Man*. Texas University, June 1976.
1213. Stephen, A. & Silter, G., Full-scale tornado-missile impact tests. 4th International Conference on Structural Mechanics in Reactor Technology (SMiRT), Paper J10/1, Berlin, 1977.
1214. Jonas, W. & Rudiger, E., Experimental and analytical research on the behaviour of reinforced concrete slabs subjected to impact loads. 4th International Conference on Structural Mechanics in Reactor Technology (SMiRT), Paper J7/6, Berlin, 1977.
1215. Beriaud, C. *et al.*, Local behaviour of reinforced concrete walls under hard missile impact. 4th International Conference on Structural Mechanics in Reactor Technology (SMiRT), Paper J7/9, Berlin, 1977.
1216. Gupta, Y. & Seaman, L., Local response of reinforced concrete to missile impacts. 4th International Conference on Structural Mechanics in Reactor Technology (SMiRT), Paper J10/4, Berlin, 1977.
1217. Bethe, H. A., Attempt of a theory of armor penetration. Frankford Arsenal Rep., Philadelphia, 1941.
1218. Taylor, G. I., Notes on H. A. Bethe's theory of armor penetration, II. Enlargement of a hole in a flat sheet at high speeds. Home Ministry Security Report, R.C. 280 UK, 1941.
1219. Taylor, G. I., The formation and enlargement of a circular hole in a thin plastic sheet. *Quart. J. Mech. Appl. Math.*, **1**, 1948.
1220. Kineke, J. H. & Vitali, R., Transient observations of crater formation in semi-infinite targets, in *Proceedings of the Sixth Symposium on Hypervelocity Impact*. Vol. II, Part 2, Springer-Verlag, Berlin, Germany, 1963.
1221. Eichelberger, R. J. & Kineke, J. H., Hypervelocity impact, in *High Speed Physics*, K. Vollrath and G. Thomer (eds), Springer-Verlag, Wien, New York, 1967.
1222. Turpin, W. C. & Carson, J. M., Hole growth in thin plates perforated by hypervelocity pellets. *Air Force Materials Lab. Rep. 70–83*, Wright-Patterson AFB, Ohio, 1970.
1223. Allison, F. E., Mechanics of hypervelocity impact, in *Proceedings of the Seventh Hypervelocity Impact Symposium*, Vol. I, Springer-Verlag, Berlin, Germany, 1965.
1224. Wilkins, M. & Guinan, M., Impact of cylinders on a rigid boundary. *J. Appl. Phys.*, **44** (1973).
1225. Bischoff, P. H. & Perry, S. H., Compressive behaviour of concrete at high strain rates. *Materials and Structures*, **24** (1991) 425–50.
1226. ACI, *Building Code Requirements for Reinforced Concrete* (ACI 318-89). 1st Printing, November, Detroit, USA, 1989.
1227. ACI-ASCE Committee 426, Shear and diagonal tension. *ACI J. Proc.*, **59**, No. 1, January, pp. 1–30 and No. 2, February, pp. 277–340 and No. 3, March, pp. 353–96, 1962.
1228. ASCE-ACI Committee, The shear strength of reinforced concrete members, by the joint ASCE-ACI Committee on Shear and Diagonal, 1973.
1229. Tension of the Committee on Masonry and Reinforced Concrete of Structural Division, Chmn., MacGregor, J. G. *J. Struct. Div., ASCE*, **99**, No. ST6, Proc., Paper 9791, June, pp. 1091–1187.

1230. Ali, G., Study of stress distribution in deep reinforced concrete beams. Master's thesis, Department of Civil Engineering, BUET, Dhaka, September 1984.

1231. Kabir, A., Shear strength of deep reinforced concrete beam. Master's thesis, Department of Civil Engineering, BUET, Dhaka, October 1982.

1232. Mau, S. T. & Hsu, T. C., Formula for the shear strength of deep beams. *ACI Struct. Journal*, **86**, No. 5 (1989) 516–23.

1233. Ramakrishnan, V. & Ananthananrayana, Y., Ultimate strength of deep beams in shear. *J. American Concr. Inst., Proc.*, **65**, No. 2 (1968) 87–98.

1234. Selvam, V. K. M. & Thomas, K., Shear strength of concrete deep beams. *Indian Concrete Journal*, **61** (1987) 219–22.

1235. Slao, W. B., Strut-and-tie model for shear behaviour in deep beams and pile caps falling in diagonal splitting. *ACI Struct. J.*, **90**, No. 4 (1993) 356–63.

1236. Slao, W. B., Deep beams revisited. *ACI Struct. J.*, **92**, No. 1 (1995) 95–102.

1237. Singh, R., Ray, S. P. & Reddy, C. S., Some tests on reinforced concrete deep beams with and without opening in the web. *Indian Concrete Journal*, **54** (1980) 189–94.

1238. Winter, G. & Nilson, A. H., *Design of concrete structures*. McGraw-Hill Book Company, 1989, 10th edn, 2nd printing, pp. 138–46.

1239. Kong, F. K. *et al.*, Instability and buckling of reinforced concrete deep beams. *Struct. Engr.*, **64B** (1986) 49.

1240. Kotsovos, M. D., Design of reinforced concrete deep beams. *Struct. Engr.*, **66** (1988) 2, 28.

1241. Kubik, L. A., Predicting the strength of reinforced concrete deep beams with web openings. *Proc. Instn Civ. Engrs*, Part 2, **69** (1980) 939.

1242. Kumar, P., Collapse load of deep reinforced concrete beams. *Mag. Concr. Res.*, **28** (1976) 94, 30.

1243. Leonhardt, F. & Walther, R., Deep beams. Bulletin 178, Deutscher Ausschuss für Stahlbeton. Berlin, 1966 (English translation: CIRIA, London, 1970).

1244. Mansur, M. A. & Alwis, W. A. M., Reinforced fibre concrete deep beams with web openings. *Int. J. Cement Composites Lightwt, Concr.*, **6** (1984) 263.

1245. Mau, S. T. & Hsu, T. T. C., Shear strength prediction for deep beams with web reinforcement. *Am. Concr. Inst. Struct. J.*, **84** (1987) 6, 513.

1246. Rasheeduzzafar, M. H. & Al-Tayyib, A. H. J., Stress distribution in deep beams with web openings. *J. Struct. Engng., Am. Soc. Civ. Engrs*, **112** (1986) 1147.

1247. Ray, S. P., Behaviour and ultimate shear strength of reinforced concrete deep beams with and without opening in web. PhD thesis, Indian Institute of Technology, Kharagpur, India, 1980.

1248. Regan, P. E. & Hamadi, Y. K., *Concrete in the oceans, Part 1: reinforced concrete deep beams with thin webs*. Cement and Concrete Associations, London, 1981.

1249. Robins, P. J. & Kong, F. K., Modified finite element method applied to reinforced concrete deep beams. *Civ. Engng. Publ. Works Review*, **68** (1973) 963.

1250. Rogowsky, D. M., MacGregor, J. G. & Ong, S. Y., Tests of reinforced concrete deep beams. *Proc. Am. Concr. Inst.*, **83** (1986) 614.

1251. Shanmugan, N. E. & Swaddiwudhipong, S., Strength of fibre reinforced deep beams containing openings. *Int. J. Cement Composites and Lightwt Concr.*, **10** (1988) 53.

1252. Smith, K. N. & Vantsiotis, A. S., Shear strength of deep beams. *Proc. ACI*, **79** (1982) 201.

1253. Subedi, N. K., Reinforced concrete deep beams: a method of analysis. *Proc. Instn Civ. Engrs*, Part 2, **85** (1988) 1.

1254. Swaddiwudhipong, S. & Shanmugan, N. E., Fibre-reinforced concrete deep beams with openings. *J. Struct. Engng. Am. Soc. Civ. Engrs.*, **111** (1985) ST8, 1679.

1255. Tang, C. W. J., Reinforced concrete deep beams: behaviour, analysis and design. PhD thesis, University of Newcastle upon Tyne, UK, 1987.

1256. Wong, H. H. A., Buckling and stability of slender reinforced concrete deep beams. PhD thesis, University of Newcastle upon Tyne, UK, 1987.

1257. Kong, F. K. & Robins, P. J., Web reinforcement effects on lightweight concrete deep beams. *Proc. Am. Concs. Inst.*, **68** (1971) 514.
1258. Kong, F. K. & Sharp, G. R., Shear strength of lightweight reinforced concrete deep beams with web openings. *Struct. Engr.*, **51** (1973) 267.
1259. Kong, F. K. & Sharp, G. R., Structural idealisation for deep beams with web openings. *Mag. Concr. Res.*, **29** (1977) 81.
1260. Kong, F. K. & Singh, A., Diagonal cracking and ultimate loads of lightweight concrete deep beams. *Proc. Am. Concr. Inst.*, **69** (1972) 513.
1261. Kong, F. K. & Singh, A., Shear strength of lightweight concrete deep beams subjected to repeated loads, in *Shear in Reinforced Concrete*. ACI Publication SP42, American Concrete Institute, Detroit, 1974, p. 461.
1262. Kong, F. K. *et al.*, Deep beams with inclined web reinforcement. *Proc. Am. Concr. Inst.*, **69** (1972a) 3.
1263. Kong, F. K. *et al.*, Shear analysis and design of reinforced concrete deep beams. *Struct. Engr.*, **50** (1972b) 405.
1264. Portland Cement Association, *Design of deep girders*. Publication ISO79.01D, PCA, Skokie, Illinois, 1980.
1265. Rogowsky, D. M., MacGregor, J. G. & Ong, S. Y., Tests of reinforced concrete deep beams. *Am. Conc. Inst.*, **83**(4) (1986) 614.
1266. Rogowsky, D. M. & MacGregor, J. G., The design of reinforced concrete deep beams. *Concr. Internat.: Design & Construction*, **8** (1986) 49.
1267. Schlaich, J., Schaefer, K. & Jennewein, J., Toward a consistent design of structural concrete. *J. Prestressed Concr. Inst.*, **32**(3) 74.
1268. Albritton, G. E., Review of literature pertaining to the analysis of deep beams. Technical Report 1-701, US Army Engineers Waterways Experiment Station, Vicksburg, Miss., 1965.
1269. American Concrete Institute, Building Code Requirements for Reinforced Concrete. ACI 318-83 (1983), American Concrete Institute, Detroit, MI.
1270. Cement and Concrete Association, Bibliography on deep beams. Library Bibliography Ch. 71 (3169), Cement and Concrete Association, London, 1969.
1271. Cheng, D. H. & Pei, M. L., Continuous deep beams. *Proc. Am. Soc. Civ. Engrs.*, **450**(80) (1954) Separate.
1272. Chow, L., Conway, H. D. & Winter, G., Stresses in deep beams. *Trans. Am. Soc. Civ. Engrs.*, **118** (1953) 686.
1273. Comité Euro-International du Bêton/Fédération Internationale de la Précontrainte, *Model code for concrete structures*. English edition, Cement and Concrete Association, London, 1978.
1274. Construction Industry Research and Information Association, *The design of deep beams in reinforced concrete*. Guide 2. Ove Arup and Partners and Construction and Industry Research and Information Association, London, 1977.
1275. Dischinger, F., Beitrag zur Theorie der Halbscheibe und des wandartigen Trägers. *Internat. Assoc. of Bridge and Struct. Engng*, **1** (1932) 69.
1276. El-Behairy, S., Spannungszustand wandartiger Träger mit Inneren angreifenden Einzelkraften. *Beton und Stahlbeton*, No. 10 (1968) 228.
1277. Forster, W. & Stegbauer, A., *Wandartige Träger: Tafeln zur Ermittlung der Bewehrung*. Werner-Verlag, Dusseldorf, 1974.
1278. Gogate, A. B., Finite element analysis of deep reinforced concrete beams with static, short term loading. PhD dissertation, Ohio State University, 1977.
1279. Gogate, A. B. & Sabnis, G. M., Design of thick pile caps. *Am. Concr. Inst. J.*, **77**(1) (1980).
1280. Leonhardt, F. & Walther, R., *Wandartige Träger*. Heft 178. Deutscher Ausschuss für Stahlbeton, Berlin, 1966.
1281. Nylander, H., Hoga Balkar I. Inverkan av Klyvnings-forhindrahindrande Armering pa spanningsfordelningen under Koncentrerad last. *Nordisk Betong*, **1** (1967) 53.

1282. Nylander, H., Hoga Balker III, Hallfasthet vid Upplag. *Nordisk Betong*, No. 1 (1967) 79.
1283. Nylander, H., Hoga Balker IV, Balker belastade I underkant. *Nordisk Betong*, No. 2 (1967) 173.
1284. Nylander, H., Hoga Balker V, Sammanfattande Synaspunkter for dimensionering. *Nordisk Betong*, No. 2 (1967) 195.
1285. Nylander, H. & Nylander, J. O., Hoga Balker II, Moment-og Spanningsfordelning I Kontinuerling hog balk. *Nordisk Betong*, No. 1 (1967) 65.
1286. Paul, I. S., Behaviour of indirectly loaded reinforced concrete thin wall ribbed panels. Eng. thesis, Department of Civil Engineering, Concordia University, Montreal, 1978.
1287. Regan, P. E. & Hamadi, Y. D., Behaviour of concrete caisson and tower members. Technical Report 4, Cement and Concrete Association for Concrete in the Ocean Management Committee, Wexham Springs, Slough, 1981.
1288. Robins, P. J. & Kong, F. K., Modified finite element method applied to reinforced concrete deep beams. *Civ. Engng Publ. Works Rev.*, **69** (1973) 963.
1289. Robinson, J. R. & Demorieux, J. M., Essais de poutres en double te en bêton armée. *Annales I.T.B.T.P.*, No. 335, Jan. 1976, p. 65.
1290. Subedi, N. K., The behaviour of reinforced concrete flanged beams with stiffeners. *Mag. Concr. Res.*, **35**, No. 122 (1983) 40.
1291. Hugenschmidt, H., Epoxy adhesive for concrete and steel. Proc. 1st Int. Congress on Polymers in Concrete, London, May 1975. The Construction Press Ltd, Hornby, 1976, pp. 195–209.
1292. Iino, T. & Otokawa, K., Application of epoxy resins in strengthening of concrete structures. Proc. 3rd Int. Congress on Polymers in Concrete, Koriyama, Japan, May 1981, Vol. II, pp. 997–1011.
1293. Ryback, M., Reinforcement of bridges by glueing of reinforcing steel. *RILEM Materials and Structures*, **16**(91) (1981) 13–17.
1294. Van Gemert, D. A., Repairing of concrete structures by externally bonded steel plates. Proc. ICP/RILEM/IBK Int. Symp. on Plastics in Material and Structural Engineering, Prague, June 1981. Elsevier Scientific Publishing Co., 1982, pp. 519–26.
1295. Lerchental, H., Bonded sheet metal reinforcement for concrete slabs. RILEM Int. Symp., Resins in Building Construction, Paris, 1967, pp. 165–73.
1296. Kajfasz, S., Concrete beams with external reinforcement bonded by fluing. RILEM Int. Symp., Synthetic Resins in Building Construction, Part 2, Paris, 1967, pp. 142–51.
1297. Irwin, C. A. K., The strengthening of concrete beams by bonded steel plates. TRRL Supp. Report 160 UC, Transport and Road Research Laboratory, Dept. of the Environment, Crowthorne, 1975, p. 8.
1298. MacDonald, M. D., The flexural behaviour of concrete beams with bonded external reinforcement. TRRL Supp. Report 415, Transport and Road Research Laboratory, Department of the Environment, Crowthorne, 1978.
1299. Raithby, K. D., External strengthening of concrete bridges with bonded steel plates. TRRL Supp. Report 612, Transport and Road Research Laboratory, Department of the Environment, Crowthorne, 1960.
1300. Jones, R. *et al.*, Composite behaviour of concrete beams with epoxy bonded external reinforcement. *Int. J. Cement Composites*, **2**, No. 2 (1980) 91–107.
1301. Swamy, R. N. & Jones, R., Behaviour of plated reinforced concrete beams subjected to cyclic loading during glue hardening. *Int. J. Cement Composites*, **2**, No. 4 (1980) 233–34.
1302. Swamy, R. N., Jones, R. & Ang, T. H., Under and over-reinforced concrete beams with glued steel plates. *Int. J. Cement Composites and Lightweight Concrete*, **4**, No. 1 (1982) 19–32.
1303. Jones, R. & Swamy, R. N., In situ strengthening of concrete structural members using epoxy-bonded steel plates. Proc. 4th Int. Congress on Polymers in Concrete, Darmstadt, September 1984, pp. 251–55.

1304. Jones, R., Swamy, R. N. & Salman, F. A. R., Structural implications of repairing by epoxy-bonded steel plates. Proc. Int. Conf. on structural faults and repair, London, April–May 1985. Engineering Technics Press, pp. 75–80.
1305. Jones, R., Swamy, R. N. & Bloxham, J., Crack control of reinforced concrete beams through epoxy bonded steel plates. Proc. Int. Conf. on adhesion between polymers and concrete. Aix-en-Provence, Sept. 1986, pp. 542–55.
1306. Ziraba, Y. N., Non-linear finite element analysis of reinforced beams repaired by plate bonding. PhD thesis, King Fahd University of Petroleum and Minerals, 1993.
1307. Jones, R. *et al.*, Composite behaviour of concrete beams with epoxy bonded external reinforcement. *Int. J. of Cement Composites and Lightweight Concrete*, **2**, No. 2 (1980), 91–107.
1308. Charif, A., Structural behaviour of reinforced concrete beams strengthened by epoxy bonded steel plates. PhD thesis, University of Sheffield, 193.
1309. Munawar, H., Flexural and shear behaviour of damaged RC beams repaired by plate bonding. MSc thesis, King Fahd University of Petroleum and Minerals, 1992.
1310. Calder, A. J. J., Exposure tests on externally reinforced concrete beams — first two years. Supplementary Report 529, Transport and Road Research Laboratory, Department of Environment, Department of Transport, 1979.
1311. He, J. H., Pilakoutas, K. & Waldron, P., RC beams strengthened externally by CFRP plates. International Conference, Rehabilitation and Development of Civil Engineering Infrastructure Systems, Beirut, Lebanon, 9–11 June, 1997 (submitted).
1312. He, J. H., Pilakoutas, K. & Waldron, P., Strengthening of reinforced concrete beams with CFRP laminates. 3rd Int. Symp. on non-metallic (FRP) reinforcement for concrete structures, Sapporo, Japan, 14–16 October 1997.
1313. Pilakoutas, K., Waldron, P. & He, J. H., Strengthening of Concrete Beams with CFRP Laminates. Composites in Infrastructure Design and Rehabilitation, November 1995.
1314. Pilakoutas, K., He, J. H. & Waldron, P., Behaviour of reinforced concrete beams strengthened with epoxy bonded carbon fibre reinforced laminates. Confidential Report, The Centre for Cement and Concrete, Department of Civil and Structural Engineering, The University of Sheffield, May 1994.
1315. Rehimi, H. & Hutchinson, A. R., Strengthening of concrete structures with externally bonded fibre reinforced plastics. Proc. 3rd Int. Conf. on Structural Adhesives in Engng, University of Bristol, July 1992.
1316. Peshkam, V. & Leeming, M. B., Application of composites to strengthening of bridges: Project ROBUST. Proc. 19th Int. BPF Composites Congress, Birmingham, 1994.
1317. Hutchinson, A. R. & Rahimi, H., Flexural strengthening of concrete beams with externally bonded FRP reinforcement. 2nd Int. Conf. on Advanced Composite Materials in Bridges and Structures, Montreal, Aug. 1996.
1318. Hutchinson, A. R. & Rahimi, H., Behaviour of reinforced concrete beams with externally bonded fibre reinforced plastics. 5th Int. Conf. on Structural Faults and Repair, Edinburgh, June/July 1993, pp. 221–28.
1319. Holzenkampfer, P., Ingenieurmodell des Verbunds geklebter Bewehrung für Betonbauteile. Dissertation TU Braunschweig, 1994.
1320. Peshkam, V., Leeming, M. B. & Fashole-Luke, P. S., The role of finite element analysis in composite plate bonding. Conf. on Infrastructure Design and Rehabilitation, NAFEMS Research Working Group, London, 1st Nov. 1995, NAFEMS.
1321. Char, M. S., Saadatmanesh, H. & Ehsani, M. R., Concrete girders externally prestressed with composite plates. *PCI Journal* (1994) 40–51.
1322. Deuring, M., Verstärken von Stahlbeton mit gespannten Faserverbundwerkstoffen. Report No. 224, Swiss Federal Laboratories for Materials Testing and Research, 1993.
1323. Dussek, I. J., Strengthening of bridge beams and similar structures by means of epoxy-resin bonded external reinforcement. *Transportation Research Record*, No. 785 (1980) 21–24.

1324. Garden, H. N., Hollaway, L. C. & Thorme, A. M., A preliminary evaluation of carbon fibre reinforced polymer plates for strengthening reinforced concrete members. *Structures and Buildings*, Proc. Instn Civ. Engrs, **123** (1997) 127–42.

1325. Kaiser, H., Bewehren von Stahlbeton mit kohlenstoffaserverstarkten Epoxidharzen. Doctoral thesis, Federal Technical High School of Zurich, 1989.

1326. Lie, T. T. & Irwin, R. J., Method to calculate the fire resistance of reinforced concrete columns with rectangular cross-section. *ACI Struct. J.*, **90** (1993) 52–60.

1327. Becker, J. & Bresler, B., FIRES — RC, a computer program for the fire response of structure reinforced concrete frames. Report No. UCB FRG 74-3, Fire Research Group, University of California, Berkeley, 1974.

1328. Bizri, H., Structural capacity of reinforced concrete columns subjected to fire induced thermal gradients. Report No. 73-I, Structural Engineering Laboratory, University of California, Berkeley, 1973.

1329. Lie, T. T. & Celikkol, B., Method to calculate the fire resistance of circular reinforced concrete columns. *ACI Mat. J.*, **88** (1991) 84–91.

1330. Ellingwood, B. & Lin, T. D., Flexure and shear behaviour of concrete beams during fires. *J. Struct. Engng*, **117** (1991) 440–58.

1331. Weeks, N. J., Lateral instability of slender reinforced concrete columns in a fire environment. PhD thesis, Aston University, 1985.

1332. *Finite element analysis of reinforced concrete*, American Society of Civil Engineers, New York, 1982, 545 pp.

1333. Nizamuddin, Z. T., Thermal and structural analysis of reinforced concrete slabs in fire environment. PhD thesis, University of California, Berkeley, 1976.

1334. Mukaddam, M., Creep analysis of concrete at elevated temperatures. *ACI J.*, **71** (1974) 72–79.

1335. Huang, Z., Platten, A. & Roberts, J., Nonlinear finite element model to predict temperature histories within reinforced concrete in fires. *Building and Environment*, **31**, No. 2 (1996) 109–18.

1336. Harmathy, T. Z. & Stanzak, W. W., Elevated-temperature tensile and creep properties of some structural and prestressing steels. ASTM Special Technical Publication No. 464, 1969, pp. 186–208.

1337. Bangash, M. Y. H., *Prototype building structures — analysis & design*. Thomas Telford, London, 1999.

1338. Lin, T. D. *et al.*, Flexural and shear behaviour of reinforced concrete beams during fire tests. Report No. NBS-GCR-87-536, Centre for Fire Research, National Bureau of Standards, Washington DC, 1987, 98 pp.

1339. Statistical analysis of earthquake ground motion parameters. Report No. NUREG/CR-1175 prepared for the U.S. Nuclear Regulatory Commission by Shannon & Wilson, Inc. and Agbabian Associates, December 1979.

1340. Abdel-Ghaffar, A. M. & Trifunac, M. D., Anti-plane dynamic soil/bridge interaction for incident plane SH-waves (draft). California Institute of Technology, Pasadena, CA, 1976.

1341. Bogdanoff, J. L., Goldberg, J. E. & Schiff, A. J., The effect of ground transmission time on the response of long structure. *Bull. Seism. Soc. Am.*, **55** (1965) 627–40.

1342. Boncheva, H., Soil amplification factor of surface waves. Proc. 6th World Conf. Earthq. Engrg, Vol. 6, New Delhi, 1977, pp. 6–189.

1343. Chandra, U., Angles of incidence of S-waves. *Bull. Seism. Soc. Am.*, **62**, No. 4 (1972) 903–15.

1344. Chen, J.-C. & Lysmer, J., Doctoral research at the University of California, Berkeley, 1979.

1345. Desai, C. S. & Christian, J. T., *Numerical methods in geotechnical engineering*. McGraw-Hill Book Co., 783 pp., 1976.

1346. Ewing, W. M., Jardetzky, W. S. & Press, F., *Elastic waves in layered media*. McGraw-Hill, 1957.

1347. Gomez-Masso, A. *et al.*, Soil–structure interaction in different seismic environments. Report No. UCB/EERC-79/18, Earthq. Engrg Res. Ctr., University of California, Berkeley, 1979.

1348. Hannon, W. J., An application of the Haskell-Thomson matrix method to the synthesis of the surface motion due to dilatational waves. *Bull. Seis. Soc. Am.*, **54** (1964) 2067–79.

1349. Haskell, N. A., The dispersion of surface waves in multi-layered media. *Bull. Seism. Soc. Am.*, **43**, No. 1 (1953) 17–34.

1350. Haskell, N. A., Crustal reflection of plane SH-waves. *J. Geophysical Res.*, **65** (1960) 4147–50.

1351. Haskell, N. A., Crustal reflection of plane P and SV-waves. *J. Geophysical Res.*, **67**, No. 12 (1962) 4751–67.

1352. Hayasni, S., Tsucnida, H. & Korata, E., Average response spectra for various subsoil conditions. Joint Meeting, U.S.–Japan Panel on Wind and Seismic Effects, UJNR, Tokyo, 2–12 May, 1971.

1353. Hays, W. W., Rogers, A. M. & King, K. W., Empirical data about local ground response. Proc. 2nd U.S. Nat. Conf. on Earthq. Engrg, Earthq. Engrg Res. Inst., Stanford University, August 1979.

1354. Idriss, I. M., Characteristics of earthquake motions — a state-of-the-art review. Proc. ASCE Geotech., Eng. Div. Spec. Conf. on Earthq. Engrg and Soil Dyns., Vol. III, Pasadena, CA, June 1978.

1355. Idriss, I. M. *et al.*, Behaviour of soft clays under earthquake loading conditions. Proc. Offshore Tech. Conf., Paper No. OTC 2671, Dallas, May 1975.

1356. Johnson, N. E. & Galletley, R. D., The comparison of response of a highway bridge to uniform and moving ground excitation. *Shock and Vibration Bull.*, **42** (1972) 2.

1357. Joyner, W. B., A method for calculating nonlinear seismic response in two dimensions. *Bull. Seism. Soc. Am.*, **65**, No. 5 (1975) 1337–357.

1358. Joyner, W. B., A FORTRAN program for calculating nonlinear seismic ground response. Open File Report No. 77-671, U.S. Geol. Survey, Menlo Park, CA, 1977.

1359. Joyner, W. B. & Chen, A. T. F., Calculation of nonlinear ground response in earthquake. *Bull. Seism. Soc. Am.*, **65** (1975) 1335–36.

1360. Laysmer, J., Lumped mass methods for Rayleigh waves. *Bull. Seism. Soc. Am.*, **60**, No. 1 (1970) 89–104.

1361. Lysmer, J. & Kuhlemeyer, R. L., Finite dynamic model for infinite media. *J. Engrg Mech. Div., ASCE*, **95**, No. EM4 (1969) 859–77.

1362. Lysmer, J. & Waas, G., Shear waves in plane infinite structures. *J. Engrg Mech. Div., ASCE*, **98**, No. EM1 (1972) 85–105.

1363. Martin, P. P., Nonlinear methods for dynamic analysis of ground response. PhD thesis, University of California, Berkeley, June 1973.

1364. Martin, P. P. & Seed, H. B., Simplified procedure for effective stress analysis of ground response. *J. Geotech. Engrg Div., ASCE*, **105**, No. GT6 (1979) 739–58.

1365. Nair, G. P. & Emery, J. J., Spatial variations in seismic motions. 2nd Canadian Conf. on Earthq. Engrg Reprint No. 5, Hamilton, Ontario, 1975.

1366. Chsaki, Y. & Haqiwara, T., On effects of soil and foundation upon earthquake inputs to buildings. Res. Paper No. 41, Bldg. Res. Inst. Ministry of Construction, Japan, 1970.

1367. Randall, M. J., Revised travel time table for S. *Geophys. J.*, **22** (1971) 229–34.

1368. Richart, F. E. Jr., Hall, J. R. Jr. & Woods, R. D., *Variations of soils and foundations*. Prentice-Hall, Inc., 1979.

1369. Scanlan, R. H., Seismic wave effects on soil–structure interaction. *Earthq. Engrg and Struct. Dyns.*, **4** (1976) 379–88.

1370. Schnabel, P. B., Lysmer, J. & Seed, H. B., SHAKE — a computer program for earthquake response analysis of horizontally layered sites. Report No. EERC 72-12, Earthq. Engrg Res. Centre, University of California, Berkeley, December 1972.

1371. Seed, H. B. & Idriss, I. M., The influence of soil conditions on ground motions during earthquakes. *J. Soil Mech. and Found. Div., ASCE*, **94**, No. SM1 (1969) 99–137.

1372. Seed, H. B. *et al.*, Relationships between maximum acceleration, maximum velocity, distance from source and local site conditions for moderately strong earthquakes. *Bull. Seism. Soc. Am.*, **66**, No. 4 (1976b) 1323–42.

1373. Seed, H. B., Uqas, C. & Lysmer, J., Site dependent spectra for earthquake resistant design. *Bull. Seism. Soc. Am.*, **66**, No. 1 (1976a) 221–43.

1374. Silva, W., Body waves in a layered anelastic solid. *Bull. Seism. Soc. Am.*, **66**, No. 5 (1976) 1539–54.

1375. Udka, T., Analysis of response of large embankments to travelling base motions. PhD dissertation, University of California, Berkeley, December 1975.

1376. Wass, G., Analysis method for footing vibrations through layered media. PhD dissertation, University of California, Berkeley, December 1972. Also published under the title: Earth vibration effects and abatement for military facilities. *U.S. Army Engineer*, WES, Vicksburg, MS, September.

1377. Wong, H. L., Dynamic soil/structure interaction. PhD dissertation, California Institute of Technology, Pasadena, CA, May 1975.

1378. Wong, H. L. & Luco, J. E., Dynamic response of rigid foundations of arbitrary shape. *Earthq. Engrg and Struct. Dyns.*, **4** (1976) 579–87.

1379. Zienkiewica, O. C., Chang, C. T. & Hinton, E., Nonlinear seismic response and liquefaction. Submitted for publ. in *Int. J. Numerical and Analytical Methods in Geomechanics*, 1978.

1380. Berger, E. & Mahin, S. A., Simplified method for evaluating soil–structure interaction effects. Proc. 9th Ann. Offshore Technology Conf., Houston, Texas, 1977.

1381. Bryant, L. M. & Matoloc, H., Three-dimensional analysis of framed structure with nonlinear pile foundations. Proc. 9th Ann. Offshore Technology Conf., Houston, Texas, 1977.

1382. Arnold, P. *et al.*, A study of soil–pile–structure systems in severe earthquakes. Proc. 9th Ann. Offshore Technology Conf., Houston, Texas, 1977.

1383. Weaver, W., Brandow, G. E. & Hoeg, K., Three-dimensional soil–structure response to earthquakes. *Bull. Seism. Soc. Am.*, **63** (1973) 1041–56.

1384. Desai, C. S. & Appel, G. C., 3-D analysis of laterally loaded structures. Proc. 2nd Int. Conf. Hum. Methods in Geomechanics, Aachen, Germany, 1976, pp. 405–18.

1385. Novak, M., Foundation and soil–structure interaction. Proc. 6th World Conf. on Earthquake Engineering, New Delhi, India, 1977, pp. 1421–48.

1386. Veletsos, A. S., Soil–structure interaction for buildings during earthquakes. Proc. 2nd Int. Conf. on microzonation for safer construction research and application, San Francisco, California, 1978, pp. 111–33.

1387. Novak, M. & Nogami, T., Soil–pile interaction in horizontal vibration. *Int. J. Earthquake Eng. Struc., Dynamics*, **5** (1977) 263–81.

1388. Kuhlemeyer, R. L., Static and dynamic laterally loaded piles. *J. Geol. Eng. Div., Am. Soc. Civ. Engrs*, **105** (1979) 289–304.

1389. Kuhlemeyer, R. L., Vertical vibration of piles. *J. Geotech. Eng. Div., Am. Soc. Civ. Engrs*, **105** (1979) 273–87.

1390. Khandoker, J. U. & Batta, R. C., On a curve description model for soils. Proc. ASCE-EMD Speciality Conference, Purdue University, Vol. II, 1983, pp. 1016–19.

1391. Brandov, G. E., Soil–foundation–structure interaction during earthquake excitations. PhD dissertation, Stanford University, 1971.

1392. Weaver, W. & Nelson, M. F., Three-dimensional analysis of a tier building. *J. Struc. Div., ASCE*, No. ST6 (1966) 385–404.

1393. Becker, E. B., Carey, G. F. & Oden, J. T., *Finite elements — an introduction, Vol. I*. Prentice-Hall, Inc., NY, 1981.

1394. Khandoker, J. U. & Batra, R. C., A curve description model for soils subjected to cyclic loads, pending publication.
1395. Jackson, J. G. Jr., Factors that influence the development of soil constitutive relations. U.S. Army Waterways Expt. Stan., Misc. Paper No. 4-980, 1968.
1396. Kawamura, S., Umemura, H. & Osawa, Y., Earthquake motion measurement of pile-supported building on reclaimed ground. Proc. 6th World Conf. on Earthquake Engng, New Delhi, India, **2** (1977) 1563–69.
1397. Sugimura, Y., Earthquake observation and dynamic analysis of a building supported on long piles. Proc. 6th World Conf. on Earthquake Engng, New Delhi, India, **2** (1977) 1570–75.
1398. Pezien, J., Scheffey, C. F. & Parmalee, R. A., Seismic analysis of bridges on long piles. *J. Eng. Div., Am. Soc. Civ. Engrs*, **90** (1954) (EM3), 223–54; also see Penzien, J., Soil–pile foundation interaction. In *Earthquake engineering*, R. L. Wiegel (ed.), Prentice-Hall Inc., Englewood Cliffs, NJ, 1970.
1399. Bielak, J. & Palancia, V. J., Dynamic behaviour of structure with pile-supported foundations. Proc. 6th World Conf. on Earthquake Engng, New Delhi, India, **2** (1977) 1576–82.
1400. Novak, M., Effect of soil on structural response to wind and earthquake, in *Earthquake Engineering and Structural Dynamics*, Vol. 3, 1974, pp. 79–96.
1401. Novak, M. & Beredugo, Y. O., Vertical vibrations of embedded footings. *J. ASCE Soil Mech. Found. Div.*, **98** (1972) (SM12) 1291–310.
1402. Novak, M., Dynamic stiffness and damping of piles. *Canadian Geotech. J.*, **11** (1974) 574–98.
1403. Parmelee, R. A., Perelman, D. S. & Lee, S. L., Seismic response of multistorey structure on flexible foundation. *Bull. Seism. Soc. Am.*, **29** (1969) 1070.
1404. Poulos, H. G. & Davis, E. H., *Elastic solutions for soil and rock mechanics*. New York, Wiley and Sons, 1974.
1405. Rainer, J. H., Simplified analysis of dynamic structure–ground interaction. *Canadian J. Civ. Engng*, **2**, No. 3 (1975) 345–56.
1406. Rainer, J. H., Damping in dynamic structure–foundation interaction. *Canadian Geotech. J.*, **12** (1975) 13–22.
1407. Richart, F. E., Hall, J. R. Jr & Woods, R. D., *Vibrations of soils and foundations*. Prentice-Hall, Englewood Cliffs, New Jersey.
1408. Rutenberg, A., Simplified P–Δ analysis for asymmetric structure. *J. Struct. Div., ASCE*, **108** (1982).
1409. Seed, H. B., Lysmer, J. & Hwang, R., Soil–structure interaction analysis for seismic response. *J. ASCE Geotech. Engng. Div.*, **101** (1974) 439–57.
1410. Seed, H. B., Lysmer, J. & Hwang, R., Soil moduli and damping factors for dynamic response analysis. Report EERC70-10, Berkeley, University of California, 1970.
1411. Seed, H. B., Witman, R. V. & Lysmer, J., Soil–structure interaction effects in the design of nuclear power plants. In *Structural and Geotechnical Mechanics*, a volume honoring N. M. Newmark, W. J. Hall (ed.), Prentice-Hall, Englewood Cliffs, New Jersey, 1977.
1412. Skinner, R. I., Robinson, W. H. & McVerry, G. H., *An introduction to seismic isolation*. Wiley and Sons, 1993.
1413. Skinner, R. I. *et al.*, Structures Congress '91. Proc. 9th ASCE Structures Congress, 1991.
1414. Stokoe, K. H. II & Woods, R. D., In situ shear wave velocity by cross-hole method. *J. ASCE Soil Mech. Found. Div.*, **98** (1972) 443–60.
1415. Structural Engineers Association of California, 1968, 1973, 1974, 1976 and 1987. Recommended lateral force requirements of commentary. San Francisco, SEAOC.
1416. Thomson, W. T., *Vibration theory and application*. Prentice-Hall, New York, 1965.
1417. Tasi, N. C., Modal damping for soil–structure interaction. *J. ASCE Engng Mech. Div.*, **100** (1974) 323–41.

1418. Vaish, A. K. & Chopra, A. K., Earthquake finite element analysis of structure–foundation systems. *J. ASCE Engng Mech. Div.*, **100** (1974) 1011–16.

1419. Veletsos, A. S. & Nair, V. V., Seismic interaction on structures on hysteretic foundations. *J. ASCE Struct. Div.*, **101** (1975) 109–29.

1420. Veletsos, A. S., Dynamics of structure–foundation systems, in *Structural and Geotechnical Mechanics* — a volume honoring N. M. Newmark, W. J. Hall (ed.), Prentice-Hall, Englewood Cliffs, NJ, 1977, pp. 333–61.

1421. Veletsos, A. S. & Meek, J. W., Dynamic behaviour of building foundation systems. *Earthquake Engng and Struct. Dynams*, **3**(2) (1974) 121–38.

1422. Veletsos, A. S. & Verbic, B., Vibration of viscoelastic foundations. *Earthquake Engng and Struct. Dynams*, **2**(1) (1973) 87–102.

1423. Veletsos, A. S. & Wei, Y. T., Lateral and rocking vibration of footings. *J. ASCE Soil Mech. Found. Div.*, **97** (1971) 1227–48.

1424. Veletsos, A. S. & Verbic, B., Basic response functions for elastic foundations. *J. ASCE Engng Mech. Div.*, **100** (1974) 189–202.

1425. Wiegel, R. L. (ed.), *Earthquake engineering*. Prentice-Hall, New York, 1970.

1426. Wilson, E. L. & Habibullah, A., Static and dynamic analysis of multi-storey buildings including P–Δ effects. *Earthquake Spectra*, **3**, No. 2 (1987).

1427. Wilson, E. L., Suharwardy, M. I. & Habibullah, A., *A seismic analysis method which satisfies the 1988 lateral force requirements*. Computers and Structures, Inc., Berkeley, California, 1989.

1428. Wilson, E. L., Yuan, M. & Dickens, J., Dynamic analysis by direct superposition of Ritz vectors. *Earthquake Engng and Struct. Dynamics* (1982) 10.

1429. Hoen, D., *MARCO — a program system for estimation of structural system modal parameters from vibration measurements*. Theory manual, SINTER Report STF71 F87057, SINTER, Trondheim, 1987.

1430. Hen, C., System identification of structures excited by stochastic load processes. Dr. in thesis, 1991: 37, Division of Marine Structures, The Norwegian Institute of Technology, Trondheim, 1991.

1431. Jefferys, E. R., Private communications. London/Stavanger, November 1993.

1432. Krostad, J. R. *et al.*, A new non-slender ringing load approach verified against measurements. OMAE '96 15th Int. Conf. on Offshore Mechanics and Arctic Engineering, Florence, June 1996.

1433. Myrvoll, F., Instrumentation of the skirt piled Gullfaks C gravity platform for performance monitoring. BOSS '92 Int. Conf. on Behaviour of Offshore Structures, London, July 1992.

1434. Spidsoe, N. & Hilmarsen, B., Measured dynamic behaviour of North Sea gravity platforms under extreme environmental conditions. OTC — 15th Offshore Technology Conf., Houston, May 1983.

1435. Spidsoe, N. & Langen, I., Damping of fixed offshore platform. BOSS '85 Int. Conf. on Behaviour of Offshore Structures, Delft, July 1985.

1436. Tjelta, T. I., Janbu, N. & Grande, L., Observations on drainage control effects on Gullfaks C gravity structure. BOSS '92 Int. Conf. on Behaviour of Offshore Structures, London, July 1992.

1437. USNRC, *Earthquake engineering for concrete dams: design performance and research needs*. National Academy Press, Washington, DC, 1990, p. 43.

1438. Bangash, M. Y. H., *Prototype Building Structures*, Thomas Telford, London, 1999.

1439. Cheung, Y. I. & Nag, D. K., Plates and beams on elastic foundations, linear and nonlinear behaviour. *Géotechnique*, **18** (1969).

1440. Clough, G. W. & Taeri, Y., Static analysis of earth retaining structures, Chapter 15. *Numerical methods in geotechnical engineering*, C. S. Desai and J. T. Christian (eds), McGraw-Hill Book Co., New York, 1977.

1441. Desai, C. S., Solution of stress-deformation problems in soil and rock mechanics using FEM. PhD dissertation, University of Texas, 1968.
1442. Desai, C. S. & J. F., *Introduction to finite element method for engineering analysis*. Van Nostran Reinhold Co., New York, 1972.
1443. Desai, C. S. & Appal, G. C., 3-D analysis of laterally loaded structures. Proc. 2nd Int. Conf. Num. Methods Geomech. Blacksburg, VA, 1976.
1444. Desai, C. S. & Patil, U. K., Finite element analysis of some interaction problems. Proc. Int. Symp. on Soil–Structure Interaction, University of Roorkes, 1979.
1445. Hooper, J. A., Foundation interaction analysis. Chapter 5, *Developments in Soil Mechanics — 1*, C. R. Scott (ed.), Allied Science Publishers Ltd, London, 1978.
1446. Jain, M. B., Analysis of curved rafts on winkler bed using FEM. M-Tech. thesis, ITT Delhi, 1981.
1447. Roy, S. K., Analysis of axially loaded piles by FEM. M-Tech. thesis, ITT Delhi, 1983.
1448. Varadarajan, A. & Arora, K. R., An interactive study of strip footing — sand bed system by FEM. 3rd Int. Conf. on Numerical Methods in Geomechanics, pp. 1041–51.
1449. Inagaki, H. *et al.*, Performance of caisson type walls at Kobe Port. *Soils and Foundations*. Special Issue, January 1966, pp. 119–36.
1450. Okabe, S., General theory of earth pressure. *J. Japan Civ. Engng Soc.*, **12**, No. 1 (1924).
1451. Monobe, N. & Matsuo, H., On the determination of earth pressure during earthquakes. Proc. World Engng Congress, 1929, pp. 177–85.
1452. Seed, H. B. & Whitman, R. V., Design for earth retaining structures for dynamic loads. ASCE Special Conf. — lateral stress in the ground and design of earth retaining structures, Cornell, 1970, pp. 108–47.
1453. Krinttzsky, E. L., Gould, J. P. & Edinger, P. H., Fundamentals of earthquake resistant construction.
1454. Ebeling, R. M. & Morrison, E. E., The seismic design of waterfront retaining structure. Technical report ITL-92-11, USAE, Waterways Experiment Station, Corps of Engineers, Vicksburg, 1992.
1455. Steedman, R. S. & Zeng, X., Hydrodynamic pressures on a flexible quay wall. Proc. Eur. Conf. Structural Dynamics, Bochum, Balkema, 1990, pp. 843–50.
1456. Westergaard, H. M., Water pressure on dams during earthquakes. *Trans. Am. Soc. Civ. Engrs*, **98** (1993) 418–72.
1457. Matsuzawa, H., Ishibashi, I. & Kawamura, M., Dynamic soil and water pressures of submerged soils. *Proc. ASCE, J. Geotech. Engng*, **111** (1985) No. 10.
1458. Amano, R., Azuma, H. & Ishii, Y., Aseismic design of walls in Japan. Proc. 1st World Conf. Earthquake Engng, Paper 32, 1956, pp. 1–17.
1459. Matsuo, H. & Ohara, S., Lateral earth pressure and stability of quay walls during earthquakes. Proc. 2nd World Conf. Earthquake Engng, Tokyo, 1960.
1460. Whitman, R. V. & Ting, N. H., in *Experimental results for tilting wall with saturated backfill*.
1461. Arulanandan & Scott (eds), Proc. Conf. Verification of Numerical Procedures for the Analysis of Soil Liquefaction Problems, Balkema, 1994, pp. 1515–28.
1462. Bolton, M. D., Geotechnical stress analysis for bridge abutment design. Contractor report 270. Transport and Road Research Laboratory, Crowthorne, Berks, 1991.
1463. Murphy, V. A., The effect of ground characteristics on the aseismic design of structures. Proc. 2nd World Conf. Earthquake Engng, Tokyo, 1960, pp. 231–47.
1464. Bolton, M. D. & Steedman, R. S., Modelling the seismic resistance of retaining structures. Proc. XI Int. Conf. Soil Mech. Fndn Engng, San Francisco, 4, Balkema, 1985, pp. 1845–48.
1465. Bolton, M. D. & Steedman, R. S., Centrifugal testing of microconcrete retaining walls subjected to base shaking. Proc. Conf. Soil Dynamics and Earthquake Engng, Southampton, Balkema, 1982, pp. 311–29.
1466. Steedman, R. S., Modelling the behaviour of retaining walls in earthquakes. Cambridge University, PhD thesis, 1984.

1467. Steedman, R. S. & Zeng, X., The influence of phase on the calculation of pseudostatic earth pressure on retaining walls. *Géotechnique*, **40**, No. 1 (1990) 103–12.
1468. Japan Ministry of Transport, Report of a survey of the damage caused to port facilities etc. by Great Hanshin/Awaji earthquake. Ministry of Transport, Ports and Harbours Bureau, PHRI, No. 3, Port Construction Bureau, May 1995 (in Japanese).
1469. Earthquake Engineering Research Centre, Geotechnical reconnaissance of the effects of the January 17, 1995 Hyogoken-Nambu earthquake, Japan. Report No. UCB/EERC-95/01, Berkeley, August 1995.
1470. EQE International, The January 17, 1995 Kobe earthquake: on EQE summary report. EQE International, San Francisco, 1995.
1471. Japan Ministry of Transport, Technical standards for port and harbour facilities in Japan. Ports and Harbours Bureau, 1991. English version by Overseas Coastal Area Development Institute of Japan, Tokyo.
1472. Newmark, N. M., Effect of earthquakes on dams and embankments. 5th Rankine Lecture. *Géotechnique*, **15**, No. 2 (1965) 139–60.
1473. Richards, R. & Elms, D. G., Seismic behaviour of gravity retaining walls. *Proc. ASCE, J. Geotech. Engng Div.*, **105** (1979) 449–64.
1474. Franklin, A. G. & Chang, F. K., Earthquake resistance of earth and rockfill dams. Report 5: Permanent displacements of earth dams by Newmark analysis. Misc. paper S71-17, USAE Waterways Experiment Station, November 1977.
1475. Shitman, R. V. & Liao, S., Seismic design of gravity retaining walls. Misc. paper GL85-1, USAE Waterways Experiment Station, Vicksburg, 1985.
1476. Nadim, F., Tilting and sliding of gravity retaining walls. Dept. Civil Engng MIT, SM thesis, 1980.
1477. Steedman, R. S. & Zeng, X., Rotation of large gravity walls on rigid foundations under seismic loading. Proc. ASCE annual convention, Washington, November 1996.
1478. Veletsos, A. S. & Younan, A. H., Dynamic response of cantilever retaining walls. *Proc. ASCE J. Geotech. Geoenviron. Engng* (1997) 161–72.
1479. Steedman, R. S. & Zeng, S., On the behaviour of quay walls in earthquakes. *Géotechnique*, **43** No. 3 (1993) 417–31.
1480. Cole, K., Building over abandoned shallow land mines. Paper 1: Considerations of risk and reliability. *Ground Engineering*, **26**, No. 1, Jan./Feb. 1993.
1481. Matsuo, H., Experimental study on the distribution of earth pressure acting on a vertical wall during earthquakes. *J. Japanese Soc. Civ. Engrs*, **27**, No. 2 (1941).
1482. Nadim, F., Tilting and sliding of gravity retaining walls. MS thesis, Department of Civil Engineering, Massachusetts Institute of Technology, Cambridge, Massachusetts, 1980.
1483. Nadim, F. & Whitman, R. V., Coupled sliding and tilting of gravity retaining walls during earthquakes. Proc. 8th World Conf. on Earthquake Engng, 1984, pp. 477–84.
1484. Newmark, N. M., Effects of earthquakes on dams and embankments. *Géotechnique*, **14**, No. 2 (1965) 139–60.
1485. NYSDOT, Bridge Inventory Manual. Bridge Inventory and Inspection System, Albany, NY, 1991.
1486. Okabe, S., General theory of earth pressure. *J. Japanese Soc. Civ. Engrs*, **12**, No. 1 (1926).
1487. Richards, R. & Elms, D., Seismic behaviour of gravity retaining walls. *J. Geotech. Engng Div., ASCE*, **105**, No. 4 (1979).
1488. Richards, R. & Elms, D. G., Seismic behaviour of tied back walls. Report 87-8, Dept. of Civil Engineering, University of Canterbury, Christchurch, New Zealand, 1987.
1489. Richards, R., Elms, D. G. & Budhu, M., Dynamic fluidization of soils. *J. Geotech. Engng, ASCE*, **116**, No. 5 (1990).
1490. Richards, R. & Shi, Z., Seismic bearing capacity of shallow foundations. Pacific Conf. on Earthquake Engineering, 20–23 November 1991, New Zealand.

1491. Richards, R., Budhu, M. & Elms, D. G., Seismic fluidization and foundation behaviour. Proc. 2nd Int. Conf. on Recent Advances in Geotechnical Earthquake Engineering and Soil Dynamics, S. Prakash (ed.), University of Missouri, Rollo, 1991.
1492. Richards, R., Elms, D. G. & Budhu, M., Seismic bearing capacity and settlements of foundations. *J. Geotech. Engng, ASCE*, **119**, No. 4 (1993).
1493. Richards, R. & Elms, D. G., Seismic passive resistance of tied back walls. *J. Geotech. Engng, ASCE*, **118**, No. 7 (1992) 996–1011.
1494. Richards, R., Fishman, K. L. & Divito, R. C., Threshold accelerations for rotation or sliding of bridge abutments. *J. Geotech. Engng, ASCE*, scheduled to appear September 1996.
1495. Shi, X., Plastic analysis for seismic stress fields. PhD dissertation, State University of New York at Buffalo, New York, 1993.
1496. Siddharthan, R., Ara, S. & Anderson, J., Seismic displacement of rigid retaining walls. Proc. 4th U.S. National Conference on Earthquake Engineering, 1990, pp. 673–80.
1497. Siddharthan, R., Gowda, P. & Norris, G., Displacement based design of retaining walls. Proc. 2nd Int. Conf. on Recent Advances in Geotechnical Earthquake Engineering and Soil Dynamics, 1999, pp. 657–61.
1498. Siddharthan, R., Ara. S. & Norris, G., Simple rigid plastic model for seismic tilting of rigid walls. *J. Struct. Engng, ASCE* (1992) 469–87.
1499. Steedman, R. S., Modelling the behaviour of retaining walls in earthquakes. PhD thesis, Engineering Department, Cambridge University, Cambridge, England, 1984.
1500. Uwabe, T. & Moriya, M., Shaking table tests of sliding gravity-type retaining walls during earthquake. Proc. 9th World Conf. on Earthquake Engineering, Tokyo, Japan, 1988, Vol. III, pp. 685–90.
1501. Whitman, R. V., Seismic design and behaviour of gravity retaining walls, design and performance of earth retaining structures. Proc. Conf., ASE, 1990, pp. 817–41.
1502. Whitman, R. V., Predicting earthquake-induced tilt of gravity retaining walls. *Retaining structures, Proc. Conf., Inst. Civ. Engrs* (1992), pp. 750–58.
1503. Wood, J. M., Earthquake induced pressures on rigid wall structures. *Bull. New Zealand Soc. for Earthquake Engng*, **8**, No. 3 (1975) 175.
1504. Younkins, J., Seismic vulnerability of bridge abutments in New York State. Master of Engineering Project, submitted to State University of New York at Buffalo, 1994.
1505. Younis, C. J. & Tadibanksh, I. G., Response of sliding rigid structure to base excitation. *J. Engng Mech., ASCE*, **110** (1984) 417–32.
1506. Li, L., Base isolation measure for a seismic building in China. Proc. 8th World Conf. Earthquake Engng, San Francisco, 1984, pp. 791–98.
1507. Malushte, S. R. & Singh, M. P., A study of seismic response characteristic of structures with friction damping. *Earthquake Engng Struct. Dyn.*, **18** (1989) 767–83.
1508. Skinner, R. I., Robinson, W. H. & McWerry, G. H., *An introduction to seismic isolation*. Wiley, New York, 1993.
1509. Kelly, J. M., A seismic base isolation: review and bibliography. *Soil Dyn. Earthquake Engng*, **5** (1986) 202–16.
1510. Kelly, J. M. & Tsai, H. C., Seismic response of light internal equipment in base-isolated structure. *Earthquake Engng Struct. Dyn.*, **13** (1985) 711–32.
1511. Su, L., Ahmadi, G. & Tadjbakhsh, I. G., A comparative study of performances of various base isolation systems. Part II: sensitivity analysis. *Earthquake Engng Struct. Dyn.*, **19** (1990) 21–34.
1512. Wu Lin, Tsung & Hone, Chan-Chi, Base isolation by free rolling rods under basement. *Earthquake Engng Struct. Dyn.*, **22** (1993) 261–73.
1513. Zayas, V. & Low, S., A simple pendulum technique for achieving seismic isolation. *Earthquake Spectra*, **6** (1990) 317–33.
1514. Mokha, A. *et al.*, Experimental study of friction pendulum isolation system. *J. Struct. Engng ASCE*, **117** (1981) 1201–18.

1515. Zayas, V. & Al-Hussaini, T., Summary of testing of the friction pendulum seismic isolation system, 1986–1993. Proc. ATC.17.1, Seismic Isolation, Passive Energy Dissipation and Active Control, Vol. 1, San Francisco, 1993, pp. 377–88.
1516. Amin, N., Mokha, A. & Fatehi, H., Seismic isolation retrofit of the U.S. Court of Appeals Building. Proc. ATC-17-1 Seismic Isolation Passive Energy Dissipation of Active Control, Vol. 1, San Francisco, 1993, pp. 185–96.
1517. Jingwen, Xue, *Tribology and the technique of lubrication*. Weapon Industry Press, 1992 (in Chinese).
1518. Xiangyun Huang, Test study on structure base isolated by ball system with restoring property. Thesis of the Master's Degree, Xian University of Architecture & Technology, Xian, China, 1996 (in Chinese).
1519. Qiang Zhou, Experimental study and numerical analysis on structure base isolated by ball system with restoring property. Thesis of the Master's Degree, Xian University of Architecture & Technology, Xian, China, 1997 (in Chinese).
1520. Lau, S. L. & Cheung, Y. K., Amplitude incremental variational principle for nonlinear vibration of elastic system. *J. Appl. Mech. ASME*, **48** (1981) 959–64.
1521. Cheung, Y. K. & Lau, S. L., Incremental time-space finite strip method for nonlinear structural vibrations. *Earthquake Engng Struct. Dyn.*, **10** (1982) 239–53.
1522. Cheung, Y. K., Chen, S. H. & Lau, S. L., Application of the incremental harmonic balance method to cubic non-linearity systems. *J. Sound Vib.*, **140** (1990) 273–86.
1523. Pierre, C., Ferri, A. A. & Dowell, E. H., Multi-harmonic analysis of dry friction damped system using an incremental harmonic balance method. *J. Appl. Mech. ASME*, **52** (1985) 958–64.
1524. Westergaard, H. M., Water pressures on dams during earthquakes. *Trans. ASCE*, **98** (1933) 418.
1525. Zanger, C. N., Hydrodynamic pressures on dams due to horizontal earthquake effects. Engineering Monograph No. 11, US Bureau of Reclamation, May 1952.
1526. Brahtz, J. H. A. & Heilbron, C. H. Jr., Discussion on water pressures on dams during earthquakes by H. M. Westergaard. *Trans. ASCE*, **98** (1933) 452.
1527. Hatano, T., Seismic force effect on gravity dam. *Trans. JSCE*, **3**, No. 1 (1949) 80; *Trans. JSCE*, **3**, No. 2 (1949) 174; *Trans. JSCE*, **5**, No. 3 (1950) 83 (in Japanese).
1528. Chakarabarti, P. & Chopra, A. K., Hydrodynamic effects in earthquake response of gravity dams. *J. Struct. Div., ASCE*, **100** (ST6) (1974), Paper 10616, p. 1211.
1529. Chopra, A. K., Earthquake response of concrete gravity dams. *J. Eng. Mech. Div., ASCE*, **96** (1970) (EM4), Paper 7485, p. 448.
1530. Hatano, T., Takahashi, T. & Tsutsumi, T., Dynamical study of Tsukabaru Gravity Dam. Technical Report C5801, Central Res. Inst. Electric Power Industry, June 1958.
1531. Hatano, T., An examination on the resonance of hydrodynamic pressure during earthquakes due to the elasticity of water. Technical Report C-65001, Central Res. Inst. Electric Power Industry, Nov. 1965.
1532. Nakagawa, T. T. & Hatano, T., Analytical solution of hydrodynamic pressure with reflective condition at reservoir bottom during earthquakes. *Proc. JSCE*, **229** (1974) 119.
1533. Chopra, A. K., Hydrodynamic pressures on dams during earthquakes. *J. Eng. Mech. Div., ASCE*, **93** (1967) (EM6), Paper 5695, p. 205.
1534. Nakagawa, T. T., *The near-field finite difference solution for wave propagation problems in infinite media*. Studies in mathematics and its applications. Kinokuniya/North-Holland Publishing Co., **18** (1986) 561.
1535. Nakagawa, T. T., Numerical solution in the near-field to linear hyperbolic systems with application to an elastic foundation problem. *Comput. Methods in Appl. Sci. Engng V*, North-Holland Publishing Co., 1982, 207.
1536. Nakagawa, T. T. & Shojiri, H., A new method of time domain analysis for structure on a semi-infinite foundation. Proc. 8th WCEE, Prentice-Hall Inc., New Jersey, III, 1984, 761.

1537. Chopra, A. K., Earthquake response analysis of concrete dams, in *Advanced Dam Engineering for Design, Construction and Rehabilitation*, Van Nostrand Reinhold, New York, 1988.

1538. Zienkiewicz, O. C. & Taylor, R. L., *The finite element method*, Vol. 1, McGraw-Hill, London, 1989.

1539. Ahmad, S., Irons, B. & Zienkiewicz, O. C., Analysis of thick and thin shell structures by curved finite elements. *Int. J. Num. Meth. Engng*, **2** (1970).

1540. Yogt, F., *Ueber die Berchnung der Fundamentdeformation*. Det Norske Videnskaps Akademi, Oslo, 1925.

1541. *Treatise on dams*. United States Department of the Interior Bureau of Reclamation (USBR), Denver, Colorado, 1956.

1542. Clough, R. W. & Penzien, J., *Dynamics of structures*. McGraw-Hill, New York, 1975.

1543. Newmark, N. M., A method of computation for structural dynamics. *J. Engng Mech. Div., ASCE*, **85** (1959) 67–91.

1544. Paz, M., *Structural dynamics — theory and computation*. Van Nostrand Reinhold, New York, 1985.

1545. Delgado, R. M. & Faria, R. M., Dynamic analysis of Caldeirao Dam (in Portuguese). Technical Report, Faculadade de Engenharia da Universidade do Porto, March 1987.

1546. Delgado, R. M. & Faria, R. M., The foundation influence in the earthquake response of arch dams. *Numerical methods in geomechanics*, A. A. Balkema, Rotterdam, 1989.

1547. Westergaard, H., Water pressure on dams during earthquakes. *Trans. ASCE*, 1933.

1548. Priscu, R. *et al.*, *Earthquake engineering for large dams*. 2nd edn, Editura Acadamiei and John Wiley and Sons, 1985.

1549. Delgado, R. M., The finite element method in the dynamic analysis of dams including the solid–fluid interaction (in Portuguese). PhD thesis, Faculdade de Engenharia da Universidade do Porto, 1984.

1550. White, N., Valliappan, S. & Lee, I. K., Unified boundary for finite dynamic models. *J. Engng Mech. Div., ASCE*, **103** (1977) 949–64.

1551. Marques, J. M. M. C., Development of a finite element package for the solution of nonlinear dynamic problems with application to the seismic analysis of dams. MSc thesis, Faculdade de Engenharia da Universidade do Porto, 1985.

1552. Guedes, J. P. M. *et al.*, Infinite domains in dynamic problems. Internal Report, Faculdade de Engenharia da Universidade do Porto, 1992.

1553. Lysmer, J. *et al.*, FLUSH — a computer program for approximate 3D analysis of soil–structure interaction. Report UCB-EERC, University of California, Berkeley, California, USA, **75**, No. 30 (1975).

1554. Vucetic, M. & Dobry, R., Effects of soil plasticity on cyclic response. *J. Geotech. Engng Div., ASCE*, **117** (1991) (GT1).

1555. Seed, H. B. *et al.*, Moduli and damping factors for dynamic analyses of cohesionless soils. Report UCB-EERC, University of California, Berkeley, California, USA, **84**, No. 14 (1984).

1556. Hardin, B. O., The nature of stress–strain behaviour for soils. *J. Geotech. Engng Div., ASCE*, Special Conf. Earthquake Engng Soil Dynamics, Pasadena, **1** (1978) 1–90.

1557. Anderson, D. G. & Woods, R. D., Time dependent increase in shear modulus of clay. *J. Geotech. Engng Div., ASCE*, **102** (1976) (GT5).

1558. Zeghal, M. & Abdel-Ghaffar, A. M., System indentification of earth dams using seismic records. *Proc. ASCE, Struct. Congr. Seismic Engng*, San Francisco, 1989, pp. 179–88.

1559. Sagaseta, C., Cuellar, V. & Pastor, M., Modelling stress–strain time behaviour of natural soils — cyclic loading. Proc. 10th Eur. Conf. on Soil Mech. and Found. Engng, Firenze, 26–30 May 1991.

1560. Joyner, W. B. & Boore, B. M., Peak horizontal acceleration and velocity from strong-motion records including records from the 1979 Imperial Valley, California, Earthquake. *Bull. Seism. Soc. Am.*, **71** No. 6 (1981) 2011–38.

1561. Sabetta, F. & Pugliese, A., Attenuation of peak horizontal acceleration and velocity from Italian strong-motion records. *Bull. Seism. Soc. Am.*, **77**, No. 5 (1987) 1481–1513.
1562. Mononobe, H. A., Seismic stability of the earth dam. Proc. 2nd Congr. on Large Dams, Washington DC, USA, **4**, 1936.
1563. Ambraseys, N. N., On the seismic behaviour of earth dams. Proc. 2nd World Conf. Earthquake Engng, Tokyo, **1** (1960) 331.
1564. Gazetas, G., A new dynamic model for earth dams evaluated through case histories. *Soils and Foundations*, Map. Soc. SMFE, **21**, No. 1 (1981) 67–78.
1565. Clough, R. W., *The finite element method in structural mechanics*. John Wiley and Sons, London, Chapter 7, 1965.
1566. Wilson, E. L., A computer program for the dynamic stress analysis of underground structures. Struct. Engng Lab. Report No. 68-1, University of California, Berkeley, California, USA, 1968.
1567. Zienkiewicz, O. C., *The finite element method in engineering science*. McGraw-Hill, London, 1971.
1568. Rayleigh, L., *Theory of sound*. Vol. 1, Dover Publications, New York, 1945.
1569. Idriss, I. M. *et al.*, QUAD 4 — a computer program for evaluating the seismic response of soil structures by variable damping finite element procedures. Rep. EERC, University of California, Berkeley, California, USA, **73**, No. 16 (1973).
1570. Romo Organista, M. P. *et al.*, PLUSH — a computer program for probabilistic finite element analysis of seismic soil structure interaction. Report UCB-EERC, University of California, Berkeley, California, USA, **77**, No. 1 (1977).
1571. Terzaghi, K., Mechanisms of landslides. *Geol. Survey America, Engng Geol.*, 1950.
1572. Seed, H. B. & Martin, G. R., The seismic coefficient in earth dam design. *J. Soil Mech. Found. Div., ASCE*, **104** (1966) (GT7).
1573. Ambraseys, N. N. & Sarma, S. K., The response of earth dams to strong earthquakes. *Géotechnique*, **17**, No. 3 (1967) 181–213.
1574. Sarma, S. K., Seismic stability of earth dams and embankments. *Géotechnique*, **25**, No. 4 (1975).
1575. Makdisi, F. I. & Seed, H. B., Simplified procedure for estimating dam and embankment earthquake induced deformations. *J. Geotech. Engng Div., ASCE*, **104** (1978) (GT7).
1576. Goodman, R. E. & Seed, H. B., Earthquake-induced displacement in sand embankments. *J. Soil Mech. Found. Div., ASCE*, **92** (1966) (SM2), 125–46.
1577. Paoliani, P., A method for evaluating pore pressure rise during dynamic loadings of sands. 2nd Int. Conf. on Soil Dynamics and Earthquake Engineering on board the QE2, 1985.
1578. Ruggeri, G. *et al.*, The seismic response of El Infiernillo embankment dam: equivalent linear approach and Newmark method for evaluation of permanent displacements. 1st ICOLD Benchmark Workshop, Bergamo, Italy, 1991.
1579. Dakoulas, P., Nonlinear response of dams on alluvial deposits in Narrow Canyon. *J. Dyn. and Earthq. Engng*, **9**, No. 6 (1990).
1580. Elgamai, A. W. M., Scott, R. F. & Succarieh, M. F., La Villita Dam response during five earthquakes including permanent deformations. *J. Geotech. Engng Div., ASCE*, **116** (1990) (GT10).
1581. De Alba, P. A. *et al.*, Analyses of dam failure in 1985 Chilean earthquake. *J. Geotech. Engng Div., ASCE*, **114** (1987) (GT12).
1582. ICOLD, Earthquake analysis procedures for dams — state of the art. Bulletin 52, 1986.
1583. Lee, K. L., Seismic permanent deformations in earth dams. Rep. UCLA-Eng-7497, University of California, Berkeley, California, USA, 1975.
1584. Griffiths, D. V. & Prevost, J. H., Two and three-dimensional dynamic finite element analysis of the Long Valley dam. *Géotechnique*, **38**, No. 3 (1988).

1585. Seed, H. B. *et al.*, Re-evaluation of the slide in the lower San Fernando dam in the earthquake of 9 Feb. 1971. Report UCB-EERC, University of California, Berkeley, California, USA, **88**, No. 4 (1988).
1586. Lai, S. S. & Seed, H. B., Dynamic response of Long Valley dam in the Mammoth Lake earthquake series of 25–27 May 1980. Report UCB-EERC, University of California, Berkeley, California, USA, **85**, No. 12, (1985).
1587. Kagawa, T. *et al.*, FLUSH — A computer program for the three-dimensional analysis of earth dams. Report UCT-EERC, University of California, Berkeley, California, USA, **81**, No. 14, (1981).
1588. Baldovin, E. & Paoljani, P., Seismic stability of earth and rockfill dams. Proc. of the 6th Int. Conf. on Num. Methods in Geomechanics, Innsbruck, 11–15 April 1988.
1589. Banerjee, N. G., Seed, H. B. & Chan, C. K., Cyclic behaviour of dense coarse-grained materials in relation to the seismic stability of dams. Report UCB-EERC, University of California, Berkeley, California, USA, **79**, No. 13, (1979).
1590. Ishihara, K., Evaluation of soil properties for use in earthquake response analysis. *Num. Mod. Geom.*, R. Dungar, G. N. Pande and J. A. Studer (eds), 1982.
1591. Marcuson, W. F. & Krinitsky, E. L., Dynamic analysis of Fort Peck dam. *US Army Eng. Wat. Exp. Station, Techn. Rep.*, Vicksburg, Miss., USA, S76(1), 1976.
1592. CDMG, Strong motion records from the 2 May 1983 Coalings, California, earthquake. Cal. Div. of Mines and Geology, Rep. OSMS, California, USA, **83**, No. 5.2, (1983).
1593. Cornell, C. A., Engineering seismic risk analysis. *Bull. Seism. Soc. Am.*, **58** (1968).
1594. Finn, W. D. L., *Estimating how embankment dams behave during earthquakes*. Water Power and Dam Construction, April 1991.
1595. Lin, J. S. & Chao, B. K., Estimation of shear modulus and damping factors of earth dam materials. *J. Earthq. Engng and Struct. Dyn.*, **19**, 1990.
1596. Gazetas, G., Seismic response of earth dams: some recent developments. *J. Soil Dyn. and Earthq. Engng*, **6**, No. 1 (1987).
1597. Newmark, N. M., Effects of earthquake on dams and embankments. *Géotechnique*, **15**, No. 2 (1965).
1598. Seed, H. B., Considerations on the earthquake-resistant design of earth and rockfill dams. *Géotechnique*, **29**, No. 3 (1979).
1599. Sherard, J. L., Earthquake considerations in earth dam design. *J. Soil Mech. Div., ASCE*, **93** (1967) (SMA4).
1600. Dakoulas, P. & Gazetas, G., Seismic shear vibration of embankment dams in semi-cylindrical valleys. *J. Earthq. Eng. and Struct. Dyn.*, **14** (1986).
1601. Åtkinson, J. H. & Sallfors, G., Experimental determination of stress–strain–time characteristics in laboratory and in-situ tests. Proc. 10th Eur. Conf. on Soil Mech. and Found. Eng., Firenze, 26–30 May 1991.
1602. Bangash, M. Y. H., Prototype building structures — analysis & design, Section A3, Thomas Telford, London, 1999.
1603. Kelly, J. M., Implementation of base isolation in the United States. Proc. 6th Symp. on Seismic, Shock and Vibration Isolation. ASME Pressure Vessel and Piping Conference, Denver, Colorado, 1993, 159–70.
1604. Kelly, J. M., Base isolation in Japan. Report No. UCB/EERC-88/20, Earthquake Engineering Research Center, University of California, Berkeley, California, 1988.
1605. Allen, E. W. & Bailey, J., Seismic isolation retrofitting of the Salt Lake City and County Buildings using base isolation. Proc. 9th World Conf. on Earthquake Engineering, Tokyo-Kyoto, Japan, 1988, pp. 633–38.
1606. Kelly, J. M., The application of seismic isolation for the retrofit of historic buildings. Proc. Colloquium on Seismic Stabilization of Historic Buildings, The Getty Conservation Institute, Quito, Ecuador, 1993.

1607. Elsesser, E. *et al.*, Repair of five historic buildings damaged by the Loma Prieta Earthquake. Proc. Conf. on the seismic retrofitting of historical buildings. Western Region of the National Park Service, Federal Emergency Management Agency, San Francisco, California, 1991, pp. 4-1–4-40.

1608. Sparacio, R. *et al.*, Retrofitting of the St. Pietro Church at Frigento by use of seismic isolation, as a first experience in view of applications to artistic and historical buildings in Italy. Proc. Int. Post-SMiRT Conference Seminar, Napoli, Italy, 1993, pp. 287–93.

1609. Kelly, J. M. *et al.*, Mechanisms of energy absorption in special devices for use in earthquake resistant structures. *Bull. NZ Nat. Soc. Earthq. Engng*, **5**, No. 3 (1972), 63–88.

1610. Robinson, W. H. & Greenbank, L. R., An extrusion energy absorber suitable for the protection of structures during an earthquake. *J. Earthquake Engrg Struct. Dyn.*, **4**, No. 3 (1976) 251–59.

1611. Skinner, R. I. *et al.*, *An introduction to seismic isolation.* John Wiley & Sons Ltd, West Sussex, England, 1993.

1612. Bridgestone Corporation, Rubber technology for seismic isolation. Proc. 11th Int. Conf. on Struct. Mech. in Reactor Technology, SMiRT, Tokyo, Japan, 1991, pp. 45–56.

1613. Kelly, J. M. & Chalhoub, M. S., Earthquake simulation testing of a combined sliding bearing and rubber bearing isolation system. Report No. UCB/EERC-92/11, Earthquake Engineering Rssearch Center, University of California, Berkeley, California, 1992.

1614. Mokha, A. *et al.*, Teflon bearings in aseismic base isolation: experimental studies and mathematical modeling. Report No. NCEER-88-0038, National Center for Earthquake Engineering, SUNY, Buffalo, NY, 1988.

1615. Wilson, E. L., An efficient computational method for base isolation and energy dissipation analysis. Proc. ATC17-1 seminar on seismic isolation, passive energy dissipation and active control, San Francisco, California, 1993, **1**, 365–76.

1616. Kelly, J. M., Base isolation: linear theory and design. *Earthquake Spectra*, **6**, No. 2 (1990) 223–44.

1617. Winters, C. W. & Constantinou, M. C., Evaluation of static and response spectrum analysis. Report No. NCEER-93-0004, National Center for Earthquake Engineering Research, SUNY, Buffalo, NY, 1993.

1618. Nagarajaiah, S. *et al.*, Nonlinear dynamic analysis of three-dimensional base isolated structures 3D-BASIS. Report No. NCEER-89-0019, National Center for Earthquake Engineering Research, SUNY, Buffalo, NY, 1989.

1619. Wilson, E. L., SADSAP-Sta. and Dyn. Struct. Anal. Program with Nonlinear Damping and Isolation Elements, El Cerrito, Califormia, 1992.

1620. Bangash, M. Y. H., *Impact and explosion.* Blackwell Science, Oxford, 1993.

APPENDIX 1

Relevant element library

Appendix IA: Material and structural matrices

Table AIA.1. [D] — Variable Young's modulus and constant Poisson's ratio

$D_{11} = \dfrac{E_1(E')^3 - E_{cr}}{\nu''}$	$D_{12} = \dfrac{\nu E_1 E_2 (E')^2 + E_{cr}}{\nu''}$	$D_{13} = \dfrac{\nu E_1 E_3 (E')^2 + E_{cr}}{\nu''}$	$D_{14} = 0$	$D_{15} = 0$	$D_{16} = 0$
	$D_{22} = \dfrac{E_2 E_3 (E')^2 + E_{cr}}{\nu''}$	$D_{23} = \dfrac{\nu E_2 E_3 (E')^2 + E_{cr}}{\nu''}$	$D_{24} = 0$	$D_{25} = 0$	$D_{26} = 0$
		$D_{33} = \dfrac{E_3(E')^3 - E_{cr}}{\nu''}$	$D_{34} = 0$	$D_{35} = 0$	$D_{36} = 0$
			$D_{44} = G_{12}$	$D_{45} = 0$	$D_{46} = 0$
				$D_{55} = G_{23}$	$D_{56} = 0$
					$D_{66} = G_{31}$

$$E_{cr} = \nu^2 E_1 E_2 E_3 E'$$

$$E' = (E_1 + E_2 + E_3)/3$$

$$\nu'' = (E')^3 - 2E_1 E_2 E_3 \nu^2 - E'\nu^2(E_1 E_2 + E_1 E_3 + E_2 E_3)$$

$$G_{12} = E_{12}/2(1 + \nu)$$

$$E_{12} = (E_1 + E_2)/2$$

$$G_{23} = E_{23}/2(1 + \nu)$$

$$E_{23} = (E_2 + E_3)/2$$

$$G_{31} = E_{31}/2(1 + \nu)$$

$$E_{31} = (E_3 + E_1)/2$$

Table AIA.2. [D] — Variable Young's modulus and Poisson's ratio

$$\begin{bmatrix} D_{11} = \dfrac{(1-\nu_{23}\nu_{32})}{\bar{\nu}}E_1 & D_{12} = \dfrac{(\nu_{12}+\nu_{12}\nu_{32})}{\bar{\nu}}E_2 & D_{13} = \dfrac{(\nu_{13}+\nu_{12}\nu_{23})}{\bar{\nu}}E_3 & D_{14}=0 & D_{15}=0 & D_{16}=0 \\ D_{21} = \dfrac{(\nu_{21}+\nu_{23}\nu_{31})}{\bar{\nu}}E_1 & D_{22} = \dfrac{(1-\nu_{13}\nu_{31})}{\bar{\nu}}E_2 & D_{23} = \dfrac{(\nu_{23}+\nu_{13}\nu_{21})}{\bar{\nu}}E_3 & D_{24}=0 & D_{25}=0 & D_{26}=0 \\ D_{31} = \dfrac{(\nu_{31}+\nu_{21}\nu_{32})}{\bar{\nu}}E_1 & D_{32} = \dfrac{(\nu_{32}+\nu_{12}\nu_{31})}{\bar{\nu}}E_2 & D_{33} = \dfrac{(1-\nu_{12}\nu_{21})}{\bar{\nu}}E_3 & D_{34}=0 & D_{35}=0 & D_{36}=0 \\ D_{41}=0 & D_{42}=0 & D_{43}=0 & D_{44} & D_{45}=0 & D_{46}=0 \\ D_{51}=0 & D_{52}=0 & D_{53}=0 & D_{54}=0 & D_{55} & D_{56}=0 \\ D_{61}=0 & D_{62}=0 & D_{63}=0 & D_{64}=0 & D_{65}=0 & D_{66} \end{bmatrix}$$

$$\bar{\nu} = 1 - \nu_{12}\nu_{21} - \nu_{13}\nu_{31} - \nu_{23}\nu_{32} - \nu_{12}\nu_{23}\nu_{31} - \nu_{21}\nu_{13}\nu_{32}$$

Table AIA.2. continued

Due to symmetry of compliances, the following relations can be written:

$$E_1\nu_{21} = E_2\nu_{12} \qquad D_{55} = G_{23}$$
$$E_2\nu_{32} = E_3\nu_{23} \qquad D_{66} = G_{13}$$
$$E_3\nu_{13} = E_1\nu_{31}$$

The values of G_{12}, G_{23} and G_{13} are calculated in terms of modulus of elasticity and Poisson's ratio as follows:

$$G_{12} = \frac{1}{2}\left[\frac{E_1}{2(1+\nu_{12})} + \frac{E_2}{2(1+\nu_{21})}\right] = \frac{1}{2}\left[\frac{E_1}{2(1+\nu_{12})} + \frac{E_1}{2\left(\dfrac{E_1}{E_2} + \nu_{12}\right)}\right]$$

$$G_{23} = \frac{1}{2}\left[\frac{E_2}{2(1+\nu_{23})} + \frac{E_3}{2(1+\nu_{32})}\right] = \frac{1}{2}\left[\frac{E_2}{2(1+\nu_{23})} + \frac{E_2}{2\left(\dfrac{E_2}{E_3} + \nu_{23}\right)}\right]$$

$$G_{13} = \frac{1}{2}\left[\frac{E_3}{2(1+\nu_{31})} + \frac{E_1}{2(1+\nu_{13})}\right] = \left[\frac{E_3}{2(1+\nu_{31})} + \frac{E_2}{2\left(\dfrac{E_3}{E_1} + \nu_{31}\right)}\right]$$

For isotropic cases;

$$E_1 = E_2 = E_3 = E$$
$$\nu_{12} = \nu_{13} = \nu_{23} = \nu_{21} = \nu_{31} = \nu_{32} = \nu$$

[D] — Constant Young's modulus and Poisson's ratio

$$[D] = \frac{E}{(1+\nu)(1-2\nu)}\begin{bmatrix} 1-\nu & \nu & \nu & 0 & 0 & 0 \\ \nu & 1-\nu & \nu & 0 & 0 & 0 \\ \nu & \nu & 1-\nu & 0 & 0 & 0 \\ 0 & 0 & 0 & \dfrac{1-2\nu}{2} & 0 & 0 \\ 0 & 0 & 0 & 0 & \dfrac{1-2\nu}{2} & 0 \\ 0 & 0 & 0 & 0 & 0 & \dfrac{1-2\nu}{2} \end{bmatrix}$$

Bulk and shear moduli

$$[D] = \begin{bmatrix} K+\frac{4}{3}G & K-\frac{2}{3}G & K-\frac{2}{3}G & 0 & 0 & 0 \\ K-\frac{2}{3}G & K+\frac{4}{3}G & K-\frac{2}{3}G & 0 & 0 & 0 \\ K-\frac{2}{3}G & K-\frac{2}{3}G & K+\frac{4}{3}G & 0 & 0 & 0 \\ 0 & 0 & 0 & G & 0 & 0 \\ 0 & 0 & 0 & 0 & G & 0 \\ 0 & 0 & 0 & 0 & 0 & G \end{bmatrix}$$

Table AIA.2. continued

For plane stress

$$[D] = \frac{E}{1-\nu^2}\begin{bmatrix} 1 & \nu & 0 \\ \nu & 1 & 0 \\ 0 & 0 & \dfrac{1-\nu}{2} \end{bmatrix}$$

For plane strain

$$[D] = \frac{E(1-\nu)}{(1+\nu)(1-2\nu)}\begin{bmatrix} 1 & \dfrac{\nu}{1-\nu} & 0 \\ \dfrac{\nu}{1-\nu} & 1 & 0 \\ 0 & 0 & \dfrac{1-2\nu}{2(1-\nu)} \end{bmatrix}$$

For axisymmetric cases

$$[D] = \frac{E}{(1+\nu)(1-2\nu)}\begin{bmatrix} 1-\nu & \nu & \nu & 0 \\ \nu & 1-\nu & \nu & 0 \\ \nu & \nu & 1-\nu & 0 \\ 0 & 0 & 0 & \dfrac{1-2\nu}{2} \end{bmatrix}$$

Table AIA.3. Two-dimensional elastic beam (courtesy STRUCOM, London)

Element matrices and load vectors

The element stiffness matrix in element co-ordinates is:

$$[K_1] = \begin{bmatrix} \dfrac{AE}{L} & 0 & 0 & -\dfrac{AE}{L} & 0 & 0 \\ 0 & \dfrac{12EI}{L^3(1+\phi')} & \dfrac{6EI}{L^2(1+\phi')} & 0 & -\dfrac{12EI}{L^3(1+\phi')} & \dfrac{6EI}{L(1+\phi')} \\ 0 & \dfrac{6EI}{L^2(1+\phi')} & \dfrac{EI(4+\phi')}{L(1+\phi')} & 0 & -\dfrac{6EI}{L^2(1+\phi')} & \dfrac{EI(2-\phi')}{L(1+\phi')} \\ -\dfrac{AE}{L} & 0 & 0 & \dfrac{AE}{L} & 0 & 0 \\ 0 & -\dfrac{12EI}{L^3(1+\phi')} & \dfrac{6EI}{L^2(1+\phi')} & 0 & \dfrac{12EI}{L^3(1+\phi')} & -\dfrac{6EI}{L^2(1+\phi')} \\ 0 & \dfrac{6EI}{L^2(1+\phi')} & \dfrac{EI(2-\phi')}{L(1+\phi')} & 0 & -\dfrac{6EI}{L^2(1+\phi')} & \dfrac{EI(4+\phi')}{L(1+\phi')} \end{bmatrix}$$

where

A = cross-sectional area

E = Young's modulus

L = element length

I = moment of inertia

$\phi' = \dfrac{12EI}{GA^{s}L^{2}}$

G = shear modulus

$A^{s} = \dfrac{A}{F^{s}}$ = shear area

F^{s} = shear deflection constant

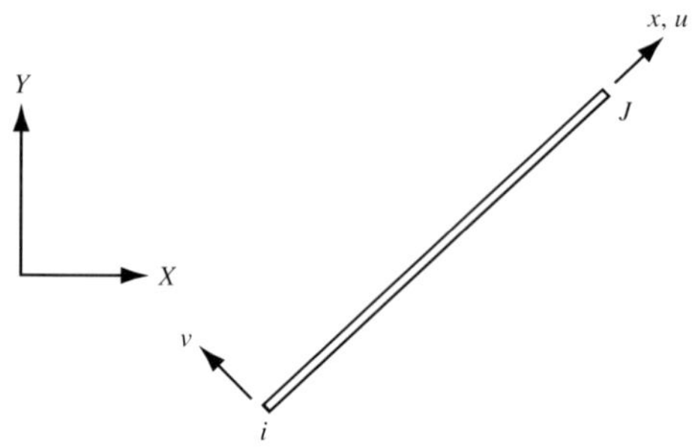

Fig. AIA.1. Two-dimensional beam element centroidal axis

For uniform lateral pressure

$$P_1 = P_4 = 0$$

$$P_2 = P_5 = -\frac{PL}{2}$$

$$P_3 = -P_6 = -\frac{PL^2}{12}$$

P = uniform applied pressure (units = force/length)

Stress calculations

The centroidal stress at end i is:

$$\sigma_i^{\text{dir}} = \frac{F_{x,i}}{A}$$

where

σ_i^{dir} = centroidal stress

$F_{x,i}$ = axial force

The bending stress is

$$\sigma_i^{\text{bnd}} = \frac{M_i t}{2I}$$

where

σ_i^{bnd} = bending stress at end i

M_i = moment at end i

t = thickness of beam in element z direction

Table AIA.4. Three-dimensional elastic beam (courtesy STRUCOM, London)

Element matrices and load vectors

All element matrices and load vectors are generated in the element co-ordinate system and must subsequently then be converted to the global co-ordinate system. The element stiffness matrix is:

$$[K_1] = \frac{AE}{L}\begin{bmatrix} C_1 & 0 & 0 & -C_1 & 0 & 0 \\ 0 & 0 & 0 & 0 & 0 & 0 \\ 0 & 0 & 0 & 0 & 0 & 0 \\ -C_1 & 0 & 0 & C_1 & 0 & 0 \\ 0 & 0 & 0 & 0 & 0 & 0 \\ 0 & 0 & 0 & 0 & 0 & 0 \end{bmatrix}$$

where

A = element cross-sectional area

E = Young's modulus

L = element length

C_1 = value given in the table below.

The element mass matrix is the same as the element stress stiffness matrix:

$$[S_1] = \frac{F}{L}\begin{bmatrix} 0 & 0 & 0 & 0 & 0 & 0 \\ 0 & C_2 & 0 & 0 & -C_2 & 0 \\ 0 & 0 & C_2 & 0 & 0 & -C_2 \\ 0 & 0 & 0 & 0 & 0 & 0 \\ 0 & -C_2 & 0 & 0 & C_2 & 0 \\ 0 & 0 & -C_2 & 0 & 0 & C_2 \end{bmatrix}$$

Value of stiffness coefficient (C_1)

Previous iteration resulted in a tensile stress	Previous iteration resulted in a compressive stress
1.0	0.0
1.0	1.0×10^{-6}
0.0	1.0
1.0×10^{-6}	1.0

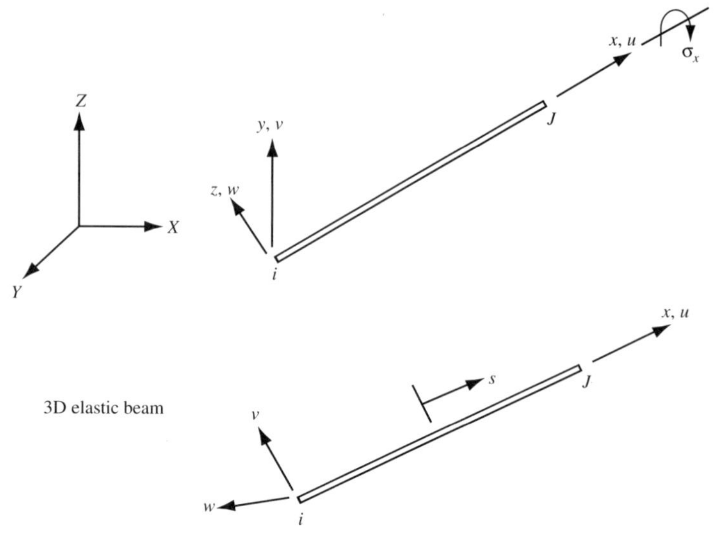

Fig. AIA.2. Three-dimensional elastic beam.

Table AIA.5. Stiffness and mass matrices (courtesy STRUCOM, London)
Orders of degrees of freedom

The stiffness matrix in element co-ordinates is:

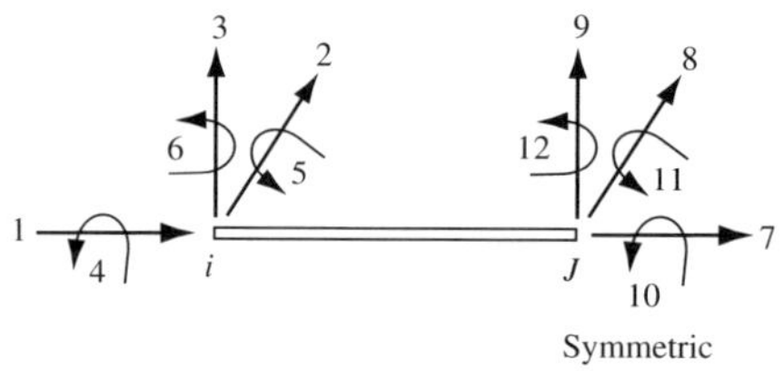

Fig. AIA.3. Order of degrees of freedom.

$$[K_1] = \begin{bmatrix} AE/L & & & & & & & & & & & \\ 0 & a_z & & & & & & & & & & \\ 0 & 0 & a_y & & & & & & & & & \\ 0 & 0 & 0 & GJ/L & & & & & \text{Symmetric} & & & \\ 0 & 0 & d_y & 0 & e_y & & & & & & & \\ 0 & c_z & 0 & 0 & 0 & e_z & & & & & & \\ -AE/L & 0 & 0 & 0 & 0 & 0 & AE/L & & & & & \\ 0 & b_z & 0 & 0 & 0 & d_z & 0 & a_z & & & & \\ 0 & 0 & b_y & 0 & c_y & 0 & 0 & 0 & a_y & & & \\ 0 & 0 & 0 & -GJ/L & 0 & 0 & 0 & 0 & 0 & GJ/L & & \\ 0 & 0 & d_y & 0 & f_y & 0 & 0 & 0 & c_z & 0 & e_y & \\ 0 & c_z & 0 & 0 & 0 & f_z & 0 & d_y & 0 & 0 & 0 & e_z \end{bmatrix}$$

where

$A =$ cross-sectional area

$E =$ Young's modulus

$L =$ element length

$G =$ shear modulus

$r_y = \sqrt{\dfrac{I_{yy}}{A}} =$ radius of gyration

$r_z = \sqrt{\dfrac{I_{zz}}{A}} =$ radius of gyration

$$[M_\mathrm{l}] = M_\mathrm{t}$$

$$\begin{bmatrix}
1/3 & & & & & & & & & & & \\
0 & A_z & & & & & & & & & & \\
0 & 0 & A_y & & & & & & & & & \\
0 & 0 & 0 & J_x/3A & & & & & \text{Symmetric} & & & \\
0 & 0 & -C_y & 0 & E_y & & & & & & & \\
0 & C_z & 0 & 0 & 0 & E_z & & & & & & \\
1/6 & 0 & 0 & 0 & 0 & 0 & 1/3 & & & & & \\
0 & B_z & 0 & 0 & 0 & D_z & 0 & A_z & & & & \\
0 & 0 & B_y & 0 & -D_y & 0 & 0 & 0 & A_y & & & \\
0 & 0 & 0 & J_x/6A & 0 & 0 & 0 & 0 & 0 & J_x/3A & & \\
0 & 0 & D_y & 0 & F_y & 0 & 0 & 0 & C_y & 0 & E_y & \\
0 & -D_z & 0 & 0 & 0 & F_z & 0 & -C_z & 0 & 0 & 0 & E_z
\end{bmatrix}$$

where

$M_\mathrm{t} = (\rho A + m)L(1 - \varepsilon^{\mathrm{in}})$

$\rho =$ density

$m =$ added mass

$\varepsilon^{\mathrm{in}} =$ prestrain

$A_z = A(r_z, \phi'_y)$

$A_y = A(r_y, \phi'_z)$

$B_z = B(r_z, \phi'_y)$

$\vdots$

$F_z = F(r_z, \phi'_y)$

$F_y = F(r, \phi'_z)$

and where

$$A(r,\phi') = \frac{13/35 + 7/10\phi' + 1/3\phi'^2 + 6/5(r/L)^2}{(1+\phi')^2}$$

$$B(r,\phi') = \frac{9/70 + 3/10\phi' + 1/6\phi'^2 - 6/5(r/L)^2}{(1+\phi')^2}$$

$$C(r,\phi') = \frac{(11/210 + 11/120\phi' + 1/24\phi'^2 + (1/10 - 1/2\phi')(r/L)^2)L}{(1+\phi')^2}$$

$$D(r,\phi') = \frac{(13/420 + 3/40\phi' + 1/24\phi'^2 - (1/10 - 1/2\phi')(r/L)^2)L}{(1+\phi')^2}$$

$$E(r,\phi') = \frac{(1/105 + 1/60\phi' + 1/120\phi'^2 + (2/15 + 1/6\phi' + 1/3\phi'^2)(r/L)^2)L^2}{(1+\phi')^2}$$

$$F(r,\phi') = -\frac{(1/140 + 1/60\phi' + 1/120\phi'^2 + (1/30 + 1/6\phi' - 1/6\phi'^2)(r/L)^2)L^2}{(1+\phi')^2}$$

$$J = \text{torsional moment of inertia} = \left\{\begin{matrix} J_x & \text{if } I_x = 0 \\ I_x & \text{if } I_x \neq 0 \end{matrix}\right\}$$

$$I_x = \text{input as } IXX$$

$$J_x = \text{polar moment of inertia} = I_y + I_z$$

The element mass matrix in element co-ordinates is:

$$[M_1] = (\rho A + m)L(1 - \varepsilon^{\text{in}})$$

$$\times \begin{bmatrix} 1/3 & 0 & 0 & 1/6 & 0 & 0 \\ 0 & A(r,\phi') & C(r,\phi') & 0 & B(r,\phi') & -D(r,\phi') \\ 0 & C(r,\phi') & E(r,\phi') & 0 & D(r,\phi') & -F(r,\phi') \\ 1/6 & 0 & 0 & 1/3 & 0 & 0 \\ 0 & B(r,\phi') & D(r,\phi') & 0 & A(r,\phi') & -C(r,\phi') \\ 0 & -D(r,\phi') & -F(r,\phi') & 0 & -C(r,\phi') & E(r,\phi') \end{bmatrix}$$

where

$$\rho = \text{density}$$

$$m = \text{added } m$$

$$\varepsilon^{\text{in}} = \text{prestrain}$$

$$A(r,\phi') = \frac{13/35 + 7/10\phi' + 1/3\phi'^2 + 6/5(r/L)^2}{(1+\phi')^2}$$

$$B(r,\phi') = \frac{9/70 + 3/10\phi' + 1/6\phi'^2 - 6/5(r/L)^2}{(1+\phi')^2}$$

$$C(r,\phi') = \frac{(11/210 + 11/120\phi' + 1/24\phi'^2 + (1/10 - 1/2\phi')(r/L)^2)L}{(1+\phi')^2}$$

$$D(r,\phi') = \frac{(13/420 + 3/40\phi' + 1/24\phi'^2 - (1/10 - 1/2\phi')(r/L)^2)L}{(1+\phi')^2}$$

$$E(r,\phi') = \frac{(1/105 + 1/60\phi' + 1/120\phi'^2 + (2/15 + 1/6\phi' + 1/3)(r/L)^2)L^2}{(1+\phi')^2}$$

$$F(r,\phi') = \frac{(1/140 + 1/60\phi' + 1/120\phi'^2 + (1/30 + 1/6\phi' + 1/6\phi'^2)(r/L)^2)L^2}{(1+\phi')^2}$$

$$r = \sqrt{\frac{I}{A}} = \text{radius of gyration}$$

The element pressure load vector in element coordinates is:

$$\{F_1^{pr}\} = \lfloor P_1 \quad P_2 \quad P_3 \quad P_4 \quad P_5 \quad P_6 \rfloor^T$$

Value of stress stiffness coefficient (C_2)

Previous iteration resulted in a tensile stress	Previous iteration resulted in a compressive stress
1.0	0.0
1.0	$\frac{AE}{F \times 10^6}$
0.0	1.0
$\frac{AE}{F \times 10^6}$	1.0

$F = $ for the first iteration: $AE\varepsilon^{in}$

for all subsequent iterations: the axial force in the element as computed in the previous stress pass of the element (output quantity) FORC

C_2 = value given in the table above.

The matrix for the tension-only or compression-only spar is given by:

$$[M_l] = \frac{M_t}{2}\begin{bmatrix} 1 & & & & & & & & & & & \\ 0 & 1 & & & & & & & & & & \\ 0 & 0 & 1 & & & & & & & & & \\ 0 & 0 & 0 & 0 & & & & & & & & \\ 0 & 0 & 0 & 0 & 0 & & & \text{Symmetric} & & & & \\ 0 & 0 & 0 & 0 & 0 & 0 & & & & & & \\ 0 & 0 & 0 & 0 & 0 & 0 & 1 & & & & & \\ 0 & 0 & 0 & 0 & 0 & 0 & 0 & 1 & & & & \\ 0 & 0 & 0 & 0 & 0 & 0 & 0 & 0 & 1 & & & \\ 0 & 0 & 0 & 0 & 0 & 0 & 0 & 0 & 0 & 0 & & \\ 0 & 0 & 0 & 0 & 0 & 0 & 0 & 0 & 0 & 0 & 0 & \\ 0 & 0 & 0 & 0 & 0 & 0 & 0 & 0 & 0 & 0 & 0 & 0 \end{bmatrix}$$

Appendix IB: Element types, shape functions, derivatives, stiffness matrices

Table AIB.1. [K] — Shear and torsion included for line element

$$[D] = \begin{bmatrix}
\frac{EA}{L} & & & & & & & & & & & \\
0 & \frac{12EI_\xi}{L^3(1+\bar{\tau}_\eta)} & & & & & & & & & & \\
0 & 0 & \frac{12EI_\xi}{L^3(1+\bar{\tau}_\xi)} & & & & & & & & & \\
0 & 0 & 0 & \frac{GJ}{L} & & & & & & & & \\
0 & 0 & \frac{-6EI_\eta}{L^2(1+\tau_\xi)} & 0 & \frac{(4+\bar{\tau}_\xi)EI_\eta}{L(1+\bar{\tau}_\eta)} & & & & & & & \\
0 & \frac{6EI_\xi}{L^2(1+\bar{\tau}_\eta)} & 0 & 0 & 0 & \frac{(4+\bar{\tau}_\eta)EI_\xi}{L(1+\bar{\tau}_\eta)} & & & & & & \\
\frac{-EA}{L} & 0 & 0 & 0 & 0 & 0 & \frac{AE}{L} & & & & & \\
0 & \frac{-12EI_\xi}{L^3(1+\bar{\tau}_\eta)} & 0 & 0 & 0 & \frac{-6EI_\xi}{L^2(1+\bar{\tau}_\eta)} & 0 & \frac{12EI_\xi}{L^3(1+\bar{\tau}_\eta)} & & & & \\
0 & 0 & \frac{-12EI_\eta}{L^3(1+\bar{\tau}_\xi)} & 0 & \frac{6EI_\eta}{L^2(1+\bar{\tau}_\xi)} & 0 & 0 & 0 & \frac{12EI_\eta}{L^3(1+\bar{\tau}_\xi)} & & & \\
0 & 0 & 0 & \frac{-GJ}{L} & 0 & 0 & 0 & 0 & 0 & \frac{GJ}{L} & & \\
0 & 0 & \frac{-6EI_\eta}{L^2(1+\bar{\tau}_\xi)} & 0 & \frac{(2-\bar{\tau}_\xi)EI_\eta}{L(1+\bar{\tau}_\xi)} & 0 & 0 & 0 & \frac{6EI_\eta}{L^2(1+\bar{\tau}_\xi)} & 0 & \frac{(4+\bar{\tau}_\xi)EI_\eta}{L(1+\bar{\tau}_\xi)} & \\
0 & \frac{6EI_\xi}{L^2(1+\bar{\tau}_\eta)} & 0 & 0 & 0 & \frac{(2-\bar{\tau}_\eta)EI_\xi}{L(1+\bar{\tau}_\eta)} & 0 & \frac{-6EI_\xi}{L^2(1+\bar{\tau}_\eta)} & 0 & 0 & 0 & \frac{(4+\bar{\tau}_\eta)EI_\xi}{L(1+\bar{\tau}_\eta)}
\end{bmatrix}$$

where

$$\bar{\tau}_\eta = \frac{12EI_\xi}{GA_{s\eta}L^2} = 24(1+\nu)\frac{A}{A_{s\eta}}\left(\frac{T_\xi}{L}\right)^2 \qquad \bar{\tau}_\xi = \frac{12EI_\eta}{GA_{s\xi}L^2} = 24(1+\nu)\frac{A}{A_{s\xi}}\left(\frac{T_\eta}{L}\right)^2$$

A_s = shear area T = torsional moment of inertia ρ = density ξ, η local axes are parallel to Z and Y axes

Table AIB.2. Chain rule

$$\begin{bmatrix} \frac{\partial u}{\partial X} \\ \frac{\partial u}{\partial Y} \\ \frac{\partial u}{\partial Z} \\ \frac{\partial v}{\partial X} \\ \frac{\partial v}{\partial Y} \\ \frac{\partial v}{\partial Z} \\ \frac{\partial w}{\partial X} \\ \frac{\partial w}{\partial Y} \\ \frac{\partial w}{\partial Z} \end{bmatrix} = \frac{1}{\det J} \begin{bmatrix} C_{11} & C_{12} & C_{13} & 0 & 0 & 0 & 0 & 0 & 0 \\ C_{21} & C_{22} & C_{23} & 0 & 0 & 0 & 0 & 0 & 0 \\ C_{31} & C_{32} & C_{33} & 0 & 0 & 0 & 0 & 0 & 0 \\ 0 & 0 & 0 & C_{11} & C_{12} & C_{13} & 0 & 0 & 0 \\ 0 & 0 & 0 & C_{21} & C_{22} & C_{23} & 0 & 0 & 0 \\ 0 & 0 & 0 & C_{31} & C_{32} & C_{33} & 0 & 0 & 0 \\ 0 & 0 & 0 & 0 & 0 & 0 & C_{11} & C_{12} & C_{13} \\ 0 & 0 & 0 & 0 & 0 & 0 & C_{21} & C_{22} & C_{23} \\ 0 & 0 & 0 & 0 & 0 & 0 & C_{31} & C_{32} & C_{33} \end{bmatrix} \begin{bmatrix} \frac{\partial u}{\partial \xi} \\ \frac{\partial u}{\partial \eta} \\ \frac{\partial u}{\partial \zeta} \\ \frac{\partial v}{\partial \xi} \\ \frac{\partial v}{\partial \eta} \\ \frac{\partial v}{\partial \zeta} \\ \frac{\partial w}{\partial \xi} \\ \frac{\partial w}{\partial \eta} \\ \frac{\partial w}{\partial \zeta} \end{bmatrix}$$

where

$$C_{11} = \frac{\partial Y}{\partial \xi}\frac{\partial Z}{\partial \xi} - \frac{\partial Z}{\partial \eta}\frac{\partial Y}{\partial \xi} \qquad C_{12} = \frac{\partial Z}{\partial \xi}\frac{\partial Y}{\partial \zeta} - \frac{\partial Y}{\partial \xi}\frac{\partial Z}{\partial \zeta}$$

$$C_{13} = \frac{\partial Y}{\partial \xi}\frac{\partial Z}{\partial \zeta} - \frac{\partial Z}{\partial \xi}\frac{\partial Y}{\partial \zeta} \qquad C_{21} = \frac{\partial Z}{\partial \xi}\frac{\partial X}{\partial \zeta} - \frac{\partial X}{\partial \xi}\frac{\partial Z}{\partial \zeta}$$

$$C_{22} = \frac{\partial X}{\partial \xi}\frac{\partial Z}{\partial \zeta} - \frac{\partial Z}{\partial \xi}\frac{\partial X}{\partial \zeta} \qquad C_{23} = \frac{\partial Z}{\partial \xi}\frac{\partial X}{\partial \eta} - \frac{\partial X}{\partial \xi}\frac{\partial Z}{\partial \eta}$$

$$C_{31} = \frac{\partial X}{\partial \eta}\frac{\partial Y}{\partial \zeta} - \frac{\partial Y}{\partial \eta}\frac{\partial X}{\partial \zeta} \qquad C_{32} = \frac{\partial Y}{\partial \xi}\frac{\partial X}{\partial \zeta} - \frac{\partial X}{\partial \xi}\frac{\partial Y}{\partial \zeta}$$

$$C_{33} = \frac{\partial X}{\partial \xi}\frac{\partial Y}{\partial \eta} - \frac{\partial Y}{\partial \xi}\frac{\partial X}{\partial \eta}$$

$\det [J]$ = the determinant of the Jacobian matrix.

Table AIB.3. Solid isoparametric elements
Eight-noded solid element

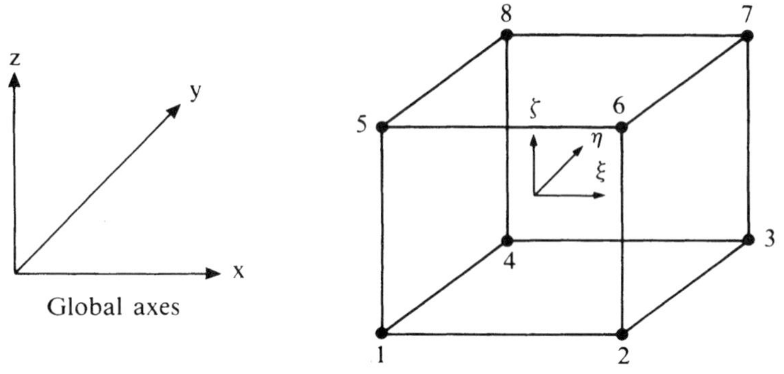

Node i	Shape functions $N_i(\xi, \eta, \zeta)$	Derivatives		
		$\dfrac{\partial N_i}{\partial \xi}$	$\dfrac{\partial N_i}{\partial \eta}$	$\dfrac{\partial N_i}{\partial \zeta}$
1	$\frac{1}{8}(1-\xi)(1-\eta)(1-\zeta)$	$-\frac{1}{8}(1-\eta)(1-\zeta)$	$-\frac{1}{8}(1-\xi)(1-\zeta)$	$-\frac{1}{8}(1-\eta)(1-\xi)$
2	$\frac{1}{8}(1+\xi)(1-\eta)(1-\zeta)$	$\frac{1}{8}(1-\eta)(1-\zeta)$	$-\frac{1}{8}(1+\xi)(1-\zeta)$	$-\frac{1}{8}(1+\eta)(1-\eta)$
3	$\frac{1}{8}(1+\xi)(1+\eta)(1-\zeta)$	$\frac{1}{8}(1+\eta)(1-\zeta)$	$\frac{1}{8}(1+\xi)(1-\zeta)$	$-\frac{1}{8}(1+\eta)(1+\eta)$
4	$\frac{1}{8}(1-\xi)(1+\eta)(1-\zeta)$	$-\frac{1}{8}(1+\eta)(1-\zeta)$	$\frac{1}{8}(1-\xi)(1-\zeta)$	$-\frac{1}{8}(1-\eta)(1+\eta)$
5	$\frac{1}{8}(1-\xi)(1-\eta)(1+\zeta)$	$-\frac{1}{8}(1-\eta)(1+\zeta)$	$-\frac{1}{8}(1-\xi)(1+\zeta)$	$\frac{1}{8}(1-\eta)(1-\eta)$
6	$\frac{1}{8}(1+\xi)(1-\eta)(1+\zeta)$	$\frac{1}{8}(1-\eta)(1+\zeta)$	$-\frac{1}{8}(1+\xi)(1+\zeta)$	$\frac{1}{8}(1+\eta)(1-\eta)$
7	$\frac{1}{8}(1+\xi)(1+\eta)(1+\zeta)$	$\frac{1}{8}(1+\eta)(1+\zeta)$	$\frac{1}{8}(1+\xi)(1+\zeta)$	$\frac{1}{8}(1+\eta)(1+\eta)$
8	$\frac{1}{8}(1-\xi)(1+\eta)(1+\zeta)$	$-\frac{1}{8}(1+\eta)(1+\zeta)$	$\frac{1}{8}(1-\xi)(1+\zeta)$	$\frac{1}{8}(1-\eta)(1+\eta)$

Table AIB.4. Twenty-noded solid element

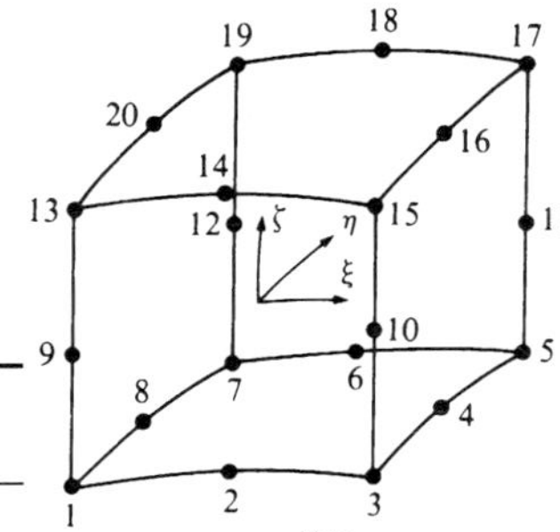

Node i	Shape functions $N_i(\xi,\eta,\zeta)$	Derivatives $\dfrac{\partial N_i}{\partial \xi}$	$\dfrac{\partial N_i}{\partial \eta}$	$\dfrac{\partial N_i}{\partial \zeta}$
1	$\frac{1}{8}(1-\xi)(1-\eta)(1-\zeta)(-\xi-\eta-\zeta-2)$	$\frac{1}{8}(1-\eta)(1-\zeta)(2\xi+\eta+\zeta+1)$	$\frac{1}{8}(1-\xi)(1-\zeta)(2\eta+\xi+\zeta+1)$	$\frac{1}{8}(1-\xi)(1-\eta)(2\zeta+\eta+\xi+1)$
2	$\frac{1}{4}(1-\xi^2)(1-\eta)(1-\zeta)$	$-\frac{1}{2}(1-\eta)(1-\zeta)\xi$	$-\frac{1}{4}(1-\xi^2)(1-\zeta)$	$-\frac{1}{4}(1-\xi^2)(1-\eta)$
3	$\frac{1}{8}(1+\xi)(1-\eta)(1-\zeta)(\xi-\eta-\zeta-2)$	$\frac{1}{8}(1-\eta)(1-\zeta)(2\xi-\eta-\zeta-1)$	$\frac{1}{8}(1+\xi)(1-\zeta)(2\eta-\xi+\zeta+1)$	$\frac{1}{8}(1+\xi)(1-\eta)(2\zeta-\xi+\eta+1)$
4	$\frac{1}{4}(1+\xi)(1-\eta^2)(1-\zeta)$	$\frac{1}{4}(1-\eta^2)(1-\zeta)$	$-\frac{1}{2}(1+\xi)(1-\zeta)\eta$	$-\frac{1}{4}(1-\eta^2)(1+\xi)$
5	$\frac{1}{8}(1+\xi)(1+\eta)(1-\zeta)(\xi+\eta-\zeta-2)$	$\frac{1}{8}(1+\eta)(1-\zeta)(2\xi+\eta-\zeta-1)$	$\frac{1}{8}(1+\xi)(1-\zeta)(2\eta+\xi-\zeta-1)$	$\frac{1}{8}(1+\xi)(1+\eta)(2\zeta-\xi-\eta+1)$
6	$\frac{1}{4}(1-\xi^2)(1+\eta)(1-\zeta)$	$-\frac{1}{2}(1+\eta)(1-\zeta)\xi$	$\frac{1}{4}(1-\xi^2)(1-\zeta^2)$	$-\frac{1}{4}(1-\xi^2)(1+\eta)$
7	$\frac{1}{8}(1-\xi)(1+\eta)(1-\zeta)(-\xi+\eta-\zeta-2)$	$\frac{1}{8}(1+\eta)(1-\zeta)(2\xi-\eta+\zeta+1)$	$\frac{1}{8}(1-\xi)(1-\zeta)(2\eta-\xi-\zeta-1)$	$\frac{1}{8}(1-\xi)(1+\eta)(2\zeta-\eta+\xi+1)$
8	$\frac{1}{4}(1-\xi)(1-\eta^2)(1-\zeta)$	$-\frac{1}{4}(1-\eta^2)(1-\zeta)$	$-\frac{1}{2}(1-\xi)(1-\zeta)$	$-\frac{1}{4}(1-\eta^2)(1-\xi)$
9	$\frac{1}{4}(1-\xi)(1-\eta)(1-\zeta^2)$	$-\frac{1}{4}(1-\zeta^2)(1-\eta)$	$-\frac{1}{4}(1-\xi)(1-\zeta^2)$	$-\frac{1}{2}(1-\xi)(1-\eta)\zeta$
10	$\frac{1}{4}(1+\xi)(1-\eta)(1-\zeta^2)$	$\frac{1}{4}(1-\eta)(1-\zeta^2)$	$-\frac{1}{4}(1+\xi)(1-\zeta^2)$	$-\frac{1}{2}(1+\xi)(1-\eta)\zeta$
11	$\frac{1}{4}(1+\xi)(1+\eta)(1-\zeta^2)$	$\frac{1}{4}(1+\eta)(1-\zeta^2)$	$\frac{1}{4}(1+\xi)(1-\zeta^2)$	$-\frac{1}{2}(1+\xi)(1+\eta)\zeta$
12	$\frac{1}{4}(1-\xi)(1+\eta)(1-\zeta^2)$	$-\frac{1}{4}(1+\eta)(1-\zeta^2)$	$\frac{1}{4}(1-\xi)(1-\zeta^2)$	$-\frac{1}{2}(1-\xi)(1+\eta)\zeta$
13	$\frac{1}{8}(1-\xi)(1-\eta)(1+\zeta)(-\xi-\eta+\zeta-2)$	$\frac{1}{8}(1-\eta)(1+\zeta)(2\xi+\eta-\zeta+1)$	$\frac{1}{8}(1-\xi)(1+\zeta)(2\eta+\xi-\zeta+1)$	$\frac{1}{8}(1-\xi)(1-\eta)(2\zeta-\eta-\xi-1)$
14	$\frac{1}{4}(1-\xi^2)(1-\eta)(1+\zeta)$	$-\frac{1}{2}(1-\eta)(1+\zeta)\xi$	$-\frac{1}{4}(1-\xi^2)(1+\zeta)$	$\frac{1}{4}(1-\xi^2)(1-\eta)$
15	$\frac{1}{8}(1+\xi)(1-\eta)(1+\zeta)(\xi-\eta+\zeta-2)$	$\frac{1}{8}(1-\eta)(1+\zeta)(2\xi-\eta+\zeta-1)$	$\frac{1}{8}(1+\zeta)(1+\xi)(2\eta-\xi-\zeta+1)$	$\frac{1}{8}(1-\eta)(1+\xi)(2\zeta+\xi-\eta-1)$
16	$\frac{1}{4}(1+\xi)(1-\eta^2)(1+\zeta)$	$\frac{1}{4}(1-\eta^2)(1+\xi)$	$-\frac{1}{2}(1+\xi)(1+\zeta)\eta$	$\frac{1}{4}(1+\xi)(1-\eta^2)$
17	$\frac{1}{8}(1+\xi)(1+\eta)(1+\zeta)(\xi+\eta+\zeta-2)$	$\frac{1}{8}(1+\eta)(1+\zeta)(2\xi+\eta+\zeta-1)$	$\frac{1}{8}(1+\xi)(1+\zeta)(2\eta+\xi+\zeta-1)$	$\frac{1}{8}(1+\xi)(1+\eta)(2\zeta+\eta+\xi-1)$
18	$\frac{1}{4}(1-\xi^2)(1+\eta)(1+\zeta)$	$-\frac{1}{2}\xi(1+\eta)(1+\zeta)$	$\frac{1}{4}(1-\xi^2)(1+\zeta)$	$\frac{1}{4}(1-\xi^2)(1+\eta)$
19	$\frac{1}{8}(1-\xi)(1+\eta)(1+\zeta)(-\xi+\eta+\zeta-2)$	$\frac{1}{8}(1+\eta)(1+\zeta)(2\xi-\eta-\zeta-1)$	$\frac{1}{8}(1-\xi)(1+\zeta)(2\eta-\xi+\zeta-1)$	$\frac{1}{8}(1-\xi)(1+\eta)(2\zeta-\xi+\eta-1)$
20	$\frac{1}{4}(1-\xi)(1-\eta^2)(1+\zeta)$	$-\frac{1}{4}(1-\eta^2)(1+\zeta)$	$-\frac{1}{2}(1-\xi)(1+\zeta)\eta$	$\frac{1}{4}(1-\xi)(1-\eta^2)$

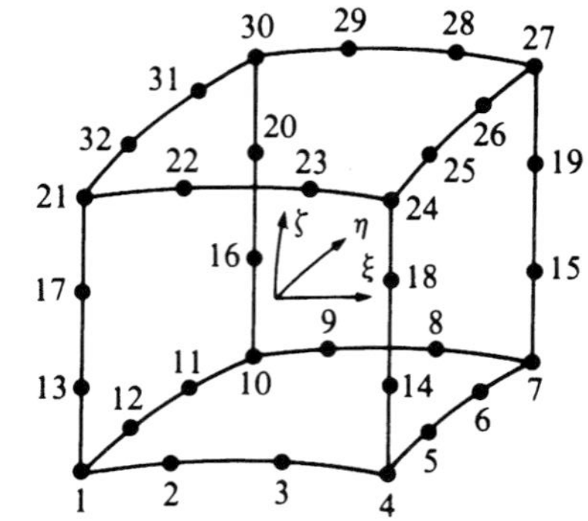

Table AIB.5. Thirty-two-noded solid element

Node i	Shape functions $N_i(\xi,\eta,\zeta)$	Derivatives $\frac{\partial N_i}{\partial \xi}$	$\frac{\partial N_i}{\partial \eta}$	$\frac{\partial N_i}{\partial \zeta}$
23	$\frac{9}{64}(1-\xi^2)(1+3\xi)(1-\eta)(1+\zeta)$	$\frac{9}{64}(3-2\xi-9\xi^2)(1-\eta)(1+\zeta)$	$-\frac{9}{64}(1-\xi^2)(1+3\xi)(1+\zeta)$	$\frac{9}{64}(1-\xi^2)(1+3\xi)(1-\eta)$
24	$\frac{9}{64}(1+\xi)(1-\eta)(1+\zeta)[\xi^2+\eta^2-\frac{19}{9}+\zeta^2]$	$\frac{9}{64}(1+\zeta)(1-\eta)(2\xi+3\xi^2+\zeta^2+\eta^2-\frac{19}{9})$	$\frac{9}{64}(1+\xi)(1+\zeta)(2\eta^2-3\eta-\xi^2-\zeta^2+\frac{19}{9})$	$\frac{9}{64}(1+\xi)(1-\eta)(2\zeta+3\zeta^2+\xi^2+\eta^2-\frac{19}{9})$
25	$\frac{9}{64}(1-\eta^2)(1-3\eta)(1+\xi)(1+\zeta)$	$\frac{9}{64}(1-\eta^2)(1-3\eta)(1+\zeta)$	$\frac{9}{64}(9\eta^2-2\eta-3)(1+\xi)(1+\zeta)$	$\frac{9}{64}(1-\eta^2)(1-3\eta)(1+\xi)$
26	$\frac{9}{64}(1-\eta^2)(1+3\eta)(1+\xi)(1+\zeta)$	$\frac{9}{64}(1-\eta^2)(1+3\eta)(1+\zeta)$	$\frac{9}{64}(3-2\eta-9\eta^2)(1+\xi)(1+\zeta)$	$\frac{9}{64}(1-\eta^2)(1+3\eta)(1+\xi)$
27	$\frac{9}{64}(1+\xi)(1+\eta)(1+\zeta)(\xi^2+\eta^2+\zeta^2-\frac{19}{9})$	$\frac{9}{64}(1+\xi)(1+\zeta)(2\xi+3\xi^2+\eta^2+\zeta^2-\frac{19}{9})$	$\frac{9}{64}(1+\xi)(1+\zeta)(2\eta+3\eta^2+\xi^2+\zeta^2-\frac{19}{9})$	$\frac{9}{64}(1+\xi)(1+\eta)(2\zeta^2+3\zeta+\xi^2+\eta^2-\frac{19}{9})$
28	$\frac{9}{64}(1-\xi^2)(1+3\xi)(1+\eta)(1+\zeta)$	$\frac{9}{64}(1+\eta)(1+\zeta)(3-2\xi-9\xi^2)$	$\frac{9}{64}(1-\xi^2)(1+3\xi)(1+\zeta)$	$\frac{p}{64}(1-\xi^2)(1+3\xi)(1+\eta)$
29	$\frac{9}{64}(1-\xi^2)(1-3\xi)(1+\eta)(1+\zeta)$	$\frac{9}{64}(1+\eta)(1+\zeta)(9\xi^2-2\xi-3)$	$\frac{9}{64}(1-\xi^2)(1-3\xi)(1+\zeta)$	$\frac{9}{64}(1-\xi^2)(1-3\xi)(1+\eta)$
30	$\frac{9}{64}(1-\xi)(1+\eta)(1+\zeta)(\zeta^2+\eta^2+\xi^2-\frac{19}{9})$	$\frac{9}{64}(1+\eta)(1+\eta)(2\xi-3\xi^2-\eta^2-\zeta^2+\frac{19}{9})$	$\frac{9}{64}(1-\xi)(1+\zeta)(2\eta+3\eta^2+\zeta^2+\xi^2-\frac{19}{9})$	$\frac{9}{64}(1-\xi)(1+\eta)(2\zeta+3\zeta^2+\xi^2+\eta^2-\frac{19}{9})$
31	$\frac{9}{64}(1-\eta^2)(1+3\eta)(1-\xi)(1+\zeta)$	$-\frac{9}{64}(1-\eta^2)(1+3\eta)(1+\zeta)$	$\frac{9}{64}(3-2\eta-9\eta^2)(1-\xi)(1+\zeta)$	$\frac{9}{64}(1-\eta^2)(1+3\eta)(1-\xi)$
32	$\frac{9}{64}(1-\eta^2)(1-3\eta)(1-\xi)(1+\zeta)$	$-\frac{9}{64}(1-\eta^2)(1-3\eta)(1+\zeta)$	$\frac{9}{64}(9\eta^2-2\eta-3)(1-\xi)(1+\zeta)$	$\frac{9}{64}(1-\eta^2)(1-3\eta)(1-\xi)$

Table AIB.6. Isoparametric membrane elements

Four-noded membrane element

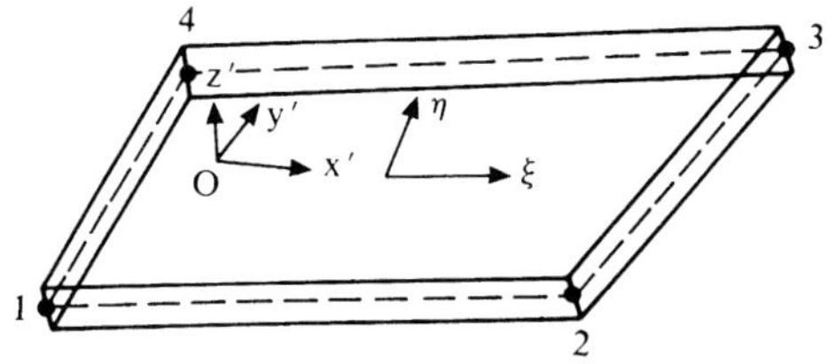

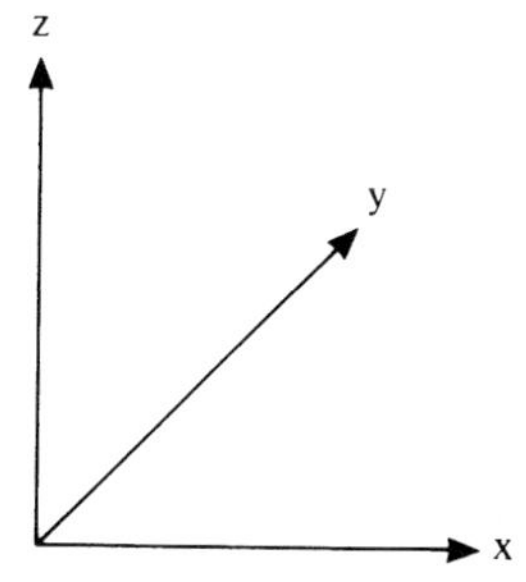

Global cartesian system

Node i	Shape functions $N_i(\xi, \eta)$	Derivatives $\frac{\partial N_i}{\partial \xi}$	$\frac{\partial N_i}{\partial \eta}$
1	$\frac{1}{4}(1-\xi)(1-\eta)$	$-\frac{1}{4}(1-\eta)$	$-\frac{1}{4}(1-\xi)$
2	$\frac{1}{4}(1+\xi)(1-\eta)$	$\frac{1}{4}(1-\eta)$	$-\frac{1}{4}(1+\xi)$
3	$\frac{1}{4}(1+\xi)(1+\eta)$	$\frac{1}{4}(1+\eta)$	$\frac{1}{4}(1+\xi)$
4	$\frac{1}{4}(1-\xi)(1+\eta)$	$-\frac{1}{4}(1+\eta)$	$\frac{1}{4}(1-\xi)$

Eight-noded membrane element

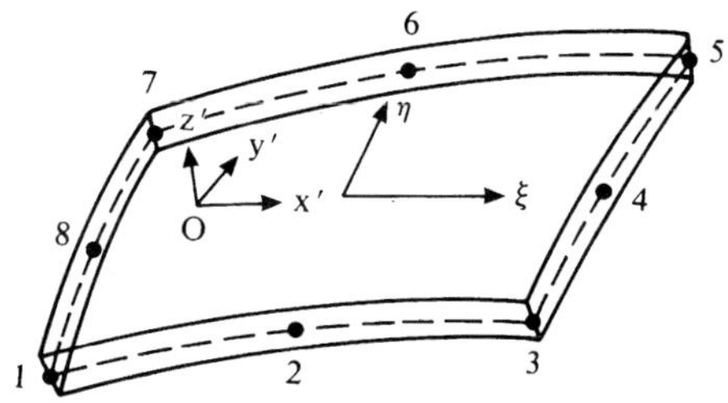

Node i	Shape functions $N_i(\xi, \eta)$	Derivatives $\frac{\partial N_i}{\partial \xi}$	$\frac{\partial N_i}{\partial \eta}$
1	$\frac{1}{4}(1-\xi)(1-\eta)(-\xi-\eta-1)$	$\frac{1}{4}(1-\eta)(2\xi+\eta)$	$\frac{1}{4}(1-\xi)(2\eta+\xi)$
2	$\frac{1}{2}(1-\xi^2)(1-\eta)$	$-\xi(1-\eta)$	$-\frac{1}{2}(1-\xi^2)$
3	$\frac{1}{4}(1+\xi)(1-\eta)(\xi-\eta-1)$	$\frac{1}{4}(1-\eta)(2\xi-\eta)$	$\frac{1}{4}(1+\xi)(2\eta-\xi)$
4	$\frac{1}{2}(1-\eta^2)(1+\xi)$	$\frac{1}{2}(1-\eta^2)$	$-\eta(1+\xi)$
5	$\frac{1}{4}(1+\xi)(1+\eta)(\xi+\eta-1)$	$\frac{1}{4}(1+\eta)(2\xi+\eta)$	$\frac{1}{4}(1+\xi)(2\eta+\xi)$
6	$\frac{1}{2}(1-\xi^2)(1+\eta^2)$	$-\xi(1+\eta)$	$\frac{1}{2}(1-\xi^2)$
7	$\frac{1}{4}(1-\xi)(1+\eta)(-\xi+\eta-1)$	$\frac{1}{4}(1+\eta)(2\xi-\eta)$	$\frac{1}{4}(1-\xi)(2\eta-\xi)$
8	$\frac{1}{2}(1-\eta^2)(1-\xi)$	$-\frac{1}{2}(1-\eta^2)$	$-\eta(1-\xi)$

Table AIB.7. Twelve-noded membrane element

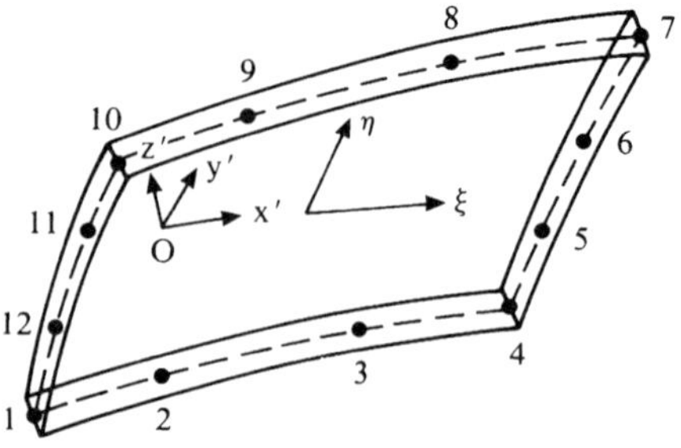

Node i	Shape functions $N_i(\xi, \eta)$	Derivatives $\frac{\partial N_i}{\partial \xi}$	Derivatives $\frac{\partial N_i}{\partial \eta}$
1	$\frac{9}{32}(1-\xi)(1-\eta)[(\xi^2+\eta^2-\frac{10}{9}]$	$\frac{9}{32}(1-\eta)[2\xi-3\xi^2-\eta^2+\frac{10}{9}]$	$\frac{9}{32}(1-\xi)[2\eta-3\eta^2-\xi^2+\frac{10}{9}]$
2	$\frac{9}{32}(1-\xi)(1-\xi^2)(1-\eta)$	$\frac{9}{32}(1-\eta)(3\xi^2-2\xi-1)$	$-\frac{9}{32}(1-\xi)(1-\xi^2)$
3	$\frac{9}{32}(1-\eta)(1-\xi^2)(1+\xi)$	$\frac{9}{32}(1-\eta)(1-2\xi-3\xi^2)$	$-\frac{9}{32}(1-\xi^2)(1+\xi)$
4	$\frac{9}{32}(1+\xi)(1-\eta)[\xi^2+\eta^2-\frac{10}{9}]$	$\frac{9}{32}(1-\eta)[2\xi+3\xi^2+\eta^2-\frac{10}{9}]$	$\frac{9}{32}(1+\xi)[2\eta-3\eta^2-\xi^2-\frac{10}{9}]$
5	$\frac{9}{32}(1+\xi)(1-\eta^2)(1-\eta)$	$\frac{9}{32}(1-\eta^2)(1-\eta)$	$\frac{9}{32}(1+\xi)(3\eta^2-2\eta-1)$
6	$\frac{9}{32}(1+\xi)(1-\eta^2)(1+\eta)$	$\frac{9}{32}(1-\eta^2)(1+\eta)$	$\frac{9}{32}(1+\xi)(1-2\eta-3\eta^2)$
7	$\frac{9}{32}(1+\xi)(1+\eta)[\xi^2+\eta^2-\frac{10}{9}]$	$\frac{9}{32}(1+\eta)[2\xi+3\xi^2+\eta^2-\frac{19}{9}]$	$\frac{9}{32}(1+\xi)[2\eta+3\eta^2+\xi^2-\frac{10}{9}]$
8	$\frac{9}{32}(1+\eta)(1-\xi^2)(1+\xi)$	$\frac{9}{32}(1+\eta)(1-2\xi-3\xi^2)$	$\frac{9}{32}(1-\xi^2)(1+\xi)$
9	$\frac{9}{32}(1+\eta)(1-\xi^2)(1-\xi)$	$\frac{9}{32}(1+\eta)(3\xi^2-3\xi-1)$	$\frac{9}{32}(1-\xi^2)(1-\xi)$
10	$\frac{9}{32}(1-\xi)(1+\eta)[\xi^2+\eta^2-\frac{10}{9}]$	$\frac{9}{32}(1+\eta)[2\xi-3\xi^2-\eta^2+\frac{10}{9}]$	$\frac{9}{32}(1-\xi)[2\eta+3\eta^2+\xi^2-\frac{10}{9}]$
11	$\frac{9}{32}(1-\xi)(1-\eta^2)(1+\eta)$	$-\frac{9}{32}(1+\eta)(1-\eta^2)$	$\frac{9}{32}(1-\xi)(1-2\eta-3\eta^2)$
12	$\frac{9}{32}(1-\xi)(1-\eta^2)(1-\eta)$	$-\frac{9}{32}(1-\eta)(1-\eta^2)$	$\frac{9}{32}(1-\xi)(3\eta^2-2\eta-1)$

Table AIB.8. Three-dimensional reinforced concrete solid element

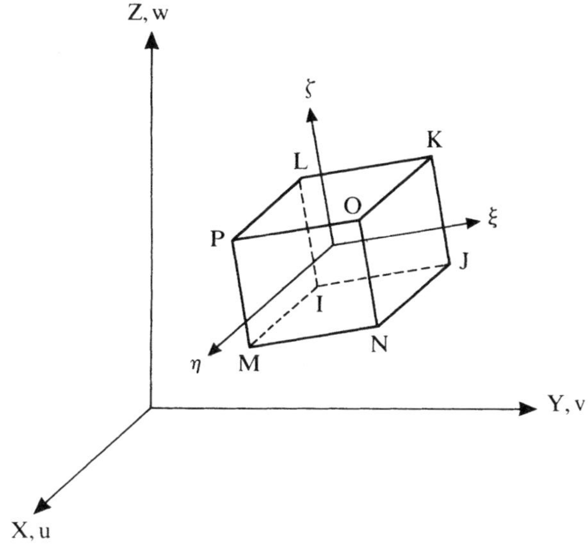

	Shape functions
Stiffness matrix $[K]$	$u = \frac{1}{8}[u_I(1-\xi)(1-\zeta)(1-\eta)$ $+u_J(1+\xi)(1-\zeta)(1-\eta)$ $+u_K(1+\xi)(1+\zeta)(1-\eta)$ $+u_L(1-\xi)(1+\zeta)(1-\eta)$ $+u_M(1-\xi)(1-\zeta)(1+\eta)$ $+u_N(1+\xi)(1-\zeta)(1+\eta)$ $+u_O(1+\xi)(1+\zeta)(1+\eta)$ $+u_P(1-\xi)(1+\zeta)(1+\eta)]$ $+u_1(1-\xi^2)$ $+u_2(1-\zeta^2)$ $+u_3(1-\eta^2)$ $v = \frac{1}{8}(v_I(1-\xi)\ldots$ (similar to u) $w = \frac{1}{8}(w_I(1-\xi)\ldots$ (similar to u)

Table AIB.9. Crack tip solid element

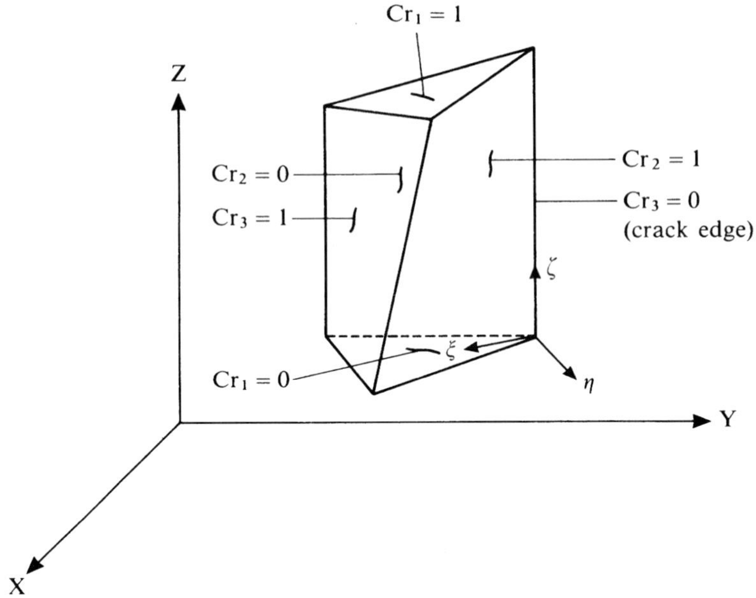

	Shape functions
Stiffness matrix $[K]$	6-node option $\xi = \alpha_1 + \alpha_2\sqrt{Cr_3} + \alpha_3 Cr_2\sqrt{Cr_3} + \alpha_4 Cr_1 + \alpha_5 Cr_1\sqrt{Cr_3} + \alpha_6 Cr_2 Cr_1\sqrt{Cr_3}$ $\eta = \alpha_7 + \alpha_8\sqrt{Cr_3} + \alpha_9 Cr_2\sqrt{Cr_3} + \alpha_{10} Cr_1 + \alpha_{11} Cr_1\sqrt{Cr_3} + \alpha_{12} Cr_2 Cr_1\sqrt{Cr_3}$ $\zeta = \alpha_{13} + \alpha_{14} Cr_3 + \alpha_{15} Cr_2 Cr_3 + \alpha_{16} Cr_1 + \alpha_{17} Cr_1 Cr_3 + \alpha_{18} Cr_2 Cr_1 Cr_3$

Table AIB.10. Shape function for a prism element

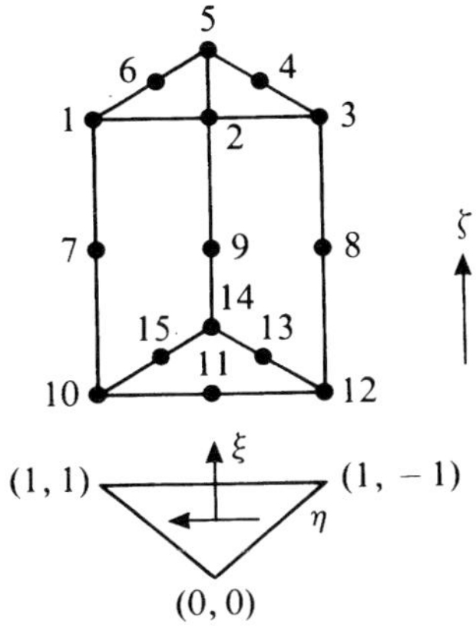

Fifteen nodes

$$N_1(\xi,\eta,\zeta) = -\tfrac{1}{2}(1+\zeta)(1-\xi^{1/2})(\sqrt{2}\xi^{1/2}-\zeta)$$

$$N_2(\xi,\eta,\zeta) = +\tfrac{1}{2}(1+\sqrt{2})(1+\zeta)(1-\xi^{1/2})(\xi^{1/2}+\eta\xi^{-1/2})$$

$$N_3(\xi,\eta,\zeta) = +\tfrac{1}{4}(1+\zeta)(\xi^{1/2}+\eta\xi^{-1/2})[(2+\sqrt{2})(\xi^{1/2}-1)-(\xi-\eta-\zeta)]$$

$$N_4(\xi,\eta,\zeta) = +\tfrac{1}{2}(1+\zeta)(\xi^{3/2}-\eta^2\xi^{-1/2})$$

$$N_5(\xi,\eta,\zeta) = +\tfrac{1}{4}(1+\zeta)(\xi^{1/2}-\eta\xi^{-1/2})[(2+\sqrt{2})(\xi^{1/2}-1)-(\xi+\eta-\zeta)]$$

$$N_6(\xi,\eta,\zeta) = +\tfrac{1}{2}(1+\sqrt{2})(1+\xi)(\xi^{1/2}-\eta\xi^{-1/2})(1-\xi^{1/2})$$

$$N_7(\xi,\eta,\zeta) = (1-\zeta^2)(1-\xi^{1/2})$$

$$N_8(\xi,\eta,\zeta) = +\tfrac{1}{2}(1-\zeta^2)(\xi^{1/2}+\eta\xi^{-1/2})$$

$$N_9(\xi,\eta,\zeta) = +\tfrac{1}{2}(1-\zeta^2)(\xi^{1/2}-\eta\xi^{-1/2})$$

$$N_{10}(\xi,\eta,\zeta) = -\tfrac{1}{2}(1-\zeta)(1-\xi^{1/2})(\sqrt{2}\xi^{1/2}+\zeta)$$

$$N_{11}(\xi,\eta,\zeta) = +\tfrac{1}{2}(1+\sqrt{2})(1-\zeta)(\xi^{1/2}+\eta\xi^{-1/2})(1-\xi^{1/2})$$

$$N_{12}(\xi,\eta,\zeta) = +\tfrac{1}{4}(1-\zeta)(\xi^{1/2}+\eta\xi^{-1/2})[(2+\sqrt{2})(\xi^{1/2}-1)-(\xi-\eta+\zeta)]$$

$$N_{13}(\xi,\eta,\zeta) = +\tfrac{1}{2}(1-\zeta)(\xi^{3/2}-\eta^2\xi^{-1/2})$$

$$N_{14}(\xi,\eta,\zeta) = +\tfrac{1}{4}(1-\zeta)(\xi^{1/2}-\eta\xi^{-1/2})[(2+\sqrt{2})(\zeta^{1/2}-1)-(\xi+\eta+\zeta)]$$

$$N_{15}(\xi,\eta,\zeta) = +\tfrac{1}{2}(1+\sqrt{2})(1-\zeta)(1-\xi^{1/2})(\xi^{1/2}-\eta\xi^{-1/2})$$

Table AIB.11. Right tetrahedral element

Four-noded

Co-ordinates:

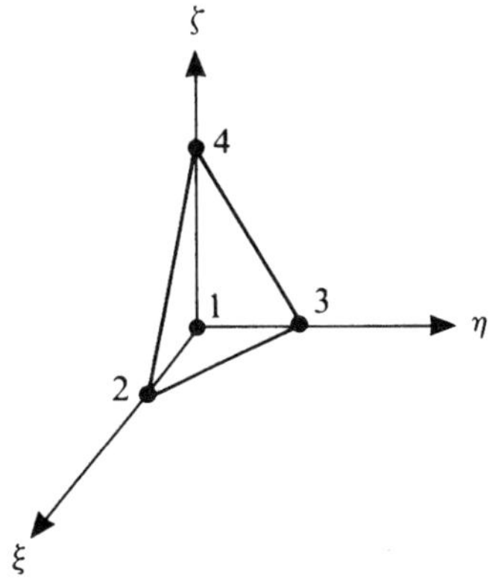

Nodal no.	ξ_i	η_i	ζ_i
1	0	0	0
2	1	0	0
3	0	1	0
4	0	0	1

$$N_1(\xi,\eta,\zeta) = (1-\xi-\eta-\zeta)$$

$$N_2(\xi,\eta,\zeta) = +\xi$$

$$N_3(\xi,\eta,\zeta) = +\eta$$

$$N_4(\xi,\eta,\zeta) = +\zeta$$

Ten-noded

Co-ordinates:

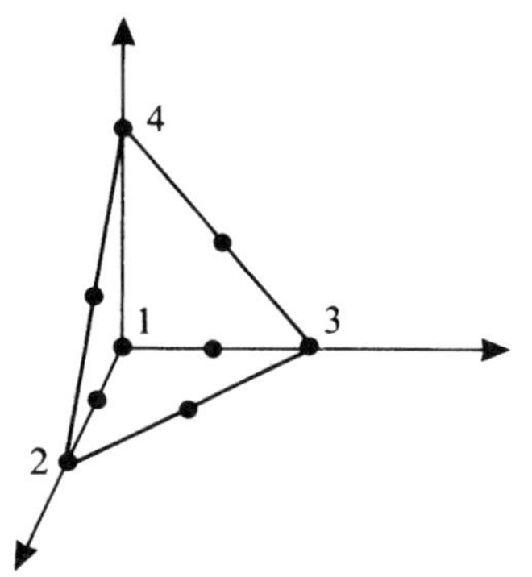

Nodal no.	ξ_i	η_i	ζ_i
1	0	0	0
2	1	0	0
3	0	1	0
4	0	0	1
5	$\frac{1}{2}$	0	0
6	$\frac{1}{2}$	$\frac{1}{2}$	0
7	0	$\frac{1}{2}$	0
8	0	0	$\frac{1}{2}$
9	$\frac{1}{2}$	0	$\frac{1}{2}$
10	0	$\frac{1}{2}$	$\frac{1}{2}$

$$N_1(\xi,\eta,\zeta) = 2(1-\xi-\eta-\zeta)^2 - (1-\xi-\eta-\zeta) \qquad N_2(\xi,\eta,\zeta) = (2\xi-1)\xi$$

$$N_3(\xi,\eta,\zeta) = (2\eta-1)\eta \qquad N_4(\xi,\eta,\zeta) = (2\zeta-1)\zeta$$

$$N_5(\xi,\eta,\zeta) = 4\xi(1-\xi-\eta-\zeta) \qquad N_6(\xi,\eta,\zeta) = 4\xi\eta$$

$$N_7(\xi,\eta,\zeta) = 4\eta(1-\xi-\eta-\zeta) \qquad N_8(\xi,\eta,\zeta) = 4\zeta(1-\xi-\eta-\zeta)$$

$$N_9(\xi,\eta,\zeta) = 4\xi\zeta \qquad N_{10}(\xi,\eta,\zeta) = 4\eta\zeta$$

Table AIB.12. Shell element

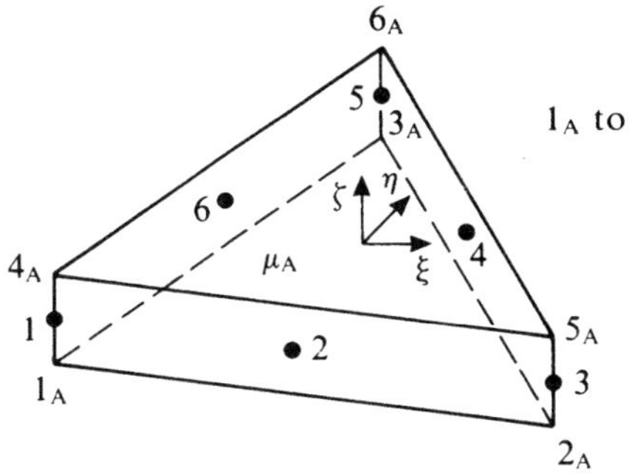

μ_A = area coordinate
1_A to 6_A = modified nodal system

Shape functions

$N_{1_A} = \frac{1}{2}(1-\mu_A)\xi \qquad N_{2_A} = \frac{1}{2}(1-\mu_A)\eta$

$N_{3_A} = \frac{1}{2}(1-\mu_A)\zeta \qquad N_{4_A} = \frac{1}{2}(1+\mu_A)\xi$

$N_{5_A} = \frac{1}{2}(1+\mu_A)\eta \qquad N_{6_A} = \frac{1}{2}(1+\mu_A)\zeta$

Triangular element
Three-noded

Co-ordinates:

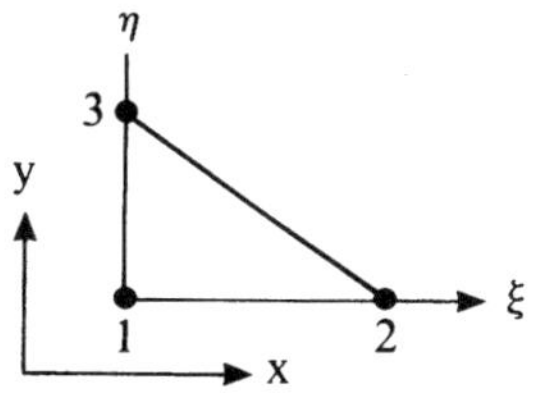

Nodal no.	ξ_i	η_i
1	0	0
2	1	0
3	0	1

$N_1(\xi,\eta) = 1-\xi-\eta \qquad N_2(\xi,\eta) = \xi \qquad N_3(\xi,\eta) = \eta$

$$\frac{\partial N_1}{\partial \xi} = -1 \qquad \frac{\partial N_1}{\partial \eta} = -1 \qquad \frac{\partial N_2}{\partial \xi} = 1$$

$$\frac{\partial N_2}{\partial \eta} = 0 \qquad \frac{\partial N_3}{\partial \xi} = 0 \qquad \frac{\partial N_3}{\partial \eta} = 1$$

Six-noded

Co-ordinates:

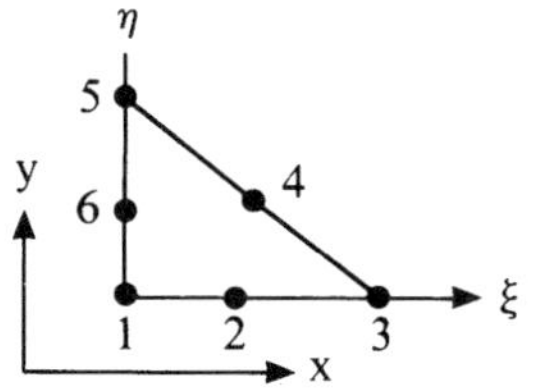

Nodal no.	ξ_i	η_i
1	0	0
2	$\frac{1}{2}$	0
3	1	0
4	$\frac{1}{2}$	$\frac{1}{2}$
5	0	1
6	0	$\frac{1}{2}$

$N_1(\xi,\eta) = 1-3\xi-3\eta+2\xi^2+2\eta^2+4\xi\eta \qquad N_4(\xi,\eta) = 4\xi\eta$

$N_2(\xi,\eta) = 4\xi-4\xi^2-4\xi\eta \qquad N_5(\xi,\eta) = -\eta+2\eta^2$

$N_3(\xi,\eta) = -\xi+2\xi^2 \qquad N_6(\xi,\eta) = 4\eta-4\xi-4\eta^2$

Table AIB.13. Boom elements

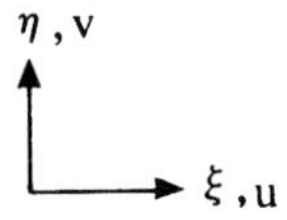

Two nodes

Displacement functions

$u = a_1 + b_1\xi$

$v = a_2 + b_2\xi$

Degrees of freedom = 4

Three nodes

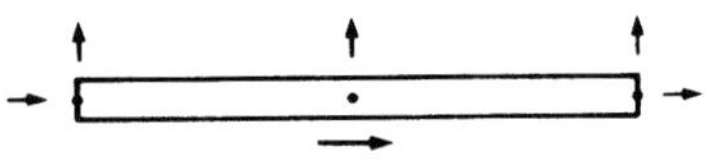

Displacement functions

$u = a_1 + b_1\xi + c_1\xi^2$

$v = a_2 + b_2\xi + c_2\xi^2$

Degrees of freedom = 6

Two nodes

Displacement functions

$u = u_1L_1(\xi) + u_2L_2(\xi)$

$v = v_1L_1(\xi) + v_2L_2(\xi)$

Degrees of freedom = 4

Beam elements

Two nodes

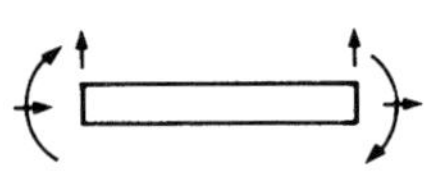

Displacement functions

$u = a_1 + b_1\xi$

$v = a_2 + b_2\xi + c_2\xi^2 + d_2\xi^3$

Degrees of freedom = 6

Three nodes

Displacement functions

$u = a_1 + b_1\xi + c_1\xi^2$

$v = a_2 + b_2\xi + c_2\xi^2 + d_2\xi^3 + e_2\xi^4 + f_2\xi^5$

Degrees of freedom = 9

Two nodes

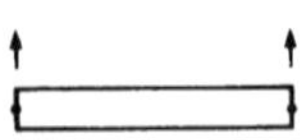

Displacement functions

$u = a_1 + b_1\xi + c_1\xi^2 + d_1\xi^3$

$v = a_2 + b_2\xi + c_2\xi^2 + d_2\xi^3 + e_2\xi^4 + f_2\xi^5$

Degrees of freedom = 10

Table AIB.14. Linear and quadratic two-dimensional isoparametric elements

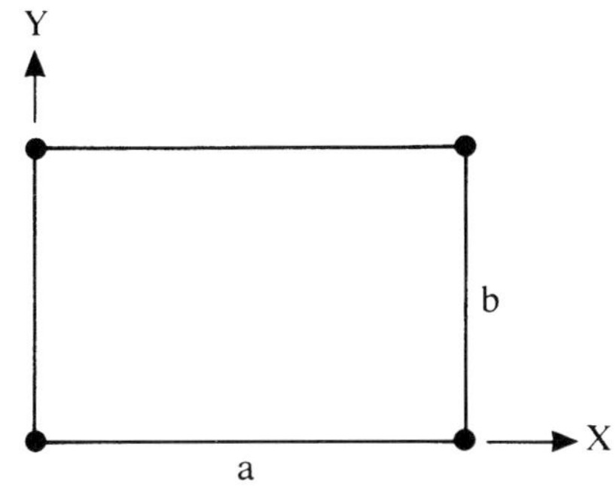

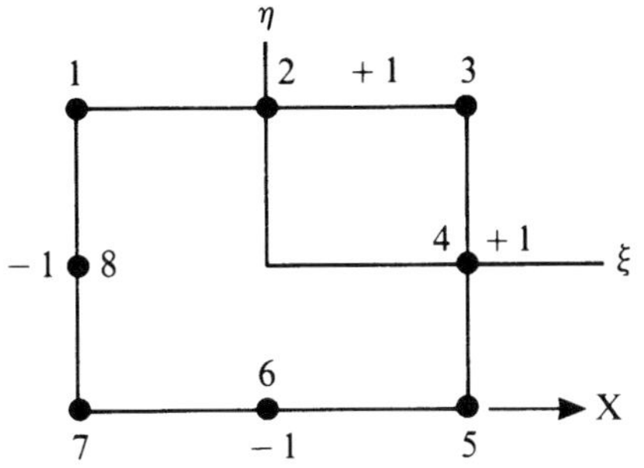

Linear shape functions:

$$N_1 = \tfrac{1}{4}(1-\xi)(1+\eta) \qquad N_2 = \tfrac{1}{4}(1+\xi)(1+\eta) \qquad N_3 = \tfrac{1}{4}(1+\xi)(1-\eta) \qquad N_4 = \tfrac{1}{4}(1-\xi)(1-\eta)$$

$$\left.\begin{array}{l} \text{mid-side nodes } 2, 6 \quad (\xi_r = 0) \\ N_r = \tfrac{1}{2}(1-\xi^2)(1+\eta_0) \\ \text{mid-side nodes } 4, 8 \quad (\eta_r = 0) \end{array}\right\} \text{Quadratic}$$

Quadratic general shape function for the corner nodes:

$$N_r = \tfrac{1}{4}(1+\xi_0)(1+\eta_0)(\xi_0+\eta_0+1) \quad (\xi_0 = \xi\xi_r, \eta_0 = \eta\eta_r)$$

$$[B] = \sum_{r=1}^{4}[B]_r = \sum_{r=1}^{4}\begin{bmatrix} \dfrac{\partial N_r}{\partial y} & 0 \\ 0 & \dfrac{\partial N_r}{\partial y} \\ \dfrac{\partial N_r}{\partial y} & \dfrac{\partial N_r}{\partial x} \end{bmatrix} \qquad [J] = \begin{bmatrix} \dfrac{\partial X}{\partial \xi} & \dfrac{\partial Y}{\partial \xi} \\ \dfrac{\partial X}{\partial \eta} & \dfrac{\partial Y}{\partial \eta} \end{bmatrix} = \begin{bmatrix} \dfrac{a}{2} & 0 \\ 0 & \dfrac{b}{2} \end{bmatrix}$$

$$[B] = \begin{bmatrix} \dfrac{-(1+\eta)}{2a} & 0 & \dfrac{(1+\eta)}{2a} & 0 & \dfrac{(1-\eta)}{2a} & 0 & \dfrac{-(1-\eta)}{2a} & 0 \\ 0 & \dfrac{1-\xi}{2b} & 0 & \dfrac{(1+\xi)}{2b} & 0 & \dfrac{-(1+\xi)}{2b} & 0 & \dfrac{-(1-\xi)}{2b} \\ \dfrac{1-\xi}{2b} & \dfrac{-(1+\eta)}{2a} & \dfrac{1+\xi}{2b} & \dfrac{1+\eta}{2a} & \dfrac{-(1+\xi)}{2b} & \dfrac{(1+\eta)}{2a} & \dfrac{-(1-\xi)}{2b} & \dfrac{-(1-\eta)}{2a} \end{bmatrix}$$

$$[K] = \int B^{T''} D B \,\mathrm{d\,vol} = t\iint [B]^{T''}[D][B]|J|\,\mathrm{d}\xi\,\mathrm{d}\eta$$

Table AIB.15. Linear strain triangular element (six-noded)

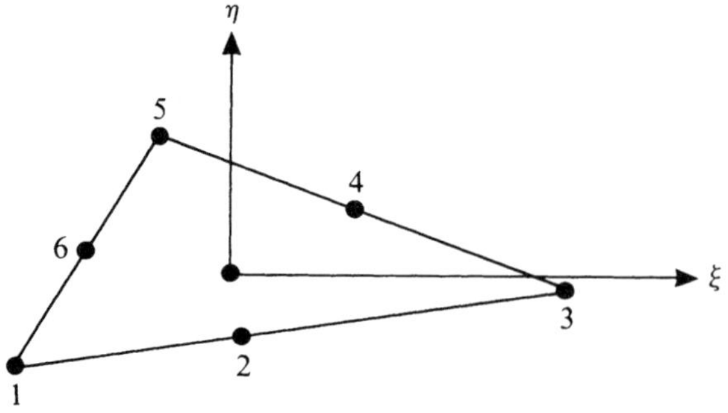

$$[B] = \begin{bmatrix} 0 & 1 & 0 & 2\xi & \eta & 0 & 0 & 0 & 0 & 0 & 0 & 0 \\ 0 & 0 & 0 & 0 & 0 & 0 & 0 & 0 & 1 & 0 & \xi & 2\eta \\ 0 & 0 & 1 & 0 & \xi & 2\eta & 0 & 1 & 0 & 2\xi & \eta & 0 \end{bmatrix}$$

$$[K] = [B]^{T''}[D][B]\,\mathrm{d}A = \frac{\text{Area} \times E}{1 - \nu^2}$$

$$= \begin{bmatrix} 0 & 0 & 0 & 0 & 0 & 0 & 0 & 0 & 0 & 0 & 0 \\ 0 & 1 & 0 & 0 & 0 & 0 & 0 & \nu & 0 & 0 & 0 \\ 0 & 0 & d' & 0 & 0 & 0 & d' & 0 & 0 & 0 & 0 \\ 0 & 0 & 0 & 4\overline{\xi\xi} & 2\overline{\xi\eta} & 0 & 0 & 0 & 0 & 2\nu\overline{\xi\xi} & 4\nu\overline{\xi\eta} \\ 0 & 0 & 0 & 2\overline{\xi\eta} & d'\overline{\xi\xi} + \overline{\eta\eta} & 0 & 0 & 0 & G'_1 & (\nu + d')\overline{\xi\eta} & 2\nu\overline{\eta\eta} \\ 0 & 0 & 0 & 0 & 2d'\overline{\xi\eta} & G' & 0 & 0 & G''_2 & 2d'\overline{\eta\eta} & 0 \\ 0 & 0 & 0 & 0 & 0 & G'' & 0 & 0 & 0 & 0 & 0 \\ 0 & 0 & d' & 0 & 0 & 0 & 0 & 0 & 0 & 0 & 0 \\ 0 & \nu & 0 & 0 & 0 & 0 & 0 & 1 & 0 & 0 & 0 \\ 0 & 0 & 0 & 0 & 2d'\overline{\xi\xi} & G''_2 & 0 & 0 & G''_1 & 2d'\overline{\xi\eta} & 0 \\ 0 & 0 & 0 & 2d'\overline{\xi\xi} & (\nu + d')\overline{\xi\eta} & G'_1 & 0 & 0 & G' & \overline{\xi\xi} + d'\overline{\eta\eta} & 2\overline{\xi\eta} \\ 0 & 0 & 0 & 4d'\overline{\xi\eta} & 2\nu\overline{\eta\eta} & 0 & 0 & 0 & 0 & 2\overline{\xi\eta} & 4\overline{\eta\eta} \end{bmatrix}$$

$d' = (1 - \nu)/2 \qquad \overline{\xi\xi} = \frac{1}{12}(\xi_1^2 + \xi_3^2 + \xi_5^2) \qquad \overline{\xi\eta} = \frac{1}{12}(\xi_1\eta_1 + \xi_3\eta_3 + \xi_5\eta_5)$

$\overline{\eta\eta} = \frac{1}{12}(\eta_1^2 + \eta_3^2 + \eta_5^2) \qquad G' = 2d'\overline{\xi\eta} \qquad G'' = 4d'\eta\eta$

$G'_1 = 2d'\overline{\xi\xi} \qquad G'_2 = 4d'\overline{\xi\xi} \qquad G''_1 = 4d'\overline{\xi\xi} \qquad G''_2 = 4d'\overline{\xi\eta}$

Table AIB.16. Constant stress/strain [K] matrix

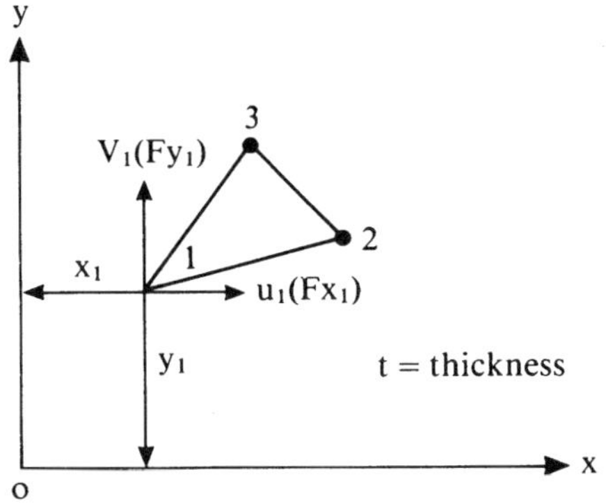

Triangular

Plane stress:

$$\bar{d}_{11} = \bar{d}_{22} = E/(1 - \nu^2)$$

$$\bar{d}_{12} = \bar{d}_{21} = \nu E/(1 - \nu^2)$$

$$\bar{d}_{33} = E/2(1 + \nu)$$

Plane strain:

$$\bar{d}_{11} = \bar{d}_{22} = E(1 - \nu)/(1 + \nu)(1 - 2\nu)$$

$$\bar{d}_{12} = \bar{d}_{21} = \nu E/(1 + \nu)(1 - 2\nu)$$

$$\bar{d}_{33} = E/2(1 + \nu)$$

$$
[K]=\frac{t}{4\Delta}\begin{bmatrix}
\begin{array}{l}\bar{d}_{11}(y_2-y_3)^2\\+\bar{d}_{33}(x_3-x_2)^2\end{array} &
\begin{array}{l}\bar{d}_{12}(x_3-x_2)(y_2-y_3)\\+\bar{d}_{33}(x_3-x_2)(y_2-y_3)\end{array} &
\begin{array}{l}\bar{d}_{11}(y_2-y_3)(y_3-y_1)\\+\bar{d}_{33}(x_3-x_2)(x_1-x_3)\end{array} &
\begin{array}{l}\bar{d}_{12}(x_1-x_3)(y_2-y_3)\\+\bar{d}_{33}(x_3-x_2)(y_3-y_1)\end{array} &
\begin{array}{l}\bar{d}_{11}(y_1-y_2)(y_2-y_3)\\+\bar{d}_{33}(x_2-x_1)(x_3-x_2)\end{array} &
\begin{array}{l}\bar{d}_{12}(x_2-x_1)(y_2-y_3)\\+\bar{d}_{33}(x_3-x_2)(y_1-y_2)\end{array} \\
\begin{array}{l}\bar{d}_{21}(x_3-x_2)(y_2-y_3)\\+\bar{d}_{33}(x_3-x_2)(y_2-y_3)\end{array} &
\begin{array}{l}\bar{d}_{22}(x_3-x_2)^2\\+\bar{d}_{33}(y_2-y_3)^2\end{array} &
\begin{array}{l}\bar{d}_{12}(y_3-y_1)(x_3-x_2)\\+\bar{d}_{33}(x_1-x_3)(y_2-y_3)\end{array} &
\begin{array}{l}\bar{d}_{22}(x_3-x_2)(x_1-x_3)\\+\bar{d}_{33}(y_2-y_3)(y_3-y_1)\end{array} &
\begin{array}{l}\bar{d}_{21}(x_3-x_2)(y_1-y_2)\\+\bar{d}_{33}(x_2-x_1)(y_2-y_2)\end{array} &
\begin{array}{l}\bar{d}_{22}(x_2-x_1)(x_3-x_2)\\+\bar{d}_{33}(y_1-y_2)(y_2-y_3)\end{array} \\
\begin{array}{l}\bar{d}_{11}(y_2-y_3)(y_3-y_1)\\+\bar{d}_{33}(x_1-x_3)(x_3-x_2)\end{array} &
\begin{array}{l}\bar{d}_{12}(x_3-x_2)(y_3-y_1)\\+\bar{d}_{33}(x_1-x_3)(y_2-y_3)\end{array} &
\begin{array}{l}\bar{d}_{11}(y_3-y_1)^2\\+\bar{d}_{33}(x_1-x_3)^2\end{array} &
\begin{array}{l}\bar{d}_{12}(x_1-x_3)(y_3-y_1)\\+\bar{d}_{33}(x_1-x_3)(y_3-y_1)\end{array} &
\begin{array}{l}\bar{d}_{11}(y_1-y_2)(y_3-y_1)\\+\bar{d}_{33}(x_1-x_3)(x_2-x_1)\end{array} &
\begin{array}{l}\bar{d}_{12}(x_2-x_1)(y_3-y_1)\\+\bar{d}_{33}(x_1-x_3)(y_1-y_2)\end{array} \\
\begin{array}{l}\bar{d}_{21}(x_1-x_3)(y_2-y_3)\\+\bar{d}_{33}(x_3-x_2)(y_3-y_1)\end{array} &
\begin{array}{l}\bar{d}_{22}(x_1-x_3)(x_3-x_2)\\+\bar{d}_{33}(y_2-y_3)(y_3-y_1)\end{array} &
\begin{array}{l}\bar{d}_{12}(x_1-x_3)(y_3-y_1)\\+\bar{d}_{33}(x_1-x_3)(y_3-y_1)\end{array} &
\begin{array}{l}\bar{d}_{22}(x_1-x_3)^2\\+\bar{d}_{33}(y_3-y_1)^2\end{array} &
\begin{array}{l}\bar{d}_{21}(x_1-x_3)(y_1-y_2)\\+\bar{d}_{33}(x_2-x_1)(y_3-y_1)\end{array} &
\begin{array}{l}\bar{d}_{22}(x_1-x_3)(x_2-x_1)\\+\bar{d}_{33}(y_1-y_2)(y_3-y_1)\end{array} \\
\begin{array}{l}\bar{d}_{11}(y_1-y_2)(y_2-y_3)\\+\bar{d}_{33}(x_2-x_1)(x_3-x_2)\end{array} &
\begin{array}{l}\bar{d}_{12}(x_3-x_2)(y_1-y_2)\\+\bar{d}_{33}(x_2-x_1)(y_2-y_3)\end{array} &
\begin{array}{l}\bar{d}_{11}(y_1-y_2)(y_3-y_1)\\+\bar{d}_{33}(x_1-x_3)(x_2-x_1)\end{array} &
\begin{array}{l}\bar{d}_{12}(x_1-x_3)(y_1-y_2)\\+\bar{d}_{33}(x_2-x_1)(y_3-y_1)\end{array} &
\begin{array}{l}\bar{d}_{11}(y_1-y_2)^2\\+\bar{d}_{33}(x_2-x_1)^2\end{array} &
\begin{array}{l}\bar{d}_{12}(x_2-x_1)(y_1-y_2)\\+\bar{d}_{33}(x_2-x_1)(y_1-y_2)\end{array} \\
\begin{array}{l}\bar{d}_{21}(x_2-x_1)(y_2-y_3)\\+\bar{d}_{33}(x_3-x_2)(y_1-y_2)\end{array} &
\begin{array}{l}\bar{d}_{22}(x_2-x_1)(x_3-x_2)\\+\bar{d}_{33}(y_1-y_2)(y_2-y_3)\end{array} &
\begin{array}{l}\bar{d}_{12}(x_2-x_1)(y_3-y_1)\\+\bar{d}_{33}(x_1-x_3)(y_1-y_2)\end{array} &
\begin{array}{l}\bar{d}_{22}(x_1-x_3)(x_2-x_1)\\+\bar{d}_{33}(y_1-y_2)(y_3-y_1)\end{array} &
\begin{array}{l}\bar{d}_{21}(x_2-x_1)(y_1-y_2)\\+\bar{d}_{33}(x_2-x_1)(y_1-y_2)\end{array} &
\begin{array}{l}\bar{d}_{22}(x_2-x_1)^2\\+\bar{d}_{33}(y_1-y_2)^2\end{array}
\end{bmatrix}
$$

Table AIB.17. Linear strain rectangular element

Rectangular four-noded element

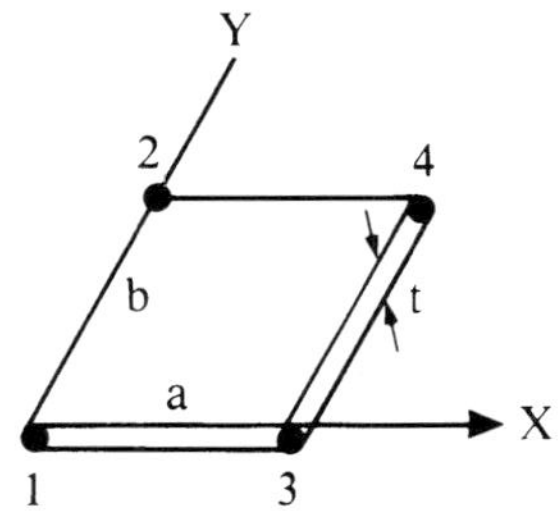

$$[K] = \frac{t}{12}\begin{bmatrix}
\begin{array}{l}4\bar{d}_{11}/(a/b)\\+4\bar{d}_{33}(a/b)\end{array} & & & & & & & \\
3\bar{d}_{21}+3\bar{d}_{33} & \begin{array}{l}4\bar{d}_{22}(a/b)\\+4\bar{d}_{33}/(a/b)\end{array} & & & & \text{symmetric} & & \\
\begin{array}{l}2\bar{d}_{11}/(a/b)\\-4\bar{d}_{33}(a/b)\end{array} & 3\bar{d}_{21}-3\bar{d}_{33} & \begin{array}{l}4\bar{d}_{11}/(a/b)\\+4\bar{d}_{33}(a/b)\end{array} & & & & & \\
-3\bar{d}_{21}+3\bar{d}_{33} & \begin{array}{l}-4\bar{d}_{22}(a/b)\\+2\bar{d}_{33}/(a/b)\end{array} & -3\bar{d}_{21}-3\bar{d}_{33} & \begin{array}{l}4\bar{d}_{22}(a/b)\\+4\bar{d}_{33}/(a/b)\end{array} & & & & \\
\begin{array}{l}-4\bar{d}_{11}/(a/b)\\+2\bar{d}_{33}(a/b)\end{array} & -3\bar{d}_{21}+3\bar{d}_{33} & \begin{array}{l}-2\bar{d}_{11}/(a/b)\\-2\bar{d}_{33}(a/b)\end{array} & 3\bar{d}_{21}+3\bar{d}_{33} & \begin{array}{l}4\bar{d}_{11}/(a/b)\\+4\bar{d}_{33}(a/b)\end{array} & & & \\
3\bar{d}_{21}-3\bar{d}_{33} & \begin{array}{l}2\bar{d}_{22}(a/b)\\-4\bar{d}_{33}/(a/b)\end{array} & 3\bar{d}_{21}+3\bar{d}_{33} & \begin{array}{l}-2\bar{d}_{22}(a/b)\\-2\bar{d}_{33}/(a/b)\end{array} & -3\bar{d}_{21}-3\bar{d}_{33} & \begin{array}{l}4\bar{d}_{22}(a/b)\\+4\bar{d}_{33}/(a/b)\end{array} & & \\
\begin{array}{l}-2\bar{d}_{11}/(a/b)\\-2\bar{d}_{33}(a/b)\end{array} & -3\bar{d}_{21}-3\bar{d}_{33} & \begin{array}{l}-4\bar{d}_{11}/(a/b)\\+2\bar{d}_{33}(a/b)\end{array} & 3\bar{d}_{21}-3\bar{d}_{33} & \begin{array}{l}2\bar{d}_{11}/(a/b)\\-4\bar{d}_{33}(a/b)\end{array} & -3\bar{d}_{21}+3\bar{d}_{33} & \begin{array}{l}4\bar{d}_{11}/(a/b)\\+4\bar{d}_{33}(a/b)\end{array} & \\
-3\bar{d}_{21}-3\bar{d}_{33} & \begin{array}{l}-2\bar{d}_{22}(a/b)\\-2\bar{d}_{33}/(a/b)\end{array} & -3\bar{d}_{21}+3\bar{d}_{33} & \begin{array}{l}2\bar{d}_{22}(a/b)\\-4\bar{d}_{33}/(a/b)\end{array} & 3\bar{d}_{21}-3\bar{d}_{33} & \begin{array}{l}-4\bar{d}_{22}(a/b)\\+2\bar{d}_{33}/(a/b)\end{array} & 3\bar{d}_{21}+3\bar{d}_{33} & \begin{array}{l}4\bar{d}_{22}(a/b)\\+4\bar{d}_{33}/(a/b)\end{array}
\end{bmatrix}$$

Table AIB.18. The rectangular finite element for plate flexure

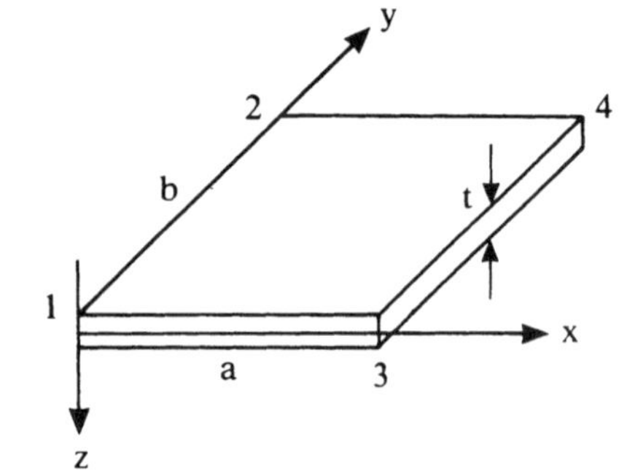

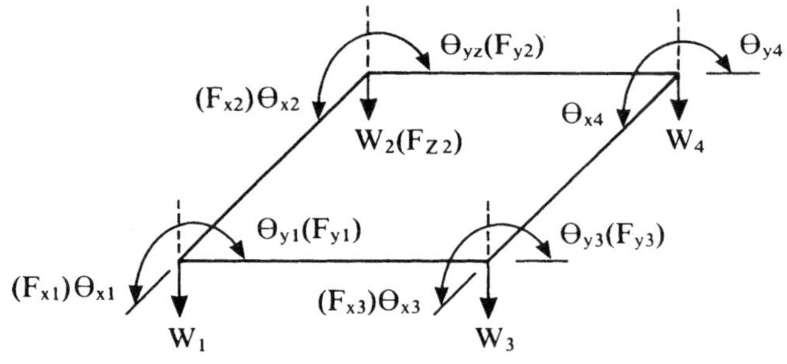

$$w = \text{displacement}$$
$$= a_1 + a_2x + a_3y + a_4x^2 + a_5xy + a_6y^2$$
$$+ a_7x^3 + a_8x^2y + a_9xy^2 + a_{10}y^3$$
$$+ a_{11}x^3y + a_{12}xy^3$$

$$\theta_x = -\,\partial w/\partial y \qquad \theta_y = \partial w/\partial x$$

$$\{\delta_1\} = \begin{Bmatrix} \theta_{x_1} \\ \theta_{y_1} \\ w_1 \end{Bmatrix} \text{ etc.}$$

$$\{F_1\} = \begin{Bmatrix} F_{x_1} \\ F_{y_1} \\ F_{z_1} \end{Bmatrix} \text{ etc.}$$

$$[K] = \begin{bmatrix} L^{\mathrm{I}} & & & \\ L^{\mathrm{II}} & L^{\mathrm{III}} & & \\ L^{\mathrm{IV}} & L^{\mathrm{V}} & L^{\mathrm{VI}} & \\ L^{\mathrm{VII}} & L^{\mathrm{VIII}} & L^{\mathrm{IX}} & L^{\mathrm{X}} \end{bmatrix}$$

$$[L^{\mathrm{I}}] = \begin{bmatrix} A & & \\ -B & C & \\ -D & E & F \end{bmatrix} \qquad [L^{\mathrm{II}}] = \begin{bmatrix} G & H & I \\ J & K & L \\ M & N & O \end{bmatrix}$$

$$[L^{\mathrm{III}}] = \begin{bmatrix} A & & \\ B & C & \\ D & E & F \end{bmatrix} \qquad [L^{\mathrm{IV}}] = \begin{bmatrix} P & Q & R \\ S & T & S' \\ X & Y & Z \end{bmatrix}$$

$$[L^{\mathrm{V}}] = \begin{bmatrix} G' & H' & I' \\ J' & K' & L' \\ M' & N' & O' \end{bmatrix} \qquad [L^{\mathrm{VI}}] = \begin{bmatrix} A & & \\ B & C & \\ -D & -E & F \end{bmatrix}$$

$$[L^{\mathrm{VII}}] = \begin{bmatrix} G' & H' & -I' \\ J' & K' & L' \\ M' & -N' & O' \end{bmatrix} \qquad [L^{\mathrm{VIII}}] = \begin{bmatrix} P & Q & -R \\ S & T & S' \\ -X & Y & Z \end{bmatrix}$$

$$[L^{\mathrm{IX}}] = \begin{bmatrix} G & H & I \\ J & K & -L \\ -M & -N & O \end{bmatrix} \qquad [L^{\mathrm{X}}] = \begin{bmatrix} A & & \\ -B & C & \\ D & -E & F \end{bmatrix}$$

$H = 0 = J \qquad Q = O = S \qquad H' = 0 = J'$

$$A = 20a^2 D_y + 8b^2 D_{xy}$$

$$B = 15abD_1$$

$$C = 20b^2 D_x + 8a^2 D_{xy}$$

$$D = 30\frac{a^2}{b} D_y + 15bD_1 + 6bD_{xy}$$

$$E = 30\frac{b^2}{a} D_x + 15aD_1 + 6aD_{xy}$$

$$F = 60\frac{b^2}{a^2} D_x + 60\frac{b^2}{a^2} D_y + 30D_1 + 84D_{xy}$$

$$G = 10a^2 D_y - 2b^2 D_{xy}$$

$$I = -30\frac{a^2}{b} D_y - 6bD_{xy}$$

$$K = 10b^2 D_x - 8a^2 D_{xy}$$

$$L = 15\frac{b^2}{a} D_x - 15aD_1 - 6aD_{xy}$$

$$O = 30\frac{b^2}{a} - 60\frac{a^2}{b^2} D_y - 30D_1 - 84D_{xy}$$

$$P = 10a^2 D_y - 8b^2 D_{xy}$$

$$R = -15\frac{a^2}{b} D_y + 15bD_1 + 6bD_{xy}$$

$$G' = 5a^2 D_y + 2b^2 D_{xy}$$

$$M' = 15\frac{a^2}{b} D_y - 6bD_{xy}$$

$$T = 10b^2 D_x - 2a^2 D_{xy}$$

$$Y = 30\frac{b^2}{a} D_x + 6aD_{xy}$$

$$K' = 5b^2 D_x + 2a^2 D_{xy}$$

$$N' = 15\frac{b^2}{a} D_x - 6aD_{xy}$$

$$Z = 60\frac{b^2}{a^2} D_x + 30\frac{a^2}{b^2} D_y - 30D_1 - 84D_{xy}$$

$$O' = -30\frac{b^2}{a^2} D_x - 30\frac{a^2}{b^2} D_y + 30D_1 + 84D_{xy}$$

Table AIB.19. The Semiloof shell element

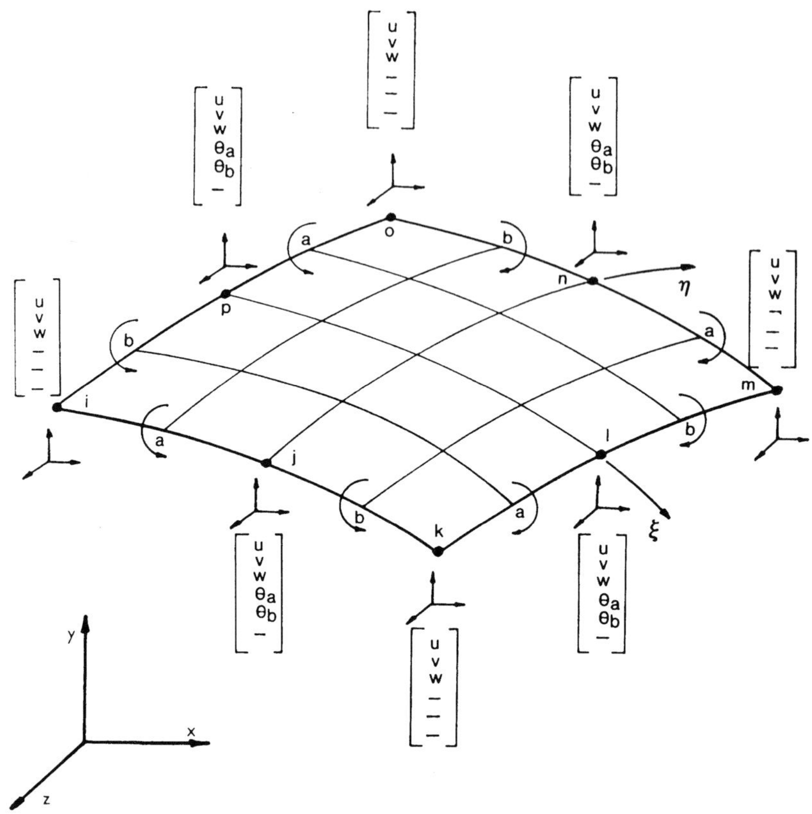

Loof nodal degrees of freedom

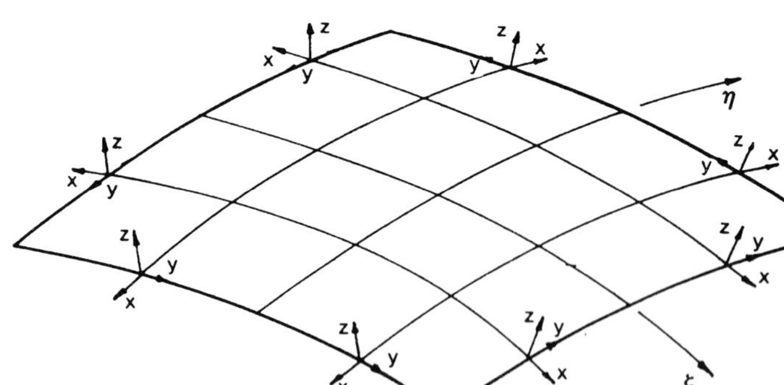

Bending strain nomenclature

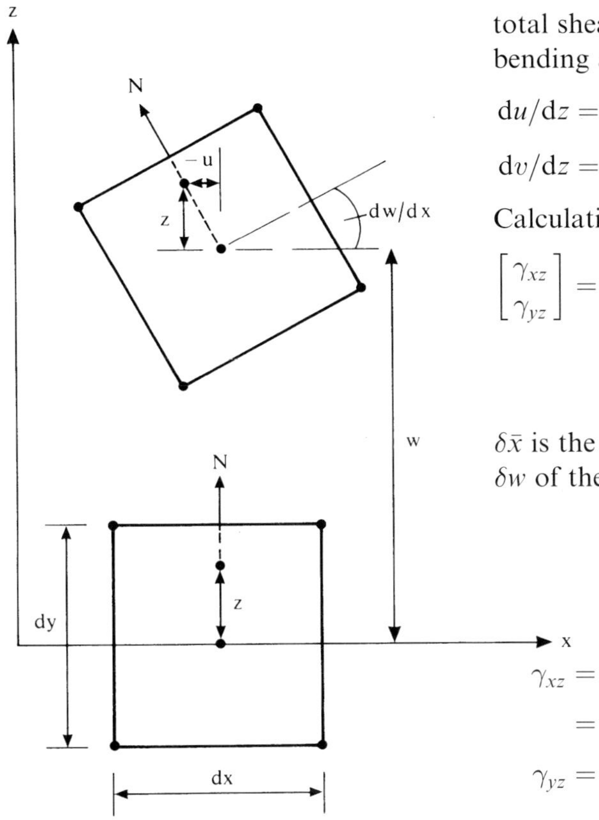

total shear strain
bending and membrane

$$du/dz = -z(d^2w/dx^2) + du/dx$$

$$dv/dz = -z(d^2w/dy^2) + dv/dy$$

Calculations of transverse shear strains

$$\begin{bmatrix} \gamma_{xz} \\ \gamma_{yz} \end{bmatrix} = \begin{bmatrix} w_x \\ w_y \end{bmatrix} + \frac{\delta}{T}\begin{bmatrix} \bar{x} \\ \bar{y} \end{bmatrix} - \frac{1}{T}\begin{bmatrix} ux & uy \\ vx & vy \end{bmatrix}\begin{bmatrix} R \\ S \end{bmatrix}$$

$\delta\bar{x}$ is the difference in x displacement.
δw of the top and bottom surfaces.

$$\gamma_{xz} = du/dz + dw/dx$$
$$= \text{shear strain on } xz \text{ face}$$
$$\gamma_{yz} = dv/dz + dw/dy$$
$$= \text{shear strain on } xy \text{ face}$$
$$du/dz = (\delta\bar{x}/T) - (R/T)(du/dx) - (S/T)(du/dy)$$

Shear strain nomenclature

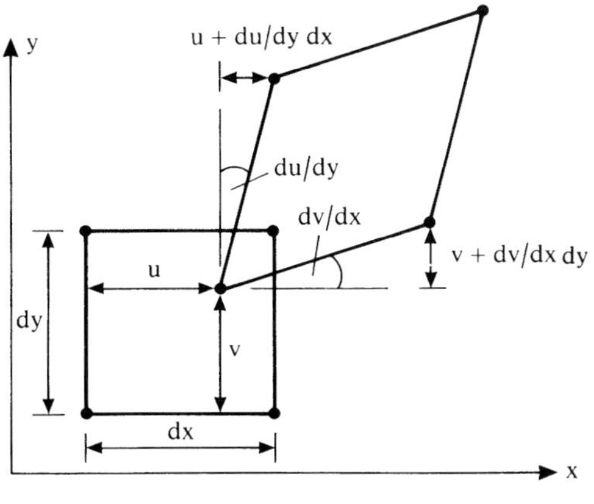

Note: x, y axes are the same as X, Y in the text.

Calculation of equivalent rotational second derivatives

$$\begin{bmatrix} u_{xz} & v_{xz} \\ u_{yz} & v_{yz} \end{bmatrix} = \frac{1}{T}\begin{bmatrix} \delta_x \\ \delta_y \end{bmatrix}\overline{\begin{matrix} \bar{x} & \bar{y} \end{matrix}} - \frac{1}{T}\begin{bmatrix} R_x & S_x \\ R_y & S_y \end{bmatrix}\begin{bmatrix} u_x & v_x \\ u_y & v_y \end{bmatrix} + \frac{1}{T}\begin{bmatrix} T_x \\ T_y \end{bmatrix}\overline{\begin{matrix} w_x & w_y \end{matrix}}$$

$$u_{xz} = -w_{xx} = -\mathrm{d}^2w/\mathrm{d}x^2$$

therefore $z(u_{xz})$ = direct tensile strain in the y direction at height z, due to rotations about the x axis.

$$v_{yz} = -w_{yy} = -\mathrm{d}^2w/\mathrm{d}y^2$$

$z(v_{yz})$ = direct tensile strain in the x direction at height z, due to rotations about the y axis.

$$u_{yz} + v_{xz} = -2w_{xy} = -2(\mathrm{d}^2w/\mathrm{d}x\,\mathrm{d}y)$$

$z(u_{yz} + v_{xz})$ = shear strain in the xy plane at height z, due to rotations about the x and y axes.

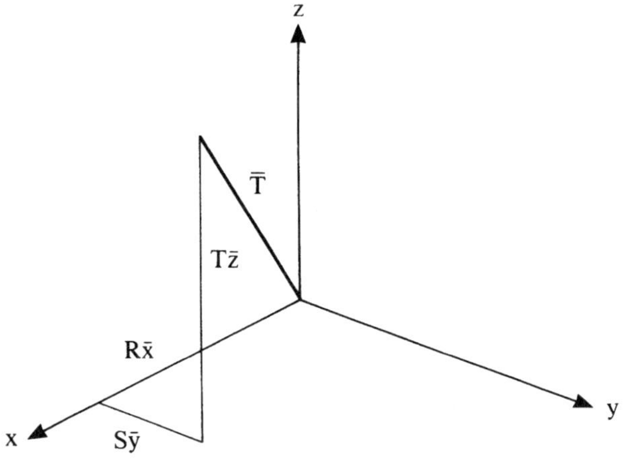

An orthogonal vector system is defined where R, S and T represent slope components of the vector thickness $\bar{T}$

$$\bar{T} = R_{\bar{x}} + S_{\bar{y}} + T_{\bar{z}}$$

where T is the resolute thickness in the z direction, R is the resolute thickness in the x direction, and S is the resolute thickness in the y direction

$$\bar{T} = \sum_{j=1}^{9} \bar{T}jL_j$$

where L_j is a Loof mapping function.

Table AIB.20. Material compliance matrix

$$[D] = \frac{Et}{(1-\nu^2)} \begin{bmatrix} 1 & \nu & 0 & 0 & 0 & 0 \\ \nu & 1 & 0 & 0 & 0 & 0 \\ 0 & 0 & 1/2(1-\nu) & 0 & 0 & 0 \\ 0 & 0 & 0 & t^2/12 & \nu t^2/12 & 0 \\ 0 & 0 & 0 & \nu t^2/12 & t^2/12 & 0 \\ 0 & 0 & 0 & 0 & 0 & t^2(1-\nu)/24 \end{bmatrix}$$

APPENDIX II

Transformation matrices

Table AII.1. Stress and strain transformation matrices

$\{T''_\sigma\}\{T''_\varepsilon\}$

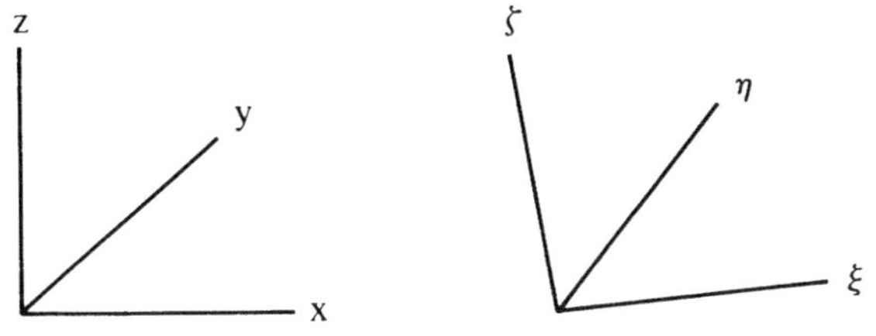

Global axis Local axis

	x	y	z
ξ	l_1	m_1	n_1
η	l_2	m_2	n_2
ζ	l_3	m_3	n_3

Direction cosines of the two axes are given by:

$$l_1 = \cos(\xi, x) \qquad m_1 = \cos(\xi, y) \qquad n_1 = \cos(\xi, z)$$
$$l_2 = \cos(\eta, x) \qquad m_2 = \cos(\eta, y) \qquad n_2 = \cos(\eta, z)$$
$$l_3 = \cos(\zeta, x) \qquad m_3 = \cos(\zeta, y) \qquad n_3 = \cos(\zeta, z)$$

The following relationships can be written for local and global strain and stress vectors:

$$\{\varepsilon'_x\} = [T_\varepsilon]\{\varepsilon_x\} \qquad \{\sigma_x\} = [T_\varepsilon]^T\{\sigma'_x\}$$

and also

$$\{\sigma'_x\} = [T_\sigma]\{\sigma_x\} \qquad \{\varepsilon_x\} = [T_\sigma]^T\{\varepsilon'_x\}$$

$$[T''_\varepsilon] = \begin{bmatrix} l_1^2 & m_1^2 & n_1^2 & l_1m_1 & m_1n_1 & l_1n_1 \\ l_2^2 & m_2^2 & n_2^2 & l_2m_2 & m_2n_2 & l_2n_2 \\ l_3^2 & m_3^2 & n_3^2 & l_3m_3 & m_3n_3 & l_3n_3 \\ 2l_1l_2 & 2m_1m_2 & 2n_1n_2 & l_1m_2 + l_2m_1 & m_1n_2 + m_2n_1 & l_1n_2 + l_2n_1 \\ 2l_2l_3 & 2m_2m_3 & 2n_2n_3 & l_2m_3 + l_3m_2 & m_2n_3 + n_2m_3 & l_2n_3 + l_3n_2 \\ 2l_1l_3 & 2m_1m_3 & 2m_1n_3 & l_1m_3 + m_1l_3 & m_1n_3 + m_3n_1 & l_1n_3 + n_1l_3 \end{bmatrix}$$

$$[T''_\sigma] = \begin{bmatrix} l_1^2 & m_1^2 & n_1^2 & 2l_1m_1 & 2m_1n_1 & 2l_1n_1 \\ l_2^2 & m_2^2 & n_2^2 & 2l_2m_2 & 2m_2n_2 & 2l_2n_2 \\ l_3^2 & m_3^2 & n_3^2 & 2l_3m_3 & 2m_3n_3 & 2l_3n_3 \\ l_1l_2 & m_1m_2 & n_1n_2 & l_1m_2 + l_2m_1 & m_1n_2 + n_1m_2 & l_1n_2 + l_2n_1 \\ l_2l_3 & m_2m_3 & n_2n_3 & l_2m_3 + l_3m_2 & m_2n_3 + n_2m_3 & l_2n_3 + l_3n_2 \\ l_1l_3 & m_1m_3 & n_1n_3 & l_1m_3 + l_3m_1 & m_1n_3 + m_3n_1 & l_1n_3 + n_1l_3 \end{bmatrix}$$

Table AII.2. Algorithm for principal stresses

Principal stresses and direction cosines D1, D2, D3 are the direction cosines of principal stresses PS1, PS2, PS3

```
      IF (X5 .GE. X6 .AND. X6 .GE. X7) GOTO 430
      IF (X5 .GE. X7 .AND. X7 .GE. X6) GOTO 431
      IF (X6 .GE. X5 .AND. X5 .GE. X7) GOTO 432
      IF (X6 .GE. X7 .AND. X7 .GE. X5) GOTO 433
      IF (X7 .GE. X5 .AND. X5 .GE. X6) GOTO 434
      IF (X7 .GE. X5 .AND. X6 .GE. X5) GOTO 435
  430 X1=X5
      X2=X6
      X3=X7
      GOTO 438
  431 X1=X5
      X2=X7
      X3=X6
      GOTO 438
  432 X1=X6
      X2=X5
      X3=X7
      GOTO 438
  433 X1=X6
      X2=X7
      X3=X5
      GOTO 438
  434 X1=X7
      X2=X5
      X3=X6
      GOTO 438
  435 X1=X7
      X2=X6
      X3=X5
  438 CONTINUE

      PRINCIPAL STRESSES

      PS1(IPT)=X1
      PS2(IPT)=X2
      PS3(IPT)=X3
      DO 440 IS=1,3
      GOTO (443,445,447),IS
  443 AS1=G1-X1
      AS2=G2-X1
      AS3=G3-X1
      GOTO 444
  445 AS1=G1-X2
      AS2=G2-X2
      AS3=G3-X2
      GOTO 444
  447 AS1=G1-X3
      AS2=G2-X3
      AS3=G3-X3
  444 CONTINUE
      AK=G4
      BK=G5
      CK=G6
      YAP1=AS2*CK-BK*AK
      YAP2=AK*AK-AS1*AS2
      IF (YAP1 .EQ. 0.0) YAP1=1.0
      IF (YAP2 .EQ. 0.0) YAP2=1.0
      BJM1=(BK*BK-AS2*AS3)/YAP1
      BJM2=(AS1*BK-AK*CK)/YAP2
      BJ1=BJM1*BJM1
      BJ2=BJM2*BJM2
      ZIP=DSQRT(BJ1+BJ2+1.0)
```

APPENDIX III

Material and cracking matrices

Table AIII.1. Material matrices [*D*] & [*D**] cracking matrices

Uncracked material matrix [D]

$$[D] = \begin{bmatrix} D_{11} & D_{12} & D_{13} & 0 & 0 & 0 \\ D_{21} & D_{22} & D_{23} & 0 & 0 & 0 \\ D_{31} & D_{32} & D_{33} & 0 & 0 & 0 \\ 0 & 0 & 0 & D_{44} & 0 & 0 \\ 0 & 0 & 0 & 0 & D_{55} & 0 \\ 0 & 0 & 0 & 0 & 0 & D_{66} \end{bmatrix}$$

Cracked material matrices [D*]

Cracks in principal direction 'one'

$$D_{11}^* = D_{12}^* = D_{13}^* = D_{21}^* = D_{31}^* = 0$$

$$D_{22}^* = D_{22} - D_{12}D_{12}/D_{11}$$

$$D_{33}^* = D_{33} - D_{13}D_{13}/D_{11}$$

$$D_{23}^* = D_{23} - D_{12}D_{13}/D_{11}$$

$$D_{44}^* = \beta' D_{44}$$

$$D_{55}^* = D_{55}$$

$$D_{66}^* = \beta' D_{66}$$

Cracks in principal direction 'two'

$$D_{22}^* = D_{12}^* = D_{21}^* = D_{23}^* = D_{32}^* = 0$$

$$D_{11}^* = D_{11} - D_{21}D_{21}/D_{22}$$

$$D_{33}^* = D_{33} - D_{23}D_{23}/D_{22}$$

$$D_{13}^* = D_{13} - D_{12}D_{23}/D_{22}$$

$$D_{31}^* = D_{13}^*$$

$$D_{44}^* = \beta' D_{44}$$

$$D_{55}^* = \beta' D_{55}$$

$$D_{66}^* = D_{66}$$

Cracks in principal direction 'three'

$$D^*_{33} = D^*_{13} = D^*_{31} = D^*_{23} = D^*_{31} = 0$$
$$D^*_{11} = D_{11} - D_{13}D_{13}/D_{33}$$
$$D^*_{22} = D_{22} - D_{23}D_{23}/D_{33}$$
$$D^*_{12} = D_{12} - D_{13}D_{23}/D_{33}$$
$$D^*_{21} = D^*_{12}$$
$$D^*_{44} = D_{44}$$
$$D^*_{55} = \beta' D_{55}$$
$$D^*_{66} = \beta' D_{66}$$

Cracks in principal directions 'one' and 'two'

$$D^*_{11} = D^*_{22} = D^*_{12} = D^*_{21} = 0$$
$$D^*_{13} = D^*_{31} = D^*_{23} = D^*_{32} = 0$$
$$D^*_{33} = D_{33} - D_{13}D_{13}/D_{11} - D_{23}D_{23}/D_{22}$$
$$D^*_{44} = \beta' D_{44}$$
$$D^*_{55} = \beta' D_{55}$$
$$D^*_{66} = \beta' D_{66}$$

Cracks in principal directions 'two' and 'three'

$$D^*_{22} = D^*_{33} = D^*_{23} = D^*_{32} = 0$$
$$D^*_{13} = D^*_{31} = D^*_{12} = D^*_{21} = 0$$
$$D^*_{11} = D_{11} - D_{21}D_{21}/D_{22} - D_{13}D_{13}/D_{33}$$
$$D^*_{44} = \beta' D_{44}$$
$$D^*_{55} = \beta' D_{55}$$
$$D^*_{66} = \beta' D_{66}$$

Cracks in principal directions 'three' and 'one'

$$D^*_{11} = D^*_{33} = D^*_{12} = D^*_{21} = 0$$
$$D^*_{13} = D^*_{31} = D^*_{23} = D^*_{31} = 0$$
$$D^*_{22} = D_{22} - D_{12}D_{12}/D_{11} - D_{23}D_{23}/D_{33}$$
$$D^*_{44} = \beta' D_{44}$$
$$D^*_{55} = \beta' D_{55}$$
$$D^*_{66} = \beta' D_{66}$$

Table AIII.2. Computer program ISOPAR: material models subroutines

Cracks in all three principal directions

$$[D^*] = [0]$$

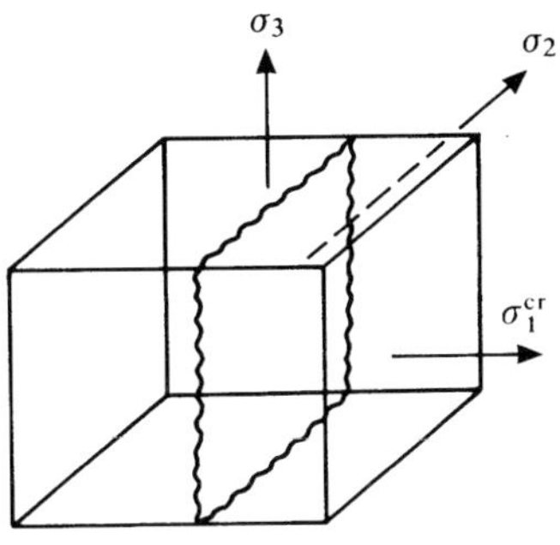

(a) One crack

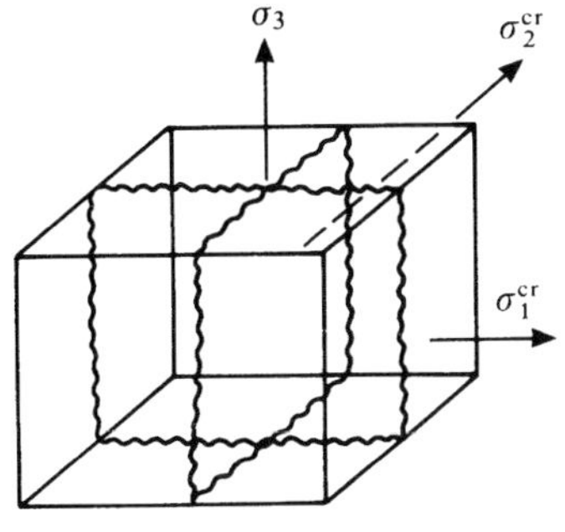

(b) Two cracks

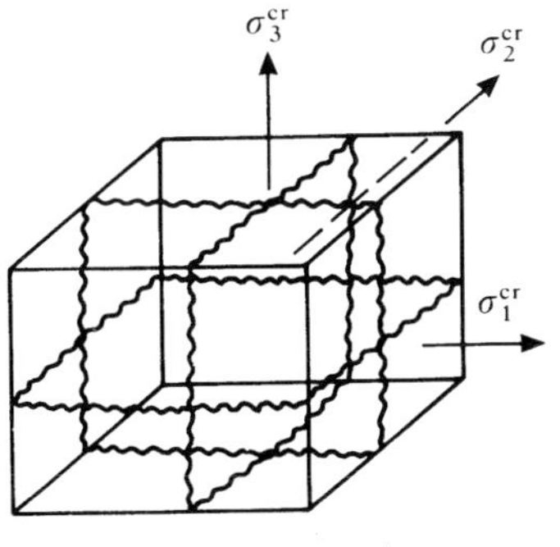

(c) Three cracks

Material matrix for reinforcement

```
      IMPLICIT REAL*8(A – H,O – Z)
      COMMON /MTMD3D/ D(6,6),STRESS(6),STRAIN(6),IPT,NEL
      DIMENSION PROP(1),DS(6,6),SIG(1),EPS(1),NCK(1),PS1(1),PS2(1),
     1               PS3(1),DC1(1),DC2(1),DC3(1)
      DO 111 II = 1,6
      DO 111 JJ = 1,6
  111 DS(II,JJ) = 0.0
      DS(1,1) = PROP(9)/PROP(6)*PROP(2)
      DS(2,2) = PROP(10)/PROP(7)*PROP(2)
      DS(3,3) = PROP(11)/PROP(8)*PROP(2)
      CALL TESTCK (PROP,SIG,EPS,NCK,PS1,PS2,PS3,DC1,DC2,DC3)
      IF (NCK(1) .EQ. 1 .OR. NCK(2) .EQ. 1 .OR. NCK(3) .EQ. 1)
     @   GOTO 220
      CALL DMAT(PROP,NCK)
  220 DO 222 III = 1,6
      DO 222 JJJ = 1,6
  222 D(III,JJJ) = D(III,JJJ) + DS(III,JJJ)
      RETURN
   END
```

Orthotropic variable-modulus model for concrete

```
      IMPLICIT REAL*8(A – H,O – Z)
      DIMENSION E(3),G(3,3),D(6,6),PROP(1)
      DO 222 II = 1,6
      DO 222 JJ = 1,6
 222  D(II,JJ) = 0.0
      AA = (1.0 – PROP(S))/(1.0 + PROP(5))*(1.0 – 2.0*PROP(5))
      BB = PROP(5)/(1.0 – PROP(5))
      E(1) = PROP(12)*PROP(1)*PROP(6) + PROP(2)*PROP(9)
      E(2) = PROP(12)*PROP(1)*PROP(7) + PROP(2)*PROP(10)
      E(3) = PROP(12)*PROP(1)*PROP(8) + PROP(2)*PROP(11)
      DO 7100 J = 1,3
      DO 7100 K = 1,3
7100  G(J,K) = 0.25*(AA*(E(J) + E(K))) – 2.0*AA*BB*DSQRT(E(J)*E(K))
      D(1,1) = AA*E(1)
      D(1,2) = AA*BB*DSQRT(E(1)*E(2))
      D(1,3) = AA*BB*DSQRT(E(1)*E(3))
      D(2,1) = D(1,2)
      D(2,2) = AA*E(2)
      D(2,3) = BB*DSQRT(E(2)*E(3))
      D(3,1) = D(1,3)
      D(3,2) = D(2,3)
      D(3,3) = AA*E(3)
      D(4,4) = G(1,2)
      D(5,5) = G(1,3)
      D(6,6) = G(2,3)
      RETURN
      END
```

Ottoson model

```
      IMPLICIT REAL*8(A – H,O – Z)
      COMMON /MTMD3D/ DEP(6,6),STRESS(6),STRAIN(6),IPT,NEL
      DIMENSION PAR(3,5),FS(6,6),FSTPOS(6,6),PROP(1),SIG(1),
     @          DVI1DS(6),DVJ2DS(6),DVJ3DS(6),DVTHDS(6)
      OPEN (UNIT = 5,FILE = 'PARAMETERS',STATUS = 'OLD')
      READ (5,*,END = 3700)((PAR(IF,JF),JF = 1,5),IF = 1,3)
3700  CLOSE (5)
      PK = PROP(3)/PROP(4)
      IP = 0
      JP = 0
      IF (PK .LE. 0.08) IP = 1
      IF (PK .EQ. 0.10) IP = 2
      IF (PK .GE. 0.12) IP = 3
      IF (PK .LT. 0.10) JP = 1
      IF (PK .GT. 0.10) JP = 2
      IF (IP .EQ. 0) GOTO 3800
      A = PAR(IP,2)
      B = PAR(IP,3)
      PK1 = PAR(IP,4)
      PK2 = PAR(IP,5)
      GOTO 3900
3800  SUB1 = PK – PAR(JP,1)
      SUB2 = PAR(JP + 1,1) – PAR(JP,1)
      A = SUB1*(PAR(JP + 1,2) – PAR(JP,2))/SUB2 + PAR(JP,2)
      B = SUB1*(PAR(JP + 1,3) – PAR(JP,3))/SUB2 + PAR(JP,3)
      PK1 = SUB1*(PAR(JP + 1,4) – PAR(JP,4))/SUB2 + PAR(JP,4)
```

```
      PK2 = SUB1*(PAR(JP + 1,5) - PAR(JP,5))/SUB2 + PAR(JP,5)
 3900 VARI1 = SIG(1) + SIG(2) + SIG(3)
      VARJ2 = 1.0/6.0*((SIG(1) - SIG(2))**2 + (SIG(2) - SIG(3))**2 +
     @          (SIG(3) - SIG(1))**2) + SIG(4)**2 + SIG(5)**2 + SIG(6)**2
      VARI13 = VARI1/3.0
      VI131 = SIG(1) - VARI13
      VI132 = SIG(2) - VARI13
      VI133 = SIG(3) - VARI13
      VARJ3 = VI131*(VI132*VI133 - SIG(5)**2) - SIG(4)*(SIG(4)*VI133
     @          - SIG(5)*SIG(5)) + SIG(6)*(SIG(4)*SIG(5) - SIG(6)*VI132)
      VAR3TH = 1.5*3.0**(0.5)*VARJ3/VARJ2**1.5
      IF (VAR3TH .GE. 0.0) GOTO 4000
      ALAM = 22.0/21.0 - 1.0/3.0*ACOS(- PK2*VAR3TH)
      TOTLAM = PK1*COS(ALAM)
      DFD3TH = PK1*PK2*VARJ2**0.5*SIN(ALAM)/(3.0*PROP(4)*
     @          SIN(ACOS(- PK2*VAR3TH)))
      GOTO 4100
 4000 ALAM = 1.0/3.0*ACOS(PK2*VAR3TH)
      TOTLAM = PK1*COS(ALAM)
      DFD3TH = PK1*PK2*VARJ2**0.5*SIN(ALAM)/(3.0*PROP(4)*
     @          SIN(ACOS(PK2*VAR3TH)))
 4100 DFDI1 = B/PROP(4)
      DFDJ2 = A/PROP(4)**2 + TOTLAM/(PROP(4)*VARJ2**0.5)
      DVI1DS(1) = 1.0
      DVI1DS(2) = 1.0
      DVI1DS(3) = 1.0
      DVI1DS(4) = 0.0
      DVI1DS(5) = 0.0
      DVI1DS(6) = 0.0
      DVJ2DS(1) = 1.0/3.0*(2.0*SIG(1) - SIG(2) - SIG(3))
      DVJ2DS(2) = 1.0/3.0*(2.0*SIG(2) - SIG(1) - SIG(3))
      DVJ2DS(3) = 1.0/3.0*(2.0*SIG(3) - SIG(1) - SIG(2))
      DVJ2DS(4) = 2.0*SIG(4)
      DVJ2DS(5) = 2.0*SIG(5)
      DVJ2DS(6) = 2.0*SIG(6)
      DVJ3DS(1) = 1.0/3.0*(VI131*(- VI132 - VI133)) + 2.0*VI132*VI131
     @          - 2.0*SIG(5)**2 + SIG(4)**2 + SIG(6)**2
      DVJ3DS(2) = 1.0/3.0*(VI132*(- VI131 - VI133)) + 2.0*VII31*VI133
     @          - 2.0*SIG(6)**2 + SIG(4)**2 + SIG(5)**2
      DVJ3DS(3) = 1.0/3.0*(VI133*(- VI131 - VI132)) + 2.0*VI131*VI132
     @          - 2.0*SIG(4)**2 + SIG(5)**2 + SIG(6)**2
      DVJ3DS(4) = - 2.0*VI133*SIG(4) + 2.0*SIG(5)*SIG(6)
      DVJ3DS(5) = - 2.0*VI131*SIG(5) + 2.0*SIG(4)*SIG(6)
      DVJ3DS(6) = - 2.0*VI132*SIG(6) + 2.0*SIG(4)*SIG(5)
      CONVJ2 = 3.0*3.0**0.5/(2.0*VARJ2*1.2)
      VJ3J2 = VARJ3/VARJ2**0.5
      DVTHDS(1) = CONVJ2*(- 0.5*VJ3J2*(2.0*SIG(1) - SIG(2) - SIG(3)) + DVJ3DS(1))
      DVTHDS(2) = CONVJ2*(- 0.5*VJ3J2*(2.0*SIG(2) - SIG(1) - SIG(3)) + DVJ3DS(2))
      DVTHDS(3) = CONVJ2*(- 0.5*VJ3J2*(2.0*SIG(3) - SIG(1) - SIG(2)) + DVJ3DS(3))
      DVTHDS(4) = CONVJ2*(- 3.0*VJ3J2*SIG(4) + DVJ3DS(4))
      DVTHDS(5) = CONVJ2*(- 3.0*V33J2*SIG(5) + DVJ3DS(5))
      DVTHDS(6) = CONVJ2*(- 3.0*VJ3J2*SICA6) + DVJ3DS(6))
      DO 4200 IS = 1,6
      FS(IS,1) = DFDI1*DVI1DS(IS) + DFDJ2*DVJ2DS(IS) + DFD3TH*DVTHDS(IS)
 4200 FSTPDS(1,IS) = FS(IS,1)
      RETURN
      END
```

Table AIII.3. Finite element thermal analysis of concrete: Wickstrom program

U. L. F. Wickstrom program on thermal analysis

Reproduced courtesy of U. L. F. Wickstrom and the Lund Institute of Technology, Sweden.

```
      SUBROUTINE PROG2(IX,IY,NN,NE,NR,N,KTOP,NODFL,MNODFL,X,Y,T,TT,TMAX,
     1 ELA,EV4,A,MAX,P,W,EN,F,FLOW,AXIAL,NODCPL,NODINT,DTA)
C-----------------------------------------------------------------------
C-----
C-----                      *** TASEF ***
C-----
C----- TEMPERATURE ANALYSIS OF STUCTURES EXPOSED TO FIRE
C-----
C----- FINITE ELEMENT PROGRAM FOR ANALYSIS OF TRANSIENT NONLINEAR
C----- HEAT TRANSFER PROBLEMS
C-----
C----- PROGRAMMED BY
C----- ULF WICKSTROM
C----- LUND INSTITUTE OF TECHNOLOGY
C----- MARCH 1979
C-----
C-----------------------------------------------------------------------
C----- THIS IS THE MAIN CONTROL ROUTINE
C-----
C----- DEFINITIONS OF VARIABLES
C----- IX, IY      NUMBER OF X- AND Y- LINES
C----- NN          NUMBER OF NODES IN BASE STRUCTURE
C----- NE          NUMBER OF ELEMENTS IN BASE STRUCTURE
C----- NR          NUMBER OF REGIONS
C----- N           VECTOR OF REGION NUMBERS
C----- KTOP        NODES ADJACENT TO EACH ELEMENT
C----- NODEL       ELEMENTS ADJACENT TO EACH NODE
C----- MNODEL      NUMBER ELEMENTS ADJACENT TO EACH NODE
C----- X, Y        NODE COORDINATES
C----- T           CURRENT NODAL TEMPERATURES
C----- TT          MAXIMUM NODAL TEMPERATURE5
C----- TMAX        TRUE IF MAXIMUM NODAL TEMPERATURE OBTAINED
C----- ELA,EV4     DUMMY GEOMETRICAL CONSTANTS
C----- A           HEAT CONDUCTION MATRIX
C----- P           HEAT CAPACITY VECTOR
C----- W           NODAL VOLUME VECTOR
C----- EN          MODAL ENTHALPY VECTOR
C----- F           INTERNAL NODAL HEAT FLOW VECTOR
C----- FLOW        EXTERNAL NODAL HEAT FLOW VECTOR
C----- AXIAL       TRUE IF AXISYMMETRIC PROBLEM
C----- NODCPL      INDICATES COUPLED NODES
C----- NODINT      INDICATES INTERFACE NODES
C----- DTA         DUMMY VECTOR FOR CRITICAL TIME INCREMENT CALCULATION
C-----
C----- PARAMETER CONSTANT5
C-----
C----- NB          MAXIMUM NUMBER OF NODE GROUPS

      SUBROUTINE OUTMA2(IX,IY,NN,NE,X,Y,TIME,KTIME,T,TT,TMAX,FLOW,AXIAL)
C----- THIS ROUTNE PRINTS MAXIMUM CALCULATED NODAL TEMPERATURES
```

```
          COMMON/FIRE/TIM(5C),TB(50),TITFIR
          INTEGER TITFIR(18)
          LOGICAL TMAX,AXIAL
          DIMENSION X(NN),Y(NN),T(NN),TT(NN),TMAX(NN),FLOW(NN)
          PRINT 200,TITFIR,X(NN),Y(NN)
          IDUM1 = 1-IY
          DO 10 I = 1,IX
          IDUM1 = IDUM1 + IY
          IDUM2 = IDUM1 + IY-1
          IF (IY.LE.7) PRINT 210, (J,TT(J),J = IDUM1,IDUM2)
          IF (IY.GT.7) PRINT 230, (TT(J),J = IDUM1,IDUM2)
   10     CONTINUE
          PRINT 220,TIME,KTIME
  200     FORMAT(/////1X,75(1HF)/2H,FI/*F     MAXIMAL TEMPERATURES'/' F*,
          1 18A4/' F XMAX = ',F8.3,1CX,'YMAX = ',F8.3/' F')
  210     FORMAT('F',13(I5,F5.0))
  220     FORMAT(2H F/2H F/' F MAX-TIME',F7.2,10X,'NUMBER OF '
          1          'TIME INCREMENTS',15/2H F/2H F/2H F,75(1HF))
  230     FORMAT(' F',18F7.C)
          RETURN
          END

          SUBROUTINE OUT2(IXIY,NN,NE,X,Y,TIME,KTIME,DELTI.T.TT,TMAX,FLOW,
          1      TFIRE,NODT,AXIAL)
C - - - - - THIS ROUTINE PRINTS NODAL TEMPERATURES AND VOID AIR
          TEMPERATURES
          DIMENSION X(NN),Y(NN),T(NN),TT(NN),TMAX(NN),FLOW(NN)
          LOGICAL TMAX,AXIAL,LDUM,LEN
          COMMON/ENCON/H(50),TAIR(2)
          COMMON/ENCLOS/LEN,NENC,NENCNG(2),IGREN(2,4),NNODEN(2),
          1      INODEN(100),XSYM(7),YSYM(2)
          COMMON/TOUT/II,TOUT(100),TIMMAX,DTMAX,TIMFAC,KTMAX,KUPDA
          TIME1 = TIME-DELTI
          DO 5 IJ = 1,NN
          IF(TMAX(IJ)) GOTO 5
C - - - - - IF THE NODAL TEMPERATURE DECREASES SET TMAX = .TRUE. AND PRINT
C - - - - - MAX TEMPERATURE TT
          IF(TT(IJ).GT.1.001*T(IJ))
          1      PRINT 200,1J,TT(IJ),TIME1,DELTI
          IF(TT(IJ).GT.1.001*T(IJ)) TMAX(IJ) = .TRUE.
          TT(IJ) = AMAX 1 (TT(IJ),T(IJ))
    5     CONTINUE
C - - - - - IF TIME = TOUT PRINT ALL TEMPERATURES
          IF((TIME-TOUT(II)).LT.-1.E-4) GOTO 70
          PRINT 100,TIME,KTIME,TFIRE,NODT
          IF(.NOT.LEN) GOTO 30
          PRINT 300
          DO 20 I = 1,NENC
   20     PRINT 310,I,TAIR(I)
   30     CONTINUE
          II1 = II + 1
          IDUM1 = 1-IY
          LDUM = IY.LT.7
          DO 10 I = 1,IX
          IDUM1 = IDUM1 + IY
          IDUM2 = IDUM1 + IY - 1
```

```
      SUBROUTINE BRBCA(BR,BC,EPSIG,BET,BAR,NUMI,N3,ING1)
C - - - - - FORM BOUNDARY RADIATION AND CONVECTION MATRICES
      DIMENSION BR(NUMI,2),BC(NUMI,2),BAR(NB,NUMI)
      BR(1,1)=0.
      BR(1,2)=.33333333*BAR(ING1,2)
      NUM1=NUMI-1
      IF(NUM1.EQ.1) GOTO 20
      DO 10 I=2,NUM1
      BR(I,1)=.16666667*BAR(ING1,I)
      BR(I,2)=.3333333*(BAR(ING1,I)+BAR(ING1,I+1))
   10 CONTINUE
   20 CONTINUE
      BR(NUM1,1)=.16666667*BAR(ING1,NUMI)
      BR(NUM1,2)=.33333333*BAR(ING1,NUMI)
C
      DO 30 I=1,NUMI
      DO 30 J=1,2
      BC(I,J)=BET*BR(I,J)
      BR(I,J)=EPSIG*BR(I,J)
   30 CONTINUE
      RETURN
      END

      SUBROUTINE BRBCB(BR,BC,TR,TC,TRD,TCD.NUMI,DTA,NN,MAX,FLOW,TG,
      1    T,ING1)
C - - - - - THIS ROUTINE CALCULATES EXTERNAL HEAT FLOW BY RADIATION AND
C - - - - - CONVECTION AND ADDS THE CORRESPONDING CONTRIBUTIONS TO THE
C - - - - - VECTOR DATA FOR CALCULATION OF CRITICAL TIME INCREMENT
      DIMENSION BR(NUMI,2),BC(NUMI,2),DTA(NN),FLOW(NN),T(NN)
      1, TR(NUMI),TC(NUMI),TRD(NUMI),TCD(NUMI)
      PARAMETER NB=10,NNP=30,NNB2=2*NNB
      COMMON/BNOD/NUMB(NB),NBOUND(NB,NNB).BAREA(NB,NNB),
      1    EPSG(NB),BETA(NB),CPG(NB),FA(NB)
      LOGICAL FA
C
C - - - - - FIRST NODE
C
      NODE=NBOUND(ING1,1)
      TR2=TR(1)
      TC2=TC(1)
      TR3=TR(2)
      TC3=TC(2)
      TRCD2=TRD(1)
      TCD2=TCD(1)
      TRD3=TRD(2)
      TCD3=TCD(2)
      BR2=BR(1,2)
      BC2=BC(1,2)
      BR3=BR(2,1)
      BC3=BC(2,1)
      FLW=BR2*TR2+BC2*TC2
      FLW=FLW+BR3*TR3
      FLW=FLW+5C3*TC3
      DA=BR2*TRD2+BC2*TCD2
      DA=DA+BR3*TRD3
```

```
      SUBROUTINE PTBNDB (T,TFIRE)
C - - - - - SET PRESCRIBED NODAL BOUNDARY TEMPERATURE
      DIMENSION T(1)
      PARAMETER NB = 10,NNB = 30,NNB2 = 2*NNB
      COMMON/PTB/NPTNG,NPTG(NB)
      COMMON/BNOD/NUMB(NP),NBOUND(NB,NNB),BAREA(NB,NNB),TH(NB),
     1    EPSG(NS),BETA(NB),CPG(NB),FA(NB)
      COMMON/UNIT/SIGMA,TABS,TINIT,TAMB,TAMB4
      LOGICAL FA
C - - - - -
      IF(NPTNG.EQ.0) RETURN
C - - - - - EACH PRESCRIBED TEMPERATURE BOUNDARY NODE GROUP
      DO 10 IB = 1,NPTNG
      TG = TAMB
      ING1 = NPTG(IB)
      IF(FA(ING1)) TG - TFIRE
      NUMI  = NUMB(ING1)
      DO 10 I = 1,NUMI
      NODE = NBOUND(ING1,I)
      T(NODE)  = TG
 10   CONTINUE
      RETURN
      END

      SUBROUTINE RADVEC (F,ETA,N,Q)
C - - - - - THIS ROUTINE FORMS THE LOCAL ENCLOSURE SURFACE RADIATION HEAT
C - - - - - EXCHANGE VECTOR Q = E*ETA
      DIMENSION Q(1),ETA(1),E(N,N)
      QTOT = 0.
      DO 20 I = 1,N
      QT = 0.
      DO 10 J = 1,N
 10   QT = QT + E(I,J)*ETA(J)
      QTOT = QTOT + QT
 20   Q(I) = QT
      RETURN
220   FORMAT (/' TOTAL RADIATION HEAT EXCHANGE',E11.3)
      END

      SUBROUTINE REG2(NN,NE,NR,N,KTOP,X,Y,NODEL,MNODEL)
C - - - - - THIS SUBROUTINE FORMS VECTOR OF REGION NUMBERS N OF EACH
      ELEMENT
      DIMENSION X(NN),Y(NN),N(NE),KTOP(4,NE),NODEL(4,NN),MNODEL(NN)
      PARAMETER MNR = 10
      COMMON/RGEO/ELFICT(MNR),ET(MNR),SRDIAC(4,MNR)
      LOGICAL ELFICT
      EPS = 1.E - 7
      DO 10 I = 1,NE

      N(I) = 1
      IF(NR.EQ.1) GOTO 10
      ND1 = KTOP(1,I)
      ND2 =  KTOP(4,I)
      DO 5 J = 2,NR
      IF((X(ND1) - SRDIAC(3,J)).GT. - EPS) GOTO 5
      IF((Y(ND1) - SRDIAC(4,J)).GT. - EPS) GOTO 5
      IF((X(ND2) - SRDIAC(1,J)).LT.EPS) GOTO 5
```

```
      IF((Y(ND2) – SRDIAC(2,J)).LT.EPS) GOTO 5
      N(I) = J
  5   CONTINUE
 10   CONTINUE
      DO 40 T = 1,NN
      II = 0
 20   DO 30 IE = 1,NE
      N1 = N(IE)
      IF(ELFICT(N1)) GOTO 30
      DO 20 J = 1,4
      IF(KTOP(J,IF).NF.I) GOTO 20
      II = II + 1
      NODEL(II,I) = IE
      IF(II.EQ.4) GOTO 30
 20   CONTINUE
 30   CONTINUE
      MNODEL(I) = II
 40   CONTINUE
      RETURN
      END

      SUBROUTINE ENCON1
C - - - - - THIS ROUTINE FORMS CONVECTION ARRAY H
      PARAMETER NB = 10,NNB = 30,NNB2 = 2*NNB
      COMMON/BNOD/NUMB(NB),NBOUND(NB,NNB),BAREA(NB,NNB),TH(NB),
      1     EPSG(NB),BETA(NB),CPG(NB),FA(NB)
      COMMON/ENCLOS/LEN,NENC,NENCNG(2),IGREN(2,4),NNODEN(2),
      1     INODEN(100),XSYM(2),YSYM(2)
      COMMON/ENCON/H(50),TAIR(2)
      COMMON/DUMMY/HZ(25),DUM2(25)
      LOGICAL LEN
      LOGICAL XSYM,YSYM,SYM
      INTEGER EN
      IND = 1
C - - - - - FORM ZONE CONVECTION ARRAY
C - - - - - EACH VOID
      DO 150 EN = 1,NENC
      SYM = XSYM(EN).OR.YSYM(EN)
      IN = 0
      NENG = NENCNG(EN)
C - - - - - EACH NODE GROUP
      DO 10 IG = 1,NENG
      I1 = IGREN(EN,IG)
      NUMI = NUMB(I1)
      BE = BETA(I1)
C - - - - - EACH ZONE
      DO 10 I = 2,NUMI
      IN = IN + 1
 10   HZ(IN) = BE*BAREA(I1,I)
C - - - - - FORM NODE CONVECTION ARRAY
      CALL HTRANS(HZ,H(IND),IN,SYM)
      N = IN
      IF(SYM) N = N + 1
      IND = IND + N
150   CONTINUE
      RETURN
      END
```

```
      SUBROUTINE TIME
C - - - - - READ TIME INTEGRATION CONTROL DATA
      COMMON/TOUT/II,TOUT(100),TIMMAX,DTMAX,TIMEAC,KTMAX,KUPDA
      PRINT 200
C - - - - -
      READ 100,NT,TIMMAX,DTMAX,TIMFAC,KTMAXKUPDA
C - - - - -
      IF(DTMAX.EQ.0) DTMAX = TIMMAX
      IF(TIMFAC.EQ.0) TIMFAC = .8
      IF(KTMAX.EQ.0) KTMAX = 1000
      IF(KUPDA.EQ.0) KUPDA = 1
C - - - - -
      READ 100,(TOUT(I),I = 1,NT)
C - - - - -
      PRINT 210,TIMMAX,DTMAX,TIMFAC,KTMAX,KUPDA
      PRINT 220,(TOUT(I),I = 1,NT)
100   FORMAT( )
220   FORMAT(' PRINT OUT TIMES',3X,8G7.21(19X,8G7.2))
200   FORMAT(//' TIME'/'****'/)
210   FORMAT('MAXIMUM TIME = ',G8.31' MAXIMUM TIME INCREMENT = ',G8.3/
      1     ' CRITICAL TIME INCREMENT FACTOR = ',G8.3/
      2     ' MAXIMUM NUMBER OF TIME INCREMENTS = ',15/
      3     ' NUMBER OF STEPS BETWEEN UPDATING OF CONDUCTION
             MATRIX = ',I5)
      RETURN
      END

      SUBROUTINE ENRAD1(X,Y)
C - - - - - FORM RADIATION MATRICES FOR EACH VOID AND STORE THEM IN
C - - - - - THE VECTOR E.
C - - - - - CALCULATE VIEW-FACTOR MATRIX VIEW AND ZONE AREA VECTOR D
      DIMENSION X(1),Y(1),A(25,25),B(25,25)
      PARAMETER NB = 10,NNB = 30,NNB2 = 2*NNB
      COMMON/BNOD/NUMB(NB),NBOUND(NB,NNB),BAREA(NB,NNB),
      1   EPSG(NB),BETA(NB),CPG(NB),FA(NB)
      COMMON/ENCLOS/LEN,NENC,NENCNG(2),IGREN(2,4),NNODEN(2),
      1   INODEN(100),XSYM(2),YSYM(2)
      COMMON/ENRAD/E(1000)
      COMMON/UNIT/SIGMA,TABS
      COMMON/DIM/MAXNG,MAXNOD
      COMMON/DUMMY/D(25),DUM2(25)
      DIMENSION VIEW(25,25)
      EQUIVALENCE (A(1),VIEW(1))
      DATA IND,IE/O,1/
      LOGICAL LEN
      LOGICAL XSYM,YSYM.SYM
      INTEGER EN
C - - - - - EACH VOID
      DO 150 EN = 1,NENC
      CALL VIEWFC(X,Y,D,EN,VIEW,MAXNOD)
C - - - - - FORM THE MATRICES A AND B
      NENG = NENONG(EN)
      IN = 0
C - - - - - EACH NODE GROUP
      DO 120 IG = 1,NENG
      I1 = IGREN(EN,IG)
```

```
            NUMI = NUMB(I1)
C - - - - - EACH ZONE
            DO 120 I = 2,NUMI
            IN = IN + 1
            JN = 0
            DO 120 JG = 1,NENG
            J1 = IGREN(EN,JG)
            NUMJ = NUMB(J1)
            EPSJ = EPSG(J1)
            DO 120 J = 2,NUMJ
            JN = JN + 1
            B(IN,JN) = VIEW(IN,JN)*SIGMA
            A(IN,JN) = – VIEW(IN,JN),(1. – EPSJ)/EPSJ/D(JN)
            IF B(IN.NE.JN) GOTO 120
            B(IN,JN) =  – SIGMA + B(IN,JN)
            A(IN,JN) = 1./EPSJ/D(JN) + A(IN,JN)
  120       CONTINUE
            N = IN
C - - - - - INVERT A AND STORE RESULT IN A
            CALL INVER(A,N,MAXNOD)
C - - - - - MULTIPLY A AND B AND STORE RESULT IN A
            CALL MULT(A,B,N,MAXNOD)
            SYM = .FALSE.
            IF(XSYM(EN).OR.YSYM(EN)) SYM = .TRUE.
            NZ = N
            IF(SYM) N = N + 1
C - - - - - TRANSFORM THE LOCAL RADIATION MATRICE A AND STORE THE RESULT IN
C - - - - - VECTOR E
C - - - - - B IS EMPLOYED AS A DUMMY MATRIX
            CALL ETRANS(A,B,E(IE),N,NZ,SYM,MAXNOD)
            IE = IE + N*N

            SUBROUTINE FQBNDB(T,FLOW,DTA,NN,MAX,TFIRE)
C - - - - - THIS ROUTINE PREPARES CALCULATION OF PRESCRIBED BOUNDARY FLOW
            DIMENSION T(NN),DTA(NN),FLOW(NN)
            PARAMETER NB = 10,NNB = 30,NNB2 = 2*NNB
            COMMON/FQB/NFQNG,NFQG(NB),TR(NNB),TC(NNB)
            1     ,BR(NNB2),BC(NNB2),TRD(NNB),TCD(NNB)
            COMMON/BNOD/NUMB(NB),NBOUND(NB,NNB),BAREA(NB,NNB),
            1     EPSG(NB),BETA(NB).CPG(NB),FA(NB)
            COMMON/UNIT/SIGMA,TABS,TINIT,TAMB.TAMB4
            LOGICAL FA
C - - - - - NULL FLOW VECTOR
            DO 777 1 = 1,NN
  777       FLOW(I) = 0.
C - - - - - RETURN IF NO PRESCRIBED BOUNDARY FLOW
            IF(NFQNG.EQ.0). RETURN
            TF4 = (TFIRE + TABS)**4
            IND = 1
C - - - - - EACH BOUNDARY FLOW NODE GROUP
            DO 30 IB = 1,NFQNG
            TG4 = TAMB4
            TG = TAMB
            ING1 = NFQG(IB)
            IF(FA(ING1)) TG = TFIRE
            IF(FA(ING1)) TG4 = TF4
            NUMI = NUMB(ING1)
```

```
      CP=CPG(ING1)
      DO 20 I=1,NUMI
      NODE=NBOUND(ING1,I)
      TNODE=T(NODE)
      TNABS=TNODE+TABS
C - - - - - RADIATION
      TRD(I)=4.*TNABS**3
      TR(I)=TG4-TNABS**4
C - - - - - CONVECTION
      DUM=TG-TNODE
      TCD(I)=CP*ABS(DUM)**(CP-1.)
```

APPENDIX IV

AIV.1. Gaussian quadrature

AIV.2. Hourglassing

AIV.1. The Gauss–Legendre quadrature

Integration points, co-ordinates and weighting factors

Typical integral

$$\int_a^b F(x)\,\mathrm{d}x = \int_{-1}^{+1} F(x)\,\mathrm{d}x = \sum_{i=1}^{n} A_i f(x_i)$$

n	*Point* ξ_i	*Weights* w_i
1-point formula	0.000 000 000 0	2×10^{10}
2-point formula	±0.577 350 269 2	1×10^{10}
3-point formula	0.000 000 000 0	0.888 888 888 9
	±0.774 596 669 2	0.555 555 555 5
4-point formula	±0.339 981 043 5	0.652 145 154 8
	±0.861 136 311 6	0.347 854 845 1
5-point formula	0.000 000 000 0	0.568 888 888 9
	±0.538 469 310 1	0.478 628 670 5
	±0.906 179 845 9	0.236 926 885 0
6-point formula	±0.238 619 186 1	0.467 913 934 6
	±0.661 209 386 5	0.360 761 573 0
	±0.932 469 514 2	0.171 324 492 4

Typical case: 20-noded isoparametric element

$$\begin{aligned}\int_{-1}^{+1}\int_{-1}^{+1}\int_{-1}^{+1} F(\xi,\eta,\zeta)\,\mathrm{d}\xi\,\mathrm{d}\eta\,\mathrm{d}\zeta = {} & B_6\{F(-b,0,0)+F(b,0,0)+F(0,-b,0)\\ & + F(0,b,0)+F(0,0,b)+F(0,0,b)\}\\ & + C_8\{F(-c,-c,-c)+F(c,-c,-c)\\ & + F(c,c,-c)+F(c,c,c)+F(-c,c,-c)\\ & + F(-c,-c,c)+F(-c,c,c)+F(c,-c,c)\}\end{aligned}$$

$$B_6 = 0.886\,426\,593$$

$$C_8 = 0.335\,180\,055$$

$$b = 0.795\,822\,426$$

$$c = 0.758\,786\,911$$

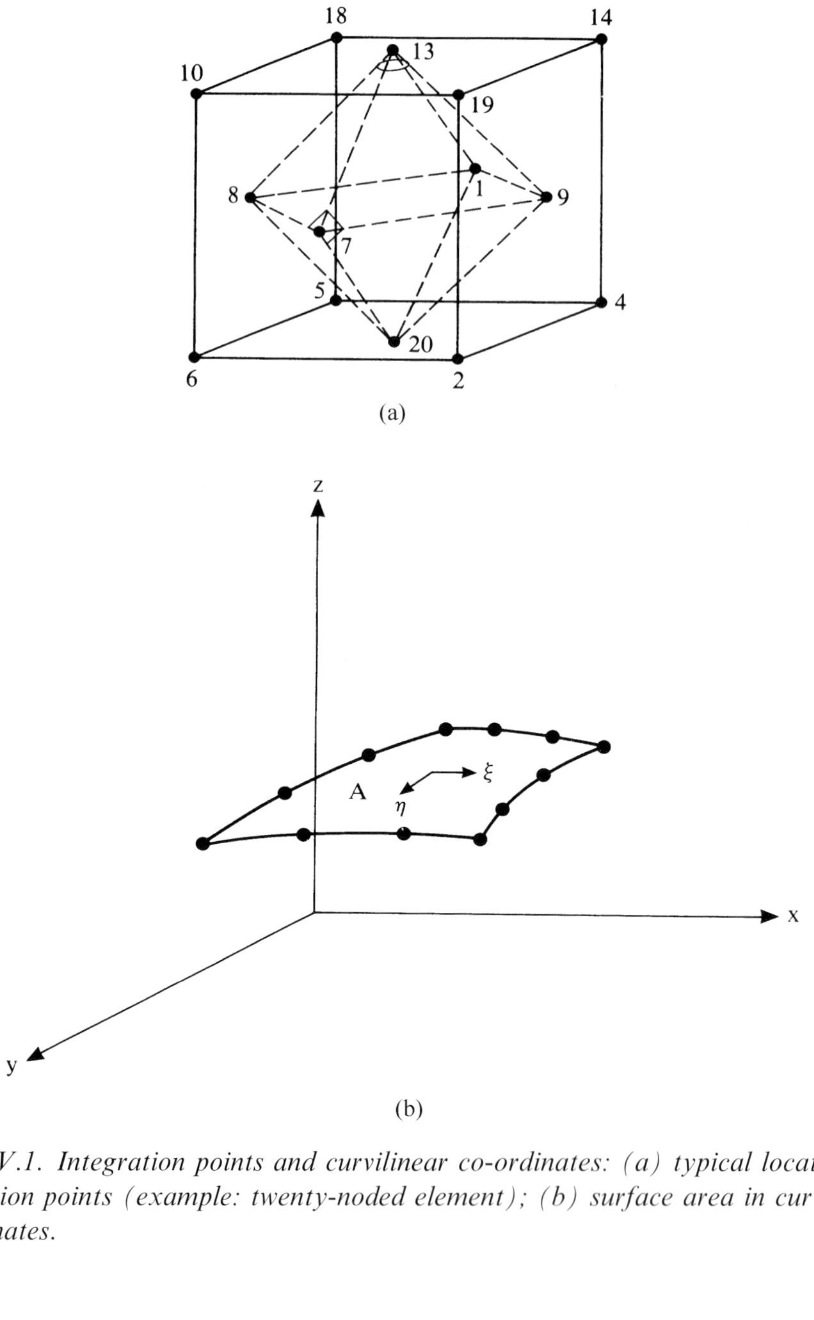

Fig. AIV.1. Integration points and curvilinear co-ordinates: (a) typical locations of integration points (example: twenty-noded element); (b) surface area in curvilinear co-ordinates.

AIV.2. Hourglassing

Hourglassing is a numerical phenomenon, which occurs when using one Gauss point integration. Deformation occurs at the nodes of each element without incurring strain at the centroid of the element. The deformation of the elements forms mode shapes and in a typical 8-noded brick, twelve hourglass modes exist. The shapes of the hourglass modes are illustrated in Fig. AIV.1 and, in the case of a solid model, have the appearance of a saw-tooth pattern. These modes are also referred to as zero energy modes because no energy dissipation occurs during hourglassing [A1]. Hourglassing occurs in a time period less than the time period of the structural response. Reference is made to other researchers [A3–A5] for theoretical work in this field.

The cause of hourglassing is due to the velocities of diagonally opposite nodes being equal. The calculation of strain rates in elements involves the difference in velocities of diagonally opposite corners, but when the velocities of these opposite nodes are equal the result is a strain rate equal to zero. Consequently, no increase in strain occurs and hence the deformation is not resisted. Figure AIV.3 illustrates the deformation of a typical quadrilateral shell element caused by hourglassing.

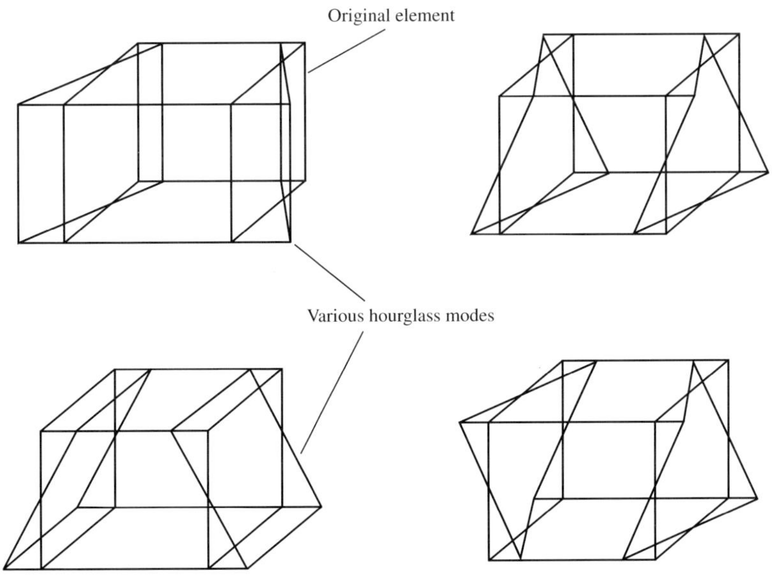

Fig. AIV.2. Hourglass mode shapes of an 8-noded solid element.

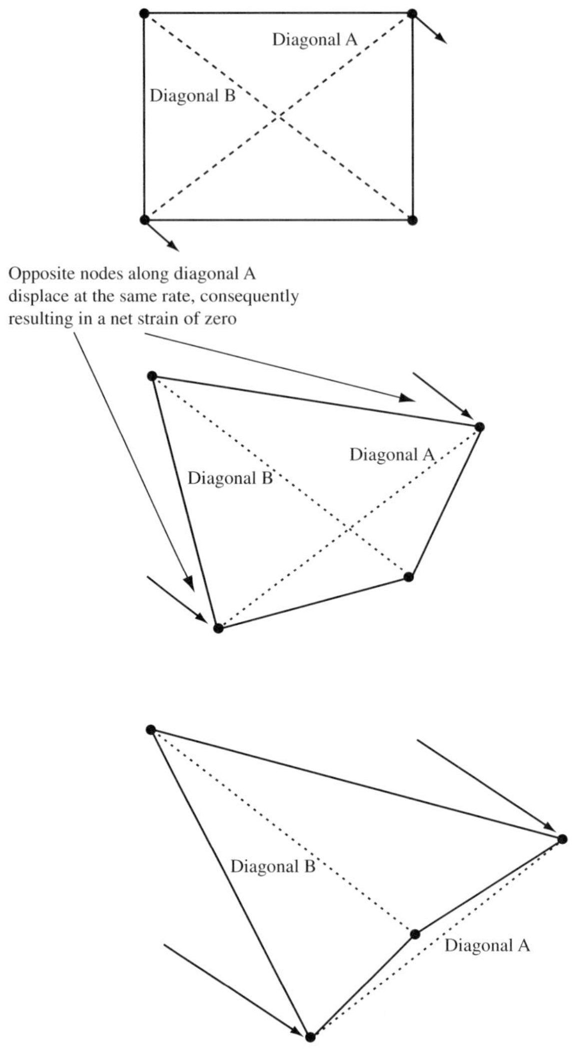

Fig. AIV.3. Hourglassing of a quadrilateral shell element.

References

A1. Monks, G. W., ed. *DYNA3D Users Guide — Issue 2 Sept 1990*, Doc. No. 1237H, DYNA3D Users Group, 1990.

A2. Hancock, S. L., An hourglass subtraction procedure, TCAM Technical Memo 73-6, Applied Mechanics Department, Physics International Company, San Leandro, CA, USA, 1973.

A3. MSC, MSC/Patran version 7.5, a pre-processor for use with ABAQUS and DYNA FE codes, MacNeal-Schwendler Corporation, Los Angeles, CA, USA, 1998.

A4. LSTC, LS-TAURUS, a post processor visualisation code for LS-DYNA and LS-NIKE, Livermore Software Technology Corporation Inc., Livermore, CA, USA, 1995.

A5. Bathe, K. J., Finite element procedures in engineering analysis, Prentice-Hall Inc., New Jersey, USA, 1982.

APPENDIX V

Acceleration and convergence

AV.1. Criteria for convergence and acceleration

Convergence criteria

To ensure convergence to the correct solution by finer sub-division of the mesh, the assumed displacement function must satisfy the convergence criteria given below:

(a) Displacements must be continuous over element boundaries.
(b) Rigid body movements should be possible without straining.
(c) A state of constant strain should be reproducible.

Euclidean norm $\psi_i/R_i \leq C$. The term ψ_i represents the unbalanced forces and the norm of the residuals. With the aid of the iterative scheme described above, the unbalanced forces due to the initial stresses $\{\sigma_0\}$ become negligibly small. As a measure of their magnitude, the norm of the vector $||\psi_i||$ is used. The Euclidean norm and the absolute value of the largest component of the vector are written as

$$||\psi_i|| = (|\psi_1|^2 + \cdots + |\psi_n|^2)^{1/2} \tag{AV.1}$$

$$||R_i|| = (|\{R_i\}^T\{R_i\}|)^{1/2}$$

the convergence criterion adopted is

$$||\psi|| = \max_i |\psi_i| < C = 0.001 \tag{AV.2}$$

Uniform acceleration

Various procedures are available for accelerating the convergence of the modified Newton–Raphson iterations. Figure AV.1 shows the technique of computing individual acceleration factors, δ_1 and δ_2 are known. Then, assuming a constant slope of the response curve, and from similar triangles, the value of δ_3 is computed.

$$\frac{\delta_1}{\delta_2} = \frac{\delta_2}{\delta_3} \qquad \delta_3 = \delta_2\frac{\delta_2}{\delta_1} \tag{AV.3}$$

When δ_3 is added to δ_2, then the accelerated displacement δ_2' is expressed as

$$\delta_2' = \delta_2 + \delta_3 = \delta_2\left(1 + \frac{\delta_2}{\delta_1}\right) = \alpha\delta_2 \tag{AV.4}$$

where the acceleration factor α is

$$\alpha = 1 + \frac{\delta_2}{\delta_1} \tag{AV.5}$$

Generally the range of α is between 1 and 2. The value of $\alpha = 1$ for zero acceleration, and the value of α reaches the maximum value of 2 when the slope of the δ–R curve approaches zero.

The acceleration factor α is computed individually for every degree of freedom of the system. The displacement vector obtained from the linear stiffness matrix $[k_0]$ is then multiplied by the $[\alpha]$ matrix having the above constants on its diagonals. The remaining components of $[\alpha]$ are zero. The accelerated displacement vector is then expressed as follows:

$$\{\Delta u_i'\} = [a_{i-1}]\{\Delta u_i\} \quad \text{(AV.6)}$$

From these accelerated displacements $\{\Delta u_i'\}$, the initial stresses $\{\sigma_0\}$ are found and they are equilibrated with the forces $\{\psi_i\}$. They are then used for the next solution

$$\{\Delta \bar{u}_i\} = [k_0]^{-1}\{\psi_i\} \quad \text{(AV.7)}$$

which results in a new set of acceleration factors. Now an estimate for the displacement increment is made in order to find the incremental stresses and total stresses.

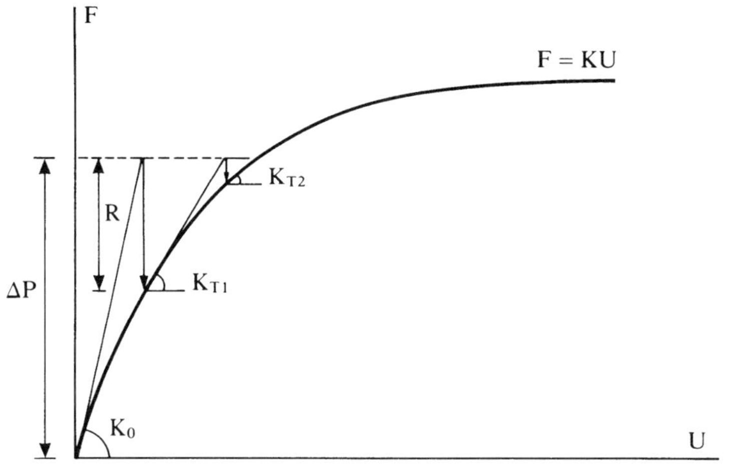

Fig. AV.1a. Newton–Raphson method.

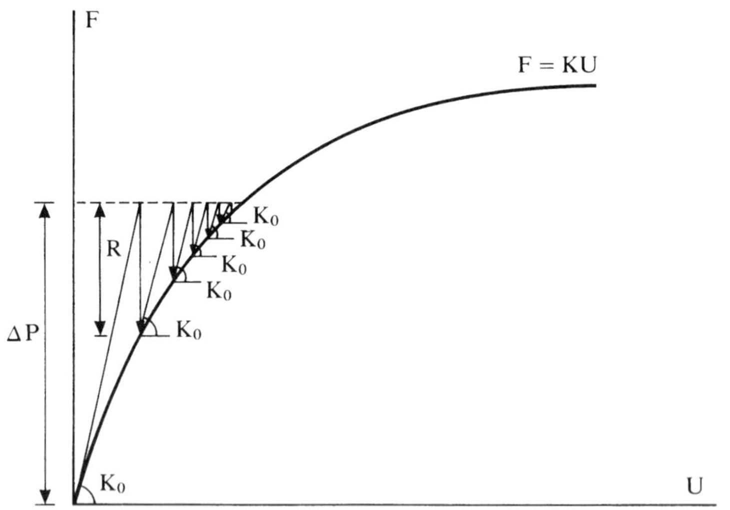

Fig. AV.1b. Initial stress method.

Note: ΔP is a specific value of F.

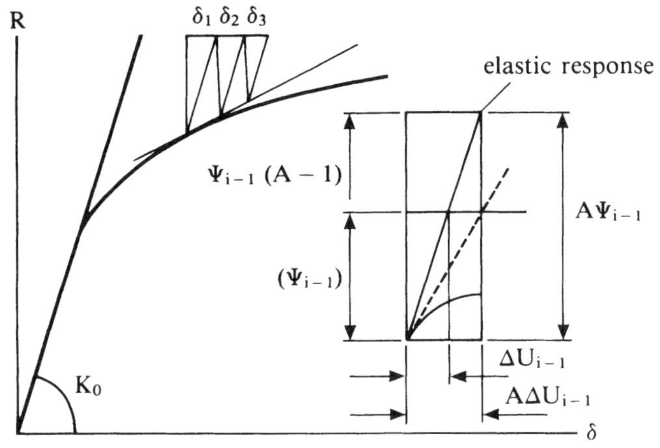

Fig. AV.1c. Technique of computing acceleration factors.

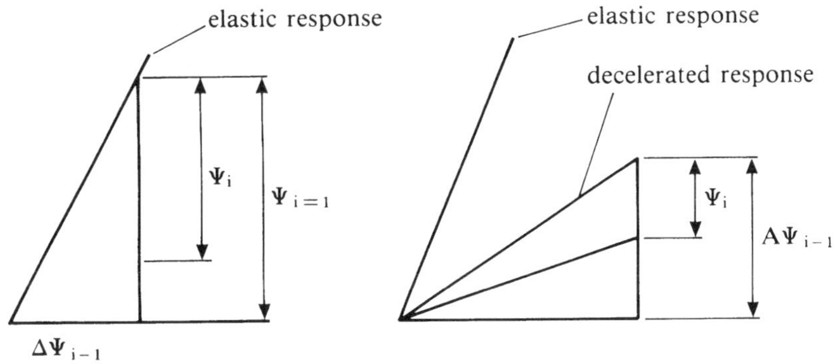

Fig. AV.1d. Graphical representation.

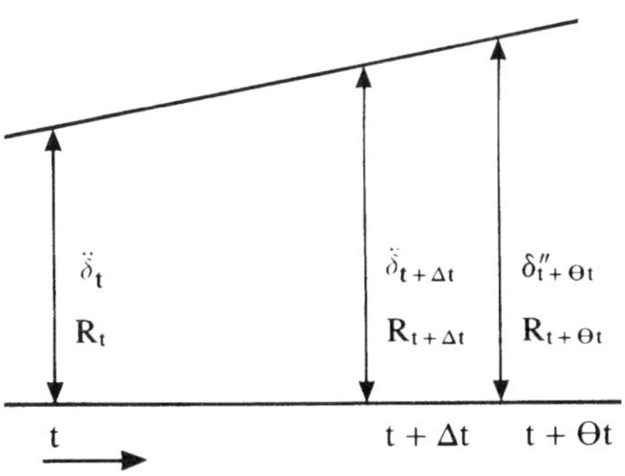

Fig. AV.1e. Linear acceleration and load assumptions of the Wilson θ method.

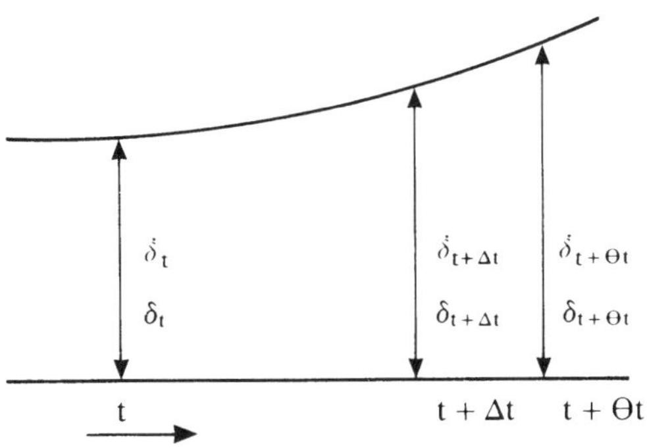

Fig. AV.1f. Quadratic and cubic variation of velocity and displacement assumptions of the Wilson θ method.

The residual forces needed to re-establish equilibrium can now easily be evaluated

$$\{\hat{\psi}_i\} = \int_v [B]^T \{\sigma_{0_T}\} \, dV - \{R_i\} \tag{AV.8}$$

where $\{R_i\}$ represents the total external load; dV is the volume.

A new displacement now results from

$$\{\Delta u_{i+1}\} = -[k_0]^{-1} \{\hat{\psi}_i\} \tag{AV.9}$$

In order to carry out these iterative steps, numerical integration is required. First of all the evaluation of $\{\hat{\psi}_i\}$ from the initial stresses is required, and this requires integration over the elastic–plastic region only. The value of $\{\hat{\psi}_i\}$ is computed by carrying out the integration over the entire domain of the analysis. Since these kinds of accelerated steps unbalance the equilibrium, therefore it has to be re-established by finding the residual forces $\{\hat{\psi}_i\}$. Since the state of stress produced by the accelerated displacements is not in balance with the residual forces of the previous iteration, the new residual forces $\{\hat{\psi}_i\}$ of eqn (AV.9) must balance $\{\sigma_T\}$ and $\{R_i\}$. Here the acceleration scheme is needed to preserve equilibrium, which will eventually make the equivalent forces over the whole region unnecessary. This is achieved by applying a uniform acceleration, i.e. the same acceleration factor $\bar{A}$ to all displacements, found by averaging the individual factors α_i

$$\bar{A} = \frac{1}{n} \sum_{i=1}^{n} \alpha_i \tag{AV.10}$$

The force–displacement equation is then written by multiplying both sides with the scalar quantity $\bar{A}$ without disturbing the equilibrium.

$$\bar{A}\{\Delta u_i\} = [k_0]^{-1} \bar{A}\{\psi_i\} \tag{AV.11}$$

Now to evaluate $\{\psi_{i+1}\}$, the previous value of $\{\psi_i\}$ must be multiplied by $\bar{A}$, and the previously accelerated forces from the initial stresses $\{\sigma_0\}$ must be included such that

$$\{\psi_{i+1}\} = \int_V [B]^T \{\sigma_0\} \, dV - (A - 1)\{\psi_{i-1}\} \tag{AV.12}$$

APPENDIX VI

Concrete cubes and steel bars — experimental and finite element results

AVI.1. Preparation of concrete

Cement	43 lb	1:2:4 mix (cement, sand, aggregate)
Sand	86 lb	
Concentrated aggregate	172 lb	
Water	26 lb	

The concrete is mixed in a laboratory batch mixer.

AVI.2. Testing of concrete

Concrete is poured into test cubes and cylindrical moulds to test the strength.

Results

7 days:	Cube 1		ultimate loads and strength
	Cube 2		
28 days:	Cube 3	Dry	
	Cube 2		
	Cube 3	Wet	
	Cube 4		

Concrete cube strengths and the yield stress of steel reinforcement are as shown in Table AVI.1.

Two air-dried and two water-cured test cubes were tested for strength.

Table AVI.1

	Test cube			
	Air-dried		Water-cured	
	1	2	3	4
Concrete strength (kN)	1197	1150	1291	1286
Average strength (kN)	(1090[a]) 1173.5		(1270[a]) 1288.5	

[a] Results from finite element incremental analysis.

Size of test cube = $150 \times 150 \times 150$ mm

Area of test cube = $150 \times 150 = 22\,500\,\text{mm}^2$

Average concrete strength $f_{cu} = 54.71\,\text{N/mm}^2$

Characteristic of reinforcement $f_y = 410\,\text{N/mm}^2$

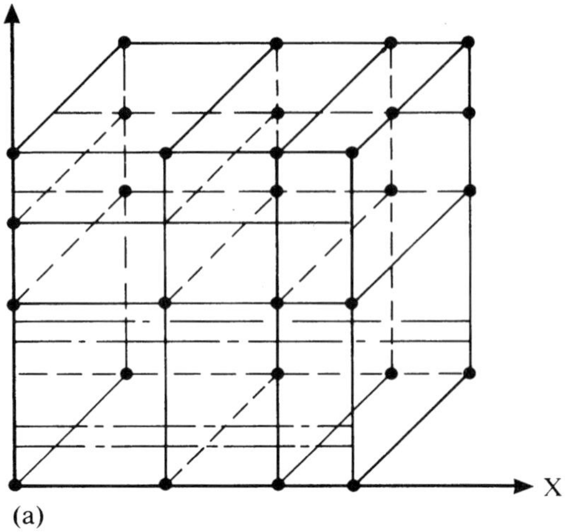

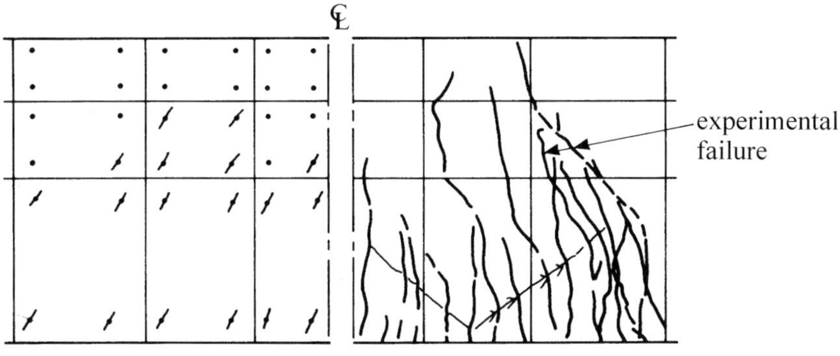

Fig. AVI.1. (a) Finite element mesh; (b) computed crack patterns.

AVI.3. Tension strength of steel bars

Area of 10 mm diameter bar
Area $A = \pi(10^2/4) = 78.5\,\text{mm}^2$

Bar diameter (mm)	Area (mm)	Yield load (kN)	Failure load (kN)	Yield stress (N/mm^2)
10	78.5	29 (31)[a]	39 (42)[a]	369.4 (410)

[a] Results from finite element incremental analysis.

Figure AVI.1 shows the finite element mesh and computed crack patterns. For cases for comparison refer to Figs. I.5, I.10, I.16–I.19.

APPENDIX VII

Stress–strain curves for reinforcements and prestressing tendons (data for the program isopar)

AVII.1. Data calculation for ISOPAR on reinforcing steel

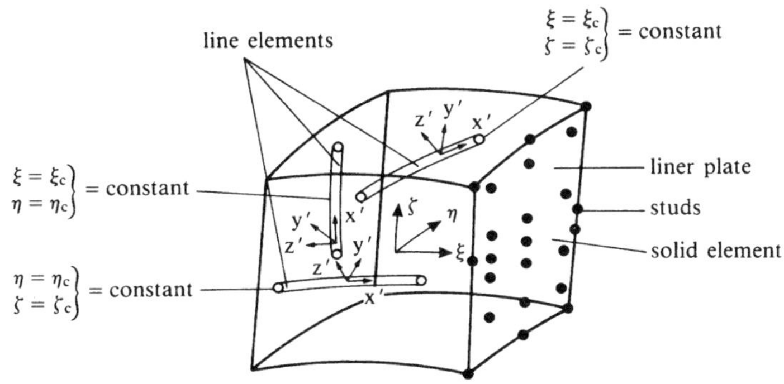

Fig. AVII.1. Finite element modelling of liner–stud interaction.

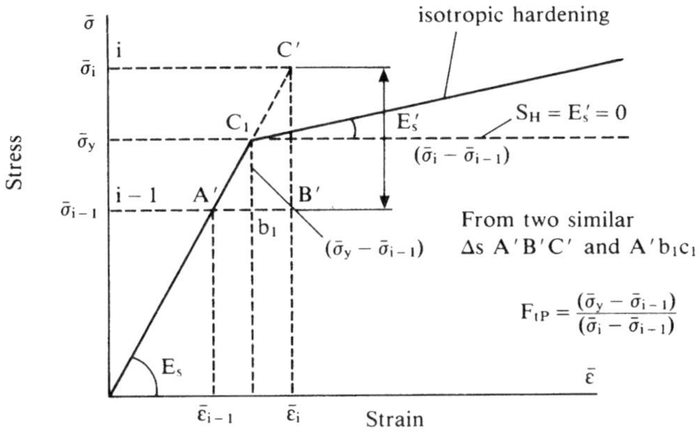

S_{H} = strain hardening

$$S_{\mathrm{H}} = \frac{\mathrm{d}\sigma}{\mathrm{d}\varepsilon_{\mathrm{p}}} = \frac{E_{\mathrm{s}}E_{\mathrm{s}}'}{E_{\mathrm{s}} + E}$$

$$E_{\mathrm{s}} = K_0$$

Fig. AVII.2. Equivalent stress–strain curve for steel. (E_s = initial yield modulus; E_s' = post-yield modulus.)

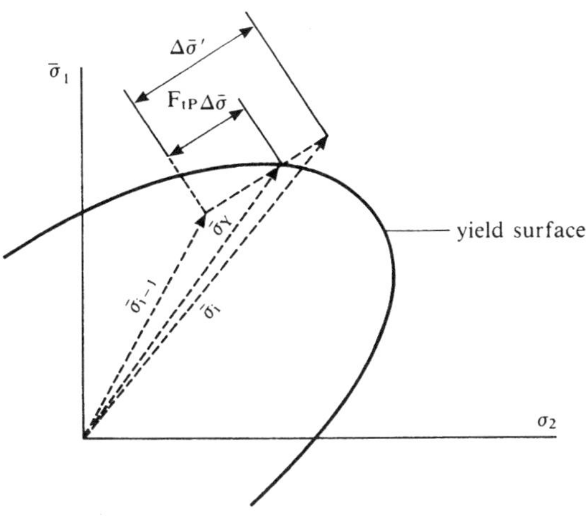

Fig. AVII.3. Failure surface.

Transitional factor, elastic to plastic. From the figure $\bar{\sigma}_{i-1} + F_{\mathrm{TR}}\,\Delta\bar{\sigma}' = \bar{\sigma}_{\mathrm{Y}}$

$$F_{\mathrm{tP}} = \frac{\bar{\sigma}_{\mathrm{Y}} - \bar{\sigma}_{i-1}}{\Delta\bar{\sigma}'} = \frac{\bar{\sigma}_{\mathrm{Y}} - \bar{\sigma}_{i}}{\bar{\sigma}_{i} - \bar{\sigma}_{\mathrm{Y}}}$$

$$\bar{\sigma} = \sigma_{\mathrm{eq}}$$

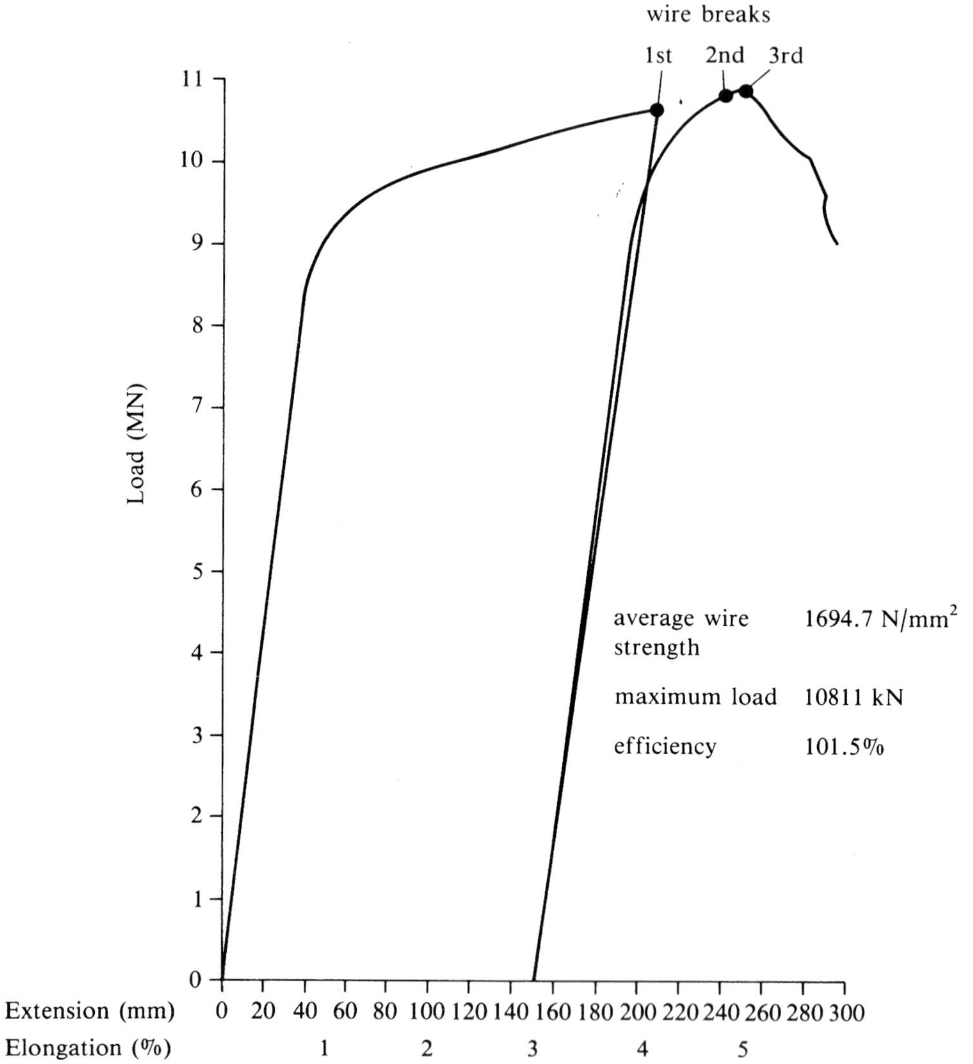

Fig. AVII.4. BBRV straight tendons 163/7 system (stress–strain test no. 11 (courtesy of Simon-Cave, Manchester, 1989)).

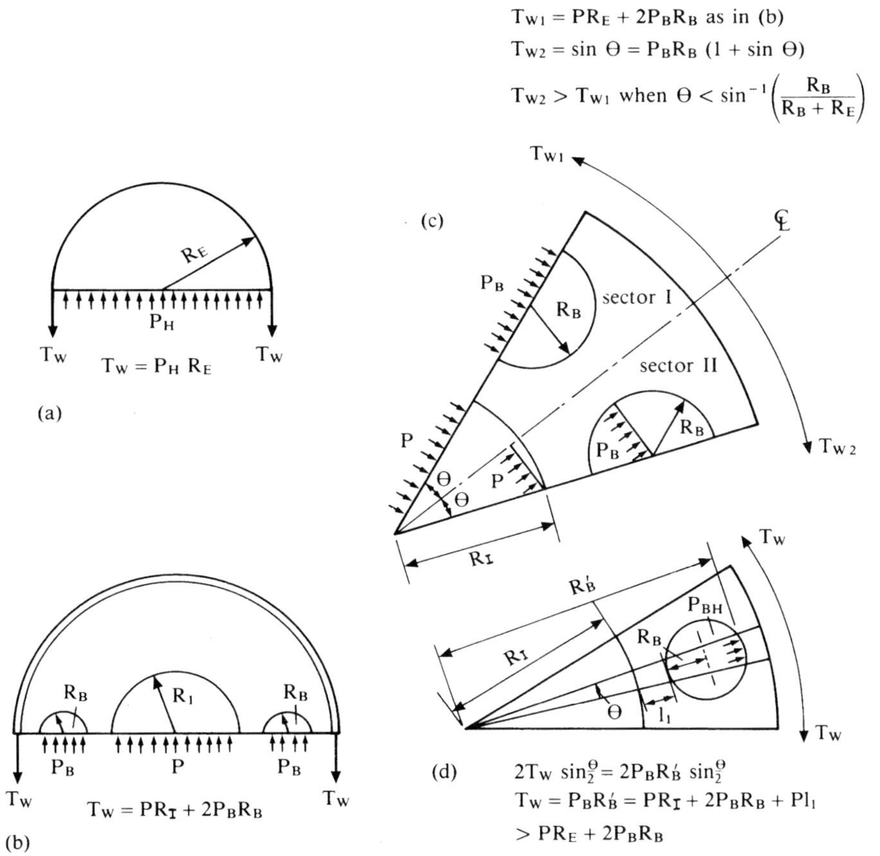

Fig. AVII.4. Wire/strand load versus uniform pressure: (a) Single cavity; (b) half main cavity with two boiler holes; (c) two sectors of multi-cavity; (d) a sector of main cavity with one boiler hole.

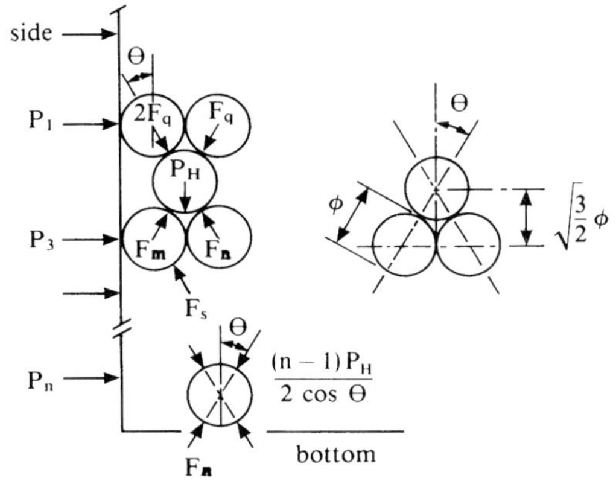

Fig. AVII.5. Contact forces from wires and strands.

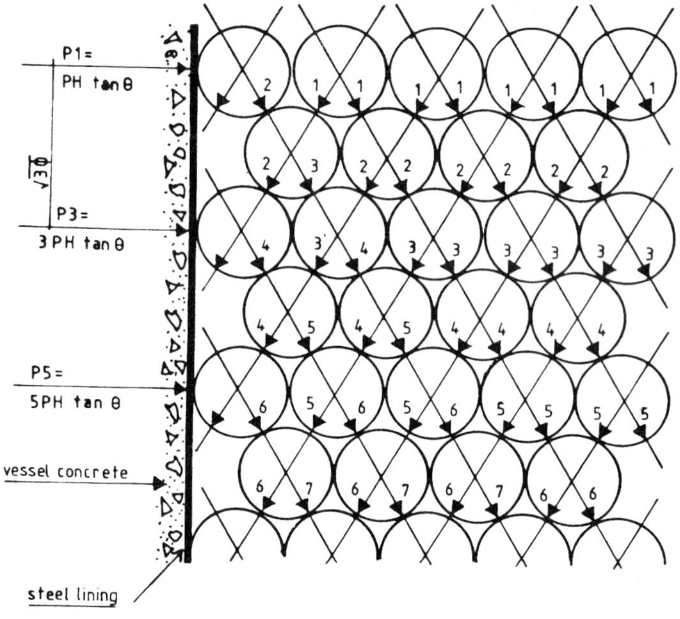

Fig. AVII.6. Wire/strand winding — odd number of layers — values × F_q (excluding bottom layer).

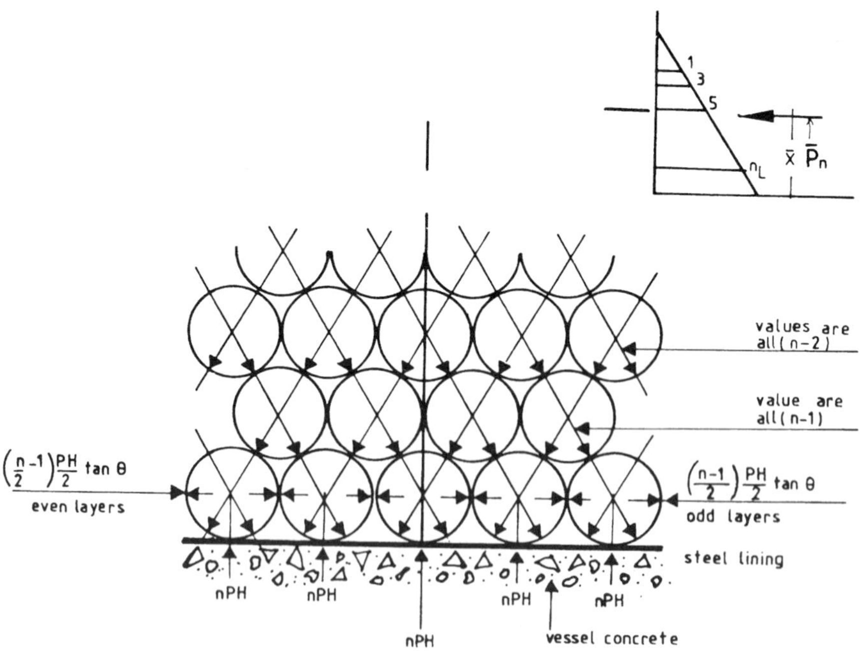

Fig. AVII.7. Wire/strand winding — even and odd layers (centre zone at the base).

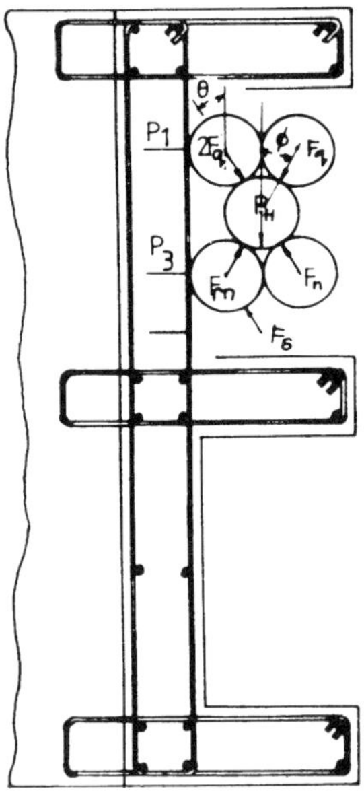

Wires

Even wires

No. of Layers $= \frac{1}{2}\left[n_L \cdot l/\phi + n_L(l/\phi - 1)\right]$

$P_H = \sum P_H$ Circumferential force

$= \frac{nT_W}{l}\left[1/R_1 + n\phi\right]$

P_n = nTh layer pressure

$= nP_H \tan\theta$

For example

$F_q = 1/2\, P_H \sec\theta$

$P_3 = 2F_q \sin\theta + 2P_H \tan\theta$

$F_m = F_q + 1/2\, P_H \sec\theta$

$F_s \quad F_q + 3/2\, P_H \sec\theta$

Strands

$$P_H = -\left[\frac{\pi E \phi^4 \nu \sin\alpha \cos^4\alpha}{64R^3(1+\nu)} - \frac{\pi\phi^4\sigma_t(\sin\alpha\cos\alpha + n_1 \sin\alpha\cos^4\alpha)}{4R}\right]$$

$$T_W = \sigma_t \frac{\pi\phi^2}{4}(1 + n_1 \cos\alpha \times (1 + n_1\cos^3\alpha)$$

$$\sin\phi_1 = \frac{1}{\sqrt{\left[1 + \frac{\sin^2\alpha}{\tan^2(\frac{\pi}{2} - \frac{\pi}{n})}\right]}}$$

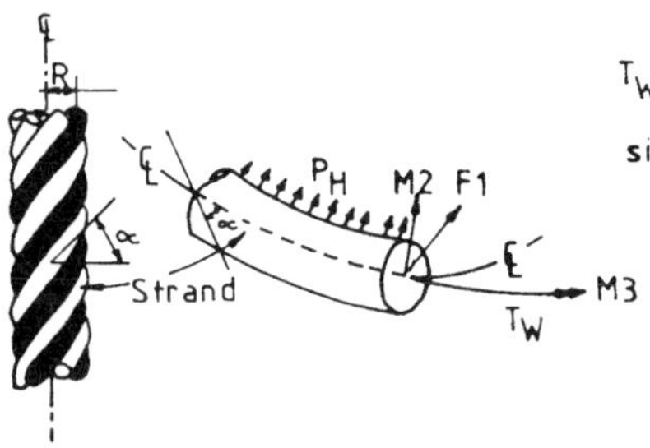

Strength Reduction & Ultimate Tensile Stresses

$\sigma ut_1 = \bar{k}\{\sigma_1 - \nu(\sigma_2 - \sigma_3)\}$

$\sigma ut_1 = \bar{k}\{\sigma_3 - \nu(\sigma_1 - \sigma_3)\}$

$\sigma ut_1 = \bar{k}\{\sigma_3 - \nu(\sigma_1 - \sigma_3)\}$

$\bar{k}$ = Strength Reduction factor issued later on with equations for wires / strands

Fig. AVII.8. Load from wires/strands.

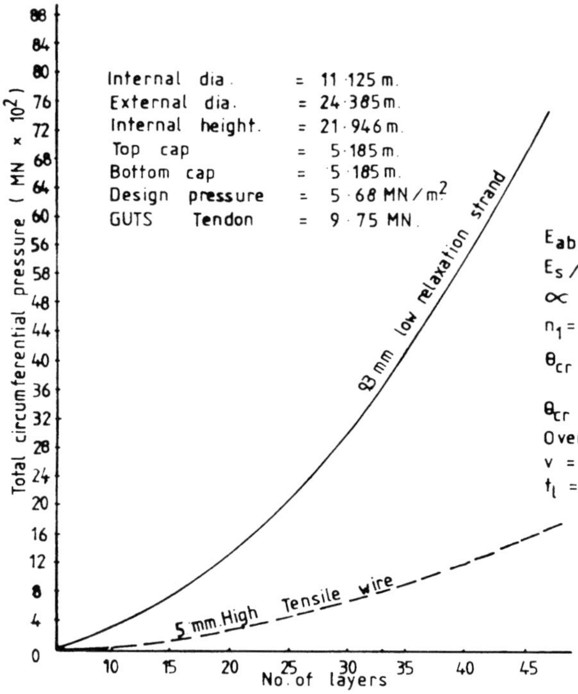

Fig. AVII.9. Circumferential pressure versus numbers of layers (HTCGR pressure vessel parameters).

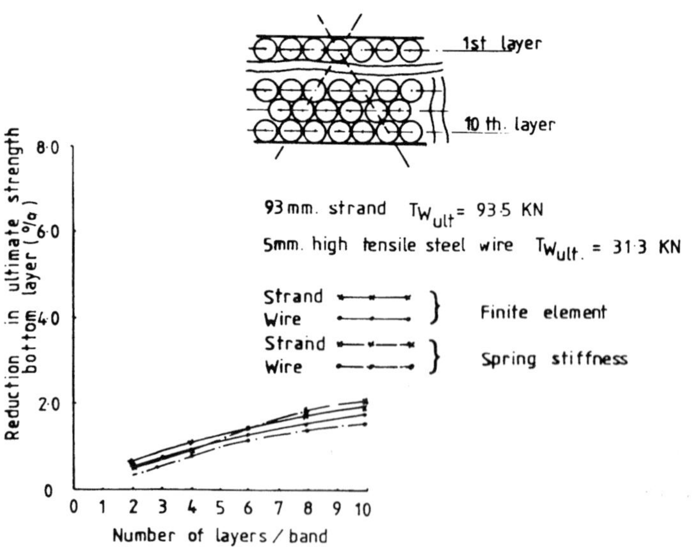

Fig. AVII.10. Band layer and strength reduction.

Table AVII.1. Summary of circumferential and vertical tendon losses excluding friction

Type of loss	Walls tons (MN)					Cap tons (MN)	
	Circumferential				Vertical	Circumferential	
	2	3	4	7	1	5	6
Relaxation	22.0 (0.219)	30.8 (0.306)	28.6 (0.284)	25.0	29.4 (0.249)	19.1 (0.190 3)	22.8 (0.227)
Elastic	18.4 (0.183)	27.0 (0.269)	12.7 (0.126)	10.1 (0.10)	19.1 (0.19)	10.3 (0.102)	13.2 (0.131)
Creep	28.5 (0.283)	65.3 (0.65)	52.5 (0.523)	29.5 (0.293)	104.9 (1.045)	25.3 (0.252)	43.3 (0.431)
Pressure	−19.5 (−0.128)	−15.9 (−0.158)	−0.4 (−0.003 9)	−1.3 (−0.012 9)	−9.1 (−0.09)	−5.3 (−0.052)	−9.5 (−0.094)
Thermal	−12.9 (−0.128)	+14.6 (−0.145)	+9.1 (−0.09)	+2.2 (−0.021)	−28.9 −0.289	−10.6 (−0.105)	−4.0 (−0.039)
Total tons	36.5 (0.363)	121.8 (1.213)	102.5 (1.021)	65.5 (0.652)	116.4 (1.159)	38.8 (0.386)	65.3 (0.650)
% loss	5.0	16.5	14.0	8.9	15.9	5.3	9.0
Average % loss/ group of tendons	9.0				15.9	7.8	

Geometry C_5-type tendons

Total tendon length $= 2(l_1 + l_2' + l_3')$

$$\cos\theta = \frac{R_2}{R_E} \qquad \theta = 50^\circ\, 49'$$

$$\sin\gamma = \frac{l_2}{R_E} = 0.5801 \qquad \gamma = 35^\circ\, 27'$$

$$\cos\theta_i = \frac{R_2 - l'''}{R_E} = 0.6004 \qquad \theta_1 = 53^\circ\, 6'$$

$$l_1 + l_2' = R_E \sin\theta_1 \qquad 36.1864\,\text{ft}\ (11.036\,\text{m})$$

$$l_5' = \frac{2\pi \times R_2(90^\circ + \theta_2)}{360^\circ} = 46.7611\,\text{ft}\ (14.26\,\text{m})$$

$$\theta = 90^\circ - (\theta + \gamma) = 90^\circ - \left\{\cos^{-1}\frac{R_2}{R_E} + \sin^{-1}\frac{R_2}{R_E}\right\} = 3^\circ\, 44'$$

Total tendon length $= 165.8950\,\text{ft}\ (50.59\,\text{m}) = (2(l_1 + l_3' + l_5'))$

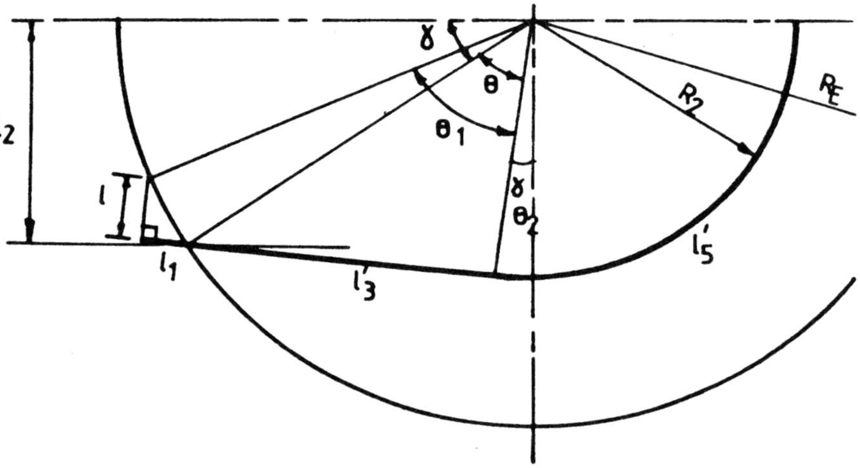

Fig. AVII.11. Geometry of circumferential tendons or cables.

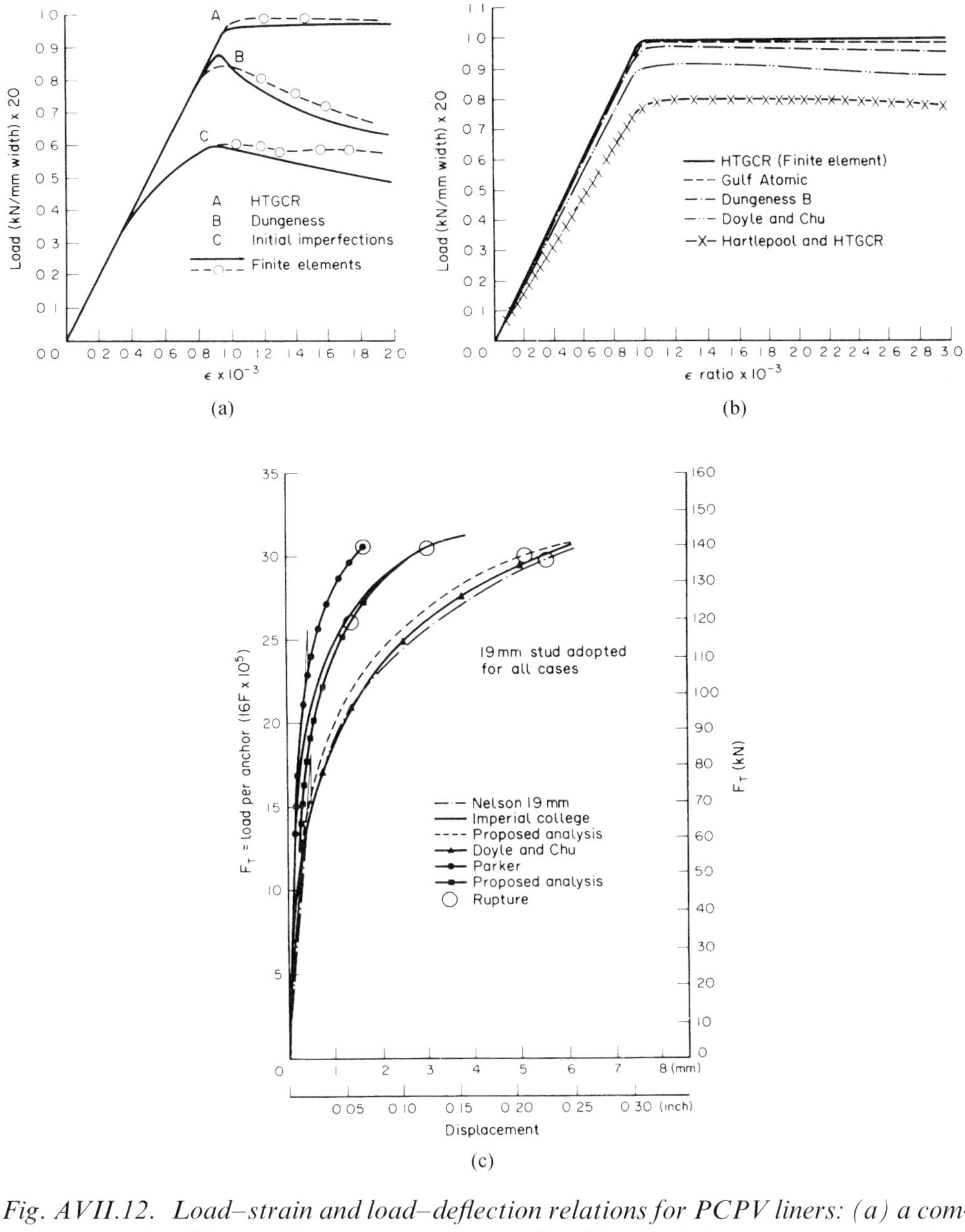

Fig. AVII.12. Load–strain and load–deflection relations for PCPV liners: (a) a comparative study of liners; (b) a comparative study of liners; (c) a comparative study of stud loads to failure.

APPENDIX VIII

Concrete cracking — step-by-step formulation and computer sub-routines

AVIII.1. General steps of crack flow and crack calculations

(1) The load increment $\{\Delta P_n\}$ is applied where n is the load increment.

(2) The total is accumulated as $\{P_n\} = \{P_{n-1}\} + \{\Delta P_n\}$, and $\{R\} = \{\Delta P_n\}$, where $\{R\}$ is the residual load vector.

(3) Incremental displacement is computed as $\{\Delta U_i\} = [K]_e^{-1}\{R\}$, where i is the iteration.

(4) Total displacements are now accumulated in the following form: $\{U_i\} = \{U_{i-1}\} + \{\Delta U_i\}$.

(5) Strain increments are calculated from step 4 as $\{\Delta\varepsilon_i\} = [B]\{\Delta U_i\}$. The accumulated strains at this stage would then be written as

$$\{\varepsilon_i\} = \{\varepsilon_{i-1}\} + \{\Delta\varepsilon_i\}$$

(6) The stress increments are calculated using the current nonlinear constitutive matrices of various models described earlier: $\{\Delta\sigma_i\} = \{f(\sigma)\}\{\Delta\varepsilon\}$. The accumulated stresses are computed as $\{\sigma_i\} = \{\sigma_{i-1}\} + \{\Delta\sigma_i\}$. In order to differentiate stresses at elastic and plastic conditions, a stress point indicator I_p is introduced.

$$\begin{aligned} I_p &= 0 \text{ (elastic point)} \\ &= 1 \text{ (plastic point)} \\ &= 2 \text{ (unloading from plastic state)} \end{aligned}$$

(7) The stress increment is calculated using the elastic material matrix as $\{\Delta\sigma_i'\} = [D']\{\Delta\varepsilon\}$. Total stresses are given as $\{\sigma_i'\}_T = \{\sigma_{i-1}\} - \{\Delta\sigma_i'\}$.

(8) The stress $\{\sigma_i\}$ is now calculated using step 7: $\{\sigma_i\} = \{f(\sigma_i')\}$, $\{\sigma_{i-1}\} = \{f(\sigma_{i-1}\}$ — any yield criterion required.

(9) If a plastic point is obtained, step 11 should be considered.

(10) If $\sigma_i \geq \sigma_y$ — plastic point ($I_p = 1$), transition from elastic to plastic, calculate factor F_{TP} using Figs VI.3 and VI.4:

$$F_{TP} = \left(\frac{\sigma_y - \bar{\sigma}_{i-1}}{\sigma_i - \sigma_{i-1}}\right)$$

The stress at the yield surface $\{\sigma_i\}_y^* = \{\sigma_{i-1}\} + F_{TP}^*\{\Delta\sigma_i\}$. Elasto-plastic stress increments are calculated as $\{\Delta\sigma_i\} = [D]_{ep}\{\sigma_i\}^*(1 - F_{TP})\{\Delta\varepsilon\}$. Total stress $\{\sigma_i\}_T = \{\sigma_i\} + \{\Delta\sigma_i\}$

(11) Plastic point from steps 9 and 10, check for unloading, i.e. if $\bar{\sigma} \geq \sigma_y$ it is necessary to proceed to step 12. For the unloading case at this point, set $I_p = 2$, total stress $\{\sigma_i\}_T$ then given by $\{\sigma_i\} = \{\sigma_{i-1}\} + \{\Delta\sigma_i\}$. Set $\{\sigma_y\} = \{\bar{\sigma}_{i-1}\}$, and the procedure is repeated for the additional increments.

(12) Loading at this point $\{\Delta\sigma_i\} = [D]_{ep}\{\bar{\sigma}_{i-1}\}\{\Delta e\}$. Total stress $\{\sigma_i\}_T = \{\bar{\sigma}_{i-1}\} + \{\Delta\varepsilon\}$.

(13) Sometimes it is necessary to correct stresses from the equivalent stress–strain curve: $\sigma_{corr} = \sigma_{i-1} + S_H\,\Delta\varepsilon_p$, where $\Delta\varepsilon_p = \sqrt{\frac{2}{3}\Delta\varepsilon_{ij}^p\,\Delta_{ij}^p}$ = equivalent plastic strain increment. S_H is the strain hardening parameter, such that $\Delta\bar{\varepsilon}_p = \lambda$.

Equivalent stress, calculated from the current stress state, is given by $\{\sigma_i\} = F_{TP}\{(\sigma_i)\}$.The correct stress state, which is on the yield surface, will therefore be given as $\{\sigma_i\} = F_{TP}\{\sigma_i\}$. The total stresses are converted into equivalent nodal loads from $\int_v [B]^T \{\sigma_i\}\,\mathrm{d\,vol}$, and the residual load vector is calculated from $\{R\} = \{F_n\} - \int_v [B]^T \{\sigma_i\}\,\mathrm{d\,vol}$.

(14) Check for convergence. For detailed information see Appendix V.

$$(||R||/||F||) \leq \mathrm{TOL} \qquad (||\Delta U||/||U||) \leq \mathrm{TOL}$$

where TOL is chosen from 0.01 to 0.001; $||R|| = \sqrt{R_i^{T''} R_i}$ is the Euclidean norm of the residuals; $||F||\sqrt{P^T P}$ is the Euclidean norm of the externally applied load; and $||\Delta U|| = \sqrt{\Delta U_i^{T''} U_i}$ is the Euclidean norm.

If convergence is not achieved, step 3 is invoked and all the steps repeated for the next iteration. If convergence is achieved, then proceed with the next load increment.

AVIII.2. Flow chart on cracking phenomenon

Crack indicators NCK (1), NCK (2), NCK (3)

NCK (1) — crack normal to the principal stress '1'
NCK (2) — crack normal to the principal stress '2'
NCK (3) — crack normal to the principal stress '3'

NCK (1) = 0, NCK (2) = 0, NCK (3) = 0 — no cracks

NCK (1) = 1, NCK (2) = 1, NCK (3) = 1 — cracks open

NCK (1) = 2, NCK (2) = 2, NCK (3) = 2 — cracked closed

$\underline{\sigma}$, $\underline{\varepsilon}$ — stress strain state at integrated point
ε_i — principal strains
σ_i — principal stresses, i = 1, 2, 3
σ_t — limiting tensile strength of concrete
T_ε, T_σ — transformation matrix

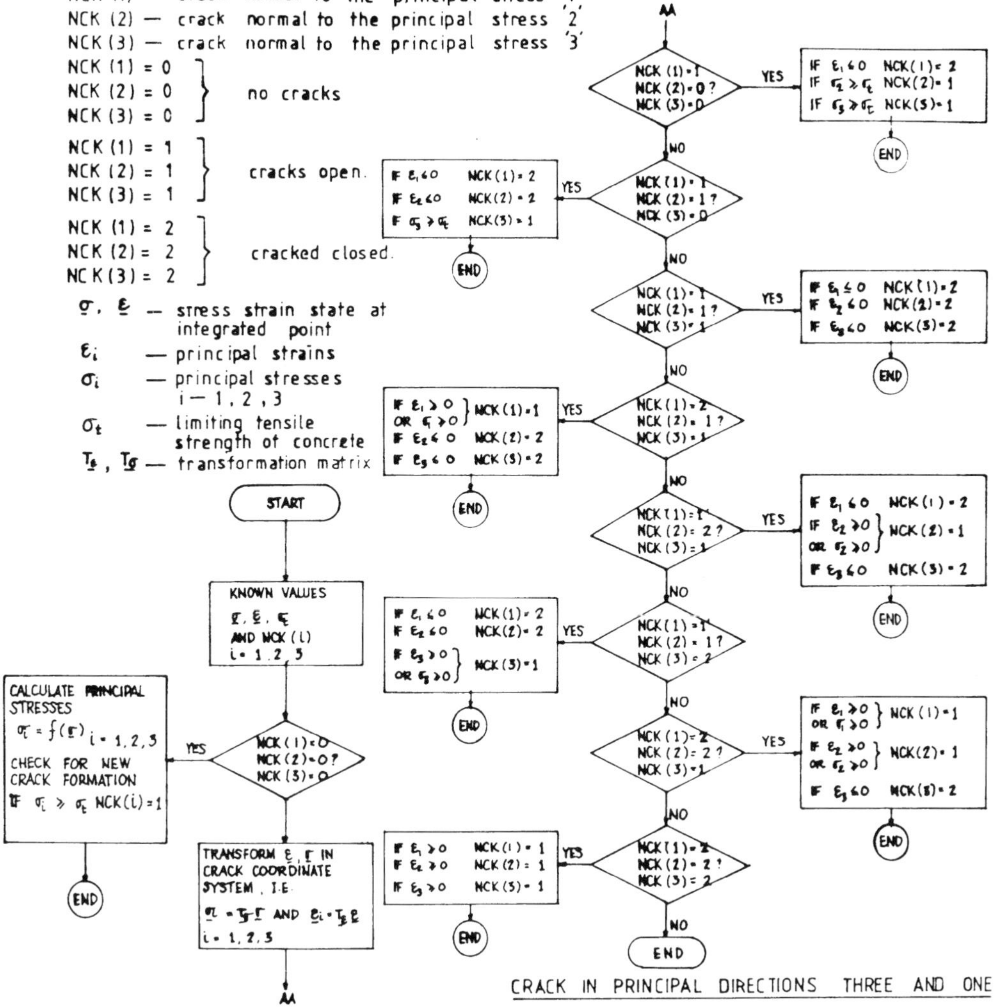

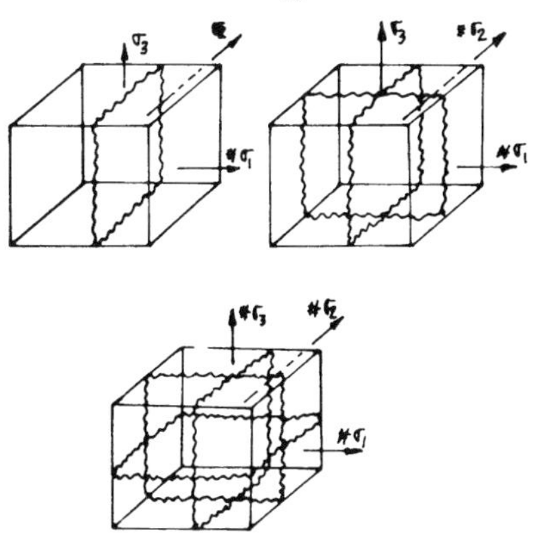

CRACK IN PRINCIPAL DIRECTIONS THREE AND ONE

$$D^*_{11} = D^*_{33} = D^*_{12} = D^*_{21} = 0$$

$$D^*_{13} = D^*_{31} = D^*_{23} = D^*_{31} = 0$$

$$D^*_{22} \quad D_{22} - D_{12}\frac{D_{12}}{D_{11}} - D_{23}\frac{D_{23}}{D_{33}}$$

$$D^*_{44} = \beta' D_{44}$$

$$D^*_{55} = \beta' D_{55}$$

$$D^*_{66} = \beta' D_{66}$$

CRACKS IN ALL THREE PRINCIPAL DIRECTIONS

$$[D^*] = [0]$$

AVIII.3. Program ISOPAR to print displacements and (if ISTAT.NE.0) velocities and accelerations

(*Jointly developed by J. Tang and the author.*)

```
      IND.EQ.0, PRINT DISPL/VEL/ACC AT ALL NODES
      IND.NE.0, PRINT DISPL/VEL/ACC ONLY AT NODES
                CONTAINED IN PRINTOUT BLOCKS

      COMMON /EL/ IND,ICOUNT,NPAR(20),NUMEG,NEGL,NEGNL,IMASS,IDAMP,ISTAT,
     1            NDOFDM,KLIN,IEIG,IMASSN,IDAMPN
      COMMON /PRCON/ NPB,IDC,IVC,IAC,IPC,IPNODE(2,8)
      DIMENSION DISP(NEQ),VEL(NEQ),ACC(NEQ),ID(NDOF,1)
      DIMENSION D(6)

      READ ID ARRAY INTO CORE

      REWIND 8
      NDBLK = NUMNP
      READ (8) ((ID(I,J),I = 1,NDOF),J = 1,NUMNP)

      PRINT DISPLACEMENTS

      IC = 4
      IF (IND.EQ.0) GO TO 10
      IF (IND.EQ.0) GO TO 180
   10 WRITE (6,2000)
      WRITE (1,2000)
      IC = IC + 5
      DO 150 IB = 1,NPB
      NODE1 = IPNODE(1,IB)
      IF (NODE1.EQ.0) GO TO 150
      NODE2 = IPNODE(2,IB)
      IF (IND.EQ.0) NODE1 = 1
      IF (IND.EQ.0) NODE2 = NUMNP

      DO 100 II = NODE1,NODE2
      IC = IC + 1
      IF (IC.LT.56) GO TO 105
      WRITE (6,2045)

      WRITE (1,2045)
      IC = 4
  105 DO 110 I = 1,6
  110 D(I) = 0.
      DO 120 I = 1,NDOF
      KK = ID(I,II)
      IL = I
      IF (NDOF.EQ.2) IL = I + 1
  120 IF (KK.NE.0) D(IL) = DISP(KK)
      WRITE (1,2010) II,D
  100 WRITE (6,2010) II,D

      IF (IND.EQ.0) GO TO 180
      IF (IC.GE.55) GO TO 150
      IC = IC + 1
      WRITE (6,2050)
      WRITE (1,2050)
  150 CONTINUE
  180 IF (ISTAT.EQ.0) RETURN
      PRINT VELOCITIES
```

```
      IF (IND.EQ.0) GO TO 201
      IF (IVC.EQ.0) GO TO 280
  201 IC=IC+5+IDC
      IF (IDC.NE.0) WRITE (6,2050)
      IF (IDC.NE.0) WRITE (1,2050)
      IF (IC.GE.54) GO TO 205
      WRITE (6,2020)
      WRITE (1,2020)
      GO TO 206
  205 WRITE (6,2022)
      WRITE (1,2022)
      IC=4
  206 DO 250 IB=1,NPB
      NODE1=IPNODE(1,IB)
      IF (NODE1.EQ.0) GO TO 250
      NODE2=IPNODE(2,IB)
      IF (IND.EQ.0) NODE1=1
      IF (IND.EQ.0) NODE2=NUMNP

      DO 200 II=NODE1,NODE2
      IC=IC+1
      IF (IC.LT.56) GO TO 207
      WRITE (6,2022)
      WRITE (1,2022)
      IC=4
  207 DO 210 I=1,6
  210 D(I)=0.
      DO 220 I=1,NDOF
      KK=ID(I,II)
      IL=1
      IF (NDOF.EQ.2) IL=I+1
  220 IF (KK.NE.0) D(IL)=VEL(KK)
      WRITE (1,2010) II,D
  200 WRITE (6,2010) II,D

      IF (IND.EQ.0) GO TO 280
      IF (IC.GE.55) GO TO 250
      IC=IC+1
      WRITE (6,2050)
      WRITE (1,2050)

  250 CONTINUE

      PRINT ACCELERATIONS

  280 IF (IND.EQ.0) GO TO 290
      IF (IAC.EQ.0) RETURN
      IF (IDC.EQ.0 .AND. IVC.EQ.0) GO TO 305
  290 IC=IC+6
      IF (IC.GE.54) GO TO 303
      WRITE (6,2050)
      WRITE (1,2050)
      WRITE (6,2030)
      WRITE (1,2030)
      GO TO 308
  303 WRITE (6,2032)
      WRITE (1,2032)
      IC=4
      GO TO 308
```

```
305 IC = IC + 5
    WRITE (6,2030)
    WRITE (1,2030)
308 DO 350 IB = 1,NPB
    NODE1 = IPNODE(1,IB)
    IF (NODE1.EQ.0) GO TO 350
    NODE2 = IPNODE(2,IB)
    IF (IND.EQ.0) NODE1 = 1
    IF (IND.EQ.0) NODE2 = NUMNP

    DO 300 II = NODE1,NODE2
    IC = IC + 1
    IF (IC.LT.56) GO TO 307
    WRITE (6,2032)
    WRITE (1,2032)
    IC = 4
307 DO 310 I = 1,6
310 D(I) = 0.
    DO 320 I = 1,NDOF
    KK = ID(I,II)
    IL = I
    IF (NDOF.EQ.2) IL = I + 1
320 IF (KK.NE.0) D(IL) = ACC(KK)
    WRITE (1,2010) II,D
300 WRITE (6,2010) II,D

    IF (IND.EQ.0) RETURN
    IF (IC.GE.55) GO TO 350
    IC = IC + 1
    WRITE (6,2050)
    WRITE (1,2050)
350 CONTINUE

    RETURN
```

Subroutine CRACKD (PROP,NCK,SIG,EPS,PS1,PS2, PS3,DC1,DC2,DC3) IMPLICIT REAL*8(A – H,O – Z)

```
    SET UP MATERIAL MATRICES FOR CRACKED CONCRETE

    COMMON /MTMD3D/ D(6,6),STRESS(6),STRAIN(6),IPT,NEL
    DIMENSION DD(6,6),PROP(1),NCK(1),SIG(1),EPS(1),
  @           PS1(1),PS2(1),PS3(1),DC1(1),DC2(1),DC3(1)
    CALL PRINCL (IPT,STRESS,PS1,PS2,PS3,DC1,DC2,DC3)
    CALL RCMOD (PROP,D)
    CALL DMAT (PROP,NCK)
    DO 222 I = 1,6

    DO 222 J = 1,6
222 DD(I,J) = 0.0
    JJJ = 1
    LL = 0
    IF (NCK(1).EQ.1) LL = 1
    IF (NCK(2).EQ.1) LL = 2
    IF (NCK(3).EQ.1) LL = 3
    IF (NCK(1).EQ.1.AND.NCK(2).EQ.1) LL = 4
    IF (NCK(2).EQ.1.AND.NCK(3).EQ.1) LL = 5
    IF (NCK(1).EQ.1.AND.NCK(3).EQ.1) LL = 6
    IF (NCK(1).EQ.1.AND.NCK(2).EQ.1.AND.NCK(3).EQI) LL = 7
```

```
      IF (LL.EQ.7) GOTO 99
      IF (JJJ.EQ.0) GOTO 200
      IF (LL.EQ.0) GOTO 999
      GOTO (113,114,115,116,117,118),LL

      ONLY ONE DIRECTION CRACKED

  113 CONTINUE

      CRACK IN DIRECTION 1

      DD(1,1)=0.0
      DD(1,2)=0.0
      DD(1,3)=0.0
      DD(2,1)=0.0
      DD(2,2)=D(2,2) - D(1,2)*D(1,2)/D(1,1)
      DD(2,3)=D(2,3) - D(1,3)*D(1,2)/D(1,1)
      DD(3,1)=0.0
      DD(3,2)=DD(2,3)
      DD(3,3)=D(3,3) - D(1,3)*D(1,3)/D(1,1)
      DD(4,4)=PROP(12)*D(4,4)
      DD(5,5)=DD(5,5)
      DD(6,6)=PROP(12)*D(6,6)
      GOTO 121
  114 CONTINUE

      CRACK IN DIRECTION 2

      DD(1,1)=D(1,1) - D(2,1)*D(2,1)/D(2,2)
      DD(1,2)=0.0
      DD(1,3)=D(1,3) - D(1,2)*D(2,3)/D(2,2)
      DD(2,1)=0.0
      DD(2,2)=0.0
      DD(2,3)=0.0
      DD(3,1)=DD(1,3)
      DD(3,2)=0.0
      DD(3,3)=D(3,3) - D(2,3)*D(2,3)/D(2,2)
      DD(4,4)=PROP(12)*D(4,4)
      DD(5,5)=PROP(12)*D(5,5)
      DD(6,6)=D(6,6)
      GOTO 121
  115 CONTINUE

      CRACK IN DIRECTION 3

      DD(1,1)=D(1,1) - D(1,3)*D(1,3)/D(3,3)
      DD(1,2)=D(1,2) - D(1,3)*D(2,3)/D(3,3)
      DD(1,3)=0.0
      DD(2,1)=DD(1,2)
      DD(2,2)=D(2,2) - D(2,3)*D(2,3)/D(3,3)
      DD(2,3)=0.0
      DD(3,1)=0.0
      DD(3,2)=0.0
      DD(3,3)=0.0
      DD(4,4)=D(4,4)
      DD(5,5)=D(5,5)*PROP(12)
      DD(6,6)=D(6,6)*PROP(12)
      GOTO 121
  116 CONTINUE
      CRACKS IN TWO DIRECTIONS
      CRACKS IN DIRECTIONS 1 & 2
```

```
      DENOM = D(1,1)*D(2,2) - D(1,2)*D(2,1)
      DD(1,1) = 0.0
      DD(1,2) = 0.0
      DD(1,3) = 0.0
      DD(2,1) = 0.0
      DD(2,2) = 0.0
      DD(2,3) = 0.0
      DD(3,1) = 0.0
      DD(3,2) = 0.0
      DD(3,3) = D(3,3)
     1          - D(3,1)*(D(2,2)*D(1,3) - D(1,2)*D(2,3))/DENOM
     2          - D(3,2)*(D(1,1)*D(2,3) - D(2, l)*D(3,1))/DENOM
      DD(4,4) = PROP(12)*D(4,4)
      DD(5,5) = PROP(12)*D(5,5)
      DD(6,6) = PROP(12)*D(6,6)
      GOTO 121
  117 CONTINUE

      CRACKS IN DIRECTIONS 3 & 2

      DENOM = D(2,2)*D(3,3) - D(2,3)*D(3,2)
      DD(1,1) = D(1,1)
     1          - D(1,2)*(D(3,3)*D(2,1) - D(3,1)*D(2,3))/DENOM
     2          - D(1,3)*(D(2,2)*D(3,1) - D(2,1)*D(3,2))/DENOM
      DD(1,2) = 0.0
      DD(1,3) = 0.0
      DD(2,1) = 0.0
      DD(2,2) = 0.0
      DD(2,3) = 0.0
      DD(3,1) = 0.0
      DD(3.2) = 0.0
      DD(3,3) = 0.0
      DD(4,4) = PROP(12)*D(4,4)
      DD(5,5) = PROP(12)*D(5,5)
      DD(6,6) - PROP(12)*D(6,6)
      GOTO 121
  118 CONTINUE

      CRACKS IN DIRECTIONS 1 & 3

      DENOM = D(1,1)*D(3,3) - D(3,2)*D(1,3)
      DD(1,1) = 0.0
      DD(1,2) = 0.0
      DD(1,3) = 0.0
      DD(2,1) = 0.0
      DD(2,2) = D(2,2)
     1          - D(2,1)*(D(3,3)*D(1,2) - D(3,2)*D(1,3))/DENOM
     2          - D(2,3)*(D(1,1)*D(3,2) - D(3,1)*D(1,2))/DENOM
      DD(2,3) = 0.0
      DD(3,1) = 0.0
      DD(3,2) = 0.0
      DD(3,3) = 0.0
      DD(4,4) = PROP(12)*D(4,4)
      DD(5,5) = PROP(12)*D(5,5)
      DD(6,6) = PROP(12)*D(6,6)
  121 CONTINUE
      GOTO 99
  200 CONTINUE
      IF (LL. .EQ. 0) GOTO 999
```

```
      GOTO (1,2,3,4,5,6),LL
    1 CONTINUE
      DD(2,2) = D(2,2)
      DD(2,3) = D(2,3)
      DD(3,2) = DD(2,3)
      DD(3,3) = D(3,3)
      DD(4,4) = PROP(12)*D(4,4)
      DD(5,5) = PROP(12)*D(5,5)
      DD(6,6) = PROP(12)*D(6,6)
      GOTO 99
    2 CONTINUE
      DD(1,1) = D(1,1)
      DD(1,3) = D(2,3)
      DD(3,1) = D(1,3)
      DD(3,3) = D(3,3)
      DD(4,4) = PROP(12)*D(4,4)
      DD(5,5) = PROP(12)*D(5,5)
      DD(6,6) = D(6,6)
      GOTO 99
    3 CONTINUE
      DD(1,1) = D(1,1)
      DD(2,2) = D(2,2)
      DD(1,2) = D(1,2)
      DD(3,3) = D(3,3)
      DD(2,1) = DD(1,2)
      DD(4,4) = D(4,4)
      DD(5,5) = PROP(12)*D(5,5)
      DD(6,6) = PROP(12)*D(6,6)
      GOTO 99
    4 CONTINUE
      DD(3,3) = D(3,3)
      DD(4,4) = PROP(12)*D(4,4)
      DD(5,5) - PROP(12)*D(5,5)
      DD(6,6) = PROP(12)*D(6,6)
      GOTO 99
    5 CONTINUE
      DD(1,1) = D(1,1)
      DD(4,4) = PROP(12)*D(4,4)
      DD(5,5) = PROP(12)*D(5,5)
      DD(6,6) = PROP(12)*D(6,6)
      GOTO 99
    6 CONTINUE
      DD(2,2) = D(2,2)
      DD(4,4) = D(4,4)
      DD(5,5) = D(5,5)
      DD(6,6) = D(6,6)
   99 CONTINUE

      CRACKS IN ALL THREE DIRECTIONS
      TRANSFER DD TO D

      DO 101 J = 1,6
      DO 101 K = 1,6
      D(J,K) = DD(J,K)
  101 CONTINUE
  999 CONTINUE
      RETURN
      END
```

APPENDIX IX

Data and results for concrete offshore structures

AIX.1. Concrete offshore structures

Brent C production platform (Fig. AIX.1)
The caisson of the Brent C platform is made up of 36 square concrete cells. The external walls of the outer cells are semi-circular so as to resist any external over-pressure, and the roofs of the cells are in the shape of truncated pyramids. Four of these cells rise to form four towers where the utilities are kept. The base slab is 4 m thick to give stability to the platform when it is being towed out.

Brent D production platform
This platform consists of a base with three concrete towers. Some utilities are kept in these towers as well as in the depth of the deck truss, the rest are contained in the modules mounted on the steel deck truss assembly. No anchors or piles are needed when on location on the seabed.

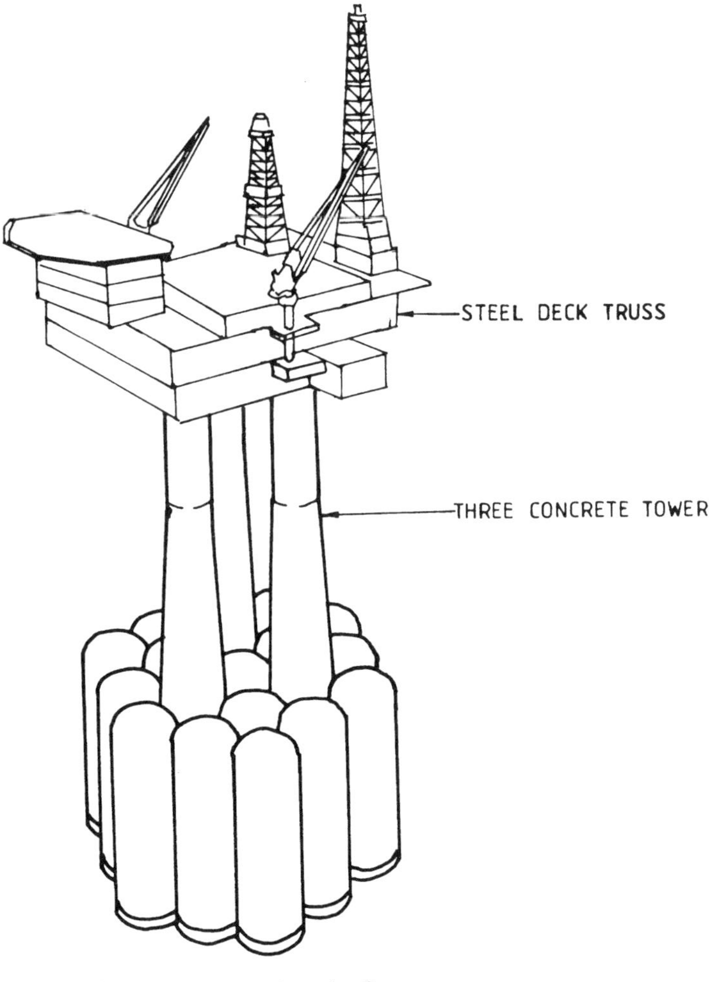

Fig. AIX.1. Brent C concrete gravity platform.

Table AIX.1. Basic data on two offshore platforms.

	Brent D	Brent C
Environment		
Water depth	142 m	141 m
Substructure		
Caisson shape	hexagon	square
Caisson base area	6360 m^2	10 340 m^2
Caisson height	58 m	57 m
Number of legs	3	4
Weight of sub-structure in air	175 000 tons	283 000 tons
Superstructure		
Height of deck above water	25.2 m	23.5 m
Area of deck	3400 m^2	4000 m^2
Type of deck construction	plate girder	lattice girder
Deck weight	3200 tons	6400 tons
Weight of equipment within deck	1864 tons	3420 tons
Weight of modules and equipment on deck	—	9860 tons

AIX.2. Environmental loads

Wave loads, current loads and wind loads are the main loads acting on gravity platforms. The other loads imposed are due to ice and tidal fluctuations. Stokes non-linear fifth order wave theory is adopted for wave crest and incidental wave potential as shown in Fig. AIX.2.

The potential function assumed is

$$\begin{aligned}\frac{\beta_{w\phi}}{(CB)} = \frac{2(\pi)(\phi)}{L(CB)} &= (\lambda A_{11} + \lambda^3 A_{13} + \lambda^5 A_{15})(\cosh(\beta_w(z)))\sin\theta \\ &+ (\lambda^2 A_{22} + \lambda^4 A_{24})(\cosh(2\beta_w(z)))\sin(2\theta) \\ &+ (\lambda^3 A_{33} + \lambda^5 A_{35})(\cosh(3\beta_w(z)))\sin(3\theta) \\ &+ (\lambda^4 A_{44})(\cosh(4\beta_w(z)))\sin(4\theta) \\ &+ (\lambda^5 A_{55})(\cosh(5\beta_w(z)))\sin(5\theta) \qquad \text{(AIX.1)}\end{aligned}$$

where $\beta_w = 2\pi/\lambda$; λ is a parameter; and A_{ij} are constants depending on the ratio of water depth to wave length.

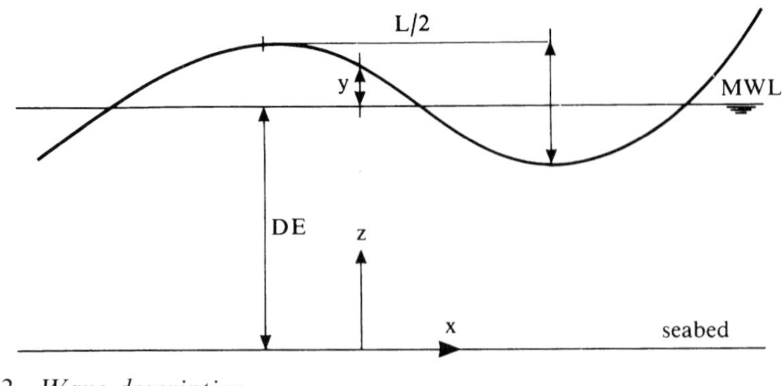

Fig. AIX.2. Wave description.

The final velocity and acceleration components adopted are given below.

Velocity components
Horizontal

$$\begin{aligned} U_1 = (CB)((\lambda A_{11} + \lambda^3 A_{13} + \lambda^5 A_{15}) \cosh(\beta_w P_N) \cos\theta \\ + 2(\lambda^2 A_{22} + \lambda^4 A_{24}) \cosh(2\beta_w P_N) \cos 2\theta \\ + 3(\lambda^3 A_{33} + \lambda^5 A_{35}) \cosh(3\beta_w P_N) \cos 3\theta \\ + 4(\lambda^4 A_{44}) \cosh(4\beta_w P_N) \cos 4\theta \\ + 5(\lambda^5 A_{55}) \cosh(5\beta_w P_N) \cos 5\theta) \end{aligned} \tag{AIX.2}$$

Vertical

$$\begin{aligned} W_1 = (CB)(\lambda A_{11} + \lambda^5 A_{15}) \sinh(\beta_w P_N) \sin\theta \\ + 2(\lambda^2 A_{22} + \lambda^4 A_{24}) \sinh(2\beta_w P_N) \sin(2\theta) \\ + 3(\lambda^3 A_{33} + \lambda^5 A_{35}) \sinh(3\beta_w P_N) \sin(3\theta) \\ + 4(\lambda^4 A_{44}) \sinh(4\beta_w P_N) \sin(4\theta) \\ + 5(\lambda^5 A_{55}) \sinh(5\beta_w P_N) \sin(5\theta) \end{aligned} \tag{AIX.3}$$

Acceleration Components
The acceleration components can be found by differentiating the expressions for the horizontal and vertical velocities. These are given below.

Horizontal

$$\begin{aligned} U_2 = \beta_w (CB)^2 (\lambda A_{11} + \lambda^3 A_{13} + \lambda^5 A_{15}) \cosh(\beta_w P_N) \sin\theta \\ + 4(\lambda^2 A_{22} + \lambda^4 A_{24}) \cosh(2\beta_w P_N) \sin(2\theta) \\ + 9(\lambda^3 A_{33} + \lambda^5 A_{35}) \cosh(3\beta_w P_N) \sin(3\theta) \\ + 16(\lambda^4 A_{44}) \cosh(4\beta_w P_N) \sin(4\theta) \\ + 25(\lambda^5 A_{55}) \cosh(5\beta_w P_N) \sin(5\theta) \end{aligned} \tag{AIX.4}$$

Vertical

$$\begin{aligned} W_1 = (-\beta_w (CB)^2 (\lambda A_{11} + \lambda^3 A_{13} + \lambda^5 A_{15}) \sinh(\beta_w P_N) \cos\theta \\ + 4(\lambda^2 A_{22} + \lambda^4 A_{24}) \sinh(2\beta_w P_N) \cos(2\theta) \\ + 9(\lambda^3 A_{33} + \lambda^5 A_{35}) \sinh(3\beta_w P_N) \cos(3\theta) \\ + 16(\lambda^4 A_{44}) \sinh(4\beta_w P_N) \cos(4\theta) \\ + 25(\lambda^5 A_{55}) \sinh(5\beta_w P_N) \cos(5\theta) \end{aligned} \tag{AIX.5}$$

The equations for the vertical and horizontal components of velocity and acceleration will be used together with current loads and Morrison's formula, taking into account diffraction of waves on a large base to calculate the wave forces.

Modified Morrison's equation

The Morrison equation, expressing force as the sum of a velocity-dependent term (drag) and an acceleration-dependent term (inertia), is given by

$$F_{\mathrm{M}} = \tfrac{1}{2}(P_{\mathrm{w}}(CD)(D_0)(U_1)|U_1| + (P_{\mathrm{w}})(CM)\pi(D_0)^2(U_2)/4 \qquad \text{(AIX.6)}$$

The appropriate criterion for the neglect of the drag is $D_0/H > 0.2$. Diffraction is added if the ratio of the diameter of the main sections of the platform to the design wave length is >0.2. For the Condeep platform having three small circular towers and a large hexagonal-shaped base formed from caissons, diffraction is assumed.

The diffraction coefficient CH is given by

$$CH = (F_{\mathrm{u}} + F_{\mathrm{d}})/F_{\mathrm{u}} \qquad \text{(AIX.7)}$$

where F_{u} = contribution to the undisturbed incident wave, known as the Froude–Krylor component (Morrison's equation); and F_{d} = contribution due to disturbance.

$$F = \text{inertia coefficient} = (CH)F_{\mathrm{u}} \qquad \text{(AIX.8)}$$

The program ISOPAR includes two diffraction-coefficient formulae and they are given below.

Maccamy and Fuchs

$$CH = \frac{4(DL)^2}{\pi^3(DIA)^2[j_1'(x)^2 + Y_1'(x)^2]} \qquad \text{(AIX.9)}$$

when

$$X = (\pi)(DL)(DIA) \qquad j_1' = \tfrac{1}{2} \qquad Y_1' = \frac{2}{\pi x^2}$$

Hobgen and Standing

$$CH = 1 + (0.75)\left(\frac{h_{\mathrm{c}}}{2R_{\mathrm{e}}}\right)^{1/3}(1 - 0.3)\left(\frac{2\pi}{DL}\right)^2 R_{\mathrm{e}}^2 \qquad \text{(AIX.10)}$$

where h_{c} = caisson height; R_{e} = equivalent radius of the caisson base; and DL = design wave length.

Current loads

Currents in conjunction with the wave loads normally result in stresses and overturning moments on the gravity platforms. The forces due to currents acting on the structure are obtained by the formula:

$$F_{DC} = (\tfrac{1}{2})(p_w(CD)(CU)^2(D_0) \qquad \text{(AIX.11)}$$

where D_0 *is a specific value of* ϕ.

Due to lack of observational data on the existing offshore structure, a conservative approach is adopted by defining the current velocity (CU) at any elevation (RL) above the seabed as:

For RL less than 15 m

$$CU = 0.24 + (0.3)(RL)/15\,(\text{m/s}) \qquad \text{(AIX.12a)}$$

For RL greater than or equal to 15 m

$$CU = 0.54 + (0.84)(RL - 15)/155\,(\text{m/s}) \qquad \text{(AIX.12b)}$$

In addition velocity is 1.38 m/s at the water surface and 0.24 m/s at the bottom.

Wind loads

The wind load adopted is based on Refs [1], [2] and [19].

Storm wind

The maximum one-minute sustained wind velocity of 125 mph is used in conjunction with a storm wind.

Operating winds and gusts

A maximum instantaneous gust velocity of 160 mph is used on the deck.

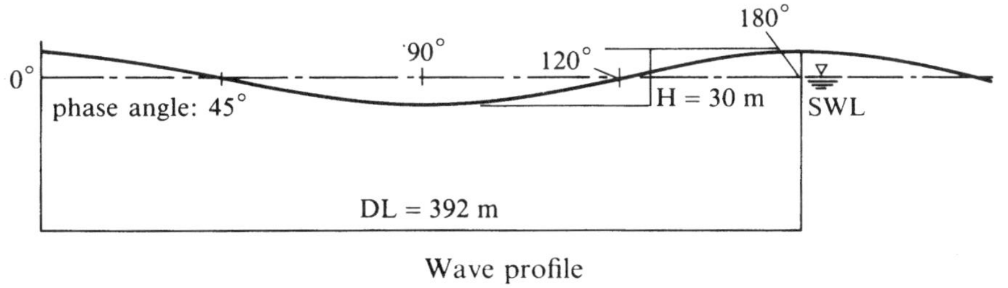

Fig. AIX.2. A wave profile.

AIX.3. Airy wave theory

This is a programme to analyse the effects of surface wave on offshore structures by means of Airy wave theory, which is related to the surface data for the water velocity, acceleration, and pressure beneath the wave.

Notation

H	Wave height
D	Water depth
X	Horizontal direction
Y	Vertical direction

N	Wave profile
U	Horizontal velocity
V	Vertical velocity
W	Frequency
DL	Wave length
K	Wavenumber
T	Period
G	Gravitational acceleration
DEN	Density of water
AX	Horizontal acceleration
AY	Vertical Acceleration
P	Gauge pressure
C	Wave speed
FWAV	Wave force
FD	Frictional force from drag
FI	Inertia force due to water particle acceleration
CD	Drag coefficient
CM	Coefficient of virtual mass
A	Projected area of the structure
DIA	Diameter of the structure
MOM	Total moment
MD	Moment arising from drag force
MI	Moment arising from inertia force
CI	Inertia coefficient
CU	Current velocity
FCU	Current force
FENV	Environmental force

AIX.4. Computer subroutines for concrete offshore structures

```
WRITE (6,'(A)') 'PLEASE INPUT THE WATER DEPTH (m):'
READ (5,*)D
WRITE (6,'(A)') 'PLEASE INPUT THE WAVE HEIGHT (m):'
READ (5,*)H
WRITE (6,'(A)') 'PLEASE INPUT THE PERIODIC TIME (sec):'
```

This is a sub-routine to determine the wave force
by Morrison's equation

```
SUBROUTINE WAVEFO (Y,DEN,DIA,U,PI,AX,FWAV)
IMPLICIT REAL (A – Z)
DIMENSION FD(10000),FI(10000),FWAV(10000),U(10000),AX(10000)
CD = 0.60
Cm = 2.0
FD(Y + 1) = (0.5*CD*DEN*DIA*U(Y + 1)*ABS(U(Y + 1)))/1000.0
FI(Y + 1) = (0.25*CM*DEN*PI*DIA**2.0*AX(Y + 1))/1000.0
FWAV(Y + 1) = FD(Y + 1) + FI(Y + 1)
RETURN
END
```

This is a sub-routine to determine the moments
arising from the drag and inertia forces

```
SUBROUTINE MOMENT (J,K,Y,D,DEN,DIA,W,H,THETA,MOM)
IMPLICIT REAL (A – Z)
DIMENSION MOM(10000)
CD = 0.60
CI = 2.0
IF (J.EQ.Y) THEN
  MOM(Y + 1) = 0.0
  GOTO 1234
     ENDIF
     Q1A = 2.0*(K*(J – Y))*SINH(2.0*(K*(J – Y)))
     Q1B = COSH(2.0*K*(J – Y)) – 2.0*(K*(J – Y))**2.0 – 1.0
     Q1C = (SINH(K*D))**2.0
     Q1 = (Q1A – Q1B)/Q1C
- - - - - - - - - - - - - - - - - - - - - - - - - - - - - - - - - - - -
     Q2A = K*Y*SINH(K*(J – Y))
     Q2B = COSH(K*(J – Y)) – 1.0
     Q2C = SINH(K*D)
     Q2 = (Q2A – Q2B)/Q2C
- - - - - - - - - - - - - - - - - - - - - - - - - - - - - - - - - - - -
     MCOM = DEN*DIA*W**2.0*H/(8.0*K**2.0)
     MD = MCOM*CD*H*Q1*ABS(COS(THETA))*COS(THETA)/8.0
     MD = (DEN*CD*DIA)/(64.0*K**2.0)*(W*H)**2.0*Q1
  *     *ABS(COS(THETA))*COS(THETA)
     MI = (DEN*CI*PI*DIA**2.0)/(2.0*K*2.0*4.0)
  *     *W**2.0*H*Q2*SIN(THETA)
     MOM(Y + 1) = (MD + MI)/1000.0
1234 RETURN
     END
```

This is a sub-routine to find the current velocity

```
SUBROUTINE CURVEL (Y,CU)
      IMPLICIT REAL (A – Z)
      DIMENSION CU(10000)
      CU(Y + 1) = 0.54 + (0.84)*(Y – 15.0)/155.0
      RETURN
      END
```

This is a sub-routine to determine the current force

```
SUBROUTINE CURRFO (Y,DEN,CU,DIA,FCU)
      IMPLICIT REAL (A – Z)
      DIMENSION FCU (10000), CU(10000)
      CD = 0.60
      FCU(Y + l) = (0.5*DEN*CD*CU(Y + l)**2*DIA)/1000
      RETURN
      END
```

This is a sub-routine to find the environmental force

```
      SUBROUTINE ENVOFO (Y,FWAV,FCU,FENV)
      IMPLICIT REAL (A – Z)
      DIMENSION FENV(10000),FWAV(10000),FCU(10000)
      FENV(Y + 1) = FWAV(Y + 1) + FCU(Y + 1)
      RETURN
      END
```

Sub-routine for drilling tower with caisson base

```
      SUBROUTINE DTOWER (DIA,Y)
      IMPLICIT REAL (A – Z)
      IF (Y.LT.65.0)                      DIA = 110.0
      IF ((Y.GE.65.0).AND.(Y.LT.72.0))    DIA = 2.0*11.758
      IF ((Y.GE.72.0).AND.(Y.LT.82.0))    DIA = 2.0*11.067
      IF ((Y.GE.82.0).AND.(Y.LT.92.0))    DIA = 2.0*10.108
      IF ((Y.GE.92.0).AND.(Y.LT.102.0))   DIA = 2.0* 9.150
      IF ((Y.GE.102.0).AND.(Y.LT.112.0))  DIA = 2.0* 8.192
      IF ((Y.GE.112.0).AND.(Y.LT.122.0))  DIA = 2.0* 7.233
      IF (Y.GE.122.0)                     DIA = 2.0* 6.275
      RETURN
      END
```

Sub-routine for utility tower with caisson base

```
      SUBROUTINE UTOWER (DIA,Y)
      IMPLICIT REAL (A – Z)
      IF (Y.LT.65.0)                      DIA = 110.0
      IF ((Y.GE.65.0).AND.(Y.LT.72.0))    DIA = 2.0*11.708
      IF ((Y.GE.72.0).AND.(Y.LT.82.0))    DIA = 2.0*10.967
      IF ((Y.GE.82.0).AND.(Y.LT.92.0))    DIA = 2.0* 9.908
      IF ((Y.GE.92.0).AND.(Y.LT.102.0))   DIA = 2.0* 8.850
      IF ((Y.GE.102.0).AND.(Y.LT.112.0))  DIA = 2.0* 7.792
      IF ((Y.GE.112.0).AND.(Y.LT.122.0))  DIA = 2.0* 6.735
      IF ((Y.GE.122.0)                    DIA = 2.0* 5.675
      RETURN
      END
```

A subroutine to find the diffraction coefficient by Maccamy and Fuchs

```
SUBROUTINE MACFUC (Y,DIA,PI,DL,CHF)
IMPLICIT REAL (A - Z)
DIMENSION CHF(10000)
DIJ = 0.5
DIY = 2.0/PI/(PI*DIA*DL)**2
CHF(Y + 1) = 4.0*DL**2/(PI**3*DIA**2*(DIJ**2 + DIY**2)**0.5)
RETURN
END
```

A subroutine to find the diffraction coefficient by Hobgen and Standing

```
SUBROUTINE HOBSTA (Y,PI,DL,DIA,HC,CHH)
IMPLICIT REAL (A - Z)
DIMENSION CHH(10000)
HC = 60.0
RC = DIA/2.0
PW = 1.0/3.0
CHH(Y + 1) = (1.0 + 0.75*(HC/2.0/RC)**PW*
*    (1.0 - 0.3*(2.0*PI/DL)**2*RC**2))
RETURN
END
```

A subroutine to find the total wave force by Morrison's equation and Maccamy and Fuchs coefficient

```
SUBROUTINE TOTFOM (Y,FWAV,CHF,FOM)
IMPLICIT REAL (A - Z)
DIMENSION FOM(10000),FWAV(10000),CHF(10000)
FOM(Y + 1) = FWAV(Y + 1)*CHF(Y + 1)
RETURN
END
```

A subroutine to find the total wave force by Morrison's equation and Hobgen and Standing coefficient

```
SUBROUTINE TOTFOH (Y,FWAV,CHH,FOH)
IMPLICIT REAL (A - Z)
DIMENSION FOH(10000),FWAV(10000),CHH(10000)
FOH(Y + 1) = FWAV(Y + 1)*CHH(Y + 1)
RETURN
END
```

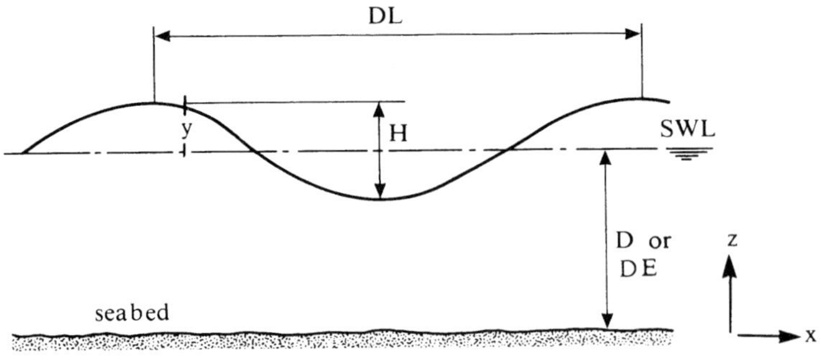

Fig. AIX.3. Wave definition sketch.

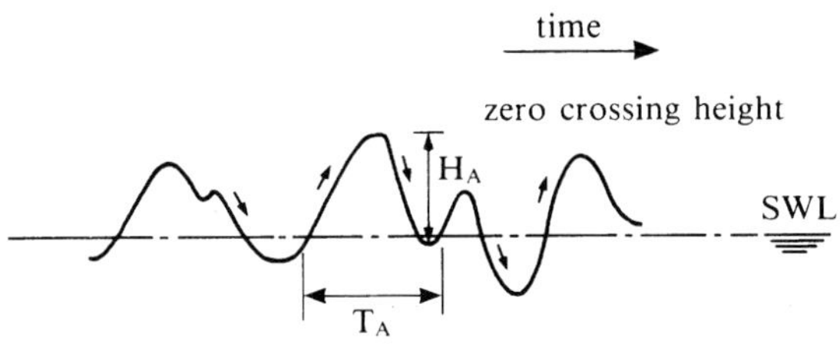

Fig. AIX.4. Typical sea surface record.

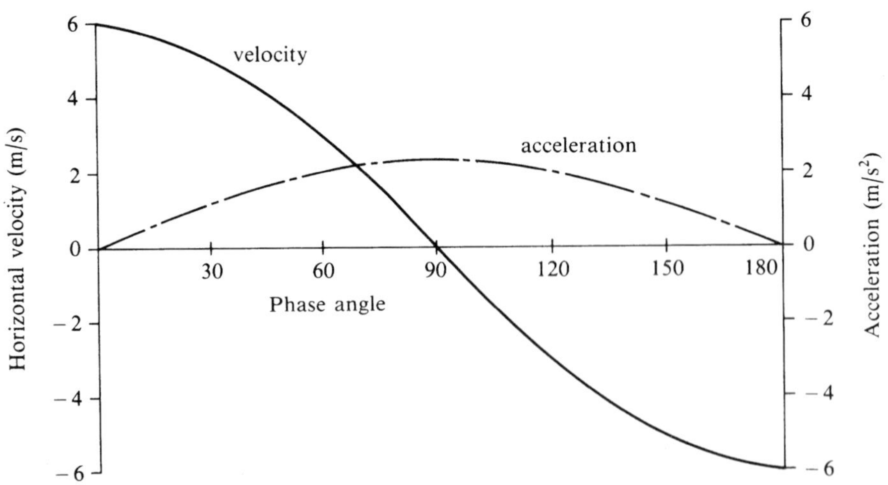

Fig. AIX.5. The variation of horizontal velocity and acceleration with respect to phase angle at a water depth of 145 m.

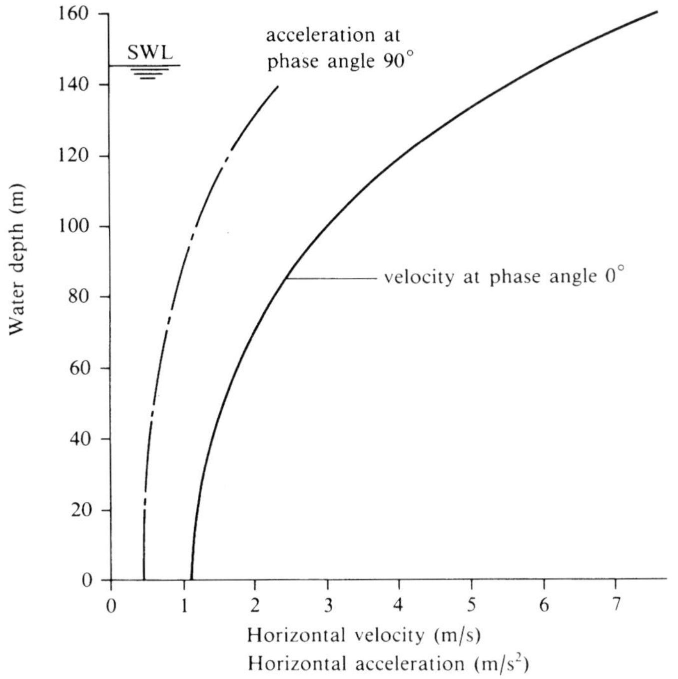

Fig. AIX.6. Variation of horizontal velocity and acceleration with water depth.

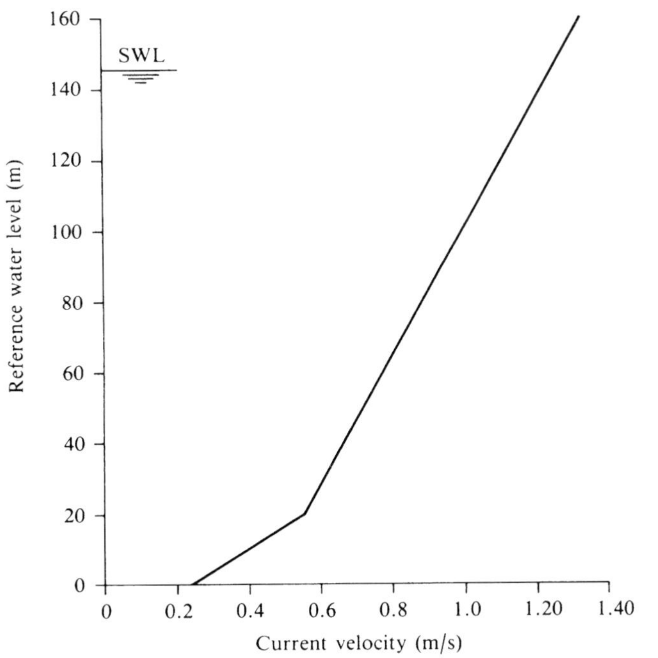

Fig. AIX.7. Current velocity profile.

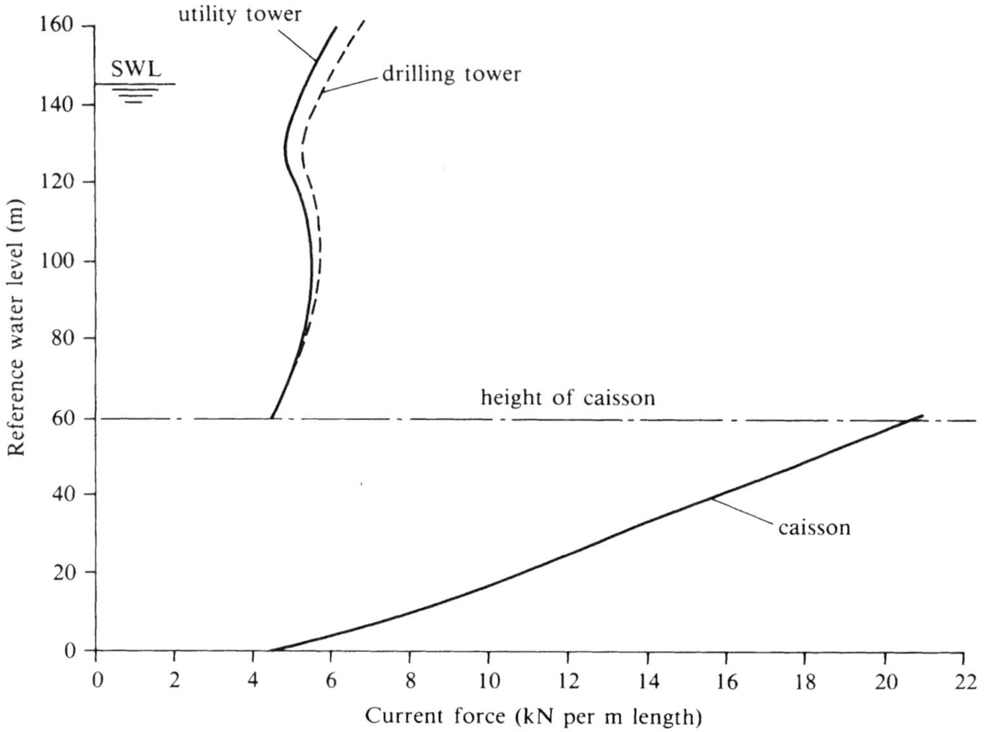

Fig. AIX.8. The distribution of current force/unit length with water level.

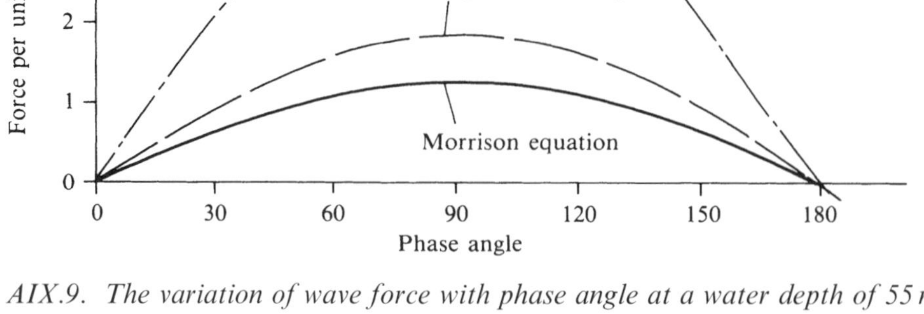

Fig. AIX.9. The variation of wave force with phase angle at a water depth of 55 m.

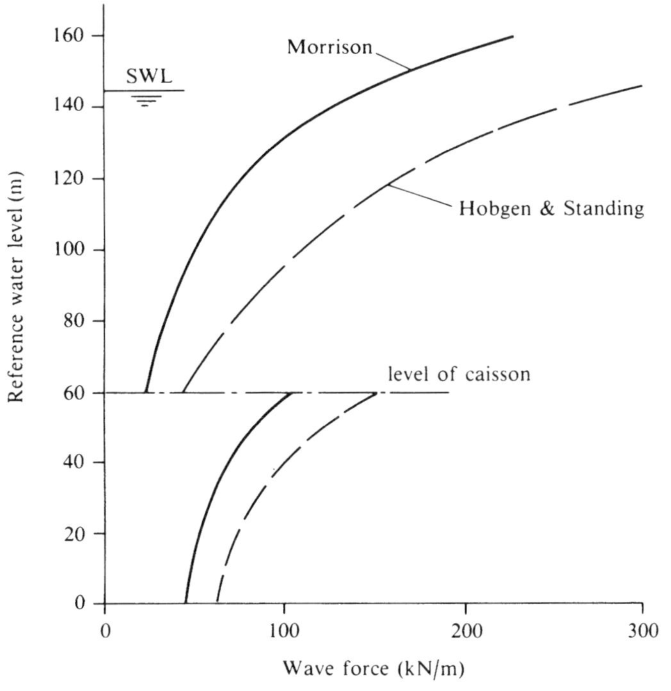

Fig. AIX.10. Wave force profile at phase angle 0°.

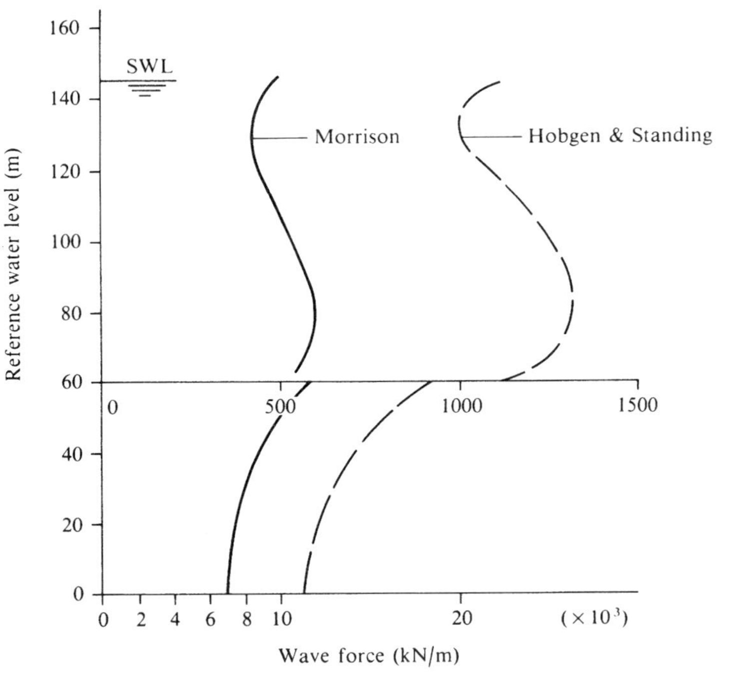

Fig. AIX.11. Wave force profile at phase angle 90°.

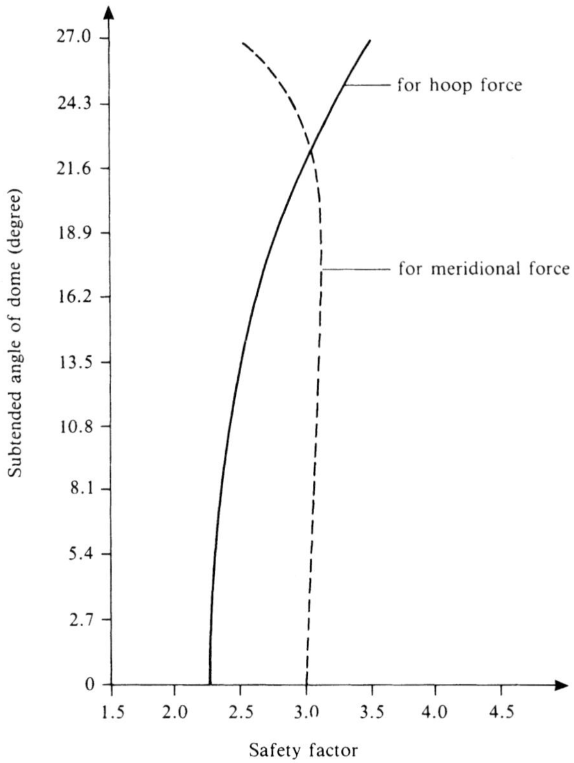

Fig. AIX.12. Angle of the tower dome versus safety factor.

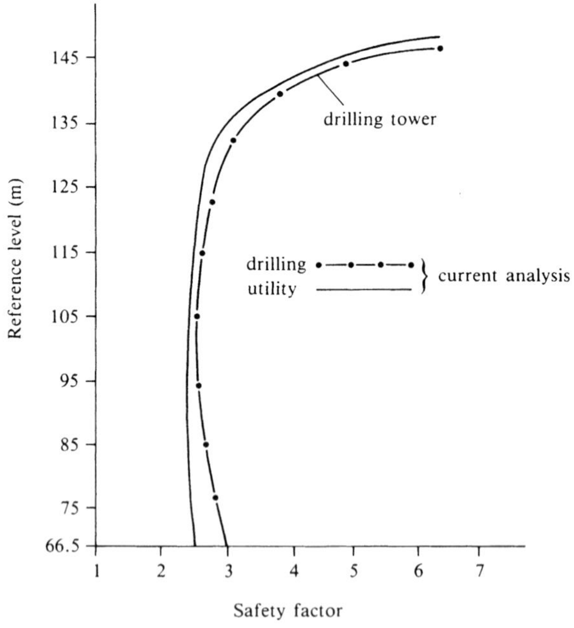

Fig. AIX.13. Drilling and utility towers' heights versus safety factors.

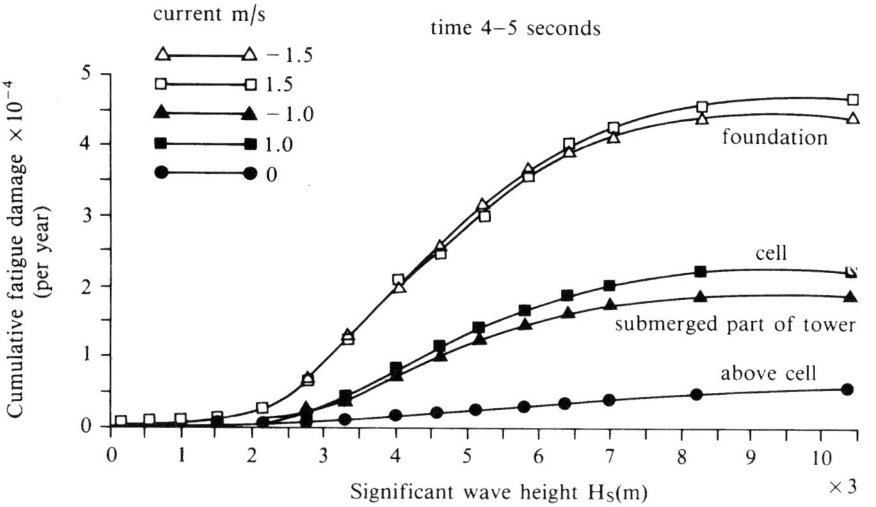

Fig. AIX.14. Cumulative damage with significant wave height in the presence of current.

AIX.5. Olson IMPLOSION analysis [1029]

General equations have been reproduced for cylinders from Olson's work [1029]. They are then modelled in a special subprogram IMPLOSION given in this section by the author and A. Aziz. This work is then compared in the main text using the ISOPAR program.

If P_i and P_0 denote the internal and external pressure respectively, the conditions at the outer and inner surfaces of the cylinder are (Fig. AIX.15)

$$(\sigma_r)_{r=R_0} = -P_0 \qquad (\sigma_r)_{r=R_0-t} = P_i = -P_0 \tag{AIX.13}$$

The sign on the right-hand side of each equation is negative because the normal stress is taken as positive for tension. Therefore

$$P_{\text{imp}} = \frac{t(2R_0 - t)}{2R_0^2}\sigma_{\max} \tag{AIX.14}$$

In terms of outer diameter D_0

$$P_{\text{imp}} = \frac{2t(D_0 - t)}{D_0^2}\sigma_{\max} \tag{AIX.15}$$

In terms of mean radius R

$$P_{\text{imp}} = 4\frac{t}{\left(2 + \frac{t}{R}\right)^2}\sigma_{\max} \tag{AIX.16}$$

where

$$R = \frac{D_0 - t}{2} \tag{AIX.17}$$

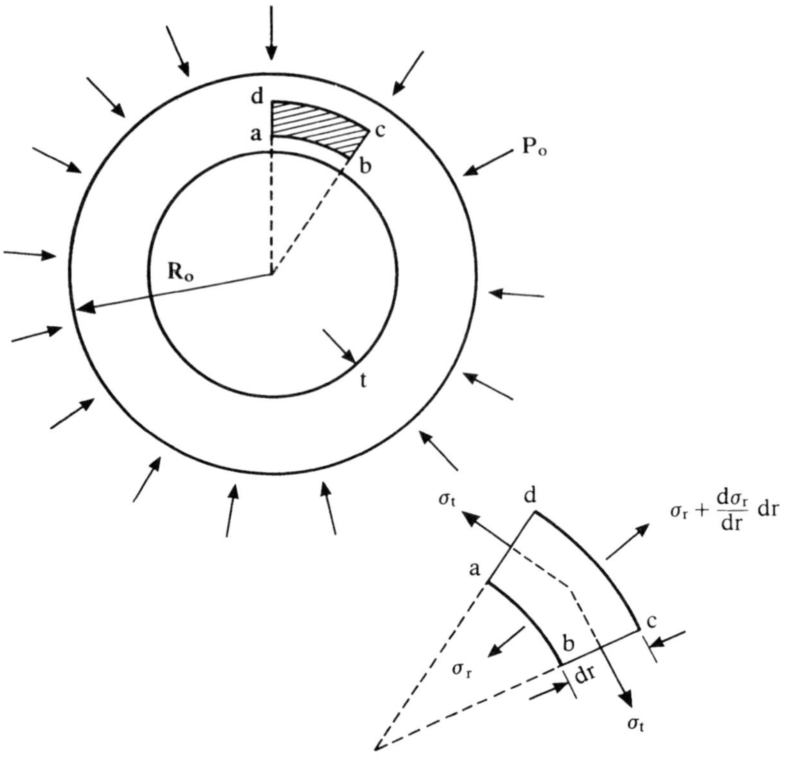

Fig. AIX.15. Cylinder under loads.

The separation point between thin-walled and thick-walled cylinders is selected at $t/D_0 = 0.063$, while the separation point for short, moderately long and long thin-walled cylinders is given in Fig. AIX. 16.

A thick-walled cylinders of all lengths
B thin-walled cylinders
B_1 short cylinder
B_2 moderately long cylinder
B_3 long cylinder

Thick-walled cylinders

The implosion pressure is developed by using an average wall hoop stress equation and a strength increase factor K to account for the difference in concrete length under multi-axial loading. For a cylindrical structure the ratio of hoop to axial stress is nearly 2:1. From studies of biaxial loading of concrete, for the loading case where principal stresses are in this ratio, the strength increase factor $K = 1.25$. The expression to predict implosion strength is developed as follows.

The average wall stress equation is

$$\sigma_{\text{imp}} = \frac{P_{\text{imp}} \times R_0}{t} \qquad \text{(AIX.18)}$$

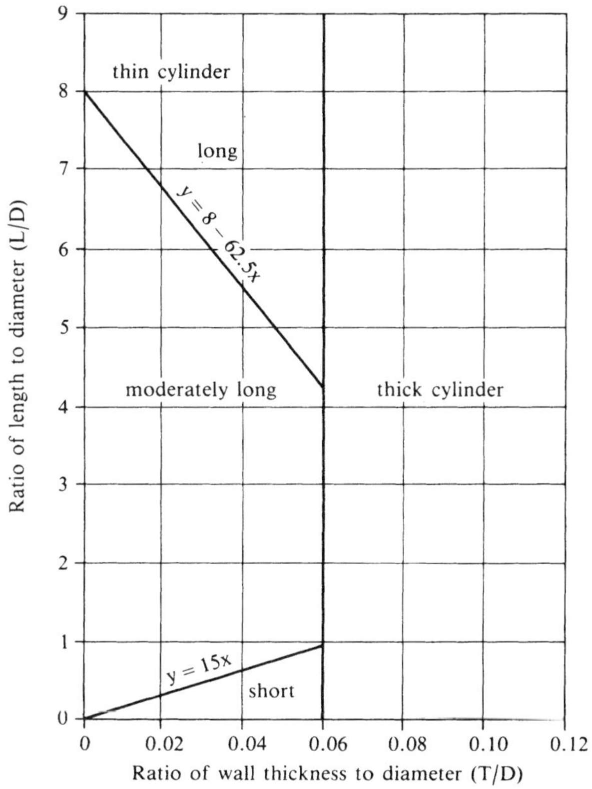

Fig. AIX.16. Cylindrical structures of different geometries [1029].

where

$$\sigma_{imp} = Kf'_c \quad \text{and} \quad R_0 = \frac{D_0}{2} \tag{AIX.19}$$

therefore

$$P_{imp} = 2K\frac{t}{D_0}f'_c \tag{AIX.20}$$

Thin-walled cylinders

Short cylinders

Short thin-walled cylinders are prevented from buckling by the influence of end closures. When the cylinder wall stress in the hoop direction equals the ultimate strength of the concrete, a material failure will cause implosion.

$$\sigma_{max} = f'_c \qquad P_{imp} = 4\left[\frac{\frac{t}{R}}{\left(2+\frac{t}{R}\right)}\right]f'_c \tag{AIX.21}$$

Moderately long cylinders

For thin-walled cylinders of moderate length, end closures influence the behaviour of the cylindrical shell by restraining radial displacement. An expression to predict the implosion pressure of moderately long cylinders can be written

$$P_{\text{imp}} = 2330\eta \left[\frac{\left(\frac{t}{R}\right)^{5/2} \left(\frac{R}{L}\right)}{\left(2 + \frac{t}{R}\right)^2} \right] f_c' \tag{AIX.22}$$

A factor β is introduced in this equation depending on the type of end condition. In the computer program the value of β is taken as 1.2 assuming the cylinders are fixed at the end.

$$P_{\text{imp}} = 2330 \times \beta \times \eta \left[\frac{\left(\frac{t}{R}\right)^{5/2} \left(\frac{R}{L}\right)}{\left(2 + \frac{t}{R}\right)^2} \right] f_c' \tag{AIX.23}$$

η = plasticity reduction factor; E_i = initial tangential modulus of elasticity.

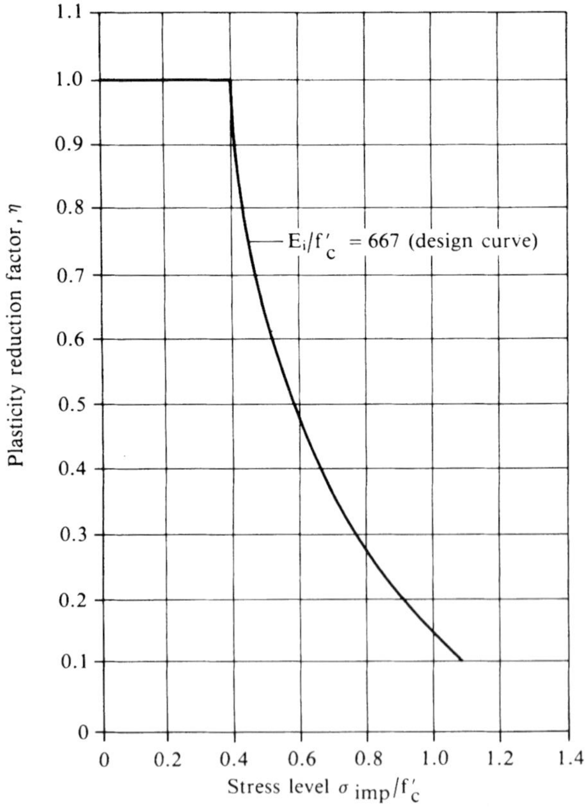

Fig. AIX.17. η curve versus stress levels [1029].

The plasticity reduction factor accounts for the inelastic material behaviour in the elastic buckling analysis. σ_{max}/f'_c is the ratio of wall stress at implosion (experimental) to wall stress at implosion (analytical).

Because the plasticity reduction factor is a function of modulus of elasticity, the η curve was selected at $E_i/f'_c = 667$ (Fig. AIX.17). This value was obtained from the relationship

$$E_i = \frac{2f'_c}{\varepsilon_{ult}} \qquad \varepsilon_{ult} = 0.003 \tag{AIX.24}$$

Long cylinders

A long thin-walled cylinder has its mid-length portion unaffected by the end closure conditions. For this case the elliptical out-of-round shape, where the number of buckles is two, will be the geometry at implosion.

Using the same substitution $E_i/f'_c = 667$ and $\nu = 0.18$ the equation obtained is

$$\frac{\sigma_{max}}{f'_c} = 172\left(\frac{t}{R}\right)^2 \eta$$

The implosion pressure of a long cylinder can be written as

$$P_{imp} = 686 \times \eta \times \left[\frac{\left(\frac{t}{R}\right)^3}{\left(2+\frac{t}{R}\right)^2}\right] \times f'_c \tag{AIX.25}$$

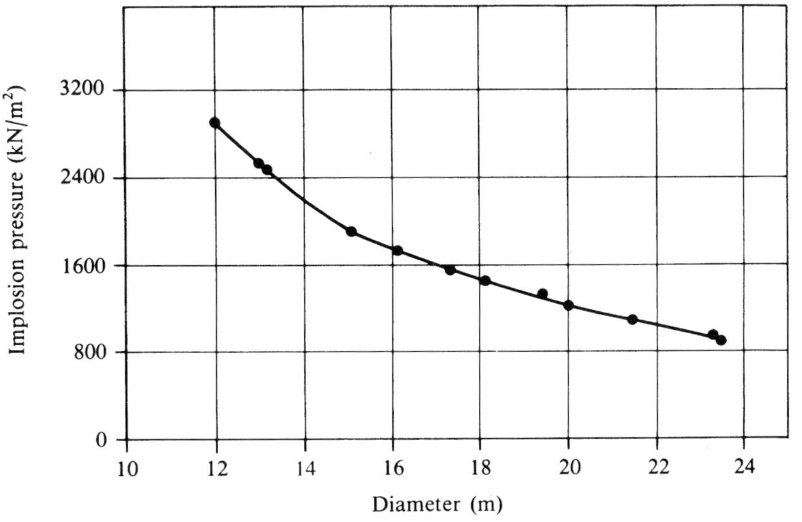

Fig. AIX.18. Implosion pressure versus diameter (utility and drilling tower). Results from ISOPAR *(Bangash). (See Implosion program in this Section.)*

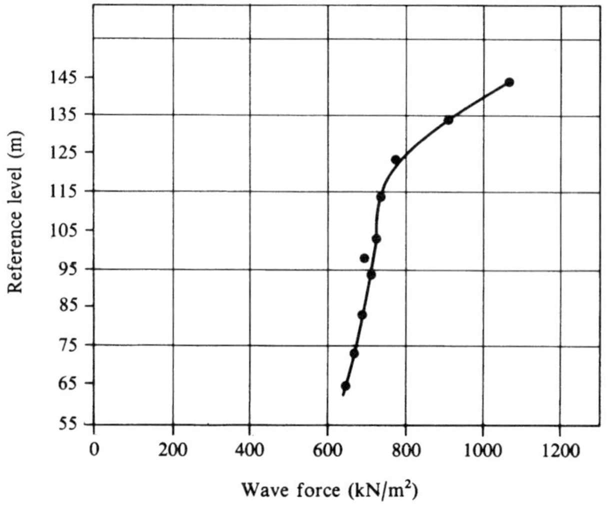

Fig. AIX.19. Reference level versus wave force (utility tower). Results from ISOPAR *(Bangash). (See Implosion program in this Section.)*

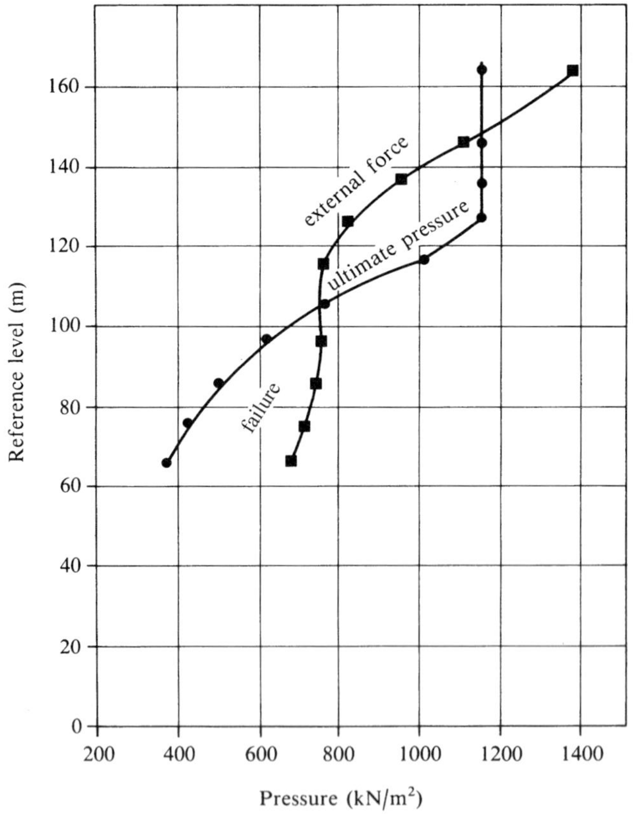

Fig. AIX.20. Reference level versus pressure (utility tower). Results from ISOPAR *using Olson's equations (Bangash). (See Implosion program in this Section.)*

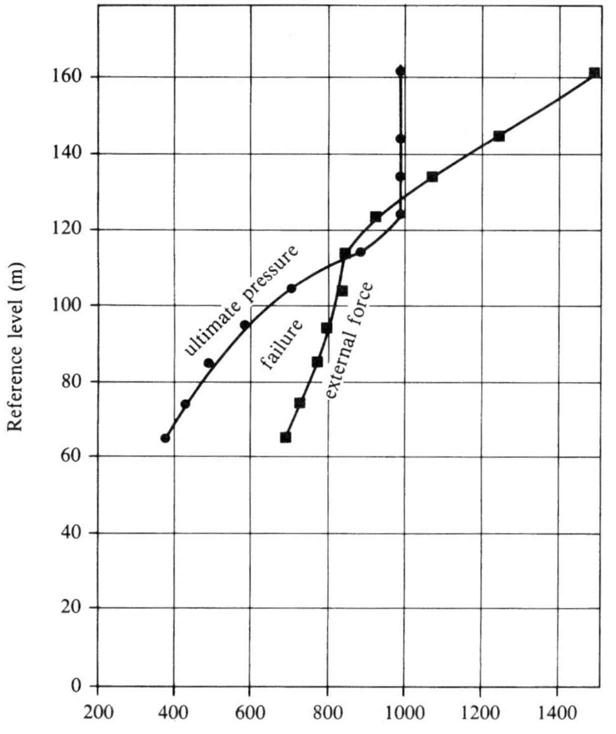

Fig. AIX.21. Reference level versus pressure (drilling tower). Results from ISOPAR *using Olson's equations (Bangash). (See* IMPLOSION *program in this Section.)*

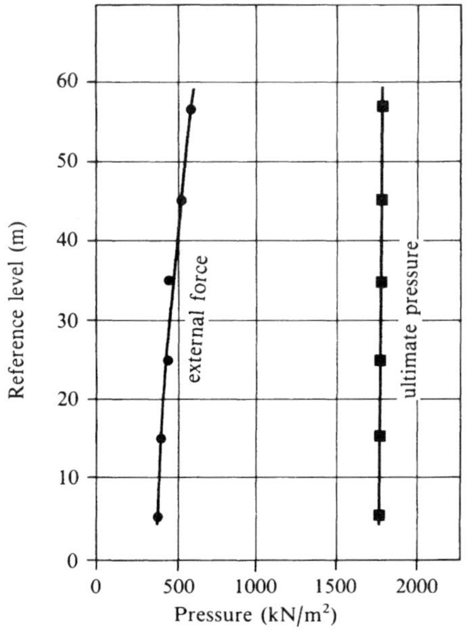

Fig. AIX.22. Reference level versus pressure (caisson base). Results from ISOPAR *using Olson's equations (Bangash). (See* IMPLOSION *program in this Section.)*

Input data and output of computer program: Condeep and 100 years wave data

Condeep

The parameters given are as follows.

(a) Utility tower and drilling tower

Length = 172.0 m
Thickness = 0.65 m
Concrete stress $f_c = 40\,000\,\text{kN/m}^2$ ($40\,\text{N/mm}^2$)
Long-term loading factor $\lambda = 0.7$
Material factor $\gamma_m = 1.5$
Load factor $\gamma_f = 1.2$

(b) Caisson base

Length = 56.5 m
Thickness = 1.2 m

The other data are similar to those for the tower.

One hundred years wave data

Wave period = 16 s
Water depth (from mean water level) = 145 m
Wave height = 30 m
Density of seawater = $10.1\,\text{kN/m}^3$
$C_D = 0.6$
$C_I = 2.0$

AIX.6.

```
C             **********************************
*             *       PROGRAM TO PREDICT IMPLOSION         *
*             *         IN GRAVITY PLATFORM                *
*             **********************************

*    ==================================================================
*    I                                                                I
*    I  NOMENCLATURE:                                                 I
*    I  =============                                                 I
*    I    UT   = UTILITY TOWER            LW    = WAVE LENGTH         I
*    I    DT   = DRILLING TOWER           K     = WAVE NUMBER         I
*    I    C    = CAISSON BASE             W     = WAVE FREQUENCY      I
*    I    D    = DIAMETER                 FD    = DRAG FORCE          I
*    I    R    = MEAN RADIUS              FI    = INERTIA FORCE       I
*    I    PIMP = IMPLOSION PRESSURE       FDI   = DRAG & INERTIA FORCE I
*    I    PUL  = ULT. LIMIT PRESSURE      FCU   = CURRENT FORCE       I
*    I    ITA  = PLASTICITY FACTOR        FDIFF = DIFFRACTION FORCE   I
*    I    L    = LENGTH                   CI    = COEFFICIENT OF INERTIA I
*    I    RL   = REDUCED LEVEL            CD    = COEFFICIENT OF DRAG I
*    I    FM   = MAX FORCE                P     = DENSITY             I
*    I                                                                I
*    ==================================================================

      PARAMETER (N=9, M=6)
      REAL DUT(N),DDT(N),DC(N),ITA(N),PIMP(N),PIMP1(N),PIMP2(N),
     *PIMP3(N),PIMP4(N),PIMP5(N),PIMP6(N),PIMP7(N),PIMP8(N),PIMP9(N),
     *PIMP10(N),PIMP11(N),PIMP12(N),RLUT(N),RLDT(N),RLC(N),YUT(N),
     *YDT(N),YC(N),XUT(N),XDT(N),XC(N),DUTI(N),DDTI(N),DC1(N),RUT1(N),
     *RUT2(N),RUT3(N),RDT1(N),RDT2(N),RDT3(N),RC1(N),RC2(N),RC3(N),
     *RLUT1(N),RLUT2(N),RLUT3(N),RLUT4(N),RLDT1(N),RLDT2(N),RLDT4(N),
     *RLCI(N),RLC2(N),RLC3(N),RLC4(N),RL1(N),RL2(N),RL3(N),RL4(N),
     *A(N),B(N),D(N),D11(N),RC(N),RDT(N),RUT(N),YUT1(N),YUT2(N),
     *YUT3(N),YDT1(N),YDT2(N),YDT3(N),LUT,LDT,LC,YC1(N),YC2(N),YC3(N),
     *X2(N),Y(N),RL(N),RLDT3(N),ITA1(N),ITA2(N),FA1(9),FA2(9),FA3(9),
     *FA4(9),FA5(9),FA6(9),Z(9),GA1(9),GA2(9),GA3(9),GA4(9),GA5(9),
     *GA6(9),GA7(9),GA8(9),GA9(9),FI(N),FD(N),Y4(N),CIV(N),C(N),E(N),
     *E1(N),DIFFRC(N),DIFFRU(N),DIFFRD(N),F1(N),RATIO(N),FCU(14),CU(N),
     *ANGLEU(N),ANGLED(N),ANGLEC(N),T1(N),F14(N),F23(N),
     *V1(N),V2(N),V3(N),V4(N),V5(N),U1(N),U2(N),U3(N),U4(N),U5(N),R1(N),
     *THITA(180),FD1(180),FI1(180),ANGLE(180),F32(N),F41(N),
     *R2(N),R3(N),R4(N),R5(N),K,K1,KO,LW,NITA,NITAC,NITAT,Y5(N),Y6(N),
     *FDIFF(N),FDIFFU(N),FDIFFD(N),FDIFFC(N),AUT(N),BUT(N),CUT(N),
     *D11UT(N),ADT(N),BDT(N),CDT(N),D11DT(N),AC(N),BC(N),CC(N),
     *D11C(N),ITA3(N),ANGLE1(N),YL(N),X(N),FD2(180),FD3(180),FD4(180)
      REAL FD5(180),F(180),FTOTAL(N),FT(N),PULA(N),PULB(N),PULC(N),
     *FIUT(N),FDDT(N),FIC(N),FIDT(N),FDUT(N),FDC(N),FCUUT(N),FCUDT(N),
     *FCUC(N),PULD(N),FM1(180),FM2(180),F1A(180),F2A(180),F3A(180),
     *F4A(180),F1B(180),F1C(180),PUL1(N),PUL2(N),PUL3(N),PUL4(N),
     *PUL5(N),PUL6(N),PUL7(N),PUL8(N),PUL9(N),PUL10(N),PUL11(N),
     *FDI(N),FDIUT(N),FDIDT(N),FDIC(N),FDI1(180),D1(N),PUL12(N)
      DATA TDT,TUT,TC,LDT,LUT,LC/0.65,0.65,1.2,172.0,172.0,56.5/
      DATA FC,GAMA,GAMAM,GAMAF,LU,LD,LC1/40000.0,0.7,1.5,1.2,1,2,3/
      DATA DIUT,DIDT,DIC/12.0,13.0,0.0/
      OPEN(20,FILE='DIAMRL',STATUS='OLD')
      READ(20,*)(RLUT(I),DUT(I),RLDT(I),DDT(I),I=1,N)
```

```
      OPEN(20,FILE='DIARLC')
      READ(20,*)(RLC(I),DC(I),I=1,M)
      OPEN(20,FILE='SLUT')
      OPEN(20,FILE='SLC')
      READ(20,*)(Y6(I),I=1,M)
      CALL CALC (DUT,LUT,TUT,RUT,YUT1,YUT2,YUT3,XUT)
      CALL CALC (DDT,LDT,TDT,RDT,YDT1,YDT2,YDT3,XDT)
      CALL CALC (DC,LC,TC,RC,YC1,YC2,YC3,XC)
      CALL CALC1(LU,LUT,YUT1,YUT2,YUT3,XUT,DUT,RUT,RLUT,NORLU1,
     *    NORLU2,NORLU3,NORLU4,RUT1,RUT2,RUT3,DUT1,A1UT,B1UT,
     *        C1UT,D12UT,AUT,BUT,CUT,D11UT,RLUT1,RLUT2,RLUT3,RLUT4)
      CALL CALC1(LD,LDT,YDT1,YDT2,YDT3,XDT,DDT,RDT,RLDT,NORLD1,NORLD2,
     *        NORLD3,NORLD4,RDT1,RDT2,RDT3,DDT1,A1DT,B1DT,C1DT,D12DT,
     *                   ADT,BDT,CDT,D11DT,RLDT1,RLDT2,RLDT3,RLDT4)
      CALL CALC1(LC1,LC,YC1,YC2,YC3,XC,DC,RC,RLC,NORLC1,NORLC2,NORLC3,
     *    NORLC4,RC1,RC2,RC3,DC1,A1C,B1C,C1C,D12C,AC,BC,CC,D11C,
     *          RLC1,RLC2,RLC3,RLC4)
      PRINT1
      PRINT2
      PRINT5
      PRINT4, (RLUT(I),DUT(I),RLDT(I),DDT(I),I=1,N)
      PRINT7
      PRINT6,(RLC(I),DC(I),I=1,M)
      PRINT50
   50 FORMAT(/9X,'IMPLOSION PRESSURE AND ULTIMATE PRESSURE :',/9X,42
     *        ('='))
      PRINT8
      CALL THICK(DUT1,NORLU1,LUT,TUT,FC,GAMA,GAMAM,GAMAF,PUL1.PIMP1)
      CALL THIN1(RUT1,NORLU2,LUT,TUT,FC,CAMA,GAMAM,GAMAF,PUL4,PIMP4)
      CALL THIN2(RUT2,NORLU3,LUT,TUT,FC,GAMA,GAMAM,GAMAF,ITA,
     *            PUL7,PIMP7)
      CALL THIN3(RUT3,NORLU4,LUT,TUT,FC,GAMA,GAMAM,GAMAF,ITA2,
     *            PUL10,PIMP10)
      IF(AUT(1).EQ.1)THEN
      PRINT23
      PRINT27
      PRINT11,(RLUT1(I),PIMP1(I),PUL1(I),I=1,NORLU1)
      ELSEIF(BUT(1).EQ.1)THEN
      PRINT24
      PRINT27
      PRINT14,(RLUT2(I),PIMP4(I),PUL4(I),I=1,NORLU2)
      ELSEIF(CUT(1).EQ.1)THEN
      PRINT25
      PRINT27
      PRINT17,(RLUT3(I),PIMP7(I),PUL7(I),ITA(I),I=1,NORLU3)
      ELSEIF(D11UT(1).EQ.1)THEN
      PRINT26
      PRINT27
      PRINT20,(RLUT4(I),PIMP10(I),PUL10(I),ITA2(I),I=I,NORLU4)
      END IF
      IF(A1UT.EQ.1.AND.AUT(1).NE.1)THEN
      PRINT23
      PRINT11,(RLUT1(I),PIMP1(I),PUL1(I),I=I,NORLU1)
      ELSEIF(B1UT.EQ.1.AND.BUT(1).NE.1)THEN
      PRINT24
      PRINT14,(RLUT2(I),PIMP4(I),PUL4(I),I=1,NORLU2)
      ELSEIF(C1UT.EQ.1.AND.CUT(1).NE.1)THEN
```

```
      PRINT25
      PRINT17,(RLUT3(I),PIMP7(I),PUL7(I),ITA(I),I=1,NORLU3)
      ELSEIF(D12UT.EQ.1.AND.D11UT(1).NE.1)THEN
      PRINT26
      PRINT20,(RLUT4(I),PIMP10(I),PUL10(I),ITA2(I),I=1,NORLU4)
      ENDIF
      PRINT10
      CALL THICK(DDT1,NORLD1,LDT,TDT,FC,GAMA,GAMAM,GAMAF.PUL2.PIMP2)
      CALL THIN1(RDT1,NORLD2,LDT,TDT,FC,GAMA,GAMAM,GAMAF,PUL5,PIMP5)
      CALL THIN2(RDT2,NORLD3,LDT,TDT,FC,GAMA,GAMAM,GAMAF,ITA,
     *          PUL8,PIMP8)
      CALL THIN3(RDT3,NORLD4,LDT,TDT,FC,GAMA,GAMAM,GAMAF,ITA,
     *          PUL11,PIMP11)
      IF(ADT(1).EQ.1)THEN
      PRINT23
      PRINT27
      PRINT12,(RLDT1(I),PIMP2(I),PUL2(I),I=1,NORLD1)
      ELSEIF(BDT(1).EQ.1)THEN
      PRINT24
      PRINT27
      PRINT15,(RLDT2(I),PIMPS(I),PUL5(I),I=1,NORLD2)
      ELSEIF(CDT(1).EQ.1)THEN
      PRINT25
      PRINT27
      PRINT18,(RLDT3(I),PIMP8(I),PUL8(I),ITA(I),I=1,NORLD3)
      ELSEIF(D11DT(1).EQ.1)THEN
      PRINT26
      PRINT27
      PRINT21,(RLDT4(I),PIMP11(I),PUL11(I),ITA(I),I=1,NORLD4)
      END IF
      IF(A1DT.EQ.1.AND.ADT(1).NE.1)THEN
      PRINT23
      PRINT12,(RLDT1(I),PIMP2(I),PUL2(I),I=1,NORLD1)
      ELSEIF(B1DT.EQ.1.AND.BDT(1).NE.1)THEN
      PRINT24
      PRINT15,(RLDT2(I),PIMP5(I),PUL5(I),I=1,NORLD2)
      ELSEIF(C1DT.EQ.1.AND.CDT(1).NE.1)THEN
      PRINT25
      PRINT18,(RLDT3(I),PIMP8(I),PUL8(I),ITA(I),I=1,NORLD3)
      ELSEIF(D12DT.EQ.1.AND.D11DT(1).NE.1)THEN
      PRINT26
      PRINT21,(RLDT4(I),PIMP11(I), PUL11(I),ITA(I),I=1,NORLD4)
      END IF
      PRINT9
      CALL THICK(DC1,NORLC1,LC,TC,FC,GAMA,GAMAM,GAMAF,PUL3,PIMP3)
      CALL THIN1(RC1,NORLC2,LC,TC,FC,GAMA,GAMAM,GAMAF,PUL6,PIMP6)
      CALL THIN2(RC2,NORLC3,LC,TC,FC,GAMA,GAMAM,GAMAF,ITA1,
     *          PUL9,PIMP9)
      CALL THIN3(RC3,NORLC4,LC,TC,FC,GAMA,GAMAM,GAMAF,ITA3,
     *          PUL12,PIMP12)
      IF(AC(1).EQ.1)THEN
      PRINT23
      PRINT27
      PRINT13,(RLC1(I),PIMP3(I),PUL3(I),I=1,NORLC1)
      ELSEIF(BC(1).EQ.1)THEN
      PRINT24
      PRINT27
```

```
      PRINT16,(RLC2(I),PIMP6(I),PUL6(1),I=1,NORLC2)
      ELSEIF(CC(1).EQ.1)THEN
      PRINT25
      PRINT27
      PRINT19,(RLC3(I),PIMP9(I),PUL9(I),ITA1(I),I=1,NORLC3)
      ELSEIF(D11C(1).EQ.1)THEN
      PRINT26
      PRINT27
      PRINT22,(RLC4(I),PIMP12(I),PUL12(I),ITA3(I),I=1,NORLC4)
      END IF
      IF(A1C.EQ.1.AND.AC(1).NE.1)THEN
      PRINT23
      PRINT13,(RLC1(I),PIMP3(I),PUL3(I),I=1,NORLC1)
      ELSEIF(B1C.EQ.1.AND.BC(1).NE.1)THEN
      PRINT24
      PRINT16,(RLC2(I),PIMP6(I),PUL6(I),I=1,NORLC2)
      ELSEIF(C1C.EQ.1.AND.CC(1).NE.1)THEN
      PRINT25
      PRINT19,(RLC3(I),PIMP9(I),PUL9(I),ITA1(I),I=1,NORL3)
      ELSEIF(D12C.EQ.1.AND.D11C(1).NE.1)THEN
      PRINT26
      PRINT22,(RLC4(I),PIMP12(I),PUL12(1),ITA3(I),I=1,NORLC4)
      END IF
    1 FORMAT(4(/),19X,'5.3 RESULT OF PROGRAM'/19X,22('='))
    2 FORMAT(/12X,'UTILITY TOWER',8X,'DRILLING TOWER'/12X,13('='),8X,
     *       14('='))
    5 FORMAT(11X,'RL',9X,'D',11X,'RL',9X,'D'/11X,2('='),9X,1('='),
     *       11X,2('='),9X,1('='))
    4 FORMAT(9X,F6.2,4X,F6.2,7X,F6.2,4X,F6.2)
    7 FORMAT(/12X,'CAISSON BASE'/12X,12('='),/11X,'RL',9X,'D',/11X,
     *         2('='),9X,1('='))
    6 FORMAT(9X,F6.2,4X,F6.2)
    8 FORMAT(/18X,'UTILITY TOWER'/18X,13('='))
   10 FORMAT(/18X,'DRILLING TOWER'/18X,14('='))
    9 FORMAT(///18X,'CAISSON BASE'/18X,12('='))
   11 FORMAT(9X,F6.2,4X,F7.2,4X,F7.2)
   12 FORMAT(9X,F6.2,4X,F7.2,4X,F7.2)
   13 FORMAT(9X,F6.2,4X,F7.2,4X,F7.2)
   14 FORMAT(9X,F6.2,4X,F7.2,4X,F7.2)
   15 FORMAT(9X,F6.2,4X,F7.2,4X,F7.2)
   16 FORMAT(9X,F6.2,4X,F7.2,4X,F7.2)
   17 FORMAT(9X,F6.2,6X,F7.2,10X,F7.2,7X,F5.2)
   18 FORMAT(9X,F6.2,6X,F7.2,10X,F7.2,7X,F5.2)
   19 FORMAT(9X,F6.2,6X,F7.2,10X,F7.2,7X,F5.2)
   20 FORMAT(9X,F6.2,6X,F7.2,10X,F7.2,7X,F5.2)
   21 FORMAT(9X,F6.2,6X,F7.2,10X,F7.2,7X,F5.2)
   22 FORMAT(9X,F6.2,6X,F7.2,10X,F7.2,7X,F5.2)
   23 FORMAT(9X,'THICK-WALLED CYLINDER'/9X,21('='))
   24 FORMAT(9X,'THIN-WALLED SHORT CYLINDER'/9X,26('='))
   25 FORMAT(9X,'THIN-WALLED MODERATELY LONG'/9X,27('='))
   26 FORMAT(9X,'THIN-WALLED LONG CYLINDER'/9X,25('='))
   27 FORMAT(10X,'RL(m)',4X,'PIMP(kn/m~2)',4X,'ULT.PRES(kn/m~2)',3X,
     *       'ITA',/IOX,5('='),4X,12('='),4X,16('='),3X,3('='))
      CALL WAVE(DUT,DIUT,LUT,Y4,FDIUT,FIUT,ANGLEU,RATIOU,DIFFRU,
     *      FMU,FDIFFU,NODIFU,NODFFU,NODU,NOIU,NOINTU,NODRAU,YUT7)
      CALL WAVE(DDT,DIDT,LDT,Y5,FDIDT,FIDT,ANGLED,RATIOD,DIFFRD,FMD,
     *         FDIFFD,NODIFD,NODFFD,NODD,NOID,NOINTD,NODRAD,YDT7)
```

```
      CALL WAVE(DC,DIC,LC,Y6,FDIC,FIC,ANGLEC,RATIOC,DIFFRC,FMC,
     *          FDIFFC,NODIFC,NODFFC,NODC,NOIC,NOINTC,NODRAC,YC7)
      PRINT30
 30   FORMAT(/9X,'PRESSURE LOADING DUE TO WAVE :',/9X,30('='))
      PRINT31
 31   FORMAT(/18X,'UTILITY TOWER',/18X,13('='))
      CALL PRINTF(RLUT,NODFFU,FDIFFU,NODIFU,NOINTU,NODU,NOIU,FIUT,FDUT,
     *            ANGLEU)
      PRINT100,FMU,YUT7
100   FORMAT(/7X,'MAX FORCE =',F7.2,'(KN/M~2) OCCUR @ R. LEVEL=',F6.2,
     *       /7X,51('='))
      PRINT32
 32   FORMAT(/18X,'DRILLING TOWER',/18X,14('='))
      CALL PRINTF(RLDT,NODFFD,FDIFFD,NODIFD,NOINTD,NODD,NOID,FIDT,FDDT,
     *            ANGLED)
      PRINT102,FMD,YDT7
102   FORMAT(/7X,'MAX FORCE =',F7.2,'(KN/M~2) OCCUR @ R. LEVEL=',F6.2,
     *        /7X,51('='))
      PRINT33
 33   FORMAT(14(/),18X,'CAISSON BASE',/18X,12('='))
      CALL PRINTF(RLC,NODFFC,FDIFFC,NODIFC,NOINTC,NODC,NOIC,FIC,FDC,
     *            ANGLEC)
      PRINT34
      CALL FCUP(DUT,LUT,Y4,FCUUT)
      CALL FCUR(DDT,LDT,Y5,FCUDT)
      CALL FCUR(DC,LC,Y6,FCUC)
 34   FORMAT(/9X,'CURRENT FORCE:',/9X,14('='))
      PRINT35
 35   FORMAT(/18X,'UTILITY TOWER',/18X,13('='))
      CALL PRINTC(RLUT,FCUUT,LUT)
      PRINT36
 36   FORMAT(/18X,'DRILLING TOWER',/18X,14('='))
      CALL PRINTC(RLDT,FCUDT,LDT)
      PRINT37
 37   FORMAT(/18X,'CAISSON BASE',/18X,12('='))
      CALL PRINTC(RLC,FCUC,LC)
      PRINT38
 38   FORMAT(5(/),9X,'TO CHECK WHETHER THE STRUCTURE IS SAFE',
     *        /9X,38('='))
      PRINT39
 39   FORMAT(/18X,'UTILITY TOWER',/18X,13('='))
      DO 70 I=1,N
      IF(NODFFU.NE.1)FDIFFU(I)=0.0
      IF(NOINTU.NE.1)FIUT(I)=0.0
      IF(NODRAU.NE.1)FDIUT(I)=0.0
      FTOTAL(I)=FDIUT(I)+FIUT(I)+FDIFFU(I)+FCUUT(I)
 70   CONTINUE
      CALL FTOTL(LUT,RLUT,FTOTAL,PUL1,PUL4,PUL7,PUL10)
      PRINT40
 40   FORMAT(/18X,'DRILLING TOWER',/18X,14('='))
      DO 80 I=1,N
      IF(NODFFD.NE.1)FDIFFD(I)=0.0
      IF(NOINTD.NE.1)FIDT(I)=0.0
      IF(NODRAD.NE.1)FDIDT(I)=0.0
      FTOTAL(I)=FDIDT(I)+FIDT(I)+FDIFFD(I)+FCUDT(I)
 80   CONTINUE
      CALL FTOTL(LDT,RLDT,FTOTAL,PUL2,PULS,PUL8,PUL11)
```

```
      PRINT41
   41 FORMAT(/18X,'CAISSON BASE',/18X,12('=')) 
      DO 60 I=1,M
      IF(NODFFC.NE.1)FDIFFC(I)=0.0
      END IF
      END IF
      RETURN
      END

C ===========================================================================
*               ********************                                        *
*               * 1. THICK-WALLED CYLINDER *                                *
*               *------------------*                                        *
*               ********************                                        *
* ===========================================================================

      SUBROUTINE THICK(D1,NORL1,L,T,FC,GAMA,GAMAM,GAMAF,PUL,PIMP)
      PARAMETER (N=9)
      REAL D1(N),PIMP(N),PUL(N),K,L
      K=1.25
      DO 3 I=1,NORL1
      PIMP(I)=2*K*(T/D1(I))*FC
      PUL(I)=GAMA/(GAMAM*GAMAF)*PIMP(I)
    3 CONTINUE
      RETURN
      END

C ===========================================================================
*               2. THIN-WALLED CYLINDER                                     *
* ===========================================================================
*                                                                           *
*               ********************                                        *
*               * 2.1 SHORT        *                                        *
*               *------------------*                                        *
*               ********************                                        *
* ===========================================================================

      SUBROUTINE THIN1(R1,NORL2,L,T,FC,GAMA,GAMAM,GAMAF,PUL,PIMP)
      PARAMETER (N=9)
      REAL R1(N),PIMP(N),PUL(N),L
      DO 1 I=1,NORL2
      PIMP(I)=4*FC*(T/R1(I))/(2+T/R1(I))**2
      PUL(I)=GAMA/(GAMAM*GAMAF)*PIMP(I)
    1 CONTINUE
      RETURN
      END

C ===========================================================================
*               ********************                                        *
*               * 2.2 MODERATELY LONG *                                     *
*               *------------------*                                        *
*               ********************                                        *
* ===========================================================================

      SUBROUTINE THIN2(R,NORL,L,T,FC,GAMA,GAMAM,GAMAF,ITA,PUL,PIMP)
      PARAMETER (N=9)
      REAL R(N),PIMP(N),ITA(N),Y(N),PUL(N),T,L
```

```
      CALL CALCI(NORL,R,T,L,ITA,Y)
      BETA = 1.2
      DO 1 I = 1,NORL
      PIMP(I) = 2330*ITA(I)*FC*BETA*((T/R(I))**2.5*(R(I)/L)/
     *((2 + T/R(I))**2))
      PUL(I) = GAMA/(GAMAM*GAMAF)*PIMP(I)
    1 CONTINUE
      RETURN
      END

C ==========================================================================
*               * * * * * * * * * * * * * * * * * * *                      *
*               *  2.3 LONG                         *                      *
*               * - - - - - - - - - - - - - - - - - *                      *
*               * * * * * * * * * * * * * * * * * * *                      *
* ==========================================================================

      SUBROUTINE THIN3(R,NORL,L,T,FC,GAMA,GAMAM,GAMAF,ITA,PUL,PIMP)
      PARAMETER (N = 9)
      REAL R(N),PIMP(N),ITA(N),Y(N),PUL(N),T,L
      CALL CALCI1(NORL,R,T,L,ITA)
      DO 1 I = 1,NORL
      PIMP(I) = 686*ITA(I)*FC*(T/R(I))**3/(2 + T/R(I))**2
      PUL(I) = GAMA/(GAMAM*GAMAF)*PIMP(I)
    1 CONTINUE
      RETURN
      END

C ==========================================================================
*               * * * * * * * * * * * * * * * * * * * * * * * * * * * * * * * * * *   *
*               *  SUBROUTINE TO CALCULATE ITA [ PLASTICITY FACTOR ] *   *
*               *          FOR MODERATELY LONG CYLINDER              *   *
*               *  - - - - - - - - - - - - - - - - - - - - - - - - - *   *
*               * * * * * * * * * * * * * * * * * * * * * * * * * * * * * * * * * *   *
* ==========================================================================

      SUBROUTINE CALCI(NORL,R,T,L,ITA,Y)
      PARAMETER (N = 9)
      REAL R(N),ITA(N),X2(N),Y(N),T,L
      DO 1 I = 1,NORL
      DO 2 X1 = 0.4,1.10,1D - 5
      ITA(I) = 1.0 - 0.35*(10*X1 - 4) + 0.15/2*(10*X1 - 4)*(10*X1 - 5)
     *        - 0.05/6*(10*X1 - 4)*(10*X1 - 5)*(10*X1 - 6)
     *        - 0.03/24*(10*X1 - 4)*(10*X1 - 5)*(10*X1 - 6)*(10*X1 - 7)
     *        + 0.1/120*(10*X1 - 4)*(10*X1 - 5)*(10*X1 - 6)*(10*X1 - 7)*(10*X1 - 8)
     *        - 0.15/720*(10*X1 - 4)*(10*X1 - 5)*(10*X1 - 6)*(10*X1 - 7)*(10*X1 - 8)
     *      *(10*X1 - 9)+0.15/5040*(10*X1 - 4)*(10*X1 - 5)*(10*X1 - 6)*(10*X1 - 7)
     *      *(10*X1 - 8)*(10*X1 - 9)*(10*X1 - 10)
      X2(I) = 583*(T/R(I))**1.5*(R(I)/L)*ITA(I)
      Y(I) = INT(X2(I)) - INT(X0)
      IF(ABS(Y(I)).EQ.0.0)GOTO 1
    2 CONTINUE
    1 CONTINUE
      RETURN
      END
```

APPENDIX X

Geometry and forces for domes for offshore cells, reactor containment and LNG tanks

AX.1. Geometry and forces of elliptical domes

Meridional thrust due to wave/current loads

$$MTL = \frac{wla^2}{2b}Q \tag{AX.1}$$

Hoop thrusts due to wave/current loads

$$HTL = \frac{wla^2}{2b}\frac{\{(2(y/b)^2 - 1)\}}{Q} \tag{AX.2}$$

where

W_v = wave/current load on dome surface per unit area

$$Q = \sqrt{1 - (a^2 - b^2)^2/a^2)(1 - (y/b)^2)} \tag{AX.3}$$

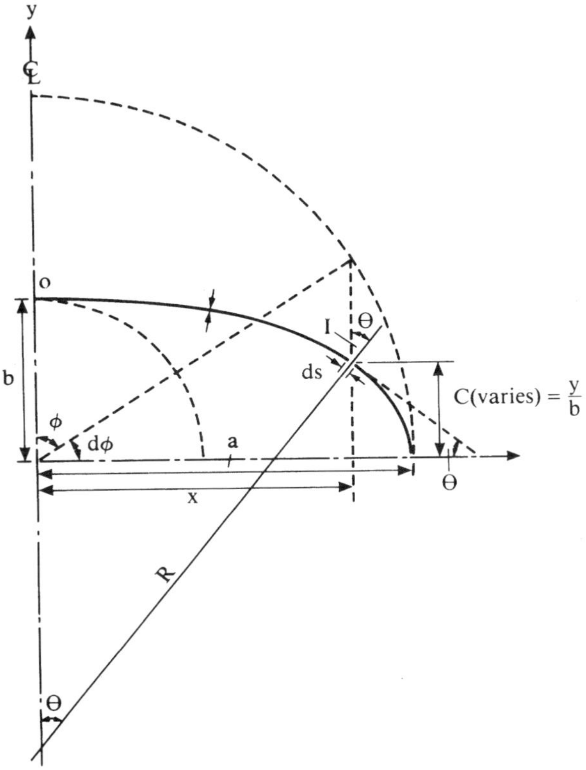

Fig. AX.1. Geometry of dome shapes.

External applied forces

$$H = N_\phi = \frac{W_v r_0}{2\sin\phi} = \frac{W_v r_0}{2} \tag{AX.4}$$

$$T = N_\theta = W_v\left(r_2 - \frac{r_2^2}{2r_1}\right) \tag{AX.5}$$

where

$$r_1 = \frac{a^2}{(a^2 \sin^2 \phi + b^2 \cos^2 \phi)^{1/2}} \tag{AX.6}$$

Resisting forces

Meridional and hoop forces

$$M_{\text{TD}} = \text{meridional thrust due to deep load} = \frac{2\pi a^2 W_s C}{2\pi \sin\theta} \tag{AX.7}$$

$$\sin\theta = \frac{b(1-(y/b)^2)^{1/2}}{a(1-(a^2-b^2)^2(1-(y/b)^2))^{1/2}} \tag{AX.8}$$

Hoop thrust due to dead load

$$HTD = \frac{-WD}{2\pi R \sin^2\theta} + wa^2 y((y/b)^2 + \frac{b^2}{a^2}(1-(y/b)^2)) \tag{AX.9}$$

where $WD = 2\pi a^2 W_s C$.

APPENDIX XI

The superelement and substructuring, and large strain formulations

AXI.1. Superelement and substructuring

In general terms, such formulations are described by the following:

$$\left[\begin{array}{c|c} K & K_{\mathrm{R}} \\ \hline K_{\mathrm{R}}^{\mathrm{T}''} & K_{\mathrm{RR}} \end{array}\right]\left\{\frac{U}{U_{\mathrm{R}}}\right\}=\left\{\frac{F}{F_{\mathrm{R}}}\right\} \quad \text{(AXI.1)}$$

The subscript R represents reaction forces. The top half of Eq. (AXI.1) is used to solve for $\{U\}$:

$$\{U\} = -[K]^{-1}[K_{\mathrm{R}}]\{U_{\mathrm{R}}\} + [K]^{-1}\{F\} \quad \text{(AXI.2)}$$

The reaction forces $\{F_{\mathrm{R}}\}$ are computed from the bottom half of the equation as

$$\{F_{\mathrm{R}}\} = [K_{\mathrm{R}}]^{\mathrm{T}''}\{U\} + \{K_{\mathrm{RR}}\}\{U_{\mathrm{R}}\} \quad \text{(AXI.3)}$$

Equation (AXI.2) must be in equilibrium with eqn (AXI.3).

For large structures with complicated features, a substructure (superelement) may be adopted on the lines suggested in eqn (AXI.1). This superelement may then be used as a reduced element from the collection of elements. If subscripts γ and γ' represent the retained and removed degrees of freedom of the equations partitioned into two groups, then the expressions in eqn (AXI.1) can be written as

$$\left[\begin{array}{c|c} K_{\gamma\gamma} & K_{\gamma\gamma'} \\ \hline K_{\gamma'\gamma} & K_{\gamma'\gamma'} \end{array}\right]\left\{\frac{U_{\gamma}}{U_{\gamma'}}\right\}=\left\{\frac{F_{\gamma}}{F_{\gamma'}}\right\} \quad \text{(AXI.4)}$$

Equation (AXI.4) when expanded assumes the following form:

$$\{F_{\gamma}\} = [K_{\gamma\gamma}]\{U_{\gamma}\} + [K_{\gamma\gamma'}]\{U_{\gamma'}\} \quad \text{(AXI.5)}$$

$$\{F_{\gamma'}\} = [K_{\gamma'\gamma}]\{U_{\gamma}\} + [K_{\gamma'\gamma}]\{U_{\gamma'}\} \quad \text{(AXI.6)}$$

When a dynamic analysis is carried out, the subscript γ (retained) represents the dynamic degrees of freedom.

When eqn (AXI.6) is solved, the value of $U_{\gamma'}$ is then written, similarly to eqn (AXI.2),

$$\{U_{\gamma'}\} = [K_{\gamma'\gamma'}]^{-1}\{F_{\gamma'}\} - [K_{\gamma'\gamma'}]^{-1}[K_{\gamma'\gamma}]\{U_{\gamma}\} \quad \text{(AXI.7)}$$

Substituting $\{U_{\gamma'}\}$ into eqn (AXI.5) gives:

$$[\![K_{\gamma\gamma}] - [K_{\gamma'\gamma}][K_{\gamma\gamma'}]^{-1}[K_{\gamma'\gamma}]]\!]\{U_{\gamma}\} = [\{F_{\gamma}\} - [K_{\gamma\gamma'}][K_{\gamma\gamma'}]^{-1}\{F_{\gamma'}\}] \quad \text{(AXI.8)}$$

or

$$[\bar{K}]\{\bar{U}\} = \{\bar{F}\} \quad \text{(AXI.9)}$$

where

$$[\bar{K}] = [K_{\gamma\gamma}] - [K_{\gamma\gamma'}][K_{\gamma'\gamma'}]^{-1}[K_{\gamma'\gamma}] \quad \text{(AXI.10)}$$

$$\{\bar{F}\} = \{F_{\gamma}\} - [K_{\gamma\gamma'}][K_{\gamma'\gamma'}]^{-1}\{F_{\gamma'}\} \quad \text{(AXI.11)}$$

$$\{\bar{U}\} = \{U_{\gamma}\} \quad \text{(AXI.12)}$$

and $[\bar{K}]$ and $\{\bar{F}\}$ are generally known as the substructure stiffness matrix and load vector, respectively.

In the above equations, the load vector for the substructure is taken as a total load vector. The same derivation may be applied to any number of independent load vectors. For example, one may wish to apply thermal, pressure, gravity and other loading conditions in varying proportions. Expanding the right-hand sides of eqns (AXI.5) and (AXI.6) gives:

$$\{F_\gamma\} = \sum_{i=1}^{n} \{F_{\gamma i}\} \tag{AXI.13}$$

$$\{F_{\gamma'}\} = \sum_{i=1}^{n} \{F_{\gamma' i}\} \tag{AXI.14}$$

where n = the number of independent load vectors.

Substituting into eqn (AXI.12)

$$\{\bar{F}\} = \sum_{i=1}^{n} \{F_{\gamma\gamma'}\} - [K_{\gamma\gamma'}][K_{\gamma'\gamma'}]^{-1} \sum_{i=1}^{n} \{F_{\gamma' i}\} \tag{AXI.15}$$

AXI.2. Large strain

When the strains in a material exceed more than a few per cent, the changing geometry due to this deformation can no longer be neglected. Analyses which include this effect are called large strain, or finite strain, analyses. A large strain analysis is performed in a static and dynamic transient analysis.

The theory of large strain computations can be addressed by defining a few basic physical quantities (motion and deformation) and the corresponding mathematical relationship. The applied loads acting on a body make it move from one position to another. This motion can be defined by studying a position vector in the 'deformed' and 'undeformed' configuration. Say the position vectors in the 'deformed' and 'undeformed' state are represented by $\{\boldsymbol{x}\}$ and $\{\boldsymbol{X}\}$ respectively, then the motion (displacement) vector $\{\boldsymbol{u}\}$ is computed by

$$\{\boldsymbol{u}\} = \{\boldsymbol{x}\} - \{\boldsymbol{X}\} \tag{AXI.16}$$

The deformation gradient $[F_\mathrm{d}]$ is defined as:

$$[F_\mathrm{d}] = \frac{\partial\{\boldsymbol{x}\}}{\partial\{\boldsymbol{X}\}} + [\mathbf{I}] = [\boldsymbol{R}][\boldsymbol{u}] \tag{AXI.17}$$

where $[\mathbf{I}]$ = identity matrix.

The volume change at a point

$$\frac{\mathrm{d}V}{\mathrm{d}V_0} = \det[F] \tag{AXI.18}$$

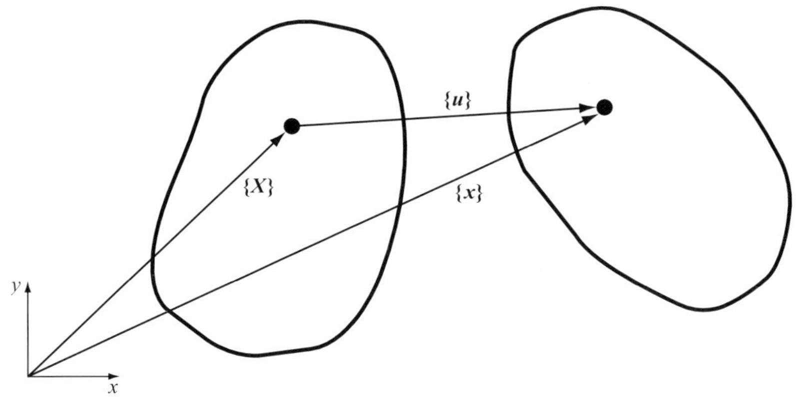

Fig. AXI.1. Position vectors and motion of a deforming body.

where

V_0 = original volume

V = current volume

$[\mathbf{u}]$ = right stretch shape change matrix

Largestrain $\{\varepsilon\}$ is written as

$$\{\varepsilon\} = \ln[\mathbf{u}] = \sum \ln(\lambda_i)\{e_i\}\{e_i\}^{T''} \quad \text{(AXI.19)}$$

where

λ_i = eigenvalues of $[\mathbf{u}]$

e_i = eigenvectors of $[\mathbf{u}]$

$\mathbf{R}$ = rotation matrix $([\mathbf{R}]^{T''}[\mathbf{R}] = [\mathbf{I}])$

Element formulation using Lagrangian formulation technique

$$[\bar{\mathbf{K}}_i]\Delta u_i = [F^a] - [F_i^{nr}] \quad \text{(AXI.20)}$$

where

$[\bar{\mathbf{K}}_i]$ = tangent matrix = $[\mathbf{K}_i] + [S_i]$

$[\mathbf{K}_i]$ = usual stiffness matrix

$$= \int [B_i]^{T''}[D_i][B_i]\,\mathrm{d(vol)} \quad \text{(AXI.21)}$$

$[\mathbf{S}_i]$ = stress stiffness contribution or geometric stiffness

$$= \int [\mathbf{G}_i]^{T}[\boldsymbol{\tau}_i][\mathbf{G}_i]\,\mathrm{d(vol)} \quad \text{(AXI.22)}$$

where

$[\mathbf{G}_i]$ = matrix of shape function derivatives

$[\boldsymbol{\tau}_\mathbf{i}]$ = matrix for the current Cauchy stresses i.e. true stresses σ_i

$$[\mathbf{F}_i^{nr}] = \int [\mathbf{B}_i]^{T''} \{\sigma_i\} \, d(\text{vol}) \tag{AXI.23}$$

The increment of large strain $[\Delta\varepsilon_n]$ is then written as

$$[\Delta\varepsilon_n] = [\boldsymbol{R}_{1/2}]^{T''} [\Delta\bar{\varepsilon}_n][\boldsymbol{R}_{1/2}] \tag{AXI.24}$$

where

$[\boldsymbol{R}_{1/2}]$ = rotational matrix from the polar decomposition of the deformation gradient evaluated at the midpoint configuration

$$[F_{1/2}] = [R_{1/2}][U_{1/2}] \tag{AXI.25}$$

where

$$[F_{1/2}] = [I] + \frac{\partial\{U_{1/2}\}}{\partial\{X\}} \tag{AXI.26}$$

Large deformation and large strains are obtained where these formulations are linked to the main analyses, static dynamic and blast type of loading are encountered.

APPENDIX XII

Bridge decks

AXII.1. Finite element analysis of bridge decks

The finite element method employs an assemblage of discrete two- or three-dimensional members as well. The elements are connected at nodal points which possess an appropriate number of degrees of freedom. Material properties can be incorporated which will truly represent various components of the bridge deck. Static and dynamic problems can easily be tackled by this single method, making it a most powerful technique.

The bridge is idealised as an assemblage of discrete parts, known as elements. The next step is to evaluate the element properties. This is followed by the structural analysis of the element assemblage. Forces, stresses, moments, displacements, strains, etc. are calculated. Where a computed program is interactive, results are plotted which will show the exact behaviour of various components of the bridge.

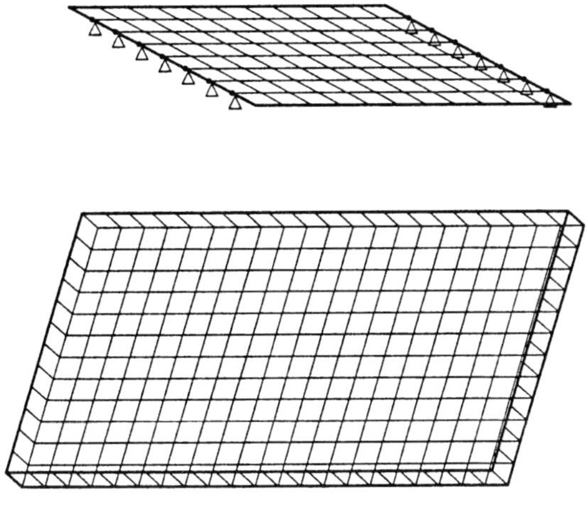

Fig. AXII.1. Finite element mesh scheme for the concrete grillage.

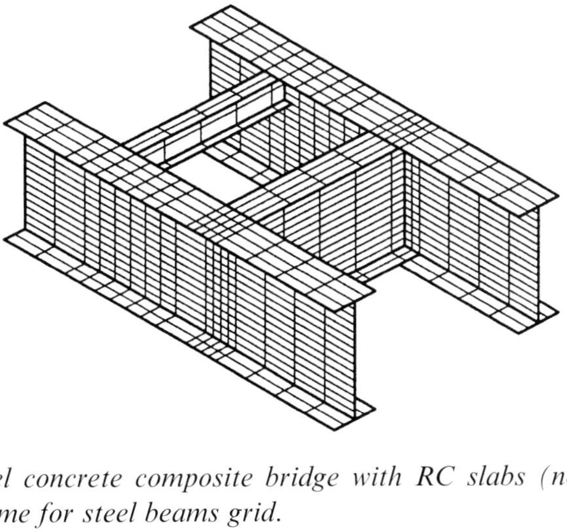

Fig. AXII.2. Steel concrete composite bridge with RC slabs (not shown) — finite element mesh scheme for steel beams grid.

Again, the purpose of the analysis using the finite element method has to be clearly established. Where dynamic problems are investigated, the results must provide displacements, velocities, accelerations, frequencies, etc.

Where plasticity and material/geometrical non-linearity is to be investigated, the analysis should be thoroughly justified. Material failure criteria must be included, together with the solution and acceleration procedures. A convergence criterion should be established. Normally, suitable graphic techniques are associated with any finite element program package, which are used to provide meaningful results. Typical finite element meshes for bridge decks are shown in Figs AXII.1 to AXII.5.

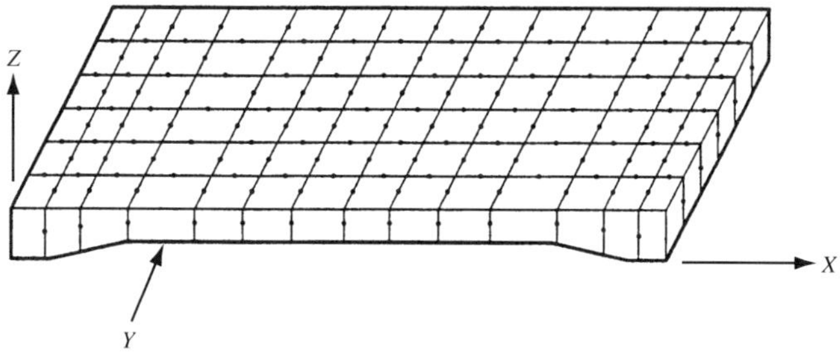

Fig. AXII.3. Finite element mesh scheme for the deck slab.

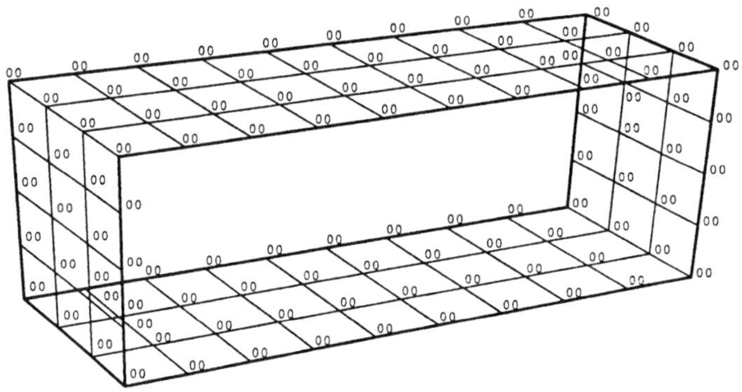

Fig. AXII.4. Finite element mesh scheme of a box culvert.

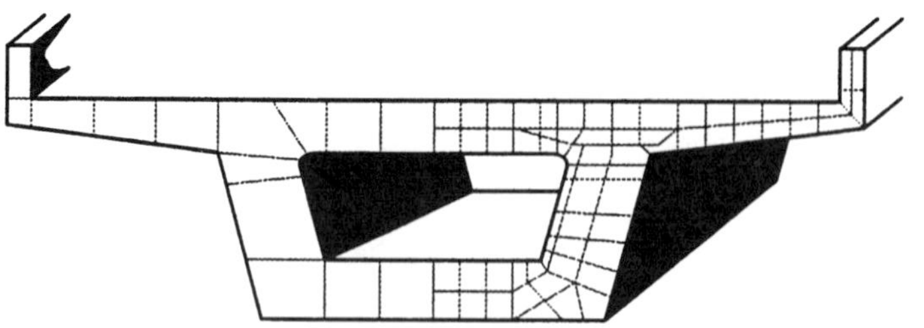

Fig. AXII.5. Finite element mesh scheme for a three-dimensional box girder.

APPENDIX XIII

Interface or gap elements in current computer programs

AXIII.1. Introduction

Potential of interface elements in the engineering disciplines

In soil–structural interaction problems, interface elements play a very important role in bringing about close relations between various structural and soil/rock elements. In concrete mechanics examples can be cited such as the bond between concrete and reinforcements, softening and aggregate interlock in discrete cracks, friction in concrete connections, bedding of reinforced/plain/prestressed concrete structures in soils/rocks and so on. This Appendix reviews three methods which are recommended for both static and dynamic analysis.

AXIII.2. Hallquist *et al.* method

Hallquist *et al.* developed a useful concept of master and slave nodes sliding on each other. As shown in Fig. AXIII.1 slave nodes are constrained to slide on master segments after impact occurs and must remain on a master segment until a tensile interface force develops. The zone in which a slave segment exists is called a slave zone. A separation between the slave and the master line is known as void. The following basic principles apply at the interface:

(a) update the location of each slave node by finding its closest master node or the one on which it lies,
(b) for each master segment, find out the first slave zone that overlaps,
(c) show the existence of the tensile interface force.

Constraints are imposed on global equations by a transformation of the nodal displacement components of the slave nodes along the contact interface. Such a transformation of the displacement components of the slave nodes will eliminate

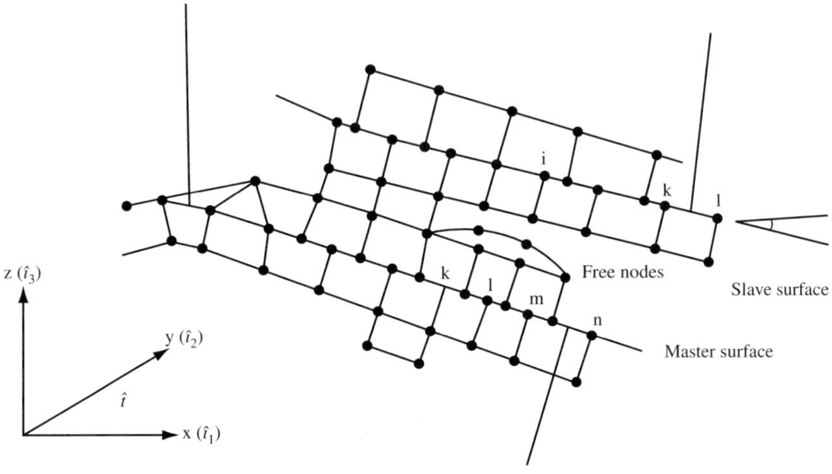

k_n — Interface normal stiffness m D, material stiffness matrix
Values varied as 10^3 to 10^6 N/mm^3

Fig. AXIII.1. Hallquist contact method (modified by Bangash).

their normal degrees of freedom and distribute their normal force components to the nearby master nodes. This is done using explicit time integration, as described in the finite element solution procedures. Thereafter impact and release conditions are imposed. The slave and master nodes are shown in Fig. AXIII.1. Hallquist *et al.* gave a useful demonstration of the identification of the contact point which is the point on the master segment to the slave node n_s and which finally becomes non-trivial during the execution of the analyses. When the master segment $\hat{t}$ is given the parametric representation and t is the position vector drawn to the slave node n_s, the contact point co-ordinate must satisfy the following equations:

$$\begin{aligned} &\frac{\partial \hat{r}}{\partial \xi}(\xi_c, \eta_c) \times [\hat{t} - \hat{r}(\xi_c, \eta_c)] = 0 \\ &\frac{\partial \hat{r}}{\partial \eta}(\xi_c, \eta_c) \times [\hat{t} - \hat{r}(\xi_c, \eta_c)] = 0 \end{aligned} \qquad \text{(AXIII.1)}$$

where (ξ_c, η_c) are the co-ordinates on the master surface segment S_i. Where penetration through the master segment S_i occurs, the slave node n_s (containing its contact point) can be identified using the interface vector f_s

$$f_s = -lk_i n_i \qquad \text{if } l < 0 \qquad \text{(AXIII.2)}$$

to the degree of freedom corresponding to n_s, and

$$f_m^i = N_i(\xi_c \eta_c) f_s \qquad \text{if } l < 0 \qquad \text{(AXIII.3)}$$

where

$$l = \hat{n}_i \cdot [\hat{t} - \hat{r}(\xi_c, \eta_c)] < 0 \qquad \text{(AXIII.4)}$$

A unit normal

$$\hat{n}_i = \hat{n}_i(\xi_c, \eta_c); \qquad \hat{t}_i = \hat{n}_i \sum_{j=1}^{n} N_j (F_1)^j(t) \qquad \text{(AXIII.5)}$$

$$k_i = f_{si} K_i A_i^2 / V_i \qquad \text{(AXIII.6)}$$

where

$(F_1)^j(t)$ = impact at the jth node

K = stiffness factor

K_i, V_i, A_i = bulk modulus, volume and face area, respectively

f_{si} = scale factor normally defaulted to 0.10

$N_i = \frac{1}{4}(1 + \xi\xi_i)(1 + \eta\eta_i)$ for a 4-node linear surface

Bangash extended this useful analysis for other shape functions, such as N_i for 8-noded and 12-noded elements. On the basis of this theory and owing to the non-availability of the original computer source, a new sub-program CONTACT was written in association with the program ISOPAR. The subprogram CONTACT is in three dimensions.

Gap element

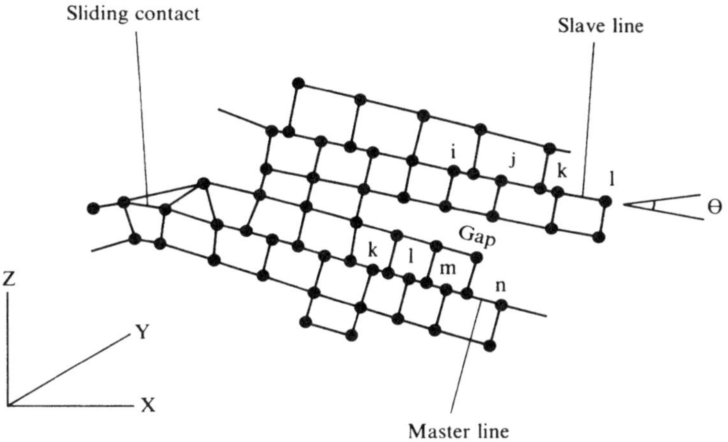

$$\{\hat{\vec{\mathbf{F}}}_{\mathrm{N}}\} = [\mathrm{K}_{\gamma'\gamma'}]_i\{U_{\gamma'}\}_i = [\Sigma k]\{\Sigma\Delta_i, \Delta_j \ldots\} = \{\hat{\vec{\mathbf{F}}}_{i,j}, \ldots\} + \{\pm\mu\hat{\vec{\mathbf{F}}}_{\mathrm{n}} \ldots \pm k\Delta_i \ldots\}$$

$$\Delta_{\mathrm{sl}} = \text{distance of sliding}$$

$$= (\Delta_j - \Delta_i) - \frac{\mu|\hat{\vec{\mathbf{F}}}_{\mathrm{n}}|}{[\mathrm{K}_{\gamma'\gamma'}]}$$

$$\mu = \text{friction}$$

$$\{\hat{\vec{\mathbf{F}}}_{\mathrm{SN}}\} \leq \mu\{\hat{\vec{\mathbf{F}}}_{\mathrm{N}}\} \quad \text{no sliding}$$

$$\geq \mu\{\hat{\vec{\mathbf{F}}}_{\mathrm{N}}\} \quad \text{sliding}$$

$$= 0 \quad \text{contact broken}$$

$$\theta = \cos^{-1}\frac{X}{\gamma} \quad \text{or} \quad \sin^{-1}\frac{Y}{\gamma}$$

AXIII.3. DELFT interface fraction type element

This is suited to isoparametric quadrilateral two-dimensional elements as shown in Fig. AXIII.2. This gap or contact element is dealt with in detail by Rots [1207] and Schellekens [1208] using traction profile.

Consider the general case of an isoparametric two-dimensional interface element that connects two isoparametric solid elements. Figure AXIII.2 shows one example. The element geometry x and the element displacement field u are interpolated as

$$x = Hx_i \tag{AXIII.7}$$

$$u = Hu_i \tag{AXIII.8}$$

where x_i is the nodal geometry vector in global co-ordinates, u_i is the nodal displacement vector in global co-ordinates and H is a logical grouping of the shape functions

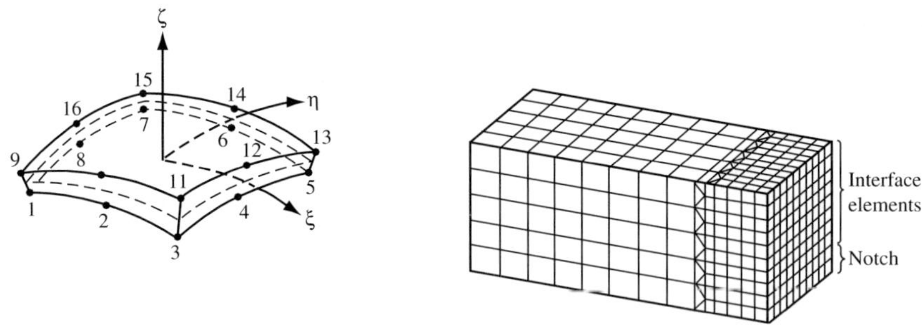

Fig. AXIII.2. Quadratic two-dimensional quadrilateral interface element.

N which are expressed in the natural co-ordinates ξ and η along the surface. The same interpolations are adopted for top and bottom sides of the interface, which implies that the nodes should overlap. If the nodes do not overlap, interpolations should be set up with reference to mid-plane.

The relative displacement field Δu is defined as

$$\Delta u = Lu \tag{AXIII.9}$$

with L being a 3×6 matrix, filled with 1, -1 and 0 such that the relative displacement is the displacement at the top side minus the displacement at the bottom side. Here, the interface elements deviate from continuum elements, where L contains differential operators.

The interface constitutive behaviour is described in terms of a traction-relative displacement law in the local co-ordinate system at integration points. This requires the global relative displacements Δu to be transformed to local relative displacements u as

$$\Delta \bar{u} = \Theta^{\mathrm{T}} \Delta u \tag{AXIII.10}$$

Θ contains the normalised local axes n, s and t as columns. These local axes are set up according to eqn (AXIII.7).

Substitution of eqn (AXIII.8) into eqn (AXIII.9) and of the result thereof into eqn (AXIII.10) yields

$$\Delta \bar{u} = B u_i \tag{AXIII.11}$$

where the notation $B = \Theta^{\mathrm{T}} LH$ has been introduced for the local relative displacement–nodal displacement matrix. Introducing the local traction-relative displacement law as

$$\hat{t} = D \Delta \bar{u} \tag{AXIII.12}$$

the stiffness matrix can be written in standard form:

$$K = \int_{\xi=-1}^{\xi=+1} \int_{\eta=-1}^{\eta=+1} B^{\mathrm{T}} D B \det(J) \, \mathrm{d}\xi \, \mathrm{d}\eta \tag{AXIII.13}$$

with $\det J$ being the determinant of the Jacobian operator.

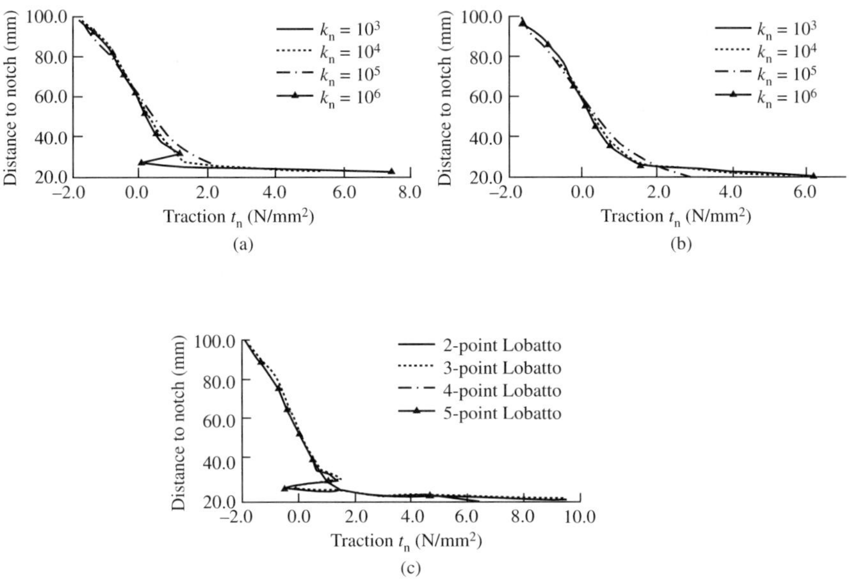

Fig. AXIII.3. Interface element — behaviour checked with other researchers based on DELFT.

In the case of three-dimensional analysis, Fig. AXIII.3 shows the interface element for three-dimensional elastic analysis of a beam. Element interaction schemes are described in the text. This element has been tested with other methods of analysis [1207, 1208] by Newton–Cotes, Lobatto, Gauss and node-lumping schemes. The comparative study of results is given in Fig. AXIII.3.

AXIII.4. ANSYS contact elements

CONTAC52 — three-dimensional point to point contact element

CONTAC52 represents two surfaces which may maintain or break physical contact and may slide relative to each other. The element is capable of supporting only compression in the direction normal to the surfaces and shear (Coulomb friction) in the tangential direction. The element has three degrees of freedom at each node: translation in the nodal x, y and z directions as shown in Fig. AXIII.4.

The element may initially be preloaded in the normal direction or it may be given a gap specification. A specified stiffness acts in the normal and tangential directions when the gap is closed and not sliding.

The geometry node locations and the co-ordinate system for this element are shown in Fig. AXIII.4. The element is defined by two nodes, two stiffnesses (KN and KS), an initial gap or interference (GAP) and an initial element status (START). The orientation of the interface is defined by the node locations. The interface is assumed to be perpendicular to the I–J line. The element co-ordinate system has its origin at node I and the x-axis is directed towards node J. The interface is parallel to the element y–z plane.

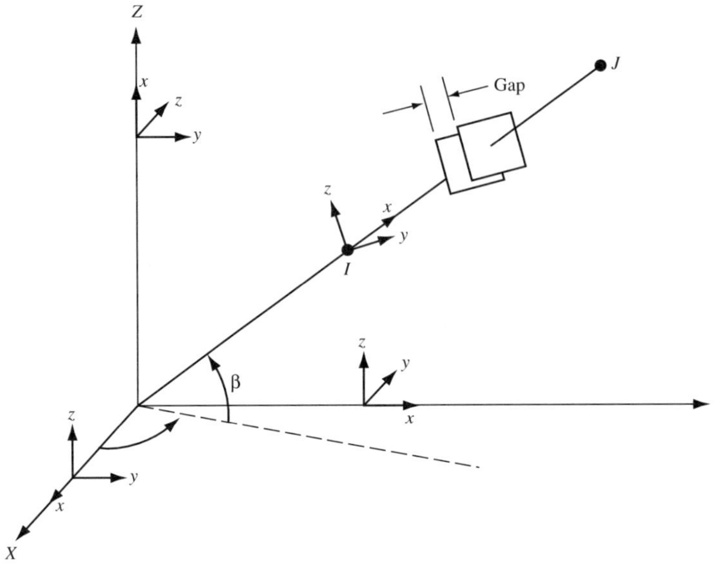

Fig. AXIII.4. CONTAC52 three-dimensional point-to-point contact element.

The normal stiffness KN should be based on the stiffness of the surfaces in contact. The sticking stiffness KS represents the stiffness in the tangential direction when elastic Coulomb friction is selected ($\mu > 0.00$ and KEYOPT(1) = 0). The coefficient of friction μ is input as material property MU and is evaluated at the average of the two node temperatures. Stiffness may also be computed from the maximum expected force divided by the maximum allowable surface displacement. KS defaults to KN.

The initial gap defines the gap size (if positive) or the displacement interference (if negative). An interference causes the nodes to separate. The only material property used is the interface coefficient of friction μ. A zero value should be used for frictionless surfaces.

The force–deflection relationships for the interface element can be separated into the normal and tangential (sliding) directions as shown in Fig. AXIII.5. The element condition at the beginning of the first sub-step is determined from the START

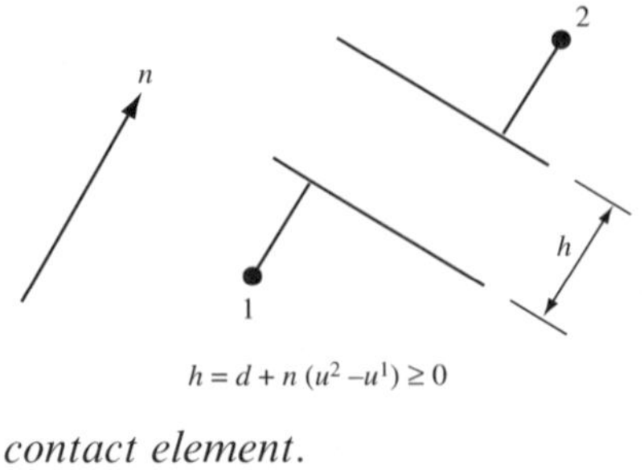

Fig. AXIII.5. GAPUNI contact element.

Table AXIII.1. CONTAC52 input summary

Element name	CONTAC52
Nodes	I, J
Degree of freedom	UX, UY, UZ
Real constants	KN, GAP, START, KS A negative value of GAP assumes an initial interference condition If START = 0.0 or blank, initial status of element is determined from gap input If START = 1.0, gap is initially closed and not sliding (if MU ≠ 0.0), or sliding (if MU = 0.0) If START = 2.0 gap is initially closed and sliding If START = 3.0, gap initially open
Material properties	MU
Surface loads	None
Body loads	Temperatures: T(I), T(J)
Special features	Nonlinear, adaptive descent
KEYOPT(1)	Used only with MU > 0.0 0 Elastic Coulomb friction (KS used for sticking stiffness) 1 Rigid Coulomb friction (resisting force only)
KEYOPT(4)	0 Gap size based on gap real constant 1 Gap size determined from initial node locations (ignore gap real constant)
KEYOPT(7)	Used with automatic loading to control contact time predictions 0 Predictions are made to achieve the minimum time (or load) increment whenever a change contact status occurs 1 Predictions are made to maintain a reasonable time (or load) increment (recommended)

parameter. If the interface is closed and sticking, KN is used in the gap resistance and KS is used for sticking resistance. If the interface is closed but sliding, KN is used in the gap resistance and the constant friction force μFN is used for the sliding resistance.

In the normal direction when the normal force (FN) is negative, the interface remains in contact and responds as a linear spring. As the normal force becomes positive, contact is broken and no force is transmitted.

In the tangential direction for FN < 0 and the absolute value of the tangential force (FS) less than $\mu|\mathrm{FN}|$, the interface sticks and responds as a linear spring. For FN < 0 and FS $= \mu|\mathrm{FN}|$ sliding occurs. If contact is broken, FS $= 0$.

A summary of the element input is given in a step-by-manner in Table AXIII.1.

Output data

The solution output associated with the element is in two forms:

(a) nodal displacements included in the overall nodal solution, and
(b) additional element output as shown in Table AXIII.2.

The force displacement curves are generally illustrated.

Table AXIII.2. Output for gap element

Name	Definition	O	R
UT(Y, Z)	Displacement (node J-node 1) in element *y* and *z* directions	2	2
FS	Tangential (friction) force (vector sum)	2	2
ANGLE	Principal angle of friction force in elementary-*z* plane	2	2

1 — if the value of START is:
2 — sliding contact
3 — gap open (not given in Table AXIII.2)
2. If MU > 0.0

AXIII.5. ABAQUS gap contact element

The following overview underlines various steps and options for gap elements.

Overview

Gap elements:

- allow for contact between two nodes;
- allow for the nodes to be in contact (gap closed) or separated (gap open) with respect to particular directions and separation conditions;
- are always defined in three dimensions but can also be used in two-dimensional and axisymmetric models;
- allow contact to be defined on any type of element, including superelements and user-defined elements;
- provide the contact forces, not stresses between two points in a local basis system;
- require that some limitations be placed on the rotation of the contact direction (GAPUNI and GAPCYL elements only) and
- can be used to stimulate an inextensible cable (by using GAPSPHER elements).

Choosing and defining a gap element

(a) GAPUNI elements model contact between two nodes when the contact direction is fixed in space.
(b) GAPCYL elements model contact between two nodes when the contact direction is fixed in a plane.
(c) GAPSPHER elements model contact between two nodes when the contact direction is arbitrary in space.
(d) Gap elements are defined by specifying the two nodes forming the gap and providing geometric data defining the initial state of the gap.

GAPUNI elements

The behaviour of the interface being modelled with a GAPUNI element is defined by the initial distance (clearance) d of the gap and the contact direction, n.

Clearance between GAPUNI nodes
ABAQUS defines the current clearance between two nodes of the gap h as

$$h = d + n(u^2 - u^1)$$

where u^1 and u^2 are the total displacements at the first and the second node forming the GAPUNI element. Figure AXIII.5 shows the configuration of the GAPUNI element. When h becomes the gap contact element it is closed only when the constraint $h = 0$ is imposed.

Gap element library
This section defines the gap elements available in ABAQUS.

References
- GAP contact elements
- *GAP.

Element types

GAPUNI	unidirectional gap between two nodes
GAPCYL	cylindrical gap between two nodes
GAPSHER	spherical gap between two nodes

Active degrees of freedom — 1, 2, 3(u_x, u_y, u_z).
Additional solution variables — three additional variables relating to the contact and friction forces.

Nodal co-ordinates required
GAPUNI: if the contact direction n is specified on the *GAP option, the nodal co-ordinates are not relevant; however, it is useful to define the co-ordinates of the two nodes for plotting purposes.

GAPCYL and GAPSPHER: X, Y, Z

Element property definition
Use the *GAP option to specify the initial clearance or overclosure in GAPUNI elements; a negative value indicates an initial overclosure. For GAPCYL and GAPSPHER elements, specify the maximum separation as a positive number or the minimum separation as a negative number.

Element loading
None.

Element output

S11	force in the gap
S12	first friction force normal to the gap direction
S13	second friction force normal to the gap direction

E11 current opening h of the gap element

E12 relative displacement (slip) in the first direction orthogonal to the contact direction

E13 relative displacement (slip) in the second direction orthogonal to the contact direction.

The increments of shear slip are the relative displacement increments projected onto the two local directions that are orthogonal to the contact direction.

In two-dimensional or axisymmetric models when the contact direction is along the first (X or r), the active slip direction is E13 and the active shear force is S13. In any other two-dimensional or axisymmetric case the active slip direction is E12 and the active shear force is S12.

Nodes associated with the element

Two nodes: the ends of the gap.

The behaviour of a GAPCYL element is defined by the initial separation distance between nodes, d; the current positions of the element's nodes, and the axis of the GAPCYL element. The axis of the GAPCYL element defines the plane in which the contact direction n lies. Specify d and the direction cosines of the GAPCYL element axis on the data line of the *GAP option.

The value $d = 0$ is not allowed: it would enforce the distance between the nodes to be exactly zero at all times, which does not correspond to a contact problem.

Case 1. Defining the gap clearance when d is positive

If d is positive, the GAPCYL element models contact between two rigid tubes of different diameter where the smaller tube is located inside the larger tube. In this case, d is the maximum allowable separation. Each tube is represented by a node on its axis with the axes connected by the GAPCYL element and d corresponds to the difference between the radius of the tubes. The gap between the tubes closes when the two nodes become separated by more than d in any direction in the plane defined by the axis of the GAPCYL element.

ABAQUS defines the current gap opening, h in GAPCYL elements for Case 1 as

$$h = d - |\bar{x}^2 - \bar{x}^1| \qquad \text{with } \bar{x}^N = x^N - a(ax^N) \tag{AXIII.14}$$

where x^N is the current position of node N; d is the specified initial separation and a is the axis of the GAPCYL element.

If the initial position of the tube axes is such that the distance between them is less than d the GAPCYL element is open initially. If the distance is equal to d the element is closed initially and if the distance is greater than d an initial overclosure (interference) is defined.

Case 2. Defining the gap clearance when d is negative

If d is negative, the GAPCYL element models external contact between two parallel rigid cylinders. In this case, d is the minimum allowable separation of the nodes.

Each cylinder is represented by a node on its axis connected by the GAPCYL element and d corresponds to the sum of the radii of the cylinders. The gap closes when the two nodes approach each other to within d in any direction in the plane defined by the axis of the GAPCYL element.

ABAQUS defines the current gap opening h in GAPCYL elements for Case 2 as

$$h = |x^2 - x^1| - |d| \qquad \text{with } x^{\mathrm{N}} = x^{\mathrm{N}} - a(ax^{\mathrm{N}}) \tag{AXIII.15}$$

if the initial position of the cylinder axes is such that the distance between them is greater than $|d|$ the GAPCYL element is open initially. If the distance is equal to $|d|$ the element is closed initially; and if the distance is less than $|d|$ an initial over-closure (interference) is defined. Details about modelling interference fit problems with GAP elements are discussed below.

Local basis system for GAPCYL element output

ABAQUS reports the force across the gap and the shear forces that are orthogonal to the contact direction as element output for GAPCYL elements. It also reports the current clearance in the gap, h, and the relative motions of the elements' nodes that are orthogonal to the contact direction. The relative motions and the shear forces are reported in local surface directions that are formed using the standard ABAQUS convention for defining directions on surfaces in space. The contact direction defines a surface in space on which the local axes are formed and the slip is calculated from the relative motions in the surface directions.

ABAQUS updates the contact direction for GAPCYL elements based on the motion of the nodes forming the elements. However, the orientation of a is not updated during the analysis.

GAPSPHER elements

GAPSPHER elements can be used to model two very different contact situations: contact between two rigid spheres where the smaller sphere is inside the larger, hollow sphere and contact between two rigid spheres along their external surfaces.

The behaviour of a GAPSPHER element is defined by the minimum or maximum separation distance between the nodes d and the current positions of the element's nodes. Specify the minimum or maximum separation distance d on the data line of the *GAP option. The contact direction is defined by the current position of the nodes.

The value $d = 0$ is not allowed: it would enforce the distance between the nodes to be exactly zero at all times, which does not correspond to a contact problem.

Case 1. Defining the gap clearance when d is positive

If d is positive, the GAPSPHER element models contact between a rigid sphere inside another (larger) hollow rigid sphere. In this case d is the maximum allowable separation of the nodes forming the gap. Each sphere is represented by a node at its centre, with the centres connected by the GAPSPHER element; and d corresponds to the difference between the radii of the spheres. The gap closes when the two nodes become separated by more than d.

ABAQUS defines the current gap opening h for Case 1 as

$$h = d - |x^2 - x^1| \tag{AXIII.16}$$

with x^N the current position of node N and d the specified separation.

If the initial position of the tube axes is such that the distance between them is less than d, the GAPSPHER element is open initially. If the distance is equal to d, the element is closed initially, and if the distance is greater than d an initial overclosure (interference) is defined.

Case 2. Defining the gap clearance when d is negative

If d is negative, the GAPSPHER element models extend contact between two rigid spheres. In this case d is the minimum allowable separation of the nodes forming the gap. Each sphere is represented by a node at its centre connected by the GAPSPHER element and $|d|$ corresponds to the sum of the radii of the spheres. The gap closes when the two nodes approach each other to within $|d|$.

ABAQUS defines the current gap opening h for Case 2 as

$$h = |x^2 - x^1| - |d| \tag{AXIII.17}$$

If the initial position of cylinder axes is such that the distance between them is greater than $|d|$, the GAPSPHER element is open initially. If the distance is equal to $|d|$ the element is closed initially and if the distance is less than $|d|$ an initial overclosure (interference) is defined.

Local basis system for GAPSPHER element output

ABAQUS reports the force across the gap and the shear forces that are orthogonal to the contact h and the relative motions of the elements node that are orthogonal to the contact direction. The relative motions and the shear forces are reported in local surface directions that are formed using the standard direction that formed using the standard ABAQUS convention for defining direction on surfaces in space. The contact direction defines a surface in space on which the local axes are formed, and the slip is calculated from the relative motion in the surface directions.

ABAQUS updates the contact direction for GAPSPHER elements based on the motion of the nodes forming the elements.

Defining nondefault mechanical interactions with gap elements

The default mechanical surface interaction model for problems modelled with gap elements is hard frictionless contact. Optional mechanical surface interaction models can be assigned by using the appropriate options in conjunction with the *GAP option defining the elements properties. The following mechanical surface interaction models are available:

(a) friction
(b) modified 'hard' contact, softened contact and viscous damping.

Modelling large initial interference with gap elements
Specifying a large negative initial overclosure (interference) may lead to convergence problems as ABAQUS tries to resolve the overclosure in a single increment. Use the *CONTACT INTERFERENCE, TYPE = ELEMENT option to allow ABAQUS to resolve the overclosure gradually.

Usage: *CONTACT INTERFERENCE, TYPE = ELEMENT

Specify d on the data line of the *GAP option. If a positive value is provided the gap is open initially. If $d = 0$ the gap is initially closed. If d is negative, the gap is considered overclosed at the start of the analysis and an initial interference fit problem is defined.

Specifying the contact direction
The user can specify the contact direction on the data line of the *GAP option. Otherwise ABAQUS will calculate the gap direction n by using the initial positions of the two nodes forming the element, X^1 and X^2:

$$n = (X^2 - X^1)/|X^2 - X^1| \qquad \text{(AXIII.18)}$$

An error message is issued if $X^3 = X^1$ (if the two gap element nodes have the same initial co-ordinates). In this situation the user must define n on the *GAP option. The normal n usually points from the first node of the element to the second, unless the gap is overclosed at the start of the analysis. In that case, specify n so that the correct contact direction is used for the gap element.

If the user specifies the gap direction n rather than allowing ABAQUS to calculate it, the contact calculations consider only the displacements of the gap elements nodes and the ordering of the nodes in the element definition. The initial co-ordinates of the nodes play no role in the calculations.

The output from the GAPUNI gives force across the gap and shear forces orthogonal to the contact direction.

GAPCYL elements and GAPSPHER elements are treated as outside the scope of this book since contact examples between rigid steel tubes internally or externally contacted have not been included in the text. For an in-depth study the reader is referred to analytical and users' manuals of ANSYS, available from Swanson's Incorporation of USA.

AVIII.6. LS-DYNA gap/contact elements

Introduction
The gap elements or contact elements are based on the Hallquist method given above. The number of slave segments and number of master segments are defined by interface control cards. IREAD flag is to read additional control cards. Ties

interface contact definitions are based on offset option which can be used with rigid bodies. Typical letters and numbers are given below as:

Type number (the letters 'a', 'm', 'o' and 'p' must be in col. 13)

1 — sliding without penalties
p1 — symmetric sliding with penalties (recommended)
2 — tied
02 — tied with offsets permitted, see note 1 below
3 — sliding, impact, friction
a3 — sliding, impact, friction, no segment orientation
m3 — sliding, impact, friction-metal forming option
4 — single surface contact
5 — discrete nodes impacting surface
a5 — discrete nodes impacting surface, no segment orientation
m5 — discrete nodes impacting surface-metal forming option
6 — discrete nodes tied to surface, see note 1 below
06 — discrete nodes tied to surface with offsets permitted
7 — shell edge tied to shell surface, see note 1 below
07 — shell edge tied to shell surface with offsets permitted
8 — nodes spot welded to surface
9 — tiebreak interface
10 — one way treatment of sliding, impact, friction
a10 — one way treatment, no segment orientation
m10 — one way treatment of sliding, impact, friction-metal forming option
11 — box/material limited automatic contact for shells
12 — automatic contact for shells (no additional input required)
13 — automatic single surface with beams and arbitrary orientations
a13 — like above but with extra searching for airbag contact
14 — surface to surface eroding contact
15 — single surface eroding contact
16 — node to surface eroding contact
17 — surface to surface symmetric/asymmetric constraint method
18 — node to surface constraint method
19 — rigid body to rigid body contact with arbitrary force/deflection curve (this option may be used with deformable bodies)
20 — rigid nodes to rigid body contact with arbitrary force deflection curve (this option may be used with deformable bodies)
21 — rigid body to rigid body contact with arbitrary force/deflection curve; unlike option 19 this is a one-way treatment (this option may be used with deformable bodies)
22 — single edge treatment for shell surface edge to edge treatment
23 — simulated draw bead
25 — force transducer contact for penalty based contact types — not for types 2, 6, 7, 17 and 18, see type 27 below for the constraint type

26 — automatic single surface, beams to beams, beam to shell edge
27 — force transducer contact for constraint based contact types; applies to types 2, 6, 7, 17 and 18 only.

Static coefficient of friction μ_s
EQ.-1: part based friction coefficients are used. Applies to contact types a3, a5, a10, 13, 15 and 26 only.

Tying will only work if the surfaces are near each other. The criteria used to determine whether a slave node is tied down is that it must be 'close'. For shell elements, 'close' is defined as distance δ, less than:

$$\begin{aligned}\delta_1 &= 0.60^{*}(\text{thickness\_slave\_node} + \text{thickness\_master\_segment})\\ \delta_2 &= 0.05^{*}\min(\text{master\_segment\_diagonals})\\ \delta &= \max(\delta_1, \delta_2)\end{aligned} \tag{AXIII.19}$$

If a node is further away it will not be tied and a warning message will be printed.

For thermal control card

h_{rad} = radiation conductance

l_{gap} = length or thickness of gap between sliding surfaces

h = heat transfer conductance

$$h = h_{cont}, \text{ if the gap thickness is } 0 \leq l_{gap} \leq l_{min} \tag{AXIII.20}$$

$$h = h_{cond} + h_{rad}, \text{ if the gap thickness is } l_{min} \leq l_{gap} \leq l_{max} \tag{AXIII.21}$$

$$h = 0, \text{ if the gap thickness is } l_{gap} > l_{max} \tag{AXIII.22}$$

APPENDIX XIV

AXIV.1. Data on the Kobe earthquake

AXIV.2. Computer program subroutine for soil–structure interaction

AXIV.1. The Japan Meteorological Agency (JMA) intensity scale summary

0 = No feeling: registered by seismographs only
1 = Slight: felt by some individuals at rest
2 = Weak: felt by most people
3 = Rather strong: shaking of houses and buildings
4 = Strong: strong shaking of houses and buildings and unstable objects fall over
5 = Very strong: cracking of plaster walls, heavy objects such as tombstones and stone lanterns fall over, and damage to masonry chimneys and mud-plastered warehouses
6 = Disastrous: collapse of up to 30% of wooden houses, numerous landslides and embankment failures, fissures on flat ground
7 = Very disastrous: collapse of more than 30% of wooden houses, falling of objects, wavy deformation observed in horizon

Source of the approximate comparison of earthquake intensity scales

M. D. Trifunac and A. G. Brady, The correlation of seismic intensity scales with the peaks of recorded strong ground motion, *Bulletin of Seismological Society of America*, Vol. 65, 1976.

The Modified Mercalli (MM) and the Japan Meteorological Agency (JMA) intensity scales are compared in the table below, along with the definitions of the JMA scale. It is of note that initially the Japanese authorities rated the worst hit area of Kobe as 6 on the JMA scale, but later increased that ranking to 7.

Seismological details for the Hyogo-ken Nanbu earthquake of 20 h 46 m 51.0 s UT, 16 January 1995 (5 h 46 m 51.0 s local time on 17 January 1995) are as follows:

M_L (local or Richter magnitude)	7.2
M_w (moment magnitude)	6.9
M_s (surface wave magnitude)	6.9
Epicentre location	34.527N, 135.005E
Hypocentral depth	16 km

AXIV.2. Buildings and bridges — soil–structure interaction

The gap element developed by ANSYS can be used for soil–structure interaction. In addition, various springs/dampers are used to model soil with the structure. The models for raft and separate footings foundations are shown in the figures below.

Comparison of the Japan Meteorological Agency (JMA) and the modified Mercalli (MM) intensity scales

JMA	0	1	2		3		4	5		6	7	
MM	I	II	III	IV	V	VI	VII	VIII	IX	X	XI	XII

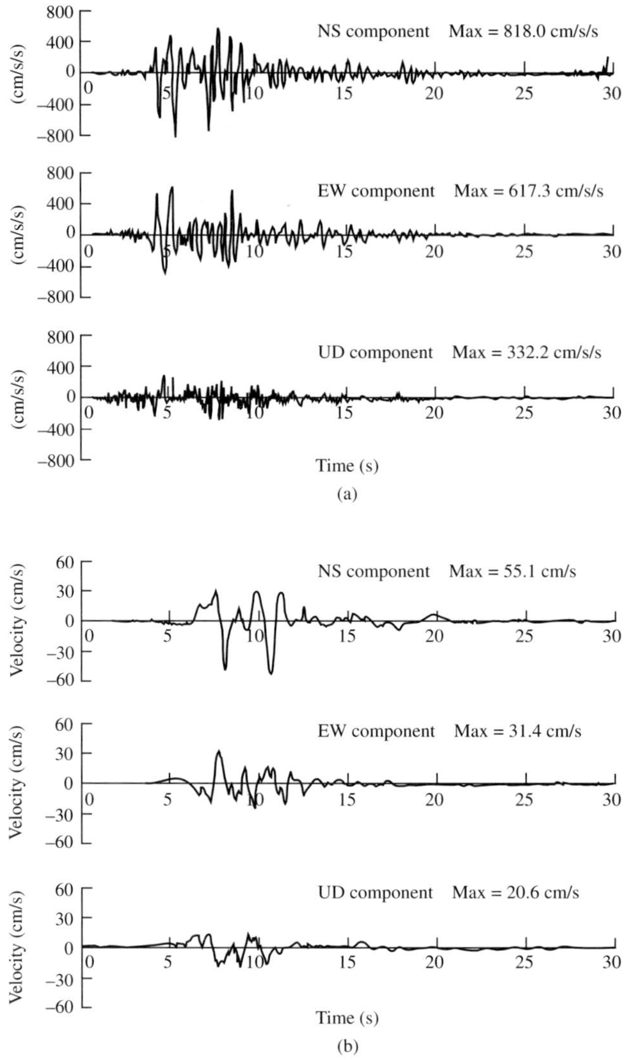

Fig. AXIV.1. Examples of recorded motions: (a) JMA Kobe Station (acceleration). (b) Kobe University (velocity).

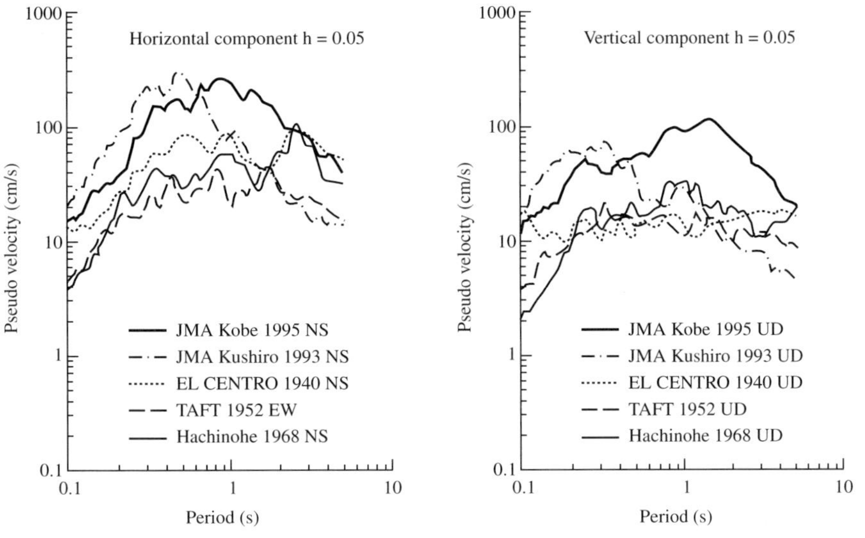

Fig. AXIV.2. Comparison of velocity response spectra.

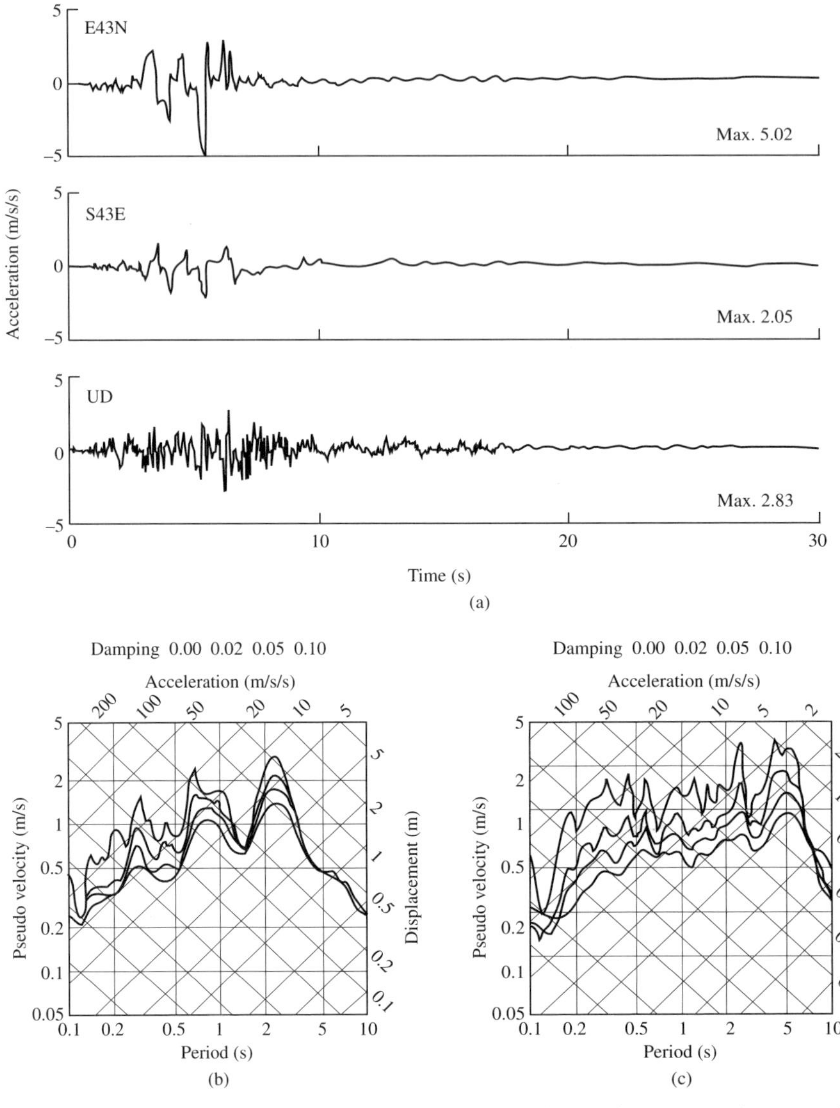

Fig. AXIV.3. Representative strong-motion data from the Hyogo-ken Nanbu main-shock: (a) Accelerogram recorded at the Kobe Port Construction Office (note that the E43N component was clipped at 5.02 m/s/s), (b) response spectrum for site Kobe Motoyama, (c) response spectrum for site Sakai. (Both of the Kobe sites were less than 1 km from the 'fault zone', while the Sakai site was south of Osaka and 27 km from the 'fault zone'.)

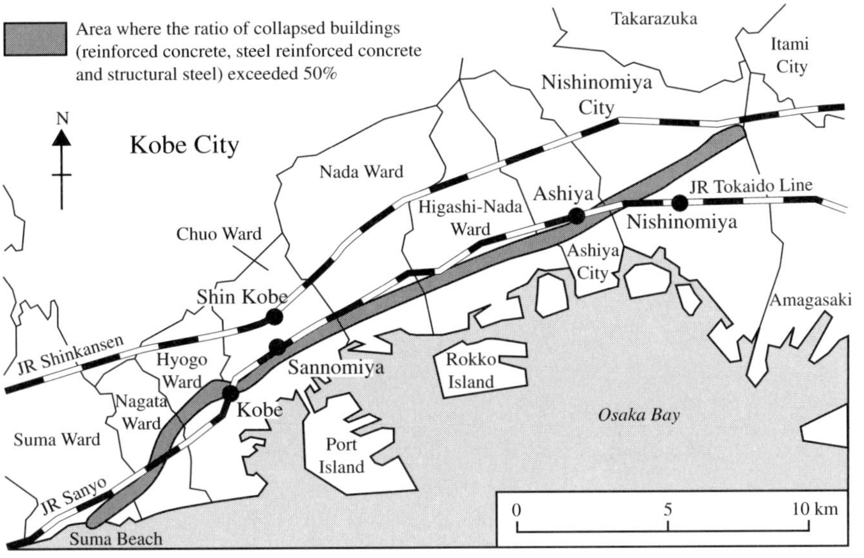

Fig. AXIV.4. Area of intense damage to buildings ('collapse' means either total collapse or loss of one or more storeys due to damaged columns and/or walls).

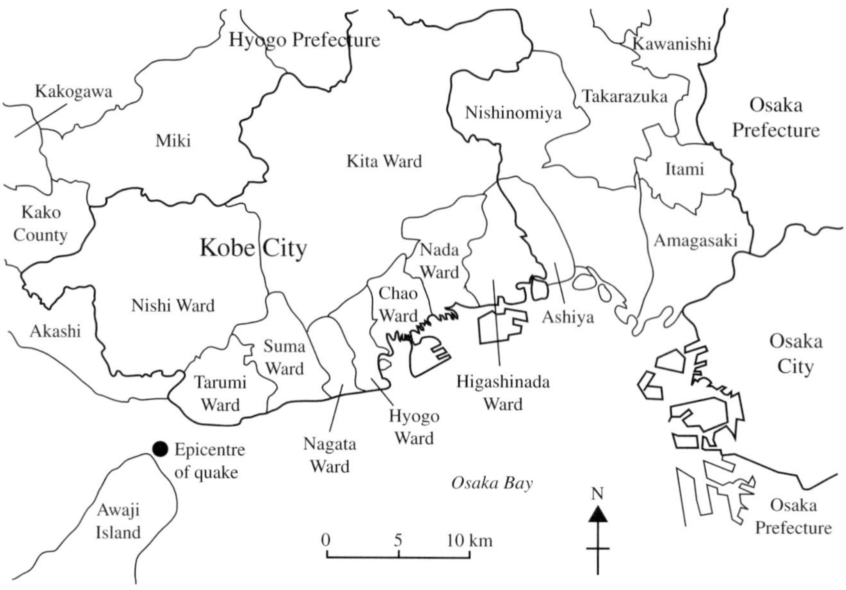

Fig. AXIV.5. The Hyogo-ken Nanbu earthquake (the Great Hanshin earthquake) of 17 January 1995 — Kobe City and the surrounding areas.

Table AXIV.1. Peak ground accelerations from sites within 10 km of the zone of intense destruction, the 'fault zone', of the Hyogo-ken Nanbu earthquake

Site name	PGA (g)	Estimated distance from 'fault zone' (km)
Nishinomiya	0.81	0
Takatori Station	0.63	0
Kobe Motoyama	0.79	≤ 1
Shin-Kobe Station	0.57	1
Kobe Port Construction Office	>0.5	1
Fukiai	0.85	2
Kobe University	0.31	2
Amagasaki	0.33	3
Amagasaki (port)	0.48	4
Takarazuka Station	0.61	6
Mishi-Akashi Station	0.49	8
Shin-Osaka Station	0.25	10

Fig. AXIV.6. The Hyogo-ken Nanbu Earthquake.

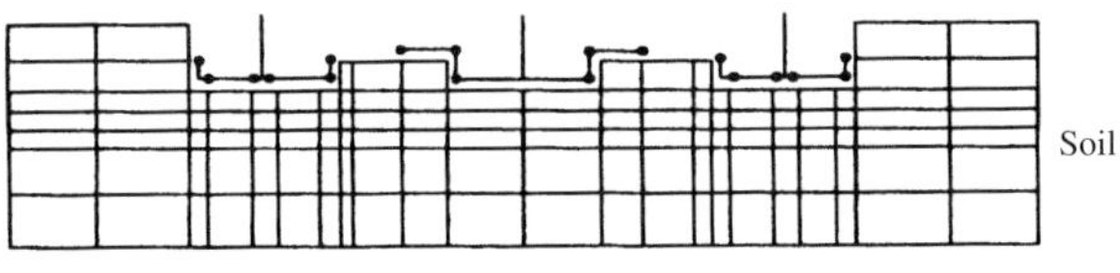

Fig. AXIV.7. Soil–structure model for a nuclear reactor (the nuclear reactor and heat exchanger models have been simplified; excitation is applied at the base of the model as a time history).

Table AXIV.2. Foundation interaction: spring and damper values for circular bases

Motion	Equivalent spring constant	Equivalent damping coefficient
Horizontal	$k_x = \dfrac{32(1-\nu)GR}{7-8\nu}$	$c_x = 0.576k_x R(\rho/G)^{1/2}$
Rocking	$k_\psi = \dfrac{8GR^3}{3(1-\nu)}$	$c_\psi = \dfrac{0.30}{1+B_\psi} k_\psi R(\rho/G)^{1/2}$
Vertical	$k_z = \dfrac{4GR}{1-\nu}$	$c_z = 0.85k_z R(\rho/G)^{1/2}$
Torsion	$k_1 = 16GR^3/3$	$c_z = \dfrac{(k_t I_t)^{1/2}}{1+2I_t/\rho R^5}$

ν is Poisson's ratio for the foundation medium, G is the shear modulus for the foundation medium, R is the radius of the circular basemat, ρ is the mass density of the foundation medium, $B_\psi = 3(1-\nu)I_0/8\rho R^5$, I_0 is the total mass amount of inertia of the structure and basemat about the rocking axis at the base and I_t is the polar mass moment of inertia of the structure and basemat.

Table AXIV.3. Foundation interaction: spring and damper values for rectangular bases

Motion	Equivalent spring constant	Equivalent damping coefficient
Horizontal	$k_x = 2(1+\nu)G\beta_x(BL)^{1/2}$	Use the results for a circular base with the following equivalent radius R
Rocking	$k_\psi = \dfrac{G}{1-\nu}\beta_\psi BL^2$	$R = (BL/\pi)^{1/2}$ for translation $R = (BL^3/3\pi)^{1/4}$ for rocking
Vertical	$k_z = \dfrac{G}{1-\nu}\beta_z(BL)^{1/2}$	
Torsion	$R = [BL(B^2+L^2)/6\pi]^{1/4}$	

B is the width of the basemat perpendicular to the direction of horizontal excitation, L is the length of the

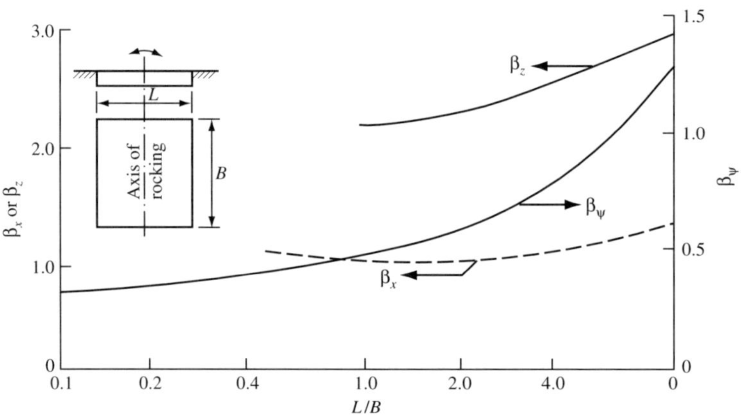

Fig. AXIV.8. Constants β_x, β_ψ, β_z for rectangular bases (after Whitman and Richart (1967)).

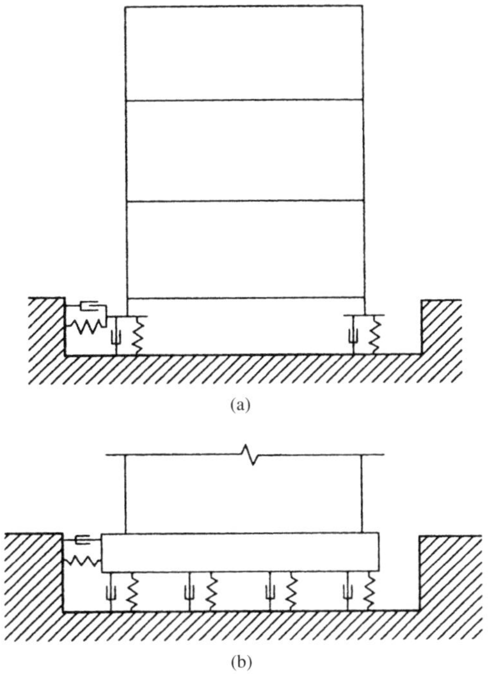

Fig. AXIV.9. Modelling of springs and dampers under foundations: (a) separate footings with tie; (b) raft foundation.

AXIV.3

```
REFERENCE - *FLUSH - A COMPUTER PROGRAM FOR APPROXIMATE 3-D ANALYSIS OF SOIL-
STRUCTURE INTERACTION PROBLEMS*
*****************************************************************************
      COMPLETE PROGRAM FOR COMPLEX RESPONSE ANALYSIS OF SOIL-STRUCTURE
      SYSTEMS BY THE FINITE ELEMENT METHOD.
      THE EQUIVALENT LINEAR METHOD IS USED TO CONSIDER STRAIN-COMPATIBLE
      SOIL PROPERTIES IN THE SYSTEM.
      PLANE TRIANGULAR OR QUADRILATERAL SOLID ELEMENTS, BEAM ELEMENTS AND
      TRANSMITTING BOUNDARIES MAY BE USED. APPROXIMATED 3-D SIMULATION BY
      VISCOUS BOUNDARIES.
*****************************************************************************
C     SET CONSTANTS AND INITIALIZE ARRAY FOR MAX. ACCEL. AND TIME
C     ***********************************************************************
C
      IFIN=0
      IEMR=1
      IARY(1)=0
      NT=KEY
      REWIND NT
      ICRE=1
      DDT=DT
      NCUT4=NCUT4E(NKT)+KINT(1)-1
      KST2=KST/2
      IF (NCUT4.GT.KST2) NCUT4=KST2
      NCUT7=2*NCUT4+1
      DDW=DW(1)/KINT(1)
      W=DDW
      BB(1)=BB(1)/32.2
      DD(1)=CMPLC(C.,C.)
      DO 1 KTS=2,NCUT4
      BB(KTS)=BB(KTS)/32.2
      DD(KTS)=-BB(KTS)*W*W
      W2(KTS)=-386.4/W/W
      W=W+DDW
1     CONTINUE
      CALL FSET (FSTRT,FLAST,NINT,ND,LENG,TT,FR)
C
      DO 2 N=1,NOUT
      TIME1(N)=C.
2     ACCMAX(N)=0.
C
C     ***********************************************************************
C     CHECK COMMON SIZE FOR RESPONSE IN FREQUENCY DOMAIN
C     ***********************************************************************
C
      NAAA=NBLKCN/NCUT
      IF (NAAA.LT.NCUT) GO TO 4
C
      J1=1
      J2=NOUT
      DO 3 KQ=1,NCUT
      READ (NT) (AAAC(I), I=J1,J2)
      J1=J1+NCUT
      J2=J2+NCUT
3     CONTINUE
```

```
      NA=1
      NB=NOUT
      NC=NOUT
      GO TO 8
C
4     NA=1
      NB=NAAA
      NC=NAAA
5     KF=0
      DO 7 KQ=1,NCUT
      READ (NT) (ACCC(I), I=1,NOUT)
      KU=NA
      DO 6 KZ=1,NO
      KF=KF+1
      AAAC(KF)=ACCC(KU)
      KU=KU+1
6     CONTINUE
7     CONTINUE
C
C
8     CONTINUE
      PRINT 48
      KX=0
C
C     **********************************************************************
C     FIND OUTPUT REQUIREMENTS AT NODAL POINTS
C     **********************************************************************
C
      DO 35 KY=NA,NB
      B3=HOR
      ITYPE=0
      KK=KOUTP(KY)
      KTPP=AINT(KEYSPC(KY))
      KSPEC=IABS(KTPP)
      KV=KK/2
      IF (KEYSPC(KY).LT.0) GO TO 1009
      IF (KV*2.NE.KK) GO TO 10
      B3=VERT
      ITYPE=1
      DO 9 KQ=1,7
      IF (KEY.EQ.9) GO TO 90
      HEAD(KQ)=VERTI(KQ)
      HEAD1(KQ)=VDISP(KQ)
      HEAD2(KQ)=VACCE(KQ)
      GO TO 9
   90 CONTINUE
      HEAD(KQ)=VERL(KG)
      HEAD1(KQ)=VDIL(KG)
      HEAD3(KQ)=VACL(KG)
9     CONTINUE
      GO TO 12
10    DO 11 KQ=1,7
      IF (KEY.EQ.9) GO TO 80
      HEAD1(KQ)=HDISP(KQ)
      HEAD(KQ)=HCRIZ(KQ)
      HEAD2(KQ)=HACCE(KQ)
      GO TO 11
```

```
   80 CONTINUE
      HEAD1(KQ)=HDIL(KQ)
      HEAD(KQ)=HORL(KG)
      HEAD2(KQ)=HACL(KQ)
11    CONTINUE
      GO TO 1006
1009  B3=0.
      N=N-1
      I=N
      J=N+NN
      M=MVB(N)
      DO 160 L=J,M,NN
      I=I+1
      DL1=REAL(A(L))
      DL2=AIMAG(A(L))
      DL3=REAL(V2(I))
      DL4=AIMAG(V2(I))
      DLR=DL1*DL3-DL2*DL4
      DLI=DL1*DL4*DL2*DL3
      DL1=REAL(V2(N))
      DL2=AIMAG(V2(N))
      DL3=DL1-DLR
      DL4=DL2-DLI
      X3=SNGL(DL3)
      X4=SNGL(DL4)
  160 V2(N)=CMPLX(X3,X4)
C
C**** COMPUTES RAYLEIGH QUOTIENT
C
      CE=0.
      DO 170 N=1,NN
      V1(N)=(V2(N)-V1(N)*CD)/EV
      OE=OE-U1(N)*V1(N)+U2(N)*V2(N)
      U1(N)=C1(N)*V1(N)
  170 U2(N)=A1(N)*V2(N)
      CE=CE*CD
C
C**** COMPUTES ELEMENTS OF R.H.S. FOR NEW RAYLEIGH QUOTIENT
C
      DO 180 N=3,NN
      L=N-2
      U1(L)=U1(L)+C3(L)*V1(N)
      U1(N)=U1(N)+C2(L)*V1(L)
      U2(L)=U2(L)+A3(L)*V2(N)
  180 U2(N)=U2(N)+A3(L)*V2(L)
C
C**** NEW ESTIMATE FOR EIGENVALUE
C
      CD=0.
      DO 190 N=1,NN
  190 CD=CD-U1(N)*V1(N)+U2(N)*V2(N)
      DEV=CE/CD
      IF (IT.GT.15) DEV=DEV*0.5
      EV=EV+DEV
      CF=CD
      CD=CSQRT(2./CD)
C
```

```
C**** CHECK FOR CONVERGENCE
C
      C=DEV/EV
      X=ABS(REAL(C))*AES(AIMAG(C))
      IF (IK.EQ.1.AND.X.LT.EPS2) GO TO 300
      IF (X.LT.EPS1) IK=1
  200 CONTINUE
      PRINT 2020
      PRINT 202,X
      STOP
C
C**** SELECT PROPER EIGENVALUES
      RAMD=PQ/(1.-2.*FO)
      GG(1,1)=G/H
      GG(1,3)=-GG(1,1)
      GG(2,2)=2.*G*(1.+RAMD)/H
      GG(2,4)=-GG(2,2)
      GG(3,1)=GG(1,3)
      GG(3,3)=GG(1,1)
      GG(4,2)=GG(2,4)
      GG(4,4)=GG(2,2)
      WRITE(NRAP) GG
C
C     ***********************************************************************
C     FORM LAYER MASS MATRIX
C     ***********************************************************************
C
      CM(1,1)=H*RO/3.*RT+(1.-RT)*.5*H*RO
      CM(1,3)=H*RO/6.*RT
      CM(2,2)=CM(1,1)
      CM(2,4)=CM(1,3)
      CM(3,1)=CM(1,3)
      CM(3,3)=CM(1,1)
      CM(4,2)=CM(2,4)
      CM(4,4)=CM(1,1)
      WRITE(NTAP) CM
C
      RETURN
      END
      SUBROUTINE MOORE (NHC)
      NNN=(NHC-1)/512
      NNN=(NNN+1)*512
      CALL SETFLS(NNN)
      RETURN
      END
      SUBROUTINE MOTION (TIME1,DD,W2,ACC,ACCC,ACCMAX,TT,FR,RD,PRV,PAA,SAA,SV,
      T,FOUAM,IDEN,AAAC,XORD,YORD,KEYSPC,KOUTP,DAMP,BB,INV,S,NCUT,BU2GMAX,NDP,
      KST,NBLKCM,KEY,NOUT,NMBR)
C
C     ***********************************************************************
C
C     COMPUTE ACCELERATION, DISPLACEMENT AND RESPONSE SPECTRUM
C     ARGUMENTS
C     TIME1  = ARRAY FOR TIME AT MAX. ACCELERATION
C     DD     = ARRAY FOR EARTHQUAKE RECORD TIMES FREQUENCY SQUARE
                TO FIND ACCELERATION FROM DISPLACEMENT
C     W2     = ARRAY FOR INVERSE OF FREQUENCY SQUARE
```

```
C     ACC    = ARRAY FOR OUTPUT IN REAL VALUES
C     ACCMAX = ARRAY FOR MAX. ACCELERATION
C     TT     = ARRAY FOR PERIOD IN RESPONSE SPECTRUM ANALYSIS
C     FR     = ARRAY FOR FREQUENCY IN RESPONSE SPECTRUM ANALYSIS
C     RD     = ARRAY FOR RELATIVE DISPLACEMENT IN RESPONSE SPECTRUM
C     PRV    = ARRAY FOR PSUDRO RELATIVE VELOCITY IN RESPONSE SPECTRUM
C     PAA    = ARRAY FOR PSUDRO ABSOLUTE ACC. RESPONSE SPECTRUM
C     SA     = ARRAY FOR ABSOLUTE ACC. RESPONSE SPECTRUM
C     SV     = ARRAY FOR RELATIVE VELOCITY RESPONSE SPECTRUM
C     T      = ARRAY FOR TIME IN PLOTTING
C     IDEN   = IDENTIFICATION FOR PLOT
C     AAAC   = ARRAY FOR READING DATA FROM TAPE
C     XORD   = ARRAY FOR X COORDINATE OF NODAL POINT
C     YORD   = ARRAY FOR Y COORDINATE OF NODAL POINT
C     KEYSPC = ARRAY FOR KEY NUMBER OF OUTPUT CONTROL
C     KOUTP  = ARRAY FOR NODAL POINT NUMBER WHERE OUTPUT IS REQUIRED
C     DAMP   = ARRAY FOR DAMPING USED IN RESPONSE SPECTRUM COMPUTATION
C     BB     = ARRAY FOR INPUT MOTION IN FREQUENCY DOMAIN
C     INV    = ARRAY FOR FAST FOURIER TRANSFORMATION
C     S      = ARRAY FOR FAST FOURIER TRANSFORMATION
C     NCUT   = TOTAL FREQUENCY STEP IN THIS ITERATION
C     U2GMAX = MAX. ACCELERATION OF INPUT MOTION
C     NDP    = TOTAL NUMBER OG DAMPING USED IN RESPONSE SPECTRUM
C     KST    = TOTAL NUMBER OF FREQUENCY COMPONENTS OF INPUT MOTION UP TO TOTFR
C
C
C     ***********************************************************************
C
      COMMON/CONST1/TITLE,EQN,ENAME.ITER,KERRO,TIM1,DT,NELM,KINTS,MATYP
      COMMON/CONST2/IFLAG,MFFT,N2,N3,KDOUB(5),LENGTH,LENG,NNDAM
      COMMON/CONST4/ND/NDAM,DW(5)
      COMMON/CONST5/HOR,VERT
      COMMON/CONST9/FSTRT,FLAST,NINT,NSKIP
      COMMON/CONS15/NCUTK(6),NCUT4E(5),KINT(5),FRKINT(5),MCUT2(5),
     1   NKT,FRKOD(5),RFR(5)
      COMMON/AMPIL/IARY(100),IEMR,IFIN
      COMMON/BLINE/LINE
C
      DIMENSION IDDN(10,1)
      DIMENSION FOUAM(1)
      DIMENSION TITLE(12),EQN(12)
      DIMENSION ABSIS(10),HDISP(7),VDISP(7)
      DIMENSION XORD(1),YORD(1),KEYSPC(1),KOUTP(1),DAMP(1),BB(1)
      DIMENSION IVN(1),S(1),TIME1(1),DD(1),W2(1),ACC(1),ACCC(1)
      DIMENSION ACCMAX(1),TT(1),FR(1),RD(1),PRV(1),PAA(1),T(1),
     A   IDEN(10,1)
      DIMENSION AAAC(1),SA(NDP,1),SV(NDP,1)
C
      INTEGER UNIT(2),UNIT1,UNIT2,UNIT3
      INTEGER HEAD(7),HEAD1(7),HEAD2(7)
      INTEGER ROTAT(7),RACCE(7),RODIS(7),HORIZ(7),VERTI(7),HACCE(7)
     1   VACCE(7)
      INTEGER VERL(7),VDIL(7),VACL(7),HDIL(7),HORL(7),HACL(7)
      REAL KEYSP
      COMPLEX ACCC,AAAC,BB,DD
C
      DATA UNIT/6H RAC/S,6HEC**2 /
```

```
      DATA UNIT1/6H G /
      DATA UNIT2/6H IN. /
      DATA UNIT3/6H /
      DATA ROTAT/6H ROTAT,6HIONAL,6HMOTION,6H AT NO,6HDAL PO,6HINT,6H /
      DATA RACCE/6H ROTAT,6HIONAL,6HACCEL.,6H AT NO,6HDAL PO,6HINT,6H /
      DATA RODIS/6H ROTAT,6HIONAL,6HDISPLA,6HCEMENT,6H AT NO,6HDAL PO,6HINT /
      DATA HORIZ/6H HORIZ,6HONTAL,6HMOTION,6H AT NO,6HDAL PO,6HINT,6H /
      DATA VERTI/6H VERTI,6HCAL MO,6HTION A,6HT NODA,6HL POIN,6HT,6H /
      DATA HACCE/6H HORIZ,6HONTAL,6HACCEL.,6H (6) A,6HT NODA,6HL POIN,6HT /
      DATA VACCE/6H VERTI.6HCAL AC,6HCEL.,6H (6) A,6HT NODA.6HL POIN,6HT /
      DATA ABSIS/6H TIME,6HIN SEC,6HONDS,7*6H /
      DATA HDISP/6H HORIZ,6HONTAL,6HDISPLA,6HCEMENT,6H AT NO,6HDAL PO,6HINT /
C     14B-1. ELEMENT PROPERTY CARD. SAME AS FOR SOLID ELEMENTS
C
C     14B-2. ADDITIONAL PROPERTY FOR BEAM ELEMENT (3F10.2)
C        1-10 ACR(N)  CROSS SECTION AREA OF BEAM PER UNIT WIDTH (FT)
C       11-20 RIN(N)  MOMENT OF INERTIA OF BEAM PER UNIT WIDTH (FT**4)
C       31-30 EFA(N)  RATIO BETWEEN EFFECTIVE SHEAR AREA AND TOTAL
C                     CROSS SECTIONAL AREA.
C                     IF 0, NO SHEAR DEFORMATION IS CONSIDERED.
C
C  14C. VOID ELEMENTS. MDLM CARDS (615).
C                      SKIP IF NO VOID ELEMENTS
C        1- 5 M        VOID ELEMENT NUMBER. NUMBERING STARTS FROM 1.
C        6-10 MDPI(M)  NODAL POINT I
C       11-15 MDPJ(M)  NODAL POINT J
C       16-20 MDPK(M)  NODAL POINT K
C       21-25 MDPL(M)  NODAL POINT L
C       26-30 MLAY(M)  LAYER NUMBER AT WHICH THE VOID ELEMENT IS LOCATED.
C
C  15.   MATERIAL CURVES. MATYP GROUPS OF THREE CARDS.
                         NOTE***SKIP THESE CARDS IF MATYP=0.
C
C  15A. IDENTIFICATION CARD FOR MATERIAL CURVES (I5,12A6).
C         1- 5 N         MATERIAL TYPE NUMBER.
C         6-77 TTL       MATERIAL IDENTIFICATION
C
C  15B.  SHEAR MODULUS REDUCTION FACTORS - G FACTOR (11F5.3).
C         1- 5 MODULUS REDUCTION FACTOR(=1.0) AT 10**(-4.0) PER CENT STRAIN
C         6-10 MODULUS REDUCTION FACTOR AT 10**(-3.5) PER CENT STRAIN
C        11-15 MODULUS REDUCTION FACTOR AT 10**(-3.0) PER CENT STRAIN
C        16-20 MODULUS REDUCTION FACTOR AT 10**(-2.5) PER CENT STRAIN
C        21-25 MODULUS REDUCTION FACTOR AT 10**(-2.0) PER CENT STRAIN
C        26-30 MODULUS REDUCTION FACTOR AT 10**(-1.5) PER CENT STRAIN
C        31-35 MODULUS REDUCTION FACTOR AT 10**(-1.0) PER CENT STRAIN
C        36-40 MODULUS REDUCTION FACTOR AT 10**(-0.5) PER CENT STRAIN
C        41-45 MODULUS REDUCTION FACTOR AT 10**( 0.0) PER CENT STRAIN
C        46-50 MODULUS REDUCTION FACTOR AT 10**( 0.5) PER CENT STRAIN
C        51-55 MODULUS REDUCTION FACTOR AT 10**( 1.0) PER CENT STRAIN
C
C  15C.  FRACTIONS OF CRITICAL DAMPING, IN PER CENT - D FACTOR C11F5.3).
C         1- 5 FRACTION OF CRITICAL DAMPING AT 10**(-4.0) PER CENT STRAIN
C         6-10 FRACTION OF CRITICAL DAMPING AT 10**(-3.5) PER CENT STRAIN
C        10-15 FRACTION OF CRITICAL DAMPING AT 10**(-3.0) PER CENT STRAIN
C        16-20 FRACTION OF CRITICAL DAMPING AT 10**(-2.5) PER CENT STRAIN
C        21-25 FRACTION OF CRITICAL DAMPING AT 10**(-2.0) PER CENT STRAIN
C        26-30 FRACTION OF CRITICAL DAMPING AT 10**(-1.5) PER CENT STRAIN
```

```
C        31-35 FRACTION OF CRITICAL DAMPING AT 10**(-1.0) PER CENT STRAIN
C        36-40 FRACTION OF CRITICAL DAMPING AT 10**(-0.5) PER CENT STRAIN
C        41-45 FRACTION OF CRITICAL DAMPING AT 10**( 0.0) PER CENT STRAIN
C        46-50 FRACTION OF CRITICAL DAMPING AT 10**( 0.5) PER CENT STRAIN
C        51-55 FRACTION OF CRITICAL DAMPING AT 10**( 1.0) PER CENT STRAIN
C
C
C    16. NODAL POINT CARDS. NDPT CARDS (I5, 2F10.3,I5,F10.4).
C         1- 5 M          NODAL POINT NUMBER
C         6-15 XORD(M)    X-COORDINATE,FT.
C        16-25 YORD(M)    Y-COORDINATE,FT.
C        26-30 KEYBC(M)   KEY FOR DISPLACEMENT BOUNDARY CONDITIONS
C                         RELATIVE TO RIGID BOUNDARY
```

*** ELEMENT DATA ***

SOLID ELEMENTS

ELM	NPI	NPJ	NPK	NPL	LAY	M.TYP	DENSITY (PCF)	POISSON R	G-MAX (KSF)	G-USE (KSF)	DAMP-USE (FRACTION)	VS-USE (FPS)	AREA (FT2)
1	1	8	10	2	1	1	1.2000E+02	3.5000E-01	3.0000E+03	1.5300E+03	4.0000E-02	6.4074E+02	3.7500E+01
2	2	10	12	3	2	2	1.3500E+02	3.0000E-01	3.3000E+03	2.1300E+03	5.8000E-02	7.1277E+02	5.0000E+01
3	3	12	30	31	3	2	1.3500E+02	3.0000E-01	3.3000E+03	1.7500E+03	8.2000E-02	6.4607E+02	5.0000E+01
4	1	6	8	8	1	1	1.2000E+02	3.5000E-01	3.0000E+03	2.0000E+03	7.0000E-02	7.3258E+02	1.2500E+01
5	10	16	17	12	2	0	1.4000E+02	2.0000E-01	6.0000E+03	6.0000E+03	5.0000E-02	1.1747E+03	5.0000E+01
6	12	17	29	30	3	2	1.3500E+02	3.0000E-01	3.3000E+03	2.3000E+03	7.0000E-02	7.4067E+02	5.0000E+01
7	27	26	22	16	1	-1	1.4000E+02	2.0000E-01	6.0000E+03	6.0000E+03	5.0000E-02	1.1747E+03	2.5000E+01
8	16	22	24	17	2	0	1.4000E+02	2.0000E-01	6.0000E+03	6.0000E+03	5.0000E-02	1.1747E+03	5.0000E+01
9	17	24	28	29	3	2	1.3500E+03	3.0000E-01	3.3000E+03	2.3000E+03	7.0000E-02	7.4067E+02	5.0000E+01

BEAM ELEMENTS

ELM	NPI	NPJ	NPK	NPL	BEAM-LENGTH (FT)	DENSITY (PCF)	POISSON R	G-USE (KSF)	DAMP-USE (FRACTION)	CROSS-AREA (FT2)	INER-MO. (FT4)	EFA
10	4	5	6	7	1.0000E+01	1.5000E+02	3.0000E-01	8.0000E+03	2.0000E-02	1.0000E+00	8.3300E-02	6.5000E-01
11	6	7	8	9	2.5000E+00	1.5000E+02	3.0000E-01	8.0000E+03	2.0000E-02	1.0000E+00	8.3300E-02	6.5000E-01
12	8	9	10	11	2.5000E+00	1.5000E+02	3.0000E-01	8.0000E+03	2.0000E-02	1.0000E+00	8.3300E-02	6.5000E-01
13	10	11	12	13	5.0000E+00	1.5000E+02	3.0000E-01	8.0000E+03	2.0000E-02	1.0000E+00	8.3300E-02	6.5000E-01
14	18	19	4	5	2.0000E+01	1.5000E+02	3.0000E-01	8.0000E+03	2.0000E-02	1.0000E+00	8.3300E-02	6.5000E-01
15	14	15	6	7	1.0000E+01	1.5000E+02	3.0000E-01	8.0000E+03	2.0000E-02	1.0000E+00	8.3300E-02	6.5000E-01
16	20	21	14	15	1.0000E+01	1.5000E+02	3.0000E-01	8.0000E+03	2.0000E-02	1.0000E+00	8.3300E-02	6.5000E-01
17	18	19	20	21	1.0000E+01	1.5000E+02	3.0000E-01	8.0000E+03	2.0000E-02	5.0000E-01	4.1500E-02	6.5000E-01
18	20	21	22	23	5.0000E+00	1.5000E+02	3.0000E-01	8.0000E+03	2.0000E-02	5.0000E-01	4.1600E-02	6.5000E-01
19	22	23	24	25	5.0000E+00	1.5000E+02	3.0000E-01	8.0000E+03	2.0000E-02	5.0000E-01	4.1600E-02	6.5000E-01

VOID ELEMENTS

ELM	NPI	NPJ	NPK	NPL	LAY	AREA (FT2)
1	20	22	16	14	1	5.0000E+01
2	14	15	10	6	1	5.0000E+01

AREA COVERED BY SOLID AND VOID ELEMENTS = 4.7500E+02

TOTAL BEAM LENGTH = 8.0000E+01

*** NODAL POINT DATA ***

N.P.	XORD (FT)	YORD (FT)	MASS (KIPS/FT)	
1	30.000	15.000	-0.	
2	30.000	10.000	-0.	
3	30.000	5.000	-0.	
4	20.000	25.000	-0.	
5	20.000	25.000	-0.	ROTATIONAL NODE
6	20.000	15.000	-0.	
7	20.000	15.000	-0.	ROTATIONAL NODE
8	20.000	12.500	-0.	
9	20.000	12.500	-0.	ROTATIONAL NODE
10	20.000	10.000	-0.	
11	20.000	10.000	-0.	ROTATIONAL NODE
12	20.000	5.000	-0.	
13	20.000	5.000	-0.	ROTATIONAL NODE
14	10.000	15.000	-0.	
15	10.000	15.000	-0.	ROTATIONAL NODE
16	10.000	10.000	-0.	
17	10.000	5.000	-0.	
18	0.	25.000	-0.	FIXED IN Y DIRECTION
19	0.	25.000	-0.	ROTATIONAL NODE
20	0.	15.000	10.000	FIXED IN Y DIRECTION
21	0.	15.000	10.000	ROTATIONAL NODE
22	0.	10.000	-0.	FIXED IN Y DIRECTION
23	0.	10.000	-0.	ROTATIONAL NODE
24	0.	5.000	-0.	FIXED IN Y DIRECTION
25	0.	5.000	-0.	ROTATIONAL NODE
26	0.	12.500	-0.	FIXED IN Y DIRECTION
				NODE OF INTERNAL STRUCTURE
27	10.000	12.500	-0.	NODE OF INTERNAL STRUCTURE
28	0.	0.	-0.	RIGID BASE
29	10.000	0.	-0.	RIGID BASE
30	20.000	0.	-0.	RIGID BASE
31	30.000	0.	-0.	RIGID BASE

1

*** INDENTIFICATION - CONTROL MOTION DT=0.02 SEC. 256 POINTS ***

0 MAX. ACCEL. PRIOR TO SCALING=.079866 G MAX. ACCEL. AFTER SCALING=.300000 G

ACCELERATION VALUES PRIOR TO SCALING (DT=.0200 SEC.)

								CARD NO.
.024666	.033454	.029644	.057712	.056018	.046083	.021503	-.006433	1
-0.14804	-.023711	-.034450	-.047430	-.042368	-.021383	-.009957	-.016338	2
-.011724	-.014001	-.012644	-0.24034	-.025163	-.027820	-.008109	-.023016	3
-.008375	.009219	.009809	.013743	.008303	.011212	.013405	.021812	4
.008682	.014134	-.016334	-.003920	-.008478	-.034596	-.034986	-.021345	5
-.017234	-.007731	.004235	.008909	.025632	.039188	.049076	.025019	6
.016487	.025882	.036617	.027130	.002505	-.012823	-.016477	-.023853	7
-0.20339	-.044898	-.052331	-.033881	-.021102	.000444	-.006793	-.009941	8
-.006113	-.003944	-.002391	.004444	.016259	-.003173	.034998	.040554	9
.053448	.042974	.037406	.015052	-.004224	-.031040	-.027064	.003969	10
-.003005	-.004459	-.035501	-.042812	-.056992	-.043319	-.018590	.006093	11
.017767	.011630	.035536	.045457	.047950	.037972	.007507	-.003016	12
.012080	-.009642	-.018144	-.020529	-.031665	-.039970	-.036507	-.017702	13
-.027351	-.018155	.001120	-.007774	.015006	.034004	.052183	.036515	14
.053162	.055285	.064441	.037363	-.001313	-.014732	-.044701	-.051041	15
-.035211	-.062378	-.075229	-.071437	-.049989	-.007210	.023328	.041704	16
.079866	.064498	.055579	.028429	.012166	.018566	.034832	.038080	17
.019443	.011823	-.002465	-.024867	-.030910	-.047033	-.039485	-.018358	18
-.012618	.002066	.013092	.000748	-.009420	-.019380	-.025121	-.018908	19
-.017959	.001018	.011918	-.010157	-.014632	.001636	.003841	.006743	20
.001684	.001678	.001666	.001659	.001645	.001638	.001622	.001614	21
.001597	.001578	.001569	.001559	.001539	.001527	.001506	.001494	22
.001471	.001458	.001434	.001419	.001394	.001379	.001352	.001336	23
.001307	.001290	.001260	.001242	.001211	.001192	.001159	.001139	24
.001105	.001084	.001048	.001026	.000989	.000966	.000927	.000904	25
.000863	.000839	.000797	.000772	.000728	.000703	.000657	.000631	26
.000584	.000557	.000508	.000480	.000429	.000401	.000348	.000319	27
.000265	.000236	.000179	.000150	.000091	.000061	0.	-.000030	28
-.000093	-.000123	-.000188	-.000218	-.000286	-.000316	-.000387	-.000416	29
-.000490	-.000519	-.000596	-.000623	-.000704	-.000730	-.000815	-.000839	30
-.000929	-.000950	-.001045	-.001062	-.001165	-.001176	-.001289	-.001291	31
-.001417	-.001404	-.001553	-.001514	-.001701	-.001605	-.001895	-.001581	32

0.	0.	0.	0.	0.	0.	0.	0.	33
0.	0.	0.	0.	0.	0.	0.	0.	34
0.	0.	0.	0.	0.	0.	0.	0.	35
0.	0.	0.	0.	0.	0.	0.	0.	36
0.	0.	0.	0.	0.	0.	0.	0.	37
0.	0.	0.	0.	0.	0.	0.	0.	38
0.	0.	0.	0.	0.	0.	0.	0.	39
0.	0.	0.	0.	0.	0.	0.	0.	40
0.	0.	0.	0.	0.	0.	0.	0.	41
0.	0.	0.	0.	0.	0.	0.	0.	42
0.	0.	0.	0.	0.	0.	0.	0.	43
0.	0.	0.	0.	0.	0.	0.	0.	44
0.	0.	0.	0.	0.	0.	0.	0.	45
0.	0.	0.	0.	0.	0.	0.	0.	46
0.	0.	0.	0.	0.	0.	0.	0.	47
0.	0.	0.	0.	0.	0.	0.	0.	48
0.	0.	0.	0.	0.	0.	0.	0.	49
0.	0.	0.	0.	0.	0.	0.	0.	50
0.	0.	0.	0.	0.	0.	0.	0.	51
0.	0.	0.	0.	0.	0.	0.	0.	52
0.	0.	0.	0.	0.	0.	0.	0.	53
0.	0.	0.	0.	0.	0.	0.	0.	54
0.	0.	0.	0.	0.	0.	0.	0.	55
0.	0.	0.	0.	0.	0.	0.	0.	56
0.	0.	0.	0.	0.	0.	0.	0.	57
0.	0.	0.	0.	0.	0.	0.	0.	58
0.	0.	0.	0.	0.	0.	0.	0.	59
0.	0.	0.	0.	0.	0.	0.	0.	60
0.	0.	0.	0.	0.	0.	0.	0.	61
0.	0.	0.	0.	0.	0.	0.	0.	62
0.	0.	0.	0.	0.	0.	0.	0.	63
0.	0.	0.	0.	0.	0.	0.	0.	64

ACTUAL MAXIMUM ACCELERATION USED IN LAST ITERATION OF THIS RUN = .286207 (G)

TIME REQUIRED FOR INPUT AND INITIALIZATION = .412 SEC

```
**********************************************
* SOIL PROFILE AND PROPERTIES FOR FREE FIELD *
**********************************************
```

FREQUENCY NUMBER	FREQUENCY (HZ)	SOLUTION TIME (SEC)
0	0.	.012
1	.09766	.023
5	.48828	.034
9	.87891	.045
13	1.26953	.059
17	1.66016	.073
21	2.05078	.087
25	2.44141	.101
29	2.83203	.115
33	3.22266	.129
37	3.61328	.143
41	4.00391	.157
45	4.39453	.170
49	4.78516	.184
57	5.56641	.198
65	6.34766	.207
73	7.12891	.221
81	7.91016	.235
89	8.69141	.249
97	9.47266	.263
105	10.25391	.277
113	11.03516	.291
121	11.81641	.304
129	12.59766	.318
137	13.37891	.329
145	14.16016	.341
153	14.94141	.355
169	16.50391	.369
185	18.06641	.383
201	19.62891	.397

MATERIAL PROPERTIES OBTAINED BY USING SHEAR STRAIN ESTIMATED IN FREQUENCY DOMAIN

LAYER	G-USED (KSF)	G-NEW (KSF)	DIF-G (PERCENT)	DAMP-USED (FRACTION)	DAMP-NEW (FRACTION)	DIF-DAMP (PERCENT)	VS-USED (FPS)	VS-NEW (FPS)	DIFF-VS (PERCENT)	VP-USED (FPS)
1	1493	1445	3.4	.040	.041	3.0	633	623	1.7	1318
2	2100	2053	2.3	.060	.063	4.7	708	700	1.1	1324
3	1670	1625	2.8	.087	.090	3.1	631	623	1.4	1181

MATERIAL PROPERTIES OBTAINED USING EFFECTIVE SHEAR STAIN COMPUTED IN TIME DOMAIN

LAYER	G-USED (KSF)	G-NEW (KSF)	DIF-G (PERCENT)	DAMP-USED (FRACTION)	DAMP-NEW (FRACTION)	DIF-DAMP (PERCENT)	VS-USED (FPS)	VS-NEW (FPS)	DIFF-VS (PERCENT)	VP-USED (FPS)
1	1493	1456	2.6	.040	.041	2.3	633	625	1.3	1318
2	2100	2056	2.1	.060	.063	4.4	708	700	1.1	1324
3	1670	1611	3.7	.087	.091	4.1	631	620	1.8	1181

81	7.91016	.233
89	8.69141	.247
97	9.47266	.260
105	10.25391	.273
113	11.03516	.287
121	11.81641	.301
129	12.59766	.315
137	13.37891	.329
145	14.16016	.342
153	14.94141	.356
169	16.50391	.370
185	18.06641	.384
201	19.62891	.397

MATERIAL PROPERTIES OBTAINED USING EFFECTIVE SHEAR STRAIN ESTIMATED IN FREQUENCY DOMAIN

LAYER	G-USED (KSF)	G-NEW (KSF)	DIF-G (PERCENT)	DAMP-USED (FRACTION)	DAMP-NEW (FRACTION)	DIF-DAMP (PERCENT)	VS-USED (FPS)	VS-NEW (FPS)	DIFF-VS (PERCENT)	VP-USED (FPS)
1	1530	1475	5.0	.040	.041	2.5	641	625	2.5	1334
2	2130	2064	3.2	.058	.062	6.9	713	702	1.6	1333
3	1750	1656	5.7	.082	.088	6.6	646	629	2.8	1209

MATERIAL PROPERTIES OBTAINED USING EFFECTIVE SHEAR STRAIN COMPUTED IN TIME DOMAIN

LAYER	G-USED (KSF)	G-NEW (KSF)	DIF-G (PERCENT)	DAMP-USED (FRACTION)	DAMP-NEW (FRACTION)	DIF-DAMP (PERCENT)	VS-USED (FPS)	VS-NEW (FPS)	DIFF-VS (PERCENT)	VP-USED (FPS)
1	1530	1493	2.5	.040	.040	.2	641	633	1.2	1334
2	2130	2100	1.4	.058	.060	3.4	713	708	.7	1333
3	1750	1670	4.8	.082	.087	5.7	646	631	2.4	1209

MAXIMUM SHEAR COMPUTED FROM ITS TIME HISTORY

COMPARISON BETWEEN MAXIMUM SHEAR STRAINS

LAYER	SIG-X (PSF)	SIG-Y (PSF)	SIG-XY (PSF)	EFF. SHEAR STRN (PERCENT)	TIME DOMAIN MAX. SHEAR STRN (PERCENT)	FREQ DOMAIN MAX. SHEAR STRN (PERCENT)
1	0.	0.	1.1898E+02	5.0549E-03	7.7767E-03	8.4719E-03
2	0.	0.	3.6445E+02	1.1122E-02	1.7110E-02	1.8153E-02
3	0.	0.	6.0509E+02	2.2475E-02	3.4576E-02	3.5387E-02

TIME REQUIRED FOR COMPUTING STRESS AND STRAIN = .526 SEC

```
*************************************************************************
* SOLUTIONS OBTAINED BY GAUSSIAN ELIMINATION FOR FINITE ELEMENT MODEL *
*************************************************************************
```

FREQUENCY NUMBER	FREQUENCY (HZ)	SOLUTION TIME (SEC)
0	0.	.039
1	.09766	.076
5	.48828	.114
9	.87890	.152
13	1.26953	.189
17	1.66015	.226
21	2.05078	.264
25	2.44140	.302
29	2.83202	.339
33	3.22265	.376
37	3.61327	.410
41	4.00390	.447
45	4.39452	.482
49	4.78515	.520
57	5.56639	.553
65	6.34764	.592
73	7.12889	.630
81	7.91014	.667
89	8.69139	.703
97	9.47263	.741
105	10.25388	.779
113	11.03513	.817
121	11.81638	.855
129	12.59763	.895
137	13.37887	.930
145	14.16012	.967
153	14.94137	1.005
169	16.50387	1.043
185	18.06636	1.082
201	19.62886	1.117

MATERIAL PROPERTIES OBTAINED USING EFFECTIVE SHEAR STRAIN ESTIMATED IN FREQUENCY DOMAIN

ELM	G-USED (KSF)	G-NEW (KSF)	DIF-G (PERCENT)	DAMP-USED (FRACTION)	DAMP-NEW (FRACTION)	DIF-DAMP (PERCENT)	VS-USED (FPS)	VS-NEW (FPS)	DIFF-VS (PERCENT)	VP-USED (FPS)
1	1530	1027	49.0	.040	.055	27.0	641	525	22.1	1334
2	2130	2167	1.7	.058	.056	3.8	713	719	-.9	1333
3	1750	1721	1.7	.082	.084	2.1	646	641	.8	1209
4	2000	675	196.2	.070	.074	5.5	733	426	72.1	1525
5	6000	6000	0.	.050	.050	0.	1175	1175	0.	1918
6	2300	1761	30.6	.070	.081	13.9	741	648	14.3	1386
7	6000	6000	0.	.050	.050	0.	1175	1175	0.	1918
8	6000	6000	0.	.050	.050	0.	1175	1175	0.	1918
9	2300	1785	28.9	.070	.080	12.3	741	652	13.5	1386
10	8000	8000	0.	.020	.020	0.	1310	1310	0.	2452
11	8000	8000	0.	.020	.020	0.	1310	1310	0.	2452
12	8000	8000	0.	.020	.020	0.	1310	1310	0.	2452
13	8000	8000	0.	.020	.020	0.	1310	1310	0.	2452
14	8000	8000	0.	.020	.020	0.	1310	1310	0.	2452
15	8000	8000	0.	.020	.020	0.	1310	1310	0.	2452
16	8000	8000	0.	.020	.020	0.	1310	1310	0.	2452
17	8000	8000	0.	.020	.020	0.	1310	1310	0.	2452
18	8000	8000	0.	.020	.020	0.	1310	1310	0.	2452
19	8000	8000	0.	.020	.020	0.	1310	1310	0.	2452

MATERIAL PROPERTIES OBTAINED USING EFFECTIVE SHEAR STRAIN COMPUTED IN TIME DOMAIN

ELM	G-USED (KSF)	G-NEW (KSF)	DIF-G (PERCENT)	DAMP-USED (FRACTION)	DAMP-NEW (FRACTION)	DIF-DAMP (PERCENT)	VS-USED (FPS)	VS-NEW (FPS)	DIFF-VS (PERCENT)	VP-USED (FPS)
1	1530	1000	53.0	.040	.056	28.4	641	518	23.7	1334
2	2130	2185	2.5	.058	.055	5.2	713	722	1.3	1333
3	1750	1733	1.0	.082	.083	1.2	646	643	.5	1209
4	2000	792	152.5	.070	.065	8.3	733	461	58.9	1525
5	6000	6000	0.	.050	.050	0.	1175	1175	0.	1918
6	2300	1781	29.1	.070	.080	12.5	741	652	13.6	1386
7	6000	6000	0.	.050	.050	0.	1175	1175	0.	1918
8	6000	6000	0.	.050	.050	0.	1175	1175	0.	1918
9	2300	1804	27.5	.070	.079	10.9	741	656	12.9	1386
10	8000	8000	0.	.020	.020	0.	1310	1310	0.	2452
11	8000	8000	0.	.020	.020	0.	1310	1310	0.	2452
12	8000	8000	0.	.020	.020	0.	1310	1310	0.	2452
13	8000	8000	0.	.020	.020	0.	1310	1310	0.	2452
14	8000	8000	0.	.020	.020	0.	1310	1310	0.	2452
15	8000	8000	0.	.020	.020	0.	1310	1310	0.	2452
16	8000	8000	0.	.020	.020	0.	1310	1310	0.	2452
17	8000	8000	0.	.020	.020	0.	1310	1310	0.	2452
18	8000	8000	0.	.020	.020	0.	1310	1310	0.	2452
19	8000	8000	?	?Y4Y	?Y4Y	Y.	242X	131X	C?	4674

SOLID ELEMENTS

MAXIMUM SHEAR COMPUTED FROM ITS TIME HISTORY					COMPARISON BETWEEN MAXIMUM SHEAR STRAINS	
					TIME DOMAIN	FREQ DOMAIN
ELEM.	SIG-X (PSF)	SIG-Y (PSF)	SIG-XY (PSF)	EFF. SHEAR STRN (PERCENT)	MAX. SHEAR STRN (PERCENT)	MAX. SHEAR STRN (PERCENT)
1	2.9732E+02	1.1126E+02	4.0317E+02	1.7376E-02	2.6733E-02	2.4799E-02
2	2.3346E=02	6.1902E+01	2.9643E+02	9.5950E-03	1.4762E-02	1.5308E-02
3	4.8045E+01	2.2133E+01	5.4579E+02	2.0277E-02	3.1195E-02	3.1808E-02
4	1.2243E+03	1.6759E+02	8.1245E+02	3.0828E-02	4.7427E-03	7.1088E-02
5	5.2061E+02	1.6515E+02	5.1990E+02	5.8637E-03	9.0211E-03	8.2671E-03
6	7.8058E+01	1.4631E+02	6.6334E+02	1.8750E-02	2.8846E-02	2.9798E-02
7	7.8955E+01	1.6150E+01	6.0225E+01	6.9622E-04	1.0711E-03	1.4274E-03
8	7.7656E+01	3.7634E+01	3.2279E+02	3.5001E-03	5.3848E-03	5.8396E-03
9	6.1234E+01	6.5539E+01	6.3912E+02	1.8062E-02	2.7788E-02	2.8671E-02

APPENDIX XV

Concrete fatigue analysis data

AXV.1. Introduction

Fatigue strength is defined as a fraction of the static strength that can be supported repeatedly for a given number of cycles. It is represented by the stress–fatigue life curves, known as the *S–N* curves. A number of fatigue models are given in Chapter V. This appendix gives additional models for fatigue in concrete with comparative study of input data and results. Figures AXV.1 to AXV.4 show relations of failure cycles to modulus of rupture, maximum stress and strain and strain rates. Figure AXV.5 gives stress level versus fatigue life when concrete is in compression. Figure AXV.6 shows results for the stress levels with and without lateral confining pressures. It is important to notice a comparative study with the finite element fatigue analysis of the model samples. The *S–N* curves are usually plotted for a given constant minimum stress level (S_{min} = constant) or for a maximum stress level (S_{max} = constant). Fatigue tests usually exhibit a large scatter. The S_{max} fatigue strength depends on both these stress levels in the cycle. A Goodman diagram or a Smith diagram represents the mean. The finite element analysis gives a true curve. In this diagram a 20-noded isoparametric element as a solid element for concrete has been used. For reinforcement, conventional and pre-stressing three-dimensional 3-node elements in the isoparametric element families have been adopted and they match with the nodes of the solid elements. In rare cases, the reinforcing bar elements are left in the body of the solid elements or a smear method for reinforcement is adopted for the fatigue analysis.

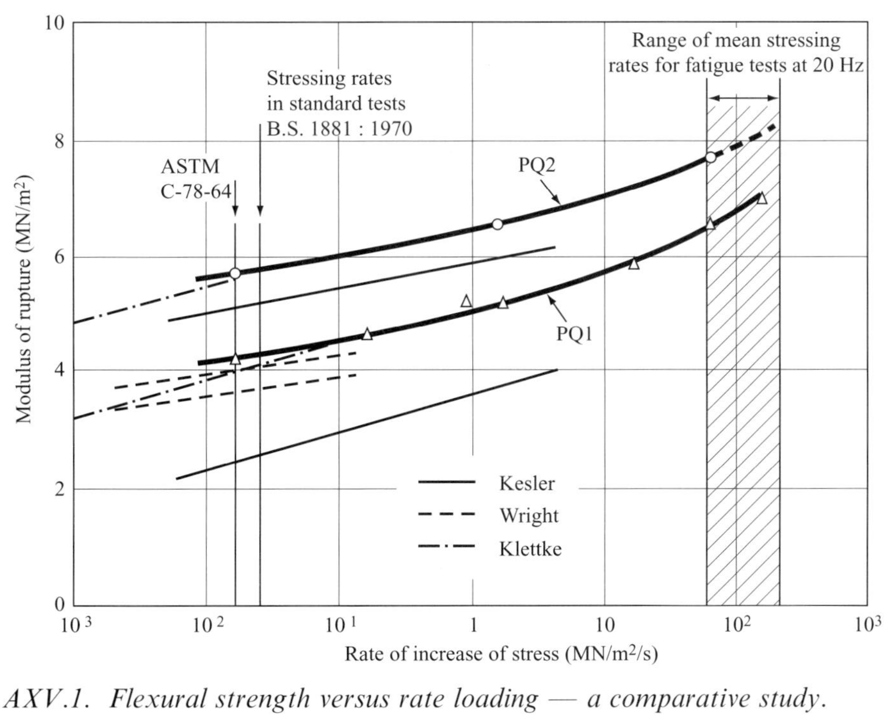

Fig. AXV.1. Flexural strength versus rate loading — a comparative study.

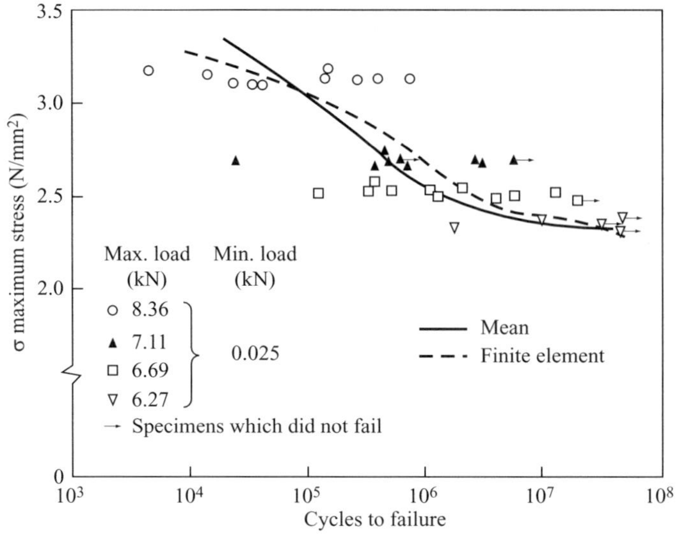

Fig. AXV.2. S–N curve for saturated concrete (Galloway, Harding and Raithby, 1979).

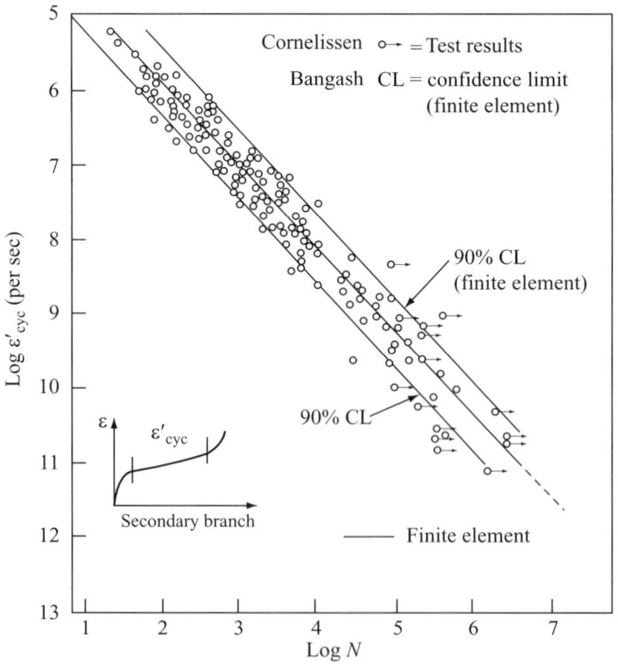

Fig. AXV.3. Correlation between tensile cycle strain rate and fatigue life (Cornelissen & Reinhardt, 1984).

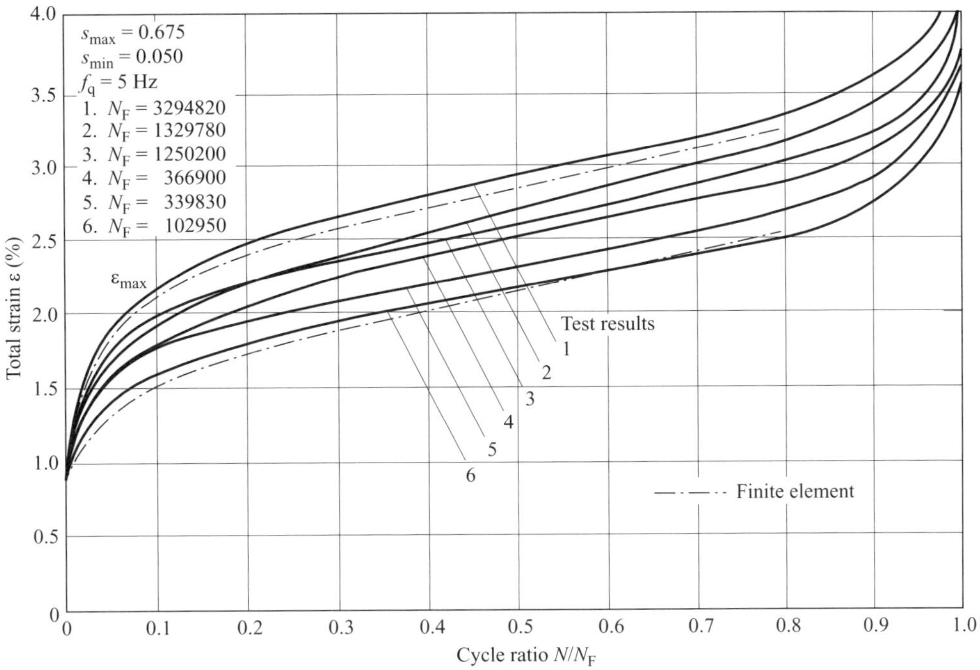

Fig. AXV.4. Measured and calculated variation of total maximum strain with the cycle ratio.

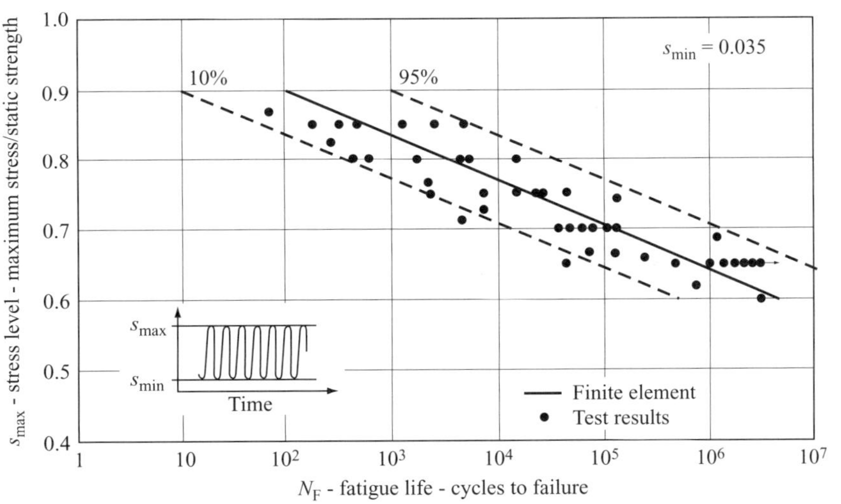

Fig. AXV.5. Stress level versus fatigue life (S–N curves) for concrete in compression.

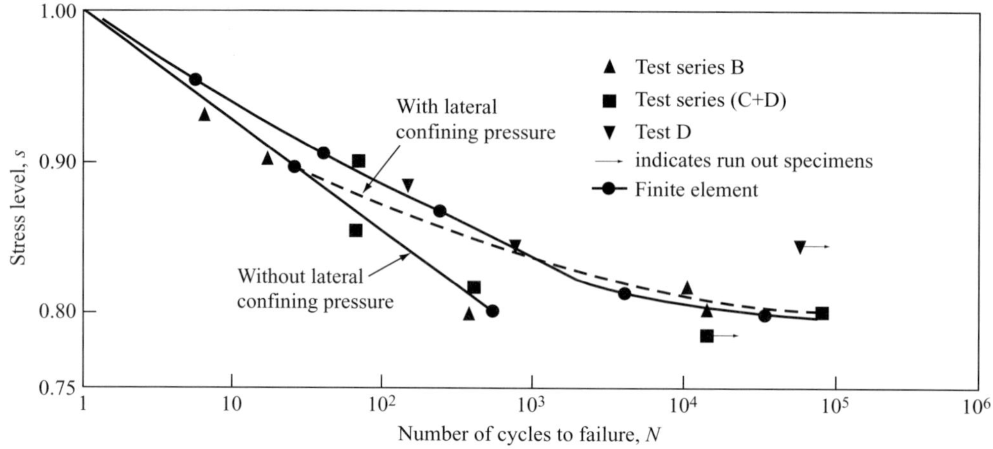

Fig. AXV.6. S–N curves for concrete specimens with and without lateral confining pressure.

AXV.2. Bibliography

Abeles, P. W., Brown, E. I. & Hu, C. H., Fatigue resistance of under-reinforced prestressed concrete beams subjected to different stress ranges; Miner's hypothesis. *Abeles Symposium on Fatigue, ACI Convention, Hollywood, Florida*, 1972.

Abeles, P. W., Brown, I. E. & Slepetz, J. M., Fatigue resistance of partially prestressed concrete beams to large range loading. *AIBSE Congress, New York*, 1968.

Abeles, P. W., Resistance of prestressed concrete to dynamic loading, its fatigue resistance; Miner's hypothesis. *IABSE Symposium, Lisbon*, September 1973.

ACI Committee 215, Considerations for design of concrete structures subjected to fatigue loading. *ACI Journal*, March 1974, 97–121.

Award, M. E. & Hilsdorf, H. K., Strength and deformation characteristics of plain concrete subjected to high repeated and sustained loads. *Abeles Symposium of Fatigue on Concrete*, SP-41, ACI, 1974, 1–13.

Borges, J. F., Report Theme V. *IABSE Symposium, Lisbon*, September 1973.

Freudenthal, A. M. & Heller, R. A., *Journal of Aero/Space Science*, **26**(7), July 1959, 431–442.

Hilms, J. B. & Ekberg, C. E., Statistical analysis of fatigue characteristics of under-reinforced prestressed concrete flexural members. Report, Iowa State University, May 1965.

Hilsdorf, H. K. & Kesler, C. E., Fatigue strength of concrete under varying flexural stresses. *ACI Journal*, **63**(10), October 1966, 1059–1076.

Hu, C. H., Repeated overloads on partially prestressed concrete beams. *PhD Dissertation*, Department of Civil Engineering, Duke University, Durham, N.C. 1970.

Leeuwen, J. & Siemes, A. J. M., Fatigue of concrete, part 2. Memo 78-80-bg-SIM/PEM, IBBC-TNO, Institute TNO for Building Materials and Building Structures, Delft, The Netherlands, 1978.

Ople, F. S. & Hulsbos, C. L., Probable fatigue life of plain concrete with stress gradient. *ACI Journal*, **63**(1), January 1966, 59–82.

Raithby, K. D. & Galloway, J. W., Effect of moisture conditions, age, and rate of loading on fatigue of plain concrete. *Abeles Symposium of Fatigue on Concrete*, SP-41, ACI, 1974, 14–34.

Takhar, S. S., Jordaan, I. J. & Gamble, B. R., Fatigue of concrete under lateral confining pressure. *Abeles Symposium of Fatigue on Concrete*, SP-41, ACI, 1974, 59–69.

Tepfers, R, Gørlin, J. and Samuelsson, T., Concrete subjected to pulsating load and pulsating deformation of different pulse waveforms. *Nordisk Betong*, No. 4, 1973, 27–36.

Tide, R. H. R. & Van Horn, D. A., A statistical study of the static and fatigue properties of high strength prestressing strand. *Fritz Engineering Laboratory Report* 309.2, Lehigh University, June 1966.

Venuti, W. J., A statistical approach to the analysis of fatigue failure of prestressed concrete beams. *ACI Journal*, **62**(11), November 1965, 1375–1392.

Warner, R. L. & Hulsbos, C. L., Fatigue properties of prestressing strand. *PCI Journal*, February 1966, 32–52.

Westerberg, B., Utmatting av Betongkonstruksjoner. *Föredrag vid Dansk Selskap for Bygningsstatik*, January 1975.

APPENDIX XVI

Ductility of beams

AXVI.1. Relation between displacement ductility and curvation ductility

The displacement ductility is defined as

$$\mu_d = \frac{\delta_m}{\delta_e} \tag{AXVI.1}$$

The curvation ductility is defined as

$$\mu_c = \frac{\kappa_m}{\kappa_e} \tag{AXVI.2}$$

where

δ_m = the maximum displacement

δ_e = the maximum elastic displacement

κ_m = the maximum curvation

κ_e = the maximum elastic curvation

The maximum elastic displacement δ_e can be defined as

$$\begin{aligned}\delta_e &= \frac{5}{384} \times \frac{q \times l^4}{E \times I}\\ &= \frac{5}{384} \times \frac{8 \times M_e \times l^2}{E \times I}\\ &= \frac{5}{48} \times \kappa_e \times l^2 \end{aligned} \tag{AXVI.3}$$

The maximum displacement δ_m is defined as

$$\begin{aligned}\delta_m &= \theta \times \tfrac{1}{2}\\ &= \kappa_m \times l_s \times \tfrac{1}{2}\end{aligned} \tag{AXVI.4}$$

where

l_s = the length of the plastic hinge.

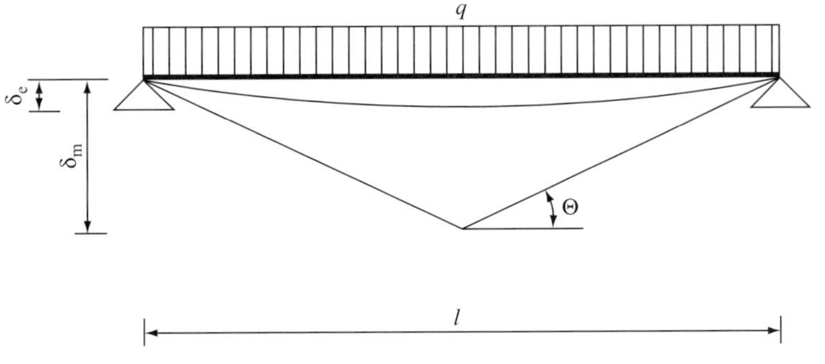

Fig. AXVI.1. Simply supported beam.

The total length of the plastic hinge can be stated to $2 \times d$ so the effective length of the plastic hinge for the rotation θ is d.

$$\delta_{\mathrm{m}} = \frac{\kappa_{\mathrm{m}} \times d \times l}{2} \tag{AXVI.5}$$

So the relation between the displacement ductility and the curvation ductility can be found

$$\mu_{\mathrm{d}} = \frac{48 \times \kappa_{\mathrm{m}} \times d \times l}{10 \times \kappa_{\mathrm{e}} \times l^2}$$

$$= 4.8 \times \frac{d}{l} \times \mu_{\mathrm{c}} \tag{AXVI.6}$$

AXVI.2. Curvature ductility

General

The ductility of a concrete cross-section will be defined as the ratio of the ultimate curvation and the yield curvation. To determine the ductility it is necessary to calculate the curvations of a reinforced cross section.

In the moment–curvation diagram three points are of importance:

1. The point where it is assumed that the concrete will crack in the tension zone.
2. The point where the ultimate moment is reached, $\varepsilon_{\mathrm{cu}} = 0.003$.
3. The ultimate limit is reached.

Tension–yield moment

As a result of equilibrium of internal forces the concrete strain in the extreme compressive fiber, $\varepsilon_{\mathrm{cm}}$, and the depth of the compressive zone, c, can be calculated when the tension reinforcement reaches its yield strain.

The calculations for the yield moment and the ultimate moment is based on the assumption of the stress–strain figure as is shown in Figure AXVI.2. Also, the concrete stress distributions over the depth of the compressive zone are given.

The stress–strain relation is defined by the following relations where $f_{\mathrm{cl}} = f_{\mathrm{c}}'$:

$$\varepsilon_{\mathrm{c}} \leq 0.002 \qquad f_{\mathrm{c}} = f_{\mathrm{cl}} \times \left(\frac{2 \times \varepsilon_{\mathrm{c}}}{0.002} - \left(\frac{\varepsilon_{\mathrm{c}}}{0.002} \right)^2 \right) \tag{AXVI.7}$$

$$0.002 < \varepsilon_{\mathrm{c}} < \varepsilon_{20\mathrm{c}} \qquad f_{\mathrm{c}} = f_{\mathrm{cl}} \times (1 - z \times (\varepsilon_{\mathrm{c}} - 0.002)) \tag{AXVI.8}$$

where

$$z = \frac{0.5}{\dfrac{3 + 0.002 \times f_{\mathrm{cl}}}{f_{\mathrm{cl}} - 1000} - 0.002}$$

The compressive concrete force then is defined as

$$N'c = \alpha \times f_{\mathrm{cl}} \times b \times c$$

$$\varepsilon_{\mathrm{cm}} \leq 0.002 \qquad \alpha = \frac{\varepsilon_{\mathrm{cm}}}{0.002} \times \left(1 - \frac{\varepsilon_{\mathrm{cm}}}{0.006} \right) \tag{AXVI.9}$$

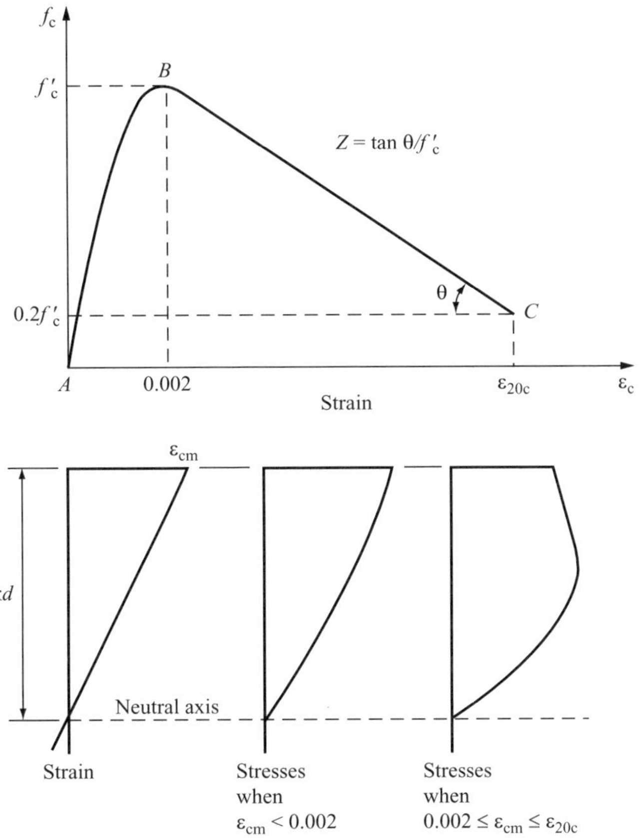

Fig. AXVI.2. Assumed stress–strain relation for concrete.

$0.002 < \varepsilon_c \le \varepsilon_{20c}$

$$\alpha = \frac{0.004}{3 \times \varepsilon_{cm}} + \left(1 - \frac{0.002}{\varepsilon_{cm}}\right) \times \left(1 - \frac{z}{2} \times (\varepsilon_{cm} - 0.002)\right) \qquad \text{(AXVI.10)}$$

The depth of the compression zone and the concrete strain can be calculated from the requirement of equilibrium of internal forces:

$$0 = N + N_{sb} + N_{sp} - N_{st} - N'_c \qquad \text{(AXVI.11)}$$

where

$$N_{sb} = A_{sb} \times f_y$$

$$N_{sp} = t \times b \times f_p$$

$$N_{st} = \frac{c - a}{c} \times \varepsilon_c \times E_s \times A_{st} < A_{st} \times f_y$$

$$N'_c = \alpha \times c \times b \times f'_{cu}$$

By knowing ε_{cm} and c, the yielding curvation can be calculated with the following formula

$$\kappa_{yield} = \frac{\varepsilon_c}{c} \qquad \text{(AXVI.12)}$$

Ultimate moment

The ultimate moment will appear when the ultimate concrete strain, ε_{cu}, is reached. The ultimate concrete strain is 0.3%.

The depth of the compression zone can be calculated from the requirement of equilibrium of internal forces

$$0 = N + N_{sb} + N_{sb} - N_{st} - N'_c \qquad \text{(AXVI.13)}$$

where

$$N_{sb} = A_{sb} \times f_y$$

$$N_{sp} = t \times b \times f_p$$

$$N_{st} = \frac{c-a}{c} \times \varepsilon_{cu} \times E_s \times A_{st} < A_{st} \times f_y$$

$$N'_c = \alpha \times c \times b \times f'_{cu}$$

The curvation in the ultimate limit state can be calculated as follows

$$\kappa_{uls} = \frac{\varepsilon_{cu}}{c}$$

The curvation ductility for the ultimate limit state is defined as

$$\text{ductility} = \frac{\kappa_{uls}}{\kappa_{yield}}$$

Sample calculations for curvation ductility are given in Table AXVI.1.

AXVI.3. Bibliography

ACI Manual of Concrete Practice, 1984.

Blume, J. A., Newmark, N. M. & Corning, L. H., *Design of multistory reinforced concrete buildings for earthquake motions*. PCA, Illinois, 1961.

CEB-FIP model code 1990, Paris, Sept. 1990.

Crawford, E. A., *Protection from non-nuclear weapons*. AFWL-TR-70-127, 1971.

Dilger, W. H., Koch, R. & Kowalczyk, R., Ductility of plain and confined concrete under different strain rates. *ACI Journal*, Jan.–Feb. 1984.

Fafititis, A. & Shah, S. P., Prediction of ultimate behavior of confined columns subjected to large deformations. *ACI Journal*, Jul.–Aug. 1985.

Park, R. & Dai Ruitong, Ductility of doubly reinforced concrete beam sections. *ACI Structural Journal*, March–April 1988.

Table AXVI.1. Curvation ductility

Moment–curvation relationship of a reinforced concrete cross-section

Cross-section input	*Material input*
$h = 950\,\text{mm}$	Concrete:
$b = 1000\,\text{mm}$	$f_{\text{cl}} = 24\,\text{MPa}$
$a = 70\,\text{mm}$	$f_{\text{ifc}} = 1.2\,\text{MPa}$
$A_{\text{st}} = 3272\,\text{mm}^2$	
$A_{\text{sb}} = 5362\,\text{mm}^2$	Reinforcement:
$t = 4\,\text{mm}$	$f_{\text{y}} = 460\,\text{MPa}$
$N = 0\,\text{N}$	$d_{\text{ifs}} = 1.2$
	$E_{\text{s}} = 200\,000\,\text{MPa}$
	Scabbing plate:
	$f_{\text{p}} = 275\,\text{MPa}$
	$d_{\text{ifp}} = 1.1$

Calculated properties

$A_{\text{c}} = b \times h$	$f_{\text{c}} = f_{\text{cl}} \times d_{\text{ifc}}$
$d = h - a$	$f_{\text{y}} = f_{\text{y}} \times d_{\text{ifs}}$
$A_{\text{sp}} = t \times b$	$E_{\text{sy}} = \dfrac{f_{\text{y}}}{E_{\text{s}}}$
	$f_{\text{p}} = f_{\text{p}} \times d_{\text{ifp}}$
$A_{\text{c}} = 9.5 \times 10^5\,\text{mm}^2$	$f_{\text{c}} = 28.8\,\text{MPa}$
$d = 880\text{mm}$	$f_{\text{y}} = 552\,\text{MPa}$
$A_{\text{sp}} = 4 \times 10^2\,\text{mm}^2$	$E_{\text{sy}} = 0.00276$
	$f_{\text{p}} = 302.5\,\text{MPa}$

$$z = \frac{0.5}{\left[\dfrac{3 + 0.002 \times f_{\text{c}} \times 145}{f_{\text{c}} \times 145 - 1000}\right] - 0.002}$$

$$z = 318$$

$$E_{20\text{c}} = \frac{0.8}{z} + 0.002$$

$$E_{20\text{c}} = 0.00452$$

$$\alpha_{20\text{c}} = \frac{0.004}{2E_{20\text{c}}} + \left[1 + \frac{0.002}{E_{20\text{c}}}\right]\left[1 - \frac{z}{2}(E_{20\text{c}} - 0.002)\right]$$

$$\alpha_{20\text{c}} = 0.63$$

Table AXVI.1. continued

Yielding moment

Equilibrium can be found for the following condition
$E_{c_1} = 0.00118$ and $c_1 = 264$ mm so:

$N_{sb} = A_{sb} \times f_y$ $\quad N_{sb} = 2.96 \times 10^6$ N

$N_{sp} = A_{sp} \times f_p$ $\quad N_{sp} = 1.21 \times 10^6$ N

$$N_{st_1} = \left[\frac{E_{c_1} + E_{sy}}{d}(d-a) - E_{sy}\right] E_s \times A_{st}$$

$N_{st_1} = \text{if}\,[N_{st_1} < A_{st} \times f_y, N_{st_1}, A_{st} \times f_y]$ $\quad N_{st_1} = 5.667 \times 10^5$ N

$$\alpha_1 = \frac{E_{c_1}}{0.002}\left[1 - \frac{E_{c_1}}{0.006}\right]$$

$N'_{c_1} = \alpha_1 \times f_c \times c_1 \times b$ $\quad N'_{c_1} = 3.602 \times 10^6$ N

$N'_{c_1} + N_{st_1} - N_{sb} - N_{sp} - N = -1.863 \times 10^{-9}$

Yielding moment and curvation

$$M_1 = N_{sp}\left[a - \frac{t}{2}\right] + N'_{c_1}\left[d - \frac{c_1}{3}\right] + N_{st_1}(d-a) - N\left[d - \frac{h}{2}\right]$$

$$\kappa_{yield} = \frac{E_{c_1}}{c_1}$$

$M_1 = 3.395 \times 10^9$ Nmm

$\kappa_{yield} = 4.478 \times 10^{-6}\ \text{mm}^{-1}$

$\kappa_1 = \kappa_{yield}$

$$\text{ductility}_1 = \frac{\kappa_1}{\kappa_{yield}}$$

Ultimate moment

According to ductility of doubly reinforced concrete beam sections, the maximum strain in the concrete is 0.003. As a result of equilibrium, the depth of the compressive zone can be calculated.

$c_2 = 150$ mm $\quad E_{c_2} := E_{cu}$ $\quad E_{c_2} = 0.003$

$$\alpha_2 = \frac{0.004}{3E_{cu}} + \left[1 - \frac{0.002}{E_{cu}}\right] \times \left[1 - \frac{z}{2}(E_{cu} - 0.002)\right]$$

$\alpha_2 = 0.724$

$N'_{c_2} = \alpha_2 \times f_c \times c_2 \times b$ $\quad N'_{c_2} = 3.125 \times 10^6$ N

Table AXVI.1. continued

$$N_{st_2} = \frac{c_2 - a}{c_2} E_{cu} \times E_s \times A_{st}$$

$$N_{st_2} = \text{if}\,[N_{st_2} < A_{st} \times f_y, N_{st_2}, A_{st} \times f_y] \qquad N_{st_2} = 1.045 \times 10^6\,\text{N}$$

$$N'_{c_2} + N_{st_2} - N_{sb} - N_{sp} - N = 0$$

Ultimate moment and curvation

$$M_2 = N_{sp}\left[a - \frac{t}{2}\right] + N'_{c_2}[d - 0.45c_2] + N_{st_2}(d - a) - N\left[d - \frac{h}{2}\right]$$

$$\kappa_2 = \frac{E_{cu}}{c_2}$$

$$M_2 = 3.468 \times 10^9\,\text{Nmm}$$

$$\kappa_2 = 2.004 \times 10^{-5}\,\text{mm}^{-1}$$

$$\text{ductility}_2 := \frac{\kappa_2}{\kappa_{yield}} \qquad \text{ductility}_2 = 4.5$$

Ultimate limit

The strain in the tension reinforcement:

$$E_{sb_3} = 0.08$$

This strain will be reached for the following condition

$$c_3 = 161\,\text{mm}$$

$$E_{c_3} = 0.01792 < E_{c_{max}} = 0.018$$

The requirement for equilibrium of internal forces

$$N_{st_3} = E_{c_3}\frac{c_3 - a}{c_3} E_s \times A_{st}$$

$$N_{st_3} = \text{if}\,[N_{st_3} < f_y \times A_{st}, N_{st_3}, f_y \times A_{st}] \qquad N_{st_3} = 1.806 \times 10^6\,\text{N}$$

$$\alpha_3 = 1 - \frac{0.001c_3}{E_{c_3}[c_3 - a]}$$

$$N'_{c_3} = \alpha_3[c_3 - a]b \times f_c \qquad N'_{c_3} = 2.364 \times 10^6\,\text{N}$$

$$N'_{c_3} + N_{st_3} - N - N_{sb} - N_{sp} = 4.66 \times 10^{-10}$$

The ultimate limit of curvation and moment

$$M_3 = N_{sp}\left[a - \frac{t}{2}\right] + N'_{c_3}[d - c_3 + 0.55[c_3 - a]] + N_{st_3}(d - a) - N\left[d - \frac{h}{2}\right]$$

Table AXVI.1. continued

$$\kappa_3 = \frac{E_{c_3}}{c_3}$$

$$M_3 = 3.363 \times 10^9 \text{ Nmm}$$

$$\kappa_3 = 1.113 \times 10^{-4} \text{ mm}$$

The ductility is

$$\text{ductility}_3 = \frac{\kappa_3}{\kappa_{\text{yield}}} \qquad \text{ductility}_3 = 24.8$$

APPENDIX XVII

Blast effects on concrete structures

AXVII.1. Introduction

Bangash has dealt in detail with the blast effects on building structures, and air blast and ground shock have been thoroughly assessed [A1].

Here a summary is given on the numerical aspect of concrete elements.

Blast wave scaling laws

The Hopkinson–Cranz scaling law is commonly known as 'cube-root' scaling. Two bombs of weight W_1 and W_2 having diameters d_1 and d_2, respectively, are made up of the same explosive materials. The charge weights W_1 and W_2 are related as

$$\frac{d_1}{d_2} = \left(\frac{W_1}{W_2}\right)^{1/3}$$

since W_1 and W_2 are proportional to d_1^3 and d_2^3, respectively.

d_1 and d_2 are related to W_1 and W_2. Table AXVII.1 gives a list of some explosive loads.

Table AXVII.1. Terrorist attacks with explosives, scaled distances

Site/country	Explosion load
Oklahoma City/USA	4800 lbf (33.1 ton of TNT)
The World Trade Centre, NY/USA	1200 lbf (0.6 ton of TNT)
The Jewish Community Centre, Buenos Aires/Argentina	275 kg of TNT
St. Mary Axe, London/UK	350 kg of TNT
Apartments, Israeli Embassy, London/UK	350 kg of TNT
Staples Corner, London/UK	300 kg of TNT

AXVII.2. Methods of analysis and design

Methods of analysis have been fully explained in seismic and fire engineering sections of this text. However, there is an additional proviso that whatever methods are used they require adequate resolution in space and time to account for high intensity and short duration loading. The following methods are recommended:

(a) The modelling of components such as beams, slabs, columns and walls can be carried out by a single or two-degree-of-freedom system. The response of such components can be found by using standard charts and handbooks. Limit state analysis and a plastic method of analysis can be useful tools.

(b) For complex structural components, or for the entire building, non-linear finite element analysis using explicit or implicit or mixed hybrid modelling on micro- or main-frame computers should be carried out.

AXVII.3. Structural response analysis to blast loading

The structural response depends on a number of factors which include the layout, structural detailing, material properties, damping characteristics, explosive charge source and its range and degrees of freedom of the structure. For rapid blast analyses, graphs have been produced to calculate various parameters needed for concrete structures by the U.S. Corps of Engineers.

AXVII.4. Practical design examples based on British practice

(a) A reinforced concrete wall is loaded by a blast from a vehicle bomb of 100 kg actual mass. The wall is rigidly connected at the foundation and free at the top. Using the following data, calculate the required reinforcement for the wall:

Wall height = 4 m

Vertical reinforcement $\rho_{SU} = 0.5\%$ at each face

$R = R' =$ range $= 4.0$ m

Hemispherical charge factor = 1.8

f_y = Static yield stress of the reinforcement = 460 N/mm^2

f_{dy} = Dynamic yield stress of the reinforcement = $1.2f_y$

Concrete grade = 40

Type III category rotation of the base allowed during explosion: $\theta = 12°$. Justify the impulsive load analysis.

(b) If the same wall is fixed at the top as well as at the bottom and is subjected to a quasi-static load of a triangular shape shown in Fig. AXVII.1 and adopting BS 8110 and other relevant criteria, check the reinforcement and the wall thickness while behaving as a single-degree elastoplastic system.

(c) If the wall is subject to vehicle bombs of 50, 100, 150, 200, 250 and 300 kg of TNT at a random range R' of 1 m to 10 m from the explosion to the centre of the wall, draw curves for

$$\frac{x}{L}\left(\frac{\text{wall deflection}}{\text{wall span}}\right) \text{ versus } R'$$

for various vehicle bombs given. The wall is assumed damaged when

$$\frac{x}{L} = \frac{1}{60}$$

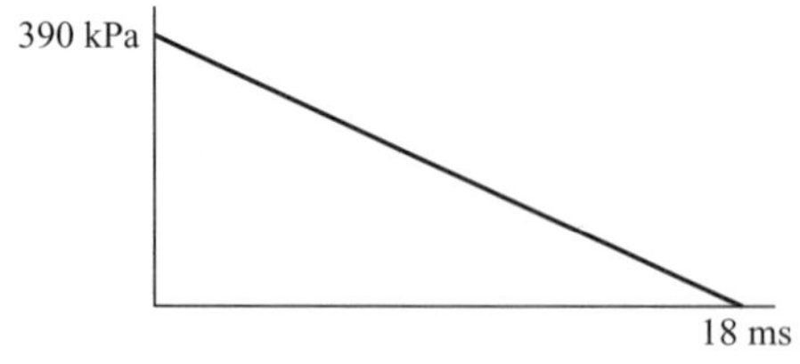

Fig. AXVII.1. Triangular pressure-pulse function.

Calculations

The vehicle bomb mass produces a hemispherical charge of mass $1.8 \times 100\,\text{kg} = 180\,\text{kg}$ where 100 kg is the actual mass.

$$R' = \text{range} = 4.0\,\text{m}$$

$$Z = \text{the scaled distance} = \frac{4.0}{180^{1/3}} = 0.705\,\text{m/kg}^{1/3}$$

$$i_r = I_r = \text{reflected overpressure}$$

impulse = 5095 kPa-ms against 180 kg hemispherical charge.

For category type III behaviour, the material properties are:

$$f_y = \text{reinforcement yield stress} = 460\,\text{N/mm}^2 \quad \text{(static value)}$$

$$f_{dy} = 1.2f_y = 1.2 \times 460 = 522\,\text{N/mm}^2 \approx f_{ds} \quad \text{(dynamic value)}$$

$$\frac{I_r^2}{2K_{LM}\rho d_c} = \left(\frac{2}{H}\rho_v f_{ds} d_c^2\right)\tan\theta$$

$$\theta = \text{rotation of the base of the wall} = 12^\circ \quad \text{(Type III category)}$$

$$\tan\theta = 0.2126$$

$$\rho_v = 0.5\%, \quad \text{i.e. reinforcement on both sides is equal}$$

$$K_{LM} = \text{load–mass factor} = 0.66$$

$$\frac{(5095)^2}{2 \times 0.67 \times 2400 d_c} = \frac{2}{4.0} \times \frac{0.5}{100} \times 552 \times 10^6 d_c^2 \times 0.2126$$

$$d_c = 0.302\,\text{m} = 302\,\text{mm}$$

$$T_c = d = \text{overall thickness}$$

$$= d_c + 2 \times \text{cover of } 40 + 2 \text{ assumed-size bar halves}$$

$$= 302 + 80 + 25$$

$$= 407\,\text{mm}$$

Adopt 425 mm or 17 in RC wall (T_c) (see Fig. AXVII.2);

$$A_S = 0.005 \times 302 \times 1000 = 1510\,\text{mm}^2/\text{m} = \text{width of wall}$$

The reinforcement should be generously placed to account for the cover variations of the material and blast load and is a principal design assumption. A total of 25% steel on top is recommended.

Adopt T25-250 where $A_{S_{prov}} = 1963\,\text{mm}^2/\text{m}$ = width of wall

When checking for shear:

$$v = \frac{r_u(4.0 - 0.302)}{0.302} = 12.245 r_u$$

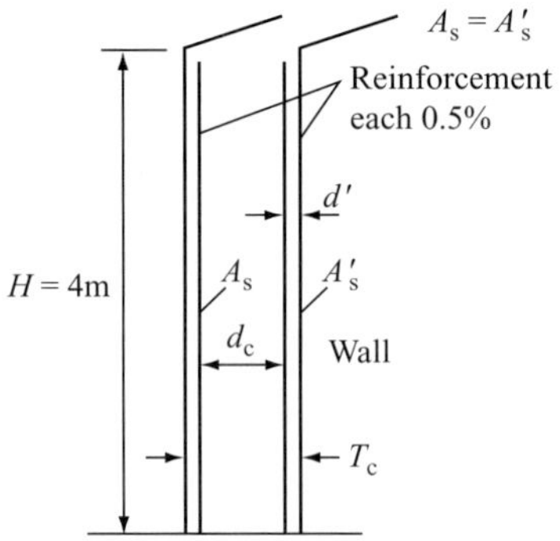

Fig. AXVII.2. Reinforced concrete wall.

$$r_u = \frac{2}{(4)^2} \times 0.005 \times 552 \times 10^6 (0.302)^2$$

$$= 31\,465.38$$

$$v = 0.3853\ \text{N/mm}^2$$

$$\frac{100A_S}{bd} = 0.65 \qquad f_{cu} = 40\ \text{N/mm}^2 \qquad v_c = 0.59 \quad \text{(BS 8110, Table 3.9)}$$

Using Table 13 of BS 8110

$$v < 0.5v_c \not< \frac{0.59}{2} = 0.295\ \text{N/mm}^2$$

A minimum shear link is needed when

$$\frac{0.59}{2} < v < v_c + 0.4$$

$$0.295 < 0.3853 < 0.7853\ \text{N/mm}^2$$

$$A_{Sv} = \frac{0.4 b_v s_v}{0.87 f_y}$$

$$= 45\ \text{mm}^2$$

Take

$$f_{ds} = 1.1 \times 250$$

$$= 275\ \text{N/mm}^2$$

$$b_v = 150$$

$$s_v = 300$$

$$\text{Then, the ultimate shear} = 31\,465.38 \times 4 = 125\,861.52\,\text{N/m}$$

$$= 126\,\text{N/mm}^2$$

$$f_{ds} = 1.1 \times 460 = 506\,\text{N/mm}^2$$

$$b = 150\,\text{mm} \quad \text{to match up with stirrups}$$

$$A_d = \text{area of the diagonal bars @}45^\circ$$

$$= \frac{126 \times 150}{506} \approx 37.35\,\text{mm}^2$$

Therefore, 1 row of R8-150 bars needed.

For the delay time $= t_d = T_s = 50\,\text{ms}$:

$$\frac{t_m}{t_d} = \frac{I}{r_u t_d} = \frac{5095 \times 1000}{31\,465.38 \times 50} = 3.24 > 3.0 \quad \text{QED}$$

Hence the impulsive load analysis is valid.

For the instance where a quasi-static load is applied to the wall restricted top and bottom, it follows:

$$r_u = \frac{8(M_{Ps} + M_{Pm})}{H^2}$$

$$M_P = M_{Ps} + M_{Pm}$$

$$A_S = A'_S = 1963\,\text{mm}^2/\text{m}$$

$$x = \text{depth of neutral axis}$$

$$= \frac{d - z}{0.45} = \frac{d - 0.95d}{0.45}$$

$$0.05d = 0.45x$$

$$x = \frac{0.05 \times 302}{0.45} = 33.6\,\text{mm}$$

Ignoring the contribution of the compression reinforcement ($K = K' = 0.156$)

$$M_P = 1963 \times 555(302 - 0.45 \times 33.6) \times 10^{-3}$$

$$= 310\,856.28\,\text{Nm/m}$$

Hence

$$r_u = \frac{(M_{Ps} + M_{Pm}) \times 8}{(4)^2} = \frac{2 \times 310\,856.28 \times 8}{4^2}$$

$$= 310\,856\,\text{N/m}^2$$

$$\rho_s = 0.005 \qquad n = \frac{200\,\text{GN}}{30\,\text{GM}} = 6.67$$

$$I = Fbd^3$$

$$F = 0.0246 \qquad T = 2\pi\sqrt{\frac{K_{LM}M_d}{K_E}}$$

$$M_d = 2400 \times 0.425 = 1020\,\mathrm{kg/m^2}$$

$$K_{LM} = 0.66$$

K_E (for both ends fixed) = equivalent stiffness

$$= \frac{307EI}{H^4} = 1.51 \times 10^6\,(\mathrm{N/m^2})/\mathrm{m}$$

$$T = 0.025 = 20\,\mathrm{ms}$$

For a single-degree-of-freedom system (elastoplastic system)

$$\frac{r_u}{P_r} = \frac{310\,856}{390 \times 10^3} = 0.797 \approx 0.8$$

$$t_d = 18\,\mathrm{ms} \qquad \frac{t_d}{T} = \frac{18}{20} = 0.9$$

$$\frac{X_m}{X_E} = 3.0 \qquad X_E = \frac{r_u}{K_E} = 0.206\,\mathrm{m \ or \ 206\,mm}$$

$$X_m = 3.0 \times 206 = 618\,\mathrm{mm} < 4000 \times \tan 12^\circ = 850\,\mathrm{mm} \quad \mathrm{QED}$$

Similar analyses have been carried out for various ranges or stand-off distances and explosive weights. The final results are plotted in Fig. AXVII.3.

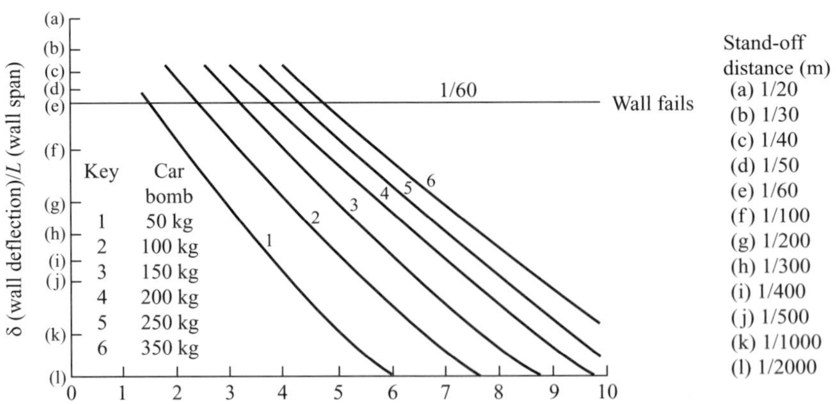

Fig. AXVII.3. δ/L versus stand-off damage assessment.

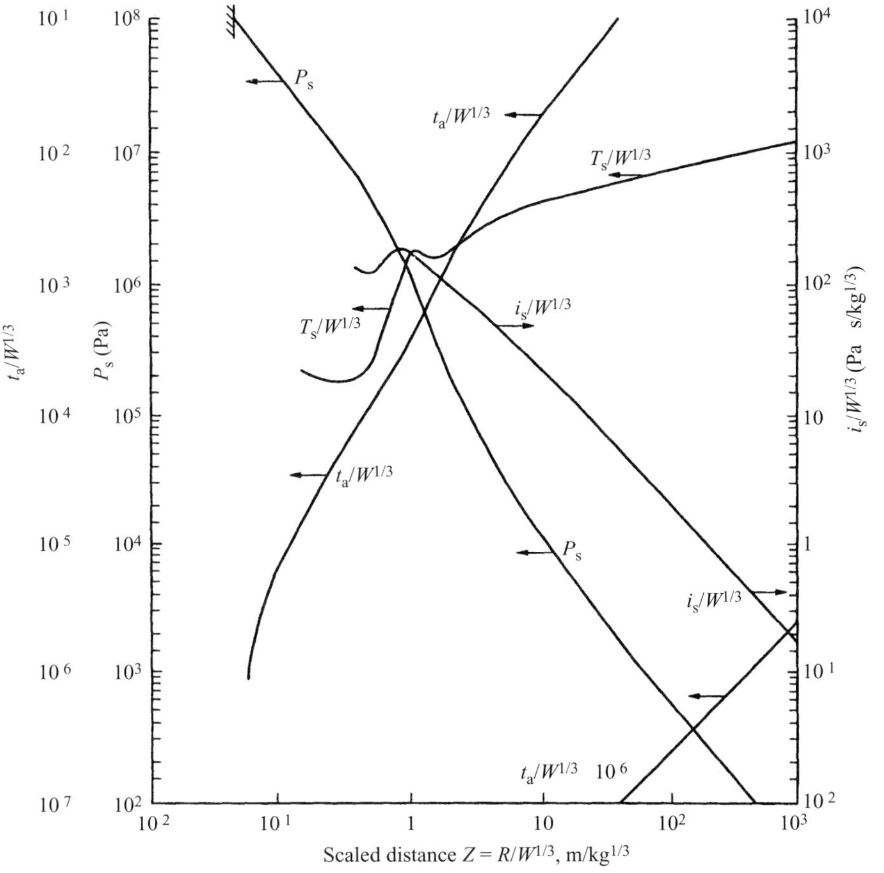

Fig. AXVII.4. Side-on blast parameters for TNT.

Calculate scaled distance Z for the following stand-off distances R, using the charts (Figs AXVII.4 to AXVII.6) prepared by Baker *et al.* for 27 kg, 100 kg and 500 kg spherical TNT explosive charges:

Charge (kg)	$R = 10$ m	$R = 6$ m	$R = 4$ m	$R = 2$ m
27	3.33	1.998	1.332	0.666
100	2.155	1.293	0.862	0.431
500	1.26	0.756	0.504	0.252

$R = 10$ m

$R = 6$ m

$R = 4$ m

$R = 2$ m

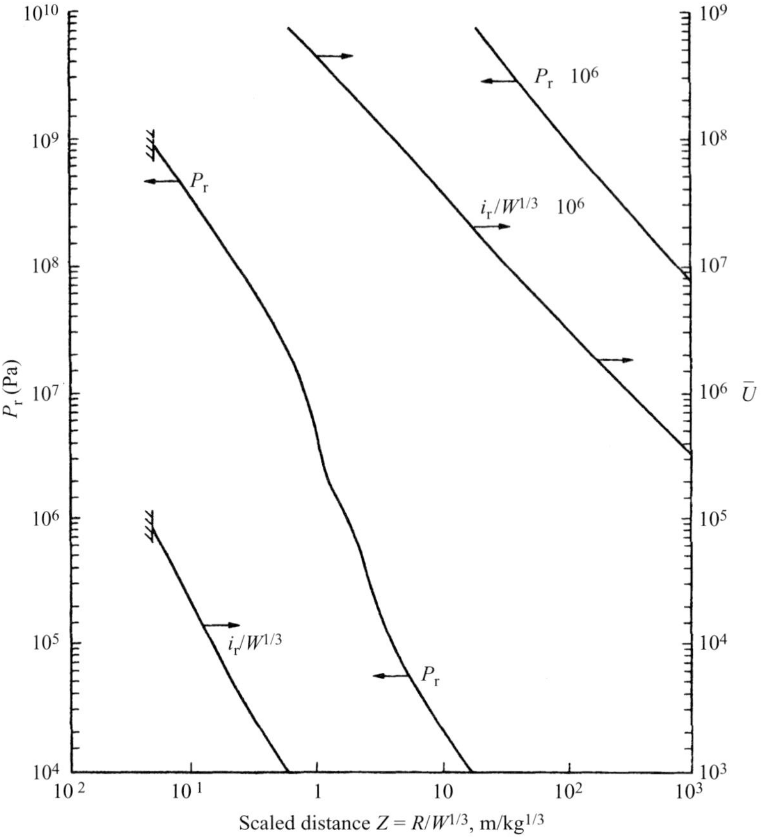

Fig. AXVII.5. Normally reflected blast parameters for TNT.

$$R = 10\,\text{m} \qquad W = 100\,\text{kg}$$

$$Z = \frac{10}{100^{1/3}} = 2.155\frac{\text{m}}{\text{kg}_{\text{TNT}}^{1/3}}$$

$$R = 10\,\text{m} \qquad W = 500\,\text{kg}$$

$$Z = \frac{10}{500^{1/3}} = 1.26\frac{\text{m}}{\text{kg}_{\text{TNT}}^{1/3}}$$

Compute the reflected pressure P_r and the corresponding impulse i_r and the Hopkinson–Cranz blast parameters for the stand-off distance of 6 m using these charts and explosive charge of 27 kg of TNT. Determine P_r and i_r values and time T_s for an explosive force of 100 kg of TNT for the stand-off $R = 10$ m

$$Z = \frac{R}{W^{1/3}}\left(\frac{\text{m}}{\text{kg}^{1/3}}\right)$$

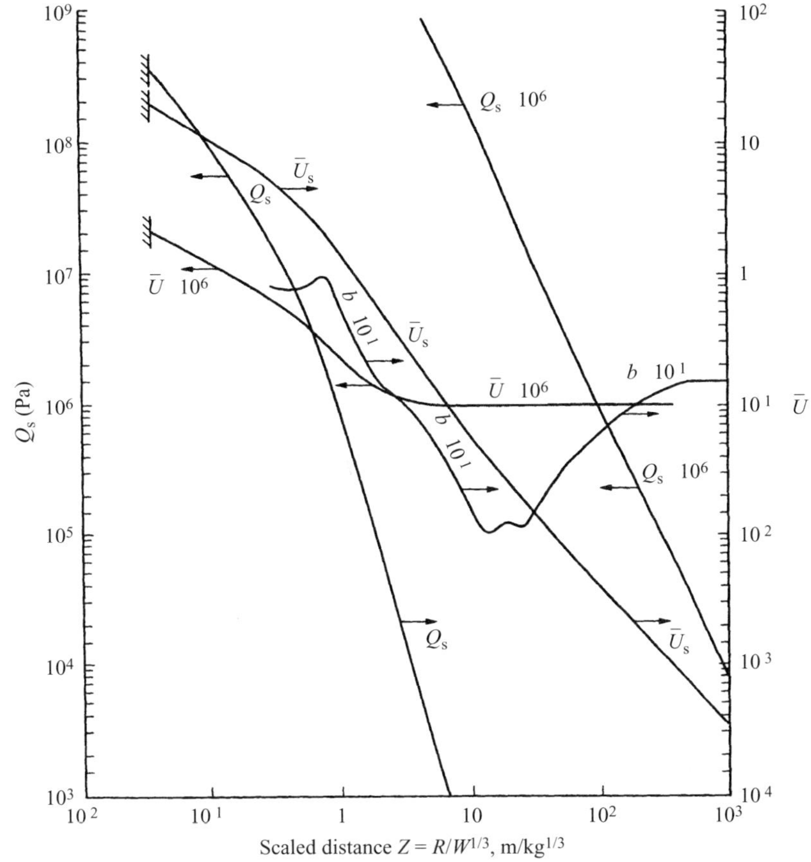

Fig. AXVII.6. Additional side-on blast parameters for TNT.

For

$$R = 10\,\text{m} \qquad W = 27\,\text{kg}$$

$$Z = \frac{10}{27^{1/3}} = 3.33\frac{\text{m}}{\text{kg}_{\text{TNT}}^{1/3}}$$

Consider the calculations below.

$$W = 27\,\text{kg of TNT}$$

$$Z = 1.998\,\text{m}/\text{kg}_{\text{TNT}}^{1/3}$$

$$P_\text{s} = \text{the incidence overpressure} = 1.649 \times 10^5\,\text{Pa}$$

$$\frac{t_\text{a}}{W_{\text{TNT}}^{1/3}} = 1.8455 \times 10^{-2}\frac{\text{s}}{\text{kg}_{\text{TNT}}^{1/3}}$$

where

$$t_a = \text{arrival time (sec)}$$

$$t_a = 1.8455 \times 10^{-2} \frac{\text{s}}{\text{kg}_{\text{TNT}}^{1/3}} \times 3\,\text{kg}_{\text{TNT}}^{1/3} = 5.434 \times 10^{-2}\,\text{s}$$

$$\frac{T_s}{W_{\text{TNT}}^{1/3}} = 1.8455 \times 10^{-2} \frac{\text{s}}{\text{kg}_{\text{TNT}}^{1/3}} \times 3\,\text{kg}_{\text{TNT}}^{1/3} = 5.434 \times 10^{-2}\,\text{s} = t_a$$

where

$$T_s = \text{time in seconds}$$

Figure AXVII.5 now gives reflected pressure P_r and scaled reflected specific impulse $i_r/W^{1/3}$:

$$P_r = 7.291 \times 10^5\,\text{Pa}$$

$$\frac{i_r}{W_{\text{TNT}}^{1/3}} = 2.795 \times 10^2 \frac{\text{Pa} \cdot \text{s}}{\text{kg}_{\text{TNT}}^{1/3}}$$

$$i_r = \frac{2.795 \times 10^2 \cdot \text{Pa} \cdot \text{s}}{\text{kg}_{\text{TNT}}^{1/3}} \times 3\,\text{kg}_{\text{TNT}}^{1/3} = 8.385 \times 10^2$$

AXVII.5. Finite element analysis of blast effects on multistory concrete buildings

Figure AXVII.7 shows a typical concrete building. The three-dimensional finite element mesh scheme is shown in Fig. AXVII.8 for the same building.

Dharancepathy *et al.* [A2] have given the concept of critical distance versus mach stem to illustrate the height the blast reaches. Figure AXVII.8 gives such a height and the building substructure is decided accordingly.

Here d_0 is the ground zero, H_m mach stem, H_b is the height of burst, d is the distance. These phenomena help in the finite element analysis to decide whether to apply airburst (regular reflection) or surface burst (Mach reflection) conditions in calculating the blast parameters differently for different elements of the building frames or floors.

A CRAY-3 machine was used to analyse this building. The time taken to produce results (such as nodal forces, velocities, accelerations, displacements, damage etc.) was 5 h 3.35 min, with $4\frac{1}{2}$ million lines of results.

Data

W = vehicle bomb force = 5000 kg × 3

placed at three different levels at x, y and z axes

R' = range or d critical distance = 15 m

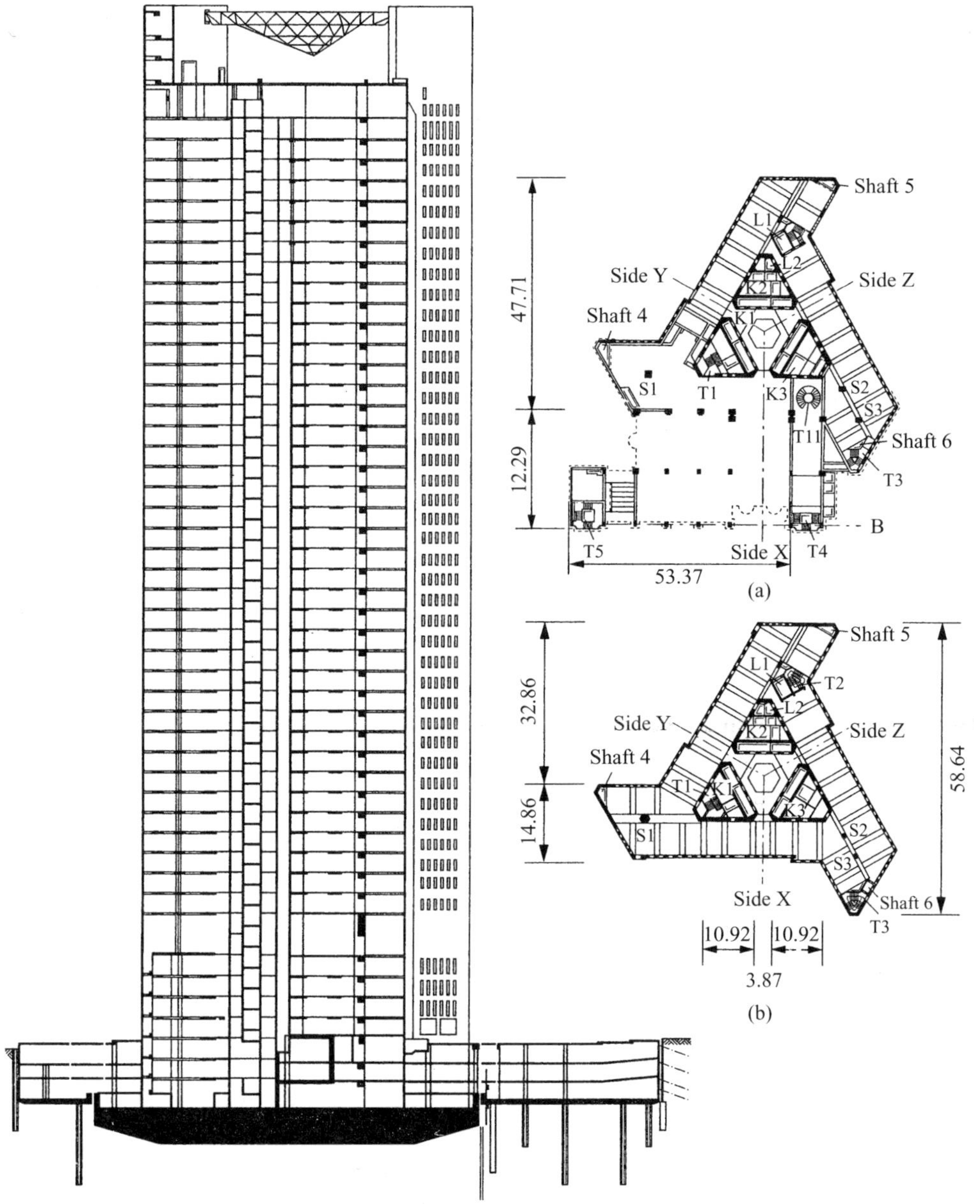

Fig. AXVII.7. Cross-section of the Trianon Building, Frankfurt: a multi-storey building parallel to the z-axis.

Fig. AXVII.8. Finite element mesh scheme for the Trianon Building, Frankfurt, Germany.

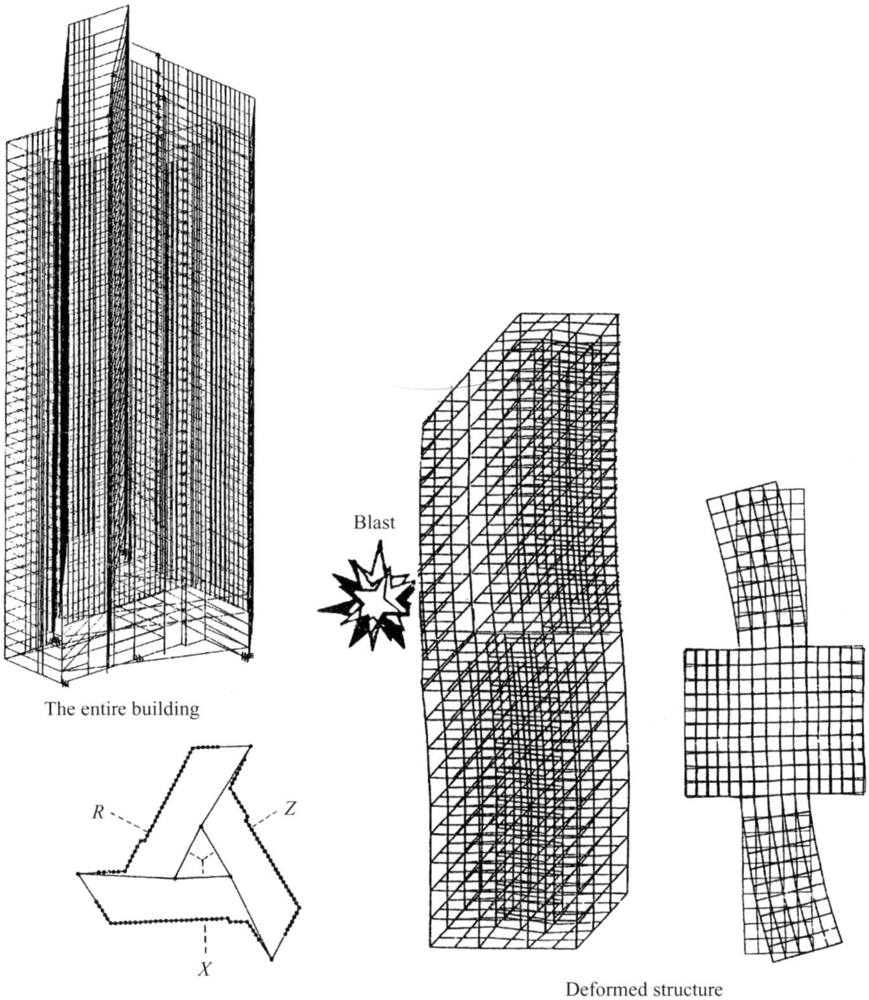

Fig. AXVII.9. Finite element mesh scheme for deformation and damage.

Material properties

Concrete:

$$f_c' = 36\,\mathrm{MN/m^2}$$

$$E_c = 21.5 \times 10^3\,\mathrm{MN/m^2}$$

$$= \text{Poisson's ratio} = 0.17$$

$$\frac{\text{weight}}{\text{density}} = 2400\,\mathrm{kg/m^2}$$

$$f_c' = 0.87 f_{cu}$$

$$f_{cu} = 41.7\,\mathrm{NM/m^2}$$

Fig. AXVII.10. Damaged zones and damage scenario.

Steel bars:

$$f_y = 280\,\mathrm{MN/m^2}$$

$$E = 200\,\mathrm{GN/m^2}$$

Endochronic theory: as damage theory.

Element data

Outer shell 155 300 elements
Line elements 6 degrees of freedom
Solid elements 8 noded, 20 noded and 4 noded
Solid elements each floor = 15 840 elements
Substructures 3 NO each 18 storey = H_m
Total number of storeys = 54
Total number of bars/floors 12,672 bars — sprayed by volume 4%
Lacings/stirrups = 0.1 × main steel for each element
All verticals, beams etc. are assumed to have 0.5% of the gross area at the joints.

Results

It is difficult to include the entire data, which comprise the size of a big room. The blast shows deformations for a single substructure of 18 storeys within the building. The value of H_m computed was 18 storeys and hence deformation and damage is indicated as in Fig. AXVII.9 under a blast of 200 kg burst at $H_b = 2.0$ m from the ground level where the estimated height of the truck is assumed. Most columns and floors are cracked and displaced particularly at the X and Y sections.

Excessive deflections of damaged walls and frames are observed along the Z-axis of the building. Windows up to the 18 storey level have been damaged and annealed glass has been damaged. The entire scenario is shown in Fig. AXVII.10.

AXVII.6. References

A1. Bangash, M. Y. H., *Prototype building structures*. Thomas Telford Limited, London, 1999.
A2. Dharancepathy, M. V. *et al.*, Critical distance for blast-resistant design. *Computer & Structures*, **54**(4), 1995, 587–595.

Index

Note: Figures and Tables are indicated (in this index) by *italic page numbers*